Mineral Nutrition of Higher Plants
Second Edition

Horst Marschner

Institute of Plant Nutrition
University of Hohenheim
Germany

ACADEMIC PRESS
Harcourt Brace & Company, Publishers
London   San Diego   New York
Boston   Sydney   Tokyo   Toronto

ACADEMIC PRESS LIMITED
24–28 Oval Road
LONDON NW1 7DX

U.S. Edition Published by
ACADEMIC PRESS INC.
San Diego, CA 92101

This book is printed on acid free paper

A catalogue record for this book is available from the British Library

ISBN 0-12-473542-8 (HB)
ISBN 0-12-473543-6 (PB)

Typeset by Paston Press Ltd, Loddon, Norfolk
Printed in Great Britain by The University Printing House, Cambridge

Preface to First Edition

Mineral nutrients are essential for plant growth and development. Mineral nutrition of plants is thus an area of fundamental importance for both basic and applied science. Impressive progress has been made during the last decades in our understanding of the mechanisms of nutrient uptake and their functions in plant metabolism; at the same time, there have also been advances in increasing crop yields by the supply of mineral nutrients through fertilizer application. It is the main aim of this textbook to present the principles of the mineral nutrition of plants, based on our current knowledge. Although emphasis is placed on crop plants, examples are also presented from noncultivated plants including lower plants in cases where these examples are considered more suitable for demonstrating certain principles of mineral nutrition, either at a cellular level or as particular mechanisms of adaptation to adverse chemical soil conditions.

Plant nutrition as a subject is closely related to other disciplines such as soil science, plant physiology and biochemistry. In this book, mineral nutrients in soils are treated only to the extent considered necessary for an understanding of how plant roots acquire mineral nutrients from soils, or how roots modify the chemical soil properties at the soil–root interface. Fundamental processes of plant physiology and biochemistry, such as photosynthesis and respiration, are treated mainly from the viewpoint of how, and to what extent, they are affected or regulated by mineral nutrients. Crop physiology is included as an area of fundamental practical importance for agriculture and horticulture, with particular reference to source–sink relationships as affected by mineral nutrients and phytohormones.

Mineral nutrition of plants covers a wide field. It is therefore not possible to treat all aspects with the detail they deserve. In this book, certain aspects are covered in more detail, either because they have recently become particularly important to our understanding of mineral nutrition, or because many advances have been made in a particular area in the last decade. Naturally, personal research interests and evaluation are also factors which have influenced selection. Particular emphasis is placed on short- and long-distance transport of mineral elements, on source–sink relationships, and on plant–soil relationships. It is also the intention of this book to enable the reader to become better acquainted with the mechanisms of adaptation of plants to adverse chemical soil conditions. The genetical basis of mineral nutrition is therefore stressed, as well as the possibilities and limitations of "fitting crop plant to soils", especially in the tropics and subtropics.

I have written this textbook for graduate students and researchers in the various fields of agricultural, biological and environmental sciences, who already have a profound knowledge of plant physiology, biochemistry and soil science. Instead of extensive explanations of basic processes, emphasis is placed on representative examples—tables, figures, schematic presentations—illustrating the various aspects of mineral nutrition. In a textbook of such wide scope, generalizations cannot be avoided, but relevant literature is cited for further and more detailed studies. In the literature, preference has been given to more recent publications. Nevertheless, representative examples of classical contributions are also cited in the various sections.

Although this book is written by one person, it is nevertheless the product of cooperation at various levels. My interest in plant nutrition and my scientific career in this field are due to the inspiration of Dr. G. Michael. The book as it is presented here would not have been accomplished without the excellent support of two colleagues, Dr. V. Römheld and Mr. Ernest A. Kirkby. I am very much indebted to both of them. Dr. Römheld not only prepared the drawings but also gave highly valuable advice regarding the arrangement of the tables and improvements to the text. My old friend Ernest A. Kirkby corrected the English and improved the first draft considerably by valuable suggestions and stimulating criticism. My colleagues in the institute, Dr. P. Martin, Dr. W. J. Horst and Dr. B. Sattelmacher helped me greatly, both by valuable discussions in various subject areas treated in this book and by keeping me free for some time from teaching and administrative responsibilities. Many colleagues were kind enough to supply me with their original photographs, as indicated in the legend of the corresponding figures.

The preparation of such a manuscript requires skilful technical assistance. I would especially like to thank Mrs. H. Hoderlein for typing the manuscript.

Last but not least, I have to thank my family for encouraging me to write the book and for their assistance and patience throughout this time-consuming process.

Stuttgart-Hohenheim   Horst Marschner
August 1985

Preface to Second Edition

As mentioned in the first edition the main aim of this textbook is to present the principles of the mineral nutrition of higher plants, based on current knowledge. This ambitious aim requires that the content of the book has to be updated regularly to take into account new developments in the subject as has been done in this second edition. The structure of the textbook has not been altered and the subject matter and number of chapters remains the same. The contents of the chapters, however, have been revised and on average about half the figures, tables and references replaced. The introduction of these more recent findings was based on the principle that newer examples and references are given priority, provided the quality of the information is at least similar to that which is being replaced. In Part I more emphasis has been placed on root–shoot interactions, stress physiology, water relations, and functions of micronutrients. In view of the worldwide increasing interest in plant–soil interactions, Part II has been considerably altered and extended. This is particularly true for Chapter 14 on the effects of external and internal factors on root growth, and Chapter 15 on the root–soil interface (root exudates, rhizosphere microorganisms, mycorrhizae).

The second edition would not have been accomplished without the support of many colleagues, friends and co-workers. Of these colleagues I am particularly grateful to Dr. Ismail Cakmak, Dr. Albrecht Jungk, Dr. Volker Römheld and Dr. Alexander Hansen. And again my old friend Ernest A. Kirkby took the most difficult task not only of correcting the English but also of improving the presentation by valuable suggestions and detailed, constructive criticism. I am also highly indebted to Dr. Eckhard George and his team for skilfully drawing the figures, to my daughter Petra and Dr. Ulrich Grauer for critically reading the text and the proofs, and to Mrs. Helga Hoderlein for the high quality of her technical assistance, especially in preparing the manuscript.

Stuttgart-Hohenheim
December 1993

Horst Marschner

One year after the second edition of his *Mineral Nutrition of Higher Plants* was published, Horst Marschner died in September 1996. He had contracted malaria during a visit to agricultural research projects in West Africa.

Horst Marschner was born in 1929. After an apprenticeship on a farm, he studied Agriculture and Chemistry at Jena University, and worked at Hohenheim University and the Technical University in Berlin. From 1977, he was a Professor at the Institute of Plant Nutrition, Stuttgart-Hohenheim. In addition, Horst Marschner went to California and to Australia for sabbatical terms, was responsible for field projects in many developing countries, and was visited in Hohenheim by guest scientists from all over the world who enjoyed his enthusiasm. In his research, he was interested in many subjects from soil science to plant physiology. He contributed specifically to the understanding of uptake and utilization of mineral nutrients by plants, rhizosphere effects, environmental aspects of plant nutrition, and plant adaptation to low nutrient supply and adverse soil conditions. He became a leading figure in plant mineral nutrition, and believed that science and rational thinking should be used to improve human living conditions.

In addition to writing several hundred scientific publications, Horst Marschner was a dedicated teacher and mentor of young scientists. We know of many people who enjoyed his contributions at meetings and conferences. Research in plant nutrition was fascinating to him, and he transmitted this fascination to those around him. His close involvement with practical experimentation was the basis for the clarity in his communications.

When asked to publish a book on mineral nutrition of plants, Horst Marschner knew the risks and challenges of writing as the single author of a general textbook. He was gratified by the success of *Mineral Nutrition of Higher Plants*, the very positive comments of his colleagues, and the world-wide appreciation by scientists and students. Close to retirement from his official duties as Head of the Institute, he was eagerly looking forward to spending more time discussing and presenting progress in his field, the science of plant nutrition. Now, after his unexpected death, this textbook has to serve as a legacy to an energetic, kind, constructive, and stimulating man.

Stuttgart-Hohenheim
March 1997

Eckhard George
Volker Römheld

Contents

Glossary of Plant Species

Botanical names of plant species frequently cited in this book.

Alder	*Alnus glutinosa* (L.) Gaertn.
Alfalfa	*Medicago sativa* L.
Almond	*Amygdalus communis* L.
Apple	*Malus sylvestris* Mill.
Apricot	*Prunus armeniaca* L.
Artichoke	*Cynara scolymus* L.
Aubergine	*Solanum melongena* L.
Barley	*Hordeum vulgare* L.
Bean (Common bean, snap bean)	*Phaseolus vulgaris* L.
Beech	*Fagus sylvatica* L.
Black gram	*Vigna mungo* (L.) Hepper
Black walnut	*Juglans nigra* L.
Blue lupin	*Lupinus angustifolius* L.
Broccoli	*Brassica oleracea* L. convar. *botrytis* var. *italica* Plenck
Brussels sprout	*Brassica oleracea* var. *gemmifera* L.
Cabbage	*Brassica oleracea* L. var. *capitata*
Carrot	*Daucus carota* L. ssp. *sativus* (Hoffm.) Arcang.
Cassava	*Manihot esculenta* Crantz
Castor bean	*Ricinus communis* L.
Cauliflower	*Brassica oleracea* L. convar. *botrytis* var. *botrytis*
Chickpea	*Cicer arietinum* L.
Chinese cabbage	*Brassica chinensis* L.
Celery	*Apium graveolens* L. var. *rapaceum* (Mill.)
Coffee	*Coffea* ssp.
Common bean	*Phaseolus vulgaris* L.
Cotton	*Gossypium hirsutum* L.
Couchgrass	*Agropyron repens* L.
Cowpea	*Vigna unguiculata* (L.) Walp.
Cucumber	*Cucumis sativus* L.

Douglas fir	*Pseudotsuga menziesii* (Mirb.) Franco
Faba bean	*Vicia faba* L.
Grapevine	*Vitis vinifera* L. ssp. *vinifera*
Groundnut	*Arachis hypogaea* L.
Italian ryegrass	*Lolium multiflorum* Lam.
Kallar grass	*Leptochloa fusca* L. Kunth.
Kentucky bluegrass	*Poa pratensis* L.
Leek	*Allium porrum* L.
Lentil	*Lens culinaris* L.
Lettuce	*Lactuca sativa* L.
Leucaena	*Leucaena leucocephala*
Loblolly pine	*Pinus taeda* L.
Maize	*Zea mays* L.
Mango	*Mangifera indica* L.
Maritime pine	*Pinus pinaster* Soland in Ait.
Melon	*Cucumis melo* L.
Mung bean	*Phaseolus mungo* L.
Norway spruce	*Picea abies* (L.) Karst.
Oat	*Avena sativa* L.
Oil palm	*Elaeis guineensis* Jacq.
Onion	*Allium cepa* L.
Pea	*Pisum sativum* L.
Peach	*Prunus persica* L. Batsch.
Peanut	*Arachis hypogaea* L.
Pearl millet	*Pennisetum glaucum* (L.) *R. Br. s. l.*
Pigeon pea	*Cajanus cajan* L. Huth.
Poppy	*Papaver somniferum* L.
Potato	*Solanum tuberosum* L.
Pumpkin	*Cucurbita pepo* L.
Rape	*Brassica napus* L. var. *napus*
Red beet	*Beta vulgaris* L. ssp. *vulgaris* var. *conditiva* Alef.
Red clover	*Trifolium pratense* L.
Red pepper	*Capsicum annuum* L.
Reed	*Phragmites communis* Trinius
Rice	*Oryza sativa* L.
Rhodes grass	*Chloris gayana* Kunth.
Rye	*Secale cereale* L.
Ryegrass	*Lolium perenne* L.
Scots pine	*Pinus sylvestris* L.
Sesbania	*Sesbania sesban*
Shortleaf pine	*Pinus echinata* Mill.
Sitka spruce	*Picea stichensis* Bong. Carr.
Soybean	*Glycine max* (L.) Merr.
Stinging nettle	*Urtica dioica* L.
Strawberry	*Fragaria vesca*
Sugar beet	*Beta vulgaris* L. ssp. *vulgaris*

Sugarcane	*Saccharum officinarum* L.
Sorghum	*Sorghum bicolor* (L.) Moench
Spinach	*Spinacea oleracea* L.
Squash	*Cucurbita maxima* Duch.
Subterranean clover	*Trifolium subterraneum* L.
Sunflower	*Helianthus annuus* L.
Sweet potato	*Ipomea batatas* L. Lam.
Taniers	*Xanthosoma* sp.
Tea plant	*Thea sinensis* L.
Teosinte	*Zea mays* L. ssp. *mexicana*
Timothy	*Phleum pratense* L.
Tobacco	*Nicotiana tabacum* L.
Tomato	*Lycopersicon esculentum* L.
Turnip	*Brassica rapa* var. *rapa* L.
Watercress	*Chenopodium quinoa*
Water melon	*Citrullus vulgaris* Schrad.
Welsh onion	*Allium fistulosum* L.
Wheat	*Triticum aestivum* L.
White clover	*Trifolium repens* L.
White lupin	*Lupinus albus* L.
White mustard	*Sinapis alba* L.
Winter radish	*Raphanus sativus* L. var. *niger* (Mill.) S. Kerner
Yams	*Dioscorea alata* L.
Yellow lupin	*Lupinus luteus* L.

Part I

Nutritional Physiology

1

Introduction, Definition, and Classification of Mineral Nutrients

The beneficial effect of adding mineral elements (e.g., plant ash or lime) to soils to improve plant growth has been known in agriculture for more than 2000 years. Nevertheless, even 150 years ago it was still a matter of scientific controversy as to whether mineral elements function as nutrients for plant growth. It was mainly to the credit of Justus von Liebig (1803–1873) that the scattered information concerning the importance of mineral elements for plant growth was compiled and summarized and that the mineral nutrition of plants was established as a scientific discipline. These achievements led to a rapid increase in the use of mineral fertilizers. By the end of the nineteenth century, especially in Europe, large amounts of potash, superphosphate, and, later, inorganic nitrogen were used in agriculture and horticulture to improve plant growth.

Liebig's conclusion that the mineral elements nitrogen, sulfur, phosphorus, potassium, calcium, magnesium, silicon, sodium, and iron are essential for plant growth was arrived at by observation and speculation rather than by precise experimentation. The fact that the 'mineral element theory' was based on this unsound foundation was one of the reasons for the large number of studies undertaken at the end of the nineteenth century. From these and other extensive investigations on the mineral composition of different plant species growing on various soils, it was realized as early as the beginning of this century that neither the presence nor the concentration of a mineral element in a plant is a criterion for essentiality. Plants have a limited capability for the selective uptake of those mineral elements which are essential for their growth. They also take up mineral elements which are not necessary for growth and which may even be toxic.

The mineral composition of plants growing in soils cannot therefore be used to establish whether a mineral element is essential. Once this fact was appreciated, both water and sand culture experiments were carried out in which particular mineral elements were omitted. These techniques made possible a more precise characterization of the essentiality of mineral elements and led to a better understanding of their role in plant metabolism. Progress in this research was closely related to the development of analytical chemistry, particularly in the purification of chemicals and methods of estimation. This relationship is reflected in the time scale of the discovery of the essentiality of micronutrients (Table 1.1).

Table 1.1
Discovery of the Essentiality of Micronutrients for Higher Plants

Element	Year	Discovered by
Iron	1860	J. Sachs
Manganese	1922	J. S. McHargue
Boron	1923	K. Warington
Zinc	1926	A. L. Sommer and C. B. Lipman
Copper	1931	C. B. Lipman and G. MacKinney
Molybdenum	1938	D. I. Arnon and P. R. Stout
Chlorine	1954	T. C. Broyer et al.
Nickel	1987	P. H. Brown et al.

The term *essential mineral element* (or mineral nutrient) was proposed by Arnon and Stout (1939). These authors concluded that, for an element to be considered essential, three criteria must be met:

1. A given plant must be unable to complete its life cycle in the absence of the mineral element.

2. The function of the element must not be replaceable by another mineral element.

3. The element must be directly involved in plant metabolism – for example, as a component of an essential plant constituent such as an enzyme – or it must be required for a distinct metabolic step such as an enzyme reaction.

According to this strict definition those mineral elements which compensate for the toxic effects of other elements or which simply replace mineral nutrients in some of their less specific functions, such as maintenance of osmotic pressure, are not essential, but can be described as 'beneficial' elements (Chapter 10). It is still difficult to generalize when discussing which mineral elements are essential for plant growth. This is particularly obvious when higher and lower plants are compared (Table 1.2). For higher plants the essentiality of 14 mineral elements is well established, although the known requirement for chlorine and nickel is as yet restricted to a limited number of plant species.

Table 1.2
Essentiality of Mineral Elements for Higher and Lower Plants

Classification	Element	Higher plants	Lower plants
Macronutrient	N, P, S, K, Mg, Ca	+	+ (Exception: Ca for fungi)
Micronutrient	Fe, Mn, Zn, Cu, B, Mo, Cl, Ni	+	+ (Exception: B for fungi)
Micronutrient and 'beneficial' element	Na, Si, Co I, V	± —	± ±

Because of continuous improvements in analytical techniques, especially in the purification of chemicals, this list might well be extended to include mineral elements that are essential only in very low concentrations in plants (i.e., that act as micronutrients). This holds true in particular for sodium and silicon, which are abundant in the biosphere. The essentiality of these two mineral elements has been established for some higher plant species (Chapter 10). Most micronutrients are predominantly constituents of enzyme molecules and are thus essential only in small amounts. In contrast, the macronutrients either are constituents of organic compounds, such as proteins and nucleic acids, or act as osmotica. These differences in function are reflected in the average concentrations of mineral nutrients in plant shoots that are sufficient for adequate growth (Table 1.3). The values can vary considerably depending on plant species, plant age, and concentration of other mineral elements. This aspect is discussed in Chapters 8 to 10.

Table 1.3
Average Concentrations of Mineral Nutrients in Plant Shoot Dry Matter that are Sufficient for Adequate Growth[a]

Element	Abbreviation	μmol g^{-1} dry wt	mg kg^{-1} (ppm)	%	Relative number of atoms
Molybdenum	Mo	0.001	0.1	—	1
Nickel[b]	Ni	~0.001	~0.1	—	1
Copper	Cu	0.10	6	—	100
Zinc	Zn	0.30	20	—	300
Manganese	Mn	1.0	50	—	1 000
Iron	Fe	2.0	100	—	2 000
Boron	B	2.0	20	—	2 000
Chlorine	Cl	3.0	100	—	3 000
Sulfur	S	30	—	0.1	30 000
Phosphorus	P	60	—	0.2	60 000
Magnesium	Mg	80	—	0.2	80 000
Calcium	Ca	125	—	0.5	125 000
Potassium	K	250	—	1.0	250 000
Nitrogen	N	1000	—	1.5	1 000 000

[a]From Epstein (1965).
[b]Based on Brown et al. (1987b).

2

Ion Uptake Mechanisms of Individual Cells and Roots: Short-Distance Transport

2.1   General

As a rule there is a great discrepancy between the mineral nutrient concentration in the soil or nutrient solution, on the one hand, and the mineral nutrient requirement of plants, on the other. Furthermore, soil and also in some cases nutrient solutions may contain high concentrations of mineral elements not needed for plant growth. The mechanisms by which plants take up nutrients must therefore be selective. This selectivity can be demonstrated particularly well in algal cells (Table 2.1), where the external and internal (cell sap) solutions are separated by only two membranes: the plasma membrane and the tonoplast.

In *Nitella* the concentration of potassium, sodium, calcium, and chloride ions is higher in the cell sap than in the pond water, but the concentration ratio differs considerably between the ions. In *Valonia* growing in highly saline seawater, on the other hand, only potassium is much more concentrated in the cell sap, whereas the sodium and calcium concentrations remain at a lower level in the cell sap than in the seawater.

Although usually less dramatic, selectivity of ion uptake is also a typical feature of

Table 2.1

Relationship between Ion Concentration in the Substrate and in the Cell Sap of *Nitella* and *Valonia*[a]

Ion	*Nitella* concentration (mM)			*Valonia* concentration (mM)		
	A, Pond water	B, Cell sap	Ratio B/A	A, Seawater	B, Cell sap	Ratio B/A
Potassium	0.05	54	1080	12	500	42
Sodium	0.22	10	45	498	90	0.18
Calcium	0.78	10	13	12	2	0.17
Chloride	0.93	91	98	580	597	1

[a]Modified from Hoagland (1948).

Table 2.2
Changes in the Ion Concentration of the External (Nutrient) Solution and in the Root Press Sap
of Maize and Bean

Ion	External concentration (mM)			Concentration in the root press sap (mM)	
	Initial	After 4 days[a]		Maize	Bean
		Maize	Bean		
Potassium	2.00	0.14	0.67	160	84
Calcium	1.00	0.94	0.59	3	10
Sodium	0.32	0.51	0.58	0.6	6
Phosphate	0.25	0.06	0.09	6	12
Nitrate	2.00	0.13	0.07	38	35
Sulfate	0.67	0.61	0.81	14	6

[a]No replacement of water lost through transpiration.

higher plants. When plants are grown in a nutrient solution of limited volume, the external concentration changes within a few days (Table 2.2). The concentrations of potassium, phosphate, and nitrate decline markedly, whereas those of sodium and sulfate can even increase, indicating that water is taken up faster than either of these two ions. Uptake rates, especially for potassium and calcium, differ between the two plant species (maize and bean). The ion concentration in the root press sap is generally higher than that in the nutrient solution; this is most evident in the case of potassium, nitrate, and phosphate.

The results obtained from both lower and higher plants demonstrate that ion uptake is characterized by the following:

1. *Selectivity*. Certain mineral elements are taken up preferentially, while others are discriminated against or almost excluded.

2. *Accumulation*. The concentration of mineral elements can be much higher in the plant cell sap than in the external solution.

3. *Genotype*. There are distinct differences among plant species in ion uptake characteristics.

These results pose many questions. For example, how do individual cells and higher plants regulate ion uptake? Is ion uptake a reflection of demand or are ions that play no role in plant metabolism or that are even toxic also taken up? For discussion of regulation of ion uptake at a cellular level it is necessary to follow the pathway of solutes (ions, charged and uncharged molecules) from the external solution through the cell wall and the plasma membrane into the cytoplasm and vacuole.

2.2 Pathway of Solutes from the External Solution into the Cells

2.2.1 Influx into the Apoplasm

Movement of low-molecular-weight solutes (e.g. ions, organic acids, amino acids, sugars) from the external solution into the cell walls of individual cells or roots (the *free*

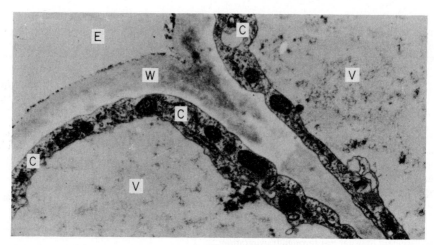

Fig. 2.1   Cross section of two rhizodermal cells of a maize root. V, vacuole; C, cytoplasm; W, cell wall; E, external solution. (Courtesy of C. Hecht-Buchholz.)

space) is a nonmetabolic, passive process, driven by diffusion or mass flow (Fig. 2.1). Nevertheless, the cell walls can interact with solutes and thus may facilitate or restrict further movement to the uptake sites of the plasma membrane of individual cells or roots.

Primary cell walls consist of a network of cellulose, hemicellulose, (including pectins) and glycoproteins, the latter may represent between 5% and 10% of the dry weight of the cell walls (Cassab and Varner, 1988). This network contains pores, the so-called interfibrillar and intermicellar spaces, which differ in size. For root hair cells of radish a maximum diameter of 3.5–3.8 nm (35–38 Å) has been calculated; maximum values for the pores in plant cell walls are in the range of 5.0 nm (Carpita *et al.*, 1979). In comparison hydrated ions such as K^+ and Ca^{2+} are small as shown below being only in the order of 10–20% of the cell wall pore size. The pores themselves would thus not be expected to offer any restriction to movement of ions in the free space.

	Diameter (nm)
Rhizodermal cell wall (maize; Fig. 2.1)	500–3000
Cortical cell wall (maize)	100–200
Pores in cell wall	<5.0
Sucrose	1.0
Hydrated ions	
K^+	0.66
Ca^{2+}	0.82

In contrast to mineral nutrients and low-molecular-weight organic solutes, high-molecular-weight solutes (e.g., metal chelates, fulvic acids, and toxins) or viruses and other pathogens are either severely restricted or prevented by the diameter of pores from entering the free space of root cells.

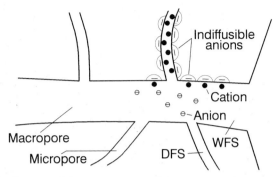

Fig. 2.2 Schematic diagram of the pore system of the apparent free space. DFS, Donnan free space; WFS, water free space.

In this network, a variable proportion of the pectins consist of polygalacturonic acid, originating mainly from the middle lamella. Accordingly both in roots and in the cell wall continuum of other plant tissue, the so-called *apoplasm*, the carboxylic groups ($R \cdot COO^-$) act as cation exchangers. In roots therefore cations from the external solution can accumulate in a nonmetabolic step in the *free space*, whereas anions are 'repelled' (Fig. 2.2).

Because of these negative charges the apoplasm does not provide a *free space* for the movement of charged solutes, and Hope and Stevens (1952) introduced the terms *apparent free space* (AFS). This comprises the *water free space* (WFS), which is freely accessible to ions and charged and uncharged molecules, and the *Donnan free space* (DFS), where cation exchange and anion repulsion take place (Fig. 2.2). Ion distribution within the DFS is characterized by the typical Donnan distribution which occurs in soils at the surfaces of negatively charged clay particles. Divalent cations such as Ca^{2+} are therefore preferentially bound to these cation-exchange sites. Plant species differ considerably in cation-exchange capacity (CEC), that is, in the number of cation-exchange sites (fixed anions; $R \cdot COO^-$), located in cell walls, as shown in Table 2.3.

As a rule, the CEC of dicotyledonous species is much higher than that of monocotyledonous species. The effective CEC decreases as the external pH falls (Allan and Jarrell, 1989), and is usually much lower in intact roots than the values shown in Table 2.3.

Table 2.3
Cation Exchange Capacity of Root Dry Matter of
Different Plant Species[a]

Plant species	Cation exchange capacity meq $(100 \text{ g})^{-1}$ dry wt
Wheat	23
Maize	29
Bean	54
Tomato	62

[a]Based on Keller and Deuel (1957).

Table 2.4
Uptake and Translocation of Zinc by Barley Plants[a]

Zinc supplied as[b]	Rate of uptake and translocation (μg Zn g^{-1} dry wt per 24 h)	
	Roots	Shoots
ZnSO$_4$	4598	305
ZnEDTA	45	35

[a]Based on Barber and Lee (1974).
[b]Concentration of zinc in nutrient solution: 1 mg l^{-1}.

Because of spatial limitations (Casparian band and exodermis; Section 2.5.1) only part of the exchange sites of the AFS are directly accessible to cations from the external solution. Nevertheless, the differences shown are typical of those that exist between plant species.

Exchange adsorption in the AFS of the apoplasm is not an essential step for ion uptake or transport through the plasma membrane into the cytoplasm. Nevertheless, the preferential binding of di- and polyvalent cations increases the concentration of these cations in the apoplasm of the roots and thus in the vicinity of the active uptake sites at the plasma membrane. As a result of this indirect effect, a positive correlation can be observed between the CEC and the ratio of Ca^{2+} to K^+ contents in different plant species (Crooke and Knight, 1962; Haynes, 1980). Effective competition between H^+ or mono- and polyvalent aluminium species or both with magnesium for binding sites in the apoplasm of roots is obviously a main factor responsible for the depression in magnesium uptake and appearance of magnesium deficiency in annual (Rengel, 1990; Tan et al., 1991) and forest tree species (Marschner, 1991b) grown on acid mineral soils (Section 16.3).

The importance of cation binding in the AFS for uptake and subsequent shoot transport is also indicated by experiments with the same plant species but with different binding forms of a divalent cation such as zinc (Table 2.4). When zinc is supplied in the form of an inorganic salt (i.e., as free Zn^{2+}), the zinc content not only of the roots but also of the shoots is several times higher than when zinc is supplied as a chelate (ZnEDTA), that is, without substantial binding of the solute in the AFS. In addition, restricted permeation of the chelated zinc within the pores of the AFS may be a contributing factor. Using these differences in uptake rate between metal cations like Zn^{2+} (and also Cu^{2+} and Mn^{2+}) and their complexes with synthetic chelators in so-called *chelator-buffered solutions* calculations can be made of the concentrations of free metal cations in the external solution required for optimal plant growth (Bell et al., 1991; Laurie et al., 1991). According to these calculations, extremely low external concentrations at the plasma membrane of root cortical cells appear to be adequate to meet plant demand for these micronutrient cations (Section 2.5.4).

With heavy-metal cations in particular, binding in the apoplasm can be quite specific. Copper, for example, may be bound in a nonionic form (coordinative binding) to nitrogen-containing groups of either glycoproteins or proteins of ectoenzymes, such as phosphatases or peroxidases, in the cell wall (Harrison et al., 1979; Van Cutsem and

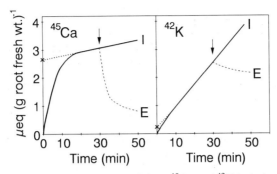

Fig. 2.3 Time course of influx (I) and efflux (E) of ^{45}Ca and ^{42}K in isolated barley roots. After 30 min (arrow) some of the roots were transferred to solutions with nonlabelled Ca^{2+} and K^+. The proportion of the exchangeable fraction in the apparent free space is calculated by extrapolation to zero time ($\times$).

Gillet, 1982). This cation binding in the apoplasm can contribute significantly to the total cation content of roots, as shown by studies of the uptake of polyvalent cations such as copper, zinc, and iron. This is also demonstrated by the data in Table 2.4. When supplied in non-chelated form high contents of polyvalent cations in the roots compared to the shoots therefore not necessarily reflect immobilization in the cytoplasm or vacuoles but may result from preferential binding in the apoplasm of the root cortex.

The root apoplasm may also serve as transient storage pool for heavy metals such as iron and zinc which can be mobilized, for example by specific root exudates such as phytosiderophores, and translocated subsequently into the shoots (Zhang *et al.*, 1991b, c). The size of this storage pool for iron probably plays a role for genotypical differences in sensitivity to iron deficiency in soybean (Longnecker and Welch, 1990). On the other hand, excessive uptake of calcium may be restricted by precipitation as calcium oxalate in the cell walls of the cortex (Fink, 1992b).

2.2.2 Passage into the Cytoplasm and the Vacuole

Despite some selectivity for cation binding in the cell wall (Section 2.2.1), the main sites of selectivity in the uptake of cations and anions as well as solutes in general are located in the *plasma membrane* of individual cells. The plasma membrane is an effective barrier against the diffusion of solutes either from the apoplasm into the cytoplasm (influx) or from the cytoplasm into the apoplasm and the external solution (efflux). The plasma membrane is also the principal site of active transport in either direction. The other main barrier to diffusion is the *tonoplast* (vacuolar membrane). In most mature plant cells the vacuole comprises more than 80–90% of the cell volume (Leigh and Wyn Jones, 1986; Wink, 1993; also see Fig. 2.1) acting as central storage compartment for ions, but also for other solutes (e.g. sugars, secondary metabolites).

It can be readily demonstrated that the plasma membrane and the tonoplast function as effective barriers to diffusion and exchange of ions, as shown for example, for K^+ and Ca^{2+} (Fig. 2.3). Most of the Ca^{2+} (^{45}Ca) taken up within 30 min (influx) is still readily exchangeable (efflux) and is almost certainly located in the AFS. In contrast only a minor fraction of the K^+ (^{42}K) is readily exchangeable within this 30-min period, most of the K^+ having already been transported across the membranes into the cytoplasm and vacuoles ('inner space').

Although the plasma membrane and the tonoplast are the main biomembranes directly involved in solute uptake and transport in roots, it must be borne in mind that compartmentation by biomembranes is a general prerequisite for living systems (Leigh and Wyn Jones, 1986). Solute transport into organelles such as mitochondria and chloroplasts must also therefore be regulated by membranes which separate these organelles from the surrounding cytoplasm. An example of solute transport across the outer chloroplast membrane is given in Section 8.4 for phosphorus and sugars.

The capability of biomembranes for solute transport and its regulation is closely related to their chemical composition and molecular structure. Before the mechanisms of solute transport across membranes are discussed in more detail (Sections 2.4 and 2.5), it is therefore appropriate to consider some fundamental aspects of the composition and structure of biomembranes.

2.3   Structure and Composition of Membranes

The capacity of plant cell membranes to regulate solute uptake has fascinated botanists since the nineteenth century. At that time the experimental techniques available limited the investigation of the process. Nevertheless, even by the early years of the twentieth century some basic facts of solute permeation across the plasma membrane and tonoplast had been established, as for example of the inverse relationship between membrane permeation and the diameter of uncharged molecules and the rates at which they permeate membranes. These ultrafilter-like properties of membranes have been confirmed more recently, at least in principle (Table 2.5).

Thus, in addition to the cell walls (Section 2.2.1) cell membranes are effective barriers to solutes of high molecular weight. Most synthetic chelators such as EDTA (see also Table 2.4) and microbial siderophores as specific chelators of iron (Section 16.5) are of high molecular weight and their rate of permeation is restricted through the

Table 2.5
Reflection Coefficient (δ) of Some Nonelectrolytes
at the Cell Membranes of *Valonia utricularis*[a]

Compound	δ[b]	Molecule radius (nm)
Raffinose	1.00	0.61
Sucrose	1.00	0.53
Glucose	0.95	0.44
Glycerol	0.81	0.27
Urea	0.76	0.20

[a]Based on Zimmermann and Steudle (1970).
[b]1.00 indicates that the membranes are impermeable to the solute; 0 indicates that the membranes are freely permeable to the solute.

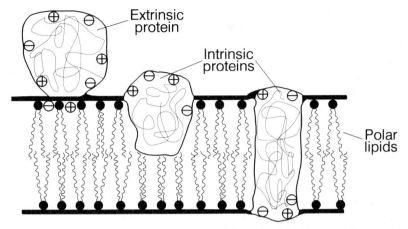

Fig. 2.4 Model of a biomembrane with polar lipids and with either extrinsic or intrinsic, integrated proteins. The latter can cross the membrane to form 'protein channels'.

plasma membrane of root cells. It is possible, therefore, to use high-molecular-weight organic solutes such as polyethyleneglycol at high external concentrations as effective osmotica in order to induce water deficiency (drought stress) in plants.

Molecules which are highly soluble in organic solvents, i.e. with lipophilic properties, penetrate membranes much faster than would be predicted on the basis of their size. The solubility of these molecules in the membrane and their ability to diffuse through the lipid core of the membranes are presumably the main factors responsible for the faster permeation.

Membranes are typically composed of two main classes of compounds: proteins and lipids. Carbohydrates comprise only a minor fraction of membranes. The relative abundance of proteins and lipids can be quite variable depending on whether the membrane is a plasma, mitochondrial, or chloroplast membrane (Clarkson, 1977). Membranes also differ in diameter, for example in spinach from 10.5 nm (plasma membrane) to 8.1 nm (tonoplast) and 6.3 nm (endoplasmic reticulum; Auderset *et al.*, 1986). However, all biomembranes have some common basic structure as shown in a model in Fig. 2.4.

Polar lipids (e.g. phospholipids) with the hydrophilic, charged head regions (phosphate, amino, and carboxylic groups) are oriented towards the membrane surface. Protein molecules can be attached (extrinsic proteins), for example, by electrostatic binding to the surfaces as membrane-bound enzymes. Other proteins may be integrated into membranes (intrinsic proteins), or traverse the membranes to form 'protein channels' (transport proteins) which serve to function in membrane transport of polar solutes such as ions (Section 2.4).

Three polar lipids represent the major lipid components of membranes: phospholipids, glycolipids, and less abundant, sulfolipids (except in the thylakoid membranes of chloroplasts, where they occur in substantial amounts). Examples of these polar lipids are shown below:

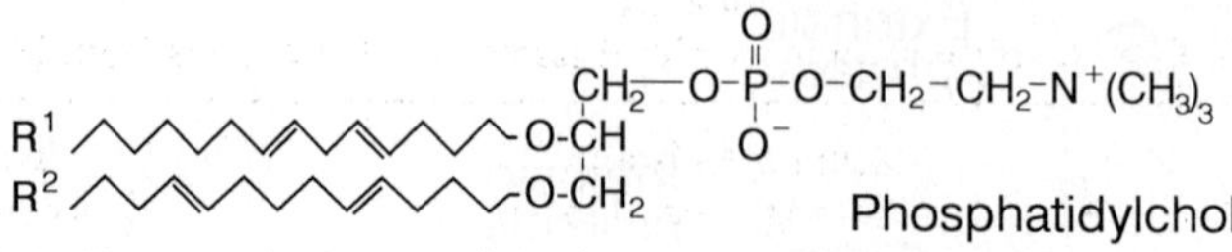

Phosphatidylcholine (lecithin)

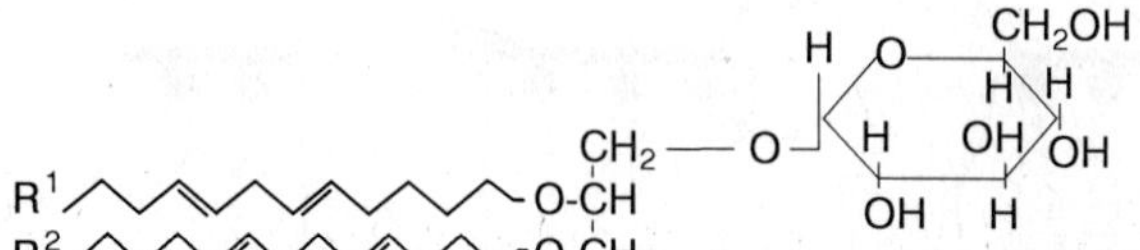

Monogalactosyl diglyceride

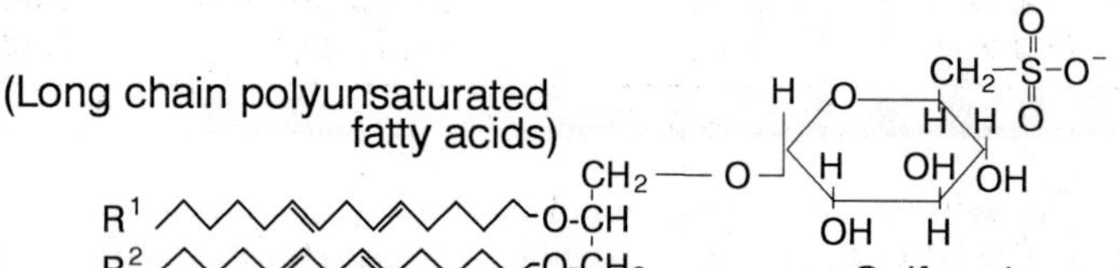

(Long chain polyunsaturated fatty acids)

Sulfoquinovosyl diglyceride

Another important group of membrane lipids consists of sterols, for example β-sistosterol:

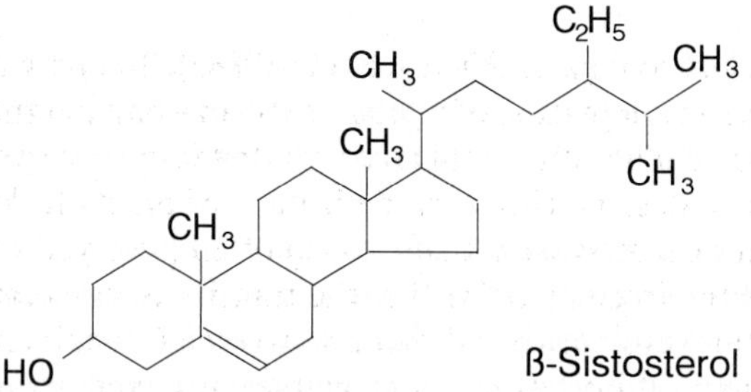

ß-Sistosterol

Through their structural role in membranes sterols may indirectly affect transport processes such as the activity of the proton pumping ATPase in the plasma membrane (Sandstrom and Cleland, 1989). In agreement with this assumption the sterol content is very low in endomembranes (e.g. endoplasmic reticulum) but may make up more than 30% of the total lipids in the plasma membrane (Brown and DuPont, 1989) and also in the tonoplast (Table 2.6). Despite these differences in lipids, the fatty acid composition of the phospholipids is similar in both membranes. The long-chain fatty acids in polar membrane lipids vary in both the length and degree of unsaturation (i.e. number of double bounds) which influence the melting point (Table 2.6).

Lipid composition not only differs characteristically between membranes of individual cells but also between cells of different plant species (Stuiver et al., 1978), it is also strongly affected by environmental factors. In leaves, for example, distinct annual variations in the levels of sterols occur (Westerman and Roddick, 1981) and in roots both phospholipid content and the proportion of highly unsaturated fatty acids decrease

Table 2.6

Lipid and Fatty Acid Composition of Plasma Membranes and Tonoplasts from Mung Bean[a]

Lipids	Plasma membrane μmol mg^{-1} protein	Tonoplast μmol mg^{-1} protein
Phospholipids	1.29	1.93
Sterols	1.15	1.05
Glycolipids	0.20	0.80

Fatty acid	Chain length	Fatty acid composition of the phospholipids Melting point (°C)	Plasma membrane (% of total)	Tonoplast (% of total)
Palmitic acid	C$_{16}$	+62.8	35	39
Stearic acid	C$_{18}$	+70.1	6	6
Oleic acid	C$_{18:1}$[b]	+13.0	9	9
Linoleic acid	C$_{18:2}$[b]	−5.5	21	22
Linolenic acid	C$_{18:3}$[b]	−11.1	19	20
Others	—	—	10	4

[a]Based on Yoshida and Uemura (1986). Reprinted by permission of the American Society of Plant Physiologists.
[b]Numeral to the right of the colon indicates the number of double bounds.

under zinc deficiency (Cakmak and Marschner, 1988c). In many instances changes in lipid composition reflect adaptation of a plant to its environment through adjustment of membrane properties. Generally, highly unsaturated fatty acids predominate in plants that grow in cold climates. During acclimatization of plants to low temperatures an increase in highly unsaturated fatty acids is also often observed (Bulder *et al.*, 1991). Such a change shifts the freezing point (i.e. the transition temperature) of membranes to a lower temperature and may thus be of importance for maintenance of membrane functions at low temperatures. It is questionable, however, to generalize about the effect of temperature on lipid composition of membranes. In rye, for example, which is a cold-tolerant plant species, the proportion of polyunsaturated fatty acids in the roots decreased rather than increased as the roots were cooled (White *et al.*, 1990b).

During acclimatization of roots to low temperatures synthesis of new membrane proteins is also enhanced (Mohapatra *et al.*, 1988) and phospholipids increase considerably (Kinney *et al.*, 1987). Since phospholipids probably act as receptors for phytohormones such as gibberellic acid, increasing responsiveness of membranes to gibberellic acid at low temperatures may be related to these changes (Singh and Paleg, 1984).

The property of membranes in ion selectivity and lipid composition are often highly ·orrelated as for example between chloride uptake and sterols (Douglas and Walker, ₁983) and galactolipids (Section 16.6). Also the crop plant species bean, sugar beet and barley differ not only in the fatty acid composition of root membranes (Stuiver *et al.*, 1978) but also considerably in the uptake of sodium (Section 10.2).

Alterations in the lipid composition of root membranes are also typical responses to changes in the mineral nutrient supply or exposure to salinity (Kuiper, 1980). Of the mineral nutrients, calcium plays the most direct role in the maintenance of membrane integrity, a function which is discussed in Section 2.5.2. In soybean roots, changes in calcium and nitrogen supply affect the ratio of saturated to unsaturated fatty acids as well as the uptake rate of certain herbicides (Rivera and Penner, 1978). An increase in membrane permeability can be observed in roots suffering from phosphorus deficiency (Ratnayake *et al.*, 1978) and zinc deficiency (Welch *et al.*, 1982; Cakmak and Marschner, 1988b, 1990). In the case of phosphorus deficiency, a shortage of phospho-lipids in the membranes has been assumed to be the responsible factor. In the case of zinc deficiency, autoxidation in the membranes of highly unsaturated fatty acids is presumably involved in membrane leakiness (Section 9.4).

The dynamic nature of membranes is clearly demonstrated, for example, by the rapid decrease in efflux of low-molecular-weight solutes (potassium, sugars, amino acids) after resupplying of zinc to zinc-deficient roots (Cakmak and Marschner, 1988b). Another example is the rapid incorporation of externally supplied membrane constitu-ents such as phospholipids into the membrane structure. For the plasma membrane turnover rates seem to be in the order of only a few hours (Steer, 1988a). Such high turnover rates indicate that certain subunits (e.g. with intrinsic proteins; Fig. 2.4) are already synthesized and transported to the plasma membrane via secretory vesicles as for example of the Golgi apparatus (Coleman *et al.*, 1988).

The incorporation also of externally supplied compounds, however, renders mem-branes more sensitive to injury. The incorporation of antibiotics such as nystatin induces the formation of pores ('holes') in the membranes and a corresponding rapid leakage of low-molecular-weight solutes such as potassium. Monocarboxylic acids such as acetic acid and butyric acid, also induce membrane injury. The undissociated species of these acids are readily taken up and lead to a sharp rise in membrane leakiness, as indicated by the leakage of potassium and nitrate from the root tissue (Lee, 1977). The capacity of monocarboxylic acids to induce membrane leakiness increases with the chain length of the acids [C_2 (acetic acid) $\rightarrow C_8$ (caprylic acid)] and hence with increased lipophilic behaviour, as well as with a lowering of the external pH ($R \cdot COO^- + H^+ \rightarrow R \cdot COOH$). Undissociated monocarboxylic acids may increase membrane leakiness by changing the fatty acid composition of membranes, particularly by decreasing the proportion of polyunsaturated fatty acids such as linolenic acid (Jackson and St. John, 1980). The effect of monocarboxylic acids on the membrane permeability of roots is of considerable ecological importance, since these acids accumulate in waterlogged soils (Section 16.4).

These examples demonstrate that composition, structure and integrity of membranes are affected by a range of environmental factors. In the last decade in particular increasing evidence has accumulated that a range of environmental stress factors such as high light intensity, drought, chilling, air pollutants, and also mineral nutrient deficiencies are harmful to plants by impairment of membrane integrity, and that elevated levels of toxic oxygen species are causally involved in this impairment (Elstner, 1982; Hippeli and Elstner, 1991).

As shown in a model in Fig. 2.5, these toxic oxygen species are either radicals such as superoxide ($O_2^{\cdot-}$) or hydroxyl ($OH^{\cdot}$), or the molecule hydrogen peroxide (H_2O_2). All are formed in various reactions and metabolic processes where oxygen is involved, for example photosynthesis (Asada, 1992) and respiration, including oxidation of NADPH

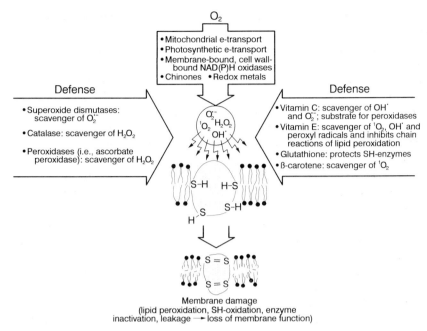

Fig. 2.5 Model of generation of, and membrane lipid peroxidation by, oxygen radicals and hydrogen peroxide, and systems for scavenging and detoxification. (I. Cakmak, unpublished.)

or NADH at the plasma membrane–cell wall interface (Werf *et al.*, 1991; Vianello and Macri, 1991). Toxicity by activated oxygen species and its derivates is caused, for example by oxidation of thiol groups (-SH) of enzymes and peroxidation of polyunsaturated fatty acids in membranes (Halliwell, 1978; Pukacka and Kuiper, 1988). Aerobic organisms including plants possess a range of defense systems (Fig. 2.5) for detoxification of oxygen radicals and hydrogen peroxide, including superoxide dismutase ($O_2^{\cdot-} \rightarrow H_2O_2$) and peroxidases/catalase ($H_2O_2 \rightarrow H_2O$).

Mineral nutrition of plants may affect at various levels both, generation of toxic oxygen species and hydrogen peroxide, and mechanisms for detoxification. These may be summarized as follows:

1. As a component of detoxifying enzymes (e.g. zinc, copper, manganese or iron in superoxide dismutases; iron in peroxidases and catalase);

2. By the accumulation of precursors for radical formation (e.g. phenols and quinones) under nutrient deficiency (e.g. boron deficiency);

3. Through a decrease in sink activity (i.e. demand) and accumulation of photosynthates and correspondingly elevated levels of toxic oxygen species in source leaves under mineral nutrient deficiencies (e.g. potassium and magnesium).

These various aspects are discussed in more detail in the relevant sections on photosynthesis (Chapter 5) and functions of mineral nutrients (Chapters 8 and 9).

2.4   Solute Transport Across Membranes

2.4.1   Transport and Energy Demand

Intact membranes are effective barriers to the passage of ions and uncharged molecules. On the other hand, they are also the sites of selectivity and transport against the concentration gradient of solutes. In the experiment recorded in Table 2.2, for example, the potassium concentration in maize root press sap (which is approximately equal to the potassium concentration of the vacuoles) rose to 80 times the value of that in the external solution. In contrast, the sodium concentration in the root press sap remained lower than that in the external solution. It is generally agreed that such selectivity and accumulation requires both, an energy supply as a driving force and specific binding sites, *carriers* or *permeases* in the membranes, most likely proteins such as the sulfate-binding protein isolated from microorganisms (Pardee, 1967) or the sulfate permease in roots (Hawkesford and Belcher 1991; Section 2.5.6). A direct coupling of carrier-mediated selective ion transport and consumption of energy-rich phosphates in the form of ATP was proposed in this process. In this model the ATP was supplied via respiration (oxidative phosphorylation in the mitochondria; Section 8.4.3) and required for activation of the carrier, the ion binding to the carrier, the membrane transport of the carrier–ion complex, or the release of the ion from the carrier at the internal surface of the membrane.

Strong support for the involvement of ATP in carrier-mediated ion transport was first presented by Fisher *et al.* (1970). Studying the uptake of K^+ by roots of various plant species, these workers demonstrated a close relationship between K^+ uptake and ATPase activity (Fig. 2.6). Furthermore, $Mg \cdot ATPases$ (Section 8.5) of the plasma membrane are strongly stimulated by K^+ (Fisher *et al.*, 1970; Briskin and Poole, 1983) so that ions such as K^+, when added to the external solution, trigger their own transport across the plasma membrane.

The energy demand for ion uptake by roots is considerable, especially during rapid vegetative growth (Table 2.7). Of the total respiratory energy cost, expressed as ATP consumption, up to 36% is required for ion uptake. With increase in plant age this

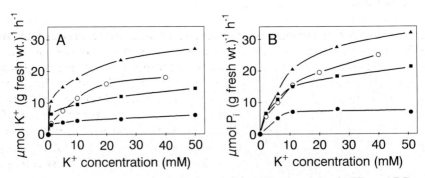

Fig. 2.6   (A) Potassium ion uptake (influx) and (B) ATPase activity (ATP → ADP + P_i) in isolated roots of different plant species. Key: ▲, barley; ○, oat; ■, wheat; ●, maize. (After Fisher *et al.*, 1970.)

Table 2.7
Respiratory Energy Costs for Ion Uptake in Roots of *Carex diandra*[a]

Proportion of total ATP demand required for	Plant age (days)		
	40	60	80
Ion uptake	36	17	10
Growth	39	43	38
Maintenance of biomass	25	40	52

[a]Based on Werf *et al.* (1988).

proportion declines in favour of ATP demand for growth and maintenance of biomass. In principle, similar results have been found with maize (Werf *et al.*, 1988).

These calculations at a whole plant level on ATP demand for ion uptake by roots have to be interpreted with care with respect to energy demand for membrane transport of ions in root cells. Firstly, these calculations include energy demand for radial transport through the roots and secretion into the xylem (Sections 2.7 and 2.8). Secondly, a relatively large proportion of carbohydrates supplied from the shoot to the roots are oxidized via the nonphosphorylating mitochondrial electron transport chain ('alternative pathway'; Section 5.3) yielding less ATP synthesized per molecule of carbohydrate oxidized. Taking this shift in respiratory pathway into account, a requirement of one molecule ATP per ion transported across the plasma membrane has been calculated (Lambers *et al.*, 1981). Such calculations are based on net uptake and include energy requirement for re-uptake ('retrieval') of ions from the apoplasm of the root ('efflux costs') which are assumed to be in the range of 20% of the influx costs (Bouma and De Visser, 1993). Thirdly, a direct coupling of ATP consumption and ion transport across membranes is the exception rather than the rule. As discussed below, ATP-driven pumps at the plasma membrane and the tonoplast also have functions other than transport of mineral nutrients and organic solutes across the membranes.

2.4.2 Active and Passive Transport: Electrogenic Pumps, Carriers, Ion Channels

Solute transport across membranes is not necessarily an active process. Solutes may be more concentrated on one side of the membrane (i.e. they may possess more free energy) and thus diffuse from a higher to a lower concentration (or chemical potential). This 'downhill' transport across a membrane is, in thermodynamic terms, a passive transport with the aid of carriers, or across aqueous pores (Clarkson, 1977). In cells, such downhill transport of ions across the plasma membrane may be maintained by a lowering of the ion activity in the cytoplasm, for example, due to adsorption at charged groups (e.g., $R \cdot COO^-$ or $R \cdot NH_3^+$) or to incorporation into organic structures (e.g., phosphate into nucleic acids). This is particularly true in meristematic tissues (e.g., root tips).

In contrast, membrane transport against the gradient of potential energy ('uphill') must be linked directly or indirectly to an energy-consuming mechanism, a 'pump' in

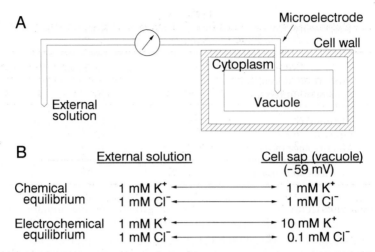

Fig. 2.7 A. Schematic presentation of the system for measuring electropotentials in plant cells. B. Example of the calculation of ion distribution at chemical and electrochemical equilibrium assuming an electropotential of −59 mV.

the membrane. To determine whether an ion is actively transported across a membrane, however, both the activity or concentration of the ion on either side of the membrane (i.e., the chemical potential gradient) and the electrical potential gradient (i.e., differences in millivolts) across the membrane must be known. By means of microelectrodes inserted into the vacuoles, a strongly negative electrical potential can be measured between the cell sap and the external solution (Fig. 2.7). The first measurements of this kind were made in cells of giant algae such as *Chara*, where strongly negative electrical potentials of between −100 and −200 mV were found. The same method was used by Higinbotham *et al.* (1967) and Glass and Dunlop (1979) to demonstrate the existence of similar electrical potential gradients in cells of higher plants.

The concentration at which cations and anions on either side of a membrane are in electrochemical equilibrium or which ions in the external solution are in equilibrium with those in the vacuole can be calculated according to the Nernst equation:

$$E \text{ (mV)} = -59 \log \frac{\text{concentration inside (vacuole)}}{\text{concentration outside (external solution)}}$$

According to this equation, at a negative electropotential of −59 mV, monovalent cations such as K^+ or anions such as Cl^- would be in electrochemical equilibrium if their concentration in the vacuole were 10 times higher (K^+) or 10 times lower (Cl^-) than in the external solution (Fig. 2.7). For divalent cations or anions the difference between chemical and electrochemical equilibrium differs by even more than the factor of 100. In cells of higher plants the electrical potential differences between vacuoles and the external solution are generally higher than −59 mV (Table 2.8). Thus, as a rule, in terms of electrophysiology, only anion uptake into the vacuoles would always require an active transport process. This is indicated in Table 2.8 in the differences between the

Table 2.8
Experimentally Determined and Calculated Ion Concentration (mM) According to the Electrical
Potential Differences in Roots of Pea and Oat[a]

Ion	Pea roots (-110 mV)		Oat roots (-84 mV)	
	Experimental	Calculated	Experimental	Calculated
Potassium	75	74	66	27
Sodium	8	74	3	27
Calcium	2	10 800	3	1400
Chloride	7	0.014	3	0.038
Nitrate	28	0.027	56	0.076

[a]Composition of the external solution: 1 mM KCl, 1 mM $Ca(NO_3)_2$, and 1 mM NaH_2PO_4. Based on Higinbotham *et al.* (1967).

anion concentrations in the vacuoles based on calculations according to the electrochemical equilibrium and the anion concentrations in the vacuoles actually found in the experiment.

In this example, the only cation that would require active transport for its uptake is K^+ in oat roots. At low external K^+ concentration, active transport is usually required (Cheeseman and Hanson, 1979). For Na^+ and Ca^{2+} in particular, the equilibrium concentration in the cell sap (i.e., the calculated ones) would be much higher than that found experimentally in the steady state (Table 2.8). A possible explanation for this discrepancy is that the plasma membrane strongly restricts permeation by these ions or that the ions are pumped (transported) back into the external solution. For Na^+ such an efflux pump at the plasma membrane of root cells has been established for various plant species (Jacoby and Rudich, 1985; Schubert and Läuchli, 1988). For Ca^{2+} active extrusion (Ca^{2+} efflux pump) at the plasma membrane also exists in root cells (Olbe and Sommarin, 1991). The general importance of this Ca^{2+} efflux pump at both plasma membrane and tonoplast for the functioning of cells is discussed in Section 8.6. Since the Ca^{2+} concentrations in soil solutions are usually higher than 1 mM, a Ca^{2+} efflux pump to prevent Ca^{2+} transport along the electrochemical gradient (e.g., Table 2.8) would have a considerable energy requirement. It is likely, therefore, that additional physicochemical factors such as the size and charge of Ca^{2+} strongly restrict permeation along the electrochemical potential gradient across the plasma membrane.

In recent years impressive progress has been made in understanding both the mechanisms leading to the formation of electropotentials across membranes and the importance of these potentials for cell growth and functioning. Progress was possible by new techniques allowing measurements of membrane potentials and ion fluxes in isolated membrane vesicles or in sections of membranes (patch clamp technique, Hedrich and Schroeder, 1989). Some of the principles of ion transport in membranes are shown in Fig. 2.8. An ATP-driven H^+ pump ('proton motive force'; Poole, 1978) transports H^+ through the membrane from the internal to the external surface, for example in the plasma membrane from the cytoplasm to the apoplasm, thereby creating a gradient in pH and electropotential. Transport of cations and anions along the gradient is mediated either by ion selective carriers or channels. This model also takes

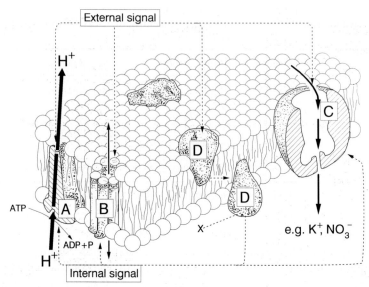

Fig. 2.8 Principal mechanisms of ion transport in plasma membranes. (A) H^+ pumping ATPase; (B) ion channel; (C) carrier; (D) coupling proteins for signal perception and transduction. (Modified from Hedrich *et al.*, 1986; with permission from Trends in Biochemical Sciences.)

into account another important membrane function, namely that of a receptor of external and internal signals (e.g. pH, ion concentration) and the transformation of these signals into membrane transport processes.

There is also general agreement that in plants proton pumps are located in both plasma membrane and tonoplast, and that their primary function is the regulation of the pH in the cytoplasm (Kurdjian and Guern, 1989). These pumps transport H^+ from the cytoplasm either through the plasma membrane into the apoplasm or through the tonoplast into the vacuole leading to the typical pH differences between these compartments. As a rule, the pH of the cytoplasm (cytosol) is 7.3–7.6, the vacuole 4.5–5.9 and the apoplasm about 5.5 (Kurdjian and Guern, 1989). Accordingly, in plant cells extrusion of protons from the cytoplasm across the plasma membrane and tonoplast is the primary energized process and provides the driving force for the secondary energized transport process of cations and anions along the electrochemical gradient. The functioning and location of proton pumps and of transport of cations and anions across the plasma membrane and the tonoplast are summarized in a model in Fig. 2.9. The membrane transport of amino acids and sugars follows the same principles, i.e., it is driven by membrane-bound proton pumps (Section 5.4.1).

In the model shown in Fig. 2.9 the plasma membrane-bound H^+-ATPase plays a key role in both the regulation of cytoplasmic pH and the driving force for cation and anion uptake. It is therefore considered as the 'master enzyme' (Serrano, 1990). Considerable progress had been made in understanding how the activity of this enzyme is regulated, also at a gene level (Palmgren, 1991). The activity of the plasma membrane H^+-ATPase is particularly high in root hairs where it consumes as much as 25–50% of the cellular ATP (Felle, 1982). Accordingly, low carbohydrate levels in roots not only decrease the

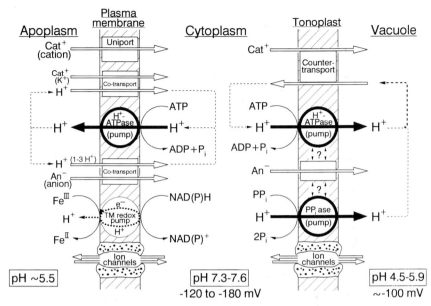

Fig. 2.9 Model for the functioning and location of electrogenic proton pumps (H$^+$-ATPase, PP$_i$ase), transmembrane redox pump (NAD(P)oxidase), ion channels, and transport of cations and anions across the plasma membrane and tonoplast.

net extrusion of protons into the external medium but also decrease the cytosolic pH (Schubert and Mengel, 1986). The plasma membrane H$^+$-ATPase is stimulated by monovalent cations in the order K$^+$ > Na$^+$ and is relatively insensitive to anions (O'Neill and Spanswick, 1984). An example of the stimulation by K$^+$ has been given in Fig. 2.6. Calcium in general and the cytosolic free Ca^{2+} concentration in particular play a key role in the stimulation of this enzyme. An increase in cytosolic free Ca^{2+} strongly increases the activity of this plasma membrane-bound enzyme, maximal stimulation occurring even at values lower than 7 μM free Ca^{2+} (Schaller and Sussman, 1988). Activation is presumably achieved by effects of free Ca^{2+} on a protein kinase associated with the internal surface of the plasma membrane (Section 8.6.7).

Cations are transported 'downhill' along the electrical potential gradient (about -120 to -180 mV) across the plasma membrane into the cytoplasm in a *uniport*, mediated by specific structures (carriers, permeases) in the membrane (Fig. 2.9). For potassium (K$^+$), however, at low external concentrations (<1 mM) the high affinity uptake system operates against the prevailing electrochemical potential difference (e.g. 10 μM K$^+$ outside; ~80 mM K$^+$ activity in the cytoplasm). Thus, an energized K$^+$ transport is required either as carrier-mediated K$^+$-proton antiport (counter-transport) or, more likely as an 1:1 K$^+$-*proton symport*, or *cotransport* (Kochian et al., 1989; Maathuis and Sanders, 1993).

For anion transport across the plasma membrane a *proton–anion cotransport* (or *symport*) operates, using the steep electrical (potential difference) and chemical (pH difference) gradient for protons as the driving force. Evidence for proton–anion

Table 2.9
Some Characteristics of the Proton Pumps in the Tonoplast

	ATPase	PP$_i$ase
H$^+$ pumping activity[a] (mol H$^+$ m^{-2} s^{-1})	214	95
Activity affected[b]		
Stimulated by	Mg^{2+}, Cl$^-$	Mg^{2+}, K$^+$, NO$_3^-$
Not affected or inhibited by	K$^+$, NO$_3^-$	Cl$^-$

[a]From Hoffmann and Bentrup (1989).
[b]Compiled data from Bennet et al. (1984); Marquardt and Lüttge (1987) and Pugliarello et al. (1991).

cotransport at the plasma membrane of root cells has been presented for chloride (Ullrich and Novacky, 1990), phosphate (Dunlop, 1989) and nitrate (McClure et al., 1990a), as well as for amino acids (Petzold et al., 1989). The stoichiometry of this cotransport is not yet clear; more than one proton may be transported per negative charge of the anion (Lass and Ullrich-Eberius, 1984; McClure et al., 1990b).

There are also other views on the mechanisms of anion transport across the plasma membrane. According to Liu (1979), phosphate uptake by corn roots is mediated by OH$^-$/phosphate countertransport in which the downhill transport of OH$^-$ from the high electrochemical potential in the cytoplasm (high pH and strong negative charge) into the apoplasm is coupled with a countertransport of phosphate anions into the cytoplasm. However, experimental evidence for exchange or transport of OH$^-$ or HCO$_3^-$, across membranes as driving force for anion transport is weak, as specific inhibitors of OH$^-$ and HCO$_3^-$ transport are not available (Kurdjian and Guern, 1989).

At the tonoplast the existence of two functionally and physically distinct proton pumps is now established, an H$^+$-ATPase and an inorganic pyrophosphatase, PP$_i$ase (Fig. 2.9). Both proton pumps are phosphohydrolases using either ATP or inorganic pyrophosphate as energy source (Section 8.4). Magnesium is essential for both pumps, indicating that Mg · ATP and Mg · PP$_i$ are the substrates. Inorganic pyrophosphate is generated in several major biosynthetic pathways (Rea and Sanders, 1987) such as starch synthesis (Section 8.4) or activation of sulfate (Section 8.3.2). The concentration of PP$_i$ in the cytosol is assumed to be in the range of 50–390 μM which is adequate to drive this proton pump (Chanson et al., 1985).

The relative contribution of the PP$_i$ase to the total proton pumping activity at the tonoplast seems to be lower than that of the ATPase, and both pumps are quite differently affected by inorganic cations and anions (Table 2.9). Except for the essentiality of Mg^{2+} for both pumps, the Mg · ATPase is stimulated by chloride and inhibited (or not affected) by potassium and nitrate whereas the reverse is true for the Mg · PP$_i$ase. The H$^+$ pumping Mg · PP$_i$ase exhibits an obligatory dependency on the presence of K$^+$ and the K$^+$ activation occurs at the cytoplasmic phase (Davies et al., 1991b). The implications of this difference in capacity and sensitivity of both pumps for ion accumulation in vacuoles or radial transport of ions across the roots (Section 2.7) are not clear. Different proportions of activities of both pumps in the tonoplast of roots may well be involved in the genotypical differences between plants in chloride translocation into the shoots (Chapter 16).

The proton pumps at the tonoplast are required for maintenance of a high cytosolic pH and simultaneously provide the driving force for cation transport into the vacuole as *countertransport* (or *antiport*, Fig. 2.9). This countertransport is not only important, for example, for turgor regulation (high vacuolar K^+ concentrations, Section 8.7) but also for the maintenance of low cytosolic concentrations of sodium (Garbarino and DuPont, 1989; Section 16.6.5) and calcium (Chanson, 1991; Section 8.6). For anions the situation is less clear. Whereas in leaf cells of plants with crassulacean acid metabolism (CAM, Section 5.2.4) a stoichiometric proton–malate anion cotransport into the vacuoles has been demonstrated, for roots the evidence for coupling of such a proton–anion transport is weak. Transport of anions from the cytoplasm into the vacuole may follow the gradient in electropotential which is less negative in the vacuole as compared with the cytoplasm (Fig. 2.9).

Whereas the existence of the two proton pumps at the tonoplast (ATPase; PP_iase) and of the H^+-ATPase at the plasma membrane is well established, it is still a matter of controversy whether a second system of proton translocation across the plasma membrane from the cytoplasm to the apoplasm is involved in the formation and maintenance of the transmembrane electropotential and pH (Fig. 2.9). This second system is linked to a redox chain with NAD(P)H as electron donor. There is good evidence for the involvement of this system in auxin-induced enhanced growth (Morré et al., 1988), in antimicrobial activity (Dahse et al., 1989), or in the reduction of Fe(III) at the surface of the root plasma membranes (Bienfait and Lüttge, 1988). However, so far mainly ferricyanide or other artificial compounds have to be used as electron acceptors to achieve a substantial transmembrane proton transport capacity (Luster and Buckhout, 1988). The role of this redox system for transmembrane proton gradients and ion transport remains questionable until a physiological electron acceptor can be found (Serrano, 1989). In view of the effectivity of ascorbate (ascorbic free radical) as electron acceptor for the transmembrane redox pump (Gonzales-Reyes et al., 1992) and the relatively high ascorbate concentrations in the apoplasm of leaves (Polle et al., 1990) a role of this system in leaves in plasma membrane transport of ions and other solutes cannot be excluded (Chapter 3).

More recently, the existence of *ion channels* also in plant cell membranes has been established and, thus, channels are included in current models of membrane transport of ions (Figs. 2.8 and 2.9). Ion channels are unique among transport proteins in their ability to regulate or 'gate' ion flux subject to the physical–chemical environment of the channel protein (Blatt and Thiel, 1993). These channels permit rapid passive permeation (uniport) of ions through membranes. Open channels catalyze the permeation of 10^6 to 10^8 ions per second which is at least three (Tester, 1990) or even five (Bentrup, 1989) orders of magnitude faster than carrier-mediated transport of ions. However, ion channels are closed most of the time, and their number per cell seems to be rather small. For example, in the plasma membrane of leaf cells about 200 K^+ channels per cell are assumed (Kourie and Goldsmith, 1992). So far, specific channels for K^+, Ca^{2+}, H^+ and Cl^- have been identified, and for NO_3^- a channel is postulated at the tonoplast (Tyerman, 1992).

There are many assumptions about the function of these ion channels in plant cells.

They are important for osmoregulation, for example in guard cells of leaves (Schauf and Wilson, 1987) and in seismonastic and nyctinastic movements of leaves (Schroeder and Hedrich, 1989), i.e., in processes where rapid transport of low molecular solutes like K^+ or Cl^- between cell compartments is required as a response to environmental signals (Section 8.7). Selective Ca^{2+} channels in the plasma membrane and tonoplast leading to rapid increase in cytosolic free Ca^{2+} concentration are considered of key importance for signal transduction and functioning of Ca^{2+} as a secondary messenger in the cytoplasm by modulating enzyme activities (Briskin, 1990; Section 8.6.7).

Besides these specific functions of ion channels their role in ion uptake, for example by roots, is not very clear. For the uptake of divalent cations in general and Ca^{2+} in particular, opening of the channels facilitates rapid influx into the cytoplasm of root cells. For K^+ uptake an inward rectifying channel is proposed which opens upon hyperpolarization of the plasma membrane and facilitates the K^+ influx in presence of high external concentrations (>1 mм K^+; Maathuis and Sanders, 1993; White, 1993). Another high conductance channel in the plasma membrane of root cells is permeable for both monovalent and divalent cations, opens upon depolarization (i.e. drop in membrane potential) and permits rapid influx of cations such as Ca^{2+} but facilitates rapid efflux of K^+ along the electrochemical potential gradient (outward rectifying K^+ channel) at low (<1 mм) external K^+ concentrations (White, 1993). Thus, cation channels in the plasma membrane of root cells presumably play an important part in uptake of divalent cations even at low external concentrations but for monovalent cations and K^+ in particular only at high external concentrations.

Although channels in membranes allow rapid, passive fluxes of solutes, the dimensions of these channels are not suited for permeation of macromolecules. Nevertheless, macromolecules like proteins (e.g. insulin; Horn *et al.*, 1990) or ferritin particles (Joachim and Robinson, 1984; Section 8.4.5) are also taken up by plant cells. Most probably, *endocytosis* (*pinocytosis*) is the responsible mechanism in which plasma membrane vesicles mediate the permeation. Interestingly, the uptake of insulin in plant cells via endocytosis requires a coupling of the protein with the vitamin biotin (Horn *et al.*, 1990). This capacity for uptake of macromolecules reflects the dynamic properties of the membrane structures (Section 2.4.1). The importance of this uptake capacity in plants should not be overestimated, however, as the pores in the cell walls strongly restrict the permeation of macromolecules (Section 2.2.1).

2.4.3 The Kinetics of Transport

As a rule ion uptake by plant cells and roots has features of a saturation kinetics. This is in accordance with the assumption of control, as for example by the number of binding sites of ions (carrier, permeases), or the capacity of the proton efflux pumps, in the plasma membrane and tonoplast (Section 2.4.2). The pioneering work of Emanuel Epstein and his group in the early 1950s contributed fundamentally to the better understanding of ion uptake and its regulation in plants by regarding the kinetics of ion transport through membranes of plant cells as formally equivalent to the relationship between an enzyme and its substrate, using terms of enzymology (Fig. 2.10). Comparing a carrier to an enzyme molecule and the ion to the substrate for the enzyme, the transport rate of an ion is dependent on the following two factors:

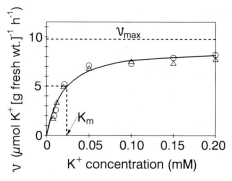

Fig. 2.10 Rate of K^+ uptake (v) as a function of the external concentration of KCl (○) or K_2SO_4 (△); $K_m = 0.023$ mM. (After Epstein, 1972.)

V_{max} A capacity factor denoting the maximal transport rate when all available carrier sites are loaded, that is, the maximal transport rate.

K_m The Michaelis constant, equal to the substrate ion concentration giving half the maximal transport rate.

When the concentration range is low, the uptake is often well described by the Michaelis–Menten kinetics, as shown in an example in Fig. 2.10 for potassium uptake by barley roots. It is evident from this figure that the uptake isotherm of potassium is the same whether the source of potassium is KCl or K_2SO_4. As we shall see later, however, when the substrate concentrations are higher, the accompanying anion has an effect on the uptake rate of the cation and vice versa. The K_m value reflects the affinity of the carrier sites for the ion, just as in enzymatic reactions where it indicates the affinity of the enzyme for the substrate.

As a first approximation in the low concentration range Michaelis–Menten kinetics may also be applied to describe uptake rates of sulfate (Deane-Drummond, 1987) and phosphate (Seeling and Claassen, 1990). As summarized by Jensen *et al.* (1987), however, and shown in a few examples cited below, the formal application of the Michaelis–Menten kinetics is too restrictive for theoretical reasons (Section 2.4.2) and often not in accordance with experimental results.

The original concept of a single carrier-mediated mechanism of ion transport (one carrier system for each ion) did not sufficiently describe the kinetics of uptake when wide concentration ranges were tested. At K^+ concentrations above 1 mM, for example, the kinetics differ considerably from those at lower concentrations (Epstein *et al.*, 1963). The selectivity of the binding sites is lower (Na^+ competes with K^+) and the accompanying anion has an effect on the uptake rate (Section 2.5.5). These differences led to the hypothesis of a *dual system*, System I with higher selectivity and System II with lower selectivity, either located at the same or at different membranes (i.e. plasma membrane, tonoplast, respectively). For further details see Epstein (1972).

Deviations in the uptake isotherm, when considered over a wide concentration range, especially at high external concentrations were interpreted as indicative of 'multiphasic' carrier systems (Nissen, 1991) or as allosteric inhibition (negative feedback regulation) of the carrier sites at high cellular concentrations (Glass, 1983). In

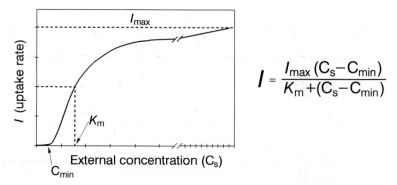

$$I = \frac{I_{max}(C_s - C_{min})}{K_m + (C_s - C_{min})}$$

Fig. 2.11   Schematic presentation of the relationships between uptake rates (net influx = I) of ions and their external concentrations; C_{min} = net uptake zero (influx = efflux).

view of the usually very low concentrations particularly of phosphorus and potassium in soil solutions (Section 13.2.2), and results of ion uptake studies in the low concentration range ($<10 \mu M$) the term (C_{min}) was introduced defining the concentration at which net uptake of ions ceases before the ions are completely depleted (Fig. 2.11). The C_{min} concentration is an important factor in ion uptake from soils, because it is the lowest concentration at which roots can extract an ion from the soil solution (Section 13.2). C_{min} concentrations differ considerably between plant species (Asher, 1978). For phosphorus, for example, a value of 0.12 μM has been found in tomato (Itoh and Barber, 1983a), 0.04 μM in soybean (Silberbush and Barber, 1984) and 0.01 μM in ryegrass (Breeze et al., 1984). For potassium, the corresponding values were 2 μM in maize (Barber, 1979) and 1 μM in barley (Drew et al., 1984). C_{min} concentrations for nitrate may vary between more than 50 μM and less than 1 μM depending not only on the plant species but also the environmental conditions (Deane-Drummond and Chaffey, 1985; Marschner et al., 1991); for ammonium the C_{min} concentrations decrease from 30 to 1.5 μM as the root zone temperature increases (Marschner et al., 1991).

In order to interpret uptake kinetics of ions by plant roots, using the Michaelis–Menten equation not only the term C_{min} has been introduced but also the term I for influx which replaces velocity (v) (Fig. 2.11). Very often though only the *net uptake* of ions is determined which is the resultant of influx and *efflux*. Particularly at low external concentrations efflux can become similar in magnitude to that of influx and thus be an important component in determining net uptake as shown in Table 2.10 for phosphorus. At a given external concentration, the efflux of phosphorus is also 5–8 times higher in roots of phosphorus sufficient as compared with phosphorus-deficient plants (McPharlin and Bieleski, 1989). In agreement with this, the C_{min} concentrations are 0.34 μM in the phosphorus-sufficient and 0.08 μM in the phosphorus-deficient plants.

Efflux of ions and other solutes is not only affected by the transmembrane electro-potential and the integrity of the plasma membrane but also by the concentration and activity of the respective ions in the cytoplasm. In pea, for example, the high net uptake of sulfate in sulfur-deficient roots drops to about 30% within 1 h due to a marked increase in sulfate efflux, despite of an even slight increase in influx (Deane-Drummond, 1987). For nitrate and ammonium too the efflux component may account

Table 2.10

Effect of low Phosphorus Concentrations on Influx and Efflux of Phosphorus in Maize Roots[a]

P concentration supplied (μM)	P flux (nmol P g^{-1} root fresh wt min^{-1})		Efflux (%)
	Influx	Efflux	
0.2	0.21	0.15	71
2.0	4.40	0.32	7

[a]Elliott et al. (1984).

Table 2.11

Influence of the Phosphorus Nutritional Status of Soybean Plants on Short-term Uptake Parameters of Phosphorus[a]

Plants grown at P concentration (μM)	P content (% dry wt)		I_{max} (mol cm^{-1} s^{-1} $\times 10^{-14}$)	K_m (μM)
	Shoot	Root		
0.03	0.22	0.23	17.6	1.6
0.3	0.34	0.30	16.9	1.7
3.0	0.59	0.56	6.5	1.2
30.0	0.66	0.90	3.7	1.0

[a]Based on Jungk et al. (1990).

for a high proportion – almost 40–50% of the influx – most probably relating to the high concentrations of nitrate and ammonium in the cytoplasm (Lee and Clarkson, 1986; Jackson et al., 1993). The rapid exchange between ions in the external solution and in the cytoplasm is reflected in the half-time for exchange ($t_{1/2}$) which is for sulfate in the range of 10–20 min (Deane-Drummond, 1987) and for nitrate between 4 min (Lee and Clarkson, 1986) and 107 min (Macklon et al., 1990). These rates of exchange with the cytoplasmic pool are usually orders of magnitude higher than the rates of exchange with the ions in the vacuole (e.g. about 700 h for nitrate; Macklon et al., 1990). Because of both the high exchange rates and the small volume of the cytoplasmic compartment (~5% of the total cell volume in differentiated cells), the role of efflux for the net uptake can only be measured in short-term studies, usually with radioisotopes (e.g. ^{13}N; ^{32}P). In view of the current models on structure and functioning of the plasma membrane (Fig. 2.9), the relatively high efflux rates of ions may be related to either ion channels or proton-mediated transport from the cytoplasm into the apoplasm by cotransport (anions) or countertransport (cations).

The parameters of ion-uptake kinetics are also strongly affected by the nutritional status of the plants. This holds true not only for C_{min} but also for K_m and particularly I_{max}. An example for this is given in Table 2.11 for phosphorus. With increasing phosphorus content in the plants, K_m slightly decreases and I_{max} rapidly decreases, indicating an effective feed-back regulation. As I_{max} values were based on net uptake in this experiment, the contribution of increased efflux with higher internal phosphorus concentrations cannot be evaluated. At least for nitrate, however, measurements of influx clearly indicate that besides efflux other mechanisms contribute to the decline in net uptake when internal concentrations are high. In barley roots with increasing internal nitrate concentrations both I_{max} and K_m decrease by a factor of 4–5, indicating an effective feedback regulation of the influx component (Siddiqi *et al.*, 1990).

Evaluations of the kinetic parameters of nitrate uptake are complicated by the fact that there are obviously two uptake and transport systems located in the plasma membrane.One is constitutive, with a low capacity system, and the other is a system that is inducible by nitrate and which has both, a higher affinity (lower K_m) and higher transport capacity (higher I_{max}) for nitrate (Behl *et al.*, 1988; Siddiqi *et al.*, 1990). Accordingly, both, K_m and I_{max} are quite different in noninduced as compared with induced plants (Wieneke, 1992). Inhibitors of protein synthesis strongly depress the formation of this inducible transport system (Wieneke, 1992) in which arginine groups seem to play an essential role in binding or transport of nitrate or both (Ni and Beevers, 1990). Some of the implications of the feedback regulation for mineral nutrient uptake are discussed in Section 2.5.6.

In the high concentration range a linear relationship is often found between external concentrations (>1mM) and influx rate of ions, as for example rubidium (Kochian and Lucas, 1982) and nitrate (Siddiqi *et al.*, 1990). It is quite likely that the linear relationships (formerly also defined as System II, Epstein, 1972) are reflections of passive ion fluxes in ion channels through the plasma membrane along the gradient in ion concentrations or activities. In view of the usually low ion concentrations in soil solutions, the ecological importance of these mechanisms operating in the high concentration range for the mineral nutrition of soil-grown plants may be questioned for potassium, but not for divalent cations (Section 2.4.2). Non-energized (passive) influx of ions through channels in the plasma membrane certainly plays an important role in saline soils (Section 16.6), in mineral element transport in plants, especially xylem loading in the roots (Section 2.8) and xylem unloading, i.e. uptake in cells along the pathway and in leaf cells (Section 3.2).

2.5 Characteristics of Ion Uptake by Roots

2.5.1 Influx into the Apoplasm

Before reaching the plasma membrane of root cells ions have to pass through the cell walls. In general, movement of ions and other low-molecular-weight solutes by diffusion or mass flow is not restricted to the external surface of the roots, that is the rhizodermal cells (Fig. 2.1). The cell walls and water-filled intercellular spaces of the root cortex are also, at least to a certain extent, accessible to these solutes from the external solution.

The main barrier to solute flux in the apoplasm of roots is the endodermis, the innermost layer of cells of the cortex (Fig. 2.12). In the radial and transverse walls of the endodermis, hydrophobic incrustations (suberin) – the Casparian band – constitute an

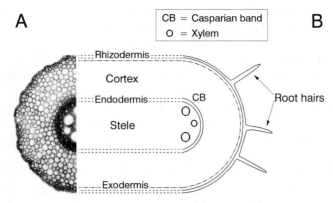

Fig. 2.12 A. Cross section of a differentiated root zone of maize. B. Schematic presentation of cross section.

effective barrier against passive solute movement into the stele. In most species of the angiosperms suberin lamellae are also found in the *hypodermis*, or *exodermis* (cell layer below rhizodermis; Peterson, 1988; Enstone and Peterson, 1992). The exodermis (Fig. 2.12) may also function as a barrier to protect the inner cortex from colonization by microorganisms, for example of sorghum roots by the endophyte *Polymyxa* sp. (Galamay *et al.*, 1992).

Compared with the Casparian band the formation of this *exodermis* along the root axis is delayed, particularly in fast-growing roots (Peterson, 1988). There are different views on the effectiveness of the exodermis as a barrier against passive solute flux into the root apoplasm (Clarkson *et al.*, 1987; Peterson, 1988; Section 2.7). In plants adapted to submerged conditions the exodermis serves another function, namely as an effective barrier against oxygen diffusion (leakage) from the root aerenchyma into the rhizosphere (Section 14.3).

The volume of root tissue accessible for passive solute flux – the *free space* – represents only a small fraction ~5% in maize (Shone and Flood, 1985) of the total root volume. The presence of this free space enables individual cortex cells to take up solutes directly from the external solution. For a given volume of the free space the extent of solute flux into the free space depends on various factors such as the rate of transpiration, solute concentration and root hair formation. There has been a tendency to overestimate the importance of the free space for ion uptake by roots. As shown more than 25 years ago by Vakhmistrov (1967), at low external concentration root hair formation is usually extensive and uptake, of potassium and phosphorus for example, is limited mainly to the rhizodermal cell layer. This is particularly true for roots growing in soil (see Section 2.10). More recently, also in solution culture the particular role of the rhizodermal cells for uptake of nitrate (Deane-Drummond and Gates, 1987) and sulfate (Holobrada and Kubica, 1988) has been demonstrated and also the predominant location of plasma membrane-bound H^+ pumping ATPases in the rhizodermal cells of roots (Felle, 1982; Parets-Soler *et al.*, 1990). Other examples for the key role of rhizodermal cells in ion uptake are given in Section 2.7.

Table 2.12
Relationship Between Ion Radius and Uptake of Alkali
Cations in Barley Roots[a]

Cation	Ion radius[b] (nm)	Uptake rate (μmol g^{-1} fresh wt per 3 h)
Lithium	0.38	2
Sodium	0.36	15
Potassium	0.33	26
Cesium	0.31	12

[a]Cations were supplied at pH 6.0 as bromide salts, 5 mM. Based on
Jacobson *et al.* (1960).
[b]Hydrated ion; data from Conway (1981).

2.5.2 Role of Physicochemical Properties of Ions and Root Metabolism

Despite the interactions of ions with charged groups in the cell walls of the apoplasm, the characteristics of ion uptake by roots are determined primarily by the transport across membranes, and the plasma membrane in particular (Section 2.3). There are some physicochemical properties of ions and other solutes, e.g. ion diameter and valency, which to a considerable extent determine their rates of membrane transport, despite the dynamic properties of membranes.

2.5.2.1 Ion Diameter

For ions with the same valency there is often a negative correlation between the uptake rate and the ion radius. An example of this for monovalent cations is given in Table 2.12. The inverse relationship between ion radius and uptake rate is observed only when Li^+, Na^+, and K^+ are compared. Despite its smaller diameter, Cs^+ is taken up at a much lower rate than K^+. Obviously, factors other than ion diameter are involved in the regulation of uptake; one of these is the affinity of membrane-bound carriers, or channels, for ions of a given valency.

2.5.2.2 Molecule versus Ion Uptake and the Role of Valency

Membrane constituents, particularly phospho- and sulfolipids and proteins, contain electrically charged groups, and ions interact with these groups. As a general rule, the strength of this interaction increases in the following order:

$$\text{Uncharged molecules} < Cat^+, An^- < Cat^{2+}, An^{2-} < Cat^{3+}, An^{3-}$$

Conversely, the uptake rate often decreases in this order. The increase in the diameter of a hydrated ion with valency is certainly an additional factor responsible for this order. A few examples will illustrate this general pattern.

As shown in Fig. 2.13 the uptake rate of boron falls dramatically when the external pH is increased. This pattern of behaviour is closely related to the shift in the ratio of boric acid to borate anion. This is in complete agreement with the results obtained with

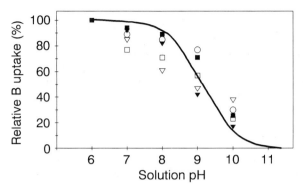

Fig. 2.13 Relative uptake of boron by barley roots as a function of the external solution pH. Uptake at pH 6 = 100 at each supply concentration. Solid line: percentage of undissociated H_3BO_3. Key for boron concentrations mg l^{-1}: $\triangledown$, 1.0; $\square$, 2.5; $\bigcirc$, 5.0; $\blacktriangledown$, 7.5; $\blacksquare$, 10.0. (Reproduced from Oertli and Grgurevic, 1975, by permission of the American Society of Agronomy.)

other weak acids such as acetic acid (Toulon *et al.*, 1989). The ratio of dissociated to undissociated species of acids – and thus the solution pH – also determines the membrane permeation of certain phytohormones such as abscisic acid (ABA); only the undissociated species readily permeates cell membranes, a factor which is closely related to the action of ABA in cells (Kaiser and Hartung, 1981; Cowan *et al.*, 1982).

In the case of weak acids, the high membrane permeation of the uncharged species is certainly not merely the result of less interaction with the charged groups in the membranes. In addition, downhill transport might be involved along the pH gradient across the plasma membrane (Fig. 2.9), which also reflects a corresponding chemical gradient toward the cytoplasm for the uncharged species [e.g., H_3BO_3 (outside) $\rightarrow$ $H_2BO_3^-$ (cytoplasm)].

A special case of pH-dependent uptake of a molecular species is observed when ammonium nitrogen is supplied (Section 2.5.3). At high external pH the uptake increases sharply, probably owing to an increase in the proportion of the molecular species (NH_3 and NH_4OH).

In the case of phosphorus in the pH range of 5–8 two ionic species occur together

$$H_2PO_4^- \underset{+H^+}{\overset{-H^+}{\rightleftharpoons}} HPO_4^{2-}$$

Consequently, at low pH, $H_2PO_4^-$ dominates, whereas the reverse is true at high pH. As shown in Fig. 2.14, in the pH range of 8.5–5.6 there is a striking positive correlation between the proportion of $H_2PO_4^-$ in the external solution and the uptake rate of phosphate. In contrast, there is a much smaller effect on the uptake rate of sulfate, since in this pH range only the divalent anion SO_4^{2-} occurs.

The increase in phosphate uptake, however, cannot be explained solely in terms of the fall in pH from 8.5 to 5.5 with the associated shift in valency. As we shall see later (Section 2.5.3) pH-induced changes in proton–anion cotransport are presumably the main factor responsible for this enhanced anion uptake.

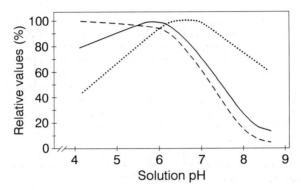

Fig. 2.14 Relationship of solution pH, proportion of $H_2PO_4^-$ (– – –), and uptake rate of phosphate (——) and sulfate (·····) by bean plants; relative values. (After Hendrix, 1967.)

2.5.2.3 *Metabolic Activity*

In order for ions and other solutes to accumulate against a concentration gradient, an expenditure of energy is required, either directly or indirectly. The main source of energy in nonphotosynthesizing cells and tissues (including roots) is respiration. Thus, all factors which affect respiration may also influence ion accumulation. The few examples below demonstrate this connection.

Oxygen. As oxygen tension decreases, the uptake of ions such as potassium and phosphate falls, particularly at very low oxygen tensions (Table 2.13). Consequently, oxygen deficiency is one of the factors which may restrict plant growth in poorly aerated substrates (e.g., waterlogged soils; Section 16.4).

Carbohydrates: The main energy substrate for respiration are carbohydrates. Therefore, in roots and other nonphotosynthesizing tissues, under conditions of limited carbohydrate supply from a *source* (e.g. leaves) a close correlation can often be found between carbohydrate content and the uptake of ions, e.g. potassium (Mengel, 1962). In roots within a few hours after excision with depletion of carbohydrate content

Table 2.13

Effect of Oxygen Partial Pressure around Roots on Uptake of Potassium and Phosphate by Barley Plants[a]

Oxygen partial pressure (%)	Uptake[b]	
	Potassium	Phosphate
20	100	100
5	75	56
0.5	37	30

[a]Based on Hopkins *et al.* (1950).
[b]Data represent relative values.

Table 2.14
Effect of Root Excision on Sugar Content, Respiration (O_2 Uptake),
and Nitrogen Uptake in Barley Roots[a]

Time (h) after excision	Sugar content (μmol g^{-1} dry wt)	Net uptake (μmol g^{-1} dry wt min^{-1})		
		O_2	NH_4^+	NO_3^-
0	82	4.5	1.8	1.5
3	51	3.3	1.1	1.0

[a]Recalculated from Bloom and Caldwell (1988).

respiration and nitrogen uptake also decrease (Table 2.14). These relationships are of particular ecological importance, for example, when leaves are removed (grazing, cutting) or in dense plant stands when light supply to the basal leaves is limited, since basal leaves are the main source of carbohydrates for the roots. Root carbohydrate content may also become a growth-limiting factor at high supply of ammonium nitrogen in combination with high root zone temperatures (Kafkafi, 1990; see also Section 8.2.4).

Diurnal fluctuations in net proton excretion (Vogt *et al.*, 1987) and in ion uptake rate by roots may also be at least in part causally related to corresponding fluctuations in carbohydrate supply to the roots. Distinct diurnal patterns in uptake rates (maxima during the day, minima during the night) can be observed for nitrate (Clement *et al.*, 1978b) and nitrate and potassium (Le Bot and Kirkby, 1992). Root carbohydrate content may act as a coarse control for ion uptake and is one of the factors responsible for the diurnal fluctuations in ion uptake. However, in maize roots, for example, diurnal fluctuations in nitrate uptake were only loosely related to the root carbohydrate content (Fig. 2.15). In contrast, the relations were close between root carbohydrate content and nitrate reductase activity (NRA).

There is good evidence that the root carbohydrate content as a regulating factor for diurnal fluctuations in ion uptake is superimposed by internal factors such as nutrient demand or retranslocation of nutrients or both (Section 2.5.6). In soybean growing under short-day conditions, the typical diurnal fluctuations of nitrate uptake could be reversed by an intervening 3 h period of low light (i.e. imitating long-day conditions, repressed flower initiation); uptake rates of nitrate were then twice as high during the night as compared with the day (Raper *et al.*, 1991).

Temperature. Whereas physical processes such as exchange adsorption of cations in the AFS are only slightly affected by temperature (Q_{10}[*] $\approx$ 1.1–1.2), chemical reactions are much more temperature dependent. An increase in temperature of 10°C usually enhances chemical reactions by a factor of 2 ($Q_{10} \approx 2$). For biochemical reactions, Q_{10} values considerably higher than 2 are quite often observed. Also, for the uptake of ions such as potassium, Q_{10} is often much higher than 2, at least within the physiological

[*]Q_{10}, or quotient $_{10}$ refers to the change in the rate of a reaction or process (i.e., rate of membrane transport) imposed by a change in the temperature of 10°C.

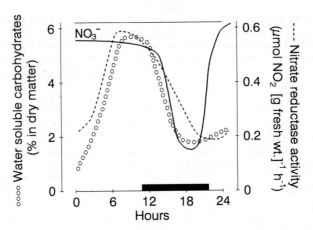

Fig. 2.15 Diurnal fluctuations in nitrate uptake (——), nitrate reductase activity and content of water soluble carbohydrates in maize roots. Nitrate uptake: relative values, uptake at the end of the light period = 100. (Based on Keltjens and Nijenstein, 1987; by courtesy of Marcel Dekker Inc.)

temperature range (Fig. 2.16). A comparison of the Q_{10} values for ion uptake and respiration reveals that ion uptake is more temperature dependent, especially below 10°C.

This may possibly indicate that in chilling-sensitive plant species like maize, restricted ion uptake at low temperatures is primarily the result of low membrane fluidity and strongly depressed activity of the plasma membrane-bound proton pump (Kennedy and Gonsalves, 1988). At supraoptimal temperatures root respiration further increases whereas ion uptake declines (Fig. 2.16), indicating again that respiration and membrane transport of ions are not directly coupled.

In studies on temperature effects on ion uptake two major problems arise: short-term

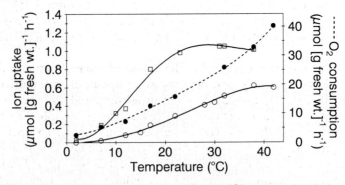

Fig. 2.16 Effect of temperature on rates of respiration (●) and uptake of phosphorus (○) and potassium (□) (supply of 0.25 mM potassium and 0.25 mM phosphorus) by maize root segments. (After Bravo and Uribe, 1981.)

Table 2.15

Long-term Effects of the Root Zone Temperature (RZT) and of the Temperature at the Stem Base (Shoot Growing Zone Temperature, SGT) on Shoot and Root Growth and Uptake of Nitrate and Potassium by Maize Plants; Temperature Treatment for Eight Days[a]

	Temperature treatment		(SGT°C/RZT°C)
	24/24	12/12	24/12
Shoot growth (g fresh wt day^{-1})	1.91	0.32	1.34
Root growth (g fresh wt day^{-1})	0.85	0.20	0.26
Nitrate uptake (pmol g^{-1} fresh wt h^{-1})	6.4	4.2	7.6
Potassium uptake (pmol g^{-1} fresh wt h^{-1})	2.5	1.2	3.1

[a]After Engels and Marschner (1992b).

effects of root zone temperature (e.g. decreases from 20°C to 10°C, and vice versa, within minutes or less than 1 h) reflect direct temperature effects on the uptake system, but are of limited ecological relevance. On the other hand, long-term effects of root zone temperature (e.g., growth for several days or weeks at different root zone temperatures) are of more ecological significance as they include root adaptation, for example changes in membrane properties (Section 2.3). However, in the long-term, different root zone temperatures may affect root and shoot growth quite differently and, thus, the root–shoot dry weight ratio (Clarkson *et al.*, 1986, 1992; see also Section 14.7). Accordingly, long-term effects of different root zone temperatures on ion uptake include feedback regulations via demand, an example for this is shown for maize in Table 2.15.

As is to be expected for a chilling sensitive plant species, low root zone temperatures (12°C) decrease shoot and root growth and uptake rates of nitrate and potassium. However, the depression in ion uptake at low temperature was not a temperature effect on the roots per se, but reflection of a feedback regulation via lower demand. This was shown by lifting the shoot growing zone (stem base) out of the cooling zone (24/12°C). Shoot growth was greatly increased (i.e. the demand for nutrients) and so were the uptake rates of nitrate and potassium per unit root weight (Table 2.15). In chilling-tolerant graminaceous species, as a rule, poor growth at low root zone temperatures, is also obviously not due to limited uptake of nutrients such as nitrogen (Clarkson *et al.*, 1992).

Low root zone temperatures may affect the uptake of mineral nutrients differently, phosphorus uptake being usually more depressed than uptake of other nutrients (Engels and Marschner, 1992a). The uptake rate of nitrate as compared with ammonium seems to be particularly depressed at low root zone temperatures in chilling sensitive plant species like cucumber (Tachibana, 1987) whereas in chilling-tolerant species (barley, ryegrass) the high preference for ammonium compared with nitrate uptake is not much affected by the root zone temperature (MacDuff and Jackson, 1991; Clarkson *et al.*, 1992). Compared with calcium and magnesium, uptake rates of potassium are often more affected by root zone temperatures. In winter wheat a corresponding increase in K/(Ca + Mg) ratios in the shoots with increasing root zone temperature is considered as a causal factor in tetany for grazing beef cattle on winter wheat forage (Miyasaka and Grunes, 1990).

In contrast to plants grown in solution culture, in soil-grown plants spatial availability of mineral nutrients is also of great importance for mineral nutrient uptake (Engels and Marschner, 1990). Root zone temperature may thus affect mineral nutrient uptake primarily via modulation of root growth and morphology (Section 14.3).

2.5.3   Interactions between Ions

In the preceding sections, for the sake of simplicity the transport of a particular ion was treated as a singular process, regulated only by the physicochemical properties of the ion and the metabolic activities of individual cells and roots. In reality, however, in the external solution (soil or nutrient solution) both cations and anions are present in different concentrations and forms. Various interactions between ions during their uptake are therefore to be expected. Some of these general interactions are discussed in this section.

2.5.3.1   Competition

For ion transfer from the external solution to the cytoplasm binding at transport sites (e.g. carriers) in the plasma membrane is an important step (Section 2.4; Fig. 2.9). Competition between ions of the same electrical charge can thus be expected, assuming that the number of binding sites is small in relation to the concentration of competing ions or the capacity of the electrogenic proton pump is limited or both. Such competition occurs particularly between ions with similar physicochemical properties (valency and ion diameter), for example, between the alkali cations potassium (K^+) and rubidium (Rb^+). Since the radius of hydrated Rb^+ is similar to that of hydrated K^+, the binding sites at the plasma membrane of root cells do not seem to distinguish between these two cations (Erdei and Trivedi, 1991), although rubidium cannot replace potassium in its functions in plant metabolism. Radioactive rubidium (^{86}Rb) is often used as a tracer for K^+ uptake studies, although this can give misleading results under certain circumstances as is also true for the flux rates (Behl and Jeschke, 1982). Compared with potassium and rubidium, the affinity of the binding sites for cesium is low and thus, in general cesium uptake is distinctly depressed by potassium (Erdei and Trivedi, 1991). In contrast, uptake of potassium is hardly affected by cesium.

Of the monovalent cations competition between potassium (K^+) and ammonium (NH_4^+) is difficult to explain simply by competition for binding sites at the plasma membrane (Table 2.16). Whereas NH_4^+ is quite effective in competing with K^+, the converse (inhibition of NH_4^+ uptake by K^+) is not observed. This seems quite surprising, but Mengel et al. (1976) obtained similar results with rice. These authors suppose that at least a substantial proportion of ammonium nitrogen is not taken up in the form of NH_4^+, but that NH_3 also permeates the plasma membrane after deprotonation, leaving H^+ in the external solution. Studies with lower plants indicate that deprotonation before uptake (i.e., permeation as NH_3) may become increasingly important at higher substrate concentrations of NH_4^+ (Bertl et al., 1984). Inhibition of uptake of K^+ and particularly Ca^{2+} and Mg^{2+} by NH_4^+ as compared with NO_3^- (Shaviv

Table 2.16
Interaction between Uptake of NH_4^+ and K^+ in Maize Roots[a,b]

| | Contents in roots (μmol g^{-1} fresh wt) | | | |
| | Ammonium (NH_4^+) | | Potassium | |
$(NH_4)_2SO_4$ (mM)	$-K^+$	$+K^+$	$-K^+$	$+K^+$
0	6.9	6.7	8.2	53.7
0.15	7.3	7.1	6.7	48.4
0.50	17.1	13.5	8.9	41.1
5.00	29.4	31.5	9.3	27.1

[a]Based on Rufty *et al.* (1982a).
[b]Duration of the experiment: 8 h; +K indicates addition of 0.15 mM K^+; calcium concentration constant at 0.15 mM.

Table 2.17
Effect of K^+ and Ca^{2+} on the Uptake of Labeled Mg^{2+} (^{28}Mg) by Barley Seedlings[a]

| | Mg^{2+} Uptake (μeq Mg^{2+}(10 g)$^{-1}$ fresh wt (8 h)$^{-1}$) | | |
	$MgCl_2$	$MgCl_2 + CaSO_4$	$MgCl_2 + CaSO_4 + KCl$
Roots	165	115	15
Shoots	88	25	6.5

[a]Concentration of each cation: 0.25 meq l^{-1}. Based on Schimansky (1981).

et al., 1987) would then be merely a reflection of competition for negative charges within individual cells, or in the whole plant (Engels and Marschner, 1993) that is of cation–anion relationships (Section 2.5.4).

Of the mineral nutrients that are taken up as cations, the binding strength is rather low of the highly hydrated Mg^{2+} at the exchange sites in the cell walls (Section 2.2.1) and presumably also at the binding sites at the plasma membrane. Other cations such as K^+ and Ca^{2+} therefore compete quite effectively with Mg^{2+} and strongly depress the uptake rate of Mg^{2+} (Table 2.17). This strong competition is in agreement with observations of magnesium deficiency induced in crop plants by extensive application of potassium and calcium fertilizers.

A particular effective competition on Mg^{2+} uptake is exerted by Mn^{2+} (Table 2.18). Inhibition of Mg^{2+} uptake by far exceeds the 1:1 competition for specific binding sites at the plasma membrane of root cells. Presumably, Mn^{2+} not only competes much more effectively but also in some way blocks the binding or transport sites for Mg^{2+} or both. In contrast, the uptake of K^+ is only slightly depressed by increasing Mn^{2+} concentrations (Heenan and Campbell, 1981).

Competition and limited selectivity of binding sites at the plasma membrane are also observed for anions. Some well-known examples are competition between sulfate and

Table 2.18

Effect of Increasing Manganese Concentrations in the Substrate on Uptake Rates of Manganese and Magnesium in Roots of Soybean Plants[a]

Nutrient	Manganese supply (μM)		
	1.8	90	275
Manganese	0.5	3.1	4.8
Magnesium	121.8	81.1	20.2

[a]Data represent micromoles of nutrient taken up per gram of root dry weight. Based on Heenan and Campbell (1981).

molybdate, sulfate and selenate, and phosphate and arsenate. Increasing sulfate concentrations in the rooting medium strongly depress molybdenum uptake which is of beneficial effect for plant growth and animal nutrition on soils with toxic levels of molybdenum (Pasricha *et al.*, 1977; Chatterjee *et al.* 1992) but may become a critical factor on low molybdenum soils (Section 8.3). Antagonistic interactions between selenate and sulfate are quite distinct and of considerable practical importance in view of both, the selenium requirement of humans and animals, and the increasing concern on excessive selenate levels in certain soils (Tanji *et al.*, 1986). Increasing sulfate levels very effectively decrease selenate uptake and selenium content in plants (Mikkelsen and Wan, 1990), irrespectively of the selenium tolerance of the plant species (Section 10.5).

Arsenate and phosphate are taken up by the same transport system in both, lower and higher plants, leading to excessive uptake and toxicity of arsenate in plants growing on soils with high arsenate levels. In *Holcus lanatus* L. arsenate-tolerant and nontolerant genotypes exist, and in the tolerant genotypes the arsenate uptake is much lower (Meharg and Macnair, 1992). This low arsenate uptake is achieved by suppression of the phosphorus deficiency-induced high affinity uptake system (Section 2.5.6) in the tolerant plants. This is of advantage in respect to restriction of arsenate uptake and thus, arsenate tolerance. It might, however, have consequences for phosphorus nutrition, unless other mechanisms of phosphorus acquisition of more importance than the high affinity system come into play in plants acquiring phosphorus from the soil (Meharg and Macnair, 1992; see also Section 15.4).

The examples of strong competition between K^+ and Rb^+ and between anions as SO_4^{2-} and SeO_4^{2-} demonstrate that the selectivity of the binding sites in root plasma membranes is not a reflection of the role of a given mineral element in plant metabolism, but merely a reflection of the physicochemical similarities between ions that are plant nutrients (e.g. K^+ and SO_4^{2-}) and ions which have no function in metabolism (e.g. Rb^+ and SeO_4^{2-}). Plants are thus unable to exclude unneeded ions from uptake. This aspect has important practical implications in, for example, the channeling of certain heavy metals into the food chain via their uptake by plants (Marschner, 1983).

Another distinct type of anion competition occurs between chloride and nitrate. The chloride content in plants, particularly in roots, can be strongly depressed by increasing

Table 2.19

Effects of Nitrate Concentrations in the Nutrient Solution on
Chloride Contents in Roots and Shoots of Barley Plants[a]

Concentation in nutrient solution (mM)		Chloride content (μmol g^{-1} fresh wt)	
Cl$^-$	NO$_3^-$	Roots	Shoot
1	0	52	93
1	0.2	26	73
1	1.0	13	54
1	5.0	9	46

[a]Based on Glass and Siddiqi (1985).

nitrate concentrations (Table 2.19). This depression seems to be the result of both, negative feedback effects from nitrate stored in the vacuoles of root cells as well as chloride influx at the plasma membrane (Glass and Siddiqi, 1985). It is still a matter of controversy whether nitrate and chloride are transported by the same (Pope and Leigh, 1990) or different (Dhugga et al., 1988) carrier systems. The net influx of nitrate is decreased by chloride, and the chloride already accumulated in the vacuoles seems to be particularly effective in this respect (Cram, 1973).

Competition between nitrate and chloride during uptake is of great importance for crop production. The competing effect of chloride can be used to decrease the nitrate content of such plant species as spinach which tend to accumulate large amounts of nitrate and to use it mainly as an osmoticum (Section 8.2). On the other hand, in saline soils the competing effect of chloride on nitrate uptake may impair nitrogen nutrition of the plants (Bernal et al., 1974). Under these conditions increasing nitrate supply can be an effective means to improve the nitrogen nutritional status of the plants and simultaneously prevent chloride toxicity in sensitive plant species (Section 16.6.3).

A particular case of competition is that exerted by ammonium on nitrate uptake. In almost all cases external ammonium strongly suppresses net uptake of nitrate. In contrast, externally supplied nitrate generally has little or no effect on net uptake of ammonium (Breteler and Siegerist, 1984). Accordingly by supplying NH$_4$NO$_3$, ammonium is usually taken up very much preferentially than nitrate, and at higher external ammonium concentrations, uptake of nitrate is suppressed until the ammonium concentration is considerably lowered. In Norway spruce such a threshold value for ammonium depletion is about 100 μM NH$_4^+$ (Marschner et al., 1991). In short-term experiments with barley external ammonium inhibited net influx of nitrate within 3 min, and on removing the external ammonium net influx of nitrate resumed within 3 min (Lee and Drew, 1989). Rapid ammonium influx into the cytoplasm and a decrease in transmembrane potential are considered as factors possibly involved in the rapid suppression of net nitrate influx (H$^+$–NO$_3^-$ symport), an explanation that would also account for the inhibitory effect of ammonium on cation uptake, for example potassium (Table 2.16).

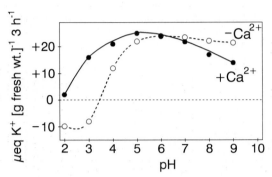

Fig. 2.17 Net uptake of K^+ from 5 mM KBr by barley roots as a function of the pH of the external solution and without ($-Ca^{2+}$) or with ($+5$ mM Ca^{2+}) calcium supply. (Modified from Jacobson *et al.*, 1960.)

2.5.3.2   Role of pH

The interaction between protons (H^+) and other cations and between hydroxyl (OH^-) or bicarbonate (HCO_3^-) and other anions is of general importance for plant mineral nutrition. Because pH values below about 7.5 are more common in soil solutions than higher values, the role of H^+ has attracted more attention. A typical pattern of the influence of the external pH on the net uptake rate of a cation is shown in Fig. 2.17 for K^+. As the H^+ concentration increases (i.e. the pH falls), in the absence of Ca^{2+} in the external solution the net uptake of K^+ sharply declines below pH 5; below pH 4 (i.e. above 10^{-4} M H^+) there is a considerable net efflux of K^+ from the roots. The addition of Ca^{2+} is quite effective in reducing the H^+-induced depression in net K^+ uptake and also in preventing net K^+ efflux at low pH.

The pH effects on cation uptake are in accordance with the key role of the plasma membrane-bound proton efflux pump as the driving force for ion uptake (Section 2.4.2; Fig. 2.9). At high substrate H^+ concentrations the efficiency of the H^+ efflux pump decreases whereas the downhill transport of H^+ from the apoplasm into the cytoplasm is enhanced. In agreement with the latter supposition the electropotentials of root cells decrease from -150 mV at pH 6 to -100 mV at pH 4 (Dunlop and Bowling, 1978). Accordingly, cation uptake in general is inhibited at low substrate pH. The ameliorating effect of Ca^{2+} on net cation uptake at low pH (e.g. Fig. 2.17) is most likely related to its function in the maintenance of plasma membrane integrity (Section 8.6.5). The enhancing effect of Ca^{2+} on the activity of the plasma membrane-bound proton pump (Section 2.4.3) may be a contributing factor, facilitating the potassium uptake at low external pH through the operation of a K^+–H^+ cotransport (Fig. 2.9).

In contrast to cation uptake the uptake of anions is either not affected or stimulated at low pH. In short-term experiments with maize roots decreasing the external pH from 8 to 4 increased the nitrate influx by a factor of about 10 (McClure *et al.*, 1990b) and phosphate by a factor of about 3 (Sentenac and Grignon, 1985). For phosphate this increase at low pH was also observed when the concentration of the monovalent species $H_2PO_4^-$ was kept constant, indicating that a shift in the proportion of $H_2PO_4^-$ to HPO_4^{2-} (Section 2.5.2.2) is not a major factor responsible for the decrease in phosphorus uptake

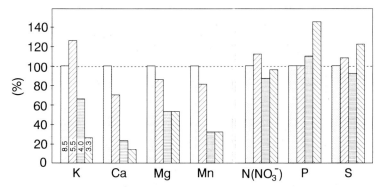

Fig. 2.18 Effect of pH of external solution on mineral element content (percentage of dry weight) of shoots of bean (*Phaseolus vulgaris*); pH 8.5, 5.5, 4.0, and 3.3, respectively, as indicated in the columns for potassium. Relative values, pH 8.5 = 100. (Data recalculated from Islam *et al.*, 1980.)

at high pH. A decrease in anion uptake with increase in external pH is to be expected in view of the importance of proton–anion cotransport from the apoplasm into the cytoplasm (Fig. 2.9). At higher substrate pH, a shift in the buffer system $CO_2/H_2CO_3/HCO_3^-$ in favour of HCO_3^- becomes an important additional factor impairing the efficiency of the proton–anion cotransport by consumption of protons in the apoplasm (Toulon *et al.*, 1989).

In principle, the differences in pH effects on cation and anion uptake are well-known phenomena, for example, in rice (Zsoldos and Haunold, 1982) and soybean (Rufty *et al.*, 1982b). In the latter case, a decrease in the pH from 6.1 to 5.1 resulted in an increase in the ratio of anion to cation uptake from about 1.0 to 1.25. In long-term growth experiments too the different effects of the external pH on uptake of cations and anions are reflected in the mineral element content in the plants (Fig. 2.18). Cation contents are more affected than the anion contents. Various reasons may be put forward to account for this. One of which is that ion uptake is also regulated by other factors, for example the internal concentration, i.e. the nutritional status of the plants (Section 2.5.6).

The effect of external pH on nitrogen uptake depends on whether nitrogen is supplied as NH_4^+ or NO_3^-. As is to be expected lowering the external pH increases the uptake of NO_3^- but decreases uptake of NH_4^+ (Zsoldos and Haunold, 1982). The interpretation of results with both forms of nitrogen at different external pH values is, however, complicated by side effects, such as distinct changes in cation-anion balance in the plants (Section 2.5.4) and on root metabolism and function (Section 8.2.4).

2.5.3.3 Ion Synergism and the Role of Ca²⁺

Synergism, like competition, is a feature of ion interaction during uptake. Stimulation of cation uptake by anions and vice versa is often observed and is mainly a reflection of the necessity of maintaining charge balance within the cells. Synergism in uptake can also be the result of an increase in metabolic activity of the roots when mineral nutrients

Table 2.20
Effect of Ca^{2+} on the Rates of K^+ and Cl^- Uptake in Barley Roots. External pH 5.0

| External | Uptake rate (μeq g^{-1} dry wt (2 h)$^{-1}$) | | | |
solution (mM)	K^+ Influx	K^+ Net uptake	Cl^- Influx	Cl^- Net uptake
0.1 KCl	116 ± 3	117 ± 6	35 ± 1	34 ± 4
0.1 KCl + 1.0 CaSO$_4$	137 ± 2	140 ± 7	53 ± 3	52 ± 4

Table 2.21
Effect of Ca^{2+} on the K^+/Na^+ Selectivity of Roots

| External solution | Uptake rate (μeq g^{-1} fresh wt (4 h)$^{-1}$) | | | | | |
| | Maize | | | Sugar beet | | |
NaCl + KCl (10 meq l^{-1} each)	Na^+	K^+	$Na^+ + K^+$	Na^+	K^+	$Na^+ + K^+$
− Calcium	9.0	11.0	20.0	18.8	8.3	27.1
+ Calcium[a]	5.9	15.0	20.9	15.4	10.7	26.1

[a] 0.5 mM CaCl$_2$.

are supplied after a period of deprivation. In long-term experiments involving different growth rates, when 'concentration' or 'dilution' of mineral nutrients in the dry matter plays an important role, the interpretation of mutual effects of ions during uptake is rather difficult and should be undertaken with care.

An example of synergism is Ca^{2+}-stimulated cation and anion uptake, first discovered by Viets (1944). As has been shown in Fig. 2.17 Ca^{2+} stimulates net uptake of K^+ at low pH, mainly by counteracting the negative effects of high H^+ concentrations on plasma membrane integrity and functioning of the proton efflux pump. Accordingly at low external pH Ca^{2+} not only enhances net influx of K^+ (countertransport) but also of anions as Cl^- (proton–anion cotransport). An example for this is given in Table 2.20 for 'low salt' barley roots (i.e. with low internal concentrations of K^+ and Cl^-).

Due to its stabilizing effect on the plasma membrane Ca^{2+} also plays an important role in selectivity of ion uptake, the K^+/Na^+ selectivity of roots in particular (Table 2.21).

Despite the genotypical differences in the K^+/Na^+ uptake ratio between the 'natrophobic' maize and the 'natrophilic' sugar beet (Section 10.2) in both plant species the presence of Ca^{2+} shifts the uptake ratio in favour of K^+ at the expense of Na^+. In this experiment only net uptake rates were measured. Calcium may have exerted this influence on K^+/Na^+ selectivity via stimulation of the Na^+ efflux pump, either as countertransport (antiport) of K^+/Na^+ (Jeschke and Jambor, 1981) or of H^+/Na^+ (Mennen et al., 1990), or via general effects on plasma membrane integrity, e.g. affecting ion channels.

Calcium as divalent cation stabilizes biomembranes and favours a high transmem-

Table 2.22
Effect of the Accompanying Ion on the Rate of K^+ and Cl^-
Uptake by Maize Plants[a]

Concentration (meq l^{-1})	Uptake rate (μeq g^{-1} fresh wt h^{-1})			
	K^+ from		Cl^- from	
	KCl	K_2SO_4	KCl	$CaCl_2$
0.2	1.6	1.6	0.8	0.7
2.0	2.7	1.9	2.0	1.0
20.0	5.7	2.2	4.3	2.1

[a]Recalculated from Lüttge and Laties (1966).

brane electropotential by reacting with negatively charged groups (e.g. of phospholipids) thereby influencing the physicochemical properties and functioning of biomembranes. This function of Ca^{2+} is reflected, for example, in higher efflux rates of low-molecular-weight solutes from calcium deficient roots and membrane vesicles from those roots (Matsumoto, 1988) as well as lower selectivity in ion uptake when it is absent in the external solution (Table 2.21). Calcium can be removed fairly readily from its binding sites at the outer surface of the plasma membrane, by chelators, for example (van Steveninck, 1965), or can be exchanged by high concentrations of H^+ or metal cations including Na^+ (Lynch et al., 1987). Accordingly, for fulfilling its function in the plasma membrane of roots, the requirement of Ca^{2+} in the external solution is dependent on these environmental factors. High Ca^{2+} concentrations are particularly required, for example, in saline substrates, for maintenance of high K^+/Na^+ selectivity in uptake and for the salt tolerance of plants (Section 16.6).

2.5.4 Cation-Anion Relationships

Because cation uptake and anion uptake are regulated differently (Fig. 2.9), direct interactions between cations and anions do not necessarily occur. For instance, at low external concentrations the uptake rate of a cation is not affected by the accompanying anion and vice versa, as shown in Table 2.22 for K^+ and Cl^-. At high external concentrations, however, an ion which is taken up relatively slowly can depress the uptake of an oppositely charged more mobile ion: for example, SO_4^{2-} depresses K^+ uptake, Ca^{2+} depresses Cl^- uptake.

Different uptake rates of cations and anions require both regulation of cellular pH and compensation of electrical charges. Obviously, at high external concentrations this regulation becomes a limiting factor for the uptake of K^+ when accompanied by SO_4^{2-} and for Cl^- when accompanied by Ca^{2+} (Table 2.22). Under these conditions, nonspecific competition between ions of the same charge can also occur. For example, cations such as K^+ which are rapidly transported across the plasma membrane may depress the uptake rate of cations with slower transport rates such as Mg^{2+} and Ca^{2+}, not by competition for binding sites at the plasma membrane, but by nonspecific

Table 2.23
Relationship between Cation–Anion Uptake and Organic Acid Content in Isolated
Barley Roots[a]

External solution (meq l^{-1})	Uptake (μeq g^{-1} fresh wt)		Change in organic acid (μeq g^{-1} fresh wt)	$^{14}CO_2$ Fixation (relative)
	Cations	Anions		
2 K_2SO_4	17	1	+15.1	145
1 KCl	28	29	−0.2	100
1 $CaCl_2$	1	15	−9.7	60

[a]Based on Hiatt (1967a, b) and Hiatt and Hendricks (1967).

competition for 'native' anions in the cytoplasm or the vacuole, if the rate of synthesis of these anions becomes limiting.

In principle, different rates in the cation–anion uptake ratio by roots are an important cause for intracellular pH perturbations. The stabilization of cytosolic pH in the range 7.3–7.6 (Section 2.4.2) is achieved by the so-called *pH stat* which consists of two components, the *biophysical pH stat*, characterized by proton exchange through the plasma membrane or tonoplast (Fig. 2.9), and the *biochemical pH stat* which mainly involves production and consumption of protons, achieved by the formation and removal of carboxylic groups (Davies, 1986; Raven, 1986). The functioning of the biochemical pH stat is reflected in Table 2.23 in the net changes in organic acid content in roots in relation to the imbalance in cation–anion uptake ratio. When K_2SO_4 is supplied, the excess cation uptake is compensated by an equivalent net synthesis of organic acid anions, and the excess of inorganic anion uptake with $CaCl_2$ supply by a decrease in organic acid anions. These changes in carboxylation and decarboxylation of organic acids are also reflected in the different rates of CO_2 fixation in the roots (dark fixation).

The main reactions involved in the biochemical pH stat in relation to different cation–anion uptake ratios are shown schematically in Fig. 2.19. Excessive cation uptake (Fig. 2.19A) is correlated with pH increase in the cytoplasm and induces enhanced synthesis of organic acids, thereby providing anions (R · COO$^-$) for pH stabilization and charge compensation and subsequent transport of cations and anions either into the vacuole or the shoot. In contrast, excessive anion uptake (Fig. 2.19B) is correlated with pH decrease in the cytoplasm (e.g. proton–anion cotransport, Fig. 2.9). Maintenance of high cytoplasmic pH requires enhanced decarboxylation of organic acids from the storage pool (i.e. the vacuoles). As consequences of this biochemical pH stat, imbalance in cation–anion uptake ratio increases or decreases root concentrations of organic acid anions and also the pH in the root apoplasm and external solution. In the experiment recorded in Table 2.23 when K_2SO_4 was supplied, the net efflux of H$^+$ was 4.3 μeq g^{-1} root fresh weight per 2 h, leading to a decrease in external pH from 5.60 to 5.12 (Hiatt and Hendricks, 1967). The cation–anion balance in plants and the consequences for rhizosphere pH and mineral nutrition of plants has been reviewed recently by Haynes (1990) and is discussed further in Section 15.3.

In the cytoplasm the equilibrium between carboxylation (CO_2 fixation) and decarboxylation is regulated mainly by the pH sensitivity of two enzymes, PEP carboxylase

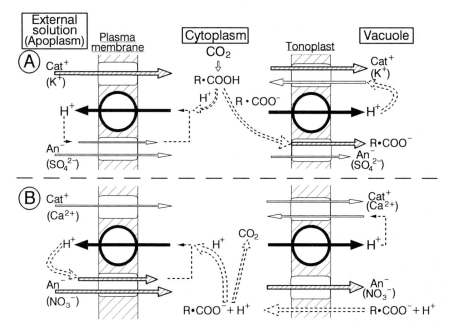

Fig. 2.19 Model for internal pH stabilization and for charge compensation at different ratios of cation–anion uptake from the external solution. A. Excessive uptake of cations (Cat^+) when, for example, K_2SO_4 is supplied. B. Excessive uptake of anions (An^-) when, for example, $Ca(NO_3)_2$ is supplied. (For further details see Fig. 2.9.)

and the malic enzyme (Fig. 2.20). An increase in pH activates the enzyme PEP carboxylase [reaction (1)], and both the rate of CO_2 fixation and the synthesis of oxaloacetate are increased. After oxaloacetate is reduced to malate by the enzyme malate dehydrogenase, the malate can be directly translocated into the vacuoles [reaction(2)], where it acts as a counterion for cations (Fig. 2.19A). Alternatively, the malate can be incorporated into the cytoplasmic pool of the organic acids of the Krebs cycle (Section 5.3), and another organic acid from this pool (e.g. citric acid) can be translocated into the vacuole. In plant species which accumulate large amounts of oxalate (e.g. members of the Chenopodiaceae), an oxalate-based biochemical pH stat may play an important role (Davies, 1986).

If, however, anions are taken up in excess and thus proton–anion cotransport predominates (Fig. 2.19B), the pH of the cytoplasm decreases and the malic enzyme [reaction(4)] is activated, leading to decarboxylation of malate and the production of CO_2. As a result of these reactions, the cytoplasmic pH is stabilized and the cation–anion ratio in the cells maintained. This biochemical pH stat responds rapidly to supply of K_2SO_4 and PEP carboxylase activity is increased by 70% within 20 min (Chang and Roberts, 1992).

The form of nitrogen supply (NH_4^+; NO_3^-; N_2 fixation) plays a key role in the cation–anion relationships in plants. About 70% of the cations and anions taken up by plants are represented by either NH_4^+ or NO_3^- (van Beusichem *et al.*, 1988). Thus, in principle, NH_4^+-fed plants are characterized by a high cation–anion uptake ratio and, in contrast,

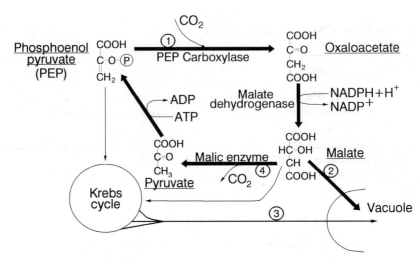

Fig. 2.20 Model of the pathways of CO_2 fixation ('dark fixation') and decarboxylation. Reactions (1)–(4) are explained in the text.

NO_3^--fed plants by a high anion–cation uptake ratio. However, the corresponding effects on biochemical pH stat and on organic acid content are different from those to be expected from Fig. 2.19, as the nitrogen assimilation in roots is correlated with production or consumption of protons. Assimilation of NH_4^+ produces protons in about equimolar ratio ($3\ NH_4^+ \rightarrow 3\ R \cdot NH_2 + 4\ H^+$, Raven, 1986). As higher plants have rather limited capacity in the shoots for disposal of protons (Raven, 1986), NH_4^+ assimilation has to take place in the roots (Engels and Marschner, 1993) and the protons are excreted into the external solution in about equimolar ratio to the NH_4^+ uptake (Marschner et al., 1991) or the excess of cation over anion uptake (van Beusichem et al., 1988). Thus, despite a high total cation uptake by NH_4^+-fed plants, the pH in the cytoplasm decreases during NH_4^+ assimilation and has to be stabilized by both, enhanced proton excretion and decarboxylation of organic acids (Fig. 2.19B).

By comparison, legumes dependent on biological nitrogen (N_2) fixation are characterized by cation–anion uptake ratio > 1 and, thus, elevated levels of organic acid anions and enhanced net excretion of protons in amounts lower than in NH_4^+-fed plants ($1.5\ N_2 fix \rightarrow 1\ H^+$; Raven, 1986; Allen et al., 1988).

In contrast, in NO_3^--fed plants, assimilation of NO_3^- is correlated with an approximately equimolar production of OH^- – or consumption of H^+ ($3\ NO_3^- \rightarrow 3\ R \cdot NH_2 + 2\ OH^-$; Raven, 1986). Depending whether NO_3^- reduction and assimilation take place in the roots or shoots, or carboxylates are retranslocated in the phloem back to the roots (Section 8.2), a varied amount of organic acid synthesis is required either in the shoots or the roots for the biochemical pH stat and charge compensation. Accordingly, the form of nitrogen supply considerable affects both the mineral element composition and the organic acid content of plants (Table 2.24).

As a consequence of the differences in cation–anion uptake ratio, nitrogen assimilation and cellular pH stabilization, also the pH in the external solution is strongly

Table 2.24
Influence of the Form of Nitrogen Supply on the Ionic Balance in the Shoots of Castor Bean Plants[a]

Form of N supply	Cations				Anions					
	K^+	Ca^{2+}	Mg^{2+}	Total	NO_3^-	$H_2PO_4^-$	SO_4^{2-}	Cl^-	Organic acids[b]	Total
NO_3^-	99	85	28	212	44	18	11	2	137	212
NH_4^+	55	43	22	120	0	23	33	5	59	120

[a]Van Beusichem *et al.* (1988); data in meq $(100\ g)^{-1}$ dry wt.
[b]Calculated from the difference of Cations − Anions.

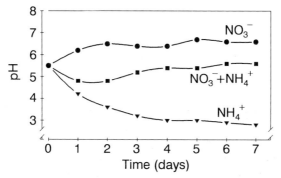

Fig. 2.21 Time course of external solution pH when sorghum plants were supplied with only NO_3^-, only NH_4^+, or both at a ratio of $8\ NO_3^-$ to $1\ NH_4^+$. Total nitrogen concentration, 300 mg l^{-1}. (Redrawn from Clark, 1982b, by courtesy of Marcel Dekker.)

affected whether nitrogen is supplied as NH_4^+ or NO_3^- (Fig. 2.21). In NO_3^--fed plants with preferential NO_3^- reduction in the roots such as sorghum, the external pH usually increases considerably with time, and with mixed supply, after preferential NH_4^+ uptake and depletion of external NH_4^+ the pH decrease is transient followed by pH increase as typically for NO_3^--fed plants.

Interestingly, in NO_3^--fed plants the typical pH increase may also be reversed to a drastic pH decrease under conditions where NO_3^- uptake and assimilation are impaired and cation–anion uptake ratio increased considerably. This situation is typical in many plant species for phosphorus deficiency (Schjorring, 1986), zinc deficiency (Cakmak and Marschner, 1990) and iron deficiency in dicots (Section 2.5.5).

In NH_4^+-fed plants net proton excretion and thus pH stat in the roots becomes increasingly difficult at low external pH (Fig. 2.21). Particularly at high levels of NH_4^+ supply, the pH in the root tissue in general (Findenegg *et al.*, 1982) and also in the cytoplasm of root cells (Gerendás *et al.*, 1990) may decrease substantially. Poor growth of NH_4^+-fed plants at low external pH (Findenegg *et al.*, 1982) is most probably related at least in part to the difficulty of pH stat regulation in the cytoplasm (Section 8.2).

Maintenance of pH stat involves costs in terms of photosynthates and water (Raven,

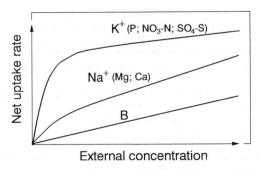

Fig. 2.22   Schematic presentation of uptake rates of potassium, sodium, and boron in barley roots with increasing supply of KCl + NaCl and boron. Uptake rates of other mineral elements in brackets.

1985). This is particularly true in relation to nitrogen nutrition. When both NH_4^+ and NO_3^- are supplied, pH stat may be achieved by similar rates of H^+ production (NH_4^+ assimilation) and H^+ consumption (NO_3^- assimilation) and thus have a very low energy requirement (Raven, 1985; Allen *et al.*, 1988). This may explain at least in part that optimal growth for most plant species is usually obtained with mixed supply of NH_4^+ and NO_3^- (Section 8.2.4).

2.5.5   External Concentrations

As discussed in Section 2.4.3 the uptake rate of ions such as K^+ is usually governed in the low concentration range by saturation kinetics. It is also highly selective and closely coupled to metabolism. In contrast at high external concentrations the uptake rate is more or less linear, is not very selective, not particularly sensitive to metabolic inhibitors (Glass *et al.*, 1990) and probably reflects ion transport through channels. In Fig. 2.22 the typical relationships between external concentration and uptake rate of K^+ are presented schematically, without consideration of C_{min} (Section 2.4.3). In principle similar uptake isotherms have been obtained for phosphorus (Loneragan and Asher, 1967), nitrate (Glass *et al.*, 1990) and sulfate (Clarkson and Saker, 1989).

In contrast to K^+ the uptake rate, for example, of Na^+ is much more concentration dependent, reflecting less specific binding sites in the plasma membrane – or a higher efficiency of a K^+-Na^+ coupled Na^+ efflux pump. A similar uptake isotherm as for Na^+ is often found for Mg^{2+} and Ca^{2+}.

The question arises as to the ecological importance of these differences in uptake isotherms. Compared with the requirement for optimal growth the concentrations in soil solutions are usually low for K^+ ($\ll 1$ mm) and phosphate (<0.05 mm); on the other hand, the concentrations of Ca^{2+} and Mg^{2+} are often considerably higher (Section 13.2). To satisfy the different requirements for these nutrients, plants have binding sites in the plasma membrane of root cells which differ in affinity (K_m) and probably in number for the various mineral nutrients.

As already stressed, however, in Section 2.5 the uptake isotherms and their parameters (K_m; I_{max}) for a given nutrient can also be strongly modulated by environmental (e.g., temperature) and internal factors (e.g., nutritional status). Another

Table 2.25

Influx of Nitrate in Barley Roots without and with Induced High Capacity Nitrate Uptake System[a]

External conc. (mM NO_3^-)	NO_3^- influx (μmol g^{-1} fresh wt h^{-1})		
	non-induced	induced 1 day[b]	induced 4 days[b]
0.02	0.10	2.75	1.54
0.30	0.39	5.27	2.86
20.0	11.54	20.87	8.02

[a]Based on Glass *et al.* (1990).
[b]Pretreatment with nitrate for one or four days.

Table 2.26

Effect of Increasing Boron Supply on Boron Content in Shoots and Shoot Dry Weight of Two Barley Genotypes[a]

	Boron supply (μM)			
	0	2.5	7.5	15
Boron content (mg kg^{-1} dry wt)				
Schooner	5.6	10.0	22.1	46.4
Sahara 3771	2.5	5.5	7.8	11.7
Shoot dry weight mg per plant				
Schooner	129	140	132	121
Sahara 3771	74	84	92	107

[a]Based on Nable *et al.* (1990a).

modulating factor is the induction of an uptake system by external supply of this nutrient. In plants grown in the absence of nitrate (noninduced) roots possess only a low capacity (constitutive) uptake system for nitrate. After nitrate supply, however, within 20 min a high capacity uptake system is formed for which protein synthesis is required (Mäck and Tischner, 1990). An example showing this is given in Table 2.25 for barley roots. The uptake capacity of the non-induced system is very low and with increasing external nitrate concentrations the uptake isotherm resembles features similar to that of Na^+ (Fig. 2.22), i.e. a large contribution of a passive, concentration-dependent response with a low response to metabolic inhibitors (Glass *et al.*, 1990). In contrast, in the induced roots (Table 2.25) the affinity and the uptake capacity for nitrate are much higher and the uptake isotherm resembles that shown in Fig. 2.22 for potassium and phosphate. Interestingly, after 4 days' induction the influx rate of nitrate is much lower, indicating a feedback regulation of the internal concentration (Section 2.5.6).

Compared with other mineral nutrients the isotherm of boron is unique. In both individual cells (Seresinhe and Oertli, 1991) and plants (Fig. 2.22) the uptake rate of boron is linearly related to the external concentration (Table 2.26). Despite this linear relationship there are distinct genotypical differences in total boron uptake. The genotype Sahara 3771 has a much lower uptake and therefore a higher requirement of external boron for optimal growth. This is of disadvantage on low boron soils but an effective mechanism in avoiding boron toxicity when grown on soils with high boron levels (Nable, 1991). Boron uptake is controlled by a nonmetabolic process (Nable *et al.*, 1990b; Paull *et al.*, 1992a), and the genotypical differences in boron uptake are under strict genetic control (Paull *et al.*, 1991, 1992a). The mechanisms responsible for the genotypical differences in boron uptake are still unknown, but apparently the same which govern silicon uptake (Nable *et al.*, 1990b).

This example of boron uptake demonstrates the difficulty in defining an external concentration required for optimal growth even for a given plant species. Genotypical differences in the required external concentration of other mineral nutrients are well documented, for example in different plant species for phosphorus (Jungk *et al.*, 1990) or for iron (Bell *et al.*, 1991). In view of the various external and internal factors affecting the uptake kinetics of ions, it is more appropriate to define a concentration range required for optimal growth rather than a threshold concentration.

As a rule, for mineral nutrients like potassium, phosphorus and also nitrogen at constant supply in the external solution optimal growth can be achieved at concentrations in the range of the high affinity system (Asher and Edwards, 1983), i.e. at concentrations much lower than in many standard nutrient solutions. For the micronutrient cations zinc, copper, manganese and iron the concentration of the free metal species (Zn^{2+}, Cu^{2+}, Mn^{2+}, Fe^{3+}) in the external solution required for optimal growth have been defined by introducing a chelate-buffered nutrient solution system (Section 2.2.1). According to these results adequate concentrations in the external solution are extremely low and in the order of 10^{-11} to 10^{-12} M for Zn^{2+} and Cu^{2+}, and even several orders of magnitude lower for Fe^{3+} (Bell *et al.*, 1991; Laurie *et al.*, 1991).

These results indicate once more that under optimal conditions in which nutrient supply is maintained constant (e.g., in nutrient solutions), only very low concentrations of mineral nutrients are required in the external solution for optimal growth. At higher supply, higher uptake rates reflect what is known as 'luxury uptake'. In soil-grown plants in general, however, and under field conditions in particular the conditions in the root environment are far from optimal, and the maintenance of a constant nutrient supply to the root surface is unlikely to occur for spatial reasons (Section 13.2). Higher external concentrations and 'luxury uptake' in preceding periods can therefore be important because they provide an internal reserve pool in periods of high demand or interrupted root supply. This also holds true for natural vegetation during transient nutrient flushes under favourable weather conditions (Rorison, 1987).

2.5.6   Internal Concentrations and Nutritional Status

For a given external concentration the uptake rate of many mineral nutrients depends on the growth rate of the plant, i.e., the nutrient demand for growth (Clement *et al.*, 1978a). In principle, as the internal concentration of a particular mineral nutrient (as an ion or metabolized, e.g. amino-N) increases, its uptake rate declines, and *vice versa*. An example for this feedback regulation is shown in Table 2.27 for potassium influx in barley roots. Similar relationships between internal concentrations and influx rate are well documented for phosphorus, and an example has been given in Table 2.11.

Table 2.27
Relationship between Content and Influx of
Potassium in Barley Roots[a]

K Content (μmol g^{-1} fresh wt)	K$^+$ Influx (μmol g^{-1} fresh wt h^{-1})
20.9	3.05
32.1	2.72
47.9	2.16
57.8	1.61

[a]From Glass and Dunlop (1979).

The mechanisms which might be involved in this feedback regulation are summarized in Fig. 2.23. Decrease in net uptake rates with increasing internal concentrations may be related to higher efflux rates [Fig. 2.23 (2)] as is well documented, for example, for nitrate (Breteler and Nissen, 1982) or phosphate (Schjorring and Jensen, 1984; McPharlin and Bieleski, 1987) and is discussed in Section 2.4.3. However, in many instances (see above) the influx rate is also affected by the internal concentrations [Fig. 2.23 (1)].

Increase in influx rate of phosphorus (Table 2.11) and of potassium (Wild *et al.*, 1979) was found to be primarily due to an increase in I_{max} rather than an increase in affinity of the binding sites (K_m), indicating that the capacity for uptake of these nutrients

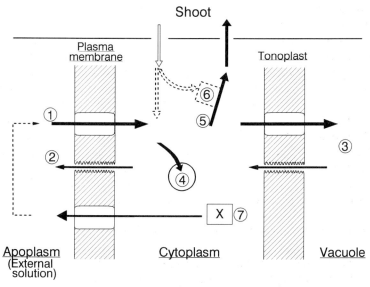

Fig. 2.23 Model of feedback regulation mechanisms of the internal concentration (nutritional status) on ion uptake by roots. (1) Influx rate, binding sites, number, affinity; channel activity; (2) efflux rate, concentration in cytosol, ions and their metabolites; channel activity; (3) vacuolar concentration; (4) transformation, incorporation; (5) xylem transport to the shoot; (6) feedback regulation from the shoot; (7) nutrient deficiency-enhanced excretion of organic solutes (e.g., organic acids; phytosiderophores).

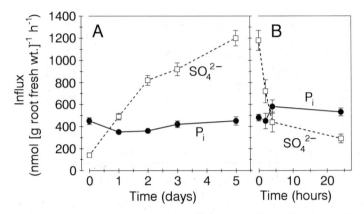

Fig. 2.24   Time course of the influx of sulfate (SO_4^{2-}) and phosphate (P_i) in roots of barley plants deprived of external sulfate supply for up to 5 days (A) and then resupply of sulfate for up to one day (B). (Redrawn from Clarkson and Saker, 1989.)

increased, either by increase in the turnover rate or number of the binding sites in the plasma membrane. Support for the latter mechanism is the increase in a 25 kDa polypeptide in the plasma membrane of roots deprived of phosphorus (Hawkesford and Belcher, 1991).

The dynamics of this feedback regulation of the internal concentration on the uptake rate is shown in Fig. 2.24 for sulfate. Without external sulfate supply the capacity of barley roots for sulfate uptake (influx) increased rapidly within 3–5 days, but decreased drastically within a few hours and was lost within one day when sulfate was resupplied. In contrast, the influx rate of phosphorus was not affected by the sulfur nutritional status. Indications for an increase in number of specific binding sites for sulfate at the plasma membrane of root cells of sulfur deficient plants had been found in wheat (Persson, 1969). Induction of the enhanced transport system for sulfate requires protein synthesis and sulfhydryl groups seem to be involved in this induced system (Clarkson and Saker, 1989). In roots deprived of sulfate supply a 36 kDa polypeptide increases in the plasma membrane which might be a component of the plasma membrane sulfate transport system (sulfate permease; Hawkesford and Belcher, 1991). The mechanism, or the chemical nature, of the negative feedback signal on sulfate uptake might be the sulfate stored in the vacuoles (Cram, 1983), i.e. mechanism (3) in Fig. 2.23, or reduced sulfur compounds such as the amino acids cysteine and methionine (Brunold and Schmidt, 1978) or reduced glutathione (Rennenberg et al., 1989).

The uptake rate of nitrogen is also closely related to the nitrogen nutritional status of the plants. For example, the uptake rate of NH_4^+ is negatively related with the root concentrations of NH_4^+ and amino acids, glutamine and asparagine in particular (Causin and Barneix, 1993). Accordingly, the uptake rate of NH_4^+ rises rapidly within a few days after withdrawal of nitrogen supply (Lee and Rudge, 1986). A decrease in NH_4^+ efflux is involved but this is not the major factor responsible for the increase in NH_4^+ uptake in N-starved plants (Morgan and Jackson, 1988). For nitrate uptake the mechanism of feedback regulation is more complex because of the apparently opposing effects of NO_3^-, namely induction of a high capacity uptake system (Table 2.25) and

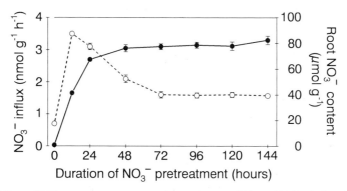

Fig. 2.25 Effect of NO_3^--pretreatment on nitrate content (●) and NO_3^- influx (○) in roots of barley plants. (Siddiqi *et al.*, 1989; reprinted by permission of the American Society of Plant Physiologists.)

negative feedback regulation by increasing internal concentrations (Fig. 2.25). In noninduced plants, NO_3^- supply increases rapidly both the influx and the content of NO_3^- in the roots. With further increase in root content, however, the influx rate is depressed. This negative feedback regulation may be caused by high NO_3^- concentrations in the vacuoles [Fig. 2.23 (3)] and expression of a turgor-regulated event (Cram, 1973; Glass, 1983). However, negative feedback regulation by elevated levels of reduced nitrogen in the form of the amino acids glutamine and asparagine (Lee *et al.*, 1992) or of NH_4^+ is more likely [Fig. 2.23 (4)] as it is known from the inhibitory effects of NH_4^+ supply on NO_3^- uptake (Section 2.5.4), or from reduced sulfur compounds on sulfate uptake (see above). Of course, transformation of a mineral nutrient in the root cells can also act as a positive feedback signal, for example, if combined with incorporation into macromolecular structures (proteins, nucleic acids) in growing cells.

Similar to the examples shown for sulfate and nitrate, after phosphorus supply is withheld from the external solution, the uptake (influx) rate of phosphorus increases after resupplying phosphorus. However, those enhanced influx rates are transient and may last only for a few hours (Lefebvre and Glass, 1982). In this feedback regulation the internal concentration of inorganic phosphorus (P_i) plays a key role, in root cells probably its concentration in the vacuoles, rather than in the cytoplasm (Lee and Ratcliffe, 1983) where the P_i concentration is usually kept quite constant (Lee and Ratcliffe, 1993). In agreement with this, in intact plants the feedback regulation of elevated internal levels can be delayed for several days, because rapid phosphorus translocation into the shoot [Fig. 2.23 (5)] prevents a marked increase in phosphorus content within the roots (Table 2.28). Resupplying phosphorus after a period of deficiency can therefore lead to greatly increased phosphorus content in the shoots and also to phosphorus toxicity (Green and Warder, 1973). Such rapid changes in phosphorus supply are unlikely to occur in soil-grown plants. In nutrient solution culture, however, this factor has to be considered, especially after replacement of solutions.

Although these deficiency-induced new binding sites are specific for each nutrient they also have a limited selectivity in terms of functional requirement of a mineral element as plant nutrient. For example in barley, in sulfur-deficient plants selenate

Table 2.28

Contents of Phosphorus in Barley Plants after Growth without
Phosphorus or Resupply of Phosphorus[a]

	Phosphorus content (μmol P g^{-1} dry wt)[b]		
	8 days $-$ P[c]	7 days $-$ P + 1 day + P[d]	7 days $-$ P + 3 days + P[e]
Shoot total	49 (20)	151 (61)	412 (176)
Youngest leaves	26 (5)	684 (141)	1647 (493)
Roots	43 (24)	86 (48)	169 (94)

[a]Based on Clarkson and Scattergood (1982).
[b]Numerals in parentheses are relative values; 100 represents control with
continuous phosphorus supply of 150 μM P throughout the experiment.
[c]Eight days of growth without phosphorus.
[d]Seven days of growth without phosphorus and 1 day of growth upon
addition of phosphorus (150 μM).
[e]Seven days of growth without phosphorus and 3 days of growth upon
addition of phosphorus (150 μM).

uptake rate is also increased, and in phosphorus-deficient plants the uptake rate of
arsenate is enhanced (Lee, 1982). Unusual and unexpected responses also sometimes
occur. In tomato roots, for example, cadmium uptake rate increases with increase in
cadmium content of the roots (Petit *et al.*, 1978). This might reflect the induction of
synthesis of compounds such as metallothioneins, or phytochelatins, which have a
specific binding affinity for heavy metals (Section 8.3). A similar mechanism might be
involved in the surprising differences in the rate of copper uptake in copper-sufficient
and copper-deficient plants: on resupplying copper to deficient plants the uptake rate is
much lower than in copper-sufficient plants (Jarvis and Robson, 1982).

The relationships between influx rate and internal concentration of a particular
mineral nutrient cannot always be explained satisfactorily by consideration of the roots
alone. Positive and negative feedback control by the shoot [Fig. 2.23 (6)] can markedly
affect the uptake rate of the roots. An example for this has been given for nitrate and
potassium in Table 2.15. Such a feedback control is essential for the coordination of
nutrient uptake depending on the demand of the plants for growth. A varied shoot
supply of sugars to the roots (Pitman, 1972b), or different rates of root export of mineral
nutrients in the xylem into the shoot [Fig. 2.23 (5)] can be considered as a coarse
feedback control. However, there are also fine controls specific for particular nutrients.
For example, the uptake rates of phosphorus (Drew *et al.*, 1984) or potassium (Table
2.29) may be more closely related to the concentrations of these nutrients in the shoots
than in the roots. There is good evidence that retranslocation of mineral nutrients from
shoot to roots may play an important role as a feedback regulation signal from the shoot
for ion uptake by the roots depending on shoot demand. This is the case for potassium
and phosphorus as well as iron (Maas *et al.*, 1988), nitrogen in form of amino acids
(Simpson *et al.*, 1982), and sulfur in the form of glutathione (Rennenberg, 1989).

In tomato plants about 20% of the K$^+$ flux in the xylem was cycling, i.e., originated
from the shoots (Armstrong and Kirkby, 1979a), and in young wheat and rye plants the

Table 2.29
Rate of Potassium Uptake by Maize Roots in Relation to the Potassium Contents of the Roots and Shoot[a]

K⁺ Uptake (pmol cm⁻¹ s⁻¹)	K Content (% dry wt)	
	Roots	Shoot
15.8	5.85	8.00
28.0	5.55	6.45
33.8	4.99	4.35
36.8	5.51	4.13

[a]From Barber (1979).

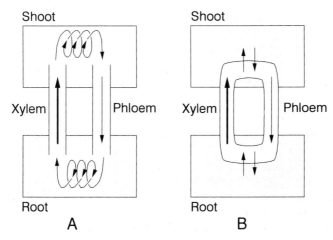

Fig. 2.26 Model for circulation of mineral nutrients between shoots and roots as a mechanism for regulating uptake by roots. (A) Nutrients are uniformly mixed with the pool in the bulk tissue. (B) Only limited exchange of the cycling nutrients with the bulk tissue. (Modified from Cooper and Clarkson, 1989.)

corresponding fraction of the amino-N flux in the xylem was over 60% (Cooper and Clarkson, 1989). Drew and Saker (1984) proposed a model for the cycling of mineral nutrients in which potassium or phosphate delivered to the shoot in excess of its demand are translocated back to the roots to convey information concerning the potassium or phosphorus nutritional status of the shoot. In order to act as a finely tuned signal, this fraction of cycling nutrients obviously exchanges only to a very limited extent with the corresponding nutrients (e.g., potassium, or amino-N) in the bulk tissue of shoots and roots (Fig. 2.26B).

At least for potassium there is good evidence that the rate of transport to the shoot according to the shoot demand is primarily regulated at the level of xylem loading in the roots (Engels and Marschner, 1992b), and a higher shoot demand is reflected by lower

K$^+$ concentrations in the sieve tubes and early maturing metaxylem in the stele of the roots (Drew et al., 1990). Thus, the shoot demand of a particular nutrient seems to be primarily and finely regulated by the shoot-derived concentration of that nutrient at the phloem–xylem vascular complex in the stele [Fig. 2.23 (6)], thereby affecting the root level in general. As shown in field-grown maize extended vascular connections exist between branch and nodal roots (McCully and Mallet, 1993). These sites of direct phloem–xylem connections are suggested as the sites of transfer not only of mineral nutrients (e.g. potassium, amino-N) but also sugars from the phloem into the xylem. An altered root level of a nutrient delivered from the shoot in turn may act as a more coarse feedback signal for the uptake at the plasma membrane for increasing or decreasing influx or efflux rates of this nutrient [Fig. 2.23 (1), (2)]. Additional mechanisms, however, may also be induced at low root contents [Fig. 2.23 (7)] as for example under phosphorus deficiency an increase in the release of extracellular phosphatases (Goldstein et al., 1988; Lee, 1988) or organic acids (see Section 15.4.2), or phytosiderophores under iron deficiency. With the exception of phytosiderophores (see below), neither extracellular phosphatases (Lefebvre et al., 1990) nor organic acids are directly coupled with enhanced uptake rates of mineral nutrients. The role of these deficiency-induced root responses for mineral nutrient acquisition is discussed in Section 15.4.

Also for a given mineral nutrient, the feedback regulation signal on the nutritional status of the shoot to the roots may lead to basically different responses of the uptake system in different plant species as discussed below for iron.

Depending on their response to iron deficiency, plants can be classified into two categories or strategies (Strategy I and Strategy II). For reviews see Marschner et al. (1986a,b), Römheld (1987a,b), Chaney (1988), and Bienfait (1989). In both strategies the responses are confined to the apical zones of growing roots and are fully repressed within about one day after resupply of iron.

Strategy I is typically for dicots and non-graminaceous monocots, and characterized by at least two distinct components of iron deficiency responses: increased reducing capacity and enhanced net excretion of protons (Fig. 2.27). In many instances also the release is enhanced of reducing and/or chelating compounds, mainly phenolics (Olsen et al., 1981; Marschner et al., 1986a). These root responses are often related to changes in root morphology and anatomy, particularly in the formation of transfer cell-like structures in rhizodermal cells (Section 9.1.6).

The most sensitive and typical response is the increase in activity of a plasma membrane bound reductase in the rhizodermal cells [Fig. 2.27 (R)]. The supposed existence of two distinct reductases, a constitutive (basic) reductase with low capacity, and an iron deficiency-induced high capacity reductase ('Turbo'; Bienfait, 1985) has not been confirmed by more recent studies (Brüggemann et al., 1991; Holden et al., 1991). The iron (chelate) reductase of the plasma membrane of deficient and sufficient roots have similar characteristics (including K_m), indicating that under iron deficiency isoforms of the constitutive reductase (most likely the 'Transmembrane redox pump', Fig. 2.9) are increased in their expression and lead to higher V_{max} (Holden et al., 1991). This increase in reductase activity is correlated with the synthesis of a new 70 kDa protein in the plasma membrane (Valenti et al., 1991). In agreement with the model in Fig. 2.9, for the iron deficiency-enhanced reductase either NADPH (Lubberding et al., 1988) or NADH (Schmidt and Janiesch, 1991) also act as an electron donor.

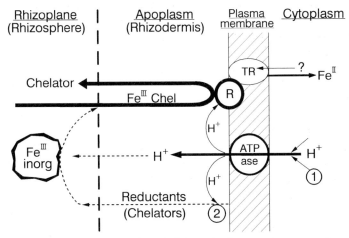

Fig. 2.27 Model for root responses to iron deficiency in dicots and non-graminaceous monocots; Strategy I: (R) inducible reductase; [TR] transporter or channel for Fe(II)?; (1) stimulated proton efflux pump; (2) increased release of reductants/chelators. (Modified from Marschner *et al.*, 1986a; Römheld, 1987a,b.)

Table 2.30
Effect of Iron Deficiency in Cucumber (Strategy I) on Proton Excretion (pH), Reducing Capacity of the Roots and Iron Uptake Rate[a]

Fe nutritional status (preculture)	Chlorophyll (mg g^{-1} dry wt)	H$^+$ excretion (pH solution)	Reducing cap. (μmol Fe(II) g^{-1} root dry wt (4 h)$^{-1}$)	Fe uptake (μmol g^{-1} root dry wt (4 h)$^{-1}$)
+Fe[b]	12.2	6.2	3.2	0.03
−Fe	7.8	4.8	96.8	2.6

[a]Compiled data from Römheld and Kramer (1983) and Römheld and Marschner (1990).
[b]Supply of 1×10^{-6} M FeEDDHA, pH 6.2.

Although the transmembrane redox pump may contribute to the net excretion of protons (Fig. 2.9) the strongly enhanced net excretion of protons under iron deficiency is most probably the result of higher activity of the plasma membrane proton efflux pump [Fig. 2.27 (1)] and not of the reductase (Bienfait *et al.*, 1989; Alcantara *et al.*, 1991). The activity of the reductase (R) which is strongly stimulated by low pH, i.e., enhanced proton excretion by the ATPase is important for the efficiency in Fe(III) reduction. Accordingly, high concentrations of HCO$_3^-$ counteract this response system in Strategy I plants (Section 16.5.3).

An example for root responses to iron deficiency and the corresponding enhanced uptake rates of iron are shown in Table 2.30 for cucumber as a Strategy I plant. The much higher reduction rates of Fe(III) at the outer surface of the plasma membrane of root rhizodermal cells are correlated with distinctly enhanced uptake rates of iron. It is not clear whether the reductase itself or an attached protein (Fig. 2.27 [TR]) mediates the transport of the reduced iron [Fe(II)] into the cytoplasm (Young and Terry, 1982; Grusak *et al.*, 1990). It is also not clear how the iron nutritional status of the cells is transformed into a 'signal' for the response mechanisms at the plasma membrane (Bienfait, 1989). The involvement of nicotianamine as chelator for Fe(II) and a sensor for iron nutritional status has recently been discussed (Pich and Scholz, 1991). Both root responses, increase in reduction rates of Fe(III) and iron uptake, are under strict genetic control (Alcantara *et al.*, 1990; Romera *et al.*, 1991) and, in a given plant species (chickpea) the responses seem to be controlled by only one dominant gene (Saxena *et al.*, 1990).

Strategy II is confined to *graminaceous plant species* (grasses) and characterized by an iron deficiency-induced enhanced release of non-proteinogenic amino acids, so-called phytosiderophores (Takagi *et al.*, 1984). The release follows a distinct diurnal rhythm and is rapidly depressed by resupply of iron (Fig. 2.28). The diurnal rhythm in release of phytosiderophores in iron-deficient plants is inversely related with the volume of a particular type of vesicles in the cytoplasm of root cortical cells (Nishizawa and Mori, 1987).

Phytosiderophores such as mugineic acid (Fig. 2.29) form highly stable complexes with Fe(III), the stability constant in water is in the order of 10^{33} (Murakami *et al.*, 1989). As a second component of Stragegy II a highly specific constitutive transport system (Translocator (Tr), Fig. 2.29) is present in the plasma membrane of root cells of grasses (Römheld and Marschner, 1990) which transfers the Fe(III) phytosiderophores into the cytoplasm. In plant species with Strategy I this transport system is also lacking. Although phytosiderophores form complexes also with other heavy metals such as zinc, copper and manganese (Fig. 2.29), the translocator in the plasma membrane has only a low affinity to the corresponding complexes (Marschner *et al.*, 1989; Ma *et al.*, 1993). Nevertheless, release of phytosiderophores may indirectly enhance the uptake rate of these other metals by increasing their mobility in the rhizosphere and in the root apoplasm (Zhang *et al.*, 1991a,b,c).

Under iron deficiency not only the release of phytosiderophores is increased but also

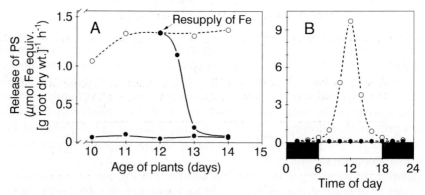

Fig. 2.28 Release of phytosiderophores (PS) from barley roots as affected by the iron nutritional status (A), and diurnal rhythm in release of phytosiderophores (B). Fe sufficient (●) and Fe deficient (○) plants. (Based on Römheld, 1987a,b and A. Walter, personal communication.)

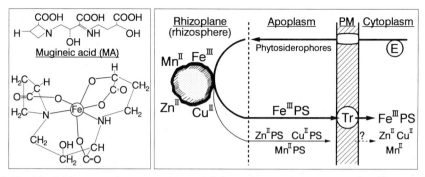

Fig. 2.29 Model for root responses to iron deficiency in graminaceous species; Strategy II: (E) enhanced synthesis and release of phytosiderophores; (Tr) translocator for Fe(III) phytosiderophores in the plasma membrane; structure of the phytosiderophore mugineic acid and its corresponding Fe(III) chelate (Modified from Nomoto *et al.* 1987; Römheld, 1987b.)

the uptake rate of the Fe(III) phytosiderophore complexes (Table 2.31) indicating a higher transport capacity (I_{max}) due either to an increase in number or the turnover rate of the translocator. As mechanism of phytosiderophore release a cotransport with either protons or potassium is discussed, and for the uptake of Fe(III) phytosiderophores a proton–anion cotransport across the plasma membrane (Mori *et al.*, 1991).

Although this phytosiderophore system resembles features of the siderophore system in microorganisms (Winkelmann, 1986) its affinity to phytosiderophores is two to three orders of magnitude higher than for siderophores such as ferrioxamine B (Bar-Ness *et al.*, 1991, 1992; Crowley *et al.*, 1992), or for synthetic iron chelates such as FeEDDHA (Römheld and Marschner, 1990; Bar-Ness *et al.*, 1991, 1992).

For the biosynthesis of phytosiderophores, methionine is the common precursor and nicotianamine an intermediate (Shojima *et al.*, 1990):

Avenic acid

↕

Methionine → Nicotianamine → 2′-Deoxymugineic acid →

Mugineic acid → 3-Hydroxymugineic acid

This pathway is under strict genetic control and the chromosomes have already been

Table 2.31
Release of Phytosiderophores (PS; mugineic acid) and Uptake of
Iron-Phytosiderophores in Iron-Sufficient and Iron-Deficient Barley
Plants[a]

Fe nutritional status (preculture)	Chlorophyll content (mg g^{-1} dry wt)	PS release (μmol g^{-1} root dry wt (4h)$^{-1}$)	Fe uptake (μmol g^{-1} root dry wt (4h)$^{-1}$)
+ Fe	12.8	0.4	0.4
− Fe	7.5	8.2	3.4

[a]Römheld and Marschner (1990). Reprinted by permission of Kluwer Academic Publishers.

identified which are responsible for the regulation, for example, of the transformation of 2'-deoxymugineic acid into mugineic acid (Mori and Nishizawa, 1989).

In graminaceous species, as well as between genotypes within a species, distinct differences exist in their 'Fe-efficiency' (e.g. sensitivity to iron-deficiency chlorosis). These differences are, as a rule, closely related to the capacity for the release of phytosiderophores under iron deficiency (Clark *et al.*, 1988; Römheld and Marschner, 1990). In some 'Fe-inefficient' mutants, however, also the uptake and utilization of Fe(III) phytosiderophores is impaired (Jolley and Brown, 1991a). The progress in identification of the control mechanism for phytosiderophore biosynthesis on the chromosome and gene level may offer a chance not only for better understanding of the biosynthetic pathway but also for gene transfer in graminaceous species for increasing their 'Fe-efficiency'.

In view of this genetic control of biosynthesis of phytosiderophores under iron deficiency, enhanced release of the same phytosiderophores also under zinc deficiency was unexpected (Zhang *et al.*, 1989). It is not yet clear whether this indicates a common control mechanism of the biosynthetic pathway by both, iron and zinc, or a zinc deficiency-induced impairment of iron metabolism at a cellular level leading to the expression of iron deficiency responses in Strategy II plants (Cakmak *et al.*, 1994).

2.5.7 Maintenance of Constant Internal Concentrations

The common approach of studying mineral nutrient uptake in relation to the external or internal concentration, as well as of the relationships between nutrient supply, concentrations in plants and growth rate has been questioned by Ingestad and coworkers. For review see Ingestad and Ågren (1992). It is argued that even in flowing nutrient solution cultures particularly during the exponential plant growth the low external concentrations required for optimal growth are difficult to maintain constant. This would even be more true for the relationships between growth-limiting nutrient supply, uptake rates and internal concentrations. Examples demonstrating have been given in Section 2.5.6.

In order to overcome these difficulties and also to define the productivity of a mineral nutrient in terms of biomass production per unit of nutrient at different internal concentrations, Ingestad and his coworkers use a different theoretical and experimental approach. In principle, this approach is based on relative values. For control of a constant relative uptake, the relative addition rates of nutrients (i.e. the supply) are divided by the amount of the nutrient already in the plant. Accordingly in this approach only the amount of nutrients supplied counts, and not the external concentration, and information on the latter are therefore also not given. Using this approach a range of different but constant relative growth rates can be achieved at different degrees of nutrient limitations. The root–shoot dry weight ratio of the nutrient-limited plants is often extremely large but, nevertheless, visual deficiency symptoms are absent.

This highly formal concept on mineral nutrition of plants is an interesting variable to the common approach in which the role of external and internal concentrations on uptake rates on growth responses and changes in various physiological and biochemical parameters (e.g. photosynthesis) are studied. The Ingestad concept allows the study of these effects of mineral nutrition at suboptimal but steady-state conditions. However, these steady-state conditions in which the relative nutrient supply is adjusted to the relative growth rate are not typical of the conditions to which field-grown plants are exposed. For field-grown plants fluctuations in nutrient supply to the roots are as typical as is true for other growth parameters such as irradiation, temperature or water supply. To cope with these fluctuations plants respond by invoking a range of mechanisms to nonoptimal (stress) conditions. Fluctuations in nutrient supply are compensated for by different uptake rates (Section 2.5.6), changes in root morphology and physiology (Chapters 14–16) or storage and remobilization of mineral nutrients (Chapters 3 and 6). All these parameters are not appreciated in Ingestad's concept.

2.6   Uptake of Ions and Water Along the Root Axis

Growing roots vary both anatomically and physiologically along their longitudinal axes. This has to be borne in mind when models for mechanisms of 'the' behaviour of root tissue and root cells are based on uptake studies with isolated roots or roots of intact plants. In the apical zone nonvacuolated cells dominate. These cells differ in many respects from vacuolated cells in the basal zones. The apical root zones have higher respiration rates (Thomson et al., 1989b) which rapidly drop when carbohydrate supply to the roots is interrupted, for example after isolation (Brouquisse et al., 1991). Nitrate reductase activity may also be much higher in apical than in basal root zones (Granato and Raper, 1989) as well as the selectivity in K^+/Na^+ uptake (Jeschke and Stelter, 1976).

In general there is a tendency for the rate of ion uptake per unit root length to decline as the distance from the apex increases. However, this tendency very much depends on factors such as the type of ion (mineral nutrient), the nutritional status and the plant species. In seminal roots of maize supplied with either potassium or calcium to different zones along the axis (Table 2.32) the uptake rate of potassium is even slightly lower in

Table 2.32

Uptake and Translocation of Potassium (^{42}K) and Calcium (^{45}Ca) Supplied to Different Zones of the Seminal Roots of Maize[a]

Nutrient (1 meq l^{-1})	Accumulation and translocation	Root zone supplied (distance from tip, cm)		
		0–3	6–9	12–15
Potassium	Transloc. to shoot	3.8	14.6	15.6
	Accum. in zone of supply	11.5	3.8	1.9
	Transloc. to root tip	—	4.3	2.0
	Total	15.3	22.7	19.5
Calcium	Transloc. to shoot	2.4	2.2	2.4
	Accum. in zone of supply	4.1	1.6	0.4
	Transloc. to root tip	—	—	—
	Total	6.5	3.8	2.8

[a] Data expressed as μeq (24 h)$^{-1}$ per 12 plants. Based on Marschner and Richter (1973).

Table 2.33
Effect of Phosphorus Nutritional Status on the Rate of
Phosphorus Uptake by Various Root Zones of Barley
Plants[a]

Pretreatment for 9 days	Root zone (distance from root tip, cm)		
	1	2	3
With phosphorus	2019	1558	970
Without phosphorus	3150	4500	4613

[a]Uptake rate expressed as pmol mm^{-3} of root segment in 24 h.
Based on Clarkson *et al.* (1978).

the apical zone, despite the high potassium demand for growth in the apical root zones. The high potassium concentration in root apical meristematic cells of about 200 mM (Huang and Van Steveninck, 1989a) is obviously maintained not only by uptake from the external solution but also delivery via the phloem either from basal root zones (Table 2.32) or from the shoot (Section 2.5.6). In perennial plant species like Norway spruce the uptake rates of potassium are also lower in apical than basal zones of nonmycorrhizal longroots (Häussling *et al.*, 1988).

In contrast to potassium the uptake of magnesium and particularly calcium is much higher in apical than in basal root zones (Ferguson and Clarkson, 1976; Häussling *et al.*, 1988). This is also shown in Table 2.32. Because of lack of phloem mobility root tips have to meet their calcium demand for growth by direct uptake from the external solution. Root apical zones take up high amounts of calcium not only for their own demand but also for delivery to the shoot (Table 2.32; Clarkson, 1984). The calcium uptake in basal root zones is usually low but may increase in basal zones where lateral roots are formed and penetrate the cortex (Häussling *et al.*, 1988).

The decline in phosphorus uptake along the root axis is much less than for calcium (Ferguson and Clarkson, 1975). In soil-grown maize this decline is mainly related to a decrease in root hair viability and, thus, in absorbing root surface area (Ernst *et al.*, 1989). The gradient in phosphorus uptake along the root axis also depends on the phosphorus nutritional status of the plant and may be reversed under deficiency in favour of the basal zones (Table 2.33). This is probably related to lower internal phosphorus concentrations in the basal zones which deliver phosphorus to apical zones under deficiency conditions. The situation is different under iron deficiency in plants with Strategy I (Section 2.5.6) where the apical, but not the basal, root zones increase their capacity for iron uptake by a factor of up to 100 (Römheld and Marschner, 1981b).

In contrast to dicotyledonous plants and perennial plants, in graminaceous species like wheat and barley the rhizodermis and cortex cells of basal (older) regions of the roots collapse and die. There are also reports (Lascaris and Deacon, 1991a,b) of early progressive senescence of rhizodermis and cortex cells behind the root hair zone in otherwise healthy looking roots ('root cortical death', RCD). This would be detrimental for the uptake of nutrients and water through the older root zones. However, the

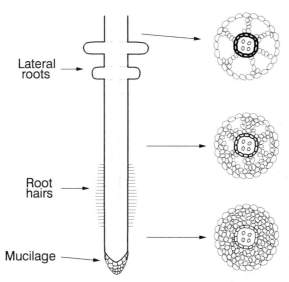

Fig. 2.30 Schematic representation of anatomical changes along the axis of a maize nodal root. In basal zones there is degeneration of cortical cells and formation of tertiary endodermis.

methods used for characterizing RCD had been questioned by Wenzel and McCully (1991).

Formation of cortical gas spaces (*aerenchyma*) particularly in more basal root zones can often be observed (Fig. 2.30). Aerenchyma formation is a typical response to oxygen deficiency in plant species adapted to wetland conditions (Section 16.4). However, it can also be induced, for example, in maize roots under fully aerobic conditions by temporary deprivation of nitrogen or phosphorus supply (He *et al.*, 1992). Despite these anatomical changes (Fig. 2.30) the basal root zones still have a considerable capacity for ion uptake (Drew and Fourcy, 1986) and also for radial transport, indicating that the strands of cells bridging the cortex maintain sufficient ion transport capacity from the rhizodermis up to the endodermis (Drew and Saker, 1986).

Gradients in uptake of water along the root axis may affect the gradients in ion uptake either indirectly via solute supply to the root surface (Section 13.2) or via direct effects on radial transport in the cortex. Water uptake rates are usually higher in the apical zones and decline thereafter sharply (Sanderson, 1983; Häussling *et al.*, 1988). This decline in basal zones is caused mainly by formation of suberin in the rhizodermis, of the exodermis and the secondary and tertiary endodermis as efficient barriers against apoplasmic solute flow. In perennial species, in particular, water uptake may increase again in basal zones where lateral roots penetrate the cortex and temporarily disrupt these barriers (Häussling *et al.*, 1988; MacFall *et al.*, 1991).

2.7 Radial Transport of Ions and Water Across the Root

There are two parallel pathways or movement of ions (solutes) and water across the cortex towards the stele: one passing through the apoplasm (cell walls and intercellular

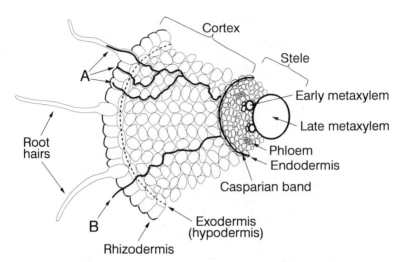

Fig. 2.31 Part of transsection of a maize root showing the symplasmic (A) and apoplasmic (B) pathway of ion transport across the root.

spaces, Section 2.2.1) and another passing from cell to cell in the *symplasm* through the *plasmodesmata*. In the symplasmic pathway ions, but not water, bypass the vacuoles (Fig. 2.31). As a rule the apoplasmic pathway of ions is constrained by the Casparian band in the walls of the endodermal cells. This band has hydrophobic properties and completely surrounds each cell. Additionally in basal root zones the cell walls of the endodermis become thicker and even lignified (tertiary endodermis). Recently, the key role of the endodermis and the Casparian band as effective barrier also for radial movement of water has been questioned. At least in young roots, despite the hydrophobic properties of the Casparian band, water seems to permeate this band readily (Peterson *et al.*, 1993; Schreiber *et al.*, 1994).

Depending on the plant species and the root zone the apoplasmic pathway may already be constrained or blocked by the exodermis (Fig. 2.31) or suberization of the rhizodermis. Formation of an exodermis is found, for example, in *Zea mays*, *Allium cepa*, or *Helianthus annuus*, but not in *Vicia faba* or *Pisum sativum* (Enstone and Peterson, 1992). However, there are somewhat different views on the function of the exodermis as effective barrier for transport of water and solutes in the apoplasm of the root cortex (Section 2.5.1). Termination of the apoplasmic pathway at the exodermis as suggested by Enstone and Peterson (1992) would confine in basal root zones both water influx and ion transport across the plasma membrane into the symplasm to the rhizodermal cells. Although rhizodermal cells play a key role in mineral nutrient uptake in general (Grunwald *et al.*, 1979) and at low external concentrations of potassium and phosphate in particular (Drew and Saker, 1986), it is not possible to generalize on the relative importance of the two pathways in ion and water transport in the root cortex. This depends on: (a) the external concentration compared with the capacity and affinity of the transport system at the plasma membrane for a given ion (e.g., $K^+ > Na^+$; $NO_3^- > H_3BO_3$; Section 2.5.5); (b) the root zone considered: depending on the growth rate of the root, the exodermis may develop between 2 cm and 12 cm proximal to the root tip (Perumalla and Peterson, 1986) and may possess 'passage cells' (Storey and Walker,

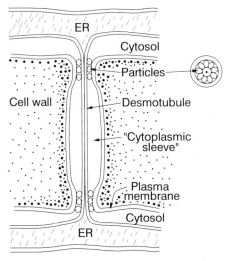

Fig. 2.32 Diagram of a plasmodesma including substructural components. ER = endoplasmic reticulum. (Modified from Beebe and Turgeon, 1991.)

1987), and (c) the hydraulic conductivity of the root zone considered (see below) and the transpiration rate. For water, calculations on the proportions of the apoplasmic pathway in radial transport in roots vary between less than 1% (Hanson *et al.*, 1985) and 76–98% (Zhu and Steudle, 1991).

The endodermis is also not a perfect barrier against the apoplasmic pathway of ions from the cortex into the stele (Fig. 2.31). In addition to the 'passage cells' in some plant species, at least at two sites along the root axis this barrier may be 'leaky'. At the root apex the Casparian band is not yet fully developed and, is thus, a zone allowing apoplasmic transport into the stele (Huang and Van Steveninck, 1989b) most likely also of calcium (Section 2.6). However, for certain solutes such as apoplasmic dyes (Enstone and Peterson, 1992) and polyvalent cations like aluminium the apoplasmic pathway might be restricted in the apical root zone by *mucilage* formed at the external surface of the rhizodermal cells (Section 15.4). The apoplasmic pathway into the stele is certainly of general importance in basal root zones where lateral roots emerge from the pericycle in the stele and where structural continuity of the endodermis is ruptured transiently during penetration of the lateral roots through the cortex. At these sites the apoplasmic pathway into the stele can be important, for example, for calcium (Häussling *et al.*, 1988), aluminium (Rasmussen, 1968) and water (Häussling *et al.*, 1988; Wang *et al.*, 1991). This 'bypass-flow' becomes particularly important for shoot water supply at high transpirational demand (Sanderson, 1983) and for uptake and shoot transport of sodium under saline conditions (Yeo *et al.*, 1987). In a given genotype environmental factors may strongly influence apoplasmic pathway also via anatomical changes. In sorghum drought stress decreases drastically root hydraulic conductivity because both, accelerated deposition of suberin and lignin in the exodermis and endodermis, and persistence of nonconducting late metaxylem vessels (Cruz *et al.*, 1992).

For mineral nutrients the symplasmic pathway plays the key role, either beginning at the rhizodermis and root hairs, at the exodermis or at the endodermis. Radial transport in the symplasm from cell to cell requires bridges, the *plasmodesmata*, across the cell walls that connect the cytoplasm of neighbouring cells (Fig. 2.32). Plasmodesmata have a complex structure, the most common type being a tube of endoplasmic reticulum (ER) running through the pore, the *desmotubule*. The transport of solutes and water between cells may occur in the 'cytoplasmic sleeve', i.e. the cytosol between the desmotubule and the plasma membrane (Fig. 2.32). At the neck region large particles (proteins) determine the molecular exclusion for the solutes at about 1000 Da (Beebe and Turgeon, 1991) and act as a kind of gate mechanism for the symplasmic pathway triggered by internal and external signals (Pitman *et al.*, 1981). Thus, plasmodesmata represent another mechanism of internal control of ion fluxes and cell to cell communication.

High intracellular Ca^{2+} concentrations induce closure of the plasmodesmata (Tucker, 1990). Also the rapid and drastic decrease in root hydraulic conductivity by oxygen deficiency is attributed to blocking of the plasmodesmata and, thus, of the symplasmic component of the radial water transport (Zhang and Tyerman, 1991). Radial root hydraulic conductivity also decreases under phosphorus and nitrogen deficiency (Radin and Mathews, 1989), and blocking of the plasmodesmata seems to be a more likely explanation than lower hydraulic conductivity of the plasma membrane. Such a decrease in radial hydraulic conductivity should not only depress delivery of water and mineral nutrients to the shoot but also limit leaf growth rate by increasing the water deficit in expanding leaves (Radin and Eidenbock, 1984).

The number of plasmodesmata per cell may vary considerably between plant species and cell type (Table 2.34). Rhizodermal cells which have developed into root hairs have more plasmodesmata than the remaining rhizodermal cells. The relatively small number of plasmodesmata in *Raphanus* raises the question as to whether the root hairs are of major importance for symplasmic radial transport in this plant species. However, not only the number but also the diameter of the individual plasmodesmata must be taken into account (Tyree, 1970) and duration of opening (see above).

In the endodermis of young barley roots, on average 20 000 plasmodesmata per cell have been found (Helder and Boerma, 1969). In the tertiary (lignified) endodermis of older zones of barley roots, there are far fewer plasmodesmata, but sufficient in number to permit considerable radial transport of both water and ions through the endodermis (Clarkson *et al.*, 1971).

Table 2.34
Intracellular K^+ Activity and Number of Plasmodesmata in Tangential Walls of Hair and Hairless Cells of the Root Epidermis[a]

Plant species	Cell type	K^+ Activity (mM)	Number of plasmodesmata Per μm^2	Per cell junction
Trianea bogotensis	Hair	133	2.06	10 419
	Hairless	74	0.11	693
Raphanus sativus	Hair	129	0.16	273
	Hairless	124	0.07	150

[a]From Vakhmistrov (1981).

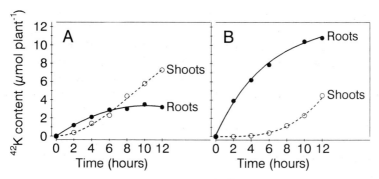

Fig. 2.33   Accumulation and translocation rate of K^+ (^{42}K) from 1 mM KCl (+0.5 mM CaSO$_4$) in barley plants (A) after preculture with 1 mM KCl or (B) without KCl.

   The mechanism of symplasmic transport of solutes seems to be chiefly diffusional, facilitated by radial water flux. Inhibition of cytoplasmic streaming may have no effect on the radial transport of potassium (Glass and Dunlop, 1979). On the other hand, in radial transport of phosphate, various metabolic steps such as incorporation into organic compounds (ATP, sugar phosphates) are involved before it is released as inorganic phosphate into the xylem (Sasaki *et al.*, 1987). During radial transport of ions competition occurs between accumulation in the vacuoles and transport in the symplasm. This competition depends on the mineral nutrient and its concentration in the vacuoles of root cells along the pathway. At low internal concentrations ('low-salt' roots), in short-term experiments this competition is reflected by a typical high accumulation rate in the roots and a delay in the translocation of the ions from the roots to the shoots of plants which originally were low in internal concentrations of the ion being investigated (Fig. 2.33). As a result of this competition, when the supply of a mineral nutrient is suboptimal the roots usually have higher contents of the particular nutrient than the shoot ('restricted translocation'). In long-term studies, this regulation mechanism is in part responsible for the often observed shift in the relative growth rates of roots and shoots in favor of the roots (Section 14.7).

   Along the symplasmic pathway preferential accumulation of certain ions like Na^+ may take place, accounting to some extent, for example, to the restricted shoot translocation of Na^+ in natrophobic plant species (Chapter 3). Preferential accumulation in roots is also a mechanism by which certain genotypes of maize ('shoot excluder') strongly restrict translocation of cadmium to the shoots (Florijn and Van Beusichem, 1993). On the other hand, symplasmic transport of phosphate and translocation into the shoots is enhanced in expense of accumulation in vacuoles of the roots not only at high vacuolar concentrations of phosphate (see above) but also of nitrate (Lamaze *et al.*, 1987). The exchange rate between ions in the vacuoles of cortex cells and those in the symplasm depends on the ion species ($K^+ > Na^+$; $NO_3^- > SO_4^{2-}$), the $t_{1/2}$ for exchange is in the order of at least a few days (Section 2.4.2).

   Radial transport of water and root hydraulic conductivity in general, and presumably also radial transport of ions, is strongly affected by maturation of xylem vessels along the root axis. For example, in graminaceous species like maize, growing in soil two types of roots are found; 'sheathed' with strong persisting soil sheath, and 'bare' roots

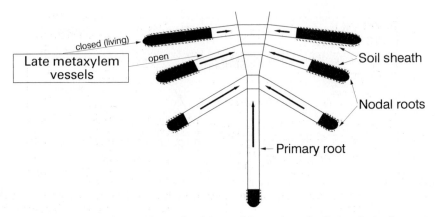

Fig. 2.34   Model of root hydraulic conductivity and formation of soil sheaths in the root system of maize and other C4 grasses. (Modified from Wenzel *et al.*, 1989; reprinted by permission of the American Society of Plant Physiologists.)

with no closely adhering soil (Fig. 2.34). This 'root dimorphism' is caused by differences in the maturation of the metaxylem (LMX) vessels (Fig. 2.31). In the sheathed zones the LMX vessels are still alive and nonconducting (McCully and Canny, 1988). Accordingly, the hydraulic conductivity of bare roots is about 100 times greater than that of sheathed roots (Wenzel *et al.*, 1989). This difference leads directly to high moisture contents in the rhizosphere soil of the sheathed zones and low moisture of the bare zones (Fig. 2.34). Living LMX vessels persist up to 20–30 cm proximal to the root tip in maize (Wenzel *et al.*, 1989), and up to 17 cm proximal to the root tip in soybean (Kevekordes *et al.*, 1988). This delay in LMX maturation not only affects hydraulic conductivity of the roots and plant water relations (Wang *et al.*, 1991) but also the mechanism of release of ions into the xylem and their transport to the shoot.

2.8   Release of Ions into the Xylem

After radial transport in the symplasm into the stele, most of the ions and organic solutes (amino acids, organic acids) are released into the xylem. This release into fully differentiated non-living xylem vessels therefore represents a retransfer from the symplasm into the apoplasm. Crafts and Broyer (1938) postulated uphill transport in the symplasm across the cortex to the endodermal cells and, in the stele, a 'leakage' into the xylem. The distinctly lower oxygen tension in the stele than in the cortex of intact roots (Fiscus and Kramer, 1970) seemed to support this view of an oxygen deficiency-induced leakage. Also electrophysiological studies apparently indicate ion movement from the symplasm into the xylem along the electrochemical gradient (Bowling, 1981).

In contrast, Pitman (1972a) put forward a two-pump model for ion transport from the external solution into the xylem, one located at the plasma membrane of root cortical cells and the other at the symplasm–xylem interface (apoplasm) in the stele (Fig. 2.35).

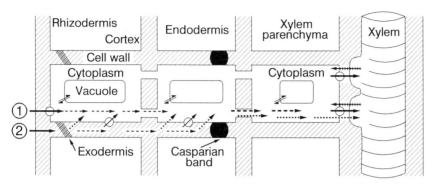

Fig. 2.35 Model for symplasmic (1) and apoplasmic (2) pathways of radial transport of ions across the root into the xylem. Key: ⊖→, active transport; ←, resorption. (Modified from Läuchli, 1976a.)

In this model the xylem parenchyma cells play a key role in ion secretion. High concentrations of ions such as K^+ in these cells, together with transfer cell-like structures (Kramer *et al.*, 1977) support this model. These cells seem to be involved also in reabsorption of ions from the xylem sap along the pathway to the shoot (Section 2.9).

The release of ions and organic solutes (*'xylem loading'*) is not well understood. The key role of a respiratory-dependent proton pump at the plasma membrane of the parenchyma cells is now well established. Protons are pumped into the xylem (DeBoer *et al.*, 1983; Mizuno *et al.*, 1985) and acidify the xylem sap which has pH values between about 5.2 and 6.0 depending, for example, on plant species and source of nitrogen supply (Arnozis and Findenegg, 1986). Similarly to the pump at the plasma membrane of cortical cells (Fig. 2.9) the proton pump at the plasma membrane of xylem parenchyma cells transfers protons into the apoplasm of the xylem vessels and may thereby act indirectly by reabsorption as a driving force for the secretion of cations (antiport). Anions may be secreted either by cotransport with the protons or along the electrical potential gradient formed by the proton pump (e.g. transport into the vacuole, Fig. 2.9). As the oxygen tension is usually lower in the stele than in the cortex, the xylem loading pump is more quickly inhibited at decreasing oxygen tensions in the root environment (DeBoer *et al.*, 1983).

Recently this concept of xylem loading as an energized process has been questioned by Wegner and Raschke (1994). By using isolated parenchyma cells from barley roots and measuring plasma membrane-potential related fluxes of cations and anions these authors suggest that similar to guard cells at closing (Section 8.7.6.2) also release of ions into the xylem sap occurs through ion channels in a process which is thermodynamically passive.

Irrespective of the different views on the mechanism there is general agreement that xylem loading is separately regulated from the ion uptake in cortical cells. This separate regulation step offers the plant the possibility of control selectivity and rate of long-distance transport to the shoot, for example as a feedback regulation depending on shoot demand [Fig. 2.23 (6)]. For example, preferential xylem loading of nitrate

Table 2.35
Root Uptake and Translocation to the Shoot of Phosphate and Sulfate in two
Genotypes of *Arabidopsis thaliana*[a]

	Phosphate[b]		Sulfate	
Genotype	Root uptake (nmol g^{-1} h^{-1})	Translocation to shoot (%)	Root uptake (nmol g^{-1} h^{-1})	Translocation to shoot (%)
Wildtype	1593	35	291	25
Mutant	1559	0.9	367	12

[a]Poirier *et al.* (1991). Reprinted by permission of the American Society of Plant Physiologists.
[b]Supply of 8 μM P$_i$.

compared with the amino acid glutamine may play an important role in partitioning of different forms of nitrogen between roots and shoots (Schobert and Komor, 1990).

Selective inhibitors of protein synthesis strongly impair xylem loading of mineral nutrients such as potassium without affecting their accumulation in the roots (Läuchli *et al.*, 1978; Morgan *et al.*, 1985). An example of separate genetic control of the xylem loading step is also shown in Table 2.35. Compared with the wildtype the mutant of *Arabidopsis* requires very high external phosphate supply for normal growth. At low phosphate supply the mutant becomes severely phosphorus deficient because of impaired shoot transport of phosphorus whereas the uptake in the roots is not different from the wildtype (Table 2.35). In contrast to phosphate sulfate transport to the shoot is similar in the mutant and the wildtype. This defect in the mutant is caused by a single recessive gene locus which obviously regulates the xylem loading of phosphorus (Poirier *et al.*, 1991). From results obtained by Sasaki *et al.* (1987) one may speculate on regulation of the enzyme glucose-6-phosphatase. In maize the activity of this enzyme and the concentration of its substrate glucose-6-phosphate are particularly high in xylem parenchyma cells. Inhibition of this enzyme by glucosamine severely depresses the xylem loading of phosphorus but not accumulation in the roots. Evidence for a particular fine regulation of phosphorus loading into the xylem is also the inability of maize plants to meet the phosphorus demand of the shoot at low root zone temperatures (Engels and Marschner, 1992a; Section 2.5.2).

The discovery of the abundance of living LMX vessels (see above) comprising more than half the total root length in maize up to 50 days old (Wenzel *et al.*, 1989) renewed the view of leakage as a mechanism of ion release into the xylem (McCully *et al.*, 1987). The concentrations of K$^+$, for example, in the vacuoles of living LMX vessels are up to 400 mM. These high levels of K$^+$, together with the other solutes in the vacuoles, are released into the transpiration stream at maturation of the LMX vessels. According to McCully and Canny (1988) this leakage from maturing xylem vessels can account for about 10% of the shoot demand of growing maize plants. Thus, at least part of the solutes in the xylem sap (including proteins) may derive not from active xylem loading but leakage from maturing xylem vessels.

2.9   Factors Affecting Ion Release into the Xylem and Exudation Rate

The permeability of plant membranes to water is much higher than that to ions. Plant cells or roots therefore behave as osmometers. Ion release (secretion, Section 2.8) into the apoplasm of the stele decreases both the osmotic potential and the water potential (they become more negative), and a corresponding net flux of water from the external solution is induced. As a result of this water flux, the hydrostatic pressure increases. As the endodermis with the Casparian band 'seal' the apoplasm of the stele, the hydrostatic pressure in the stele induces a volume flow of water and solutes (ions, molecules) in the non-living xylem vessels towards the shoot. Because of this 'root pressure' droplets are sometimes released on the tips and margins of leaves, a process known as *guttation*. This is particularly apparent in seedlings and young plants at night and in the early morning (under conditions of high relative air humidity and low transpiration). Exudation from the stumps of cut plants (e.g., freshly mown grass) is also the result of root pressure.

Root pressure and corresponding volume flow in the xylem are of particular importance for the long-distance transport of calcium into low-transpiring organs such as fruits and for nutrient cycling in plants (Section 3.4). Volume flow and composition of the xylem exudate also provide important information on the influence of external and internal factors on root metabolism and activity, mineral nutrient uptake, assimilation in the roots, release into the xylem and on cycling of mineral nutrients and organic solutes in plants.

For technical reasons, it is difficult to measure directly ion release into the xylem as the final step of the radial transport (Fig. 2.35). Thus most experimental evidences on this step are based on studies of xylem exudate, or xylem sap in general, obtained from isolated roots or decapitated plants. Because of reabsorption along the pathway (Section 3.2), and a certain contribution of solutes from maturing LMX vessels, the concentration of ions at the collecting sites can be different from that at the sites of secretion into the nonliving xylem vessels. In interpreting results on xylem exudate it has also to be kept in mind that at least two separately regulated membrane transport processes ('pumps' or 'channels') are involved in radial transport of mineral nutrients from the external solution into the xylem (Fig. 2.35), and that xylem sap volume flow is also determined by the root hydraulic conductivity. Some examples are given in the following on rates and composition of xylem exudates.

2.9.1   External and Internal Factors

As a rule, an increase in the external ion concentration leads to an increase in the concentration of ions in the xylem exudate. However, the relative concentration falls as the external concentration is increased (Table 2.36). This concentration gradient ('Concentration factor') between the external solution and the xylem exudate decreases and can even fall below 1 in the case of calcium. The exudation volume flow shows a somewhat different pattern, and is maximal at 1.0 mM external concentration (Table 2.36). At 0.1 mM this flow is limited by the ion concentration in the xylem. In contrast, at 10.0 mM, the flow is limited by the water availability (i.e., the low water potential in the external solution) and the small concentration gradient between the external solution and the xylem. The increase in the exudate concentration of the mineral nutrients, with

Table 2.36

Relationship between External Concentration, Exudate Concentration, and Exudate Volume Flow in Decapitated Sunflower Plants

External solution KNO_3 + $CaCl_2$ (mM each)	Exudate (mM)			'Concentration factor'			Exudation volume flow (ml $(4 h)^{-1}$)
	K^+	Ca^{2+}	NO_3^-	K^+	Ca^{2+}	NO_3^-	
0.1	7.3	2.8	7.4	73	28	74	4.0
1.0	10.0	3.2	10.7	10	3.2	10.1	4.5
10.0	16.6	4.2	10.3	1.7	0.4	1.0	1.6

the rise in external concentration from 1.0 to 10.0 mM, does not compensate for the decrease in the exudation volume flow. Thus, in contrast to the accumulation in roots (hyperbolic function of the external concentration, see e.g., Fig. 2.11), the rate of root pressure-driven shoot transport of mineral nutrients can decline at high external concentrations.

An increase in the root zone temperature has a much greater effect on the exudation volume flow than on the ion concentration in the exudate (Table 2.37). This is consistent with the expectation that a root behaves as an osmometer: Temperature determines the rate of ion transport in the symplasm (plasmodesmata, Section 2.7) and the release into the xylem, and water moves accordingly along the water potential gradient. There are, however, distinct differences between a root and a simple osmometer. An increase in the root temperature results in an increase in the potassium concentration but a decrease in the calcium concentration of the exudate. This shift in the potassium/calcium ratio might reflect temperature effects either on membrane selectivity or on the relative importance of the apoplasmic pathway of radial transport of calcium and water (Engels et al., 1992). Similar shifts in the potassium/calcium translocation ratio are also observed at different soil temperatures (Walker, 1969). This temperature effect may have important implications for the calcium nutrition of plants and might explain the enhancement of calcium deficiency symptoms in lettuce at elevated root temperatures, despite a slight increase in the calcium concentration of the leaf tissue (Collier and Tibbits, 1984).

Table 2.37

Temperature Effect on Exudation Volume Flow and on Potassium and Calcium Concentration in the Exudate of Decapitated Maize Plants[a]

Temperature (°C)	Exudation volume flow (ml $(4 h)^{-1}$)	Exudate concentration (mM)		Ratio K^+/Ca^{2+}
		K^+	Ca^{2+}	
8	5.3	13.4	1.5	8.9
18	21.9	15.2	1.0	15.2
28	31.7	19.6	0.8	24.5

[a]Concentration of KNO_3 and $CaCl_2$ in the external solution: 1mM each.

Table 2.38
Effect of Root Respiration on Exudation Volume Flow and Ion
Concentration in the Exudate of Decapitated Maize Plants[a]

Treatment[b]	Exudation volume flow (ml (3 h)$^{-1}$)	Exudate concentration (mM)	
		K$^+$	Ca^{2+}
O$_2$	26.5	16.6	1.8
N$_2$	5.7	15.2	1.7

[a]Concentration of KNO$_3$ and CaCl$_2$ in the external solution: 0.5 mM each.
[b]Respiration treatment consisted of bubbling oxygen or nitrogen through
the external (nutrient) solution.

The rate of release of ions into the xylem is closely related to root respiration (Table 2.38). A lack of oxygen strongly depresses the exudation volume flow but not the concentrations of potassium and calcium in the exudate. Oxygen deficiency seems to affect ion release into the xylem and root hydraulic conductivity to the same degree.

As in ion accumulation in root cells, maintenance of the cation–anion balance is necessary in the xylem exudate (Allen *et al.*, 1988; Findenegg *et al.*, 1989). Therefore, the accompanying ion may affect the transport rate even at low external concentrations (Table 2.39). When KNO$_3$ is supplied, the exudation flow rate is almost twice as high as the flow rate when an equivalent concentration of K$_2$SO$_4$ is added. Since the potassium concentration in the exudate is similar in both treatments, the transport rate of potassium supplied as K$_2$SO$_4$ is only about half the rate of potassium supplied as KNO$_3$.

In contrast to the potassium concentration, the concentrations of nitrate and sulfate in the exudate exhibit a large difference (18.1 and 0.6 meq l^{-1}, respectively) between the treatments. The corresponding difference in negative charges in the exudate is approximately compensated for by elevated concentrations of organic acid anions. In the K$_2$SO$_4$ treatment, however, the rate-limiting factor is probably the capacity of the

Table 2.39
Flow Rate and Ion Concentration in the Xylem Exudate of Wheat
Seedlings[a]

Parameter	Treatment	
	KNO$_3$	K$_2$SO$_4$
Exudation flow rate (μl h^{-1} per 50 plants)	372	180
Ion concentration (μeq ml^{-1})		
Potassium	23.3	24.5
Calcium	9.1	9.5
Nitrate	18.1	0.0
Sulfate	0.2	0.8
Organic acids	9.6	25.8

[a]Seedlings were supplied with either KNO$_3$ (1 mM) or K$_2$SO$_4$ (0.5 mM) in
the presence of 0.2 mM CaSO$_4$. From Triplett *et al.* (1980).

Table 2.40

Relationship between Photoperiod, Carbohydrate Content of Roots, and Uptake and Translocation of Potassium in Decapitated Maize Plants[a].

	Photoperiod (h)	
	12/12[b]	24/0
Carbohydrate in roots (mg)	122 (48)	328 (226)[c]
Total potassium uptake (meq)	1.3	5.0
Potassium translocation in exudation volume flow (meq)	1.0	3.5
Exudation volume flow (ml $(8\ h)^{-1}$)	30.3	88.5
Relative decline in flow rate within 8 h (%)	60	12

[a]Data per 12 plants.
[b]Hours of light/hours of darkness. This pretreatment with different day lengths was for one day (i.e., the day prior to decapitation).
[c]Numbers in parentheses denote carbohydrate content after 8 h (decapitation).

roots to maintain the cation–anion balance by organic acid synthesis; this leads to a decrease in the rate of potassium and calcium release into the xylem and a corresponding decrease in exudation flow rate.

Release of ions into the xylem and the corresponding changes in root pressure are also closely related to the carbohydrate status of the roots (Table 2.40). Variation in the length of the photoperiod for one day before decapitation affects the carbohydrate status of the roots and correspondingly the rate and duration of exudation volume flow after decapitation. Both the uptake and translocation rate of potassium in roots with a high carbohydrate content are considerably greater than in roots that are low in carbohydrate. The higher translocation rate is closely related to the exudation volume flow. In roots with a low carbohydrate content, reserves are rapidly depleted after decapitation and there is a corresponding decline in the rate of exudation volume flow within 8 h. This depletion of carbohydrates in the roots of decapitated plants is one of the factors which limits studies on exudation volume flow.

However, release of ions into the xylem and exudation volume flow are not necessarily related to the carbohydrate status of the roots but can show endogenously regulated, distinct diurnal fluctuations which are also maintained in plants transferred to continuous darkness (Ferrario *et al.*, 1992a,b). Hormonal effects may be involved in this endogenous rhythm as, for example, abscisic acid (ABA) strongly enhances exudation volume flow (Fournier *et al.*, 1987).

2.9.2 Xylem Exudate, Root Assimilation and Cycling of Nutrients

Analyses of xylem exudates provide valuable information about assimilation of mineral nutrients in the roots, for example, or the capacity of roots for nitrate reduction or in legumes for N_2 fixation. In soybean and other tropical legumes the proportion of ureides (Chapter 7) to total nitrogen in the xylem exudate reflects the nodule activity and is also a suitable indicator in field-grown plants of the relative contribution of N_2 fixation to the nitrogen nutrition of legumes (Peoples *et al.*, 1989). In non-legumes, analysis of the forms of nitrogen in the xylem sap can provide valuable information on the assimilation of nitrogen in the root and the importance of the various organic and inorganic fractions in long distance transport (van Beusichem, 1988). This also holds true for analysis of the binding forms of heavy metals in the xylem exudate (Cataldo *et al.*, 1988). Analysis of xylem exudate also provides insight into transport of hormones (e.g., ABA, cytokinins) or related compounds like polyamines (Friedman *et al.*, 1986) as signals from roots to the shoots (Chapter 5). The discovery of unexpectedly high concentrations of sugars in xylem exudates of annual species (Cataldo *et al.*, 1988; Canny and McCully, 1989) may represent sugars leaked from the phloem and swept into the xylem stream before retrieval into the phloem (McCully and Mallett, 1993). These results demonstrate the potential of xylem sap analysis for modeling not only mineral nutrient cycling but also the nitrogen and carbon economy of plants.

There are, however, several factors to be considered in the interpretation of xylem exudate analyses. Xylem sap collected from decapitated plants (exudate) represents only the root pressure component of xylem volume flow. For evaluation of the transpirational component exudates have to be collected either under vacuum at the cut stump, or by increasing the external pressure in the root zone (pressure chamber). With both methods xylem volume flow is increased and mineral nutrient concentration usually decreased. However,with these methods calculated transport rates to the shoot might be quite different from the results found in intact plants (Salim and Pitman, 1984; Allen *et al.*, 1988). Furthermore, irrespective of the collection method, the xylem sap also contains shoot-derived nutrients recycled in the phloem and reloaded (or leaked) in the roots into the xylem (Section 2.5.6). Recycling of water may also have to be considered in interpreting of xylem sap analysis (Tanner and Beevers, 1990; Chapter 3). The fractions of recycled nutrients can be particularly high in the case of potassium, nitrogen and sulfur, and may lead to misinterpretations, for example, on the capacity of roots for reduction of nitrate and sulfate. The proportion of recycled nutrients in the xylem sap depends on various factors such as plant species and nutritional status in general and shoot demand in particular as shown for potassium in Table 2.41.

When shoot demand is high potassium transport in the xylem exudate strongly increases and net translocation is also higher (sequential harvests), but net uptake is unaffected (Table 2.41). Accordingly, root content of potassium is about 20% higher in the plants with low shoot demand. The differences between net translocation and xylem

Table 2.41
Role of Shoot Demand on Net Uptake, Net Translocation and Flux of Potassium in the Xylem Exudate of Maize[a]

Shoot demand[b]	Potassium (μmol g^{-1} root fresh wt h^{-1})		
	Net uptake 0–3 days	Net translocation 0–3 days	Xylem exudate day 3
High	2.26	1.83	8.55
Low	2.28	1.17	2.46

[a]Engels and Marschner (1992b).
[b]Shoot demand altered by the shoot base temperature, see Table 2.15.

transport of potassium reflect differences in recirculation of potassium which, against expectation, were higher in plants with high shoot demand and presumably related to the role of potassium in xylem transport of nitrate (Section 3.2). This is an example of the important information xylem sap analysis can provide for regulation of xylem loading and recycling of mineral nutrients, if combined with other methods like phloem sap analysis (Allen *et al.*, 1988; Van Beusichem *et al.*, 1988) or measurements of net uptake and net changes in nutrient contents in roots and shoots.

3

Long-Distance Transport in the Xylem and Phloem and Its Regulation

3.1 General

The long-distance transport of water and solutes – mineral elements and low-molecular-weight organic compounds – takes place in the vascular system of xylem and phloem. Long-distance transport from the roots to the shoots occurs predominantly in the nonliving xylem vessels. Coniferous trees lack the continuous system of xylem vessels, and depend on tracheides which are non-living conducting cells ranging in length from 2 to 6 mm (Tyree and Ewers, 1991). In annual plant species too long-distance transport in the xylem vessels may be interrupted by tracheides, for example at the root–shoot junction (Aloni and Griffith, 1991) or in the nodes of the stem. These structures pose an internal resistance to xylem volume flow but simultaneously permit an intensive xylem–phloem solute transfer (Section 3.3.4).

Xylem transport is driven by the gradient in hydrostatic pressure (root pressure) and by the gradient in the water potential. As a reference the water potential of pure free water is defined as having a water potential of zero. Accordingly, values for water potential are usually negative. The gradient in water potential between roots and shoots is quite steep particularly during the day when the stomata are open. Values become less negative in the following sequence: atmosphere $\gg$ leaf cells $>$ xylem sap $>$ root cells $>$ external solution. Solute flow in the xylem from the roots to the shoots is therefore unidirectional (Fig. 3.1). However, under certain conditions in the shoots a

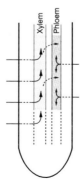

Fig. 3.1 Direction of long-distance transport of mineral elements in non-living xylem vessels and in the phloem in roots.

Table 3.1
Accumulation and Long-Distance Transport of ^{45}Ca, ^{22}Na
and ^{42}K in Maize Seedlings[a,b]

Plant part	Content (μeq per 12 plants $(24\,h)^{-1}$)		
	^{45}Ca	^{22}Na	^{42}K
Shoot	2.20	0.01	9.07
Endosperm	0.18	0.04	2.38
24–27 cm root	0.01	0.06	0.35
21–24 cm root	0.01	0.09	0.85
18–21 cm root	0.01	0.18	1.30
15–18 cm root	0.01	0.46	1.58
12–15 cm zone of supply	0.40	1.28	1.93
9–12 cm root	0	0.03	0.40
6– 9 cm root	0	0.02	0.38
3– 6 cm root	0	0.02	0.45
0– 3 cm root	0	0.01	0.75
Total	2.82	2.20	19.44

[a]Based on Marschner and Richter (1973).
[b]Each seedling was supplied with 1 meq l^{-1} of labeled nutrient solution to the root zone 12–15 cm from the root tip. The remainder of the root system was supplied with the same solution in which the nutrients were not labelled.

counterflow of water in the xylem may also occur, for example, from low-transpiring fruits back to the leaves (Lang and Thorpe, 1989; Section 3.4).

In contrast to the xylem, long-distance transport in the phloem takes place in the living sieve tube cells and is bidirectional. The direction of transport is determined primarily by the nutritional requirements of the various plant organs or tissues and occurs, therefore, from source to sink (Chapter 5). In addition, phloem transport is an important component in cycling of mineral nutrients between shoots and roots (Section 3.4) and for signal conductance of the nutritional status of the shoots (Section 2.5.6). Mineral elements can also enter the phloem in the roots and thus be translocated bidirectionally. The translocation of different mineral elements taken up by a particular zone of the root varies markedly during long-distance transport from the zone of supply, as shown in Table 3.1 for maize seedlings. For the reasons already mentioned, long-distance transport from the zone of supply to the root tip must take place in the phloem. Whereas ^{45}Ca is rapidly translocated to the shoot, the translocation of ^{22}Na toward the shoot is severely restricted. The steep gradient in the ^{22}Na content of the root sections in the direction of the shoot (basipetal) reflects resorption by the surrounding root tissue and is a typical feature of so-called natrophobic plant species (Section 10.2). Some ^{22}Na has also been translocated via the phloem to the root tip. In contrast, ^{42}K is quite mobile both in the xylem and in the phloem, and a markedly high proportion of the potassium taken up in more basal root zones is translocated via the phloem toward the root tip, which acts as a sink for this mineral nutrient.

Table 3.2
Xylem Volume Flow (Pressurized Exudation at 100 kPa) and Mineral Element Concentrations in the Xylem Sap of Soil-Grown Nodulated Soybean During Reproductive Stage[a]

Parameter	Plant development stages			
	Full pod extension	Early-mid podfill	Late podfill	Early leaf yellowing
Sap volume (ml (50 min)$^{-1}$ per plant)	1.43	1.13	0.94	0.43
Mineral element concentration				
K (mM)	6.1	5.0	4.0	2.4
Mg (mM)	3.8	2.6	1.9	1.2
Ca (mM)	4.8	3.9	3.9	2.2
P (mM)	2.5	1.6	0.9	0.4
S (mM)	1.8	1.6	2.1	1.5
B (mM)	1.0	1.5	1.6	3.2
Zn (μM)	23.0	29.0	32.0	42.0
Cu (μM)	2.7	3.6	2.8	6.9

[a]Based on Noodén and Mauk (1987).

During long-distance transport, mineral elements and organic solutes are transferred between the xylem and phloem by extensive exchange processes, referred to as loading and unloading. The transfer is mediated by specific cells called *transfer cells* (Pate and Gunning, 1972). Despite this interchange, and internal cycling, mineral nutrients, such as phosphorus, supplied to only one part of the root system (lateral or seminal roots) are transported preferentially to those parts of the shoots that have direct vascular connections with particular root zones (Stryker *et al.*, 1974). This distribution pattern is especially important for the mineral nutrition of trees that are supplied with fertilizer in a localized area of the root system.

3.2 Xylem Transport

3.2.1 Composition of the Xylem Sap

The composition and concentration of mineral elements and organic solutes in the xylem sap depend on factors such as plant species, mineral element supply to the roots, assimilation of mineral nutrients in the roots and nutrient recycling. Composition and particularly concentration of solutes are also strongly influenced by the degree of dilution by water (Section 2.9) and are therefore dependent on the transpiration rate and the time of day. Composition and concentration of xylem sap also change typically during the ontogenesis of plants (Table 3.2). In soybean during the reproductive stage xylem sap volume flow declines and the concentrations of some mineral nutrients in the sap decrease and of others increase. This decline in mineral nutrient concentrations can be reverted by depodding the plant, reflecting an easing in sink competition for photosynthates between pods and roots, thus, leading to higher uptake and xylem loading of mineral nutrients (Noodén and Mauk, 1987).

In perennial species in temperate climates composition of the xylem sap changes typically during the season not only in organic solutes (e.g. remobilized in spring) but also in nitrate concentrations and pH (Glavac et al., 1991). Polyvalent heavy metal cations in the xylem sap exist mainly in organic form complexed with organic acids, amino acids and peptides (White et al., 1981a,b). Both number and distribution of the complexes vary with plant age in annual species (Cataldo et al., 1988).

The form and proportion of the various nitrogen fractions in the xylem sap depend on the form of nitrogen supply (NO_3^-; NH_4^+; N_2 fixation), the predominant site of nitrate reduction (roots or shoots) and the proportion of recycled nitrogen (Section 3.4.4). Except at very high external NH_4^+ supply the concentration of NH_4^+ in the xylem is very low (Van Beusichem et al., 1988), in maize in the range of 1 mM, irrespective of whether nitrogen is supplied as NH_4^+ or NO_3^- (Engels and Marschner, 1993). The concentration of organic acids in the xylem sap depends primarily on the cation–anion uptake ratio by the roots (Table 2.39) and the form of nitrogen supply (Arnozis and Findenegg, 1986). In the xylem sap of annual species high concentrations of sugars may also occur, for example, in maize in which up to 5 mM may be present (Canny and McCully, 1989), and in soybean sugars may account for about 15% of the total organic carbon in the sap (Cataldo et al., 1988). Xylem sap may also contain enzymes like peroxidases (Biles and Abeles, 1991) which probably derive from maturing xylem elements (Section 2.8).

Phytohormones are a normal constituent of xylem sap, particularly cytokinins which are mainly synthesized in the roots (Section 5.6). Recently the concentration of abscisic acid (ABA) in the xylem sap has attracted wide interest as a possible nonhydraulic chemical root signal to the shoot of the root water status and also on the strength of the soil (Passioura and Gardner, 1990). As the soil dries out stomatal conductance decreases prior to decrease in leaf turgor, and inverse relationships have been shown to occur between stomatal conductance and xylem sap ABA concentrations (Zhang and Davies, 1989, 1990). Under field conditions too, for example, in maize, stomatal conductance has been found to be closely related to the ABA concentration of the xylem sap but not the current leaf water status nor ABA concentrations in the bulk leaf (Tardieu et al., 1992). There is a substantial body of evidence that high concentrations of ABA, or of 'inhibitors' other than ABA (Munns, 1992), in the xylem sap are also causally involved in a decrease in cell extension and division and, thus, also in leaf elongation in response to drying or compacting soil (Randall and Sinclair, 1988; Gowing et al., 1990; Blum et al., 1991). As soil dries out both the ionic composition and pH of the xylem sap increases (Gollan et al., 1992), and this may also alter the partitioning of ABA in the leaf cells and lead to preferential transport of ABA to the guard cells (Hartung et al., 1988; Hartung and Slovik, 1991; Section 5.6).

The increase in ABA concentrations in the xylem sap of nitrogen deficient plants and its consequences for plant water relations and leaf growth are discussed in Chapter 6. The root-derived hormonal signals in the xylem sap also distinctly affect long-distance transport of mineral nutrients, for example, via the volume flow rate in the xylem, the rate of xylem–phloem transfer (Section 3.3.4) and the mineral nutrient distribution within the shoot (Section 3.2.4).

3.2.2   Interactions Along the Pathway, Xylem Unloading

Along the pathway of solute transport in the nonliving xylem vessels (i.e., in the apoplasm), from the roots to the leaves important interactions take place between solutes and both the cell walls of the vessels and the surrounding xylem parenchyma cells. The major interactions are exchange adsorption of polyvalent cations in the cell walls, and resorption (uptake) and release of mineral elements and of organic solutes by surrounding living cells (xylem parenchyma and phloem).

3.2.2.1   Exchange Adsorption and Resorption

The interactions between cations and the negatively charged groups in the cell walls of the xylem vessels (and tracheides) are similar to those in the AFS of the root cortex (Fig. 2.2). The long-distance transport of cations in the xylem can be compared with ion movement in a cation exchanger with a corresponding decline in the translocation rate of cations such as Ca^{2+} (Jacoby, 1967) and Cd^{2+} (Senden and Wolterbeek, 1990), relative to that of water (Thomas, 1967) or anions such as phosphate (Ferguson and Bollard, 1976). This cation-exchange adsorption is not restricted to the xylem vessels; in addition the cell walls of the surrounding tissue take part in these exchange reactions (Wolterbeek et al., 1984).

The degree of retardation of cation translocation depends on the valency of the cation ($Ca^{2+} > K^{+}$), its own activity (McGrath and Robson, 1984), the activity of competing cations, the charge density of the negative groups (dicots > monocots), the pH of the xylem sap which may vary between 5 and 7 and the diameter of the xylem vessels. The translocation rate in the xylem of heavy metal cations is much enhanced when the ions are complexed, for example, in the case of copper (Smeulders and van de Geijn, 1983), zinc (McGrath and Robson, 1984) or cadmium (Senden and Wolterbeek, 1990).

Solutes can also be resorbed ('scavenged' or 'unloaded', see below) from the xylem (apoplasm) into the living cells (transport in the cytoplasm and vacuole) along the pathway of the xylem sap from the roots to the leaves. At least for resorption of potassium, the proton efflux pump in the plasma membrane of xylem parenchyma cells is the driving force for the H^{+}/K^{+} antiport (De Boer et al., 1985). With increasing path length, also the concentration of amino acids in the xylem sap may decrease for example, in nodulated legumes (Pate et al., 1964). However, the opposite may also occur, namely an increase in the solute concentrations by a factor of two or more, as has been found for calcium and magnesium in castor bean (Jeschke and Pate, 1991b).

Resorption from the xylem can be the result either of transient or permanent storage in the xylem parenchyma and other stem tissue, or of xylem–phloem transfer, mediated by specialized cells (xylem parenchyma transfer cells, Section 3.3.4). In some plant species, the resorption of certain mineral elements from the xylem sap is very pronounced and can have important consequences for the mineral nutrition of these plants. This is most evident in so-called natrophobic plant species (Section 10.2). In these plant species (e.g., *Phaseolus vulgaris*) Na^{+} is retained mainly in the roots and

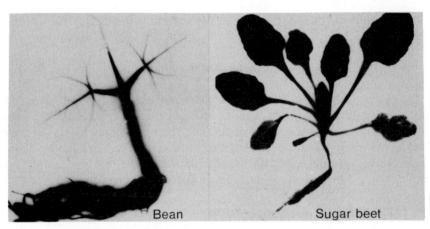
Bean   Sugar beet

Fig. 3.2   Distribution of sodium in bean (*Phaseolus vulgaris* L.) and sugar beet (*Beta vulgaris* L.)
24 h after 5 mM ^{22}NaCl was supplied to the roots. Autoradiogram.

lower stem, whereas in natrophilic species (e.g. sugar beet) translocation into the leaves
readily occurs (Fig. 3.2).

This restricted upward Na$^+$ translocation is caused by selective Na$^+$ accumulation in
the xylem parenchyma cells (Rains, 1969; Drew and Läuchli, 1987) together with
retranslocation into the roots (Fig. 3.10). In castor bean these two components led to a
depletion in the Na$^+$ concentration from 0.8 to 0.2 mM in the upward-moving xylem
stream (Jeschke and Pate, 1991b).

Resorption of Na$^+$ from the xylem sap is therefore an effective mechanism of
restricting translocation to the leaf blades. This mechanism, however, is not necessarily
advantageous for the salt tolerance of plants (Drew and Läuchli, 1987; Jeschke and
Pate, 1991b; see also Section 16.6) and is also a disadvantage in forage plants. For
animal nutrition the sodium content of the forage should be at least 0.2%. As shown in
Table 3.3, in *Lolium perenne* and *Trifolium repens*, Na$^+$ is readily translocated to the

Table 3.3

Effect of Sodium Fertilizer on the Sodium Content of Roots and Shoots of Pasture
Plants[a]

	Na Content (% dry wt)			
	Without Na fertilizer		With Na fertilizer	
Plant species	Roots	Shoots	Roots	Shoots
Lolium perenne	0.03	0.26	0.06	1.16
Phleum pratense	0.10	0.04	0.28	0.38
Trifolium repens	0.27	0.22	0.77	1.96
Trifolium hybridum	0.45	0.03	0.77	0.22

[a]Based on Saalbach and Aigner (1970).

Table 3.4
Distribution of Molybdenum in Bean and Tomato
Plants Supplied with Molybdenum in the Nutrient
Solution[a]

	Molybdenum content $(\mu g\ g^{-1}$ dry wt)	
Plant parts	Bean	Tomato
Leaves	85	325
Stems	210	123
Roots	1030	470

[a]Concentration of molybdenum in solution: 4 mg l^{-1}.
Based on Hecht-Buchholz (1973).

shoots, whereas in *Phleum pratense* and *Trifolium hybridum* this translocation is rather restricted. It is evident that in order to increase the sodium content of forage selection of suitable plant species is more important than the application of sodium fertilizers.

Resorption from the xylem sap in roots and stems can also be a determining factor in the distribution of micronutrients in plants. In certain species, such as bean and sunflower, molybdenum is preferentially accumulated in the xylem parenchyma of the roots and stems. In these species a steep gradient occurs in the molybdenum concentrations from the roots to the leaves (Table 3.4). In contrast, in other species, such as tomato, molybdenum is readily translocated from the root to the leaves. In accordance with this finding, when the molybdenum supply in the nutrient medium is high, toxicity occurs much earlier in tomato than in bean or sunflower (Hecht-Buchholz, 1973).

3.2.2.2 Release or Secretion

The composition of the xylem sap along the transport pathway can also be changed by the release or secretion of solutes from the surrounding cells. For example, in nonlegumes supplied with nitrate, the nitrate concentration in the xylem sap decreases as the path length increases, whereas the concentration of organic nitrogen, glutamine in particular, increases (Pate *et al.*, 1964). In nodulated legumes (where N_2 fixation occurs), on the other hand, the ratio of amides to amino acids is shifted in favor of the amino acids (Pate *et al.*, 1979).

Besides these specific aspects of nitrogen translocation, the release or secretion of mineral nutrients from the xylem parenchyma (and stem tissue in general) is of major importance for the maintenance of a continuous nutrient supply to the growing parts of the shoots. In periods of ample supply to the roots, mineral nutrients are resorbed from the xylem sap, whereas in periods of insufficient root supply they are released into the xylem sap. Changes in the potassium and nitrate contents of the stem base reflect this functioning of the tissues along the xylem in response to changes in the nutritional status of a plant. From this information a rapid test for nitrate in the stem base has been developed as a means for recommending rates of nitrogen fertilizer application (Section 12.3.8).

3.2.2.3 Xylem Unloading in Leaves

Despite resorption along the pathway in the stem most of the solutes and water are transported in the xylem vessels into the leaves. Here water is preferentially transported in the major veins to sites of rapid evaporation such as leaf margins, or from the vein endings mainly via symplasmic movement towards the stomata (Canny, 1990). Although the bundle sheath walls of the veins are suberized in grass leaves of C_3 and C_4 species, they do not provide a barrier against apoplasmic flux of water and solutes (Eastman *et al.*, 1988). Depending on the concentration and composition of solutes in the xylem sap entering the leaf, and the rate of water loss by transpiration, along its stream through the leaf, the solute concentration may be enriched manyfold at predictable sites, as for example, the leaf edges. This is particularly true when mineral element concentrations are high in the root medium (e.g. saline substrates) and for mineral elements such as boron and silicon (Section 3.2.4). Unless some of this excessive solute accumulation at the terminal sites of the transpiration stream is not removed, for example, by loading into the phloem, by guttation, as has been shown for boron (Oertli, 1962) or in epidermal glands in halophytes (Fitzgerald and Allaway, 1991), necrosis on the tips or margins of leaves occur (Fig. 3.7). This is a reflection of insufficient resorption of solutes along the pathway of the transpiration stream in the leaf.

Prevention of excessive solute accumulation in the leaf apoplasm by mechanisms other than uptake by the leaf cells can be achieved by the formation of salts of low solubility in the apoplasm. This strategy seems to be used particularly for the removal of soluble calcium in gymnosperms (Fink, 1991a). Calcium oxalate crystals are abundant in the needles of various gymnosperms in the cell walls of the mesophyll and particularly of the phloem and in the outer wall of the epidermis (Fig. 3.3). This mechanism of precipitation seems to be a safe way of coping with a continuous xylem import of calcium which is scarcely exported in the phloem (Section 3.4.3) and where the ionic concentrations in the symplasm have to be kept very low. The origin of oxalic acid in the apoplasm is unknown. Oxalic acid may be released from the cytoplasm or be formed in

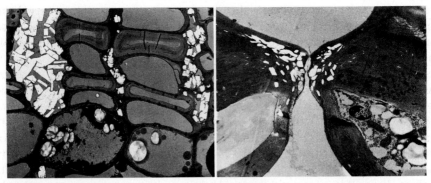

Fig. 3.3 Calcium oxalate crystals in the apoplasm of needles. (*Left*) Micrograph from the phloem of a needle from *Juniperus chinensis*; (*right*) micrograph of a stomatal pore of a needle from *Picea abies* (L.) Karst. (Courtesy of S. Fink, 1991a,c.)

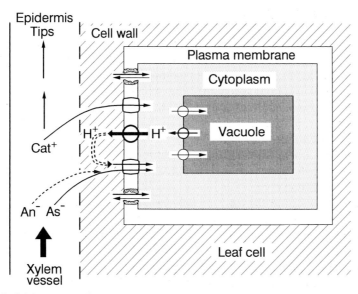

Fig. 3.4 Model for scavenging solutes from the xylem sap ('xylem unloading') in leaf cells. As⁻, amino acids.

the apoplasm during oxidative degradation of ascorbic acid which serves as antioxidant and is present in the apoplasm in fairly high concentrations (Polle *et al.*, 1990)

Of course, the xylem import of solutes into leaves and the evaporation of water does not necessarily lead to the accumulation of the solutes in the leaf apoplasm. In fast-growing plants with only low nutrient supply, the solute concentration in the xylem sap declines sharply from the roots to the leaves and within a leaf blade from the base to the tip. For example, in barley the concentration of magnesium in the xylem sap decreased from the base to the tips from 1.1 to 0.1 mm and for potassium from 18.0 to 8.0 mm (Wolf *et al.*, 1990b). Similarly, in tomato water released by guttation from the leaf tips was virtually free of inorganic solutes (Klepper and Kaufmann, 1966).

It is only recently that there has been any great interest in studying the mechanisms involved in the uptake of solutes from the apoplasm in leaves. This 'xylem unloading' or 'scavenging' of solutes from the leaf apoplasm is now reasonably well understood for the uptake of amino acids in legumes (Canny, 1988) and a number of broad-leaved trees (Wilson *et al.*, 1988). The cells of the bundle sheath are sites of intensive net proton excretion which acidifies the apoplasm (Fig. 3.4), the proton gradient across the plasma membrane acting as the driving force for cotransport of amino acids and ureides. The activity of the proton pump in legumes is high before pod filling and disappears during pod formation (Canny, 1988). In the plasma membrane of leaf cells the existence of at least two separate transport systems for aliphatic amino acids seems to be established (Li and Bush, 1991).

Although no information is available on mechanisms by which mineral nutrients are scavenged from the leaf apoplasm, a proton–anion cotransport for inorganic anions (e.g., NO_3^-) is also likely (Fig. 3.4) as is established for root cells (Fig. 2.9). Uptake of mineral nutrient cations may be mediated by a proton–antiport mechanism as has been shown for the resorption of K^+ from the xylem in the stem tissue (De Boer *et al.*, 1985). In view of the often high concentrations of mineral nutrients in the xylem sap and in the leaf apoplasm, for example between 5 and 18 mM K^+ (Wolf *et al.*, 1990b; Long and Widders, 1990) ion channels in the plasma membrane of leaf cells (Fig. 3.4) may more likely play a key role in 'xylem unloading' than in ion uptake by root rhizodermal cells (Section 2.4.2). Various methods are available for obtaining apoplasmic fluid from leaves (Meinzer and Moore, 1988) and application of these methods not only for determination of phytohormone concentrations in the leaf apoplasm (Hartung *et al.*, 1992) but also of mineral nutrients will improve our knowledge of the mechanism of this final step of the long-distance transport in the xylem from the roots into the leaves.

3.2.3 Effect of Transpiration Rate on Uptake and Translocation

The rate of water flux across the root (short-distance transport) and in the xylem vessels (long-distance transport) is determined by the root pressure and the rate of transpiration. An increase in the transpiration rate may, or may not, enhance the uptake and translocation of mineral elements in the xylem. Enhancement can be achieved in various ways, as shown in Fig. 3.5. Scheme A is true for mineral elements such as boron and silicon, except in the case of wetland rice (Section 10.3.2). Scheme C may be important for soil-grown plants (Section 15.2), particularly in saline substrates (Section 16.6). Whether or not transpiration affects uptake and translocation rate of mineral elements depends predominantly on the following factors:

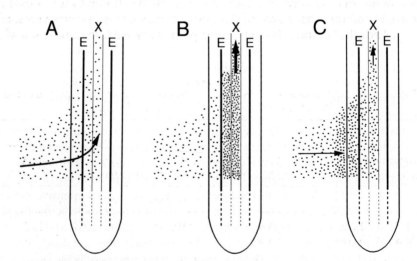

Fig. 3.5 Model for possible enhancement effects of high transpiration rates on the uptake and translocation of mineral elements in roots. A. 'Passive' transport of mineral elements through the apoplasm into the stele. B. More rapid removal of mineral elements released into the xylem vessels (Emmert, 1972). C. Increase in the mass flow of the external solution to the rhizoplane and eventually into the apparent free space of the cortex, favoring active uptake into the symplasm. E, Endodermis; X, xylem; arrow, water flux (A to C see text).

1. *Plant age.* In seedlings and young plants with a low leaf surface area, enhancement effects of transpiration are usually absent; water uptake and solute transport in the xylem to the shoots are determined mainly by the root pressure. As the age and size of the plants increase, the relative importance of the transpiration rate, particularly for the translocation of mineral elements increases.

2. *Time of day.* In leaves up to 90% of the total transpiration is stomatal. During the light period, transpiration rates and thus the potential enhancement of uptake and translocation of mineral elements are much higher than during the dark period. Short-term transient falls in the translocation rates of mineral elements at the onset of the dark period reflect the change from transpiration-mediated to root pressure-mediated xylem volume flow (Crossett, 1968). A consistent and synchronous diurnal pattern in transpiration rate and uptake rate of potassium and nitrate (Le Bot and Kirkby, 1992) is probably caused by changes in carbohydrate availability in the roots or feedback control of uptake.

Nodulated legumes show a particular diurnal pattern in shoot transport of fixed nitrogen. The sharp decrease in transpiration-driven xylem volume flow during the dark period is compensated for by a sharp increase in the concentration of fixed nitrogen (as ureides, see Chapter 7) in the xylem sap, thus keeping the total xylem transport rate of fixed nitrogen constant throughout the light/dark cycle (Rainbird *et al.*, 1983).

3. *External concentration.* It is well known that an increase in the concentration of mineral elements in the nutrient medium may enhance the effect of transpiration rate on the uptake and translocation of mineral elements. This is most likely the result of transport as shown in schemes A and C in Fig. 3.5. Usually, translocation rates are more responsive to different transpiration rates than are uptake rates, as shown for sodium in Table 3.5. The effect of transpiration on potassium is negligible in comparison with that

Table 3.5
Effect of Transpiration Rate of Sugar Beet Plants on Uptake and Translocation of Potassium and Sodium from Nutrient Solutions[a,b]

External concentration (mM)	Potassium		Sodium	
	Low transpiration	High transpiration	Low transpiration	High transpiration
	Uptake rate (μmol per plant $(4\ h)^{-1}$)			
$1\ K^+ + 1\ Na^+$	4.6	4.9	8.4	11.2
$10\ K^+ + 10\ Na^+$	10.3	11.0	12.0	19.1
	Translocation rate (μmol per plant $(4\ h)^{-1}$)			
$1\ K^+ + 1\ Na^+$	2.9	3.0	2.0	3.9
$10\ K^+ + 10\ Na^+$	6.5	7.0	3.4	8.1

[a]Based on Marschner and Schafarczyk (1967) and W. Schafarczyk (unpublished).
[b]Transpiration in relative values: low transpiration = 100; high transpiration = 650.

on sodium. This difference corresponds to the differences in the uptake isotherms of these elements at increasing external concentrations (Fig. 2.22). At low external concentrations the nitrate flux in the xylem of maize plants is also not affected by decreasing the transpiration rate to 50%, and a reduction in transpiration rate to 20% being required before a major decline in nitrate flux becomes apparent (Shaner and Boyer, 1976).

4. *Type of mineral element*. Under otherwise comparable conditions (e.g. plant age and external concentration), the effect of transpiration rates on the uptake and transport follows a typically defined ranking order of mineral elements. It is usually absent or only low for potassium, nitrate and phosphate but may become significant for sodium (Table 3.5) or calcium. As a rule, transpiration enhances the uptake and translocation of uncharged molecules to a greater extent than that of ions. There is a close relationship between transpiration rate and uptake rates of certain herbicides (Shone *et al.*, 1973). The uptake and translocation of mineral elements in the form of molecules is of great importance in the cases of boron (boric acid; Fig. 2.13) and silicon (monosilisic acid; Jones and Handreck, 1965; but see Section 10.3). A close correlation between transpiration and the uptake of silicon is shown for oat plants in Table 3.6.

There is perfect agreement between silicon content measured in plants and that predicted from the transpiration values (water loss times silicon concentration in the soil solution). Silicon accumulation in the shoot dry matter may therefore be a suitable parameter for calculations of the water use efficiency (WUE; kg dry matter produced/ kg water transpired) in field-grown cereals grown under rainfed conditions (Walker and Lance, 1991). However, this parameter is unsuitable, for example, in plants grown at different irrigation regimes (Mayland *et al.*, 1991) as well as in plants grown in nutrient solution (Jarvis, 1987), or when different genotypes within a species such as barley are compared (Nable *et al.*, 1990b).

Even in plants where close correlations between transpiration and silicon uptake are found, roots are not freely permeable to the radial transport of silicon. In wheat in endodermal cells large silicon depositions are found increasing from apical to basal root zones (Hodson and Sangster, 1989c) and such large deposition in the endodermis is typical for field-grown cereals (Bennett, 1982).

Table 3.6
Measured and Calculated Silicon Uptake in Relation to Transpiration (Water Consumption) of Oat Plants[a]

Harvest after days	Transpiration (ml per plant)	Measured uptake (mg per plant)	Calculated Si uptake[b] (mg per plant)
44	67	3.4	3.6
58	175	9.4	9.4
82	910	50.0	49.1
109	2785	156.0	150.0

[a]From Jones and Handreck (1965).
[b]Silicon concentration in the soil solution: 54 mg l^{-1}.

The absence of effects of reduced transpiration rates on the root to shoot transport of mineral nutrients may indicate a high proportion of xylem to phloem transfer in the stem tissue, or a corresponding increase in xylem sap concentrations of the mineral nutrients. Alternatively, the involvement of a nontranspirational component of xylem transport, namely the so-called *Münch-counterflow* has been stressed recently (Tanner and Beevers, 1990). This component is based on recycling of water derived from solute volume flow in the phloem from shoots to roots (Section 3.4.4) and the release of this water, and some of the solutes (recycled fraction) into the xylem. According to Tanner and Beevers (1990) the calculated amounts of water recycled in this manner vary between 9% (at high transpiration rates) and 30% (at low transpiration rates) of the total water uptake by the roots of maize plants. Values in this high order have been questioned for various reasons by Smith (1991). Nevertheless, recycling of water in plants must be taken into account in relation to recycling of mineral nutrients in general and distribution of calcium within the shoot in particular (Section 3.4.3).

3.2.4 Effect of Transpiration Rate on Distribution within the Shoot

The long-distance transport of a mineral element exclusively in the xylem should be expected to give a distinct distribution pattern in the shoot organs that depends on both transpiration rates (e.g., ml g^{-1} dry weight each day) and duration of transpiration (e.g. age of the organ). This is true, for example, for manganese (McCain and Markley, 1989) where at the same plant (maple tree) and similar leaf age the 'sun leaves' (high transpiration rates) have much higher manganese contents in their dry matter than 'shade leaves' (low transpiration rates). Both the distribution and content of silicon usually reflect the loss of water from the various organs. The silicon content increases with leaf age and is particularly high in spikelets of cereals such as barley. Even within a certain tissue, silicon distribution resembles the pathway of transpiration flow in the apoplasm. Silicon is deposited in the walls of the epidermis cells (Hodson and Sangster, 1988) or in the pericarp and outer aleurone layer of grass seeds such as *Setaria italica* (Hodson and Parry, 1982).

The distribution of boron is also related to the loss of water from the shoot organ, as shown by the boron distribution in shoots of rape in response to an increasing boron supply (Fig. 3.6). The typical gradient in the transpiration rates in the shoot organs (leaves > pods ≫ seeds) corresponds to the gradient in boron content.

Even for a particular leaf, an excessive supply of boron creates a steep gradient in the boron content: petioles < middle of the leaf blade < leaf tip (Oertli and Roth, 1969). Necrosis on the margins or leaf tips is therefore a typical symptom of boron toxicity (Fig. 3.7). In salt-affected plants the visible symptoms of toxicity (e.g., by chloride) are often quite similar, reflecting the transpiration-mediated distribution pattern within the shoot and its organs.

Frequently, a close positive correlation is observed between calcium distribution and the transpiration rates of shoot organs. This is shown, for example, by the low calcium content in the dry matter of low transpiring fleshy fruits (<0.3% calcium) as compared with that of the leaves (3–5% calcium) in the same plant. A lowering of the transpiration rate further decreases the calcium content of fruits (Table 3.7). The effect of transpiration on magnesium is much lower than its effect on calcium, and that on potassium is

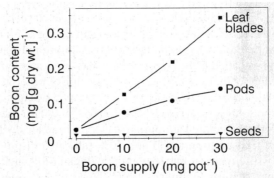

Fig. 3.6 Effect of increasing boron application to the soil on the distribution of boron in the shoots of rape. (Recalculated from Gerath *et al.*, 1975.)

negligible. Despite the correlations shown in Table 3.7, the interactions are much more complex between the rates of water and calcium influx into a plant organ (Section 3.4.3).

The fact that transpiration rates are higher and leaf water potentials are lower in mistletoe than in the host plant presumably explains why xylem parasites such as *Loranthus* can compete effectively with the host for mineral nutrients, nitrogen in particular, in the xylem fluid (Schulze *et al.*, 1984), and thereby also maintain a high influx of root-derived phytohormones such as cytokinins.

The influence of transpiration on the distribution differs not only between mineral elements but also between the various forms of the same element, as shown in Fig. 3.8 for nitrogen. Whereas the distribution within the shoot of ^{15}N from ammonium is independent of the transpiration rates (water loss) of the leaves and is translocated preferentially to the shoot apex, which acts as a sink for reduced nitrogen, ^{15}N from nitrate follows the transpiration pattern quite closely. The decrease in xylem flux of water and nitrate into older leaves of plant species such as bean is causally related to an endogenously regulated decrease in hydraulic conductivity caused by plugging of the xylem vessels at the pulvinal junction (Neumann, 1987). This plugging may be considered as a primary step of a programmed sequence leading to a decrease in xylem import of mineral nutrients and phytohormones into the leaf and, thus, to leaf senescence (Neumann, 1987).

3.3 Phloem Transport

3.3.1 Principles of Transport and Phloem Anatomy

Long-distance transport in the phloem takes place in living cells, the sieve tubes (Fig. 3.9). The principles of the transport mechanism in the phloem were proposed as early as 1930 by Münch in a *pressure flow hypothesis* (*Druckstromtheorie*) based on the principle of the osmometer. This has already been discussed in Section 2.8 for the root pressure. Münch suggested that solutes such as sucrose are concentrated in the phloem of leaves

Fig. 3.7   Boron toxicity in the leaves of lentil. (*Left*) control; (*right*) boron toxicity.

(i.e., *phloem loading*) and the water is sucked into the phloem, creating a positive internal pressure. This pressure induces a mass flow in the phloem to the sites of lower positive pressure caused by removal of solutes from the phloem. Flow rate and direction of flow are therefore closely related to the release or *unloading* at the sink. This type of pressure-driven mass flow in the phloem differs from that in the xylem in three important ways: (a) Organic compounds are the dominant solutes in the phloem sap; (b) transport takes place in living cells; and (c) the unloading of solutes at the sink plays an important role.

For mineral nutrients the main sites (sources) for phloem loading are located in the stem (Section 3.3.4) and the leaves as components of either mineral nutrient supply to growth sinks (shoot apices, fruits, roots) or of nutrient recycling (Section 3.4.4). An example for a primarily source–sink regulated transport of a mineral nutrient is shown in Fig. 3.10 for phosphorus. After application to one of the two mature primary leaves, the labelled phosphorus is transported to the shoot apex and the roots whereas transport to the other primary leaf is negligible. In contrast, sodium (Fig. 3.10) is not transported to the shoot apex but exclusively moves downwards (basipetally) to the

Table 3.7
Effect of Transpiration Rates of the Shoots of Red Pepper during Fruit Growth on the Mineral Element Content of the Fruits[a]

Transpiration rate (relative)	Mineral element content (mg g^{-1} dry wt)			Fruit dry wt (g per fruit)
	K	Mg	Ca	
100	91.0	3.0	2.75	0.62
35	88.0	2.4	1.45	0.69

[a]From Mix and Marschner (1976b).

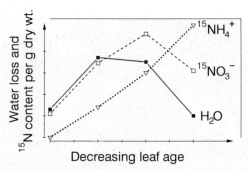

Fig. 3.8   Transpiration rates and distribution of labeled nitrogen (^{15}N) in different leaves of bean after $^{15}NO_3^-$ and $^{15}NH_4^+$ are supplied to the root. (Redrawn from Martin, 1971.)

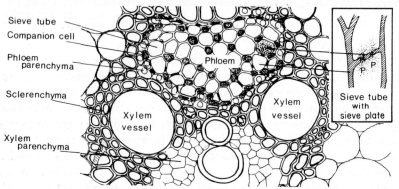

Fig. 3.9   Cross-sectional area of a vascular bundle from the stem of maize. Inset: sieve tube with sieve plate pores and 'P-protein'. (From Eschrich, 1976.)

roots where it is confined to the basal zones. From here a considerable net efflux of sodium takes place (Lessani and Marschner, 1978). This example also reflects the role of phloem transport in cycling of mineral elements and particularly in prevention of sodium accumulation in the shoots of natrophobic plant species. The capacity of bidirectional, ion specific long-distance transport is causally related to the physiology and anatomy of the phloem and its elements.

Within the phloem the sieve tube elements are associated with companion cells and parenchyma cells (Fig. 3.9). Some of these individual sieve tube elements are stretched end to end in a long series, forming the sieve tubes which are connected by conspicuous pores (inset, Fig. 3.9) called sieve plate pores. The sieve tubes are highly specialized vascular systems for the long-distance transport of solutes. The sieve tube cells contain a thin layer of cytoplasm, which forms transcellular filaments (the so-called P-protein) that pass through the sieve plate pores. The anatomical features of long-distance transport in the sieve tube across the sieve plate pores are similar to those of short-distance transport in the symplasm across the plasmodesmata.

In most plant species the sieve plate pores are lined with callose, a highly hydrated

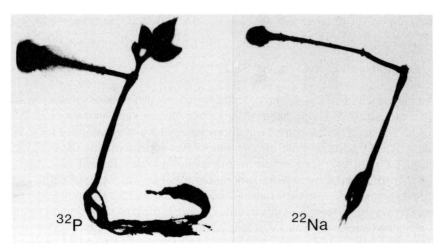

Fig. 3.10 Retranslocation of labeled phosphorus (^{32}P) and sodium (^{22}Na) after application to the tip of a primary leaf of bean. Autoradiogram, 24 h after application.

polysaccharide. There is good evidence that callose can swell rapidly and fill the pores, thus blocking long-distance transport in the sieve tubes. Callose formation is strongly enhanced by Ca^{2+} even at a concentration of a few μM (Kauss, 1987). This means that only very low concentrations of free Ca^{2+} can be present in the phloem sap for the normal functioning of long-distance transport. Plugging sieve tube pores is also induced by such factors as heat treatment or mechanical perturbation of the stem (Jaffe et al., 1985), as well as by mechanical injury of the sieve tubes, by incision, for example. Incision causes a sudden fall in the high internal pressure of the sieve tubes (>10 bars), which presumably triggers the mechanism of plugging the sieve tube plates. Considering the role of sieve tubes as food-connecting channels, this process can be thought of as performing the same function as a 'security valve' that prevents 'bleeding' when the system is injured. For experimental studies on long-distance transport this plugging mechanism is both an advantage and a disadvantage. It is an advantage in that very soon after decapitation of a plant, only xylem exudate is obtained at the stump of the root or stem; it is a disadvantage, in that, with a few exceptions – for example, the inflorescent stalks of certain palm tree species – it is very difficult to collect phloem exudate and thus to conduct extended studies on the mineral element composition of the phloem sap. There are some plant species (e.g., *Ricinus* and *Lupinus* spp.) from which small amounts of phloem exudate can be collected relatively easily by careful incision. However, with the incision technique there is always a possibility of contaminating the phloem sap by cut parenchyma cells and by substances from the apoplasm. Another method is to use sucking insects such as aphids and plant hoppers. In the process of feeding, these insects insert their stylet into the phloem tissue and sieve tubes. If the stylet is severed, for example with a laser beam (Hayashi and Chino, 1990) it remains in the tissue and the high internal pressure within the sieve tubes forces the phloem sap out of the open end of the stylet. This technique, of course, is very difficult, and the amounts of exudate obtained are quite small. For these reasons our knowledge of long-distance transport based on phloem sap analysis is rather limited, particularly for mineral elements.

3.3.2 Composition of the Phloem Sap

Phloem sap has a high pH (7–8) and contains high concentrations of solutes, on average 15–25% dry matter. A comprehensive analysis of phloem sap composition is shown in Table 3.8. The main component is usually sucrose, which may comprise up to 90% of the solids. The proportion of sucrose to other solutes depends on the site of phloem sap collection, it is very high, for example near the ear of cereals (Hayashi and Chino, 1990; Section 5.4). In addition to sucrose, among the other organic solutes amino acids are usually present in high concentrations (Table 3.8); the amides glutamine and aspara- gine may represent up to 90% of this fraction, whereas the concentrations of nitrate and ammonium are usually very low (Van Beusichem *et al.*, 1988). Organic acid anions such as citrate and malate are also abundant in the phloem sap, and in white lupin succinate concentrations may reach the same orders of magnitude as the concentration of total amino-N (Jeschke *et al.*, 1986). A whole range of other organic compounds are also found in phloem sap, for example vitamins, hormones, proteins and ATP.

Of the mineral elements, potassium is usually present in by far the highest concen-

Table 3.8
Comparison of the Levels of Organic and Inorganic Solutes in the Phloem and Xylem Exudates of *Nicotiana glauca*[a]

Substance	Phloem exudate (stem incision) pH 7.8–8.0 ($\mu g\ ml^{-1}$)[b]	Xylem exudate (tracheal) pH 5.6–5.9 ($\mu g\ ml^{-1}$)[b]	Concentration ratio phloem/xylem
Dry matter	170–196[c]	1.1–1.2[c]	155–163
Sucrose	155–168[c]	ND	—
Reducing sugars	Absent	NA	—
Amino compounds	10 808.0	283.0	38.2
Nitrate	ND	NA	—
Ammonium	45.3	9.7	4.7
Potassium	3673.0	204.3	18.0
Phosphorus	434.6	68.1	6.4
Chloride	486.4	63.8	7.6
Sulfur	138.9	43.3	3.2
Calcium	83.3	189.2	0.44
Magnesium	104.3	33.8	3.1
Sodium	116.3	46.2	2.5
Iron	9.4	0.60	15.7
Zinc	15.9	1.47	10.8
Manganese	0.87	0.23	3.8
Copper	1.20	0.11	10.9

[a]From Hocking (1980b).
[b]ND. Not present in a detectable amount; NA, data not available.
[c]$mg\ ml^{-1}$.

tration, followed by phosphorus, magnesium and sulfur (Table 3.8). Sulfur occurs in both the reduced form (glutathione > methionine > cysteine; Rennenberg *et al.*, 1979; or cysteine > glutathione; Schupp *et al.*, 1991) and as sulfate (Smith and Lang, 1988). Sulfate concentrations in the phloem sap can be as high as those of phosphate (Van Beusichem *et al.*, 1988). Chloride and sodium may also be present at considerably high concentrations (Table 3.8), but this depends strongly on the external supply and the plant species (Jeschke and Pate, 1991b). In contrast, the concentration of calcium in the phloem sap is very low.

Reliable data on micronutrient concentrations in the phloem sap are rare (Table 3.8). In phloem sap collected from broccoli by stem incision fairly high boron concentrations ($6-13\ \mu g\ ml^{-1}$) were found (Shelp, 1987). There are no data on the molybdenum and nickel concentrations in phloem sap.

With the exception of calcium, the concentration of all solutes is usually several times greater in the phloem exudate than in the xylem exudate (Table 3.8). The data in Table 3.8 on phloem sap composition are in fairly good agreement with those obtained from analyses of stems of castor bean (Van Beusichem *et al.*, 1988), white lupin (Jeschke *et al.*, 1986) and rice (Chino *et al.*, 1982). For a comprehensive review of the composition of phloem sap, the reader is referred to Ziegler (1975).

3.3.3   Mobility in the Phloem

All mineral nutrients have been found in the phloem sap, except for molybdenum and nickel where no data are available so far. The question arises, however, as to whether the phloem sap, particularly the exudate collected by incision, fully reflect the *in vivo* mobility of mineral elements in long-distance transport in the phloem from source to sink. Another approach to the study of phloem mobility is the use of labeled elements (radioactive or stable isotopes) to follow long-distance transport after application, for example, to the tip of a leaf blade (Fig. 3.10). Because of the gradient in xylem water potential, retranslocation from the leaf tips and out of the treated leaf must take place in the phloem. On the basis of such studies and in consideration of the data on phloem sap composition, mineral nutrients can be classified depending on their phloem mobility (Table 3.9). Sodium has been included as it is a mineral nutrient for some plant species,

Table 3.9
Characteristic Differences in Mobility of Mineral Nutrients in the Phloem

High Mobility	Intermediate Mobility	Low Mobility
Potassium	Iron	Calcium
Magnesium	Zinc	Manganese
Phosphorus	Copper	
Sulfur	Boron	
Nitrogen (amino-N)	Molybdenum	
Chlorine		
(Sodium)		

and its phloem mobility is of particular importance for plants growing in saline substrates.

The classification in Table 3.9 is, of course, only a first approximation as certain factors are ignored, for example, genotypical differences or the nutritional status of plants. However, for the macronutrients, except calcium (but see Section 3.4.3), phloem mobility is generally high, and for the micronutrients it is at least intermediate with the exception of manganese. For molybdenum fairly high phloem mobility has been established from both indirect (Wood *et al.*, 1986) and direct (Kannan and Ramani, 1978) measurements. Studies on boron mobility in the phloem have been carried out by following boron translocation with time into developing fruits such as peanut (Campbell *et al.*, 1975) and particularly with the aid of boron isotopes (Chamel *et al.*, 1981; Changzhi *et al.*, 1990). Such investigations considered together with the fairly high concentrations of boron occurring in phloem exudates (Section 3.3.2) clearly reveal the phloem mobility of boron. Substantial amounts of boron are translocated in the phloem to growth sinks, for example, flower buds, after foliar application (Hanson, 1991a). Thus, boron may at least be classified as of intermediate phloem mobility.

Although some long-distance transport in the phloem can be demonstrated with labeled manganese (Nable and Loneragan, 1984; El-Baz *et al.*, 1990), the mobility is generally very low. The same is true for calcium. Although substantial calcium concentrations may be found in the phloem sap (Table 3.8) it is nevertheless appropriate to classify calcium as a mineral nutrient of very low phloem mobility. If it is supposed that the phloem sap delivered to a growth sink (e.g. shoot apex, young fruit) reflects the demand of that sink for mineral nutrients, the observed ratio of calcium/potassium of about 1/100 in the phloem sap in Table 3.8 (Jeschke and Pate, 1991b) is about 5 to 10 times too low to cover this demand. A similar conclusion may be drawn from other phloem sap analysis and thus, most of the calcium demand of growth sinks has to be covered by import via the xylem (Section 3.4.3).

3.3.4   Transfer between the Xylem and Phloem

In the vascular bundles, phloem and xylem are separated by only a few cells (Fig. 3.9). In the regulation of long-distance transport, exchange of solutes between the two conducting systems is very important. From the concentration differences shown in Table 3.8 it is evident that a transfer from phloem to xylem can occur downhill, through the plasma membrane of the sieve tubes, if an adequate concentration gradient exists. In contrast, for most organic and inorganic solutes a transfer from xylem to phloem is usually an uphill transport against a steep concentration gradient between the apoplasm (xylem) and the symplasm of the surrounding xylem parenchyma cells and the cells of the phloem (Fig. 3.9). The xylem-to-phloem transfer is of particular importance for the mineral nutrition of plants, because xylem transport is directed mainly to the sites (organs) of highest transpiration, which are usually not the sites of highest demand for mineral nutrients. This transfer of organic and inorganic solutes can take place all along the pathway from roots to shoot, and the stem plays an important role in this respect (McNeil, 1980; Van Bel, 1984), most likely via transfer cells (Fig. 3.11; Kuo *et al.*, 1980; Jeschke and Pate, 1991a). In stems, sites of intensive xylem-to-phloem transfer are the nodes, which function in cereals, for example, for mineral nutrients such as potassium

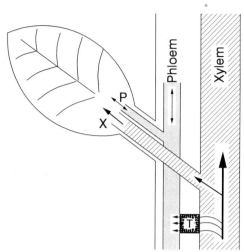

Fig. 3.11 Long-distance transport in xylem (X) and phloem (P) in a stem with a connected leaf, and xylem-to-phloem transfer mediated by a transfer cell (T).

(Haeder and Beringer, 1984a,b) and in soybean for amino acids (Da Silva and Shelp, 1990). In the transport of amino acids in soybean from roots to shoot, between 21 and 33% of the total xylem-to-phloem transfer occured in the stem and between 60 and 73% in the leaf blades.

The proportion of xylem-to-phloem transfer in the stem is influenced by the rate of xylem volume flow, i.e., also by the transpiration rate. In tomato doubling the volume flow rate reduced the transfer of amino acids in the stem distinctly in favour of a higher proportion transported to the older leaves at the expense of the shoot apex (Van Bel, 1984). A certain diurnal rhythm in the partitioning of solutes between mature leaves and shoot apex or fruits is thus also to be expected for this reason, unless it is not compensated for by a higher xylem-to-phloem transfer in the leaf blades.

Information is scarce about the opposite process, phloem-to-xylem transfer. In wheat, after anthesis retranslocation in the phloem from the flag leaf to the stem is followed by a considerable release of phosphorus, magnesium, and nitrogen, but not of potassium, into the xylem. These mineral nutrients are subsequently transported in the xylem into the ears (Martin, 1982). In white lupin at least in some regions of the stem, in the transport to the shoot apex phloem-to-xylem transfer seems to be of greater importance than transfer in the opposite direction (Jeschke et al., 1987).

3.4 Relative Importance of Phloem and Xylem for Long-Distance Transport of Mineral Nutrients

3.4.1 General

Precise quantitative assessments on the relative importance of solute transport in the phloem and xylem into parts or organs of plants are difficult to make. For such assessments not only the concentrations of solutes are required, but also the velocity of transport and the cross-sectional area of the conducting vessels, according to the following relationship:

$$\text{Specific mass transfer} = \text{velocity} \times \text{concentration}$$
$$(\text{g h}^{-1}\,\text{cm}^{-2}) \qquad (\text{cm h}^{-1}) \qquad (\text{mg ml}^{-1})$$

The velocity of transport in the xylem and phloem varies enormously. On average, velocities between 10 and 100 cm h^{-1} are often found, and the rates in the phloem are usually much lower than in the xylem. In fruit stalks of white lupins, maximal velocities of 22 cm h^{-1} in the phloem and 147 cm h^{-1} in the xylem have been reported (Pate *et al.*, 1978).

Our present knowledge on the relative importance of xylem and phloem import and export of mineral elements into plant parts or organs is mainly based on detailed analyses of phloem and xylem sap in different shoot parts of individual plants and the corresponding mineral element contents in the shoot parts at sequential harvests (Jeschke *et al.*, 1987; Jeschke and Pate, 1991a,b, 1992).

3.4.2   Mineral Nutrients with High Phloem Mobility

For mineral nutrients with high phloem mobility such as potassium, phosphorus or nitrogen as amino-N the relative importance of phloem and xylem transport into an organ mainly depends on the stage of development of the organ as shown in Table 3.10 for amino-N during the life of an individual leaf.

Throughout the life of the leaf of the nitrate-fed castor bean plant, nitrogen import by the xylem sap was high and continued at a high rate and only declined at the onset of senescence. Additional nitrogen import by the phloem during rapid leaf expansion was followed by a steep increase in phloem export so that export was greater than import. Nitrate import presented only a small fraction of the xylem import of nitrogen. The rates of phloem export of nitrogen closely matched the net rates of CO_2 fixation by the lamina (Jeschke and Pate, 1992).

In principle, similar data on the time course of import and export in xylem and phloem during the life of individual leaves have been obtained in barley for potassium

Table 3.10

Import and Export of Nitrogen During the Life of a Leaf of Nitrate-fed Castor Bean Plants[a]

Days after leaf emergence	Nitrogen (nmol per leaf)		
	Xylem (NO_3)[b]	Phloem	Net change
1–12	+2.7 (0.23)	+1.4	+4.10
13–20	+2.5 (0.43)	−1.1	+1.36
21–40	+2.8 (0.63)	−3.7	−0.87
41–60	+1.4 (0.48)	−4.0	−2.63

[a]Based on Jeschke and Pate (1992).
[b]NO_3-N import in the xylem sap.

(Greenway and Pitman, 1965) and for phosphorus (Greenway and Gunn, 1966). A lack of change in the net contents of highly phloem-mobile mineral elements in fully expanded leaves is therefore a reflection either of cessation of the import or, more likely, of an equilibrium between import and export (retranslocation).

3.4.3   Mineral Nutrients of Low Phloem Mobility; Example Calcium

Because of its low concentrations in the phloem sap (Section 3.3.2) the import of calcium into growth sinks such as shoot apices, young leaves or fruits takes place nearly exclusively in the xylem, whereas the import in the phloem is negligible as shown for castor bean in Table 3.11. This is in marked contrast to potassium of which most (terminal bud) and at least half (youngest leaves) of the total net import takes place in the phloem. For magnesium phloem import contributes to 25 and 40% of the total import, respectively.

In order to cover the relatively high calcium demand of growth sinks, particularly in dicotyledonous plant species which have a high cation exchange capacity in the apoplasm (Section 8.6), a high rate of xylem volume flow into these organs is required. Developing fruits of peanut (Hallock and Garren, 1968) and potato tubers (Krauss and Marschner, 1975) are exceptions as they can cover part of their calcium demand by direct uptake from the soil solution. Shoot apices, young leaves, particularly those enclosed by mature leaves (e.g. cabbage), and fleshy fruits are characterized by low rates of transpiration and inherent low rates of xylem volume flow. Calcium deficiency and the so-called calcium deficiency-related disorders, such as tipburn in lettuce, blossom end rot in tomato, and bitter pit in apple, are therefore widespread. For comprehensive reviews on this subject, see Bangerth (1979), Kirkby (1979) and Marschner (1983).

To increase the calcium content of growing leaves or fruits, increasing the transpiration rates of the fruits is more effective than increasing the calcium supply in the substrate (Table 3.12). As expected, potassium because of its high phloem mobility is not affected by these treatments. Furthermore, there is a distinct negative correlation between the growth rate and the calcium content in the dry matter of growing fruits, whereas this is again not observed with potassium. High growth rates are based on high

Table 3.11

Xylem and Phloem Import of Potassium, Magnesium and Calcium into the Terminal Bud and Youngest Leaves of Castor Bean[a]

	Terminal bud			Youngest leaves		
	K	Mg	Ca	K	Mg	Ca
Xylem	3.9	8.0	4.2	20.6	5.2	2.4
Phloem	20.4	2.0	0.03	19.3	2.0	0.03

[a]Data in μmol per plant per 9 days. Data recalculated from Jeschke and Pate (1991b).

Table 3.12

Effect of Environmental Factors and Growth Rate
on Calcium and Potassium Content of Red Pepper
Fruits[a]

Treatment and growth rate	Content in fruits (μmol g^{-1} dry wt)	
	Ca	K
Calcium supply to the roots		
0.5 mM	26.9	1315
5.0 mM	33.2	1228
Relative humidity in the fruit environment		
90%	32.7	1892
40%	55.4	1918
Growth rate of the fruits (mg dry wt per day)		
20.1	28.2	1772
29.9	20.7	1846
38.5	17.2	1813

[a]From Marschner (1983).

solute volume inflow via the phloem and thus correlated with high potassium, but a very low inflow of calcium. In addition, in organs with low transpiration rates, such as fleshy fruits, a high phloem solute volume flow either strongly depresses, or even reverses the direction of the xylem volume flow (Mix and Marschner, 1976c). This counter-flow of water in the xylem can be substantial, for example in grape berries (Lang and Thorpe, 1989) and lead to the export from fruits of both calcium (Mix and Marschner, 1976c) and organic solutes (Hamilton and Davies, 1988).

High transpiration rates of the whole shoot, however, often decrease rather than increase the calcium influx into low-transpiring organs such as rosettes of cauliflower (Krug et al., 1972). Under these conditions the xylem volume flow is directed to the high-transpiring outer leaves at the expense of the inner leaves or the rosettes. Inhibition of transpiration (by high relative humidity or during the dark period) usually favours the direction of the xylem volume flow towards low-transpiring organs. For example, in Chinese cabbage, an increase in relative humidity during the night increased the calcium concentrations in the inner leaves by 64% and decreased the proportion of tipburn heads by 90% (Berkel, 1988). In potato plants subjected to soil drying, calcium deficiency-related tuber necrosis could be significantly reduced by foliar spray with antitranspirants and a corresponding alteration of the leaf–tuber water potential gradients (Win et al., 1991). Diurnal shrinking (during the light period) and swelling (during the dark period) of the rosettes of cauliflower (Krug et al., 1972) or of cabbage (Wiebe et al., 1977) are closely correlated with corresponding changes in the xylem volume flow and calcium flux into various parts of the shoots.

Under conditions of low transpiration, the rate of xylem volume flow from the roots to the shoots is determined by the root pressure. The import of water and calcium via

Table 3.13
Relationship between Root Pressure (Guttation) and Calcium Transport into Expanding Strawberry Leaves[a]

Concentration of the nutrient solution[b]		Guttation (relative) (0–3)[c]	Tip necrosis (0–5)[d]	Content (μg per leaf)	
Day	Night			Ca	Mg
Concentrated	Concentrated	0.3	3.0	7	57
Concentrated	Diluted	2.4	0.3	25	77
Diluted	Concentrated	0.8	1.3	16	74
Diluted	Diluted	2.3	0.0	62	78

[a]Based on Guttridge *et al.* (1981).
[b]Root pressure varied by the concentration of the nutrient solution: Concentrated = 6.5 atm; diluted = 1.6 atm.
[c]0 = none; 3 = high.
[d]0 = none; 5 = very severe.

the xylem into low-transpiring organs therefore strongly depends on the root pressure. Water availability in the rooting medium, particularly during the dark period, is thus crucial for the long-distance transport of calcium into low-transpiring organs with high calcium demand for growth. In agreement with this, high osmotic potential of the soil solution (e.g. soil salinity) decreases both root pressure and calcium influx into young leaves or fruits and induces calcium deficiency symptoms (Bradfield and Guttridge, 1984; Mizrahi and Pasternak, 1985; Berkel, 1988).

These relationships are shown in Table 3.13 for expanding strawberry leaves. High root pressure, as indicated by the intensity of guttation, is closely correlated with an increased concentration of calcium in expanding leaves and either the absence of, or only mild symptoms of, calcium deficiency (tip necrosis). Magnesium, which is highly phloem mobile, is only slightly affected by root pressure. Root pressure also strongly depends on root respiration and oxygen supply to the roots. Interruption of the aeration of the nutrient solution during the night had no effect on calcium accumulation in the roots of tomato but reduced the calcium transport into the stem by 42% and into the leaves by 82% (Tachibana, 1991). An increase in blossom end rot in tomato by poor aeration of the rooting medium is well documented.

Calcium import rate into growth sinks is also dependent on the formation of new cation exchange sites in the apoplasm (cell walls) acting as sinks for calcium and thereby removing it from the exchange sites at the end of the xylem vessels (Marschner and Ossenberg-Neuhaus, 1977). This may contribute to the well-documented sharp decrease in the calcium import rate per unit of transpired water after termination of the extension growth of leaves (Marschner and Ossenberg-Neuhaus, 1977) and fruits (Mix and Marschner, 1976a).

In the fine regulation of the import rate of calcium into growth sinks, a nonvascular component, presumably cell-to-cell transport linked with a countertransport of auxin (IAA), plays an important role (De Guzman and De la Fuente, 1984). The basipetal (downward) polar transport of IAA in young tissue takes place most likely by H^+-IAA

Table 3.14

Effect of IAA (Auxin) and TIBA (triiodobenzoic acid) on Calcium Efflux From Sunflower Hypocotyl Segments[a]

	Calcium efflux (nmol per 20 segments h^{-1})	
	Apical end	Basal end
Control	22.4	15.4
IAA[b]	37.1	25.6
TIBA[b]	25.4	17.2

[a]De Guzman and De la Fuente (1984).
[b]Applied to the respective ends of the segments.

symport at one end of the cell and a Ca^{2+}-regulated IAA efflux at the other (Allan and Rubery, 1991). Accordingly, Ca^{2+} is required for the polar basipetal IAA transport. Simultaneously, the polar IAA transport is linked with an acropetal (upward) polar transport of calcium. Table 3.14 shows that in hypocotyl segments of sunflower calcium transport is higher towards the apical (younger) relative to the basal end of the segments, and distinctly enhanced by IAA application. This enhancing effect can be diminished by TIBA, an inhibitor of polar IAA transport (De Guzman and De la Fuente, 1986)

This causal relationship between IAA and calcium transport in a tissue can also be demonstrated in young leaves of lettuce (Banuelos *et al.*, 1988), in different shoot organs of mango trees (Cutting and Bower, 1989) and in young tomato fruits (Table 3.15). Pretreatment with TIBA inhibited polar basipetal IAA transport from the young fruit and simultaneously depressed the polar acropetal calcium transport into the fruit. In contrast, the acropetal transport rates of $^{86}Rb(K)$ and water were not affected by the TIBA treatment.

Increasing calcium import into fruits and leaves is, however, not advantageous under all circumstances. In tomato, for example, environmental factors which enhanced the

Table 3.15

Effect of TIBA Pretreatment (24 h) on Basipetal IAA Transport and Acropetal Transport of ^{45}Ca; $^{86}Rb(K)$ and $^{3}H_2O$ in 10-Day Old Tomato Fruits[a]

		Acropetal transport (Bq per fruit)		
Treatment	IAA (pmol per fruit)	^{45}Ca	^{86}Rb	$^{3}H_2O$
Control	18	33	17	39
TIBA	12	18	15	38

[a]Based on Banuelos *et al.* (1987).

calcium import into fruits increased the incidence of 'gold specks' as a physiological disorder caused by an excess rather than a deficiency of calcium in the tissue (De Kreij *et al.*, 1992). In the gold speck tissue high calcium levels were found together with many calcium oxalate crystals. Abundant formation of calcium oxalate crystals in the apoplasm of needles in gymnosperms (Fig. 3.3) is another example for excessive calcium import into an organ, and is particularly evident in trees growing on calcareous soils (Fink, 1991b).

3.4.4 Retranslocation and Cycling of Nutrients

With exception of calcium and presumably also manganese, import of nutrients in the xylem and export (retranslocation) in the phloem is a normal feature throughout the life of an individual leaf. Several pieces of evidence indicate a rapid xylem-to-phloem transfer in the leaf blades and involvement of only a small fraction of the total leaf content ('cycling' fraction; Fig. 2.26) in this process. A considerable portion of these nutrients is retranslocated in the phloem from the shoots back to the roots and may thereby serve several functions. They may be used to convey information about the nutritional status of the shoots and, via feedback regulation, also control uptake by the roots (Section 2.5.6). In natrophobic plant species retranslocation in the phloem is an important component in maintaining low sodium contents in the leaves (Fig. 3.10). This also holds true for some natrophilic, salt tolerant species such as reed (Matsushita and Matoh, 1992) but not for others such as barley (Munns *et al.*, 1987).

In plant species for which the shoot provides the site for most nitrate reduction (Section 8.2) retranslocation of nitrogen in reduced form in the phloem from shoot to the roots is required to meet root demand of reduced nitrogen for growth. Frequently, however, considerable amounts of retranslocated nutrients are again loaded into the xylem of the roots to be transported back to the shoot, i. e. they cycle in the plant. For potassium it has been convincingly demonstrated that at least in certain plant species cycling is an important process for maintenance of charge balance in shoots and roots of nitrate-fed plants (see below). In more general terms cycling of nutrients might be a useful means of smoothing out fluctuations in external supply to match a more consistent demand (Cooper and Clarkson, 1989). Cycling of mineral nutrients may also be important to compensate, at least in part, for the heterogeneous distribution of mineral nutrients in the rooting zone, for example in the case of zinc (Loneragan *et al.*, 1987; Webb and Loneragan, 1990), but not in the case of iron (Romera *et al.*, 1992). In general, cycling of mineral nutrients should not be considered in every case as a specific regulatory mechanism for a particular mineral nutrient. In many instances cycling could well be the consequence of the mechanism and the direction of the phloem transport governed by the sugar transport from leaves as the source to roots as a sink.

Comprehensive studies on mineral nutrient cycling have been made in white lupin and castor bean by Jeschke and Pate (1991b). Some of their data are summarized in Table 3.16, which give mineral nutrient import and export from leaf laminae, retranslocation through the phloem and cycling through the roots. As has already been shown for reduced nitrogen (Table 3.10), and is also the case for potassium, sodium and magnesium, export through the phloem can comprise a major fraction of the import through the xylem. Phloem export of calcium is negligible in castor bean but unexpec-

Table 3.16
Partitioning, Translocation and Cycling of Mineral Elements in White Lupin and Castor Bean[a]

	Proportion of total uptake (%)							
	White lupin				Castor bean			
Parameter	K	Na	Mg	Ca	K	Na	Mg	Ca
Import (leaf laminae) through xylem	96	45	33	29	138	11	51	39
Export (leaf laminae) through phloem	72	33	25	12	93	9	13	2
Phloem transport to the roots	59	33	20	9	85	9	15	1
Cycling through the roots	39	—[b]	10	—[b]	78	—[b]	7	—[b]

[a]Based on Jeschke and Pate (1991b).
[b]Could not be quantified.

tedly high in white lupin. This high calcium export probably relates to the exceptionally high concentrations of organic acids (mainly succinate) in the phloem sap of white lupin (Jeschke *et al.*, 1986) and is indicative of the strong chelation of calcium. Between 82 and 100% of the exported mineral elements are retranslocated in the phloem back to the roots, and a high proportion of the potassium and magnesium cycle, i.e., they are again loaded into the xylem and transported to the shoots (Table 3.16). For calcium and sodium no precise data can be given but cycling is of minor importance. For cycling of potassium, corresponding data for other plant species are 20% in tomato (Armstrong and Kirkby, 1979a) and 30% in wheat and rye (Cooper and Clarkson, 1989).

Nutrient cycling is of particular importance for the nitrogen nutrition of plants. In nitrate-fed barley plants, of the nitrogen translocated in the xylem to the shoots (100%), up to 79% was retranslocated in the phloem as reduced nitrogen back to the roots; of this 79%, about 21% was incorporated into the root tissue and the remainder cycled back in the xylem to the shoots (Simpson *et al.*, 1982). In young wheat and rye plants over 60% of the reduced nitrogen in the xylem sap represents a cycling fraction (Cooper and Clarkson, 1989). In wheat throughout ontogenesis 10–17% of the nitrogen and 12–33% of the sulfur in the xylem sap derived from the fraction recycled in the phloem from shoots to roots (Larsson *et al.*, 1991). Accordingly, in nitrate-fed plants the proportion of nitrate in the total nitrogen in the xylem sap can be used as an indicator for nitrate reduction in the roots only in plant species in which nitrate reduction is merely confined to the roots. For plant species which reduce nitrate both in roots and shoots, however, the situation is more complicated (Van Beusichem *et al.*, 1988). In castor bean, for example, about half the nitrate reduction occurs in the roots. Most of the nitrogen reduced in the roots is translocated in the xylem to the shoots, of which a considerable portion is retranslocated in the phloem to the roots and cycles back in the xylem to the shoots (Jeschke and Pate, 1991a). Thus, at any given time of sampling xylem sap, a substantial proportion of the reduced nitrogen will have already cycled at least once through the plant. This may also hold true for reduced sulfur (Schupp *et al.*, 1991).

The predominant site of nitrate reduction in plants (roots or shoots) may also have an

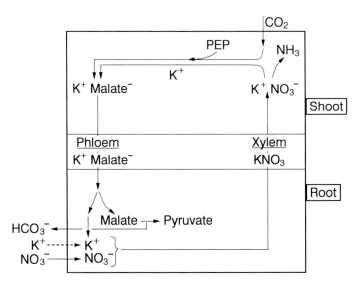

Fig. 3.12   Model for the circulation of potassium between root and shoot in relation to nitrate and malate transport (PEP, phosphoenol pyruvate). (Based on Ben-Zioni *et al.*, 1971 and Kirkby and Knight, 1977.)

important impact on potassium cycling (Fig. 3.12). Potassium plays an important role as counterion for nitrate transport in the xylem (Van Beusichem *et al.* 1988; Section 2.9). After nitrate reduction in the shoot, charge balance has to be maintained by corresponding net increase in organic acid anions (Section 2.5.4). As an alternative to storage in the leaf cell vacuoles, the organic acid anions (mainly malate) and potassium as accompanying cation can be retranslocated in the phloem to the roots. After decarboxylation of the organic acids, potassium may act again as counterion for nitrate transport in the xylem to the shoot. Strong support for this model has been provided by Touraine *et al.* (1990) in soybean, which reduce about 90% of the nitrate in shoots. In these plants close correlations were found between nitrate reduction in the shoot, retranslocation of potassium and organic acid anions (mainly malate) in the phloem, decarboxylations in the roots and release of bicarbonate. As would be predicted from the model, subjecting the shoot to light results in enhanced release of bicarbonate from the roots, and stem-feeding of potassium malate induces increase in net uptake of nitrate and the net consumption of protons by the roots (i.e., bicarbonate release).

3.5   Remobilization of Mineral Nutrients

3.5.1   General

Import and export of mineral nutrients occur simultaneously during the life of plant organs such as leaves (Table 3.16). As a rule, ageing (senescence) is associated with higher rates of export of mineral nutrients than rates of import and, thus, decrease in net content or, more precisely, in amount per organ such as a leaf (Table 3.10). In the literature the term *redistribution* and *retranslocation* are often used to describe this process. In view of the dynamics of import and export and cycling of mineral nutrients, these terms may lead to some confusion. In the following discussion therefore, this decrease in net content is denoted by the term *remobilization*.

Remobilization is based on a range of different physiological and biochemical processes: utilization of mineral nutrients stored in vacuoles (potassium, phosphorus, magnesium, amino-N, etc.), breakdown of storage proteins (e.g., in vacuoles of the paraveinal mesophyll cells of legumes; Klauer *et al.*, 1991; Section 3.2.2.3), or, finally, breakdown of cell structures (e.g., chloroplasts) and enzyme proteins thereby transforming structurally bound mineral nutrients (e.g., magnesium in chlorophyll, micronutrients in enzymes) into a mobile form.

Remobilization of mineral nutrients is important during the ontogenesis of a plant at the following stages: seed germination; periods of insufficient supply to the roots during vegetative growth; reproductive growth; and, in perennials, the period before leaf drop.

3.5.2 Seed Germination

During the germination of seeds (or storage organs such as tubers) mineral nutrients are remobilized within the seed tissue and translocated in the phloem or xylem, or both, to the developing roots or shoots. As a rule, seedlings will grow for at least several days without an external supply of mineral nutrients. In seeds many mineral nutrients (e.g., potassium, magnesium, calcium) are usually bound to phytic acid as phytate; thus, remobilization of these mineral nutrients and also of phosphorus is correlated to the phytase activity. In legume seeds, a much higher proportion of the mineral nutrients (including calcium) stored in the cotyledons is remobilized (Hocking, 1980a) than, for example, in cucumber (Ockenden and Lott, 1988a,b).

3.5.3 Vegetative Stage

During vegetative growth, nutrient supply to the roots is often either permanently insufficient (as in the case of low soil nutrient content) or temporarily interrupted (when, for example, there is a lack or excess of soil moisture). Remobilization of mineral nutrients from mature leaves to areas of new growth is thus of key importance for the completion of the life cycle of plants under these experimental conditions. This behaviour (strategy) is typical for fast-growing crop species whereas for many wild species cessation of growth occurs under adverse environmental conditions and, thus, redistribution of mineral nutrients plays a lesser important role (Chapin, 1983).

The extent to which remobilization takes place, however, also differs between mineral nutrients and this is reasonably well reflected in the distribution of visible deficiency symptoms in plants. Deficiency symptoms which predominantly occur in young leaves and apical meristems reflect insufficient remobilization. In the latter case, either the phloem mobility is insufficient (Section 3.3.3) or only a relatively small fraction of the mineral nutrients can be transformed into a mobile form in the fully expanded older leaves.

The extent of remobilization is also important for diagnosis of the nutritional status of plants (Chapter 12). Leaves and other plant organs which respond to an insufficient supply of a particular mineral nutrient to the roots by a rapid increase in remobilization of that nutrient, are more suitable for foliar (plant) analysis than less responsive leaves or other organs. However, discrepancies do exist in this respect. For example, Scott and Robson (1991) have shown that, despite the normally high mobility of magnesium in plants, interruption of magnesium supply to the roots of young wheat plants resulted in a faster decline in the concentration of magnesium in the fully expanded young leaves than in the older leaves. Such a sudden interruption of a mineral nutrient supply to the roots under otherwise optimal growth conditions, however, may well be expected to lead to a somewhat different pattern of nutrient remobilization than would occur under field conditions. In accord with this view the so-called critical deficiency concentrations of mineral nutrients in shoots of young plants (Chapter 12) obtained by the procedure of sudden interruption of root supply are also much higher (Burns, 1992) than those from field-grown plants.

3.5.4   Reproductive Stage

Remobilization of mineral nutrients is particularly important during reproductive growth when seeds, fruits, and storage organs are formed. At this growth stage root activity and nutrient uptake generally decrease, mainly as a result of decreasing carbohydrate supply to the roots ('sink competition', Chapter 5). Therefore, the mineral nutrient contents of vegetative parts quite often decline sharply during the reproductive stage (Fig. 3.13).

   The extent of this remobilization depends on various factors, including (a) the specific requirement of seeds and fruits for a given mineral nutrient; (b) the mineral nutrient status of the vegetative parts; (c) the ratio between vegetative mass (source size) and number and size of seeds and fruits (sink size); and (d) the nutrient uptake rate by the roots during the reproductive stage. Cereal grains, for example, are

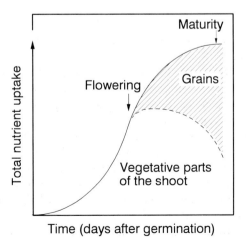

Fig. 3.13   Schematic representation of the mineral nutrient distribution in cereal plants during ontogeny.

Table 3.17

Remobilization of Mineral Nutrients in a Pea Crop between Flowering and Ripening[a]

	Mineral nutrients in leaves and stems (kg ha^{-1})				
	N	P	K	Mg	Ca
Harvest time					
June 8 (flowering)	64	7	53	5	31
June 22	87	10	66	8	60
July 1	60	7	61	8	69
July 12 (ripening)	32	3	46	9	76
Percentage of increase or decrease after June 22	−63	−73	−30	+10	+21
Percentage in seeds in relation to the total content of shoots	76	82	29	26	4

[a]Based on Garz (1966).

characterized by a high concentration of nitrogen and phosphorus, and a low concentration of potassium, magnesium, and calcium, whereas fleshy fruits (e.g. tomatoes) or storage organs (e.g. potato tubers) are high in potassium but relatively low in nitrogen and phosphorus.

A typical example of the differences in the extent of remobilization of these mineral nutrients from vegetative shoots is shown in Table 3.17 for pea plants grown under field conditions. The percentage of remobilization of nitrogen and phosphorus is very high, whereas there is a lack of remobilization of magnesium and calcium; instead, a net increase in these nutrients takes place in the vegetative organs, as has also been shown for soil-grown soybean plants (Wood *et al.*, 1986). Relatively high concentrations of nutrients in the soil solution leading to a continuous uptake by the roots and the import of nutrients into leaves after anthesis are the main responsible factor for the lack of remobilization of magnesium and calcium. An inherent low capacity for remobilization of calcium is also a contributing factor.

In cereals such as wheat up to about 90% of the total phosphorus in grains can be attributed to remobilization from vegetative parts. Much lower proportions are only found when the roots are continuously well supplied with phosphorus in sand culture (Batten *et al.*, 1986). For nitrogen a comparison of remobilization in different wheat cultivars under field conditions gave an average value for remobilization of 83%, but values ranged from 51 to 91% depending on the total nitrogen uptake of the cultivars (Van Sanford and MacKown, 1987).

The remobilization of highly phloem-mobile mineral nutrients can lead to such a rapid decline in their content in the vegetative shoots that rapid senescence is induced and plants behave as 'self-destructing' systems. From experiments with soybean, remobilization of mineral nutrients as a senescence-inducing factor has been questioned (Wood *et al.*, 1986; Mauk and Noodén, 1992). However, there are various examples (see also Chapters 5 and 6) showing this phenomenon, for example, for the

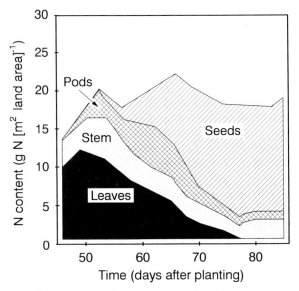

Fig. 3.14 Nitrogen partitioning in field-grown bean (*Phaseolus vulgaris* L., genotype G 5059) during reproductive growth. (Lynch and White, 1992.)

remobilization of phosphorus and senescence of the flag leaf in wheat (Batten and Wardlaw, 1987a,b), remobilization of phosphorus and disruption of carbon metabolism in source leaves of phosphorus-deficient soybean (Lauer *et al.*, 1989b), or for nitrogen remobilization and senescence in field-grown bean (Fig. 3.14). Despite the high potential for N_2 fixation of this genotype, the enhanced remobilization (and export) of nitrogen from the leaves to the pods and developing seeds soon after flowering (Fig. 3.14) strongly limited the rate of photosynthesis of the leaves and, thus, also seed yield of bean grown in the tropics (Lynch and White, 1992).

Another example of rigorous remobilization during the reproductive growth is shown in Table 3.18 for potassium in two tomato cultivars. The cultivar VF-13L was developed for mechanical harvesting and is characterized by a heavy fruit load combined with an early and uniform maturation. Severe potassium deficiency symptoms during fruit ripening occur in this cultivar even in plants growing in soils with high potassium

Table 3.18

Potassium Content of the Petioles of Two Tomato Cultivars at Various Stages of Growth[a]

Cultivar	Third cluster, full bloom	First cluster, mature green	First cluster, fruit pink	50% fruits ripe
VFN-8	5.30	6.83	3.48	0.97
VF-13L	5.24	5.86	1.80	0.40[b]

[a]Potassium content expressed as percentage of dry weight. Based on Lingle and Lorenz (1969).
[b]Severe potassium deficiency symptoms on the leaves.

Table 3.19

Mineral Element Content of Barley (cv. Palladium) Grown in a
Saline Substrate[a,b]

Plant part	Content (μmol g^{-1} dry wt)		
	K	Na	Cl
Vegetative shoot	0.22	2.27	1.52
Rachis, glume, awn	0.56	0.42	0.43
Grain	0.13	0.04	0.04

[a]Based on Greenway (1962).
[b]Substrate contained 6 mM K$^+$ and 125 mM Na$^+$ (as NaCl).

availability. Obviously, in this genotype a particularly strong sink competition for carbohydrates between fruits and roots causes a rapid decline in the root uptake of potassium during the period of high potassium demand for fruit growth. This is also an instructive example of a specific yield limitation induced by a mineral nutrient (Chapter 6) and demonstrates too some of the physiological limitations of plant breeding for higher yields.

Remobilization is highly selective for mineral nutrients. This selectivity and the corresponding discrimination against mineral elements which are either not essential or required only at very low levels is quite impressive, as shown in Table 3.19 for barley grown in a saline substrate. In the vegetative shoots the content of potassium is lower than that of sodium and chloride. During remobilization, however, potassium is highly preferred and the ratio of the three mineral elements is reversed in the ears. An additional step in the selection of nutrients takes place before their entry into the grains.

During the reproductive stage the degree of remobilization of micronutrients and of calcium is often astonishingly high compared with that during vegetative growth. In lupins (*Lupinus albus*), for example, up to 50% of the micronutrients and 18% of the calcium that originally accumulated in the leaves were retranslocated to the fruits (Hocking and Pate, 1978). Substantial remobilization of at least some of the micronutrients also occur in soil-grown plants as shown in Table 3.20 for soybean. Remobilization of molybdenum is particularly impressive, a result which has also been confirmed by Mauk and Noodén (1992).

The extent of remobilization of micronutrients strongly depends on their contents in the fully expanded leaves (Loneragan *et al.*, 1976). During grain development in wheat, for example, leaves with a high copper content lost more than 70% of their copper, whereas leaves of copper-deficient plants lost less than 20% (Hill *et al.*, 1978). This relationship between leaf nutrient status and degree of remobilization contrasts to that for the highly mobile mineral nutrients, such as nitrogen and potassium. For these nutrients a much higher proportion is remobilized in deficient plants. This inverse relationship between leaf content and the degree of remobilization of micronutrients is caused by the higher proportion of firmly bound micronutrients (structural constituents, e.g., in cell membranes and cell wall) in leaves low in nutrient content. This same relationship has been observed in fruit tree leaves after foliar application of boron (^{10}B). Whereas the foliar-applied boron was almost quantitatively exported within the

Table 3.20
Changes in Micronutrient Content of Leaf Blades of Soybean
during Podfill[a]

Parameter	Early-middle podfill (Day 64)	Late podfill (Day 88)
Fresh weight (g per 3 leaflets)	1.96	2.57
Element ($\mu g\ g^{-1}$ fresh wt)		
Fe	48.9	30.2
Zn	45.1	21.6
Mn	36.3	56.2
Cu	1.01	0.87
B	17.4	24.2
Mo	0.45	0.09

[a]Based on Wood et al. (1986).

following weeks, the content of soil-born (originally present) leaf boron remained unchanged (Hanson, 1991a,b).

The extent of remobilization of the micronutrients copper and zinc, but not manganese (Nable and Loneragan, 1984), is also closely related to leaf senescence. This is reflected, for example, in the close positive correlation that exists between the remobilization of nitrogen and that of copper (Fig. 3.15). The onset of senescence can

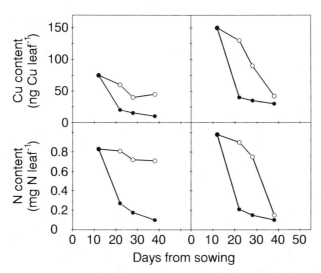

Fig. 3.15 Effects of copper supply (*left* = low; *right* = high) and shading on the copper and nitrogen content of the oldest leaf of wheat. Key: ○, unshaded; ●, shaded. (From Hill et al., 1979a.)

be accelerated by shading and this is associated with a more rapid remobilization of both nitrogen and copper; in copper-deficient plants most of the copper can then be remobilized. Nitrogen deficiency, like shading, also enhances copper remobilization (Hill *et al.*, 1978). The same is true for zinc (Hill *et al.*, 1979b). These relationships may in part be responsible for the results of field experiments showing a particularly high copper demand in plants supplied with high levels of nitrogen fertilizers and the corresponding delay in leaf senescence (Section 6.4).

The relatively high remobilization rates of micronutrients during reproductive growth as compared with vegetative growth stage are presumably the result of fruit- and seed-induced leaf senescence, in which changes in hormonal balance play an important role (Mauk and Noodén, 1992; Section 5.6). The development of sulfur-deficiency symptoms in either old or young leaves depends on the level of nitrogen supply (Loneragan *et al.*, 1976) and is most likely also related to leaf senescence.

Remobilization of mineral nutrients requires several steps: (a) mobilization within individual leaf cells; (b) short-distance transport in the symplasm to the phloem; (c) phloem loading; and (d) phloem transport. Discrepancies between high or intermediate phloem mobility (Table 3.9) and low rates of remobilization, particularly during the vegetative growth stage, are most likely caused by insufficient mobilization within the leaf cells. A large proportion of micronutrients is incorporated into cell structures and high-molecular-weight organic compounds (e.g., enzymes). During reproductive growth seed- and fruit-induced leaf senescence overcomes the most limiting step (step a) of remobilization for most micronutrients. This is most likely the reason that, despite the high to moderate phloem mobility of the micronutrients iron, zinc, copper, molybdenum and also boron, deficiency symptoms of these micronutrients during the vegetative growth first appear in young leaves and the shoot apex. These vegetative growth sinks obviously lack the capacity to produce a 'signal' strong enough to induce leaf senescence and, thus, enhanced mobilization of these mineral nutrients within leaf cells.

The extent of remobilization of mineral nutrients is attracting increasing attention in connection with the selection and breeding of genotypes of high 'nutrient efficiency'. Genotypes that grow well on soils of low nutrient availability not only may have a higher rate of uptake and translocation of a particular mineral nutrient but may also show higher nutrient efficiency at the cellular level (compartmentation, binding stage, etc.), including high rates of remobilization from older to younger leaves, seeds, and storage organs.

3.5.5 Period Before Leaf Drop (Perennials)

Remobilization of mineral nutrients (except calcium and manganese) from the leaves to woody parts is a typical feature of perennial species before leaf drop in temperate climates, and is closely related to the discoloration of leaves in the autumn. As a rule, and similar to annual species, the extent of remobilization is high for nitrogen, potassium, phosphorus, and zinc, whereas the leaf contents of calcium, boron, iron and manganese increase until leaf drop (Sanchez-Alonso and Lachica, 1987a). During this period, typical visible deficiency symptoms are often observed, indicating that during the growing period there might have been a latent deficiency of a particular mineral nutrient. In plants growing on saline substrates, preferential remobilization of certain mineral nutrients (Table 3.19) often gives rise to toxicity symptoms in leaf margins, indicating a further shift toward extreme ionic imbalance before leaf drop.

4

Uptake and Release of Mineral Elements by Leaves and Other Aerial Plant Parts

4.1 Uptake and Release of Gases and Other Volatile Compounds through Stomata

4.1.1 Uptake through Stomata

In terrestrial plants the stomata (Fig. 4.1) are the sites of exchange of gases (mainly CO_2, O_2) with the atmosphere. Their number per mm^2 of leaf surface varies between

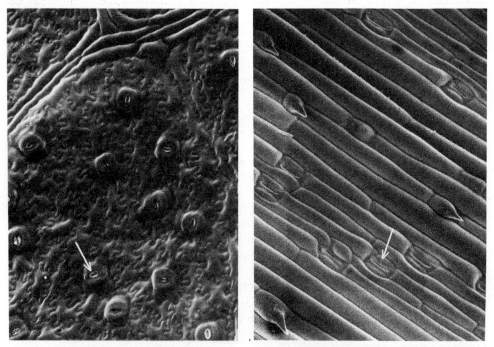

Fig. 4.1 Scanning electron micrograph of lower (abaxial) leaf surfaces of *Fagus sylvatica* (*left*) and *Puccinellia peisonis* (*right*). Arrow = stomata. (Courtesy of R. Stelzer.)

Table 4.1

Shoot Dry Weight, Nitrogen Content in the Dry Matter and Uptake of
NH_3-N from the Atmosphere in Italian Ryegrass Grown at Low Soil
Nitrate Levels and Exposed to Different NH_3 Concentrations for 33 Days[a]

NH_3 concn. ($\mu g\ m^{-3}$)	Shoot dry wt. (g per pot)	Shoot N content (% in dry matter)	Total plant N derived from NH_3 (mg per pot)
14	6.4	0.89	8
123	7.8	1.14	42
297	9.0	1.47	121
498	10.2	1.92	230
709	10.7	2.80	341

[a]Recalculated from Whitehead and Lockyer (1987).

about 20 in succulents (CAM species), 100–200 in most annual species, and more than 800 in certain tree species (e.g., *Acer montanum*). The stomata are usually more abundant (most annual species) or confined (many tree species, e.g. *Fagus sylvatica*) to the lower (abaxial) leaf surface. Mineral nutrients in the form of gases, such as SO_2, NH_3 and NO_2, also enter the leaves predominantly through the stomata and are rapidly metabolized in the leaves. In recent years foliar uptake of these gases has attracted much interest as they are major components of air pollution and their uptake can be substantial. Moreover, depending on concentration and the plant species, they can either depress or enhance plant growth. In agricultural areas the main source of ammonia (NH_3) emission is animal husbandry, and large amounts of NH_3 may be volatilized from manure and pastures grazed by cattle. As NH_3 is readily taken up by leaves, increasing ambient NH_3 concentrations increase plant growth, and the nitrogen content of the shoots was found to be linearly related to the ambient NH_3 concentrations (Table 4.1). The proportion of total plant nitrogen (shoots and roots) derived from NH_3 increased from about 4% at the lowest to about 77% at the highest concentration.

In areas of intense animal husbandry atmospheric NH_3 concentrations are, on average, between 5 and 15 $\mu g\ NH_3\ m^{-3}$, but may rise within the pasture canopy to about 85 $\mu g\ NH_3\ m^{-3}$ (Whitehead and Lockyer, 1987) and be subject to distinct diurnal fluctuation. At night a steep concentration gradient of NH_3 can occur within the canopy from the base (soil surface) to the atmosphere above the canopy (the free atmosphere above the vegetation); during the day, however, the NH_3 concentration within the canopy drops to a very low level, indicating substantial NH_3 uptake through the stomata (Lemon and Houtte, 1980). Daily uptake rates of NH_3 by leaves in a pasture have been calculated to be between 100 and 450 g nitrogen per hectare (Cowling and Lockyer, 1981), but in certain periods as much as 10–20% of the nitrogen in pasture plants can originate from gaseous NH_3 (Whitehead and Lockyer, 1987).

Ammonia emissions from animal husbandry may also be a major component in the 'dry deposition' of nitrogen (e.g. NH_4^+ in aerosols, and NH_3) in nearby agricultural, natural and forest ecosystems, and of the nitrogen uptake by the foliage, for example, of forest trees (Duyzer *et al.*, 1992). In south and east England 35–40 kg N ha^{-1} are estimated to be deposited annually on arable land, and two-thirds of this represents 'dry deposition', mainly in the form of NH_3 and HNO_3 (Goulding, 1990). The proportion of wet and dry deposition of nitrogen varies considerably between locations but as a first approximation about equal proportions of both components can be assumed for most locations in Central Europe, as is true for other mineral nutrients (Table 4.10).

The uptake of atmospheric nitrogen dioxide (NO_2) through the stomata is also linearly related to the external concentration and its metabolism is rapid (Thoene *et al.*, 1991). Long-term exposure of plants to NO_2 can contribute considerably to their nitrogen nutrition (Gupta and Narayanan, 1992).

Sulfur dioxide (SO_2) too is readily taken up through the stomata. In SO_2 fumigated Norway spruce seedlings, sulfate accumulation in the needles is a linear function of atmospheric SO_2 concentrations (Kaiser *et al.*, 1993). Short-term exposure to high concentrations (50 mg SO_2 m^{-3}) causes a long-term depression in net photosynthesis (Keller, 1981). With long-term exposure of plants to low concentrations (1.5 mg m^{-3}), however, SO_2 is similar in effect on growth, for example of tobacco, to sulfate supplied to the roots (Faller, 1972). In oats and rape plants grown under field conditions in a sulfur-deficient soil, nearly half of the total sulfur taken up over the vegetation period was found to be derived from atmospheric volatile sulfur compounds (Simán and Jansson, 1976), most probably via SO_2 foliar absorption.

Uptake of hydrogen sulfide (H_2S) by leaves which follows a distinct diurnal pattern, is closely related to the stomatal aperture. Hydrogen sulfide is toxic to sensitive plant species such as spinach even at concentrations below 0.7 mg m^{-3} (DeKok *et al.*, 1989).

4.1.2   Release through Stomata

Not only uptake of these gases by the leaves can be considerable but also losses by emission. This holds true for NH_3, H_2S and other volatile sulfur compounds. In rice, losses of volatile nitrogen compounds (mainly NH_3) through the stomata have been calculated to be as much as 15 kg N ha^{-1} over a 100-day period (Silva and Stutte, 1981). Wheat plants seem to lose NH_3 at a fairly constant rate of 60–120 ng NH_3-N m^{-2} s^{-1} before the milk ripe stage, but the rate increases to 100–200 ng NH_3-N m^{-2} s^{-1} during senescence, leading to cumulative nitrogen losses of between 2.8 and 4.4 kg ha^{-1} (Parton *et al.*, 1988). For a wheat crop losses of NH_3 by the leaves during senescence can reach about 7 kg N ha^{-1}, an equivalent of 21% of the fertilizer nitrogen applied to the soil (Harper *et al.*, 1987). In pea plants during seed filling losses of nitrogen can reach up to 30% of the total plant nitrogen, and a large proportion of this loss may be attributed to volatilization of NH_3 from aerial plant parts (Bertelsen and Jensen, 1992).

Light-dependent emission of H_2S from leaves can be considerable after exposure to high atmospheric SO_2 concentrations and is considered as an SO_2 detoxification mechanism (Sekiya *et al.*, 1982b). However, plants grown in a nonpolluted atmosphere and supplied with sulfur only in the form of sulfate in the soil also release substantial amounts of volatile sulfur compounds through the stomata (Table 4.2). The main component found in this experiment was SO_2 and its emission increased with the sulfate content in the soil. For oats and rape emissions of volatile sulfur compounds have been shown to occur within 35 days after the onset of growth and found to vary between 0.2

Table 4.2
Relation between Sulfate Content of the Soil and Sulfur Content of, and
Volatile Sulfur Emissions by, Needles of Norway Spruce[a]

	Sulfate (mg SO_4-S kg^{-1} soil)		
Parameter	97	129	181
Total S (mg g^{-1} needle dry wt)	1.0	0.9	1.2
H_2S emission (nmol mol^{-1} H_2O)	0.9	1.1	1.0
SO_2 emission (nmol m^{-2} $(2\,h)^{-1}$	4.1	8.8	10.3

[a]Based on Rennenberg *et al.* (1990).

and 2–3 kg S ha^{-1}, depending on whether the plants were grown in a soil with low or high sulfate content (Simán and Jansson, 1976).

In alfalfa plants the emissions of volatile sulfur compounds follow a distinct diurnal rhythm with maximal rates occurring around mid-day (Grundon and Asher, 1986). Both, the amounts and the spectrum of the emitted volatile sulfur compounds vary between plant species, and in the case of rape may represent up to 0.92% of the total sulfur in the plant, which under field conditions amount to between a few hundred grams and a few kilograms S per ha being emitted during the growing season (Grundon and Asher, 1988). In plants with high selenium contents, considerable release of volatile selenium occurs (Section 10.5), presumably as dimethylselenide from the breakdown of seleno-amino acids (Zayed and Terry, 1992).

4.2 Uptake of Solutes

4.2.1 Structure and Function of the Cuticular Layer

Whereas in aquatic plants the leaves and not the roots are the main sites of mineral nutrient uptake, in terrestrial plants the uptake of solutes by the surface of leaves and other aerial parts is severely restricted by the outer wall of the epidermal cells. The principle structure of the outer epidermal wall is shown schematically in Fig. 4.2, and in Fig. 4.3 two examples are given for cross sections through the outer epidermal cell walls of wheat and bean leaves.

This outer wall is covered by the cuticle (*cuticle proper*) and a layer of epicuticular waxes which are often well and typically structured (Bartloth, 1990). These waxes are excreted by the epidermal cells and consist of long-chain alcohols, ketones, and esters of long-chain fatty acids. Waxes also occur 'intracuticularly' within the cuticle and in the cutinized layer (Fig. 4.2). The cuticle consists mainly of cutin, a mixture of long-chain fatty acids. The chemical and physical properties of the cuticle differ between outer and inner surfaces, a distinct gradient occuring from the hydrophobic (lipophilic) outer surface to a more hydrophilic inner surface of the cutinized layer. The cutinized layer is normally the thickest part of the epidermal wall (Fig. 4.2) and consists of a cellulose skeleton, incrusted with cutin, wax and pectin.

The cuticle and the cutinized layer (Fig. 4.2) have diverse functions. A major function

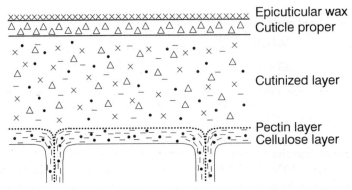

Fig. 4.2 Schematic drawing of the different layers of a typical outer epidermal wall of leaf cells.
×, wax; △, cutin; ●, pectin; −, cellulose. (Lyshede, 1982.)

is to protect the leaf from excessive water loss by transpiration. The control of water
economy in terrestrial higher plants by the stomata is dependent on the remaining
surface of the plant being very low in hydraulic conductivity. The other main function of
these structures is to protect the leaf against excessive leaching of inorganic and organic
solutes by rain (Section 4.4). It has to be kept in mind that mineral nutrients and other
solutes entering the leaves via the xylem are in the apoplasm of the leaf tissue, and a
'waterproof' barrier is required to act as an apoplasmic boundary thereby playing a
similar role to that of the Casparian band in the endodermis of the roots (Section 2.7).
The relative importance of these two main functions of the cuticle depends on climatic
conditions (arid zones versus humid tropics). In addition the cuticle is involved in
temperature control, optical properties of leaves and plays a role in defence against
pests and diseases (Chapter 11).

   Permeation of low-molecular-weight solutes (e.g., sugars, mineral elements) and
evaporation of water through the cuticle (peristomatal, or cuticular transpiration) takes
place in hydrophilic pores within the cuticle. The majority of these pores in the cuticle

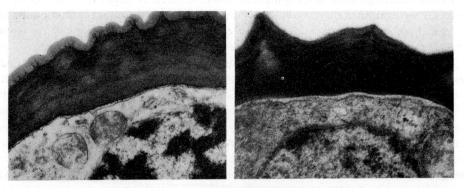

Fig. 4.3 Cross section (× 20 000) of outer epidermal cell walls of wheat (*T. aestivum*; *left*) and
bean (*P. vulgaris*; *right*) leaves. (Courtesy of Ch. Hecht-Buchholz.)

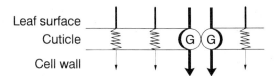

Fig. 4.4   Schematic presentation of solute penetration across the cuticular layer of leaf epidermal cells (G, guard cell).

have a diameter of less than 1 nm, and a density of about 10^{10} pores cm^{-2} has been calculated (Schönherr, 1976). These pores are readily permeable to solutes such as urea (radii 0.44 nm) but not to larger molecules such as synthetic chelates (e.g. FeEDTA). These small pores are lined with fixed negative charges (presumably mainly from polygalacturonic acids) increasing in density from the outside of the cuticle to the inside (i.e. the cutinized layer and the cell wall interface, Fig. 4.2). Accordingly, permeation of cations along this gradient is enhanced whereas anions are repulsed from this region (Tyree *et al*. 1990). Uptake of cations by leaves is thus faster than that of anions (e.g. NH_4^+ compared with NO_3^-) and is particularly fast for small, uncharged molecules such as urea. However, when applied at high concentrations as foliar sprays, differences in uptake rates of nitrogen from urea, ammonium and nitrate become negligible (Bowman and Paul, 1992).

Cuticular pore density is higher in cell walls between guard cells and subsidiary cells (Maier-Maercker, 1979). This explains the commonly observed positive correlation between number or distribution of stomata, for example, between the upper (adaxial) and the lower (abaxial) leaf surface, and the intensity of mineral nutrient uptake from foliar sprays (Levy and Horesh, 1984). Not only is the number of the cuticular pores larger around guard cells (or trichomes), but the pores also seem to have different permeability characteristics (Schönherr and Bukovac, 1978) and are most likely the sites where larger solute molecules (e.g. FeEDTA) penetrate the cuticle and are taken up by the leaf cells.

The differences in resistance to solute penetration at various parts of the cuticle are shown schematically in Fig. 4.4. It is unlikely that direct penetration of solutes from the leaf surface through open stomata into the leaf tissue plays an important role, because a cuticular layer (the internal cuticle) also covers the surface of the guard cells in stomatal cavities. Furthermore, ion uptake rates from foliar sprays are usually higher at night, when the stomata are closed, than during the day, when the stomata are open.

It has been supposed that hydrophilic microchannels ('*ectodesmata*') are present throughout the outer epidermal cell wall as pathways for water vapour and solute movement (Franke, 1967). Experimental evidence for the existence of such structures *in vivo*, however, is lacking.

4.2.2   Role of External and Internal Factors

Leaf cells, like root cells, take up mineral elements from the apoplasm. Leaf uptake is thus similarly affected by external factors, such as mineral nutrient concentration and ion valency as well as by temperature, and internal factors, such as metabolic activity

Table 4.3

Foliar Absorption and Translocation of Labeled Phosphate by Barley Plants[a,b]

	Rate of absorption and translocation (μmol P g^{-1} leaf dry wt h^{-1})	
	Control plants	Phosphorus-deficient plants
Uptake by treated leaf	5.29 ± 0.54	9.92 ± 2.17
Translocation from treated leaf	2.00 ± 0.25	5.96 ± 1.08
Translocation to roots	0.63 ± 0.04	4.38 ± 0.42

[a]From Clarkson and Scattergood (1982).
[b][^{32}P]Phosphate was supplied to the mature leaf. Duration of experiment: 3 days.

(Chamel, 1988). For a given external concentration of mineral nutrients, the rates of uptake by intact leaves are, however, much lower than the corresponding rates of uptake by roots, since the very small pores in the cuticle severely restrict diffusion from the external leaf surface into the bulk of the leaf apoplasm and hence to the plasma membrane of the leaf cells. The thickness of the cuticle differs widely between plant species (Lyshede, 1982) and is also affected by environmental factors; this is particularly evident in comparisons of plants grown under shaded and non-shaded conditions (Takeoka et al., 1983).

The rate at which leaves take up mineral nutrients supplied to their surfaces also depends on the nutritional status of a plant, as shown for phosphorus in Table 4.3. The rate of uptake by the leaves of phosphorus-deficient plants was twice as high as that of control plants well supplied with phosphorus via the roots. In addition, in the deficient plants, much more phosphorus was translocated from the leaf, particularly to the roots.

Rates of mineral element uptake by leaves usually decline with leaf age. Several factors are responsible for this decline, including a decrease in metabolic activity (sink activity), an increase in membrane permeability (i.e., an accompanying increase in leakage of ions from the vacuole and the cytoplasm into the apoplasm), and an increase in the thickness of the cuticle.

In contrast to ion uptake by root cells, uptake by green leaf cells is directly stimulated by light. This can be demonstrated either with intact leaves after vacuum infiltration of the solutes (MacDonald et al., 1975) or with leaf segments (Rains, 1968), where the resistance to solute penetration by the cuticle is minimized (Table 4.4).

During the light period, not only is the potassium uptake rate higher, but the type of energy coupling is different, as shown by the effect of 2,4-DNP, an inhibitor of oxidative phosphorylation (mitochondrial ATP synthesis). Part of the ATP necessary for the active uptake of potassium is obviously supplied by the chloroplasts, either directly via ATP from photophosphorylation or indirectly via the triosephosphate-shuttle (Yin et al., 1990; see also Section 8.3).

Mineral element uptake by intact leaves from foliar sprays, however, is often either not stimulated by light or even depressed as a result of indirect light effects. During the daytime, as the ambient temperature increases, there is usually a decrease in relative

Table 4.4
Effect of Light and Inhibitor (2,4-DNP) on Potassium
Uptake by Maize Leaf Segments[a]

	K Uptake (μmol g^{-1} h^{-1})	
	Darkness	Light
Treatment		
Control	2.3	3.7
10^{-5} M 2,4-DNP	0.2	2.0
Inhibition (%)	91	46

[a]Potassium supply 0.1 mM KCl. Based on Rains (1968).

humidity, leading to more rapid water evaporation from the foliar sprays and thus more rapid drying of the sprays at the leaf surface. Differences in rates of magnesium uptake by apple leaves from various salts [$MgCl_2 \gg Mg(NO_3)_2 > MgSO_4$] applied during light and dark periods were found to relate exclusively to differences in the solubility and hygroscopicity of these salts (Allen, 1960).

Uptake rates of mineral elements such as phosphorus by leaves increase as a hyperbolic function of increasing external concentrations as is also known for roots (Section 2.5.5). This similarity also holds true for boron, where the uptake rate by leaves is linearly related to the external concentration (Shu et al., 1991). When low concentrations of macronutrients such as potassium and phosphorus are applied to leaves, the uptake rates through leaf surfaces are particularly low. The relatively high internal concentrations of these ions in the apoplasm at the leaf tissue, for example, of potassium up to 18 mM (Section 3.2.2.3), severely restrict permeation of ions from the leaf surface into the apoplasm. This is of course not the case for deficient plants (Table 4.3) or when micronutrients are applied to the leaves.

4.3 Foliar Application of Mineral Nutrients

4.3.1 General

The foliar application of mineral nutrients by means of sprays offers a method of supplying nutrients to higher plants more rapidly than methods involving root application. The supply is more temporary, however, and several problems can occur. These include:

1. Low penetration rates, particularly in leaves with thick cuticles (e.g., citrus and coffee)
2. Run-off from the hydrophobic surfaces
3. Washing off by rain
4. Rapid drying of spray solutions
5. Limited rates of retranslocation of certain mineral nutrients such as calcium from the sites of uptake (mainly the mature leaves) to other plant parts

Table 4.5

Effect of the Urease Inhibitor PPD[a] on Urease Activity, Leaf Tip Necrosis and Content of Urea and Ammonia in Soybean Leaves after Foliar Application of 15 mg Urea per Leaf[b]

PPD (μg per leaf)	Urease activity[c]	Leaf tip necrosis (% of dry wt)	Content (% of dry wt)	
			Urea	Ammonia
0	16.1	1.3	0.10	0.031
75	5.8	5.7	0.52	0.017

[a]Phenylphosphorodiamidate
[b]Based on Krogmeier *et al.* (1989).
[c]μmol N h^{-1} g^{-1} fresh wt.

6. Limited amounts of macronutrients which can be supplied by one foliar spray (on average, 1% × 400 l ha^{-1}, an exception being urea, which can comprise 10% of a spray)

7. Leaf damage (necrosis and 'burning').

Leaf damage by high nutrient concentrations is a serious practical problem encountered in the foliar application of mineral nutrients. The damage is mainly the result of local nutrient imbalance in the leaf tissue rather than osmotic effects. This is even true for urea which is usually applied at fairly high concentrations. In soybean, for example, leaf damage (leaf tip necrosis) by foliar spray with urea could be prevented by simultaneous spraying with sucrose, despite the additional increase in the osmotic potential of the foliar spray (Barel and Black, 1979).

Leaf tip necrosis following foliar application of urea is not caused by ammonia formed through hydrolysis of urea by plant urease. As shown in Table 4.5, inhibiting plant urease and increasing the urea concentrations in the leaves correspondingly increases the incidence of leaf tip necrosis rather than decreasing it. Thus accumulation of urea in the leaf tissue is the causal factor for the leaf tip necrosis, a result which is of particular interest in view of the function of nickel as a metal component of urease (Krogmeier *et al.*, 1991; Section 9.5).

There are contradictory reports on the levels of nitrogen losses by ammonia volatilization from leaves following foliar sprays with urea. Values range from not more than 4% of the applied nitrogen in wheat (Smith *et al.*, 1991) to more than 30% in Kentucky bluegrass (Wesely *et al.*, 1987); high leaf surface moisture followed by rapid drying seems to be the major responsible factor for high losses. As a side effect of foliar application of urea, microflora populations on the leaf surface might be changed and spore germination and colony growth of pathogens such as *Erysiphe graminis* (powdery mildew) thereby reduced (Gooding and Davies, 1992).

In general, leaf damage by foliar application of nutrients is much less severe when the spray solution pH is low (Neumann *et al.*, 1983). The addition of silicon-based surfactants seems to be a means of decreasing leaf damage and simultaneously increasing the efficiency of sprays, particularly for leaves with thick cuticles (Horesh and Levy, 1981).

Table 4.6
Rates of Manganese Fertilizers (as $MnSO_4$) Required for Optimal Yield of Soybean grown in Manganese Deficient Soils[a]

Mode of manganese fertilizer application	Requirement for optimal yield (kg Mn ha^{-1})
Broadcast	14
Banded	3
Foliar sprays (2×)	0.1

[a]Based on Mascagni and Cox (1985).

4.3.2 Practical Importance of Foliar Application of Mineral Nutrients

Despite the drawbacks of supplying nutrients to plants by means of foliar application, the technique has great practical utility under certain conditions.

4.3.2.1 Low Nutrient Availability in Soils

In calcareous soils, for example, iron availability is usually very low and iron deficiency ('lime chlorosis') widespread. Foliar spraying under these conditions might be much more efficient than the application of expensive iron chelates to the soil (Horesh and Levy, 1981) and is also a method of alleviating manganese toxicity (Moraghan, 1979). Manganese deficiency is also widespread, particularly in soybean over a range of neutral and alkaline soils, and foliar sprays with manganese are again much more effective than soil applications in overcoming manganese deficiency (Table 4.6). However, since manganese is only poorly remobilized and its mobility in the phloem is low (Sections 3.3 and 3.4), two or more foliar sprays may be required within a growing season (Gettier *et al.*, 1985).

In fruit trees foliar application of boron in the autumn is a very effective procedure for increasing the boron contents in the flower buds and the fruit set in the following season (Hanson *et al.*, 1985; Hanson, 1991a,b). In swedes (*Brassica napobrassica* Mill.), brown-heart, a typical symptom of boron deficiency, can be prevented by two foliar sprays of boron (Cutcliffe and Gupta, 1987).

4.3.2.2 Dry Topsoil

In semiarid regions, a lack of available water in the topsoil and a corresponding decline in nutrient availability during the growing season are common phenomena. Even though water may still be available in the subsoil, mineral nutrition becomes the growth-limiting factor. Under these conditions, soil application of nutrients is much less effective than foliar application, as shown in Table 4.7 by the results of an experiment in which copper was applied to wheat under field conditions in a semiarid region of Australia.

Table 4.7
Effects of Soil and Foliar Application of Copper ($CuSO_4 \cdot 5H_2O$) on Growth
Parameters and Grain Yield in Wheat[a]

Treatment	Ears m^{-2}	Grains per ear	Grain yield (g dry wt m^{-2})
No application	37.0	0.14	0.03
Soil application (kg copper sulfate ha^{-1})			
2.5	28.8	2.3	1.0
10.0	58.5	2.9	2.3
Foliar application (2%; 2 kg copper sulfate ha^{-1})			
Once at stem extension	63.8	17.1	14.0
Once at stem extension and once at booting stage	127.4	52.7	79.7

[a]Based on Grundon (1980).

4.3.2.3 Decrease in Root Activity during the Reproductive Stage

As a result of sink competition for carbohydrates, root activity and thus nutrient uptake by the roots decline with the onset of the reproductive stage. Foliar sprays containing nutrients can compensate for this decline (Trobisch and Schilling, 1970). In legumes that rely on symbiotic N_2 fixation, sink competition for carbohydrates between developing seeds and root nodules may cause a marked decrease in the rate of N_2 fixation (Chapter 7). Although not always (Neumann, 1982) but quite often, foliar application of nitrogen such as urea after flowering and at pod filling can be quite effective in increasing the yield of nodulated legumes as shown in an example for soybean in Table 4.8. Both the application of urea alone and particularly in combination with a sucrose-containing surfactant (SFE), steeply increased yield and nitrogen content of the plants (Table 4.8). By labelling the urea with ^{15}N it could be demonstrated that most of this increase in nitrogen was due to enhanced N_2 fixation, probably related to a delay in leaf senescence, thus prolonging the supply of carbohydrates to the

Table 4.8
Effect of Foliar Application of Urea at Flowering and Podfill on Dry Matter Production and
Nitrogen Content in Nodulated Soybean[a]

Treatment	Dry weight (g per plant)		N content (mg per plant)		
	Seeds	Total	Seeds	Total	(from urea)
Control	4.6	21.4	234	342	(–)
1% Urea	10.2	38.1	518	680	(99)
1% Urea +0.1% SFE[b]	20.7	54.9	1204	1476	(169)

[a]Based on Ikeda et al. (1991).
[b]Sucrosemono- and diester of long-chain fatty acids.

roots and nodules (Ikeda *et al.*, 1991). Similarly, in wheat grown in phosphorus-deficient soils, foliar application of phosphorus after anthesis can considerably delay senescence of the flag leaf and, thus, increase the leaf area duration (Batten and Wardlaw, 1987).

4.3.2.4 Increase in Protein Content of Cereal Grains

In cereals such as wheat the protein content of the grains and thus their quality for certain purposes (e.g., baking, animal feeding) can be increased quite readily by the foliar application of nitrogen at later stages of growth. Nitrogen supplied at these stages is rapidly retranslocated from the leaves and directly transported to the developing grains (Section 8.2.5). Although the recovery rates of nitrogen from foliar sprays with urea are usually quite high, for example, in wheat about 70% (Powlson *et al.*, 1989), losses by volatilization of NH_3 also occur from the leaves and can be higher than with soil application (Section 4.3.1).

4.3.2.5 Increase in Calcium Content of Fruits

As shown in Chapter 3, calcium-related disorders are widespread in certain plant species. Due to the limited phloem mobility of calcium, foliar sprays are not very effective and multiple applications are required during the growing season. Nevertheless, some decrease, for example, in bitter pit in apple can be achieved by multiple application of calcium sprays, particularly if the surfaces of developing fruits are sprayed directly (Schumacher and Frankenhauser, 1968).

4.3.3 Foliar Uptake and Irrigation Methods

Foliar uptake of mineral elements can also occur as a negative side effect of sprinkler irrigation with saline water (Table 4.9).

Sprinkler irrigation leads to a greater increase in the contents of chloride and sodium in the leaves than does drip irrigation (to the soil surface only), indicating substantial,

Table 4.9
Effects of Sprinkler and Drip Irrigation with Saline Water on the Mineral Element Content of Leaves of Chili (*Capsicum fructescens* L.)[a]

Salt content of the water	Mineral content of leaves (mmol $(100 \text{ g})^{-1}$ dry wt)					
	Chloride		Sodium		Potassium	
	Sprinkler	Drip	Sprinkler	Drip	Sprinkler	Drip
Low	110	20	20	1	101	118
Medium	121	51	26	1	97	121
High	165	76	48	1	86	113

[a]Based on Bernstein and Francois (1975).

direct leaf uptake from the irrigation water. The levels of these two mineral elements in the leaves therefore become toxic quite rapidly when saline water is used for sprinkler irrigation (Francois and Clark, 1979; Maas, 1985). The effect of the two irrigation methods on the potassium content of leaves reflects another phenomenon. With sprinkler irrigation the potassium content of the leaves is lower and declines with increasing salt content of the irrigation water. This pattern indicates leaching of potassium from the leaves, and enhancement of this process by Na^+/K^+ replacement within the leaf tissue.

In general, sensitivity to foliar injury by sprinkling with saline water depends more on leaf surface properties (foliar uptake through epidermis) than on crop tolerance to salinity. Deciduous fruit trees (e.g., almond, apricot) are particularly sensitive to leaf injury when saline water is used for sprinkling, whereas, for example, cotton and sunflower are very tolerant (Maas, 1985).

4.4 Leaching of Mineral Elements from Leaves

4.4.1 Causes and Mechanisms

Leaching can be defined as the removal of inorganic and organic solutes from the aerial parts of plants by the action of aqueous solutions such as rain, irrigation, dew, and fog. These solutes are of different origin and the mechanism by which leaching occurs can differ. Four categories can be distinguished: (a) solutes actively excreted to the external surfaces, for example, salt excretion by salt glands in halophytes, or organic acids (mainly malic acid) secreted by trichomes on the stem, leaves and pods of chickpea (Lazzaro and Thomson, 1989); (b) excretion of inorganic solutes on tips and margins of leaves by guttation (root pressure-induced); (c) leaching from damaged leaf areas or (d) solutes leached from the apoplasm of intact leaf tissue. In the following discussion only the latter two possibilities are considered, since they are of general ecological importance.

The leaching rate generally increases with leaf age (Wetselaar and Farquhar, 1980). With the onset of senescence the permeability ('leakiness') of membranes increases, and there is a corresponding rise in the concentrations of organic and inorganic solutes in the apoplasm of the leaf tissue. The resulting steep concentration gradient across the cuticle favors leaching by runoff water at the leaf surface. Under field conditions in wetland rice, for example, the maximum amount of nitrogen or potassium in the shoots is reached at anthesis and declines thereafter during the rainy season by about 30% when the plants reach maturity (Tanaka and Navasero, 1964). Losses through leaching under field conditions have to be considered in studies of nutrient uptake and retranslocation throughout the growth of plants. Presumably, the losses are largely responsible for the generally lower mineral element contents in leaves of plants grown under field conditions compared with those in leaves of plants grown in the same soil, but indoors.

Mechanical leaf damage (e.g., by the wind) is quite common, tips and margins being especially susceptible. These areas of the leaf tissue are often particularly high in certain mineral elements when they are supplied in excess (Section 3.2.4). In tomato plants with a high chloride supply, for example, chloride is preferentially translocated to the leaf margin and into the leaf hairs, so that mechanical damage leads to a loss of between 10 and 70% of the total chloride taken up during the growing period (Chhabra *et al.*, 1977). Stress conditions other than mechanical damage, such as prolonged darkness, water shortage, and high temperatures, can also increase the leaching rate of mineral nutrients from leaves (Tukey and Morgan, 1963). Comparable stress is imposed on leaves or needles by air pollutants such as ozone or high acidity of the rain water or fog ('acid rain'). Leaching of solutes, mineral nutrient cations in particular, is enhanced under these conditions, either due to more rapid leaf senescence, or damaging of the cuticle, or both. On average, a decrease in the pH of rain water or fog from about 5.5 to 3.5–3.0 increases the leaching of potassium, calcium, magnesium, manganese and zinc by a factor between 2 and 10 (Mengel *et al.*, 1987; Leisen and Marschner, 1990; Turner and Tingey, 1990). Metal cations thus act as a pH buffer in the canopy and are replaced by protons from rain water.

Although leaching of cations does not usually represent much more than 1% of the total content of the leaves (Pfirrmann *et al.*, 1990), it may reach up to 10% of the annual net incorporation of cations into aerial biomass (Schuepp and Hendershot, 1989). The leaching of specific cations from aerial plant parts is compensated by higher uptake rates of these cations by the roots (Mengel *et al.*, 1987; Kaupenjohann *et al.*, 1988). As a consequence of higher cation uptake rate, rhizosphere pH may decrease (Kaupenjohann *et al.*, 1988; Leonardi and Flückiger, 1989) and, thus, part of the acid load of the canopy may indirectly be carried into the rhizosphere.

4.4.2   Ecological Importance of Uptake and Leaching of Solutes from Leaves

Uptake of solutes by, and leaching of solutes from, leaves and other aerial plant parts is important for individual plants, plant stands, and the ecosystem. This is true in particular for perennials such as forest trees where these processes can become a dominant component in mineral nutrition, influencing internal nutrient cycling, as well as the mineral element input and output of forest ecosystems and the long-term stability of these ecosystems. In Central Europe and Northern America a considerable amount of the 'wet deposition' of ammonium and nitrate nitrogen (between 3 and 10 kg N ha^{-1} per year; Brumme *et al.*, 1992) can be directly taken up by the foliage, with NH_4^+ being absorbed much more preferentially than NO_3^- (Garten and Hanson, 1990; Wilson, 1992).

Compared with nitrogen and sulfate (sulfuric acid), for most other mineral nutrients foliar uptake in natural and forest ecosystems is of lesser importance than leaching. This is also the case in areas where air pollution is not a major factor. The amount of a particular mineral nutrient that is leached depends on the type of the mineral nutrient and the amount and the intensity of rainfall. In tropical rain forests the amounts of mineral nutrients leached from the canopy are, as expected, very high, and the annual values, expressed in kilograms per hectare, are as follows: potassium, 100–200; nitrogen, 12–60; magnesium, 18–45; calcium, 25–29; and phosphorus, 4–10 (Nye and Greenland, 1960; Bernhard-Reversat, 1975). This magnitude of nutrient leaching is similar to that of the annual rate of nutrients supplied to the soil surface from the throughfall (litter) and is thus an important component of mineral nutrient recycling,

Table 4.10
Mineral Element Input in Bulk Precipitation (Wet Deposition) and Dry Deposition, the Throughfall and Leaching in a 40-year-old Scots Pine Forest[a]

Parameter	Precipitation (mm yr^{-1})	Mineral element input (kg ha^{-1} yr^{-1})				
		Ca	Mg	K	Mn	Na
Wet deposition	550	12.7	1.4	1.9	0.24	4.1
Dry deposition	—	10.2	1.1	1.5	0.19	3.7
Throughfall	397	27.8	4.3	15.1	2.50	7.4
Leaching from Canopy	—	4.9	1.8	11.7	2.07	—

[a]Based on B. Marschner *et al.* (1991).

particularly in ecosystems with low amounts of available nutrients in the soil, e.g., in highly weathered tropical soils. Reabsorption of leached mineral nutrients also offers the possibility for plants to be supplied at sites of nutrient demand (e.g., new growth) by mineral nutrients that are retranslocated within the plant to only a very limited extent (e.g., calcium and manganese).

In temperate climates losses by leaching from aerial plant parts are much lower, but still considerable (Table 4.10). Under these conditions of much lower rainfall, quantification of losses by leaching is difficult because 'dry deposition' (particulate and gaseous) may constitute a substantial portion of the mineral nutrients in the throughfall (Table 4.10).

Compared with their respective contents in the leaves, the amounts of leached calcium and particularly manganese are often very high, the data in Table 4.10 being a typical example for forest trees grown on an acid mineral soil in a temperate climate. These high amounts of calcium and manganese most likely derive from the apoplasm and reflect an advantage of leaching, namely the removal of excessive amounts of accumulated mineral nutrients which in many instances are present in toxic amounts in mature and older leaves. This holds true particularly for mineral nutrients which are not readily remobilized (calcium, manganese, as well as boron). In agreement with this, needles of Norway spruce, exposed to 'acid rain' contain fewer calcium oxalate crystals in the outer cell wall of the epidermis (Fink, 1991c), and the manganese content has a maximum in 3-year-old needles and drops drastically in older needles (Trüby and Lindner, 1990).

Besides mineral elements, substantial amounts of organic solutes can also be leached from a forest canopy reaching amounts between 25 and 60 kg organic carbon ha^{-1} per year in temperate climates (Bartels, 1990), and several hundreds of kilograms in tropical forests. As a side effect, leaching of mineral elements and organic compounds such as phenolics, organic acids, and amino acids (Tyagi and Chauhan, 1982), can affect other plant species within the canopy as well as soil microorganisms.

5

Yield and the Source–Sink Relationships

5.1 General

More than 90% of plant dry matter consists of organic compounds such as cellulose, starch, lipids, and proteins. The total dry matter production of plants, the *biological yield*, is therefore in the first place directly related to photosynthesis, the primary process of synthesis of organic compounds in green plants. In crop plants, however, yield is usually defined by the dry matter production of those plant organs for which particular crops are cultivated and harvested (e.g., grains and tubers). The term *economic yield* may be used for the harvested crop parts (Barnett and Pearce, 1983). In many crop plants therefore it is not only the total dry matter production that is of importance but also the partitioning of the dry matter. The so-called *harvest index* represents the proportion of the dry matter present in the harvested parts of the crop in relation to the total dry matter production. The partitioning of photosynthates and the source–sink relationship and its controlling mechanisms are therefore of crucial importance in crop production.

In this chapter some principles of photosynthesis are discussed, as are the related processes of photophosphorylation and photorespiration, and examples of the direct involvement of mineral nutrients are given. This discussion includes aspects of photoinhibition and photooxidation and mechanisms protecting the photosynthetic apparatus against this damage.

In higher plants the main sites of photosynthesis – the *source* (mature green leaves) – and the sites of consumption and storage – the *sink* (roots, shoot apices, seeds, and fruits) – are separated from one another. The long-distance transport of photosynthates in the phloem from source to sink is therefore essential for growth and plant yield. It is, thus, necessary to have a basic understanding of the processes of phloem loading, phloem transport and phloem unloading of photosynthates at the sink sites and the regulation of these processes, particularly in relation to the involvement of phytohormones. Finally, the source–sink relationship and the question of whether yield can be limited by source or sink are discussed.

5.2 Photosynthesis and Related Processes

5.2.1 Photosynthetic Energy Flow and Photophosphorylation

The conversion of light energy into chemical energy is brought about by a flow of electrons through pigment systems. In the chloroplasts these pigment systems are

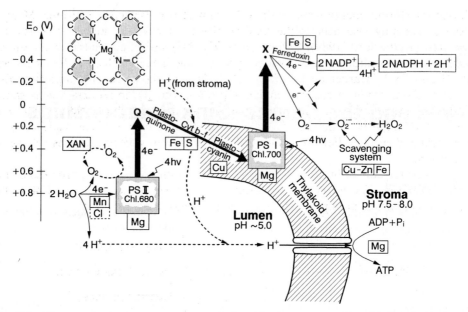

Fig. 5.1 Photosynthetic electron transport chain with photosystems II and I (PS II; PS I) and photophosphorylation. Q, Quencher; X, unknown compound; Cyt, cytochrome; XAN, xanthophyll cycle. (*Inset*) Section of the porphyrin structure of chlorophyll with the central atom magnesium.

embedded in thylakoid membranes in a distinct structural arrangement. Often, the thylakoid membranes are stacked into piles (see Fig. 5.4) which appear as grains or 'grana' under the light microscope. The principles involved in the process of electron flow are illustrated in Fig. 5.1. Light energy is absorbed by two pigment systems: photosystem II (PS II) and photosystem I (PS I). In each of these photosystems between 400 and 500 individual chlorophyll molecules and accessory pigments (e.g., carotinoids) act as centers for trapping light energy (photons), with an 'antenna' chlorophyll absorbing at 680 nm (PS II) and 700 nm (PS I). In both photosystems the absorption of light energy induces the emission and uphill transport of electrons against the electrical gradients, from $+0.8$ to -0.1 in PS II and from $+0.46$ to -0.44 V in PS I. The electrons required for this process are derived from the photolysis of water, mediated by PS II. In higher plants PS II and PS I act in series (Z scheme; for a review see Walker, 1992). At the end of the uphill transport chain, the electrons are accepted by an unknown compound X and transferred to ferredoxin, the first stable redox compound. Ferredoxin in its reduced form has a high negative potential (-0.43 V) and is able to reduce $NADP^+$ (nicotinamide adenine dinucleotide phosphate), as well as other compounds (see below).

Several mineral nutrients are directly involved in this photosynthetic electron transport chain (Fig. 5.1). In PS II and PS I chlorophyll molecules with their central magnesium atom absorb photons, thereby initiating the electron flow. The photolysis (splitting) of water is mediated by a manganese-containing enzyme complex attached to PS II. In this water-splitting system manganese clusters (Section 9.2.4) act as a device for storing energy prior to the oxidation of two molecules of water. Manganese presumably also acts as the binding site for the water molecules which are oxidized (Rutherford, 1989). Cytochromes (Cyt b-f) with their central iron atom as well as an iron–sulfur complex (Rieske protein) mediate the electron flow between PS II and PS I (Marder and Barber, 1989). One of the electron acceptors in this chain is plastocyanin, a copper-containing protein. Finally, ferredoxin acts as transmitter of electrons from a compound X as yet not well defined to $NADP^+$. This is reduced to NADPH by the ferredoxin-$NADP^+$-oxidoreductase which is anchored on the thylakoid surface. Ferredoxin is a 9 kDa iron–sulfur protein which is soluble in the stroma.

Reduced ferredoxin in the chloroplasts can also function as an electron donor for other acceptors. The ferredoxin-mediated reduction of nitrite (NO_2^-) and of sulfite (SO_3^{2-}) is of particular importance for the mineral nutrition of plants:

$$PS\ I \cdots\!\xrightarrow{e^-}\longrightarrow Fe\ \overset{S}{\underset{S}{\overset{\diagup S\diagdown}{\underset{\diagdown S\diagup}{}}}}Fe \xrightarrow[e^-]{\overset{e^-}{}} \begin{array}{l} \nearrow\text{Sulfite reductase} \\ \rightarrow NADP^+ \text{ oxireductase} \\ \searrow\text{Nitrite reductase} \end{array}$$

Both nitrite and sulfite compete within the chloroplasts with $NADP^+$ for reduction. In leaves the rates of reduction of nitrite and sulfite are much higher during the light period (Chapter 8). This coupling of nitrite and sulfite reduction with light is also an example of a more general regulatory mechanism, since photosynthesis supplies the structures (carbon skeletons) required for the incorporation of reduced nitrogen ($-NH_2$) and sulphur ($-SH$) into organic compounds such as amino acids.

The water-splitting and the passage of electrons through the transport chain in the thylakoid membrane is coupled with the pumping of protons into the thylakoid lumen (Fig. 5.1), leading to acidification to a pH of about 5. On the other hand, protons are consumed at the terminal site of the electron transport chain (formation of NADPH), raising the stroma pH to 7.5–8.0. The corresponding electrochemical potential gradient across the thylakoid membrane is used for proton-driven ATP synthesis, *photophosphorylation*. An additional component in the formation of the proton gradient is a pumping system for protons between PS II and PS I (Fig. 5.1), termed 'cyclic photophosphorylation'. The production of one ATP molecule is probably coupled with the downhill transport of three protons across the thylakoid membrane. In the stroma, ATP is required at various steps involved in CO_2 assimilation, carbohydrate synthesis as well as other ferredoxin-mediated processes (see below).

5.2.2 Photoinhibition and Photooxidation

A balance does not necessarily occur between light absorption by the PS II and PS I, corresponding electron flow, formation of reduced ferredoxin, and the consumption of electrons, for example in CO_2 assimilation. This is true under conditions of high light intensity in general and in combination with other environmental stress factors such as drought, low temperatures, or mineral nutrient deficiency in particular.

Excess excitation energy is reflected in a depression in net photosynthesis which is usually reversible (*photoinhibition*), but may also lead in the long term to irreversible damage of the photosynthetic apparatus, as indicated by chlorosis and necrosis of the leaves (photooxidation). For both symptoms the formation of toxic oxygen species is causally involved (Osswald and Elstner, 1986; Salin, 1988; see also Fig. 2.5).

Plants possess a range of protective systems which function by decreasing light absorption or by dissipating energy (e.g., change in leaf angle, light and heat reflection) or detoxification of harmful oxygen species (Fig. 2.5). The primary target for photoinhibition is PS II which produces molecular oxygen and where excessive excitation energy can be transferred from PS II to molecular oxygen forming the highly toxic singlet oxygen ($^1O_2$; Fig. 5.1). As a self-protecting mechanism, carotenoids (xanthophylls in particular) play an important role in both scavenging singlet oxygen and quenching the excited stage of PS II (Young, 1991; Demming-Adams and Adams, 1992). Furthermore, in the PS II the 32 kDa D1-polypeptide has a particularly high turnover rate as required for continuous repair of the system (Walker, 1992).

Another main site of formation of toxic oxygen species is located in the stroma of chloroplasts, where reduced ferredoxin can use molecular oxygen as an electron acceptor leading to univalent reduction of O_2 to the superoxide anion (O_2^-; Figs. 5.1 and 5.2). This reductive O_2 activation in chloroplasts is unavoidable and enhanced under conditions which give rise to an increase in the $NADPH/NADP^+$ ratio, for example low CO_2 supply or impaired CO_2 fixation, caused by a range of environmental stress factors such as low temperatures in chilling-sensitive plant species (Hodgon and Raison, 1991), salinity, drought and mineral nutrient deficiency. This is also true for low or inhibited export rates of photosythates from source leaves under mineral nutrient deficiency (Marschner and Cakmak, 1989). All these stress factors lead to elevated levels of toxic oxygen species, photoinhibition and finally photooxidation. Another factor contributing to photoinhibition is presumably set by excessive acidification of the thylakoid

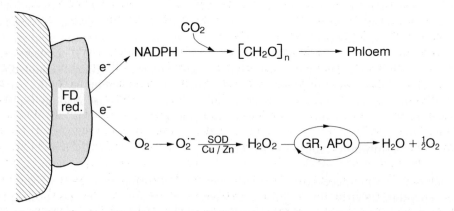

Fig. 5.2 Alternative utilization of photoreductants for CO_2 assimilation or activation of molecular oxygen and detoxification (scavenger) systems. SOD, superoxide dismutase; GR, glutathione reductase; APO, ascorbate peroxidase.

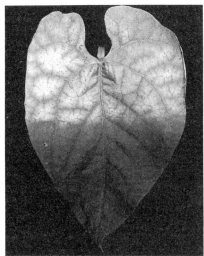

Fig. 5.3 Effect of partial shading of the leaf blades on chlorosis and necrosis in primary leaves of zinc deficient (*left*) and magnesium deficient (*right*) *Phaseolus vulgaris* plants exposed to high light intensity ($480 \, \mu E \, m^{-2} \, s^{-1}$). (From Marschner and Cakmak, 1989.)

lumen as a result of insufficient utilization of ATP in the stroma (Fig. 5.1). Under conditions of limited CO_2 assimilation a continuous high proportion of the NADPH is utilized for other processes such as scavenging H_2O_2 where ATP is not required (Walker, 1992).

In C_3 species photorespiration may be an important protective mechanism for the maintenance of a high electron flow in the carbon chain, i.e. by releasing CO_2 to chloroplasts (Wu *et al.*, 1990). Other systems, however, play a key role in detoxifying O_2^- and related compounds such as H_2O_2 (Fig. 2.5). In chloroplasts, where catalase is absent, O_2^- is detoxified by Cu–Zn superoxide dismutase producing H_2O_2 which is reduced to water by the ascorbate peroxidase–glutathione reductase cycle (Fig. 5.2). In leaves about 70–80% of the ascorbate-dependent H_2O_2-scavenging enzymes are located in the chloroplasts (Strother, 1988).

Elevated activity of the detoxifying enzymes (Fig. 5.2) and increased concentrations of their metabolites (glutathione, ascorbate) typically reflect the influence of oxidative stress, particularly under high light intensity. Such findings have been reported for pine needles during winter (Anderson *et al.*, 1990), spruce needles at noon (Schupp and Rennenberg, 1988), and in bean leaves under magnesium deficiency (Cakmak and Marschner, 1992). There is substantial evidence that toxic oxygen species are also involved in senescence of cells and organs such as leaves (Section 5.5), and that quite often the appearance of chlorosis and necrosis of leaves as visual symptoms of mineral nutrient deficiency is causally related to elevated levels of toxic oxygen species in the leaves. An example of this in bean leaves is shown in Fig. 5.3. Under zinc deficiency the level of toxic oxygen species is high (Cakmak and Marschner, 1988a,b) because of both, depressed SOD activity and lower export rates of carbohydrates as a result of low sink activity (Marschner and Cakmak, 1989). Under magnesium deficiency impaired phloem loading of carbohydrates is the main factor for the oxidative stress under high light intensity (Cakmak and Marschner, 1992). In both instances the production of photooxidants and, thus, photooxidation of leaf pigments, could almost be totally prevented by partial shading of the leaf blades (Fig. 5.3). In agreement with this, inhibited phloem loading of sucrose in genetically manipulated tobacco and tomato plants is also correlated with severe chlorosis and necrosis of the leaf blades (von Schaewen et al., 1990; Dickinson et al., 1991).

5.2.3   Carbon Dioxide Fixation and Reduction

In order to utilize the energy stored during the light reaction (as NADPH and ATP), for CO_2 reduction and the formation of photosynthates such as sugars within the chloroplasts, a CO_2 acceptor is necessary. Ribulose bisphosphate (RuBP), a C_5 compound, serves this function:

$$
\begin{array}{ccc}
\begin{array}{l}
CH_2-O-\circled{P}\\
C-OH\\
\parallel\\
C-OH\\
HC-OH\\
CH_2-O-\circled{P}
\end{array}
&
\xrightarrow[\substack{\text{RuBP Carboxylase}\\ \boxed{Mg}}]{+CO_2 \;+H_2O}
&
\begin{array}{l}
CH_2-O-\circled{P}\\
HC-OH\\
COOH\\
\\
COOH\\
HC-OH\\
CH_2-O-\circled{P}
\end{array}
\\
\text{Ribulose}\\
\text{bisphosphate (RuBP)} && 2 \times \text{Phosphoglycerate (PGA)}
\end{array}
$$

After carboxylation of RuBP two molecules of the C_3 compound phosphoglycerate (PGA) are formed; hence this route of CO_2 incorporation is referred to as the C_3 pathway. The enzyme RuBP carboxylase (Rubisco), which mediates the CO_2 incorporation, is strongly activated by Mg^{2+}. In a series of further steps PGA is reduced to glyceraldehyde-3-phosphate (GAP), a reaction using NADPH and ATP supplied from the light reaction of photosynthesis.

The principles of CO_2 fixation by the C_3 pathway in chloroplasts are illustrated in Fig. 5.4. The enzymes responsible for CO_2 fixation and carbohydrate synthesis are located in the stroma of the chloroplasts, whereas NADPH and ATP are supplied from the thylakoids. The CO_2 acceptor RuBP has to be regenerated in the Calvin–Benson cycle. The remaining carbohydrates are either utilized for transient starch formation in the chloroplasts or transferred as C_3 compounds through the chloroplast envelope into the cytoplasm for further synthesis of mono- and disaccharides. The rate of release of C_3 compounds from the chloroplasts is controlled by the concentration of inorganic phosphate (P_i) in the cytoplasm; P_i therefore has a strong regulatory effect on the ratio of starch accumulation to sugar release from the chloroplasts (Section 8.4.4).

5.2.4 C_4 Pathway of Photosynthesis and Crassulacean Acid Metabolism (CAM)

The incorporation of CO_2 into organic compounds is not restricted to the C_3 pathway described above. It has already been shown (Section 2.5.4) that an imbalance of cation–

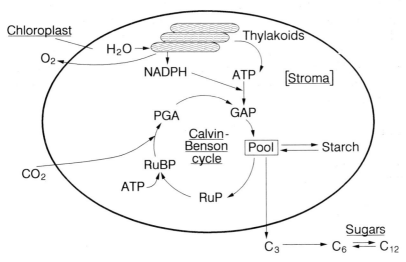

Fig. 5.4 Simplified scheme of CO_2 fixation and carbohydrate synthesis according to the Calvin–Benson cycle in C_3 plants. (Modified from Larcher, 1980.)

anion uptake in roots in favour of cations has to be compensated for by the incorporation of CO_2 via the PEP carboxylase and formation of organic acids. Another example is the reassimilation of CO_2 released by mitochondrial respiration in developing apple fruits, whereby malate accumulation steeply increases in the fruits, but CO_2 loss merely disappears (Blanke *et al.*, 1987). In principle, the same pathway of CO_2 incorporation occurs in the chloroplasts of certain plant species:

$$\begin{array}{ccccc}
\text{COO}^- & & \text{COO}^- & \text{NADPH} \quad \text{NADP}^+ & \text{COO}^- \\
| & \xrightarrow[\text{carboxylase}]{+CO_2 \;\; \text{PEP}} & | & \searrow \quad \nearrow & | \\
\text{C-O-}\textcircled{P} & & \text{C=O} & \xrightarrow[\substack{\text{Malate} \\ \text{dehydrogenase}}]{} & \text{HC-OH} \\
\| & & | & & | \\
\text{CH}_2 & & \text{CH}_2 & & \text{CH}_2 \\
& & \text{COO}^- & & \text{COO}^-
\end{array}$$

Phosphoenol-
pyruvate (PEP)   Oxaloacetate   Malate

Phosphoenol pyruvate (PEP) acts as CO_2 acceptor, forming oxaloacetate which is reduced to malate. The products of this CO_2 incorporation are C_4 compounds, either malate or the amino acid aspartate. Plant species with this C_4 pathway are therefore classified as C_4 plants (Fig. 5.5). This pathway of CO_2 incorporation, however, is confined to the chloroplasts of the mesophyll cells from where the C_4 compounds are transported to the bundle sheath cells. In these cells the C_4 compounds are decarboxylated and the CO_2 released is fixed by RuBP and channelled into the Calvin–Benson cycle. Thus in C_4 plants the final fixation and reduction of CO_2 is identical to that in C_3

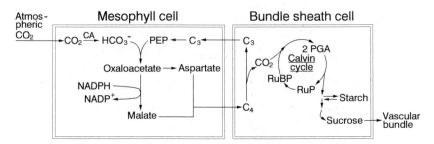

Fig. 5.5 Simplified scheme of CO_2 fixation and compartmentation in C_4 plants. CA, carbonic anhydrase.

plants. In C_4 plants, however, there is a characteristic spatial separation of these two forms of CO_2 fixation.

The remaining C_3 compounds are translocated from the bundle sheath cells back to the mesophyll cells to form PEP to act again as a CO_2 acceptor (Fig. 5.5). The relative importance of malate and aspartate in this carbon shuttle depends on plant species (Ray and Black, 1979) and nitrogen supply. In maize, under nitrogen deficiency the amount of malate operating shuttle is not affected but that of aspartate is greatly reduced (Khamis *et al.*, 1992). This shift is to be expected as nitrate reduction is confined to the mesophyll cells, and aspartate also acts as a shuttle in the transfer of reduced nitrogen into the bundle sheath cells. After transamination (Section 8.2) and decarboxylation the remaining C_3 compounds are recycled into the bundle sheath cells and the amino-N loaded into the phloem by the bundle sheath cells.

In most C_4 species the two cell types are arranged in the so-called Kranz-type leaf anatomy. The minor veins of the vascular bundles are surrounded by bundle sheath cells, forming a Kranz, or wreath. The bundle sheath cells are in turn surrounded by a layer of large mesophyll cells. Additionally, in C_4 species the chloroplasts are dimorphic, those in the bundle sheath cells being larger and possessing grana that are not as well developed as those of the mesophyll. On the other hand, the starch synthesizing enzymes are confined to the bundle sheath chloroplasts where almost all of the leaf starch is accumulated (Spilatro and Preiss, 1987).

The type of photosynthesis that takes place in C_4 plants usually occurs in plant species of tropical origin which have high photosynthetic rates and produce large amonts of dry matter (e.g. sugarcane, sorghum, maize, and various Chenopodiaceae). The C_4 mechanism for various reasons enables plants to utilize both CO_2 and water more efficiently. High CO_2 efficiency is achieved by various factors, such as the high affinity of PEP for CO_2. Additionally, in leaves of C_4 plants the sensitivity of the PEP carboxylase against negative feedback regulation of elevated malate concentrations is lowered in the light (Jiao *et al.* 1991). Furthermore, in C_4 plants the activity of carbonic anhydrase (CA) is high in the mesophyll cells but very low in the bundle sheath cells (Burnell and Hatch, 1988; Hatch and Burnell, 1990). This enzyme shifts the equilibrium $CO_2 \rightleftharpoons HCO_3^-$ in favour of HCO_3^-, which is the substrate for PEP carboxylase, and simultaneously maintains a low CO_2 partial pressure in the intercellular space of the leaves. On the other hand, in the bundle sheath cells a high CO_2 concentration is

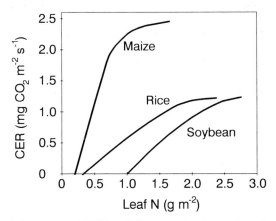

Fig. 5.6   Leaf CO_2 exchange rate (CER) at light saturation for maize, rice, and soybean plotted as a function of leaf N content per unit area. (Sinclair and Horie, 1989.)

achieved, and CO_2 is the substrate for RuBP carboxylase. The very low permeability of the bundle sheath cells for CO_2 (Furbank *et al.*, 1989) further contributes to the maintenance of localized high CO_2 partial pressure within these cells, but to a low value in the intercellular space. This is true even at high temperatures (above 25°C), where endogenous respiratory CO_2 production is correspondingly high. The temperature optimum of PEP carboxylase is also higher than that of RuBP carboxylase (Treharne and Cooper, 1969). This explains, at least in part, why at high light intensities and at high temperatures C_4 plants have considerably lower CO_2 compensation points, that is, levels of CO_2 at which CO_2 consumption and CO_2 production (respiration) are in equilibrium, than those of C_3 plants (0–20 and 50–100 ppm CO_2, respectively).

The greater efficiency of water use by C_4 plants than by C_3 plants is also related to the lower endogenous CO_2 partial pressure and the correspondingly steeper CO_2 gradient from the ambient atmosphere through the open stomata into the leaf tissue. In C_4 plants there is a relatively greater inward diffusion of CO_2 through the stomata (expressed in terms of units of water vapor lost), which can be utilized for photosynthesis and dry matter production. In addition, when stomata partially close in response to water deficit, the decrease in CO_2 influx is less in C_4 than in C_3 plants, because the internal recycling of CO_2 maintains a lower CO_2 concentration in the leaf tissue of C_4 plants. Correspondingly, the relative efficiency of water use (gram dry matter produced per gram transpired water) is around 200–300 in C_4 species compared with usually more than 500 in C_3 species.

In general, C_4 plants have a greater photosynthetic nitrogen use efficiency (NUE) than C_3 plants (Monson, 1989). An example of this is shown in Fig. 5.6, where in maize (C_4) not only the rate of CO_2 fixation (CER) is much higher, but these higher rates are achieved at a much lower leaf nitrogen content. This higher NUE occurs because in C_4 plants RuBP carboxylase operates already at CO_2 saturation. In contrast in C_3 plants due to the lower CO_2 concentration in the chloroplast RuBP carboxylase operates at only about 25% of its capacity and, thus, part of the capacity is lost by photorespiration

(Section 5.2.5). In C_3 plants, 30–60% of the soluble protein in leaves is made up of RuBP carboxylase. The comparative value is only 5–10% in C_4 plants, although additionally some 2–5% are required for the C_4 enzymes (Sage *et al.*, 1987). The higher NUE enables C_4 plants to produce more leaf area per unit nitrogen and, thus, during ontogenesis increase light utilization faster than C_3 plants, provided the higher temperature requirement of C_4 plants is met. If grown under suboptimal temperatures, despite of the higher NUE, some C_4 species may suffer more from nitrogen deficiency than C_3 species (Gebauer *et al.*, 1987).

It is not possible, however, to divide plants strictly into C_3 or C_4 categories. In C_3 plants, a substantial proportion of the CO_2 fixation occurs via the PEP carboxylase pathway, particularly in the reproductive organs, such as developing wheat grains (Wirth *et al.*, 1977) and legume fruits (Atkins *et al.*, 1977). Such observations might indicate a shift in the photosynthetic pathway toward a more efficient use of CO_2, even if the anatomical structures of typical C_4 plants are lacking.

Fixation of CO_2 via the PEP carboxylase pathway is also a characteristic feature of plant species in certain families, such as Crassulaceae and Bromeliaceae, which are particularly well adapted to dry habitats. These plants are mostly succulent; that is, they have a low surface area per unit of fresh weight. These plant species are characterized by their so-called *crassulacean acid metabolism* (CAM) and differ from C_4 plants in a number of features. (a) The stomata of CAM species are open at night. (b) Carbon dioxide enters the leaves and is fixed by PEP carboxylase in the cytoplasm with subsequent reduction to malic acid, which is stored in the vacuoles during the night. (c) During the day the stomata are closed and malic acid is released from the vacuoles; after decarboxylation, CO_2 is fixed and reduced in the chloroplasts following the C_3 pathway. Accordingly large day–night changes in vacuolar pH occur in the leaves of these plants (Lüttge, 1988), and probably both proton pumping systems (ATPase and PP_iase) are involved in the transport of malate into the vacuole (Marquardt and Lüttge, 1987).

In contrast to the spatial separation of the two steps of CO_2 fixation in C_4 plants, the separation of the three steps of CO_2 fixation in CAM species is temporal (*diurnal acid rhythm*). CAM species generally also have lower growth rates than C_4 plants. The combination of CAM and succulence is of particular advantage for adaptation to dry habitats or high soil salinity or both. Interestingly, in facultative halophytes such as *Mesembryanthenum crystallinum* a switch can be observed from C_3 photosynthesis to CAM as a physiological adaptation to drought or salinity stress, i.e. for increase in water conservation (Cushman *et al.*, 1990). The rapidity of this response increases with plant age indicating the existence of a developmental programme that regulates PEP carboxylase transcription and mRNA stability.

5.2.5 Photorespiration

In C_3 plants, light enhances not only the incorporation of CO_2 but also its evolution, a process stimulated by the presence of O_2. This light-driven evolution of CO_2 (*photorespiration*) proceeds simultaneously with the incorporation of CO_2. At high temperatures in particular, the rate of CO_2 evolution increases more than the rate of incorporation resulting in a decline in net photosynthesis. The principle of the reactions

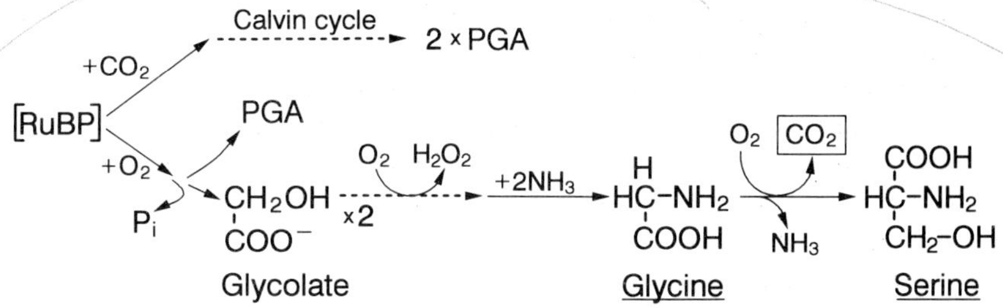

Calvin cycle → 2 × PGA

[RuBP]
+CO₂
+O₂
Pᵢ
PGA

CH₂OH / COO⁻ Glycolate ×2 O₂ H₂O₂ +2NH₃ H / HC–NH₂ / COOH Glycine NH₃ O₂ CO₂ COOH / HC–NH₂ / CH₂–OH Serine

Fig. 5.7   Photorespiration, glycolate pathway, and synthesis of the amino acids glycine and serine.

involved in photorespiration are shown in Fig. 5.7. The CO_2 released in photorespiration comes almost exclusively from glycine oxidation (Sharkey, 1988).

Photorespiration is based on the fact that RuBP carboxylase (Rubisco) can also react with O_2; that is, it can behave as an oxygenase (RuBP oxygenase). In the oxygenase reaction (Fig. 5.7) the C_5 compound RuBP is split into PGA (C_3) and glycolate, the first compound of the 'glycolate pathway'. Glycolate is released from the chloroplasts into the cytoplasm and transferred to peroxisomes in which glycolate acts as acceptor for NH_3, forming the amino acid glycine. After the translocation of glycine into the mitochondria, two molecules are converted into the amino acid serine with simultaneous liberation of CO_2 (photorespiration). In this reaction ammonia is also liberated (Fig. 5.7). In order to prevent both ammonia toxicity and losses by volatilization, reassimilation of ammonia is required via the formation of glutamine from glutamate (Section 8.2). This 'photorespiratory nitrogen cycle' has been reviewed by Wallsgrove et al. (1983).

A key factor for the rate of photorespiration is the CO_2 concentration (partial pressure) at the site of carboxylation of RuBP carboxylase. As CO_2 and O_2 compete at the sites of carboxylation reaction, photorespiration is stimulated by low CO_2 concentrations which are the ultimate consequence of CO_2 fixation in light. In C_3 plants the CO_2 concentration at the sites of carboxylation, in the stroma of chloroplasts, is similar or only about 60% of that of the ambient CO_2 concentration (Sharkey, 1988), which is around 350 ppm (0.035%). Any increase in ambient CO_2 concentrations therefore decreases the rate of photorespiration in C_3 plants. Photorespiration also takes place in C_4 plants, however, at much lower rates. In maize, for example, under ambient conditions (21% O_2, 0.035% CO_2) photorespiratory CO_2 evolution as a proportion of the net photosynthesis was about 6% as compared with 27% in wheat (De Veau and Burris, 1989). The main reason for the much lower photorespiration in C_4 plants is causally related to the higher CO_2 concentrations in the bundle sheath cells, i.e. at the sites of carboxylation of RuBP in C_4 plants. The mechanisms leading to this elevated concentrations of CO_2 have been discussed in Section 5.2.4.

The much higher rates of photorespiration in C_3 plants are the main reason why under otherwise optimal environmental conditions, high light and high temperatures in particular, rates of net photosynthesis and biomass production are considerably lower than in C_4 plants. However, photorespiration should not only be considered from the viewpoint of net CO_2 fixation. Photorespiration is an important pathway of amino acid synthesis in leaf cells (Fig. 5.7) whereby glycolate can act as primary acceptor of the NH_3 produced in the chloroplasts during the light-dependent step (NO_2^- reduction) of the nitrate reduction (Section 8.2). In the above example (De Veau and Burris, 1989) the rate of serine synthesis per unit leaf area was about twice as large in wheat as in maize. It is supposed that the photorespiratory nitrogen cycle represents the largest component of NH_3 incorporation in leaves of most C_3 plants in the light (Yu and Wo, 1991).

Photorespiration may also play a protective role against photoinhibition and photooxidation under high light by the consumption of photosynthates and making CO_2 available to chloroplasts and, thus, maintaining a higher flux rate of electrons in the electron chain of the photosystems (Fig. 5.1), and also leading to higher consumption of ATP in the stroma of chloroplasts (Wu *et al.*, 1990).

5.3 Respiration and Oxidative Phosphorylation

In nongreen tissue (e.g., roots, seeds, and tubers) or in green tissue during the dark period, respiratory carbohydrate decomposition is the main source of energy required for energy-consuming processes such as synthesis and transport. As shown in Fig. 5.8 the major steps of respiration are the decarboxylation reactions of pyruvate and of

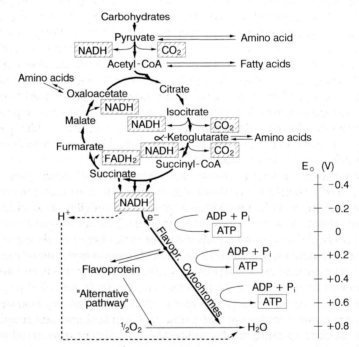

Fig. 5.8 Scheme of the tricarboxylic acid cycle (Krebs cycle) and electron transport chain with oxidative phosphorylation and 'alternative pathway' (cyanide-insensitive respiration).

organic acids in the Krebs cycle (also called the tricarboxylic acid cycle or citric acid cycle) and the oxidation of NADH (reduced nicotinamide dinucleotide). The downhill flow of electrons from the donor NADH to the acceptor O_2 takes place in the respiratory chain in the mitochondrial membranes and supplies the energy needed for ATP synthesis (oxidative phosphorylation). The principles of ATP synthesis in the mitochondria are the same as those of ATP synthesis in the thylakoids of the chloroplasts, namely, charge separation by a membrane with a corresponding proton (pH) gradient across the membrane, constituting a pump for ATP synthesis.

The NADH synthesized in the decarboxylation reactions represents a universal reducing agent in nongreen tissues and is therefore also required for various synthetic processes involving reduction, such as amino acid and fatty acid synthesis. Further-more, the various intermediates of carbohydrate decomposition are essential structures (carbon skeletons) for the synthesis of, for example, amino acids and fatty acids. The rate of respiration is therefore regulated not only by environmental factors such as temperature or by energy requirements (e.g., ATP for ion uptake in the roots), but also by the demand for reducing equivalents and intermediates.

The fact that the respiration is dependent on the requirement of ATP and NADH implies the presence of a feedback reaction in the form of endproduct inhibition if the requirement is low. Depending on the prevailing pathway of synthesis (e.g., lipid or protein synthesis) or other energy-consuming processes (e.g., transport), the require-ment of intermediates such as organic acids, amino acids or sugars for NADH (reducing agent) or for ATP (activating agent) can vary over a wide range. This variable demand is met by metabolic 'by-passes', of which the 'alternative pathway' is particularly interesting (Fig. 5.8). Depending upon the time of the day (Siedow and Berthold, 1986), phosphorus nutritional status (Theodorou and Plaxton, 1993), plant species, developmental stage and plant organ, the proportion of this alternative pathway can vary between 14–60% and 20–80% of the total respiration in the roots and leaves, respectively (Lambers *et al.*, 1983; Poorter *et al.*, 1991). In addition to the alternative pathway a large proportion of oxygen consumption and NADH oxidation may be mediated by peroxidases, which may even exceed that of the alternative pathway, for example in roots of *Brachypodium pinnatum* (Werf *et al.*, 1991). This peroxidase-mediated O_2 consumption requires activated oxygen (O_2^-) formed in reactions similar to those shown in Fig. 5.2 for chloroplasts.

In accordance with the shift in the demand for intermediates, in wheat roots the proportion of the alternative pathway also depends on the form of nitrogen supply; the proportion is very high with ammonium supply but negligible with nitrate supply (Barneix *et al.*, 1984). As will be shown later (Section 5.7), in fast-growing tissues the demand for intermediates in the synthesis of organic structures is especially high compared with the demand for NADH or ATP. A large proportion of the electrons from NADH thus bypass the cytochromes in the respiratory chain and are directly transferred from a flavoprotein to oxygen. As a consequence, less ATP is synthesized per molecule of NADH oxidized. Although this alternative pathway is less efficient in energy conversion than the cytochrome pathway, it has an important function in the regulation of metabolism and growth (Lambers, 1982). On the other hand, restricting the alternative pathway enables nongreen plant cells or tissues to increase ATP synthesis without increasing respiration or carbohydrate consumption, an aspect which demonstrates the difficulty of correlating respiration rates with active ATP-mediated processes, such as ion transport across biomembranes.

The lower efficiency of the alternative pathway in energy conversion in the form of ATP results in a higher energy dissipation in form of heat (*thermogenesis*). The capacity of several plant species to generate heat and maintain higher than ambient temperatures, for example in the inflorescences of *Arum* or of *Victoria* (Skubatz *et al.*, 1990), is causally related to a high proportion of the alternative pathway. The same holds true for ripening mango fruits, where an increase in internal temperature from 29° to about 39°C is correlated with an increase in total respiration by 57% and concomitant increase in the proportion of the alternative (cyanide insensitive) respiration from 41% to 80% (Kumar *et al.*, 1990). Plants also respond to chilling stress by an increase in the proportion of alternative respiration and an increase in their internal temperature (Ordentlich *et al.*, 1991).

5.4   Phloem Transport of Assimilates and Its Regulation

5.4.1   Phloem Loading of Assimilates

In young expanding leaves, most or all assimilates produced during photosynthesis (photosynthates) are required for growth and energy supply. Therefore, in their early growth stages green leaves, like nongreen tissues, act as sinks and require assimilates supplied by a source via the phloem. The major sources of assimilates are fully expanded leaves.

The first step in supplying young leaves with assimilates from the source leaves is short-distance transport of the assimilates from individual leaf cells to the vascular bundles, followed by loading into the phloem. As a rule sugars represent 80–90% of the assimilates exported in the phloem from the source leaves. A typical phloem sap composition has been shown in Table 3.8. In most plant species sucrose is the dominant sugar in the phloem sap, but in some plant species the trisaccharide raffinose (galactose + fructose + glucose) and sugar alcohols like mannitol or sorbitol occur. The preferential sites for phloem loading of sugars are the minor veins of a source leaf as shown in Fig. 5.9, where ^{14}C-labeled sucrose was infiltrated into the leaf.

Depending on the plant species, time of the day, and also the site of collection, the sucrose concentration in the phloem sap is in the range of 200–1000 mM. In order to achieve these high concentrations, a loading step from the leaf cells into the companion and sieve tube cells of the minor veins is required. Based on a range of experimental approaches such as application of inhibitors or uptake studies at different external pH, a *proton–sucrose cotransport* was postulated (Giaquinta 1977; Komor *et al.*, 1977). According to this model, a proton-pumping ATPase in the plasma membrane of the sieve tube cells creates a steep transmembrane potential gradient as well as a pH gradient between the lumen of the sieve tube ('symplasm') and the apoplasm (Fig. 5.10). This gradient acts as driving force for the transport of sucrose from the apoplasm into the sieve tubes in the form of proton–sucrose cotransport (symport) mediated by a specific transport protein. This transport model follows the same principle as has been

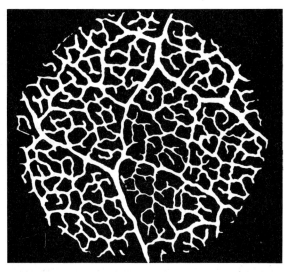

Fig. 5.9 Autoradiograph of phloem loading of [^{14}C]sucrose into source leaf tissue of bean. Sucrose concentration, 1 mM; accumulation period, 30 min. White areas = minor veins with ^{14}C. (From Giaquinta and Geiger, 1977.)

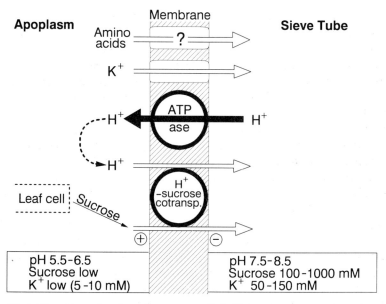

Fig. 5.10 Model for phloem loading of sucrose mediated by proton–sucrose cotransport, and for uniport of potassium and amino acids. (Based on Baker, 1978; Giaquinta, 1980.)

described in Fig. 2.9 for the proton–anion cotransport at the plasma membrane of root cells.

A range of evidence support this model. In sieve tubes a negative potential of -155 mV has been found (Wright and Fisher, 1981), and the pH in the phloem sap is between 7.5 and 8.5 compared with 5.5–6.5 in the apoplasm of leaf cells. An increase in pH of the apoplasm distinctly inhibits sucrose loading into the sieve tubes (Delrot and Bonnemain, 1981). Further evidence for proton–sucrose cotransport comes from studies with plasma membrane vesicles from sugar beet leaves demonstrating, for example, a sucrose-induced alkanization of the external medium and a 1:1 proton:sucrose transport stoichiometry (Slone and Buckhout, 1991). And finally, in transgenetic tomato and potato plants with high acid invertase activity (which hydrolyses sucrose to form hexoses) in the apoplasm the phloem loading of sucrose is drastically decreased and sucrose and other carbohydrates accumulate in the leaves (Dickinson et al., 1991; Heineke et al., 1992).

This concept of apoplasmic loading of sucrose into the sieve tubes requires release of sucrose from the individual mesophyll (or bundle sheath cells in C_4 plants) into the apoplasm. Not much is known about the sucrose concentrations in the apoplasm of leaves; in sugar beet leaves concentrations of about 20 mM have been found (Sovonick et al., 1974) and in maize leaves about 7 mM (Heyser et al., 1978).

The apoplasmic loading as a universal principle in higher plants has been questioned for various reasons, particularly by Van Bel (1989, 1993). For example, in several plant species, particularly those which translocate raffinose (e.g. Cucumis), abundant cytoplasmic connections (plasmodesmata) occur along the possible solute pathway from the mesophyll to the companion and sieve tube cells (Turgeon and Beebe, 1991). However, a symplasmic loading raises questions about the functioning of the plasmodesmata in the formation of a steep concentration gradient in sugar concentrations between leaf cells and sieve tubes (Delrot, 1987). Although both, apoplasmic and symplasmic phloem loading of sugars may exist and occur simultaneously in a given plant species (Van Bel et al., 1988), the experimental evidence for the existence of apoplasmic loading is much more conceivable, at least for sucrose. It well might be that both pathways are important along the route of sugars from the individual leaf cells to the sieve tubes, symplasmic to the region of the sieve tube–companion cell complex and their release into the apoplasm for phloem loading (Beebe and Evert, 1992).

Compared with the sugar concentrations, the amino acid concentrations in the phloem sap are usually much lower, but still in the range of 50–200 mM. Thus, loading of amino acids may be of similar importance. Results supporting the idea of a proton–amino acid cotransport similar to that for sucrose have been presented by Baker et al. (1980) showing that the loading and transport of amino acids in the phloem is also strongly depressed by raising the external pH (Table 5.1). Loading and transport rate of amino acids also depends on the type of cation present in the external solution, K^+ being much more stimulatory than Na^+. It is also well established that phloem loading of sucrose is enhanced by potassium (Peel and Rogers, 1982), unless excessive external potassium concentration lead to depolarization of the membrane potential and thus impairment of the proton–sucrose cotransport. However, it is not clear whether the stimulation by potassium is a direct effect on the loading mechanism (e.g. maintenance of the transmembrane pH gradient) or an indirect one via an increase in osmotic

Table 5.1
Effect of pH and Potassium or Sodium in the Apoplasm on the Loading and Transport of [^{14}C]Alanine in the Phloem of Castor Bean Petioles[a]

Treatment		[^{14}C]Alanine in the Phloem Exudate (^{14}C counts ml^{-1})
Ion	pH	
K$^+$	5	27 840
	8	8 441
Na$^+$	5	13 760
	8	4 090

[a]Based on Baker *et al.* (1980).

potential of phloem sap and, thus, the rate of mass flow in the sieve tubes (Section 5.4.2).

Phloem loading of amino acids is rather selective, in castor bean, for example, glutamine is loaded much more preferentially than glutamate or arginine (Schobert and Komor, 1989). In castor bean, amino acid loading is also distinctly depressed by simultaneous loading of sucrose, and vice versa. In maize leaves, asparagine is the amino acid preferentially loaded into the phloem where its concentration is about eight times higher than in the cytosol of the leaf cells (Weiner *et al.*, 1991). However, as a rule the uphill transport of sucrose into the sieve tubes is much steeper than for most of the amino acids. An example for this is shown in Table 5.2.

Sucrose was preferentially loaded into the phloem compared with the amino acids, as reflected by the ratio of amino acids/sucrose in the cytosol of about one compared with that of about 0.2 in the phloem sap (Table 5.2). Although in some instances the concentration of individual amino acids has been shown to be distinctly higher in the phloem sap than in the cytosol (Winter *et al.*, 1992) and carrier-mediated transfer into the sieve tubes may occur, the overall phloem transport of amino acids seems to depend on sucrose loading and mass flow in the phloem (Winter *et al.*, 1992). Thus, at least in barley, the phloem loading of amino acids presumably follows the concept of an uniport, similar to that for potassium (Fig. 5.10).

Table 5.2
Concentrations of Sucrose and Amino Acids in Different Cell Compartments of Barley Leaves[a]

	After 8 h light		After 5 h dark	
	Cytosol	Phloem	Cytosol	Phloem
Sucrose (mM)	~150	1030	~43	930
Amino acids (mM)	~156	186	~58	244
Ratio Aa/Suc	~1.04	0.18	~1.35	0.26

[a]Based on Winter *et al.* (1992).

5.4.2   Mechanism of Phloem Transport of Assimilates

The principles of transport in the sieve tubes, the anatomy of the phloem, and the transport direction (from source to sink) have been discussed in Chapter 3 in connection with the long-distance transport of mineral nutrients. In brief, according to the *pressure flow hypothesis* (Münch, 1930) solutes are loaded into the sieve tubes of leaves and water is sucked into the sieve tubes creating a positive internal pressure. As sucrose and other sugars are the dominant osmotically active solutes in the sieve tubes of leaves, volume flow rates are determined primarily by phloem loading of photosynthates (including amino acids) in the source and unloading at the sink. Water availability in the source leaves is also an important factor for volume flow rates in the sieve tubes (Smith and Milburn, 1980), and phloem loading associated with lateral water transport in the leaves towards the phloem (Minchin and Thorpe, 1982).

Of the mineral nutrients, potassium is usually present at the highest concentrations in the phloem sap (Section 3.3). Thus, potassium substantially contributes to the volume flow rates in sieve tubes as shown in Table 5.3 for castor bean. In plants well supplied with potassium, the concentration of potassium and the osmotic potential of the phloem sap, and particularly the volume flow rate (exudation rate), are all higher than in plants supplied with a lower level of potassium. Sucrose concentration in the phloem sap remains more or less unaffected, and a high potassium supply increases the transport rate of sucrose in the phloem by a factor of ~ 2. There could be several reasons for this enhancement of the volume flow rate by potassium, including higher rates of sucrose synthesis, enhancement of phloem loading (Fig. 5.10), or direct osmotic effects of potassium within the sieve tubes.

Along the pathway between source and sink, concentrations and composition of the phloem sap may change considerably for various reasons. Photosynthates may leak from the sieve tubes and retrieval becomes important to drive the pressure flow and to nourish the sink (Minchin and Thorpe, 1987). Along the pathway retrieval of sucrose is mediated by the same mechanism (sucrose–proton cotransport) as phloem loading in the source tissue (Grimm *et al.*, 1990). Leakage (or unloading) along the pathway may

Table 5.3
Effect of Potassium Supply to Castor Bean Plants on the Composition of Phloem Sap and Rate of Phloem Sap Exudation[a]

	Potassium supply in the growth medium	
	0.4 mM	1.0 mM
Phloem sap concentration (mM)		
Potassium	47	66
Sucrose	228	238
Osmotic potential (MPa)	1.25	1.45
Exudation rate (ml (3 h)$^{-1}$)	1.35	2.49

[a]Based on Mengel and Haeder (1977).

Table 5.4
Concentration (mM) of Various Solutes in Phloem Sap of Rice
Plants[a]

Solutes	Site of collection[b]	
	Leaf sheath (7–8 leaf stage)	Uppermost internode (one week after anthesis)
Sucrose	206	574
Amino acids	103	125
Potassium	147	40
ATP	1.63	1.76

[a]Hayashi and Chino (1990).
[b]Insect laser technique.

also serve several functions such as supply of sucrose as energy source for surrounding tissues, transient storage of starch or fructanes in stem tissues of grasses (Section 5.7.1) and adjustment of the solute composition in the sieve tubes according to the demand of the sink. In soybean, for example, sucrose concentration was found to decrease from 336 mM in the leaves to 155 mM in the roots as a growth (utilization) sink, and with a corresponding decrease in the pressure potential from 0.60 to 0.18 MPa, i.e. no compensation of osmotically active solutes in the phloem sap (Fisher, 1978). In contrast, in rice plants the solute concentration in the phloem sap increases from the source leaves towards the ear as a storage sink (Table 5.4). This increase is due to sucrose whereas the potassium concentration decreases markedly. Despite a similar total concentration of amino acids (Table 5.4), the composition differed and towards the sink the proportion of glutamine and arginine increased at the expense of glutamate and asparagine (Hayashi and Chino, 1990).

The shift in proportion of sucrose/potassium in the phloem sap (Table 5.4) reflects the demand at the sink sites. In the developing grains of cereals as starch storing organs with a low water content, the demand for sucrose is high but that for potassium low, particularly at later stages of grain filling. In agreement with this, in the peduncles of wheat ears the contribution of potassium to the total osmotic potential of the phloem sap decreases from 8% to 2% within five weeks after anthesis (Fisher, 1987). The increase in sucrose/potassium ratio (Table 5.4) is most likely the result of mobilization of carbohydrates (starch, fructans) in the stem tissue and subsequent sucrose loading into the phloem. Thus, potassium is replaced by sucrose to a considerable proportion in the phloem and presumably transferred into the vacuoles of the stem tissue, demonstrating, that the turgor of individual sieve tubes may be regulated along the pathway from source to sink, and that potassium plays an important role in this regulation (Lang, 1983). Along the pathway potassium may therefore not only fulfil the functions in phloem loading of sucrose but also represent a means of fine regulation within the coarse regulation of pressure-driven solute flow from source to sink (Martin, 1989).

Such a role of potassium is more likely than a specific function in driving a metabolically coupled solute flow through the pores at the sieve plates (Spanner, 1975). Inhibitory effects of chilling temperatures or anoxia applied along the pathway (e.g., stems, petioles) on solute volume flow rates in the phloem (Fensom *et al.*, 1984; Goeschl *et al.*, 1984) are most likely an indication of additional components as described above in turgor regulation in the sieve tubes along the pathway of an overall pressure-driven solute transport.

5.4.3 Phloem Unloading and Storage of Assimilates at Utilization and Storage Sinks

The release of solutes from the sieve tubes into the surrounding tissue at the sink sites is strongly regulated by the sink strength. Sink strength is the capacity of a tissue or organ to accumulate or metabolize photosynthates and thereby to create or maintain a concentration gradient between the solutes in the sieve tubes and the surrounding cells in the sink. Concentration gradients for sucrose and other sugars can be created by the consumption of sucrose during growth (*growth or utilization sink*) and storage of photosynthates (*storage sink*), for example as sucrose (e.g., sugar beet) or as starch (e.g., cereal grains and tubers). For a given plant part (e.g., shoot apex) or organ (e.g., wheat grain, tomato fruit), sink strength is related to the growth rate, storage capacity (e.g., number of storage cells), transport rate from the phloem to the individual storage cells, or conversion rate of photosynthates (e.g., sugars to starch, amino acids to proteins). Sink strength and, thus, phloem unloading are therefore closely related to the metabolic activity of the sink tissue or organ.

Negative feedback regulations on phloem unloading are exerted by high sucrose concentrations in a utilization sink such as growing roots (Farrar and Minchin, 1991). In seeds and grains a turgor-sensitive component is involved in phloem unloading; unloading is inhibited not only by high concentrations of sucrose in the apoplasm but also by other slowly permeating osmotica such as mannitol (Patrick, 1990, 1993).

Whether phloem unloading *per se* is an active or passive process (leakage) is controversially discussed, with arguments in favour (Van Bel and Patrick, 1985) and against (Farrar and Minchin, 1991) an active process. The phytohormone abscisic acid (ABA) seems to be involved in the unloading of sucrose (Schussler *et al.*, 1984), even low concentrations of ABA increasing the rate of sucrose efflux from phloem tissue (Vreugdenhil, 1983; Ross *et al.*, 1987). The induction of a localized increase in the membrane permeability of the phloem cells of the host seems to be the mechanism by which stem parasites such as *Cuscuta europea* acts as a sink, acquiring the assimilates and mineral nutrients they require for growth (Wolswinkel *et al.*, 1984). A particular mechanism of phloem unloading exists in *Mimosa pudica* where seismonastic responses in leaf movement are based on an action potential arising from the touched leaf, travelling through the phloem (1–10 cm s^{-1}) and leading to unloading of the phloem in the exterior region of the pulvinus, i.e. the motor cell complex (Fromm, 1991).

Utilization sinks like young leaves and roots are characterized by high activities of acid invertase in the apoplasm. This enzyme hydrolyzes sucrose to form hexoses, and thereby maintains a low sucrose concentration in the apoplasm (see also Section 5.4.1). Acid invertase-mediated hydrolysis of sucrose has therefore been considered as an important factor for enhanced phloem unloading in these types of sinks (Eschrich, 1984), a view which has recently been questioned (Möller and Beck, 1992). Strong evidence for prevailing symplasmic phloem unloading in utilization sinks, and also in

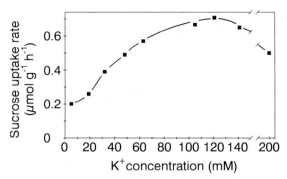

Fig. 5.11   Effect of potassium concentration on sucrose uptake rates by slices of sugar beet storage roots. Sucrose concentration, 40 mм. (From Saftner and Wyse, 1980.)

various storage sinks, comes from electron microscopic studies on the abundance of plasmodesmata connecting sieve elements with most cell types in the sink tissues such as growing leaves (Bourquin *et al.*, 1990) or potato tubers (Oparka, 1986). Symplasmic phloem unloading would not require energy (Oparka, 1990).

In starch-storing sinks, sucrose has to be hydrolyzed into hexoses in the cytosol of the storage cells, as only hexoses such as glucose-1-phosphate are transported through the envelope of the amyloplasts to serve as substrate for starch synthesis (Keeling *et al.*, 1988; Hawker *et al.*, 1991). In the stem tissue of sugar cane which acts as a sucrose storing sink, the symplasmic pathway seems to be the principal means of phloem unloading (Jacobsen *et al.*, 1992) whereas in the storage root of sugar beet, both symplasmic and apoplasmic pathways appear to be involved after phloem unloading of sucrose. Therefore, in sugar beet either sucrose (symplasmic pathway) or hexoses (apoplasmic pathway) may enter the cytosol of the individual storage cells (Lemoine *et al.*, 1988; Fieuw and Willenbrink, 1990). It is not clear whether sucrose is directly transported from the cytosol (high pH) into the storage vacuoles (low pH) via a proton–sucrose antiporter (Lemoine *et al.*, 1988) or require cleavage by sucrose synthase and subsequent synthesis by sucrose phosphate synthase (Fieuw and Willenbrink, 1990; Hawker *et al.*, 1991).

The sucrose concentration in the vacuoles of storage cells can exceed 500 mм which is about 10 times higher than in the cytosol of storage cells (Saftner *et al.*, 1983). Regardless of the mechanism involved the active step of sucrose accumulation in the vacuoles is located at the tonoplast. In storage cells of sugar beet roots the accumulation of sucrose is stimulated by potassium (Fig. 5.11). Sodium has an even greater stimulatory effect on sucrose accumulation (Saftner and Wyse, 1980; Willenbrink *et al.*, 1984). The sites of stimulation are presumably located at the tonoplast and operate, either by activation of the membrane-bound proton pumps or maintenance of high cytosolic pH required to compensate for protons of the proton–sucrose antiporter.

An example of the direct role of mineral nutrients in sucrose transport into vacuoles is shown in Table 5.5. The sucrose accumulation depends on magnesium and is further stimulated by potassium. This strongly supports the view, that a membrane-bound, magnesium-dependent proton pump is also causally involved in the sucrose transport into the vacuoles of storage cells. Activation of $Mg \cdot ATPases$ by potassium is a well-

Table 5.5
Effect of Magnesium and Potassium on Sucrose Transport
into Vacuoles Isolated from Red Beet Tissue[a]

Mg^{2+}	K^+	Uptake rate of sucrose (nmol per unit β-cyanin h^{-1})
–	–	4.9
+	–	42.3
+	+	55.3

[a]Based on Doll *et al.* (1979).

known phenomenon in ion transport at the plasma membrane of root cells (Section 2.4). However, at the tonoplast, potassium stimulates only the $Mg \cdot PP_i$ase, and not the $Mg \cdot ATPase$ (Table 2.9), indicating the involvement of a proton pump energized by PP_i.

Compared with unloading of sugars there is little information on the unloading of amino acids and their transfer to individual cells in sink tissue. For any given sink for anatomical reasons the roles of apoplasmic and symplasmic pathways in the transfer of amino acids are most likely comparable to their roles in the transfer of sugars. After unloading in roots as utilization sinks, a substantial proportion of amino acids not required for growth are recirculated in the xylem, a feature which occurs only occasionally for sugars (Section 3.2).

5.5 Shift in the Sink–Source Relationship

5.5.1 Effect of Leaf Maturation on Source Function

During its life cycle each leaf undergoes a shift in which its function as a sink changes to that of a source for both mineral nutrients and photosynthates. For mineral nutrients, this shift is correlated with a change in the prevailing long-distance transport in the phloem and xylem (Section 3.4). The long-distance transport of sugars such as sucrose, however, is restricted to the phloem, and the shift from sink to source of each leaf has to be correlated with the corresponding shift from phloem unloading (import) to phloem loading (export). As shown in Fig. 5.12, in sugar beet this transition occurs at 40–50% leaf expansion, when about half of the net photosynthetic capacity has been reached. Also in leaves of other dicotyledonous species this transition occurs when they are 30–60% fully expanded. However, there is a long transition period when bidirectional phloem transport occurs in a single leaf, with some vascular bundles importing and others exporting photosynthates (Turgeon, 1989). In graminaceous species such as sugar cane, phloem import occurs up to 90% of the final leaf length, most likely because in these species the young leaves are sheathed and, thus, shaded for a much longer time by older leaves (Robinson-Beers *et al.*, 1990).

The sink–source transition is correlated with typical changes in leaf anatomy and enzyme activities. In sugar beet leaves, during leaf maturation a shift occurs in the

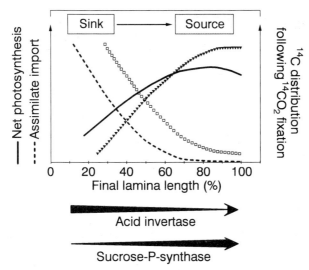

Fig. 5.12   Relationship between assimilate import, net photosynthesis, rate of sugar synthesis (▼, sucrose; □, glucose + fructose), and enzyme activities during maturation of sugar beet leaves. (Based on Giaquinta, 1978.)

incorporation of carbon into sugars, as can be demonstrated by supplying $^{14}CO_2$ to leaves of different age (Fig. 5.12). The shift in favor of sucrose synthesis is closely correlated with changes in the enzyme activities associated with carbohydrate metabolism in the leaves, namely, a decrease in acid invertase activity (sucrose hydrolysis) and a sharp increase in sucrose-P-synthase activity (sucrose synthesis). The correlation with acid invertase, a cell wall-bound enzyme, is probably a reflection of high rates of cell wall synthesis and the provision of hexoses for synthesis rather than of regulatory functions of this enzyme in phloem unloading of sucrose, which is mainly symplasmic in young leaves (Section 5.4.3). In sink leaves the activity of the cytosolic enzyme sucrose synthase (enhancing hydrolysis of sucrose) is also high and rapidly declines during sink–source transition (Turgeon, 1989). The correlation between a decrease in sucrose synthase and an increase in sucrose-P-synthase (sucrose synthesizing enzyme), and transition from sink to source (Fig. 5.12), is causal in plants where sucrose is the dominant sugar in the phloem sap, because the functioning of a leaf as a source relies on the induction and activity of this sucrose-synthesizing enzyme. Results similar to those obtained with sugar beet leaves have been found for soybean leaves during maturation (Silvius *et al.*, 1978). The same causal relationship holds true for the activity of mannose-6-P reductase in mannose-exporting leaves (Turgeon, 1989). In apple, on the other hand, the alcohol sorbitol is the dominant form of organic carbon in the phloem sap. Correspondingly, in apple leaves the activity of the enzyme responsible for sorbitol synthesis (aldose-6-P reductase) is absent in young leaves but increases markedly during leaf maturation (Loescher, 1987).

Sink–source transition is also correlated to a rapid decrease in number of plasmodesmata connecting individual leaf cells and, thus, with marked symplasmic isolation from the phloem (Bourquin *et al.*, 1990). Termination of phloem import may therefore

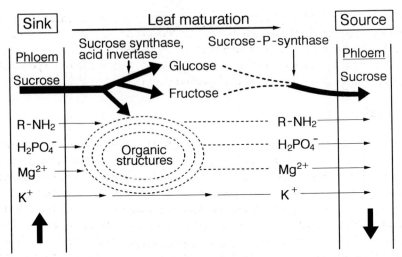

Fig. 5.13   Schematic representation of the shift from import to export of assimilates and mineral
nutrients during leaf maturation and the sink–source transition.

also be caused by interruption of symplasmic phloem unloading (Turgeon, 1989). The
photosynthetic capacity *per se* of a developing leaf is not a regulatory factor in sink–
source transition. In developing sugar beet leaves grown in light or dark, photosynthate
import (^{14}C) was markedly depressed in less than one week and stopped at about 8 days
after emergence also in the much smaller dark-grown leaves, indicating that the
transition is developmentally regulated, and severance of symplasmic connections
between the veins and the lamina may be involved (Pitcher and Daie, 1991).

   The mechanism by which the import and export of mineral nutrients are regulated
during leaf maturation is not very clear. From a consideration of both the mechanism of
phloem transport (solute volume flow) and the average composition of phloem sap in
the stem of plants during vegetative growth (Section 3.3), positive correlation would be
expected between the import rate of sugars such as sucrose into a sink leaf and the
import rate of mineral nutrients such as potassium and phosphorus, and also amino
acids (Fig. 5.13). It may be supposed that phloem unloading of these solutes is regulated
by the requirement for growth processes. However, preferential phloem transport from
source to sink can also be observed when a nonproteinogenic amino acid (α-
aminoisobutyric acid) is supplied to a source leaf (Schilling and Trobisch, 1971) or to the
stem (Van Bel, 1984). This amino acid accumulates in the sink in the soluble nitrogen
fraction, indicating that it is not the sink demand for a particular amino acid that
regulates transport from source to utilizing sinks such as developing leaves, but rather
the direction of solute volume flow in the phloem and the unloading of other solutes,
sugars in particular.

   With the onset of leaf maturation and the capacity for synthesis of sucrose and other
export sugars (e.g., mannose), the leaf becomes a new source of phloem sap as loading
of sugars begins and an increase in the volume flow rate in the phloem from the leaf is
induced. This means, of course, that the export of other solutes in the phloem such as
mineral nutrients and amino compounds can also increase. As discussed previously

(Chapter 3.4) for highly phloem mobile mineral nutrients such as potassium and phosphorus, import via the xylem and export via the phloem can be in equilibrium in mature leaves. The degree to which mature leaves also act as a source of mineral nutrients depends, however, not only on the rate of photosynthate export but also on the nutrient content of the source leaf and the demand of the sink.

5.5.2 Leaf Senescence

With the onset of leaf senescence the rates of both photosynthesis (Fig. 5.12) and export of sugars from the leaf decline. These changes are associated with an increase in membrane permeability (Poovaiah, 1979), and compartmentation is correspondingly affected. Among the other changes that occur, proteolytic enzymes such as acid proteases, previously sequestered in the vacuoles, are released into the cytoplasm and induce rapid breakdown of the proteins in the cytoplasm and chloroplasts. The decline in chlorophyll content (chlorosis) is the visible symptom of senescence. The composition of the phloem sap exported from senescing leaves also changes, the sugar concentration declines and that of low-molecular-weight organic nitrogen compounds (Pate and Atkins, 1983) and phloem mobile mineral nutrients increases (Section 3.5).

Leaf senescence can be induced when leaves are maintained in the dark, an effect which is enhanced if they are also detached. In detached leaves of *Tropaeolum* kept in darkness and sprayed with distilled water only, there is a rapid decline in chlorophyll and protein in the leaf blades within 6 days (Table 5.6). Senescence and enhanced phloem export of soluble nitrogen, potassium, and phosphorus take place, as indicated by the accumulation of these constituents at the base of the petioles. Senescence and export from the detached leaves, however, can be inhibited almost totally by spraying of the leaves with kinetin, a synthetic analogue of the phytohormone cytokinin (CYT).

In intact plants, leaf senescence is under correlative control, and root-borne CYT plays a key role in retarding leaf senescence. The apical zones of growing roots are the main sites of CYT production in plants. This CYT is transported in the xylem to the shoot and distributed with the transpiration stream, i.e., preferentially into the mature leaves. In the leaves CYT is rapidly metabolized (Noodén and Letham, 1986), and, thus a continuous root supply is required in order to prevent senescence. As a rule, in annual

Table 5.6
Inhibitory Effect of Kinetin on the Senescence of Isolated *Tropaeolum majus* Leaves in Darkness[a,b]

Treatment of leaf blades	Content in leaf blades		Content in petioles (base)			
	Chlorophyll	Protein N	Total N	Carbohydrates	K	P
Zero time	100	100	100	100	100	100
After 6 days						
+Distilled water	53	57	323	90	135	217
+Kinetin	98	87	95	47	107	117

[a] Based on Allinger *et al.* (1969)
[b] Leaf blades were sprayed daily with either distilled water or kinetin. Relative values.

Table 5.7
Changes in Xylem Sap Concentration of the Cytokinin
Dihydrozeatin Riboside during Pod Development in
Soybean and Effect of Depodding at Full Pod
Extension[a]

Developmental stage	CYT concentration (mM)	
	Control	Depodded
Pods 12 cm long	75	75
Full pod extension	13	13
Late podfill	12	50
Leaf yellowing	19	93

[a]Recalculated from Noodén et al. (1990a).

plant species with the onset of reproductive growth and preferential allocation of photosynthates into developing seeds and fruits, root growth declines (sink competition) and, thus, also export of CYT as shown in Table 5.7 for soybean plants. Pod development induces leaf senescence primarily via lowering the CYT import into the source leaves. Accordingly, depodding increases CYT export from the roots (Table 5.7) and also delays leaf senescence (Noodén et al., 1990b), as does similarly a direct supply of CYT to derooted soybean explants (Mauk et al., 1990). This general one-way pattern of CYT production in roots, transport in the xylem to the shoot and decline after onset of the reproduction phase typical for soybean for example, is not valid for all plant species. In white lupin, for example after flowering root export of CYT increases considerably, and also a large amount of CYT is transported from the leaves via the phloem into the inflorescences and developing seeds (Taylor et al., 1990). Thus, in some plant species leaf senescence may be induced not by a decline in import but by an increase in export of CYT.

The role of CYT in delaying leaf senescence can often be observed in form of 'green islands' in blades of senescing leaves of perennials in the autumn. These green islands are usually areas of fungal infection or parasitic insect attack and particularly high in CYT (Abou and Volk, 1971). Also certain strains of the gall-forming phytopathogenic bacterium *Pseudomonas savastanoi* synthesize and excrete high amounts of CYT (MacDonald et al., 1986). The production of CYT by parasites appears to be an elegant mechanism by which the parasites maintain the function of the leaf as a source at the site of infection.

Leaf senescence is not only modulated by environmental factors such as drought stress. Leaf senescence in the form of chlorosis of source leaves can be readily induced, for example, in chilling-sensitive plant species by low night temperatures (Schaffer et al., 1991) or high light in combination with zinc, magnesium or potassium deficiency (Marschner and Cakmak, 1989). In these instances, senescence is not induced by a decrease in CYT import but inhibited export of photosynthates and accumulation of large amounts of starch and sucrose in the source leaves. This type of premature leaf senescence is caused by high levels of toxic oxygen species and photooxidation of chloroplast pigments (Section 5.2.2). The rapid senescence of leaves after removal of reproductive sinks is most likely also a matter of excessive energy absorption by the source leaves and not a hormonal-regulated process.

5.6 Role of Phytohormones in the Regulation of the Sink Source Relationships

5.6.1 General

Phytohormones play an important role in the regulation of the growth and development of higher plants. This is reflected, for example, in their effect on the sink–source relationships. Both synthesis and the action of phytohormones are modulated by environmental factors, such as the mineral nutrient supply. At least some effects of deficient mineral nutrient supply on plant growth and yield are caused primarily by their influence on the phytohormone level in the plant. Some examples of these effects are given in the following sections and in Chapter 6.

Phytohormones are chemical messengers, or 'signal' substances, for which sites of synthesis and sites of action are separate in most instances. Transport either from cell to cell or from organ to organ is therefore necessary. With the exception of ethylene, phytohormones are translocated in both the phloem (Weiler and Ziegler, 1981) and the xylem. The prevailing direction of transport depends on the type of phytohormone (e.g., whether they are synthesized primarily in roots or in the shoots) and the developmental stage of the plant. Each phytohormone has a broad action spectrum; that is, the same phytohormone can affect or regulate various processes depending on its concentration and the conditions at the sites of action – the receptor sites.

5.6.2 Structure, Sites of Biosynthesis, and Main Effects of Phytohormones

The importance of the following five classes of phytohormones in higher plants is well established: cytokinins (CYT), gibberellins (GA), auxins (AUX, e.g. IAA), abscisic acid (ABA) and ethylene (ET). More recently evidence has been presented that jasmonic acid and its derivates, the jasmonates (JA), can be a new class of phytohormones (Parthier, 1991). The basic molecular structures of the six phytohormone classes are shown in Fig. 5.14, and some of their major characteristics are summarized in Table 5.8. There is a general tendency for CYT, GA and IAA to enhance growth and developmental processes, whereas ABA and JA have more antagonistic effects. The synthesis of the 'stress hormone' ABA occurs in rapid response to environmental factors such as a deficiency in water, and also the response to nitrogen deficiency is fairly rapid. Some of the actions of ABA, for example, increasing of membrane permeability (e.g., stomatal closure) or decrease in cell wall extensibility, also occur very rapidly. For ET such general classification is not possible, it may enhance growth and development, but also ripening and senescence. It is also a typical stress hormone (see below).

Cytokinins are readily mobile in plants. Although the major sites of biosynthesis are the roots, and in root to shoot transfer xylem transport prevails (Section 5.6.4), at least in *Lupinus albus* cytokinins are also mobile in the phloem and transported from source leaves into inflorescences and developing seeds (Taylor *et al.*, 1990). One of the most

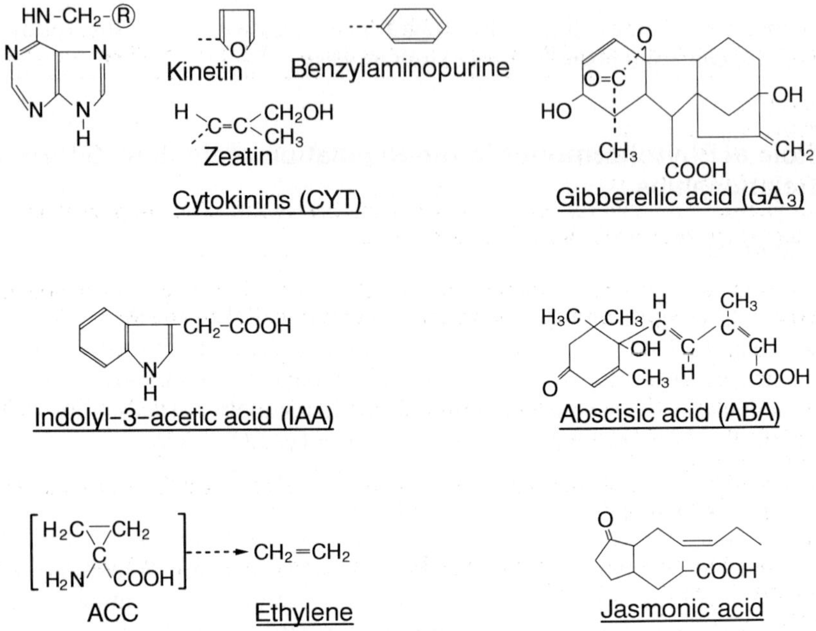

Fig. 5.14 Molecular structure of phytohormones.

distinct effects of CYT, retardation of leaf senescence, is primarily to be attributed to reduction in protein degradation rather than increase in protein synthesis (Lamattina *et al.*, 1987).

Abscisic acid, for which the carotinoids violaxanthin and neoxanthin are the precursors for biosynthesis (Parry and Horgan, 1991), is highly mobile in both xylem and phloem and also circulates in plants, as has similarly been described for some mineral nutrients (Section 3.4.4). Although the roots are also important sites of ABA biosynthesis, of the ABA in xylem sap of *Lupinus albus* only about 28% originate from biosynthesis in the roots, the remainder being derived from the shoot; under drought stress the proportion from root biosynthesis increases to about 55% (Wolf *et al.*, 1990a).

Ethylene (ET) has some pecularities compared with the other phytohormones. It is a gas and the sites of synthesis and action are located in the same tissue. ET has a range of distinct effects on plant growth and development (Table 5.8). For example, depending on its concentration it enhances or represses root growth, it is required for fruit ripening (Theologis *et al.*, 1992), induces aerenchyma formation in roots in response to flooding (He *et al.*, 1992; see also Section 16.4.3.3) and also induces the activation of defense genes under various types of environmental stresses including that induced by pathogens (Mehlhorn, 1990). The well-known effect of IAA on xylem differentiation (xylogenesis) is probably mediated by ET: IAA enhances the biosynthesis of ET (Osswald *et al.*, 1989) which in turn increases protein phosphorylation and xylogenesis (Koritsas, 1988). Enhanced biosynthesis of ET in shoots in response to O_2 deficiency (e.g., by flooding) in the rooting medium is mediated by an increase in xylem transport

Table 5.8
Pathway and Main Sites of Biosynthesis and Some Major Effects of Phytohormones

Cytokinins (CYT)
  Biosynthesis, precursors
    Purine derivates (adenine)
  Main sites of biosynthesis
    Root meristems; to some extent shoot meristems and embryo of seeds: prevailing long-distance transport via xylem from roots to shoot
  Effects
    Cell division and expansion, stimulation of RNA and protein synthesis, induction of enzymes, delay in protein degradation and senescence, apical dominance

Gibberellins (GA)
  Biosynthesis, precursors
    From mevalonic acid to the gibbane carbon skeleton; more than 80 gibberellins with this basic structure have been found
  Main sites of biosynthesis
    Expanding leaves and shoot apex; also other parts of shoots, including fruits and seeds and, presumably, roots
  Effects
    Cell expansion, breaking of dormancy of buds and seeds, induction of flowers and enzyme synthesis (especially of hydrolases)
  Inhibitors of biosynthesis
    Chlorocholine chloride (CCC), Ancymidol, Triazoles

Auxins (AUX; e.g. IAA)
  Biosynthesis, precursors
    Indol derivates of the amino acid tryptophan, the most prominent being IAA ('auxin')
  Main sites of biosynthesis
    Meristems or young expanding tissues; in dicots mainly the apical meristems and young leaves; prevailing direction of transport basipetally: polar from cell to cell, and some long distance in the vicinity of the phloem
  Effects
    Cell expansion and division (in cambial tissues), apical dominance, induction and activation of enzymes (e.g., H^+-ATPase)
  Antagonistes/inhibitors
    ABA, Coumarins, TIBA, 2,4-D, NAA and other synthetic auxins

Abscisic acid (ABA)
  Biosynthesis, precursors
    Carotenoids: violaxanthin and neoxanthin
  Main sites of biosynthesis
    Fully differentiated tissues of shoots and roots
  Effects
    Inhibits cell extension in shoot tissue, induces stomatal closure, favors abscission of leaves and fruits and enhances or induces dormancy ('dormin') of seeds and buds; inhibits DNA synthesis, activates ribonucleases, increases membrane permeability
  Antagonists/inhibitors
    IAA, CYT, GA, Fusicoccin

Ethylene (ET)
  Biosynthesis, precursors
    Methionine $\rightarrow$ 1-aminocyclopropane-1-carboxylic acid (ACC)
  Main sites of biosynthesis
    Various plant parts and organs

Continued

Table 5.8 *Continued*

Ethylene (ET) *continued*
    Effects
        Enhancement of germination, modification of root growth, aerenchyma formation,
        epinastic curvature of leaves, enhancement of flowering, ripening and senescence
    Antagonists/inhibitors
        (Co, Ag)

Jasmonic acid (JA)
    Biosynthesis, precursors
        Linolenic acid
    Main sites of biosynthesis
        Roots, shoot, fruits
    Effects
        Promotion of leaf senescence, fruit ripening, tuber formation (potato), stomata closure,
        storage protein formation; inhibition of cell growth and germination of seeds and pollen
    Antagonists
        CYT

---

of ACC (1-aminocyclopropane-1-carboxylic acid), the precursor of ET, to the shoot (Jackson, 1990a).

Jasmonates (JA) can be considered as 'stress hormones' similar to ABA, their biosynthesis is strongly enhanced under drought stress and it induces the synthesis of particular stress proteins (Parthier, 1991). In contrast to ABA, biosynthesis of JA is not enhanced by salinity stress. Jasmonates are highly phloem mobile and often more effective than ABA in inducing senescence, JA may be responsible for fruit- and seed-induced leaf senescence (Engvild, 1989), and are also very effective in inducing tuberization in stolons of potato (Pelacho and Mingo-Castel, 1991). Jasmonic acid and especially its volatile methyl ester seem to act as signaling molecules for inducing tendril coiling for example in Cucurbitaceae (Weiler, 1993).

It is still a matter of debate whether polyamines (PA) act on plant growth and development as 'second messengers' (Fig. 5.16) or can be considered as another class of phytohormones (Galston and Sawhney, 1990; Beraud *et al.*, 1992). The major polyamines are the diamine putrescine (NH_2-CH_2-CH_2-CH_2-NH_2), the triamine spermidine and the tetramine spermine. They are ubiquitous in plant cells, and highly mobile in both, xylem and phloem, and their concentrations in tissues vary between micromolar and millimolar. In cereals the main precursor of polyamine biosynthesis is the amino acid arginine, and PA biosynthesis is rapidly increased under a range of environmental stresses, drought, heat and salinity in particular (Galston and Sawhney, 1990). Polyamines also accumulate under potassium deficiency (Section 8.7), or when NH_4-N instead of NO_3-N is the source of nitrogen supplied (Gerendás and Sattelmacher, 1990). In contrast, levels are very low under nitrogen deficiency, even in combination with potassium deficiency (Altman *et al.*, 1989). The protective functions of polyamines against environmental stress factors, including ozone-induced foliar injury might be attributed to their role in the detoxification of oxygen radicals (Bors *et al.*, 1989).

In most instances PAs have a protective effect on membranes, their concentrations are very high in meristematic tissues (Palavan and Galston, 1982). In developing soybean seeds both the concentration and composition of the PAs change dramatically with time in cotyledons and embryo (Lin *et al.*, 1984). Polyamines are presumably required as polycations for stabilization of cytosolic pH, and ionic interactions with membranes and macromolecules such as DNA and RNA (Kuehn *et al.*, 1990). Polyamines delay senescence and are synergistic to CYT, they also accumulate, for example, in the 'green islands' of senescing leaves (Walters and Wylie, 1986; Section 5.6.5). Polyamines are effective inhibitors of ethylene biosynthesis; during fruit ripening a decline in PA content is correlated with a steep increase in ethylene production (Winer and Apelbaum, 1986).

Irrespective of the various effects of phytohormones on plant growth and development (Table 5.8) and the effects of environmental factors on their biosynthesis (Fig. 5.16), a typical pattern occurs in the levels of the individual phytohormones in a given organ during its growth and development. Such a pattern is shown in Table 5.9 for trifoliate leaves of bean plants. The levels of IAA, ABA and CYT are high in very young leaves and decline rapidly, particularly for ABA, during early leaf development. 'Dilution effects' by cell wall material are certainly involved in this decrease in content in the dry matter. Thereafter the contents of IAA and CYT level off and ABA rises again.

The phytohormones in developing leaves (Table 5.9) may originate from biosynthesis within the leaves, or import from other plant parts, or both. In view of the main sites of biosynthesis in plants (Table 5.8), IAA most likely originates from the leaf itself, and the gradient in levels correlates with the shift from sink to source of a leaf (Section 5.5). On the other hand, ABA is mainly synthesized in mature (source) leaves and exported with the photosynthates in the phloem to sink sites, i.e., also young leaves. As ABA increases membrane permeability not only to solutes but also to water (Glinka and Reinhold, 1971), high levels of ABA may enhance phloem unloading, but, on the other hand, may negatively affect cell extension growth. The pattern in ABA (Table 5.9) may reflect the shift from sink to source during leaf development.

For CYT the levels change much less, the high levels in the very young leaves may be

Table 5.9
Patterns of Auxin (IAA), Abscisic Acid (ABA) and Zeatin and Zeatin Riboside (CYT) Content During the Growth of Trifoliate Leaves in Bean (*Phaseolus vulgaris* L.)[a]

Area of the trifoliate leaf (cm^2)	Phytohormone content (ng g^{-1} dry wt)		
	IAA	ABA	CYT
1.3	419	568	23
6.8	336	245	19
23.4	297	146	14
57.6	217	57	11
110.0	153	106	10
191.0[b]	166	156	10

[a]From Cakmak *et al.* (1989).
[b]Fully expanded leaf.

a combination of both, biosynthesis *in situ* and phloem import, and at later stages presumably xylem import.

The levels of the various phytohormones occurring during the development of reproductive sinks such as seeds and fruits, are characterized by a distinct sequence (Fig. 5.15) which is quite different from that in developing leaves (Table 5.9). This sequence in seeds might correspond to the period of accumulation of the individual phytohormones during seed development. Maximum CYT level is reached a few days after anthesis (Jameson *et al.*, 1982) and obviously coincides with the maximum rate of cell division. In contrast, the ABA level increases much later and reaches a maximum during the period of rapid decline in the rate of dry matter accumulation. The peak in ABA level is also correlated with enhanced water loss and corresponding desiccation of the grains. A delay in increase in ABA levels in developing grains is caused by rapid degradation of the imported ABA during the early stages of grain development (Goldbach *et al.*, 1977).

The maxima of GA and IAA levels are reached when rates of dry matter accumulation are highest, that is, when both sink activity and rate of phloem unloading are greatest. The interpretation of causal relationships between sink activity and average values for phytohormone levels in seeds is complicated not only because of a lack of information about receptor sites (Fig. 5.16) but also by the differences in the levels of individual phytohormones between tissues. In soybean, for example, the levels of ABA and IAA vary dramatically and independently between embryo, cotyledons and testa during seed development (Hein *et al.*, 1984).

There is a well-established positive correlation between final grain weight and the number of endosperm cells (Singh and Jenner, 1982; Schacherer and Beringer, 1984) as well as the length of the grain-filling period (days between anthesis and maturity). In agreement with this, the single grain weight can be increased by application of CYT to the roots shortly before anthesis (Herzog and Geisler, 1977) and decreased by elevated ABA levels, induced, for example, by high leaf temperatures during the grain-filling period (Goldbach and Michael, 1976). In maize, elevated ABA levels during early

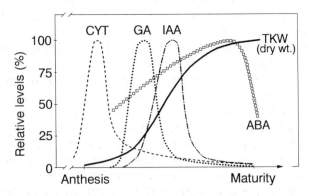

Fig. 5.15   Tentative patterns of phytohormone levels in cereal grains during grain development. CYT, cytokinins; GA, gibberellins; IAA, auxins; TKW, thousand kernel weight; ABA, abscisic acid. Relative values; maximum = 100. (Data compiled from Rademacher, 1978; Radley, 1978; Michael and Beringer, 1980; Mounla *et al.*, 1980; and Jameson *et al.*, 1982.)

kernel development decrease rate of cell division in the endosperm and, thus, storage capacity of the kernels (Myers *et al.*, 1990).

In principle, these distinct patterns in endogenous phytohormone levels also occur in fleshy fruits such as tomatoes (Desai and Chism, 1978) and grapes (Alleweldt *et al.*, 1975).

There has been much speculation about the dependency of developing seeds and fruits on the import of these phytohormones either via the xylem (e.g., cytokinins) or the phloem (ABA, GA). At least for cereals such as wheat, however, it has been convincingly demonstrated that there is no such dependency. In cultures of isolated ears, even when isolated prior to anthesis, normal kernel development can be achieved with only exogenous supply of sugars and nitrogen, but without phytohormones (Lee *et al.*, 1989; Krumm, 1991). The role of imported phytohormones into developing seeds has to be considered in terms of signals by which environmental factors modulate the seed-borne phytohormone levels and, thus, affect the growth and development of the seeds (Section 5.6.3).

Based on the knowledge of the effects of phytohormones on plant growth and development and their typical levels during organ development, it has been a promising approach from the viewpoint of crop production to alter the endogenous levels of phytohormones by the application of '*bioregulators*'. These synthetic plant hormones such as kinetin, or growth retardants such as CCC (chlorocholine chloride) and TIBA (2,3,5-triiodobenzoic acid) can regulate vegetative and reproductive growth, as well as senescence and abscission. In several cases bioregulators are used on a large scale (Jung, 1980; Nitsche *et al.*, 1985), the most successful being the 'anti-gibberellins' which interfere with the biosynthesis of GAs and inhibit cell extension, when applied at low levels, and inhibit cell division and sterol biosynthesis when applied at high levels (Grossmann, 1990). Levels of CYT and polyamines in plants may be increased as a side effect thereby delaying senescense.

During the last few years *brassinolids* have attracted much attention as bioregulators. Brassinolids are naturally occurring steroids ('brassinosteroids') with the same basic structure as the steroids such as ergosterol in the plasma membrane and tonoplast (Section 2.3). Brassinolids were first isolated from rape pollen, but more recently have been obtained from other plant species including rice, *Phaseolus* bean, *Picea* and *Pinus* species (Adam and Marquardt, 1986). Brassinolids have striking effects on plant growth and development, by increasing cell elongation and division and acting synergistically to IAA and GA. Their effects are particularly potent. For example, even at concentrations as low as 10^{-10} M, brassinolids stimulate elongation growth. They are highly lipophilic and their effects on proton excretion and membrane potential indicate that they may be capable of modifying membrane structure and function. Impressive beneficial effects on horticultural crop species have been achieved by application of brassinolids (Mandava, 1988). Information is still lacking, however, as to whether brassinolids play a role in plants as chemical messengers in the signal-conducting system. Their classification as bioregulaters is therefore more appropriate so far.

Despite the success in several areas, the application of bioregulators for manipulation of sink–source relationships and crop yield is rather limited, mainly because of the unpredictability often associated with their use. The reasons for these difficulties are discussed in the following section.

5.6.3   Phytohormones, Signal Deduction, Receptor Sites, and Gene Activation

There is often a poor correlation between the levels of endogenous phytohormones, as determined by bioassays or chemical methods, and the actions of phytohormones in plants. For example, high GA levels are found in certain dwarf mutants. The expected action of applied phytohormones is also often very much at variance with their real actions in plants. The main reasons for the poor correlation between effects of phytohormones and cellular levels are outlined in Fig. 5.16. Usually only a fraction of the total phytohormones are physiologically active, the remainder being at least temporarily inactivated either by chemical binding (e.g. CYT as zeatine ribotide or riboside; Sattelmacher and Marschner, 1978b; Koda, 1982) or by compartmentation in the cells, depending on the pH and membrane potential between the different cell compartments; this holds true particularly for ABA, GA and IAA, but not for CYT (Hartung and Slovik, 1991).

The main reason for poor correlations – or unexpected effects – however, is the requirement for receptors at the sites of phytohormone action. During cell and tissue differentiation and organ maturation not only the response (sensitivity) to given phytohormones changes; the type of action can also be different. This is demonstrated in a few examples.

The stimulation of RNA synthesis by CYT declines with increasing leaf age (Naito *et al.*, 1981), in older leaves the protecting effect against degradation of existing proteins dominating by far (Lamattina *et al.*, 1987). Inhibition of leaf elongation growth by elevated levels of ABA is confined to the responsive zone of cell elongation whereas ABA affects stomatal aperture over the whole leaf. With respect to auxin (IAA), only immature leaf tissue is responsive, mature tissue being unresponsive and neither the uptake nor the rate of metabolization of IAA are responsible for the unresponsiveness (Wernicke and Milkovits, 1987). Several IAA-binding proteins have already been identified from elongating zones of roots and shoots (Jones, 1990). These proteins are attached to the outer surface of the plasma membrane, and in maize coleoptiles more than 90% of it at the outer epidermal cell wall (Klämbt, 1990). This means that of the whole tissue in the elongation zone only one cell layer is responsive to auxin which, after attachment, induces cell wall acidification and accelerated depolymerization of the hemicellulose matrix (Wakabayashi *et al.*, 1991). In growing organs such as coleoptiles and leaves there is a strong tissue tension produced by the extending force of the inner tissue, counteracted by the contracting force of the epidermis with their thick, relatively inextensible external cell wall which is under strong tension (Kutschera, 1989; Edelmann and Kutschera, 1993). Auxin enhances elongation growth by increasing the extensibility of the outer epidermal cell wall, and ABA inhibits this IAA-induced elongation (Kutschera *et al.*, 1987).

In many instances, however, the phytohormones transported to the target tissue or cells do not directly induce the response but act via *second messengers* such as polyamines (Fig. 5.16). The role of calcium as a second messenger has also attracted much attention in the last decade. This is true particularly for the effects of environmental factors (light, temperature, mechanical damage etc.) on the modulation of cytosolic free Ca^{2+} concentration and the subsequent calmodulin-activated secretion processes (Section 8.6.7). Also auxin-regulated cell elongation seems to require Ca^{2+} as second

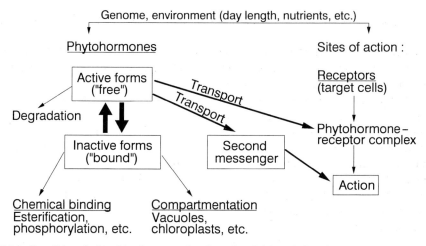

Fig. 5.16 Possible relationships between levels and activities of phytohormones, receptors, and the action of phytohormones.

messenger (Jones, 1990). Inositol triphosphate too can be considered as a typical second messenger, being released from the plasma membrane in response to a hormonal signal, and activating a protein kinase in the cytoplasm (Nishikuza, 1986).

Phytohormones or second messengers were thought to have a general regulatory role in growth and development by reaction with preexisting receptor sites thereby leading to specific physiological responses (Fig. 5.16). Although this is true in various instances, this simple concept has had to be revised for quite often the physiological response requires gene activation and the synthesis of new proteins (Fig. 5.17).

Phytohormones and second messengers have to be considered as integral parts of a signal transduction chain between environmental factors (often occurring as stress factors) and response at a cellular level. For example, low soil moisture (Section 5.6.4) or deficient nitrogen supply to the roots induces enhanced synthesis and export of ABA from the roots to the shoot. Some of this ABA in the shoot may directly bind to the receptor sites at the outer surface of the plasma membrane of guard cells inducing stomatal closure (Hartung *et al.*, 1988), or interfere with the action of IAA on cell elongation, most likely by enzyme modulation and not gene activation (Creelman *et al.*, 1990).

In its effect on seed maturation, however, ABA regulates the expression of specific genes (Skriver and Mundy, 1991). JA which increases under desiccation or osmotic stress, similarly to ABA, also induces gene expression in the direction of enhanced synthesis of (stress) proteins and in repressing the synthesis of the control proteins normally present, i.e., it reprograms gene expression (Parthier, 1991). For ET too this signal transduction via gene activation (Fig. 5.17) is well documented: in response to various environmental stress factors such as ozone, pathogens or UV light, ET levels increase and induce, or activate, defense genes and thereby increase, for example, the level of ascorbate peroxidase, a key enzyme in the detoxification of hydrogen peroxide (Mehlhorn, 1990). In its effect in inducing fruit ripening ET exerts its influence through gene expression (Theologis *et al.*, 1992).

Gene activation may also be involved in the 'memory effects' of plant organs such as leaves. Induction of storage roots and potato tubers can be achieved by environmental factors which increase the ABA/GA ratio in plants, for example, low nitrogen supply (Section 6.3) or short-day (SD) treatment. The stimulus for tuberization achieved by SD can be transmitted to noninduced plants by grafting. Recently, JA has been discussed as a potential candidate for the 'stimulus' (Vreugdenhil, 1991).

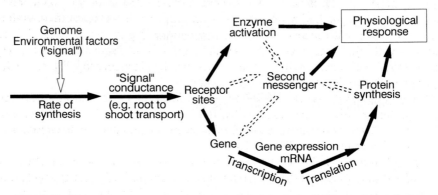

Fig. 5.17   Phytohormones as components of signal conducting systems in plants.

As phytohormones are components of a general signal transducing system (Fig. 5.17) care is required in interpretation of their endogenous levels in terms of expected effects on plant growth and development. However, the endogenous levels provide very valuable information as to whether, for example, the environmental stress factor was sufficiently severe to be transformed into a distinct hormonal signal (decrease or increase in level). Many examples exist showing not only this relationship but also the relationships between the levels of phytohormones and the physiological response, for example between ABA in xylem sap and stomatal aperture (Zhang and Davies, 1989, 1991) or CYT level and leaf senescence (Noodén *et al.*, 1990a,b).

In response to short-term changes in the environment (pH, nutrients, hormones), roots may also transmit electrical signals (action potential) to the shoot leading to a shift in foliar gas exchange (Fromm and Eschrich, 1993). According to the velocity of the signal transmission (2–5 cm s^{-1}) a route has to be involved other than the transpiration stream.

5.6.4   Effect of Environmental Factors on the Endogenous Level of Phytohormones

The synthesis, activity, and degradation of phytohormones are affected by such environmental factors as temperature, day length, and water and nutrient supply. Some of these factors are of particular ecological importance and can be varied in many instances relatively easily by agronomic and horticultural practices (e.g., the application of fertilizers). Thus, growth and development of plants and ultimately economic yield may be affected indirectly via the modulation of endogenous levels of phytohormones. In the following discussion the main attention is directed toward the mineral nutrients and water supply and their effects on phytohormone levels.

It is generally agreed that most of the CYT is synthesized in the roots (Van Staden and Davey, 1979). There is thus a close relationship between root growth in general, the number of root meristems (the sites of CYT synthesis), and CYT production in the roots in particular (Forsyth and Van Staden, 1981). Of the mineral nutrients, nitrogen has the most prominent influence on both root growth and the production and export of CYT to the shoots. Because CYT is exported mainly in the xylem, collecting xylem exudate is a simple method of obtaining information on this nitrogen effect, as shown in Table 5.10 for potato plants. When the nitrogen supply is continuous, CYT export increases with plant age, whereas when the supply is interrupted, the roots respond rapidly by a drastic decrease in CYT export. After the nitrogen supply is restored, CYT export is rapidly enhanced.

The synthesis and export of CYT are also affected by phosphorus and potassium supply, although this effect is somewhat less prominent than in the case of nitrogen (Table 5.11). Results similar to those obtained with sunflower have been obtained with perennials (Horgan and Wareing, 1980).

Although the possibility that these mineral nutrients have a direct effect on the biosynthesis of CYT cannot be dismissed, it is much more likely that they act indirectly via root growth and the induction of root primordia (see above). The close positive correlation between the number of root primordia and leaf area in tomato plants (Richards, 1981) is presumably related to CYT production.

Differences in growth responses of plant species to low root zone temperatures are also closely related to differences in synthesis and root export of CYT. In chilling sensitive plant species (e.g. *Cucumis sativus*), at low root zone temperature CYT levels fall drastically and growth declines steeply, whereas in chilling tolerant species (e.g. *Cucurbita ficifolia*) low root zone temperature markedly increases CYT levels in roots,

Table 5.10

Effect of the Supply of Nitrogen to the Roots of Potato Plants on the Export of CYT from the Roots[a]

Plant age at zero time[b] (days)	CYT exported (ng per plant $(24 \text{ h})^{-1}$)[c]	
	+N	−N
0	196	196
3	420	26
6	561	17
9[d]	—	132

[a]Based on Sattelmacher and Marschner (1978a).
[b]30 days after sprouting.
[c]The amount of CYT exported (over a 24 h period) is determined by levels in the xylem exudate. +N, continuous nitrogen supply; −N, interrupted nitrogen supply.
[d]Restoration of nitrogen supply after 6 days without nitrogen.

Table 5.11

Nutrient Supply and CYT Content of Roots and Leaves
of Sunflower Plants Grown in Nutrient Solution with
Deficient Levels of Nutrients[a]

Treatment (15 days)	CYT (kinetin equivalents, $\mu g\ kg^{-1}$ fresh wt)	
	Roots	Leaves
Control	2.38	3.36
1/10 N[b]	0.94	1.06
1/10 P	1.06	1.28
1/10 K	1.06	2.02

[a]From Salama and Wareing (1979).
[b]Indicates proportion of nutrients in relation to fully concentrated control solution.

Table 5.12

Relative Growth Rate of *Plantago major* L. as Affected by Mineral Nutrient Supply and by
Benzyladenine (BA, 10^{-8} M)[a]

Treatment[b] (Nutrient concentration	BA	Relative growth rate (mg dry wt g^{-1} day^{-1})		CYT content (pmol Z + ZR g^{-1} fresh wt)	
		Shoot	Roots	Shoot	Roots
100%	−	208	159	78	105
2%	−	49	76	21	39
100% → 2%	−	73	183	34	50
100% → 2%	+	220	163	81	110

[a]Compiled data from Kuiper (1988) and Kuiper et al. (1988).
[b]Full concentrated nutrient solution (100%) or diluted to 2%; treatments 100% → 2% for two days only.

maintain a similar CYT transport rate to the shoots and only slightly depresses growth (Tachibana, 1988).

An impressive example of the key role of CYT in modulating plant growth at high or low supply of mineral nutrients is shown in Table 5.12. When plants are grown over a long period with low nutrient supply (2%), their growth rate and CYT contents are much lower as compared with high nutrient supply (100%). Within two days after transfer from high to low nutrient levels (100 → 2%), shoot growth rate and CYT content in shoots and roots declined drastically whereas root growth rate was slightly enhanced. The decline in shoot growth rate could be prevented by amending the low nutrient supply to contain 10^{-8} M benzyladenine (CYT). Root growth was slightly depressed. During these short-term changes in growth rate, mineral nutrient contents in the shoots did not change significantly (Kuiper et al., 1988), confirming that the short-term growth responses were hormonally regulated (Kuiper et al., 1989). Most probably in the low nutrient treatment a shortage of nitrogen supply to the roots was the predominant environmental 'stress signal' (Fig. 5.16) as has been shown in *Urtica dioica* (Fetene and Beck, 1993; Fetene *et al.*, 1993). At low nitrogen supply, in *Urtica* the photosynthates were preferentially allocated to the roots whereas at high nitrogen supply, or direct supply of BA to the roots, photosynthates were preferentially allocated to the shoot apex. As an interesting side-effect, long-term supply of different nutrient levels with or without BA not only affected growth rates but also the ATPase activity of root plasma membrane vesicles isolated from these plants (Kuiper *et al.*, 1991).

Enhanced synthesis and higher levels of ABA in roots and shoots are a typical response to nitrogen deficiency, and similarly to water deficiency (drought stress) or a sudden decrease in root zone temperatures (Smith and Dale, 1988). An example of the effect of nitrogen is shown in Table 5.13. In plants well supplied with nitrogen, the ABA content of young leaves is somewhat higher than that of fully expanded or old leaves, reflecting phloem transport of ABA from the older (source) to the young (sink) leaves (Zeevaart and Boyer, 1984). Under conditions of nitrogen deficiency, the ABA content increases sharply in all parts of the shoots. In potato plants this response can be observed within 3 days, and is much more distinct in roots and xylem exudate than in shoots (Krauss, 1978a).

In crop species such as tomato and barley, short-term response to withdrawal of nitrogen supply is an immediate reduction of leaf elongation rate. Net photosynthesis, however, is not affected so that an accumulation of sugars occurs in these plants (Chapin *et al.*, 1988). This short-term response in leaf elongation rate was associated with a sharp increase in ABA content in the shoots and, thus, most likely an ABA-mediated decrease in cell wall extensibility (Section 5.6.3). Only gradually did rates of photosynthesis decline as a response of the stomata to the elevated ABA contents, and also probably to the declining CYT contents (Chapin *et al.*, 1988).

In tall fescue (*Festuca arundinacea*) low nitrogen supply decreases the number of epidermal cells as well as their elongation rate, and the duration of epidermal cell elongation is about 20 h shorter as compared with plants with high nitrogen supply (MacAdams *et al.*, 1989). As discussed in Section 5.4.3, extensibility of the epidermal

Table 5.13

Relationship between Nitrogen Supply and the ABA Content of Sunflower Plants[a]

	Plant grown in nutrient solution	
Plant part	With nitrogen	Without nitrogen (7 days)
Leaves		
Old	8.1	29.8
Fully expanded	6.8	21.0
Young	13.5	24.0
Stem	2.5	4.9

[a]Content of ABA expressed as μg/g fresh weight. Based on Goldbach *et al.* (1975).

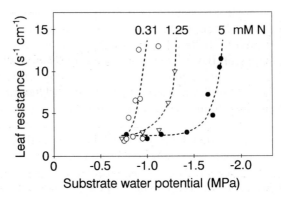

Fig. 5.18   Relationship between nitrogen supply (mM nitrate nitrogen), leaf resistance to water vapor diffusion, and substrate water potential in cotton plants. (Based on Radin and Ackerson, 1981.)

cell wall plays a key role in elongation growth of shoot organs, and the outer epidermal cell wall is most likely the target tissue of IAA and ABA action on elongation growth. In contrast to IAA and ABA the role of CYT in the short-term responses to nitrogen deficiency (e.g., Table 5.12) is not well understood. There is some evidence that CYT modulates the synthesis and the basipetal, polar IAA transport (Li and Bangerth, 1992). In potato plants too, cessation of shoot elongation is the response to withholding nitrogen supply to the roots (Krauss and Marschner, 1982). The failure to restore shoot growth under these conditions by foliar application of nitrogen (Sattelmacher and Marschner, 1979) is indicative that hormonal signals from the roots are causally involved in the cessation of shoot growth and, as discussed below (Section 6.3), also for the induction of tuber formation.

The effects of nitrogen supply on ABA levels are of importance for the water balance of plants. Under conditions of water deficiency (e.g., in dry soils or soils with a high salt content) elevated ABA levels in the leaves favor stomatal closure and thus prevent excessive water loss (Fig. 5.18). As one would expect, when plants are nitrogen deficient or are supplied with suboptimal amounts of nitrogen, they respond to a shortage of available water in the substrate (i.e., an increase in substrate water potential) by a more rapid stomatal closure (indicated by an increase in leaf resistance to water vapor diffusion) than do plants well supplied with nitrogen (Fig. 5.18). This faster stomatal response is not caused exclusively by higher ABA levels delivered by the xylem sap to the leaves but also by the different responsiveness of the leaves to elevated ABA levels (Fig. 5.19). The responsiveness of the stomata to increasing ABA concentrations provided through the xylem is higher in older than in younger leaves, and for a given leaf age, higher in leaves from nitrogen-limited than from nitrogen-sufficient plants.

The higher stomatal responsiveness to ABA in nitrogen-limited plants is also causally related to lower CYT levels. It is well documented that CYT and ABA have opposite effects on stomatal aperture (Radin *et al.*, 1982). In agreement with this, the higher stomatal responsiveness of the older, nitrogen-limited leaves to ABA (Fig. 5.19) could be at least partially reversed by a simultaneous supply of CYT (Radin and Hendrix,

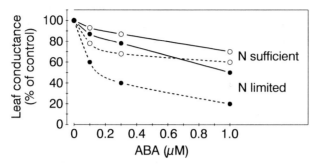

Fig. 5.19 Effect of ABA and stomatal conductance of expanded (○) and old (●) excised cotton leaves from N-sufficient (——) and N-limited (– – –) plants. (From Radin and Hendrix, 1988.)

Table 5.14
Effect of 24 h Drought Stress on Elongation Growth and ABA
Content of Hypocotyls and Roots of Soybean Seedlings[a]

Treatment	± Water	Length (mm)	ABA (μg g^{-1} dry wt)
Hypocotyl	+	40.0	0.87
Hypocotyl	−	4.4	3.60
Hypocotyl	+ +ABA[b]	29.2	3.30
Root	+	52.5	2.66
Root	−	44.7	27.56
Root	+ +ABA[b]	61.5	44.35

[a]Based on Creelman *et al.* (1990).
[b]5×10^{-6} M ABA.

1988). The higher drought resistance of low-nitrogen plants (Radin and Parker, 1979) is therefore the result not only of morphological changes in root growth (Chapter 14) or leaf anatomy (e.g., smaller leaf blades), but also of physiological changes such as an increase in the ABA/CYT ratio. Somewhat similar relationships to those described for nitrogen and stomatal response can be observed for phosphorus (Radin, 1984). In phosphorus-deficient cotton plants, more ABA is accumulated in the leaves in response to drought stress than in phosphorus sufficient plants: in the deficient plants, the stomata close at leaf water potentials of approximately −1.2 MPa, compared with −1.6 MPa in the sufficient plants. As for nitrogen-deficient plants (Fig. 5.19) the sensitivity of the stomata to ABA is increased and can be reversed by CYT under phosphorus deficiency.

One of the most sensitive and distinct plant responses to drought stress is also an increase in ABA in roots and shoots. As a rule, roots continue to grow whereas shoot growth is decreased, leading to typical increases in root/shoot length or dry weight ratio. These differences in growth response between shoots and roots to drought stress are not caused by differences in endogenous ABA levels (Table 5.14). Levels in the roots even increase to a much greater extent than in the shoots in response to drought stress. The responsiveness to ABA differs between shoots and roots, as verified by the effects of exogenous ABA on root and shoot (hypocotyl) elongation rate in plants well supplied with water (Table 5.14).

In view of the similarity of effects of drought stress and nitrogen deficiency on ABA levels in plants, the well-documented increase in root/shoot growth ratio under nitrogen deficiency can also be explained at least in part by the shift in phytohormone levels, with an increase in ABA and decrease in CYT in particular. These differences in root–shoot response to drought stress or nitrogen deficiency are in many instances advantageous to plants growing in soils limited in availability of water or nitrogen, unless root growth is restricted by mechanical impedance or soil volume.

Restriction in root growth by a small soil volume (e.g. container) strongly reduces shoot growth not because of a limitation in water or mineral nutrient supply (Robbins and Pharr, 1988). Leaf expansion and lateral stem formation are both in particular impaired. Simultaneously, carbohydrates (starch in particular) accumulate in the leaves (Robbins and Pharr, 1988) and the carboxylation efficiency of photosynthesis decreases (Thomas and Strain, 1991). These changes indicate that under conditions of root growth restriction by the soil volume root-derived phytohormones may act as a stress signal to the shoot similar to that described for drought stress or nitrogen deficiency.

The levels of GA can also be modulated by environmental factors, particularly by day length, and also by nitrogen nutrition. In potato plants, for example, interrupting nitrogen supply to the roots induces a sharp drop in the GA levels of the shoots, associated with a rapid increase in ABA levels (Krauss and Marschner, 1982). After restoration of the nitrogen supply, levels of GA and ABA respond quite rapidly in the opposite direction, GA increasing and ABA falling. Comparable changes in GA and ABA levels induced by nitrogen supply can also be observed in the tubers of potato plants (Krauss, 1978b) where the changes are correlated with distinct differences in tuber growth pattern (Chapter 6).

The effects of nitrogen on GA levels are presumably indirect. The main sites of GA synthesis are the shoot apex and the expanding leaves. Environmental factors which favour the growth rate of the shoots (e.g. high nitrogen supply, sufficient water supply) therefore, also indirectly favor GA synthesis which is also reflected in changes in plant morphology. In cereals to which high levels of nitrogen have been applied, for example, elongation of the stems is enhanced and the potential danger of lodging increases. In order to counteract these effects in cereals, high levels of nitrogen fertilizer are often applied in combination with growth retardants such as CCC which depress GA synthesis.

Not only the level but also the form of nitrogen supplied to the roots affect the level of phytohormones and second messengers such as polyamines, and thereby also plant growth and development and, ultimately, plant yield. Examples for this are given in Chapter 6.

5.6.5 Phytohormones and Sink Action

During the growth and development of a plant organ the levels of different phytohormones vary distinctly (Table 5.9; Fig. 5.15) and are usually correlated with the sink strength and, in leaves, the transition from sink to source. The level of phytohormone activity is also important for sink competition, as for example, between the reproductive and vegetative sinks of a plant (Section 5.5). In the following discussion a few examples are given of this involvement.

Expanding leaves act as a strong sink for photosynthates, and the application of ABA not only reduces leaf expansion immediately – similarly to the effect of drought stress – but simultaneously enhances the export rate of photosynthates. This is indicated by the lower dry mass gain resulting from an otherwise unaltered rate of photosynthesis (Table 5.15). This ABA-induced rapid shift from sink to source is most likely caused by a decrease in IAA activity leading to a decrease in cell wall extension (Section 5.6.3) and sink strength of the leaf (see below). It has been shown in flowers of melon, that an increase in ABA content alters the IAA metabolism and greatly decreases the content and proportion of free IAA (Dunlap and Robacker, 1990) which is considered as a main hormonal component responsible for the sink strength of an organ.

As early as 1950 Nitsch convincingly demonstrated the role of IAA in the sink action of developing strawberry fruits. Removal of the seeds from the developing fruits resulted in the immediate cessation of fruit growth. Application of IAA to the seedless fruits replaced the sink action of the seeds and restored the growth rate of the fruits. This indicates that solute volume flow via the phloem into developing strawberry fruits is mediated by IAA produced in the seeds.

Phytohormones in general and IAA in particular are also involved in the role of tissues as sinks for mineral nutrients as shown in Table 5.16 for phosphorus in bean plants. Removal of the seeds and, especially, the fruit markedly reduces the accumulation of ^{32}P in the peduncles. This can be restored to a large extent, however, by IAA application to the cut end of the stump. When IAA is applied in combination with kinetin, the accumulation of ^{32}P in the peduncle is even greater than in the control, indicating that the action of the fruit as a sink for ^{32}P can be stimulated by treatment with phytohormones. IAA similarly enhances the accumulation of ^{14}C in the peduncle after exposure of a mature leaf to $^{14}CO_2$.

Auxins also play an important role in dominance phenomena (Bangerth, 1989). Dominance, or correlative, phenomena which are widespread in the plant kingdom, are particularly common in reproductive sinks between fruits (e.g., individual tomato fruits on the same truss), seeds (e.g., grains in medial, proximal and acropetal position within an ear) and in utilization (vegetative) sinks (e.g., terminal versus lateral buds). Competition for photosynthates may offer an explanation for this phenomenon.

Table 5.15
Effect of Applied ABA (1 nmol per leaflet), Drought Stress and Low Light Intensity on Rate of Photosynthesis, Leaf Area Expansion and Change in Dry Mass in Soybean during an 8 h Light Period[a]

Parameter	Treatments			
	Control	ABA	Drought	Low light
Photosynthesis (μmol CO_2 m^{-2} s^{-1})	14.8	14.0	14.6	2.4
Area expansion (cm^2)	4.5	3.2	2.7	4.2
Dry mass gain (mg per leaflet)	33	26	27	7

[a]According to Bunce (1990).

Table 5.16

Effects of Seed or Fruit Removal and Hormone Application on the Accumulation of Leaf-Applied ^{32}P in the Peduncles of Bean[a]

Treatment	^{32}P (cpm)
Control (intact fruit)	373
Seeds removed	189
Fruit removed	34
Fruit removed and cut end treated with	
Lanolin	6 (320)[b]
Kinetin	20
IAA	235 (5520)[b]
IAA + kinetin	471

[a]Based on Seth and Wareing (1967).
[b]Numbers in parentheses indicate counts per minute (cpm)
^{14}C from $^{14}CO_2$ applied to a mature leaf.

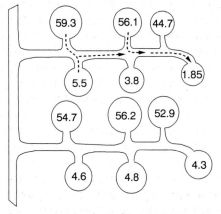

Fig. 5.20 Tomato trusses with natural sequence of fruit development (upper truss) and 'synchronized' fruit development by pollination at the same day (lower truss). Numbers in fruits represent final fruit weight in g (upper fruits) and polar IAA export (ng per fruit per day) of 10-day-old fruits (lower fruits). (Bangerth, 1989.)

However, dominance frequently occurs very early in the ontogeny of reproductive and utilization sinks when competition for the limited amount of photosynthates available is less likely and may even be discounted. A correlative dominance signal may thus account for this effect. An example of this is shown in Fig. 5.20. Earlier developed fruits dominate over later developed ones, and dominance is achieved by higher polar, basipetal transport of IAA as a 'signal' of higher sink activity. The higher IAA export from dominating sinks seems to have additional repressing effects ('autoinhibition') on those fruits which are dominated (Bangerth, 1989). The same phenomenon is also

Table 5.17

Effect of Foliar Sprays on the Growth of Carrot Plants[a]

Spray	Weight (g dry wt per plant)			Ratio Shoot/Root
	Shoot	Root	Total	
H_2O	3.2	10.9	14.1	0.29
Kinetin	7.3	8.8	16.1	0.83
GA	9.9	5.7	15.6	1.74
CCC	2.8	10.8	13.6	0.26

[a]Sprays were applied once per week for 7 weeks. From Linser *et al.* (1974).

observed in the dominance of apical versus lateral buds on a stem (Li and Bangerth, 1992).

However, in several instances more than one of the dominance mechanisms may be involved. In potato plants, for example, in the very early stages of tuber development dominance phenomena on a hormonal basis may prevail whereas competition for photosynthates operates at later stages (Engels and Marschner, 1986).

Although it is relatively easy to modify the activities of the endogenous phytohormones by environmental factors such as nitrogen supply, these changes are complex and the plant system is not an easy one to manipulate. Direct application of phytohormones for increasing sink activity appears more straightforward, but has been successful only in a few cases. Some reasons for the particular difficulties in manipulating reproductive sinks have been discussed by Michael and Beringer (1980; see also Fig. 5.15). One of the few successful examples has been the increase in seed yield in faba bean by foliar application of GA at the six-leaf stage (Belucci *et al.*, 1982). The yield increase obtained was mainly the result of an increased number of pods and seeds per plant. Faba beans are well known for the high proportion of flowers which abort and the application of GA decreases this abortion.

A similar mode of action seems to be responsible for the increase in grain yield and harvest index in maize after foliar application of CYT (Smiciklas and Below, 1992). Yield increase by CYT was achieved by a depression in kernel abortion, i.e. sink initiation. Interestingly, this effect of CYT was dependent on the form of nitrogen supplied. It occurred only with NO_3-N supply whereas with NH_4-N supply yield was similar to that in the NO_3-N + CYT as in the NH_4NO_3 − CYT treatments.

In plants with vegetative storage sinks such as tuber and root crops, manipulations of sink activity by application of phytohormones appears somewhat easier and successful results have been reported on storage roots of winter radish (Starck *et al.*, 1980) or potato tubers (Ahmed and Sagar, 1981). In all these experiments, however, phytohormones have not and could not be applied to the sink organs directly, so that their effects have to be indirect and may therefore not be as envisaged. An example demonstrating this complication is shown in Table 5.17. Both the application of kinetin (CYT) and GA as foliar sprays greatly increased the shoot growth of carrot plants, but this increase was largely at the expense of the growth of the storage root. This is a typical example of both sink competition between shoot and root and effects of phytohormones on the sink

Table 5.18
Effect of Removal of Source Leaves of White Mustard on the Photosynthesis and Assimilate Export of a Remaining Source Leaf (no. 2)[a]

Treatment	Photosynthetic rate of leaf no. 2 (%)	[14]C export from leaf no. 2 (%)
Control	100	36
Source leaves removed (nos. 3–6)	187	62

[a]Based on Römer (1971).

strength of tissues and organs. CCC inhibits shoot growth without affecting the storage root growth; that is, it supports the sink strength of the storage root and produces plants similar to the untreated control but with a somewhat larger harvest index (storage root/shoot ratio). This example again demonstrates that use of 'bioregulators' which influence phytohormone biosynthesis or the action of a phytohormone in plants are, as a rule, more effective in modifying activity and strength of sinks than direct application of phytohormones.

5.7 Source and Sink Limitations on Growth Rate and Yield

5.7.1 Shift in Limitations During Ontogenesis, Role of Environmental Factors

The growth rate of sink tissues and organs such as the roots, shoot apex, fruits, and storage organs can be limited either by supply of photosynthates from the source leaves (*source limitation*) or by a limited capacity of the sink to utilize the photosynthates (*sink limitation*). Sink limitation can be related to low rates of phloem unloading or cell division, a small number of storage cells, low conversion rate of photosynthates (e.g., sugars to starch), or low number of sinks (e.g., grains/ear). Sink–source limitations are characterized by strong genotype/environment interactions and the ratio source size (e.g., leaf area) to sink size (e.g. number of fruits/plant). For example, in crop species genotypes with a high harvest index (e.g., high fruit weight/total biomass ratio) tend more often to become source limited than genotypes with low harvest index. Within a given genotype and environment, during ontogenesis a shift from sink to source limitation is quite typical. Environmental stress factors such as drought, extreme temperatures and nutrient deficiency are particularly important in the transition phase from vegetative to reproductive growth or induction of storage organs and, thus, for sink or source limitations of the crop yield. In the following examples, both types of limitation are considered with particular emphasis on ontogenesis and environmental factors, except mineral nutrients which are discussed in detail in Chapter 6.

During vegetative growth in plants with a large number of source leaves, the growth rate is often limited by the capacity of young leaves (sinks) to utilize the photosynthetic potential of the source. In mustard plants, for example (Table 5.18), the removal of four source leaves led to an approximate doubling of both the rate of photosynthesis and the export of photosynthates from the remaining source leaf. In principle, the same results have been obtained in young plants of *Populus nigra* (Tschaplinski and Blake, 1989). This demonstrates that, in intact plants with all the leaves present, the potential photosynthetic capacity of leaves was not being fully utilized.

The decline in rate of net photosynthesis often observed in the afternoon (photoinhibition, Section 5.2.2) is the expression of a temporary sink limitation (Foyer, 1988), reflected in sugar and starch accumulation, and either partial or total closure of the stomata (Hall and Milthorpe, 1978). Under high light intensity in the long run sink limitation may even lead to source destruction (photooxidation, Section 5.2.2) unless alternative sinks are available at least as temporary sinks such as stems of cereals for carbohydrate storage (Williams and Farrar, 1988). The rate of net photosynthesis, and also of transpiration per unit leaf area, therefore often increases after fruit or seed set, the increase being dependent on the presence of transient sinks and on the former degree of photoinhibition. Evidence has been presented that ABA and CYT are involved in this regulation. Because the transport rates of sucrose and ABA from the source leaves to the sink are positively correlated, when the sink demand for photosynthates is high, the ABA concentrations in the source leaves are low and the stomata are fully open (Setter *et al.*, 1980). The high transpiration rates of the source leaves in turn enhance xylem volume flow into the leaves not only of water but also of the CYT synthesized in the roots (Carmi and Koller, 1979), a mechanism which is also important for delaying leaf senescence (Mauk *et al.*, 1990). It remains an open question whether an additional hormonal signal from the developing fruits to the source leaves as feedback-signal is involved in fruit or seed-enhanced senescence of the source leaves (Noodén *et al.*, 1990a,b).

Particularly in plants with vegetative storage organs (sinks) such as storage roots and tubers, after the induction of such sinks shoot growth rate usually declines or ceases and rate of net photosynthesis may be maintained or also fall. However, in plants (e.g. carrot, sugar beet) with such sinks which gradually change from a utilization sink (increase in storage root growth) to a storage sink (storage of carbohydrates), the situation is more complex and net rates of photosynthesis are not a suitable parameter for evaluating source–sink relationships. In early stages of growth the requirement for structural material for synthesis of proteins, membranes, and cell walls is high as compared with the energy demand (e.g., ATP). A high proportion of the sugars imported is therefore respired via the 'alternative pathway' (Section 5.3). An example of this is shown in Table 5.19 for the storage roots of carrots. Rapid growth of the storage root and sugar storage occurred at the expense of lateral root growth and started at about day 25. Thereafter the growth rate of the roots (i.e., of the main sink) was maintained at a similarly high level, despite the sharp decline in the rate of net photosynthesis and in photosynthate export to the roots. This apparent discrepancy is the result of a shift in the respiratory pathway. Before sugar storage, respiration in the tap roots was high and the alternative pathway contributed 46% to the root respiration (compared with 54% for cytochrome respiration). With the onset of sugar storage the respiration rate decreased, the decrease being particularly pronounced in the alternative respiratory pathway. Thus, despite a reduction of net photosynthesis to less than half with increasing plant age, and also a reduction in export of photosynthates to the roots by 42%, the dry matter increase in the storage roots declined by only 10% (Table

Table 5.19
Photosynthate Production and Utilization in Shoots and
Roots of Carrot[a,b]

	Plant age (d)		
	18–25	25–32	32–39
Shoot			
Photosynthesis	555	381	236
Respiration	119	55	35
Growth	236	193	85
Export to roots	200	133	116
Roots			
Alternative respiration	49	9	4
Cytochrome respiration	58	40	28
Growth	93	84	84

[a]From Steingröver (1981).
[b]Data expressed as mg organic carbon (CH_2O) g^{-1} shoot d^{-1}.

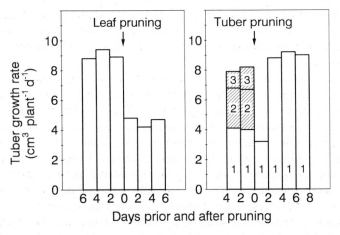

Fig. 5.21 Growth rate of potato tubers per plant as affected by reduction of the leaf area to 50% (leaf pruning) and number of tubers (tuber pruning; removal of tubers 2 and 3). (Engels and Marschner, 1987.)

5.19). Despite this increase in efficiency in carbon transformation of photosynthates, however, the carrot plants were most likely source limited at later growth stages.

Also in potato plants after tuberization a shift is rather typical from sink to source limitation (Fig. 5.21). At high tuber growth rates a decrease in source capacity by pruning half of the leaves also reduces the tuber growth rate to one half. Conversely, increasing the source–sink ratio by removal of some of the tubers increases the growth rate of the remaining tuber within 2–4 days to a level comparable to that prior to pruning (Fig. 5.21).

For a given plant species, genotypical differences in sink–source relationships and limitations are often related to differences in the ratio of source size (leaf area) to sink size (e.g., number of grains or tubers per plant). This was demonstrated for three maize genotypes, which were subjected to defoliation treatment (25% reduction of the leaf area) 2 weeks after 50% silking (Barnett and Pearce, 1983). Defoliation hardly affected grain weight in the genotype with a relatively large source size and a large source/sink ratio. In contrast, defoliation of the other two genotypes with a smaller source/sink ratio reduced grain yield. In all three genotypes, defoliation reduced the stalk weight as a consequence of the mobilization of nonstructural carbohydrates stored in the stalk (Barnett and Pearce, 1983). This reflects the role of the stem of graminaceous species as a transient storage pool for photosynthates during vegetative growth (Setter and Meller, 1984). In graminaceous species of cool and temperate climate (e.g., *Agrostis* species, or wheat), the main transient storage carbohydrates in stems are fructans (polymerized fructose with chain length mainly between 3 and 8 units; Schnyder *et al.*, 1988), accumulated prior to and also during the first weeks after anthesis, and utilized thereafter for grain filling (Dubois *et al.*, 1990). In maize, large amounts of nitrogen are stored in the stem, and nearly half the nitrogen in the grains may derive from this source (Ta, 1991). Thus, graminaceous species have a substantial internal buffer for short-term compensation in fluctuations in source capacity during early reproductive growth. In winter rape after anthesis leaf area index decreases steeply and the developing seeds are supplied mainly with carbon fixed by the pods or photosynthates from transient storage in the stem (Clasen *et al.*, 1991). In wheat under drought stress after flowering remobilization of photosynthates from the stem is of crucial importance for grain growth (Blum *et al.*, 1983).

The source capacity is not only determined by the leaf area but also leaf area duration, the latter being of crucial importance for the length of storage and often also for the final yield. In potato plants grown at different elevations and temperature regimes, at high temperatures, despite a much larger leaf area (source size) at the time of tuber induction the final tuber yield was low because of early leaf senescence (Manrique and Bartholomew, 1991). In contrast, at low temperatures (high elevation) leaf area was only a quarter of that at high temperatures, but leaf area duration much longer and, thus, also the tuber yield. In principle, similar differences are achieved with temperatures as with different levels of nitrogen supply (Chapter 6).

Root zone and canopy temperatures may affect source–sink relationships to an even greater extent in plants with reproductive sinks. Although this is true for both suboptimal and supraoptimal temperatures, the effects toward sink limitation are often more distinct for the latter as shown in an example in Table 5.20 for cotton. Total plant dry weight was not much affected by the temperature regime but dry matter allocation changed dramatically. Only at 30/20°C was a high proportion of the dry matter allocated to the reproductive sink (bolls), whereas at lower temperatures this proportion decreased greatly, and at supraoptimal temperatures (35/25°C) the plants failed to form reproductive sinks (flowers).

Failure of inducing flowers or to set seeds and fruits are a well known phenomena at supraoptimal temperatures or drought stress during anthesis. This failure might be attributed to elevated levels of ABA in the reproductive organs (Zeng and King, 1986). In wheat, for example, drought stress during meiosis of pollen mother cells decreased the proportion of fertile spikelets from 68% (well watered) to 44%, and simultaneously

Table 5.20
Effect of Day/Night Temperatures on Dry Matter Allocation in Cotton Plants[a]

Dry matter (kg m^{-2})	Temperature regime (°C day/night)			
	20/10	25/15	30/20	35/25
Total	2.26	2.84	3.11	2.68
Stems and Leaves	1.68	2.16	1.62	2.27
Bolls	0.18	0.33	1.31	0.05
Roots	0.29	0.27	0.17	0.29

[a]Based on Reddy *et al.* (1991).

increased content of ABA in the ears from 35 to 111 μg g^{-1} fresh weight (Morgan, 1980). Application of ABA to the ears of well-watered plants also decreased spikelet fertility from 68% to 37%. The inhibitory effect of ABA on fertilization is restricted to a short period before anthesis (Saini and Aspinall, 1982).

However, conclusions about the role of ABA as the decisive 'signal' of drought stress and also for depressing fertilization have to be drawn with care. In maize, short-term interruption of carbohydrate supply to the flowers during anthesis can strongly affect seed set (Table 5.21). Stem injection of a liquid medium from tissue culture with high sucrose concentrations (150 g l^{-1}) over a 5 or 7.5 day water-deficit period prevented failure of reproduction, regardless of whether the liquid medium contained the phytohormones (IAA, CYT) or not.

The mechanism of how sucrose affects seed set is not clear, and the above results cannot exclude the possibility that ABA was involved in this process; in many instances it is not only the concentration but also the cellular compartmentation of ABA as affected by exogenous and endogenous factors that is of crucial importance for its action (Radin and Hendrix, 1988; Hartung *et al.*, 1988). Nevertheless, the role of sucrose concentration in the reproductive organs during anthesis for seed set does appear to be

Table 5.21
Effect of Low Leaf Water Potential (MPa) During Anthesis and of Stem Injections of either Culture Medium (CM) only, or Hormones (HO), on Seed Set and Grain Yield in Maize[a]

Treatment	Leaf (MPa)	Seeds (no. per ear)	Seed size (mg per seed)	Total seed wt. (g per plant)
Control	−0.63	431	176	75
Low	−1.81	0	0	0
Low + HO[b]	−1.71	19	225	4
Low + CM[c]	−1.62	302	203	60

[a]Based on Boyle *et al.* (1991).
[b]CYT + 2,4-D.
[c]Culture medium (Murashige); 150 g sucrose l^{-1}; but minus hormones.

of great importance and requires more attention, and not only the phytohormones. This role of sucrose is supported, in principle, also by studies in cultures with isolated wheat ears where the number of fertile spikelets could be significantly increased by elevated concentrations of sucrose in the medium supplied (Lee et al., 1989; Krumm, 1991).

In most plant species several sinks compete for photosynthates. An example of competition between reproductive sinks has been given in Fig. 5.20. In addition, and as a rule, in the reproductive stage, the fruits, seeds and storage organs are the dominant sinks. Under conditions of sink limitation, competition is reflected by a decrease in the growth rate of the other sinks.

Table 5.22

Effect of Fruit Load on Dry Matter Production and Distribution and on Water Consumption of *Citrus madurensis* Lour[a]

	Number of fruits per plant		
	0	50	100
Dry weight (g per plant)			
Fruits	0	134	175
Vegetative shoots and flowers	457	305	118
Roots	68	49	17
Total dry wt	525	488	310
Water transpired			
l per plant	91	90	59
l kg^{-1} leaf dry wt	370	520	1030

[a]Based on Lenz and Döring (1975).

In perennials with indeterminate vegetative growth during the reproductive phase, sink competition by the developing fruits can be quite dramatic, as in the case of citrus trees (Table 5.22). With increasing fruit load the growth of vegetative shoots and roots is greatly reduced. Simultaneously the total dry weight per plant decreases, indicating that more is affected by the fruit load than the distribution of photosynthates. The effect of increasing fruit load on water consumption per plant is much lower than that on dry weight. The transpiration rate per unit leaf dry weight increases by nearly a factor of 3. The close positive correlations between the number of fruits per plant and the transpiration rate per unit leaf area is a general phenomenon (Lenz, 1970). Plants with a heavy fruit load are therfore more sensitive to inadequate supplies of water and mineral nutrients (Lenz, 1970), because their shoots place a higher demand on their small root system than do the shoots of plants without fruits or with only a small number of fruits.

In nodulated legumes characterized by N_2 fixation, the root nodules represent an additional sink for carbohydrates supplied from the leaves. As shown in Table 5.23 the removal of source leaves leads to a decrease in both nodule growth and N_2 fixation, whereas the removal of flowers and pods (competing sinks) results in an increase in both

Table 5.23

Effect of Defoliation or Removal of Flowers and Developing Pods on
Nodule Weight and Nitrogen Content of Soybean Plants[a]

Treatment	Dry weigt of root nodules (mg per plant)	Nitrogen (mg per plant)
Control	298	475
Defoliation	176	266
Removal of flowers and pods	430	548

[a]Data are for plants harvested after 60 days of growth.
Based on Bethlenfalvay *et al.* (1978).

nodule weight and N_2 fixation to values that are higher than those of untreated control
plants.

Sink competition in legumes between fruits (seeds) and root nodules for carbohydrates from the source leaves often leads to a sharp decline in N_2 fixation rates at the
onset of a rapid fruit growth. During this period, however, the demand of the fruits for
organic nitrogen is also very high. An insufficient supply of fixed N_2 from the root
nodules is then compensated for by enhanced mobilization of nitrogen from the leaves,
an effect which can accelerate senescence and cause 'self-destruction' of the system. An
example of this has been given in Fig. 3.14 for beans grown under field conditions, and
another is shown in Fig. 5.22 in more detail in a model experiment with soybean
explants. During pod-filling, supply of water only resulted in a rapid senescence and
abscission of the source leaves. Supply of mineral nutrients delayed senescence to the
level of the intact control plants. Supply of CYT only, was more effective in delaying
source leaf senescence, even compared with intact plants, and with the combined
supply of CYT and mineral nutrients source leaf senescence could be totally prevented
during the experimental time. This experiment demonstrates again that for a long leaf
area duration, maintenance of root growth and activity during the reproductive stage is
also of crucial importance, and this holds true not only for nodulated legumes.

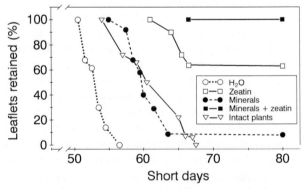

Fig. 5.22   Abscission of leaf blades in pod-bearing soybean explants and intact plants. Treatment solutions were supplied through the stem base of explants. (Mauk *et al.*, 1990.)

5.7.2   Concluding Remarks

In this section sink and source limitations in relation to yield have been discussed on a single-plant basis. Under field conditions, however, yield is determined by dry or fresh weight per unit surface area (e.g., hectares), that is, the sum of the yield of the individual plants. The number of plants and the number of reproductive or storage organs per unit surface (the plant density) are therefore important yield components. As plant density increases, however, low light conditions and competition for light and, thus, source limitation become more important. Under nearly all circumstances under which crop species are grown therefore canopy photosynthesis would benefit from an increase in maximal net photosynthesis at low, but not at high, light intensities (Austin, 1989). At least for maize, yield increase during the last three decades has been achieved mainly by selection of genotypes with higher leaf longevity (i.e., leaf area duration), and a large leaf area index in combination with more efficient light utilization by a steeper leaf angle (Dwyer *et al.*, 1991).

6

Mineral Nutrition and Yield Response

6.1 General

Various factors are required for plant growth such as light, CO_2, water and mineral nutrients. Increasing the supply of any of these factors from the deficiency range increases the growth rate and yield, although the response diminishes as the supply of the growth factor is increased. This relationship was formulated mathematically for mineral nutrients by Mitscherlich as a *law of diminishing yield increment* (Mitscherlich, 1954; Boguslawski, 1958). According to this formulation, the yield response curves for a particular mineral nutrient are asymptotic; when the supply of one mineral nutrient (or growth factor) is increased, other mineral nutrients (or growth factors) or the genetic potential of crop plants become limiting factors. Typical yield response curves for mineral nutrients are shown in Fig. 6.1. The slopes of the three curves differ. Micronutrients have the steepest and nitrogen the flattest slope, if the nutrient supply is expressed in the same mass units. The slopes reflect the different demands of plants for particular mineral nutrients.

It is now established that some of the assumptions made by Mitscherlich were incorrect. The slope of the yield response curve for a particular mineral nutrient cannot be described by a constant factor, nor is the curve asymptotic. Also when there is an abundant supply of nutrients, a point of inversion is obtained, as shown for micro-nutrients in Fig. 6.1. This inversion point also exists for other mineral nutrients such as nitrogen (e.g., in the case of grain yield depression by lodging in cereals) and is caused

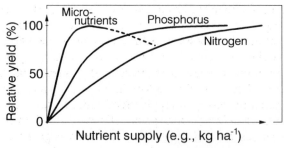

Fig. 6.1 Yield–response curves for nitrogen, phosphorus, and micronutrients.

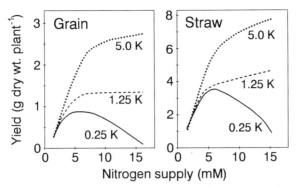

Fig. 6.2 Effect of increasing nitrogen supply at three potassium levels (mM) on grain and straw yield of barley grown in water culture. (Reproduced from MacLeod, 1969, by permission of the American Society of Agronomy.)

by a number of factors such as the toxicity of a nutrient per se or the induced deficiency of another nutrient. The effects of high nitrogen supply on the phytohormone level and thus on development processes can also be the cause of yield depression. Furthermore, distinct deviations from typical yield response curves (Fig. 6.1) can be obtained when mineral nutrients such as copper are supplied in very low quantities to a severely deficient copper fixing soil. In this case seed set is either prevented or severely inhibited (Section 6.3).

An example of the effect of interaction between mineral nutrients on yield is given in Fig. 6.2. At the lowest potassium level, the response to increasing nitrogen supply is small and at high nitrogen supply yield depression is severe. Under field conditions, however, yield depressions caused by excessive nutrient supply are usually less severe.

Yield response curves differ between grain and straw, particularly at higher potassium levels (Fig. 6.2). In contrast to the straw yield, the grain yield levels off when the nitrogen supply is high, reflecting sink limitation (e.g., small grain number per ear), sink competition (e.g., enhanced formation of tillers), or source limitation (e.g., mutual shading of leaves).

Yield response curves are strongly modulated by interactions between mineral nutrients and other growth factors. Under field conditions the interactions between water availability and nitrogen supply are of particular importance. In maize, for example, with increasing nitrogen supply and different soil moisture levels, the grain yield response curves obtained (Shimshi, 1969) are similar to those shown for different potassium levels (Fig. 6.2). The depressions in yield that accompany a large supply of nitrogen in combination with low soil moisture levels are presumably caused by several factors such as (a) delay in stomatal response to water deficiency (Chapter 5), (b) the higher water consumption of vegetative biomass and the correspondingly higher risk of drought stress in critical periods of grain formation, and (c) increase in shoot–root dry weight ratio with increasing nitrogen supply (Section 8.2.5), an effect which seems to be more prominent in C_3 than in C_4 plant species (Hocking and Meyer, 1991).

Yield response curves can differ not only between vegetative and reproductive organs (Fig. 6.2) but also between the yield components of harvested products. In most

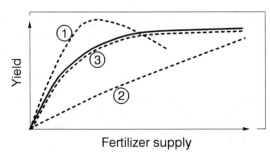

Fig. 6.3 Schematic representation of yield response curves of harvested products. Key: ———, quantitative yield (e.g., dry matter per hectare); – – –, qualitative yield (e.g., content of sugar, protein, and mineral elements; for examples (1) to (3) see following text).

crops, both quantity (e.g., dry matter yield in tons per hectare) and quality (e.g., content of sugars or protein) are important yield components. As shown schematically in Fig. 6.3 maximum quality can be obtained either before [curve(1)] or after [curve(2)] the maximum dry matter yield has been reached, or both yield components can have a synchronous pattern [curve(3)]. Examples of the behaviour described by curve (1) are nitrate accumulation in spinach and sucrose accumulation in sugar beet with increasing levels of nitrogen fertilizer. Examples of curve (2) are the change in protein content of cereals, grains of forage plants with increasing supply of nitrogen fertilizer, or the change in content of certain mineral elements with increasing mineral nutrient supply, for example, magnesium and sodium in forage plants, or zinc and iron in cereal grains for human consumption. Examples of curve (3) are common under conditions where with increase in mineral nutrient supply the number of either reproductive sinks (e.g. grains) or vegetative storage sinks (e.g. tubers) is increased.

6.2 Leaf Area Index and Net Photosynthesis

Positive yield response curves are the result of different individual processes, such as an increase in leaf area and net photosynthesis per unit leaf area (i.e., effects at the source) or an increase in fruit and seed number (i.e., effects at the sink). In this section the main emphasis is put on processes that primarily affect the source site, although feedback regulation from the sink sites are often involved, and may even dominate the source processes.

Generally, the density of a crop population is expressed in terms of the *leaf area index* (LAI), which is defined as the leaf area of plants per unit area of soil. For example, an LAI of 5 means that there is a 5 m^2 leaf area of plants growing on a soil area of 1 m^2. As a rule the crop yield increases until an optimal value in the range of 3–6 is reached, the exact value depending on plant species, light intensity, leaf shape, leaf angle and other factors. At a high LAI, mutual shading usually becomes the main limiting factor. When the water supply is limited, however, drought stress and corresponding negative effects, particularly at the sink sites (Section 6.3), can decrease the optimal LAI to values far below those resulting from mutual shading.

Table 6.1
Inhibition of Leaf Growth by Nitrogen Deficiency in Different Plant Species[a]

Plant species	Average growth inhibition (%)	
	Day	Night
Cereals (wheat, barley, maize, sorghum)	16	18
Dicotyledons (sunflower, cotton, soybean, radish)	53	8

[a]Based on Radin (1983).

When the nutrient supply is suboptimal, the leaf growth rate, and thus the LAI, can be limited by low rates of net photosynthesis or insufficient cell expansion or both these factors. This is particularly evident with suboptimal supply of nitrogen and phosphorus. In plants suffering from nitrogen deficiency, elongation rates of leaves may decline before there is any reduction in net photosynthesis (Chapin *et al.*, 1988), this decline being the result of decrease in both number and duration of extension of epidermal cells (MacAdam *et al.*, 1989). Besides hormonal effects (Section 5.6) a decrease in root hydraulic conductivity might be involved, leading to a decrease in water availability in expanding leaf blades (Radin and Boyer, 1982). The effect of nitrogen deficiency on leaf expansion differs between plant species (Table 6.1). In cereals (monocotyledons) cell expansion is inhibited to the same extent during the day and night. In dicotyledons, however, the inhibition is much more severe during the daytime. This difference in response is related to morphological differences among species and corresponding differences in competition for the water available for transpiration and for cell expansion. In dicotyledons, cell expansion occurs in leaf blades which are exposed to the atmosphere and therefore experience a high rate of transpiration during the daytime. In cereals, however, cell expansion occurs at the base of the leaf blade. This zone is protected from the atmosphere by the sheath of the preceding leaf, so that little transpiration occurs from this zone of elongation. In contrast to leaf expansion, net photosynthesis per unit leaf area is depressed to a similar extent in both groups of plants by nitrogen deficiency.

Similar results to those shown in Table 6.1 for the nitrogen effect in dicotyledons have been obtained for phosphorus in cotton plants (Radin and Eidenbock, 1984). It was found that phosphorus deficiency severely inhibited leaf growth rate only during the daytime, and had very little effect at night. These day/night differences were primarily a response to limited water availability for cell expansion during the daytime, caused by low hydraulic conductance of the root system as a result of phosphorus deficiency. The small size and often dark-green color of the leaf blades in phosphorus deficient plants are the result of impaired cell expansion and a correspondingly larger number of cells per unit surface area (Hecht-Buchholz, 1967). In addition, particularly in dicots the number of leaves might be reduced through a lower number of nodes branching (Lynch *et al.*, 1991). However, in many instances, a transient deficiency of phosphorus or nitrogen during the early growth of cereals or maize might reduce final yield not as a consequence of smaller leaf area but as the result of a lower number of spikelets per ear or grains per cob (Römer and Schilling, 1986; Barry and Miller, 1989).

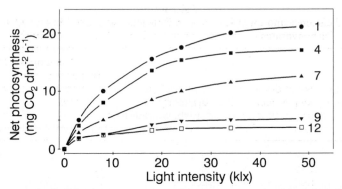

Fig. 6.4 Photosynthetic light response curves of a mature sugar beet leaf 1, 4, 7, 9 and 12 days after the plant was transferred to nitrogen-free nutrient solution. (Reproduced from Nevins and Loomis, 1970, by permission of the Crop Science Society of America.)

Mineral nutrient deficiency can also delay plant development. For example, in barley the number of days to reach the booting stage is about twice as much for manganese-deficient as for manganese-sufficient plants (Longnecker *et al.*, 1991b).

Mineral nutrition can also influence net photosynthesis in various ways (Natr, 1975; Barker, 1979). The direct involvement of some mineral nutrients in the electron transport chain in the thylakoid membranes, in detoxification of oxygen free radicals and in photophosphorylation has been summarized in Fig. 5.1 and discussed in Sections 5.2.1 and 5.2.2. Mineral nutrients are also required for chloroplast formation, either for synthesis of proteins, thylakoid membranes or chloroplast pigments. In green leaf cells, for example, up to 75% of the total organic nitrogen is located in the chloroplasts, mainly as enzyme protein. A deficiency of mineral nutrients that are directly involved in synthesis of protein or chloroplast pigments or electron transfer therefore results in the formation of chloroplasts with lower photosynthetic efficiency (Spencer and Poss-ingham, 1960), and also in a change in the fine structure of chloroplasts in a more or less specific manner (Hecht-Buchholz, 1972). In leaves of spinach about 24% of the total nitrogen is allocated to thylakoid membranes; nitrogen nutrition therefore also affects the amount of thylakoids per unit leaf area (Terashima and Evans, 1988). A mineral nutrient deficiency can also depress net photosynthesis by influencing the CO_2 fixation reaction and entry of CO_2 through the stomata. Finally, starch synthesis in the chloroplasts and transport of sugars across the chloroplast envelope into the cytoplasm are directly controlled by the concentration of inorganic phosphate (Heldt *et al.*, 1977). These functions of mineral nutrients in photosynthesis are discussed in more detail in Chapters 8 and 9.

In the range between suboptimal and optimal nutrient supply, close positive correlations are often observed between mineral nutrient content of leaves and the rate of net photosynthesis (Nátr, 1975). In principle, these correlations can also be indirectly demonstrated in mature leaf blades when the root supply of mineral nutrients such as nitrogen is withheld (Fig. 6.4).

As nitrogen becomes increasingly deficient, the light response curve of net photosyn-thesis in the mature leaf declines to low levels (Fig. 6.4), and an increasing proportion of

Table 6.2
Effect of Zinc Deficiency and Light Intensity on Shoot Growth and Content of Chlorophyll and Carbohydrates in Primary Leaves of Bean (*Phaseolus vulgaris*)[a]

Light intensity ($\mu E\ m^{-2}\ s^{-1}$)	Shoot dry wt. (g per plant)		Chlorophyll (mg g^{-1} dry wt.)		Carbohydrates (mg glucose equiv. g^{-1} dry wt)			
					Sucrose		Total[b]	
	+Zn	−Zn	+Zn	−Zn	+Zn	−Zn	+Zn	−Zn
80	1.24	1.13	19.2	17.3	10	11	40	42
230	2.38	1.13	16.6	7.8	11	54	42	124
490	3.80	1.16	11.2	4.5	17	82	77	138

[a]Based on Marschner and Cakmak (1989).
[b]Sucrose, reducing sugars, starch.

the absorbed light energy which is not used in photochemical reactions is dissipated as heat (Demming and Winter, 1988). Nevertheless, the photosynthetic efficiency per unit chlorophyll can even increase under nitrogen deficiency (Khamis *et al.*, 1990b), presumably reflecting a response to decrease in source–sink ratio. In contrast, in manganese-deficient leaves the photosynthetic efficiency per unit chlorophyll is drastically decreased and can be restored within two days after foliar application of manganese, indicating a direct effect on photosystem II (Fig. 5.1) rather than an indirect effect via source–sink relationships (Kriedemann *et al.*, 1985).

Similar changes in the light response curves shown for nitrogen deficiency (Fig. 6.4) are also found under phosphorus deficiency (Lauer *et al.*, 1989a), and deficiency of a range of other mineral nutrients. In many instances in the deficient plants, however, despite poor utilization of higher light intensities, carbohydrates accumulate in leaves (Grabau *et al.* 1986a; Rao *et al.*, 1990) and also in roots (Khamis *et al.*, 1990a) of phosphorus-deficient plants. Thus, low photosynthetic efficiency of source leaves might often be the result of feedback regulation induced by a lower demand for photosynthates at the sink sites. An example for this is shown for zinc-deficiency in Table 6.2.

With increasing light intensity plant dry weight increases in the zinc-sufficient plants but not in the zinc-deficient plants (Table 6.2). Although chlorophyll content drastically declines with increasing light intensities, particularly in the zinc deficient plants the carbohydrate content increases steeply, indicating that the lack of growth response to increasing light intensities reflects a sink and not source limitation.

Accumulation of photosynthates under high light intensity in source leaves of deficient plants not only decreases utilization of light energy but also poses a stress. This high light stress is indicated, for example, by an increase in the antioxidative defense mechanisms in the deficient leaves (Cakmak and Marschner, 1992; Fig. 5.2), photooxidation of chloroplast pigments (Table 6.2) and enhanced leaf senescence. These side effects of mineral nutrient deficiency decrease not only current photosynthesis and LAI but also *leaf area duration* (LAD), that is the length of time in which the source leaves supply photosynthates to sink sites, an aspect that is discussed in Section 6.4.

6.3   Mineral Nutrient Supply, Sink Formation, and Sink Activity

6.3.1   General

In crop species where fruits, seeds, and tubers represent the yield, the effects of mineral nutrient supply on yield response curves are often a reflection of sink limitations, imposed by either a deficiency or an excessive supply of mineral nutrients during certain critical periods of plant development, including flower induction, pollination, and tuber initiation. These effects can be either direct (as in the case of nutrient deficiency) or indirect (e.g., effects on the levels of photosynthates or phytohormones).

6.3.2   Flower Initiation

In apple trees, flower formation is affected to a much greater extent by the time or form of nitrogen application or both these factors than by the level of nitrogen supply. Compared with a continuous nitrate supply a short-term supply of ammonium to the roots was found to more than double both the percentage of buds developing inflorescences and the arginine content in the stem (Table 6.3). Arginine is a precursor of polyamines which also accumulate particularly in leaves of plants supplied with high levels of ammonium (Gerendás and Sattelmacher, 1990).

The causal involvement of polyamines in the ammonium-induced enhanced development of inflorescences in apple trees is indicated by the similar effects obtained by infiltrating polyamines into the petioles (Table 6.3). These results on the flower-inducing effects of ammonium supply confirm earlier results of Grasmanis and Edwards (1974). Since the apple trees in this study were amply supplied with nitrogen throughout the growing season, it is unlikely that these effects on flower initiation (i.e. on developmental processes) are related to a general nutritional role of nitrogen. It appears more probable that some nitrogen compounds such as polyamines may function as second messenger in flower initiation (Section 5.6.3).

Table 6.3
Effect of Ammonium-N or Polyamines on Flower Initiation and Arginine Content in Apple Trees[a]

Treatment	Percentage flowering	Stem arginine content ($\mu g\ g^{-1}$ dry wt)
Control, nitrate continuously	15	1120
NH_4-N for 24 h[b]	37	2570
NH_4-N for 1 week	40	2330
Putrescine[c]	51	—
Spermine[c]	47	—
NH_4-N for 24h[b]	50	—

[a]Based on Rohozinski *et al.* (1986).
[b]8 mM NH_4^+ in the nutrient solution.
[c]8 mM petiole infiltration.

Table 6.4
Shoot Growth, Flower Induction and CYT in Xylem Exudate of Apple Root
Stock M7 as Affected by the Form of Nitrogen Supply[a]

Form of N-supply	Shoot length (cm)	No. lateral shoots (spurs)	Flowering buds (% of emerged)	CYT (nmol $(100 \text{ g})^{-1}$ shoot fresh wt)
NO_3-N	326	6.4	7.4	0.002
NH_4NO_3	268	6.0	8.2	0.373
NH_4-N	209	8.9	20.7	0.830

[a]Based on Gao *et al.* (1992).

Most probably, changes in the phytohormone level in general and of CYT in particular are involved in this enhancing effect of ammonium supply on flowering (Buban *et al.*, 1978). In apple root stocks, ammonium supply as compared with nitrate, not only increased flower bud formation but also CYT concentration in the xylem exudate and the number of flower-bearing lateral branches (spurs), whereas the total shoot length was depressed (Table 6.4). Promotion of flower morphogenesis by CYT is well documented for various other plant species (Bruinsma, 1977; Herzog, 1981).

Flower formation in apple trees (Bould and Parfitt, 1973), tomato (Menary and Van Staden, 1976), and wheat (Rahman and Wilson, 1977) is also positively correlated with phosphorus supply. The positive correlations between the number of flowers and CYT level in tomato (Menary and Van Staden, 1976), on the one hand, and between the phosphorus supply and the CYT level on the other (Horgan and Wareing, 1980), provide additional evidence that CYT also contributes to the enhancing effect of phosphorus on flower formation. Basically similar conclusions were drawn from the effects of potassium on flower formation in *Solanum sisymbrifolium* (Wakhloo, 1975a,b). Low potassium levels in the leaves were correlated with a high proportion of sterile female flowers. This sterility did not occur in plants of either high or low potassium status when the plants had been sprayed with CYT.

These various results strongly confirm the supposition that the effects of mineral nutrient supply on flower formation are brought about by changes in phytohormone level. This is also true for the beneficial effects of nitrogen fertilizer application before anthesis in increasing grain number per ear in wheat (Herzog, 1981) or seed number per plant in sunflower (Steer *et al.*, 1984). However, seed number per plant can also be increased by high concentrations of sucrose prior to flower initiation (Waters *et al.*, 1984), high light intensity (Stockman *et al.*, 1983) or stem injection of sucrose under drought stress conditions (Boyle *et al.*, 1991; Section 5.7). The possibility can not be ignored, therefore, that mineral nutritional status may also affect flower initiation and seed set by increasing the supply of photosynthates during critical periods of the reproductive phase.

6.3.3 Pollination and Seed Development

The number of seeds or fruits or both per plant can also be directly affected by mineral nutrient supply. This is clearly the case with various micronutrients. In cereals in

Table 6.5
Effect of Increasing Copper Supply in Wheat (cv. Chatilerma)
Grown on Copper-Deficient Soil[a]

	Copper supply (mg per pot)			
	0	0.1	0.4	2.0
Number of tillers	22	15	13	10
Straw yield (g)	7.7	9.0	10.3	10.9
Grain yield (g)	0.0	0.5	3.5	11.8

[a]Total number of plants per pot: 4. Based on Nambiar (1976c).

particular, copper deficiency affects the reproductive phase (Table 6.5). The critical period in copper-deficient plants is the early booting stage at the onset of pollen formation (microsporogenesis). When the copper deficiency is severe, no grains are produced even though the straw yield is quite high owing to enhanced tiller formation (loss of apical dominance of the main stem). As the copper supply is increased, grain yield rises sharply, whereas the straw yield is only slightly enhanced. These results provide an informative example of both sink limitation on yield and deviation from the typical response curve (Fig. 6.1) between grain yield and mineral nutrient supply. Under field conditions, in copper deficient soils, especially when the top soil is dry, copper application to the soil is often much less effective than foliar application at the early vegetative growth stage in increasing the grain yield (Section 4.3).

The primary causes of failure in grain set in copper-deficient plants are inhibition of anther formation, the production of a much smaller number of pollen grains per anther, and particularly the nonviability of the pollen (Graham, 1975), in part because of lack of supply of carbohydrates to the developing pollen grains (Jewell et al., 1988).

In principle, similar results as those found for copper deficiency (Table 6.5) are obtained with zinc or manganese deficiency. In maize, zinc deficiency prior to micro-sporogenesis (~35 days after germination) did not significantly affect vegetative growth and ovule fertility but decreased pollen viability and cob dry weight by about 75% (Sharma et al., 1990). Also under manganese deficiency vegetative growth of maize is much less depressed than grain yield (Table 6.6). In the deficient plants anther

Table 6.6
Effect of Manganese Deficiency on Growth, Fertilization and Grain Yield in Maize[a]

Mn supply (μg l^{-1})	Shoot dry wt. (g per plant)	Grain wt (g per plant)	Single grain wt (mg)	Pollen (no. per anther)	Pollen germination (%)
550	82.5	69.3	302	2770	85.6
5.5	57.8	11.8	358	1060	9.4

[a]Sharma et al. (1991). Reprinted by permission of Kluwer Academic Publishers.

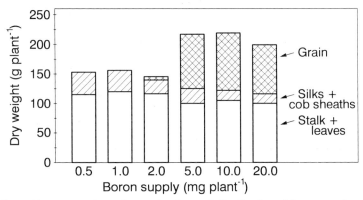

Fig. 6.5   Effect of boron supply on the production and distribution of dry matter in maize plants. (Based on Vaughan, 1977.)

development is delayed and fewer and smaller pollen grains are produced with very low germination rates. In contrast, ovule fertility was not significantly affected by manganese deficiency (Sharma *et al.*, 1991), a result, which is in agreement with the effect of copper deficiency in wheat (Graham, 1975).

The production and viability of pollen are also affected by molybdenum. In maize, a decrease in the molybdenum content of pollen was correlated with a decrease in the number of pollen grains per anther as well as a decrease in the size and viability of the pollen grains (Section 9.6). As yet no information is available as to the extent to which molybdenum deficiency also depresses fertilization and grain set. However, it is well documented that preharvest sprouting in maize (Section 9.6) and wheat (Cairns and Kritzinger, 1992), causing severe yield losses in certain areas, is very high in seeds with low molybdenum content and can be effectively decreased by molybdenum supply to the soil or as a foliar spray.

Boron is another mineral nutrient that affects fertilization. Boron is essential for pollen tube growth (Section 9.7), a role reflected under conditions of boron deficiency by a decrease in the number of grains per head in rice (Garg *et al.*, 1979) or even a total lack of fertilization in barley and rice (Ambak and Tadano, 1991). The failure of seed formation in maize suffering from boron deficiency is caused by the nonreceptiveness of the silks to the pollen (Vaughan, 1977). As the level of boron nutrition increases, vegetative growth, including the structural growth of the silks, is either not affected or is even somewhat depressed (Fig. 6.5). In contrast, grain formation is absent in severely deficient plants but increases dramatically when the boron supply is adequate. Obviously there is a minimum boron requirement, which is in the range of 3 mg boron per maize plant for fertilization and grain set. Figure 6.5 provides another example of both a strict sink limitation induced by mineral nutrient deficiency and a yield response curve quite different from the typical curve.

Low boron supply not only inhibits flowering and seed development but may also produce seeds with low boron content in plants even without visual symptoms of boron

Table 6.7
Effect of Potassium Supply to Wheat on ABA Content and Weight of Grains[a]

	ABA content (ng per grain) days after anthesis				Days from anthesis to full ripening	Weight of a single grain (mg)
K Supply	28	35	38	44		
Low (deficient)	7.7	13.4	16.5	2.2	46	16.0
High	3.7	4.4	ND[b]	9.4	75	34.4

[a]Based on Haeder and Beringer (1981).
[b]ND, Not determined.

deficiency. These low boron seeds have a low germination rate and produce a high percentage of abnormal seedlings (Bell *et al.*, 1989).

In lowland rice, grain yield might be considerably decreased by spikelet sterility induced by low temperatures (below 20°C) during anthesis. This temperature sensitivity can be drastically decreased by high supply of potassium (Haque, 1988). Increasing potassium contents in the panicles from 0.61% to 2.36% in the dry matter decreased spikelet sterility after three days at 15°C from 75% to 11%. The reasons for this protecting effect of potassium are not known, high nitrogen contents in the low potassium plants are probably involved (Haque, 1988).

In certain plant species, such as soybean, drop of flowers and developing pods is a major yield-limiting factor. Nitrogen or phosphorus deficiency during the flowering period enhances flower and pod drop and depresses seed yield correspondingly (Streeter, 1978; Lauer and Blevins, 1989). Supplying ample amounts of nitrogen or phosphorus during this critical phase is therefore quite effective in reducing flower and pod drop and in increasing final seed yield in soybean (Brevedan *et al.*, 1978; Lauer and Blevins, 1989). Although the physiological reasons for both the flower and pod drop and the decline in drop produced by nitrogen application are not yet known in detail, it is certain that phytohormones, especially CYT and ABA, are involved. An ample nitrogen supply both increases CYT and decreases ABA (Section 5.6.4) and hence decreases flower and pod drop, as would be expected from the specific role of ABA in the formation of abscission layers. Accordingly, maize kernel abortion can be reduced by either foliar application of CYT or supplying the roots with ammonium-N (Smiciklas and Below, 1992), the latter treatment being particularly effective in increasing the CYT contents in the plants (Table 6.4).

Premature ripening of fruits and seeds imposed by water or nutrient deficiency is another yield-limiting factor. In this case it is not the number of grains but the weight (size) of a single grain or fruit that is low. There is substantial evidence that elevated ABA levels are also involved in premature ripening. An example of this is shown in Table 6.7 for potassium-deficient wheat plants. In these plants, and particularly 4–6 weeks after anthesis, the levels of ABA in the grains are much higher than those in the grains of plants well supplied with potassium. Correspondingly, the grain-filling period in potassium-deficient plants is much shorter and the weight of a single grain at maturity is lower than that in control plants. As has been shown before (Section 5.6.5), high

ABA levels in grains coincide with a sharp decline in the sink activity of grains. It is quite likely that the elevated ABA levels in the flag leaves of potassium-deficient wheat plants (Haeder and Beringer, 1981), and a correspondingly higher ABA import to the developing grains, are responsible for the premature ripening and not the source limitation of a mineral nutrient *per se* (Section 6.4).

6.3.4 Tuberization and Tuber Growth Rate

In root and tuber crops such as sugar beet or potato, the induction and growth rate of the storage organ are strongly influenced by the environmental factors. In contrast to crop species in which seeds and fruits are the main storage sinks, root and tuber crops often exhibit a distinct sink competition between vegetative shoot growth and storage tissue growth for fairly long periods after the onset of storage growth. This competition is particularly evident in so-called indeterminate genotypes of crop species, for example, in potato (Kleinkopf *et al.*, 1981). In general, environmental factors (e.g., high nitrogen supply) with pronounced favorable effects on vegetative shoot growth delay the initiation of the storage process and decrease growth rate and photosynthate accumulation in storage organs – for example, of sugar beet (Forster, 1970) and potato (Ivins and Bremner, 1964; Gunasena and Harris, 1971).

A large and continuous supply of nitrogen to the roots of potatoes delays or even prevents tuberization (Krauss and Marschner, 1971). After tuberization the tuber growth rate is also drastically reduced by high nitrogen supply, whereas the growth rate of the vegetative shoot is enhanced. The effect of nitrogen supply on tuber growth rate is illustrated in Table 6.8. Resumption of the tuber growth rate to normal levels after interruption of the nitrogen supply indicates that sink competition between the vegetative shoot and tubers can readily be manipulated by means of the nitrogen supply.

In potato, cessation of tuber growth caused by a sudden increase in nitrogen supply to the roots induces 'regrowth' of the tubers, that is, the formation of stolons on the tuber apex (Krauss and Marschner, 1976, 1982). Interruption and resupply of nitrogen, therefore, can result in the production of chain-like tubers or so-called secondary growth (Fig. 6.6). After a temporary cessation of growth, resumption of the normal

Table 6.8

Growth Rate of Potato Tubers in Relation to Nitrate Supply to the Roots of Potato Plants[a]

Nitrate concentration (mM)	Nitrate uptake (meq per day per plant)	Tuber growth rate (cm³ per day per plant)
1.5	1.18	3.24
3.5	2.10	4.06
7.0	6.04	0.44
Nitrogen supply withheld for 6 days	—	3.89

[a]From Krauss and Marschner (1971).

Fig. 6.6 Secondary growth and malformation of potato tubers induced by alternating high and low nitrogen supply to the roots. (Krauss, 1980.)

growth rate is usually restricted to a certain area of the tubers (meristems or 'eyes'), leading to typical malformations and knobby tubers, which are often observed under field conditions after transient drought periods. Similar effects on cessation of tuber growth and 'regrowth' can be achieved by exposing growing tubers to high temperatures, which instantly inhibit starch synthesis and lead to the accumulation of sugars in the tubers (Krauss and Marschner, 1984; Van den Berg *et al.*, 1991), followed by a decrease in ABA levels in the tubers and 'regrowth'.

The effects of nitrogen supply on tuber growth rate and 'regrowth' are brought about by nitrogen-induced changes in the phytohormone balance both in the vegetative shoots and in the tubers. As already shown (Section 5.6.4), an interruption of the nitrogen supply results in a decrease both in CYT export from roots to shoots and in the sink strength and growth rate of the vegetative shoot. A corresponding increase in the ABA/GA ratio of the shoots seems to trigger tuberization. In agreement with this, tuberization can also be induced by the application of either ABA or the GA antagonist CCC (Krauss and Marschner, 1976) or by the removal of the shoot apices, the main sites of GA synthesis (Hammes and Beyers, 1973). On the other hand, after cessation of growth the regrowth of tubers induced by a sudden increase in the nitrogen supply is correlated with a decrease in the ABA/GA ratio not only in the vegetative shoots but also in the tubers, where the GA level increases by a factor of 2 but the ABA level drops to less than 5% of that in normal growing tubers (Krauss, 1978b).

6.4 Mineral Nutrition and the Sink–Source Relationships

In root and tuber crops, unlike grain crops, the sink–source relationship is quite labile even after the onset of the storage process. This has to be considered, for example, in the application of nitrogen fertilizer to potato. On the one hand, a high nitrogen supply is important for rapid leaf expansion and for obtaining an LAI between 4 and 6, a value considered necessary for high tuber yields (Kleinkopf *et al.*, 1981; Dwelle *et al.*, 1981). On the other hand, a high supply of nitrogen delays either tuberization or the onset of the linear phase of tuber growth. The principles of these interactions are demonstrated in Fig. 6.7. The advantage of earlier tuberization obtained by supplying a low level of nitrogen is offset by a low LAI and earlier leaf senescence, that is a short LAD and a

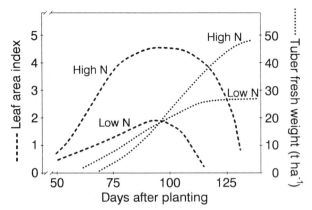

Fig. 6.7 Time course of leaf area index and fresh weight of potato tubers at two levels of nitrogen supply. (Based on Ivins and Bremner, 1964, and Kleinkopf *et al.*, 1981.)

correspondingly lower tuber yield. When the nitrogen supply is high, both LAI and LAD, and thus final tuber yield, are much higher. However, higher tuber yield induced by a large nitrogen supply can be realized only when the vegetation period is sufficiently long, that is, in the absence of early frost (Clutterbuck and Simpson, 1978) or in the absence of severe drought stress.

The early decline in LAI when the supply of nitrogen is low (Fig. 6.7) indicates that the final tuber yield is limited by the source. The question arises as to the reasons for this source limitation. In potato plants at maturity, between 60% and 80% of the total nitrogen is located in the tubers (Kleinkopf *et al.*, 1981). Therefore, when the nitrogen supply is low, exhaustion of nitrogen in the source leaves presumably plays a key role in leaf senescence and in the termination of tuber growth.

However, these simple relationships between nitrogen supply, LAI, LAD and tuber yield (Fig. 6.7) are not only modified by the length of the growing period but also the mineralization rate of soil nitrogen and temperature during tuber growth. At high nitrogen supply and high LAI, mutual shading of the basal leaves may not only drastically decrease their net photosynthesis but also the LAD by rapid leaf senescence (Firman and Allen, 1988), a process which is further enhanced at high ambient temperatures (Manrique and Bartholomew, 1991). Thus, a lower, but more continuous supply of nitrogen which allows an earlier tuberization and continuous root growth and CYT production, and which is more effective on LAD than on LAI, might often lead to higher tuber yields than a rapid establishment of a high LAI by high nitrogen supply during early growth. Interestingly, an increase in LAD is one of the major character-istics of the new, high-yielding varieties in maize and wheat developed during the last 30 years (Austin, 1989; Tollenaar, 1991).

Competition for nitrogen rather than for carbohydrates supplied from the source leaves can also be the main limiting factor for seed yield in mustard and rape plants (Trobisch and Schilling, 1969; Schilling and Trobisch, 1970). In mustard plants the developing seeds and leaves compete for nitrogen, and seed set, seed growth, and final seed yield are determined primarily by the size of the nitrogen pool in the vegetative parts. In crucifers, the flower differentiation at the auxiliary stems occurs after the onset

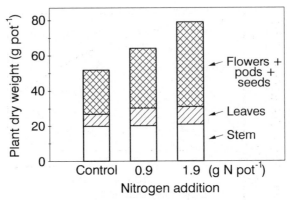

Fig. 6.8 Effect of addition of 0.9 and 1.9 g nitrogen at the onset of flowering on total dry weight and dry weight distribution in shoots of white mustard plants. (Based on Trobisch and Schilling, 1970.)

of flowering of the main stem and is strongly dependent on the availability of nitrogen during this period. Additional application of nitrogen at the onset of flowering therefore leads to an increase in seed number and yield (Fig. 6.8).

The example with mustard plants in Fig. 6.8 demonstrates that source limitation can be imposed by nitrogen rather than carbohydrates. This aspect has also to be considered in source–sink manipulations. Removing source leaves from a plant is a common procedure for evaluating source limitation in photosynthesis (Section 5.7). Of course, when source leaves are removed, nitrogen and other mineral nutrients are also removed. Therefore, shading of source leaves may have a different effect on the reduction of seed yield from that of leaf removal. Shading the source leaves of mustard plants reduced the seed yield by only 20%, whereas removal of these same leaves reduced the seed yield by 50% (Trobisch and Schilling, 1969). In plant species like mustard and rape, mutual shading of source leaves is usually much less detrimental on yield than, for example, in cereals and tuber crops, as in the crucifers after the onset of flowering the stems and pods provide a high proportion of the photosynthates required in the developing seeds (Section 5.7.1).

In principle, each of the mineral nutrients can become the dominant factor inducing source limitation on final yield of seeds, fruits and tubers, provided that it can be readily retranslocated from the source (Section 3.5). Whether such a limitation exists depends on such factors as the availability of the given nutrient in the soil, its concentration and amount (source size) in the vegetative shoot, the specific demand of the sink for the nutrient, and the growth rate of the sink. For example, in fleshy fruits or tubers, the potassium content is very high (2–3% of the dry weight), and at maturity most of the potassium is located in the fruits or tubers. Potassium-induced source limitation is therefore more likely in this type of crop. An example of this has been given in Section 5.7 for tomato genotypes. In contrast, in mature cereal plants, as much as 80% of the total amount of nitrogen or phosphorus is located in the grains, compared with less than 20% of the total potassium (see also Section 3.5.4). Thus, in cereal plants where there is a suboptimal supply of the three mineral nutrients during the vegetative stage, source limitation during grain filling is most likely induced by nitrogen or phosphorus, but not

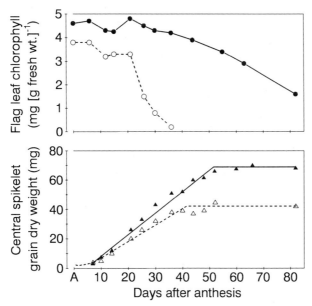

Fig. 6.9 Time course of chlorophyll content in flag leaf and dry weight accumulation in spikelet grains of phosphorus-sufficient (●, ▲) and phosphorus-deficient (○, △) wheat plants. (Modified from Batten and Wardlaw, 1987a.)

by potassium. The importance of export of mineral nutrients to developing seeds for senescence of source leaves has been discussed in Section 5.7. In soybean, for example, enhanced leaf senescence and also premature pod drop could be counteracted not only by an increase in phosphorus supply to the roots but also by stem infusion of phosphorus (Grabau *et al.*, 1986a), indicating that in this case phosphorus deficiency limited seed yield by decreasing LAD.

In moderately phosphorus-deficient cereal plants source limitation may also become the dominant factor for grain yield. In wheat these relationships are particularly prominent for the flag leaf blade which may deliver between 51% and 89% of the phosphorus found in the grains at harvest (Batten *et al.*, 1986). As shown in Fig. 6.9, in phosphorus-deficient plants the flag leaf senesced rapidly and its photosynthetic activity approached zero by the time the grains were only 60% of their final potential dry weight, leading to a reduction in final grain yield by 40% and in phosphorus content by 75%.

The results shown in Fig. 6.9 reflect source limitation of either phosphorus or photosynthates. Evidence against limitation by photosynthates had been presented by additional experiments where shading of ears increased the rate of net photosynthesis severalfold in the deficient flag leaf and also substantially delayed its senescence (Chapin and Wardlaw, 1988). In contrast, in the phosphorus-sufficient plants shading of the ear was without significant effects on flag leaf net photosynthesis. Thus, enhanced leaf senescence in the phosphorus-deficient plants is most likely caused by accumulation of photosynthates in the flag leaf and a subsequent photooxidation of the chloroplast pigments and destruction of membranes (Section 5.2.2) which, in turn, enhances remobilization and retranslocation of phosphorus from source to sink.

These examples illustrate the role of mineral nutrients as yield-limiting factors when fruits, seeds, or other organs are the dominant sink sites and mineral nutrient uptake by the roots is declining. Progress in selecting and breeding for genotypes with a high *harvest index* (ratio of economic yield to total dry matter) and short periods of fruit growth or ripening (e.g., the filling period in cereals) might be severely restricted, not because of the limited capacity of the source to supply carbohydrates, but rather because of the limited amount of mineral nutrients such as potassium, nitrogen, phosphorus, and magnesium that are available for retranslocation from source to sink.

7

Nitrogen Fixation

7.1 General

At present the terrestrial input of nitrogen from biological N_2 fixation is held to be in the range of $139–170 \times 10^6$ t N per year as compared with 65×10^6 t N per year provided by fertilizer nitrogen (Peoples and Craswell, 1992). The conversion of the inert N_2 molecule into combined nitrogen (NH_3; NO_3^-, etc.) which can be utilized as a mineral nutrient is brought about either by reduction to ammonia (NH_3) or oxidation to nitrate (NO_3^-). This conversion, also referred to as fixation, is highly energy consuming. In both industrial and biological conversion the reaction $N_2 \rightarrow 2\,NH_3$ dominates. In industrial fixation, N_2 is catalytically reduced to NH_3 by reaction with hydrogen (produced, e.g., from natural gas) in the Haber–Bosch process ($N_2 + 3\,H_2 \rightarrow 2\,NH_3$) under conditions of high temperature and pressure. The increase in both the costs of fossil energy and the worldwide demand for nitrogen fertilizer used for food production are major reasons for renewed interest in biological N_2 fixation as an alternative or at least a supplement to the use of chemical nitrogen fertilizer.

7.2 Biological Nitrogen-Fixing System

The capability of biological fixation of atmospheric nitrogen (N_2) is restricted to organisms with prokaryotic cell structure, namely, bacteria and blue–green algae (Cyanobacteria). According to present knowledge, some species in 11 of the 47 bacterial families and some species in the eight Cyanobacteria families are diazotrophs, i.e., capable of N_2 fixation (Werner, 1980). Agriculturally significant N_2 fixation is carried out by Eubacteria, many of which are heterotrophic, depending on supply of reduced carbon (e.g., *Azospirillum*). Others are autotrophic and able to reduce CO_2 (e.g., *Anabaena*). A number of species of the genera *Frankia* (Thallobacteria), *Nostoc* and *Anabaena* (Cyanobacteria) and the rhizobia (Protobacteria) are of particular importance because of their symbiotic capabilities. Based on their growth rates, two distinct groups of rhizobia exist, the fast-growing genus *Rhizobium* and the slow-growing genus *Bradyrhizobium*.

Three major strategies of N_2 fixation can be differentiated in terrestrial ecosystems,

System of N_2 fixation ($N_2 \longrightarrow NH_3$) and microorganisms involved	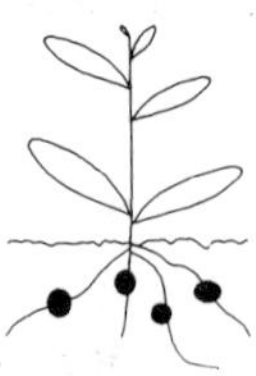 Symbiosis (e.g., *Rhizobium*, *Actinomycetes*)	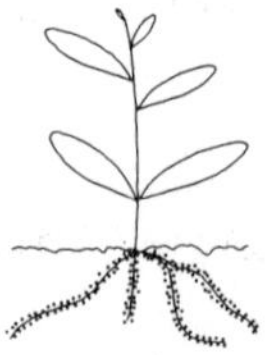 Associations (e.g., *Azospirillum*, *Azotobacter*)	 Free living (e.g., *Azotobacter*, *Klebsiella*, *Rhodospirillum*)	
Energy source (organic carbon)	Sucrose and its metabolites (from the host plant)	Root exudates from the host plant	Heterotroph: Plant residues	Autotroph: Photo-synthesis
Estimates of amounts fixed (kg N $ha^{-1}yr^{-1}$)	Legumes: 50–400 Nodulated non-legumes: 20–300	10–200	1–2	10–80

Fig. 7.1   Type, energy source, and fixation capabilities of biological N_2 fixation systems in soils. (Courtesy of K. Isermann; modified.)

symbiotic, associative, and free-living nitrogen fixing organisms, differing in both, energy source and fixation capability (Fig. 7.1). On average, symbiotic systems have the highest fixation capability since not only the energy in the form of carbohydrates is provided by the plant, but also other conditions (e.g., export of reduced N) are optimized for efficient N_2 fixation (Section 7.4). In this system the plants benefit directly since more than 90% of the fixed nitrogen is rapidly translocated from the bacteria to the plant. Nodulated legumes, such as alfalfa and soybean, in symbiosis with *Rhizobium* and *Bradyrhizobium* are among the most prominent N_2-fixing systems in agriculture. Leguminous trees such as *Leucaena leucocephala* are attracting increasing attention in agroforestry (Danso *et al.*, 1992). In forest and woodland ecosystems, however, most of the nitrogen input from biological N_2 fixation is provided by nodulated nonleguminous symbiotic systems between *Actinomycetes* (genus *Frankia*) and perennial species, for example of the genera *Alnus* and *Casuarina*.

   Some N_2-fixing systems with high host–bacteria specificity do not develop nodules. The habitat of these bacteria is the root surface and intercellular spaces of cortex cells. In these *rhizosphere associations* (Fig. 7.1) the host plant provides root exudates as energy source for N_2 fixation. However, the benefit to the plant of this type of N_2 fixation is mainly indirect, as approximatly 90% of the fixed nitrogen becomes only available to the plant after the death of the bacteria.

   The free-living N_2-fixing soil bacteria are mostly heterotrophic (e.g., *Azotobacter*) and thus often restricted in their N_2-fixation capacity by substrate limitation due to inadequate availability of organic residues (Fig. 7.1). As a result, their contribution to nitrogen fixation in terrestrial ecosystems is small, unless the bacteria are carbon autotrophic, as are cyanobacteria such as *Anabaena*.

   On average the relative contribution of symbiotic, associative and free-living N_2-fixing systems is in the order of 70% symbiotic and 30% nonsymbiotic (Peoples and

Craswell, 1992). The latter include N_2 fixation in the phyllosphere of forest trees (Section 7.6).

7.3 Biochemistry of Nitrogen Fixation

Since industrial N_2 fixation requires both high temperature and pressure, the question arises as to how this reaction can proceed in living cells growing at low temperatures and atmospheric pressure.

Biological reduction of N_2 to NH_3 is also a highly endergonic process with a minimum energy requirement of ca. 960 kJ mol^{-1} N fixed (Sprent and Raven, 1985). In all N_2-fixing microorganisms the principal steps of this reaction are the same (Fig. 7.2). The key enzyme complex, referred to as *nitrogenase*, is unique to N_2-fixing microorganisms and has been found, for example, in aerobic and anaerobic bacteria, cyanobacteria, and root nodules of legumes and nonlegumes. It consists of two oxygen-sensitive nonheme iron proteins. The smaller of the two proteins is referred to as the Fe-protein, has a molecular mass between 52 and 73 kDa and consists of two subunits and a single Fe_4S_4 unit (cluster). The larger MoFe-protein with a molecular mass of approximatly 240 kDa, consists of four subunits and contains 30 Fe and 2 Mo atoms (Thorneley, 1992). Besides this 'classic' form of nitrogenase there is in some diazotrophic bacteria a functional nitrogenase which contain vanadium (V) in place of molybdenum (Rosendahl *et al.*, 1991). Other forms of nitrogenase appear to contain neither of the two elements (Eady *et al.*, 1987).

For the nitrogenase reaction, energy in the form of ATP and reducing equivalents (electrons) are required (Fig. 7.2), supplied by respiration (ATP) and electron carriers, usually ferredoxin (see Chapter 5). Nitrogenase catalyzes the reduction of several substrates, including H^+, N_2, and C_2H_2. The principal reaction for dinitrogen reduction is as follows

$$N_2 + 8\,H^+ + 8\,e^- + 16\,Mg \cdot ATP \rightarrow 2\,NH_3 + H_2 + 16\,Mg \cdot ADP + 16\,P_i$$

The actual ATP requirement for N_2 fixation is considerably in excess of 16 mol per mol N_2 fixed, as a rule between 25 and 30 mol are required (Bergersen, 1991). According to the above reaction, evolution of H_2 is not a 'byproduct' of N_2 fixation but part of the chemistry of binding and reduction of N_2 as shown in more detail in Fig. 7.3. Protonation of Mo is the first step required for binding of N_2, during which H_2 is released. Gaseous H_2 is not only a competitive inhibitor of N_2 fixation but also

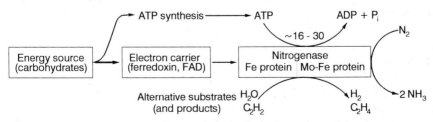

Fig. 7.2   Scheme illustrating the energy supply and principal reactions of the nitrogenase system. (Based on Evans and Barber, 1977.)

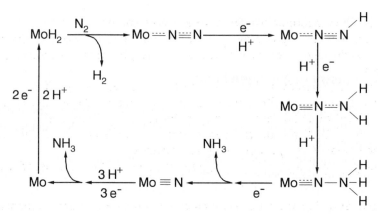

Fig. 7.3 Mechanism of protonation and dinitrogen reduction of molybdenum-nitrogenase. (Modified from Richards, 1991.)

represents a net loss of energy, unless the H_2 can be reprocessed by means of the *uptake hydrogenase (hup)*. This enzyme, which is of widespread occurrence in aerobic diazotrophic bacteria, contains FeS groups and requires nickel (Ni) for activity.

Even under optimal conditions for N_2 fixation, on average 25% of the total electron flux through nitrogenase is allocated to the formation of H_2 (Simpson, 1987). The uptake hydrogenase, splitting H_2 to $2\,H^+$ and $2\,e^-$, recycles some of the electrons for subsequent use in N_2 reduction. In legumes, selection of *Rhizobium* strains with high *hup* activity has therefore been considered as an important factor in increasing efficiency of N_2 fixation (Schubert *et al.*, 1978). The gain of energy by recycling H_2 is presumably small, protection of nitrogenase from both free oxygen (O_2) and inhibition by H_2 seeming to be the primary functions of the uptake hydrogenase (Bergersen, 1991).

The utilization of alternative substrates in the nitrogenase reaction (Fig. 7.2) offers an important tool for studies on nitrogenase activity and N_2 fixation. Most important in this context is acetylene (C_2H_2) which is reduced by nitrogenase to ethylene (C_2H_4) which can be readily measured in very low concentrations by gas chromatography. Supplying acetylene and then determining the rate of ethylene evolution (acetylene reduction assay, ARA) is a relatively simple method and has contributed very much in the past in research on biological nitrogen fixation (Bothe *et al.*, 1983). The relationship between acetylene reduction and N_2 fixation, however, is not constant and may differ considerably with plant age and particularly between legume species (Hansen *et al.*, 1987). Furthermore, in several symbioses nitrogenase activity is inhibited fairly rapidly by acetylene (Hunt and Layzell, 1993). Thus, as an indirect method the ARA is more suitable for obtaining relative data rather than for quantification of the N_2 fixation rate. For the latter purpose use of ^{15}N is much more appropriate, particularly if the problem of a suitable nonfixing 'reference plant' can be solved (Danso *et al.*, 1993).

Nitrogenase is extremely sensitive to oxygen (O_2). To protect nitrogenase from irreversible inactivation by O_2 in vivo, diazotrophic microorganisms have developed various strategies and mechanisms (Becana and Rodriguez-Barruaco, 1989). These include:

1. Living under anaerobic conditions (e.g., *Clostridium*)
2. Providing microaerophilic conditions at the enzyme site by consumption of most of the O_2 through excessive respiration, i.e., respiratory protection (e.g., *Azotobacter*)
3. Living in colonies covered by slime sheets, which restrict O_2 diffusion
4. Spatial separation of nitrogenase and sites of photosynthesis, i.e., O_2 evolution (e.g., N_2 fixation in heterocysts in cyanobacteria like *Anabaena*)
5. Controlling O_2 diffusion through mechanical barriers, and enzymatically by leghemoglobin (e.g., in root nodules of legumes)
6. Scavenging enzymes for toxic oxygen species and H_2O_2 (e.g., ascorbate peroxidase in root nodules of legumes).

The high energy demand (ATP), which can be provided in large amounts only by aerobic respiration, coupled with the necessity to protect the nitrogenase from O_2, requires a delicate system of oxygen pressure regulation at cellular level. This regulation is best realized in symbiotic systems and a major factor for their higher effectivity in N_2 fixation compared with other systems.

7.4   Symbiotic Systems

7.4.1   General

In principle two types of symbiotic systems exist in relation to the carbon supply required for N_2 fixation:

   I. Nodulated legumes and nodulated nonlegumes
  II. Symbioses with cyanobacteria (blue–green algae)

In system I the N_2-fixing microorganisms are either bacteria belonging to the genera *Rhizobium* and *Bradyrhizobium* (in legumes), or actinomycetes, of the genus *Frankia* (nonlegumes; actinorhizal symbiosis). In the tropics and subtropics about 200 plant species form actinorhizal symbioses (Peoples and Craswell, 1992). The *Casuarina* system in particular has recently attracted much attention (Torrey and Racette, 1989; Diem and Dommergues, 1990), with woody nodules that may reach a diameter of 10 cm and more. Casuarinas play an important role in agriculture associated with shifting cultivation and in aforestation of eroded land surfaces.

In system II the fixing bacteria rely mainly on their photosynthetic ability to meet the carbon (energy) requirement for N_2 fixation. Examples of system II are cyanobacteria of the genus *Nostoc* living symbiotically with fungi (e.g., lichens), free-floating freshwater ferns of the genus *Azolla* with the heterocyst-forming cyanobacteria of the genus *Anabaena*, or woody species of the genera *Cycas* and *Macrozamia* with cyanobacteria living in the coralloid roots (Pate *et al.*, 1988). In some instances the symbiotic interactions between cyanobacteria and the host invalidates a strict classification between systems I and II (see above), i.e. the microsymbiont relies on photosynthates provided by the host (Söderbäck and Bergman, 1992). For details on the various types of symbiotic interactions the reader is referred to Werner (1987).

In agricultural production systems the freshwater fern *Azolla* in symbiosis with *Anabaena azolla* has long been appreciated for its contribution to the nitrogen balance in paddy soils, which is held to be on average 50–80 kg fixed N per ha each year (Bothe *et al.*, 1983). From long-term field studies even higher values (79–103 kg N ha^{-1} yr^{-1}; App *et al.* 1984; or 40 kg N ha^{-1} within 40 days; Peoples and Craswell, 1992) have been calculated. In other agricultural production systems, however, the main suppliers of fixed N$_2$ are nodulated legumes. The following discussion of symbiotic systems therefore focuses on legumes where the knowledge of the functioning of the system is also most advanced.

7.4.2   Infection and Host Specifity

The initial basis for our understanding of the symbiotic relationship between legumes and *Rhizobium* species was provided by Hellriegel and Wilfarth in 1888, although the beneficial effect of legumes in crop rotation on the growth of the subsequent crops was appreciated and exploited even in ancient agriculture.

In both the legume and nonlegume systems, symbiosis is characterized by a more or less distinct host preference or even host specificity. Table 7.1 gives examples of the host preference in the genera *Rhizobium* and *Bradyrhizobium*. Therefore, if legumes are introduced into soils in which the same or a symbiotically related legume has not previously been grown, seed inoculation with the appropriate species is required if effective N$_2$ fixation is to be achieved. However, within a given combination (i.e. bacteria/host plant) there are large differences between strains of bacteria and host genotype in relation to both infection, nodulation and effectivity in N$_2$ fixation which may offer the potential to increase N$_2$ fixation, for example, by strain selection and genetic manipulation (Peoples and Craswell, 1992).

Depending on the plant species (Rolfe and Gresshoff, 1988) infection by the microsymbiont may occur on developing root hairs (e.g., in clover), at the junction of lateral roots through structurally altered cell walls of the root cortex (e.g., peanut, as well as also in many actinorhizal systems) or at the base of the stem. In stem-nodulating species such as *Sesbania rostrata* infection takes place preferentially at sites where

Table 7.1

Examples of Host Preference in the Genera *Rhizobium* and *Bradyrhizobium*[a]

Bacterial genus/species	Host plant
Rhizobium leguminosarum	
biovar *viciae*	*Vicia, Lens, Pisum sativum*
biovar *phaseoli*	*Phaseolus*
biovar *trifolii*	*Trifolium*
Rhizobium meliloti	*Medicago sativa, Melilotus*
Bradyrhizobium japonicum	*Glycine*
Bradyrhizobium lupinus	*Lupinus*
Bradyrhizobium arachis	*Arachis*

[a]Formerly referred to as slow-growing rhizobia.

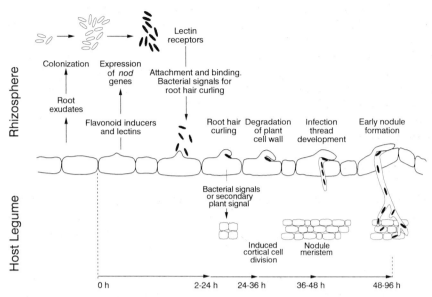

Fig. 7.4 Model of the early events of host plant infection by *Rhizobium* over a 96 h period. (From Djordjevic and Weinman, 1991.)

lateral root primordia may pierce the epidermis, in some species temporary flooding is required for nodulation to occur. Stem nodules are formed basically by the same bacteria as in the case of root nodules, and, as a rule, root and stem nodules occur on the same plant. Compared with root nodules, both, formation and activity of stem nodules are less depressed by high soil levels of mineral nitrogen (Alazard *et al.*, 1988; Becker *et al.*, 1990).

The first step of host plant infection by the microsymbiont requires recognition of the host. The principles of host plant recognition and infection are summarized in Fig. 7.4 for rhizobia which infect root hairs (e.g., *R. leguminosarum* biovar *trifolii* on *Trifolium* spp.). Distinct phenolic compounds (flavonoids) such as lutein released by the roots act at low concentrations as a signal for chemotaxis of rhizobia in the rhizosphere, and stimulate expression of nodulation (*nod*) genes in the rhizobia at higher concentrations, which are to be expected at the root surface (Bauer and Caetano-Anollés, 1990). Induction of *nod* genes is required for the production of lectins and the attachment of the bacteria on the developing root hairs. Root hair curling and cortical cell division by bacterial 'signals', probably cytokinins (Sturtevant and Taller, 1989), are the main features of the next step which is followed by development of the nodule meristem and invasion by the rhizobia, mediated by the host plant-derived infection thread (Fig. 7.4). Inside the dividing nodule cells rhizobia are released into the plant cell cytoplasm and packaged inside a plant-derived membrane (peribacteroid membrane). During this process bacteria are transformed into bacteroids which are several times larger than the original bacteria and are devoid of a cell wall. Infected root cells may contain up to 20 000 bacteroids. The transformation to bacteroids is closely related to the synthesis of hemoglobin, nitrogenase and other enzymes required for N_2 fixation. This is also the

Fig. 7.5 Sections of a nodulated root of soybean (*Glycine max*) with determinate nodules (*left*) and of *Mimosa pudica* L. with elongated indeterminate nodules (*right*). (Courtesy of A. P. Hansen.)

stage of nodule development which is strongly impaired by deficiency of certain mineral nutrients (Section 7.4.5).

The *nod* gene-inducing activity of root exudates and the adsorption of the rhizobia at the root surface is strongly affected by the calcium concentration and pH (Richardson *et al.*, 1988). In the concentration range of 0–1.5 mM the adsorption of *R. meliloti* on *Medicago sativa* is linearly related to the concentrations of calcium and magnesium, and higher calcium concentrations are required to compensate for the negative effect of low pH (i.e., high H^+ concentrations) on the adsorption of rhizobia (Caetano-Anollés *et al.* 1989). Thus, the well-known negative effects of low pH and low calcium concentrations on nodulation of alfalfa (*M. sativa*) are reflected as early as this first step of nodulation. In addition, in *Trifolium subterraneum* the *nod* gene-inducing activity of root exudates is also higher in plants grown at higher calcium concentrations (Richardson *et al.*, 1988).

The plant can restrict nodule number via a process termed 'autoregulation' or feedback regulation by which the formation of nodules on one part of the root inhibits subsequent nodule formation in other regions of the root (Sutherland *et al.*, 1990). This negative feedback reaction is exerted during the onset of cell division in the root cortex (Caetano-Anollés and Gresshoff, 1991), but is obviously absent in 'supernodulating' mutants (Day *et al.*, 1986). The form of nitrogen supply may alter the extent of autoregulation. Nitrate seems to amplify this signal (Day *et al.*, 1989) whereas ammonium interferes with its functioning (Waterer *et al.*, 1992; see Section 7.4.5).

Depending on the plant species and environmental factors, N_2 fixation starts at the earliest between 10 and 21 days after infection. Functioning nodules also differ typically between plant species in shape and size, and may be of indeterminate type (continue to grow) or of determinate type (Rolfe and Gresshoff, 1988). An example of determinate and indeterminate nodules is presented in Fig. 7.5.

During the lag-phase between infection and onset of N_2 fixation the host plant has to

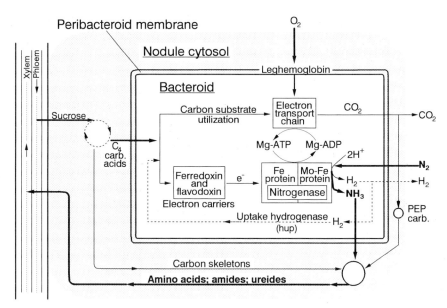

Fig. 7.6 Model of relationships between nitrogenase and related reactions in bacteroids and the cytosol of the host in legume nodules.

supply all mineral nutrients, photosynthates and amino acids required for the growth of the bacteria and the nodule tissue, without direct benefit to the host. Consequently, legumes rely also on an adequate supply of combined nitrogen during this period ('starter N') in order to establish a source leaf area large enough to supply photosynthates and the other solutes to the growing nodules.

7.4.3 Functioning of Root Nodules, Energetics of Nitrogen Fixation

The large potential of nodulated legumes for N_2 fixation is mainly based on three factors, direct supply of photosynthates to the N_2-fixing bacteroids in the nodules, effective maintenance of very low O_2 concentrations in the interior of the nodules for protection of the nitrogenase, and rapid export of the fixed nitrogen (Fig. 7.6). Sucrose is delivered by the phloem into the nodules where the carbon and nitrogen metabolism is adapted to an O_2-limited environment with free O_2 concentrations in the range of 10–40 nM (Vance and Gantt, 1992). The low O_2 concentrations are achieved by two mechanisms, existence of an O_2 diffusion barrier in the cortex of the nodules, and high respiration rates of the bacteroids for energy production (oxidative phosphorylation). Providing the bacteroids with sufficient O_2 in an O_2-limited environment requires finely tuned diffusion of O_2, mediated by leghemoglobin (Fig. 7.6), a red-coloured enzyme with a central iron atom in the porphyrin ring (identical to cytochromes; Section 8.7). The concentration of leghemoglobin is closely but not linearly correlated with the N_2-fixing capacity of root nodules (Werner *et al.*, 1981).

The high O_2 consumption in nodules for provision of energy also creates a great potential for production of toxic oxygen species (Section 5.2.2; Fig. 5.2). This is true in

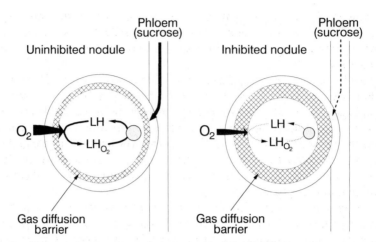

Fig. 7.7   Model of possible mechanism of inhibition of N_2 fixation rate by O_2 limitation of nitrogenase activity. (Modified from Vessey and Waterer, 1992.)

particular for leghemoglobin which is also subject to autoxidation in which O_2^- and H_2O_2 are released. For protection against this toxicity, legume nodules contain the enzymes of the ascorbate-glutathione cycle as part of a defense mechanism (Fig. 5.2) which rapidly responds to changes in ambient O_2 concentrations in the nodules (Dalton *et al.*, 1991). In soybean and alfalfa effective nodules contain a considerably higher capacity for peroxide scavenging than do ineffective nodules (Dalton *et al.*, 1993).

More recently the variable barrier in the outer root cortex restricting and modulating O_2 diffusion into the nodules has attracted great interest for various reasons (Fig. 7.7). In this barrier of one to five cell layers in thickness the intercellular spaces are filled with water. Since the diffusion coefficient for O_2 in air is about 10^4 times higher than in water, a water barrier in the cortex by imbition of the intercellular spaces represents a most effective mechanism limiting O_2 diffusion into the interior of the nodules (Blevins, 1989). In highly active nodules the intercellular spaces are mainly filled with air. However, a change in water status of the plant, for example as a result of drought or defoliation, may cause imbibition of the intercellular spaces with water and, thus, drastically decrease O_2 supply to the leghemoglobin in the host cytosol and subsequently to the bacteroids (Fig. 7.7).

Import and export of water in nodules is another factor which indirectly affects the rate of N_2 fixation. As the water supply to nodules is coupled to the solute import via the phloem (Section 5.4), any restriction of this import decreases the turgor pressure in the nodules and thereby indirectly nodule permeability for O_2 (Hunt and Layzell, 1993). The rate of N_2 fixation might also be limited by the availability of water for export of reduced nitrogen in the xylem from the nodules. Most of the water involved in this export from the nodules is derived from the phloem. A decrease in phloem solute import (e.g., shading) presumably also inhibits nitrogen export and N_2 fixation by water limitation for xylem export (Streeter, 1993). Many environmental factors which depress nitrogenase activity may therefore at least in part function via modification of both, water relations and the O_2 diffusion barrier in the nodules (Sections 7.4.4 and 7.4.5).

In root nodules of the *Frankia* symbiosis the nitrogenase is protected from free O_2 mainly by the cell walls around the microsymbiont (vesicles); these cell walls adapt to the ambient O_2 pressure by variation in thickness and incorporation of multiple lipid layers (Parsons *et al.*, 1987).

Owing to the low concentrations of free oxygen in the host cytosol of legume nodules, the degradation of carbohydrates by glycolysis dominates by way of the oxidative pathway, and malic acid is synthesized preferentially as is known from the carbon metabolism in roots of flooding-tolerant species (Section 16.4.3). Malic acid acts as a substrate for the reductive synthesis of the C_4 carboxylic acids fumarate and succinate which are transported into the bacteroids to serve as an energy substrate for nitrogenase (McKay *et al.*, 1988). Other carbon skeletons from glycolysis serve as substrates for the NH_3 assimilation in the cytosol (Fig. 7.6).

At certain stages of the growth period nodules may consume as much as 50% of the photosynthates produced by legume plants such as cowpea (Pate and Herridge, 1978). About half of these photosynthates are respired as CO_2. However, between 25 and 30% of the respired CO_2 can be reassimilated by the nodules via PEP carboxylase providing up to 25% of the carbon needed for malate and aspartate synthesis, required for the assimilation of NH_3 and export to the host plant (Fig. 7.6). On a fresh weight basis, PEP carboxylase activity in nodules is between 20 times (pea) and 1000 times (soybean) higher than in the roots (Deroche and Carrayol, 1988).

The carbon costs of N_2 fixation vary with host species, bacterial strain and plant development. They may also increase with plant age by a factor of about three, as for example in soybean, but remain almost constant in red clover (Warembourg and Roumet, 1989). Estimates on carbon costs per g N fixed vary from between 12 g C (Streeter, 1985) to 6 g C (Vance and Heichel, 1991) if the costs for nodule growth etc. are included. As average values of the different systems of N_2 fixing legumes (e.g., differing in *hup* activity or type of nitrogen compound exported), 36–39% of the carbon costs are required for nodulation, nodule growth and maintenance respiration, 42–45% for nitrogenase activity, and 16–22% for NH_3 assimilation and export (Layzell *et al.*, 1988). In *Medicago sativa* the carbon costs varied between 5.1 and 8.1 g C g^{-1} N fixed, depending on the rhizobial strain, and about 50% of these costs were required for growth of nodules and maintenance respiration (Twary and Heichel, 1991). Assuming a fixation of 150 kg N ha^{-1} on the basis of 5–8 g C g^{-1} N, the total carbon costs expressed in sucrose units would be in the range of 2–3 t ha^{-1} which the shoots have to supply for growth and functioning of the nodules. The high turnover of carbon in the nodules is reflected by a substantial requirement for newly fixed carbon continuously provided by the shoot, and by about two times higher respiration rates per unit dry weight of nodulated compared with non-nodulated legume roots (Ryle *et al.*, 1979; Hansen *et al.*, 1992).

Although the carbon costs for N_2 fixation are fairly high, these costs have to be compared with costs for assimilation of bound nitrogen, for example, nitrate. In plant species such as lupins in which nitrate is preferentially reduced in the roots, the carbon costs have been calculated to be in a similar range to that for N_2 fixation (Pate *et al.*, 1979). In cowpea about 2.3 g C are required per g NO_3-N reduced (Sasakawa and La Rue, 1986), excluding the costs for nitrate uptake, synthesis and maintenance of the reduction system, and assimilation and export of the reduced nitrogen. However, in

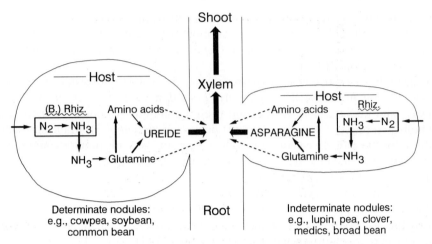

Fig. 7.8 Incorporation of ammonia into translocated solutes of nitrogen in legumes with determinate and indeterminate nodules. (Modified from Atkins, 1987; reprinted by permission of Kluwer Academic Publishers.)

plant species with preferential nitrate reduction in the leaves, these carbon costs are considerably lower (Section 8.2.1).

The nitrogen fixed in the bacteroids of root nodules is released as NH_3 to the host cytosol (Fig. 7.6), solely by simple diffusion across the peribacteroid membranes (Udvardi and Day, 1990). In the host cytosol the NH_3 is assimilated via the glutamine synthase (GS)/glutamate synthase (GOGAT) pathway (Vance and Gantt, 1992), as is also the case for NH_3 assimilation in roots and shoots of the host (Section 8.2.1). High concentrations of NH_3 as well as of the endproducts of its assimilation, glutamine, also repress the activity of the nitrogenase (Shanmugam *et al.*, 1978). This has important implications for the application of nitrogen fertilizer to legumes and other N_2 fixing systems (Section 7.4.6).

Generally, glutamine is a nitrogen compound exported in only small amounts from nodules (Fig. 7.8). In legume species with indeterminate nodules glutamine is mainly metabolized to asparagine in the nodules and delivered in the xylem to the shoots. In contrast, in legume species with determinate nodules glutamine is metabolized in the nodules to the ureides allantoin and allantoic acid (Section 8.2.3) and translocated in the xylem to the shoot.

7.4.4 Effect of Endogenous and Exogenous Factors on Nitrogen Fixation Rates

In general there is a close correlation between the N_2 fixation rate and a continuous supply of newly fixed carbohydrates to the root nodules. This is reflected, for example, in diurnal fluctuations in N_2 fixation rates (Rainbird *et al.*, 1983). However, these diurnal fluctuations in N_2 fixation rates may also reflect, at least in part, fluctuations in soil temperature and corresponding differences in O_2 diffusion rates into the nodules (Weisz and Sinclair, 1988).

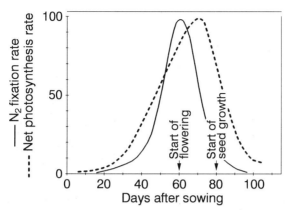

Fig. 7.9 Changes in rates of net photosynthesis and N_2 fixation during growth of cowpea. Relative values. (Based on Herridge and Pate, 1977.)

If nodulated legume plants are kept for several days under low light, not only does the N_2 fixation rate decrease (Tricot *et al.*, 1990) but also an increasing proportion of the root nodules are transformed from the N_2 fixing red, and functioning leghemoglobin-containing types to nonfixing types (Haystead and Sprent, 1981). In peanut, genotypical variations in N_2-fixation rates under field conditions were mainly caused by differences in light interception, i.e., canopy characteristics (Williams *et al.*, 1990). In mixed stands of legumes with nonlegumes, the competitiveness of legumes is, therefore, closely related to light interception and light intensity. However, legume species differ in the type of their response to a decreasing supply of photosynthates from the shoot, e.g., after shading or defoliation. Peanut plants, unlike other legumes, can sustain N_2 fixation under these conditions for prolonged time, which is likely to be attributable to the presence of stored photosynthates (as lipid bodies) in the nodules (Siddique and Bal, 1991).

In annual grain legumes, but to a lesser extent in pasture legumes such as white clover (Haystead and Sprent, 1981), a characteristic pattern occurs in the N_2 fixation rate during ontogenesis, as shown in Fig. 7.9 for cowpea. The maximum rate of N_2 fixation is reached at the beginning of the flowering period and is followed by a rapid decline. After some delay, a decline in the rate of net photosynthesis also occurs (Peoples *et al.*, 1983). The decline in N_2-fixation rate at the beginning of flowering is most likely an expression of sink competition for photosynthates between the developing pods and the root nodules. The rate of decline after flowering and pod set in N_2-fixation rates is dependent on various factors such as the source–sink relationships (e.g., size of source leaves versus number of pods; Section 5.7) and the sink strength of the nodules. The rate of decline is more rapid, for example, in cowpea (Fig. 7.9; Peoples *et al.*, 1983) than in soybean, where distinct differences between cultivars in the rate of decline also exist (Spaeth and Sinclair, 1983). Thus the beneficial effects of late application of nitrogen fertilizer on legume seed yield depend on the genotype, being high, for example, in pea but negligible in faba bean (Schilling, 1983).

Removal of flowers and developing pods may prevent this decline in N_2-fixation rates

(Table 5.23), if the decline is caused by sink competition for photosynthates. However, the presence of fruits does not necessarily depress and may even enhance N_2 fixation (Twary and Heichel, 1991; Hansen *et al.*, 1993), probably by eliminating feedback inhibition associated with excess availability of recently fixed nitrogen. Under field conditions such feedback regulations may limit N_2-fixation rates more often than limited supply of photosynthates (Fujita *et al.*, 1988). This type of feedback regulation deserves more attention in the current views on regulation mechanisms of N_2 fixation in legumes.

Defoliation of legumes by cutting or grazing induces a rapid and drastic decline in nitrogenase activity, in clover up to 95% being lost within a few hours (Hartwig *et al.*, 1987). However, similarly to the effect of shading, the effect of cutting on nodule activity also varies between species, and is, for example, very severe in *Sesbania* as compared with *Leucaena* (Fownes and Anderson, 1991). It is generally assumed that these effects of shading and defoliation are caused by deprivation of photosynthates as energy source for nitrogenase; however O_2 deprivation of nodules might be involved or even be the primary regulatory factor. For example, after cutting clover the drastic drop in nitrogenase activity could be overcome to a large extent by increasing the ambient O_2 concentration in the rooting zone (Hartwig *et al.*, 1987). In contrast, elevated O_2 concentrations depressed nitrogenase activity in the intact (undefoliated) clover plants. Similar differences in the response of nitrogenase activity to elevated ambient O_2 concentrations have been found in soybean between fully illuminated control plants and those exposed to prolonged darkness (Carroll *et al.*, 1987). Changes in turgor pressure of the nodules are likely be involved in these different responses to O_2 (Section 7.4.3). Short-term restriction of phloem supply in particular, may change the turgor in the nodules and the diffusion barrier (Fig. 7.7) thereby inducing O_2 limitation rather than carbon limitation (Walsh, 1990; Hunt and Layzell, 1993).

7.4.5 Effects of Mineral Nutrients Other Than Nitrogen

Mineral nutrients may influence N_2 fixation in legumes and nonlegumes at various levels of the symbiotic interactions: infection and nodule development, nodule function, and host plant growth. Furthermore, the relative requirement for a given mineral nutrient for host plant growth on the one hand and establishment and functioning of the symbiosis on the other differs in many instances. For reviews on this topic the reader is referred to O'Hara *et al.* (1988a), Martin (1990) and Robson and Bottomley (1991).

Inhibition of nodulation is a major limiting factor in N_2 fixation of many legume species grown in acid mineral soils. Increase in soil pH by liming is therefore very effective in increasing nodule number, for example in common bean (Buerkert *et al.*, 1990) or alfalfa (Pijnenberg and Lie, 1990). Various factors are responsible for poor nodulation in acid mineral soils, high concentrations of protons and of monomeric aluminium (Alva *et al.*, 1990) and in particular, low calcium concentrations. As shown in Fig. 7.10, nodule formation has a much greater requirement for calcium than root and shoot growth of the host plant, and also as the growth of rhizobia (Loneragan and Dowling, 1958). Additional experiments (Lowther and Loneragan, 1968) have demonstrated that after nodule initiation further nodule growth was not affected by a decrease in the calcium concentration, indicating that only the first step of infection is highly

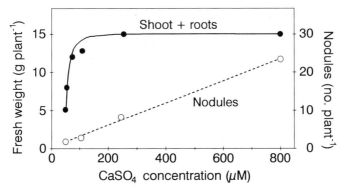

Fig. 7.10 Effect of calcium concentrations in nutrient solution (pH 5.0) on fresh weight (●) and nodule number (○) in subterranean clover. (Based on Lowther and Loneragan, 1968.)

sensitive to calcium supply. As discussed in Section 7.4.2, at low calcium concentrations, particularly in combination with high proton concentrations, the adsorption of the rhizobia at the host root surface is impaired. However, in *Medicago* there are also remarkable differences between acid-tolerant and acid-sensitive species with respect to the *nod*-gene inducing activity of root exudates. In contrast to acid-sensitive species, in the tolerant species the root exudates released at low pH and low calcium concentrations are also effective in inducing *nod*-gene activity (Howieson *et al.*, 1992), indicating a very delicate mechanism in acid tolerance of legume species.

Poor nodulation in acid soils can also be caused by low survival of certain strains of *Rhizobium* (Howieson *et al.*, 1988) or *Bradyrhizobium*, and effects on root morphology. Inhibition in root hair formation by low concentrations of calcium (Ewens and Leigh, 1985) and high concentrations of aluminum (Section 16.3) and protons (Franco and Munns, 1982) may be involved in impaired nodulation in species where root hairs are the dominant infection sites.

On acid mineral soils the net release of protons, i.e., rhizosphere acidification, inherently coupled with symbiotic nitrogen nutrition of legumes and nodulated non-legumes (Section 15.3.2) is a negative side effect. Depending on the legume species, between 37 and 49 mg H^+ are produced per gram fixed nitrogen, amounting to an annual production per hectare of 4.6 kg H^+ in sweet clover (*Melilotus alba* Medik) and 15.2 kg H^+ in alfalfa (Lui *et al.*, 1989). On average, to neutralize the acidity formed in the production of one ton of dry matter by various legume species, an equivalent of 80–96 kg of lime ($CaCO_3$) would be required (Jarvis and Hatch, 1985).

A high phosphorus supply is needed for nodulation. This requirement might be higher than for root or shoot growth of the host plant as shown in in Table 7.2 for soybean. The minimum concentration of phosphorus required for nodulation was about 0.5 μg P l^{-1} in the external solution. In particular, an increase in the concentration from 200 to 500 μg P l^{-1} resulted in a greater increase in the nodule dry weight relative to the shoot and, especially, to the root dry weight. In principle the results shown in Table 7.2 were confirmed by Israel (1987). With increasing phosphorus supply dry weight of soybean plants increased by a factor of six, nodule weight by a factor of 30,

Table 7.2

Relationship between Phosphorus Supply and Dry Weight of
Roots, Nodules and Shoots in Soybean Plants[a]

P supply $(\mu g\,l^{-1})$	Dry weight (g per plant)		
	Roots	Nodules	Shoots
0.5	0.60	0.07	1.21
20	0.76	0.10	1.55
50	0.79	0.16	1.86
200	1.23	0.35	4.22
500	1.35	0.64	6.57

[a]Based on Cassman et al. (1980).

specific nitrogenase activity doubled and nitrogen content in the plant dry matter increased by 70%.

When legumes dependent on symbiotic nitrogen receive an inadequate supply of phosphorus, they may therefore also suffer from nitrogen deficiency. In soybean grown in a soil low in available phosphorus, a similar yield increase was obtained with mineral fertilizer supply in the combination of either 90 kg P/0 kg N or 0 kg P/120 kg N per hectare (Dadson and Acquaah, 1984).

The phosphorus content per unit dry weight is usually considerably higher in the nodules than in the roots and shoots, particularly at low external phosphorus supply (Adu-Gyamfi et al., 1989; Hart, 1989). Nodules therefore provide a strong sink for phosphorus. The capability of developing nodules to compete with other vegetative sinks (root and shoot meristems) for phosphorus at limited external supply may differ between legume species and may in part be responsible for the doubts raised on suggestions of high specific requirements of phosphorus for nodulation (Jakobsen, 1985; Robson and Bottomley, 1991).

The high phosphorus requirement for nodulation is responsible, at least in part, for the interactions in many legume species between the infection of roots with VA mycorrhiza and nodulation. Mycorrhizal infection of roots increases the acquisition of phosphorus in plants grown in low phosphorus soils and, thus, nodulation might be increased either by application of phosphorus fertilizer or root infection with mycorrhiza (Section 15.7.1).

As molybdenum is a metal component of nitrogenase, all N_2-fixing systems have a specific high molybdenum requirement (Chapter 9.5). Although at low supply, molybdenum is preferentially transported into the nodules (Brodrick and Giller, 1991a), molybdenum deficiency-induced nitrogen deficiency in legumes relying on N_2 fixation is widespread, particularly in acid mineral soils of the humid and subhumid tropics. Under these conditions, seed pelleting or soil treatment with molybdenum are effective methods of obtaining high rates of N_2 fixation in legumes as shown for groundnut in Table 7.3. Seed pelleting with 100 g Mo ha^{-1} increased nitrogenase activity and leaf nitrogen content and particularly nodule dry weight, whereas mineral nitrogen decreased nodule dry weight and suppressed nitrogenase activity compared with plants supplied with phosphorus only. At maturity the plants supplied with mineral fertilizer

Table 7.3
Effect of Nitrogen Fertilizer (2×30 kg N ha^{-1} as NH$_4$NO$_3$) and Molybdenum Seed Pelleting (100 g Mo ha^{-1} as MoO$_3$) on Peanut (*Arachis hypogaea*) Grown in an Acid Sandy Soil[a]

| | Early podfilling (68 DAP) | | Maturity (90 DAP) | | | |
| | Nodule dry wt (mg per plant) | Nitrogenase (μmol C$_2$H$_4$ g^{-1} nodule fresh wt) | Leaf N content (mg g^{-1} dry wt) | Dry wt (kg ha^{-1}) | | N uptake |
Treatment				Shoots	Pods	(kg ha^{-1})
+P[b]	80	50	25	861	1570	77
+P+N	70	43	33	1817	1783	110
+P+Mo	180	60	37	1380	1948	119

[a]Based on Hafner *et al.* (1992). Reprinted by permission of Kluwer Academic Publishers.
[b]16 kg P ha^{-1} as single superphosphate.

had a higher shoot dry weight but lower pod dry weight compared with the plants supplied with molybdenum. This lower harvest index (Section 5.7) in the plants fertilized with mineral nitrogen was most likely the result of higher water consumption and a more severe drought stress during early podfilling (Hafner *et al.*, 1992). Thus, under certain ecological conditions supply of only 100 g Mo ha^{-1} might not only enhance N$_2$ fixation, total nitrogen uptake and drought tolerance but also increase pod yield more than supply of 60 kg N ha^{-1} as mineral fertilizer.

Iron is required for several key enzymes of the nitrogenase complex as well as for the electron carrier ferredoxin (Fig. 7.2) and for some hydrogenases. A particular high iron requirement exists in legumes for the heme component of hemoglobin. Therefore, in legumes iron is required in a greater amount for nodule formation than for host plant growth, for example in lupins (Tang *et al.*, 1990) and peanut (Table 7.4). Although iron deficiency did not significantly affect shoot growth it severely depressed nodule mass and particularly leghemoglobin content, number of bacteroids and nitrogenase activity, compared with those plants five days after a foliar spray of iron. In contrast to peanut, in lupin (*Lupinus angustifolius*) iron is not retranslocated into the nodules after a foliar spray, and direct iron supply at the infection sites at the roots required for effective nodulation (Tang *et al.*, 1990; 1992a).

Table 7.4
Iron Deficiency-Limited Nodule Development in Peanut (*Arachis hypogaea* cv. Tainau 9) Grown in a Calcareous Soil[a]

Treatment (foliar spray FeSO$_4$)	Foliage (g fresh wt per plant)	Nodules (mg fresh wt per plant)	Iron (μg g^{-1} nodule dry wt)	Leghem. (nmol g^{-1} fresh wt nodule)	Bacteroid content (10^6 g^{-1} fresh wt)	ARA[c] (pmol min^{-1} per plant)
$-$Fe	3.2	15	800	6	170	3
$+$Fe[b]	3.3	60	1470	86	7300	35

[a]O'Hara *et al.* (1988b).
[b]On day 10; all measurements ($\pm$Fe) at day 15.
[c]pmol acetylene reduced (C$_2$H$_2 \to$ C$_2$H$_4$).

Table 7.5

Effect of Cobalt Supply for Six Weeks on Nodule Development and Function in *Lupinus angustifolius*[a]

Parameter	Co supply (mg $CoSO_4 \cdot 7H_2O$ (6 kg)$^{-1}$ soil)				
	0	0.01	0.05	0.1	0.5
Foliage mass (g fresh wt)	5.0	6.1	7.5	9.6	14.0
Nodule mass (g fresh wt)	2.9	2.8	2.5	2.3	1.1
Bacteroid content (no. 10^9 per nodule)	6.0	12.0	12.5	20.5	22.5
Leghemoglobin content (nmol g^{-1} lateral root fresh wt)	—	1	11	20	120
Nitrogenase activity nmol C_2H_2 reduced (g^{-1} nodule fresh wt min^{-1})	10	21	58	104	172
nmol C_2H_2 reduced (nmol^{-1} leghemoglobin min^{-1})	1.1	2.5	3.7	3.8	3.2

[a]Recalculated from Riley and Dilworth (1985a).

The critical step impaired by iron deficiency is not the infection *per se* but the further division of cortical cells, i.e. early stages of nodule development (Tang *et al.*, 1992a). In dicotyledonous plant species grown on calcareous soils, high bicarbonate concentrations not only induce visual symptoms of iron deficiency (chlorosis) and decrease net photosynthesis but may particularly depress nodulation and N_2 fixation in legumes (Tang *et al.*, 1991).

Although nickel is a constituent of a variety of uptake hydrogenases, and lower hydrogenase activity has been found in bacteroids isolated from nickel-deficient soybean plants (Klucas *et al.*, 1983), evidence is lacking that under field conditions N_2 fixation of legumes is impaired by nickel deficiency.

Cobalt is required for the synthesis of leghemoglobin and, thus, for the growth of legumes relying on symbiotically fixed nitrogen, cobalt is an essential mineral nutrient (Section 10.4). Cobalt deficiency effects nodule development and function at different levels and to different degrees (Table 7.5). In lupins relying on symbiotic N_2 fixation cobalt deficiency depresses host plant growth, but not nodule mass, which even increases in the deficient plants. The most sensitive parameter for cobalt limitation is the bacteroid content of the nodules. Whereas the synthesis of leghemoglobin also responded markedly to cobalt supply, the increase in activity of the nitrogenase per unit hemoglobin was only relatively small.

7.4.6 Effect of Mineral Nitrogen

Of the mineral nutrients, very often combined nitrogen (nitrate, ammonium, urea, amino acids) both in soils and in plants has the most prominent influence on biological N_2 fixation. In legumes (and other symbiotic N_2 fixing systems) combined nitrogen can enhance or depress N_2 fixation, depending on a range of factors, and the level of supply in particular. As shown schematically in Fig. 7.11 increasing supply of combined

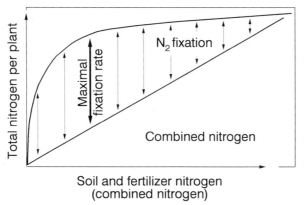

Fig. 7.11 Simplified scheme of the relationship between N_2 fixation and nitrogen uptake from soil and fertilizer in nodulated legumes.

nitrogen results in an asymptotic increase in total nitrogen per plant or per unit surface area (e.g. hectares). The contribution of N_2 fixation to the total nitrogen per plant is increased by moderate levels of soil or mineral fertilizer nitrogen but declines at high levels, reflecting the depression of N_2 fixation by the high levels of either soil or fertilizer nitrogen.

The enhancing effect of low levels of combined nitrogen on N_2 fixation in legumes is related to the lag phase between root infection and the onset of N_2 fixation. A shortage of nitrogen in the host plant during this lag phase is detrimental to the formation of a source leaf area that is sufficiently large to supply the photosynthates needed for nodule growth and activity. At zero or very low levels of combined nitrogen the enhancement effect by fertilizer nitrogen (Fig. 7.11) depends on the nitrogen reserves in the seeds, i.e. also on the seed size. As a rule, the highest nodulation and nodule activity (N_2 fixation) is therefore obtained when the seed reserves and combined nitrogen either from the soil reserves or from fertilizers is available in amounts that are sufficient for vigorous plant growth during the first weeks of legume establishment. This beneficial effect is demonstrated in Table 7.6 at a treatment level of 25 kg nitrogen as 'starter-N'.

When the levels of combined nitrogen increase, nitrogenase activity declines drastically and number of nodules also decreases (Table 7.6). Shoot growth, however, continues to increase, indicating the shift from symbiotic to inorganic nitrogen nutrition. The highest nitrogen content in the shoot dry weight coincides with the highest nitrogenase activity but not with the highest dry weight of the plants. Evidently, at the maximum of N_2 fixation under these experimental conditions dry matter production was source-limited (Section 5.7) and the additional supply of photosynthates required for N_2 fixation restricted plant growth.

In principle, similar results as shown for bean (Table 7.6) have been obtained in soybean. Low levels of mineral fertilizer supply as 'starter-N' increased nodulation and total amount of nitrogen derived from N_2 fixation, but high levels of fertilizer nitrogen drastically decreased nodulation and even prevented N_2 fixation (T. George *et al.*, 1992).

The depression of nodulation and nodule activity by high levels of combined nitrogen

Table 7.6

Influence of Nitrogen Fertilizer Supply on Nitrogenase Activity and Shoot and Root Growth in Bean[a]

Nitrate N fertilizer supply (kg ha^{-1})	Nitrogenase activity (μmol C$_2$H$_4$ produced per plant h^{-1})		N in shoots (%) day 49	Dry weight (g per plant) day 49	
	35 days	49 days		Shoot and roots	Nodules
0	1.13	0.19	1.54	2.53	0.18
25	2.26	0.33	1.82	3.35	0.28
50	0.60	0.10	1.67	3.65	0.13
100	0.14	0.03	1.69	4.35	0.11

[a]Based on Sundstrom et al. (1982).

depends very much on the plant genotype and form of nitrogen supplied. It is generally agreed that in legumes infection and nodule development are more sensitive to nitrate rather than ammonium inhibition. In pea, a continuous supply of moderately high levels of ammonium nitrogen (1 mM) may not only increase nodulation and N$_2$ fixation but even stimulate continuous formation of small nodules as is known from 'supernodulating' genotypes (Waterer et al., 1992). Although in general a continuous high nitrate supply strongly depresses nodulation (Table 7.6), marked genotypical differences in 'nitrate-sensitivity' exist. As shown by Harper and Gibson (1984), high nitrate supply inhibits nodulation much more in soybean than in chickpea or lupins, whereas the nitrogenase activity is severely inhibited in lupins and chickpea but only slightly affected in subterranean clover. In common bean, differences exist even between cultivars in the inhibition of nodulation by high nitrate supply (Martin, 1990). Genotypical differences between plant species in 'nitrate-sensitivity' may be related, at least in part, to the mode of root infection. In actinorhizal plants infection may occur at root hairs (e.g., in *Alnus glutinosa*) or at sites of lateral root (e.g., in *Elaeagnus angustifolia*). High nitrate concentrations depressed nodulation drastically only in the species with root hair infection, whereas nodulation was insensitive to high nitrate concentrations in the other type of infection (Kohls and Barker, 1989).

Recently, much attention has been paid to the inhibitory effect of nitrate on nitrogenase activity in established nodules. Nitrate might inhibit nitrogenase activity at various levels, and either indirectly or directly as brought about by:

1. Competition for photosynthates. This type of inhibition is expected to be particularly important in those plant species where nitrate reduction occurs principally in roots rather than in shoots.

2. Feedback regulation by elevated levels of reduced nitrogen compounds in the host plant in general and in the phloem sap delivered to the nodules in particular. In view of the importance of circulation of reduced nitrogen in the plants (Section 3.4.3) this aspect is often not sufficiently appreciated as an important mechanism also in the regulation of symbiotic N$_2$ fixation, particularly in field-grown plants.

3. Direct inhibition by nitrate taken up by, and reduced in the nodules. This might lead to competition for reducing equivalents and malate (Heckmann et al., 1989),

Table 7.7
Examples of Estimates of Amounts and Proportions of Plant Nitrogen Derived from Symbiotic
N_2 Fixation in Legumes in Tropical and Subtropical Production Systems[a]

Plant species	Total N ($kg\ ha^{-1}$ per crop)	N_2 fixation ($kg\ N\ ha^{-1}$ per crop)	Proportion of total N derived from N_2 fixation
Arachis hypogaea	126–319	37–206	22–92
Glycine max	33–643	17–450	14–97
Vigna unguiculata	25–100	9–39	12–70
Phaseolus vulgaris	71–183	3–57	16–71
Leucaena leucocephala	288–344	98–231	34–100
Sesbania rostrata	157–312	83–286	36–100

[a]Based on Peoples and Craswell (1992).

nitrite toxicity (Becana *et al.*, 1985) and O_2 deficiency (Vessey *et al.*, 1988). The latter two factors are considered of particular relevance. Nitrate supply induces nitrate reductase activity in the bacteroids fairly rapidly, but in some strains nitrite reductase is induced after a considerable delay, leading to the accumulation of nitrite, NO_2^- (Arrese-Igor *et al.*, 1990), which can directly inactivate leghemoglobin by the formation of nitrosylhemoglobin (Kanayama and Yamamoto, 1991).

The inhibitory effect of nitrate on nitrogenase activity may relate to the effects of O_2. In accord with the supposition of O_2 deprivation, inhibition of nitrogenase activity by nitrate can be alleviated to a considerable extent by increasing the ambient O_2 partial pressure in the rhizosphere (Vessey *et al.*, 1988). Similarly to stem girdling and defoliation, nitrate supply also increases severalfold the resistance of the nodules to O_2 diffusion (Fig. 7.7) (Vessey *et al.*, 1988), and this decrease is probably caused by the reduction in phloem import of photosynthates and other solutes and, thus, also a decrease in osmotic pressure in the nodules (Vessey and Waterer, 1992).

Irrespective of the mechanisms involved, in many soils high levels of mineral nitrogen in general and nitrate in particular are an important factor in limiting the potential of N_2 fixation in legumes as well as in other N_2-fixing systems. Breeding for 'nitrate tolerance' is therefore considered as an important line of development (Peoples and Craswell, 1992). However, in many other soils the potential of N_2 fixation is not realized because of other constraints related to mineral nutrition, namely deficiencies in molybdenum, iron, phosphorus or calcium, and high concentrations of protons and aluminum, i.e., soil acidity.

7.5 Amounts of Nitrogen Fixed in Symbiotic Systems, and Transfer in Mixed Stands

Amounts of N_2 fixed annually in symbiotic systems vary over a wide range. Inherent differences in biomass production and environmental factors such as low and high soil fertility contribute to this variability, and an example of the wide range obtained is given in Table 7.7 for some tropical and subtropical legumes. Both the total amounts and the proportions derived from N_2 fixation vary greatly, in some instances nearly all the nitrogen recovered in the plants appeared to be derived from N_2 fixation. There is a tendency for the proportion of fixed nitrogen to be higher in the tree species (*Leucaena* and *Sesbania*). However, estimates of N_2 fixation vary greatly, particularly in trees, because of serious methodological difficulties. Some of the tree species such as *Leucaena leucocephala* definitely have a much higher N_2-fixing potential than others as for example the *Acacia* species (Danso *et al.*, 1992).

Part of the fixed nitrogen remains in the soil as root residues and nodules or returns to the soil with litter fall, etc. In annual species some of the fixed nitrogen after harvest becomes available for the next crop. In mixed stands of legumes and nonlegumes (e.g., grasses), direct transfer of fixed nitrogen during the growing season is possible, for example via VA mycorrhizal hyphae (Section 15.7), although the extent to which it occurs is small and most likely in the range of 10% or less of the total nitrogen fixed (Morris *et al.*, 1990).

Legumes as sole crops usually have a lower and more variable proportion of fixed nitrogen than when grown in mixture with nonlegumes, provided the portion of legumes in the mixture dominates (Mallarino *et al.*, 1990a). One factor responsible for the higher efficiency in mixed stands is presumably the lower level of mineralized nitrogen maintained in the soil. In *Phaseolus vulgaris* genotypes nitrogen derived from N_2 fixation may vary between 5.6 and 21.1% in monoculture, and 18.2 and 56.6% when intercropped with maize (Tsai *et al.*, 1993).

In temperate climates in mixed stands of *Trifolium* ssp. with *Lolium* ssp the annual N_2 fixation has been estimated to be in the order of 232–308 kg N ha^{-1} ($\sim$75–86% of the total plant N), and the apparent clover-to-grass transfer of fixed nitrogen contributes up to 52 kg N ha^{-1} per year (Boller and Nösberger, 1987). In mixed stands of *Trifolium* ssp. and *Festuca arundinacea*, the total amounts of fixed nitrogen were between 300 and 390 kg N ha^{-1} per year, representing 50–70% of the total nitrogen of the legume (Mallarino *et al.*, 1990a). In this study the proportion of nitrogen in the grass provided by the N_2 fixation of the legumes increased from 20% in the year of sward establishment to 60% in the second year (Mallarino *et al.*, 1990b). In grazed pasture swards of *Trifolium repens* and *Lolium perenne* of the fixed nitrogen (269 kg ha^{-1} per year), in addition to the belowground transfer of 60 kg N ha^{-1} per year, an additional amount of 70 kg N ha^{-1} per year was provided to the grass by aboveground transfer via animal excrement (Ledgard, 1991).

7.6 Free-Living and Associative Nitrogen-Fixing Microorganisms

7.6.1 Free-Living Nitrogen-Fixing Microorganisms

Diazotrophic bacteria are ubiquitous in soils. They are either anaerobes (e.g., *Clostridium pasteurianum*), facultative anaerobes (e.g., *Klebsiella*) or aerobes (e.g., *Azotobacter* and *Azospirillum*). Because of carbon-limitation (Section 7.1), the actual contribution of these carbon-heterotrophic bacteria to the nitrogen input is considered to be very small, and on average less than 1 kg N ha^{-1} per year (Bothe *et al.*, 1983). Their competitiveness against other carbon-heterotrophic microorganisms can be increased by supplying carbon sources with a high C/N ratio. For example, incorporation of wheat straw strongly enhances not only CO_2 liberation but also nitrogenase activity in the soil (Roper *et al.*, 1989). Diazotrophic anaerobes living in the gastroenteric cavities of soil animals on decaying organic matter (Citernesi *et al.*, 1977) may contribute to this increase in nitrogenase activity.

The situation is different for carbon-autotrophic (photosynthetic) cyanobacteria living at the soil surface or in shallow water (wetland rice fields). From long-term field experiments in Rothamsted (UK) the annual contribution ot these microorganisms to the nitrogen balance has been calculated to be between 13 and 38 kg N ha^{-1} (Witty *et al.*, 1979). In wetland rice fields the annual N_2 fixation rates of cyanobacteria are assumed to be potentially between 42 and 150 kg N ha^{-1}, but actual values have been found to be between 10 and 80 kg N ha^{-1}, depending on availability of light and combined nitrogen (Watanabe, 1986). This actual annual contribution of cyanobacteria is higher than that estimated for rhizosphere associations in rice (10–30 kg N ha^{-1}), but less than estimated for the *Azolla–Anabaena* symbiosis (20–100 kg N ha^{-1}; Roger and Ladha, 1992).

Free-living diazotrophs are also abundant on leaf surfaces, especially on leaves of plants growing in the humid tropics (Ruinen, 1975). These microorganisms in the *phyllosphere* (the leaves of a canopy) consist of carbon-autotrophic (cyanobacteria) and carbon-heterotrophic (e.g., *Bacillus polymyxa*) species, the latter relying on photosynthates leached from leaves. Associations between plants and microbes also occur, for example, between cyanobacteria and the leaves of certain tree species, in which a direct transfer of fixed nitrogen to the host leaf is substantial and may account for 20–25% of the nitrogen in the leaf tissue (Bentley and Carpenter, 1984). Estimates of the annual fixation rates of diazotrophic bacteria in the phyllosphere of forest trees are in the range of 10–20 kg N ha^{-1} in temperate forests and up to 90 kg N ha^{-1} in tropical rain forests (Favilli and Messini, 1990).

There is some prospect of utilizing N_2 fixation in the phyllosphere in tropical agricultural systems. Spraying leaves of wetland rice several times with N_2-fixing bacteria substantially increases nitrogenase activity of the leaves, the grain yield, and the nitrogen content of the plants (Nandi and Sen, 1981). In wetland rice, and also in wheat, this may save supplying up to 50 kg N ha^{-1} as mineral fertilizer (Sen Gupta *et al.*, 1982). However, as nitrogenase activity is strongly suppressed by combined nitrogen (Section 7.4.6), in crop plants a major contribution of this type of N_2 fixation in the phyllosphere can only be expected when plants are severely nitrogen deficient (Sen Gupta *et al.*, 1982).

7.6.2 Associative Nitrogen Fixation

A considerable proportion of the carbon fixed during photosynthesis in higher plants is released into the rhizosphere in the form of root exudates or decaying root cells (Section 15.4). The population density of soil microorganisms, including diazotrophic bacteria, in the rhizosphere is thus several times higher than in the bulk soil (Section 15.5). For cases where the diazotrophic bacteria live preferentially at the root surface (rhizoplane) and also within intercellular spaces of the cortex cells, the term *rhizosphere associations* has been introduced (Döbereiner, 1983). However, the distinction between free-living associative, and symbiotic diazotrophs is not very clear. In contrast to symbiosis

Table 7.8
Effect of Combined Nitrogen (NH_4NO_3) on the Growth of Maize
Inoculated with *Azospirillum*[a]

NH_4NO_3 supply ($g\,l^{-1}$)	Nitrogenase activity (nmol C_2H_4 per plant h^{-1})	Shoot dry weight (g per plant)
0	200	0.49
0.04	156	0.97
0.08	10	1.84
0.16	0	2.93

[a]From Cohen *et al.* (1980).

where the collaboration between the different organisms (host, microbe) is mutually beneficial, in associations the partnership is more casual and the nitrogen transfer more indirect (Klucas, 1991).

Diazotrophic bacteria can also be found in the stele of roots and in the stems of graminaceous species (De-Polli *et al.*, 1982) where they also multiply (Ruppel *et al.*, 1992). They may even infect epidermal and cortical root cells (Pohlman and McColl, 1982), in which their nitrogenase activity is not only retained but is also much less depressed by external nitrate supply than in the free-living bacteria (Gantar *et al.*, 1991).

Within the last two decades six genera of diazotrophic bacteria, living in association with roots, mainly of graminaceous plants, have been identified (Boddey and Döbereiner, 1988). The most common genera in associated systems are *Azospirillum*, *Azotobacter*, *Klebsiella*, *Enterobacter* and *Pseudomonas*. *Azospirillum* and *Azotobacter* appear to be dominant in the tropics and are fairly rare in temperate regions, presumably because of their high temperature optimum which is in the range of 30°C, irrespective of their origin (Jain *et al.*, 1987; Harris *et al.*, 1989). In temperate regions the genera *Klebsiella* and *Enterobacter*, which have a lower temperature optimum, are more prominent (Jain *et al.*, 1987; Jagnow *et al.*, 1991).

Similarly to symbiotic systems (Section 7.3) nitrogenase in associative diazotrophs is extremly sensitive to free oxygen (Burris *et al.*, 1991), and their activity is suppressed by combined nitrogen (Table 7.8). With increasing nitrogen supply the nitrogenase activity is drastically depressed, but the shoot growth of the host plant increases. Similar results have been obtained with tall oatgrass, *Arrhenatherum elatior* (Martin *et al.*, 1989).

High ambient O_2 concentrations almost totally prevent associative N_2 fixation. In maize a decrease in ambient O_2 partial pressure in the rhizosphere from 10 kPa to 2 kPa increased N_2 fixation rates by a factor of 10, and a factor of 20 when additional carbon substrate (malate) was supplied (Alexander and Zuberer, 1989). In soil-grown plants these microaerophilic conditions required for high nitrogenase activity are presumably achieved in associative systems within colonies in microsites in the rhizosphere, at the rhizoplane or in the cortex, and may shift the optimum for nitrogenase activity to much higher O_2 concentrations in the rhizosphere (Christiansen-Weniger and van Veen, 1991). In lowland rice, microaerophilic conditions exist in the oxygenated zone around the roots of intact plants (Section 15.3.4); the majority of diazotrophic bacteria in the rhizosphere of rice are aerobic species (Balandreau *et al.*, 1975).

Compared with the free-living diazotrophs, associative diazotrophs have better access to root exudates as an energy substrate for nitrogenase activity. Nevertheless, these associations are also often carbon-limited as indicated by distinct diurnal (Troll-denier, 1977) and seasonal fluctuations in nitrogenase activity (Sims and Dunigan, 1984). Nitrogenase activity is low during the first weeks of host plant growth, increases sharply at the booting stage, for example in barley (Vinther, 1982), and reaches its peak at flowering, for example in rice (Sano *et al.*, 1981). However, the establishment and effectivity of specific associations depends not only on the amount but also on the composition of the carbon substrate. Associative bacteria, for example, *Azospirillum* prefer C_4 carboxylic acids, particularly malate as substrate (Alexander and Zuberer, 1989; Werner *et al.*, 1989), which is also important for chemotaxis in highly specific associations such as *Leptochloa fusca* with *Azospirillum halopraeferans* (Reinhold *et al.*, 1988).

In contrast to symbiotic and free-living diazotrophic systems, reliable data are lacking on the efficiency of energy conversion in associative systems. On average, for free-living diazotrophic bacteria the figures vary between 10 and 50 mg N fixed per gram carbon utilized, and only for *Azospirillum* have higher figures ($\sim$92 mg N g^{-1} C) been found (Klucas, 1991).

Estimates as to the amount of N fixed by these associations vary largely, from a few kilograms to several hundred kilograms of nitrogen per hectare each year. The very high N_2 fixation rates are usually obtained from short-term studies of nitrogenase activity with the ARA, and should be regarded with caution. Measurements of ARA are a poor parameter for quantification of the N_2 fixation in symbiotic systems (Section 7.4) and even less suited for associative systems, where a direct transfer of fixed nitrogen from the bacteria to the host plant is lacking. It takes several days before nitrogen fixed in the roots can be detected in the shoots (Alexander and Zuberer, 1989). For quantification of the nitrogen in plants derived from associative N_2 fixation the ^{15}N isotope dilution technique is appropriate, so far a suitable, non-associative (non-fixing) reference plant is available (Miranda *et al.*, 1990; Chalk, 1991).

In Table 7.9 examples are given of estimates of the proportion of plant nitrogen derived from associative N_2 fixation. The range is large, and low (or zero) values may reflect conditions where, for example, the supply of mineral nitrogen was high, i.e. conditions prevailed which also inhibit or prevent symbiotic N_2 fixation (Fig. 7.11). In temperate species, most studies have been done with wheat, either uninoculated or inoculated with various diazotrophic species, and most data are in the range of 10% or below unless particular combinations of host cultivars and bacterial strains, for example, of *Azospirillum* are used when in pot cultures up to 39% of the plant nitrogen may derive from N_2 fixation (Bhattarai and Hess, 1993). However, under field conditions in cereals (C_3 plants) grown in temperate climate the amount of nitrogen in plants derived from associative N_2 fixation may be at most 5–10 kg N ha^{-1} per cropping season (Idris *et al.*, 1981). Often observed growth enhancement effects by inoculation with diazotrophic bacteria are attributable to other causes as for example, hormonal effects (see below).

The situation is different, however, in sugar cane and C_4 forage grasses in the tropics

Table 7.9
Estimates of Proportions of Nitrogen Derived from Rhizosphere
Associations in Nonlegumes[a]

Plant species	Proportion of total plant nitrogen
Rice (*Oryza sativa* L.)	0–35
Wheat (*Triticum aestivum* L.)	0–47
Sugar cane (*Saccharum* sp.)	2–56
	(60–80)[b]
Forage grasses	
Brachiaria humidicola	30–40
Leptochloa fusca	2–41

[a]Compiled data from Chalk (1991).
[b]Boddey *et al.* (1991).

and subtropics (Table 7.9). It is now well established that under field conditions in uninoculated soils a fairly large proportion of the plant nitrogen in sugar cane and C_4 forage grasses derives from rhizosphere associations. These associations are characterized by high host specificity. For example, of the 31 cultivars (ecotypes) of the tropical bahia grass (*Paspalum notatum*) only one (Batateis) produced root exudates which specifically stimulated growth of *Azotobacter paspali* (Döbereiner, 1983). In *Sorghum bicolor* (*nutans*), cultivar differences in nitrogenase activity by a factor of 10 were related to differences in root exudation rates and, presumably in the microaerophilic conditions in the rhizosphere (Werner *et al.*, 1989). In 25 ecotypes of *Panicum maximum* the proportion of nitrogen in the plants derived from associative N_2 fixation varied between 16% and 39%, and in one ecotype this proportion was zero (Table 7.10).

Sugar cane is the crop plant species with the highest potential of associative N_2 fixation. In long-term studies with 10 cultivars of sugar cane between 60% and 80% of the nitrogen in the plants derived from these associations (Boddey *et al.*, 1991), in other studies proportions of 49–70% have been found (Urquiaga *et al.*, 1992). Estimates of the potential of N_2 fixation in sugar cane cropping systems are in the range of more than

Table 7.10
Estimated Proportion of Nitrogen in Different Ecotypes of
Panicum maximum JACQ. Derived from Associative Nitrogen
Fixation (ANF)[a]

Ecotype	Proportion derived from ANF
T 84	39.4
T 110	29.7
T 29	23.2
K 190 B	15.9
KK 16	0

[a]Selected data from Miranda *et al.* (1990).

200 kg ha^{-1} (Boddey *et al.*, 1991). These estimates are supported, in principle, by experience in continuous sugar cane cropping systems where 100–200 kg N ha^{-1} is removed by each cane harvest without a decline in soil nitrogen content and with only small input of fertilizer nitrogen (Chalk, 1991; Urquiaga *et al.*, 1992). The particularly high effectivity of the associative N$_2$ fixation in sugar cane may occur for various reasons: high quantity and quality (C$_4$ carboxylic acids?) of root exudates; high soil temperatures and soil moisture levels and, thus, microaerophilic conditions in the rhizosphere; and, most important, the association with a particular bacteria. In sugar cane in addition to *Azospirillum*, a new diazotrophic bacterium, *Acetobacter diazotrophicus* has been found not only in the rhizosphere but also the root cortex and the stems (Gillis *et al.*, 1989; Boddey *et al.*, 1991). The nitrogenase activity of this bacterium is resistant to fairly high ammonium concentrations and the bacterium behaves like an endophyte, for infection either damaged tissue or VA mycorrhiza as a 'biological carrier' seem to be required (Boddey *et al.*, 1991).

Many diazotrophic bacteria produce and also secrete phytohormones like auxins, cytokinins and gibberellins (Martin *et al.*, 1989; Cacciari *et al.*, 1989; Jagnow *et al.*, 1991) and thereby enhance net proton excretion in wheat roots (Bashan, 1990) and growth of roots and shoots, for example of maize (Nieto and Frankenberger, 1991), chickpea (Del Gallo and Fabbri, 1991) and *Casuarina* (Rodriguez-Barrueco *et al.*, 1991). An example of the effects of *Azospirillum brasilense* on root morphology of soil-grown wheat plants is shown in Section 14.5.3.

These hormone-induced changes in root morphology by diazotrophic bacteria may have important consequences for acquisition of sparingly soluble mineral nutrients such as phosphorus (Section 13.2), as well as on other rhizosphere microorganisms, both pathogens and growth-promoting bacteria (Jagnow, 1990; Jagnow *et al.*, 1991). These changes brought about by hormonal effects are most likely also the reason for enhanced nodulation in legumes after inoculation with *Azospirillum* (Martin *et al.*, 1989; Del Gallo and Fabbri, 1991).

During the last decade many field experiments have been conducted on the effect of inoculation particularly with *Azospirillum* and *Azotobacter* on the growth and yield of several temperate and tropical crop species (Jagnow, 1987). The results are highly variable, but increases in grain yield of up to about 40% in wheat and barley (Jagnow *et al.*, 1991), 33% in pearl millet (Wani *et al.*, 1988) or 20% in sorghum (Okon *et al.*, 1988) may be achieved by inoculation. Whether this yield increase is attributable to N$_2$ fixation, hormonal or other effects on growth and development, remains unclear (Wani *et al.*, 1988; Jagnow *et al.*, 1991). However, substantial evidence is in favour of hormonal effects (Okon *et al.*, 1988; Martin *et al.*, 1989; Saric *et al.*, 1990), supporting results from former field experiments in wheat where an increase in grain yield by inoculation with *Azospirillum* was also achieved by foliar application (Rynders and Vlassak, 1982), and was independent of the level of mineral nitrogen fertilizer supplied (Avivi and Feldman, 1982; Rynders and Vlassak, 1982).

In conclusion, diazotrophic associative bacteria may enhance host plant growth either by N$_2$ fixation and improving plant nitrogen nutrition, or by production of phytohormones and thereby altering root morphology and physiology as well as root and shoot growth and developmental processes, or by both (Jagnow *et al.*, 1991). The relative importance of the two components may depend on many factors, the plant

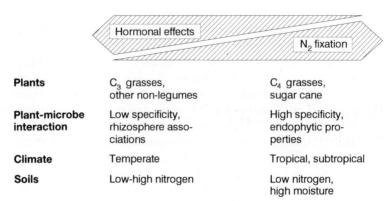

Plants	C$_3$ grasses, other non-legumes	C$_4$ grasses, sugar cane
Plant-microbe interaction	Low specificity, rhizosphere asso- ciations	High specificity, endophytic pro- perties
Climate	Temperate	Tropical, subtropical
Soils	Low-high nitrogen	Low nitrogen, high moisture

Fig. 7.12 Tentative scheme of the relative importance of hormonal effects and N$_2$ fixation in associations between diazotrophic bacteria and plant roots.

genotype within a given species in particular (Table 7.10). However, as schematically summarized in Fig. 7.12, some general predictions are possible, for example, the dominance of hormonal effects in plants grown in temperate climates and possessing mainly less specific rhizosphere associations (Martin *et al.*, 1989). In contrast, in C$_4$ forage grasses and in sugar cane with highly specific associations and grown under ecological conditions which favour associative N$_2$ fixation (high soil temperature and water content, high irradiation and root exudation) the growth enhancement by N$_2$ fixation dominates. The principles of the pattern shown in Fig. 7.12 are most likely not confined to cultivated plants but also hold true for the natural vegetation.

7.7 Outlook

In the near future, the main potential for increasing biological N$_2$ fixation in agricultural and forest ecosystems lies in the improvement of existing systems (Quispel, 1991; Bohlool *et al.*, 1992). Improvements can be achieved by overcoming limitations in symbiotic systems such as lack of proper bacterial strains or constraints imposed by mineral nutrient deficiency, soil acidity, and excessive levels of mineral fertilizers. In associative systems the potential is presumably higher in selection of proper cultivars (Table 7.10) and cropping systems, e.g., incorporation of crop residues with large C/N ratio (Kretzschmar *et al.*, 1991; Hafner *et al.*, 1993), than in inoculation with diazotro- phic bacteria. In the long run there may be the possibility to increase the efficiency of N$_2$ fixation in legumes by genetic manipulation of the rhizobia. However, the more challenging prospect is to transform associative diazotroph bacteria like *Azospirillum* into more efficient endophytes as has already been achieved in sugar cane, or to manipulate *Rhizobium* and other diazotrophs to also express their *nod*-gene activity in nonlegumes and form nodules ('paranodules') with nitrogenase activity as has been demonstrated, for example, in wheat, rape and rice (Quispel, 1991; Kennedy and Tchan, 1992).

8

Functions of Mineral Nutrients: Macronutrients

8.1 Classification and Principles of Action of Mineral Nutrients

By definition, mineral nutrients have specific and essential functions in plant metabolism. Depending on how great the growth requirement for a given nutrient, the nutrient is referred to as either a *macronutrient* or a *micronutrient*. Another classification, based on physicochemical properties, divides nutrients into *metals* (potassium, calcium, magnesium, iron, manganese, zinc, copper, molybdenum, nickel) and *nonmetals* (nitrogen, sulfur, phosphorus, boron, chlorine). Both classifications are inadequate since each mineral nutrient can perform a variety of functions, and some of these functions are only loosely correlated to either quantity of requirement or physicochemical properties. A mineral nutrient can function as a constituent of an organic structure, as an activator of enzyme reactions, or as a charge carrier and osmoregulator. In this book the more common classification of macro- and micronutrients is preferred.

The main functions of mineral nutrients such as nitrogen, sulfur, and phosphorus that serve as constituents of proteins and nucleic acids are quite evident and readily described. Other mineral nutrients, such as magnesium and the micronutrients (except chlorine), may function as constituents of organic structures, predominantly of enzyme molecules, where they are either directly or indirectly involved in the catalytic function of the enzymes. Potassium, and presumably chlorine, are the only mineral nutrients that are not constituents of organic structures. They function mainly in osmoregulation (e.g., in vacuoles), the maintenance of electrochemical equilibria in cells and their compartments and the regulation of enzyme activities. Naturally, because of their low concentrations, micronutrients do not play a direct role in either osmoregulation or the maintenance of electrochemical equilibria.

The different types of functions that mineral elements perform in enzyme reactions require further comment. Nitrogen and sulfur are integral constituents of protein structure and thus of *apoenzymes* (Fig. 8.1). For the catalytic reaction of the majority of enzymes, a nonprotein component is required, namely, a coenzyme, a prosthetic group, or a metal component. The difference between coenzymes and prosthetic groups is primarily a matter of convention. Typical coenzymes are ATP and FAD; typical prosthetic groups are chlorophyll, cytochromes, and nitrogenase, in which a

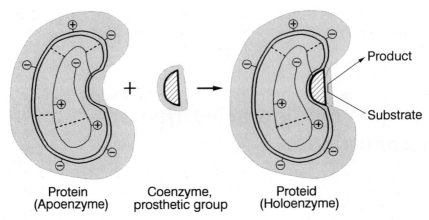

Protein	Coenzyme,	Proteid
(Apoenzyme)	prosthetic group	(Holoenzyme)

Fig. 8.1 Schematic presentation of the components of an enzyme molecule. Shaded area: hydration shells of water molecules (cluster).

metal acts as the functional group. In several enzymes the prosthetic group resembles the metal component only. Examples of this are given in Chapter 9. Most of the metal atoms integrated into metalloproteins are transition metals, which perform their catalytic function through a change in valency. This is the case for iron in cytochromes, copper in plastocyanin, and molybdenum in nitrogenase. In some enzymes, however, the metal performs its catalytic function by forming an enzyme–substrate–metal complex (e.g., magnesium in ATPase). Detailed information on the enzyme function of heavy metals can be found in Sandmann and Böger (1983).

In recent years phosphorylation of enzyme proteins (apoenzymes) for regulation of their activity has attracted great interest. A large number of enzymes, for example PEP carboxylase and H^+-ATPase, undergo reversible phosphorylation (Section 8.4), and *protein phosphorylation* is a general mechanism for regulating cellular metabolism and development (Budde and Randall, 1990). Enzymes which catalyze the phosphorylation of a protein are referred to as *protein kinases*. Effectors which regulate protein kinases are, for example, polyamines, calmodulin and, in particular, free Ca^{2+}. In plants more than 30 protein kinases have been identified and 10 of these are dependent on Ca^{2+} (Budde and Randall, 1990). This mechanism offers, at least for some mineral nutrients, the possibility of functioning as a component of a signal chain. Examples of involvement of mineral nutrients in this type of regulation of enzyme activities are given in the relevant sections.

Mineral nutrients can have a dominant role in enzyme reactions by other means also. Potassium is a typical example of a mineral nutrient that exerts its regulatory function by changing the conformation of the protein component of the enzyme. Proteins are charged macromolecules which are highly hydrated in living, metabolically active cells (Fig. 8.1). Because of intermolecular hydrogen bonding, water molecules form partial but nonpermanent associations ('structures' or 'clusters') and thus have a stabilizing effect on protein conformation. Solutes, including mineral nutrients, alter the physical properties of the solvent water through the formation of hydration shells around the ions, as well as the properties of the protein molecule through interactions, particularly with the charged groups of the macromolecule (electrostatic interaction). The hydration, stability, and conformation of enzymes or other biopolymers (e.g., membranes) are therefore affected not only by temperature and pH but also by the type (cation or anion, and their valencies) and concentration of the mineral nutrients. The conformation (spatial orientation) of an enzyme is again a crucial factor for both the affinity between the active centre of the enzyme and the substrate (K_m value) and the turnover rate of the enzyme (V_{max}; see Fig. 2.10). Potassium, as the major cytoplasmic cation (Section 8.7), has a prominent effect on the conformation of enzymes and thus regulates the activities of a large number of enzymes. For details on the interactions between inorganic ions and enzymes, see Wyn Jones and Pollard (1983).

The distribution of mineral nutrients between different types of cells within a given tissue (e.g., epidermis cells, guard cells, mesophyll cells of a leaf) also provides important informations about the functions of mineral nutrients. This is particularly true for the distribution of ions in different cellular compartments (Leigh and Wyn Jones, 1986). In the last decade much progress has been made in this respect by applying techniques such as X-ray microanalysis, NMR (nuclear magnetic resonance), ion-selective microelectrodes, or fluorescent dyes, for studies on ion distribution in the cytoplasm and the organelles contained within it (e.g., chloroplasts) and the vacuole. New insight into the functions of mineral nutrients as for example of calcium as a second messenger, is based on these studies of cellular compartmentation.

Similar progress leading to a better understanding of the functions of mineral nutrients is to be expected in the future from research not only in comparing genotypes or mutants within a given plant species but also by introducing modern approaches and the techniques of molecular biology and genetics into studies on the mineral nutrition of plants. Promising results on this line have already been obtained.

In this chapter typical examples are given of the various functions of macronutrients, while micronutrients are dealt with in Chapter 9. Mineral elements that either replace certain mineral nutrients in some of their functions (e.g., sodium replacing potassium) or stimulate growth by other means are discussed in Chapter 10.

8.2 Nitrogen

8.2.1 Assimilation of Nitrogen

Nitrate and ammonium are the major sources of inorganic nitrogen taken up by the roots of higher plants. Most of the ammonium has to be incorporated into organic compounds in the roots (see Section 8.2.1.2), whereas nitrate is readily mobile in the xylem and can also be stored in the vacuoles of roots, shoots, and storage organs. Nitrate accumulation in vacuoles can be of considerable importance for cation–anion balance (Section 2.5.3), for osmoregulation, particularly in so-called 'nitrophilic' species such as *Chenopodium album* and *Urtica dioica* (Smirnoff and Stewart, 1985), and for the quality of vegetable and forage plants. However, in order to be incorporated into organic structures and to fulfil its essential functions as a plant nutrient, nitrate has to be reduced to ammonia. The importance of the reduction and assimilation of nitrate for plant life is similar to that of the reduction and assimilation of CO_2 in photosynthesis.

8.2.1.1 Nitrate Reduction and Assimilation

Mechanism. Nitrate reduction in higher as well as in lower plants follows the reaction:

$$NO_3^- + 8\,H^+ + 8\,e^- \rightarrow NH_3 + 2\,H_2O + OH^-$$

Some bacteria use nitrate as an electron acceptor under anaerobic conditions ('nitrate respiration') and produce nitrogenous gases (N_2, N_2O and NO_x), a process which causes a considerable loss of combined nitrogen from soils by denitrification.

The reduction of nitrate to ammonia is mediated by two enzymes: *nitrate reductase* (NR), which involves the two-electron reduction of nitrate to nitrite, and *nitrite reductase* (NiR) which transforms nitrite to ammonia in a six-electron reduction (Fig. 8.2). In higher plants nitrate reductase is a complex enzyme containing two identical subunits of the one shown in Fig. 8.2, i.e. it exists as a dimer. In microorganisms tetramers may also occur, explaining the differences in molecular weights of nitrate reductases between about 200 kDa and about 500 kDa (Warner and Kleinhofs, 1992). Each subunit can function separately in reduction of nitrate, and contains three prosthetic groups, flavine adenine dinucleotide (FAD), cytochrome 557 (Cyt_c) and molybdenum cofactor (MoCo). The $FADH_2$ prosthetic group of nitrate reductase can also reduce other electron acceptors such as Fe(III) provided as citrate or cyanide, i.e., show 'diaphorase activity' (Campbell and Redinbaugh, 1984). There are, however, doubts that the $FADH_2$ of the 'inducible nitrate reductase' is involved in Fe(III) reduction (Corzo *et al.*, 1991).

Nitrite reductase is a monomeric polypeptide of about 60–63 kDa containing a siroheme prosthetic group. In contrast to nitrate reductase which is localized in the cytoplasm, nitrite reductase is localized in the chloroplasts in leaves and in the proplastids of roots and other nongreen tissue (Oaks and Hirel, 1985). In green leaves, the electron donor is reduced ferredoxin, generated in the light by photosystem I (Chapter 5). In the dark and particularly in the roots and other nongreen tissue a protein similar to ferredoxin may serve in this function (Solomonson and Barber, 1990)

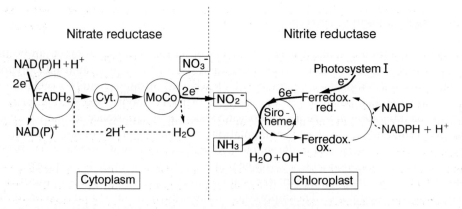

Fig. 8.2 Schematic representation of the sequence of nitrate assimilation in leaf cells. (Based on Beevers and Hageman, 1983 and Warner and Kleinhofs, 1992.)

Table 8.1

Effect of Pretreatment with Molybdenum on Nitrate Reductase Activity in Wheat Leaf Segments[a]

Molybdenum supply during plant growth (μg per plant)	Pretreatment of leaf segments (μg Mo l^{-1})	Nitrate reductase activity (μmol NO$_2^-$ g^{-1} fresh wt) after	
		24 h	70 h
0.005	0	0.2	0.3
0.005	100	2.8	4.2
5.0	0	–	8.0
5.0	100	–	8.2

[a]From Randall (1969).

and energy for production of reducing equivalents is provided by glycolysis (Bowsher *et al.*, 1989).

Despite the spatial separation of nitrate reductase and nitrite reductase (Fig. 8.2), as a rule, nitrite rarely accumulates in intact plants under normal conditions. Root nodules of legumes are obviously exceptions to this rule of synchronization of the activity of both enzymes (Section 7.4.6). Certain herbicides such as diuron strongly and selectively inhibit nitrite reductase in leaves and correspondingly increase nitrite content in the tissue (Peirson and Elliott, 1981).

In C$_4$ plants, mesophyll and bundle sheath cells differ in their functions not only in CO$_2$ assimilation (Section 5.2.4) but also in nitrate assimilation. Both, nitrate reductase and nitrite reductase are localized in the mesophyll cells and are absent in the bundle sheath cells (Vaughn and Campbell, 1988). This 'division of labour' in C$_4$ plants, whereby mesophyll cells utilize light energy for nitrate reduction and assimilation and bundle sheath cells for CO$_2$ reduction, is most probably the cause for higher photosynthetic nitrogen use efficiency (NUE) in C$_4$ compared with C$_3$ plants (Moore and Black, 1979). Because of the particular CO$_2$ concentration mechanism in the bundle sheath cells, less RuBP carboxylase (Rubisco) is required in C$_4$ than in C$_3$ plants. In C$_3$ plants nitrogen in Rubisco accounts for 20–30% of the total leaf nitrogen compared with less than 10% in C$_4$ plants, plus 2–5% nitrogen for PEP carboxylase in C$_4$ plants (Sage *et al.*, 1987).

Nitrate reductase is an enzyme that is regulated by several different modes exerted at different levels, namely, enzyme synthesis and degradation, reversible inactivation, and concentration of substrate and effectors (Solomonson and Barber, 1990). The enzyme has a half-life of only a few hours (Beevers and Hageman, 1983), and in plants receiving no nitrate it is merely absent (Li and Gresshoff, 1990). Nitrate reductase can be induced within a few hours by addition of nitrate (Oaks *et al.*, 1972) and suppressed by certain amino acids (Breteler and Smit, 1974; Oaks, 1991).

As would be expected, nitrate reductase activity is very low in molybdenum-deficient plants (Table 8.1). Incubation of deficient leaf segments in solutions containing molybdenum markedly increases enzyme activity. The enzyme can even be reactivated *in vitro* if the molybdenum-free apoenzyme is treated with molybdenum-containing complexes (Notton and Hewitt, 1979). The striking differences in the nitrate reductase

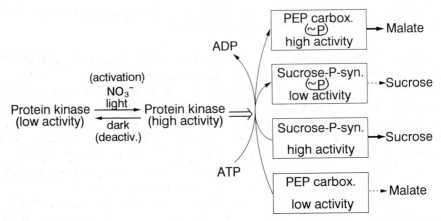

Fig. 8.3 Model of NO_3^- and light as 'signal' for protein phosphorylation and deactivation of sucrose-P-synthase and activation of PEP carboxylase. (Modified from Champigny and Foyer, 1992; reprinted by permission of the American Society of Plant Physiologists.)

activities of molybdenum-deficient and molybdenum-sufficient plants and the rapid response to molybdenum supply can be used to determine the molybdenum nutritional status of plants (Witt and Jungk, 1977; see also Chapter 12).

The high turnover rate of nitrate reductase and the distinct modulation of its activity by effectors such as light, nitrate, or phytohormones, has initiated many studies using this enzyme as a model for gene regulation by environmental factors in general (Section 5.6.3) and nitrate in particular. The well-known increase in nitrate reductase activity by cytokinin (CYT) is expressed at the level of nitrate reductase mRNA which is increased by CYT but suppressed by ABA (Lu et al., 1992). Nitrate not only induces nitrate reductase but also nitrite reductase (Aslam and Huffaker, 1989) by altering gene expression mainly by enhancing transcription of the respective genes (Redinbaugh and Campbell, 1991). Decrease in the use efficiency of the nitrate reductase transcript for production of the nitrate reductase protein is obviously the main factor responsible for the much lower activity of nitrate reductase in old compared with young leaves (Kenis et al., 1992; Fig. 8.5). The decrease in CYT import into older leaves (Section 5.6.5) might be involved in this age-dependent decline in transcription.

In addition to its function in inducing synthesis of nitrate reductase, nitrate, together with light, might act as a 'signal' altering the partitioning of photosynthetic carbon flow in leaves (Fig. 8.3). For assimilation of ammonia there is a high demand for carbon skeletons and it can be assumed that competition occurs between sucrose synthesis and amino acid synthesis as has been shown in tomato leaves (Hucklesby and Blanke, 1992). The carbon flow between both pathways seems to be regulated via cytosolic protein kinases which modulate the activity of two key enzymes, sucrose-P synthase and PEP carboxylase, by phosphorylation. Nitrate itself functions as a signal metabolite (Fig. 8.3). However, both enzymes respond to phosphorylation in an inverse fashion, sucrose-P-synthase is inactivated and PEP carboxylase activated, and thereby there is photo-synthate partitioning from preferential sucrose synthesis towards amino acid synthesis. A high PEP carboxylase activity is required to replenish tricarboxylic acids provided by the TCA for ammonia assimilation, according to the principle shown for assimilation of excessive cations taken up by roots (Section 2.5.4).

Localization in Roots and Shoots. In most plant species both roots and shoots are capable of nitrate reduction, and roots may reduce between 5 and 95% of the nitrate taken up. The proportion of reduction carried out in roots and shoots depends on various factors, including the level of nitrate supply, the plant species, the plant age, and has important consequences for the mineral nutrition and carbon economy of plants. In general, when the external nitrate supply is low, a high proportion of nitrate is reduced in the roots. With an increasing supply of nitrate, the capacity for nitrate reduction in the roots becomes a limiting factor and an increasing proportion of the total nitrogen is translocated to the shoots in the form of nitrate (Fig. 8.4).

Striking differences exist between plant species in both, the proportion of nitrate reduced in the roots and the response to increasing external nitrate concentrations. For example, at a supply of 1 mM nitrate the proportion of xylem sap nitrogen as nitrate was 15% in white clover (*Trifolium repens*), 33% in pea (*Pisum sativum*) and 59% in *Xanthium stumarium*, and at 10 mM nitrate these proportions increased to 66%, 41% and 69%, respectively (Andrews, 1986b). However, as discussed in Section 3.4.4, recycling of reduced nitrogen can be an important component of reduced nitrogen in the xylem sap and, thus, the percentage of the total N as nitrate in the xylem sap is not a very reliable parameter of the proportion of root reduction of nitrate. Nevertheless, there is a general pattern between plant species in partitioning of nitrate reduction and assimilation between roots and shoots (Andrews, 1986a). In temperate perennial species such as maritime pine (Scheromm and Plassard, 1988) or peach (Gojon *et al.*, 1991a), as well as in temperate annual legumes, virtually all (perennials) or most (legumes) of the nitrate is reduced in the roots when the external concentration is not much higher than 1 mM. In contrast, as suggested by Andrews (1986a), tropical and subtropical annual and perennial species tend to reduce a fairly large proportion of the nitrate in the shoots, even at low external supply, and the proportion between root and shoot reduction remains similar with increase as the external concentration is increased. This suggestion cannot be generalized, however, as for example in Australian open-

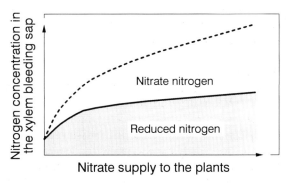

Fig. 8.4 Schematic representation of the effect of the level of nitrate supply in the rooting medium of noninoculated field pea (*Pisum arvensis* L.) on the nitrogen compounds in the xylem sap of decapitated plants. (Data recalculated from Wallace and Pate, 1965.)

forest plants (Stewart *et al.*, 1990) or woody plants growing in cerrado and forest communities in Brazil (Stewart *et al.*, 1992), at least in some under- and overstorey species the capacity for nitrate reduction in the leaves is low compared to the roots.

For a given species, the proportion of nitrate reduced in the roots increases with temperature (Theodorides and Pearson, 1982) and plant age (Hunter *et al.*, 1982). The uptake rate of the accompanying cation also affects this proportion. With potassium as the accompanying cation, translocation of both potassium and nitrate to the shoots is rapid; correspondingly, nitrate reduction in the roots is relatively low. In contrast, when calcium or sodium is the accompanying cation, nitrate reduction in the roots is considerably higher (Rufty *et al.*, 1981; Förster and Jeschke, 1993).

The preferential site of nitrate reduction, roots or shoots, may have an important impact on carbon economy of plants, and probably also have ecological consequences for the adaptation of plants to low-light and high-light conditions (Smirnoff and Stewart, 1985). Reduction and assimilation of nitrate have a high energy requirement and are costly processes when carried out in roots (Section 8.2.1.1). When expressed in ATP equivalents, this requirement represents 15 mol ATP for the reduction of one mole of NO_3^- and an additional of 5 mol ATP for ammonia assimilation (Salsac *et al.*, 1987). In barley, where a high proportion of nitrate reduction occurs in the roots, up to 23% of the energy from root respiration is required for absorption (5%), reduction (15%) and assimilation of the reduced nitrogen (3%), as compared with only 14% for assimilation when ammonium nitrogen is supplied (Bloom *et al.*, 1992). In contrast, for nitrate reduction in leaves reducing equivalents can be directly provided by photosystem I and ATP from phosphorylation. Under low-light conditions or in fruiting plants (Hucklesby and Blanke, 1992) this may lead to competition between CO_2 and nitrate reduction. On the other hand, under high-light conditions and excessive light absorption (photoinhibition, photooxidation; Section 5.2.2) nitrate reduction in leaves may not only use energy reserves but also alleviate high-light stress.

Leaf Age. During the ontogenesis of an individual leaf, a typical pattern is observed in nitrate reductase activity (Fig. 8.5). Maximum activity occurs when the rate of leaf expansion is maximal. Thereafter, the activity declines rapidly. Thus, in fully expanded leaves, nitrate reductase activity is usually very low, and often the nitrate levels are correspondingly high (Santoro and Magalhaes, 1983). This age-dependent pattern in nitrate reductase activity is also typical for cell cultures (Maki *et al.*, 1986), and reasons for this have been discussed above. In roots, nitrate reductase activity is high in expanding cells of the apical zones and declines rapidly towards the basal root zones (Dudel and Kohl, 1974).

Because of low phloem mobility of nitrate (Chapter 3), in fully expanded leaves with low nitrate reductase activity (Fig. 8.5), high nitrate contents are of limited use for the nitrogen metabolism of plants. Furthermore, in the individual cells nitrate is stored nearly exclusively in the vacuoles (Martinoia *et al.*, 1981). Although the rate of release of nitrate from the vacuoles in leaf cells is increased by decreasing nitrate import into the leaf (Gojon *et al.*, 1991b), the release from the vacuole into the cytoplasm can become a rate limiting step for nitrate reduction (Rufty *et al.*, 1982c), and thus for the utilization of stored nitrate nitrogen in growth processes (Clement *et al.*, 1979). Interruption of the nitrate supply to the roots may therefore lead to a drop in both

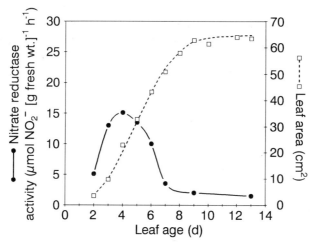

Fig. 8.5 Time course of nitrate reductase activity and leaf area development during the ontogeny of the first trifoliate leaf of soybean. (Modified from Santoro and Magalhaes, 1983.)

nitrate reductase activity in the leaves and shoot growth rate, despite a still high nitrate content in the shoot (Blom-Zandstra and Lampe, 1983). These results have important consequences for the timing of nitrate fertilizer supply.

Light. In green leaves a close correlation exists between light intensity and nitrate reduction. For example, there is a distinct diurnal pattern of reduction in the shoots but not in the roots (Fig. 8.6). The daytime proportion of nitrate reduction in shoots and roots therefore differs from the proportion at night (Aslam and Huffaker, 1982).

The rate of nitrate reduction in leaves is affected by light in various ways. Diurnal rhythm in nitrate reduction (e.g. Fig. 8.6) may reflect fluctuations in carbohydrate level

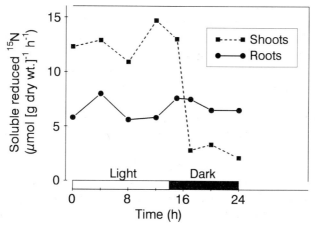

Fig. 8.6 Accumulation of soluble reduced ^{15}N in maize during a 24 h period of $^{15}NO_3$-supply to the roots. (Based on Pearson *et al.*, 1981.)

Table 8.2
Time Course of Nitrate Content in Spinach Leaves during the
Light Period from 9:00 to 18:00[a]

Time of day	Concentration of nitrate N ($mg\ kg^{-1}$ fresh wt)	
	Leaf blade	Petioles
8:30	228.2	830.2
light 9:30	166.6	725.1
light 13:30	100.8	546.0
light 17:30	91.0	504.0
18:30	106.4	578.2

[a]Steingröver *et al.* (1982).

(Aslam and Huffaker, 1984) and in the corresponding supply of reducing equivalents and carbon skeletons. However, besides these coarse regulations various mechanisms of fine regulation exist, on the level of enzyme modulation in carbon partitioning (Fig. 8.3), or direct modulation of the nitrate reductase by enzyme phosphorylation (Kaiser and Spill, 1991). In the light–dark transition this inactivation of nitrate reductase occurs within a few minutes and, thus, accumulation of nitrite is prevented (Riens and Heldt, 1992).

In plants with preferential nitrate reduction in the leaves, diurnal fluctuations in nitrate reductase activity may lead to a distinct decrease in nitrate content during the light period (Table 8.2). Irrespective of this light effect the nitrate content of petioles of spinach like that of many other nitrate-storing plant species, is higher than that of the leaf blades. Plants cultivated permanently under low-light conditions (e.g., in glasshouses during winter) may contain nitrate concentrations that are several times higher than those of plants grown under high-light conditions (e.g., in an open field during the summer). Shortage in reducing equivalents and carbon skeletons, as well as feedback regulation from amino acids accumulated as a result of low demand for growth, are additional factors. This is particularly evident in certain vegetables such as spinach and other members of the Chenopodiaceae which have a high preference for nitrate accumulation in the shoots and which obviously use nitrate in vacuoles for osmoregulation. In these species, under low-light conditions, nitrate concentrations in the leaves can reach more than 6000 $mg\ kg^{-1}$ fresh wt, that is, ~100 mM (Wedler, 1980). Under these low-light conditions nitrate can fully replace sugars in their osmotic function, and the same is true for nitrate compensation by chloride (Veen and Kleinendorst, 1985).

Nitrate Assimilation and Osmoregulation. In those plant species where most or all nitrate assimilation occurs in the shoots, organic acid anions are synthesized in the cytoplasm and stored in the vacuole (Fig. 8.7) in order to maintain both cation–anion balance and intracellular pH. This might lead to osmotic problems if nitrate reduction were to continue after the termination of leaf cell expansion (Raven and Smith, 1976). However, several mechanisms exist for the removal of excess osmotic solutes from the shoot tissue:

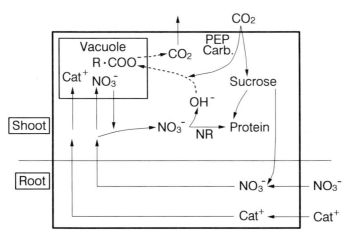

Fig. 8.7   Model of nitrate assimilation in shoots.

1. Precipitation of excess solutes in an osmotically inactive form. Synthesis of oxalic acid for charge compensation in nitrate reduction and precipitation as calcium oxalate are common in plants, including sugar beet (Egmond and Breteler, 1972).

2. Retranslocation of reduced nitrogen (amino acids and amides) together with phloem-mobile cations, such as potassium and magnesium, to areas of new growth.

3. Retranslocation of organic acid anions, preferentially malate, together with potassium into the roots and release of an anion (OH^- or HCO_3^-) after decarboxylation. In exchange for the released anions, nitrate can be taken up without cations, because the endogenous potassium acts as a counterion for long-distance transport, a mechanism discussed in Chapter 3 (Fig. 3.12).

8.2.1.2   Assimilation of Ammonium

Whereas nitrate can be stored in vacuoles without detrimental effect, ammonium and in particular its equilibrium partner ammonia

$$NH_3 \text{ (dissolved in water)} \rightleftharpoons NH_4^+ + OH^-$$

are toxic at quite low concentrations. The formation of amino acids, amides and related compounds is the main pathway of detoxification of either ammonium ions taken up by the roots or ammonia derived from nitrate reduction or N_2 fixation. Whereas ammonium (NH_4^+) concentrations in the cytoplasm are usually below 15 μM there is evidence that substantial amounts of ammonium may be stored in the vacuoles where the low pH prevents formation of ammonia (Roberts and Pang, 1992).

The principal steps in the assimilation of ammonium ions supplied to the roots are uptake into the root cells and incorporation into amino acids and amides with a simultaneous release of protons for charge compensation (Fig. 8.8). The permeation of ammonia across the plasma membrane, with proton liberation occurring before the permeation, has been discussed as an alternative model (Mengel *et al.*, 1976). As shoots have a rather limited capacity for disposal of protons, nearly all of the ammonium taken

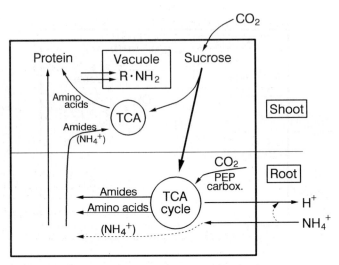

Fig. 8.8   Model of ammonium assimilation in roots.

up has to be assimilated in the roots (Raven, 1986), and the assimilated nitrogen transported in the xylem as amino acids and amides to the shoot (Fig. 8.8). In some plant species, such as lowland rice, a substantial proportion of the ammonium (NH_4^+) may be transported to, and assimilated in, the shoots (Magalhäes and Huber, 1989).

Ammonium assimilation in roots also has a large requirement for carbon skeletons for amino acid synthesis. These carbon skeletons are provided by the tricarboxylic acid cycle (TCA), and the removed intermediates have to be replenished by increased activity of PEP carboxylase (Fig. 8.8). With NH_4-N compared to NO_3-N supply the net carbon fixation in roots is up to threefold higher in rice and tomato (Ikeda *et al.*, 1992) and about fivefold higher in maize (Cramer *et al.*, 1993). The principles of this 'dark fixation' of CO_2 have been discussed in connection with excessive cation uptake (Fig. 2.20), N_2 fixation in nodulated legumes (Fig. 7.6) and nitrate assimilation (Fig. 8.7).

In order to minimize the carbon costs for root-to-shoot transport, the bulk of the nitrogen assimilated in the roots is transported in form of nitrogen-rich compounds with N/C ratios > 0.4 (Streeter, 1979). One, rarely two or more, of the following compounds dominate in the xylem exudate of the roots: the amides glutamine (2N/5C) and asparagine (2N/4C); the amino acid arginine (4N/6C); and the ureides allantoin and allantoic acid (4N/4C). In agreement with this model of carbon economy, in phloem transport to developing fruits, which are nonphotosynthetic sinks, amino acids with an N/C ratio of > 0.4 are the predominant transport forms of nitrogen (Pate, 1975).

The low-molecular-weight organic nitrogen compounds used predominantly for long-distance transport or for storage in individual cells differ among plant families (Table 8.3). In legumes in general and in soybean in particular, the majority of the fixed nitrogen transported in the xylem of nodulated roots is incorporated into the ureides allantoin and allantoic acid (Blevins, 1989).

Despite the different sites of ammonia assimilation (roots, root nodules, and leaves) the key enzymes involved are in each case glutamine synthetase and glutamate synthase

Table 8.3

Low-Molecular-Weight Organic Nitrogen Compounds which are Important Storage and Long-Distance Transport Forms

Compound	Plant family
Glutamine, asparagine	Gramineae
Glutamine	Ranunculaceae
Asparagine	Fagaceae
Arginine, glutamine	Rosaceae
Proline, allantoin	Papilionaceae
Betaine	Chenopodiaceae

Allantoin (ureide)

Betaine

(Fig. 8.9). Both enzymes have been found in roots, in chloroplasts, and in N_2-fixing microorganisms, and assimilation of most if not all ammonia derived from ammonium uptake, N_2 fixation, nitrate reduction, and photorespiration (Chapter 5) is mediated by the glutamine synthetase-glutamate synthase pathway (Blevins, 1989).

In this pathway the amino acid glutamate acts as the acceptor for ammonia, and the amide glutamine is formed (Fig. 8.9). Glutamine synthetase has a very high affinity for ammonia (low K_m value) and is thus capable of incorporating ammonia even if present in very low concentrations. Glutamine synthetase is activated by high pH and high

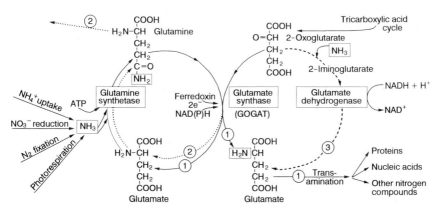

Fig. 8.9 Model of ammonia assimilation pathways (1,2) Glutamine synthetase–glutamate synthase pathway, with low NH_3 supply (1) and with high NH_3 supply (2). (3) Glutamate dehydrogenase pathway. GOGAT, Glutamine–oxoglutarate aminotransferase.

concentrations of both magnesium and ATP, and all three factors are increased in the chloroplast stroma upon illumination (see Section 8.5.4).

In chloroplasts, light-stimulated nitrate reduction and enhanced ammonia assimilation are efficiently coordinated through import of 2-oxoglutarate from the stroma and export of glutamate from the stroma of chloroplasts into the cytoplasm (Woo *et al.*, 1987). Efficient coordination is required to prevent high ammonia levels that might uncouple photophosphorylation (Krogmann *et al.*, 1959). Ammonia toxicity may be related to the rapid permeation of ammonia across biomembranes. For example, ammonia, but not ammonium (NH_4^+), diffuses rapidly across the outer membranes of chloroplasts (Heber *et al.*, 1974).

The other enzyme in ammonia assimilation, glutamate synthase (GOGAT), catalyzes the transfer of the amide group ($-NH_2$) from glutamine to 2-oxoglutarate, which is a product of the tricarboxylic acid cycle (Fig. 8.9). For this reaction either reduced ferredoxin (from photosystem I) or NAD(P)H (from respiration) is required. The reaction results in the production of two molecules of glutamate, of which one is required for the maintenance of the ammonia assimilation cycle and the other can be transported from the sites of assimilation and utilized elsewhere for biosynthesis of proteins, for example. As an alternative, when the ammonia supply is large, both glutamate molecules can act as an ammonia acceptor, and one molecule of glutamine leaves the cycle (pathway (2), Fig. 8.9).

Another enzyme, glutamate dehydrogenase (Fig. 8.9) which has a low affinity for ammonia (high K_m value), only becomes important in ammonia assimilation at very high supply of ammonium in combination with low pH of the nutrient solution, related to high contents of free ammonia in the root tissue (Magalhães and Huber, 1989).

8.2.2 Amino Acid and Protein Biosynthesis

The organically bound nitrogen of glutamate and glutamine can be used for the synthesis of other amides, as well as ureides, amino acids, amines, peptides, and high-molecular-weight peptides such as proteins. Although plants may contain up to 200 different amino acids, only about 20 of them are required for protein synthesis. The peptide chain of each protein has a genetically fixed amino acid sequence. The carbon skeletons for these different amino acids are derived mainly from intermediates of photosynthesis, glycolysis, and the tricarboxylic acid cycle (Fig. 8.10).

The transfer of the amino group from amino acids to other carbon skeletons – the transamination reaction – is catalyzed by aminotransferases, which are also referred to as transaminases:

$$
\underset{\text{Amino acid (A)}}{\underset{\displaystyle R_A}{\overset{\displaystyle NH_2}{H-C-COOH}}} + \underset{\text{2-oxo acid (B)}}{\underset{\displaystyle R_B}{\overset{\displaystyle O}{C-COOH}}} \;\underset{\text{transferases}}{\overset{\text{Amino-}}{\rightleftharpoons}}\; \underset{\text{2-oxo acid (A)}}{\underset{\displaystyle R_A}{\overset{\displaystyle O}{C-COOH}}} + \underset{\text{Amino acid (B)}}{\underset{\displaystyle R_B}{\overset{\displaystyle NH_2}{H-C-COOH}}}
$$

The prosthetic group of the transaminases is pyridoxal phosphate, a vitamin B_6 derivative. Higher plants contain a complete set of transaminases capable of shuttling amino groups between appropriate acceptors. Monogastric animals and humans rely on

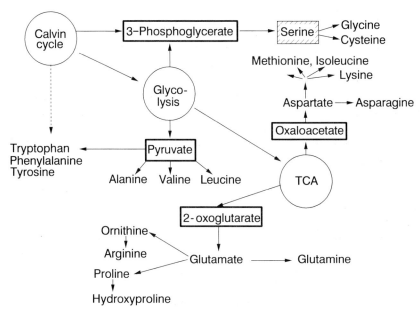

Fig. 8.10 Amino acid biosynthesis from various intermediates of the Calvin cycle, glycolysis and tricarboxylic acid cycle (TCA).

an external supply in the diet of both the transaminase prosthetic group (i.e., on vitamin B_6) and certain amino acids which cannot be synthesized and which therefore are 'essential' in the diet (e.g., valine, leucine, lysine, methionine, and tryptophan).

In protein synthesis the individual amino acids are coupled by peptide bonds (R_1-CO-NH-R_2) in a condensation reaction:

$$\text{Amino acid} \xrightleftharpoons[+H_2O]{-H_2O} \text{Dipeptide} \xrightleftharpoons[+H_2O]{-H_2O} \text{Polypeptide/protein}$$

Proteins are polypeptides formed from more than 100 individual amino acids, and their sequence is determined by the genetic information carried by double stranded molecules (Fig. 8.11). Expression of the genetic information starts in the nucleus of the cell with the synthesis of messenger ribonucleic acid (mRNA) which is a single stranded copy of an activated DNA fragment (transcription). The mRNA diffuses into the cytoplasm and becomes attached by ribosomes, the 'factories' of protein biosynthesis. By their distinct spatial structure the ribosomes allow two molecules of aminoacyl-tRNA to bind to the mRNA according to their codon at the top. Now the two amino acid residues are in position to be enzymatically linked by the formation of a peptide bonding. Subsequently, the ribosome moves on along the mRNA, releases the first tRNA and allows the next aminoacyl-tRNA to bind to the mRNA resulting in an elongation of the peptide. This translocation process is terminated when the ribosome reaches the stop signal demanding its breakdown and the release of the peptide chain.

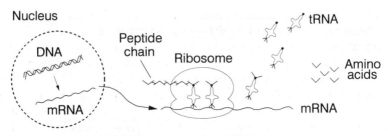

Fig. 8.11   Main steps of protein biosynthesis.

Mineral nutrients have an important role in these processes. The maintenance of the highly ordered ribosome structure requires divalent cations, Mg^{2+} in particular. Magnesium is also necessary for the activation of amino acids by ATP, and potassium is involved in the step of chain elongation. Zinc is the metal component of RNA polymerase, and iron is required in some way for ribosome integrity. These effects of individual mineral nutrients are discussed in more detail in later sections.

Synthesis and degradation of proteins occur simultaneously in cells and tissues. The turnover rates of proteins, i.e., their half-life time, differs widely, from a few hours to several days and weeks. Particularly high turnover rates are found in some enzyme proteins, for example, nitrate reductase and the D1 protein in photosystem II (Section 5.2.2). For a given protein, rate of synthesis and degradation are strongly modulated by endogenous and exogenous factors, including stage of development (e.g., leaf age), or phytohormones as components for transmitting environmental 'signals' to the level of gene expression and protein synthesis (Section 5.6; Fig. 5.17). Mineral nutrients have a strong influence on both synthesis and degradation, either directly (e.g., as a component of the enzyme) or indirectly via altering the membrane structure and, thus, compartmentation (e.g., calcium) or the phytohormone balance, which is particularly true for nitrogen.

8.2.3   Role of Low-Molecular-Weight Organic Nitrogen Compounds

In higher plants low-molecular-weight organic nitrogen compounds (Fig. 8.12) not only act as intermediates between the assimilation of inorganic nitrogen and the synthesis –

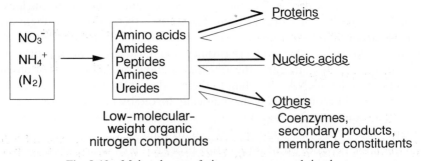

Fig. 8.12   Major classes of nitrogen compounds in plants.

or degradation – of high-molecular-weight compounds. They are also important for various other reasons as shown in the following examples. In contrast to lower plants, animals, and humans, higher plants are not capable of excreting substantial amounts of organically bound nitrogen, for example, as urea. Although plants can store large amounts of nitrate, they are not capable of reoxidizing organically bound nitrogen to nitrate, which would be a safe storage form, for example, in periods of enhanced protein degradation (e.g., senescing leaves). Of the low-molecular-weight nitrogen compounds mainly amino acids and amides serve this function as buffer and transient storage, in addition to their function in long distance transport of reduced nitrogen.

In contrast to amino acids and amides, the ureides allantoin and allantoate (allantoic acid) primarily serve the function of transport of nitrogen fixed in root nodules of certain legume species such as soybean (Fig. 7.8). It is not yet clear whether ureides are also synthesized in roots and leaves of nonnodulated legumes (Rice *et al.*, 1990). In the shoots, most of the ureides are degraded in the leaves and only a small proportion is directly transported to the developing pods and degraded there before being transported into the cotyledons (Coker and Schaefer, 1985). The pathways of degradation of ureides are shown in Fig. 8.13. In soybean the liberation of ammonia and CO_2 from degradation of ureides takes place without involvement of urea as an intermediate, i.e. via the allantoate amidohydrolase pathway (Fig. 8.13). A higher requirement of nickel as metal component of the urease (Section 9.5) would therefore not be expected in ureide-transporting nodulated legumes.

Another important class of low-molecular-weight organic nitrogen compounds are amines and polyamines, their biosynthesis being mediated by decarboxylation of amino acids. Amines are components of the lipid fraction of biomembranes (Section 2.3), the amino component ethanolamine, for example, is synthesized by decarboxylation of the

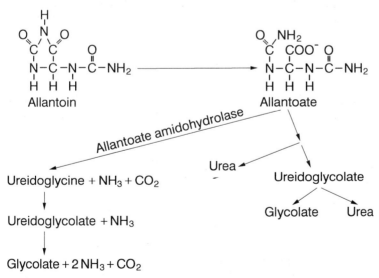

Fig. 8.13 Pathway of degradation of the ureides allantoin and allantoate in shoots of nodulated soybean. (Modified from Winkler *et al.*, 1988; reprinted with permission from Trends in Biochemical Sciences.)

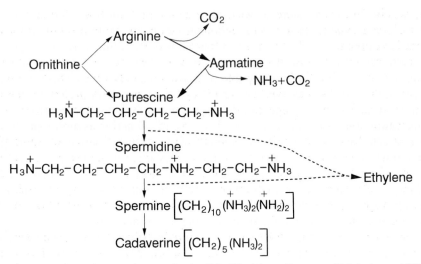

Fig. 8.14 Pathway of polyamine synthesis. (Modified from Evans and Malmberg, 1989.)

amino acid serine. More recently polyamines have attracted attention as secondary messengers (Section 5.6.3) and in the protection of membranes (Evans and Malmberg, 1989). In plants the amino acid arginine is the main precursor of polyamine synthesis (Fig. 8.14). Putrescine which is usually the dominant polyamine in plants, may constitute up to 1.2% of the plant dry matter (Smith, 1988). The polyamine content is particularly high in meristematic tissues, in plants supplied with high levels of ammonium (Gerendás and Sattelmacher, 1990), and under potassium deficiency (Klein *et al.*, 1979; Section 8.7).

Polyamines seem to serve many functions in plants, either in free or bound form, for example with various phenolics (Evans and Malmberg, 1989). They are causally involved in cell division, embryogenesis, and floral initiation and development, an example of the latter has been given in Table 6.3. Polyamines are also quite effective in delaying senescence of leaves by inhibiting the activity of acid proteinases (Shih *et al.*, 1982). Polyamines are also potentially involved in ethylene biosynthesis (Fig. 8.14). Many of the effects and functions of polyamines, including inhibition of ethylene synthesis (Miyazaki and Yang, 1987), are probably related to their capacity to act as scavengers of oxygen free radicals and thereby inhibit peroxidation of membrane lipids (Section 2.3) and, thus, stabilize biomembranes (Blevins, 1989; Evans and Malmberg, 1989).

Another low-molecular-weight organic nitrogen compound important for stabilization of cell structures and osmoregulation is betaine (also referred to as glycine betain, Table 8.3). Under salt or drought stress, the synthesis of betaine and its accumulation particularly in the cytoplasm is greatly enhanced where it acts as 'compatible solute' to counteract the osmotic perturbation caused by high vacuolar concentrations of inorganic ions such as chloride and sodium, which are incompatible with cytoplasmatic metabolism. Glycine betaine protects enzymes from inactivation by high NaCl concentrations (Fig. 16.24) and membranes against heat destabilization (Jolivet *et al.*, 1982). Low-molecular-weight organic nitrogen compounds are therefore important for the adaptation of plants to saline substrates (Section 16.6).

In response to exposure of plants to high concentrations of certain heavy metals, cadmium in particular, the synthesis of polypeptides is induced which have a very high content of the sulfur-containing amino acid cysteine and which bind large amounts of heavy metals. These 'phytochelatins' may play an important role in heavy metal detoxification (Section 8.3).

Peptides had also been considered to be involved in long-distance transport of heavy metals in the xylem. However, most of the heavy metals in the xylem are presumably present either as free cations or complexed with organic acids (White *et al.*, 1981a,b; Kochian, 1991). Copper is an exception, it is transported in the xylem exclusively in complexed form (Graham, 1979), most likely with organic nitrogen ligands, amino acids in particular (Kochian, 1991).

Not much is known of the role of the large number of nonproteinogenic amino acids in plants. However, at least some of them are important for mineral nutrition of plants. *Nicotianamine* is an effective chelator of Fe(II) and presumably plays a role in iron homeostasis, i.e., maintenance of a beneficial physiological stability of the iron level within cells (Stephan and Grün, 1989) and in phloem transport of iron and other heavy metals (Stephan and Scholz, 1993). In addition, nicotianamine is a precursor of a group of other nonproteinogenic amino acids, the so-called *phytosiderophores*, which are of particular importance for iron acquisition of graminaceous plant species (Section 9.1.6).

8.2.4 Ammonium versus Nitrate Nutrition

Whether ammonium or nitrate as sole source of nitrogen supply is better for plant growth and yield formation depends on many factors (Kirkby, 1981). An important factor is the plant species, with the general pattern that *calcifuges* – plants adapted to acid soils – and plants adapted to low soil redox potential (e.g., wetland rice) have a preference for ammonium (Ismunadji and Dijkshoorn, 1971). In contrast, *calcicoles* – plants with a preference for calcareous, high pH soils – utilize nitrate preferentially (Kirkby, 1967). However, as a rule, highest growth rates and plant yields are obtained by combined supply of both, ammonium and nitrate, for reasons summarized in the following.

As ammonium or nitrate comprise about 80% of the total cations and anions taken up by plants, the form of nitrogen supply has a strong but inverse impact on the uptake of other cations and anions, on cellular pH regulation and on rhizosphere pH (Section 2.5.4). The ammonium assimilation in roots produces about one proton per molecule ammonium taken up (Section 2.5.4) which has to be excreted into the external medium. At low external pH net excretion of protons is impaired and cytosolic pH may also fall (Gerendás *et al.*, 1990), explaining the relationship between growth retardation and pH decline in ammonium-fed plants (Findenegg *et al.*, 1982). Increases in content of polyamines at high ammonium supply and low external pH (Smith and Wilshire, 1975; Gerendás and Sattelmacher, 1990) are most likely a refection of pH homeostasis, as their synthesis is also stimulated by low pH (Smith and Sinclair, 1967). When

Table 8.4
Effect of pH and Nitrogen Source in the Nutrient Solution on the Assimilation and Transpiration Rate of Cucumber Plants[a]

pH	Nitrogen source (mM)		Ammonia[b]	Assimilation rate ($mg\ CO_2\ dm^{-2}\ h^{-1}$)	Transpiration rate ($g\ H_2O\ dm^{-2}\ h^{-1}$)
	Nitrate N	Ammonium N			
6.50	3	0	0	6.15	2.00
7.75	3	0	0	6.55	2.18
6.50	3	5	0.01	6.60	1.80
7.75	3	5	0.16	4.48	1.39

[a]Based on Schenk and Wehrmann (1979).
[b]Calculated NH_3 concentration in the aqueous solution.

rhizosphere pH is stabilized, using $CaCO_3$, polyamine contents remain low even with ammonium nutrition (Smith and Wilshire, 1975).

The high demand of carbon skeletons for ammonium assimilation in roots is reflected not only in higher activities of PEP carboxylase as compared with nitrate-fed plants (Arnozis *et al.*, 1988) but also in the approximate doubling of the rates of O_2 consumption per unit root weight (Matsumoto and Tamura, 1981). As a result, compared with nitrate-fed plants, the sugar content of those supplied with ammonium is lower in the roots and also decreases much more as the root zone temperature is increased. Accordingly, plant growth, particularly root growth, is poor in ammonium-fed plants when both root zone temperature and ammonium concentration are high (Kafkafi, 1990). These negative effects of sole ammonium nutrition are particularly distinct in combination with a very low supply of potassium (Lips *et al.*, 1990).

In a given plant species the suitability of ammonium for achieving high growth rates and yield therefore depends on root zone temperature (Clarkson and Warner, 1979) and other factors which determine carbohydrate supply to the roots such as, for example, light intensity (Lavoie *et al.*, 1992). In ammonium-fed plants not only low but also high substrate pH greater than 7 can become critical because of an increase in free ammonia concentrations in the substrate leading to ammonia toxicity (Table 8.4).

Compared with ammonium, nitrate has the advantage of also being a storage form in plants with no necessity to be assimilated in the roots. In addition, nitrate nutrition induces an increase rather than a decrease in rhizosphere pH (Section 2.5.4), and there is no risk of toxicity at alkaline pH. The advantage of lower demand of carbon skeletons in the roots however is lost in plant species which exclusively reduce and assimilate nitrate in the roots. Furthermore, an increase in rhizosphere pH resulting from nitrate nutrition might have negative side effects on mineral nutrient acquisition in alkaline soils (Section 16.5) and also on availability within the plants, for example, of iron (Alloush *et al.*, 1990).

The relative advantage of both forms of nitrogen on plant growth also strongly depends on the external concentrations (Fig. 8.15). At low concentrations the growth differences between ammonium and nitrate supply are small, and in plant species such as potato, adapted to low pH and, thus, ammonium nutrition, growth of ammonium-

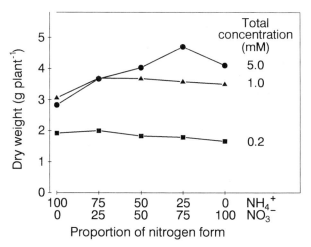

Fig. 8.15   Effect of form and concentration of nitrogen supplied on dry weight of 30-day-old maize plants. (Based on Xu *et al.*, 1992.)

fed plants might be better, particularly root growth, than of nitrate-fed plants (Gerendás and Sattelmacher, 1990). However, with increasing external concentrations, the advantage of nitrate as sole source of nitrogen increases (Gerendás and Sattelmacher, 1990) and the relative growth depression by sole ammonium supply becomes more distinct (Fig. 8.15). However, such high concentrations of ammonium, often applied in solution culture studies, are far beyond the concentrations occurring in soil solutions (Section 13.2.2). Thus, results from solution culture studies with ammonium concentrations much higher than about 1 mM are of rather limited relevance for soil-grown plants. In sand cultures supplied with nutrient solutions, this critical concentration range of ammonium is higher due to the development of depletion zones around roots (Section 13.2).

With a few exceptions, highest growth rates are achieved with mixed supply of both forms of nitrogen, the optimal proportions depend strongly on the total concentrations supplied (Fig. 8.15). This holds true in principle also for tissue cultures (Grimes and Hodges, 1990). When both forms of nitrogen are supplied, it is easier for the plant to regulate intracellular pH and to store some of the nitrogen at low energy costs. In higher plants a major factor contributing to the higher rates of vegetative and particularly reproductive growth are certainly related to the effects of ammonium on the phytohormone balance in plants (Section 6.3). In field-grown crop plants this particular effect of ammonium on reproductive growth can be achieved at least to some extent by repeated injection of ammonium fertilizer (Sommer and Six, 1982) or, more easily, by application of ammonium fertilizer together with nitrification inhibitors (Adriaanse and Human, 1990; Smiciklas and Below, 1992). Nitrification inhibitors may include naturally occurring compounds such as neem cake, obtained from *Pougamia glabra* (Thomas and Prasad, 1983), or synthetic ones, such as N-Serve or dicyanodiamide (Amberger, 1989).

Urea, another nitrogen source, can be classified according to its effect on plant metabolism and growth somewhere between nitrate and ammonium (Kirkby and Mengel, 1967). Urea can be taken up directly by the roots or aerial parts (Chapter 4); after being taken up by the roots (Hartel, 1977), it is rapidly hydrolyzed by urease either within the roots (e.g., in soybean) or after translocation to the shoots (e.g., in maize). In soils the hydrolysis of urea usually takes place before root uptake.

8.2.5 Nitrogen Supply, Plant Growth, and Plant Composition

Depending on the plant species, development stage, and organ, the nitrogen content required for optimal growth varies between 2 and 5% of the plant dry weight. When the supply is suboptimal, growth is retarded; nitrogen is mobilized in mature leaves and retranslocated to areas of new growth. Typical nitrogen-deficiency symptoms, such as enhanced senescence of older leaves, can be seen. An increase in the nitrogen supply not only delays senescence and stimulates growth but also changes plant morphology in a typical manner (Fig. 8.16), particularly if the nitrogen availability is high in the rooting medium during the early growth. An increase in shoot-root dry weight ratio with increase in nitrogen supply takes place in both perennial and annual plant species (Levin *et al.* 1989; Olsthoorn *et al.* 1991), and is more distinct in wheat than maize (Hocking and Meyer, 1991). This increase in shoot-root ratio might even be larger in terms of shoot and root length (Klemm, 1966), a shift which is unfavorable for acquisition of nutrients and water from soil at later growing stages.

Typical nitrogen-induced changes in leaf morphology are shown in Table 8.5 for rice. The length, width and area of the leaf blades increase, but the thickness decreases. In addition, the leaves become increasingly droopy, an effect that interferes with light interception.

In cereals the enhancement of stem elongation by nitrogen increases the susceptibility to lodging. This change in shoot morphology is less distinct with ammonium than

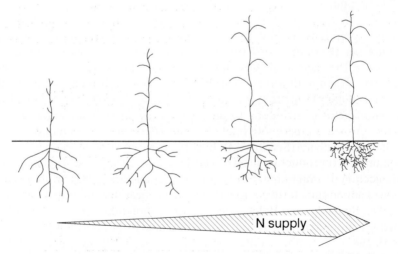

Fig. 8.16   Schematic representation of the effect of increasing levels of nitrogen supply to the roots during early growth stages on the root and shoot growth of cereal plants.

Table 8.5
Effect of Increasing Nitrogen Supply as NH_4NO_3 on Leaves of Rice

N supply $(mg\ l^{-1})$	Leaf blade			
	Length (cm)	Width (cm)	Area (cm^2)	Thickness $(mg\ cm^{-2})$
5	49.0	0.89	30.6	4.9
20	56.1	1.13	47.8	4.1
200	60.3	1.25	56.1	3.8

[a]Based on Yoshida et al. (1969).

with nitrate nutrition (Sommer and Six, 1982) and is presumably related to nitrogen-induced changes in the phytohormone balance (Chapter 5). In cereals this side effect of nitrogen on stem elongation can become the dominant yield limiting factor when high doses of nitrogen fertilizers are applied. Short stem length, which is obtained by breeding in most high-yielding cultivars, can also be induced by the application of growth retardants such as chlorocholine chloride (CCC; trade names: Chlormequat or Cycocel), counteracting the negative side effects of a large nitrogen supply (Table 8.6). The increase in grain yield obtained by CCC application is mainly, but not exclusively, the result of its counteractive effect on lodging. Other effects on development (tillering, flowering, etc.) are involved, as would be expected from the application of bioregulators that interfere with the phytohormone balance in plants by inhibiting the biosynthesis of gibberellins. In contrast to its inhibitory effect on shoot elongation, CCC is without effect on root elongation (Steen and Wünsche, 1990).

Nitrogen alters plant composition much more than any other mineral nutrient. For example, as shown in Table 8.7, the dry matter production of ryegrass is increased by nitrogen in a typical yield response curve. Simultaneously, the total nitrogen content increases, but the contents of the two main storage carbohydrates in grasses, polyfructosans and starch, decrease drastically. In contrast to the carbohydrate content, lignin

Table 8.6
Interacting Effects of Nitrogen Fertilizer Supply and Growth Regulation by Chlorocholine Chloride (CCC) on Lodging and Grain Yield of Winter Wheat[a]

N supply $(kg\ ha^{-1})$	Degree of lodging[b]		Grain yield $(t\ ha^{-1})$	
	−CCC	+CCC	−CCC	+CCC
0	2.4	1.0	3.97	4.18
80	4.8	1.2	4.71	5.13
120	5.8	1.8	4.67	5.13
160	6.3	1.7	4.80	5.31

[a]Based on Jung and Sturm (1966).
[b]1 = no lodging, 9 = total lodging.

Table 8.7
Effect of Increasing Nitrogen Supply as NH_4NO_3 on Dry
Matter Production and Composition of Ryegrass[a]

	Nitrogen supply (g per pot)			
	0.5	1.0	1.5	2.0
Dry matter (g per pot)	14.9	23.2	26.2	26.0
Composition (% dry wt)				
Total nitrogen	2.0	2.8	3.6	4.2
Sucrose	7.7	7.3	7.1	6.3
Polyfructosans	10.0	4.3	1.8	1.1
Starch	6.1	3.4	2.1	1.4
Cellulose	14.4	13.9	13.9	17.6

[a]From Hehl and Mengel (1972).

content in grasses may increase rather than decrease with high nitrogen supply (Kaltofen, 1988) as the amino acids phenylalanine and tyrosine are precursors of lignin synthesis. However, part of this enhancement effect on lignin content in mature tissues may be counteracted by enhanced formation of new leaves with their typically lower lignin content.

Whether these changes in plant composition represent an increase or decrease in quality depends on the further use of the plant material. The increase in total nitrogen content has to be interpreted with care. Total nitrogen and 'crude protein' (total nitrogen content multiplied by a factor between about 5.7 and 6.25, depending on plant species and source of crude protein) are the sum of both protein and soluble nitrogen, the latter including, for example, amino acids and amides as well as nitrate. In general, the proportion of soluble nitrogen increases with elevated levels of nitrogen supply and is higher in leaves and storage organs with high water content, but low in grains and seeds.

The shift in plant composition with increasing nitrogen supply (Table 8.7) reflects a competition for photosynthates among the various metabolic pathways. This competition is modulated by internal and external factors. Figure 8.17 summarizes the general pattern of variable nitrogen supply in relation to the composition of vegetative shoots. When the nitrogen supply is suboptimal, ammonia assimilation increases both the protein content and leaf growth and correspondingly the leaf area index (LAI). As long as the increase in LAI is correlated with an increase in net photosynthesis, the requirement of carbon skeletons for ammonia assimilation does not substantially depress other biosynthetic pathways related to carbohydrates (sugars, starch, cellulose, etc.), storage lipids, or oils. In this nitrogen concentration range, the plant composition does not change substantially, but the total production of plant constituents per unit surface area (e.g., per hectare) increases. As the nitrogen supply is further increased, however, a higher proportion of the assimilated nitrogen is sequestered in storage pools, as amides, for example, but in C_3 plants it is also stored in the chloroplasts in form of Rubisco (Millard, 1988).

Owing to mutual shading a further increase in LAI has no effect on the rates of net

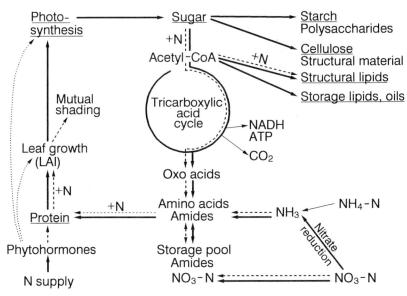

Fig. 8.17 Model of the effects of nitrogen supply on leaf growth and on various plant constituents. Key: ⟶, suboptimal to optimal nitrogen supply; −−−→, high to excessive nitrogen supply.

photosynthesis. Above this transition point, additional ammonia assimilation merely increases the nitrogen content at the expense of the content of most of the other major plant constituents. An interpretation of the effects of nitrogen on the growth and composition of plants requires, of course, a consideration of the effects of nitrogen supply on the phytohormone balance (Chapters 5 and 6).

In contrast to the content of storage lipids and oils, the content of lipids in green leaves is closely related to the nitrogen supply (Beringer, 1966). This apparent contradiction reflects a different type of regulation (Fig. 8.17). In leaves the majority of lipids are galactolipids, acting as structural components of the chloroplasts. Correspondingly, an enhancement of protein synthesis and chloroplast formation leads to an increase in the lipid content of leaves, as well as to an increase in chloroplast constituents such as chlorophyll and carotene, the provitamin A (Schulze, 1957). An increase in the protein content of vegetables or cereal grains is often positively correlated with an increase in the content of B vitamins: riboflavin, thiamine, and nicotinic acid (Dressel and Jung, 1979), indicating that a high proportion of this protein represents enzyme proteins in which these B vitamins act as prosthetic group (Section 8.1). The enhancing effect of nitrogen supplied to the roots on the formation and content of nicotine in tobacco (Schmid, 1967) is also indirect. The alkaloid nicotine is not only structurally closely related to cytokinins (adenine derivates) but is also synthesized only in the apical zones of growing roots. The enhancement of root growth in tobacco by nitrogen therefore increases the synthesis and export to the shoots of both cytokinins and nicotine.

Elevated nitrate contents in certain plant species and plant organs indicate imbalance between supply and demand for growth. Such elevated contents in plants at harvest are

Table 8.8
Effect of Additional Nitrogen Supply as NH_4NO_3 at Anthesis on Wheat Grains[a]

Protein fraction	Percentage of lysine N of total protein N		Amount of protein fractions (mg protein $(10\ g)^{-1}$ grain dry wt)	
	9.5^b	19.1^c	9.5^b	19.1^c
Albumin	4.05	4.00	13.6	13.3
Globulin	2.20	2.25	12.5	16.7
Prolamine	0.40	0.40	41.6	123.0
Glutelin	1.90	1.85	36.5	86.1

[a]From Ewald (1964).
[b]Percentage of crude protein in the dry weight (control).
[c]Percentage of crude protein in the dry weight (additional nitrogen supply).

uneconomical in terms of nitrogen utilization and also undesirable nutritionally. Nitrite might be formed from nitrate during either the storage or processing of vegetables. Infants fed these nitrite-containing foods run the serious risk of developing methemo-globinemia. In addition, nitrite increases the risk of formation of the carcinogenic nitrosamine (Tannenbaum et al., 1978; Heyns, 1979). High nitrate content (>1–2% in the dry weight) in forage can be toxic to ruminants (Prins, 1983) and in canned vegetables high nitrate contents cause detinning of containers within a few months of storage (Farrow et al., 1971). Effective procedures for eliminating high nitrate contents in plants at harvest are the adjustment of the nitrogen supply to the prevailing light conditions during the growing period, partial replacement of nitrate by chloride (Alt and Stüwe, 1982) or ammonium (Hähndel and Wehrmann, 1986; Van der Boon et al., 1990), better control of the nitrogen supply from both soil reserves and fertilizers, and, in general, utilization of the knowledge available on nitrate accumulation in plants (Breimer, 1982).

In cereals a high grain protein content is necessary for processing (bread-baking quality) and for nutrition. Increasing nitrogen supply to the roots, however, affects the nitrogen content of grains mainly indirectly via retranslocation from vegetative growth and might lead to mutual shading and lodging. These complications can be overcome by late application of nitrogen until anthesis, either as foliar or soil application. Also after soil application most of this nitrogen bypasses the leaves and is directly translocated as amides and amino acids from the roots to the developing grains (Michael et al., 1960). In wheat or barley, however, the corresponding increase in the protein content of the grains is correlated with a decrease in the content of the essential amino acid lysine in the grain protein, that is, with a decrease in the nutritional quality of the protein. As shown in Table 8.8 the decrease in the lysine content of grain protein brought about by the late application of nitrogen results from a shift in the proportions of the various grain protein fractions, mainly in favor of the endosperm protein prolamine, which has a very low lysine content. In agreement with the regulation of the synthesis of individual proteins (albumin, prolamine, etc.) according to the genetic code (Fig. 8.11), the proportion of lysine in the various protein fractions is not altered by the nitrogen supply.

A similar relationship occurs in barley between nitrogen supply, protein content, and lysine content. In oat and rice, on the other hand, the main endosperm storage protein is glutelin. Increasing the grain protein content in these species by late nitrogen application therefore has only negligible effects on lysine content and protein quality.

8.3 Sulfur

8.3.1 General

Although atmospheric SO_2 is taken up and utilized by the aerial parts of higher plants (Chapter 4), the most important source of sulfur is sulfate taken up by the roots. In the physiological pH range, the divalent anion SO_4^{2-} is taken up by the roots at relatively low rates, and long-distance transport of sulfate occurs mainly in the xylem (Chapter 3). In several respects, sulfur assimilation has many common features with nitrate assimilation. For example, reduction is necessary for the incorporation of sulfur into amino acids, proteins, and coenzymes, and in green leaves ferredoxin is the reductant for sulfate. Unlike nitrate nitrogen, however, sulfate can also be utilized without reduction and incorporated into essential organic structures such as sulfolipids in membranes or polysaccharides such as agar. Also in contrast to nitrogen, reduced sulfur can be reoxidized in plants. In this oxidation reaction the reduced sulfur of cysteine is converted into sulfate (Sekiya *et al.*, 1982a), the 'safest' storage form of sulfur in plants. The oxidation of reduced sulfur compounds also seems to play an important role as negative feedback signal for sulfate reduction (Schmidt, 1986).

8.3.2 Sulfate Assimilation and Reduction

For a comprehensive recent review the reader is referred to Schmidt (1992). In higher plants and in green algae, the first step of sulfur assimilation is the activation of the sulfate ion by ATP (Fig. 8.18). In this reaction the enzyme ATP sulfurylase catalyzes the replacement of two phosphate groups of the ATP by the sulfuryl group, which leads to the formation of adenosine phosphosulfate (APS) and pyrophosphate (Fig. 8.18). This enzyme is regulated by various external (e.g., light) and internal (e.g., reduced sulfur compounds) factors. The activated sulfate, adenosine phosphosulfate (APS), can serve as substrate for the synthesis of sulfate esters (pathway (1)) or sulfate reduction (pathway (2)). For the synthesis of sulfate esters such as sulfolipids the enzyme APS kinase catalyzes the formation of phosphoadenine phosphosulfate (PAPS) in an ATP-dependent reaction (Fig. 8.18). From PAPS the activated sulfate can be transferred to an hydroxyl group forming a sulfate ester.

For the sulfate reduction (pathway (2)) the activated sulfate of APS (Fig. 8.18) as well as that of PAPS (Schmidt and Jäger, 1992), is transferred by APS or PAPS sulfotransferases to a thiol (R-SH) group of a suitable carrier. It is not clear whether glutathione (Fig. 8.19) acts *in vivo* as such a carrier. This transfer of sulfate to thiol groups, mediated by sulfotransferases, is associated with a reduction of sulfate to sulfite (SO_3^{2-}). The subsequent reduction of carrier-bound sulfite to sulfide (S^{2-}) may involve sulfite reductase or organic thiosulfate reductase (Brunold, 1993). For both types of reductases in chloroplasts reduced ferredoxin is the electron donor (Fig. 8.18).

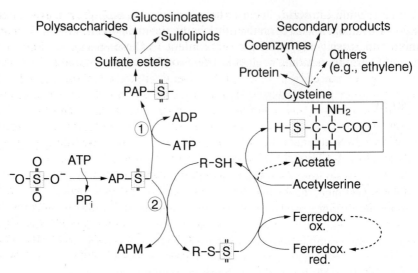

Fig. 8.18 Pathways of sulfur assimilation in higher plants and green algae. (1) Synthesis of sulfate esters; (2) sulfate reduction according to the APS pathway. (Based on Schiff, 1983 and Schmidt and Jäger, 1992.)

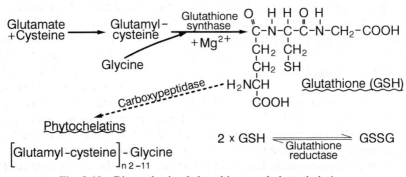

Fig. 8.19 Biosynthesis of glutathione and phytochelatins.

The newly formed —SH group is transferred to acetylserine, which is split into acetate and the amino acid cysteine. Cysteine, the first stable product of the assimilatory sulfate reduction, acts as a precursor for the synthesis of all other organic compounds containing reduced sulfur, as well as for other biosynthetic pathways, such as the formation of ethylene (Miyazaki and Yang, 1987).

Assimilatory sulfate reduction is regulated at various levels (Stulen and De Kok, 1993) by:

1. modulation of the activity of ATP sulfurylase;
2. the availability of sulfate *in situ* at the site of ATP sulfurylase;
3. change in the level of APS sulfotransferase;
4. the availability of acetylserine for cysteine synthase.

High concentrations of cysteine inhibit APS sulfotransferase (Brunold and Schmidt,

1978), the enzyme catalyzing the transfer APS $\rightarrow$ R-S-SO$_3^-$ (Fig. 8.18). With ammonium as compared with nitrate nutrition the activity of this enzyme is increased (Brunold and Suter, 1984). At high cellular levels of either cysteine (Sekiya *et al.*, 1982a) or SO$_2$ (Sekiya *et al.*, 1982b), the evolution of hydrogen sulfide (H$_2$S) from green cells is strongly enhanced by light. The light-dependent SO$_2$ reduction coupled with H$_2$S release from green leaves (Chapter 4) is considered an important mechanism for the detoxification of SO$_2$ in leaves and needles (Sekiya *et al.*, 1982b). This type of sulfate reduction may be considered a modification of the dissimilatory sulfate reduction pathway in prokaryotic anaerobes such as *Desulfovibrio* which use sulfate as an oxidant in the formation of ATP and sulfide during respiration (Schiff, 1983).

In higher plants the enzymes of the assimilatory sulfate reduction are localized in the chloroplasts (Fankhauser and Brunold, 1978) but can also be found, at a much lower level, in the plastids of roots (Fankhauser and Brunold, 1979). In C$_4$ plants the bundle sheath chloroplasts are the main sites of sulfate assimilation (Schmutz and Brunold, 1984), whereas the mesophyll chloroplasts are the sites of nitrate assimilation (Section 8.2.1). However, mesophyll chloroplasts contain at least a sulfite reductase and a cysteine synthase (Schmidt, 1986).

In general, sulfate reduction is several times higher in green leaves than in roots, and in leaves the reaction is strongly stimulated by light (Willenbrink, 1964; Fankhauser and Brunold, 1978). Because of the requirement for ferredoxin as a reductant for the carrier-bound sulfite this light enhancement is to be expected. The enhancement of sulfate reduction by light might also be related to higher levels of serine (acetylserine; Fig. 8.18) synthesized during photorespiration (Fig. 5.7). Sulfate reduction in the leaves leads to export in the phloem of reduced sulfur compounds, mainly as glutathione (Rennenberg, 1989), to sites of demand for protein synthesis (e.g., in the shoot apex, fruits, but also roots) and probably also for regulation of sulfate uptake by roots (Rennenberg, 1989; Section 3.4). During leaf development, the pattern of sulfate reduction is similar to that of nitrate reduction; that is, it is maximal during leaf expansion but declines rapidly after leaf maturation (Schmutz and Brunold, 1982). Compared with nitrate reduction, the reduction of sulfate seems to be under a more strict negative feedback control, as the accumulation of reduced nitrogen compounds such as amino acids is a common feature, but not that of high contents of reduced sulfur compounds. Secondary plant products are an exception (Section 8.3.4).

8.3.3 Metabolic Functions of Sulfur

Sulfur is a constituent of the amino acids cysteine and methionine and hence of proteins. Both of these amino acids are precursors of other sulfur-containing compounds such as coenzymes and secondary plant products. Sulfur is a structural constituent of these compounds (e.g., R^1—C—S—C—R^2) or acts as a functional group (e.g., R—SH) directly involved in metabolic reactions. About 2% of the organic reduced sulfur in the plant is present in the water-soluble thiol (—SH) fraction, and under normal conditions the tripeptide glutathione accounts for more than 90% of this fraction (De Kok and Stulen, 1993). Glutathione serves many functions in plants and its metabolism has been recently reviewed by Bergmann and Rennenberg (1993). The synthesis of glutathione occurs in two steps (Fig. 8.19). In the first step glutamylcysteine is produced from glutamate and cysteine, and in the second step glycine is coupled to glutamylcysteine, mediated by glutathione synthase, an enzyme with an absolute requirement of magnesium for activity (Hell and Bergmann, 1988). In some leguminous species, in the second step alanine rather than glycine is used by glutathione synthase, forming homoglutathione which functions analogously to glutathione (Rennenberg and Lamoureux, 1990).

In plants the glutathione content is usually higher in leaves than in roots, and in leaves more than 50% of it is localized in the chloroplasts where it may reach millimolar concentrations (Rennenberg and Lamoureux, 1990). Also in root apical zones, for example of maize, glutathione being the main low-molecular-weight thiol compound, the content is in the range of 0.7 mmol kg^{-1} fresh weight, about four times higher than that of cysteine (Nieto-Sotelo and Ho, 1986). Glutathione is readily water soluble and a powerful antioxidant in plants, and is most probably of much more importance than the cysteine–cystine redox system. Particularly in the chloroplasts the antioxidants glutathione and ascorbate both play a key role in detoxification of oxygen radicals and hydrogen peroxide, for example in the ascorbate peroxidase – glutathione reductase cycle (Section 5.2). In the cells, glutathione is largely maintained in its reduced form by the enzyme glutathione reductase (Fig. 8.19). The antioxidative role of glutathione is reflected, for example, in the increase in glutathione reductase activity at high light intensities in magnesium-deficient plants (Cakmak and Marschner, 1992), or in response to other oxidative stresses such as ozone or sulfur dioxide (Smith *et al.*, 1990b). Conjugation of reduced glutathione to a number of xenobiotics such as atrazine (used for weed control) is also the mechanism of detoxification and, thus, of resistance of some plant species to certain xenobiotics (Schröder *et al.*, 1990).

Glutathione may also function as transient storage pool of reduced sulfur (Schütz *et al.*, 1991) and thereby maintain a certain cellular cysteine concentration (Schmidt and Jäger, 1992). Glutathione is also the precursor of *phytochelatins* (Fig. 8.19) which function in detoxifying certain heavy metals in higher plants (Grill *et al.*, 1987; Rauser, 1990). Cells respond to exposure to high concentrations of heavy metals such as cadmium and zinc, by the synthesis of polypeptides high in cysteine content, termed 'metallothioneins'. In plants, unlike mammals, these polypeptides have much lower molecular weight and might be termed as 'Class III metallothioneins' or phytochelatins (Rauser, 1990).

Phytochelatins consist of repetitive glutamylcysteine units (between 2 and more than 10) with a terminal glycine, and are synthesized by degradation of glutathione mediated by a carboxypeptidase (Fig. 8.19). Phytochelatins are capable of binding heavy metal cations via thiol coordination and thereby detoxifying them (Grill *et al.*, 1987). The synthesis of phytochelatins in roots is most effectively stimulated by cadmium, less so by zinc and copper, and negligibly by nickel (Tuckendorf and Rauser, 1990). An example for cadmium is shown in Table 8.9. Synthesis of phytochelatins is greatly increased by exposure of the roots to 3 μM cadmium, and this increase is accompanied by a rapid decline in the glutathione content. This inverse relationship is evident after only 1–2 h exposure to cadmium. Phytochelatin synthesis is induced by exposure to 0.05 μM cadmium, and synthesis by far exceeds the amount required for detoxification of the heavy metal (Tukendorf and Rauser, 1990).

Differences between ecotypes of *Silene vulgaris* in cadmium tolerance are presum-

Table 8.9

Contents of Free Cysteine, Total Glutathione, and Phytochelatins and Cadmium in the Apical 10 cm of Maize Roots Exposed to 0 or $3 \mu M \, Cd^{2+}$ for 24 h[a]

Cd^{2+} (μM)	Thiol (nmol g^{-1} fresh wt)			Cd in roots (nmol g^{-1} fresh wt)
	Cysteine	Glutathione	Phytochelatins	
0	43	421	3	n.d.
3	44	156	230	13.1

[a]Based on Tukendorf and Rauser (1990). ND, Not determined.

ably related to differences in phytochelatin synthesis (Verkleij *et al.*, 1990). However, a general key role of phytochelatins in heavy metal tolerance of plants, for example zinc tolerance, is questionable (Rauser, 1990).

Thioredoxins are another important family of thiols in higher plants, besides glutathione and its related compounds. Thioredoxins are low-molecular-weight proteins of about 12 kDa with two well conserved cysteine residues which form a redox-active, intermolecular disulfide bridge. Plant cells contain two different systems capable of reducing thioredoxins: in chloroplasts the ferredoxin/thioredoxin system, and in the cytoplasm the NADP/thioredoxin system (Schürmann, 1993). In chloroplasts, thioredoxins function primarily as regulatory proteins in carbon metabolism. In the reduced form thioredoxins activate, for example, fructose-1,6-bisphosphatase and several enzymes of the Calvin cycle, and thus act as a regulatory link between provision of reducing equivalents (PS II) and assimilation of CO_2.

Reduced sulfur is a structural constituent of several coenzymes and prosthetic groups such as ferredoxin (Section 9.1), biotin (Vitamin H), and thiamine pyrophosphate (Vitamin B_1). In many enzymes and coenzymes such as urease, sulfotransferases (Fig. 8.18) and coenzyme A, the —SH groups act as functional groups in the enzyme reaction. In the glycolytic pathway, for example, decarboxylation of pyruvate and the formation of acetyl coenzyme A are catalyzed by a multienzyme complex involving three sulfur-containing coenzymes: thiamine pyrophosphate (TPP), the sulfhydryl-disulfide redox system of lipoic acid, and the sulfhydryl group of coenzyme A:

The acetyl group (—CO—CH₃) of the coenzyme A is then transferred to the tricarboxylic acid cycle or to the fatty acid synthesis pathway (Section 8.2, Fig. 8.17). The coupling of C_2 units in the synthesis of long-chain fatty acids requires transient carboxylation, which is mediated by the sulfur-containing coenzyme biotin and activated by manganese.

Also as structural component of the proteins, cysteine has particular effects on structure and function of the proteins. The reversible formation of disulfide bonds between two adjacent cysteine residues (cysteinyl moiety) in the polypeptide chain is of fundamental importance for the tertiary structure and thus the function of enzyme proteins. This bond may form a permanent (covalent) cross-link between polypeptide chains or a reversible dipeptide bridge, comparable with the redox functions of glutathione (Fig. 8.18). During dehydration, the number of disulfide bonds in proteins increases at the expense of the —SH groups, and this shift is associated with protein aggregation and denaturation (Tomati and Galli, 1979). The protection of —SH groups in proteins from the formation of disulfide bridges is considered to be of great importance for providing cellular resistance to dehydration (caused by drought and heat) and frost damage (Levitt, 1980).

The most important sulfur-containing compounds of the secondary metabolism are the *alliins* and *glucosinolates*. These are of particular relevance for horticulture and agriculture (Schnug, 1993). Alliin is the common name for *S*-alk(en)ylcysteine sulfoxides which are the characteristic compounds of the genus *Allium*:

| Alliins | | Allicins |

More than 80% of the total sulfur in *Allium* species may be bound to such compounds, in onion (*Allium cepa*) for example as *S*-propylcysteine sulfoxide (R = —CH$_2$—CH$_2$—CH$_3$). Enzymatic cleavage of alliins is mediated by alliinase. Loss in cellular compartmentation by mechanical damage of the tissue greatly enhances enzyme activity through increased availability of the substrate and leads to the formation of allicins as precursors of a large number of volatile substances such as mono- and disulfides with a characteristic odor.

Glucosinolates are characteristic compounds of the secondary metabolism of at least 15 dicotyledonous taxa, including the Brassicaceae (Schnug, 1993). Glucosinolates contain sulfur both as a sulfhydryl and a sulfo group, the side chain R varies between plant species:

Glucosinolates are stored in vacuoles and their hydrolysis is catalyzed by the cytosolic enzyme myrosinase which is present in only a very small number of cells in a given organ such as a leaf or seed (Höglund *et al.*, 1991; Wink, 1993). Hydrolysis leads to the liberation of glucose, sulfate and volatile compounds such as isothiocyanates in *Brassica napus*. Similar to the alliinase also myrosinase activity in cells is greatly enhanced by mechanical damage of cells.

The role of secondary sulfur compounds is not fully understood. They definitely act as defense substances (phytocids, feeding deterrents) although the importance of this defense mechanism has presumably been overestimated in the past (Ernst, 1993). This is certainly true for glucosinolates which serve important functions as sulfur storage in plants. During periods of low sulfur supply to the roots but high plant demand (e.g. rapid vegetative growth, seed formation) glucosinolates are degraded by myrosinase and both sulfur molecules are reutilized through the normal sulfur assimilation pathway (Schnug, 1993).

Sulfur in its nonreduced form, i.e. as sulfate ester, is a component of sulfolipids and is thus a structural constituent of all biological membranes. In sulfolipids the sulfo group is coupled by an ester bond to a C_6 sugar, for example, glucose:

Sulfolipids are particularly abundant in the thylakoid membranes of chloroplasts, about 5% of the chloroplast lipids are sulfolipids (Schmidt, 1986). Sulfolipids may also be involved in the regulation of ion transport across biomembranes. Sulfolipid levels in roots have been shown to be positively correlated with plant salt tolerance, the higher the level the greater the tolerance (Erdei *et al.*, 1980; Stuiver *et al.*, 1981).

8.3.4   Sulfur Supply, Plant Growth, and Plant Composition

Sulfur requirement for optimal growth varies between 0.1 and 0.5% of the dry weight of plants. For the families of crop plants, the requirement increases in the order Gramineae < Leguminosae < Cruciferae and this is also reflected in corresponding differences in the sulfur content (percentage of dry weight) of their seeds: 0.18–0.19, 0.25–0.3, and 1.1–1.7, respectively (Deloch, 1960). The sulfur content of protein also varies considerably both between the protein fractions of individual cells (Table 8.10) and among plant species. On average, proteins from legumes contain less sulfur than proteins from cereals, the N/S ratios being 40:1 and 30:1, respectively (Dijkshoorn and Wijk, 1967).

As with nitrogen deficiency, also under sulfur deficiency shoot growth is more

Table 8.10
Effect of Sulfur Deficiency on Leaf Composition in Tomato[a]

Treatment	Content in leaves (mg $(100\ g)^{-1}$ dry wt)			Sulfur content of protein ($\mu g\ mg^{-1}$ protein)	
	Chlorophyll	Protein	Starch	Cytoplasm	Chloroplast
Control ($+SO_4^{2-}$)	5.8	48.0	2.8	13.5	6.5
S-Deficiency	0.9	3.5	27.0	3.8	5.2

[a]Based on Willenbrink (1967).

depressed than root growth, leading, for example in tomato, to a decrease in shoot–root dry weight ratio from 4.4 in sulfur sufficient to 2.0 in sulfur deficient plants (Edelbauer, 1980). Interruption of sulfur supply within a few days decreases root hydraulic conductivity, stomatal aperture and net photosynthesis (Karmoker et al., 1991). The reduced leaf area in sulfur-deficient plants is the result of both smaller size and, particularly number of leaf cells (Burke et al., 1986). The number of chloroplasts per mesophyll cell might or might not be affected, for example, in wheat (Burke et al., 1986) or distinctly decreased, for example, in spinach (Dietz, 1989).

A drastic decrease in chlorophyll content of leaves is a typical feature of sulfur deficiency (Burke et al., 1986; Dietz, 1989). This is to be expected as in leaves a high proportion of the protein is located in the chloroplasts where the chlorophyll molecules comprise prosthetic groups of the chromoproteid complex. Accordingly, under sulfur deficiency shortage of the sulfur-containing amino acids cysteine and methionine not only inhibits protein synthesis but also decreases the chlorophyll content in leaves in a similar manner (Table 8.10). In contrast, starch may accumulate as a consequence either of impaired carbohydrate metabolism at the sites of production (the source) or of low demand at the sink sites (growth inhibition). In response to sulfur deficiency, proteins of low sulfur content are synthesized, which is most evident in the cytoplasm of leaf cells (Table 8.10).

In sulfur-deficient plants, inhibition of protein synthesis is correlated with an accumulation of soluble organic nitrogen and nitrate (Table 8.11). Amides are usually present in much higher concentrations and proportions in the soluble nitrogen fraction (Freney et al., 1978; Karmoker et al., 1991). The sulfate content is extremely low in deficient plants and increases markedly when the sulfate supply is sufficient for optimal growth. The sulfate content of plants is therefore a more sensitive indicator of sulfur nutritional status than the total sulfur content, the best indicator being the proportion of sulfate sulfur in the total sulfur content (Freney et al., 1978).

Inhibition of protein synthesis during sulfur deficiency leads to chlorosis, just as it does during nitrogen deficiency. Unlike nitrogen, however, sulfur is more uniformly distributed between old and new leaves and its content is similarly affected in old and young leaves by the level of sulfate supply (Freney et al., 1978). Furthermore, the distribution of sulfur in sulfur-deficient plants is also affected by the nitrogen supply;

Table 8.11
Effect of Sulfate Concentration in the Nutrient Solution on Plant Fresh Weight and Sulfur and Nitrogen Content of Cotton Leaves[a]

Supply ($mg\ SO_4^{2-}\ l^{-1}$)	Leaf dry wt (g per plant)	Sulfur or Nitrogen (% dry wt)				
		Sulfate S	Organic S	Nitrate N	Soluble organic N	Protein N
0.1	1.1	0.003	0.11	1.39	2.23	0.96
1.0	2.4	0.003	0.12	1.37	2.21	1.28
10.0	3.4	0.009	0.17	0.06	1.19	2.56
50.0	4.7	0.10	0.26	0.00	0 51	3.25
200.0	4.7	0.36	0.25	0.10	0.45	3.20

[a]Based on Ergle and Eaton (1951).

sulfur-deficiency symptoms may occur either in young (ample nitrogen) or in old (low nitrogen) leaves (Robson and Pitman, 1983), indicating that the extent of remobilization and retranslocation from older leaves depends on the rate of nitrogen deficiency-induced leaf senescence, a relationship which is also evident for the micronutrients copper and zinc (Section 3.5). In legumes, during the early stages of sulfur deficiency, nitrogenase activity in the root nodules is much more depressed than photosynthesis (DeBoer and Duke, 1982); symptoms of sulfur deficiency in symbiotically grown legumes are therefore indistinguishable from nitrogen-deficiency symptoms (Anderson and Spencer, 1950). In root nodules of sulfur-deficient legumes the bacteroids are still well supplied with sulfur (O'Hara *et al.*, 1987). The high sensitivity of nitrogenase activity to sulfur deficiency therefore reflects either impaired host plant metabolism or one of the various 'signals' modulating nitrogenase activity (Section 7.4).

In sulfur-deficient plants not only the protein content decreases but also the sulfur content in the proteins (Table 8.10), indicating that proteins with lower proportion of methionine and cysteine but higher proportions of other amino acids such as arginine and aspartate are synthesized (Table 8.12). Changes in protein composition brought about by changes in mineral nutrient supply have been shown already in relation to the nitrogen supply (Section 8.2.5). The decrease in sulfur-rich proteins under sulfur deficiency is not confined to wheat grains (Table 8.12) but can also be found in other cereals and legumes (Randall and Wrigley, 1986) and in maize. Under sulfur deficiency, in wheat a low-molecular-weight sulfur-rich polypeptide fraction decreases (Castle and Randall, 1987), and in maize the proportion of the major storage protein zein, which has a low sulfur content, increases by about 30% whereas the sulfur-rich glutelin decreases by 36 to 71% (Baudet *et al.*, 1986). In soybean seeds one of the two main storage proteins, the 7S globulins, are devoid of methionine. Stem infusion of methionine increased methionine content in soybean seeds by 23% (Grabau *et al.*, 1986b).

The lower sulfur content of proteins influences nutritional quality considerably. Methionine is an essential amino acid in human nutrition and often a limiting factor in

Table 8.12
Effect of Sulfur Fertilization on the Amino Acid
Composition of Endosperm Protein from Wheat[a]

Amino acid	Amino acid content (nmol $(16 \text{ g})^{-1}$ protein N)	
	Control[b]	Sulfur deficiency[c]
Methionine	11	5
Cysteine	21	7
Arginine	27	34
Aspartate	33	93

[a]Based on Wrigley *et al.* (1980).
[b]0.25% total S in dry wt.
[c]0.10% total S in dry wt.

diets in which seeds are a major source of protein (Arora and Luchra, 1970). Furthermore, a decrease in the cysteine content of cereal grains reduces the baking quality of flour, since disulfide bridging during dough preparation is responsible for the polymerization of the glutelin fraction (Ewart, 1978).

In Brassicaceae the content of glucosinolates and their volatile metabolites is closely related to sulfate supply. Their contents in plants can be increased beyond the level at which sulfate supply affects growth (Table 8.13). From the qualitative viewpoint this increase can be considered favorable (e.g., because it enhances the taste of vegetables, making them spicier) or unfavourable (e.g., because it decreases acceptability as animal feed). During the last decade much progress has been made in selection and breeding of new cultivars, for example, in rape with much lower glucosinolate contents (Mietkowski and Beringer, 1990). When grown at sites with low sulfur supply these new cultivars are, however, more sensitive to sulfur deficiency than the traditional cultivars with high glucosinolate content (Schnug, 1993). This higher sensitivity to sulfur deficiency may at least in part be explained by the role of glucosinolates as transient storage compounds of sulfur (Section 8.3.3).

In highly industrialized areas the sulfur requirement of plants is often met fully or to a substantial degree by atmospheric SO_2 pollution. In Northern Europe, however,

Table 8.13
Effect of Sulfate Supply on Yield and Mustard Oil Content of
the Shoots of *Brassica juncea*[a]

Sulfate supply (mg S per pot)	Shoots (g fresh wt)	Mustard oil content (mg $(100 \text{ g})^{-1}$ fresh wt)
1.5	80	2.8
15.0	208	8.1
45.0	285	30.7
405.0	261	53.1
1215.0	275	52.1

[a]Based on Marquard *et al.* (1968).

industrial SO_2 emissions have been drastically decreased in the last decade. Thus, sulfur deficiency is becoming more widespread in northern Europe in agricultural areas with high production levels of rape seed crops in particular (Schnug, 1993). Worldwide, sulfur deficiency in crop production is quite common in rural areas, particularly in high rainfall areas, for example in the humid tropics, and also in temperate climates (Murphy and Boggan, 1988) and in highly leached soils. Under these conditions, the application of nitrogen fertilizers in the form of urea is ineffective unless sulfur is applied simultaneously (Wang *et al.*, 1976).

8.4 Phosphorus

8.4.1 General

Unlike nitrate and sulfate, phosphate is not reduced in plants but remains in its highest oxidized form. After uptake – at physiological pH mainly as $H_2PO_4^-$ – either it remains as inorganic phosphate (P_i) or is esterified through a hydroxyl group to a carbon chain (C—O—℗) as a simple phosphate ester (e.g., sugar phosphate) or attached to another phosphate by the energy-rich pyrophosphate bond ℗~℗ (e.g., in ATP). The rates of exchange between P_i and the ℗ in ester and the pyrophosphate bond are very high. For example, P_i taken up by roots is incorporated within a few minutes into organic ℗ but thereafter is released again as P_i into the xylem (Section 2.8). Another type of phosphate bond is characterized by the relative high stability of its diester state (C—℗—C). In this association phosphate forms a bridging group connecting units to more complex or macromolecular structures.

8.4.2 Phosphorus as a Structural Element

The function of phosphorus as a constituent of macromolecular structures is most prominent in nucleic acids, which, as units of the DNA molecule, are the carriers of genetic information and, as units of RNA, are the structures responsible for the translation of the genetic information (Fig. 8.11). In both DNA and RNA, phosphate forms a bridge between ribonucleoside units to form macromolecules:

(Section of RNA molecule)

Phosphorus is responsible for the strongly acidic nature of nucleic acids and thus for the exceptionally high cation concentration in DNA and RNA structures. The proportion of phosphorus in nucleic acids to total organically bound phosphorus differs between tissues and cells; it is high in meristems and low in storage tissues.

The bridging form of phosphorus diester is also abundant in the phospholipids of biomembranes (Section 2.3). There it forms a bridge between a diglyceride and another molecule (amino acid, amine, or alcohol). In biomembranes the amine choline is often the dominant partner, forming phosphatidylcholine (lecithin):

$$\text{~~~~~~~~~~~~~~~~~~~O-CH}_2$$
$$\text{~~~~~~~~~~~~~~~~~~~O-CH~~~~~~O}^-$$
$$\text{~~~~~~~~~~~~~~~~~~H}_2\text{C-O-}\textcircled{P}\text{-O-C-C-N}^+(\text{CH}_3)_3$$
$$\text{~~~~~~~~~~~~~~~~~~~~~~~~~O}$$

The functions of phospholipids (and also of sulfolipids; see Section 8.3) are related to their molecular structure. There is a lipophilic region (consisting of two long-chain fatty acid moieties) and a hydrophilic region in one molecule; at a lipid–water interface, the molecules are oriented so that the boundary layer is stabilized (Fig. 2.4). The electrical charges of the hydrophilic region play an important role in the interactions between biomembrane surfaces and ions in the surrounding medium.

8.4.3 Role in Energy Transfer

Although present in cells only in low concentrations, phosphate esters (C—$\textcircled{P}$) and energy-rich phosphates ($\textcircled{P} \sim \textcircled{P}$) represent the metabolic machinery of the cells. Up to 50 individual esters formed from phosphate and sugar alcohols have been identified, about 10 of which, including glucose-6-phosphate and phosphoglyceraldehyde, are present in relatively high concentrations in cells. The common structure of phosphate esters is as follows:

$$\textcircled{R}\text{-C-C-O-}\textcircled{P}\text{-OH}$$

Most phosphate esters are intermediates in metabolic pathways of biosynthesis and degradation. Their function and formation are directly related to the energy metabolism of the cells and to energy-rich phosphates. The energy required, for example, for biosynthesis of starch or for ion uptake is supplied by an energy-rich intermediate or coenzyme, principally ATP:

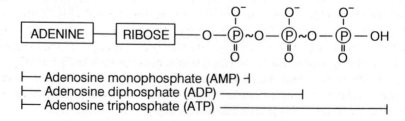

Energy liberated during glycolysis, respiration (Section 5.3), or photosynthesis (Section 5.2) is utilized for the synthesis of the energy-rich pyrophosphate bond, and on hydrolysis of this bond ~30 kJ per mole ATP are released. This energy can be transmitted with the phosphoryl group in a phosphorylation reaction to another compound, which results in the activation (priming reaction) of this compound:

$$
\begin{array}{l}
\text{Adenosine} - \text{P} \sim \text{P} \sim \text{(P)} \qquad\qquad \text{HO} - \boxed{\text{R}} \\
\qquad\quad [\text{ATP}] \\
\\
\text{Adenosine} - \text{P} \sim \text{P} \longleftarrow \qquad\longrightarrow \text{(P)} - \text{O} - \boxed{\text{R}} \dashrightarrow \\
\qquad\quad [\text{ADP}]
\end{array}
$$

ATP is the principal energy-rich phosphate required for starch synthesis. The energy-rich pyrophosphate bonds of ATP can also be transmitted to other coenzymes which differ from ATP only in the nitrogen base, for example, uridine triphosphate (UTP) and guanosine triphosphate (GTP), which are required for the synthesis of sucrose and cellulose, respectively. The activity of ATPases, mediating the hydrolysis and, thus, energy transfer is affected by many factors including mineral nutrients such as magnesium (Section 8.5), calcium (Section 8.6) and potassium (Section 8.7; Chapter 2). In some phosphorylation reactions the energy-rich inorganic pyrophosphate (PP_i) is liberated and the adenosine (or uridine) moiety remains attached to the substrate:

$$
\begin{array}{l}
\text{Adenosine} - \text{P} \sim \boxed{\text{P} \sim \text{P}} \qquad \text{(P)} - \text{O} - \boxed{\text{R}} \\
\qquad\quad [\text{ATP}] \\
\\
\qquad\quad \boxed{\text{P} \sim \text{P}_i} \longleftarrow \qquad\longrightarrow \text{Adenosine} - \text{P} - \text{(P)} - \text{O} - \boxed{\text{R}} \dashrightarrow \\
\\
\qquad [\text{Pyrophosphate}]
\end{array}
$$

Liberation of PP_i takes place in all of the major biosynthetic pathways, for example, acylation of CoA in fatty acid synthesis, formation of APS in sulfate activation (Fig. 8.18), of starch in chloroplasts and of sucrose in the cytosol (Fig. 8.20). Various enzymes can make use of PP_i, for example the UDP-glucose phosphorylase (Fig. 8.20) and the proton pumping inorganic pyrophosphatase at the tonoplast (Section 2.4.2; Fig. 2.9). The cellular concentrations of PP_i are in the range of 100–200 nmol g^{-1} fresh weight and, thus, in a similar range as ATP (Duff *et al.*, 1989). In leaves, PP_i concentrations are similar in the cytosol and stroma of chloroplasts and remain rather stable during the light–dark cycle (Eberl *et al.*, 1992).

Table 8.14
Turnover Times and Rates of Synthesis of Organic Phosphate Fractions in *Spirodela*[a]

Phosphorus fraction	Amount (nmol g^{-1} fresh wt)	Turnover (min)	Synthesis rate (nmol P g^{-1} fresh wt/min^{-1})
ATP	170	0.5	340
Glucose 6-phosphate	670	7	95
Phospholipids	2700	130	20
RNA	4900	2800	2
DNA	560	2800	0.2

[a]Based on Bieleski and Ferguson (1983).

In actively metabolizing cells, energy-rich phosphates are characterized by extremely high rates of turnover. From pulse-labeling experiments with ^{32}P, the turnover rates of various phosphorus compounds can be calculated (Table 8.14). It is impressive that such a small amount of ATP can satisfy the energy requirement of plant cells. It has been calculated, for example, that 1 g of actively metabolizing maize root tips synthesize about 5 g ATP per day (Pradet and Raymond, 1983). The amount of phospholipids and RNA is much higher compared with ATP, but they are much more stable compounds and have a relatively low rate of synthesis (Table 8.14).

Phosphorylation of enzyme proteins by ATP, GTP or ADP is another mechanism by which energy-rich phosphates can modulate enzyme activities:

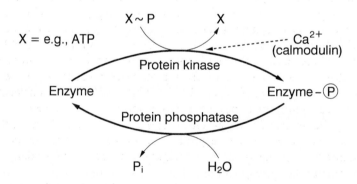

This regulatory phosphorylation is mediated by protein kinases and can result in activation, inactivation and/or changes in the allosteric properties of the target molecule (Budde and Chollet, 1988). Protein kinases normally phosphorylate serine and threonine residues of the polypeptide chain (Ranjeva and Boudet, 1987). Dephosphorylation is generally a hydrolytic reaction catalyzed by protein phosphatases. Protein phosphorylation is considered as a key factor in signal-transduction, for example, in phytochrome-mediated responses of plants (Singh and Song, 1990). An example of this has been given for the light-stimulated enhancement of nitrate assimilation in leaves (Fig. 8.3). PEP carboxylase is one of the key enzymes regulated by phosphorylation, both in C_3 and C_4 plants. In C_4 plants and in CAM plants (Section 5.2.4) phosphorylation increases activity of the PEP carboxylase and simultaneously the enzyme becomes less sensitive to negative feedback control by high malate concentrations (Budde and Chollet, 1988).

8.4.4 Compartmentation and Regulatory Role of Inorganic Phosphate

In many enzyme reactions, P_i is either a substrate or an end product (e.g., $ATP \rightarrow ADP + P_i$). Furthermore, P_i controls some key enzyme reactions. Compartmentation of P_i is therefore essential for the regulation of metabolic pathways in the cytoplasm and chloroplasts. In fruit tissue of tomato, for example, P_i released from the vacuoles into the cytoplasm can stimulate phosphofructokinase activity (Woodrow and Rowan, 1979); the latter is the key enzyme in the regulation of substrate flux into the glycolytic pathway. Thus an increase in the release of P_i from vacuoles can initiate the respiratory burst correlated with fruit ripening (Woodrow and Rowan, 1979). Delayed fruit ripening in phosphorus-deficient tomato plants (Pandita and Andrew, 1967) may be related to this function of P_i.

In vacuolated cells of higher plants the vacuole acts as storage pool, or 'nonmetabolic pool', of P_i, and at adequate phosphorus supply of the plants ~85–95% of the total P_i is located in the vacuoles (Bieleski and Ferguson, 1983). In contrast, in leaves of phosphorus-deficient plants virtually all P_i is found in the cytoplasm and chloroplasts, i.e. in the 'metabolic pool' (Foyer and Spencer, 1986). In leaves, the total phosphorus content may vary by a factor of 20 without affecting photosynthesis, as the P_i concentration in the cytoplasm is regulated in a narrow range by an effective phosphate homeostasis in which the P_i in the vacuole acts as buffer (Mimura *et al.*, 1990). The same is true for roots where the cytoplasmic P_i concentration is maintained quite constant in the range of 6.0 mM (maize) and 4.2 mM (pea) also under phosphorus deficiency, unless the vacuolar pool is depleted (Lee *et al.*, 1990). Under phosphorus deficiency in leaves cytoplasmic P_i concentrations may drop from about 5 mM to less than 0.2 mM. Simultaneously the levels of energy-rich phosphates drop to 20–30% of the original level. Enhancement of alternative respiration seems to reflect metabolic adaptation allowing respiration to proceed because it negates the necessity for adenylates and P_i (Theodorou and Plaxton, 1993).

In leaves, photosynthesis and carbon partitioning in the light–dark cycle are strongly affected by the P_i concentrations in the stroma of chloroplasts and the compartmentation between chloroplasts and cytosol (Fig. 8.20). In the light, for optimal photosynthesis in chloroplasts a P_i concentration in the range of 2.0–2.5 mM seems to be required, and photosynthesis is almost totally inhibited when the P_i concentrations fall below about 1.4–1.0 mM (Robinson and Giersch, 1987; Heber *et al.*, 1989). Due to the large demand of P_i for phosphorylated intermediates of photosynthesis (Fig. 8.20), in leaves of phosphorus deficient plants (i.e. without vacuolar buffer) the P_i concentrations may drop to 50% following dark–light transition (Sicher and Kremer, 1988).

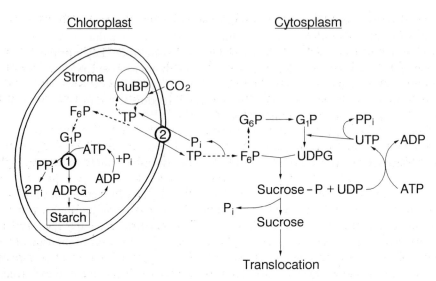

Fig. 8.20 Involvement and regulatory role of phosphate in starch synthesis and carbohydrate transport in a leaf cell. (1) ADP-glucose pyrophosphorylase: regulates the rate of starch synthesis; inhibited by P_i and stimulated by PGA. (2) Phosphate translocator: regulates the release of photosynthates from chloroplasts; enhanced by P_i. TP, Triosephosphate (glyceraldehyde-3-phosphate, GAP; dihydroxyacetone phosphate; DHAP); F_6P, fructose 6-phosphate; G_6P, glucose 6-phosphate. (Based on Walker, 1980.)

The role of P_i on carbon partitioning between chloroplasts and cytosol has been impressively demonstrated with isolated chloroplasts by Heldt *et al.* (1977). An increase in external P_i concentrations up to about 1 mM stimulated net photosynthesis (total carbon fixation), but drastically decreased incorporation of the fixed carbon into starch. The concentration of P_i in the stroma required for severe inhibition of starch synthesis was about 5 mM.

The inhibition of starch synthesis by high concentrations of P_i is caused by two separately regulated mechanisms located in the chloroplasts. The key enzyme of starch synthesis in chloroplasts, ADP-glucose pyrophosphorylase [pathway (1), Fig. 8.20], is allosterically inhibited by P_i and stimulated by triosephosphates. The ratio of P_i to triosephosphates therefore strongly influences the rate of starch synthesis in chloroplasts (Portis, 1982), at high ratios the enzyme being 'switched off'. The other mechanism regulated by P_i is the release of triosephosphates (glyceraldehyde-3-phosphate and dihydroxyacetone phosphate), the main products of CO_2 fixation leaving the chloroplasts. This release is controlled by a phosphate translocator, a specific carrier located in the inner membrane of the chloroplast envelope [pathway (2), Fig. 8.20] and facilitating the counterexchange $P_i \leftrightarrow$ triosephosphate. In spinach, this translocator protein amounts to about 15% of the total protein of the chloroplast envelope (Heldt *et al.*, 1991). In C_4 plants and CAM plants this translocator also accepts another triose, namely phosphoenolpyruvate (PEP). Via the phosphate translocator the net uptake of P_i into the chloroplasts regulates the release of photosynthates from the chloroplast. High P_i concentrations in the cytosol therefore deplete the stroma triosephosphate metabolites, which serve as both substrates for, and activators of, starch synthesis. Thus, inhibition of starch synthesis by high P_i concentrations is also the result of substrate depletion.

In guard cells of *Pisum sativum* the phosphate translocator in the chloroplast envelop [pathway (2), Fig. 8.20] transports glucose-6-phosphate, similarly as it is known for amyloplasts in storage cells (see below). This mechanism enables guard cells to synthesize starch although they lack fructose-1,6-bisphosphate synthase, the enzyme required for $C_3 \rightarrow C_6$ biosynthesis (Overlach *et al.*, 1993).

Carbon dioxide fixation in the Calvin cycle is a process in which five-sixths of the carboxylation products are required in the stroma to regenerate the CO_2 acceptor ribulose bisphosphate (RuBP). Excessive export of triosephosphates induced by high P_i concentrations leads to the depletion of these metabolites which are required for the regeneration of RuBP (Fig. 8.20). In isolated chloroplasts high external P_i concentrations therefore also inhibit total CO_2 fixation (Flügge *et al.*, 1980). However, in intact plants not high, but much more often, low P_i concentrations in the cytosol and chloroplasts prevail, for example, under phosphorus deficiency. Accumulation of large amounts of starch in the chloroplasts is one of the typical features of phosphorus deficiency (see Table 8.17), and this starch is also insufficiently remobilized at night (Qiu and Israel, 1992) or during reproductive growth (Giaquinta and Quebedeaux, 1980).

Accumulation of starch in chloroplasts of phosphorus-deficient plants is to be expected for at least two reasons (Fig. 8.20), namely the low P_i concentrations in the cytosol and, thus, low export of trioses from the chloroplast, and the increase in activity of ADPG-pyrophosphorylase (Rao *et al.*, 1990) due to low P_i concentrations in the stroma. The shift in favour of utilizing trioses for starch synthesis might even reduce Calvin cycle activity and CO_2 fixation by limiting regeneration of RuBP (Fredeen *et al.*, 1990). Accumulation of starch and sugars in leaves of phosphorus deficient plants can also be indirectly the result of lower export due to limitations in ATP for sucrose–proton cotransport in phloem loading (Section 5.4), and lower demand at the sink sites (Rao *et al.*, 1990).

Typically, under phosphorus deficiency shoot growth is much more depressed than photosynthesis. The finely tuned homeostasis of P_i in the cytosol and chloroplasts is one reason for this and a higher activity of various enzymes of carbohydrate metabolism and, thus, turnover of P_i, might be another (Rao *et al.*, 1990). Of course, with severe phosphorus deficiency various parameters of photosynthesis are also impaired such as the carboxylation efficiency (Lauer *et al.*, 1989a).

In principle, similar regulation of starch synthesis takes place in amyloplasts of storage cells. ADP-glucose pyrophosphorylase is also the key enzyme in the regulation of starch synthesis in potato tubers (Mohabir and John, 1988) and in grains, for example, maize (Plaxton and Preiss, 1987). When isolated from these storage tissues, the enzyme is severely inhibited by P_i. In contrast, starch accumulation in the endosperm of wheat grains is not affected by high P_i levels (Rijven and Gifford, 1983), which may indicate that these cells have a particularly large capability for effective P_i compartmentation.

In storage cells, the transport of phosphorylated trioses from the cytosol into the amyloplasts also proceeds by strict countertransport with P_i; however, the phosphorus translocator also accepts glucose-6-phosphate and releases P_i in a C_6–P_i shuttle (Heldt *et al.*, 1991).

Table 8.15

Effect of Phosphorus Supply on Phosphorus Fractions of Tobacco Leaves[a]

P supply $(mg \, l^{-1})$	Leaf dry wt (g per leaf)	Phosphorus fraction (mg P $(100 \, g)^{-1}$ leaf dry wt)			
		Lipid	Nucleic acid	Ester	Inorganic
2	0.82	32	74	36	33
6	1.08	83	134	91	83
8	1.10	89	133	104	123
20	1.06	91	142	109	338

[a]Based on Kakie (1969).

8.4.5   Phosphorus Fractions and the Role of Phytate

When phosphorus supply is increased from the deficiency to the sufficiency range, the major phosphorus fractions in vegetative plant organs also usually increase as shown in a typical example for leaves in Table 8.15. With further increase in supply only P_i as the major storage form of phosphorus in highly vacuolated tissue increases. However, higher and lower plants may also store phosphorus in two other major forms, namely polyphosphate and phytate.

The storage of phosphate in cells as inorganic polyphosphates is widespread among lower plants (bacteria, fungi, and green algae). It has also been found in higher plants, for example, in apple leaves (Schmidt and Buban, 1971) and in various species of Australian heath plants (Jeffrey, 1968). Polyphosphates synthesized by plants are linear polymers of P_i (up to 500 molecules, or more) with pyrophosphate linkages energetically equivalent to ATP. Polyphosphates may therefore function as energy storage compounds and as compounds controlling the P_i level in the metabolic pool of the cells. They certainly also function as cation exchangers, for example, for potassium in *Chlorella* (Peverly *et al.*, 1978) and also other cations in mycorrhizal fungi (Strullu *et al.*, 1982). Polyphosphate formation in the hyphae of mycorrhizal fungi is attracting much attention because of its proposed role in the phosphate nutrition of mycorrhizal plants. Hyphae take up P_i from the soil solution and synthesize polyphosphates; these may act as transient energy and storage pools of phosphorus in the hyphae, and are subsequently transported either as polyphosphates or P_i to the host roots (Section 15.7).

Phytate is the typical storage form of phosphorus in grains and seeds. Phytates are the salts of phytic acid, myoinositol, 1,2,3,4,5,6-hexaki*s*phosphate. Phytic acid is synthesized from the cyclic alcohol myoinositol by esterification of the hydroxyl groups with phosphate groups:

Myoinositol $+6 H_3PO_4$ $(-6 H_2O)$ Phytic acid

The sparingly soluble calcium–magnesium salt of phytic acid is termed phytin. Phytic acid also has high affinity for zinc and iron. In legume seeds and cereal grains the main phytates are the potassium–magnesium salts (Ogawa *et al.*, 1979a; Prattley and Stanley, 1982). The proportions of potassium, magnesium, and also of calcium associated with phytic acid might considerably vary between plant species and even between different tissues of a seed (Welch, 1986; Ockenden and Lott, 1988a,b). Phytate phosphorus makes up ~50% of the total phosphorus in legume seeds, 60–70% in cereal grains, and about 86% in wheat mill bran (Lolas *et al.*, 1976). In cereals and legumes phytates are deposited in electron-dense globoid crystals found inside membrane-bound intracellular protein bodies, in cereal grains mainly in the aleurone layer, and in legumes in cotyledons and embryo axes (Lott and Buttrose, 1978; Welch, 1986). In grains and seeds, phytates are also the main storage sites of potassium and magnesium, and in some instances also of calcium and zinc (Mikus *et al.*, 1992).

Phytate in the form of the potassium–magnesium–calcium salt is also the major form of phosphorus in pollen grains (Scott and Loewus, 1986) where it is deposited in form of discrete particles, degraded by phytase during pollen germination (Baldi *et al.*, 1987). Phytates are also found in roots and tubers of several crops such as carrot, artichoke and alfalfa, which contain 15–23% of the total phosphorus in the form of phytic acid (Campbell *et al.*, 1991). The high affinity of phytic acid also to zinc, iron and other heavy metals is probably of importance for heavy metal binding and thereby detoxification in the roots. In cortical root cells of zinc-tolerant ecotypes of *Deschampsia caespitosa*, up to 60% of the valencies of phytic acid were occupied by zinc (Van Steveninck *et al.*, 1987a,b). The abundance of phytic acid as storage form of phosphorus in roots, tubers, grains, seeds and pollen is probably a major source of the phytate phosphorus found in the organic phosphorus fraction in soils.

In legumes and cereals, during the early stages of seed and grain development the content of phytate is low (Fig. 8.21) but rises sharply during the period of rapid starch synthesis. In contrast, the content of P_i during early stages of seed and grain development is generally low and further declines during rapid phytate formation. When phosphorus supply to the roots is increased after anthesis, phytate is the only phosphorus fraction which increases in grains (Michael *et al.*, 1980).

Phytates are presumably involved in the regulation of starch synthesis during grain filling or tuber growth. The synthesis of phytate and a decrease in the P_i level within the grains are closely related (Fig. 8.21; Michael *et al.*, 1980). In addition, in grains and seeds in the final stage of the filling period, with the onset of desiccation phytic acid may act as a major cation trap that eliminates excessive cellular concentrations of potassium and magnesium.

Some phosphorus is associated with the starch fraction and is incorporated into the starch grains. In cereals only a small proportion is involved, but in potato tubers up to 40% of the total phosphorus may be incorporated in starch. Starch-bound phosphorus may reflect another type of compartmentation of P_i and control of its concentration at the sites of starch synthesis. It could also act as a source of phosphorus for sugar export from the amyloplasts during the sprouting of tubers.

The function of phytate in seed germination is quite obvious. In the early stages of seedling growth, the embryo has a large requirement for mineral nutrients, including magnesium (necessary for phosphorylation and protein synthesis), potassium (required

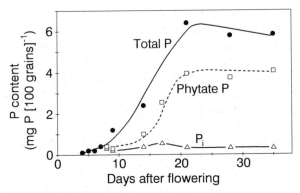

Fig. 8.21   Time course of content of inorganic phosphorus (P_i) and phytate phosphorus in rice grains during grain development. (Based on Ogawa *et al.*, 1979b.)

for cell expansion), and phosphorus (incorporated in membrane lipids and nucleic acids). In agreement with this, digestion of the globoid crystals as storage sites of phytate is one of the earliest observable changes in the protein bodies of the cotyledons during germination (Lott and Vollmer, 1973). Degradation of phytates, catalyzed by phytases, leads to a rapid decline in phytate-bound phosphorus (Table 8.16). At least in pollen grains, activation of phytase requires calcium (Scott and Loewus, 1986) which might explain in part the role of calcium in pollen germination (Section 8.6).

In germinating rice seeds (Table 8.16) within the first 24 h most of the phosphorus released from phytate is incorporated into phospholipids, indicating membrane reconstitution, which is essential for compartmentation and thus for the regulation of metabolic processes within the cells. An increase in P_i and phosphate ester levels reflects the onset of intensive respiration, phosphorylation, and related processes. The degradation of phytate continues with time, and finally the levels of DNA and RNA phosphorus increase, indicating enhancement of cell division and net protein synthesis.

Table 8.16
Changes in Phosphorus Fractions of Rice Seeds during Germination[a]

Duration of germination (h)	Phosphorus fraction (mg P g^{-1} dry wt)				
	Phytate	Lipid	Inorganic	Ester	RNA + DNA
0	2.67	0.43	0.24	0.078	0.058
24	1.48	1.19	0.64	0.102	0.048
48	1.06	1.54	0.89	0.110	0.077
72	0.80	1.71	0.86	0.124	0.116

[a]From Mukherji *et al.* (1971).

The rate of phytate degradation is also controlled by P_i; high levels of P_i repress the synthesis of phytase (Sartirana and Bianchetti, 1967). During degradation of phytic acid, various inositol phosphates with a lower phosphorus content occur as intermediates, and some of them represent a significant proportion of the phospholipid fraction of membranes. In addition inositol-1,4,5-(tri)phosphate might serve the function of a secondary messenger regulating calcium-channels in membranes of plant cells (Lehle, 1990; Section 8.6.7) as is known for cyclic AMP in animal cells and many microorganisms (Spiteri et al., 1989).

Phytates have attracted considerable attention among nutritionists. These compounds interfere with intestinal resorption of mineral elements, especially zinc as well as iron and calcium. They thereby cause nutritional deficiencies in both monogastric animals (Welch et al., 1974) and humans (Welch and House, 1984), especially children (Hambidge and Walravens, 1976). For a given supply, the amount of zinc resorbed by the intestine is determined by the zinc/phytate ratio in the diet (Lantzsch et al., 1980). In humans, on cereal diets, zinc deficiency results from both the low zinc content of the grains and the consumption of phytate-rich unleavened whole-wheat bread (Reinhold et al., 1973). This problem can be alleviated by zinc supplementation in the diet, by an increase in the zinc/phytate ratio in seeds and grains through the application of zinc fertilizers (Peck et al., 1980). Breeding for low phytic acid content might be another alternative. Such an approach is possible, in principle, but at least in wheat low phytic acid content is not only correlated with a decrease in total phosphorus but also in protein content of the grain (Raboy et al., 1991).

8.4.6 Phosphorus Supply, Plant Growth, and Plant Composition

The phosphorus requirement for optimal growth is in the range of 0.3–0.5% of the plant dry matter during the vegetative stage of growth. The probability of phosphorus toxicity increases at contents higher than 1% in the dry matter. However, many tropical food legumes are rather sensitive and toxicity may occur already at phosphorus contents in the shoot dry matter of 0.3–0.4% in pigeon pea (*Cajanus cajan*) and 0.6–0.7% in black gram (*Vigna mungo*; Bell et al., 1990). In plants suffering from phosphorus deficiency, reduction in leaf expansion and leaf surface area (Fredeen et al., 1989), and also number of leaves (Lynch et al., 1991), are the most striking effects (Table 8.17). Leaf expansion is strongly related to the extension of epidermal cells (Section 5.6), and this process might be particularly impaired in phosphorus deficient plants for various reasons, for example low phosphorus content of epidermal cells (Treeby et al., 1987) and decrease in root hydraulic conductivity (Radin, 1990). In contrast to the severe inhibition in leaf expansion the contents of protein (Rao and Terry, 1989) and of chlorophyll per unit leaf area are not much affected (Table 8.17). Often, the chlorophyll content is even increased under phosphorus deficiency (Rao and Terry, 1989), and leaves have a darker green color as cell and leaf expansion are more retarded than chloroplast and chlorophyll formation (Hecht-Buchholz, 1967). However, the photosynthetic efficiency per unit of chlorophyll is much lower in phosphorus deficient leaves (Lauer et al., 1989b).

Table 8.17
Effect of Phosphorus Deficiency on Various Growth Parameters and Contents of Phosphorus and
Carbohydrates in Soybean[a]

Parameter		Treatment	
		High P	Low P
Leaf area (dm^2)		12.1	1.8
No. primary trifoliates		7	4
Shoot/root dry wt ratio		~4.2	~1.0
Chlorophyll (mg dm^{-2})		3.02	2.80
P content leaves P_i		4.43	0.28
(mg g^{-1} dry wt) P_{org}		2.44	0.59
P content (total P mg^{-1} dry wt)			
Stem and petioles		5.84	1.14
Roots		10.54	1.29
Ratio $\dfrac{\text{Total Root P}}{\text{Total Shoot P}}$		0.54	1.57
Carbohydrates Leaves	Starch	0.4	12.8
(g m^{-2} leaf)	Sucrose	0.7	0.2
Roots	Starch	23	160
(mg g^{-1} fresh wt)	Sucrose	16	177

[a]Based on Fredeen *et al.* (1989). Reprinted by permission of the American Society of Plant Physiologists.

In contrast to shoot growth, root growth is much less inhibited under phosphorus deficiency, leading to a typical decrease in shoot–root dry weight ratio (Table 8.17). In bean (*Phaseolus vulgaris*) this ratio decreased from 5.0 in phosphorus sufficient to 1.9 in phosphorus deficient plants, which is in contrast to magnesium deficient plants where root growth is severely impaired and the ratio increases to 10.0 (Section 8.5.4, Table 8.22). As a rule, the decrease in shoot–root dry weight ratio in phosphorus deficient plants is correlated with an increase in partitioning of carbohydrates towards the roots, indicated by a steep increase particularly in sucrose content of the roots of phosphorus deficient plants (Khamis *et al.*, 1990a; Table 8.17). In bean, of the total amount of carbohydrates per plant, 22.7% were partitioned in the roots of phosphorus deficient plants compared to 15.7% in the phosphorus sufficient and only 0.18 in the magnesium deficient plants (I. Cakmak, pers. comm.). Phosphorus deficiency might even enhance elongation rate of individual root cells and of the roots (Anuradha and Marayanan, 1991). In *Stylosanthes hamata* under phosphorus deficiency, shoot growth declines rapidly but roots continue to grow, not only because of retaining most phosphorus but also of additional net translocation of phosphorus from the shoot to the roots (Smith *et al.*, 1990a). In certain plant species phosphorus deficiency-induced formation of 'proteoid roots' is another type of response (Section 14.4). Although total respiration is not much altered in phosphorus deficient roots, the proportion of alternative respiration (Section 5.3) increases from about 40–50% in phosphorus-sufficient to 80–90% in phosphorus-deficient roots, a shift which can be reversed within a few hours after resupply of phosphorus (Rychter and Mikulska, 1990).

Despite these adaptive responses in increasing phosphorus acquisition by roots, not only is shoot growth rate retarded by phosphorus limitation but also the formation of reproductive organs. Flower initiation is delayed (Rossiter, 1978), the number of flowers decreased (Bould and Parfitt, 1973), and seed formation restricted in particular (Barry and Miller, 1989). Premature senescence of leaves is another factor limiting seed yield in phosphorus-deficient plants (Section 6.4).

8.5 Magnesium

8.5.1 General

Magnesium is a small divalent cation with a hydrated ionic radius of 0.428 nm and a very high hydration energy of $1908 \, J \, mol^{-1}$. Its rate of uptake can be strongly depressed by other cations, such as K^+, NH_4^+ (Kurvits and Kirkby, 1980), Ca^{2+}, and Mn^{2+} (Heenan and Campbell, 1981), as well as by H^+, that is, by low pH (Chapter 2). Magnesium deficiency induced by competing cations is thus a fairly widespread phenomenon (Section 8.5.6).

The functions of magnesium in plants are mainly related to its capacity to interact with strongly nucleophilic ligands (e.g., phosphoryl groups) through ionic bonding, and to act as a bridging element and/or form complexes of different stabilities. Although most bonds involving magnesium are mainly ionic, some are mainly covalent, as in the chlorophyll molecule. Magnesium forms ternary complexes with enzymes in which bridging cations are required for establishing a precise geometry between enzyme and substrate (Clarkson and Hanson, 1980), RuBP carboxylase is such an example (Pierce, 1986). A substantial proportion of the total magnesium is involved in the regulation of cellular pH and the cation–anion balance.

8.5.2 Binding Form and Compartmentation

In green leaves a major function of magnesium, and certainly its most familiar function, is its role as the central atom of the chlorophyll molecule (Fig. 5.1). The proportion of the total magnesium bound to chlorophyll depends very much on the magnesium supply (Michael, 1941). In subterranean clover, this proportion ranges from 6% in leaves of plants with high magnesium supply to 35% in leaves of magnesium deficient plants (Scott and Robson, 1990a,b). Under low light conditions the proportion of total magnesium bound in chlorophyll might even be higher than 50%, for example in magnesium-deficient poplar (Dorenstouter *et al.*, 1985). An example of such a range is shown in Table 8.18. Depending on the magnesium nutritional status, between 6% and

25% of the total magnesium is bound to chlorophyll. As a rule, another 5–10% of the total magnesium in leaves and needles is firmly bound to pectate in the cell walls or precipitated as sparingly soluble salts in the vacuole (e.g. as phosphate), and the remaining 60–90% is extractable with water. In most instances, growth is depressed and visual symptoms of magnesium deficiency occur when the proportion of magnesium in the chlorophyll exceeds 20–25%.

In cells of mature leaf tissue, ~15% of the whole cell volume is occupied by the chloroplasts, the cytoplasm and the cell wall (~5% each), and the remaining 85% by the vacuole (Cowan *et al.*, 1982; Leigh and Wyn Jones, 1986). Similarly to inorganic phosphorus (P_i, Section 8.4.4), the concentration of magnesium which is not firmly bound in organic structures but located in the 'metabolic pool' has also to be strictly regulated. The concentration of magnesium in the metabolic pool of leaf cells (i.e. in the cytoplasm and chloroplasts) is assumed to be in the range of 2–10 mM (Leigh and Wyn Jones, 1986). As for P_i, for magnesium the vacuole is also the main storage pool required for maintenance of magnesium homeostasis in the 'metabolic pool'. In needles of magnesium-sufficient Norway spruce the magnesium concentrations in the vacuoles are in the range of 13–17 mM in mesophyll cells and 16–120 mM in endodermis cells. These high concentrations in endodermis cells obviously function as a buffer in maintaining magnesium homeostasis in other cells throughout the season (Stelzer *et al.*, 1990). In addition, vacuolar magnesium is also important for the cation–anion balance and turgor regulation of cells.

Also in the 'metabolic pool' the magnesium distribution between the cytosol and the chloroplast has to be well regulated. In isolated chloroplasts, photosynthesis is strongly inhibited even by 5 mM magnesium in the external solution (i.e. cytosolic site). This inhibition is caused by a decrease in potassium influx and corresponding acidification of the stroma upon illumination (Wu *et al.*, 1991; Section 8.7). Inhibition of photo-synthesis by high magnesium concentrations in the 'metabolic pool' may occur in intact plants under drought stress (Section 8.5.6).

8.5.3 Chlorophyll and Protein Synthesis

Chlorophyll and heme synthesis share a common pathway up to the level of proto-chlorophyll (Fig. 9.1). Insertion of magnesium into the porphyrin structure as the first step of chlorophyll biosynthesis is catalyzed by the magnesium-chelatase (Walker and

Table 8.18

Content and Binding Form of Magnesium in One-Year-Old Needles of Norway Spruce Grown at two Locations[a]

Location (soil)	Total Mg (mg g^{-1} dry wt)	Proportion of total Mg (%)		
		Water soluble	Pectate, phosphate	Chlorophyll
I Rendzina	1.47	91.2	2.6	6.2
II Podsol	0.31	64.8	10.0	25.2

[a]Based on Fink (1992a).

Weinstein, 1991). For activation, this enzyme also requires ATP and, thus, additional magnesium (Section 8.5.4). Chlorophyll breakdown requires two enzymes, a magnesium-dechelatase leading to pheophytin, and chlorophyllase for dephytylation of the porphyrin (Langmeier *et al.*, 1993).

Magnesium also has an essential function as a bridging element for the aggregation of ribosome subunits (Cammarano *et al.*, 1972), a process that is necessary for protein synthesis. When the level of free magnesium (Mg^{2+}) is deficient, or in the presence of excessive levels of K^+ (Sperrazza and Spremulli, 1983), the subunits dissociate and protein synthesis ceases. Magnesium is also required for RNA polymerases and hence for the formation of RNA in the nucleus. This latter role might be related to both bridging between individual DNA strands and neutralization of the acid proteins of the nuclear matrix (Wunderlich, 1978).

As shown in Fig. 8.22, net synthesis of RNA immediately stops in response to magnesium deficiency, and synthesis resumes rapidly after the addition of magnesium. In contrast, protein synthesis remains unaffected for more than 5 h, but thereafter it rapidly declines. The requirement for magnesium in protein synthesis can also be directly demonstrated in chloroplasts (Table 8.19). As free magnesium (Mg^{2+}) readily permeates the chloroplast envelope (Mg^{2+} channels?), a concentration of at least 0.25–0.40 mM Mg^{2+} is required at the cytosolic site to prevent net efflux of Mg^{2+} from the chloroplast and, thus, to maintain protein synthesis (Deshaies *et al.*, 1984).

In leaf cells at least 25% of the total protein is localized in chloroplasts. This explains why a deficiency of magnesium particularly affects the size, structure, and function of chloroplasts, including electron transfer in photosystem II (McSwain *et al.*, 1976). In magnesium-deficient plants retranslocation from mature to young leaves is enhanced and, thus, visual deficiency symptoms typically appear on mature leaves, indicating enhanced rates of protein degradation, including structural proteins of the thylakoids. This also explains why in magnesium-deficient plants the other plastid pigments are often affected in the same way as chlorophyll (Table 8.20). Regardless of this decline in chloroplast pigments, starch accumulates in magnesium-deficient chloroplasts and is mainly responsible for the increase in the dry matter content of magnesium-deficient leaves (Scott and Robson, 1990a; Table 8.20). Impaired export of photosynthates is another causal factor leading to enhanced degradation of chlorophyll in magnesium deficient source leaves (Section 8.5.5).

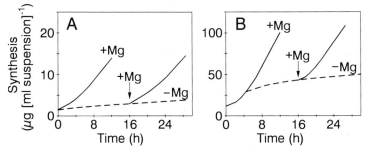

Fig. 8.22   Effect of magnesium supply on (A) RNA and (B) protein synthesis in *Chlorella pyrenoidosa* suspension culture. (Based on Galling, 1963.)

Table 8.19
Magnesium Requirement for the Incorporation of ^{14}C [Leucine] into the Protein Fraction of Isolated Wheat Chloroplasts[a]

Magnesium concentration (mM)	^{14}C Incorporation (cpm mg^{-1} chlorophyll)	Relative value
0	412	11.5
0.5	688	19.5
5.0	3550	100.0

[a]Based on Bamji and Jagendorf (1966).

8.5.4 Enzyme Activation, Phosphorylation, and Photosynthesis

There is a long list of enzymes and enzyme reactions which require or are strongly promoted by magnesium, for example, glutathione synthase (Section 8.3.3) or PEP carboxylase. For this latter enzyme in the presence of magnesium, the substrate phosphoenolpyruvate (PEP) is bound in greater quantities and more tightly (Wedding and Black, 1988). Most of the magnesium-dependent reactions can be categorized by the general type of reaction to which they conform, such as the transfer of phosphate (e.g., phosphatases and ATPases) or of carboxyl groups (e.g., carboxylases). In these reactions magnesium is preferentially bound to nitrogen and phosphoryl groups and this is also, for example, the case with ATP:

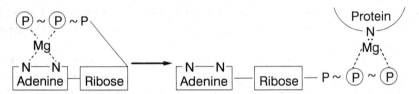

The substrate for ATPases, as well as inorganic PP$_i$ases (Rea and Sanders, 1987), is Mg·ATP rather than free ATP. The Mg·ATP complex is formed with reasonable stability above pH 6, and this complex can be utilized by the active sites of ATPases for the transfer of the energy-rich phosphoryl group (Balke and Hodges, 1975). An

Table 8.20
Magnesium Deficiency-Induced Changes in Plastid Pigments and Leaf Dry Matter in Rape[a]

Treatment	Chlorophyll (a and b) (mg g^{-1} fresh wt)	Carotenoids (mg g^{-1} fresh wt)	Leaf dry matter (%)
Control	2.33	0.21	13.6
Magnesium-deficient	1.33	0.11	17.7

[a]Based on Baszynski et al. (1980).

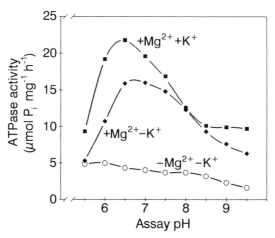

Fig. 8.23   Effect of pH, magnesium (3 mM), and potassium (50 mM) on the ATPase activity of the plasma membrane protein of maize roots. (Based on Leonard and Hotchkiss, 1976.)

example of the Mg^{2+} requirement of membrane-bound ATPases is shown in Fig. 8.23. It is evident that $Mg \cdot ATP$ rather than ATP is the substrate for plasma membrane-bound ATPases in maize roots. Maximal activity requires the presence of both Mg^{2+} and K^+. Therefore, the concentration of free Mg^{2+} strongly affects phosphorylation reactions. In meristematic cells of magnesium-sufficient roots about 90% of the cytoplasmic ATP is complexed to magnesium and the concentration of free Mg^{2+} is only about 0.4 mM as compared with total magnesium concentrations of 3.9 mM in the tissue (Yazaki *et al.*, 1988).

Also the synthesis of ATP (phosphorylation: $ADP + P_i \rightarrow ATP$) has an absolute requirement for magnesium as a bridging component between ADP and the enzyme. As shown in Table 8.21 ATP synthesis in isolated chloroplasts (photophosphorylation, Section 5.2.1) is increased considerably by external supply of magnesium. Because the endogenous magnesium content of chloroplasts is still relatively high even in the control (no added cations), the additional magnesium supply can only further stimulate

Table 8.21
Effect of Cations in the Incubation Medium on the Photophosphorylation of Isolated Pea Chloroplasts[a]

Cation in incubation medium[b]	Photophosphorylation rate (μmol ATP formed mg^{-1} chlorophyll h^{-1})
None	12.3
5 mM Mg^{2+}	34.3
5 mM Ca^{2+}	4.3

[a]Based on Lin and Nobel (1971).
[b]Incubation medium contained ADP, P_i, and cation as indicated.

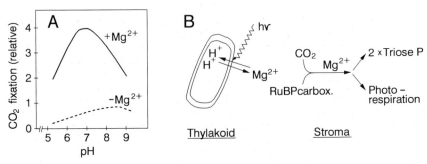

Fig. 8.24   A. Activation of ribulose-1,5-bisphosphate (RuBP) carboxylase from spinach leaves by magnesium. (Modified from Sugiyama *et al.*, 1969.) B. Model for light-induced magnesium transport from intrathylakoid space into the stroma of chloroplasts with subsequent activation of the RuBP carboxylase/oxygenase.

photophosphorylation. The addition of Ca^{2+} severely inhibits photophosphorylation. It is also for this reason, that a low calcium concentration also has to be maintained *in vivo* within the chloroplasts at the sites of photophosphorylation (Section 8.6.7).

Another key reaction of magnesium is the modulation of RuBP carboxylase in the stroma of chloroplasts (Pierce, 1986). The activity of this enzyme is highly dependent on both magnesium and pH (Fig. 8.24A). Binding of magnesium to the enzyme increases its affinity (K_m) for the substrate CO_2 and the turnover rate V_{max} (Sugiyama *et al.*, 1968). Magnesium also shifts the pH optimum of the reaction towards the physiological range (below 8). In chloroplasts the light-triggered activation of RuBP carboxylase is related to increases both in pH and in the magnesium concentration of the stroma. On illumination, protons are pumped from the stroma into the intrathylakoid space, creating a proton gradient required for ATP synthesis (Fig. 8.24B; Section 5.2.1). The light-induced transport of protons from the stroma is counterbalanced by transport of Mg^{2+} (and H^+) from the intrathylakoid space into the stroma which becomes more alkaline (Oja *et al.*, 1986), in wheat leaf chloroplasts the stroma pH may increase from about 7.6 in the dark to about 8.0 in the light (Heineke and Heldt, 1988). This light-triggered reaction increases the magnesium concentration of the stroma from ~2 mM in the dark to ~4 mM in the light (Portis and Heldt, 1976; Portis, 1981). Changes of this magnitude in both pH and magnesium concentration are sufficient to increase the activity of RuBP carboxylase and of other stromal enzymes which depend on high magnesium concentrations and which have a pH optimum above 6.

One of the key enzymes with a high magnesium requirement and high pH optimum is fructose-1,6-bisphosphatase which, in chloroplasts, regulates assimilate partitioning between starch synthesis and export of triosephosphates (Gerhardt *et al.*, 1987). Another key enzyme with similar magnesium requirements is glutamine synthetase (O'Neal and Joy, 1974). A light-induced increase in nitrite reduction and thus NH_3 production requires a simultaneous increase in the activity of enzymes such as glutamine synthetase regulating ammonia assimilation within the chloroplasts. Thus the model of regulation for CO_2 fixation and reduction (Fig. 8.24B) also holds true, in principle, for nitrite reduction and ammonia assimilation.

Table 8.22
Shoot and Root Dry Weight and Carbohydrate Content in Primary Leaves and Roots of Magnesium and Phosphorus Deficient *Phaseolus vulgaris*[a]

| | Dry weight (g per plant) | | | | Carbohydrate (mg g^{-1} dry wt)[b] | | | |
| | | | | | Leaves | | Roots | |
Treatment	Shoot	Roots	S/R	Chlorophyll (mg g^{-1} dry wt)	Starch	Sugars	Starch	Sugars
Control	2.5	0.50	5.0	11	10	27	4	51
−Mg	1.5	0.15	10.0	4	77	166	4	11
−P	0.9	0.48	1.9	12	43	34	8	35

[a]I. Cakmak (unpublished).
[b]mg glucose equivalents.

8.5.5   Carbohydrate Partitioning

The accumulation of nonstructural carbohydrates (starch, sugars) is a typical feature in source leaves of magnesium-deficient plants (Fischer and Bussler, 1988; Table 8.22) and is mainly responsible for the higher dry matter content of these leaves (Table 8.20), indicating that photosynthesis *per se* is less impaired than starch degradation in chloroplasts, sugar metabolism within the cells and/or phloem loading of sucrose. Accumulation of starch is also found in phosphorus-deficient leaves but associated with high chlorophyll content of the leaves (Tables 8.17 and 8.22). Accumulation of carbohydrates in source leaves of magnesium-deficient plants, for example, in *Phaseolus vulgaris*, is correlated with a distinct decrease in carbohydrate content at sink sites such as the pods (Fischer and Bussler, 1988) and the roots (Table 8.22). Limitation of carbohydrate supply to the roots strongly impairs root growth and increases the shoot–root dry weight ratio (Scott and Robson, 1990a), this is just the opposite of what is observed under phosphorus deficiency (Table 8.22).

Most probably inhibition of phloem loading of sucrose in magnesium-deficient source leaves is responsible for the shift in carbohydrate partitioning. The key role of a proton pumping ATPase for phloem loading of sucrose (proton–sucrose cotransport) was discussed in Section 5.4. For optimal activity this enzyme seems to require a magnesium concentration of about 2 mM (Williams and Hall, 1987), and in deficient leaves the concentration of magnesium is most likely much lower in the 'metabolic pool' in general and at the plasma membrane of sieve tube cells in particular. In agreement with this assumption, phloem loading of sucrose can be restored within one day after resupply of magnesium to deficient plants (I. Cakmak, unpublished). In needles of magnesium deficient conifers the accumulation of starch seems to be the result not only of inhibited phloem loading but also destruction of the phloem (Fink, 1991d, 1992a).

Accumulation of photosynthates in leaves exerts a feedback regulation on RuBP carboxylase/oxygenase in favour of the oxygenase reaction and, thus, enhanced O_2 activation (Section 5.2). Accordingly, in magnesium-deficient leaves the formation of superoxide radicals (O_2^-) and hydrogen peroxide (H_2O_2) is enhanced and, in response to this, the content of antioxidants like ascorbate, and the activity of the superoxide

Table 8.23

Effect of Magnesium Deficiency on Content of Chlorophyll, Antioxidants and Activity of Oxygen Radical and H_2O_2 Scavenging Enzymes in Primary Leaves of *Phaseolus vulgaris*[a]

Mg supply (μM)	Chlorophyll (mg g^{-1} dry wt)	Ascorbate (μmol g^{-1} fresh wt)	Soluble thiol (SH) (nmol g^{-1} fresh wt)	Enzyme activities (relative values)		
				SOD[b]	AsPo[c]	GR[d]
1000	11.3	0.9	0.6	100	100	100
20	5.3	6.2	2.3	229	752	310

[a]Cakmak and Marschner (1992).
[b]Superoxide dismutase.
[c]Ascorbate peroxidase.
[d]Glutathione reductase.

radical and H_2O_2 scavenging enzymes (Table 8.23). Magnesium-deficient leaves and needles are therefore highly photosensitive, and symptoms of chlorosis and necrosis strongly increase with the light intensity to which the leaves are exposed (Marschner and Cakmak, 1989; Fig. 5.3).

8.5.6 Magnesium Supply, Plant Growth and Composition

The magnesium requirement for optimal plant growth is in the range of 0.15–0.35% of the dry weight of the vegetative parts. Chlorosis of fully expanded leaves is the most obvious visible symptom of magnesium deficiency. In accordance with the function of magnesium in protein synthesis, the proportion of protein nitrogen is depressed and that of nonprotein nitrogen increased in magnesium-deficient leaves. As calculated on both unit leaf area and unit chlorophyll, the rate of photosynthesis is lower in leaves of magnesium-deficient plants (negative feedback regulation) and carbohydrates accumulate. The appearance of slight and transient magnesium-deficiency symptoms during the vegetative growth stage, however, is not necessarily associated with a depression of the final yield unless irreversible changes, such as a reduction in grain number per ear in cereals, occur (Forster, 1980). At permanently insufficient root supply, remobilization of magnesium from mature leaves reduces leaf area duration as indicated, for example, in perennials such as Norway spruce where content of magnesium and chlorophyll as well as rate of photosynthesis of the older needles decrease in spring when the new flush develops (Lange *et al.*, 1987).

In the last decade increasing evidence has been presented that magnesium deficiency is widespread in forest ecosystems in Central Europe (Liu and Hüttl, 1991), accentuated by other stress factors, in particular air pollution (Schulze, 1989) and soil acidification (Marschner, 1992). Impairment of root growth which is also typical for declining spruce stands (Roberts *et al.*, 1989) under magnesium deficiency has a considerable impact on acquisition not only of magnesium but also of other mineral nutrients and of water and, thus, on drought resistance and adaptation to nutrient-poor sites.

When magnesium is deficient and the export of carbohydrates from source to sink sites impaired, there is also a decrease in the starch content of storage tissues such as potato tubers (Werner, 1959) and the single-grain weight of cereals (Beringer and Forster, 1981). In cereal grains, however, magnesium might play an additional role in the regulation of starch synthesis through its effect on the level of P_i and formation of Mg-K-phytate. As has been shown in Section 8.4.4, high P_i levels inhibit starch synthesis. In magnesium-deficient wheat grains, twice as much phosphorus remains as P_i, and there is a correspondingly smaller proportion of phytate phosphorus, compared with the grains adequately supplied with magnesium (Beringer and Forster, 1981).

Increasing the magnesium supply beyond the growth-limiting level results in additional magnesium being stored mainly in the vacuoles, acting as buffer for magnesium homeostasis in the 'metabolic pool' and as a criterion for charge compensation and for osmoregulation in the vacuole. However, high magnesium contents in the leaves (e.g. 1.5% of the leaf dry matter) might become critical under drought stress. As the leaf water potential falls, the magnesium concentrations in the 'metabolic pool' increase from about 3–5 mM to 8–13 mM in sunflower. Such high concentrations, for example, in the stroma of chloroplasts, inhibit photophosphorylation and photosynthesis (Rao *et al.*, 1987). In pea under drought stress the magnesium concentrations in the chloroplasts have been reported to increase to a concentration as high as 24 mM (Kaiser, 1987).

In most instances, elevated magnesium contents improve the nutritional quality of plants. For example, hypomagnesaemia (grass tetany) is a serious disorder of ruminants and low magnesium content of feed and reduced efficiency of magnesium resorption are major causative factors (Grunes *et al.*, 1970). Increase in magnesium content of forage grasses by magnesium fertilizer supply is relatively easy to achieve, breeding of cultivars with high magnesium content, for example, in Italian ryegrass, is an alternative (Moseley and Baker, 1991). Insufficient magnesium intake in the human diet leading to a magnesium-deficiency syndrome has also attracted considerable attention.

8.6 Calcium

8.6.1 General

Calcium is a relatively large divalent cation with a hydrated ionic radius of 0.412 nm and a hydration energy of 1577 J mol^{-1}. In the apoplasm, part of the calcium is firmly bound in structures, another part is exchangeable at the cell walls and at the exterior surface of the plasma membrane. A high proportion of calcium might be sequestered in vacuoles whereas its concentration in the cytosol is extremly low. The same is true for the mobility of calcium in the symplasm from cell to cell and in the phloem. Most of the functions of calcium as a structural component of macromolecules are related to its capacity for coordination, by which it provides stable but reversible intermolecular linkages, predominantly in the cell walls and at the plasma membrane. Calcium can be supplied at high concentrations and might reach more than 10% of the dry weight, for example in mature leaves, without symptoms of toxicity or serious inhibition of plant growth, at least in calcicole plant species. The functions of calcium in plants have been comprehensively reviewed by Hanson (1984) and Kirkby and Pilbeam (1984). In recent years calcium has attracted much interest in plant physiology and molecular biology because of its function as a second messenger in the signal conduction between environmental factors and plant responses in terms of growth and development. This function of calcium is causally related to its strict compartmentation at the cellular level (Section 8.6.7).

8.6.2 Binding Form and Compartmentation

In contrast to the other macronutrients, a high proportion of the total calcium in the plant tissue is often located in the cell walls (apoplasm). This unique distribution is mainly the result of an abundance of binding sites for calcium in the cell walls (Table 8.24) as well as of the restricted transport of calcium into the cytoplasm. In the middle lamella it is bound to $R \cdot COO^-$ group of the polygalacturonic acids (pectins) in a more or less readily exchangeable form (Section 2.2). In dicotyledons such as sugar beet, which have a large cation-exchange capacity, and particularly when the level of calcium supply is low, up to 50% of the total calcium can be bound as pectates (Table 8.24; Armstrong and Kirkby, 1979b). In storage tissue of apple fruits, the cell wall-bound fraction of calcium can make up as much as 90% of the total (Faust and Klein, 1974).

With increase in calcium supply, in many plant species the proportion of calcium oxalate increases (Table 8.24). In some instances oxalate-bound calcium might represent the dominant binding form of calcium, for example, in mature sugar beet leaves (Egmond and Breteler, 1972) or needles of Norway spruce (Table 8.25). Although for the average of the needles the calcium bound to pectate is extremly low, high local contents of calcium pectate are typical for certain fractions of the needle tissue, the thick walls of the sieve tube cells in the phloem in particular (Fink, 1991b).

Whereas in most angiosperms calcium oxalate crystals are confined to the vacuoles of leaf cells, sometimes as 'crystal cells' arranged in a species-specific pattern (Borchert, 1990), in Pinaceae such as Norway spruce, most of the calcium oxalate crystals occur in the apoplasm, in the cell walls or in the intercellular spaces (Fink, 1991a–c; Fig. 3.3). In

Table 8.24
Relationship between Calcium Supply and Proportion of Total Calcium in Various Binding Forms in Young Sugar Beet Plants[a]

Binding form of calcium	Calcium supply (meq l^{-1})	
	0.33	5.0
Water soluble	27	19
Pectate	51	31
Phosphate	17	19
Oxalate	4	25
Residue	1	6

[a]Based on Mostafa and Ulrich (1976).

Table 8.25

Content and Binding Form of Calcium in One-Year-Old Needles
of Norway Spruce Grown at Two Locations[a]

Location (soil)	Total Ca (mg g⁻¹ dry wt)	Proportion of total Ca		
		Water soluble	Pectate, phosphate	Oxalate
I Rendzina	7.85	7.4	0.8	91.8
II Podsol	1.60	33.2	2.6	64.2

[a]Based on Fink (1992a).

species with a very low capacity for calcium binding in the pectate fraction of the cell walls (Table 8.25) the precipitation of calcium oxalate in the apoplasm is another mechanism for binding and compartmentation of calcium, as an alternative to the formation of calcium oxalate in the vacuoles. Depending on the plant species and families, calcium in the vacuoles may be also bound to pectin-type polyanions, or, in the apoplasm, precipitated as calcium carbonate (Kinzel, 1989).

A typical distribution of calcium in cells of fully expanded tissue with high cation exchange capacity of the cell walls is shown in Fig. 8.25. There are distinct areas and compartments with high or very low calcium concentrations. High calcium concentrations are found in the middle lamella of the cell wall, at the exterior surface of the plasma membrane, in the endoplasmic reticulum (ER), and in the vacuole. Most of the water soluble calcium in a plant tissue is located in the vacuoles, accompanied by organic anions (e.g., malate) or inorganic anions (e.g., nitrate, chloride). The binding form of calcium in the ER is not clear. In contrast to the cell wall, ER and vacuole, the concentration of calcium in the cytosol is extremly low and maintained in the range of 0.1–0.2 μM of free Ca^{2+} (Felle, 1988; Evans *et al.*, 1991). Such low concentrations are essential for various reasons, such as prevention of precipitation of P_i, competition with

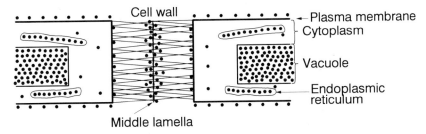

Fig. 8.25 Schematic representation of two adjacent cells with a typical distribution of calcium (●).

Mg^{2+} for binding sites and, last but not least, a prerequisite for the function of calcium as second messenger (Section 8.6.7).

The low cytosolic free Ca^{2+} concentrations are achieved by a generally low constitutive permeability of membranes for calcium, and by the action of membrane transporters removing calcium from the cytosol and expelling it to the apoplasm or accumulating it in intracellular stores such as the endoplasmic reticulum (ER), the chloroplasts, and the vacuole (Fig. 8.26). In contrast to animal cells, in plant cells the mitochondria are presumably of minor importance for storage of calcium (Evans *et al.*, 1991).

The major Ca^{2+} transporter at the plasma membrane and presumably also at the ER (Jones *et al.*, 1993) is a calcium pumping ATPase (Ca^{2+}/H^+ antiporter; Kasai and Muto, 1990). The Ca^{2+} transport at the tonoplast is also achieved by a Ca^{2+}/H^+ antiporter, energized by the proton-motive force of proton-pumping ATPase and PP_iase (Section 2.4.2). On average, these antiporters maintain a concentration difference of a factor up to 10^5 between free Ca^{2+} in the vacuole and in the cytosol (Schumaker and Sze, 1990). Also the chloroplasts may store large amounts of calcium (6.5–15 mM total calcium, mostly bound to thylakoid membranes), but in the stroma the concentration of free Ca^{2+} is only in the range of 2.4–6.3 μM (Kreimer *et al.*, 1988). On illumination calcium is transported along the electrochemical potential gradient from the cytosol into the stroma of the chloroplasts (Fig. 8.26).

An example of the importance of low cytosolic free Ca^{2+} concentrations for the functioning of certain key enzymes is shown in Table 8.26. The cytosolic fructose-1,6-bisphosphatase regulates sucrose synthesis from triosephosphates delivered by the chloroplasts (Fig. 8.20). A Ca^{2+} concentration as low as 1 μM severely inhibits the activity of this enzyme, even in the presence of 1000 times higher concentrations of magnesium (1 mM). The inhibitory effect of Ca^{2+} on Mg^{2+} is competitive, and the *in vivo* concentration of free Mg^{2+} in the cytosol is most probably lower than 4 mM (Section 8.5).

These results also demonstrate the important regulatory function of the calcium

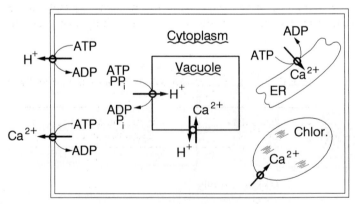

Fig. 8.26   Calcium transport processes in cell membranes for maintenance of low cytosolic free Ca^{2+}. (Modified from Evans *et al.*, 1991.)

Table 8.26
Effect of Calcium Concentration (Free Ca^{2+}) on the
Activity of Cytosolic Fructose-1,6-bisphosphatase[a]
from Spinach Leaves[b]

Mg concn	Ca^{2+} concentration (μM)				
(mM)	0	0.1	1.0	10	100
1.0	300	250	80	20	—
4.0	760	760	710	620	250

[a]Enzyme activity measured as nmol mg^{-1} protein min^{-1}.
[b]Recalculated from Brauer *et al.* (1990).

transport from the cytosol into the chloroplasts on illumination. Interestingly, chloro-plastic fructose-1,6-bisphosphatase is stimulated rather than inhibited by a rise in free Ca^{2+} in the stroma (Kreimer *et al.*, 1988).

8.6.3 Cell Wall Stabilization

Calcium bound as pectate in the middle lamella is essential for strengthening of the cell walls and of plant tissues. This function of calcium is clearly reflected in the close positive correlation between cation exchange capacity of cell walls and calcium content in plant tissues required for optimal growth (Section 8.6.8). The degradation of pectates is mediated by polygalacturonase, which is drastically inhibited by high calcium concentrations (Table 8.27). In agreement with this, in calcium-deficient tissue polyga-lacturonase activity is increased (Konno *et al.*, 1984), and a typical symptom of calcium deficiency is the disintegration of cell walls and the collapse of the affected tissues, such as the petioles and upper parts of the stems (Bussler, 1963).

In the leaves of plants receiving high levels of calcium during growth or grown under conditions of high light intensity, a large proportion of the pectic material exists as calcium pectate. This makes the tissue highly resistant to degradation by polygalacturo-nase (Cassells and Barlass, 1976). The proportion of calcium pectate in the cell walls is also of importance for the susceptibility of the tissue to fungal and bacterial infections

Table 8.27
Effect of Calcium on the Hydrolysis of Sodium Pectate by
Polygalacturonase[a]

Ca^{2+} Concentration (mg l^{-1})	Amount of galacturonic acid released (μmol $(4 h)^{-1}$)
0	3.5
40	2.5
200	0.6
400	0.2

[a]Based on Corden (1965).

(Chapter 11) and for the ripening of fruits. As shown by Rigney and Wills (1981) in experiments with tomato pericarp tissue during fruit development, the calcium content of the cell walls increases to the fully grown immature stage, but this is followed by a drop in the content and change in binding form of calcium in the tissue just before ripening ('softening' of the tissue). In agreement with this, in a nonripening mutant of tomato the content of bound calcium remains high and the polygalacturonase activity low throughout the fruit growth (Poovaiah, 1979). Increasing the calcium content of fruits, for example, by spraying several times with calcium salts during fruit development or by post-harvest dipping in $CaCl_2$ solution, leads to an increase in the firmness of the fruit (Cooper and Bangerth, 1976) and delays or even prevents fruit ripening (Wills *et al.*, 1977). However, correlations between the decrease in bound calcium and fruit ripening might not always occur as, for example, during degradation of the middle lamella by methylesterase also new binding sites for calcium are formed (Burns and Pressey, 1987).

8.6.4 Cell Extension, Secretory Processes

In the absence of an exogenous calcium supply, root extension ceases within a few hours (Fig. 8.27). This effect is more distinct in a calcium-free nutrient solution than in distilled water, an observation consistent with the role of calcium in counterbalancing the harmful effects of high concentrations of other cations at the plasma membrane. Although calcium is also involved in cell division (Schmit, 1981), the cessation of root growth in the absence of exogenous calcium is primarily the result of inhibited cell extension. Calcium provides cell wall rigidity by cross-linking the pectic chains of the middle lamella. On the other hand, for extension growth cell wall loosening is required, a process in which auxin-induced acidification of the apoplasm and replacement of calcium from the cross-links of the pectic chain are involved, although this is only one component of the event (Cleland *et al.*, 1990). Auxin also activates calcium channels in the plasma membrane and thereby leads to a transient increase in cytosolic free Ca^{2+} concentrations (Felle, 1988) which, in turn, stimulates the synthesis of cell wall precursors in the cytosol and its secretion into the apoplasm (Brummell and Hall, 1987). In accordance with this, pollen tube growth is also dependent on the presence of

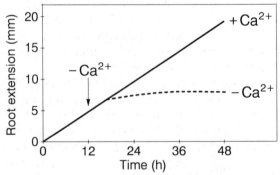

Fig. 8.27 Relationship between the extension of primary roots of bean and the calcium concentration (±2 mM) in the nutrient solution. (Based on Marschner and Richter, 1974.)

calcium in the substrate, and the direction of its growth is chemotropically controlled by the extracellular calcium gradient (Mascarenhas and Machlis, 1964). Pollen tube growth is characterized by secretion of cell wall material from the cytosol into the apoplasm at the apex of the pollen tube.

In shoots elongation growth, for example, of coleoptiles is mainly determined by the extensibility of the epidermal layer which is not only the target of auxin action (Section 5.6) but is also characterized by much higher calcium concentrations than the bulk of the other tissue (Hodson and Sangster, 1988).

Calcium is required for the formation of secretory vesicles and their fusion with the plasma membrane leading to *exocytosis*, for example, of cellulose precursors for cell wall formation, as well as of formation of mucilage or callose. The secretion is triggered by a rise in cytosolic free Ca^{2+} concentrations from about 0.1 to 1.0 μM or even higher (Steer, 1988b). In some instances, inflow of calcium from the apoplasm is restricted to a certain part of the plasma membrane, which then acts as focus for exocytosis and thereby establishes a cellular polarity (e.g., pollen tube or root hair growth). Polarity in calcium influx into the cytosol and action of calcium efflux pumps at the opposite site of the cell are also realized in the calcium–auxin counterflow in shoot apices or young fruits (Section 3.4).

In root caps, secretion of mucilage depends on the extracellular calcium concentration. Removal of extracellular (apoplasmic) calcium drastically reduces secretory activity of the root cap cells but also blocks the gravitropic response in plant roots (Bennet *et al.*, 1990). Also the auxin transport inhibitors like TIBA inhibit both calcium redistribution and gravitropism (Lee *et al.*, 1984). The root cap is the site of perception of the gravity signal (Poovaiah *et al.*, 1987) leading to redistribution of apoplasmic calcium towards the lower site of the root caps (Bjorkman and Cleland, 1991) and enhanced mucilage secretion at this site. This in turn permits higher rates of basipetal transport of growth inhibitors (ABA?) in the mucilage to the root extension zone at the lower site of the root (Bennet *et al.*, 1990). In principle, redistribution of calcium is also involved in gravitropic responses of coleoptiles (Slocum and Roux, 1983).

Callose formation is another example of calcium-induced enhanced secretory processes (Fig. 8.29). At the plasma membrane–cell wall interface normally cellulose (1,4 β-glucan units) is synthesized. In response to injury such as mechanical damage and parasitic infection (Kauss, 1987; Kauss *et al.*, 1991) or high aluminum concentrations (Wissemeier *et al.*, 1987; Jorns *et al.*, 1991), however, a switch to callose (1,3 β-glucan units) production can occur (Lerchl *et al.*, 1989). This switch is also triggered by an increase in cytosolic free Ca^{2+} by a factor of about 10 (Kauss, 1987). Callose deposition in the stigma in response to incompatible pollination also appears to be a calcium-dependent process (Singh and Paolillo, 1990).

Stimulation of α-amylase activity in germinating cereal seeds is one of the few examples of enzyme stimulation by high (mM) calcium concentrations. This stimulation is also mediated through a calcium-induced enhanced synthesis and secretion of α-amylase in aleurone cells (Mitsui *et al.*, 1984; Bush *et al.*, 1986). Calcium is a constituent of α-amylase which is synthesized on the rough ER. Transport of Ca^{2+} through the ER membranes is enhanced by GA and inhibited by ABA, leading to the typical stimulation (GA) and inhibition (ABA) of α-amylase activity in aleurone cells (Bush *et al.*, 1993).

8.6.5 Membrane Stabilization

The fundamental role of calcium in membrane stability and cell integrity is reflected in various ways. It can be demonstrated most readily by the increased leakage of low-molecular-weight solutes from cells of calcium-deficient tissue (e.g., tomato fruits; Goor, 1966) and, in severely deficient plants, by a general disintegration of membrane structures (Hecht-Buchholz, 1979) and a loss of cell compartmentation (Fig. 8.28).

Calcium stabilizes cell membranes by bridging phosphate and carboxylate groups of phospholipids (Caldwell and Haug, 1981) and proteins, preferentially at membrane surfaces (Legge *et al.*, 1982). There can be an exchange between calcium at these binding sites and other cations (e.g. K^+, Na^+, or H^+). The exchange of plasma

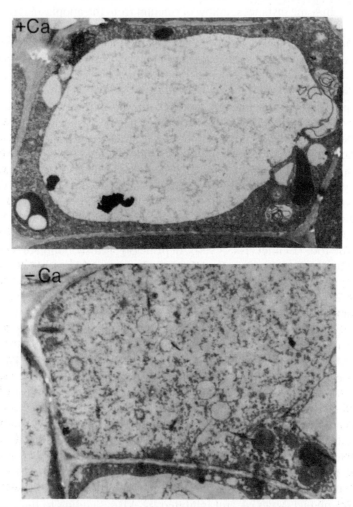

Fig. 8.28   Calcium nutritional status and fine structure of cells of potato sprouts. (*Top*) Calcium-sufficient. (*Bottom*) Calcium-deficient; loss of compartmentation. (Courtesy of Ch. Hecht-Buchholz.)

Table 8.28
Influence of Calcium on Carbohydrate Loss from Cotton Roots[a]

Treatment			Carbohydrate loss (μg per seedling $(4\,h)^{-1}$)
Aeration	Temperature (°C)	Solution	
O_2	31	Distilled water	18
O_2	5	Distilled water	57
O_2	5	10^{-5} M Ca^{2+}	7
N_2	31	Distilled water	89
N_2	31	10^{-5} M Ca^{2+}	7

[a]Based on Christiansen *et al.* (1970).

membrane-bound calcium, for example, by high external concentrations of sodium is a main factor involved in salinity stress (Lynch *et al.*, 1987; Rengel, 1992b; Section 16.6). Replacement of plasma membrane-bound calcium – or blocking of calcium channels – is also discussed as causative factors in aluminum toxicity (Huang *et al.*, 1992b; Rengel and Elliott, 1992; Rengel, 1992a).

To fulfil its functions at the plasma membrane, therefore, calcium must always be present in the external solution, where it regulates the selectivity of ion uptake (Section 2.5) and prevents solute leakage from the cytoplasm. The membrane-protecting effect of calcium is most prominent under stress conditions such as low temperature (Zsoldos and Karvaly, 1978) and anaerobiosis (Table 8.28). Calcium also alleviates the damage of tissues caused by freezing–thawing stress. During thawing of frozen tissues such as onion bulbs, the efflux of low-molecular-weight solutes is increased, particularly of potassium ions. During this efflux, plasma membrane-bound calcium is also replaced leading to further increase in membrane permeability and solute efflux. Accordingly, an increase in external calcium concentration drastically reduces solute efflux and freezing injury (Arora and Palta, 1988). The ameliorating effect of external calcium supply on low temperature stress or anaerobiosis (Table 8.28) is most likely based on a similar mechanism as in the case of freezing stress.

In calcium-deficient tissues the impairment of membrane integrity leads to increased respiration rates which are related to enhanced leakage of respiratory substrates from the vacuole to the respiratory enzymes in the cytoplasm (Bangerth *et al.*, 1972). Calcium treatment of deficient tissues therefore decreases respiration rates; it also enhances the net rate of protein synthesis (Faust and Klein, 1974). These features of calcium deficiency are similar to those related to senescence. In agreement with this, in leaf segments net degradation of protein and chlorophyll (Table 8.29) can be deferred by addition of cytokinins (benzyladenine, BA), or calcium. The effects of both substances are additive. Senescence is closely related to peroxidation of membrane lipids through elevated levels of free oxygen radicals (Section 2.3), and the protecting effect of both calcium and BA on peroxidation of membrane lipids is indicated by the distinctly lower lipoxygenase activity (Table 8.29).

Calcium not only protects membrane lipids but is also involved in enhanced breakdown. This is true for phospholipids and well documented during breakdown of

Table 8.29
Influence of Calcium and Benzyladenine (BA) on Chlorophyll Content[a] of Cowpea Leaf Disks Incubated in the Dark[b]

Treatment	Incubation time (h)				Lipoxygenase activity after 72 h[c]
	0	24	48	72	
Control (distilled H_2O)	2.50	1.84	1.59	1.01	2.5
15 mM $CaCl_2$	—	2.05	2.00	1.95	1.5
15 μM BA	—	1.92	1.85	1.62	1.5
15 μM BA + 15 mM $CaCl_2$	—	2.67	2.52	2.31	0.9

[a]Measured as mg chlorophyll g^{-1} fresh wt.
[b]Based on Swamy and Suguna (1992).
[c]μmol O_2 consumed mg^{-1} protein min^{-1}.

membranes of lipid bodies in cotyledons during seed germination, for example, of *Ricinus*. At least two lipophytic enzymes associated with protein bodies are stimulated by calcium or calmodulin (Paliyath and Thompson, 1987), and 30 μM free Ca^{2+}, which may derive from phytate hydrolysis (Section 8.4.5), stimulate the activity of phospholipase by a factor of about 40 (Hills and Beevers, 1987).

Calcium stimulates a range of membrane-bound enzymes (Rensing and Cornelius, 1980), particularly ATPases at the plasma membrane of roots of certain plant species (Kuiper and Kuiper, 1979). Because the activities of many membrane-bound enzymes are modulated by membrane structure, calcium presumably enhances the activity of those enzymes even though it is bound to noncatalytic sites at the membranes (Clarkson and Hanson, 1980). In agreement with this assumption, this stimulation requires millimolar concentrations of free Ca^{2+}. In contrast, for stimulation or inhibition of cytosolic or chloroplastic enzymes only micromolar concentrations of calcium are required, indicating another mechanism of regulation (Section 8.6.7).

8.6.6   Cation–Anion Balance and Osmoregulation

In vacuolated cells of leaves in particular, a large proportion of calcium is localized in the vacuoles, where it might contribute to the cation–anion balance by acting as a counterion for inorganic and organic anions (Kinzel, 1989). In plant species which preferentially synthesize oxalate in response to nitrate reduction, the formation of calcium oxalate in the vacuoles is important for maintenance of a low cytosolic free Ca^{2+} concentration (Kinzel, 1989). The same holds true for plant species with preferential formation of calcium oxalate in the apoplasm (Section 8.6.3).

The formation of sparingly soluble calcium oxalate is also important for the osmoregulation of cells and provides a means of salt accumulation in vacuoles of nitrate-fed plants without increasing the osmotic pressure in the vacuoles (Osmond, 1967). In mature sugar beet leaves, for example, up to 90% of the total calcium is bound to oxalate (Egmond and Breteler, 1972).

Although indirectly, in its function as second messenger calcium also plays a key role in osmoregulation. Stomatal movement or nyctinastic and seismonastic movements are

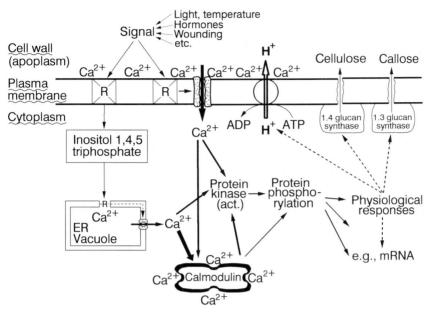

Fig. 8.29 Model of the role of calcium as second messenger in the signal transduction in plant cells. R = receptor sites, e.g., phytochrome, or binding sites of hormones, e.g., ABA, IAA.

typically turgor-regulated processes brought about by turgor changes in individual cells (guard cells) or tissues (e.g., motor cells of pulvini). These turgor changes are driven by fluxes mainly of potassium (Section 8.7.6), chloride and malate (Section 9.8) as osmotic active components. It is now well established that for transduction of the signals (e.g., light, touch) to the physiological response in terms of membrane-bound ion pumps a transient change of cytosolic free Ca^{2+} is required (Roblin *et al.*, 1989; Mansfield *et al.*, 1990). The action of ABA on stomatal closure depends on the calcium concentrations in the leaf epidermis, which are usually much higher than in the other cells (Atkinson *et al.*, 1990). ABA-induced activation of calcium channels and sharp increase in cytosolic free Ca^{2+} concentrations seem to block proton pumps and open anion channels, and both events lead to loss in turgor of the guard cells and stomatal closure (Atkinson *et al.*, 1990).

8.6.7 Calcium as Second Messenger

The function of calcium as second messenger is based on the very low cytosolic free Ca^{2+} concentrations in the range of 0.1–0.2 μM (Trewavas and Gilroy, 1991) and the high concentrations in various adjacent compartments (Fig. 8.26). Environmental signals can activate calcium channels in the membranes of these calcium pools and thereby increase calcium influx and cytosolic free Ca^{2+} concentrations (Fig. 8.29). Such an increase in cytosolic free Ca^{2+} is induced, for example, by ABA (Tester, 1990), IAA (Felle, 1988), light (Roblin *et al.*, 1989), pathogenic infection (Atkinson *et al.*, 1990), and mechanical stress or injury (Rincon and Hanson, 1986). An increase in cytosolic free Ca^{2+} concentration is also involved in *thigmorphogenesis*, that is the long-term response to mechanical stimulation (e.g., wind), expressed as altered growth pattern, for example, shorter stem length (Roberts and Harmon, 1992).

It well might be, that many of these environmental signals induce a depolarization of the plasma membrane (potential difference in the range of -120 to -180 mV; Section 2.4), for example, a sudden temperature drop (Minorsky and Spanswick, 1989) and thereby activate calcium channels in the plasma membrane (Davies, 1987). It is not yet clear how environmental signals received by the plasma membrane are transmitted to the calcium pools in the ER and vacuole (Fig. 8.29). Inositol-1,4,5-triphosphate released from the plasma membrane may act as second messenger (Trewavas and Gilroy, 1991).

In the cytosol the principal targets of calcium signals are calcium-binding proteins known as *calcium-modulated proteins*, the most well known of which are *calmodulins* (CaM), and *calcium-dependent* but *CaM-independent* protein kinases (Roberts and Harmon, 1992). Calcium dependent protein kinases are directly stimulated by calcium, and these enzymes phosphorylate other enzymes (Section 8.4.4). Such enzymes are located, for example, in the plasma membrane. The proton pumping ATPase is such an enzyme which is phosphorylated by the action of a calcium-dependent protein kinase (Roberts and Harmon, 1992). In other instances calcium is bound with high selectivity, but reversibly, to CaM, a polypeptide with four binding sites for calcium (Fig. 8.29). Binding of calcium changes the conformation and activity of CaM. A relatively large number of CaM-dependent enzymes are known, including NAD kinase which catalyses the conversion of NAD^+ into NADP, the terminal acceptor of electrons in the chloroplasts (Fig. 5.1). Also the calcium-translocating ATPase, which is an important component in calcium homeostasis in the cytosol (Fig. 8.26) is stimulated by CaM.

The concentrations of CaM are much higher in meristematic tissues and in root cap cells than in more mature zones (Roberts and Harmon, 1992). Le Gales *et al.* (1980) found high levels of calcium-binding proteins (calmodulin?) in the cytoplasm of calcicole plant species but only low levels in calcifuge species. This might reflect a mechanism for the adaptation of plants to calcareous soils (Section 16.5). By such a mechanism the plants may maintain a low cytosolic free Ca^{2+} concentration in root cells. In view of the very low cytosolic free Ca^{2+} concentrations and the often high calcium concentrations in the vacuoles, a shuttle system for calcium transfer between the plasma membrane and the tonoplast is likely and either CaM (Marmé, 1983) or the ER (Kinzel, 1989) might be involved. The regulatory function of cytosolic free Ca^{2+} also implies that the concentration of free Ca^{2+} in the apoplasm of growing tissues (e.g. shoot and root apical zones) has to be strictly regulated, and explains why root growth is so sensitive to environmental factors which decrease the calcium concentration in the apoplasm.

8.6.8 Calcium Supply, Plant Growth, and Plant Composition

The calcium content of plants varies between 0.1 and $>5.0\%$ of dry wt depending on the growing conditions, plant species, and plant organ. The calcium requirement for optimum growth is much lower in monocotyledons than in dicotyledons (Loneragan *et al.*, 1968; Loneragan and Snowball, 1969), as shown in Table 8.30. In well-balanced,

Table 8.30
Effect of Calcium Concentration in the Nutrient Solution on
Relative Growth Rates of Plants and Calcium Content in the
Shoots[a]

Plant species	Ca^{2+} concentration (μM)				
	0.8	2.5	10	100	1000
Relative growth rate					
Ryegrass	42	100	94	94	93
Tomato	3	19	52	100	80
Calcium content (mg g^{-1} dry wt)					
Ryegrass	0.6	0.7	1.5	1.7	10.8
Tomato	2.1	1.3	3.0	12.9	24.9

[a]Based on Loneragan *et al.* (1968) and Loneragan and Snowball (1969).

flowing nutrient solutions with controlled pH, maximal growth rates were obtained at calcium supply levels of 2.5 μM (ryegrass) and 100 μM (tomato), i.e., differing by a factor of 40. This difference is primarily a reflection of the calcium demand at the tissue level, which is lower in ryegrass (0.7 mg) than in tomato (12.9 mg). Genotypical differences in calcium requirement are closely related to the binding sites in the cell walls, that is, the cation-exchange capacity. This also explains why the calcium requirement of some algal species is in the micronutrient range or even difficult to demonstrate at all (O'Kelley, 1969).

The differences between monocotyledons and dicotyledons in calcium demand shown for ryegrass and tomato (Table 8.30) have been confirmed, in principle, for a large number of plant species (Islam *et al.*, 1987). However, *Lupinus angustifolius* had a calcium requirement both in supply and tissue content, which was comparable to monocotyledons, and the growth of this species was severely depressed at higher calcium contents in the tissue. Such typical calcifuge behaviour might be related to insufficient capacity for compartmentation and/or physiological inactivation of calcium (e.g., precipitation as calcium oxalate).

Another factor determining the calcium requirement for optimum growth is the concentration of other cations in the external solution. Because of its replacement by other cations from its binding sites at the exterior surface of the plasma membrane, the calcium requirement increases with increasing external concentrations of heavy metals (Wallace *et al.*, 1966), aluminum (Section 16.3), sodium chloride (LaHaye and Epstein, 1971), or protons. At low compared to high pH the Ca^{2+} concentration in the external solution has to be several times higher in order to counteract the adverse effect of high H^+ concentrations on root elongation (Table 8.31). A similar relationship exists between external pH and the calcium requirement for nodulation of legumes (Chapter 7). In order to protect roots against the adverse effects of high concentrations of various other cations in the soil solutions the Ca^{2+} concentrations required for optimal growth has to be much higher in soil solutions than in balanced flowing nutrient solutions (Asher and Edwards, 1983).

An increase in the concentration of Ca^{2+} in the external solution leads to an increase

Table 8.31

Effects of Calcium Concentration and Solution pH on the
Growth Rate of Seminal Roots of Soybean[a]

Ca^{2+} concentration $(mg\ l^{-1})$	Root growth rate $(mm\ h^{-1})$	
	pH 5.6	pH 4.5
0.05	2.66	0.04
0.50	2.87	1.36
2.50	2.70	2.38

[a]Based on Lund (1970).

in the calcium content in the leaves, but not necessarily in low-transpiring organs such as fleshy fruits and tubers supplied predominantly via the phloem. Plants have developed mechanisms for restricting the transport of calcium to these organs by maintaining low calcium concentrations in the phloem sap (Chapter 3), or by precipitation of Ca^{2+} as oxalate along the sieve tubes (Liegel, 1970; Fink, 1991b) or in the seed coat (Mix and Marschner, 1976c). The physiological reasons for this restriction are now much better understood (Section 8.6.7). Dilution of the calcium content in the tissue by growth is another way of maintaining a low calcium level, which is necessary in fruits and storage tissues for rapid cell expansion and high membrane permeability (Mix and Marschner, 1976a). High growth rates of low-transpiring organs, however, increase the risk that the tissue content of calcium falls below the critical level required for cell wall stabilization and membrane integrity, and perhaps also for functioning as second messenger. In fast-growing tissues and organs so-called calcium deficiency-related disorders are wide-spread, such as tipburn in lettuce, blackheart in celery, blossom end rot in tomato or watermelon, and bitter pit in apple (Shear, 1975; Bangerth, 1979).

Table 8.32

Effects of Calcium Sprays during the Growing Season on the
Calcium Content and Percentage of Wastage of 'Cox' Apples
during Storage[a,b]

	Wastage (%)	
	Unsprayed	Sprayed
Calcium content (mg $(100\ g)^{-1}$ fresh wt)	3.35	3.90
Storage disorders		
Lentical blotch pit	10.4	0
Senescence breakdown	10.9	0
Internal bitter pit	30.0	3.4
Gloesporium rots	9.1	1.7

[a]From Sharpless and Johnson (1977).
[b]Sprays containing 1% calcium nitrate were applied four times during the growing season. The apples were stored for 3 months at 3.5°C.

Low tissue contents of calcium in fleshy fruits also increase the losses caused by enhanced senescence of the tissue and by fungal infections (Table 8.32). Even a relatively small increase in the calcium level of the fruits can be effective in preventing or at least drastically decreasing the economical losses caused by various storage disorders including rotting due to *Gloesporium* infection.

8.7 Potassium

8.7.1 General

Potassium is a univalent cation with a hydrated ionic radius of 0.331 nm and a hydration energy of 314 J mol^{-1}. Its uptake is highly selective and closely coupled to metabolic activity (Section 2.4). It is characterized by high mobility in plants at all levels – within individual cells, within tissues, and in long-distance transport via the xylem and phloem. Potassium is the most abundant cation in the cytoplasm and K$^+$ and its accompanying anions make a major contribution to the osmotic potential of cells and tissues of glycophytic plant species. For various reasons (see below) K$^+$ has an outstanding role in plant water relations (Hsiao and Läuchli, 1986). Potassium is not metabolized and it forms only weak complexes in which it is readily exchangeable (Wyn Jones *et al.*, 1979). Therefore, K$^+$ does not strongly compete for binding sites requiring divalent cations (e.g., Mg^{2+}). On the other hand, due to its high concentrations in the cytosol and chloroplasts it neutralizes the soluble (e.g., organic acid anions and inorganic anions) and insoluble macromolecular anions and stabilizes the pH between 7 and 8 in these compartments, the optimum for most enzyme reactions. For example, a decrease in pH from 7.7 to 6.5 almost completely inhibits nitrate reductase activity (Pflüger and Wiedemann, 1977). The various functions of potassium in plants have been reviewed by Läuchli and Pflüger (1978).

8.7.2 Compartmentation and Cellular Concentrations

In almost all cases, cytosolic potassium concentrations are maintained in the range 100–200 mM (Leigh and Wyn-Jones, 1984), and this is also true for chloroplasts (Schröppel-Meier and Kaiser, 1988). In its functions, in these compartments ('metabolic pool') K$^+$ is not replaceable by other inorganic cations such as Na$^+$ (Section 10.2). In contrast, the vacuolar K$^+$ concentrations might vary between 10 and 200 mM (Hsiao and Läuchli, 1986) or may even reach 500 mM in the guard cells of stomata (Outlaw, 1983). The functions of K$^+$ in cell extension and other turgor-driven processes are related to the K$^+$ concentration in the vacuoles. In its osmotic functions in the vacuoles K$^+$ is replaceable to a varying degree by other cations (Na$^+$, Mg^{2+}, Ca^{2+}) or organic solutes (e.g. sugars). In contrast to calcium the concentrations of potassium in the apoplasm are usually low, with the exception of specialized cells or tissues (stomata, pulvini) where the apoplasmic potassium concentrations might transiently rise up to 100 mM.

For the fast transport of K$^+$ between different cell compartments and cells within a tissue, K$^+$ channels are required in the membranes, also in thylakoids for the rapid K$^+$ transport from the intrathylakoid space to the stroma upon illumination of chloroplasts

(Tester and Blatt, 1989). The channels in the membranes open and close ('gate') at different sequence and length in response to environmental signals and changes in electropotential of the membranes and permit permeation rates of ions such as K^+ which are at least three orders of magnitude faster than that catalyzed by pumps and carriers (Tester, 1990). Although the K^+ channels are, in principle, similar to the Ca^{2+} channels (Section 8.6.7), their function is totally different. Potassium ions act directly as solutes changing the osmotic potential in the compartments and thereby turgor, and as carriers of charges also the membrane potential.

8.7.3   Enzyme Activation

A large number of enzymes is either completely dependent on or stimulated by K^+ (Suelter, 1970). Potassium and other univalent cations activate enzymes by inducing conformational changes in the enzyme protein. All macromolecules are highly hydrated and stabilized by firmly bound water molecules forming an electrical double layer. At univalent salt concentrations of about 100–150 mM, maximum suppression of this electrical double layer and optimization of the protein hydration occur (Wyn Jones and Pollard, 1983). This concentration range agrees well with the K^+ concentrations in the cytosol and in the stroma of plants well supplied with K^+ (Leigh and Wyn Jones, 1984). In general, K^+-induced conformational changes of enzymes increase the rate of catalytic reactions, V_{max}, and in some cases also the affinity for the substrate, K_m (Evans and Wildes, 1971).

In potassium-deficient plants some gross chemical changes occur, including an accumulation of soluble carbohydrates, a decrease in content of starch, and an accumulation of soluble nitrogen compounds. These changes in carbohydrate metabolism are presumably related to the high K^+ requirement of certain regulatory enzymes, particularly pyruvate kinase and phosphofructokinase (Läuchli and Pflüger, 1978). As shown in Fig. 8.30, the activity of starch synthase is also highly dependent on univalent cations, and of these K^+ is the most effective. The enzyme catalyzes the transfer of glucose to starch molecules:

$$ADP\text{-glucose} + starch \rightleftharpoons ADP + glucosyl\text{-starch}$$

Potassium similarly activates starch synthase isolated from a variety of other plant species and organs (e.g., leaves, seeds, and tubers), the maximum lying in the range of 50–100 mM K^+ (Nitsos and Evans, 1969). Higher concentrations, however, may have inhibitory effects (Preusser et al., 1981).

Another function of K^+ is the activation of membrane-bound proton-pumping ATPases (Section 2.4). This activation not only facilitates the transport of K^+ from the external solution across the plasma membrane into the root cells but also makes potassium the most important mineral element in cell extension and osmoregulation.

Tissues of potassium-deficient plants exhibit a much higher activity of certain hydrolases or of oxidases such as polyphenol oxidase than do tissues of normal (sufficient) plants. It is not clear whether these changes in enzyme activity are caused by direct or indirect effects of K^+ on the synthesis of the enzymes. An instructive example of indirect effects is the accumulation of the diamine putrescine in potassium-deficient plants by a factor of 80–100 (Houman et al., 1991; Tachimoto et al., 1992). The enzymes

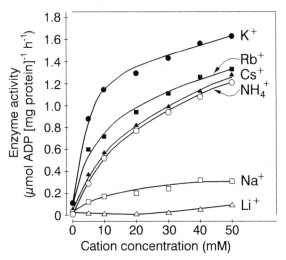

Fig. 8.30 Effect of univalent cations (as chlorides) on the activity of ADP-glucose starch synthase from maize. (Nitsos and Evans, 1969.)

which catalyze the synthesis of putrescine from arginine via agmatine (Fig. 8.14) are inhibited by high K^+ concentrations (Reggiani *et al.*, 1993) and stimulated by low cellular pH. Considering the dominant role of K^+ in the maintenance of high cytoplasmic pH, it appears that enhanced synthesis of putrescine as divalent cation is a reflection of a homeostasis of the cytosol pH; in potassium-deficient plants the putrescine concentrations may account for up to 30% of the deficit in K^+ equivalents (Murty *et al.*, 1971). In agreement with this compensatory function of putrescine, external supply of putrescine to potassium-deficient plants enhanced growth and prevented visual symptoms of potassium deficiency (Tachimoto *et al.*, 1992).

8.7.4 Protein Synthesis

Potassium is required for protein synthesis in higher concentrations than for enzyme activation which approaches its maximum at about 50 mM K^+ (e.g., Fig. 8.30). In cell-free systems the rate of protein synthesis by ribosomes isolated from wheat germ has an optimum at 130 mM K^+ and ~2 mM Mg^{2+} (Wyn Jones *et al.*, 1979). It is probable that K^+ is involved in several steps of the translation process, including the binding of tRNA to ribosomes (Wyn Jones *et al.*, 1979). In green leaves the chloroplasts account for about half of both leaf RNA and leaf protein. In C_3 plants the majority of the chloroplast protein is RuBP carboxylase (Section 8.2). Accordingly, the synthesis of this enzyme is particularly impaired under potassium deficiency and responds rapidly to resupply of K^+ as shown in Table 8.33. The maximum activation was obtained at K^+ concentrations in the external solution of 10 mM. This concentration was obviously sufficient to obtain a more than 10-fold higher K^+ concentration in the chloroplasts which is required for high rates of protein synthesis.

The role of K^+ in protein synthesis is not only reflected in the accumulation of soluble nitrogen compounds (e.g., amino acids, amides, and nitrate) in potassium-deficient

Table 8.33

Effect of Potassium on the Incorporation of [^{14}C]Leucine into RuBP Carboxylase in the Leaves of Potassium-Deficient Alfalfa Plants[a]

Preincubation medium (mM KNO$_3$)	[^{14}C]Leucine incorporation (dpm mg^{-1} RuBP carboxylase (24 h)$^{-1}$)
0.0	99
0.01	167
0.10	220
1.00	274
10.00	526
Control (K-sufficient plants)	656

[a]Potassium-deficient leaves were preincubated in light for 20 h with potassium. From Peoples and Koch (1979).

plants (Mengel and Helal, 1968), but can also be demonstrated directly following the incorporation of ^{15}N-labeled inorganic nitrogen into the protein fraction. For example, within 5 h in potassium-sufficient and potassium-deficient tobacco plants, 32% and 11%, respectively, of the total ^{15}N taken up had been incorporated into the protein nitrogen (Koch and Mengel, 1974). From studies by Pflüger and Wiedemann (1977) it seems highly probable that K$^+$ not only activates nitrate reductase but is also required for its synthesis. On the other hand, under potassium deficiency also the expression of a number of polypeptides is increased, especially of a 45 kDa membrane-bound polypeptide (Fernando *et al.*, 1990). It is not clear to what extent this enhanced synthesis is a response to cytosolic pH decrease.

8.7.5 Photosynthesis

In higher plants potassium affects photosynthesis at various levels. Potassium is the dominant counterion to the light-induced H$^+$ flux across the thylakoid membranes (Tester and Blatt, 1989) and for the establishment of the transmembrane pH gradient necessary for the synthesis of ATP (photophosphorylation), in analogy to ATP synthesis in mitochondria.

The role of potassium in CO$_2$ fixation can be demonstrated most clearly with isolated chloroplasts (Table 8.34). An increase in the external K$^+$ concentration to 100 mM, that is, to about the K$^+$ concentration in the cytosol of intact cells, stimulates CO$_2$ fixation more than threefold. On the other hand, the ionophore valinomycin, which makes biomembranes 'leaky' for passive K$^+$ flux, severely decreases CO$_2$ fixation. The effect of valinomycin can be compensated for by high external K$^+$ concentrations.

For maintenance of high pH in the stroma, upon illumination an additional influx of K$^+$ from the cytosol is required and mediated by a H$^+$/K$^+$ counterflow through the chloroplast envelope (Wu *et al.*, 1991). This counterflow is impaired under drought stress. During dehydration isolated chloroplasts lose large amounts of their K$^+$, and photosynthesis decreases; this decrease can be overcome by high concentrations of

Table 8.34
Influence of the Antibiotic Valinomycin and Potassium on the Rate of Carbon
Dioxide Fixation in Isolated Intact Spinach Chloroplasts[a]

Treatment	Rate of CO_2 fixation (μmol mg^{-1} chlorophyll h^{-1})	Percentage of control
Control	23.3	100
100 mM K$^+$	79.2	340
1 μM valinomycin	11.0	47
1 μM valinomycin + 100 mM K$^+$	78.4	337

[a]From Pflüger and Cassier (1977).

extrachloroplastic K$^+$ (Pier and Berkowitz, 1987), similarly as in case of valinomycin (Table 8.34). Also in intact plants the decrease in photosynthesis under drought stress is much less severe at high K$^+$ supply (Fig. 8.31). Supply of 2 mM K$^+$ supports maximal photosynthesis in well-watered plants but not under drought stress. The depression in photosynthesis under drought stress is, however, much less severe in plants supplied with 6 mM K$^+$ (Fig. 8.31). This ameliorating effect of potassium was associated with distinctly higher leaf contents of potassium (Pier and Berkowitz, 1987). The higher potassium requirement in the leaves of plants exposed to drought or salinity stress (Chow *et al.*, 1990) is primarily caused by the necessity to maintain high stromal K$^+$ concentrations under these conditions (Sen Gupta *et al.*, 1989).

In the absence of drought or salinity stress, with inadequate K$^+$ supply and correspondingly lower leaf contents, the rate of photosynthesis is lower (Fig. 8.31) and is closely related to the potassium contents in the leaves (Table 8.35). In deficient plants various parameters of CO_2 exchange are affected. An increase in potassium content in the leaves increases rate of photosynthesis and RuBP carboxylase activity, as well as photorespiration, probably due to a stronger depletion of CO_2 at the catalytic sites of

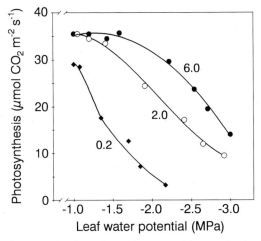

Fig. 8.31 Effect of K$^+$ supply (mM) to wheat plants on photosynthesis of leaves at declining leaf water potentials. (Based on Sen Gupta *et al.*, 1989.)

Table 8.35
Relationship between Potassium Content in Leaves, Carbon Dioxide Exchange, and RuBP Carboxylase Activity in Alfalfa[a]

Leaf potassium (mg g^{-1} dry wt)	Stomatal resistance (s cm^{-1})	Photosynthesis (mg CO$_2$ dm^{-2} h^{-1})	RuBP carboxylase activity (μmol CO$_2$ mg^{-1} protein h^{-1})	Photorespiration (dpm dm^{-2})	Dark respiration (mg CO$_2$ dm^{-2} h^{-1})
12.8	9.3	11.9	1.8	4.0	7.6
19.8	6.8	21.7	4.5	5.9	5.3
38.4	5.9	34.0	6.1	9.0	3.1

[a]From Peoples and Koch (1979).

the enzyme (Section 5.2). With increase in potassium content dark respiration decreases (Table 8.35). Higher respiration rates are a typical feature of potassium deficiency (Botrill *et al.*, 1970). It is apparent from Table 8.35 that potassium nutritional status may also affect photosynthesis in leaves via its function in stomatal regulation.

8.7.6 Osmoregulation

In Chapter 3 it was shown that a high osmotic potential in the stele of roots is a prerequisite for turgor-pressure-driven solute transport in the xylem and for the water balance of plants. In principle, at the level of individual cells or in certain tissues, the same mechanisms are responsible for cell extension and various types of movement. Potassium, as the most prominent inorganic solute, plays a key role in these processes (Hsiao and Läuchli, 1986).

8.7.6.1 Cell Extension

Cell extension involves the formation of a large central vacuole occupying 80–90% of the cell volume. There are two major requirements for cell extension: an increase in cell wall extensibility, and solute accumulation to create an internal osmotic potential (Fig. 8.32). In most instances cell extension is the consequence of the accumulation in the

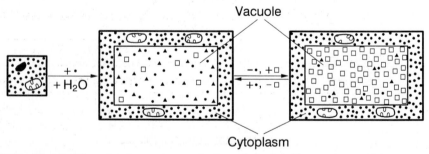

Fig. 8.32 Model of the role of potassium and other solutes in cell extension and osmoregulation. Key: ●, K$^+$; □, reducing sugars; sucrose, Na$^+$; ▲, organic acid anions.

cells of K^+, which is required for both stabilizing the pH in the cytoplasm and increasing the osmotic potential in the vacuoles. In *Avena* coleoptiles, IAA-stimulated H^+ efflux is electrochemically balanced by a stoichiometric K^+ influx; in the absence of external K^+, IAA-induced elongation declines and ceases after a few hours (Haschke and Lüttge, 1975). In cucumber cotyledons K^+ supply enhances extension by a factor of ~4 in response to the application of cytokinins (Green and Muir, 1979). Similarly, cell extension in leaves is closely related to their potassium content. In expanding leaves of bean plants suffering from potassium deficiency, turgor, cell size, and leaf areas were significantly lower than in expanding leaves well supplied with potassium (Mengel and Arneke, 1982). This inverse relationship between potassium nutritional status of plants and cell size also holds true for storage tissues such as carrots (Pfeiffenschneider and Beringer, 1989).

As shown in Table 8.36 the enhancement of stem elongation by GA is also dependent on the K^+ supply. Potassium and GA act synergistically, the highest elongation rate being obtained when both GA and K^+ are applied. The results in Table 8.36 seem to indicate further that K^+ and reducing sugars act in a complementary manner to produce the turgor potential required for cell extension. In the plants with a low K^+ supply, however, GA-stimulated growth was correlated with a marked increase in K^+ concentration in the elongation zone to a level similar to that of the reducing sugars (Guardia and Benloch, 1980). As K^+ was supplied together with Cl^- (KCl), a substantial proportion of the effects on plant growth and sugar concentrations are presumably combined effects of both K^+ and Cl^- on osmotic pressure.

These and other data from the literature demonstrate that K^+, in association with either inorganic anions or organic acid anions, is the main solute required in the vacuoles for cell extension. The extent to which sugars and other low-molecular-weight organic solutes contribute to the osmotic potential and turgor-driven cell expansion depends on the potassium nutritional status of plants, as well as on plant species and specific organs. In plant species such as tall fescue, in the elongation zone of leaf blades about half the imported sugars are used for accumulation of osmotically active fructanes in the vacuoles (Schnyder *et al.*, 1988).

After completion of cell extension, for maintenance of the cell turgor K^+ can be fairly readily replaced in the vacuoles by other solutes such as Na^+ or reducing sugars (Fig. 8.32). Inverse relationships between tissue concentrations of K^+ and sugars, reducing

Table 8.36
Effects of Potassium and Gibberellic Acid (GA) on Plant Height and Concentrations of Sugars and Potassium in the Shoots of Sunflower Plants[a]

Treatment		Plant height (cm)	Concentration (μmol g^{-1} fresh wt)		
KCl (mM)	GA (mg l^{-1})		Reducing sugars	Sucrose	Potassium
0.5	0	7.0	19.1	5.0	10.2
0.5	100	18.5	38.5	5.4	13.2
5.0	0	11.5	4.6	4.1	86.5
5.0	100	26.0	8.4	2.5	77.8

[a]Based on de la Guardia and Benlloch (1980).

sugars in particular, are widespread phenomena (Pitman *et al.*, 1971) and can also be observed during the growth of storage tissues. As shown by Steingröver (1983) the osmotic potential of the press sap from the storage roots of carrots remains constant throughout growth. Before sugar storage begins, K^+ and organic acids are the dominant osmotic substances. During sugar storage, however, an increase in the concentration of reducing sugars is compensated for by a corresponding decrease in the concentration of K^+ and organic acid anions. In storage roots of sugar beet, in principle, the same holds true for the concentrations of sucrose and K^+ (Beringer *et al.*, 1986).

8.7.6.2 Stomatal Movement

In most plant species K^+, associated with an anion, has the major responsibility for turgor changes in the guard cells during stomatal movement. An increase in the K^+ concentration in the guard cells increases their osmotic potential and results in the uptake of water from the adjacent cells and a corresponding increase in turgor in the guard cells and thus stomata opening as shown for faba bean in Table 8.37.

The accumulation of K^+ in the guard cells of open stomata can also be shown by X-ray microprobe analysis (Fig. 8.33). Closure of the stomata in the dark is correlated with K^+ efflux and a corresponding decrease in the osmotic pressure of the guard cells.

Light-induced accumulation of K^+ in the guard cells is driven by a plasma membrane-bound proton pumping ATPase (Fig. 8.34) as it is known for K^+ uptake in root cells (Section 2.4). Accordingly, stomatal opening is preceded by a decrease in pH of the guard cell apoplasm (Edwards *et al.*, 1988). The accumulation of K^+ in the vacuoles has to be balanced by a counteranion, mainly Cl^- or malate^{2-}, depending on the plant species and concentrations of Cl^- in the vicinity of the guard cells. In epidermal cells of wheat leaves, the Cl^- concentrations are often much higher than in the mesophyll cells (Hodson and Sangster, 1988). The transport of Cl^- into the guard cells is mediated by a Cl^-/H^+ symport at the plasma membrane (Fig. 8.34A). In order to achieve the high membrane transport rates of ions, channels are the principal pathways (Raschke *et al.*, 1988; Tester, 1990).

At low Cl^- availability, or in plant species which do not use Cl^- as accompanying anion for K^+ in guard cells (Fig. 8.34B), the H^+-driven K^+ influx activates PEP carboxylase activity, similarly as, for example, in root cells in response to cat$^+$ > an$^-$ uptake (Section 2.5.4). The newly formed malate in the guard cells serves as the

Table 8.37
Relationship between Stomatal Aperture and Characteristics of Guard Cells of Faba Bean[a]

	Stomatal aperture (μm)	Amount per stoma (10^{-14} mol)		Guard cell volume (10^{-12} l per stoma)	Guard cell osmotic pressure (MPa)
		K^+	Cl^-		
Open stoma	12	424	22	4.8	3.5
Closed stoma	2	20	0	2.6	1.9

[a]From Humble and Raschke (1971).

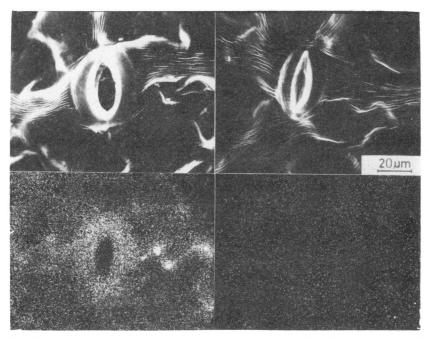

Fig. 8.33   Electron probe analyzer picture (*upper*) and corresponding X-ray microprobe images of the potassium distribution (*lower*) in open and closed stomata of faba bean. (Courtesy of B. Wurster.)

accompanying anion for K^+ in the vacuole, and as energy source for ATP synthesis in the mitochondria (Fig. 8.34B). The C_3 compound phosphoenolpyruvate (PEP) required for malate synthesis is supplied by starch degradation in the guard cell chloroplasts (Schnabl, 1980). In plant species, such as onion, that lack starch in the guard cell chloroplasts, the action of Cl^- as a counterion for K^+ influx is of fundamental importance, at least for stomatal regulation (Schnabl, 1980), an aspect which is discussed in Section 9.8.

Closure of the stomata is induced by darkness or ABA and is associated with a rapid efflux of K^+ and accompanying anions from the guard cells. The ABA, inducing stomatal closure may derive from the roots via the xylem as a 'non-hydraulic' signal (Davies and Meinzer, 1990; Section 5.6), perhaps amplified by simultaneously low CYT concentrations in the xylem sap (Meinzer *et al.*, 1991). However, endogenous ABA from guard cells may also serve this function, ABA concentrations in the guard cells being about 2.5 mM compared with about 0.9 mM in other epidermal cells in faba bean (Brinckmann *et al.*, 1990). As discussed in Section 8.6.7, ABA also activates Ca^{2+} channels in the plasma membrane of guard cells (Atkinson *et al.*, 1990). An increase in cytosolic free Ca^{2+} depolarizes the plasma membrane and thereby activates voltage-dependent anion channels and, thus, shifts the plasma membrane from a K^+-conducting state to an anion-conducting state which further decreases membrane potential and enhances K^+ efflux (Hedrich *et al.*, 1990). This effect of Ca^{2+} holds true, in principle, also for the tonoplast.

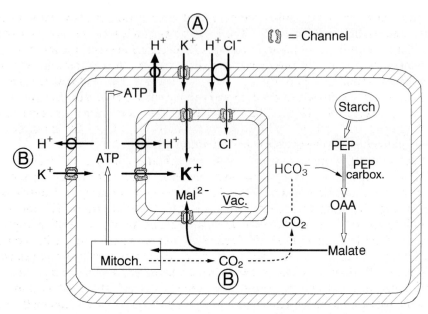

Fig. 8.34 Model of stomatal opening driven by proton pumps and K^+ + Cl^- transport (A) or K^+ + malate transport (B) into guard cell vacuole. (Modified from Raschke *et al.*, 1988.)

Stomatal closure is associated with a steep increase in the concentrations of K^+ and Cl^- in the apoplasm of guard cells of *Commelina communis*, for example from 3 mM K^+ and 4.8 mM Cl^- in open stomata to 100 mM K^+ and 33 mM Cl^- in closed stomata (Bowling, 1987). Interestingly, in the root and shoot of angiosperm parasites, such as *Striga* and *Loranthus*, stomata remain open permanently and do not respond to darkness, ABA or drought stress. This anomalous behaviour is caused by exceptionally high K^+ concentrations in the leaves of these parasites (which lack a phloem) and the lack of capability to dispose K^+ from the guard cells, required for stomatal closure (Smith and Stewart, 1990).

Sugars have been discussed as an alternative osmotic solute for stomatal opening (Tallman and Zeiger, 1988). However, the rate of sugar production in guard cells is insufficient to meet the high requirement for rapid stomatal opening (Reckmann *et al.*, 1990). Nevertheless, under potassium deficiency sugars may contribute substantially to osmoregulation in guard cells (Poffenroth *et al.*, 1992), this however, makes them very slow to respond. This slow response of sugar-loaded guard cells results in incomplete opening and closure of stomata in potassium-deficient plants (Hsiao and Läuchli, 1986).

8.7.6.3 Light-Driven and Seismonastic Movements

In leaves of many plants, particularly in the *Leguminosae*, leaves re-orientate their laminae photonastically in response to light signals, either to nondirectional light signals (*circadian rhythm*, e.g., leaf blades folded in the dark and unfolded in the light), or directional light signals (e.g., leaf blades re-orientate towards the light source).

These photonastic responses either increase light interception or allow avoidance of damage by excess light (Koller, 1990). These movements of leaves, and also of leaflets, are brought about by reversible turgor changes in specialized tissues, the motor organs (or pulvini). The turgor changes cause shrinking and swelling of cells in opposing regions (extensor and flexor) of the motor organ, and K^+, Cl^- and malate^{2-} are the major solutes involved in osmoregulation and volume change and, thus, leaf and leaflet movement (Satter et al., 1988). The principles of the mechanisms responsible for stomatal movement also hold true for leaf and leaflet movement, only the scales are different, individual cells versus specialized tissues.

In leaflet movement, the driving force of K^+ influx is also a plasma membrane-bound H^+-ATPase (Satter et al., 1988) and, thus, leaflet movement can be prevented by anaerobiosis or vanadate (Antowiak et al., 1992). In the primary leaf pulvinus of *Phaseolus vulgaris* during circadian leaf movement the concentration of H^+ and K^+ in the apoplasm of the extensor change in opposite directions: at swelling (upward movement of the leaf lamina) the pH decreases from 6.7 to 5.9 and the K^+ concentration from 50 mM to 10 mM, and vice versa, when the extensor cells shrink (Starrach and Mayer, 1989). In the extensor cell walls the cation exchange capacity is particularly high and, thus, an important reservoir of K^+ and H^+ (Starrach et al., 1985). In the leaf movements too, the environmental signals (e.g. light) activate Ca^{2+} channels in the membranes and thereby increase cytosolic free Ca^{2+} concentrations (Roblin et al., 1989). However, in contrast to guard cells, in the motor organ the extensor and the flexor regions respond to these signals in opposite ways.

A similar mechanism is responsible for the movement of leaves and other plant parts in response to a mechanical stimulus, for example in insectivorous plants or in *Mimosa*. In response to such *seismonastic* signals, for example in *Mimosa pudica*, the leaflets fold within a few seconds and reopen after about 30 min (Campbell and Thomson, 1977). This turgor-regulated response is correlated with a redistribution of K^+ within the motor organ (Allen, 1969). In the seismonastic reactions a rapid long-distance transport of the 'signal' from the touched leaflet to other leaflets also takes place. This 'signal' is an action potential, traveling in the phloem ($1–10$ cm s^{-1}) to the motor organs and inducing phloem unloading of sugars (sucrose) in the motor organ (Fromm, 1991). High local concentrations of sugars might contribute to the ion channel-mediated changes in turgor pressure in a given region of the motor organ required for leaflet movement.

8.7.7 Phloem Transport

Potassium has important functions in both loading of sucrose, and in the rate of massflow-driven solute transport in the sieve tubes (Section 5.4). This function of K^+ is related to the necessity to maintain a high pH in the sieve tubes for sucrose loading and the contribution of K^+ to the osmotic potential in the sieve tubes and, thus, the transport rates of photosynthates from source to sink. This is demonstrated in Table 8.38 for sugar cane. In potassium-sufficient plants ($+K$) within 90 min about half of the ^{14}C-labeled photosynthates are exported from the source leaf to other organs, and about 20% of it to the stalk as the main storage organ in sugar cane. In contrast, in the potassium-deficient plants ($-K$) the export rates were much lower, even after four hours.

Table 8.38

Effect of Potassium Nutritional Status of Sugarcane Plants on the Translocation of ^{14}C Labeled Photosynthates after Feeding $^{14}CO_2$ to a Leaf Blade[a]

	^{14}C distribution (%)			
	90 min		4 h	
Plant part	+K	−K	+K	−K
Fed leaf blade	54.3	95.4	46.7	73.9
Sheath of fed leaf blade	14.2	3.9	6.8	8.0
Joint of fed leaf and leaves and joint above fed leaf	11.6	0.7	17.0	13.6
Stalk below fed leaf	20.1	<0.1	29.5	4.6

[a]Total label = 100. Based on Hartt (1969).

Similarly, in legumes with an adequate (compared with an inadequate) K^+ supply, the root nodules have a greater supply of sugars, which increases their rates of N_2 fixation and export of bound nitrogen (Mengel *et al.*, 1974; Collins and Duke, 1981).

8.7.8 Cation–Anion Balance

In charge compensation, K^+ is the dominant cation for counterbalancing immobile anions in the cytoplasm, the chloroplasts, and quite often also for mobile anions in the vacuoles, the xylem and phloem. The accumulation of organic acid anions in plant tissues is often the consequence of K^+ transport without an accompanying anion into the cytoplasm (e.g., root or guard cells). The role of K^+ in the cation–anion balance is also reflected in nitrate metabolism, in which K^+ is often the dominant counterion for NO_3^- in long-distance transport in the xylem as well as for storage in vacuoles. As a consequence of NO_3^- reduction in leaves, the remaining K^+ requires the stoichiometric synthesis of organic acids for charge balance and pH homeostasis; part of this newly formed potassium malate may be retranslocated to the roots for subsequent utilization of K^+ as a counterion for NO_3^- within the root cells and for xylem transport (Chapter 3). In nodulated legumes, this recirculation of K^+ may serve a similar function in the xylem transport of amino acids (Jeschke *et al.*, 1985).

8.7.9 Potassium Supply, Plant Growth and Plant Composition

Next to nitrogen, potassium is the mineral nutrient required in the largest amount by plants. The potassium requirement for optimal plant growth is in the range 2–5% of the plant dry weight of vegetative parts, fleshy fruits, and tubers. In natrophilic species, however, the requirement for K^+ can be much lower (Section 10.2). When K^+ is deficient, growth is retarded, and net retranslocation of K^+ is enhanced from mature leaves and stems, and under severe deficiency these organs become chlorotic and necrotic, depending on the light intensity to which the leaves are exposed (Marschner and Cakmak, 1989). Lignification of vascular bundles is also impaired (Pissarek, 1973), a factor which might contribute to the higher susceptibility of potassium-deficient plants to lodging.

When the soil water supply is limited, loss of turgor and wilting are typical symptoms of potassium deficiency. The lower sensitivity of potassium-sufficient plants to drought stress is related to several factors (Lindhauer, 1985): (a) the role of K^+ in stomatal regulation, which is the major mechanism controlling the water regime of higher plants, and (b) the importance of K^+ for the osmotic potential in the vacuoles, maintaining a high tissue water content even under drought conditions. Lower sensitivity to drought stress in terms of biomass production and yield might also be the result of higher K^+ concentrations in the stroma and correspondingly higher rates of photosynthesis (Fig. 8.31), or of lower levels of ABA in the plants (Section 5.6).

Plants receiving an inadequate supply of potassium are often more susceptible to frost damage (Larsen, 1976), which, at the cellular level, is related in some respect to water deficiency. An example of this effect is shown in Table 8.39 for potato plants. Frost damage is inversely related to the potassium content of leaves, at least when the increase in potassium is still correlated with an increase in tuber yield. Inadequate potassium supply is therefore one factor leading to an increase in the risk of frost damage.

The changes in enzyme activity and organic compounds that take place during potassium deficiency are in part responsible for the higher susceptibility of potassium-deficient plants to fungal attack (Chapter 11). These changes in composition also affect the nutritional and technological (processing) quality of harvested products. This is most obvious in fleshy fruits and tubers with their high potassium requirement. In tomato fruits, for example, the incidence of so-called ripening disorders ('greenback') increases with inadequate potassium supply (van Lune and van Goor, 1977), and in potato tubers a whole range of quality criteria are affected by the potassium content in the tuber tissue (Table 8.40).

In several cases the relationships between potassium concentrations and gross changes in tuber tissue (e.g. osmoregulation and cation–anion balance) are quite obvious. In other cases, quality disorders are related directly to the citric acid content

Table 8.39

Relationship in Potato between Potassium Supply, Tuber Yield, Potassium in Leaves, and Percentage of Leaves Damaged by Frost[a]

Potassium supply (kg ha^{-1})	Tuber yield (tons ha^{-1})	Potassium content of leaves (mg g^{-1} dry wt)	Percentage of foliage damaged by frost
0	2.39	24.4	30
42	2.72	27.6	16
84	2.87	30.0	7

[a]Average values of 14 locations. Based on Grewal and Singh (1980).

Table 8.40
Effect of Increasing Potassium Contents in Potato Tubers on the Composition and Quality of the Tubers

Type of change	Effect of increasing K^+	Responsible mechanism	References[a]
Water content	Increase	Osmoregulation	1, 2, 3
Reducing sugars	Decrease	Osmoregulation	3
Citric acid	Increase	Cation–anion balance	4
Starch	Decrease	?	1,2,3
Black spot disorder	Decrease	Lower polyphenol oxidase activity?	2, 5
Darkening of press sap	Decrease	High citric acid/low polyphenol oxidase	4
Discoloration after cooking	Decrease	High citric acid/low chlorogenic acid	5, 6
Storage loss	Decrease	Lower respiration and fungal diseases	2

[a]Key to references: 1, Beringer *et al.* (1983). 2, Mirswa and Ansorge (1981). 3, A. Krauss and H. Marschner (unpublished). 4, Welte and Müller (1966). 5, Vertregt (1968). 6, Hughes and Evans (1969).

and thus only indirectly to potassium. Although differences between cultivars may modify the relationships, they do not eradicate them.

By increasing the potassium supply to plant roots it is relatively easy to increase the potassium content of various organs except grains and seeds, which maintain a relatively constant potassium content of 0.3% of the dry weight. When the potassium supply is abundant 'luxury consumption' of potassium often occurs, which deserves attention for its possible interference with the uptake and physiological availability of magnesium and calcium.

9

Functions of Mineral Nutrients: Micronutrients

9.1 Iron

9.1.1 General

In aerated systems maintained in the physiological pH range, the concentrations of ionic Fe^{3+} and Fe^{2+} are below 10^{-15} M. Chelates of Fe(III) and occasionally of Fe(II) are therefore the dominant forms of soluble iron in soil and nutrient solutions. As a rule, Fe(II) is taken up preferentially compared with Fe(III), but this also depends on the plant species (Strategy I and II, Section 2.5.6). In long-distance transport in the xylem, there is a predominance of Fe(III) complexes (Section 3.2).

Iron as a transition element is characterized by the relative ease by which it may change its oxidation state

$$Fe^{3+} \underset{-e^-}{\overset{+e^-}{\rightleftharpoons}} Fe^{2+}$$

and by its ability to form octahedral complexes with various ligands. Depending on the ligand, the redox potential of Fe(II/III) varies widely. This variability gives its special importance in biological redox systems. The high affinity of iron for various ligands (e.g., organic acids or inorganic phosphate) makes it unlikely that ionic Fe^{3+} or Fe^{2+} is of any importance in short- or long-distance transport in plants. Furthermore, in aerobic systems many low-molecular-weight iron chelates, and free iron in particular (either Fe^{3+} or Fe^{2+}) is very effective in producing oxygen and hydroxyl radicals (Halliwell and Gutteridge, 1986) and related compounds, for example

$$O_2 + Fe^{2+} \rightarrow O_2^- + Fe^{3+}$$

or in the Fenton reaction

$$H_2O_2 + Fe^{2+} \rightarrow Fe^{3+} + OH^- + OH^{\cdot}$$

These radicals are mainly responsible for peroxidation of polyunsaturated fatty acids of membrane lipids (Section 2.3). To prevent oxidative damage iron has either to be

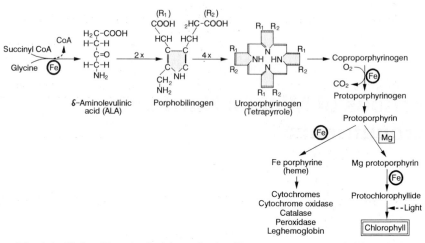

Fig. 9.1 Role of iron in the biosynthesis of heme coenzymes and chlorophyll.

tightly bound or incorporated into structures (e.g., heme and nonheme proteins) which allow controlled reversible oxidation–reduction reactions

$$Fe(II) \underset{+e^-}{\overset{-e^-}{\rightleftharpoons}} Fe(III)$$

including those in antioxidant protection.

9.1.2 Iron-containing Constituents of Redox Systems

9.1.2.1 Heme Proteins

The most well-known heme proteins are the cytochromes, which contain a heme iron–porphyrin complex (Fig. 9.1) as a prosthetic group. Cytochromes are constituents of the redox systems in chloroplasts (Fig. 5.1) in mitochondria, and also a component in the redox chain in nitrate reductase (Fig. 8.2). The particular role of iron in leghemoglobin and nitrogenase has been discussed in Chapter 7. There is some evidence that small amounts of leghemoglobin are also present in the roots of plants which are not capable of forming root nodules (Appleby *et al.*, 1988). This leghemoglobin may act as a signal molecule indicating O_2 deficiency and initiating a metabolic shift towards fermentation.

Other heme enzymes are catalase and peroxidases. Under conditions of iron deficiency, the activity of both types of enzymes declines. This is particularly the case for catalase activity in leaves (Table 9.1). The activity of this enzyme is therefore an indicator of the iron nutritional status of plants (Section 12.4).

Catalase facilitates the dismutation of H_2O_2 to water and O_2 according to the reaction

$$H_2O_2 \rightarrow H_2O + 1/2\ O_2$$

The enzyme plays an important role in association with superoxide dismutase (Section 9.2.3), as well as in photorespiration and the glycolate pathway (Fig. 5.4).

Table 9.1

Effect of Iron Deficiency on Chlorophyll Content and Enzyme Activity in Tomato Leaves[a]

Treatment	Iron in leaves (μg g^{-1} fresh wt)	Chlorophyll (mg g^{-1} fresh wt)	Enzyme activity (relative)	
			Catalase	Peroxidase
+Fe	18.5	3.52	100	100
−Fe	11.1	0.25	20	56

[a]Based on Machold (1968).

Peroxidases of various types (isoenzymes) are abundant in plants. They catalyze the following reactions:

$$XH_2 + H_2O_2 \rightarrow X + 2H_2O$$

and

$$XH + XH + H_2O_2 \rightarrow X - X + 2H_2O$$

An example of the first type of reaction has been given in Fig. 5.2, showing the role of ascorbate peroxidase in the detoxification of H_2O_2 in chloroplasts. In the second type of reaction, cell wall-bound peroxidases catalyze the polymerization of phenols to lignin. Peroxidases are abundant particularly in cell walls of the epidermis (Hendricks and van Loon, 1990) and rhizodermis (Codignola *et al.*, 1989) and required for biosynthesis of lignin and suberin. Both synthetic pathways require phenolic compounds and H_2O_2 as substrates. The formation of H_2O_2 is catalyzed by the oxidation of NADH at the plasma membrane/cell wall interface (Mäder and Füssl, 1982). The principles of these reactions are as follows:

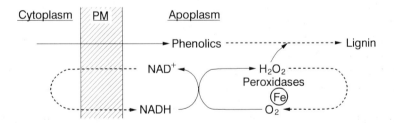

In iron-deficient roots, peroxidase activity is much depressed (Sijmons *et al.*, 1985) and phenolics accumulate in the rhizodermis, for example, in sunflower (Römheld and Marschner, 1981a). Phenolics are also released at much higher rates from the roots of iron-deficient compared with iron-sufficient plants (Hether *et al.*, 1984; Marschner *et al.*, 1986a). Certain phenolics such as caffeic acid, are very effective in chelation and reduction of inorganic Fe(III), and a component of Strategy I in iron acquisition (Section 2.5.6). It might well be that the alterations in cell wall formation of rhizodermal cells under iron deficiency (see Fig. 9.4) are causally related to the impaired peroxidase activity.

9.1.2.2 Iron–Sulfur Proteins

In these nonheme proteins iron is coordinated to the thiol group of cysteine or to inorganic sulfur as clusters, or to both. The most well-known one is ferredoxin, which acts as an electron transmitter in a number of basic metabolic processes according to the principle:

$$
\xrightarrow{\ e^-\ }
\begin{array}{c}
-\mathrm{Cys-S} \\
\mathrm{Fe} \\
-\mathrm{Cys-S}
\end{array}
\begin{array}{c}
\mathrm{S} \\
\mathrm{S}
\end{array}
\begin{array}{c}
\mathrm{Fe} \\
\end{array}
\begin{array}{c}
\mathrm{S-Cys-} \\
\mathrm{S-Cys-}
\end{array}
\xrightarrow{\ e^-\ }
\begin{array}{l}
\mathrm{NADP^+\ (photosynthesis)} \\
\mathrm{Nitrite\ reductase} \\
\mathrm{Sulfite\ reductase} \\
\mathrm{N_2\ reduction} \\
\mathrm{GOGAT}
\end{array}
$$

Details of the function of ferredoxin in these processes have been discussed in the relevant sections. In iron-deficient leaves, the ferredoxin content is decreased to a similar extent as the chlorophyll content (Table 9.2), and the fall in ferredoxin level is associated with a lower nitrate reductase activity (NRA). Both, ferredoxin content and NRA can be restored on resupplying iron. In view of the involvement of iron at various steps in nitrate reduction (Fig. 8.2) positive correlations between iron supply, ferredoxin content and nitrate reduction are to be expected.

Another example of iron–sulfur proteins are those isoenzymes of superoxide dismutase (SOD) which contain iron as a metal component of the prosthetic group (FeSOD). Superoxide dismutases detoxify superoxide anion free radicals (O_2^-) by formation of H_2O_2 (Fig. 5.1) and may contain Cu, Zn, Mn or Fe as metal components (Section 9.2.2; Fridovich, 1983; Sevilla *et al.*, 1984). In chloroplasts FeSOD is the typical isoenzyme of SOD (Kwiatowsky *et al.*, 1985), but it may also occur in mitochondria and peroxisomes in the cytoplasm (Droillard and Paulin, 1990).

Aconitase is an iron–sulfur protein (Brouquisse *et al.*, 1986) which catalyzes the

Table 9.2
Effect of Iron Deficiency on Content of Chlorophyll and Ferredoxin and Nitrate Reductase Activity in Citrus Leaves[a]

Fe content (μg g^{-1} dry wt)	Chlorophyll (mg g^{-1} dry wt)	Ferredoxin (mg g^{-1} dry wt)	Nitrate reductase (nmol NO_2 g^{-1} fresh wt h^{-1})
96	1.80	0.82	937
62	1.15	0.44	408
47	0.55	0.35	310
47 → 81[b]	—	0.63	943

[a]Based on Alcaraz *et al.* (1986).
[b]40 h after infiltration of intact iron-deficient leaves with 0.2% FeSO$_4$.

Table 9.3
Relationship between Iron Supply, Chlorophyll Content in Leaves, and Organic Acid Content in Roots of Oat[a]

Treatment	Chlorophyll content (relative)	Organic acid content $(\mu g \ (10 \ g)^{-1}$ fresh wt)			
		Malic	Citric	Other	Total
+Fe	100	39	11	23	73
−Fe	12	93	67	78	238

[a]Based on Landsberg (1981).

isomeration of citrate to isocitrate in the tricarboxylic acid cycle (Fig. 5.8). Iron as the metal component of the prosthetic group is required for both, stability and activity of the enzyme (Hsu and Miller, 1968), and the iron cluster of the enzyme is responsible for the spatial orientation of the substrates (citrate and isocitrate); valency changes are not involved in the reaction (Beinert and Kennedy, 1989). In iron-deficient plants aconitase activity is lower (De Vos et al., 1986), and reactions in the tricarboxylic acid cycle are disturbed leading to organic acids, particularly citric and malic acid (Table 9.3). In roots of iron-deficient tomato plants an increase in organic acid content is closely correlated with enhanced CO_2 dark fixation and net excretion of H^+, i.e. acidification of the rhizosphere (Miller et al., 1990). The causal relationships between lower aconitase activity and organic acid accumulation in roots of iron deficient plants are still a matter of controversy (De Vos et al., 1986; Pich et al., 1991).

Riboflavin also accumulates in most dicotyledenous plant species under iron deficiency, and its release from roots may be enhanced by a factor of 200 in iron deficient plants (Welkie and Miller, 1989). Accumulation of riboflavin is presumably the result of alterations in purine metabolism due to impairment of xanthine oxidase (Schlee et al., 1968), another enzyme with iron–sulfur clusters as a prosthetic group. In microorganisms such as cyanobacteria, but not in higher plants, such changes in purine metabolism are expressions of a shift in electron transfer in the respiratory chain from iron-dependent to iron-independent enzymes, for example, flavodoxin (Huang et al., 1992a).

9.1.3 Other Iron-Requiring Enzymes

There are a number of less well-characterized enzymes in which iron acts either as a metal component in redox reactions or as a bridging element between enzyme and substrate. In iron-deficient plants, the activities of some of these enzymes are impaired (perhaps owing to a lower affinity for iron) and may often be responsible for gross changes in metabolic processes.

For biosynthesis of ethylene, methionine is the principal precursor. Along this biosynthetic pathway in the conversion of 1-aminocyclopropane-1-carboxylic acid (ACC) into ethylene, a two-step one-electron oxidation takes place, catalyzed by Fe(II) (see Fig. 9.5). Accordingly, ethylene formation is very low in iron deficient cells and can be restored immediatly on resupplying iron, and without the involvement of protein synthesis (Bouzayen *et al.*, 1991).

Lipoxygenases are enzymes containing one atom iron per molecule (Hildebrand, 1989), and catalyze the peroxidation of linolic and linolenic acid, i.e., of long-chain polyunsaturated fatty acids (Section 2.3). High lipoxygenase activity is typical for fast-growing tissues and organs, and may become critical for membrane stability. Lipoxygenase-mediated lipid peroxidation is involved in senescence of cells and tissues and in the early events of hypersensitivity in incompatible host–pathogen combinations and, thus, in disease resistance (Nagarathna *et al.*, 1992). In leaves of iron deficient plants, lipoxygenase activity and chlorophyll content are positively and closely correlated (Boyer and van der Ploeg, 1986) indicating the possibility of a close association of the enzyme with thylakoid membranes.

Low chlorophyll content (chlorosis) of young leaves is the most obvious visible symptom of iron deficiency. Various factors are responsible for this decrease, the most direct one being the role of iron in the biosynthesis of chlorophyll (Fig. 9.1). The common precursor of chlorophyll and heme synthesis is δ-aminolevulinic acid (ALA), and the rate of ALA formation is controlled by iron (Pushnik and Miller, 1989). Iron is also required for the formation of protochlorophyllide from Mg-protoporphyrin (Fig. 9.1). Feeding ALA to iron-deficient leaf tissue leads to an increase in the Mg-protoporphyrin level, whereas the protochlorophyllide and chlorophyll levels remain low compared with the levels in leaf tissue adequately supplied with iron (Spiller *et al.*, 1982; Pushnik *et al.*, 1984). Coproporphyrinogen oxidase (Fig. 9.1) is also an iron-containing protein (Chereskin and Castelfranco, 1982).

9.1.4 Chloroplast Development and Photosynthesis

As a rule, iron deficiency has much less effect on leaf growth, cell number per unit area, or number of chloroplasts per cell than on the size of the chloroplasts and protein content per chloroplast (Table 9.4). Only with severe iron deficiency is cell division also inhibited (Abbott, 1967) and, thus, leaf growth reduced. Iron is required for protein synthesis, and the number of ribosomes – the sites of protein synthesis – decrease in

Table 9.4
Effect of Iron Deficiency on Leaves and Chloroplasts of Sugar Beet[a]

| Parameter | Chlorophyll (mg cm^{-2}) | | |
	Control >40	Mild deficiency 20–40	Severe deficiency <20
Soluble protein (mg cm^{-2} leaf area)	0.57	0.56	0.53
Mean leaf cell volume (10^{-8} cm^3)	2.64	2.78	2.75
Chloroplasts (no. per cell)	72	77	83
Average chloroplast volume (μm^3)	42	37	21
Protein N (pg per chloroplast)	1.88	1.34	1.24

[a]From Terry (1980).

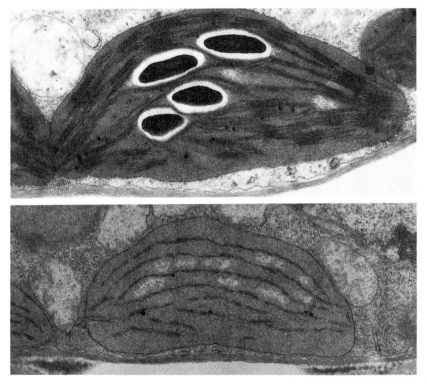

Fig. 9.2   Fine structure of chloroplasts from iron-sufficient (*upper*) and iron-deficient (*lower*) soybean (*Glycine* max. L.) plants (×24 000). (Courtesy of Ch. Hecht-Buchholz.)

iron-deficient leaf cells (Lin and Stocking, 1978). However, under iron deficiency, protein synthesis in chloroplasts is much more impaired than in the cytoplasm (Shetty and Miller, 1966). In iron-deficient maize leaves, for example, the total protein content decreases by 25% but that of the chloroplasts by 82% (Perur *et al.*, 1961), most probably because of a particularly high iron requirement of chloroplastic mRNA and rRNA (Spiller *et al.*, 1987).

In the thylakoid membranes about 20 iron atoms are directly involved in the electron transport chain per unit of PS II and PS I (Terry and Abadia, 1986; Rutherford, 1989). This high iron requirement for the structural and functional integrity of the thylakoid membranes, and the additional iron requirement for ferredoxin and the biosynthesis of chlorophyll explain the high sensitivity of chloroplasts in general and the thylakoids in particular to iron deficiency (Fig. 9.2).

In iron-deficient leaves, however, not all photosynthetic pigments and components of the electron transport chain are decreased to the same extent (Table 9.5). The activity of PS I is much more depressed than of PS II. Resupplying iron to chlorotic leaves increases the function of PS I as an electron transmitter much more than that of PS II. The individual components of PS I, P 700, cytochromes, and protein, increase in a similar manner, indicating that iron is involved in the regulation of PS I development and assembling the subunits in the thylakoid membranes (Pushnik and Miller, 1989). If

Table 9.5
Effect of Iron Nutritional Status of Tobacco Leaves on Contents of Chlorophyll and Photosystem I (PS I) Components and Photosynthetic Electron Transport Capacity of PS II and PS I[a]

| Fe treatment | Fe (μg cm^{-2} leaf) | Chlorophyll (μg cm^{-2} leaf) | PS I components | | | e$^-$-Transport capacity[b] | |
			P700	Cytochromes (pmol cm^{-2})	Protein (μg cm^{-2})	PS II	PS I
+Fe	1.44	89	545	599	108	56	840
−Fe	0.25	26	220	201	38	30	390
−Fe + Fe[c]	1.16	24	430	474	79	36	764

[a]Recalculated from Pushnik and Miller (1989).
[b]μeq cm^{-2} leaf h^{-1}.
[c]10 days after foliar application of iron.

iron deficiency becomes more severe, the activity of PS II also drops drastically and is much more difficult to restore (Table 9.4; Morales *et al.*, 1991). In contrast to the impairment of photosynthetic electron transport, respiratory activity in iron-deficient leaves is usually unaffected (Pushnik and Miller, 1989), presumably because terminal oxidation by cytochrome oxidase in mitochondria is catalyzed by copper and not by iron (Section 9.3).

In iron-deficient leaves, contents of chlorophyll and β-carotene decline to the same extent (Morales *et al.*, 1990), whereas certain xanthophylls may even increase (Quilez *et al.*, 1992). This increase has been observed particularly in zeaxanthin, violaxanthin and antheroxanthin (Abadia *et al.*, 1991) which are involved in the so-called *xanthophyll-cycle*, a system which is important for non-radiative (i.e., heat-generating) energy dissipation under high light intensity (Section 5.2.2). This shift in chloroplast pigment composition under iron deficiency is probably related to structural changes (Fig. 9.2) as most xanthophylls are located in the chloroplast envelope, and not in the thylakoids (Terry and Abadia, 1986).

Iron-deficient leaves are characterized by low contents of starch and sugars (Arulanathan *et al.*, 1990). This is to be expected in view of the low chlorophyll content, impairment of photosynthetic electron transport and low ferredoxin content and depressed regeneration of reduced ferredoxin. An additional factor contributing to low carbohydrate content is the retarded regeneration of ribulosebisphosphate which acts as substrate for the CO_2 in the Calvin cycle. This may limit photosynthesis in iron-deficient leaves (Arulanthan *et al.*, 1990) and also possibly in part account for the lower CO_2 fixation rate per unit chlorophyll in iron-deficient as compared with iron-sufficient leaves (Sharma and Sanwal, 1992).

9.1.5 Localization and Binding State of Iron

When plants are grown under controlled conditions, about 80% of the iron is localized in the chloroplasts in rapidly growing leaves, regardless of iron nutritional status (Fig. 9.3). Under iron deficiency a shift in the distribution of iron occurs only within the chloroplasts, whereby the lamellar iron content increases at the expense of the stromal iron.

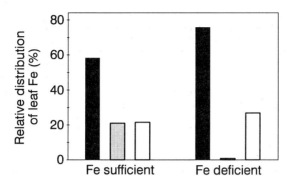

Fig. 9.3 Intracellular distribution of iron in leaf blades of iron-sufficient and iron-deficient sugar beet plants. Solid bars, lamellar iron; stippled bars, stromal iron; open bars, extrachloroplastic iron. (Redrawn from Terry and Low, 1982, by courtesy of Marcel Dekker, Inc.)

Iron can be stored in plant cells in the stroma of plastids as phytoferritin (Fe content 12–23% dry wt). It consists of a protein shell, and in the interior up to 5000 atoms of iron as Fe(III) can be stored. Phytoferritin often has a well-defined crystalline form with the proposed formula $(FeO \cdot OH)_8 \cdot (FeO \cdot OPO_3H_2)$ (Seckbach, 1982). Its content is high in dark-grown leaves (up to 50% of the total iron), but it rapidly disappears during regreening (Mark *et al.*, 1981) and remains very low in green leaves. After resupplying iron to deficient plants, however, the uptake rate is exceptionally high (Section 2.5.6) and phytoferritin leaf content may transiently increase dramatically (Lobreaux *et al.*, 1992) and make up as much as 30% of that of the total leaf iron (Van der Mark *et al.*, 1982; Platt-Aloia *et al.*, 1983). The localization of phytoferritin is not confined to chloroplasts: it can also be detected in the xylem and phloem (Smith, 1984). Phytoferritin is abundant too in seeds, as for example, in legumes. During germination phytoferritin is rapidly degraded, probably catalyzed by the released Fe^{2+} and generation of hydroxyl radicals which destroy the protein shell (Bienfait, 1989; Lobreaux and Briat, 1991). Phytoferritin may also act as storage for iron in nodules of legumes, both for heme synthesis during nodule development and heme degradation during senescence (Ko *et al.*, 1987).

If plants are grown under controlled conditions (e.g., in nutrient solutions), there is a close positive correlation between the total iron content of the leaves and the chlorophyll content when the supply of iron (as chelates) is suboptimal (Römheld and Marschner, 1981a; Terry and Abadia, 1986). This correlation, however, is often poor or absent in plants grown in calcareous soils, when there is a large supply of phosphorus (Mengel *et al.*, 1984a), or when different forms of nitrogen are supplied (Machold, 1967; Mengel and Geurtzen, 1988). Under these conditions, the content of iron in chlorotic leaves may be similar to or even higher than that in green leaves. These discrepancies are related in part to the localization and binding state of iron in leaves. A proportion of iron might be precipitated in the apoplasm of leaves (Mengel and Geurtzen, 1988) and not physiologically available. As shown by Mössbauer spectrometry, most of the iron in plants is in the ferric (FeIII) form (Terry and Abadia, 1986) and of particular physiological importance is the fraction which undergoes reversible Fe(II)/Fe(III) oxidoreduction (Machold *et al.*, 1968). Extraction of leaves with dilute acids or chelators to characterize the so-called 'active iron' often improves the correlation between iron and chlorophyll content in leaves of field-grown plants (Mengel and Bübl, 1983; Abadia *et al.*, 1989). Neither the composition nor the location of this extracted 'active iron', however, are known (Abadia, 1992).

9.1.6   Root Responses to Iron Deficiency

In leaves of all plant species the major symptom of iron deficiency is inhibition of chloroplast development. For roots, however, both morphological and physiological changes brought about by the deficiency and responses to this lack of iron depend upon plant species (Strategy I and II, Section 2.5.6). In both dicots and monocots, with the exception of the grasses (graminaceous species), iron deficiency is associated with inhibition of root elongation, increase in the diameter of apical root zones, and abundant root hair formation (Römheld and Marschner, 1981a; Chaney *et al.*, 1992b).

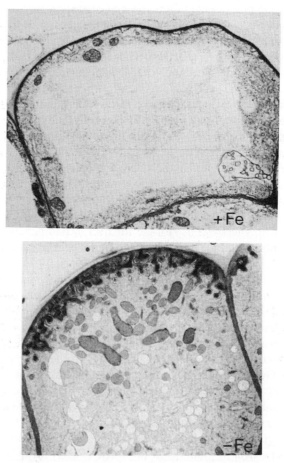

Fig. 9.4 Sections of rhizodermal cells of sunflower. (*Top*) Iron sufficient. (*Bottom*) Iron deficient. (Courtesy of D. Kramer.)

These morphological changes are often associated with the formation of cells with a distinct wall labyrinth typical of transfer cells (Fig. 9.4). These transfer cells may be induced either in the rhizodermis (Fig. 9.4) or in the hypodermis (Landsberg, 1989). The iron deficiency-induced formation of rhizodermal transfer cells (Kramer *et al.*, 1980) is part of a regulatory mechanism for enhancing iron uptake. These transfer cells are presumably the sites of iron deficiency-induced root responses of Strategy I, namely enhanced net excretion of protons and reducing capacity as well as of release of phenolic compounds. After the supply of iron is restored, not only do the physiological root responses disappear, but also the transfer cells degenerate within 1 to 2 days. When iron supply is suboptimal [e.g., when the concentrations of Fe(III) chelates are low or sparingly soluble inorganic Fe(III) compounds are supplied], rhythmic changes in root morphology and root-induced changes in the substrate pH and uptake rate of iron are observed. The growth rate of the shoots and the chlorophyll content, however, remain unaffected (Römheld and Marschner, 1981b).

In perennial and annual dicotyledenous species such as *Ficus benjamina* (Rosenfield *et al.*, 1991) and *Lupinus cosentinii* (White and Robson, 1989), the formation of proteoid roots (Section 15.5) is enhanced not only in response to phosphorus deficiency but also to iron deficiency. Proteoid roots are characterized by particularly high capacity to reduce Fe(III) and excrete protons (Marschner *et al.*, 1986a,b; Rosenfield *et al.*, 1991). In these properties proteoid roots are thus similar to apical root zones containing transfer cells.

In graminaceous species (Strategy II) these iron deficiency-induced morphological and physiological changes are absent. Instead, roots release phytosiderophores (PS) as chelators for Fe(III) (Section 2.5.6). The pathway of PS biosynthesis is understood reasonably well (Fig. 9.5). L-Methionine is the dominant precursor (Mori and Nishizawa, 1987), and three molecules of it are used to form one molecule of nicotianamine which, after deamination and hydroxylation, is converted into 2-deoxymugineic acid

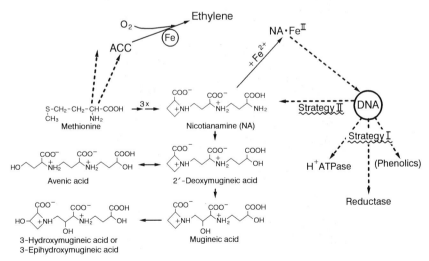

Fig. 9.5 Model of phytosiderophore biosynthesis and of some other iron-related factors in roots. (Based on Shojima *et al.*, 1989 and Scholz *et al.*, 1992.)

and further to other PS (Fig. 9.5), depending upon plant species (Römheld and Marschner, 1990). In contrast to the biosynthetic pathway, the regulation of gene expression of PS synthesis by iron deficiency is much less understood, but some progress has been made (Okumura et al., 1991).

Nicotianamine (NA) is not only a precursor of PS biosynthesis but is also a strong chelator of Fe(II), but not of Fe(III) (Scholz et al., 1988). It is also essential for the proper functioning of Fe(II) dependent processes (Pich et al., 1991). It seems to play an important role in iron homeostasis within cells and cellular compartments (Fig. 9.5), perhaps by regulation of a repressor protein (Bienfait, 1989; Scholz et al., 1992). Nicotianamine might be the link between the two strategies of iron deficiency-induced root responses, perhaps reflecting differences in codon usage in genes of dicots in comparison with monocots (Campbell and Gowri, 1990).

9.1.7 Iron Deficiency and Iron Toxicity

The critical deficiency content of iron in leaves is in the range of 50–150 mg Fe kg^{-1} dry wt. The content refers to total iron and is, therefore, only of limited value for characterization of the iron nutritional status of field-grown plants (Section 9.1.6). In general, C_4 species require a higher iron supply than C_3 species, but their critical deficiency contents are similar, namely about 72 mg Fe kg^{-1} in C_3 species and about 66 mg Fe kg^{-1} in C_4 species (Smith et al., 1984). In fast-growing meristematic and expanding tissues, for example shoot apices, the critical deficiency contents are much higher, presumably in the range of 200 mg Fe kg^{-1} dry wt for total iron, and 60–80 mg Fe kg^{-1} dry wt for 'active iron' (Häussling et al., 1985). In legumes iron demand for nodule development is particularly high (Section 7.4.5).

Iron deficiency is a worldwide problem in crop production on calcareous soils. It is the major factor responsible for so-called lime-induced chlorosis (Section 16.5). Iron deficiency might also limit CO_2 fixation of the phytoplankton in oceans such as the Pacific (Greene et al. 1992).

On the other hand, iron toxicity ('bronzing') is a serious problem in crop production on waterlogged soils, it is the second most severe yield-limiting factor in wetland rice (Section 16.4). The critical toxicity contents are above 500 mg Fe kg^{-1} leaf dry wt, but very much dependent on other factors such as content of other mineral nutrients (Yamauchi, 1989). Iron toxicity may also play a role under dryland conditions and is probably an early event of drought-induced damage in photosynthetic tissue caused by iron-catalyzed formation of oxygen free radicals in the chloroplasts (Price and Hendry, 1991).

9.2 Manganese

9.2.1 General

Manganese can exist in the oxidation states 0, II, III, IV, VI and VII. In biological systems, however, it mainly occurs in oxidation states II, III, and IV with Mn(II) and

Mn(IV) being fairly stable and Mn(III) unstable (Hughes and Williams, 1988). In plants, Mn(II) is by far the dominant form, but it can readily be oxidized to Mn(III) and Mn(IV). Manganese therefore plays an important role in redox processes. Manganese (II) forms only relatively weak bounds with organic ligands. The ionic radius of Mn^{2+} (0.075 nm) lies between Mg^{2+} (0.065 nm) and Ca^{2+} (0.099 nm) and it can therefore substitute or compete, in various reactions involving either of these ions. The binding strength of all three ions for ligands based on oxygen donors are roughly the same (Hughes and Williams, 1988) or higher for Mn^{2+}, as for example, by a factor of about four in case of ATP (Burnell, 1988). This has important consequences for the compartmentation of Mn^{2+} in cells and interactions between manganese and magnesium nutrition.

9.2.2 Manganese Containing Enzymes

Although a relatively large number of enzymes are activated by Mn^{2+} (Section 9.2.3), to date the existence of only two manganese-containing enzymes is well established, namely the manganese-protein in photosystem II (PS II) and the manganese-containing superoxide dismutase (MnSOD). Former reports on a manganese-containing acid phosphatase (Uehara et al., 1974) have not been confirmed. This enzyme contains two atoms of iron per molecule and requires iron but not manganese for its activity (Hefler and Averill, 1987).

Superoxide dismutases (SOD) are present in all aerobic organisms and play an essential role in the survival of these organisms in the presence of oxygen (Elstner, 1982; Fridovich, 1983). They protect tissues from the deleterious effect of the oxygen free radical O_2^- formed in various enzyme reactions in which a single electron is transmitted to O_2:

$$O_2 + e^- \longrightarrow O_2^- \text{ (Superoxide)}$$

$$O_2^- + O_2^- + 2H^+ \xrightarrow{\text{Superoxide-dismutase (SOD)}} H_2O_2 \text{ (Hydrogen peroxide)} + O_2$$

$$2H_2O_2 \longrightarrow 2H_2O + O_2$$

The conversion of O_2^- is catalyzed by SOD, and the subsequent dismutation of H_2O_2 into H_2O and O_2 facilitated by either peroxidases, catalase (Elstner, 1982) or, in chloroplasts, by an ascorbate specific peroxidase (Fig. 5.2; Polle et al., 1992). In illuminated green cells the chloroplasts are the organelles with the highest rate of oxygen turnover, including the formation of O_2^- and H_2O_2. Accordingly, in green leaves more than 90% of the SOD is located in the chloroplasts and only 4–5% is in the mitochondria (Jackson et al., 1978).

The isoenzymes of SOD differ in their metal component which might be either iron (FeSOD, Section 9.1.2.2), manganese (MnSOD) or copper+zinc (CuZnSOD). The FeSOD is mainly confined to chloroplasts. CuZnSOD is also found in chloroplasts but it occurs too in the cytosol and mitochondria (Palma et al., 1986). MnSOD is not so widely distributed in various families of higher plants (Sandmann and Böger, 1983). Within cells it too is mainly located in mitochondria and peroxisomes. There are controversal reports concerning the occurrence of MnSOD in chloroplasts. In pea, for example, it is absent (Palma *et al.*, 1986) whereas in tobacco it is most likely present (Bowler *et al.*, 1991). In transgenic plants of tobacco with elevated levels of MnSOD both chlorophyll degradation in light and solute leakage from chloroplasts and mitochondria were much lower than in the non-transgenic control plants with low MnSOD levels (Bowler *et al.*, 1991). Both free-living and symbiotic rhizobia (bacteroids) possess only MnSOD which is also found in the cytosol of nodules, whereas CuZnSOD is found only in the nodule cytosol (Becana and Salim, 1989).

The most well-known and best-documented example of a manganese-containing enzyme is the 33 kDa polypeptide (protein) of the water-splitting system associated with PS II (Section 5.2.1). In this system four manganese atoms arranged as a cluster act as a device for storing positive charges prior to the four-electron oxidation of two molecules of water:

$$2\,H_2O \quad \diagup\!\!\!\diagdown \quad 4\,Mn^{III} \quad \diagdown \quad$$
$$\text{Enzyme} \qquad 4\,e^- \quad \diagup\!\!\!\!\searrow hv$$
$$4\,H^+ + O_2 \quad \diagdown\!\!\!\diagup \quad 4\,Mn^{II} \qquad \boxed{\text{Pigment 680}}^{(+)}$$

The functioning of the manganese atoms in both transient electron storing and electron transmitting is coupled with fluctuations in the oxidation state of manganese between Mn(II) and Mn(IV) (Rutherford, 1989). In photosynthesizing cells this role in PS II is the most sensitive function of manganese to be impaired by manganese deficiency (Section 9.2.4).

9.2.3 Manganese Dependent or Activated Enzymes

Manganese acts as cofactor, activating about 35 different enzymes (Burnell, 1988). Most of these enzymes catalyze oxidation–reduction, decarboxylation, and hydrolytic reactions. Manganese has a primary role in the tricarboxylic acid cycle (TCA) in oxidative and nonoxidative decarboxylation reactions, as for example, the NADPH specific decarboxylating malate dehydrogenase, malic enzyme, and isocitrate dehydrogenase:

$\boxed{\text{Malic enzyme}}$ catalyzing the reaction :

$$\text{Malate} + NADP^+ \xrightarrow[Mn^{2+},\ Mg^{2+}]{} \text{Pyruvate} + NADPH + H^+ + CO_2$$

$\boxed{\text{Isocitrate dehydrogenase}}$ catalyzing the reaction :

$$\text{Isocitrate} + NADP^+ \xrightarrow[Mn^{2+},\ Mg^{2+}]{} \text{Oxalosuccinate} + NADPH$$

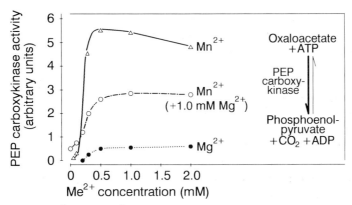

Fig. 9.6   Effect of Mn^{2+} and Mg^{2+} on the activity of PEP carboxykinase from *Urochloa panicoides*. ATP concentration was kept constant at 0.25 mM. (Burnell, 1986.)

Most of the studies on manganese activation of enzymes have been carried out *in vitro* and in many instances Mn^{2+} can be replaced by Mg^{2+}, or vice versa. Since the concentration of Mg^{2+} in the cells is, on average, about 50–100 times higher than that of Mn^{2+}, activation of enzymes by Mn^{2+} *in vivo* is presumably only important for those enzymes where Mn^{2+} is much more effective than Mg^{2+}. An example of a much higher effectivity of Mn^{2+} is the chloroplast RNA polymerase where for activation about 10 times higher concentrations of Mg^{2+} are required than of Mn^{2+} (Ness and Woolhouse, 1980). An absolute requirement of Mn^{2+} occurs in the bundle sheath chloroplasts of those C_4 plants in which oxaloacetate acts as the carbon shuttle (Section 5.2.4) and where the decarboxylation is catalyzed by PEP carboxykinase. This enzyme has an absolute requirement for Mn^{2+} which cannot be replaced by Mg^{2+} (Fig. 9.6). Indeed, Mg^{2+} inhibits the activity. Maximum activity occurs at a Mn/ATP ratio of one, suggesting that the substrate for the enzyme is the Mn·ATP complex (Burnell, 1986) and not Mg·ATP, as in most other reactions (Section 8.5).

Manganese activates several enzymes of the shikimic acid pathway, and subsequent pathways, leading to the biosynthesis of aromatic amino acids, such as tyrosine, and various secondary products, such as lignin, flavonoids, as well as IAA (Burnell, 1988; Hughes and Williams, 1988). For example, Mn^{2+} affects phenylalanine ammonia-lyase (PAL) and stimulates peroxidases required for lignin biosynthesis. In manganese-deficient leaves the IAA oxidase activity is exceptionally high, as is also the case for leaves suffering from manganese toxicity (Morgan *et al.*, 1976). The role of manganese in IAA oxidase activity is still obscure (Horst, 1988). In the biosynthetic pathway of isoprenoids producing carotenoids, sterols and GA, manganese-dependent enzymes have also been found, as for example, a phytoene synthetase (Wilkinson and Ohki, 1988).

In nodulated legumes such as soybean which mainly transport nitrogen in the form of allantoin and allantoate to the shoot (Section 7.4) the degradation of these ureides in the leaves (Winkler *et al.*, 1985) and in the seed coat (Winkler *et al.*, 1987) is catalyzed by the enzyme allantoate amidohydrolase which appears to have an absolute requirement of Mn^{2+} (Winkler *et al.*, 1985). Arginase is another manganese-dependent enzyme in nitrogen metabolism (Burnell, 1988).

A role of Mn^{2+} in nitrate reductase activity was formerly supposed because of an increase in nitrate content in manganese-deficient leaves. This accumulation of nitrate, however, is the consequence of a shortage of reducing equivalents in the chloroplasts and of carbohydrates in the cytoplasm, as well as of negative feedback regulation resulting from lower demand of reduced nitrogen in the new growth of the deficient plants. There is no evidence of a direct role of Mn^{2+} in nitrate reductase activity (Leidi and Gomes, 1985).

Manganese (Mn^{2+}) can readily displace Mg^{2+} from ATP since Mn^{2+} binds ATP four times more strongly than Mg^{2+}. At high concentrations of Mn^{2+} the ATP in the cytoplasm is readily saturated by Mn^{2+} (Pfeffer et al., 1986). In order for normal functioning of $Mg \cdot ATP$ as the main energy-transmitting system concentrations of Mn^{2+} in the cytosol and in the stroma of chloroplasts have therefore to be maintained at a low level. In agreement with this, most of the Mn^{2+} is sequestered in vacuoles (Pfeffer et al. 1986; Clarkson, 1988) or in other cell compartments such as Golgi vesicles (Hughes and Williams, 1988). Depression of net photosynthesis in leaves high in manganese is caused by inhibition of the RuBP carboxylase reaction (Houtz et al., 1988), most likely owing to replacement of Mg^{2+} by Mn^{2+} and, is thus, a reflection of inadequate compartmentation of Mn^{2+}.

9.2.4 Photosynthesis and Oxygen Evolution

The particular role of manganese in photosynthesis was discovered in green algae (Pirson, 1937). In *Chlorella* the manganese requirement for optimal growth is about 1000 times lower under heterotrophic (darkness and external supply of carbohydrates) than under autotrophic conditions, i.e. carbon supply via photosynthesis (Eyster *et al.*, 1958). Also in higher plants photosynthesis in general and photosynthetic O_2 evolution in PS II, in particular, are the processes which respond most sensitively to manganese deficiency (Fig. 9.7). A decrease in manganese content of young leaves has only a small effect on chlorophyll content (Fig. 9.7) or leaf dry weight (Nable *et al.*, 1984) but photosynthetic O_2 evolution drops by more than 50%. Resupplying Mn^{2+} to deficient

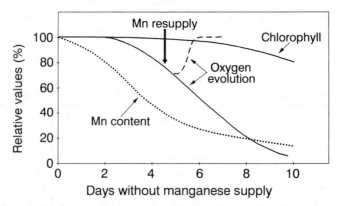

Fig. 9.7 Effect of withdrawal of manganese supply and resupply of manganese on content of manganese and chlorophyll and photosynthetic O_2 evolution in young leaves of *Trifolium subterraneum*. (Recalculated from Nable *et al.*, 1984.)

leaves restores photosynthetic O_2 evolution within one day to the levels in leaves adequately supplied with manganese. Similar results have been obtained in wheat (Kriedemann *et al.*, 1985). Manganese deficiency-induced alterations in O_2 evolution are correlated with changes in the ultrastructure of thylakoid membranes, namely the loss of certain particles (PS II functional units) associated with the stacked areas of thylakoid membranes. Resupplying manganese restores the number of the particles in the thylakoid membranes (Simpson and Robinson, 1984).

When manganese deficiency becomes more severe, the chlorophyll content also decreases and the ultrastructure of the thylakoids is drastically changed. These ultrastructural alterations are either very difficult to restore, or irreversible and are presumably caused by inhibition of biosynthesis of lipids (Section 9.2.5) and carotenoids. They are not brought about by enhanced photooxidation (lipid peroxidation) of the thylakoids and chlorophyll (Polle *et al.*, 1992).

9.2.5   Proteins, Carbohydrates, and Lipids

Although Mn^{2+} activates RNA polymerase (Ness and Woolhouse, 1980), protein synthesis is obviously not specifically impaired in manganese-deficient tissues. The protein content of deficient plants is either similar to (Table 9.6) or somewhat higher than that of plants adequately supplied with manganese (Lerer and Bar-Akiva, 1976). The accumulation of soluble nitrogen is a reflection of a shortage in reducing equivalents and carbohydrates for nitrate reduction, as well as a lower demand for reduced nitrogen. Manganese deficiency has the most severe effect on the content of nonstructural carbohydrates, as shown in Table 9.6 for the soluble (sugar) fraction. This decrease in carbohydrate content is particularly evident in the roots and most likely a key factor responsible for the depression in root growth of the deficient plants (Table 9.6; Marcar and Graham, 1987).

The role of manganese in lipid metabolism is more complex. In manganese-deficient leaves not only the chlorophyll content is lower but even more so, the content of typical thylakoid membrane constituents such as glycolipids and polyunsaturated fatty acids. These are depressed in content by up to 50% (Constantopoulus, 1970). This depression in lipid content in chloroplasts can be attributed to the role of Mn^{2+} in biosynthesis of fatty acids (coupling of C_2 units; Section 8.3) and of carotenoids and related compounds (Section 9.2.4).

Table 9.6
Effect of Manganese Deficiency on the Growth and Composition of Bean Plants[a]

	Leaves		Stems		Roots	
Parameter	+Mn	−Mn	+Mn	−Mn	+Mn	−Mn
Dry wt (g per plant)	0.64	0.46	0.55	0.38	0.21	0.14
Protein nitrogen (mg g^{-1} dry wt)	52.7	51.2	13.0	14.4	27.0	25.6
Soluble nitrogen (mg g^{-1} dry wt)	6.8	11.9	10.0	16.2	17.2	21.7
Soluble carbohydrates (mg g^{-1} dry wt)	17.5	4.0	35.6	14.5	7.6	0.9

[a]From Vielemeyer *et al.* (1969).

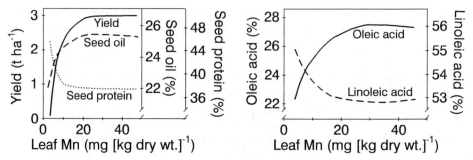

Fig. 9.8 Relationships between leaf manganese content and seed yield and seed composition of soybean. (Reproduced from Wilson *et al.*, 1982, by permission of the Crop Science Society of America.)

Distinct changes are observed in the lipid content and composition of the seeds of manganese-deficient plants (Fig. 9.8). In the deficiency range the manganese content of the leaves and both the seed yield and oil content are positively correlated. In contrast, the seed protein content is negatively correlated with the manganese content of the leaves, which in this case is an expression of a typical 'concentration effect' resulting from inhibited seed growth under manganese deficiency. The fatty acid composition of the oil is also markedly altered, the content of linoleic acid (Fig. 9.8) and certain other fatty acids increasing (Wilson *et al.*, 1982). This is counteracted by a decrease in oleic acid content. The lower oil content in the seeds of deficient plants probably results mainly from lower rates of photosynthesis and thus a decreased supply of carbon skeletons for fatty acid synthesis. In addition a direct involvement of manganese in the biosynthesis of the fatty acids might be a contributing factor. The reasons for the changes in fatty acid composition in relation to manganese supply are obscure.

Lower lignin contents in the manganese-deficient plants (Table 9.7) are a reflection of the requirement for manganese in various steps of lignin biosynthesis. The decrease is particularly evident in the roots, and an important factor responsible for the lower resistance of manganese deficient plants to root infecting pathogens (Chapter 11).

Table 9.7

Relation Between Manganese and Lignin Content in Shoots and Roots of Young Wheat Plants[a]

	Mn content (mg kg^{-1} dry weight)			
	4.2	7.8	12.1	18.9
Lignin (% of dry wt)				
Shoots	4.0	5.8	6.0	6.1
Roots	3.2	12.8	15.0	15.2

[a]Recalculated from Brown *et al.* (1984).

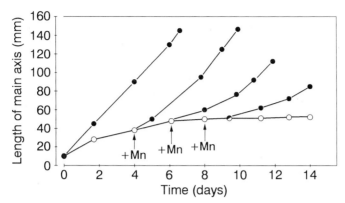

Fig. 9.9 Effect of transfer from manganese-deficient to complete medium on growth of main axis of excised tomato roots. Key: ○, manganese absent; ●, manganese present. (Based on Abbott, 1967.).

9.2.6 Cell Division and Extension

Inhibition of root growth in manganese-deficient plants is caused by shortage in carbohydrates as well as by a direct requirement for growth (Campbell and Nable, 1988). The rate of elongation seems to respond more rapidly to manganese deficiency than does the rate of cell division. As shown in Fig. 9.9 with isolated tomato roots in sterile culture and an ample supply of carbohydrates, but without manganese, there is a rapid decline in extension of the main axis within less than 2 days. Resupplying manganese rapidly restores the growth rate to normal levels if the deficiency is not too severe. In manganese-deficient plants, the formation of lateral roots ceases completely (Abbott, 1967). There is a greater abundance of small nonvacuolated cells in these roots than in control roots, which indicates that cell elongation is impaired to a greater extent by manganese deficiency than is cell division, an observation supported by tissue culture experiments (Neumann and Steward, 1968).

9.2.7 Manganese Deficiency and Toxicity

Manganese deficiency is abundant in plants growing in soils derived from parent material inherently low in manganese, and in highly leached tropical soils. It is also common on soils of high pH containing free carbonates, particularly when combined with a large organic matter content (Farley and Draycott, 1973). Manganese deficiency can be corrected by soil or foliar application of $MnSO_4$ (Reuter *et al.*, 1988), but this latter method has its limitations (Section 4.3.2). A high manganese content in the seeds, either supplied naturally from the parent plants or artificially by soaking the seeds in $MnSO_4$, can considerably improve plant growth and seed yield on manganese-deficient soils as has been shown for barley (Longnecker *et al.*, 1991a). Plant species and cultivars within a species differ considerably in susceptibility to manganese deficiency when grown on low manganese soils. For example, oat, wheat, soybean or peaches are very susceptible whereas maize and rye are not susceptible (Reuter *et al.*, 1988). In contrast, the critical deficiency contents of manganese in plants are similar, varying between 10

Table 9.8

Critical Toxicity Contents of Manganese in the Shoots of Various Plant Species[a,b]

Species	Manganese content $(mg\ kg^{-1}\ dry\ wt)$
Maize	200
Pigeon pea	300
Soybean	600
Cotton	750
Sweet potato	1380
Sunflower	5300

[a]Based on Edwards and Asher (1982).
[b]Critical toxicity contents are associated with a 10% reduction in dry matter production.

and 20 mg Mn kg^{-1} dry weight in fully expanded leaves, regardless of plant species or cultivar or prevailing environmental conditions. Only *Lupinus angustifolius* has a critical deficiency content which is twice as high as that of other plant species (Hannam and Ohki, 1988). Below the critical deficiency content dry matter production (Ohki *et al.*, 1979), net photosynthesis, and chlorophyll content decline rapidly, whereas rates of respiration and transpiration remain unaffected (Ohki *et al.*, 1981). Manganese-deficient plants are more susceptible to damage by freezing temperatures (Buntje, 1979) and require twice as long to reach booting stage than manganese-sufficient plants (Longnecker *et al.*, 1991b). The fall occurring in grain number and grain yield in manganese-deficient plants is presumably a combination of both low pollen fertility (Sharma *et al.*, 1991) and shortage in carbohydrate supply for grain filling (Longnecker *et al.*, 1991b).

In dicotyledons, interveinal chlorosis of the younger leaves is the most distinct symptom of manganese deficiency, whereas in cereals, greenish gray spots on the more basal leaves ('gray speck', '*Dörrfleckenkrankheit*') are the major symptoms. In legumes, manganese-deficiency symptoms on the cotyledons are known as 'marsh spot' in peas or as 'split seed' disorder in lupins; the latter disorder includes discoloration, splitting, and deformity of seeds (Campbell and Nable, 1988).

In contrast to the narrow range of critical deficiency content in leaves, the critical toxicity content varies widely between plant species and environmental conditions. An example of the differences between crop species is given in Table 9.8. Even within a species the critical toxicity content can vary largely between cultivars (Edwards and Asher, 1982; Horst, 1988).

Of the environmental factors affecting critical toxicity contents, temperature and the presence of silicon are of particular importance. At high temperatures the critical toxicity contents in the leaves are often much higher than those at low temperatures (Heenan and Carter, 1977; Rufty *et al.*, 1979). The effect of silicon is comparable to that of high temperatures; that is, to increase tissue tolerance to manganese (Section 10.3). There are conflicting reports on the effect of high light intensity on manganese toxicity, for example, of increasing the severity of toxicity symptoms (Horiguchi, 1988b; Nable *et al.*, 1988) or lessening toxicity symptoms (Wissemeier and Horst, 1992). The diversity of manganese toxicity symptoms is a major reason for these contradictory results.

In many plant species symptoms of manganese toxicity are characterized by brown speckles on mature leaves, and these symptoms are much less distinct in high light compared with low light conditions (Wissemeier and Horst, 1992). Although these brown speckles contain oxidized manganese, the brown colour derives not from manganese (e.g., MnO_2) but from oxidized polyphenols (Wissemeier and Horst, 1992). The formation of brown speckles is preceded by enhanced callose formation in the same area, indicating toxic effects of Mn^{2+} on the plasma membrane and enhanced Ca^{2+} influx (Wissemeier and Horst, 1987) as a 'signal' for callose formation (Section 8.6). Intensity of formation of brown speckles can be used as a simple and rapid method for screening different cultivars for manganese tolerance (Wissemeier and Horst, 1991).

In leaves of manganese-tolerant plant species like sunflower or stinging nettle, at high manganese content brown spots are also often found around the base of trichomes (Blamey *et al.*, 1986; Hughes and Williams, 1988) and deposition of manganese oxides in these sites is considered a tolerance mechanism.

In many instances manganese toxicity symptoms occur as interveinal chlorosis and necrosis (Nable *et al.*, 1988; Horiguchi *et al.*, 1988b). Particularly in dicots such as bean (Horst and Marschner, 1978b) and cotton (Foy *et al.*, 1981) these symptoms are combined with deformations of the young leaves ('crinkle leaf'), which is a typical symptom of calcium deficiency. In these instances, induced deficiencies of other mineral nutrients such as iron, magnesium and calcium dominate, or are at least involved (Horst, 1988). Induced deficiency in iron and magnesium is caused by both inhibited uptake and competition (or imbalance) at the cellular level. Examples of the inhibitory effect of Mn^{2+} on Mg^{2+} uptake have been given in Section 2.5.3, and the competition at a cellular level discussed above. Accordingly, manganese toxicity can often be counteracted by a high supply of magnesium (Löhnis, 1960).

In contrast to iron and magnesium, induction of calcium-deficiency symptoms ('crinkle leaf') by high tissue contents of manganese is most likely an indirect effect on the calcium transport to expanding leaves. Acropetal calcium transport is mediated by a basipetal countertransport of IAA (Section 3.4.3), and high IAA oxidase activity, or polyphenoloxidase activity in general, is a typical feature in tissues with high manganese contents (Horst, 1988). Calcium-deficiency symptoms induced by manganese toxicity are therefore most likely caused by enhanced degradation of IAA, a process which is aggravated, for example, by high light intensity (Horst, 1988). Loss in apical dominance and enhanced formation of auxillary shoots ('witches' broom') constitute another symptom of manganese toxicity (Kang and Fox, 1980), further supporting the hypothesis of a relationship between impaired basipetal IAA transport and manganese toxicity.

9.3 Copper

9.3.1 General

Copper is a transition element and shares similarities with iron, such as the formation of highly stable complexes and easy electron transfer:

$$Cu^{2+} \underset{-e^-}{\overset{+e^-}{\rightleftharpoons}} Cu^+$$

Divalent copper is readily reduced to monovalent copper, which is unstable. Most of the functions of copper as a plant nutrient are based on the participation of enzymatically bound copper in redox reactions. In the redox reactions of the terminal oxidases copper enzymes react directly with molecular oxygen. Terminal oxidation in living cells is therefore catalyzed by copper and not by iron.

Copper has a high affinity for peptide and sulfhydryl groups, and thus, to cysteine-rich proteins in particular, as well as for carboxylic and phenolic groups. Therefore, in the soil solution (Hodgson *et al.*, 1966) as well as in roots (press sap) and in the xylem sap (Graham, 1979), more than 98–99% of the copper is present in complexed form. This is most likely also the case in the cytoplasm and its organelles where the concentration of free Cu^{2+} and Cu^+ is extremly low.

9.3.2   Copper Proteins

According to Sandmann and Böger (1983) three different forms of proteins exist in which copper is the metal component (Cu-proteins): (a) *blue proteins* without oxidase activity (e.g., plastocyanin), which function in one-electron transfer; (b) *non-blue proteins*, which represents peroxidases and oxidize monophenols to diphenols; and (c) *multicopper proteins* containing at least four copper atoms per molecule, which act as oxidases (e.g., ascorbate oxidase and diphenol oxidase) and catalyze the reaction

$$2AH_2 + O_2 \rightarrow 2A + 2H_2O$$

Cytochrome oxidase is a mixed copper–iron protein catalyzing the terminal oxidation in mitochondria (Section 9.1.2).

Under copper deficiency the activity of these copper enzymes decreases fairly rapidly, and in most, but not all (Section 9.3.2.3), cases these decreases are correlated with distinct metabolic changes and inhibition of plant growth.

9.3.2.1   *Plastocyanin*

In general, more than 50% of the copper localized in chloroplasts is bound to plastocyanin. This Cu-protein has a molecular weight of $\sim$10 kDa and contains one copper atom per molecule. Plastocyanin is a component of the electron transport chain of photosystem I (Fig. 5.1). A proportion of 3 to 4 molecules of plastocyanin per 1000 molecules of chlorophyll appears to be the rule (Sandmann and Böger, 1983).

Under copper deficiency there is a close relationship between the copper content of leaves and the plastocyanin content and, thus, the activity of PS I whereas the chlorophyll content is only slightly affected (Table 9.9).

Compared with PS I, the activity of PS II is usually less depressed by copper deficiency (Table 9.10; Henriques, 1989). Lower activities of PS II in copper-deficient plants are related to other functions of copper in chloroplasts. Copper is a component of other chloroplast enzymes (see below) and is required for the synthesis of quinones; the decrease in plastoquinone in the chloroplasts (Table 9.10) may reflect this function of

Table 9.9
Relationship between Copper Content and Some Chloroplast Constituents and Activities of Copper-containing Enzymes in Pea Leaves[a]

Cu (μg g^{-1} dry wt)	Chlorophyll (μmol g^{-1} dry wt)	Plastocyanin (nmol μmol^{-1} chlorophyll)	Photosynthetic e$^-$ transport PS I (relative)	Enzyme activities		
				Diamine oxidase	Ascorbate oxidase	CuZnSOD (EU mg^{-1}
				(μmol g^{-1} protein h^{-1})		protein)[b]
6.9	4.9	2.4	100	0.86	730	22.9
3.8	3.9	1.1	54	0.43	470	13.5
2.2	4.4	0.3	19	0.24	220	3.6

[a]Based on Ayala and Sandmann (1988a)
[b]EU = enzyme unit

copper. In copper-deficient chloroplasts the inhibition of electron transport is further accentuated by the lack of two polypeptides in the chloroplast membrane, which are probably necessary to maintain appropriate membrane fluidity to ensure the mobility of plastoquinone molecules to transport electrons between the two photosystems (Droppa et al., 1984).

9.3.2.2 Superoxide Dismutase

The various types of SOD isoenzymes and their requirement for the detoxification of superoxide radicals (O_2^-) have been discussed in Section 9.2. The copper–zinc SOD (CuZnSOD) has a molecular weight of ~32 kDa, and at the active site probably one copper and one zinc atom are closely connected by a common histidine nitrogen (Sandmann and Böger, 1983). The copper atom in the CuZnSOD is directly involved in the mechanism of detoxification of O_2^- generated in photosynthesis (Elstner, 1982).

The CuZnSOD is located in the cytoplasm in mitochondria (Duke and Salin, 1985) and glyoxysomes (Sandalio and del Rio, 1987), but occurs also in the chloroplasts (Palma et al., 1986; Kröniger et al., 1992), together with the FeSOD (Section 9.1). In the glyoxysomes CuZnSOD is presumably involved in the control of peroxidation of membrane lipids and, thus, in senescence (Sandalio and del Rio, 1987).

Table 9.10
Effect of Copper Deficiency in Spinach on Chloroplast Pigments and Photosynthetic Electron Transport in Photosystems II and I[a]

Treatment	Chloroplast pigment content (μg g^{-1} leaf fresh wt)			Plastocyanin (10^{-9} mol mg^{-1} chlorophyll)	Photosystem activity (relative)	
	Chlorophyll	Carotenoids	Plastoquinone		PS II	PS I
+Cu	1310	248	106	5.16	100	100
−Cu	980	156	57	2.08	66	22

[a]Based on Baszynski et al. (1978).

Under copper deficiency, CuZnSOD activity also declines drastically in leaves (Table 9.9). This decline is true both, for chloroplastic and cytosolic CuZnSOD, and is accompanied by a simultaneous corresponding increase in activity of the MnSOD (Ayala and Sandmann, 1988a). As the MnSOD is exclusively located in the cytoplasm (Section 9.2), there is at least in this compartment a coordinated compensation mechanism for the formation of SOD isoenzymes. It is not known whether this is also the case for a corresponding compensation of CuZnSOD by FeSOD in the chloroplasts. The dramatic changes in the ultrastructure of chloroplasts with severe copper deficiency (disintegration of intergrana lamellae and swollen grana stocks; Henriques, 1989; Casimiro et al., 1990) are typical of oxidative damage and most likely indicative of inadequate detoxification of O_2^- in copper-deficient chloroplasts.

9.3.2.3 Cytochrome Oxidase

This terminal oxidase of mitochondrial electron transport chain (Fig. 5.5) contains two copper atoms and two iron atoms in the heme configuration. The activity of the enzyme can be blocked by cyanide; the remaining respiratory O_2 consumption of cells is then mediated by the quinal oxidase known as the 'alternative oxidase' (in the 'alternative pathway', see Section 5.3). This enzyme contains copper but no heme iron. It is therefore unlikely that in copper-deficient cells the alternative respiration can function to compensate low cytochrome oxidase activity. As respiration rates either remain unaffected or are only moderately decreased by copper deficiency, cytochrome oxidase seems to be present in large excess in the mitochondria (Blingny and Douce, 1977; Ayala and Sandmann, 1988b).

9.3.2.4 Ascorbate Oxidase

Ascorbate oxidase catalyzes the oxidation of ascorbic acid to dehydroascorbic acid according to the equation:

Ascorbic acid   Dehydro-
                                              ascorbic acid

The enzyme contains at least four copper atoms per molecule which operate a four-electron reduction of O_2 to water. The enzyme occurs in cell walls and the cytoplasm, and may act as a terminal respiratory oxidase, as shown above, or in combination with polyphenol oxidases (Section 9.3.2.6). Ascorbate oxidase activity decreases in copper-deficient plants (Table 9.9) and is a sensitive indicator of the copper nutritional status of

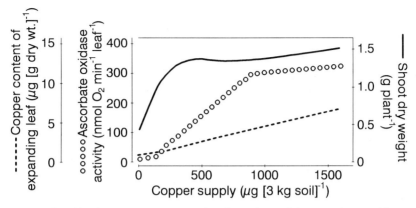

Fig. 9.10 Relationships between copper supply, shoot dry weight, ascorbate oxidase activity, and copper content of subterranean clover. (Modified from Loneragan *et al.*, 1982a.)

a plant (Fig. 9.10). Although in this case no direct relationship was found between a decrease in enzyme activity and plant growth, there is a close positive correlation in the suboptimal concentration range between the copper content of leaf tissue and its ascorbate oxidase activity (Fig. 9.10).

On the basis of this correlation a rapid and simple colorimetric field test for ascorbate oxidase activity has been developed for the diagnosis of copper deficiency. The results of this test agree closely with the diagnosis based on chemical analysis of the copper content of leaves (Delhaize *et al.*, 1982).

Resupplying copper to deficient plants can restore the activity of ascorbate oxidase only in very young, but not in mature leaves (Table 9.11), suggesting that the active holoenzyme can only be synthesized in leaf blades during their very early development.

Table 9.11
Effect of Copper Nutritional Status of Subterranean Clover on Ascorbate Oxidase Activity (AOA) and Protein Content in Very Young and Mature Leaves[a]

Leaf age	Parameter	Cu nutritional status (Cu supply)		
		−Cu	−Cu + Cu[b]	+Cu
Very young	μg Cu g^{-1} dry wt	<0.5	17.9	13.4
	AOA[c]	10	245	240
	protein (mg g^{-1} dry wt)	17.6	38.4	40.7
Mature	μg Cu g^{-1} dry wt	1.0	7.9	10.0
	AOA[c]	5.0	5.0	34.0
	protein (mg g^{-1} dry wt)	36.6	43.9	40.0

[a]Based on Delhaize *et al.* (1985).
[b]Resupply of Cu to deficient plants.
[c]nmol O_2 consumed per leaf min^{-1}.

This is in contrast to plastocyanin the activity of which can also be restored in mature leaves on resupplying copper (Droppa *et al.*, 1984).

9.3.2.5 Diamine Oxidase

Polyamine oxidases are flavoproteins which catalyze the degradation of polyamines as for example spermidine (Section 9.2) to form putrescine, H_2O_2 and NH_3. Polyamine oxidases preferentially degrade tri- and tetra-amines which are the main forms present in graminaceous species (Federico *et al.*, 1990). The degradation of putrescine (diamine) and, to some extent also of spermidine (triamine), is mediated by diamine oxidase, a copper-containing enzyme. Diamine oxidase is widespread in different plant species, particularly in legumes. Its activity decreases in copper-deficient plants (Table 9.9) and can be restored by resupplying copper (Delhaize *et al.*, 1985). Similarly to ascorbate oxidase (Table 9.11), this restoration of activity is confined to very young leaves. In copper-deficient leaves the apoenzyme of diamine oxidase is absent, copper obviously being required to modulate the level of mRNA coding for the enzyme. This latter process is confined to the very early stages of leaf development (Delhaize *et al.*, 1986).

Diamine oxidase is mainly located in the apoplasm of the epidermis and the xylem of mature tissues where it presumably functions as an H_2O_2-delivery system for peroxidase activity in the process of lignification and suberization (Angelini *et al.*, 1990). In agreement with this, diamine oxidase activity increases in response to wounding and is closely correlated with the lignification of the wounded area (Scalet *et al.*, 1991).

9.3.2.6 Phenol Oxidases

These enzymes catalyze oxygenation reactions of plant phenols. Phenol oxidases are abundant in cell walls but are also located in the thylakoid membranes of chloroplasts, where they are presumably required for the synthesis of plastocyanin, a constituent of the photosynthetic e⁻ transport chain (Section 5.2.1). These enzymes have two distinct functions: (a) they hydroxylate monophenols to diphenols, resembling, for example, tyrosinase activity, and (b) they oxidize diphenols to o-quinones, for example, resembling dihydroxyphenylalanine (DOPA) oxidase activity:

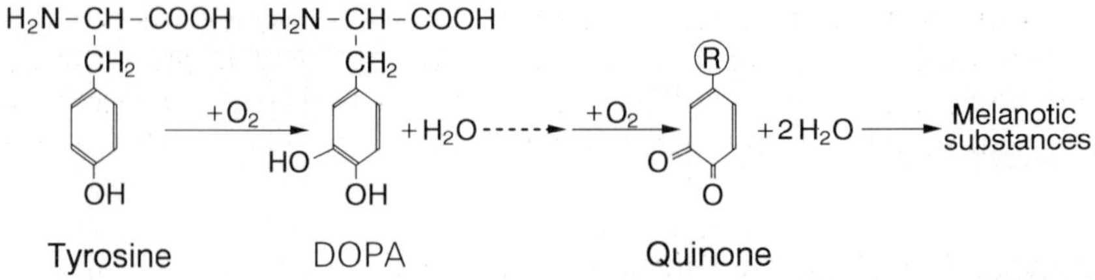

Both reactions need molecular oxygen. They are coupled to each other, if monophenols are the substrates. They are named according to their most important substrates as monophenol oxidases, polyphenol oxidases, phenolases, DOPA oxidases, tyrosinases etc. Their specificity is rather low.

Table 9.12

Effect of Copper Deficiency on Flowering and Enzyme Activities in *Chrysanthemum morifolium*[a]

Treatment	Copper content (μg g^{-1} leaf dry wt)	No. of flowering shoots per plant	No.of open flowers per plant	Enzyme activity in leaves (relative)		
				Polyphenol oxidase	IAA oxidase	Peroxidase
Cu sufficient	7.9	14.2	13.1	100	100	100
Cu deficient	2.4	8.3	0.5	26	52	41

[a]Based on Davies *et al.* (1978).

Polyphenol oxidases are involved in the biosynthesis of lignin (see Section 9.3.4) and alkaloids and in the formation of brown melanotic substances, which are sometimes formed when tissues are wounded (e.g., in apple and potatoes). These substances are also active as phytoalexins, which inhibit spore germination and fungal growth. Under copper deficiency, the decrease in polyphenol oxidase activity is quite severe (Table 9.12) and is correlated with an accumulation of phenolics (Judel, 1972) and a decrease in the formation of melanotic substances. The latter effect is reflected, for example, in the close correlation between the color of spores of *Aspergillus niger* and copper nutritional status. With an ample copper supply the spores are black; with mild deficiency they are light brown; and with severe deficiency they are white.

Polyphenol oxidase activity is almost absent in copper-deficient leaves of subterranean clover (Delhaize *et al.*, 1985) or soybean. In this latter species there is an almost linear relationship between polyphenol oxidase activity and copper supply (Marziah and Lam, 1987). Polyphenol oxidase activity in soybean leaves is only lowered by copper deficiency and not by any of the other micronutrient deficiencies.

A decline in polyphenol oxidase activity with copper deficiency may be at least indirectly responsible for the delay in flowering and maturation often observed in the copper-deficient plants (Reuter *et al.*, 1981) and shown for the flowering of *Chrysanthemum* in Table 9.12. Copper deficiency led to a decrease in the number of flowering shoots, but mainly prevented the opening of flowers. As would be expected, the polyphenol oxidase activity was much lower in copper-deficient plants, but the activities of IAA oxidase and peroxidase were also lower. On the other hand, in tissue cultures regeneration of plants is often severely impaired by high activities of polyphenol oxidase. Accordingly, the percentage of shoot-regenerating explants is inversely correlated with the copper content of the stock plants, and best results on regeneration are achieved with explants from severely copper deficient stock plants (Schum *et al.*, 1988).

9.3.3 Carbohydrate, Lipid and Nitrogen Metabolism

In plants suffering from copper deficiency the content of soluble carbohydrates is considerably lower than normal during the vegetative stage (Brown and Clark, 1977; Mizuno *et al.*, 1982). However, after anthesis, when grains have developed as a dominant sink, copper deficient plants have only a few grains (Section 6.3.3), remain

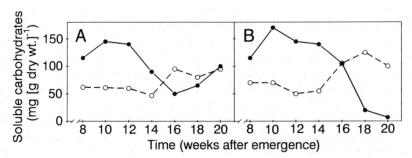

Fig. 9.11 Concentrations of soluble carbohydrates in flag leaves (A) and roots (B) of wheat plants grown at two copper levels as a function of plant age. Key: ●, +Cu; ○, −Cu. (Modified from Graham, 1980a.)

green (i.e., actively photosynthesizing) and build up high contents of soluble carbohydrates both in the leaves and in the roots (Fig. 9.11). The leaves of deficient plants may even release droplets of honeydew-like substances (Graham, 1980a).

In view of the role of copper in PS I low rates of photosynthesis and low contents of carbohydrates are to be expected, at least during the vegetative stage. However, in severely copper deficient plants a drop in net CO_2 fixation to about 50% expressed both in terms of unit chlorophyll (Bottrill *et al.*, 1970) or leaf area (Casimiro *et al.*, 1990) cannot be attributed solely to a much lower activity of PS I. Lower activity of PS II must also be a contributing factor, probably due to impaired synthesis of carotenoids in PS II (Ayala *et al.*, 1992), of quinones, and of disintegration of thylakoid membranes (Henrique, 1989). In copper-sufficient plants 11 copper atoms per 1000 chlorophyll molecules are located in the PS II complex (Ayala *et al.*, 1992). Under severe copper deficiency alterations in polypeptides of PS II occur (Droppa *et al.*, 1984) and the lipid composition changes in favour of the less unsaturated fatty acids, for example, $18:3 \rightarrow 18:2$ (Ayala *et al.*, 1992). These changes in fatty acid composition in the thylakoids and in the PS II complex are probably related to functions of copper in the desaturation of long-chain fatty acids (e.g., $18:2 \rightarrow 18:3$). Another Cu-protein, the alternative oxidase in the glyoxysomes is involved in this desaturation (Wahle and Davies, 1977).

Low carbohydrate contents in copper-deficient plants are involved in impaired pollen formation and fertilization (Section 9.3.5), and are certainly the main reason for depressed nodulation and N_2 fixation in copper-deficient legumes (Cartwright and Hallsworth, 1970). Symptoms of nitrogen deficiency occur which can be overcome by the application of mineral nitrogen (Snowball *et al.*, 1980). In nonleguminous copper-deficient plants the protein content might be somewhat lower or similar (Table 9.11) or even higher (Bottrill *et al.*, 1970) than in copper-sufficient plants. In some instances too there is an accumulation of amino acids and nitrate in the deficient plants (Brown and Clark, 1977). Evidence is lacking, however, that copper is directly involved in the biosynthesis of proteins, except of the copper-containing proteins (Sections 9.3.2.4 and 9.3.2.5).

It has been shown repeatedly that nitrogen application accentuates copper deficiency, and when nitrogen supply is high, the application of copper fertilizers is required for maximum yield (Thiel and Finck, 1973; Robson and Reuter, 1981). In addition to unspecific interactions (e.g., growth enhancement by nitrogen), nitrogen has specific effects on copper availability and mobility, including (a) sequestration of a higher proportion of copper complexed to amino acids and proteins in mature tissue and (b) a decrease in the rate of retranslocation of copper from old leaves to areas of new growth. Retranslocation of copper is closely related to leaf senescence (Section 3.5). Because high nitrogen supply delays senescence, it also retards copper retranslocation (Hill *et al.*, 1978). Impaired retranslocation of copper into new growth is also involved in copper deficiency (stem deformations) in *Pinus radiata* stands established on fertile pasture sites (Hopmans, 1990). In agreement with this, the critical deficiency levels of copper in the dry matter of the whole shoot required for maximum growth increase with increasing nitrogen supply (Thiel and Finck, 1973).

9.3.4 Lignification

Impaired lignification of cell walls is the most typical anatomical change induced by copper deficiency in higher plants. This gives rise to the characteristic distortion of young leaves, bending and twisting of stems and twigs (stem deformation and 'pendula' forms in trees, Oldenkamp and Smilde, 1966; Hopmans, 1990) and an increase in the lodging susceptibility of cereals, particularly in combination with a high nitrogen supply (Vetter and Teichmann, 1968).

As shown in Table 9.13 copper has a marked effect on the formation and chemical composition of cell walls. In deficient leaves the ratio of cell wall material to the total dry matter decreases; simultaneously the proportion of α-cellulose increases whereas the lignin content is only about half that of leaves adequately supplied with copper. This effect on lignification is even more distinct in the sclerenchyma cells of stem tissue (Fig. 9.12). In plants suffering from severe copper deficiency the xylem vessels are also insufficiently lignified. A decrease in lignification occurs even with mild copper deficiency and is thus a suitable indicator of the copper nutritional status of a plant (Rahimi and Bussler, 1974; Pissarek, 1974).

Lignification responds rapidly to copper supply; transition periods of copper deficiency during the growth period can be readily identified by variations in the degree of lignification in stem sections (Bussler, 1981b).

The inhibiton of lignification in copper-deficient tissue is related to a direct role of at least two copper-enzymes in lignin biosynthesis. Polyphenol oxidase catalyzes the

Table 9.13

Effect of Copper Nutritional Status on Cell Wall Composition of the Youngest Fully Emerged Leaves of Wheat[a]

Treatment	Cu content ($\mu g\ g^{-1}$ dry wt)	Cell wall content (% of dry matter)	Percentage of cell walls			Percentage of dry wt	
			α-Cellulose	Hemicellulose	Lignin	Total phenolics	Ferulic acid
+Cu	7.1	46.2	46.8	46.7	6.5	0.73	0.50
−Cu	1.0	42.9	55.3	41.4	3.3	0.82	0.69

[a]From Robson *et al.* (1981).

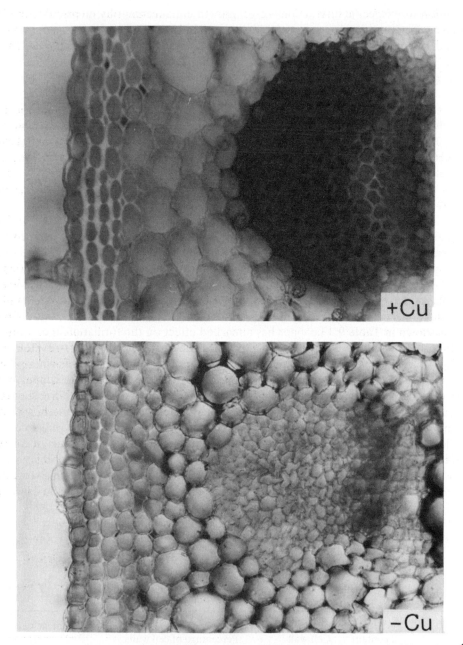

Fig. 9.12 Stem sections of sunflower plants grown with sufficient copper supply (50 μg Cu l^{-1}) and without copper supply. (*Top*) Copper sufficient; walls of the sclerenchyma cells are thick and lignified. (*Bottom*) Copper deficient; walls of the sclerenchyma cells are thin and nonlignified. (Rahimi and Bussler, 1974.)

oxidation of phenolics as precursors of lignin, and diamine oxidase provides the H_2O_2 required for oxidation by peroxidases. Accordingly, in copper-deficient tissues not only is the activity of both enzymes lower but also phenolics accumulate (Table 9.13).

9.3.5 Pollen Formation and Fertilization

Copper deficiency affects grain, seed, and fruit formation much more than vegetative growth (see also Section 6.3.3). A typical example is shown in Table 9.14. Supplying 0.5 μg of copper produced maximum dry matter yields of roots and shoots but flower formation was impaired, and no fruits were formed. For fruit formation a much greater copper supply was required. The decline in root, leaf and stem dry weights with a 1.0 and 5.0 μg copper supply reflects sink competition (Section 5.7). With a 10 μg supply, toxicity occurred.

As discussed in Chapter 6, the main reason for the decrease in the formation of generative organs is the nonviability of the pollen from copper-deficient plants (Graham, 1975). When the copper supply is adequate, the anthers containing pollen and the ovaries have the highest copper content in the flowers (Knight *et al.*, 1973) and obviously also the highest copper demand. The critical stage of copper deficiency-induced pollen sterility is microsporogenesis (Graham, 1975). Dell (1981) found that reduced seedset in copper-deficient plants might be the result of the inhibition of pollen release from the stamina, since lignification of the anther cell walls is required to rupture the stamina and the subsequent release of the pollen. In copper-deficient plants lignification of the anther cell wall (tapetum) is reduced or absent (Dell, 1981); the tapetum becomes expansionary instead of supplying the developing pollen with nutrients. This alteration in copper-deficient plants is also reflected in the lack of the massive starch reserves in the pollen grains which is typical for copper-sufficient plants (Jewell *et al.*, 1988).

After grain set in wheat (Hill *et al.*, 1979c) and seed set in subterranean clover (Reuter *et al.*, 1981), further grain and seed growth, surprisingly, are not influenced by copper nutritional status of the plants, even though at maturity the copper content of wheat grains in plants adequately supplied with copper is five to six times higher than in

Table 9.14
Relationship between Copper Supply and Growth and Dry Matter Distribution in Red Pepper[a]

Copper supply (μg per pot)	Dry weight (g per plant)			
	Roots	Leaves and stem	Buds and flowers	Fruits
0.0	0.8	1.7	0.16	None
0.5	1.6	3.3	0.28	None
1.0	1.5	3.2	0.38	0.87
5.0	1.4	3.0	0.36	1.81
10.0	1.2	2.0	0.28	1.99

[a]From Rahimi (1970).

deficient plants. This result further emphasizes the importance of adequate copper supply during fertilization for final seed and fruit yield.

9.3.6 Copper Deficiency and Toxicity

9.3.6.1 Copper Deficiency

Copper deficiency is often observed in plants growing on soils either inherently low in total copper (e.g., ferrallitic and ferruginous coarse textured soils, or calcareous soils derived from chalk) and on soils high in organic matter where copper is complexed to organic substances (Alloway and Tills, 1984). High nitrogen availability can also accentuate copper deficiency (Section 9.3.3).

The critical deficiency level of copper in vegetative plant parts is generally in the range of $1-5 \mu g \, g^{-1}$ dry weight, depending on plant species, plant organ, developmental stage, and nitrogen supply (Thiel and Finck, 1973; Robson and Reuter, 1981). In general, the critical deficiency level in the youngest emerged leaf is less affected by environmental factors (Section 12.3.3). Plant species differ considerably in sensitivity to copper deficiency, wheat, oat, and spinach being much more sensitive than, for example, pea, rye, and rape (Alloway and Tills, 1984). Stunted growth, distortion of young leaves, necrosis of the apical meristem, and bleaching of young leaves ('white tip' or 'reclamation disease' of cereals grown in organic soils), and/or 'summer dieback' in trees, in addition to those already discussed, are typical visible symptoms of copper deficiency (Rahimi and Bussler, 1973a). Enhanced formation of tillers in cereals and of auxiliary shoots in dicotyledons are secondary symptoms caused by necrosis of the apical meristem. Wilting in young leaves, also characteristic of copper-deficient plants is either the result of impairment of water transport due to insufficient lignification of the xylem vessels (Rahimi and Bussler, 1973b; Pissarek, 1974) or of structural weaknesses in the cell wall system rather than the result of a low water content per se (Graham 1976). Inhibition of calcium transport to areas of new growth may occur in copper-deficient plants (Brown, 1979), but it is probably a secondary symptom related to the alteration in phenol metabolism and, thus, acropetal calcium transport ($Ca^{2+}/$ IAA countertransport, Section 3.4.3).

Foliar applications of copper in the form of inorganic salts, oxides, or chelates are required as means of rapidly correcting copper deficiency in soil-grown plants. Soil applications of inorganic copper salt, oxides, or slow-release metal compounds are more appropriate for long-term effects. Selecting genotypes which are highly efficient in copper uptake and, particularly, efficient in translocating copper from the roots to the shoots and retranslocation within the shoot, is a promising long-term approach to the prevention of copper deficiency under certain ecological conditions (e.g. top soil drying).

9.3.6.2 Copper Toxicity

For most crop species, the critical toxicity level of copper in the leaves is above 20–30 $\mu g \, g^{-1}$ dry wt (Hodenberg and Finck, 1975; Robson and Reuter, 1981). There are, however, marked differences in copper tolerance between plant species (e.g., bean is

Table 9.15
Relationship between Copper Supply (Nutrient Solution), Dry Weight, and Copper Content of Tomato Plants[a]

Cooper supply (μg l^{-1})	Dry weight (g per plant)		Copper content (mg kg^{-1} dry wt)		
	Roots	Shoots	Roots	Stems and petioles	Leaf blades
0	0.3	2.6	4.0	2.8	3.0
2.5	2.5	9.4	3.8	2.1	3.2
5.0	3.2	11.2	6.4	2.4	4.1
50.0	3.4	12.0	64.0	4.3	14.6
250.0	1.6	9.7	360.0	6.2	20.3

[a]From Rahimi and Bussler (1974).

much more tolerant than maize); these differences are directly related to the copper content of the shoots (Bachthaler and Stritesky, 1973). In certain copper-tolerant species ('metallophytes') of the natural vegetation the copper content in leaves can be as high as 1000 μg g^{-1} dry wt (Morrison et al., 1981). Copper toxicity may induce iron deficiency (Taylor and Foy, 1985; Bergmann, 1988), depending on the source of iron supply (Rahimi and Bussler, 1973a; Taylor and Foy, 1985). Chlorosis can also be a direct result of the action of high copper concentrations on lipid peroxidation and thus the destruction of membranes (Sandmann and Böger, 1983; Mattoo et al., 1986).

A large copper supply usually inhibits root growth before shoot growth (Lexmond and Vorm, 1981). This does not mean, however, that roots are more sensitive to high copper concentrations; rather, they are the sites of preferential copper accumulation when the external copper supply is large, as shown in Table 9.15 for tomato plants. In plants receiving a large supply, the copper content of the roots rises proportionally to the concentration of copper in the external medium, whereas transport to the shoot is still highly restricted. Without analysis of roots, critical toxicity contents of copper in the shoots do not necessarily therefore provide an appropriate indication of the copper tolerance of plants. This is an especially important consideration when genotypes are being compared. Even at high supply, up to 60% of the total copper in roots might be bound to the cell wall fraction and the cell wall–plasma membrane interface (Iwasaki et al., 1990).

In nontolerant plants, inhibition of root elongation and damage of the plasma membrane of root cells, as reflected by enhanced potassium efflux, are immediate responses to a high copper supply (Baker and Walker, 1989a,b; De Vos et al., 1991). Certain changes in root morphology such as inhibited elongation and enhanced lateral root formation (Savage et al., 1981), might be related to the sharp decrease in IAA oxidase activity in roots exposed to high copper concentrations (Coombes et al., 1976).

For various reasons there is increasing concern about copper toxicity in agriculture (Tiller and Merry, 1981). These include the high copper contents in soils caused by the long-term use of copper-containing fungicides (e.g., in vineyards), industrial and urban activities (air pollution, city waste, and sewage sludge), and the application of pig and poultry slurries high in copper. Mechanisms of copper tolerance in plants are therefore of interest for crop production on copper-polluted soils.

9.3.6.3 Mechanisms of Copper Tolerance

Genotypical differences in tolerance to copper and other heavy metals are well known in certain species and ecotypes of natural vegetation (Ernst, 1982; Woolhouse, 1983). It has been known for centuries that, in mining areas particularly, a special flora (*metallophytes*) that is highly tolerant to these metals develops on outcrops. In some species cotolerance to several heavy metals exists (Baker and Walker, 1989a,b). Heavy metal tolerance might be constitutional in some species whereas in others heavy metal tolerance is mainly inducible (Baker and Walker, 1989a,b). In perennials root infection with ectomycorrhiza may play an important role in heavy metal tolerance of the host plant (Section 15.8).

The mechanism of (heavy) metal tolerance, including copper tolerance, in higher plants can be grouped into various mechanisms as summarized in Fig. 9.13. The relative importance of the various mechanisms is dependent on plant species and populations (ecotypes) within a species, and whether they belong to the 'excluder' or 'includer' types (Baker, 1987). In *Silene cucubalus* the high copper tolerance of one population is closely related to its capability of restricting copper uptake and thereby preventing damage to the root cell plasma membranes (De Vos *et al.*, 1991), i.e. by a combination of mechanisms (1), (2), and (5) in Fig. 9.13.

Although immobilization in the cell walls might also play a role in copper tolerance

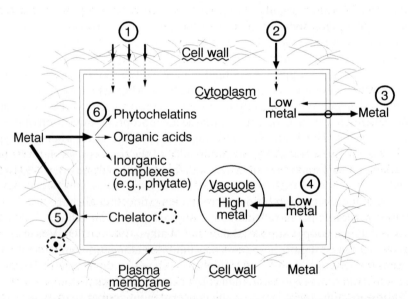

Fig. 9.13 Possible mechanisms of heavy metal tolerance of plants. (Modified from Tomsett and Thurman, 1988.) (1) Binding to cell wall. (2) Restricted influx through plasma membrane. (3) Active efflux. (4) Compartmentation in vacuole. (5) Chelation at the cell wall–plasma membrane interface. (6) Chelation in the cytoplasm.

(Turner, 1970) the capacity of this mechanism is limited (Woolhouse, 1983). Compartmentation, either as soluble or insoluble complexes, within the cytoplasm and in the vacuole is an important mechanism of copper tolerance (Wu *et al.*, 1975; Fig. 9.13, (4) and (6)). This holds true in particular for low-molecular-weight polypeptides such as phytochelatins (Section 8.3). From a copper tolerant *Agrostis* species a phytochelatin has been isolated which has 20.8% cysteine residues and binds 54 mg copper per gram protein (Curvetto and Rauser, 1979). In maize the synthesis of the cysteine-rich phytochelatins is induced by supplying high concentrations of cadmium, and also, to a lesser extent, by copper (Tukendorf and Rauser, 1990). In pea cultivars differing in copper tolerance, high copper supply induces enhanced synthesis of copper (Cu(II)) binding polypeptides more in the tolerant than in the sensitive cultivar; contrary to expectation, in both instances these polypeptides were low in cysteine but rich in leucine and isoleucine, i.e., they represented a different type of phytochelatin (Palma *et al.*, 1990). In bean leaves, high copper supply enhances the synthesis of proteins which are almost identical to plastocyanin (Nicholson *et al.*, 1980).

Enhanced synthesis of phytochelatins in response to high external copper supply is, however, not necessarily correlated with high copper tolerance. The opposite might be true as shown in *Silene cucubalus* (De Vos *et al.*, 1992). In the tolerant population, both copper uptake and phytochelatin synthesis remained low whereas in the sensitive population both increased steeply, correlated with a depletion in the glutathione pool, i.e., the precursor of phytochelatins of the cysteine type. Depletion of the glutathione pool led to impaired detoxification of H_2O_2 and related compounds and, thus, oxidative damage of the cells in the copper-sensitive population. Chlorosis in leaves and lipid peroxidation of thylakoid membranes (Sandmann and Böger, 1983), or other cell membranes (Matoo *et al.*, 1986), induced by high copper supply, might reflect such oxidative damage caused by depletion of the glutathione pool.

9.4 Zinc

9.4.1 General

Zinc is taken up predominantly as a divalent cation (Zn^{2+}); at high pH, it is presumably also taken up as a monovalent cation ($ZnOH^+$). In long-distance transport in the xylem, zinc is either bound to organic acids or occurs as the free divalent cation (Section 3.2.1). In the phloem sap the zinc concentrations are fairly high (Section 3.3.2) with zinc probably complexed to low-molecular-weight organic solutes (Kochian, 1991). In plants as in other biological systems, zinc exists only as Zn(II), and does not take part in oxidoreduction reactions. The metabolic functions of zinc are based on its strong tendency to form tetrahedral complexes with N-, O-, and particularly S-ligands and it thereby plays both a functional (catalytic) and structural role in enzyme reactions (Vallee and Auld, 1990). In the last decade the role of zinc in protein molecules involved in DNA replication and in the regulation of gene expression has attracted much interest (Coleman, 1992). Changes in metabolism brought about by zinc deficiency are quite complex. Nevertheless, some of the changes are typical and can be rather well explained by the functions of zinc in specific enzyme reactions or steps in particular metabolic pathways.

9.4.2 Zinc-Containing Enzymes

There is a large number of enzymes in which zinc is an integral component of the enzyme structure (zinc-enzymes). In these enzymes zinc has three functions: catalytic, cocatalytic (coactive), or structural (Vallee and Auld, 1990; Vallee and Falchuk, 1993). In enzymes with catalytic zinc functions (e.g., carbonic anhydrase and carboxypeptidase) the zinc atom is coordinated to four ligands, three of which are amino acids, with histidine (His) being the most frequent, followed by glutamine (Glu) and asparagine (Asp); a water molecule is the fourth ligand at all catalytical sites (model I):

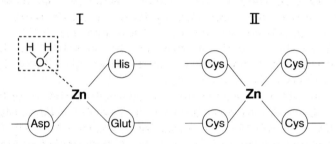

In enzymes with structural zinc functions (e.g., alcohol dehydrogenase, and the proteins involved in DNA replication and gene expression) the structural zinc atoms are coordinated to the S-groups of four cysteine residues (model II) forming a tertiary structure of high stability (Coleman, 1992). Most of the zinc-enzymes have only one zinc atom per molecule, the alcohol dehydrogenase being an exception.

9.4.2.1 Alcohol Dehydrogenase

This enzyme contains two zinc atoms per molecule, one with catalytic and the other with structural functions (Coleman, 1992). The enzyme catalyzes the reduction of acetaldehyde to ethanol

$$\text{Pyruvate} \xrightarrow{\quad CO_2 \quad} \text{Acetaldehyde} \xrightarrow[\text{NADH}]{\quad NAD^+ \quad} \text{Ethanol}$$

In higher plants under aerobic conditions, ethanol formation takes place mainly in meristematic tissues, such as root apices. In zinc-deficient plants, alcohol dehydrogenase activity decreases, but the consequences for plant metabolism are not known. The situation is different in plants grown in waterlogged or submerged soils. In lowland rice flooding stimulates the activity of root alcohol dehydrogenase twice as much in zinc-sufficient compared with zinc-deficient plants, and the lower activity of this key enzyme in anaerobic metabolism might impair root functions of submerged rice considerably (Moore and Patrick, 1988).

9.4.2.2   Carbonic Anhydrase

This enzyme contains a single zinc atom which catalyzes the hydration of CO_2

$$CO_2 + H_2O \rightleftharpoons HCO_3^- + H^+$$

Carbonic anhydrase (CA) from dicotyledons consists of six subunits and has a molecular weight of 180 kDa, and six zinc atoms per molecule (Sandmann and Böger, 1983). The enzyme is localized both in the chloroplasts and in the cytoplasm (Fig. 9.14).

In evaluating the role of CA, and particularly of that in chloroplasts, a distinction is essential between C_3 and C_4 plants and, in C_4 plants between mesophyll and bundle sheath chloroplasts.

As a rule in C_3 plants there is a lack of a direct relationship between CA activity and photosynthetic CO_2 assimilation (Ohki, 1976) or growth (Dell and Wilson, 1985) of plants with different zinc nutritional status (Fig. 9.15). The activity is closely related to the zinc content, but CO_2 assimilation per unit leaf area (Fig. 9.15) or dry matter production (Dell and Wilson, 1985) are affected only by very low CA activities. With extreme zinc deficiency CA activity is absent, but even when CA activity is low, maximum net photosynthesis can occur (Fig. 9.15).

In C_4 plants, however, the situation is different (Burnell and Hatch, 1988; Hatch and Burnell, 1990). A high CA activity is required in the mesophyll chloroplasts to shift the equilibrium in favour of HCO_3^-, the substrate for PEP carboxylase (Fig. 9.14) which forms C_4 compounds (e.g. malate) for the shuttle into the bundle sheath chloroplasts (Section 5.2.4). Here CO_2 is released and serves as a substrate for RuBP carboxylase. In agreement with this, despite similar total activities in leaves of C_3 and C_4 plants, in C_4 plants of the total CA activity only 1% is located in the bundle sheath chloroplasts (Burnell and Hatch, 1988), but 20–60% is associated with the plasma membrane (Utsunomiya and Muto, 1993).

Evidence has been presented that at least in C_4 plants the *in vivo* activity of CA might just be sufficient to prevent the rate of conversion of CO_2 to HCO_3^- from limiting photosynthesis (Hatch and Burnell, 1990). Accordingly, zinc deficiency may have a more dramatic effect on the rate of photosynthesis in C_4 compared with C_3 plants

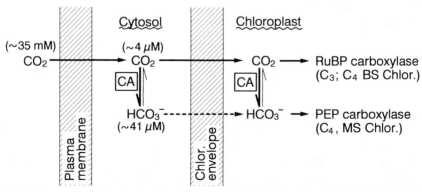

Fig. 9.14. Functioning of carbonic anhydrase (CA) in leaf cells of C_3 and C_4 plants. BS = bundle sheath chloroplasts; MS = mesophyll chloroplasts. (Based on Edwards and Walker, 1983 and Hatch and Burnell, 1990.)

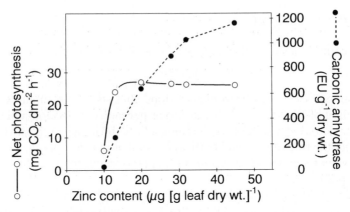

Fig. 9.15 Relationship between the zinc content of leaf blades and net photosynthesis and carbonic anhydrase activity in cotton. EU, enzyme units. (Modified from Ohki, 1976.)

(Burnell *et al.*, 1990). Carbonic anhydrase is an adaptive enzyme, its synthesis is rapidly increased at low ambient CO_2 concentrations (Dionisio-Sese *et al.*, 1990) and its activity drops within a few days in the dark, or under nitrogen deficiency. Changes in activity occur always in parallel with PEP carboxylase (Burnell *et al.*, 1990), indicating that the expression of the genes coding for the two proteins may be closely related.

9.4.2.3 CuZn-Superoxide Dismutase

In this isoenzyme zinc is associated with copper (CuZnSOD). The localization and role of CuZnSOD have been discussed in Section 9.3. Most likely the copper atom represents the catalytic metal component and zinc the structural. With zinc deficiency SOD activity is much lower but can be restored *in vitro* by resupplying zinc to the assay medium (Vaughan *et al.* 1982), indicating that the zinc atom is an essential structural component for the normal functioning of CuZnSOD.

The decrease in SOD activity which takes place under zinc deficiency is particularly critical, as a simultaneous increase in the rate of O_2^- generation occurs (Table 9.16). The correspondingly much higher level of the toxic O_2^- radicals, and related oxidants (Section 2.3) are major factors responsible for peroxidation of membrane lipids and an increase in membrane permeability (Cakmak and Marschner, 1988c).

9.4.2.4 Other Zinc-Containing Enzymes

Zinc is the metal component in a number of other enzymes (Coleman, 1992).

1. Alkaline phosphatase.
2. Phospholipase: Both these enzymes contain three zinc atoms each, and at least one of these atoms has catalytical functions.
3. Carboxypeptidase, which hydrolyses peptide cleavages, and contains a single zinc atom with catalytical functions.
4. RNA polymerase which contains two zinc atoms per molecule, and is inactive if zinc is removed (Prask and Plocke, 1971; Falchuk *et al.*, 1977).

Table 9.16

Effect of Zinc Deficiency on Generation of Superoxide Radicals (O_2^-), and Activity of Superoxide Dismutase (SOD) in Roots of Cotton[a]

Zinc supply during growth	Dry weight (g per 4 plants)		Activities mg^{-1} protein	
	Shoots	Roots	O_2^- generation (nmol min)$^{-1}$	SOD[b] (EU)
+ Zn	3.1	0.8	1.3	75
− Zn	1.8	0.5	3.7	35

[a]Cakmak and Marschner (1988a).
[b]EU = enzyme unit.

nutritional status has been studied extensively in bacteria, animals and man. There is little information, however, on this relationship in higher plants.

9.4.3 Zinc-Activated Enzymes

In higher plants zinc is either required for, or at least modulates, the activity of a large number of various types of enzymes, including dehydrogenases, aldolases, isomerases and transphosphorylases. Some examples are given below.

Inorganic pyrophosphatases (PP$_i$ases) are important components of the proton-pumping activity in the tonoplast (Section 2.4). Besides the well-known Mg^{2+}-dependent enzyme (Mg · PP$_i$ase), in leaves a PP$_i$ase isoenzyme exists which is Zn^{2+} dependent (Zn · PP$_i$ase). In rice leaves the activity ratios of Mg · /Zn · PP$_i$ase varies between 3 and 6, and both pyrophosphatases are probably different isoenzymes (Lin and Kao, 1990).

The role of zinc in DNA and RNA metabolism, in cell division, and protein synthesis has been documented for many years, but only recently has a new class of zinc dependent protein molecules (zinc metalloproteins) been identified which is involved in DNA replication, transcription and, thus, regulation of gene expression (Coleman, 1992; Vallee and Falchuk, 1993). For transcription, zinc is required in these proteins for binding to specific genes by forming tetrahedral complexes with amino acid residues of the polypeptide chain (Fig.9.16).

By this means the polypeptide chain forms a loop or 'finger' of usually 11–13 amino acid residues which bind the specific DNA sequences ('zinc finger motif'). In these DNA-binding metalloproteins zinc is therefore directly involved in the translation step of gene expression and activation or repression of DNA elements.

9.4.4 Protein Synthesis

The rate of protein synthesis and the protein content of zinc-deficient plants are drastically reduced whereas amino acids accumulate (Table 9.17). On resupplying zinc to deficient plants protein synthesis is resumed quite rapidly. Besides the function of zinc described above, at least two other functions of zinc in protein metabolism are mainly responsible for these changes. Zinc is a structural component of the ribosomes

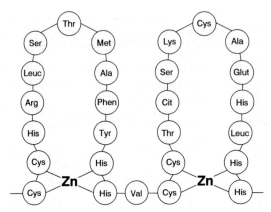

Fig. 9.16 Schematic presentation of the role of zinc in tertiary structure of the peptide chain in replication proteins ('zinc finger').(Based on Coleman, 1992 and Vallee and Falchuk, 1993.)

and essential for their structural integrity. The zinc content of ribosomal RNA in zinc-sufficient cells of *Euglena* is in the range 650–1280 μg g^{-1} RNA, whereas that in zinc-deficient cells is 300–380 μg g^{-1} RNA (Prask and Plocke, 1971). In the absence of zinc, ribosomes disintegrate, but reconstitution takes place after resumption of zinc supply.

In shoot meristems of rice, disintegration of the 80S ribosomes (soluble fraction in the cytoplasm) takes place when the zinc content drops below 100 μg g^{-1} dry weight. In contrast to this rather high zinc content required for ribosome integrity, protein content starts to decrease at considerably lower zinc content (Fig. 9.17). In tobacco tissue culture cells the corresponding contents were 70 μg Zn for the decrease in 80S ribosomes and 50 μg Zn for the decrease in protein content (Obata and Umebayashi, 1988).

A specific high requirement of zinc at sites of protein synthesis has been also shown in pollen tubes where the zinc content at the growing tip was about 150 μg g^{-1} dry weight compared with about 50 μg in more basal regions (Ender *et al.*, 1983).

In the shoot meristems, and presumably also in other meristematic tissues, a zinc content of at least 100 μg g^{-1} dry weight is essential for the maintenance of protein

Table 9.17

Effect of Zinc Supply on Shoot Dry Weight and Composition of Shoot Apical Parts (Young Leaves and Shoot Apex) of Bean Plants[a]

		Content in young leaves and shoot apex				
Zn supply	Shoot dry wt (g per 3 plants)	Zn (μg g^{-1} dry wt)	Free amino acids (μmol g^{-1} dry wt)	Protein (mg g^{-1} fresh wt)	Tryptophan (μmol g^{-1} dry wt)	IAA (ng g^{-1} fresh wt)
+Zn (1 μM)	8.24	52	82	28	0.37	239
−Zn	3.66	13	533	14	1.32	118
−Zn, +Zn[b]	4.53	141	118	30	0.27	198

[a]From Cakmak *et al.* (1989).
[b]Resupply of 3 μM Zn for 3 days.

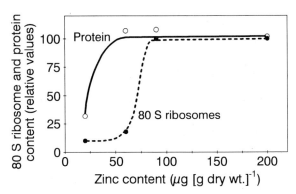

Fig. 9.17 Relationship between content of zinc, 80S ribosomes and protein in the soluble fraction of rice shoot meristematic tissue. (From Kitagishi *et al.*, 1987.)

synthesis. As shown in Table 9.18 this is about 5–10 times more than the content in mature leaf blades considered to be in the adequate range. For other mineral nutrients this gradient is usually less steep, and for some the gradient might be reversed, depending on the nutritional status (Chapter 12). To meet the high zinc demand in the shoot meristem, most of the root-supplied zinc is preferentially translocated to the shoot meristem, mediated by xylem–phloem transfer in the stem (Kitagishi and Obata, 1986).

Low protein contents and high amino acid contents in zinc-deficient plants are not only the result of reduced transcription and translation but also of enhanced rates of RNA degradation. Higher rates of RNase activity are a typical feature of zinc deficiency (Sharma *et al.*, 1982). A clear inverse relationship exists between zinc supply and RNase activity, and also between RNase activity and protein content (Table 9.19).

An increase in RNase activity is observed even before symptoms of zinc deficiency such as stunted growth and changes in leaf anatomy become apparent (Dwivedi and Takkar, 1974).

9.4.5 Carbohydrate Metabolism

Many zinc-dependent enzymes are involved in carbohydrate metabolism in general and of leaves in particular. Besides its function in the carbonic anhydrase reaction, zinc is

Table 9.18
Mineral Element Content in Meristematic Tissue of the Youngest
Leaf and of Mature Leaf Blades of Zinc-sufficient Rice Plants[a]

	Mineral element content in dry matter				
	Zn (μg g^{-1})	Mn (μg g^{-1})	Mg (%)	Ca (%)	K (%)
Meristem	204	188	0.42	0.23	3.01
Leaf blade	18	540	0.89	0.60	1.28

[a]Based on Kitagishi and Obata (1986).

Table 9.19
Effect of Zinc Supply on Fresh Weight, RNase Activity, and
Protein Nitrogen in Soybean (*Glycine wighii*)[a]

Zinc supply (mg l^{-1})	Fresh weight (g per plant)	RNase activity (%)[b]	Protein nitrogen (% fresh wt)
0.005	4.0	74	1.82
0.01	5.1	58	2.25
0.05	6.6	48	2.78
0.10	10.0	40	3.65

[a]Based on Johnson and Simons (1979).
[b]% hydrolysis of RNA substrate.

required, for example, for the activity of two other key enzymes, fructose 1,6-bisphosphatase and the aldolase.

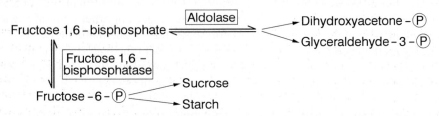

Both enzymes are located in the chloroplasts as well as in the cytoplasm. Fructose-1,6-bisphosphatase is a key enzyme in the partitioning of C_6 sugars in the chloroplasts and the cytoplasm. Aldolase regulates the transfer of C_3 photosynthates from the chloroplasts into the cytoplasm, and within the cytoplasm the flow of metabolites via the glycolytic pathway. Under zinc deficiency aldolase activity is drastically and rather specifically reduced so that the activity of this enzyme may serve as an indicator of the zinc nutritional status of plants (O'Sullivan, 1970).

As a rule, in leaves suffering from zinc deficiency, a sharp decline in carbonic anhydrase activity is the most sensitive and obvious change in activity of enzymes of the

Table 9.20
Effect of Increasing Zinc Deficiency on the Activities of Enzymes in
Leaves of Maize Plants Grown without Zinc Supply[a]

Enzyme	Percentage decrease in activity after days without zinc supply		
	5	10	15
Fructose-1,6-bisphosphatase	36	50	65
Carbonic anhydrase	84	76	84
PEP Carboxylase	<1	5	34
RuBP Carboxylase	9	41	38
Malic enzyme	<1	22	37

[a]Relative values; zinc-sufficient plants = 100. Based on Shrotri *et al.* (1983).

Table 9.21
Effect of Zinc Deficiency and Resumption of Zinc Supply on Zinc and
Carbohydrate Contents of Cabbage Leaves[a]

Parameter	Zinc supply (μM)		
	1.0	0.001	$0.001 + 2.0^b$
Zinc content (mg kg^{-1} dry wt)	21	14	30
Sugars (mg g^{-1} fresh wt)	4.2	9.1	5.0
Starch (mg g^{-1} fresh wt)	7.5	24.6	19.2
Hill reaction activity (relative)	100	48	66

[a]Based on Sharma *et al.* (1982).
[b]24 h after resumption of 2.0 μM zinc supply.

carbohydrate metabolism (Table 9.20). The activity of fructose-1,6-bisphosphatase also declines fairly rapidly; that of other enzymes is affected to a much lesser extent, however, particularly with mild zinc deficiency.

Despite a decrease in enzyme activities and in the rate of photosynthesis (as indicated by the activity of the Hill reaction) in most instances sugars and starch accumulate in zinc-deficient plants (Table 9.21). As early as 24 h after the zinc supply is restored, the sugar levels and the Hill reaction activity are again comparable to those of the adequately supplied control plants continuously receiving 1.0 μM zinc. The accumulation of carbohydrates in zinc-deficient leaves increases with light intensity (Marschner and Cakmak, 1989) and is an expression of impaired new growth, particularly of the shoot apices, i.e., of lower sink activity.

In conclusion, most experimental evidence obtained with green plants supports the view that zinc deficiency-induced changes in carbohydrate metabolism are not primarily responsible for either growth retardation or the visible symptoms of zinc deficiency.

9.4.6 Tryptophan and Indoleacetic Acid Synthesis

The most distinct zinc deficiency symptoms – stunted growth and 'little leaf' – are presumably related to disturbances in the metabolism of auxins, indoleacetic acid (IAA) in particular. The mode of action of zinc in auxin metabolism is still obscure. In zinc-deficient tomato plants retarded stem elongation has been shown to be correlated with a decrease in IAA level, and resumption of both stem elongation and IAA level takes place after zinc is resupplied. The responses of increasing levels of IAA to the zinc treatment was more rapid than that of elongation growth (Tsui, 1948). Low levels of IAA in zinc-deficient plants might be the result of inhibited synthesis or enhanced degradation of IAA (Cakmak *et al.*, 1989). Tryptophan is most likely the precursor for the biosynthesis of IAA.

There are conflicting reports as to the requirement of zinc for the synthesis of tryptophan. The increase in tryptophan content in the dry matter of rice grains by zinc fertilization of the plants growing on a calcareous soil (Singh, 1981) would support such an assumption. However, such an increase might be an expression of a general increase in protein content in the grains and, thus, the result is inconclusive in terms of zinc requirement for tryptophan synthesis.

In leaves of zinc deficient plants tryptophan might increase rather than decrease (Cakmak et al. 1989; Domingo et al., 1992), most likely as a result of impaired protein synthesis as shown in Table 9.17. In zinc deficient leaves tryptophan content increases similarly to other amino acids (Cakmak et al., 1989). Although the lower IAA content in zinc-deficient leaves might indicate a role for zinc in the biosynthesis of IAA from tryptophan (as postulated by Salami and Kenefick, 1970), lower IAA contents are more likely the result of enhanced oxidative degradation of IAA (Fig. 9.18).

9.4.7 Membrane Integrity

Zinc is required for maintenance of integrity of biomembranes. It might bind to phospholipid and sulfhydryl groups of membrane constituents or form tetrahedral complexes with cysteine residues of polypeptide chains (Section 9.4.2) and thereby protect membrane lipids and proteins against oxidative damage. Zinc might also control the generation of toxic oxygen radicals by interfering with the oxidation of NADPH as well as by scavenging O_2^- in its function as a metal component in CuZn SOD (Cakmak and Marschner, 1988a,b). Accordingly, under zinc deficiency there is a typical increase in plasma membrane permeability, for example in roots (Welch et al., 1982) indicated by leakage of low-molecular-weight solutes, a decrease in phospholipid content and in the degree of unsaturation of fatty acids in membrane lipids (Table 9.22). As soon as 12 h after resupplying zinc some restoration of membrane integrity can be observed. Plasma membrane vesicles, isolated from zinc-deficient roots, also have a higher passive permeability than vesicles from zinc-sufficient roots (Pinton et al., 1993).

Table 9.22
Effect of Zinc Nutritional Status of Cotton Plants on Leakage of Low-Molecular-Weight Solutes (Root Exudates) and Lipid Composition of Roots[a]

Treatment	Roots Zn content (μg g^{-1} dry wt)	Root exudates (g^{-1} dry wt (6 h)$^{-1}$)				Lipid content	
		Amino acids (μg)	Sugars (μg)	Phenolics (μg)	Potassium (mg)	Phospholipids (μg g^{-1} fresh wt)	Fatty acids sat/unsat
+Zn	258	48	375	117	1.68	2230	0.79
−Zn	16	165	751	161	3.66	1530	0.90
−Zn+Zn[b]	121	94	652	130	2.32	ND	ND

[a]Based on Cakmak and Marschner (1988c).
[b]Resupply of Zn to deficient plants for 12 h; ND = not determined.

Table 9.23
Relation Between Zinc Content in Roots and Shoots, Chlorophyll Content, Superoxide Generation and NADPH Oxidation in Root Extracts (48 000 g Supernatant) of Bean Plants[a]

Treatment	Zinc content ($\mu g\ g^{-1}$ dry wt)		Chlorophyll (mg g^{-1} dry wt)	O_2^- generation (nmol mg^{-1} protein min^{-1})	NADPH oxidation (nmol mg^{-1} protein min^{-1})
	Roots	Shoots			
+Zn	44	37	7.4	2.2	18.3
−Zn	11	10	3.6	6.6	61.0
−Zn + Zn[b]	69	71	4.1	4.3	40.0

[a]Cakmak and Marschner (1988a).
[b]Resupply of zinc to deficient plants for 2 days.

Increased membrane permeability in zinc-deficient plants is brought about by higher rates of O_2^- generation (Table 9.16) as a result of raised activity of an NADPH-dependent O_2^- generating oxidase (Table 9.23). Higher activity of this oxidase is either a reflection of a direct role of zinc in regulation of enzyme activity, or an indirect result of the alterations in structure and composition of the membranes (Table 9.22).

Most likely, many of the most obvious symptoms of zinc deficiency such as leaf chlorosis and necrosis, inhibited shoot elongation and increased membrane permeability are expressions of oxidative stress brought about by higher generation of O_2^- and a simultaneously impaired detoxification system in zinc-deficient plants. These events are summarized schematically in Fig. 9.18.

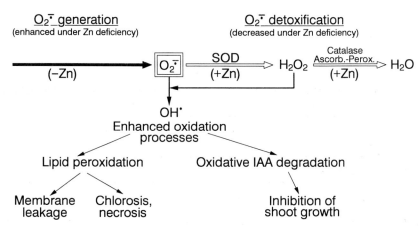

Fig. 9.18 Involvement of zinc in the generation and detoxification of superoxide radicals, and effects of oxygen-free radicals on membrane function and IAA metabolism. (Compiled from Cakmak and Marschner, 1988a,b, and Cakmak *et al.*, 1989.)

9.4.8 Phosphorus-Zinc Interactions

Large applications of phosphorus fertilizers to soils low in available zinc can induce zinc deficiency ('P-induced Zn deficiency'; Robson and Pitman, 1983), by altering either soil or plant factors. High phosphorus contents in soils can decrease solubility of zinc in soils (Marschner and Schropp, 1977; Loneragan et al., 1979), although such effects do not always occur (Pasricha et al., 1987). High phosphorus supply is often associated with a reduction in root growth and a lesser degree of infection of roots with VA mycorrhizae (Section 15.6). Both these factors are important for the acquisition of zinc. A decrease in zinc content in shoots and an induction of zinc deficiency symptoms by high phosphorus supply is often also the result of enhanced shoot growth and, thus, 'dilution' of zinc in the plants (Loneragan et al., 1979; Neilsen and Hogue, 1986). There are, however, additional physiological interactions between phosphorus and zinc within the plants involved, as phosphorus-induced zinc deficiency can also be readily demonstrated in nutrient solution culture. With increasing phosphorus contents in the shoot dry matter zinc-deficiency symptoms become more severe, although the zinc content in the dry matter is not decreased. However, the physiological availability of zinc is decreased as indicated, for example, in lower proportions of water-extractable zinc and lower SOD activities in leaves (Cakmak and Marschner, 1987). High phosphorus contents in the shoot might therefore decrease solubility and mobility of zinc both within the cells and in long-distance transport to the shoot apex.

In solution culture at high phosphorus but low zinc supply, the phosphorus-induced zinc deficiency is often associated with an exceptional high phosphorus content and symptoms of phosphorus toxicity in mature leaves (Loneragan et al., 1979; Cakmak and Marschner, 1986), which might occasionally be mistaken for evidence of accentuation of zinc deficiency because of the large P/Zn ratio. As shown in Table 9.24, zinc uptake is not affected by increasing phosphorus concentrations in the external solution. In the absence of zinc, or with low external concentrations, however, the content of phosphorus in the shoot dry matter is very high, leading to toxicity symptoms. In general, a phosphorus content of greater than ~2% in the leaf dry weight can be considered as toxic.

Table 9.24

Effects of Zinc and Phosphorus Concentrations in the Nutrient Solution on the Growth and Zinc and Phosphorus Content of the Shoots of Ochra (*Abelmoschus esculentum* L.)[a]

Zn supply (μM)	Dry weight (g per plant)[b]		Zinc content (μg g^{-1} dry wt)[b]		Phosphorus content (mg g^{-1} dry wt)[b]	
	P_1	P_2	P_1	P_2	P_1	P_2
0	8.3	9.5	15	15	11.0	24.1
0.25	9.6	9.9	27	27	9.6	20.2
1.0	9.8	11.6	54	57	8.7	11.8

[a]Based on Loneragan et al. (1982b).
[b]P_1, 0.25 mM phosphorus; P_2, 2.0 mM phosphorus.

Table 9.25
Effect of Micronutrient Deficiency on Dry Weight and Phosphorus
Content in Roots and Shoots of Cotton Plants[a]

Treatment	Dry weight (g per plant)		P content in dry matter (%)	
	Roots	Shoots	Roots	Shoots
Control	0.18	1.21	1.03	1.10
−Zn	0.13	0.70	1.15	2.65
−Fe	0.16	0.98	1.00	0.90
−Mn	0.15	0.93	0.96	1.20
−Cu	0.16	1.00	1.38	1.40

[a]Cakmak and Marschner (1986).

The much higher phosphorus content in the shoot dry matter of zinc-deficient plants supplied with high phosphorus concentrations (P_2) can be only to some extent attributed to a 'concentration effect' particularly in the mature leaves (Webb and Loneragan, 1988). The main reason for the high phosphorus content in the leaves is that zinc deficiency enhances both the uptake rate of phosphorus by the roots and the translocation to the shoots (Table 9.25). This enhancement effect is specific for zinc deficiency and is not observed when other micronutrients are deficient.

Zinc affects the phosphorus metabolism of roots (Loughman et al., 1982), perhaps also the loading step of phosphorus into the xylem (Section 2.8). Zinc deficiency also increases the permeability of the plasma membrane of root cells to phosphorus, as well as to chloride (Welch et al., 1982) and boron, and might even also lead to boron toxicity (Graham et al., 1987b; Singh et al., 1990b). Thus, enhanced phosphorus uptake in zinc-deficient plants can in part be an expression of higher passive permeability of the plasma membranes of root cells or impaired control of xylem loading.

The high phosphorus content in the shoots of zinc-deficient plants, however, is also the result of a specific impairment of retranslocation of phosphorus in the phloem (Table 9.26) and, thus, of an important 'signal' of the shoot control on phosphorus uptake by the roots (Section 3.4.4). The mechanism by which zinc deficiency impairs retranslocation of phosphorus from the shoots is obscure, as in the deficient plants neither the retranslocation of ^{86}Rb nor ^{36}Cl is impaired (Table 9.26).

Table 9.26
Distribution of ^{32}P, ^{86}Rb, and ^{36}Cl Between Shoots and Roots of Zinc-Sufficient and Zinc-Deficient Cotton Plants, 19 h after Stem Application, Relative Values[a]

Zn nutritional status	^{32}P		^{86}Rb		^{36}Cl	
	Shoots	Roots	Shoots	Roots	Shoots	Roots
+Zn	66	34	62	38	29	71
−Zn	92	8	66	34	32	68

[a]Marschner and Cakmak (1986).

9.4.9   Zinc Binding Forms and Bioavailability

The knowledge on binding forms of zinc in vegetative plant organs is poor, except for that in enzymes (see above) and at the toxic concentration range (Section 9.4.10). In lettuce leaves extracted with ammonium bicarbonate a low-molecular-weight fraction (~1.25 kDa) was isolated which contained sulfur, reducing sugars, amino-nitrogen and 73% of the total soluble zinc in the leaves (Walker and Welch, 1987). However, this does not necessarily imply an *in vivo* importance of this fraction as redistribution of zinc during extraction might occur.

In contrast to vegetative organs, much information is available on the localization and binding forms of zinc in seeds and grains. In grains and seeds, most of the zinc and other mineral nutrients, are localized in so-called 'protein bodies' in the form of discrete particles, the globoid crystals (Lott and Buttrose, 1978; Welch, 1986). These globoids consist mainly of phytate, i.e., salts of phytic acid (Table 9.27). In wheat seeds, similarly high zinc contents (600 μg g^{-1} dry wt) were found in protein bodies in the scutellum (Mazzolini *et al.*, 1985). In the protein bodies of maize germ the phytate is mainly a mixed K + Mg-salt. However, relatively large proportions of the micronutrients zinc, manganese and iron, but not copper, are also bound as phytate.

Zinc is very tightly bound to phytic acid and the formation of additional protein–zinc–phytic acid complexes increases the resistance to hydrolysis (Prattley *et al.*, 1982). During germination, however, phytase hydrolyses the phytate quite readily (Section 8.4). The strong binding of zinc to phytic acid is of much concern to nutritionists as it depresses the bioavailability of zinc for monogastric animals and man. A close negative correlation occurs, for example, in soybean products between phytic acid (phytate) content and zinc bioavailability for rats (Zhou *et al.*, 1992). It is possible to depress the phytate content of seeds and grains by selection and breeding, or by phosphorus deficiency, but as a rule, a lower phytate content is correlated with other negative effects as discussed in Section 8.4.

As the formation of phytate is not confined to reproductive organs (Section 8.4), decreased physiological availability of zinc in vegetative plant organs resulting from the formation of phytate might also be of importance, particularly in the context of phosphorus-induced zinc deficiency.

Table 9.27
Mineral Element Content in Germ and Protein Bodies in the Germ of Maize Kernels[a]

	Mineral element content								
	Zn	Fe	Mn	Cu	Ca	K	Mg	P_{tot}	P_{phyt}
		(μg g^{-1} dry wt)				(mg g^{-1} dry wt)			
Germ	163	186	30	12	449	27	10	30	23
Protein bodies	565	490	170	11	1645	68	44	89	88

[a]Marschner, Ehret and Haug (unpublished).

9.4.10   Zinc Deficiency and Toxicity

9.4.10.1   Zinc Deficiency

Zinc deficiency is widespread among plants grown in highly weathered acid soils and in calcareous soils. In the latter case zinc deficiency is often associated with iron deficiency ('lime chlorosis'). The low availability of zinc in calcareous soils of high pH results mainly from the adsorption of zinc to clay or $CaCO_3$ rather than from the formation of sparingly soluble $Zn(OH)_2$ or $ZnCO_3$ (Trehan and Sekhon, 1977). In addition zinc uptake and translocation to the shoot are inhibited by high concentrations of bicarbonate, HCO_3^- (Forno *et al.*, 1975; Dogar and van Hai, 1980). This effect has striking similarities to the effect of HCO_3^- on iron (Section 16.5). In contrast to iron deficiency, however, zinc deficiency in plants grown in calcareous soils can be corrected fairly readily by application to the soil of inorganic zinc salts such as $ZnSO_4$ (Nayyar and Takkar,1980).

The most characteristic visible symptoms of zinc deficiency in dicotyledons are stunted growth due to shortening of internodes ('rosetting') and a drastic decrease in leaf size ('little leaf'), as shown in Fig. 9.19. Under severe zinc deficiency the shoot apices die ('die-back') as is widely distributed, for example, in forest plantations in South Australia (Boardman and McGuire, 1990). Quite often these symptoms are combined with chlorosis, which is either highly contrasting or diffusive ('mottle leaf'). These symptoms are usually more severe in high light intensities than in partial shade (Boardman and McGuire, 1990), for reasons which have been discussed in Section 5.2.2. In cereals such as sorghum, chlorotic bands along the midrib and red, spotlike discoloration (caused by anthocyanins) on the leaves often occur (Furlani *et al.*, 1986).

Fig. 9.19   Zinc deficiency in apple with typical inhibition of internode elongation ('rosetting') and reduction in leaf size ('little leaf').

Symptoms of chlorosis and necrosis in older leaves of zinc-deficient plants are often secondary effects caused by toxicity of phosphorus or boron (Section 9.4.8), or by photooxidation resulting from impaired export of photosynthates (Section 5.2.2). In dicotyledons zinc deficiency symptoms may be similar in appearance to virus infections and may be mistaken as such, as happened in the case of 'crinkle disease' in hop (*Humulus scandens*) (Schmidt *et al.*, 1972).

Under zinc deficiency shoot growth is usually more inhibited than root growth (Zhang *et al.*, 1991a), root growth might even be enhanced at the expense of the shoot growth as, for example, in wheat (Cumbus, 1985). Under zinc deficiency root exudation of low-molecular-weight solutes is also increased. Whereas in dicotyledonous species amino acids, sugars, phenolics, and potassium dominate (Table 9.22), in graminaceous species the main solutes are phytosiderophores (Zhang *et al.*, 1991a) which are released in a distinct diurnal pattern (Zhang *et al.*, 1991b), as is typical for iron deficiency (Section 2.5.6). It is not clear whether the enhanced release of phytosiderophores under both zinc and iron deficiency are separately regulated or more likely the expression of a zinc-deficiency-induced disturbance of iron metabolism in the plants (Cakmak *et al.*, 1994).

In leaves, the critical deficiency levels are below 15–$20\,\mu g$ Zn g^{-1} dry weight (but see Section 9.4.4). Grain and seed yield are depressed to a relatively greater extent by zinc deficiency than the total dry matter production, probably due at least in part to impaired pollen fertility in deficient plants (Section 6.4). Plant species differ in their sensitivity to zinc deficiency, maize, cotton, and apple are much more sensitive than, for example, wheat, oat, or pea.

9.4.10.2 Zinc Toxicity

When the zinc supply is large, zinc toxicity can readily be induced in nontolerant plants, inhibition of root elongation being a very sensitive parameter (Godbold *et al.*, 1983; Ruano *et al.*, 1988). Quite often, zinc toxicity leads to chlorosis in young leaves. This may be an induced deficiency, for example, of magnesium or iron, because of the similar ion radius of Zn^{2+} and Fe^{2+} (Woolhouse, 1983) and Zn^{2+} and Mg^{2+} (Boardman and McGuire, 1990). Induced manganese deficiency might also be of importance as high zinc supply strongly decreases the manganese content of plants (Ruano *et al.* 1987).

In bean plants zinc toxicity inhibits photosynthesis at various steps and through different mechanisms. Depressed RuBP carboxylase activity is presumably caused by competition with magnesium (Van Assche and Clijsters, 1986a), and inhibited PS II activity by replacement of manganese in the thylakoid membranes (Van Assche and Clijsters, 1986b). Whereas in the thylakoid membranes of control plants about six atoms of both manganese and zinc are bound per 400 chlorophyll molecules, under zinc toxicity this proportion shifts to two manganese and 30 zinc atoms.

The critical toxicity levels in leaves of crop plants are from as low as $100\,\mu g$ Zn g^{-1} dry weight (Ruano *et al.*, 1988) to more than $300\,\mu g$ Zn g^{-1} dry weight, the latter values being more typical. Increasing soil pH by liming is the most effective procedure for decreasing both zinc content and zinc toxicity in plants (White *et al.*, 1979). In comparison with the genotypical differences between plants of the natural vegetation,

genotypical differences in zinc tolerance between crop plants are small, but nevertheless marked, even within the same species. In soybean genotypes zinc tolerance is
positively correlated with the zinc content of the leaves. In other words, in this example
the mechanism of tolerance is not exclusion from uptake but tolerance of the tissue to
high contents of zinc (White *et al.*, 1979).

9.4.10.3 Zinc Tolerance

As with copper tolerance, the mechanisms responsible for zinc tolerance have long
been a major topic of interest in ecophysiology (Baker and Walker, 1989a,b; Verkleij
and Schat, 1989). Zinc tolerance has also become a topic of interest in agriculture and
crop physiology, as zinc is the heavy metal found to occur in the greatest concentrations
in the majority of wastes arising in modern, industrialized communities (Boardman and
McGuire, 1990).

The principal mechanisms of heavy metal tolerance have been discussed in Section
9.3, and illustrated in Fig. 9.13. In contrast to copper, exclusion from uptake, or binding
to the cell walls, does not seem to be important for zinc tolerance (Qureshi *et al.*, 1985;
Vazquez *et al.*, 1992). However, a particular mechanism of exclusion might exist in
forest tree species such as *Pinus sylvestris* where certain ectomycorrhizal fungi retain
most of the zinc in their mycelium and, thus, strongly increase the 'zinc tolerance' of the
host plant (Colpaert and Van Assche, 1992).

In the case of zinc, tolerance is achieved mainly through sequestering zinc in the
vacuoles as shown in a typical example in Table 9.28. Whereas in the nontolerant clone
receiving an abundant supply of zinc there is a preferential accumulation of zinc in the
cytoplasm, in the tolerant clone the zinc concentration in the cytoplasm remains low;
instead, zinc is sequestered in the vacuoles. Fairly close positive correlations in tolerant
genotypes occur between accumulation of organic acids such as malate and citrate, and
accumulation of zinc, indicating that complexation of zinc with organic acids in the
vacuoles is in many cases an important mechanism of zinc tolerance (Godbold *et al.*,
1983, 1984).

Table 9.28

Effect of Zinc Supply on Concentrations of Zinc in the Cytoplasm and
Vacuoles of Roots of a Zinc Tolerant and Nontolerant Clone of
Deschampsia caespitosa[a]

External concentration (mM Zn^{2+})	Bound zinc in the cytoplasm (mM)		Soluble zinc in the vacuoles (mM)	
	Nontolerant	Tolerant	Nontolerant	Tolerant
0.10	7.1	10.6	3.7	5.3
0.75	33.4	6.2	2.1	33.4

[a]Based on Brookes *et al.* (1981).

There is general agreement that phytochelatins play no role in zinc tolerance (Grill *et al.*, 1988; Robinson, 1990; Davies *et al.*, 1991a). This raises the question as to the mechanisms of zinc tolerance in nonvacuolated, meristematic tissues such as root apices. Here, other mechanisms exist such as sequestering of zinc by binding to phytate as occurs in a zinc-tolerant ecotype of *Deschampsia caespitosa* (Van Steveninck *et al.*, 1987a, b). In addition, complexation of zinc in the cytoplasm to amino acids might be of importance. Zinc tolerance in *Deschampsia caespitosa* is increased in plants supplied with ammonium as compared with nitrate nutrition. This probably results from the greater accumulation of asparagine in the cytoplasm of ammonium-fed plants and which forms quite stable complexes with zinc (Smirnoff and Stewart, 1987).

9.5 Nickel

9.5.1 General

Nickel is chemically related to iron and cobalt. Its preferred oxidation state in biological systems is Ni(II), but it can also exist in the states Ni(I) and Ni(III) (Cammack *et al.*, 1988). Nickel forms stable complexes, for example, with cysteine and citrate (Thauer *et al.*, 1980) and in nickel-enzymes it is coordinated to various ligands.

Nickel is an essential trace element (micronutrient) for a large number of bacteria, for example, as a metal component in urease and many hydrogenases, and also for acetogenic and methanogenic bacteria (Ankel-Fuchs and Thauer, 1988). There are also a number of earlier reports on the stimulation of germination and growth of various crop species by low concentrations of nickel in the substrate (for a review, see Mishra and Kar, 1974). The first clear evidence for the function of nickel in urease in higher plants was provided by Dixon *et al.* (1975). Later, a requirement of nickel in legumes was shown, regardless of the form of nitrogen nutrition (Eskew *et al.*, 1984) and, the essentiality of nickel for nonlegumes has also now been established (Brown *et al.* 1987a). According to these studies, nickel meets the requirements for classification as mineral nutrient for higher plants. At least some of the functions of nickel are now clearly defined, and plants cannot complete their life cycle without adequate nickel supply. Nickel is therefore now included in the list of micronutrients.

9.5.2 Nickel-Containing Enzymes

In biological systems a number of enzymes are well defined in which nickel is the metal component required for the activity. In these enzymes nickel is coordinated either to N- and O-ligands (e.g., in urease), S-ligands (cysteine-residues, e.g., hydrogenase), or N-ligands of tetrapyrrol structures.

In higher plants, urease is so far the only known nickel-containing enzyme. Urease isolated from jack bean (*Canavalia ensiformis* L.) has a molecular weight of 590 kDa, and consists of six subunits (i.e., it is hexameric), each subunit containing two nickel atoms (Dixon *et al.*, 1980). In the subunits nickel is coordinated to N- and O-ligands (Alagna *et al.*, 1984), and one of the Ni-O bonds can possibly be displaced by water molecules during hydrolytic reactions (see p. 365).

$$O{=}C\overset{-NH_2}{\underset{-NH_2}{}} + H_2O \longrightarrow 2NH_3 + CO_2$$

Nickel is not required for the synthesis of the enzyme protein (Winkler *et al.*, 1983) but, as the metal component, is essential for the structure and catalytic function of the enzyme (Klucas *et al.*, 1983).

In hydrogenases from sulfate-reducing, photosynthetic, and hydrogen-oxidizing bacteria [i.e., the hydrogen uptake hydrogenases of rhizobia (Section 7.3)] the nickel is associated with Fe-S clusters and might also take part in redox reactions (Cammack *et al.*, 1988). *Rhizobium* and *Bradyrhizobium* produce hydrogen-uptake hydrogenases both when free-living and as bacteroids in the root nodules (Maier *et al.*, 1990). In free-living rhizobia, without nickel supply the hydrogenase activity is extremly low, but can be restored within three hours by resupplying of nickel (Maier *et al.*, 1990).

9.5.3 Role of Nickel in Nitrogen Metabolism

When supplied with urea as sole source of nitrogen, in the absence of nickel growth of *Lemna* (Gordon *et al.*, 1978) and of cell cultures of soybean was very poor and urease activity low (Polacco, 1977). Addition of nickel increased both growth and urease activity more than fivefold in the soybean cell culture. In low nickel plants supplied with urea, not only is the utilization of this form of nitrogen impaired but also urea toxicity occurs. Foliar application of urea is often associated with urea toxicity (Section 4.3), and the severity of toxicity symptoms closely relate to the nickel nutritional status of the plants as shown in Table 9.29 for soybean. In plants without nickel supply, urease activity in leaves is low and foliar application of urea leads to a large accumulation of urea and severe necrosis of the leaf tips. In plants supplied with nickel, urease activity was much higher and the urea accumulation and necrosis correspondingly lower.

Table 9.29
Effect of Nickel Supply in the Nutrient Solution and Foliar Application of Urea on Leaf Tip Necrosis, Urea Content, and Urease Activity in Soybean Plants[a]

Ni supply ($\mu g\,l^{-1}$)	Foliar appl. (mg urea per leaf)	Leaf tips necrosis (% of dry wt)	Urea cont. ($\mu g\,g^{-1}$ dry wt)	Urease activity ($\mu mol\,NH_3\,h^{-1}\,g^{-1}$ dry wt)
0	0	<0.1	64	2.2
	3	5.2	1038	2.7
	6	13.6	6099	2.4
100	0	0	0	11.8
	3	2.0	299	11.3
	6	3.5	1583	9.6

[a]Based on Krogmeier *et al.* (1991). Reprinted by permission of Kluwer Academic Publishers.

In nodulated legumes such as soybean, ureides are the dominant form of nitrogen transported to the shoots (Section 7.4) where they are degraded to NH_3 and CO_2 without involving urea metabolism (Section 8.2.3). Accordingly, there is no particularly high requirement for nickel to be expected in nodulated soybean, and other ureide-type legumes, as compared with soybean supplied with mineral nitrogen (Winkler *et al.*, 1988). Regardless of the form of nitrogen nutrition (urea; NH_4-N; NO_3-N; N_2 fixation) in soybean and cowpea, without nickel supply, large amounts of urea accumulate in the leaves and symptoms of leaf tip necrosis are severe (Eskew *et al.*, 1984). Table 9.30 shows that there is a large accumulation of urea (up to 3% of the dry weight) towards the tip of the leaf blade in nickel-deficient plants. Ureide contents are low and unaffected by nickel supply, which is also true for free purines and uric acid (Walker *et al.*, 1985). These results indicate that urea is a normal intermediate in nitrogen metabolism (see below).

When seeds were used from plants grown under low nickel conditions large accumulations of urea and severe leaf tip necrosis was also found in wheat, barley, and oat plants without nickel supply (Brown *et al.*, 1987b). Root and shoot growth was significantly lower in the nickel-deprived plants, which were less green, developed interveinal chlorosis and necrosis, and the terminal 2 cm of the leaves failed to unfold.

In barley seeds from plants grown at low nickel supply a close relationship occurs between the nickel content, viability, germination rate and seedling vigour (Brown *et al.*, 1987a). This relationship is shown for germination rate in Fig. 9.20.

Viability of the nickel-deficient seeds could not be restored by soaking the seeds in a solution containing nickel, demonstrating that nickel is essential for normal seed development in the maternal plants and, thus, for completing the life cycle of the barley plant (Brown *et al.*, 1987a).

At least in barley, the critical deficiency level of nickel in the shoots is in the range of $0.1\,\mu g\,g^{-1}$ dry weight, and at lower contents amino acids and nitrate accumulate (Brown *et al.*, 1990). Changes in content of organic acids and other solutes might result from secondary events of disturbances in nitrogen metabolism in nickel-deficient plants. It is not clear to what extent these various effects of nickel deficiency are directly related to the function of nickel in the urease. In any case these studies demonstrate that in nitrogen metabolism urea is a normal metabolite whose concentration has to be

Table 9.30

Effect of Nickel Supply during Growth on Content of Urea, Ureides and Nickel in Mature Leaves of Cowpea Supplied with $NH_4NO_3{}^a$

	Content in the dry weight					
	Urea (μmol g^{-1})		Ureide (μmol g^{-1})		Nickel (μg g^{-1})	
Leaf part	+Ni	−Ni	+Ni	−Ni	+Ni	−Ni
Petiole	0.11	0	ND	ND	ND	ND
Blade base	0.56	18.1	3.6	4.5	3.73	0.11
Blade tip	2.16	238.4	ND	ND	ND	ND

^{a}Based on Walker *et al.* (1985); ND = not determined.

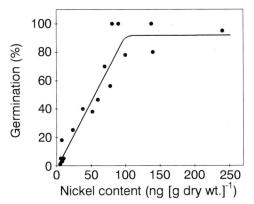

Fig. 9.20 Relationship between nickel content in seeds and germination percentage in barley. (Redrawn from Brown *et al.*, 1987a; by permission of the American Society of Plant Physiologists.)

maintained at a low level in order to prevent toxicity. Various pathways of urea biosynthesis in plants are known (Fig. 9.21). The ornithine cycle for urea biosynthesis is likely to be of general importance, as well as the higher rate of urea formation during protein degradation, for example, in mature leaves (Walker *et al.*, 1985), at onset of reproductive growth (Eskew *et al.*, 1984) and on germination of legume seeds (Horak, 1985b).

9.5.4 Nickel Content in Plants

In most plants the nickel content in the vegetative organs is in the range $1–10 \, \mu g \, g^{-1}$ dry weight, and this range mainly reflects the differences between plant species in uptake

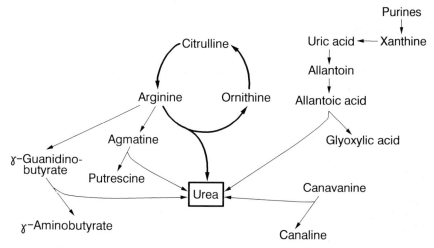

Fig. 9.21 Pathways of urea biosynthesis in plants. (Modified from Walker *et al.*, 1985.)

Table 9.31
Content of Nickel and Other Micronutrients in the Vegetative
Shoot Parts and in Seeds of Lupin (*Lupinus polyphyllus*) and Rye
(*Secale cereale*)[a]

Species		\multicolumn{6}{c}{Content (μg g^{-1} dry wt)}
		Ni	Mo	Cu	Zn	Mn	Fe
Lupin	Shoots	0.81	0.08	3.6	28	298	178
	Seeds	5.53	3.29	6.0	41	49	47
Rye	Shoots	0.62	0.17	1.6	7	16	78
	Seeds	0.28	0.33	4.4	25	27	26

[a]From Horak (1985a).

and root to shoot transport of nickel (Rebafka *et al.*, 1990). Nickel is readily mobile in the xylem and phloem (Kochian, 1991), and in some plant species preferentially translocated into the seeds. This is particularly true for the *Fabaceae* as shown in Table 9.31 in a typical example from a large number of species. Molybdenum is the only other micronutrient which is accumulated so preferentially in the legume seeds.

9.5.5 Nickel Deficiency and Toxicity

So far there is no clear evidence of nickel deficiency in soil-grown plants, or of soil bacteria (Dalton *et al.*, 1985), although in a pot experiment with a calcareous soil in wheat supplied with urea, simultaneous supply of nickel enhanced growth, even up to nickel contents in the shoot of 15–22 μg g^{-1} dry weight (Singh *et al.*, 1990a).

In general, in crop plants there is much more concern about nickel toxicity, for example, in relation to the application of sewage sludge which is often high in nickel (Marschner, 1983; Brown *et al.*, 1989). Critical toxicity levels in crop species are in the range of >10 μg g^{-1} dry weight in sensitive, and >50 μg g^{-1} dry weight in moderatly tolerant species (Bollard, 1983; Asher, 1991). In wheat the critical toxicity levels increased from 63 to 112 μg g^{-1} dry weight with increasing supply of urea (Singh *et al.*, 1990a). In sensitive species, root growth is severely inhibited even below 5 μM Ni when the Ca^{2+} concentration is low (Gabbrielli *et al.*, 1990).

9.5.6 Nickel Tolerance

Serpentine (or more precisely, ultramafic) soils are usually very high in iron, magnesium, nickel, chromium, and cobalt, but very low in calcium. The flora on these soils include many species (e.g., of the genus *Alyssum*) exhibiting hyperaccumulation of nickel, in which the nickel content in the leaves may reach 10–30 mg g^{-1} dry weight (Baker and Walker, 1989a,b; Homer *et al.*, 1991). Tolerance in these hyperaccumulators is mainly achieved by complexation of nickel with organic acids, malic and citric acid in particular; the stability of the citric acid complexes being about 150 times higher than of the malic acid complexes (Homer *et al.*, 1991). In soils underneath the canopy of hyperaccumulating trees there exists a much higher proportion of nickel resistant bacteria than beyond the canopy indicating the high rate of nickel cycling in the 'micro-ecosystem' of these trees (Schlegel *et al.*, 1991).

9.6 Molybdenum

9.6.1 General

Molybdenum is a transition element, which occurs in aqueous solution mainly as the molybdate oxyanion, MoO_4^{2-}, in its highest oxidized form [Mo(VI)]. As a metal component of enzymes it also exists as Mo(IV) and Mo(V). Due to its electron configuration, Mo(VI) shares many chemical similarities with vanadium and, particularly, tungsten. Several properties of the molybdate anion MoO_4^{2-} also resemble those of several other divalent inorganic anions, sulfate (SO_4^{2-}) and phosphate (HPO_4^{2-}) in particular, which has important implications on molybdenum availability in soils and uptake by plants (Section 9.6.6). In long-distance transport in plants molybdenum is readily mobile in xylem and phloem (Kannan and Ramani, 1978). The form in which molybdenum is translocated is unknown, but its chemical properties indicate that it is most likely transported as MoO_4^{2-} rather than in complexed form.

The requirement of plants for molybdenum is lower than that for any of the other mineral nutrients, except nickel. The functions of molybdenum as a plant nutrient are related to the valency changes it undergoes as a metal component of enzymes. In its oxidized stage it exists as Mo(VI); it is reduced to Mo(V) and Mo(IV).

In higher plants only a few enzymes have been found to contain molybdenum as a cofactor. In these enzymes molybdenum has both structural and catalytical functions and is directly involved in redox reactions. These enzymes are nitrate reductase, nitrogenase, xanthine oxidase/dehydrogenase and, presumably, sulfite reductase. The functions of molybdenum are therefore closely related to nitrogen metabolism, and the molybdenum requirement strongly depends on the mode of nitrogen supply.

9.6.2 Nitrogenase

Nitrogenase is the key enzyme complex unique to all N_2 fixing microorganisms. It consists of two iron proteins, one of which is the 240 kDa MoFe protein, composed of four subunits with each 30 Fe and 2 Mo atoms. Details of the structural arrangement and catalytical functions of molybdenum in nitrogenase have been discussed in Section 7.3. In some free-living diazotrophic bacteria (e.g., *Azotobacter chroococcum*), in addition to the Mo-nitrogenase, another nitrogenase occurs in which molybdenum is replaced by vanadium (Dilworth *et al.*, 1988).

In legumes and nonlegumes dependent on N_2 fixation, the molybdenum requirement is large, particularly in root nodules. When the external supply is low, the molybdenum content per unit dry weight of nodules is usually higher than that of leaves, whereas when the external supply is high, the content in the leaves often rises more than in the nodules (Brodrick and Giller, 1991a). When molybdenum is limiting, preferential

Table 9.32

Effect of Molybdenum on the Growth and Nitrogen Content of Alder Plants
(*Alnus glutinosa*) Grown in a Molybdenum-Deficient Soil[a]

Parameter	Molybdenum application (μg per pot)	Leaves	Stems	Roots	Nodules
Dry weight (g per pot)	0	1.79	0.59	0.38	0.007
	150	5.38	2.20	1.24	0.132
Nitrogen content (%)	0	2.29	0.92	1.79	2.77
	150	3.58	1.17	1.83	3.26

[a]From Becking (1961).

accumulation in root nodules may lead to a considerably lower molybdenum content in the shoots and seeds of nodulated legumes (Ishizuka, 1982). However, the relative allocation of molybdenum to the various plant organs varies considerably not only between plant species, but also between genotypes within a species, for example in *Phaseolus vulgaris* (Brodrick and Giller, 1991b).

As would be expected, the growth of plants relying on N_2 fixation is particularly stimulated by the application of molybdenum to deficient soils (Table 9.32). The response of nodule dry weight to molybdenum is spectacular and indirectly reflects the increase in the capacity for N_2 fixation brought about by molybdenum.

In soils low in molybdenum availability, the effect of application of molybdenum to legumes depends on the form of nitrogen supply. As shown in Table 9.33 molybdenum applied to both nodulating and nonnodulating soybean plants increased the nitrogen content and seed yield only in the nodulated plants without or with insufficient supply of nitrogen fertilizer. This demonstrates the greater requirement for molybdenum in N_2 fixation than in nitrate reduction. It also indicates that on soils with low molybdenum availability it is possible to replace the application of nitrogen fertilizer to legumes by the application of molybdenum fertilizer combined with proper rhizobial infection.

Table 9.33

Influence of Nitrogen and Molybdenum Fertilizer Supply on Leaf Nitrogen Content and Seed Yield of Nonnodulating and Nodulating Soybean Plants[a]

	Treatment (g Mo ha^{-1})	Nonnodulating (kg N ha^{-1})				Nodulating (kg N ha^{-1})			
		0	67	134	201	0	67	134	201
Nitrogen	0	3.1	4.6	5.3	5.6	4.3	5.1	5.4	5.6
(% leaf dry wt)	34	3.6	4.7	5.3	5.6	5.7	5.5	5.6	5.6
Seed yield	0	1.71	2.66	3.00	3.15	2.51	2.76	3.08	3.11
(t ha^{-1})	34	1.62	2.67	2.94	3.16	3.05	3.11	3.23	3.13

[a]Plants were grown in a soil of pH 5.6. Based on Parker and Harris (1977).

9.6.3   Nitrate Reductase

Nitrate reductase is a dimeric enzyme with three electron-transferring prosthetic groups per subunit: flavin (FAD), heme, and molybdenum. During nitrate reduction, electrons are transferred directly from molybdenum to nitrate. Details of this reduction process had been described in Section 8.2.

Nitrate reductase activity (NRA) is low in leaves of molybdenum-deficient plants, but can be readily induced within a few hours by infiltrating the leaf segments with molybdenum. As shown in Fig. 9.22, in nitrate-fed plants there is a close relation between molybdenum supply, NRA of the leaves (NRA − Mo), and yield of spinach. Of the leaf segments incubated for 2 h with molybdenum (NRA + Mo), only in those from deficient plants was there an increase in the NRA of the leaf tissue. 'Inducible NRA' can therefore be used as a test for the molybdenum nutritional status of plants (Shaked and Bar-Akiva, 1967).

As would be expected, molybdenum requirement for plant growth is strongly dependent on whether nitrogen is supplied as nitrate or ammonium (Table 9.34). In nitrate-fed plants not supplied with molybdenum growth is poor, the contents of both chlorophyll and ascorbic acid are low (main location in chloroplasts), but that of nitrate is high. Leaves showed typical symptoms of molybdenum deficiency ('whiptail', see Fig. 9.24). When ammonium is supplied, the response to molybdenum is much less marked but still present in terms of both, dry weight and ascorbic acid content. In the absence of molybdenum supply, ammonium-fed plants also developed whiptail symptoms.

There are conflicting reports as to whether there is any molybdenum requirement when plants are supplied with reduced nitrogen such as ammonium or urea. The results shown in Table 9.34 include this aspect as under the nonsterile culture conditions nitrification of ammonium occurred in the substrate and, thus, the uptake and accumulation of nitrate could not be prevented. In cauliflower plants growing under

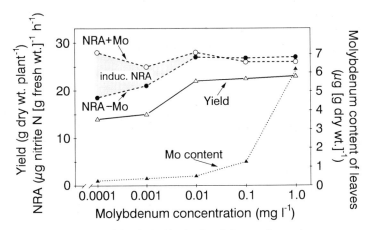

Fig. 9.22 Nitrate reductase activity (NRA) of spinach leaves from plants grown with different levels of molybdenum. Leaf segments were incubated with (NRA + Mo) or without (NRA − Mo) molybdenum for 2 h. Stippled area represents 'inducible NRA'. (Redrawn from Witt and Jungk, 1977.)

Table 9.34

Effects of Molybdenum and Source of Nitrogen on Growth and Chlorophyll, Nitrate, and Ascorbic Acid Content of Tomato[a]

Treatment[b] (nitrogen form)	Dry weight (g per plant)		Chlorophyll (mg $(100 \text{ g})^{-1}$ fresh wt)		Nitrate (mg g^{-1} dry wt)		Ascorbic acid (mg $(100 \text{ g})^{-1}$ fresh wt)	
	−Mo	+Mo	−Mo	+Mo	−Mo	+Mo	−Mo	+Mo
Nitrate	9.6	25.0	8.9	15.8	72.9	8.7	99	195
Ammonium	15.9	19.4	21.6	17.4	10.4	8.7	126	184

[a]Based on Hewitt and McCready (1956).
[b]pH of the substrate (quartz sand) buffered with $CaCO_3$.

sterile conditions (Hewitt and Gundry, 1970) those supplied with ammonium but without molybdenum did not develop any deficiency symptoms and seemed to have no molybdenum requirement, a result which is in agreement with that obtained in green algae (Ichioka and Arnon, 1955). It has been supposed (Hewitt and Gundry, 1970) that even low nitrate contents induce the synthesis of nitrate reductase (see also Section 8.2.1) and that this apoenzyme in the absence of the appropriate molybdenum cofactor may have other catalytical properties leading to metabolic disturbances similar to those induced by high levels of superoxide radicals such as peroxidation of membrane lipids (Fido *et al.*, 1977). When tungsten was applied to molybdenum-deficient plants, it was incorporated into the nitrate reductase apoenzyme and prevented the development of typical molybdenum deficiency symptoms but did not restore nitrate reductase activity. It is well known that certain metalloenzymes even within the same plant species are not absolutely metal specific. Similar metals can be incorporated and may either restore the original catalytic reaction (Sandmann and Böger, 1983), or develop a modified type of catalytic reaction.

In tobacco, replacement of molybdenum by tungsten in the apoenzyme of nitrate reductase depresses nitrate reductase activity drastically within a few hours but leads to a progressive increase not only of the apoenzyme but also the corresponding mRNA to levels that are severalfold higher than in plants supplied with molybdenum. This response suggests that tungsten inactivates nitrate reductase but simultaneously leads to overexpression of the nitrate reductase structural genes (Deng *et al.*, 1989). These genes are suppressed in plants supplied with molybdenum, probably by a higher level of reduced nitrogen.

9.6.4 Other Molybdenum-Containing Enzymes

Of the other molybdenum-containing enzymes in biological systems, most likely *xanthine oxidase/dehydrogenase* is also of general importance in higher plants. This enzyme is a dimeric metalloflavoprotein each subunit of which contains one atom of molybdenum together with one molecule of FAD and four Fe-S groups as a cluster. The apoenzymes of nitrate reductase (Section 8.2.1) and xanthine oxidase/dehydrogenase share many properties in common, including a similar molecular weight (Vunkova-Radeva *et al.*, 1988). Xanthine oxidase/dehydrogenase catalyzes the catabolic pathway of purines to uric acid:

The enzyme occurs either as a xanthine oxidase with O_2 as terminal electron acceptor, or xanthine dehydrogenase where the electrons are transferred to NAD^+ (Nguyen and Feierabend, 1978). Both forms of the enzyme have been found in higher plants, in leaves the dehydrogenase-form prevails (Nguyen and Feierabend, 1978). The enzyme is involved in the catabolism of purines and, thus, in the biosynthetic pathway of ureides which are oxidation products of purines (Section 8.5). In legumes such as soybean and cowpea, in which ureides are the most prevalent nitrogen compounds formed in root nodules (Section 7.4.3), xanthine oxidase/dehydrogenase plays a key role in nitrogen metabolism. In the cytosol of the nodules purines (e.g., xanthine) are oxidized to uric acid, the precursor of ureides. In nodulated legumes of the ureide type, under molybdenum deficiency therefore growth inhibition and low N_2 fixation rates can result from low nitrogenase activity or impaired purine catabolism in the nodules, or from both these causes.

Sulfite oxidase is another molybdenum-containing enzyme, well characterized in microorganisms, which catalyzes the oxidation of sulfite (SO_3^{2-}) to sulfate (SO_4^{2-}). It is well known, however, that this oxidation can also be brought about by other enzymes such as peroxidases and cytochrome oxidase, as well as a number of metal ions, and superoxide radicals. It is therefore not clear whether a specific sulfite oxidase is involved in the oxidation of sulfite in higher plants (DeKok, 1990), and, consequently also whether molybdenum is essential in higher plants for sulfite oxidation, as occurs for example during protein degradation and reoxidation of reduced sulfur in amino acids (Section 8.3.2).

9.6.5 Gross Metabolic Changes

In legumes dependent on N_2 fixation as a source of nitrogen, nitrogen deficiency and the corresponding metabolic changes are the most prevalent effects of molybdenum deficiency. This also often holds true for nitrate-fed plants provided the molybdenum deficiency is not severe. With severe molybdenum deficiency even the visual symptoms (e.g. whiptail; Chatterjee *et al.*, 1985; shortening of internodes and chlorosis of young leaves; Agarwala *et al.*, 1978) as well as a range of metabolic changes are distinctly different from nitrogen deficiency. These differences might relate in some way to the

Table 9.35
Effect of Molybdenum Supply to Maize Plants on Pollen Production and Viability[a]

Molybdenum supply (mg kg^{-1})	Molybdenum concentration in pollen grains (μg g^{-1} dry wt)	Pollen-producing capacity (no. of pollen grains per anther)	Pollen diameter (μm)	Pollen viability (% germination)
20	92	2437	94	86
0.1	61	1937	85	51
0.01	17	1300	68	27

[a]From Agarwala *et al.* (1979).

role of molybdenum in xanthine oxidase/dehydrogenase but many are difficult to reconcile with current knowledge of metabolic functions of molybdenum. For example, in molybdenum-deficient plants organic acids (Höfner and Grieb, 1979) and amino acids (Gruhn, 1961) accumulate, and the activity of ribonuclease is high whereas that of alanine transferase is low (Agarwala *et al.*, 1978) like the leaf contents of RNA and DNA (Chatterjee *et al.*, 1985). Molybdenum-deficient plants seem to be more sensitive to low temperature stress and waterlogging (Vunkova-Radeva *et al.*, 1988). In many studies the molybdenum deficiency has been so severe and growth so strongly depressed, that it is difficult to identify the primary metabolic functions of molybdenum.

Molybdenum deficiency also has striking effects on pollen formation in maize (Table 9.35). In deficient plants not only was tasseling delayed, but a large proportion of the flowers failed to open and the capacity of the anther for pollen production was reduced. Furthermore, the pollen grains were smaller, free of starch, had much lower invertase activity, and showed poor germination. Impaired pollen formation may also explain the failure of fruit formation in molybdenum-deficient water melon plants growing in acid soil (Gubler *et al.*, 1982).

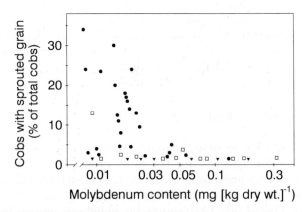

Fig. 9.23 Relationship between molybdenum content of maize grains, time of nitrogen top dressing, and percentage of sprouted cobs of maize. Top dressing with nitrogen at (▼) 30 days; (□) 40–55 days; (●) 70–85 days. (Based on Tanner, 1978.)

As shown in Fig. 9.23 the risk of premature sprouting of maize grains in standing crops greatly increases when the molybdenum content in the dry matter falls below $0.03 \, \mu g \, g^{-1}$ in the grains (Fig. 9.23), or below $0.02 \, \mu g$ in the grains and $0.10 \, \mu g$ in the leaves (Farwell *et al.*, 1991). Premature sprouting is also a serious problem in some wheat-growing areas and can be strongly depressed by foliar sprays of molybdenum (Cairns and Kritzinger, 1992). In maize the extent of premature sprouting also appears to be related to the time of nitrogen application (Tanner, 1978). Little sprouting occurred when top dressing with NH_4NO_3 took place before 60 days after germination. Sprouting of grains low in molybdenum, however, was strongly enhanced by very late nitrogen application. The causal relationships are not known. Direct stimulatory effects of high nitrate tissue levels on sprouting (Tanner, 1978) or indirect effects via other metabolic changes in deficient grains may be involved.

In wheat grains adequately supplied with molybdenum most of the molybdenum is associated with nitrate reductase, xanthine oxidase/dehydrogenase, and a 60 kDa polypeptide, all three fractions being drastically depressed under molybdenum deficiency (Vunkova-Radeva *et al.*, 1988).

9.6.6   Molybdenum Deficiency and Toxicity

Depending on plant species and source of nitrogen supply, the critical deficiency levels of molybdenum vary between 0.1 and $1.0 \, \mu g \, g^{-1}$ leaf dry weight (Gupta and Lipsett, 1981; Bergmann, 1988). In seeds the molybdenum content is highly variable (see below) but, in general, much higher in legumes than in nonlegumes (Section 9.5; Table 9.31).

In molybdenum-deficient plants, symptoms of nitrogen deficiency (e.g., legumes) and stunted growth and chlorosis in young leaves are common (Section 9.6.5). In dicotyledonous species a drastic reduction in size and irregularities in leaf blade formation (whiptail) are the most typical visual symptom (Fig. 9.24), caused by local

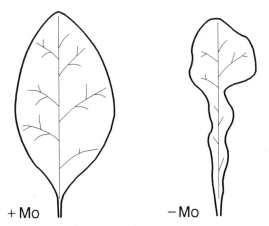

Fig. 9.24   Schematic representation of changes in leaf morphology in molybdenum-deficient cauliflower ('whiptail' symptom).

necrosis in the tissue and insufficient differentiation of vascular bundles at early stages of leaf development (Bussler, 1970).

Local chlorosis and necrosis along the main veins of mature leaves (e.g. 'Yellow spot' in citrus) and whiptail in young leaves may reflect the same type of local metabolic disturbances, occurring, however, at different stages of leaf development (Bussler, 1970b). When there is severe deficiency marginal chlorosis and necrosis on mature leaves with a high nitrate content also occur.

Molybdenum deficiency is widespread in legumes and certain other plant species (e.g. cauliflower and maize) grown in acid mineral soils with large content of reactive iron oxide hydrate and thus a high capacity for adsorbing MoO_4^{2-}. Furthermore, molybdic acid is a weak acid; with decreasing pH from 6.5 to 4.5 and below, the dissociation decreases ($MoO_4^{2-} \rightarrow HMoO_4^- \rightarrow H_2MoO_4$) and the formation of poly-anions is favoured (molybdate $\rightarrow$ tri- $\rightarrow$ hexa-molybdate) which leads to decrease in molybdenum uptake. These aspects have to be taken into account in the measurements for correction of molybdenum deficiency on acid soils.

As shown in Table 9.36, regardless whether molybdenum is supplied or not, the molybdenum content of the shoots of soybean increases when the soil pH is increased from 5.0 to 7.0 by liming. The effect of the liming treatment alone on the plant dry weight is similar to the application of molybdenum to the unlimed soil. Thus, quite often liming and molybdenum application can be seen as alternatives for stimulating legume growth on acid mineral soils. Responses of legume growth to liming therefore also strongly depend on the molybdenum status of the soils (Adams et al., 1990). A combination of both, liming and molybdenum supply often lead to distinct luxury uptake and very high molybdenum contents in the vegetative parts of the shoots and the seeds.

A high molybdenum content in seeds ensures proper seedling growth and high final grain yields in plants growing in soils low in available molybdenum (Table 9.37). Correspondingly, the effect of molybdenum application to a deficient soil on plant growth is inversely related to the seed content (Tanner, 1982) and the amount of molybdenum applied to the seed crop (Weir and Hudson, 1966).

Compared with the uptake rates of other micronutrients, the rate of molybdenum uptake by soybean plants during the first 4 weeks after germination is very low; thus the molybdenum requirement for growth has to be met mainly by retranslocation from the

Table 9.36

Relationship between Soil pH, Molybdenum Supply, and Dry Weight and Molybdenum Content of Soybean[a]

Parameter	Molybdenum supply (mg per pot)	Soil pH		
		5.0	6.0	7.0
Dry weight (g per pot)	0	14.9	18.9	22.5
	5	19.6	19.5	20.4
Molybdenum content of shoots ($\mu g \ g^{-1}$ dry wt)	0	0.09	0.82	0.90
	5	1.96	6.29	18.50

[a]Based on Mortvedt (1981).

Table 9.37
Relationship between the Molybdenum Content of Soybean Seeds and the Subsequent Seed Yield of Plants Growing in a Molybdenum-Deficient Soil[a]

Molybdenum content of seeds ($mg\ kg^{-1}$ dry wt)	Seed yield of the crop ($kg\ ha^{-1}$)
0.05	1505
19.0	2332
48.4	2755

[a]Based on Gurley and Giddens (1969).

seed (Ishizuka, 1982). Large-seeded cultivars combined with high molybdenum availability during the seed-filling period are therefore very effective in the production of seeds suitable for soils low in available molybdenum (Franco and Munns, 1981).

Seed pelleting with molybdenum is another procedure for preventing deficiency during early growth and establishing a vigorous root system for subsequent uptake from soils low in available molybdenum (Tanner, 1982). As shown in Table 9.38 seed pelleting with the relatively insoluble molybdenum trioxide at a rate of 100 g molybdenum per hectare is somewhat more effective than soil application. Seed pelleting with 100 g molybdenum in legumes such as groundnut might increase dry matter production and the amount of nitrogen in the plants more than an application of 60 kg ha^{-1} of mineral fertilizer nitrogen (Section 7.4; Table 7.3).

As molybdenum is highly phloem-mobile, foliar application is an appropriate and easy procedure for correcting acute molybdenum deficiency (Gupta and Lipsett, 1981). In legumes molybdenum applied as a foliar spray in early growth stages is preferentially translocated into the nodules (Brodrick and Giller, 1991a) and very effective in increasing final yield, for example, in soybean (Adams *et al.*, 1990) or groundnut (Table 9.39). Compared with soil application, foliar application to groundnut not only increases yield but also nitrogen uptake and the molybdenum contents of shoots, seeds and nodules.

Table 9.38
Effect of Molybdenum Trioxide Application on Dry Matter Production and Nitrogen Content of the Subtropical Pasture Legume *Desmodium intartum* Grown in a Soil of pH 4.7[a]

Molybdenum application ($g\ ha^{-1}$)	Dry weight ($kg\ ha^{-1}$)	Nitrogen content (% dry wt)
0	70	1.9
100 (soil application)	1220	3.2
100 (seed pelleting)	1380	3.4

[a]From Kerridge *et al.* (1973).

Table 9.39

Effect of Soil or Foliar Application of Molybdenum on Dry Matter Production, Nitrogen Uptake, and Molybdenum Content in Groundnut Grown on a Low Molybdenum, Acid Sandy Soil (Luvic Arenosol)[a]

Mo application (g ha^{-1})	Dry matter (kg ha^{-1})	N uptake (kg ha^{-1})	Molybdenum content (μg g^{-1} dry wt)		
			Shoots	Nodules	Seeds
0	2685	70	0.02	0.4	0.02
200 (soil)	3413	90	0.02	1.5	0.20
200 (foliar)	3737	101	0.05	3.7	0.53

[a]Based on Rebafka (1993).

A lower effectivity of soil compared with foliar applied molybdenum might reflect fixation of molybdenum in the soil, however, it is often also the result of impaired uptake by the roots. Sulfate and molybdate are strongly competing anions during uptake by the roots. In view of the wide concentration ratio of SO_4^{2-}/MoO_4^{2-} in the soil solution, sulfate containing fertilizer such as gypsum (Stout *et al.*, 1951; Pasricha *et al.*, 1977), as well as also single superphosphate (SSP) are very effective in depressing molybdenum uptake (Table 9.40). The absence of sulfate (as gypsum) in the triple superphosphate (TSP) and thus, the much higher molybdenum uptake was the main reason for the higher yield and nitrogen uptake achieved by TSP compared with SSP. In addition, with TSP seed quality was increased in terms of suitability for use in molybdenum-deficient soils.

The depression in molybdenum uptake by sulfate might also be of significance for natural ecosystems. In red cedar trees a negative relationship has been found between molybdenum and sulfur contents of tree rings, the increase in sulfur content being closely related to the historical trend in coal production and, thus, SO_2 emission in the area in which the trees were growing (Guyette *et al.*, 1989).

A unique feature of molybdenum nutrition is the wide variation between the critical deficiency and toxicity levels. These levels may differ by a factor of up to 10^4 (e.g.,

Table 9.40

Effect of Phosphate Fertilizers (13 kg P ha^{-1}) on Dry Matter Production, Nitrogen Uptake and Molybdenum Content in Groundnut Grown on a Low Molybdenum, Acid Sandy Soil (Luvic Arenosol)[a]

Phosphorus fertilizer[b]	Dry matter (kg ha^{-1})	N uptake (kg ha^{-1})	Mo content (μg g^{-1} dry wt)		
			Shoots[c]	Nodules	Seeds
−P	2000	52	0.22	4.0	1.0
+SSP	2550	62	0.09	1.5	0.1
+TSP	3150	81	0.31	8.2	3.1

[a]Based on Rebafka *et al.* (1993). Reprinted by permission of Kluwer Academic Publishers.
[b]SSP = single superphosphate; TSP = triple superphosphate.
[c]At flowering.

0.1–$1000\ \mu g$ molybdenum g^{-1} dry weight) as compared with a factor of 10 or less for boron or manganese. Under conditions of molybdenum toxicity, malformation of the leaves and a golden yellow discoloration of the shoot tissue occur, most likely owing to the formation of molybdocatechol complexes in the vacuoles (Hecht-Buchholz, 1973). Genotypical differences in molybdenum toxicity are closely related to differences in the translocation of molybdenum from roots to shoots (Chapter 3).

High but non-toxic levels of molybdenum in plants are advantageous for seed production, but such levels in forage plants are dangerous for animals, and for ruminants in particular, which are very sensitive to excessive concentrations of molybdenum. Molybdenum contents above 5–10 mg kg^{-1} dry wt of forage are high enough to induce toxicity known as molybdenosis (or 'teart'). This occurs, for example, in western parts of the United States, in Australia, and in New Zealand, often in soils with poor drainage and high in organic matter (Gupta and Lipsett, 1981), or on pastures established on retorted oil shale disposal piles (Stark and Redente, 1990). Molybdenosis is actually caused by an imbalance of molybdenum and copper in the ruminant diet, i.e. an induced copper deficiency (Stark and Redente, 1990; Miller et al., 1991). The strong depressing effect of sulfate on molybdate uptake (Table 9.40) can be used effectively in decreasing the molybdenum contents in plants to levels that are nontoxic (Pasricha et al., 1977; Chatterjee et al., 1992) either for the plants themselves or for the ruminants.

Molybdenum nutrition of plants growing in mixed pastures of legumes, herbs and grasses therefore requires special consideration. On the one hand, the relatively large requirement of legumes for N_2 fixation and for molybdenum in the seeds must be met, but at the same time toxic levels must not be allowed to accumulate in the forage of grazing animals.

9.7 Boron

9.7.1 General

Boron is a member of the metalloid group of elements which also includes silicon and germanium. These elements are intermediate in properties between metals and nonmetals, and also share many features in common in plants. The boron atom is small and has only three valencies. Boric acid is a very weak acid and in aqueous solution at pH <7, it occurs mainly as undissociated boric acid; at higher pH boric acid accepts hydroxyl ions from water thus forming a tetrahedral borate anion

$$B(OH)_3 + 2H_2O \rightleftharpoons B(OH)_4^- + H_3O^+$$

Only the monomeric species $B(OH)_3$ and $B(OH)_4^-$ are usually present in aqueous solutions at boron concentrations <25 mm; thus polymeric boron species are unlikely to occur in plants, except under boron toxicity.

Boron uptake is closely related to the pH and the external boron concentration over a wide concentration range (Section 2.5.3). Its distribution in the plants is primarily governed by the transpiration stream (Section 3.2.4) although it is also phloem-mobile and might be retranslocated in considerable amounts (Section 3.3; Shelp, 1988, 1993).

Boron is a micronutrient for vascular plants, diatoms and some species of green algae, whereas it is apparently not required by fungi and bacteria (Dugger, 1983; Loomis and Durst, 1992). An exception are the cyanobacteria in which some species require boron when depending on N_2 fixation (Section 9.7.4). According to McClendon's (1976) classification of the origins of mineral nutrient requirements, the requirement for boron is of an evolutionary nature, relating to lignification and xylem differentiation in vascular plants (Lewis, 1980a).

The role of boron in plant nutrition is still the least understood of all the mineral nutrients and what is known of boron requirement arises mainly from studies of what happens when boron is withheld or resupplied after deficiency. This poor knowledge is surprising, because on a molar basis the requirement for boron, at least for dicotyledons, is higher than that for any other micronutrient. It is quite easy in certain plant species (e.g., sunflower) very rapidly to induce a range of distinct metabolic changes and visible deficiency symptoms by withholding boron. Boron is neither an enzyme constituent nor is there convincing evidence that it directly affects enzyme activities. There is a long list of postulated roles of boron (Parr and Loughman, 1983): (a) sugar transport; (b) cell wall synthesis; (c) lignification; (d) cell wall structure; (e) carbohydrate metabolism; (f) RNA metabolism; (g) respiration; (h) indole acetic acid (IAA) metabolism; (i) phenol metabolism; (j) membranes. This long list might indicate (a) that boron is involved in a number of metabolic pathways, or (b) a 'cascade effect', as is known for the phytohormones, for example. There is increasing evidence for the latter alternative, and for a primary role of boron in the cell wall biosynthesis and structure, and for plasma membrane integrity. For a conceptional review on the role of boron in cell walls see Loomis and Durst (1992) and for a more general review see Shelp (1992).

9.7.2   Boron Complexes With Organic Structures

Boric acid has an outstanding capacity to form complexes with diols and polyols, particuarly with *cis*-diols, either as Eq (1), or Eq (2):

Polyhydroxyl compounds with an adjacent *cis*-diol configuration are required for the formation of such complexes; the compounds include a number of sugars and their derivatives (e.g., sugar alcohols and uronic acids), in particular mannitol, mannan, and polymannuronic acid. These compounds serve, for example, as constituents of the hemicellulose fraction of cell walls. In contrast, glucose, fructose, and galactose and their derivatives (e.g., sucrose) do not have this *cis*-diol configuration and thus do not form stable borate complexes. Some *o*-diphenolics, such as caffeic acid and hydroxyferulic acid, which are important precursors of lignin biosynthesis in dicotyledons

(McClure, 1976), possess the *cis*-diol configuration and hence form stable borate complexes.

The most stable borate complexes are formed with *cis*-diols on a furanoid ring, namely the pentoses ribose and apiose, the latter being a universal component of the cell walls of vascular plants (Loomis and Durst, 1992). The high boron requirement of gum-producing plants is most likely related to the function of boron in forming cross-links with the various polyhydroxy polymers such as galactomannan (Loomis and Durst, 1991). Boron forms not only stable complexes with ribose, the principal sugar component of RNA, but also with NAD^+ (Johnson and Smith, 1976). Thus, inhibited dehydrogenase activity under boron toxicity is likely to be associated with higher cytosolic boron concentrations (Section 9.7.9.2).

In higher plants at least, a substantial proportion of the total boron content is complexed in *cis*-diol configuration in the cell walls (Thellier *et al.*, 1979). The higher boron requirement in dicotyledonous compared with graminaceous species is presumably related to higher proportions of compounds with the *cis*-diol configuration in the cell walls, namely pectic substances and polygalacturonans (Loomis and Durst, 1992). It has been shown by Tanaka (1967) that the content of strongly complexed boron in the root cell walls is 3–5 μg g^{-1} dry weight in graminaceous species such as wheat, and up to 30 μg g^{-1} dry weight in dicotyledonous species such as sunflower. These differences roughly reflect the differences between the species in boron requirement for optimal growth (Section 9.7.9.1).

9.7.3 Root Elongation and Nucleic Acid Metabolism

One of the most rapid responses to boron deficiency is inhibition or cessation of root elongation, giving the roots a stubby and bushy appearance. As shown in Fig. 9.25A, inhibition of root elongation occurs as soon as 3 h after the boron supply is interrupted, becoming more severe after 6 h, and finally coming to a halt after 24 h. Twelve hours after the boron supply is restored to roots deprived of boron for the same period of time, however, root elongation again becomes rapid.

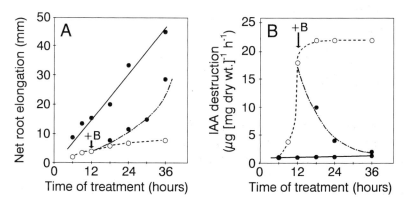

Fig. 9.25 Effect of boron deficiency on root elongation (A) and IAA oxidase activity (B) in apical 5-mm root sections of squash. Resumption of boron supply after 12 h (arrow) of boron deficiency. Key:●—●, +B; ○– –○, −B. (Redrawn from Bohnsack and Albert, 1977.)

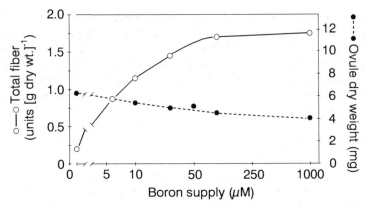

Fig. 9.26 Effect of boron supply on dry weight and fiber development of unfertilized cotton ovules cultured in the presence of IAA, gibberellic acid, and cytokinin. Total fiber units represent the ratio of fiber length to gram of dry weight. (Redrawn from Birnbaum *et al.*, 1974.)

Between 6 and 12 h after the boron supply is cut off, there is a dramatic increase in the activity of IAA oxidase in the roots (Fig. 9.25B) which falls rapidly when boron is resupplied. The similarities in the responses of root elongation and IAA oxidase activity to boron deficiency and resupply are striking. There is, however, a distinct difference in the time of response to deficiency: root elongation is inhibited ~3 h before IAA oxidase activity increases. Thus, increase in IAA oxidase activity is a secondary event of boron deficiency.

Similar responses of elongation growth to boron can be demonstrated in cotton ovules cultured *in vitro*. Epidermal cells of cotton ovules that form lint fibers begin to elongate on the day of anthesis. The degree of elongation is closely related to the external boron concentration (Fig. 9.26). It is evident that boron is necessary for fiber elongation and the prevention of callusing of the epidermal cells, as indicated indirectly by the decline in ovule dry weight. From additional observations it has been concluded that boron is required primarily for cell elongation rather than for cell division (Birnbaum *et al.*, 1974).

Root elongation is the result of various processes, including cell division, and cell elongation by loosening and re-forming of cross-links in the cell walls. The role of auxin (IAA) and calcium in elongation growth has been discussed in Section 8.6. Except for the agreement that when boron is withdrawn there is both a decrease in the rate of cell division and an inhibition in elongation growth (Kouchi, 1977; Ali and Jarvis, 1988), views conflict as to the role of boron. A decrease in DNA content and rate of DNA synthesis is a well documented phenomenon when boron is withheld (Shkol'nik, 1974). Inhibition of DNA synthesis is considered either as a primary (Krueger *et al.*, 1987) or secondary (Moore and Hirsch, 1983; Ali and Jarvis, 1988) effect of boron deficiency. The content of RNA also rapidly decreases under boron deficiency (Albert, 1965). In view of the high turnover rate of RNA this might reflect lower rates of synthesis or higher rates of degradation, for example, as a result of higher RNase activity in boron-deficient tissues (Dave and Kannan, 1980). Some support for the involvement of boron in RNA metabolism comes from the work of Birnbaum *et al.* (1977) showing that supply of certain nucleotides such as uracil can delay symptoms of boron deficiency. These and other corresponding results in the literature have been critically reviewed recently by Shelp (1993). In principle, a primary role of boron in DNA and RNA metabolism is difficult to reconcile with the much slower response of inhibition in root elongation which occurs in graminaceous species when boron is withheld. Similarly a major role of boron in nucleotide metabolism is not in keeping with the lack of boron demand in lower plants, bacteria and fungi.

9.7.4   Cell Wall Synthesis

In boron-deficient plants the cell walls are dramatically altered as reflected at both, macroscopic (e.g., 'cracked stem'; 'stem corkiness'; 'hollow stem disorder'; Bergmann, 1988, 1992; Shelp, 1988) and microscopic levels (for a review see Loomis and Durst, 1991; 1992). The cell wall diameter and the proportion of cell wall material to total dry weight are both higher in boron deficient tissues (Rajaratnam and Lowry, 1974; Hirsch and Torrey, 1980). In celery the cell wall thickness of parenchyma cells increases from 1 μm in boron-sufficient to 4 μm in boron-deficient plants (Spurr, 1957). An example of this effect of boron on cell walls is shown in Fig. 9.27. The primary cell walls of boron-deficient cells are not smooth but characterized by irregular depositions of vesicular aggregations intermixed with membraneous material (Hirsch and Torrey, 1980). There is a higher concentration of pectic substances (Rajaratnam and Lowry, 1974) and a larger proportion of glucose incorporation into β-1,3-glucan (Dugger and Palmer, 1985), the main component of callose (Section 8.6), which also accumulates in the sieve tubes of boron-deficient plants (Venter and Curries, 1977) and, thus, impairs phloem transport.

In tomato roots irregularities occur after 8 h of boron deprivation (Kouchi and Kumazawa, 1976) and indicate that boron not only complexes strongly with cell wall constituents but is required for structural integrity by forming borate-ester cross-links. These cross-links are relatively weak and therefore fulfil the function of breaking and re-forming during cell elongation, and furthermore provide negative charges for ionic interactions, for example, with Ca^{2+} (Loomis and Durst, 1992). Distinct interactions between both mineral nutrients in cell walls are reflected in the close correlations between calcium and boron content in the cell walls and demand for growth (Section 9.7.9.1), or the alterations of binding forms of calcium in the roots and impaired shoot transport of calcium in boron-deficient plants (Ramon et al., 1990). However, compared with calcium, boron is bound less firmly to the cell wall matrix, and presumably separate binding sites exist for boron and calcium (Teasdale and Richards, 1990).

Boron and germanium are chemically closely related and form similar cis-diol complexes. Substitution of germanium for boron has been shown, for example, in sunflower (McIlrath and Skok, 1966) and tomato (Brown and Jones, 1972). In both instances development of visual deficiency symptoms could be delayed for several days in the boron-deficient plants supplied with germanium. This substitution has been interpreted as a sparing effect bringing about increasing mobility of boron in plants, for example, by replacing it in the cell walls of the roots. In suspension cultures of carrot cells, however, germanium was able to substitute for boron over a long period with only somewhat lower growth rates being obtained (Loomis and Durst, 1991). This finding

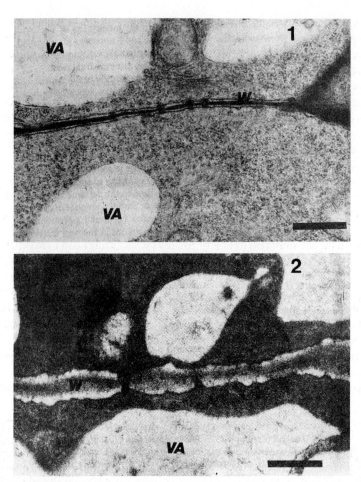

Fig. 9.27 Cell wall thickness and fine structure in parenchyma cells of leaves of *Mentha piperita* L. as affected by boron nutritional status: (1, *upper*) boron sufficient; (2, *lower*) boron deficient; W = cell wall; VA = vacuole. Bar = 1 μm. (Fischer and Hecht-Buchholz, 1985.)

suggests that the cellular binding sites of boron are relatively nonspecific and are more compatible with a structural role of boron than with a catalytical or regulatory role. In these carrot cell cultures, in the boron-sufficient cells 96–99% of the boron was present in the cell wall fraction, the corresponding value for germanium was 60% in the boron-substituted cells. In boron-deficient leaves germanium can also substitute for boron in restoring plasma membrane integrity (Section 9.7.6).

The particular role of boron for cell wall formation and functioning is also evident in pollen tube growth (Section 9.7.6) and in the heterocyst cells of cyanobacteria. In cyanobacteria boron is required only in heterocyst-forming types such as *Anabaena* when dependent on N$_2$ fixation, but not when supplied with mineral nitrogen (Mateo *et al.*, 1986; Bonilla *et al.*, 1990). In these cyanobacteria nitrogenase is located in the heterocyst cells where the envelope has the function of controlling and restricting O$_2$

diffusion and thus, protecting nitrogenase from inactivation by O_2 and toxic oxygen radicals (Garcia-Gonzales *et al.*, 1988). Under boron deficiency nitrogenase activity drops rapidly, and this decrease is correlated with dramatic morphological changes in the heterocyst cells and their envelope. The envelope comprises layers of glycolipids and polysaccharides with terminal mannose residues, i.e. potential borate complexing groups. By interacting with these groups boron modifies and stabilizes the envelope structure to function as a barrier for O_2 diffusion (Bonilla *et al.*, 1990; Garcia-Gonzales *et al.* 1991).

9.7.5 Phenol Metabolism, Auxin (IAA), and Tissue Differentiation

Boron deficiency is associated with a range of morphological alterations and changes in differentiation of tissues, similar to those induced by either suboptimal or supraoptimal levels of IAA. As lignification and xylem differentiation are unique to vascular plants, which is also true, in principle, for the boron demand, a key role of boron in IAA metabolism and in regulation of lignin biosynthesis and xylem differentiation had been proposed (Lewis, 1980a; Lovatt, 1985).

In root tips, for example, boron deficiency results in a reduction in elongation growth associated with changes in cell division from a normal longitudinal to a radial direction (Robertson and Loughman, 1974). Enhanced cell division in a radial direction with a distinct proliferation of cambial cells and impaired xylem differentiation are also features typical of the subapical shoot tissue of boron-deficient plants (Bussler, 1964; Fig. 9.28).

Enhancement of cell division in cambial stem tissue and impaired xylem differen-

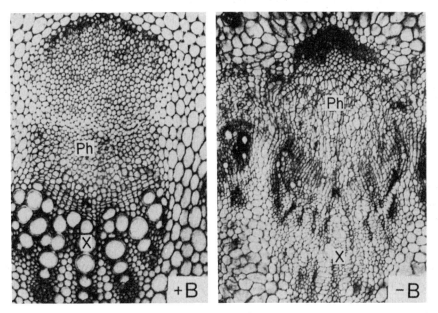

Fig. 9.28   Cross section of a vascular bundle of an upper internode of a boron-sufficient (*left*) and a boron-deficient (*right*) sunflower plant. X, xylem; Ph, phloem. (From Pissarek, 1980.)

tiation are not, however, direct effects of boron deficiency. Similar morphological changes can be obtained in boron-sufficient plants by mechanical destruction of the apical meristem of the shoot (Krosing, 1978). It can be concluded, therefore, that inhibition or even a lack of xylem differentiation is only indirectly related to boron nutrition. Furthermore, first symptoms of boron deficiency are modifications in the structure of the primary cell walls and not xylem differentiation.

The relationships between boron nutrition, auxin level, differentiation, and lignification are not clear. In boron-deficient plants auxin levels are often much higher than normal (Coke and Whittington, 1968), and an exogenous supply of IAA induces anatomical changes in the root tips similar to those caused by boron deficiency. This has led to the interpretation that boron-deficiency symptoms are a reflection of increased auxin levels (Robertson and Loughman, 1974). However, the ultrastructural changes brought about by boron deficiency and excessive IAA levels are quite different (Hirsch and Torrey, 1980). Furthermore, typical symptoms of boron deficiency may occur without any increase in the IAA level of the same tissue (Hirsch et al., 1982). In early stages of deficiency there is even a tendency for lower IAA levels in apical tissues (Fackler et al., 1985). Also no significant correlations between IAA level and boron-deficiency symptoms were found by Smirnov et al. (1977) in a comparison of different plant species or plant organs.

In view of the mechanisms of phytohormone action (Section 5.6) it is not likely that determinations, for example, of total IAA in tissues of boron-sufficient and -deficient plants will provide an adequate approach to unravel the role of boron in IAA metabolism. More specific information has been provided by Tang and de la Fuente (1986) showing that in boron-deficient hypocotyl segments basipetal IAA transport is inhibited. This is similar to what occurs under calcium deficiency and indicative of impaired membrane integrity under boron deficiency. A similar effect on basipetal IAA transport is achieved by certain flavonoids such as quercetin (Jacobs and Rubery, 1988). It might well be that interactions between boron and IAA and tissue differentiation are secondary events caused by primary effects of boron on phenol metabolism (Lewis, 1980a; Pilbeam and Kirkby, 1983). Certain phenolics are not only effective inhibitors of root elongation but also simultaneously enhance radial cell division (Svensson, 1971); that is, they induce anatomical changes that are similar to those caused by IAA.

Many contradictory results on relationships between boron, IAA, and phenol metabolism are caused by different experimental conditions. For example, high IAA levels may occur only in those plant species that, in response to boron deficiency, accumulate certain phenolics such as caffeic acid, which is an effective inhibitor of IAA oxidase activity (Birnbaum et al., 1977). Phenol contents in leaves are also strongly dependent on light intensity (Table 9.41). There is a close relationship between increasing light intensity and increasing phenol content in both, boron-sufficient and boron-deficient leaves. The gradient in the deficient leaves, however, is much steeper and this is correlated with an increase in polyphenol oxidase activity and particularly potassium efflux as an indicator of impairment of the plasma membrane integrity.

Accumulation of phenols is a typical feature of boron deficient plants and most likely related to the function of boron in the formation of cis-diol complexes with certain sugars and phenols (Fig. 9.29). Under boron deficiency the substrate flux is shifted

Table 9.41
Effect of Boron Supply (10^{-5} M = sufficient; 10^{-7} M = deficient) to Sunflower Plants Grown Under Different Light Intensities on Phenol Content, Polyphenol Oxidase Activity and Potassium Efflux from Leaf Segments[a]

Light intensity (μE m^{-2} s)	Phenol content (μg caffeic acid equiv. per 6 segments		Polyphenol oxidase activity (relative)		K$^+$ efflux (μg K per 6 segments) (2 h)$^{-1}$)	
	10^{-5} M	10^{-7} M	10^{-5} M	10^{-7} M	10^{-5} M	10^{-7} M
100	30	35	1.0	1.4	10	23
250	45	90	0.8	2.1	12	63
580	75	265	0.6	4.2	25	238

[a] I. Cakmak, personal communication.

towards the pentose phosphate cycle and, thus, enhanced phenol biosynthesis. Formation of borate complexes with certain phenols are probably involved in the regulation of the level of free phenols and the rate of synthesis of phenol alcohols as precursors of lignin biosynthesis (Pilbeam and Kirkby, 1983). Accordingly, under boron deficiency phenols accumulate and polyphenol oxidase activity is increased (Table 9.41). A high proportion of phenols together with the corresponding enzyme systems are located in the cell walls, of the epidermis in particular (Hendricks and van Loon, 1990).

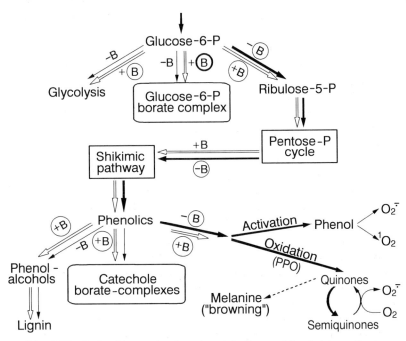

Fig. 9.29   Role of boron in phenol metabolism and lignin biosynthesis.

The accumulation of phenols, and an increase in polyphenol oxidase activity (Table 9.41) leads to highly reactive intermediates such as caffeic quinone in the cell walls (Shkol'nik et al., 1981). These quinones as well as (light) activated phenols are very effective in producing superoxide radicals (Fig. 9.29) potentially capable of damaging membranes by lipid peroxidation. Differences between dicotyledonous and graminaceous species in phenol metabolism and in the pathway of lignin biosynthesis and therefore probably also in the risk of oxidative damage of the plasma membrane, might be at least in part responsible for the differences in boron demand between these groups of plants (McClure, 1976).

Alterations in phenol metabolism in boron-deficient plants are also reflected in changes in content of certain flavonoids such as leucocyanidin, a key defense substance against sucking insects in oil palms. In palms suffering from boron deficiency the content of leucocyanidin is drastically decreased and insect damage increases accordingly (Rajaratnam et al., 1971; Rajaratnam and Hock, 1975).

9.7.6 Membrane Function

There is a range of evidence in support of a role of boron in membrane integrity and functioning. So far this has been convincingly demonstrated for the plasma membrane, but not for a role of boron in the function of other cell membranes such as tonoplast or chloroplast envelope. It has been shown (Tanada, 1978) that the formation and maintenance of membrane potentials induced by infrared light or by gravity require the presence of boron. Boron also influences the turgor-regulated nyctinastic movements of leaflets of Albizzia (Tanada, 1982) and enhances both ^{86}Rb influx and stomatal opening in Commelina communis (Roth-Bejerano and Itai, 1981).

Uptake rates of phosphorus are much lower in the root tips of boron deficient as compared with sufficient bean and maize plants (Table 9.42). However, boron pretreatment of the root tips for only 1 h markedly enhances phosphorus uptake in both boron sufficient and deficient roots and nearly restores the uptake rate of the originally boron deficient maize root and somewhat less effectively that of the boron deficient bean roots. As shown in the same study (Pollard et al., 1977) the effect of boron pretreatment on the uptake rates of chloride and rubidium was similar to that on phosphate uptake

Table 9.42
Effect of Boron Pretreatment on the Subsequent Uptake of Phosphorus
by Root Tip Zones of Faba Bean and Maize[a]

Pretreatment of root tips for 1 h	Phosphorus uptake (nmol g^{-1} h^{-1})			
	Faba bean grown with or without B		Maize grown with or without B	
	+B	−B	+B	−B
No boron	112	52	116	66
10^{-5} M $B(OH)_3$	152	108	190	171

[a]Root tip zones were 0–2 cm from the apex. From Pollard et al. (1977).

rates. Furthermore, membrane-bound ATPase activity, which was low in boron-deficient maize roots, was restored to the same level as that in boron-sufficient roots within 1 h.

It is now quite clear that these effects of boron on uptake of ions and also glucose (Goldbach, 1985), are mediated by direct or indirect effects of boron on the plasma membrane-bound H^+ pumping ATPase (Section 2.4.2). In boron-deficient cells and roots this activity is much lower and can be restored within 20–120 min after resupply of boron (Blaser-Grill *et al.*, 1989; Schon *et al.*, 1990). This restoration is indicated by both short-term changes in net excretion of protons and hyperpolarization of the membrane potential (which becomes more negative). Interestingly, in suspension-cultured tobacco cells this effect of boron on H^+-ATPase requires the presence of IAA, and vice versa, boron is required for the enhanced H^+ excretion induced by IAA (Goldbach *et al.*, 1990). The particular role of boron for plasma membrane integrity and H^+ pumping activity can also be demonstrated in vitro with membrane vesicles from boron-sufficient and boron-deficient sunflower roots (Roldan *et al.*, 1992; Ferrol *et al.*, 1993).

Although boron might act directly on the plasma membrane-bound H^+-ATPase, it is more likely that these effects are indirectly mediated as, for example, by complexing *cis*-diol groups of plasma membrane constituents such as glycoproteins or glycolipids at the cell wall–plasma membrane interface and thereby acting as a stabilizing and structural factor required for the integrity and functioning of the plasma membrane. A specific high boron content of isolated plasma membranes as compared with the rest of the protoplast of mungbean hypocotyls (Tanada, 1983) is in accordance with a particular role of boron in the plasma membrane.

Further support for the role of boron in plasma membrane integrity and function is shown in Fig. 9.30 on the potassium efflux from expanding sunflower leaves of boron-sufficient and boron-deficient plants. The leaves were isolated and immersed in either distilled water or increasing concentrations of boron. Compared with the boron-sufficient leaves potassium efflux was very high in the boron-deficient leaves, but could

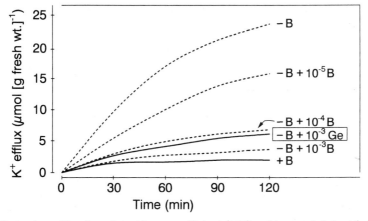

Fig. 9.30 Potassium efflux from intact boron-sufficient (+B) and boron-deficient (−B) expanding sunflower leaves and effect of external supply (10^{-5}–10^{-3} M) of boron or germanium (Ge) at zero time (−B + B; −B + Ge treatment). (Cakmak and Kurz, unpublished.)

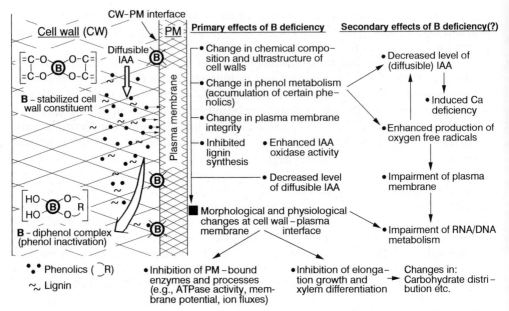

Fig. 9.31 Proposed role of boron in cell wall metabolism and related primary and secondary effects of boron deficiency. (Modified from Römheld and Marschner, 1991.)

be markedly decreased by external boron supply during the efflux period. The decrease was dependent on the external boron concentration and evident even after less than 30 min. Similarly to the potassium efflux the efflux of sugars, amino acids and phenols was also much higher in the boron-deficient leaves and could be decreased by external boron supply comparable with the potassium efflux. Interestingly, a similar decrease in potassium efflux could be achieved when the external boron supply was replaced by germanium (Fig. 9.30), indicating a substitution of boron by germanium not only in cell wall stabilitiy and functions (Section 9.7.4) but also plasma membrane integrity.

From the published evidence as to the role of boron in cell wall biosynthesis, phenol metabolism, and plasma membrane integrity, it can be concluded that in higher plants boron exerts its primary influence in the cell wall and at the plasma membrane–cell wall interface, as summarized in the model in Fig. 9.31. Changes in the cell wall and at this interface are considered as primary effects of boron deficiency leading to a cascade of secondary effects in metabolism, growth and plant composition. It should be remembered that changes in plasma membrane potential act as a signal for many changes in the cytoplasm, and also for a shift in excretion of cell wall material (Section 8.6.7).

9.7.7 Pollen Germination and Pollen Tube Growth

The particular role of boron in cell wall synthesis and plasma membrane integrity can also be shown in pollen tube growth. After germination, pollen tubes extend by tip growth, i.e. deposition of new cell wall material at the growing point rather than by overall cell wall extension. In growing pollen tubes removal of external boron leads to abnormal swelling or even burst in the tip region within 2–3 min of removal (Schmucker, 1934).

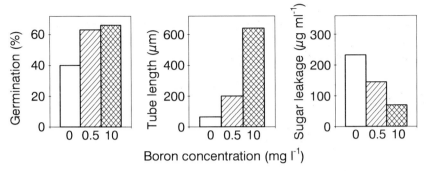

Fig. 9.32 Effect of boron concentrations on lily (*Lilium longiflorum* L.) pollen germination, tube growth, and leakage of sugar to the medium. (Redrawn from Dickinson, 1978.)

As shown in Fig. 9.32 it is not the germination but the tube length that is mostly affected by external boron supply. In addition the plasma membrane integrity is impaired when boron is deficient in the external medium as indicated by the leakage of sugars.

In flowers the boron demand for pollen tube growth has to be provided by the stigma or the silk. In maize a minimum boron content of $3\,\mu g\,g^{-1}$ dry wt in the silk is required for pollen germination and fertilization (Vaughan, 1977). The critical deficiency levels in the stigma may, however, vary considerably between cultivars and species. In grapevine (*Vitis vinifera*), which is known for its high boron requirement, with sufficient boron supply the boron content of the stigma is $50–60\,\mu g\,g^{-1}$ dry wt and even at contents of $8–20\,\mu g\,g^{-1}$ dry wt fertilization is impaired (Gärtel, 1974). According to Lewis (1980b) high boron levels in the stigma and style are required for physiological inactivation of callose from the pollen tube walls by the formation of borate–callose complexes. When the boron contents are low, callose synthesis increases and induces the synthesis of phytoalexins (including phenols) in the stigma and style, as a defense mechanism similar to that in response to microbial infection.

The particular role of boron in pollen tube growth is a major factor responsible for the usually higher demand of boron supply for seed and grain production than that needed for vegetative growth alone. This has been shown to be the case, for example, for maize (Vaughan, 1977; Chapter 6) or white clover (Johnson and Wear, 1967). In mango, irregular and periodic fruit set caused by suboptimal temperatures during pollination can at least in part be compensated by raising the boron content in the pistil and pollen grains (De Wet *et al.*, 1989). Boron also affects fertilization by increasing the pollen-producing capacity of the anthers and pollen grain viability (Agarwala *et al.*, 1981). Indirect effects might also be important such as increase in amount and composition of sugars in the nectar, whereby the flowers of species that rely on pollinating insects become more attractive to insects (Smith and Johnson, 1969; Eriksson, 1979).

9.7.8 Carbohydrate and Protein Metabolism

It had been proposed that boron plays a key role in higher plants by facilitating the short- and long-distance transport of sugars via the formation of borate–sugar complexes (for a review see Dugger, 1983). However, such a proposal is unacceptable because sucrose, the prevalent sugar transported in the phloem forms only weak complexes with boron, and in the mechanisms of phloem loading of sucrose boron is not involved (Section 5.4.1). Whereas boron facilitates sugar uptake by leaves, the export of photosynthates from the leaves is either unaffected (Weiser *et al.*, 1964) or impaired by callose formation in the sieve tubes (Venter and Currier, 1977) or by the lack of sink activity of roots and shoot apices in plants suffering from severe boron deficiency. It has also been shown (Yih and Clark, 1965) that in boron-deficient plants the rate of root elongation decreases regardless of the sugar content of the root tips. Thus, effects of boron on sugar metabolism are merely secondary effects, except on intermediates which influence the partitioning of carbohydrate fluxes between glycolysis and the pentose phosphate cycle (Fig. 9.29).

Reports that boron fertilization increases the carbohydrate content of roots and carbohydrate exudation by roots, thus leading to higher colonization of roots with VA- or ectomycorrhizae (Atalay *et al.*, 1988; Dixon *et al.*, 1989) are interesting, but difficult to interpret as direct effects of boron on sugar translocation. It is more likely that boron acts via alterations in phenol metabolism and thereby specifically on host–microbe recognition and establishment of the symbioses (Section 7.4 and 15.6). Different levels of IAA in mycorrhizal roots of fertilized and nonfertilized plants (Mitchell *et al.*, 1986) would indicate such more specific effects of boron.

There is no convincing evidence of a direct effect of boron on nitrogen metabolism, for example, nitrate reduction, amino acid and protein content, which might be higher or lower in boron-deficient plants, depending upon severity of the deficiency, plant age, and plant organ. Changes are most likely secondary effects, for example, caused by different sink activities and, thus, demand for nitrogen (Shelp, 1993). Leaf metabolism and composition might be affected by boron deficiency indirectly via its effect on cytokinin synthesis in the root tips: when the supply of boron is withheld, both the production and export of cytokinins into the shoots decrease (Wagner and Michael, 1971). In tobacco the same is true for nicotine alkaloids (Scholz, 1958).

9.7.9 Boron Deficiency and Boron Toxicity

9.7.9.1 Boron Deficiency

Boron deficiency is a widespread nutritional disorder. Under high rainfall conditions boron is readily leached from soils as $B(OH)_3$. Boron availability to plants decreases with increasing soil pH, particularly in calcareous soils and soils with a high clay content, presumably as a result of the formation of $B(OH)_4^-$ and anion adsorption. Availability also decreases sharply under drought conditions, probably because of both a decrease in boron mobility by mass flow to the roots (Kluge, 1971) and polymerization of boric acid.

Plant species differ characteristically in their capacity to take up boron when grown in the same soil (Table 9.43), which generally reflects typical species differences in the requirement of boron for growth. For example, the critical deficiency range, expressed as mg boron kg^{-1} dry weight increases from about 5–10 mg in graminaceous species (e.g., wheat) to 20–70 mg in most dicotyledonous species (e.g., clover) to 80–100 mg in gum-

Table 9.43
Boron Content of the Leaf Tissue of Plant Species
from the Same Location[a]

Plant species	Boron content $(\text{mg kg}^{-1} \text{ dry wt})$
Wheat	6.0
Maize	8.7
Timothy	14.8
Tobacco	29.4
Red clover	32.2
Alfalfa	37.0
Brussels sprouts	50.2
Carrots	75.4
Sugar beet	102.3

[a]Based on Gupta (1979).

bearing plants such as poppy (Bergmann, 1988, 1992). For evaluation of critical deficiency levels of boron, the elongation rate of the youngest leaf is a much more suitable parameter than, for example, shoot dry weight (Kirk and Loneragan, 1988). High light intensities increase sensitivity to boron deficiency (MacInnes and Albert, 1969) by raising the requirement for boron in the tissue (Tanaka, 1966). This higher boron requirement is presumably related to elevated phenol contents often observed in plants exposed to high light intensities (Table 9.41) the additional boron being needed to detoxify these compounds by complexation.

The distinct differences in boron demand particularly between graminaceous and dicotyledonous species is most likely causally related to the differences in their cell wall composition. In graminaceous species the primary cell walls contain very little pectic material and have also a much lower calcium requirement (Section 8.6). Interestingly, these two plant groups also differ typically in their capacity for silicon uptake which is usually inversely related to the boron and calcium requirement (Loomis and Durst, 1992). All three elements are mainly located in the cell walls. Although reports on calcium/boron interactions are often inconclusive (Gupta, 1979; Bergmann, 1988, 1992) these interactions are likely to have a physiological basis. Examples include similar structural functions in the cell walls and at the cell wall–plasma membrane interface, interactions in uptake and shoot transport (Section 9.7.4) and in IAA transport (Section 9.7.6). These common features also explain certain similarities in calcium- and boron-deficiency symptoms, for example, in peanut seeds (Cox and Reid, 1964) and lettuce (Crisp *et al.*, 1976).

Symptoms of boron deficiency in the shoots are noticeable at the terminal buds or youngest leaves, which become discoloured and may die. Internodes are shorter, giving the plants a bushy or rosette appearance. Interveinal chlorosis on mature leaves may occur, as might misshaped leaf blades. An increase in the diameter of petioles and stems is particularly common and may lead to symptoms such as 'stem crack' in celery, or 'hollow stem disorder' in broccoli (Shelp, 1988). Drop of buds, flowers, and developing fruits is also a typical symptom of boron deficiency. In the heads of vegetable crops

Fig. 9.33 Boron deficiency in sugar beet. (*Left*) Severe boron deficiency (heart and crown rot). (*Middle*) Mild boron deficiency (heart rot). (*Right*) Boron sufficient. (Courtesy of W. Bussler.)

(e.g., lettuce), water-soaked areas, tipburn, and brown- or blackheart occur. In storage roots of celery or sugar beet, necrosis of the growing areas leads to heart rot (Fig. 9.33). With severe deficiency the young leaves also turn brown and die, subsequent rotting and microbial infections of the damaged tissue being common. In boron-deficient fleshy fruits, not only is the growth rate lower, but the quality may also be severely affected by malformation (e.g., 'internal cork' in apple) or, in citrus, by a decrease in the pulp/peel ratio (Foroughi *et al.*, 1973).

Boron deficiency-induced reduction or even failure of seed and fruit set are well known (see Section 6.3). However, even when seed yield is not depressed in plants grown in a low boron soil, the seeds produced might have a lower quality in terms of viability as shown in Table 9.44 for black gram. Despite the same seed dry weight, the seeds with the lower boron content had a lower viability and produced a high percentage of abnormal seedlings. A boron content of 6 mg kg^{-1} seed dry weight has been considered as critical for growth of normal seedlings in black gram.

Table 9.44

Effect of Boron Fertilization on Yield, Seed Boron Content, Seed Viability and Germination of Black Gram (*Vigna mungo* L.)[a]

Treatment	Seed yield (g dry wt per plant)	B content (mg kg^{-1} seed)	Percentage of seedlings		
			Normal	Weak/abnormal	Non viable
−B	5.0	3.4	57	40	3
+B	5.1	7.4	92	6	2

[a]Based on Bell *et al.* (1989).

For the application of boron either to the soil or as a foliar spray, different sodium borates, including borax or sodium tetraborate, can be used. Boric acid or sodium borate are effective as foliar sprays, for example, to increase flower and fruitset in fruit trees (Hanson, 1991b) or in soybean. In the latter case foliar application also simultaneously lowers the risk of boron toxicity (Schon and Blevins, 1990). The amount of boron applied varies from 0.3 to 3.0 kg ha^{-1} depending on the requirement and sensitivity of the crop to boron toxicity. The narrow concentration range between boron deficiency and toxicity requires special care in the application of boron fertilizers.

9.7.9.2   Boron Toxicity and Tolerance

Boron toxicity is most common in arid and semiarid regions in plants growing on soils formed from parent material of marine origin, or related to the use of irrigation water high in boron (Nable and Paull, 1991). Boron toxicity might also occur when large amounts of municipal compost are applied (Purves and Mackenzie, 1974). Plant species, and to some extent also cultivars within a species, differ much in their boron tolerance. For example, the critical toxicity contents in leaves are in the range of (mg kg^{-1} dry wt): maize, 100; cucumber, 400; squash, 1000; and 100–270 in wheat genotypes (El-Sheikh et al., 1971; Paull et al. 1988), and about 100 in snap bean and over 330 in cowpea (Francois, 1989).

Typical boron toxicity symptoms on mature leaves are marginal or tip chlorosis or both and necrosis. They reflect the distribution of boron in shoots, following the transpiration stream (Section 3.2.4). Visual symptoms of boron toxicity on leaves might occur at much lower boron contents than required for depression of grain yield, for example, in wheat (Kluge, 1990). The boron content in wheat grains can also be raised more than twentyfold without negative effects on seed germination and seedlings growth (Nable and Paull, 1990).

Critical toxicity contents of boron in leaves have to be interpreted with reservation for various reasons. As has been shown (Section 3.2.4) there is a steep gradient in boron content within a leaf blade. In barley this gradient from the base to the tip of the leaf blade is from about 80 to 2500 μg B g^{-1} dry weight, but the average for this leaf is 208 μg g^{-1} (Nable et al., 1990c). Furthermore, the critical toxicity contents are often lower in field-grown plants compared with plants grown in a greenhouse. This difference is partially related to leaching of boron from leaves by rain (Nable et al., 1990c).

The physiology of boron tolerance and boron toxicity is not well understood. As a first approximation it appears that there is a close positive correlation between critical deficiency and toxicity contents for a wide range of plant species. Species with high demand might also have a higher capacity to sequester boron in the cell walls (Fig.9.31). When supply is excessive, inactivation as soluble complexes seems to be less important, with the exception of certain halophytes which might use compatible solutes (Section 16.6.4.4) such as sorbitol for this purpose (Rozema et al., 1992). At very high sugar concentrations fructose may also form borate complexes and thereby alleviate boron toxicity on growth of pollen tubes or tobacco cells (Yokota and Konishi, 1990). If these detoxification mechanisms become limiting, boron concentration in the cytosol might rise and cause metabolic disturbances by formation of complexes with NAD$^+$, or ribose of RNA, for example (Loomis and Durst, 1992), or specifically inhibit the ureide metabolism in the leaves of nodulated soybean (Lukaszewski *et al.*, 1992).

Within species such as barley, wheat, annual medics (*Medicago* spp.) and field peas (*Pisum sativum* L.), large genotypical differences exist in capacity to tolerate high boron concentrations in the soil or nutrient solution (Nable and Paull, 1991; Paull *et al.* 1992b). These differences are based on restrictions in boron uptake by the roots and transport into the shoot and not on high boron tolerance of the tissue (Nable, 1991). This is obviously a different mechanism compared with tomato, where root-to-shoot transport of boron rather than uptake was the most distinct difference between genotypes (Brown and Jones, 1971). Interestingly, in barley, genotypical differences in restriction of uptake by roots and transport of boron into the leaves are closely correlated with similar restrictions in uptake and transport of silicon in the respective genotypes (Nable *et al.*, 1990b).

In barley the differences in capacity to reduce boron uptake are already well defined genetically (Paull *et al.*, 1988; Nable and Paull, 1991), and obviously based on restricted passive permeation of boron through the plasma membrane of root cells (Huang and Graham 1990), and not on differences in root anatomy or transpiration rates (Nable and Paull, 1991). The mechanism causing the restriction is unclear and not demand-regulated. In barley this restriction in uptake holds true over the whole range of applied boron concentrations (Nable *et al.*, 1990b; Section 5.3).

9.8   Chlorine

9.8.1   General

Chlorine is ubiquitous in nature, and occurs in aqueous solution as the monovalent ion chloride (Cl^-). Its salts are readily soluble, the mobility of chloride in the soil is high, and its concentration in the soil solution varies over a wide range. Chloride is readily taken up by plants and its mobility in short- and long-distance transport is high. In plants chlorine occurs mainly as a free anion or is loosely bound to exchange sites. However, in higher plants more than 130 chlorinated organic compounds have also been found (Engvild, 1986). With a few exceptions (Section 9.8.5), the importance of these compounds in terms of functional requirement of chlorine for higher plants is not known. Average contents of chlorine in plants are in the range of 2–20 mg g^{-1} dry matter which is typical of the content of a macronutrient. In most plant species the chlorine requirement for optimal plant growth, however, is in the range of 0.2–0.4 mg g^{-1} dry matter, i.e., about 10 to 100 times lower. Because chlorine is usually supplied to plants as chloride from various sources (soil reserves, irrigation water, rain, fertilizers, air pollution), on a worldwide basis there is much more concern about chlorine toxicity (Section 16.6.3) than deficiency in plants. Indeed, to induce chlorine deficiency, in most plant species particular precautions are required to reduce the 'contamination' by chlorine from seeds, chemicals, water, and air. Using these precautions Broyer *et al.* (1954) were able to demonstrate the requirement of chlorine as a micronutrient for higher plants. For a more recent tabulation of plant species where chlorine deficiency has been demonstrated the reader is referred to Flowers (1988).

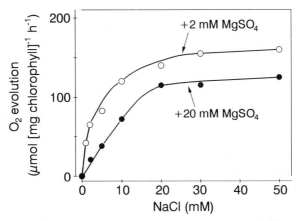

Fig. 9.34 O_2 evolution of chloride-depleted PSII particles of spinach chloroplasts as affected by varied concentrations of NaCl and $MgSO_4$. (From Itoh and Uwano, 1986.)

9.8.2 Photosynthetic O_2 Evolution

In 1946 Warburg and Lüttgens showed that the water-splitting system of photosystem II (PS II) requires chlorine, and since that time the involvement of chlorine in the splitting of water at the oxidizing site of PS II, i.e., for O_2 evolution (Section 5.2) has been confirmed in a large number of studies with chloroplast fragments. An example of this is shown in Fig. 9.34. In chlorine-depleted PS II particles of spinach chloroplasts there is a steep increase in photosynthetic O_2 evolution with increase in external chloride supply. However, other anions such as sulfate decrease the efficiency of chloride (Fig. 9.34) indicating a low selectivity of the binding sites for chloride at PS II.

In principle similar results showing chlorine-dependency of O_2 evolution have been found for the manganese-dependency, for example, by Ball *et al.* (1984) with chloroplast fragments of sugar beet. It is astonishing that besides manganese a simple anion like chloride can play such a fundamental role in the water-splitting system of PS II. Chloride might either act as a bridging ligand for stabilization of the oxidized state of manganese (Critchley, 1985), or as a structural component of the associated (extrinsic) polypeptides (Coleman *et al.*, 1987). A model of such a structural role of chloride is shown in Fig. 9.35. The various polypeptides (33 kDa, 24 kDa, 18 kDa) attached to

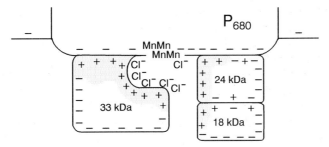

Fig. 9.35 Schematic presentation of the membrane surface of the water-splitting complex of PS II (P_{680}) with extrinsic polypeptides. (From Itoh and Uwano, 1986.)

PS II provide the charges for chloride binding, and chloride on the other hand protects the polypeptides from dissociation (Homann, 1988). The external chloride concentration required for maximal O_2 evolution varies widely, it is as high as 250 mM in pure thylakoid vesicles devoid of the three polypeptides. Resupply of the polypetides resulted in their re-arrangement and reduced the requirement of the external chloride concentration to about 5 mM (Andersson *et al.*, 1984). This is one of the reasons why the necessity for chlorine as a cofactor in the water-splitting system of PS II in vivo has been critically reviewed (Terry, 1977; Wydrzynski *et al.* 1990). It is also argued that calcium (Ca^{2+}) might play a more important structural role in PS II than chloride (Renger and Wydrzynski, 1991).

Unfortunately experiments with intact chloroplasts are inconclusive for proving the essentiality of chlorine in photosynthetic O_2 evolution, as chlorine contents are relatively high even in chloroplasts from chlorine-deficient plants. Based on their studies with halophytic and nonhalophytic plant species, including spinach, supplied with different chloride concentrations, Robinson and Downton (1985, 1986) concluded that chloroplasts maintain a rather high chloride concentration (80–90 mM) regardless of the plant species and external supply. This is in contrast to results for sugar beet (Terry, 1977) and for spinach (Van Steveninck *et al.*, 1988). By using X-ray microanalysis it was shown that the chloride concentrations in the spinach chloroplasts were closely related to the external concentration and fell as low as 1.4 mM at low external supply. This would imply that in intact chloroplasts the chloride requirement of PS II for O_2 evolution is in the range of 1 mM or lower, unless it is strictly compartmented (Fig. 9.35).

9.8.3   Tonoplast Proton-Pumping ATPase

Membrane-bound proton-pumping ATPases and PP$_i$ases are stimulated by various cations and anions. The importance of these pumps for pH regulation of the cytosol and ion uptake in roots has been discussed in Section 2.4. Whereas the proton-pumping ATPase at the plasma membrane is stimulated by monovalent cations, K^+ in particular, the proton-pumping ATPase at the tonoplast is not affected by monovalent cations but specifically stimulated by chloride (Churchill and Sze, 1984). An example showing chloride stimulation is given in Table 9.45. Bromide is somewhat less effective, sulfate has an inhibitory effect. Nitrate either stimulates the pump only slightly (Table 9.45) or even inhibits its activity (O'Neill *et al.*, 1983).

The close relationship between KCl supply and ATPase activity in roots (Section 8.7) is therefore a reflection of two different mechanisms located at different membranes:

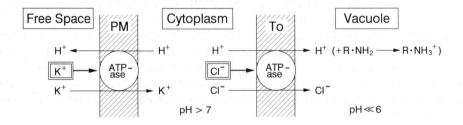

Table 9.45
Effect of Salts on the Proton-Pumping ATPase of
Tonoplast Vesicles[a]

Salt (10 mM monovalent ion)	ATPase stimulation (% of control)
No monovalent ion	10
KCl (control)	100
NaCl	102
NaBr	87
KNO$_3$	21
K$_2$SO$_4$	3

[a]Based on Mettler *et al.* (1982).

There are also striking similarities between the chloride-stimulated H$^+$-ATPase and the mechanisms regulating elongation of coleoptiles (Hager and Helmle, 1981). Severe inhibition of root elongation in chlorine-deficient plants might also be related to this function of chloride. The lack of stimulation of the tonoplast H$^+$-ATPase by nitrate is probably an important regulatory mechanism ensuring low rates of nitrate accumulation in the vacuoles of root cells and thereby rapid nitrate transport into the shoot, the main site of nitrate reduction. On the other hand the preferential transport of chloride into the vacuoles of root cells allow this ion to function as an osmotically active solute within the roots. These osmoregulatory functions require high concentrations of chlorine and are difficult to reconcile with the functions of chlorine as a micronutrient, unless these osmotic functions are confined to only particular parts of the plants or cell compartments (Section 9.8.6).

9.8.4   Stomatal Regulation

Chlorine can play an essential role in stomatal regulation. Opening and closure of stomata is mediated by fluxes of potassium and accompanying anions such as malate and chloride (Section 8.7.6.2). In plant species such as *Allium cepa* which lacks in the guard cells functional chloroplasts for malate synthesis, chloride is essential for stomatal functioning, and stomatal opening is inhibited in the absence of chloride (Schnabl, 1980). Members of the *Palmaceae* such as coconut (*Cocus nucifera* L.) and oil palm (*Elaeis guineensis* Jacq.) which might possess chloroplasts containing starch in their guard cells (Braconnier and d'Auzac, 1990) also require chloride for stomatal functioning. In coconut a close correlation occurs between potassium and chloride fluxes during stomatal opening from the subsidiary cells into the guard cells and, vice versa, during stomatal closure; in chlorine-deficient plants stomatal opening is delayed by about 3 h (Braconnier and d'Auzac, 1990). Impairment of stomatal regulation in palm trees is considered to be a major factor responsible for growth depression and wilting symptoms in chlorine deficient plants (von Uexküll, 1985; Braconnier and d'Auzac, 1990).

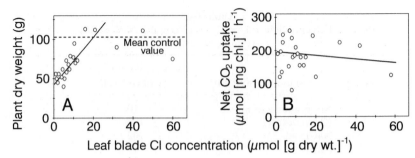

Fig. 9.36 Effect of chlorine deficiency on growth (A) and photosynthesis (B) of sugar beet. (From Terry, 1977.)

9.8.5 Chlorine Supply and Plant Growth

Besides wilting symptoms, in most plants the principal effect of chlorine deficiency is a reduction in leaf surface area and thereby plant dry weight (Fig. 9.36). This decrease in leaf area is the result of a reduction in cell division rates (Terry, 1977) and cell extension, and not of net photosynthesis per unit chlorophyll, indicating a lower chlorine requirement for photosynthetic O_2 evolution than for other chlorine-dependent processes. In sugar beet the critical deficiency content in leaf blades is in the range of 20 (Fig. 9.36) and 50 (Ulrich and Ohki, 1956) μmol Cl g^{-1} dry weight, or 0.7 and 1.7 mg Cl g^{-1} dry weight, respectively.

   Plant species plays a major role in determining the critical deficiency content of chlorine in the shoot dry weight, and growth depression when chlorine supply is interrupted as shown in Fig. 9.37 for various plant species grown in nutrient solutions under controlled environmental conditions. By withholding chlorine supply, growth was not affected in squash, but drastically reduced in lettuce. Resupplying chlorine to the deficient plants restored growth within a few days.

   Growth reduction and chlorine deficiency symptoms could be restored to 90% of the levels in plants adequately supplied with chloride by supplying bromide (Broyer, 1966).

	Relative dry weight (Control + Cl⁻ = 100)	Content of deficient plants (μg Cl [g shoot dry wt.]$^{-1}$)
	0 50 100	
Lettuce		142
Tomato		230
Sugar beet		60
Barley		142
Maize		106
Bean		181
Squash		103

Fig. 9.37 Relative shoot dry weight and chlorine content of chlorine-deficient plants. (Redrawn from Johnson *et al.*, 1957.)

Table 9.46
Effect of Chloride Supply on Chlorine Content in the Youngest Leaf
and on Growth of Kiwifruit (*Actinidia deliciosa*)[a]

Chloride supply (μM)	Content in youngest leaf (mg g^{-1} dry wt)	Total dry weight (g per plant)	Main leaf area (m^2 per leaf)
0	0.7	8	0.17
350	1.5	32	0.41
700	2.1	37	0.50
1400	4.0	34	0.43

[a]Based on Smith *et al.* (1987).

Chloride and bromide have similar physicochemical properties; for example, their hydrated ionic radii are nearly the same: 0.332 nm (Cl^-) and 0.330 nm (Br^-). Substitution of chloride by bromide is of no practical significance, however, because of the difference in their natural abundances. In the earth's crust, the sea, and the air, as well as in plants, chlorine is ~1000 times more abundant than bromine (McClendon, 1976).

Compared with most other plant species (with the exception of palm trees), kiwifruit (*Actinidia deliciosa*) has a much higher chlorine requirement (Table 9.46). In chlorine-deficient plants dry weight and leaf size are drastically reduced and interveinal chlorosis occurs on the mature leaf blades. The critical deficiency content in leaves is about 2 mg Cl g^{-1} dry weight and, thus, chlorine deficiency can readily be induced in this species. The reasons for the high chlorine requirement of kiwifruit are obscure. In the experiment shown in Table 9.46, the chlorine effects on growth were not related to changes in cation–anion balance in the plants, as increasing chloride contents in the leaves were counterbalanced by equimolar decreases in nitrate content (Smith *et al.*, 1987).

Not much is known of a specific role of chlorine as a micronutrient, for example, in cell division and extension, or in nitrogen metabolism. The contents of certain amino acids and amides are exceptionally high in chlorine-deficient cabbage and cauliflower plants (Freney *et al.*, 1959) as a result of either inhibition of synthesis or enhancement degradation of proteins. A role of chlorine in nitrogen metabolism is indicated by its stimulating effect on asparagine synthetase, which uses glutamine as a substrate:

$$\text{Glutamine} \xrightarrow[\text{Asparagine synthetase}]{\text{(NH}_3\text{)}} \text{Asparagine} + \text{Glutamic acid}$$

Either chloride or bromide enhance this transfer by a factor of ~7, whereas sulfate has an inhibitory effect. Furthermore, chloride increases the affinity of the enzyme for the substrate by a factor of 50 (Rognes, 1980). In plant species in which asparagine is the major compound in the long-distance transport of soluble nitrogen (Section 8.2), chloride might thus also play a role in nitrogen metabolism.

Some of the chlorine-containing organic compounds in plants have a biological activity as high as antibiotics and fungicides (Engvild, 1986). A particular effect of chlorine on extension growth may be supposed in some legume species such as peas and faba bean which contain substantial amounts of chlorinated IAA in their seeds. This compound enhances hypocotyl elongation at 10 times the rate of IAA itself, probably because of its higher resistance against degradation by peroxidases (Hofinger and Böttger, 1979).

9.8.6 Chlorine Supply and Osmoregulation

When expressed on the basis of the plant water content, a critical deficiency content of 0.2% in the dry weight represents 60 μmol Cl g^{-1} dry weight, or a concentration of about 6 mM Cl$^-$. This concentration is too low to be of general importance in osmoregulation of the bulk plant tissue, unless chlorine is preferentially accumulated in certain tissues (e.g., extension zones) or cells (e.g., guard cells). As a rule, however, chlorine concentrations in plants exceed this critical deficiency level by two orders of magnitude and become important in osmotic adjustment and plant water relations (Flowers, 1988) including a role in xylem volume flow and root pressure (Section 3.2). In this concentration range chloride represents the dominant inorganic anion in the vacuole. In the phloem sap chloride concentrations might be in the order of 120 mM and seem to play a role in phloem loading and unloading of sugars. This is the case for example, in barley leaves (Fromm and Eschrich, 1989), and in phloem unloading in the pulvini of *Mimosa pudica* during seismonastic leaf movement; in the latter process unloading of chloride is accompanied by potassium and sugars (Fromm and Eschrich, 1988).

Chloride, together with potassium, has a particular function in osmoregulation in the grass stigma (Heslop-Harrison and Reger, 1986). The stigma of grasses such as *Pennisetum americum* L. often extend within minutes at anthesis by cell elongation and this is mainly mediated by rapid transfer of potassium and chloride from the surrounding tissue into the stigma primordium.

In conclusion, chloride has important functions in osmoregulation at different levels. At the usually high plant contents it is a main osmoticum in the vacuoles of the bulk tissue (50–150 mM Cl$^-$), together with potassium. At low contents which are in the range of a micronutrient (~1 mM Cl$^-$ or below), these osmoregulatory functions of chloride are presumably confined to specialized tissues or cells, such as the extension zones of roots and shoots, pulvini and stigma, and guard cells, where the chloride concentrations might be much higher than the average of the bulk tissue. The stimulation by chloride of the proton-pumping ATPase at the tonoplast is in accordance with the particular role of chloride in osmoregulation.

9.8.7 Chlorine Deficiency and Toxicity

Wilting of leaves, especially at leaf margins, is a typical symptom of chlorine deficiency, even in water culture, when plants are exposed to full sunlight (Broyer *et al.*, 1954). With severe deficiency curling of the youngest leaves followed by shrivelling and necrosis might occur (Whitehead, 1985). In palm trees which have a particularly high chloride requirement of about 6 mg Cl g^{-1} leaf dry weight (Ollagnier and Wahyuni,

Table 9.47

Relationship between Chlorine Content in Leaves and Growth Disorders in
Coconut (*Cocos nucifera* L.) Trees[a]

Fertilization (kg KCl per tree)	Leaf content (% in dry matter)		Growth disorders (%)	
	K	Cl	Frond fracture	Stem cracking
0	1.61	0.07	11.6	27.0
2.25	1.64	0.41	1.7	8.1
4.50	1.66	0.51	1.2	4.5

[a]Based on von Uexküll (1985).

1986), besides wilting and premature senescence of leaves, frond fracture and stem cracking are typical symptoms of chlorine deficiency (Table 9.47).

In leaves and roots, besides cell division, cell extension is particularly impaired in deficient plants, and in roots this is associated with subapical swelling (Smith *et al.* 1987; Bergmann, 1988) and enhanced formation of short laterals, giving the roots a stubby appearance (Johnson *et al.*, 1957).

In plant species such as red clover with relatively low chloride requirements (<1 mg Cl g^{-1} leaf dry wt) the demand can be covered by a concentration of 100 μM Cl$^-$ in the nutrient solution; at 10 μM Cl$^-$ supply the shoot dry weight drops to 50% (Chisholm and Blair, 1981), indicating that the selectivity of chlorine uptake is not very high as compared, for example, with phosphorus, where the much higher requirement in the leaf dry weight (Section 8.4.6) can be covered by supply of even less than 10 μM.

The question arises as to the occurrence of chlorine deficiency under field conditions. Assuming a critical deficiency content of 1 mg Cl g^{-1} shoot dry weight, the crop requirement would be in the range 4–8 kg Cl ha^{-1}, which is about the input from rain in areas far distant from oceans, and about 10 times lower than the input from rain at sites near oceans. However, in highly leached soils with a low chlorine input from rain and other sources, chlorine deficiency might occur even in plant species with low chlorine requirements (Ozanne, 1958). The probability of chlorine deficiency and, thus response to chloride fertilizers, increases in plant species with a high chlorine requirement such as kiwifruit (Smith *et al.*, 1987; Buwalda and Smith, 1991) and palm trees in particular (Ollagnier and Wahyuni, 1986; Braconnier and d'Anzac, 1990; Table 9.46).

There are also reports of field experiments with wheat and other cereals (which have a relatively low chlorine requirement) in which increases in grain yield were obtained by chloride fertilization (e.g., KCl instead of K$_2$SO$_4$). These yield increases might be a combination of various effects, including alleviation of chlorine deficiency (Fixen *et al.*, 1986b), suppression of root rot diseases (Timm *et al.*, 1986), or a combination of suppression of diseases and improving plant water relations (Fixen *et al.* 1986a). In kiwifruit, enhanced potassium uptake rates and an improvement in potassium nutritional status might be additional side effects of chloride fertilizer application (Buwalda and Smith, 1991).

Compared with chlorine deficiency, toxicity is of much more worldwide occurrence and a general stress factor limiting plant growth particularly in arid and semiarid regions. On average, concentrations of chloride in the external solution of more than 20 mM can lead to toxicity in sensitive plant species, whereas in tolerant species the external concentration can be four to five times higher without reducing growth. These differences are related mainly to differences in the sensitivity of leaf tissue to excessive chloride contents. More than 3.5 mg Cl g^{-1} leaf dry weight ($\sim$10 mM Cl^- in the leaf water) are toxic to sensitive species, such as most fruit trees, and to bean and cotton. In contrast, 20–30 mg Cl g^{-1} leaf dry weight ($\sim$60–90 mM Cl^- in the leaf water) are not harmful to tolerant species such as barley, spinach, lettuce, and sugar beet. Genotypical differences in chlorine tolerance are closely related to salt tolerance mechanisms, which are discussed in Section 16.6.

10

Beneficial Mineral Elements

10.1 Definition

Mineral elements which either stimulate growth but are not essential (for a definition of essentiality see Chapter 1) or which are essential only for certain plant species, or under specific conditions, are usually defined as *beneficial elements*. This definition applies in particular to sodium, silicon, and cobalt. Making the distinction between beneficial and essential is especially difficult in the case of some trace elements. Developments in analytical chemistry and in methods of minimizing contamination during growth experiments may well lead in the future to an increase in the list of essential micronutrient elements and a corresponding decrease in the list of beneficial mineral elements. Nickel is the most recent example of this development.

10.2 Sodium

10.2.1 General

The sodium content of the earth's crust is ~2.8%, compared with 2.6% for potassium. In temperate regions the sodium concentration in the soil solution is on average 0.1–1 mM, thus similar to or higher than the potassium concentration. In semiarid and arid regions, particularly under irrigation, concentrations of 50–100 mM Na^+ (mostly as NaCl in the soil solution) are typical and have a rather detrimental effect on the growth of most crop plants (Section 16.6). The hydrated sodium ion (Na^+) has a radius of 0.358 nm, whereas that of the potassium ion (K^+) is 0.331 nm. Most higher plants have developed high selectivity in the uptake of potassium as compared with that of sodium, and this is particularly obvious in transport to the shoot (Chapter 3). Plant species are characterized as *natrophilic* or *natrophobic* depending on their growth response to sodium and their capacity for uptake by roots and long-distance transport of sodium to the shoots. The differences in capacity of sodium uptake by roots and translocation to the shoots are large between plant species, but also substantial between genotypes within a species. Genotypical differences in uptake by roots are related to several factors such as different activities/capacities of sodium efflux pumps (Section 2.5.3), of passive sodium permeability of the root plasma membranes (Schubert and Läuchli,

1990), but presumably not to differences in response of the root plasma membrane-bound ATPase to sodium (Mills and Hodges, 1988).

For the role of sodium in mineral nutrition of plants, three aspects have to be considered: its essentiality for certain plant species, the extent to which it can replace potassium functions in plants, and its additional growth enhancement effect.

10.2.2 Essentiality; Sodium as Mineral Nutrient

It was established in 1965 by Brownell that sodium is an essential mineral element, i.e., a mineral nutrient, for the halophyte *Atriplex vesicaria*. When contamination with sodium in the basic nutrient solution was kept to a minimum (below $0.1 \mu M$ Na^+), plants became chlorotic and necrotic and no further growth occurred, despite a high potassium content in the plants (Table 10.1). The growth response to sodium in the low concentration range (0.02 mM) was quite dramatic, although the sodium content ($\sim 0.1\%$ in the dry wt) was in a range more typical for a micronutrient. At higher supply, however, the sodium content was at a level more typical of a macronutrient, growth responses in this latter case, presumably therefore being more related to the functions of potassium such as in osmoregulation.

In further studies on various halophytes and nonhalophytes (glycophytes) responses to sodium, similar to those shown in Table 10.1 were found in species characterized by the C_4 photosynthetic pathway (Brownell and Crossland, 1972) and the CAM pathway (Brownell and Crossland, 1974). In the absence of sodium supply all C_4 species grew poorly and showed visual deficiency symptoms such as chlorosis and necrosis, or even failure to form flowers. Supply of $100 \mu M$ Na^+ enhanced growth and alleviated the visual symptoms. According to these studies and their later confirmation (Johnston *et al.*, 1988) sodium may be classified as a mineral nutrient for at least some of the C_4 species in the families Amaranthaceae, Chenopodiaceae and Cyperaceae (Brownell, 1979), the amounts of sodium required by these plant species being more typical of those for a micronutrient rather than a macronutrient. However, the conclusion of Brownell and Crossland (1972) and Brownell (1979) that sodium is essential for higher plant species in

Table 10.1

Effect of Sodium Sulfate Concentrations on the Growth and Sodium and Potassium Content of Leaves of *Atriplex vesicaria* L.[a]

Treatment (mM Na^+)	Dry wt (mg per 4 plants)	Content of leaves (mmol kg^{-1} dry wt)	
		Na	K
None	86	10	2834
0.02	398	48	4450
0.04	581	78	2504
0.20	771	296	2225
1.20	1101	1129	1688

[a]From Brownell (1965). The basic nutrient solution contained 6 mM potassium.

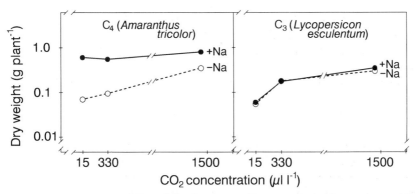

Fig. 10.1   Growth responses of C_4 and C_3 plants to sodium supply and increasing ambient CO_2 concentrations. (Based on Johnston *et al.*, 1984).

which the C_4 pathway is operative is not justified . In all these studies C_4 species such as maize or sugarcane have not been included, i.e., species which are typically natrophobic and have similar growth rates in the absence and presence of sodium supply (Hewitt, 1983). According to present knowledge sodium is essential for many, but not all C_4 species, and it is not essential for C_3 species.

Many halophytes, whether C_3 or C_4 species are distinctly enhanced in growth by high sodium concentrations in the substrate (10–100 mM Na^+). Even for extreme halophytes, however, sodium does not function as a macronutrient (Flowers et al. 1977). Growth responses of halophytes to sodium merely reflect a high salt requirement for osmotic adjustment (Flowers and Läuchli, 1983), a process in which sodium can be much more suitable than potassium (Eshel, 1985).

10.2.3   Role in C_4 Species

The principle of the C_4 photosynthetic pathway is the shuttle of metabolites between mesophyll and bundle sheath cells (Section 5.2.4) and the increase in concentration of CO_2 in the bundle sheath cells to optimize the Calvin cycle (Section 9.4). This advantage of C_4 plants over C_3 plants, becomes particularly evident therefore at low ambient CO_2 concentrations, provided the C_4 plants are supplied with sodium (Fig. 10.1). In the shoots of *A. tricolor* only sodium contents as low as 0.02% in the shoot dry weight were needed to bring about this high efficiency in CO_2 utilization at low ambient concentrations. However, in the sodium-deficient *A. tricolor* plant growth was poor and chlorosis severe at low ambient CO_2 concentration. Increasing ambient CO_2 concentrations enhanced growth as in the C_3 species tomato, where sodium effects on growth or CO_2 utilization were absent.

These different growth response curves in *A. tricolor* suggest that in sodium-deficient C_4 plants the CO_2 concentration mechanism is impaired or not operating. For the mechanism to be operative an extensive flow of metabolites between mesophyll and bundle sheath cells is required mediated through plasmodesmata and driven by the concentration gradient of the metabolites in the cytosol:

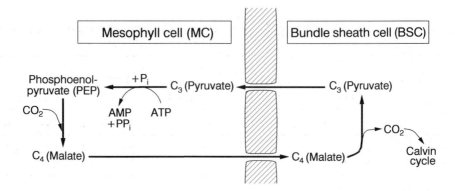

There are different types of the C_4 plants in terms of major enzymes for decarboxylation in the bundle sheath cells and in the major substrates moving between bundle sheath and mesophyll cells (Table 10.2). Hewitt (1983) suggested that the sodium requirement of C_4 plants is confined to the NAD^+-malic enzyme (ME) and PEP-carboxykinase types. This is now known not to be the case, and it is not possible to classify sodium requirement in relation to different C_4 types (Ohta *et al.* 1988; Ohnishi *et al.*, 1990).

Sodium deficiency seems particularly to impair the conversion of pyruvate into PEP which takes place in the mesophyll chloroplasts and has a high energy requirement provided by photophosphorylation. Under sodium deficiency in the C_4 species *A. tricolor*, which is an NAD^+-ME type, C_3 metabolites alanine and pyruvate were found to accumulate whereas the C_4 metabolites PEP, malate, and aspartate decrease (Table 10.3). In contrast, in tomato (a C_3 species) none of these metabolites is influenced by sodium, further suggesting that the functioning of the mesophyll chloroplasts is impaired in C_4 plants under sodium deficiency. In agreement with this, in sodium-deficient *A. tricolor* and *Kochia childsii* the activity of the PS II in the mesophyll chloroplasts was reduced and the ultrastructure drastically altered, whereas none of the parameters were affected in the bundle sheath chloroplasts (Johnston *et al.*, 1989; Grof

Table 10.2
Variations in the Biochemistry of C_4 Photosynthesis found in Specific C_4 Plants[a]

Major BSC decarboxylases	Energetics of decarbox. in BSC	Major substrates moving from[b]		Representative species
		MC→BSC	BSC→MC	
$NADP^+$ ME^c	Production 1 $NADPH/CO_2$	Malate	Pyruvate	*Zea mays* *Digitaria sanguinalis*
NAD^+ ME	Production 1 $NADH/CO_2$	Aspartate	Alanine/ pyruvate	*Atriplex spongiosa* *Portulaca oleracea*
PEP Carboxykinase	Consumption 1 ATP/CO_2	Aspartate	PEP	*Panicum maximum* *Sporobolus poiretti*

[a]From Ray and Black (1979).
[b]MC = mesophyll chloroplasts; BSC = bundle sheath chloroplasts.
[c]ME = malic enzyme.

Table 10.3
Effect of Sodium Nutrition ($-Na$ = O; $+Na$ = 0.1 mM Na^+) on Some Metabolites in Shoots of
Amaranthus tricolor (C_4) and *Lycopersicon esculentum* (C_3)[a]

	Content (μmol g^{-1} fresh weight)									
	Alanine		Pyruvate		PEPyruvate		Malate		Aspartate	
Species	$-Na$	$+Na$	$-Na$	$+Na$	$-Na$	$+Na$	$-Na$	$+Na$	$-Na$	$+Na$
A. tricolor	13.1	6.0	1.7	0.9	0.9	2.3	2.7	4.8	1.6	3.7
L. esculentum	2.5	2.6	0.1	0.1	0.2	0.2	11.3	11.3	1.9	1.9

[a]Based on Johnston *et al.* (1988).

et al., 1989). Resupplying sodium restored PS II activity and altered the metabolite level within less than three days.

The mechanism is obscure as to how sodium affects metabolism and fine structure in the mesophyll chloroplasts of responsive C_4 species. Protection from photodestruction might be involved (Grof *et al.*, 1989). In C_4 species not only the CO_2 scavenging system but also nitrate assimilation is confined to the mesophyll cells (Section 8.2.1). In agreement with this in C_4 species such as *A. tricolor* nitrate reductase activity is very low in leaves of sodium deficient plants and can be restored within less than two days after resupplying sodium (Ohta *et al.*, 1987). Sodium specifically enhances nitrate uptake by the roots and nitrate assimilation in the leaves (Ohta *et al.*, 1989). Interestingly, stimulation of nitrate reductase activity and growth enhancement were absent when NH_4-N was provided or when nitrate supply was combined with addition of tungsten, an inhibitor of nitrate reductase. Thus, in sodium-deficient C_4 species, particularly of the aspartate type, nitrogen deficiency might be an additional factor involved in impairment of the functioning of the C_4 pathway (see also Section 5.2).

A new insight into the role of sodium in mesophyll chloroplasts of different types of C_4 species has been provided (Ohnishi and Kanai, 1987; Ohnishi *et al.*, 1990) from experiments using isolated chloroplasts (Fig. 10.2). In chloroplasts of *P. miliaceum*

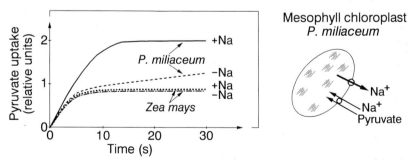

Fig. 10.2 Effect of sodium (1 mM NaCl) on pyruvate uptake into mesophyll chloroplasts of *Panicum miliaceum* (NAD^+-malic enzyme type) and *Zea mays* ($NADP^+$-malic enzyme type), and proposed Na^+/pyruvate co-transport in *P. miliaceum*. (Based on Ohnishi *et al.*, 1990.)

sodium enhanced pyruvate uptake rate with a stoichiometry of about 1:1, suggesting Na^+/pyruvate cotransport through the envelope into the chloroplast, driven by a light-stimulated Na^+ efflux pump (Fig. 10.2). In contrast, in mesophyll chloroplasts of *Zea mays* such a sodium effect on pyruvate uptake rates was absent. Based on additional evidence it is supposed that in C_4 species of the $NADP^+$-ME type (Table 10.2) such as *Zea mays* and *Sorghum bicolor* an H^+/pyruvate cotransport operates rather than the Na^+/pyruvate cotransport in the envelope of mesophyll chloroplasts (Ohnishi *et al.*, 1990). This result further stresses the necessity of differentiating between the various C_4 metabolic types in studies on the role of sodium and the need to include species for which sodium is not essential for growth and various metabolic functions in the C_4 photosynthetic pathway.

10.2.4   Substitution of Potassium by Sodium

The beneficial effects of sodium on the growth of nonhalophytes (glycophytes) are well known in agriculture and horticulture (for reviews, see, e.g., Lehr, 1953; Marschner, 1971). In general, plant species can be classified into four groups according to the differences in their growth response to sodium (Fig. 10.3).

In group A not only can a high proportion of potassium be replaced by sodium without an effect on growth; additional growth stimulation occurs which cannot be achieved by increasing the potassium content of the plants. In group B specific growth responses to sodium are observed, but they are much less distinct. Also, a much smaller proportion of potassium can be replaced without a decrease in growth. In group C substitution can only take place to a very limited extent and sodium has no specific effect on growth. In group D no substitution of potassium is possible. This classification cannot be used in a strict sense, of course, because it does not take into account, for example, differences between cultivars within a species in the substitution of potassium

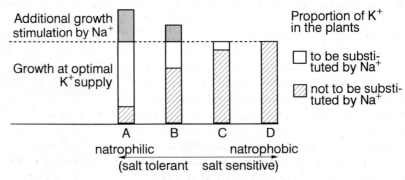

Fig. 10.3 Tentative scheme for the classification of crop plants according to both the extent to which sodium can replace potassium in plants, and additional growth stimulation by sodium. Group A: mainly members of Chenopodiaceae (e.g., sugar beet, table beet, turnip, swiss chard) and many C_4 grasses (e.g., Rhodes grass). Group B: cabbage, radish, cotton, pea, flax, wheat, and spinach. Group C: barley, millet, rice, oat, tomato, potato, and ryegrass. Group D: maize, rye, soybean, *Phaseolus* bean, and timothy.

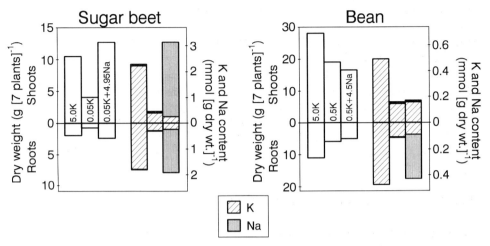

Fig. 10.4 Dry weight and potassium and sodium content of sugar beet (cv. Sharpes Klein E type) and bean (cv. Windsor Long Pod) grown in nutrient solutions with different concentrations of potassium and sodium. Concentrations in mM indicated in the columns. (Based on Hawker *et al.*, 1974.)

by sodium. These differences can be substantial, as has been shown in tomato (Figdore *et al.*, 1987, 1989).

As a rule, the differences in the growth responses of natrophilic and natrophobic species to sodium are related to differences in uptake, particularly in the translocation of sodium to the shoots (Chapter 3). An example of this is shown in Fig. 10.4 for sugar beet and bean. In sugar beet sodium is readily translocated to the shoots, where it replaces most of the potassium. This substitution increases the dry weight of the plants to values above those of potassium-deficient plants (0.05 mM K^+) and above those of plants receiving a large potassium supply (5.0 mM K^+). In contrast, the growth of potassium-deficient bean plants (0.5 mM K^+) is further depressed by sodium. The reasons for the lack of growth response in bean are quite obvious, at least for the shoots; in bean roots an effective mechanism (*exclusion mechanism*) exists for restricting sodium transport to the shoots (Chapter 3). The potential for replacement of potassium by sodium is therefore very limited or absent in group D species such as bean.

Similar differences between plant species have been found in forage grasses, such as the natrophilic ryegrass and the natrophobic timothy, in experiments under controlled conditions (Smith *et al.*, 1980; Jarvis, 1982). The different strategies for regulating sodium transport to the shoots have important consequences in pasture plants for animal nutrition and in crop plants in general for salt tolerance (Greenway and Munns, 1980). The majority of agriculturally important crops are characterized by a more or less distinct natrophobic behaviour (Groups C and D, Fig. 10.3) with correspondingly low salt tolerance. When exposed to high NaCl concentrations the exclusion mechanism in natrophobic species (referred to as *excluders*) cannot prevent massive transport of Na^+ to the shoot and inhibition of metabolic functions and growth. In soybean a close positive correlation therefore exists between the capacity to prevent sodium transport to the shoots and growth depression by high NaCl concentrations in the substrate

(Läuchli and Wieneke, 1979). In contrast, natrophilic species, especially those of group A, have a moderate to high salt tolerance and behave as *includers*. Under saline conditions they accumulate a large amount of sodium in the shoots, where it is utilized in the vacuoles of leaf cells for osmotic adjustment (Flowers and Läuchli, 1983), depending on plant species and genotypes within a species (Section 16.6.3).

In natrophilic species also, substitution of potassium by sodium in the shoots is limited. The extent of substitution differs between individual organs and between cell compartments, being very large in the vacuoles, but very limited in the cytoplasm (Leigh *et al.*, 1986). Average values for substitution for the whole shoot are therefore misleading and underestimate the essentiality of potassium for growth and metabolism. In tomato, for example, replacement of potassium by sodium takes place mainly in the petioles of expanded leaves (Besford, 1978a). In sugar beet the substitution can be very high in mature leaves, but it is much lower in expanding leaves (Lindhauer *et al.*, 1990), leading to a steep and converse gradient in the potassium/sodium ratios of leaves of different age (Table 10.4).

In old leaves nearly all the potassium can be replaced by sodium and made available for specific functions in meristematic and expanding tissues. In contrast, in young expanding leaves there is a threshold level of substitution of ~0.5 mmol potassium per gram dry weight (Table 10.4), which corresponds to a concentration of ~50 mM K^+ kg^{-1} fresh weight, and roughly with the 100–150 mM K^+ required in the cytoplasm (Leigh *et al.* 1986; Section 8.7.2).

There is evidence for the essentiality of potassium in expanding leaf tissue of sugar beet for chlorophyll formation (Marschner and Possingham, 1975), for the induction of nitrate reductase in spinach leaves (Pflüger and Wiedemann, 1977), and for the translation of mRNA on wheat germ ribosomes (Gibson *et al.*, 1984). In the latter case,

Table 10.4

Effect of the Replacement of Potassium by Sodium in the Nutrient Solution on the Potassium and Sodium Content of Sugar Beet (cv. Fia)[a]

Age and position of leaves	Supply of K^+ and Na^+ (mM)					
	5.0 K^+		0.25 K^+ + 4.75 Na^+		0.10 K^+ + 4.90 Na^+	
	K	Na	K	Na	K	Na
Whole shoot	3.0	<0.03	0.24	2.72	0.10	3.29
Old leaves (nos. 1–7)	3.43	<0.03	0.18	3.05	0.05	4.20
Middle leaves (nos. 8–15)	2.36	<0.03	0.34	2.01	0.14	2.97
Young leaves (nos. 16–22)	1.87	<0.03	0.52	1.75	0.48	1.82

[a]Sodium and potassium contents expressed as mmol g^{-1} dry weight. From Marschner *et al.* (1981b).

the specific requirement is 100–120 mM K^+ and independent of the salt tolerance of the wheat genotypes.

In natrophobic species such as maize and bean there is an absolute requirement for potassium in most of its metabolic functions (Section 8.7). Replacement of potassium by sodium may occur to some extent in the root vacuoles, whereas such substitution in the cytoplasm causes dramatic changes in the fine stucture of the cytoplasm and its organelles (Hecht-Buchholz et al., 1971).

10.2.5 Growth Stimulation by Sodium

In addition to potassium substitution (the *sparing effect* of sodium), growth stimulation by sodium is of great practical and scientific interest. It raises the possibility of applying inexpensive, low-grade potash fertilizers with a high proportion of sodium, and it increases the potential of successfully selecting and breeding for adaptation of crop plants to saline soils.

Responses to sodium differ not only between plant species but also between genotypes of a species, as shown in Table 10.5. Compared with the effect of a potassium supply only, substitution of half the potassium in the substrate by sodium led to an increase in the dry weight of the plants and the sucrose content of the storage root in all three genotypes. When 95% of the potassium in the substrate (and ~90% within the plants) was replaced by sodium, the dry weight of the plants was not affected further; instead, sucrose production per plant was severely reduced in the two genotypes Monohill and Ada. The decrease in Monohill resulted from a lower sucrose concentration (Table 10.5) and that in Ada from a shift in shoot growth at the expense of storage root growth (Marschner et al., 1981b), an effect which is typical in sugar beet at high sodium but low potassium supply (Lindhauer et al., 1990). In Fia, however, sucrose concentration and production were enhanced when more potassium was

Table 10.5

Genotypical Differences in the Response of Sugar Beet Plants to the Replacement of Potassium by Sodium in the Nutrient Solution[a]

Genotype	Treatment (mM)		Dry wt (g per plant)	Sucrose in storage root	
	K^+	Na^+		(% fresh wt)	(g per storage root)
Monohill	5.0	—	115	9.2	54.4
	2.5	2.5	133	11.9	49.6
	0.25	4.75	126	7.6	34.2
Ada	5.0	—	86	4.9	19.0
	2.5	2.5	131	7.1	43.3
	0.25	4.75	132	7.7	20.9
Fia	5.0	—	44	10.0	13.7
	2.5	2.5	65	10.4	20.3
	0.25	4.75	84	11.2	27.9

[a]Duration of the experiment 9 weeks. From Marschner et al. (1981b).

replaced by sodium. Salt tolerance differed between the three genotypes, in agreement with the general pattern of classification (Fig. 10.3). Genotype Fia tolerated up to 150 mM NaCl in the external medium without significant growth reduction, whereas growth was severely depressed at that concentration in the other two genotypes (Marschner *et al.*, 1981a).

Growth stimulation by sodium is caused mainly by its effect on cell expansion and on the water balance of plants. Not only can sodium replace potassium in its contribution to the solute potential in the vacuoles and consequently in the generation of turgor and cell expansion (Section 8.7), it may surpass potassium in this respect since it accumulates preferentially in the vacuoles (Jeschke, 1977; Nunes *et al.* 1984). The superiority of sodium can be demonstrated by the expansion of sugar beet leaf segments *in vitro* (Marschner and Possingham, 1975) as well as in intact sugar beet plants, where leaf area, thickness, and succulence are distinctly greater when a high proportion of potassium is replaced by sodium (Milford *et al.*, 1977). An example of this effect is shown in Table 10.6. The leaves are more succulent, that is they are thicker and store more water per unit leaf area. Succulence is a morphological adaptation which is usually observed in salt-tolerant species growing in saline substrates (Jennings, 1976) and is considered an important buffer mechanism against deleterious changes in leaf water potential under conditions of moderate drought stress. Better osmotic adjustment is also a major factor in growth stimulation of halophytes by high sodium supply (Flowers and Läuchli, 1983).

Sodium increases not only the leaf area but also the number of stomata per unit leaf area (Table 10.7). The chlorophyll content, however, is lower in these plants, and this may be responsible for the lower rate of net photosynthesis per unit leaf area. The higher growth rates of sugar beet plants with a high sodium but low potassium content are therefore the result, not of increased photosynthetic efficiency, but of a larger leaf area (Lawlor and Milford, 1973).

When the availability of water in the substrate is high in sugar beet sodium increases the water consumption per unit fresh weight increment (Table 10.7), that is, it decreases the water-use efficiency as has also been observed in cauliflower (Sharma and Singh, 1990). If, however, the availability of water in the substrate is lowered by the addition of mannitol, water consumption per unit fresh weight increment decreases slightly in plants supplied with sodium, but increases sharply in plants receiving a potassium

Table 10.6

Effect of Replacement of Potassium by Sodium in the Nutrient Solution on Sugar Beet Leaves (cv. Monohill)[a]

Supply (mM)	Leaf dry weight (g per plant)	Content of leaf blades (mmol g^{-1} dry wt)		Leaf area (cm^2 per leaf)	Leaf thickness (μm)	Succulence (g H$_2$O dm^{-2})
		K	Na			
5 K$^+$	7.6	2.67	0.03	233	274	3.07
0.25 K$^+$ + 4.75 Na$^+$	9.7	0.43	2.45	302	319	3.71

[a]Based on Hampe and Marschner (1982).

Table 10.7
Effect of the Replacement of Potassium by Sodium in the Nutrient Solution on Properties of Sugar Beet Leaves and on Water Consumption at Different Osmotic Potential (± Mannitol) of the Nutrient Solution[a]

Supply (mM)	Stomata lower surface (no. cm^{-2})	Chlorophyll (mg g^{-1} dry wt)	Net photosynthesis (mg CO$_2$ cm^{-2} h^{-1})	Water consumption (g H$_2$O g^{-1} fresh wt increment) −0.02 MPa[b]	−0.4 MPa[b]
5.0 K$^+$	11 807	12.1	15.2	17.7	28.2
0.25 K$^+$ + 4.75 Na$^+$	15 127	9.2	14.4	26.5	24.6

[a]Based on Hampe and Marschner (1982).
[b]Osmotic potential of the nutrient solution.

supply only. In the latter case the growth rate is much more depressed than is the transpiration rate by the decrease in osmotic potential to −0.4 MPa.

Sodium improves the water balance of plants when the water supply is limited. This obviously occurs via stomatal regulation (Fig. 10.5). With a sudden decrease in the availability of water in the substrate (*drought stress*) the stomata of plants supplied with sodium close more rapidly than plants supplied with potassium only and, after stress release, exhibit a substantial delay in opening. As a consequence, in plants supplied with sodium the relative leaf water content is maintained at a higher level even at low substrate water availability (drought periods, saline soils). Replacement of potassium by sodium in its role in stomatal opening has been shown in epidermal strips of

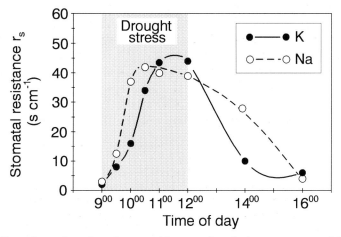

Fig. 10.5   Effect of transient drought stress (decrease in solution water potential to −0.75 MPa by the addition of mannitol) on stomatal resistance to water vapor exchange in leaves of sugar beet (cv. Monohill). Plants were grown in nutrient solutions with either 5 mM K$^+$ (●——●) or 0.25 mM K$^+$ + 4.75 mM Na$^+$ (○−−−○). (Based on Hampe and Marschner, 1982.)

Commelina species (Raghavendra *et al.*, 1976) and might therefore be a common feature of natrophilic species.

Presumably, the effects of sodium on cell expansion and water balance in plants are also mainly responsible for observations that, under field conditions, the yields of sugar beet obtained by the application of sodium fertilizers are sometimes higher than those produced by potassium fertilizers. The application of sodium fertilizers results in an increase in the leaf area index early in the growing season and a corresponding increase in light interception, thus improving water-use efficiency of leaves under conditions of moderate drought stress during the growing season (Durrant *et al.*, 1978).

Replacement at the cellular level of a high proportion of potassium by sodium might also affect the activity of enzymes which particularly respond to potassium (Section 8.7). For example, potassium is four times more effective than sodium in activating starch synthase which catalyzes the reaction of ADP-glucose to starch (Hawker *et al.*, 1974). Thus, in the leaves of plants in which a high proportion of potassium is replaced by sodium the starch content is much lower but the content of soluble carbohydrates, particularly sucrose, is much higher (Hawker *et al.*, 1974). This shift may favor cell expansion in the leaf tissue. Furthermore, sodium is more effective than potassium in stimulating sucrose accumulation in the storage tissue of sugar beet (Section 5.4). This effect of sodium on sucrose storage seems to be related to stimulation of ATPase activity at the tonoplast of beet storage cells (Willenbrink 1983). The existence of ATPases which require the presence of both potassium and sodium for maximal activity is well documented in roots of natrophilic species (Kylin and Hansson, 1971).

10.2.6 Application of Sodium Fertilizers

Given the genotypical differences in growth response to sodium and the abundance of sodium in the biosphere, one can expect the application of sodium to have beneficial effects: (a) in natrophilic plant species; (b) when soil levels of available potassium or sodium or both are low; and (c) in areas with irregular rainfall or transient drought during the growing season, or both.

The potential replacement of potassium by sodium must be taken into account when the application of fertilizers to natrophilic species is being considered. When sodium contents in the leaves are high, the potassium contents required for optimal growth decrease from 3.5 to 0.8% of the leaf dry weight in Italian ryegrass (Hylton *et al.*, 1967) and from 2.7 to 0.5% in Rhodes grass (Smith, 1974), or from 4.3 to 1.0 in lettuce (Costigan and Mead, 1987). There are also corresponding differences in optimal levels in leaves of tomato (Besford, 1978a) and sugar beet.

The sodium content of forage and pasture plants is an important factor in animal nutrition. The sodium requirement for lactating dairy cows is ~0.20% in the dry weight of the forage (Smith *et al.*, 1978; Zehler, 1981), which is distinctly higher than the average sodium content of natrophobic pasture species (Smith *et al.*, 1980). In contrast, the potassium content in these natrophobic species is usually at least adequate but often in excess of animal needs, which is in the range of 2–2.5% dry weight. The use of sodium fertilizer to increase the sodium content of forage and pasture plants is thus an important procedure in large areas of the world. A high sodium content increases the acceptability of forage to animals and enhances daily food intake (Zehler, 1981).

However, sodium fertilizers are effective only when applied to grassland or mixed pastures with a reasonably high proportion of natrophilic species.

10.3 Silicon

10.3.1 General

Silicon is the second most abundant element in the earth's crust. In soil solutions the prevailing form is monosilicic acid, $Si(OH)_4$, with a solubility in water (at 25°C) of ~2 mM (equivalent to 56 mg Si l^{-1}). On average, the concentrations in soil solutions are 14–20 mg Si l^{-1} (range about 3.5–40 mg) with a tendency to lower concentrations at high pH (>7) and when large amounts of sesquioxides are present in soils and anion adsorption is dominant (Jones and Handreck, 1967). Such conditions are widespread in highly weathered tropical soils. Concentrations of silicon in aqueous solutions higher than 56 mg Si l^{-1} indicate either supersaturation of $Si(OH)_4$ or partial polymerization of monosilicic acid.

Silicic acid, $Si(OH)_4$ has a number of similarities with boric acid, $B(OH)_3$, both are very weak acids in aqueous solutions, interact with pectins and polyphenols in the cell walls, and are mainly located in the cell walls. In contrast to boron, essentiality of silicon for higher plants has been demonstrated so far only in a few plant species, but it is beneficial for many species and, under certain circumstances, for most higher plants. It has recently been claimed that omission of silicon in nutrient solutions amounts to the imposition of an atypical environmental stress (Epstein, 1993).

10.3.2 Uptake, Content and Distribution

Higher plants differ characteristically in their capacity to take up silicon. Depending on their silicon content (SiO_2, expressed as a percentage of shoot dry weight), they can be divided into three major groups: members of the Cyperaceae, such as horsetail (*Equisetum arvense*) and wetland species of Gramineae, such as rice, 10–15%; dryland species of Gramineae, such as sugarcane and most of the cereal species, and a few dicotyledons, 1–3%; and most dicotyledons, especially legumes, <0.5%. In a survey of 175 plant species grown in the same soil, Takahashi and Miyake (1977) distinguished between *silicon accumulators* – plants in which silicon uptake largely exceeds water uptake – and *nonaccumulators* – plants in which silicon uptake is similar to or less than water uptake. Table 10.8 illustrates the relationship between plant species, silicon concentration, and silicon uptake. In this experiment, the silicon content of the shoots was both measured and calculated. In the latter case, the uptake of silicon and water were assumed to occur in the same ratio as that in the external solution. It is evident that, at low external concentrations, rice, and to a lesser extent wheat, took up more silicon than calculated, indicating an active uptake or transport mechanism. In rice but not wheat, even at the highest external concentration an active component could still be observed.

In agreement with this, in rice and other accumulator species, silicon uptake is closely related to root metabolism and not greatly affected by the transpiration rate (Okuda

Table 10.8
Measured and Calculated Silicon Content of Shoots of Plant Species Grown in Nutrient Solutions with Different Silicon Concentrations[a]

Plant species	Si concentration in nutrient solution (mg SiO_2 l^{-1})	Transpiration coefficient (l H_2O kg^{-1} dry wt)	SiO_2 content (mg kg^{-1} dry wt) Measured	SiO_2 content (mg kg^{-1} dry wt) Calculated[b]	Ratio measured/ calculated
Rice	0.75	286	10.9	0.2	54.5
	30	248	94.5	7.4	12.7
	162	248	124.0	40.2	3.1
Wheat	0.75	295	1.2	0.22	5.5
	30	295	18.4	8.9	2.1
	162	267	41.0	43.3	0.9
Soybean	0.75	197	0.2	0.15	1.3
	30	197	1.7	5.9	0.3
	162	197	4.0	31.9	0.1

[a]From Vorm (1980).
[b]Assuming 'nonselective' silicon uptake by mass flow.

and Takahashi, 1965). It is therefore most likely that in accumulator plants during radial transport across roots a substantial component of symplasmic transport of silicon is involved as is also indicated by the accumulation in the protoplast of cortical and particularly endodermal cells, for example of *Sorghum bicolor* (Hodson and Sangster, 1989a).

In contrast to wheat and particularly rice, in soybean (Table 10.8) the uptake and radial transport of silicon across the roots to the xylem vessels is much restricted at higher silicon concentrations, indicating an effective exclusion mechanism. High contents of silicon in the shoots of rice and wheat corresponded to distinctly lower transpiration coefficients (Table 10.8), a side-effect of silicon which can be considered beneficial (see Section 10.3.3).

In plant species such as wheat where a fairly close correlation exists between transpiration rate and silicon uptake, the silicon content in the shoot dry matter had been considered a suitable parameter for calculations of the water-use efficiency of a crop (Hutton and Norrish, 1974); however this approach has many limitations (Section 3.2.3). Even within a species, such as barley, cultivars differ much in their silicon uptake, in parallel to the differences in boron uptake (Nable *et al.*, 1990b).

Also in silicon accumulators such as *Sorghum bicolor* the root endodermis acts as a barrier for radial transport of silicon. Along the root axis from apical to basal zones there is a steep increase in silicon deposition in the inner wall of the endodermis cells (Hodson and Sangster, 1989a). Silification of the endodermis might act as an effective mechanical barrier against invasion of the stele by pathogens and parasites (Bennett, 1982), such as *Striga* (Maiti *et al.*, 1984; Chapter 11).

According to present knowledge, long-distance transport of silicon in plants is confined to the xylem, and relatively large amounts of silicon are deposited in the cell walls of xylem vessels (Balasta *et al.*, 1989) where it might be important in preventing compression of the vessels when the transpiration rates are high (Raven, 1983). The distribution of silicon within the shoot and shoot organs is determined by the transpiration rate of the organ (Jones and Handreck, 1967) and, for a given organ such as a leaf, is dependent on leaf age. Most of the silicon remains in the apoplasm and is deposited after water evaporation at the termini of the transpiration stream, mainly the outer walls of the epidermal cells on both surfaces of the leaves as well as in the inflorescence bracts of graminaceous species (Hodson and Sangster, 1989b). Silicon is deposited either as amorphous silica ($SiO_2 \cdot nH_2O$, 'opal') or as so-called opal phytoliths with distinct three-dimensional shapes (Parry and Smithson, 1964). The preferential deposition of silicon in the apoplasm of epidermal cells and trichomes is reflected in similarities between surface features of a leaf and structure of silicon deposits (Lanning and Eleuterius, 1989). The epidermal cell walls are impregnated with a firm layer of silicon and become effective barriers against both, water loss by cuticular transpiration and fungal infections (Chapter 11). In grasses a considerable portion of silicon in the epidermis of both leaf surfaces is also located intracellularly in so-called *silica cells* (Sangster, 1970) or 'bulliform' cells (Takeoka *et al.*, 1984).

The deposition of silicon in plant hairs on leaves, culms, inflorescence bracts and brush hairs of cereal grains such as wheat is suspected of posing a potential threat to human health (Hodson and Sangster, 1989b). There is substantial evidence that the inflorescence bracts of grasses of the genus *Phalaris* and foxtail millet (*Setaria italica*) contain sharp, elongated siliceous fibers which fall into the critical size range of fibers that have been classified as carcinogenic (Sangster *et al.*, 1983). There are striking correlations between esophageal cancer and the consumption of either foxtail millet in North China (Parry and Hodson, 1982), or of wheat contaminated with *Phalaris* in the Middle East (Sangster *et al.*, 1983).

There is increasing evidence for the necessity to modify the traditional view of silicon deposition in the cell walls as a purely physical process leading to mechanical stabilization (rigidity) of the tissue and acting as a mechanical barrier to pathogens. Silicon deposition is under rather strict metabolic and temporal control. For example, in leaf hairs of grasses the ultrastructural forms of silica deposits change particularly during transition from primary to secondary cell wall formation from sheet-like to globular silicon. These changes are obviously dictated by changes in cell wall metabolites interacting with silicic acid (formation of ester bonds) leading to bulk deposition of silicon into the mature cell wall structure (Perry *et al.*, 1987). Also in the protecting effects of silicon against insects and pathogens, in addition to a mechanical barrier, a dynamic component of redistribution of silicon is involved (Heath and Stumpf, 1986; Section 11.2.2).

10.3.3 Role in Metabolism

The essentiality of silicon in unicellular organisms such as diatoms is well documented, and many details of its metabolic functions in these organisms are known (Werner and Roth, 1983). In higher plants the essentiality of silicon is reasonably well established for *silicophile* species such as *Equisetum arvense* (Chen and Lewin, 1969) and certain wetland grass species (Takahashi and Miyake, 1977). In wetland rice lacking silicon, vegetative growth and grain production are distinctly reduced and deficiency symp-

toms, such as necrosis of mature leaves and wilting of plants might occur (Lewin and
Reimann, 1969), suggesting, but not proving, that silicon is essential for the growth of
rice. However, failure to complete the lifecycle has not yet been demonstrated. The
actual requirement of silicon for vegetative growth seems to be extremly low even for
rice, where a larger silicon requirement appears to be confined to the reproductive stage
(Table 10.9). During the reproductive stage silicon is preferentially transported into the
flag leaves, and interruption of silicon supply at this stage is detrimental for spikelet
fertility (Ma et al., 1989). Neither the sites of action (source or sink) nor the mechanism
of action are known by which silicon affects spikelet fertility.

Sugarcane is another silicon accumulator plant which strongly responds to silicon
supply. Under field conditions for optimal cane yield in the leaf dry matter at least 1%
silicon (~2.1% SiO_2) is required, and at 0.25% silicon the yield drops to about one half
(Anderson, 1991). Such drastic yield reductions are associated with typical visible
deficiency symptoms ('leaf freckling') on leaf blades directly exposed to full sunlight
(Elawad et al., 1982a,b). In contrast, under greenhouse conditions the requirement of
sugarcane for silicon seems to be extremely low (Gascho, 1977) which does not support
the assumption of essentiality of silicon as a mineral element for sugarcane.

As, even for silicon accumulator plants such as wetland rice and sugarcane, the
essentiality of silicon has yet to be demonstrated, reports of the essentiality of silicon for
tomato and cucumber (Miyake and Takahashi, 1978, 1983), soybean (Miyake and
Takahashi, 1985) and strawberry (Miyake and Takahashi, 1986) have been received
with surprise. Withholding silicon supply not only reduced fruit yield drastically but also
caused malformation of newly developed leaves, wilting, premature senescence,
impaired pollen viability, and in severe cases failure of fruit set. Lack of silicon supply
was particularly detrimental under high light conditions. On reexamination of these
results of silicon as a mineral nutrient for nonaccumulator dicotyledonous species it has
been demonstrated that these silicon effects are confined to plants supplied with high
phosphorus and insufficient zinc concentrations, and that under these conditions silicon
counteracted zinc deficiency-induced phosphorus toxicity (Marschner et al., 1990), a
disorder which was discussed in Section 9.4.8. Silicon supply increased the physiological
availability of zinc in the plants (Marschner et al., 1990) by an unknown mechanism.

Table 10.9
Effect of Silicon Supplied at Different Growth Stages on Growth and
Grain Yield of Wetland Rice[a]

| Supply[b] at vegetative stage: | −Si | +Si | −Si | +Si |
Supply[b] at reproductive stage:[c]	−Si	−Si	+Si	+Si
% SiO_2 (shoot dry wt)	0.05	2.2	6.9	10.4
Dry wt (g per pot)				
Roots	4.0	4.3	4.2	4.7
Shoots	23.5	26.5	31.0	33.6
Grain	5.3	6.6	10.3	10.8

[a]Based on Okuda and Takahashi (1965) and Takahashi and Miyake (1977).
[b]+Si, 100 mg SiO_2 l^{-1}; −Si, no silicon supply.
[c]Ear emergence.

These beneficial effects of silicon on zinc availability also deserve attention in sugar-cane.

Because of its abundance in the biosphere the essentiality of silicon as a micronutrient for higher plants is very difficult to prove. Even highly purified water contains about 20 nM silicon (Werner and Roth, 1983), and correspondingly the leaves of silicon accumulator plants that were subjected to a so-called no-silicon treatment usually contain between 1–4 mg SiO_2 g^{-1} leaf dry weight.

There have been only a few in-depth studies on metabolic changes in higher plants when silicon is omitted from the external solution or when a specific inhibitor of silicon metabolism, germanic acid, is supplied (Werner, 1967). In the absence of silicon a considerable decrease in the incorporation of inorganic phosphate into ATP, ADP, and sugar phosphates has been observed in sugarcane (Wong You Cheong and Chan, 1973); in wheat roots the proportion of lignin in the cell walls declines and that of phenolic compounds increases (Jones et al., 1978). This latter aspect deserves particular attention for various reasons. Some of the cell wall-bound silicon is presumably present as an ester-like derivative of silicic acid (R^1–O–Si–O–R^2) acting as a bridge in the structural organization of polyuronides (Jones, 1978). Furthermore, silicon seems to influence the content and metabolism of polyphenols in xylem cell walls (Parry and Kelso, 1975). As shown by Weiss and Herzog (1978) silicic acid, like boric acid, has a high affinity for o-diphenols such as caffeic acid and corresponding esters, forming mono-, di- and polymeric silicon complexes of high stability and low solubility:

Silicon may therefore affect the stability of higher plants not only as an inert deposition in lignified cell walls but also by modulating lignin biosynthesis. As stressed by Raven (1983), silicon as a structural material requires much less energy than lignin. About 2 g of glucose is required for the synthesis of 1 g of lignin; the ratio of the energy requirement for lignin to that of silicon is 20:1.

More recently, evidence has been provided that silicon not only contributes to cell wall rigidity and strengthening but might also increase cell wall elasticity during extension growth. In the primary cell walls silicon interacts with cell wall constituents such as pectins and polyphenols, and these crosslinks obviously increase cell wall elasticity during extension growth, an effect that is particularly evident in loblolly pine under drought stress (Emadian and Newton, 1989). This has an interesting parallel to the role of silicon in cotton fiber growth. During the early phase of elongation growth the silicon content of cotton fiber is fairly high (0.5% Si in the dry matter) and decreases with secondary wall thickening, i.e., cellulose deposition (Boylston, 1988). The highest silicon content has been found in cotton varieties with long fine fibers (Boylston *et al.*, 1990). This effect of silicon in the primary cell walls is the opposite of what is usually observed, for example, in leaves when large amounts of silicon are incorporated into secondary cell walls, but it has striking similarities with the function of boron in cell walls (Section 9.7.4). The relative importance of boron and silicon in primary cell walls might depend on plant species (Loomis and Durst, 1992). Graminaceous and dicotyledonous species differ very much in their cell wall composition and in their boron requirement which is opposite to their capacity of silicon uptake and growth responses to silicon supply.

10.3.4 Beneficial Effects

Silicon has a number of other, well documented and readily visible and/or measurable beneficial effects. Under field conditions, particularly in dense stands of cereals, silicon can stimulate growth and yield by several indirect actions. These include decreasing mutual shading by improving leaf erectness, decreasing susceptibility to lodging, decreasing the incidence of infections with root parasites and pathogens, leaf pathogens, and preventing manganese or iron toxicity or both.

Leaf erectness is an important factor affecting light interception in dense plant stands. For a given cultivar, leaf erectness decreases with increasing nitrogen supply (Table 10.10), for reasons discussed in Section 8.2. Silicon increases leaf erectness and thus to a large extent counteracts the negative effects of high nitrogen supply on light interception. In a similar manner, silicon counteracts the negative effects of an increasing nitrogen supply on haulm stability and lodging susceptibility (Idris *et al.*, 1975).

This effect of silicon on leaf erectness is mainly a function of the silicon depositions in the epidermal layers of the leaves (Balasta *et al.*, 1989) and, thus, over a wide range closely related to the silicon concentration supplied (Table 10.10; Balasta *et al.*, 1989). Also in dicotyledonous species, such as cucumber, silicon increases the rigidity of mature leaves, which are held more horizontally, increases their chlorophyll content and delays their senescence (Adatia and Besford, 1986).

High contents of silicon in the epidermal layers of leaves are particularly effective in

Table 10.10
Relationship between Silicon[a] and Nitrogen Supply and Leaf
Openness[b] in Rice Plants (cv. IR8) at Flowering[c]

Nitrogen supply (mg l^{-1})	Silicon supply (mg SiO_2 l^{-1})		
	0	40	200
5	23°	16°	11°
20	53°	40°	19°
200	77°	69°	22°

[a]As sodium silicate.
[b]Angle between the culm and the tip of the leaves.
[c]Based on Yoshida *et al.* (1969).

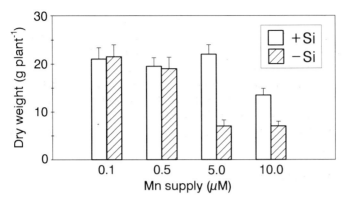

Fig. 10.6 Effect of manganese on the dry weight of bean in the absence and presence of silicon (1.55 mg SiO_2 l^{-1}). Vertical lines represent standard deviation. (Modified from Horst and Marschner, 1978a.)

increasing the resistance of tissues against attacks by fungi such as powdery mildew (Miyake and Takahashi, 1983; Adatia and Besford, 1986), blast infection in rice (Volk *et al.*, 1958) and insect pests. These aspects are discussed in more detail in Chapter 11.

Stimulation of growth by silicon can also be caused by the prevention or depression of manganese and iron toxicity (Vlamis and Williams, 1967; Ma and Takahashi, 1990a). As shown in Fig. 10.6, silicon has no effect on the growth of bean plants at low manganese concentrations. At high concentrations, however, it either prevents (5.0 μM Mn) or at least reduces (10.0 μM Mn) the severe growth depression induced by manganese toxicity. Although silicon can considerably stimulate growth at high manganese concentrations, this nevertheless has to be classified as a typical beneficial effect since it is restricted to conditions of excessive manganese supply.

The increase in manganese tolerance in bean plants does not result from lower manganese uptake but from an increase in the tolerance of the leaf tissue to high manganese contents (Horst and Marschner, 1978a). The same is true for barley (Horiguchi and Morita, 1987). In plant species such as barley and bean with low tissue tolerance to high manganese contents, and to some extent also in manganese-tolerant species such as rice (Horiguchi, 1988a), silicon alters the distribution of manganese within the leaf tissue (Williams and Vlamis, 1957). In the absence of silicon the distribution of manganese is nonhomogeneous and characterized by a spot-like accumulation (Fig. 10.7). These spots resemble the typical visible symptoms of manganese toxicity on mature leaves, namely, brown speckles surrounded by chlorotic or necrotic zones or both these symptoms. These brown speckles are mainly confined to the cell walls and contain oxidized manganese, but the brown colour derives from oxidized polyphenols (Wissemeier and Horst, 1992).

Silicon prevents the occurrence of brown speckles by bringing about a homogeneous distribution of manganese (Fig. 10.7), presumably thereby increasing tissue tolerance to high manganese contents. The critical toxicity levels of manganese in the leaf tissue of a given plant species can differ considerably, therefore, depending on the silicon content of the leaves (Horst and Marschner, 1978a). Although the severity of the

Fig. 10.7 Autoradiograph showing the effect of silicon $(0.75\ mg\ SiO_2\ l^{-1})$ on ^{54}Mn distribution in bean leaves supplied with $0.1\ mM\ ^{54}Mn$ for 6 days. Manganese content of the primary leaves: $22.0\ \mu g\ g^{-1}$ dry wt at $-Si$ and $16.7\ \mu g\ g^{-1}$ dry wt at $+Si$. (Horst and Marschner, 1978a.)

toxicity symptoms of leaves and their peroxidase activity are positively correlated, it is not clear whether silicon ameliorates manganese toxicity primarily via modulating enzyme activity or manganese distribution (Horiguchi and Morita, 1987; Wissemeier and Horst, 1992).

Table 10.11

Effect of Silicon Supply (50 mg SiO$_2$ l^{-1}) on Shoot Dry Weight, Manganese Contents in Roots and Shoots and Transpiration Rates of Rice (*Oryza sativa* L.)[a]

| Mn supply (mg l^{-1}) | Shoot dry weight (g per plant) | | Contents (mg Mn g^{-1} dry wt) | | | | Transpiration rate (mg H$_2$O g^{-1} dry wt per day) | |
| | | | Roots | | Shoots | | | |
	−Si	+Si	−Si	+Si	−Si	+Si	−Si	+Si
0.32	4.4	4.5	0.03	0.13	0.25	0.21	11.8	10.9
1.0	4.3	4.7	0.12	0.50	0.66	0.53	11.6	10.7
3.2	4.2	5.0	0.72	1.60	1.94	1.20	11.7	10.8
10.0	4.1	5.0	2.12	2.89	4.36	1.97	11.7	10.7

[a]Based on Horiguchi (1988a).

Silicon may also increase aluminum tolerance of plants as has been shown for teosinte grown in nutrient solution (Barcelo *et al.*, 1993; Section 16.3.4).

In silicon accumulators such as wetland rice, silicon increases tolerance to excessive concentrations of manganese and iron in the rooting medium mainly in another way, namely by depressing the iron and particularly manganese contents in the shoots (Ma and Takahashi, 1990a). This silicon effect on manganese contents in the shoots is only small at low external manganese concentrations but becomes much more prominent at high concentrations (Table 10.11). In contrast to the shoot contents the manganese contents of the roots (as well as the amounts per plant) are increased by silicon demonstrating that in rice and other wetland species silicon increases the 'oxidizing power' of the roots (Okuda and Takahashi, 1965). This effect of silicon is brought about by increasing the volume and rigidity of the aerenchyma (air-filled spaces in shoots and roots), thereby enhancing O$_2$ transport from the shoots to the submerged root system exposed to toxic concentrations of reduced manganese and iron (Section 16.4.3). Irrespective of the manganese content, silicon consistently decreased the transpiration rates of rice plants (Table 10.11) by decreasing the nonstomatal (cuticular) component of transpiration. In rice plants growth-enhancement by silicon supply is associated not only with lower transpiration rates but also with lower calcium uptake reflected in both lower calcium content in the shoot dry matter and amount per plant (Ma and Takahashi, 1993).

The application of silicon fertilizers such as basic slag is an effective means of meeting the high silicon demand particularly of sugarcane growing on organic soils (Elawad *et al.*, 1982a; Anderson *et al.*, 1987). In wetland rice, yield responses to silicon application can be expected at SiO$_2$ contents below 11% of the leaf dry matter (Okuda and Takahashi, 1965). Rice husks in particular and also rice straw in general are usually high in silicon content. However, silicon from rice straw is only very slowly released after incorporation of the straw into the soil and, thus, suitable only as a long-term source of silicon for rice plants (Ma and Takahashi, 1991). Desorption of phosphorus by application of silicon fertilizer is of importance in upland soils (Scheffer *et al.*, 1982) but not in paddy soils (Ma and Takahashi, 1990b). Other beneficial effects of silicon application such as reduced water loss by cuticular transpiration, and increased resistance against lodging and pests, deserve more attention in the future in crops other than rice and sugarcane (Munk, 1982). In wheat, an increase in silicon supply stimulates germination and growth, and this effect is relatively more distinct in saline substrates (Ahmad *et al.* 1992); this effect may be related to the role of silicon in elasticity of the primary cell wall (Section 10.3.3).

Silicon is an essential mineral element for animals (Nielsen, 1984), where it is a constituent of certain mucopolysaccharides in connective tissues (Jones, 1978). On the other hand, in grazing animals the uptake of a large amount of phytoliths may lead to excessive abrasion of the rumen wall, and dissolved silicon may form secondary depositions in the kidney, thereby causing serious economic loss (Jones and Handreck, 1969).

10.4   Cobalt

The role of cobalt as an essential mineral element for ruminants was discovered in 1935 in field investigations of livestock production in Australia. The requirement of cobalt for N_2 fixation in legumes and in root nodules of nonlegumes (e.g., alder) was established 25 years later (Ahmed and Evans, 1960). When *Medicago sativa* was grown under controlled environmental conditions with a minimum of cobalt contamination, plants dependent on N_2 fixation grew poorly, but growth was strongly enhanced by cobalt supply; in contrast, nitrate-fed plants grew equally well with or without supply of cobalt (Delwiche *et al.*, 1961). Kliewer and Evans (1963a) isolated the cobalamin coenzyme B_{12} from the root nodules of legumes and nonlegumes, and demonstrated the interdependence of cobalt supply, the B_{12} coenzyme content of *Rhizobium*, the formation of leghemoglobin, and N_2 fixation (Kliewer and Evans, 1963b). On the basis of these studies and later reports by other authors, it has been established that *Rhizobium* and other N_2-fixing microorganisms have an absolute cobalt requirement whether or not they are growing within nodules and regardless of whether they are dependent on a nitrogen supply from N_2 fixation or from mineral nitrogen. However, for these microorganisms the demand for cobalt is much higher for N_2 fixation than for ammonium nutrition (Kliewer and Evans, 1963a).

The coenzyme cobalamin (vitamin B_{12} and its derivatives) has Co(III) as the metal component, chelated to four nitrogen atoms at the centre of a porphyrin structure similar to that of iron in hemin. In *Rhizobium* (and *Bradyrhizobium*) species, three enzymes are known to be cobalamin-dependent and cobalt-induced changes in their activities are primarily responsible for the relationship between cobalt supply, nodulation, and N_2 fixation in legumes (Dilworth *et al.*, 1979; Dilworth and Bisseling, 1984). These enzymes are:

1. Methionine synthase. Under conditions of cobalt deficiency methionine synthesis is depressed (Table 10.12) which presumably leads to lower protein synthesis and contributes to the smaller size of the bacteroids (Table 10.12).

2. Ribonucleotide reductase. This enzyme is involved in the reduction of ribonucleotides to desoxyribonucleotides and therefore in DNA synthesis. In agreement with this, the DNA content per cell is lower (Table 10.12) and there are fewer and longer

Table 10.12
Some Characteristics of Cobalt-Sufficient and Cobalt-Deficient
Crown Nodules of *Lupinus angustifolius* L.[a]

Co status	Volume (μm^3) of bacteroids	DNA content (g $\times$ 10^{-15} per cell)	Methionine (% of total amino-nitrogen)
+	3.19	12.3	1.31
−	2.62	7.8	0.97

[a]Based on Dilworth and Bisseling (1984).

bacteroids in the root nodules of cobalt-deficient plants than in normal plants, indicating depressed rhizobial cell division (Chatel *et al.*, 1978).

3. Methylmalonyl-coenzyme A mutase. This enzyme is involved in the synthesis of heme (iron porphyrins) in the bacteroids and thus – in cooperation with the host nodule cells – in the synthesis of leghemoglobin. Therefore, under conditions of cobalt deficiency the synthesis of leghemoglobin is directly impaired, an example of this has been shown in Section 7.4.5.

Cobalt deficiency affects nodule development and function at different levels and degrees as shown in Table 10.13 for nodules of 6-week-old lupins. When lupins grown in a cobalt deficient soil are supplied with cobalt, the weight and cobalt content of the nodules increases, as does the number of bacteroids and amount of cobalamin and leghemoglobin per unit nodule fresh weight. Somewhat surprisingly but in agreement with the results of other authors, only about 12% of the total nodule cobalt was bound to cobalamin in the deficient plants (Table 10.13). Nothing is known about the location, chemical binding, and possible functions of the cobalt in nodules that is not available for cobalamin formation. Although the results in Table 10.13 suggest that the synthesis of cobalamin and leghemoglobin are the most sensitive parameters, more recent results indicate that it is the proliferation of the rhizobia which is impaired most severely under cobalt deficiency, and the low leghemoglobin contents are the consequence of this impairment (Riley and Dilworth, 1985a,b). Resupplying cobalt to deficient nodules increases bacteroid content per gram nodule fresh weight, nodule mass and N_2 fixation activity, but the degree of response declines with increasing age of the nodule tissue (Riley and Dilworth, 1985b).

Table 10.13
Effect of Cobalt on Nodule Growth and Composition in *Lupinus angustifolius* Grown in a
Cobalt-Deficient Soil and Inoculated with *Rhizobium lupini*[a,b]

Cobalt treatment	Crown nodule fresh wt (g per plant)	Cobalt content (ng g^{-1} nodule dry wt)	No. of bacteroids ($\times$ 10^9 g^{-1} nodule fresh wt)	Cobalamin (ng g^{-1} nodule fresh wt)	Leghemoglobin (mg g^{-1} nodule fresh wt)
−	0.1	45	15	5.9	0.71
+	0.6	105	27	28.3	1.91

[a]Based on Dilworth *et al.* (1979).
[b]0.19 mg cobalt as the sulfate salt was supplied per pot. Harvest after 6 weeks.

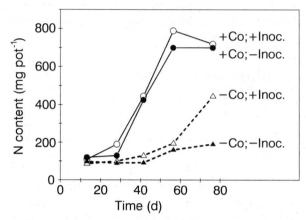

Fig. 10.8 Effect of cobalt and inoculation with *Rhizobium* on the time course of nitrogen accumulation in *Lupinus angustifolius* L. grown in a cobalt-deficient soil (eight plants per pot). (Dilworth *et al.*, 1979.)

In legumes grown in cobalt-deficient soils the nodule activity is consistently lower in plants without cobalt supply. This lower activity is reflected in either nitrogenase activity (Section 7.4.5) or nitrogen content of the plants (Fig. 10.8). In legumes dependent on N_2 fixation cobalt deficiency is therefore indicated by typical symptoms of nitrogen deficiency (Dilworth *et al.*, 1979; Robson and Snowball, 1987).

In legumes grown in cobalt-deficient soils, *Rhizobium* infection is often less extensive than in plants supplied with cobalt, and the onset of N_2 fixation, as indicated by the nitrogen accumulation in the plants, is delayed for several weeks (Fig. 10.8). Inoculation with *Rhizobium* makes little difference when cobalt has been supplied but is quite effective in deficient plants at later growth stages. This could suggest that in cobalt deficient soils limitations occur even as early as the stage of survival and infectivity of the rhizobia. However, other evidence indicates that not external but internal cobalt limitations are important, namely the inability of the host plant to provide sufficient cobalt for the rhizobia in the developing nodules (Riley and Dilworth, 1985c).

With cobalt deficiency, there is a preferential accumulation of cobalt in the nodules. On a whole-plant basis, however, the roots have the highest cobalt content. The proportion of cobalt in the shoots, nodules, and roots is 1:6:15 in cobalt-deficient and 1:3:25 in cobalt-sufficient plants (Robson *et al.*, 1979). In deficient plants the cobalt content in the nodules varies between 20 and $170 \mu g \, g^{-1}$ nodule fresh weight, depending on the plant species (Robson *et al.*, 1979). Although the cobalt content in the shoot dry matter is much higher in plants fertilized with cobalt, content is a poor indicator in the diagnosis of cobalt deficiency, at least in *Lupinus angustifolius* (Robson and Snowball, 1987).

The cobalt content of seeds of the same species varies between plants grown in different locations. In *Lupinus angustifolius* between 6 and 730 ng cobalt per gram seed weight have been found (Robson and Mead, 1980). Accordingly, when grown in cobalt-deficient soils and dependent on N_2 fixation there is a close relationship between cobalt content of the seeds, plant growth, nitrogen content and severity of the visual nitrogen

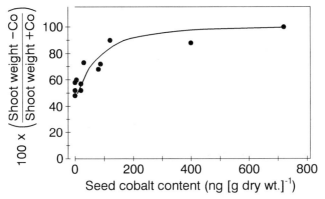

Fig. 10.9 Relation between cobalt content in seeds and response of shoot growth of *Lupinus angustifolius* L. to applied cobalt. (Based on Robson and Snowball, 1987.)

deficiency symptoms (Robson and Snowball, 1987). As shown in Fig. 10.9 the shoot growth response to increasing seed contents is very strong up to about 200 ng Co g^{-1} seed dry weight, but even at 400 ng Co g^{-1} seed dry weight there is still a 10% higher shoot growth over the soil supply of cobalt.

In large-seeded lupins about 100 ng Co g^{-1} seed dry weight is sufficient to prevent cobalt deficiency in plants grown in cobalt deficient soils (Gladstones *et al.*, 1977). Treating seeds with cobalt is therefore an effective procedure for supporting N$_2$ fixation and growth of legumes on cobalt-deficient soils (Reddy and Raj, 1975).

Field responses to cobalt fertilization of nodulated legumes are rare but have been demonstrated, for example, on poor siliceous sandy soils (Ozanne *et al.*, 1963; Powrie, 1964). Foliar sprays only are quite effective, but less so than the combination of seed treatment and foliar sprays (Table 10.14). In peanut and pigeon pea the most distinct effect of combined seed and foliar application of cobalt was on leghemoglobin content which was increased 3–4-fold in the nodule mass (Shiv Raj, 1987). The effectiveness of foliar sprays indicates a reasonable retranslocation of cobalt from leaves, as has also been shown after the application of labeled cobalt to clover and alfalfa leaves (Handreck and Riceman, 1969). In the phloem, cobalt seems to be translocated largely as a negatively charged complex (Wiersma and van Goor, 1979).

There are considerable differences in the sensitivities of legume species to cobalt deficiency. *Lupinus angustifolius* is particularly sensitive in comparison with *Trifolium*

Table 10.14
Effect of Cobalt on Peanut[a]

Cobalt treatment	No. of nodules per plant	Total nitrogen at maturity (% dry wt)	Pod yield (kg ha^{-1})
Control (−Co)	91	2.38	1232
Seed treatment	150	2.62	1687
Foliar sprays (2×)	123	3.14	1752
Seed treatment + foliar sprays (2×)	166	3.38	1844

[a]Based on Reddy and Raj (1975).

subterraneum (Gladstones *et al.*, 1977). In *L. angustifolius* cobalt increased the yield and total amount of nitrogen per plant but, unexpectedly, significantly decreased the nitrogen content as a percentage of the dry weight. Nevertheless, the cobalt-treated plants looked healthy and were dark green, whereas the cobalt-deficient plants appeared unhealthy and had yellowish leaves. One may therefore speculate that growing root nodules support plant growth not only by N_2 fixation but also by other factors such as cytokinins (Gladstones *et al.*, 1977).

There is no evidence that cobalt has any direct role in the metabolism of higher plants. A distinct growth response to cobalt was reported by Wilson and Hallsworth (1965) in clover supplied with mineral nitrogen or inoculated with ineffective *Rhizobium* strains. A similar response was noted in wheat (Wilson and Nicholas, 1967) and *Hevea* (Bolle-Jones and Mallikarjuneswara, 1957). These responses to cobalt in non-N_2-fixing higher plants are always small, and probably reflect beneficial effects of unknown nature. Reports on the occurrence of a cobalamin-dependent enzyme, leucine-2,3-aminomutase in potato (Poston, 1978) require confirmation. In contrast to higher plants, in photosynthetic lower plants such as *Euglena gracilis*, cobalamin is essential for growth and is localized in various subcellular fractions and in the thylakoids of chloroplasts (Isegawar *et al.*, 1984).

Cobalt stimulates extension growth when added to excised plant tissue or organs such as coleoptiles and hypocotyls (Bollard, 1983). Inhibition of endogenous ethylene formation by cobalt seems to be involved in this stimulation (Samimy, 1978), an effect which is also responsible for the extension of life of cut roses by cobalt application (Venkatarayappa *et al.*, 1980).

In cobalt-deficient soils cobalt application might not only enhance the N_2 fixation of legumes but also contribute to the nutritional quality of forage plants for ruminants. Cobalt is essential for ruminants because rumen microflora can synthesize sufficient vitamin B_{12} to meet the needs of the animal. Nonruminants, including humans, however have a requirement of preformed vitamin B_{12}. Cobalt deficiency is widespread in grazing ruminants on soils low in cobalt (Miller *et al.*, 1991). The critical cobalt content for ruminants is about 0.07 mg kg^{-1} dry weight of forage, that is, higher than the critical content for N_2 fixation in legumes. On average the cobalt content of plants varies between 0.05 and 0.30 mg kg^{-1} and is usually higher in legumes than in grasses (Kubota and Allaway, 1972; Kubota *et al.*, 1987).

There are contradictory reports on critical toxicity contents of cobalt, values varying from 0.4 mg kg^{-1} dry weight in clover (Ozanne *et al.*, 1963) up to a few milligrams per kilogram dry weight in bean and cabbage (Bollard, 1983). In crop and pasture species there are also distinct genotypical differences in tolerance to excessive contents of cobalt in the shoots. In some plant species (*hyperaccumulators*) adapted to metalliferous soils the cobalt content might reach $4000–10\,000$ mg kg^{-1} dry weight (Brookes and Malaisse, 1989).

10.5 Selenium

The chemistry of selenium (Se) has features in common with sulfur. Selenium, like sulfur, can exist in the $-II$ (selenide Se^{2-}), 0, $-IV$ (selenite SeO_3^{2-}) and $-VI$ (selenate

Table 10.15
Selenium Contents in Shoots of Accumulator and Non-Accumulator Species Growing on a Soil with 2–4 mg Se kg^{-1a}

Plant species	Content (mg Se kg^{-1} dry matter)
Astragalus pectinalus	4000
Stanleya pinnola	330
Gutierrezia fremontii	70
Zea mays	10
Helianthus annuus	2

[a]Based on Shrift (1969).

SeO_4^{2-}) oxidation states. From both soils and nutrient solutions plants take up selenate in strong preference to selenite (Asher *et al.*, 1977; Smith and Watkinson, 1984; Banuelos and Meek, 1989). Sulfate and selenate compete for common uptake sites in the roots and, thus, selenate uptake can be strongly decreased by high sulfate supply (Mikkelsen and Wan, 1990; Zayed and Terry, 1992). Soils also contain seleno amino acids such as selenomethionine which is readily taken up by wheat seedlings (Abrams *et al.*, 1990).

Plant species differ very much in selenium uptake and accumulation in the shoots and also in their capacity to tolerate high selenium concentrations in the rooting medium or in the shoot tissue or in both. A representative example of the differences between plant species in selenium accumulation is shown in Table 10.15. Based on these differences plants can be classified into *selenium-accumulators* and *nonaccumulators*, and those in between as *selenium-indicators*. Many species of the genera *Astragalus*, *Xylorrhiza* and *Stanleyea* are typical selenium-accumulators, and capable of growing on high selenium soils (seleniferous soils) without any detrimental effect on growth and reaching shoot selenium contents as high as 20–30 mg g^{-1} dry matter (Rosenfeld and Beath, 1964). However, within the genus *Astragalus* there are large differences between species and ecotypes in their capacity to accumulate selenium, the selenium content in accumulator types being 100–200 fold higher than in nonaccumulator types (Shrift, 1969; Davis, 1986).

Members of the Cruciferae such as black mustard and broccoli also accumulate relatively large amounts of selenium and might contain, and tolerate, several hundred μg Se g^{-1} shoot dry matter (Zayed and Terry, 1992). On the other hand, most agricultural and horticultural plant species are nonaccumulators (Shrift, 1981) and selenium toxicity can occur at contents below 100 μg Se g^{-1} (e.g., alfalfa) or less than 10 μg Se g^{-1} (e.g., wheat). Typically, the critical toxicity contents are much lower when selenite is supplied as compared with selenate (Mikkelsen *et al.*, 1989).

The large differences in selenium contents in plants first attracted attention in the 1930s when it was realized that selenium toxicity is responsible for certain disorders in animals grazing on native vegetation of seleniferous soils (Brown and Shrift, 1982; Miller *et al.*, 1991). The maximum tolerable levels of selenium in the diet depend on animal species and are in the range 1–5 μg Se g^{-1} dry matter (Miller *et al.*, 1991). Selenium toxicity can be classified into three types with decreasing severity: *acute*, which leads to rapid death; *chronic blind staggers*, manifested in blindness and paralysis, and *chronic alkali disease*, characterized by lameness and loss in vitality, the latter disease results from long-term consumption of feed containing $5\text{--}40\,\mu g$ Se g^{-1} dry matter. In agricultural areas in parts of central California high selenium contents in soils, drainage water from irrigation areas and factories have recently become of much concern as an environmental problem in general and a potential hazard for animals, and wild life in particular (Mayland *et al.* 1989; Mikkelsen *et al.*, 1989).

Earlier reports of a selenium requirement for high growth rates in accumulator species of *Astragalus* (for a review see Shrift, 1969) could not be confirmed by Broyer *et al.* (1972). These authors demonstrated that in plants grown in solution culture without selenium, toxic levels of phosphate accumulated in the leaves. The addition of selenium prevented this excessive phosphate uptake and thereby stimulated growth. At nontoxic phosphorus levels, selenium was without beneficial effect on the growth of accumulator plants of the genus *Astragalus*. This is another instructive example of the necessity of critically evaluating the so-called beneficial mineral elements. Nevertheless, in these accumulator species high selenium contents are presumably beneficial for the plants in terms of lower susceptibility to attack by a range of insects (Pate, 1983).

The large differences between plant species in tissue tolerance to selenium are causally related to differences in the detoxification of selenium. Sulfate and selenate (and selenite) have features in common not only in uptake and assimilation but also in that they compete for various enzymes in the sulfur assimilation pathway (Section 8.3.2), for example, ATP sulfurylase (Burnell, 1981), leading to the formation of selenium analogues of cysteine and methionine, namely selenocysteine and seleno-methionine (Fig. 10.10). In nonaccumulator plants, selenoamino acids are incorpor-ated into proteins which are either nonfunctioning or at least much less capable of functioning as enzyme proteins than the corresponding sulfur-containing proteins (Eustice *et al.*, 1981; Brown and Shrift, 1982). Incorporation of selenoamino acids is presumably particularly critical in enzymes with a sulfhydryl group (-SH) as catalytical site. In non-accumulator species the avoidance-strategy, i.e. restriction of selenium uptake, is therefore an important factor in tolerance to high selenium contents in the rooting medium (Wu and Huang, 1992). In contrast, in accumulator plants, the formation of selenomethionine seems to be impaired and selenocysteine transformed into nonprotein amino acids such as selenomethylcysteine (Fig. 10.10).

Similarities between sulfur and selenium metabolism in plants also exist in the production of volatile compounds released by aerial parts of plants (Section 4.1.2). The

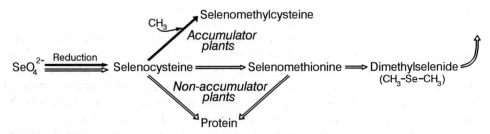

Fig. 10.10 Schematic presentation of selenium assimilation in accumulator and nonaccumulator species, and selenium volatilization. (Compiled from Burnell, 1981; Brown and Shrift, 1982; Zayed and Terry, 1992.)

main volatile selenide compound is dimethylselenide of which selenomethionine is the precursor (Fig. 10.10). The rates of selenium volatilization vary considerably between crop species. With a supply of 10 μM selenate rice, broccoli and cabbage volatilized 200–350 μg Se m^{-2} leaf area per day compared to less than 15 μg Se m^{-2} leaf area per day in sugar beet, lettuce and onion (Terry *et al.* 1992). In broccoli which accumulates up to several hundred μg Se g^{-1} dry weight the release rate of volatile selenium compounds is about seven times higher at low sulfate supply compared to high sulfate supply due to both inhibition of selenate uptake and competition within the plant at the sites of sulfur assimilation (Zayed and Terry, 1992). The sites of selenium assimilation within the plants, roots or shoots, seem to differ between supply of selenite and selenate; with selenite supply a much higher proportion is assimilated in the roots (Asher *et al.*, 1977; Gissel-Nielsen, 1987) which might explain at least in part the higher phytotoxicity of selenite, despite of its lower uptake rate compared with selenate (Smith and Watkinson, 1984).

In contrast to higher plants, selenium is an essential mineral element for animals (Miller *et al.*, 1991) and humans (Van Campen, 1991). To avoid selenium deficiency in animals the minimum requirement in the diet is in the range 0.1–0.3 μg Se g^{-1} dry matter. If this requirement is not met, in all livestock, and especially in the young, symptoms of selenium deficiency occur such as *white muscle disease, nutritional muscular dystrophy*, or *reproductive disorder* (Miller *et al.*, 1991).

In both humans and animals the only well-established function of selenium is its role as cofactor in glutathione peroxidase. This enzyme contains four atoms of selenium per molecule and is part of a multicomponent system which reduces hydrogen peroxide, lipid peroxides, and sterol peroxides and thereby protects cell constituents such as membranes from free radical and peroxide damage (Mikkelsen *et al.*, 1989; Van Campen, 1991). The function of the selenium-containing glutathione peroxidase in animals and humans is comparable to that of the glutathione reductase in higher plants, for example, in chloroplasts (Section 5.2.2). Interestingly, synthesis of a selenium-containing functioning glutathione peroxidase could be induced in the green alga *Chlamydomonas reinhardtii* by supplying selenium (Yokota *et al.*, 1988). So far, in higher plants no such studies have been conducted on the induction of a selenium-containing glutathione peroxidase. For a comprehensive recent review on selenium in higher plants the reader is referred to Läuchli (1993).

10.6 Aluminum

Aluminum is an abundant element representing about 8% of the earth crust. Aluminum concentrations in mineral soil solutions are usually much below 1 mg l^{-1} ($\sim$37 μM) at pH values higher than 5.5, but rise sharply at lower pH. The main interest in aluminum has been concerned with the ability of some plant species (accumulators) to tolerate high aluminum contents in their tissue, and the toxic effects on plant growth by high aluminum concentrations in soil or nutrient solutions (Section 16.3).

There is no convincing evidence that aluminum is an essential mineral element even for accumulator species. However, there are many reports on beneficial effects of low aluminum concentrations in the soil or nutrient solution on plant growth [for reviews see Bollard (1983) and Foy (1983)]. The aluminum concentrations at which growth stimulation has been observed vary between 71.4 μM and 185 μM in sugar beet, maize and some tropical legumes. In the tea plant, which is one of the most aluminum tolerant crop species, marked growth stimulation has been observed at aluminum concentrations as high as 1000 μM (Matsumoto et al., 1976) or even at 6400 μM (Konishi et al., 1985).

Studies with high aluminum concentrations are particularly difficult to interpret in terms of physiological responses as a high proportion, or nearly all, of the aluminum added is presumably lost by precipitation (e.g., with phosphate), or by polymerization and complexation. The nominal concentration of free aluminum is thus unknown, but certainly much lower than that applied (Asher, 1991). Supplying low concentrations of aluminum, however, has also led to distinct stimulation of root growth as observed in aluminium-tolerant genotypes, for example, of Zea mays (Clark, 1977), an effect which might be causally related to stimulation of root cap size and thereby enhanced activity of the apical meristem (Bennet and Breen, 1989; see also Section 14.3).

A general problem in most studies on the effect of low concentrations of aluminum on plant growth is the contamination of the nutrient solution with aluminum. Normally, the roots of plants growing in nutrient solutions with supposedly zero levels of aluminum contain 50–100 mg aluminum per kilogram dry matter (Bollard, 1983). In only a few experiments special care has been taken to keep the contamination as low as possible.

The nature of beneficial effects of aluminum on growth, especially of nonaccumulator species, is not clear, but there is substantial evidence that this is often a secondary effect, brought about by alleviation of toxicity caused by other mineral elements, particularly the mineral nutrients phosphorus and copper (Asher, 1991). An instructive example on this topic has been provided by Suthipradit (1991) in which aluminum activities in the nutrient solution of between 49 and 20.4 μM strongly enhanced root and shoot growth of peanut by depressing zinc uptake and shoot contents which were in the toxic range in plants without aluminum supply (Asher, 1991). As has been shown for wheat seedlings (Kinraide, 1993) alleviation of H^+ toxicity at low pH is another responsible factor for the enhancement of root growth by aluminum. This is just the opposite of what is known of the alleviation of aluminum toxicity by high H^+ concentrations (Section 16.3.4).

In conclusion, low concentrations of aluminum might have beneficial effects on growth under certain conditions, and this beneficial effect is probably a more general phenomenon in plant species with high aluminum tolerance and high capacity for aluminum uptake (accumulators). However, in nonaccumulators, negative effects of aluminum on plant growth in soils of low pH are the rule (Section 16.3).

10.7 Other Mineral Elements

The requirement for such mineral elements as iodine and vanadium is fairly well established for certain lower plant species, including marine algae (iodine) and fungi and freshwater algae (vanadium). The reports on the stimulation of growth of higher plants by iodine, titanium, and vanadium are rare and vague. An example of this is the effect of vanadium on the growth of tomato (Basiouny, 1984), or the effect of titanium on the growth (Pais, 1983), enzyme activities and photosynthesis (Dumon and Ernst, 1988) of various crop species. For further information on vanadium and iodine see Bollard (1983) and on titanium see Dumon and Ernst (1988).

More recently, interest has increased in the rare earth elements *lanthanum* (La) and *cerium* (Ce) for enhancement of plant growth. Mixtures of both elements are used on a large-scale in China as foliar sprays or seed treatment of agricultural and horticultural crop species. The amounts supplied are in the range typical for micronutrients. There are reports of substantial increases in plant growth and yield under field conditions which, however, require more careful documentation and reproduction under controlled conditions. For further informations see Asher (1991).

In the last two decades there has been a vast number of reports on the presence of heavy metals, such as cadmium, chromium, lead, and mercury, in higher plants. Most of these reports are concerned mainly with environmental pollution, the presence of heavy metals in the food chain, and genotypical differences in the critical toxicity levels of heavy metals in plants (Ernst and Joosse-van Damme, 1983). Convincing evidence of beneficial effects of these heavy metals on growth of higher plants is lacking.

11

Relationships between Mineral Nutrition and Plant Diseases and Pests

11.1 General

The effects of mineral nutrients on plant growth and yield are usually explained in terms of the functions of these elements in plant metabolism. However, mineral nutrition may also exert secondary, often unpredicted influences on the growth and yield of crop plants; by effecting changes in growth pattern, plant morphology and anatomy, and particularly chemical composition, mineral nutrients may either increase or decrease the resistance or the tolerance of plants to pathogens and pests. Whereas resistance is mainly determined by the ability of the host to limit penetration, development and/or reproduction of the invading pathogen, or limit the feeding of pests, tolerance is characterized by the ability of the host plant to maintain its own growth despite the infection or pest attack. Depending on the mineral nutrient (or beneficial mineral element), the nutritional status of the plants, host plant species and type of pathogen and pest, mineral nutrition might affect resistance or tolerance. In this section representative examples are given of the effects of mineral nutrition on both resistance and tolerance.

Considerable progress has been made in breeding and selection for increased resistance or tolerance to diseases and pests. Resistance can be increased by changes in anatomy (e.g., thicker epidermal cells and a higher degree of lignification and/or silification), and in physiological and biochemical properties (e.g., higher production of inhibitory or repelling substances). Resistance can be particularly increased by altering plant responses to parasitic attacks through enhanced formation of mechanical barriers (lignification) and the synthesis of toxins (phytoalexins). Apparent resistance can be achieved when the most susceptible growth stages of the host plant are not synchronized with the period of highest activity of parasites and pests (known as 'escape from attack', or 'outgrowing' the pathogen; Huber, 1980).

Although resistance and tolerance are genetically controlled, they are considerably influenced by environmental factors. In this context mineral nutrition of plants can be considered as an environmental factor that can be manipulated relatively easily. Although frequently unrecognized this factor has always been an important component of disease control (Huber and Wilhelm, 1988). For example, liming of soils or

Table 11.1
Nitrogen Fertilizer Supply and the Incidence of Leaf Blotch (*Rhynchosporium scalis*) in Spring Barley Cultivars[a]

Nitrogen supply (kg ha^{-1})	Flag leaf area infected by leaf blotch (%)		
	Proctor	Cambrinus	Deba Abed
0	0.4	15.4	3.6
66	1.3	21.3	20.5
132	4.5	30.5	57.3

[a]Based on Jenkyn (1976).

application of mineral fertilizers in different amounts and forms not only affects the growth and composition of the plants directly but also has profound effects on microbial activity in the soil and rhizosphere and plant resistance and tolerance to root and shoot pathogens and pests indirectly. On the other hand, symptoms of mineral nutrient deficiency in plants are often induced by soilborne root diseases or pests, impairing root growth and activity.

As a rule, the influence of mineral nutrition on plant resistance is relatively small in highly susceptible or highly resistant cultivars but very substantial in moderately susceptible or partially resistant cultivars. This is illustrated in an example in Table 11.1 on the effects of nitrogen fertilizer on leaf blotch in three barley cultivars. As the nitrogen supply is increased, the incidence of leaf blotch rises in all three cultivars. However, the absolute levels of infection, as expressed as a percentage of flag leaf affected, are different. With Proctor the increase is of no physiological or economic importance, whereas with the other two cultivars detrimental effects on photosynthesis and on grain yield would be expected.

The close correlation between nitrogen supply and leaf blotch shown in Table 11.1, however, cannot be generalized, to all fungal and parasitic diseases. Usually, a 'balanced' nutrient supply which ensures optimal plant growth is also considered optimal for plant resistance. Such an ideal situation which is, however, not the rule, is shown in Fig. 11.1 for pelargonium plants. An inverse relationship exists between nutrient supply and plant growth, on the one hand, and severity of bacterial infection, on the other. From this finding, one can conclude that plants with an optimal nutritional status have the highest resistance to diseases and that susceptibility increases as nutritional status deviates from this optimum. With respect to tolerance, there is a general pattern that in plants suffering from mineral nutrient deficiency, tolerance to diseases and pests is lower and can be increased by the supplying the deficient nutrient. Such a relationship is based on the fact that more vigorously growing plants usually have a higher capacity to compensate, for example, for losses of photosynthates or leaf and root surface area due to infection or feeding.

The interactions between higher plants and parasites and pests are complex, and to give a short outline of the role of mineral nutrients in these interactions requires considerable simplification. Nevertheless, there are some principal areas of host–parasite interactions where the roles of mineral nutrients and beneficial mineral

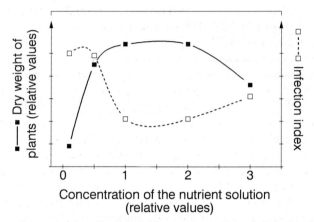

Fig. 11.1 Effect of the nutrient solution concentration on growth (noninfected plants) and on degree of infection (inoculation) with bacterial stem rot (*Xanthomonas pelargonii*) in *Pelargonium*. Relative values; water only = 0; basic nutrient solution = 1; twofold concentration of nutrient solution = 2; threefold concentration of nutrient solution = 3. (Modified from Kivilaan and Scheffer, 1958.)

elements are not only well established, but are predictable and can readily be demonstrated. It is the aim of this chapter to highlight these interactions with a few representative examples in order to demonstrate both the potential possibilities and the limitations of disease and pest control by mineral nutrition and fertilizer application. Comprehensive reviews on this subject have been presented, either at a general level (Fuchs and Grossmann, 1972; Huber, 1980, 1989; Graham, 1983), on micronutrients (Graham and Webb, 1991), and for particular mineral nutrients such as nitrogen (Huber and Watson, 1974), potassium (Perrenoud, 1977), and manganese (Huber and Wilhelm, 1988).

11.2 Fungal Diseases

11.2.1 Principles of Infection

As a rule, the germination of spores on leaf and root surfaces is stimulated by the presence of plant exudates. The flow of exudates contributes to the success or failure of infection in most fungal diseases by air- and soilborne pathogens. The rate of flow and composition of exudates depends on the cellular concentration and the corresponding diffusion gradient (Fig. 11.2). The concentrations of sugars and amino acids are high in leaves, for example, when potassium is deficient. Amino acid concentrations are also high when nitrogen supply is excessive (Section 8.2). The concentration of photosynthates in the apoplast and at the leaf surface – and also at the root surface – depends on the permeability of the plasma membrane. On average, the concentrations of amino acids and sugars in the apoplasm of both leaf and stem tissue is in the range of 1–8 mM

(Hancock and Huisman, 1981) but may rise considerably with calcium or boron deficiency (which causes increased membrane permeability) and potassium deficiency (which impairs polymer synthesis).

The concentration of soluble assimilates in the apoplasm of the host is an important factor for the growth of the parasite during the penetration and postinfection phases. Only a few groups of plant parasites are truly intracellular with direct access to assimilates in the symplast (Hancock and Huisman, 1981). Some parasites, such as powdery mildew of barley, have access only to epidermal cells. In these cases, the physical and chemical properties of the epidermal cells are of much more importance with respect to susceptibility and resistance than those of the bulk leaf tissue (Hwang *et al.*, 1983). In epidermal cells of barley, more than 90% of the soluble carbohydrates are β-cyanoglucosides which are likely to be of particular importance in resistance against powdery mildew (Pourmohseni and Ibenthal, 1991). Epidermal cells of leaves (Kojima and Conn, 1982) and stems and roots (Barz, 1977) are also characterized by much higher contents of phenolic compounds and flavonoids (i.e., substances with distinct fungistatic properties). The role of mineral nutrients in phenol metabolism is well documented, and examples of phenol accumulation have been discussed in relation to boron and copper deficiency (Chapter 9).

Most parasitic fungi and also bacteria invade the apoplasm by releasing pectolytic enzymes, which dissolve the middle lamella (Fig. 11.2). The activity of these enzymes is strongly inhibited by Ca^{2+}, which explains the close correlation between the calcium content of tissues and their resistance to fungal and bacterial diseases (see Section 11.2.4).

During infection a range of interactions occur between the hyphae and the host cells (Fig. 11.2). Inducible resistance mechanisms are associated mainly with the epidermis,

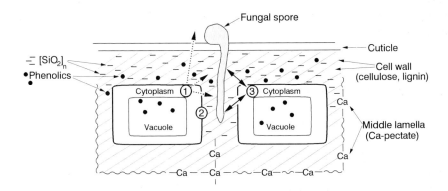

①   Diffusion of low-molecular-weight assimilates (sugars, amino acids)
②   Plasma membrane permeability
③   Interactions between fungus/epidermal cell (formation of toxins, phenolics)

Fig. 11.2   Schematic representation of the penetration of a fungal hypha on the leaf surface into the epidermal cell layer (apoplasm), and some factors which affect the penetration and growth rate of the hypha and are closely related to mineral nutrition.

the effectiveness of these mechanisms depending on the type of pathogen and the resistance of the host. Mineral nutrients and the mineral nutritional status of plants are involved in these mechanisms. Pectic enzymes of the parasite not only dissolve the middle lamella but these enzymes, or the products of pectin breakdown, also increase passive permeability of the plasma membrane and enhance K^+ efflux and H^+ influx which probably trigger hypersensitive reactions such as localized necrosis (Atkinson *et al.*, 1986). In other pathogenic fungi such as leaf spot (*Helminthosporium cynodontis* Marig.) fungal toxins induce this enhanced K^+ efflux and deplete cells and infected tissues of potassium. Accordingly, severity of disease symptoms (leaf spotting) strongly decreases with increase in potassium content of the leaves (Richardson and Croughan, 1989).

Phenolic compounds play a key role in early stages of infection (Fig. 11.2), either as phytoalexins or as precursors of lignin and suberin biosynthesis. For example, glucans of the cell wall of *Phytophthora megasperma* elicit the synthesis of isoflavones which function as phytoalexins and simultaneously lead to rapid and massive accumulation of phenolic polymers at the infection sites (Graham and Graham, 1991). Within a few hours after infection a 'signal' is transmitted to noninfected leaves which also increases their phenol synthesis (Rasmussen *et al.*, 1991). As several mineral nutrients, boron and copper in particular, have a profound influence on the biosynthesis and also binding form of phenols (Sections 9.3 and 9.7) the nutritional status of plants influences these defense responses, as has been discussed for boron by Lewis (1980b). The content of phenolics is also often high in nitrogen deficient plants and both the content of these substances (Kiraly, 1964) and their fungistatic effects (Kirkham, 1954) can decrease when the supply of nitrogen is large. A negative relationship has been observed between increasing nitrogen supply and the content of the phytoalexin stilben and resistance to downy mildew in leaves of grapevine (Bavaresco and Eibach, 1987).

In response to pathogen infection in many instances also the production of oxygen radicals (e.g. O_2^- and $OH^·$) and hydrogen peroxide (H_2O_2) is increased as a component of plant defense responses (Sutherland, 1991). This response might contribute to the hypersensitive reactions (oxidation of membrane lipids, leading to cell death), initiation of cell wall lignification, and injury of the pathogen. In view of the role of copper, zinc, iron and manganese in both generation and detoxification of oxygen radicals and hydrogen peroxide, plant resistance is markedly influenced by mineral nutrition at the level of host plant–pathogen interactions.

As tissues (particularly leaves) mature, lignification or the accumulation and deposition of silicon in the epidermal cell layer or both processes, may form an effective physical barrier to hyphal penetration (Fig. 11.2). Both lignification and silicon deposition are affected by mineral nutrition in various ways. These processes provide the main structural resistance of plants to diseases (and pests), especially in the leaves of grasses (Sherwood and Vance, 1980), and in the endodermis of roots (Section 10.3.2).

11.2.2 Role of Silicon

Grasses in general and wetland rice in particular are typical silicon accumulator plants (Section 10.3). As the silicon supply increases the silicon content of the leaves also rises, inducing a corresponding decline in susceptibility to fungal diseases such as rice blast

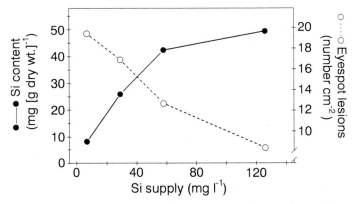

Fig. 11.3 Silicon content and susceptibility to blast fungus (*Pyricularia oryzae* Cav.) of fully expanded rice leaves. (Modified from Volk *et al.*, 1958.)

(Fig. 11.3). The increase in resistance (which manifests itself by a fall in the number of eyespots) appears to be related directly to the silicon concentration in the external solution and in the leaves.

The limitations of silicon in the control of fungal disease are also evident from Fig. 11.3. Silicon is translocated in the xylem preferentially to mature leaves (Section 3.2.4). Rice blast infection, however, occurs mainly in young leaves. As shown in Fig. 11.4, with maturation (full expansion at about day 8) and ageing of the leaves, resistance increases rapidly and becomes virtually complete whether the silicon supply is high or low. Nevertheless, the effect of silicon on the resistance of young leaves is substantial and its application should almost eliminate the effects of enhanced susceptibility to rice blast fungus when high levels of nitrogen are supplied (Osuna-Canizalez *et al.*, 1991). This profound inhibitory effect of silicon on fungal diseases is not confined to graminaceous species which are typical silicon accumulators but is also well docu-

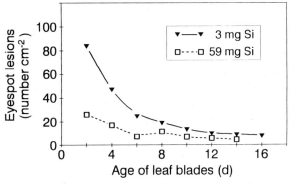

Fig. 11.4 Decline in the number of eyespot lesions (which indicates increased resistance to blast fungus) with the age of rice leaves and silicon concentration in the nutrient solution of 3 mg Si l^{-1} and 59 mg Si l^{-1}. (Modified from Volk *et al.*, 1958.)

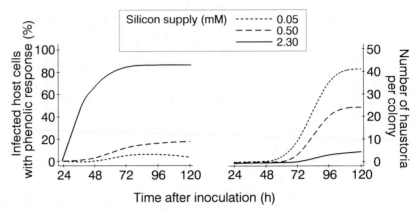

Fig. 11.5   Effect of silicon supply to leaf segments of cucumber (*Cucumis sativus*) on infection
with *Sphaerotheca fuliginea*. (Based on Menzies *et al.*, 1991.)

mented in other species in protection against powdery mildew, as for example, in
cucumber (Miyake and Takahashi, 1983; Adatia and Besford, 1986; Samuels *et al.*,
1991) or in grapevine (Bowen *et al.*, 1992).

Formation of a physical barrier in epidermal cells against the penetration of hyphae
(Fig. 11.2) or insects such as aphids (Section 11.5) is, however, not the only mechanism
by which silicon contributes to plant resistance against diseases and pests. There is also a
much more dynamic component of redistribution of silicon involved around the
infection peg (Heath and Stumpf, 1986). For example, in wheat silicon accumulates at
the sites of hyphal penetration of powdery mildew (Leusch and Buchenauer, 1988a),
and in barley within 20 h this accumulation is 3–4 times higher around the unsuccessful
infection sites than around successful ones (Carver *et al.*, 1987). This preferential
accumulation of silicon at the point of pathogen penetration requires a continuous
supply of silicon from the roots (Samuels *et al.*, 1991) or foliar sprays of silicon (Bowen
et al., 1992), indicating that after deposition and polymerization in the leaf tissue this
silicon cannot be remobilized.

Despite the close relations between silicon accumulation at the sites of penetration
and inhibition of hyphal growth and formation of haustoria, it is not the silicon itself that
has the detrimental effect but obviously phenolics which accumulate at the infection
sites (Fig. 11.5). The causal connections between silicon supply and accumulation of
phenolics at the infection sites are not clear; silicon might form weak complexes with
phenolics (Section 10.3) and thereby enhance synthesis and mobility of phenolics in the
apoplasm. Rapid deposition of phenolics or lignin at the sites of infection is a general
defense mechanism to pathogen attack and the presence of soluble silicon obviously
facilitates this resistance mechanism (Menzies *et al.*, 1991).

11.2.3   Role of Nitrogen and Potassium

There is an extensive literature on the effect of both nitrogen and potassium on parasitic
diseases because their role in modulating disease resistance is quite readily demon-

Table 11.2
Tentative Summary of the Effects of Nitrogen and Potassium Levels on the Severity of Diseases
$(+ \rightarrow ++++)$ Caused by Parasites[a]

Pathogen and disease	Nitrogen level		Potassium level	
	Low	High	Low	High
Obligate parasites				
Puccinia ssp. (rust diseases)	+	+++	++++	+
Erysiphe graminis (powdery mildew)	+	+++	++++	+
Facultative parasites				
Alternaria ssp. (leaf spot diseases)	+++	+	++++	+
Fusarium oxysporum (wilt and rot disease)	+++	+	++++	+
Xanthomonas ssp. (bacterial spots and wilt)	+++	+	++++	+

[a]Based on Kiraly (1976) and Perrenoud (1977).

strated and, furthermore, is of particular importance for fertilizer application. However, the results are often inconsistent and in some cases controversial, mainly for two reasons. (a) It is not clearly stated whether the levels of these mineral nutrients represent a deficiency, an optimal supply, or an excessive supply (see Fig. 11.1). (b) The distinction between obligate and facultative parasites in the infection pattern is not considered. As shown in Table 11.2, a high nitrogen supply increases the severity of infection by obligate parasites but has the opposite effect on diseases caused by facultative parasites, such as *Alternaria* and *Fusarium*, and most bacterial diseases, for example, *Xanthomonas* spp. In contrast to nitrogen, potassium elicits uniform responses: high concentrations increase the resistance of host plants to both obligate and facultative parasites. For soilborne pathogens the situation is more complex particularly with nitrogen and therefore the interactions between mineral nutrition and soilborne fungal and bacterial diseases are discussed in a separate section (11.4).

The principal differences in the response of obligate and facultative parasites to nitrogen are shown in Fig. 11.6. The susceptibility of wheat plants to stem rust, caused

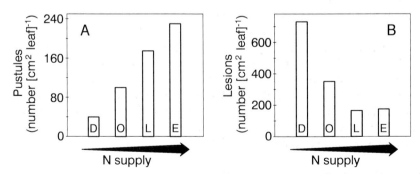

Fig. 11.6 Number of pustules of stem rust (*Puccinia graminis* ssp. *tritici*) in wheat (A) and number of necrotic lesions caused by bacterial spot (*Xanthomonas vesicatoria*) in tomato (B) grown in nutrient solutions with increasing nitrogen concentration. D, deficient; O, optimal; L, luxurious; E, excessive. (Based on Kiraly, 1976.)

by an obligate parasite, increases with increasing nitrogen supply, the nitrogen-deficient plants being the most resistant. In contrast, the susceptibility of tomato plants to bacterial leaf spot, caused by a facultative parasite, decreases with increasing nitrogen supply at levels required for optimal growth of the host plant.

These differences in response are based on the nutritional requirements of the two types of parasite. Obligate parasites rely on assimilates supplied by living cells. On the other hand, facultative parasites are semisaprophytes which prefer senescing tissue or which release toxins in order to damage or kill the host plant cells. As a rule, all factors which support the metabolic activities of host cells and which delay senescence of the host plant also increase resistance or tolerance to facultative parasites.

The role of nitrogen in increasing the susceptibility of host plants to obligate fungal parasites (Table 11.1) is a matter of concern in both agricultural and horticultural practice. This effect is related to both the nutritional requirements of the parasite and changes in the anatomy and physiology of the host plant in response to nitrogen. As discussed in Section 8.2, nitrogen in particular enhances growth rate, and during the vegetative stage of growth the proportion of young to mature tissue shifts in favor of the young tissue, which is more susceptible. In addition an increase in amino acid concentration in the apoplast and at the leaf surface seems to have a greater influence than increase in sugar concentration on the germination and growth of conidia (Robinson and Hodges, 1981). When plants are supplied with high amounts of nitrogen, the activity of some key enzymes of phenol metabolism (Matsuyama and Dimond, 1973) and the content of some phenolics (Kiraly, 1964) and of lignin may also be lower. This is because of the lower content of phenols which are precursors in the biosynthesis of lignin, together with the higher proportion of young tissue in plants grown at high nitrogen (Section 8.2.5). For example, the lignin content of rice leaves from plants supplied with high nitrogen was 500 μg per 100 g dry weight as compared with 1100 μg in low nitrogen plants (Matsuyama, 1975).

A decrease in silicon content is another change usually observed in plants in response to increasing nitrogen supply (Grosse-Brauckmann, 1957; Volk et al., 1958). This, however, is an unspecific response referred to as *dilution by growth*. The various anatomical and biochemical changes together with the increase in the content of low-molecular-weight organic nitrogen compounds as substrates for parasites are the main factors responsible for the close correlation between nitrogen supply and susceptibility to obligate parasites.

With potassium the situation is less complex. Potassium decreases the susceptibility of host plants to both types of parasites. This effect can be quite dramatic, as shown in Fig. 11.7 for stem rot in rice. In this situation the disease could be effectively controlled by the application of potassium fertilizer. In most cases, the effect of potassium is confined to the deficiency range; that is, potassium-deficient plants are more susceptible than potassium-sufficient plants to parasitic diseases of both groups. As a rule, susceptibility decreases (i.e., resistance or tolerance increase) in response to potassium in the same way as plant growth responds to increasing potassium supply (Fig. 11.8). Beyond the optimal potassium supply for growth, no further increase in resistance can be achieved by increasing the supply of potassium and its content in plants.

Results similar to those shown in Fig. 11.8 have been obtained with oil palms infected with *Fusarium* (Ollagnier and Renard, 1976) and wheat infected with stripe rust

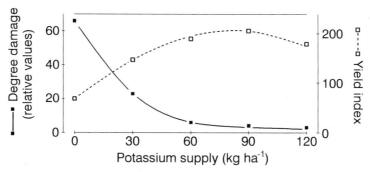

Fig. 11.7 Effect of potassium supply on grain yield of wetland rice and incidence of stem rot (*Helminthosporium sigmoideum*). Basal dressing of nitrogen and phosphorus constant at 120 and 60 kg ha^{-1}, respectively. (Based on Ismunadji, 1976.)

(Kovanci and Colakoglu, 1976). In both cases, the increase in resistance was confined to the deficiency range.

The high susceptibility of potassium-deficient plants to parasitic diseases is related to the metabolic functions of potassium (Section 8.7). In deficient plants the synthesis of high-molecular-weight compounds (proteins, starch, and cellulose) is impaired and low-molecular-weight organic compounds accumulate. In the deficiency range an increase in the potassium supply therefore leads to an increase in growth and a decrease in the content of low-molecular-weight organic compounds until growth is maximal. A further increase in supply and plant content of potassium is without substantial effect on the organic constituents of plants and thus usually on resistance or tolerance. The characteristic pattern of these changes is given in Fig. 11.9.

In the deficiency range, potassium-induced growth enhancement also causes non-specific decreases in the content of other mineral elements (dilution by growth). Beyond maximal growth there may continue to be a slight decrease in the levels of other

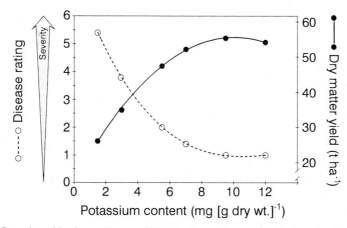

Fig. 11.8 Severity of leaf spot disease (*Helminthosporium cynodontis*) and dry matter yield in 'Coastal' bermudagrass (*Cynodon dactylon* L. Pers.) versus leaf potassium content. (Reproduced from Matocha and Smith, 1980, by permission of the American Society of Agronomy.)

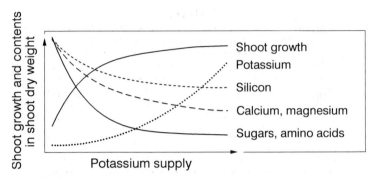

Fig. 11.9 Tentative scheme of growth response and major changes in plant composition with increasing potassium supply.

cations such as calcium and magnesium due to competition at uptake sites in the roots (Section 2.5). In plants receiving a suboptimal calcium supply the risk of both calcium deficiency-related physiological disorders and disease susceptibility may therefore increase (see Section 11.2.4).

The relationship between potassium and resistance is more complex in seeds and fruits that are supplied with potassium primarily by retranslocation from vegetative organs. In certain soybean varieties the incidence of pod blight rapidly increases late in the season in the upper pods, and this is correlated with a sharp drop in the potassium content of the pods. With an exceptionally high soil application of potassium (410–1640 kg ha^{-1}) the percentage of infected seeds can be reduced from 75% to 13% whereas the seed yield is only marginally increased (Crittenden and Svec, 1974).

11.2.4   Role of Calcium and Other Mineral Nutrients

The calcium content of plant tissues affects the incidence of parasitic diseases in two ways. First, calcium is essential for the stability of biomembranes; when the calcium content is low, the efflux of low-molecular-weight compounds (e.g., sugars) from the cytoplasm into the apoplasm is enhanced. Second, calcium polygalacturonates are required in the middle lamella for cell wall stability (Section 8.6). Many parasitic fungi and bacteria invade plant tissue by producing extracellular pectolytic enzymes such as polygalacturonase, which dissolves the middle lamella (Section 11.3.2). The activity of this enzyme is drastically inhibited by calcium (Bateman and Lumsden, 1965). The susceptibility of plants to infection with such parasites is therefore inversely related to the calcium content of the tissue, as shown in Table 11.3. In this experiment the total concentration in the nutrient solution of three cations, K^+, Ca^{2+}, and Mg^{2+}, was kept constant and only the K^+/Ca^{2+} ratio was altered. Thus a decrease in the calcium content of the plants was correlated with an increase in the potassium content. Additional experiments showed that an increase in potassium content does not necessarily lead to an increase in infection as long as the calcium content of plants is kept at a high level.

In soybean, 'twin stem' abnormality is endemic on many acid tropical soils. Necrosis of the apical meristem occurs, and the plants are simultaneously heavily infected with

Table 11.3
Relationship Between Cation Content and Severity of
Infection with *Botrytis cinerea* Pars. in Lettuce[a]

Cation content (mg g^{-1} dry wt)			Infection with *Botrytis*[b]
K	Ca	Mg	
14.4	10.6	3.2	4
23.8	5.4	4.1	7
34.2	2.2	4.7	13
48.9	1.8	4.2	15

[a]Based on Krauss (1971).
[b]Infection index: 0–5 slight infection; 6–10 moderate infection; 11–15 severe infection.

Sclerotium ssp. (Muchovej and Muchovej, 1982). Increasing the calcium supply suppresses both fungal infection and twin stem. It seems likely that the latter abnormality is the direct result of calcium deficiency (symptoms of which are apical meristem necrosis and loss of apical dominance) and that the fungal infection is a secondary event.

Various parasitic fungi preferentially invade the xylem and dissolve the cell walls of conducting vessels. This leads to plugging of the vessels and subsequent wilting symptoms (e.g., *Fusarium* wilt). In comparing the degree of infection of tomato plants with *Fusarium oxysporum* in relation to the level of calcium supply, the plants were severely infected when the calcium concentration in the xylem sap was below about 5 mM at low external supply. Nearly all plants were healthy, however, at high external calcium supply when the calcium concentration in the xylem sap was raised to about 25 mM (Corden, 1965).

Plant tissues low in calcium are also much more susceptible than tissues with normal calcium levels to parasitic diseases during storage. This is of particular concern in the case of fleshy fruits with their typically low calcium content. During storage the fruits are more susceptible not only to so-called physiological disorders (Section 8.6) but also to fungal diseases that cause fruit rotting (Fig. 11.10). Calcium treatment of fruits before storage is therefore an effective procedure for preventing losses both from physiological disorders and from fruit rotting.

There is an extensive literature on the effects of micronutrients on parasitic diseases (Graham, 1983; Graham and Webb, 1991). Of the various defense mechanisms available to plants the phenolics and lignin are the most well understood, and of the micronutrients, at least boron, manganese and copper play a key role in phenol metabolism and lignin biosynthesis (Chapter 9). Micronutrients can also affect resistance indirectly. In deficient plants not only might the defense mechanism be impaired but often plants also become a more suitable feeding substrate. With zinc deficiency, a leakage of sugars onto the leaf surface of *Hevea brasiliensis* increases the severity of infection with *Oidium* (Bolle-Jones and Hilton, 1956). In boron-deficient wheat plants the rate of infection with powdery mildew is severalfold higher than that in boron

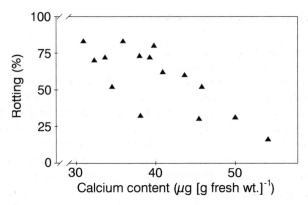

Fig. 11.10 Relationship between the calcium content of apples (cv. Cox orange) and the incidence of rotting due to *Gloesporium perennans* infection after the apples were stored for 3 months at 3°C. (Modified from Sharpless and Johnson, 1977.)

sufficient plants, and the fungus also spreads more rapidly over the plant (Schütte, 1967).

Copper has been extensively used as a fungicide, but the amounts required are at least 10–100 times higher than those required by the plants and that are used, for example, as foliar sprays to correct copper deficiency. Perhaps the best evidence of an effect of copper on host plant resistance to disease is when copper is applied to the soil and depresses leaf infections, for example of powdery mildew in wheat (Graham, 1980b), or to control stem pathogens by foliar application of copper (Table 11.4). For depression of stem and leaf pathogens, foliar application of copper is often more effective than soil application (Table 11.4) as most of the copper applied to soil is strongly bound to the soil matrix. The results in Table 11.4 also illustrate that yield increases particularly by micronutrient fertilizer application might be achieved by a combination of both improvement of the plant nutritional status and thereby metabolic activity and growth, and suppression of parasitic diseases.

Table 11.4

Effect of Copper Fertilization on Stem Melanosis (caused by *Pseudomonas cichorii*) in Wheat Grown on a Copper-Deficient Soil[a]

Treatment	Cu fertilizer (kg Cu ha^{-1})	Percent disease	Grain yield (kg ha^{-1})
Nil (control)	—	92	294
CuSO$_4$, banded	10	76	511
CuSO$_4$, incorporated	10	34	2016
CuSO$_4$, foliar spray	10	6	2116
CuChel, foliar spray	2	7	2505

[a]Based on Mahli *et al.* (1989).

11.3 Bacterial and Viral Diseases

11.3.1 Bacterial Diseases

Bacterial diseases, which are caused by various facultative parasites, can be divided into three main types: leaf spot diseases, soft rots, and vascular diseases (Grossmann, 1976). In leaf spot diseases (e.g., bacterial leaf blight, *Xanthomonas oryzae*), pathogens usually enter the host plant through the stomata. Thus the epidermal layer is a rather ineffective barrier to infection. Having entered the plant, the bacteria spread and multiply in the intercellular spaces. The effect of the mineral nutritional status of the host plant on spreading and multiplication is similar to its effect on facultative fungal parasites: for example, the multiplication and severity of the disease are enhanced when potassium and calcium contents are deficient and often (Kiraly, 1976), but not always (Fuchs and Grossmann, 1972), when nitrogen is deficient.

Spreading of bacteria within the host tissue, as in many fungal diseases, is facilitated by polygalacturonases and related pectolytic enzymes. Accordingly, the resistance of plants is closely related to their calcium content, as shown in Table 11.5 for bean plants. In infected tissue the activity of pectolytic enzymes is very high but inversely related to the calcium content of the tissue. The severity of the disease symptoms also reflects the role of calcium in resistance.

Bacterial vascular diseases spread within plants through the xylem; they lead to 'slime' formation and finally plugging of the vessels ('bacterial wilt'). The severity of such a disease, the bacterial canker in tomato is inversely correlated with the calcium content in the shoot tissue as long as calcium is in the deficiency range for the nutrition of the host plant (Table 11.6). Calcium is effective in both the susceptible and the resistant cultivar, indicating that the resistance of a cultivar is dependent on adequate calcium supply. At each level of calcium supply the resistant cultivar has higher calcium and magnesium contents, but lower potassium contents than the susceptible cultivar (Berry *et al.*, 1988). Selections of cultivars for higher potassium uptake efficiency might therefore bear a certain risk of negative effects on plant resistance to certain bacterial and fungal diseases.

Table 11.5

Relationship between the Calcium Content of Bean, the Activity of Pectolytic Enzymes in the Plant Tissue, and the Severity of Soft Rot Disease Caused by *Erwinia carotovora*[a]

Calcium content (mg g^{-1} dry wt)	Pectolytic activity (relative units)[b]				Severity of symptoms[c]
	Polygalacturonase		Pectate transeliminase		
	−	+	−	+	
6.8	0	62	0	7.2	4
16	0	48	0	4.5	4
34	0	21	0	0	0

[a]From Platero and Tejerina (1976).
[b]+, Bacterial inoculation; −, no inoculation.
[c]4 = Complete decay of plants within 6 days; 0 = no symptoms.

Table 11.6
Relationship between Calcium Supply, Calcium Content and Bacterial Canker
Disease (*Clavibacter michiganense* ssp. *michiganense* (Smith)) in a Susceptible
and a Resistant Tomato Cultivar[a]

Ca supply (mg l^{-1})	Ca content (% in shoot dry matter)		Disease development (% wilted leaves)	
	Moneymaker	Plovdiv 8/12	Moneymaker	Plovdiv 8/12
0	0.12	0.14	84	56
100	0.37	0.42	27	12
200	0.43	0.55	37	6
300	0.44	0.58	27	8

[a]Based on Berry *et al.* (1988).

The calcium nutritional status might affect plant resistance to bacterial diseases not only via stabilization of the middle lamella. Calcium is also causally involved in the hypersensitive responses to bacterial infections. In tobacco, hypersensitive reactions to *Pseudomonas syringae* requires a massive influx of Ca^{2+} from the apoplasm into the cytoplasm through Ca^{2+} channels in the plasma membrane, leading to enhanced K^+/H^+ exchange, cytoplasm acidification and death of the host cells at the infection sites (Atkinson *et al.*, 1990), comparable to the hypersensitive responses to fungal diseases (Section 11.2).

11.3.2 Viral Diseases

The multiplication of viruses is confined to living cells, and their nutritional requirements are restricted to amino acids and nucleotides. Compared with fungal diseases, bacterial diseases and pests, there are relatively few data in the literature on the effects of mineral nutrition on viral diseases. As a rule, nutritional factors which favour the growth of the host plant also favour viral multiplication. This holds true particularly for nitrogen and phosphorus (Fuchs and Grossmann, 1972; Huber, 1980), but the same tendency is observed for potassium (Perrenoud, 1977). The relationships between mineral nutrition and viral diseases are often not clear for various reasons. In deficient plants growth stimulation by mineral nutrient supply might lead to elimination of the symptoms of viral disease because the plants 'outgrow' the disease, or the symptoms are hidden. Sugar beet yellow or potato leaf roll virus symptoms might disappear when the supply of nitrogen is large even though the plants are severely infected. Furthermore, in sugar beet visual symptoms of beet mild yellowing virus (BMYV) are very similar to symptoms of manganese deficiency. When grown in manganese-deficient soils the percentage of plants with BMYV was high, and with foliar application of manganese, visual symptoms of manganese deficiency and of BMYV disappeared although the percentage of infected plants only decreased from 75% to 40% (Fritzsche and Thiele, 1984).

In many instances mineral nutritional status of the host plant might influence viral diseases indirectly via the vectors, which are mainly fungi and insects. It is assumed that about 60% of plant viruses are spread by aphids (Dreyer and Campbell, 1987), and the severity of plant attacks by aphids is strongly affected by plant nutritional status (Section 11.5). An instructive example in the role of the vector has been presented by Tomlinson and Hunt (1987). Infection of watercress with the fungal pathogen *Spongospora subterranea*, leading to crook root disease, can be effectively depressed by supraoptimal supply of zinc, i.e., levels which exceed requirements for host plant growth. This high zinc supply also very effectively suppresses watercress chlorotic leaf spot virus through control of its vector, the fungus.

11.4 Soilborne Fungal and Bacterial Diseases

The population density of microorganisms at the root surface is several times higher than that in the bulk soil. Microorganisms at the root surface include various pathogens. Competition between and repression of microorganisms, as well as chemical barriers (e.g., high concentrations of polyphenols in the rhizodermis; Barz, 1977) and physical barriers (e.g., silicon deposition at the endodermis; see Section 10.3) ensure that microbial invasion of both roots and shoots via the roots is restricted. Invasion and infection by certain microorganisms, however, is beneficial for higher plants (e.g., rhizobia or mycorrhizae).

Mineral nutrition affects soilborne fungal and bacterial diseases in various ways. For example, in manganese-deficient Norway spruce, fungistatic activity against *Fomes amosus* (Fr.) Cooks is much lower in the inner bark of roots, leading to heart root disease (Wenzel and Kreutzer, 1971). When the plant is supplied with high manganese and low nitrogen levels, both the content of these nutrients and the fungistatic activity of the inner bark increase (Alcubilla *et al.*, 1971). The incidence of infection of growing potato tubers with *Streptomyces scabies*, giving rise to common potato scab, is reduced either by lowering the soil pH or by application of manganese (Mortvedt *et al.*, 1961; McGregor and Wilson, 1964). Manganese exerts its influence not by increasing the resistance of the tuber tissue to the fungus but by directly inhibiting the vegetative growth of *S. scabies* before the onset of infection (Mortvedt *et al.*, 1963).

In peanut, preharvest pod rot is caused by severe infection with *Pythium myriotylum* and *Rhizoctonia solani*. The occurrence of this disease is closely related to the calcium content of the pod tissue and can be kept at a low level by the soil application of calcium (e.g., as gypsum) (Hallock and Garren, 1968). There are also close correlations between calcium content and rot disease in potato. The bacterial soft rot disease caused by various species of *Erwinia*, e.g., *E. carotovora* ssp. *carotovora*, can be effectively suppressed by increasing calcium contents in the peel in a range where calcium is not a limiting factor for the nutrition of the host plant (Kelman *et al.*, 1989).

In the relationships between mineral nutrition of plants and soilborne pathogens root rot disease in wheat and barley (take-all) caused by *Gaeumannomyces graminis* attracts particular attention for mainly two reasons: it seriously limits grain production in many regions of the world, and severity of the disease can be effectively controlled by mineral nutrition of the host plant (for reviews see Graham, 1983; Huber and Wilhelm, 1988; Graham and Webb, 1991). The fungus has a growth optimum at pH 7 and is very

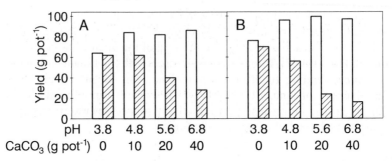

Fig. 11.11   Effect of liming and inoculation with *Gaeumannomyces graminis* var. *tritici* (take-all) on (A) straw yield and (B) grain yield of spring wheat (*Triticum sativum*). Open bars, noninoculated; striped bars, inoculated. (Modified from Trolldenier, 1981.)

sensitive to low pH. A decrease in the severity of take-all is observed even at a pH below 6.8. Liming of acid soils therefore increases the risk of root infections and yield losses by take-all. Figure 11.11 shows that, in a soil of pH 3.8, inoculation with *G. graminis* was without significant effect on growth or yield. Liming enhanced growth and increased yield in noninfected plants but had a severely depressing effect on infected plants. The incidence of take-all became more severe as the soil pH increased. Take-all is therefore a particular problem in wheat and barley production on calcareous soils.

Manganese availability in the rhizosphere and manganese content of the roots play a key role in root infection and severity of take-all, but also of other soilborne fungal diseases (Huber and Wilhelm, 1988; Graham and Webb, 1991). All factors which decrease the availability of manganese increase the severity of take-all (e.g., increase in soil pH by liming, nitrate versus ammonium fertilizers; Huber, 1980, 1989), and vice versa. The capacity of the roots to restrict penetration of fungal hyphae into the root tissue by enhanced lignification at the infection sites is impaired in manganese-deficient plants, as manganese, as well as copper, is required for biosynthesis of lignin (Graham and Webb, 1991; Chapter 9). Furthermore, *G. graminis*, like many other soilborne pathogenic fungi, is a powerful oxidizer of manganese, although considerable differences exist between isolates in their oxidation power and, thus capacity to decrease manganese availability (Wilhelm *et al.*, 1990).

Suppression of take-all by soil application of manganese fertilizers is also possible under field conditions (Brennan, 1992a) but has its limitations on calcareous soils because of rapid oxidation and immobilization of manganese in soils. Foliar sprays of manganese are not effective because of the poor phloem mobility of manganese (Section 3.3). The use of ammonium instead of nitrate nitrogen fertilizer is an effective procedure in control of take-all (Section 11.6).

In wheat zinc deficiency potentiates the severity of *Rhizoctonia* root rot, and zinc supply decreases disease score. This decrease, however, is confined to the concentration range where the zinc supply limits plant growth (Thongbai *et al.*, 1993).

Another approach is attractive as a future goal, namely the use of *Pseudomonas fluorescens* spp. as a biocontrol of take-all. Various strains of these bacteria are very effective in their capacity to reduce manganese oxides and suppress growth of *G.*

Table 11.7
Effect of Copper and Gypsum Application on Growth, Yield and on Root Infection with Take-All (*Gaeumannomyces graminis*) of Winter Wheat Grown in Copper Deficient Soil[a]

Treatment	Dry weight (g per pot)	Ears (no. per pot)	Grain (g per pot)	Infected plants (%)
Nil (control)	8.54	2.80	4.33	100
CuSO$_4$, soil	12.68	3.70	6.51	83
CuSO$_4$, foliar	12.82	3.33	6.06	100
CaSO$_4$, soil	9.81	2.70	5.41	83
CuSO$_4$ + CaSO$_4$, soil	16.96	4.67	8.98	0

[a]Based on Gardner and Flynn (1988).

graminis var. *tritici* at least *in vitro*, whereas the suppression *in vivo* is related not only to the manganese-reducing capacity of the bacterial strains (Marschner, P. *et al.*, 1991). Production of toxic substances such as cyanide by particular *P. fluorescens* strains might be involved in the suppression (Sarniguet *et al.*, 1992a, b). In soils of high pH, in particular, the suppression of *G. graminis* brought about by ammonium fertilizer application is probably not only related to rhizosphere acidification but also to quantitative and qualitative changes in *P. fluorescens* ssp. populations in favor of those which are antagonistic to *G. graminis* (Sarniguet *et al.*, 1992a,b).

The capacity of *P. fluorescens* strains to produce siderophores has also been considered as an important factor in suppression of soilborne pathogens by depriving the pathogens of iron (Kloepper *et al.* 1980: Höfte *et al.*, 1991). Production of toxins such as cyanide, however, is probably of more general importance in this respect (Schippers *et al.*, 1990).

Root infection with VA mycorrhiza is another factor which might suppress soilborne pathogens such as *Fusarium oxysporum* in tomato (Dehne and Schönbeck, 1979a,b), or wilt diseases in casuarina (Gunjal and Paril, 1992), an aspect that is discussed in more detail in Section 15.10.

The severity of take-all in wheat is not only much higher in manganese-deficient plants but also in plants suffering from nitrogen or phosphorus deficiency (Brennan, 1989, 1992b) and copper deficiency (Table 11.7). In the case of nitrogen and phosphorus deficiency the decrease in severity by supply of nitrogen and phosphorus fertilizer is most likely an expression of higher tolerance by more vigorous growth rather than increase in resistance. In contrast, in copper-deficient plants impaired biosynthesis of lignin is one of the most obvious metabolic changes and supply of copper fertilizers overcomes this impairment and thereby increases resistance. Soil and foliar applications of copper have quite different effects (Table 11.7). Foliar application increased yield but did not depress root infection with take-all indicating that despite its phloem mobility the copper concentrations at the root–soil interface and at the infection sites did not reach levels required for suppression of the pathogen. The most spectacular effect was achieved by combination of copper and calcium (gypsum) supply to the soil, probably by enhanced desorption and mobility of copper in the soil.

Table 11.8

Relationship between Mineral Nutrient Deficiencies, Number of Squash Bugs (*Anasa tristis*) per Plant, and Soluble Nitrogen Content of Squash[a]

Nutrient supply	Squash bugs (no. per plant)	Soluble nitrogen ($\mu g\ g^{-1}$ fresh wt)
Complete	1.70	32.1
−N	0.66	4.5
−P	2.11	93.7
−K	2.45	98.9
−S	3.42	143.7

[a]From Benepal and Hall (1967).

11.5 Pests

Pests are animals such as insects, mites, and nematodes which are harmful to cultivated plants. In contrast to fungal and bacterial pathogens, they have digestive and excretory systems and their dietary requirements are often less specific. Furthermore, visual factors such as color of leaves are important for 'recognition' or 'orientation'. For example, many aphid species tend to settle on yellow-reflecting surfaces (Beck, 1965). The main types of resistance of host plants to pests are: (a) physical (e.g., color, surface properties, hairs); (b) mechanical (e.g., fibers, silicon); (c) chemical/biochemical (e.g., content of stimulants, toxins, repellents). Mineral nutrition can affect all three factors to varying degrees.

Generally, young or rapidly growing plants are more likely to suffer attack by pests than are old and slow-growing plants. Therefore, there is often a positive correlation between nitrogen application and pest attack as has been shown, for example, with the whitebacked planthopper *Sogatella furcifera* (Horwáth) in rice (Salim and Saxena, 1991); in contrast, potassium-deficient plants suffered much more attack than potassium-sufficient plants. Although sugars can act as a feeding stimulant (Beck, 1965) most sucking insects such as the rice brown planthopper (Sogawa, 1982) depend much more on amino acids (Dreyer and Campbell, 1987). This is illustrated in Table 11.8 for squash bugs. Nitrogen-deficient plants showed the lowest feeding response. The number of squash bugs per plant was clearly related to the content of total soluble nitrogen in the leaves. In contrast, the protein content of the leaves did not appear to be associated with feeding preference (Benepal and Hall, 1967).

A high content of amino acids in the plants is therefore an important component for the severity of attack by sucking parasites. High contents of amino acids in plants are typical for either high nitrogen supply, or impaired protein synthesis due to certain deficiencies such as potassium or zinc. An example of nutrient imbalance in plants induced by fertilizer application is shown in Table 11.9 for oak trees attacked by cup-shield lice. Magnesium applied alone or especially when applied in combination with nitrogen and phosphorus increased the nutrient imbalance (giving rise to more severe potassium deficiency) which resulted in a corresponding increase in soluble nitrogen content (i.e. a more favorable diet for the lice). In principle, similar results had been

Table 11.9
Effects of Fertilizers Applied on a Soil Low in Available Potassium on Infestation of Oak Trees
(*Quercus pendula*) by Cup-Shield Lice (*Eulecanium refulum* Ckll.)[a]

	Fertilizer			
	K + Mg	N + P + K + Mg	Mg	N + P + Mg
No. of lice per 10-cm stem section	0.72	0.82	4.32	8.78

[a]Based on Brüning (1967).

found in citrus between the application of various mineral fertilizers and attack by purple and black scale (Chaboussou, 1976).

The close positive correlations between nitrogen supply, amino acid content and wide N/C ratio of plants, and attack by pests, which is often found in field crops is sometimes generalized for other plants and for ecosystems (Scriber and Slansky, 1981; Chapin *et al.*, 1987). However, the interactions between plants and pests are more complex and not confined to the amino acid and N/C ratio of the feed. This is particularly evident in trees where the pest attacks often depend much more on repellents or toxic compounds than on nitrogen content. For example, in *Salix dasyclados* grown at different nutrient levels and light intensities, damage of the leaves by the herbivor *Galerucella lineola* was inversely related not only to the phenol content (high light >> low light) but also to the nitrogen content of the leaves (Larsson *et al.*, 1986). In Scots pine (*Pinus sylvestris*) high nitrogen fertilization increased both the contents of nitrogen and of diterpenoids in the needles. Since these latter compounds act as allelochemicals to herbivorous insects, nitrogen fertilization did not alter the performance of the sawfly *Neodiprion sertifer* larvae on the needles (Björkman *et al.*, 1991), confirming earlier results on sawfly and caterpillars feeding on needles or leaves of forest trees (Merker, 1961).

Differences in leaf content of allelochemicals (products of secondary metabolism involved in interactions of living organisms) are obviously also responsible for the striking negative relationships between boron content and attack of leaves of oil palm seedlings by red spider mites (Table 11.10). In plants without or with very low boron supply the attack was very high but distinctly depressed as the boron supply and leaf content of the flavonoid cyanidin increased. Boron is required for the biosynthesis of cyanidin and related polyphenols which are causally involved in resistance against these insects.

Epidermal cell walls containing silicon deposits act as a mechanical barrier to the stylet and particularly the mandibles of sucking and biting insects. It has been demonstrated that the mandibles of larvae of the rice stem borer are damaged when the silicon content of rice plants is high (Jones and Handreck, 1967). The physical properties of leaf surfaces are also of great importance in regulating the severity of attack by sucking insects. Labial exploration of the surface takes place before insertion of the stylet into the tissue (Sogawa, 1982). Changes in the surface properties of leaves were presumably the main reason for the marked decrease in the attack of wheat plants by aphids when several foliar sprays containing sodium silicate were applied (Fig.

Table 11.10

Relationship between Boron Supply and Intensity of Red Spider Mite (*Tetra-nychus pieroei*) Attack on Oil Palm (*Elaeis guieensis*)[a]

			Correlation between feeding holes and leaf cyanidin content	
B supply ($mg \, l^{-1}$)	Mites (no. m^{-2})	Feeding holes (no. cm^{-2})	Feeding holes (no. cm^{-2})	Cyanidin ($\mu g \, g^{-1}$)
0	1.8	67	60–65	2–5
0.5	1.7	60	30–50	10–18
5.0	1.2	30		
50	1.0	20	10–30	20–32
500	0.9	17		
1000	0.9	12		

[a]Based on Rajaratnam and Hock (1975).

11.12). As the nitrogen supply increased, so also did the number of aphids of the species *Sitobion avenae*. However, foliar sprays with silicon reduced the number to below that in the nitrogen deficient (−N) plants. The results of this experiment also illustrate the difficulties of making generalizations about the relationship between increasing nitrogen supply and attack by sucking insects. In contrast to *S. avenae*, which is a typical ear feeder, the other aphid species, *Metoplophium dirhodum*, did not respond to increasing nitrogen supply. Differences in feeding habits and in preferences for plant organs (*M. dirhodum* prefers leaf blades) are possible reasons for the differences in response to nitrogen.

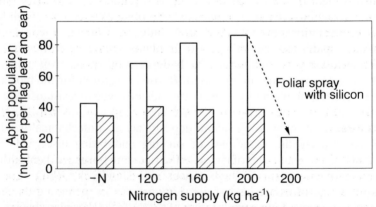

Fig. 11.12 Effect of nitrogen supply and foliar sprays containing silicon (1% Na_2SiO_2) on population density of two aphid species in winter wheat. Striped bars, *Metopolophium dirhodum*; open bars, *Sitobion avenae*; −N denotes nitrogen-deficient control plant. (Based on Hanisch, 1980.)

In wetland rice several species of leaf hoppers pose a more serious threat as vectors of viruses than as juice-sucking pests (Beck, 1965), which is another important reason for controlling sucking insects.

The striking depression of aphid populations on leaves after foliar application of silicon (Fig. 11.12) is presumably not only the result of changes in surface properties of the leaves but also of increase in soluble silicon within the leaf tissue. For the rice brown planthopper soluble silicic acid, rather than the deposited silicon in the leaves, is the effective sucking inhibitor. Silicon concentrations as low as 10 mg l^{-1} appear to be effective (Sogawa, 1982). Similarly, as with the pathogenic fungi, aphids also release pectinase into the tissue, and in resistant cultivars depolymerization of the middle lamella is slower than in nonresistant cultivars; similar differences exist between the cultivars in the rate of phytoalexin synthesis (Dreyer and Campbell, 1987). It might be that soluble silicon in the apoplasm of the leaves enhances the host response mechanisms to attack by sucking insects in a way similar to that against fungal invasion (Section 11.2.2).

Plant growth can be severely depressed by nematodes such as cereal cyst nematodes (*Heterodera avenae*) or root-lesion nematodes (*Pratylenchus penetrans*) in apple trees. Root exudates might act as signals for recognition or as repellents, but it is not clear whether mineral nutrition plays an important role in this context. There are, however, many examples showing that nematodes severely depress root growth and activity and thereby mineral nutrient uptake and the nutritional status of the plants. For example, nematodes are mainly responsible for potassium deficiency in apple replant disease (Merwin and Stiles, 1989). As shown in cotton plants, nematode attacks have much less or no effects on shoot growth at high supply of potassium but depress shoot growth severely when the potassium supply to the soil is low, although the total number of nematodes was higher in the plants receiving the high potassium supply (Oteifa and Elgindi, 1986). This is a typical example of an increase in tolerance to pests and diseases resulting from mineral nutrient supply. Such an example can also be demonstrated for micronutrients (Fig. 11.13). In barley plants grown in a manganese-deficient soil with and without manganese supply, despite a similar number of infections (immature females) in both treatments the growth of plants which were not supplied with manganese fertilizer was severely depressed, but not affected in the manganese fertilized plants. In this case the manganese application possibly offset the impaired capacity of manganese uptake caused by the nematode infection.

11.6 Direct and Indirect Effects of Fertilizer Application on Diseases and Pests

Under field conditions fertilizer application affects plant diseases and pests directly via the mineral nutritional status of the plants, and indirectly by producing dense stands and alterations in light interception and humidity within a crop stand. In addition the timing of application is an important factor, and for nitrogen in particular. This is illustrated in Table 11.11 for the relationship between time of ammonium fertilization and incidence of take-all and grain yield in wheat. Severity of take-all infection is high even without nitrogen fertilization and is increased by application of ammonium in the

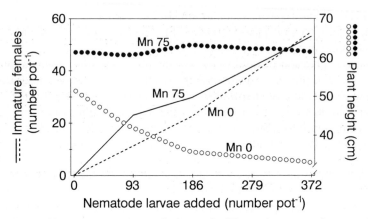

Fig. 11.13 Interaction of manganese and nematode (*Heterodera avenae*) on number of infec-tions (immature females) and height of barley plants growing in a manganese-deficient soil supplied without (Mn 0) or with 75 mg Mn $(450 \, g)^{-1}$ soil (Mn 75). (Based on Wilhelm *et al.*, 1985.)

fall, leading to severe yield depressions. In contrast, the same amount of ammonium nitrogen supplied in spring dramatically suppresses take-all, and high grain yields are obtained. The split application of nitrogen in fall and spring demonstrates that the effects of nitrogen fertilizer application on grain yield was governed more by the effects on take-all than on the nitrogen nutritional status of wheat per se. Ammonium nitrogen applied in the fall is rapidly nitrified and acts as a nitrate nitrogen source and thereby intensifies take-all on nonsuppressive soils. The use of timed ammonium fertilizer application is therefore a practical approach to suppress take-all, and variations in suppression between years and locations (Christensen *et al.*, 1987; MacNish, 1988) are probably related to the severity of the incidence and the rate of nitrification prior to the uptake by the crop.

The form of nitrogen fertilizer applied might also have other implications on pathogens in the shoots. Solubility of silicon in soils is dependent on various factors,

Table 11.11

Effect of Rate and Time of Ammonium Nitrogen Fertilizer Application on Root Infection with Take-All (*Gaeumanomyces graminis*) and Grain Yield of Winter Wheat[a]

Time of application	Rate $(kg \, N \, ha^{-1})$	Take-all index[b]	Grain yield $(kg \, ha^{-1})$
0	0	1.9	2610
Fall	83	2.8	1740
Spring	83	0.1	5290
Fall + Spring	83 + 28	1.9	2350

[a]Huber (1989).
[b]Take-all index: 0 = no infection; 4 = 100% infection.

Table 11.12
Effect of the Form of Nitrogen Fertilizer on Silicon Content and Disease Incidence of
Powdery Mildew (*Erysiphe graminis*) in Spring Wheat Grown in Soil Supplemented
with either Lime ($CaCO_3$) or Blast-Furnace Lime (BFL)[a]

Nitrogen form	Silicon content ($\%$ SiO_2 in leaf dry matter)		Disease incidence ($\%$ leaf area affected)	
	$CaCO_3$	BFL	$CaCO_3$	BFL
$Ca(NO_3)_2$	1.2	2.3	27.5	11.5
$(NH_4)_2SO_4$	2.3	7.3	18.0	2.0

[a]Recalculated from Leusch and Buchenauer (1988b).

increasing, for example, as the pH of the soil falls. In agreement with this, the silicon content in plants is not only dependent on silicon fertilization but also at least to some extent, on the form of nitrogen fertilizer applied. As shown in Table 11.12, in comparison with the application of calcium nitrate, ammonium sulfate increases silicon content in spring wheat and therefore depresses the incidence of powdery mildew, which is inversely related to the silicon content of the plants.

There are also many reports of chloride fertilizer application in macronutrient amounts leading to the suppression of various diseases, both soilborne such as take-all in wheat (Christensen *et al.*, 1987) or root rot (*Cochliobolus sativus*) in barley (Timm *et al.*1986), and of leaf diseases such as leaf rust (*Puccinia recondita*) in wheat (Fixen *et al.*, 1986a,b). The mechanism of this effect of chloride fertilizers is not clear. Chloride might act directly in the plant by improving the water balance and thereby, tolerance to diseases, or indirectly in the soil via inhibition of nitrification or enhancement of manganese mobilization (Graham and Webb, 1991).

The various effects of mineral nutritional status and of fertilizer application on diseases and pests are of direct relevance to disease and pest control by fungicides, pesticides and other chemicals. Mineral fertilizer application might thus substitute or at least reduce the demand of chemical disease control in some instances, although in other cases, however, it might also increase the demand. These interactions are illustrated in a final example in Fig. 11.14 for winter wheat naturally infected by yellow rust. In temperate climate, high levels of nitrogen applied to winter wheat early in the growing season favor abundant tillering and dense, tall crops stands and thus conditions favorable for infection. As a consequence, chemical disease control is not only most effective but usually essential with high levels of nitrogen.

The effects of both level and timing of nitrogen supply on the rust infection are striking, the most severe infection occurring with the large early single dressing (N 1.0). Split application of nitrogen decreased infection to a large extent in the early growth stages, but after the second application (at anthesis) fungal growth increased rapidly. Nevertheless, the epidemic was considerably postponed by the split application. In the plants not receiving nitrogen (N0) the infection remained low. Similar results have been obtained in wheat infected with powdery mildew (Darwinkel, 1980b).

Without chemical disease control, the infection decreased the grain yield in all

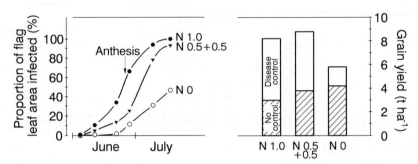

Fig. 11.14   Relationship between level and timing of nitrogen supplied to winter wheat and yellow rust infections (*Puccinia striiformis* Westend) and grain yield with and without chemical disease control. N 1.0, 160 kg N ha^{-1} as early dressing; N 0.5 + 0.5, split application: 80 kg N early and 80 kg N at anthesis; N0, no nitrogen fertilizer. (Based on Darwinkel, 1980a.)

treatments (Fig. 11.14). The extent to which the yield was decreased, however, differed between treatments, being very great in the nitrogen treatments but relatively small in the no nitrogen treatment. Thus, without chemical disease control, the highest grain yield was obtained in plants which received no nitrogen, and with disease control the highest yield was obtained in plants receiving the split application of nitrogen.

12

Diagnosis of Deficiency and Toxicity of Mineral Nutrients

12.1 Nutrient Supply and Growth Response

The well-known growth (dry matter production) versus nutrient supply curve (growth response curve) has three clearly defined regions (Fig. 12.1). In the first, the growth rate increases with increasing nutrient supply (deficient range). In the second, the growth rate reaches a maximum and remains unaffected by nutrient supply (adequate range). Finally, in the third region, the growth rate falls with increasing nutrient supply (toxic range).

In crop production, optimal nutrient supply is usually achieved by the application of fertilizer. Rational fertilizer application requires information on the nutrients that are available in the soil, on the one hand, and the nutritional status of the plants, on the other. The possibilities and limitations of using visual diagnosis and plant analysis as a basis for recommending whether or not to use fertilizer, and if so of what type and quantity, are discussed in this chapter.

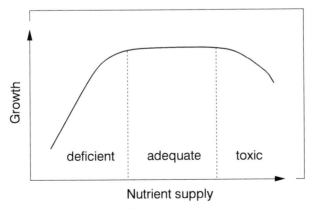

Fig. 12.1 Relationship between nutrient supply and growth.

12.2   Diagnosis of Nutritional Disorders by Visible Symptoms

As a rule, nutritional disorders that inhibit growth and yield only slightly are not characterized by specific visible symptoms. Symptoms become clearly visible when a deficiency is acute and the growth rate and yield are distinctly depressed. However, there are exceptions. For example, transient visible symptoms of magnesium deficiency in cereals which can sometimes be observed under field conditions during stem extension are without detrimental effect to the final grain yield (Pissarek, 1979). Furthermore, many annual and perennial plant species of the natural vegetation, particularly those adapted to nutrient-poor sites, adjust their growth rate to the most limiting nutrient and, thus, visible deficiency symptoms do not develop (Chapin, 1983, 1988).

Diagnosis based on visible symptoms requires a systematic approach as summarized in Table 12.1. Symptoms appear preferentially on either older or younger leaves, depending on whether the mineral nutrient in question is readily retranslocated (Section 3.5). The distribution pattern of symptoms might also be modified by the method employed to induce deficiency, i.e., permanent insufficient supply or sudden interruption of a high supply (Section 3.5).

Chlorosis or necrosis and the pattern of both are important criteria for diagnosis. As a rule, visible symptoms of nutrient deficiency are much more specific than those of nutrient toxicity, unless the toxicity of one mineral nutrient induces a deficiency of another. Visible deficiency symptoms of individual nutrients have been described

Table 12.1
Some principles of visual diagnosis of nutritional disorders

Plant part	Prevailing symptom	Disorder
		Deficiency
Old and mature leaf blades → Chlorosis	→ Uniform	N (S)
	→ Interveinal or blotched	Mg (Mn)
↘ Necrosis	→ Tip and marginal scorch	K
	→ Interveinal	Mg (Mn)
Young leaf blades and apex ↗ Chlorosis	→ Uniform	Fe (S)
	→ Interveinal or blotched	Zn (Mn)
→ Necrosis (chlorosis)		Ca, B, Cu
↘ Deformations		Mo (Zn, B)
		Toxicity
Old and mature leaf blades ↗ Necrosis	→ Spots	Mn (B)
	→ Tip and marginal scorch	B, salt (spray injury)
↘ Chlorosis, necrosis		Nonspecific toxicity

briefly in Chapters 8 and 9. For details (including colour pictures) of symptoms of nutrient disorders the reader is referred to Wallace (1961) and Bergmann (1988, 1992).

Diagnosis may be especially complicated in field-grown plants when more than one mineral nutrient is deficient or when there is a deficiency of one mineral nutrient and simultaneously toxicity of another. Such simultaneously occurring deficiencies and toxicities present difficulties in diagnosis and can be found in practice as for example, in waterlogged acid soils, where both manganese toxicity and magnesium deficiency may occur (complex symptoms). Diagnosis may be further complicated by the presence of diseases, pests, and other symptoms caused, for example, by mechanical injuries including spray damage. In order to differentiate the symptoms of nutritional disorders from these other symptoms, it is important to bear in mind that nutritional disorders always have a typical symmetric pattern: leaves of the same or similar position (physiological age) on a plant show nearly identical patterns of symptoms, and there is a marked gradation in the severity of the symptoms from old to young leaves (Table 12.1).

In order to make a more precise visual diagnosis, it is helpful to acquire additional information, including soil pH, results of soil testing for mineral nutrients, soil water status (dry/waterlogged), weather conditions (low temperature or frost) and the application of fertilizers, fungicides, or pesticides. In some instances the type and amount of fertilizer to be used can be recommended on the basis of visual diagnosis immediately. This is true of foliar sprays containing micronutrients (iron, zinc, or manganese) or magnesium. In other instances (e.g., iron deficiency chlorosis), however, visual diagnosis is an inadequate basis for making fertilizer recommendations. Nevertheless, it offers the possibility of focusing further attention on chemical and biochemical analysis of leaves and other plant parts (plant analysis) of selected mineral nutrients. This is of particular importance for annual crops, because the results are required immediately and seasonal fluctuations in the nutrient content of the plants often do not justify the high cost of carrying out a complete mineral nutrient analysis.

12.3 Plant Analysis

12.3.1 General

The use of chemical analysis of plant material for diagnostic purposes is based on the assumption that causal relationships exist between growth rate of plants and nutrient content in the shoot dry or fresh matter, or in the nutrient concentration in the tissue press sap. Mineral element composition of plant tissues is usually expressed as *content* per unit dry or fresh weight (e.g., mg g^{-1} dry wt). Although the term concentration is often used synonymously, it refers in the strict sense to a volume (e.g., mg l^{-1}). Depending on the mineral nutrient, plant species and age, the most suitable plant part or organ differs for this purpose, as well as whether or not the total content or only a certain fraction of the mineral nutrient (e.g., water extractable) should be determined. In general the nutritional status of a plant is better reflected in the mineral element content of the leaves than in that of other plant organs. Thus leaves are usually used for plant analysis. For some species and for certain mineral nutrients, nutrient contents in the dry matter may differ considerably between leaf blades and petioles, and sometimes the petioles are a more suitable indicator of nutritional status (Bouma, 1983). In fruit trees the analysis of the fruits themselves is a better indicator, especially for calcium and boron in relation to fruit quality and storage properties (Bould, 1966). Under certain climatic conditions, drought stress during seed filling in particular, in legumes seed content of zinc appears to be a more sensitive parameter to zinc fertilizer supply as, for example, foliar analysis (Rashid and Fox, 1992).

Samples from field-grown plants are often contaminated by dust or sprays and require washing procedures (Jones, 1991; Moraghan, 1991). The most severe problems in the use of plant analysis for diagnostic purposes are the often short-term fluctuations in nutrient contents (e.g., 'dilution effects' by fast growth). It is particularly difficult to establish mineral nutrient contents which are considered to reflect deficiency, sufficiency or toxicity range in relation to the effects of environmental factors as well as plant genotype and developmental stage of plants and leaves. For example, the percentage of dry matter usually increases with age of plants or organs (Walworth and Sumner, 1988) or at elevated CO_2 concentrations because of starch accumulation (Kuehny et al., 1991) with a corresponding decline in the critical deficiency concentrations of mineral nutrients. At elevated CO_2 concentrations lower critical leaf contents of nitrogen in leaves of C_3 species are also related to less Rubisco, whereas critical leaf contents of phosphorus tend to be higher for reasons which are not clear (Conroy, 1992). Strict standardization of sampling procedure and availability of suitable reference data are therefore of crucial importance for foliar analysis. The use of nutrient ratios instead of contents is another approach to meet this difficulty (Section 12.3.4). For more recent reviews on plant analysis for diagnostic purposes the reader is referred to Reuter and Robinson (1986); Martin-Prével et al. (1987); Bergmann (1988, 1992); Walworth and Sumner (1988); Jones (1991), and Heinze and Fiedler (1992).

12.3.2 Relationship between Growth Rate and Mineral Nutrient Content

A representative example on the relationships between plant growth and mineral nutrient contents in shoots is shown in Fig. 12.2 for manganese. Both under controlled environmental conditions and under field conditions the critical deficiency content

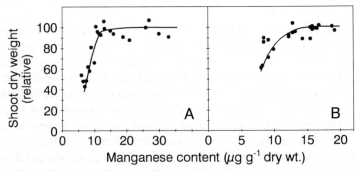

Fig. 12.2 Relationship between manganese content in youngest emerged leaf blades and shoot dry weight in barley grown in a growth chamber (A) and under field conditions (B). (From Hannam et al., 1987.)

(defined also as 'level') of manganese in the dry weight of the youngest emerged leaf blade of barley plants is in the range of 12 μg g^{-1} dry matter, if 95% of the maximal shoot dry weight (or yield) is used as reference point. It should be kept in mind that the critical deficiency contents (CDC) for example of the youngest emerged leaf blade is not necessarily identical to the CDC at the sites of new growth, the shoot meristem where the CDC might be much higher as has been shown for zinc in rice plants (Section 9.4.10).

Usually, a figure of 90% of the maximum dry matter yield is taken as a reference point in order to define the CDC of a mineral nutrient (Bouma, 1983; Ohki, 1984). In a low input system the reference point is sometimes only 80% of the maximum dry matter yield, the CDCs are correspondingly lower in the shoot dry matter, for example for phosphorus in maize and cowpea (Smyth and Cravo, 1990). Differences in reference point are often overlooked in comparisons of literature data on CDC.

The method employed for determination of the CDC is also important and contributes to different values, as has been shown for magnesium in subterranean clover, where the values of CDC for plants provided permanently with a low supply have been compared with plants where a high supply was suddenly interrupted (Scott and Robson, 1990b). It is unlikely that the latter procedure reflects the situation under field conditions. Sudden interruption of nutrient supply leads to unusual high CDCs for various nutrients (Burns, 1992), as fast growing plants become suddenly totally dependent on remobilization and retranslocation of mineral nutrients.

In view of the various factors discussed above, and also in the following sections, the use of a 'range' of values (e.g., critical deficiency range) would seem more appropriate and realistic than a single value (e.g., 12μg g^{-1}). At least it should be borne in mind that the use of a single value encompasses a range of contents and that the probability of deficiency, or sufficiency increases, with the extent of deviation from this single value. The general pattern of relationships between plant growth and mineral nutrient contents in plant tissue is shown in Fig. 12.3 in a schematic presentation. There is an ascending portion of the curve where growth either increases sharply without marked increase in nutrient content (I and II) or where increases in growth and mineral nutrient content are closely related (III). This is followed by a more or less level portion where growth is not nutrient limited and nutrient content markedly increases (IV and V) and, finally, by a portion where excessive nutrient content causes toxicity and a corresponding decline in growth (VI).

Occasionally, with an extreme deficiency of copper (Reuter *et al.*, 1981) or zinc (Howeler *et al.*, 1982b), for example, a C-shaped response curve is obtained (Fig. 12.3, region I, dashed line) in which a nutrient-induced increase in growth rate is accompanied by a decrease in its content in the dry matter, which is often referred to as the 'Piper-Steenbjerg' effect (Bates, 1971). Possible explanations for this type of response are a lack of remobilization from old leaves and stem (Reuter *et al.*, 1981) or necrosis of the apical meristems with a corresponding cessation of growth despite further uptake of small amounts of the mineral nutrient by extremely deficient plants.

Concentration and dilution effects of mineral nutrients in plants are common phenomena which must be carefully considered in interpreting nutrient contents in terms of ion antagonism or synergism, or both during uptake. This holds true in particular when the contents of mineral nutrients are in the deficiency or toxicity range

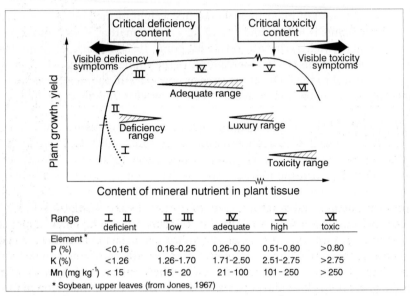

Fig. 12.3 Relationship between nutrient content (e.g., mg g^{-1} dry matter) and growth or yield (*upper*) and examples of the nutrient content in the dry matter of soybean leaves in the various nutrient supply ranges (*lower*).

(Jarrell and Beverly, 1981). If, for example, the contents of two mineral nutrients are in the deficiency range and only one of them is supplied, growth enhancement causes a 'dilution' of the other mineral nutrient (a decrease in its content in the dry matter) and severe deficiency is induced without any competition occurring in uptake or within the plant.

Central to the use of plant analysis for diagnostic purposes are the critical deficiency and toxicity contents (Fig. 12.3). Growth is maximal between the critical deficiency and toxicity contents, but for practical reasons the 90–95% of maximal growth value rather than the maximal value itself is chosen as a reference point. The mineral nutrient contents can be grouped into ranges, as shown in the lower portion of Fig. 12.3 for soybean. If nutrient contents are in the adequate range there is a high statistical probability that these nutrients are not growth-limiting factors. Certainly, contents in the luxury range further decrease the risk that these nutrients will become deficient under conditions unfavorable for root uptake (e.g., dry topsoil) or when the demand is very high (e.g., retranslocation to fruits). However, there is a greater risk of growth reduction by direct toxicity of these nutrients or by their effect in inducing a deficiency of other nutrients, i.e. nutrient imbalance (Section 12.3.5). In defining critical toxicity contents the heterogenous distribution of a nutrient within a plant organ has to be considered, for example, of boron in leaf blades (Section 9.7.9).

12.3.3 Developmental Stage of Plant and Age of Leaves

Next to the mineral nutrient supply, the physiological age of a plant or plant part is the most important factor affecting mineral nutrient content in the plant dry matter. With

Table 12.2
Critical Deficiency Content (CDC) of Copper (at Maximum Yield) in Subterranean Clover and Plant Organs and Plant Age[a]

	Age of plant (days after sowing)				
Plant part	26	40	55	98	F[b]
Whole plant tops	3.9	3.0	2.5	1.6	1.0
Youngest open leaf blade	3.2	~3	~3	~3	~3

[a]CDC expressed as mg g^{-1} dry weight. Based on Reuter *et al.* (1981).
[b]Early flowering.

the exception of calcium and sometimes iron (Sanchez-Alonso and Lachica, 1987b) and boron (Section 9.7) there is usually a fairly clear decline in mineral nutrient content in the dry matter as plants and organs age. This decline is caused mainly by a relative increase in the proportion of structural material (cell walls and lignin) and of storage compounds (e.g., starch) in the dry matter. Mineral nutrient contents corresponding to the adequate or critical deficiency range are therefore lower in old than in young plants. For example, in grain sorghum the CDC of phosphorus in the leaf dry matter has been shown to decrease from about 0.4% to 0.2% throughout the growing season (Myers *et al.*, 1987). In field-grown barley the potassium content in the shoot dry matter decreased from 5–6% in young plants to about 1% towards maturation, although the plants were well supplied with potassium (Leigh *et al.*, 1982). In this instance the decline in content was exclusively a 'dilution effect' as the potassium concentration in the tissue water (mainly representative of vacuolar sap) remained fairly constant at ~100 mM throughout the season.

Complications arising from changes in the critical deficiency content with age can be lessened by sampling tissues at specific physiological ages. For example, as shown in Table 12.2, the critical deficiency level of copper in the whole plant tops decreases drastically in clover with age but remains fairly constant at ~3 μg in the youngest leaf blades throughout the season.

The use of the youngest leaves, however, is suitable only for those mineral nutrients which either are not retranslocated or are retranslocated to only a very limited extent from the mature leaves to areas of new growth, that is, when deficiency occurs first in young leaves and at the shoot apex (Table 12.1). The situation is different for potassium, nitrogen, and magnesium; since the contents of these mineral nutrients are maintained fairly constant in the youngest expanded leaves, the mature leaves are a much better indicator of the nutritional status of a plant, as shown for potassium in Fig. 12.4. Here, the youngest leaf is not a suitable indicator because the potassium levels indicating deficiency and toxicity vary only between 3.0 and 3.5%, respectively, compared with 1.5 and 5.5% in mature leaves. This illustrates the necessity of using mature leaves to assess the nutritional status of mineral nutrients which are readily retranslocated in plants.

If young and old leaves of the same plant are analyzed separately, additional information can be obtained on the nutritional status of those mineral nutrients which are readily retranslocated. A much higher content of, say, potassium in the mature

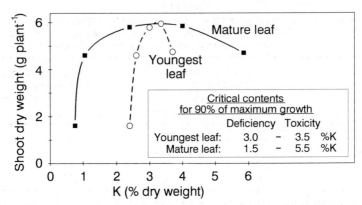

Fig. 12.4 Relationship between shoot dry weight and potassium content of mature and youngest leaves of tomato plants grown in nutrient solutions with various potassium concentrations. Inset: calculated critical contents.

leaves indicates luxury consumption or even toxicity. The reverse gradient, on the other hand, is an indicator of the transition stage between the adequate and deficient ranges; if this gradient is steep, a latent or even acute deficiency may exist. The use of gradients is particularly helpful under conditions in which relevant reference data on critical contents are lacking (e.g. for a species or cultivar) or under certain ecological conditions. If toxicity is suspected, the old leaves are the most suitable organs for plant analysis.

When choosing a given plant organ such as the most recently developed, fully expanded leaf for analysis it should be taken into account that the CDC value will decline throughout plant development, even when expressed as a concentration in the plant sap. For potassium, for example, in soybean this value falls between podset and podfilling from 65 mM to 29 mM (Bell et al., 1987). This decline throughout plant development is particularly evident for the nitrate which acts as a storage form of nitrogen in the leaves and as an indicator of the nitrogen nutritional status of the plants (Section 12.3.7). In the petioles of potato leaves the CDC of NO_3-N decreases from 2.7–3% in the dry matter at the onset of tuberization to 1.0–1.6% at later stages (Williams and Maier, 1990) and in the midribs of cauliflower from 1.1% at the 4–6 leaf stage to 0.15% at preharvest (Gardner and Roth, 1990). A similar decline holds true for sulfate as the main storage form of sulfur in plants (Huang et al., 1992c).

This decline in CDC for a given organ with age can occur for various reasons. For example, as plants become older there is a decrease in demand for nutrients for new growth. Also nutrient supply from the roots may increase although this is less likely in most field conditions in view of declining root activity. The decline in CDC appears rather to be the consequence of increase in total shoot biomass and, thus, storage capacity of mineral nutrients in the shoots, as illustrated in an example for maize in Table 12.3. Between the 4–5-leaf stage and heading both the critical deficiency concentrations of nitrate-N in the press sap and the concentrations considered as adequate (Fig. 12.3) decline during plant development. However, as the total above-ground biomass increase is more than linear, a given concentration of nitrate-N per liter

Table 12.3

Critical Deficiency and Adequate Concentrations of Nitrate Nitrogen (NO_3-N) in Press Sap of
Leaf Sheath from Basal Stem at Different Developmental Stages of Maize, and Estimated
Amounts of NO_3-N Stored in the Above-ground Biomass[a]

Developmental stage	Concentration range (mg NO_3-N l^{-1})		Estimated amounts of stored NO_3-N (kg ha^{-1})	
	Critical defic.	Adequate	Critical defic.	Adequate
4–5 leaf stage	800	1400	3.6	6.3
Onset of shooting	375	700	6.2	11.6
Shooting	250	550	7.0	15.3
Heading	250	550	10.6	23.3

[a]Based on Geyer and Marschner (1990) and Geyer (unpublished).

press sap (or per unit dry matter) represents an increasing amount of stored nitrogen
which can act as an internal buffer and maintain similar growth rates for several days
when supply from the soil declines. Using a model which takes into account changes in
growth rates and biomass as a parameter of internal demand, a single critical leaf sap
concentration of 380 mg NO_3-N l^{-1} could be identified for Brussels sprout at all growth
stages and also for various growing seasons (Scaife, 1988).

Compared with the changes in the mineral nutrient content of annual species, the
fluctuations throughout the growing season of the mineral nutrient content of leaves
and needles of trees are relatively small because of the nutrient buffering capacity of
twigs and trunk. In evergreen trees the simultaneous analysis of leaves and needles
differing in age (Table 12.4) may provide additional information on the nutritional
status of the tree. With increasing age of the needles, the content of all macronutrients
decreases, except that of calcium (Table 12.4). In Norway spruce, the silicon content
also increases with needle age (Wyttenbach et al., 1991). The decrease in nitrogen,

Table 12.4

Mineral Nutrient Content of Norway Spruce (*Picea abies* Karst)
Needles of Different Age[a]

Mineral nutrient	Age of needles (years)			
	1	2	3	4
Nitrogen	1.79	1.76	1.46	1.22
Phosphorus	0.20	0.17	0.14	0.13
Potassium	0.63	0.56	0.47	0.44
Magnesium	0.04	0.04	0.03	0.03
Calcium	0.28	0.40	0.50	0.59

[a]Contents of mineral nutrients expressed as a percentage of the dry
weight. Based on Bosch (1983).

phosphorus, potassium and magnesium with needle age (Table 12.4) might in part indicate retranslocation, but is presumably mainly caused by a dilution effect, resulting from increased lignification of the old needles. Only with calcium (and silicon) dilution is overcompensated for by a continuation of high influx into old needles. With the exception of magnesium, the data of Table 12.4 are indicative of trees well supplied with these macronutrients.

12.3.4   Plant Species

Critical deficiency contents differ between plant species even when comparisons are made for the same organs at the same physiological age. This is also true for the adequate range. These variations are mainly based on differences in the plant metabolism and plant constitution, as for example, differences in the genotypical demand of calcium and boron in cell walls. When grown under the same conditions the CDC of boron in the dry matter of the fully expanded youngest leaf is $3 \mu g\, g^{-1}$ in wheat, $5 \mu g$ in rice, but as high as $25 \mu g$ in soybean and $34 \mu g$ in sunflower (Rerkasem et al., 1988). Indigenous plant species from nutrient-rich habitats seem to have higher critical deficiency concentrations of potassium in the shoots ($\sim$100 mM) than species from nutrient-poor habitats ($\sim$50 mM; Hommels et al., 1989a). Representative data for adequate nutrient ranges of selected species are given in Table 12.5. More extensive and detailed data, including deficiency and toxicity contents, can be found in Chapman (1966); Jones (1967; 1991), Bergmann and Neubert (1976) and Drechsel and Zech (1991).

As shown in Table 12.5 the contents of macronutrients in the adequate range are of similar orders of magnitude for the various plant species; an exception is calcium, the content of which is substantially lower in monocotyledons. In all species the adequate range is relatively narrow for nitrogen, because luxury contents of nitrogen have unfavorable effects on growth and plant composition (Section 8.2.5). In apple leaves, for example, a nitrogen content of more than 2.4% often affects fruit color and storage adversely (Bould, 1966). On the other hand, the adequate range for magnesium is usually broader, due mainly to competing effects of potassium; with high potassium contents, high magnesium contents are also required to ensure an adequate magnesium nutritional status.

The contents of micronutrients in the adequate range vary by a factor of 2 or more (Table 12.5). Manganese shows the greatest variation, indicating that for manganese in particular, leaf tissue is buffering fluctuations in the root uptake of this nutrient. In plants growing in soil, manganese exhibits more rapid and distinct fluctuations in uptake than any other mineral nutrient, the rate depending on variations in soil redox potential and thus on the concentrations of Mn^{2+} (Section 16.4).

The data given in Table 12.5 are average values and offer no more than a guide as to whether a mineral nutrient is in the deficient, adequate, or toxic range. This should be borne in mind when only one or a few mineral nutrients have been analyzed and the information on possible nutrient interactions is therefore insufficient.

The critical toxicity contents of sodium and chloride are in general closely related to genotypical differences in salt tolerance. The interpretation of these contents is complicated because in saline substrates a decline in growth is often caused in the first

Table 12.5

Mineral Nutrient Contents in the Adequate Range of Some Representative Annual and Perennial Species[a]

Species (organ)	Contents (% dry wt)					Contents (mg kg^{-1} dry wt)				
	N	P	K	Ca	Mg	B	Mo	Mn	Zn	Cu
Spring wheat (whole shoot, booting stage)	3.0–4.5	0.3–0.5	2.9–3.8	0.4–1.0	0.15–0.3	5–10	0.1–0.3	30–100	20–70	5–10
Ryegrass (whole shoot)	3.0–4.2	0.35–0.5	2.5–3.5	0.6–1.2	0.2–0.5	6–12	0.15–0.5	40–100	20–50	6–12
Sugar beet (mature leaf)	4.0–6.0	0.35–0.6	3.5–6.0[b]	0.7–2.0	0.3–0.7	40–100	0.25–1.0	35–100	20–80	7–15
Cotton (mature leaf)	3.6–4.7	0.3–0.5	1.7–3.5	0.6–1.5	0.35–0.8	20–80	0.6–2.0	35–100	25–80	8–20
Tomato (mature leaf)	4.0–5.5	0.4–0.65	3.0–6.0	3.0–4.0	0.35–0.8	40–80	0.3–1.0	40–100	30–80	6–12
Alfalfa (upper shoot)	3.5–5.0	0.3–0.6	2.5–3.8	1.0–2.5	0.3–0.8	35–80	0.5–2.0	30–100	25–70	6–15
Apple (mature leaf)	2.2–2.8	0.18–0.30	1.1–1.5	1.3–2.2	0.20–0.35	30–50	0.1–0.3	35–100	20–50	5–12
Orange (*Citrus* ssp.) (mature leaf)	2.4–3.5	0.15–0.3	1.2–2.0	3.0–7.0	0.25–0.7	30–70	0.2–0.5	25–125	25–60	6–15
Norway spruce (1–2 year-old-needles)	1.35–1.7	0.13–0.25	0.5–1.2	0.35–0.8	0.1–0.25	15–50	0.04–0.2	50–500	15–60	4–10
Oak; Beech (mature leaves)	1.9–3.0	0.15–0.30	1.0–1.5	0.3–0.5	0.15–0.30	15–40	0.05–0.2	35–100	15–50	6–12

[a]Based on Bergmann (1988, 1992).
[b]Sodium content below 1.5%.

Table 12.6

Effect of Foliage Phosphorus Content on the Critical Deficiency Content (CDC) of Nitrogen and Vice Versa in *Araucaria cunninghamii*[a]

Foliage phosphorus content (% dry wt)	CDC of nitrogen (% dry wt)	Foliage nitrogen content (% dry wt)	CDC of phosphorus (% dry wt)
0.06	1.07	0.60	0.07
0.09	1.18	1.05	0.08
0.12	1.24	1.35	0.10
0.16	1.31	1.65	0.11
0.21	1.35	1.80	0.12

[a]Based on Richards and Bevege (1969).

instance by effects on the water balance of plants and not necessarily by direct toxicity of sodium or chloride or both these ions in the leaf tissue (Section 16.6).

12.3.5   Nutrient Interactions and Ratios

There is a whole range of nonspecific as well as specific interactions between mineral nutrients in plants (Robson and Pitman, 1983) which affect the critical contents. A typical example of a nonspecific interaction is shown in Table 12.6 for nitrogen and phosphorus. The CDC of nitrogen increases as the phosphorus content increases and vice versa. Similarly in maize, at low phosphorus content an increase in nitrogen content of the earleaf from 2.1 to 2.9% had little effect on yield, but at high phosphorus content yield continued to increase as earleaf nitrogen content rose well above 3% (Sumner and Farina, 1986).

Interactions between two mineral nutrients are important when the contents of both are near the deficiency range. Increasing the supply of only one mineral nutrient stimulates growth, which in turn can induce a deficiency of the other by a dilution effect. In principle, these unspecific interactions hold true for any mineral nutrients with contents at or near the critical deficiency contents. Optimal ratios between nutrients in plants are therefore often as important as absolute contents. For example, a ratio of nitrogen to sulfur of ~17 is considered to be adequate for the sulfur nutrition of wheat (Rasmussen *et al.*, 1977) and soybean (Bansal *et al.*, 1983). However, optimal ratios considered alone are insufficient because they can also be obtained when both mineral nutrients are in the deficiency range (Jarrell and Beverly, 1981), as well as in the toxicity range.

Specific interactions which affect CDC have been discussed in Chapters 8 and 9. Therefore, only two examples are reiterated here: (a) competition between potassium and magnesium at the cellular level, which usually involves the risk of potassium-induced magnesium deficiency; and (b) replacement of potassium by sodium in natrophilic species, which has to be considered in the evaluation of potassium content (see Table 12.5).

Specific interactions are also important in evaluating critical toxicity contents. The critical toxicity content of manganese, for example, differs not only between species and cultivars of a species (Section 9.2), but within the same cultivar, the difference depending on the silicon supply. In leaves the critical toxicity content of manganese can increase from 100 mg kg^{-1} dry matter in the absence of silicon to $\sim$1000 mg, i.e., by a factor of 10, in the presence of silicon in bean (Horst and Marschner, 1978a) and by a factor of 3–4 in different cowpea genotypes (Horst, 1983).

In view of the problems arising from different CDC during plant development, and of the importance of nutrient ratios in plant analysis for diagnostic purposes, a new concept had been introduced with Beaufils's Diagnosis and Recommendation Integrated System (DRIS). This system is based on the collection of as many data as possible and plant contents of mineral nutrients (so far mainly macronutrients) and use of these data for calculations of optimal nutrient ratios – nutrient indices (norm data) – for example ratios of N/P, N/K etc. (Sumner, 1977). The nutrient indices calculated through DRIS are less sensitive to changes taking place during leaf maturation and ontogenesis, but depend to some extent on location. For example, for maize earleaf tissue norm N/P ratios are on average 10.13, but 8.91 for South Africa and 11.13 for the South East of USA (Walworth and Sumner, 1988). This system requires a large number of data on both contents of different mineral nutrients in the plants of a given field, and from different locations and years. The calculated norms for ratios are thus mean values obtained from several thousand field experiments. For certain crops and under certain conditions (high yielding sites, large-scale farming) the higher analytical input might pay off by permitting a refinement in the interpretation of the data in terms of fertilizer recommendations, as has been demonstrated for sugarcane (Elwali and Gascho, 1984), maize and fruit trees (Walworth and Sumner, 1988). However, under other environmental conditions less favorable results have also been obtained with DRIS (Reuter and Robinson, 1986), and it is certainly not the method of choice in cropping systems with a diversity of annual species or for low input and small-scale farming systems.

12.3.6 Environmental Factors

Fluctuations in environmental factors such as temperature and soil moisture can affect the mineral nutrient content of leaves considerably. These factors influence both the availability and uptake of nutrients by the roots and the shoot growth rate. Their effects are more distinct in shallow-rooted annual species than in deep-rooted perennial species, which have a higher nutrient buffer capacity within the shoot. This aspect must be considered in the interpretation of both critical deficiency and toxicity contents in leaf analysis. If fluctuations in soil moisture are high, then as a rule for a given plant species the CDC of nutrients such as potassium and phosphorus are also somewhat higher in order to ensure a higher capacity for retranslocation during periods of limited root supply. The effects of irradiation and temperature on the nutrient content of leaves are described in detail by Bates (1971). For example, under high light intensity the CDC in leaves of boron and zinc are higher than under low light intensity (Sections 9.7.9). In tomato the CDC of phosphorus in mature leaves increase from 1.8 to 3.8 mg g^{-1} dry matter when the external salt concentration is increased from 10 to 100 mM (Awad *et al.*, 1990). The physiological mechanism for this higher internal requirement of phosphorus is not clear, involvement in osmotic adjustment in the mature leaves, or restricted retranslocation to expanding leaves, might be involved.

12.3.7 Nutrient Efficiency

Genotypical differences in the CDC of a nutrient can also be brought about by differences in the utilization of a nutrient. In a physiological sense, this may be expressed in terms of unit dry matter produced per unit nutrient in the dry matter (e.g., mg P g^{-1} dry matter). As an example, the difference in nitrogen efficiency between C$_3$ and C$_4$ grasses is shown in Table 12.7. Much more dry matter is produced in C$_4$ grasses than in C$_3$ grasses per unit leaf nitrogen. This is a general phenomenon which is observed in comparisons of other C$_3$ and C$_4$ species (Brown, 1985). The higher nitrogen efficiency of C$_4$ species is presumably related to the lower investment of nitrogen in enzyme proteins used in chloroplasts for CO$_2$ fixation. In C$_4$ species only 5–10% of the soluble leaf protein is found in RuBP carboxylase, compared to 30–60% in C$_3$ species (Section 5.2.4). Lower CDC in C$_4$ plants are of advantage for biomass production on nitrogen-poor sites, but not necessarily of advantage in view of the nutritional quality of forage (Brown, 1985).

Differences in the utilization of mineral nutrients are also found between cultivars, strains, and lines of a species. These differences are a component of the nutrient efficiency in general as will be discussed in detail in Section 16.2.3. In an agronomical sense, nutrient efficiency is usually expressed by the yield differences of genotypes growing in a soil with insufficient amounts of nutrients. In most instances, high nutrient efficiency is related primarily to root growth and activity, and in some instances also to the transport from the roots to the shoots (Läuchli, 1976b). Only relatively few data indicate a higher nutrient efficiency in terms of utilization within the shoots, for example, utilization of phosphorus in bean (Whiteaker et al., 1976; Youngdahl, 1990) and maize genotypes (Elliott and Läuchli, 1985), potassium in bean and tomato (Shea et al., 1967; Gerloff and Gabelman, 1983), and calcium in tomato (English and Barker, 1987; Behling et al., 1989).

In principle, higher nutrient efficiency, as reflected by lower critical deficiency contents, in one genotype compared with those in another genotype of the same species can be based on various mechanisms:

Table 12.7
Relationship between Dry Matter Production and Nitrogen Content of C$_3$ and C$_4$ Grasses[a,b]

Nitrogen supply (equivalent to kg ha^{-1})	Dry matter (g per pot)		Nitrogen content (% dry wt)	
	C$_3$	C$_4$	C$_3$	C$_4$
0	11	22	1.82	0.91
67	20	35	2.63	1.18
134	27	35	2.77	1.61
269	35	48	2.78	2.00

[a]Based on Colman and Lazemby (1970).
[b]C$_3$ Grasses: Lolium perenne and Phalaris tuberosa; C$_4$ grasses: Digitaria macroglossa and Paspalum dilatatum.

1. Higher rates of retranslocation during either vegetative or reproductive growth, for example, zinc in maize (Massey and Loeffel, 1967), nitrogen in pearl millet (Alagarswamy *et al.*, 1988) or phosphorus in bean (Youngdahl, 1990).

2. Higher nitrate reductase activity in the leaves and thus more efficient utilization of nitrogen for protein storage [e.g., in wheat grains (Dalling *et al.*, 1975) and potato tubers (Kapoor and Li, 1982)].

3. Higher proportion of replacement of potassium by sodium and thus lower CDC of potassium [e.g., in tomato (Gerloff and Gabelman, 1983)].

4. Lower proportion of nutrients which are not – or only poorly – available for metabolic processes, either due to compartmentation or chemical binding, for example of phosphorus in maize genotypes (Elliott and Läuchli, 1985). This aspect is particularly relevant for calcium, where in efficient genotypes of tomato a higher proportion of calcium is translocated to the shoot apex where more calcium also remains in the water soluble fraction (Behling *et al.*, 1989). On the other hand, a lower CDC of 0.25% Ca in the shoot dry matter of an efficient tomato genotype compared to 0.40% in an inefficient genotype was correlated to much higher potassium contents in the shoot dry matter of the inefficient genotype (English and Barker, 1987), stressing the importance of certain physiologically based nutrient ratios for the CDC.

5. Differences in the ratio of vegetative shoot growth (source) to the growth of reproductive or storage organs (sink) or both. This aspect (Section 6.4) is probably in part responsible for the general pattern in so-called modern cultivars of many crop species with a high harvest index in which the critical deficiency contents of mineral nutrients in the leaves are usually higher than those of traditional cultivars.

12.3.8 Total Analysis versus Fractionated Extraction

Most frequently it is the total dry matter content of a nutrient that is determined in plant analysis (e.g., after ashing). The determination of only a fraction of the content – for example, that which is soluble in water or in dilute acids or chelators – sometimes provides a better indication of nutritional status. In terms of plant analysis as a basis of fertilizer recommendations this is particularly true for nitrate which is a prominent storage form of nitrogen in many plant species. In those species the nitrate content is usually a much better indicator of the nitrogen nutritional status than the total nitrogen content, and the only realistic approach for a simple, rapid method of plant analysis not only as a reliable indicator of the nitrogen nutritional status of the plants but also as a basis for recommendations of top dressing of nitrogen fertilizers. This method has been successfully used, more recently, for example, in winter cereals (Wollring and Wehrmann, 1990), irrigated wheat (Knowles *et al.*, 1991), potato (Westcott *et al.*, 1991), cabbage (Gardner and Roth, 1989) and other vegetable crops (Scaife, 1988). There are only a few cases where this method has not proved to be a satisfactory predictor of responsiveness of a crop to nitrogen fertilizer (Fox *et al.*, 1989). The principal limitations of this method are set by plant species which preferentially reduce nitrate in the roots (e.g., members of the Rosaceae), or when ammonium nitrogen is supplied and taken up prior to nitrification in soils. This situation might be true for soils high in organic nitrogen with high mineralization rates during the stages of high nitrogen demand of a crop.

Table 12.8
Calcium and Oxalic Acid Content of Two Cultivars of Burley Tobacco Differing in Susceptibility to Calcium Deficiency[a]

Cultivar	Plants with Ca-deficiency symptoms (% of total)	Content of buds (meq g^{-1} dry matter)			Content of upper leaves (meq g^{-1} dry matter)		
		Ca	Oxalic acid	Ca minus oxalic acid (soluble Ca)	Ca	Oxalic acid	Ca minus oxalic acid (soluble Ca)
Ky 10	0	0.25	0.08	0.17	0.28	0.11	0.17
B 21	50	0.23	0.16	0.07	0.30	0.15	0.15

[a]Based on Brumagen and Hiatt (1966).

Determination of the arginine content in needles seems to be a better indicator than the total nitrogen content for assessing the nitrogen nutritional status and particularly of nutrient imbalances in Norway spruce stands with different levels of atmospheric nitrogen input (Ericsson *et al.* 1993).

For assessing the sulfur nutritional status of plants the content of sulfate as the main storage form of sulfur, is also a better indicator than the total sulfur content. In various legume species the CDC of SO_4-S in the dry matter of fully expanded leaves decreases during ontogenesis, for example, in alfalfa from 0.39% to 0.10% (Huang *et al.*, 1992c). In some instances the ratio of SO_4-S to total sulfur might be an even better indicator, for example in wheat (Freney *et al.*, 1978) or rice (Islam and Ponnamperuma, 1982) but not in various forage grasses and legumes.

There are conflicting reports on the suitability of using only the inorganic (or readily extractable) fraction of phosphorus, instead of total phosphorus, as a diagnostic criterion of the phosphorus nutritional status of plants. This approach seems to be suitable in the grapevine (Skinner *et al.*, 1987) but not for subterranean clover (Lewis, 1992).

Determination of only a defined fraction of a mineral nutrient might offer not only the possibility of better characterization of the reserves stored in plants (e.g., NO_3-N; SO_4-S) but also of the physiological availability of a nutrient in plant tissue. For example, extraction of leaves with diluted acids or chelators of Fe(II) for characteriz- ation of the so-called 'active iron' might improve the correlations between iron and chlorophyll content in leaves in field-grown plants (Section 9.1.5) but not necessarily so in plants grown under controlled environmental conditions in nutrient solutions (Lucena *et al.*, 1990). Determination of water-extractable zinc in leaves might reflect the zinc nutritional status of plants better than total zinc (Rahimi and Schropp, 1984), particularly in plants suffering from phosphorus-induced zinc deficiency (Cakmak and Marschner, 1987).

Another example of the importance of determination of only a defined fraction of a mineral nutrient for characterization of physiological availability is illustrated in Table 12.8. Differences in the susceptibility of tobacco cultivars to calcium deficiency were not related to the total calcium content but to the soluble fraction in the buds. These differences were caused by variations in the rate of oxalic acid synthesis and thus in the precipitation of sparingly soluble calcium oxalate. Accordingly, the critical deficiency level of total calcium was higher in B 21 than in Ky 10. Determination of only the soluble fraction would be a more appropriate method for assessing the calcium nutritional status of the two cultivars.

12.4 Histochemical and Biochemical Methods

Nutritional disorders are generally related to typical changes in the fine structure of cells and their organelles (Vesk *et al.*, 1966; Hecht-Buchholz, 1972; Niegengerd and Hecht-Buchholz, 1983) and of tissue. Light microscopic studies on changes in anatomy and morphology of leaf and stem tissue can be helpful in the diagnosis of deficiencies of copper, boron, and molybdenum (Pissarek, 1980; Bussler, 1981a). A combination of histological and histochemical methods is useful in the diagnosis of copper (Section 9.3.4) and phosphorus deficiencies (Besford and Syred, 1979).

Enzymatic methods involving marker enzymes offer another approach to assessing the mineral nutritional status of plants. These methods are based on the fact that the activity of certain enzymes is lower or higher (depending on the nutrient) in deficient than in normal tissue. Examples were given in Chapter 9 for copper and ascorbate oxidase; zinc and aldolase or carbonic anhydrase; and molybdenum and nitrate reductase. Either the actual enzyme activity is determined in the tissue after extraction or the leaves are incubated with the mineral nutrient in question to determine the inducible enzyme activities of, for example, peroxidase activity by iron (Bar-Akiva *et al.*, 1970) and nitrate reductase by molybdenum (Section 9.5.2). For assessing the manganese nutritional status, the activity of MnSOD (Section 9.3) in leaves might be used as biochemical marker (Leidi *et al.*, 1987) or, as a nondestructive method, the specific chlorophyll fluorescence (Kriedemann and Anderson, 1988).

Biochemical methods can also be used for assessing the nutritional status of plants in relation to macronutrients. The accumulation of putrescine in potassium-deficient plants (Section 8.7) is a biochemical indicator of the potassium requirement of lucerne (Smith *et al.*, 1982). Inducible nitrate reductase activity can be used as an indicator of nitrogen nutritional status (Witt and Jungk, 1974; Dias and Oliveira, 1987). Pyruvate kinase activity in leaf extracts depends on the potassium and magnesium content of the leaf tissue (Besford, 1978b). In phosphorus-deficient tissue, phosphatase activity is much higher, especially the activity of a certain fraction (Fraction B; isoenzyme) of the enzyme (Table 12.9). The increase in phosphatase activity in deficient tissue is an interesting physiological and biochemical phenomenon which might be related to enhanced turnover rates or to remobilization of phosphorus or both these factors (Smyth and Chevalier, 1984). In eucalyptus seedlings and 5-year-old plants, acid phosphatase activity is a more sensitive parameter for diagnosis of growth limitations by phosphorus than estimation of total phosphorus in leaves and stems (O'Connell and Grove, 1985), whereas in maize acid phosphatase activity increased distinctly only under severe phosphorus deficiency and appears to be suitable as a means of confirming visual diagnosis but is not sensitive enough to indicate latent phosphorus deficiency (Elliott and Läuchli, 1986).

In principle, enzymatic, biochemical and biophysical methods can be very valuable if

Table 12.9
Growth, Phosphorus Content, and Phosphatase Activity of Young Wheat Plants[a]

Phosphorus supply	Shoot dry wt (mg per plant)	Phosphorus content of shoot (%)	Phosphatase activity (μmol NPP[b] g^{-1} fresh wt min^{-1})		
			Total	Fraction A	Fraction B
High	223	0.8	5.6	4.4	0.5
Low	135	0.3	11.1	6.7	2.9

[a]From Barrett-Lennard and Greenway (1982).
[b]NPP, p-nitrophenylphosphate.

the total content or the soluble fraction of a mineral nutrient is poorly correlated with its physiological availability. Whether these enzymatic, biochemical and biophysical methods can realistically be used as alternatives to chemical analysis as a basis for making fertilizer recommendations depends on their selectivity, accuracy, and particularly whether they are sufficiently simple to provide a spot test. In the case of iron and peroxidase (Bar-Akiva *et al.*, 1978; Bar-Akiva, 1984) and copper and ascorbate oxidase (Delhaize *et al.*, 1982), enzymatic methods seem to meet these requirements. Nevertheless, calibration of the methods remains a problem when a suitable standard (nondeficient plants) is not available and there are no visible deficiency symptoms. The potential of these methods in foliar analysis for diagnostic purposes is more in solving particular problems of nutritional disorders and to supplement total and fractionated foliar analysis rather than to replace them.

12.5   Plant Analysis versus Soil Analysis

There is a long history of controversy as to whether soil or plant analysis provides a more suitable basis for making fertilizer recommendations. Both methods rely in a similar manner on calibration, that is, the determination of the relationship between contents in soils or plants and the corresponding growth and yield response curves, usually obtained in pot or field experiments using different contents of fertilizers. Both methods have advantages and disadvantages, and they also give qualitatively different results (Schlichting, 1976). Chemical soil analysis indicates the potential availability of nutrients that roots may take up under conditions favorable for root growth and root activity (Chapters 13 and 14). Plant analysis in the strict sense reflects only the actual nutritional status of plants. Therefore, in principle a combination of both methods provides a better basis for recommending fertilizer applications than one method alone. The relative importance of each method for making recommendations differs, however, depending on such conditions as plant species, soil properties, and the mineral nutrient in question.

In fruit or forest trees, soil analysis alone is not a satisfactory guide for fertilizer recommendations, mainly because of the difficulty of determining with sufficient accuracy the root zones in which deep-rooting plants take up most of their nutrients. On the other hand, in these perennial plants seasonal fluctuations in the mineral nutrient content of leaves and needles are relatively small compared with those in annual species. The nutrient content of mature leaves and needles is therefore also an accurate reflection of the long-term nutritional status of a plant. Furthermore calibrations of critical deficiency content and adequate range can be made rather precisely and refined for a specific location, plant species, and even cultivar. Therefore, in perennial species foliar analysis is in most instances the method of choice. In this instance, however, chemical soil analysis, performed at least once for a given site, is necessary for characterizing the overall level of potentially available nutrients.

In pastures, plant analysis is used more frequently than soil analysis, not only because of the peculiarities of the rooting pattern in mixed pastures (deep- and shallow-rooting species) but also because of the importance of the mineral composition of pasture and forage plants for animal nutrition.

In annual crops the short-term fluctuations of mineral nutrient contents place a severe limitation on plant analysis as a basis for making fertilizer recommendations. Chemical soil analysis is required for predicting the range of variation in plant nutrient content throughout the growing season. In annual crops a large proportion of the mineral nutrients are taken up from the topsoil, which makes soil analysis easier and increases its importance as a tool for making fertilizer recommendations (but see also Chapter 13). There is no doubt, however, that for various reasons plant analysis will also become appreciated more as an important guide to the nutrition of annual crops. Nutrient imbalances in plants, especially latent micronutrient deficiencies, present a problem particularly in intensive agriculture (Franck and Finck, 1980). This problem is also worldwide (Welch et al., 1991), though with consequences not only for plant yield but also for plant resistance to, and tolerance of, diseases and pests (Chapter 11) as well as for the quality for grazing animals (Kubota et al., 1987) and for animal and human nutrition in general (Chapter 9).

Part II

Plant–Soil Relationships

13

Nutrient Availability in Soils

13.1 Chemical Soil Analysis

The most direct way of determining nutrient availability in soils is to measure the growth responses of plants by means of field plot fertilizer trials. This is a time-consuming procedure, however, and the results are not easily extrapolated from one location to another. In contrast, chemical soil analysis – soil testing – is a comparatively rapid and inexpensive procedure for obtaining information on nutrient availability in soils as a basis for recommending fertilizer application. Soil testing has been practised in agriculture and horticulture for many years with relative success. The effectiveness of the procedure is closely related both to the extent to which the data can be calibrated with field fertilizer trials, and to the interpretation of the analysis. For a comprehensive recent review on soil testing methods for micronutrients see Sims and Johnson (1991). Quite often much more is expected from soil testing than the method allows. The reasons for this discrepancy are discussed in detail in this chapter, with special reference to phosphorus and potassium.

Soil testing makes use of a whole range of conventional extraction methods involving different forms of dilute acids, salts, or complexing agents, as well as water. Depending on the method used, quite different amounts of plant nutrients are extracted, as shown for phosphorus in Table 13.1. As a guide, 1 mg phosphorus per 100 g soil may be taken to represent ~30 kg phosphorus per hectare at a 20-cm profile depth (soil bulk density 1.5 kg l^{-1} leads to 3×10^6 kg soil ha^{-1}). Weak extractants such as water or sodium bicarbonate (Table 13.1) reflect mainly the intensity of supply (concentration in soil solution), whereas strong extractants primarily indicate the capacity of the soil to supply nutrients to the soil solution (buffer capacity). All methods used to characterize the availability of a given mineral nutrient for the plants, and thus to predict fertilizer response, must be evaluated by means of growth experiments (see Section 12.5).

Quite often a number of methods are equally suitable for soil testing for a particular mineral nutrient (Vetter et al., 1978; Bolland and Gilkes, 1992). For phosphorus for example, water extraction (Van Noordwijk et al., 1990) can be as satisfactory as extractant for determining availability as dilute acids, despite the difference in amounts of phosphorus extracted by these methods (Schachtschabel and Beyme, 1980). Typically, as is the case for phosphorus, soil testing methods provide a good indication of nutrient status of the soil, and the likelihood of fertilizer response, when the soil is

Table 13.1

Mean Content of Readily Soluble Phosphorus in 40 Soils
Extracted with Various Solutions[a]

Extraction solution	Readily soluble phosphorus $(mg\ (100\ g)^{-1}$ of air dried soil)
Neutral NH_4F (pH 7.0)	14.8
Acidic NH_4F (pH < 2)	7.4
Truog, $H_2SO_4 + (NH_4)_2SO_4$ (pH 3.0)	3.6
Acetic acid (pH 2.6)	2.5
Bicarbonate, $NaHCO_3$ (pH 8.5)	2.4
Calcium lactate (pH 3.8)	1.2

[a]Based on Williams and Knight (1963).

either acutely deficient or abundantly supplied (Bolland and Gilkes, 1992). Particularly in the upper part of the response curve relating nutrient supply to plant growth (Fig. 12.1), soil chemical analysis alone is unsatisfactory for predicting the effects of fertilizer application. Soil analysis mainly provides an indication of the capacity of a soil to supply nutrients to the plants, but does not adequately and in some cases does not at all, characterize the mobility of the nutrients in the soil. Additionally it fails to provide information about soil structure, or microbial activity, and plant factors, such as root growth and root-induced changes in the rhizosphere, which are of decisive importance for nutrient uptake under field conditions. In the following three chapters these factors are discussed, beginning with nutrient availability in relation to mobility in soils and root growth. For comprehensive treatments of this topic the reader is referred to Barber (1984), Jungk and Claassen (1989); Claassen (1990) and Jungk (1991).

13.2 Movement of Nutrients to the Root Surface

13.2.1 Principles of Calculations

The importance of the mobility of nutrients in soils in relation to availability to plants was emphasized by Barber (1962) and these ideas which were refined and further developed were summarized in a concept of 'bioavailability of nutrients' (Barber, 1984; Section 13.6). Although this concept is focused on aerated soils, its principles may also be applied to submerged soils and plant species such as lowland rice. In principle this concept may also be applied to forest trees. In mature forest stands, however, the application of this concept for the development of simulation models on nutrient delivery and uptake is considerably restricted by the high spatial heterogenity of soil and soil solution chemistry in relation to the stem distance (Koch and Matzner, 1993) and the ill-defined absorbing area of ectomycorrhizal root systems (Section 15.7).

Three components are considered in the concept: root interception, mass flow, and diffusion (Fig. 13.1). As roots proliferate through the soil they also move into spaces

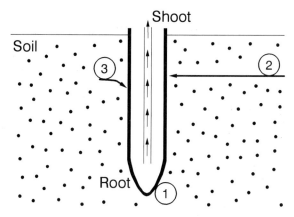

Fig. 13.1 Schematic presentation of the movement of mineral elements to the root surface of soil-grown plants. (1) Root interception: soil volume displaced by root volume. (2) Mass flow: transport of bulk soil solution along the water potential gradient (driven by transpiration). (3) Diffusion: nutrient transport along the concentration gradient. ● = Available nutrients (as determined, e.g., by soil testing).

previously occupied by soil and containing available nutrients, as, for example, adsorbed to clay surfaces. Root surfaces may thus intercept nutrients during this displacement process (Barber, 1984). Calculations of root interception are based on (a) the amounts of available nutrients in the soil volume occupied by the roots; (b) root volume as a percentage of the total soil volume – on average 1% of the topsoil volume; and (c) the proportion of the total soil volume occupied by pores, on average 50%, but very much dependent on the soil bulk density (Section 13.5). In general, only a small part of the total nutrient requirement can be met by root interception (Table 13.2).

The second component is the mass flow of water and dissolved nutrients to the root surface, which is driven by transpiration. Estimates of the quantity of nutrients supplied to plants by mass flow are based on the nutrient concentration in the soil solution and the amount of water transpired either per unit weight of shoot tissue (transpiration coefficient, e.g., 300–600 l H_2O kg^{-1} shoot dry wt) or per hectare of a crop. The contribution of diffusion, the third component relating to the supply of nutrients to the

Table 13.2
Nutrient Demand of a Maize Crop and Estimates on Nutrient Supply from the Soil by Root Interception, Mass Flow, and Diffusion[a]

Nutrient	Demand (kg ha^{-1})	Estimates on amounts (kg ha^{-1}) supplied by		
		Interception	Mass flow	Diffusion
Potassium	195	4	35	156
Nitrogen	190	2	150	38
Phosphorus	40	1	2	37
Magnesium	45	15	100	0

[a]From Barber (1984).

root surface can be calculated on the basis of the effective diffusion coefficients (Section 13.2.4). Such data are more difficult to obtain than those on mass flow. Estimates of the contribution of diffusion can also be based on differences between total uptake by plants and the sum of the amounts supplied by root interception together with mass flow. An example of such a calculation is given in Table 13.2 showing the importance of the three components for different nutrients. It is apparent that, in this soil, nitrogen and magnesium are supplied mainly by mass flow whereas the supply of potassium and phosphorus depends mainly on diffusion. Furthermore, for magnesium, supply by mass flow is greater than uptake, so that accumulation of this nutrient would be expected at the root surface as indeed is also often found. Accumulation at the root surface is particularly true for calcium (Section 15.2).

The term *root interception* has been criticized by Brewster and Tinker (1970) as it contains a diffusion component. Current concepts of solute movement in the soil–root system consider only mass flow and diffusion, and include root interception in the diffusion component (Barber, 1984; Claassen, 1990; Jungk, 1991). As will be shown in Chapter 15, however, conditions at the soil–root interface are sometimes considerably different in a number of aspects from those at a distance from the roots. These conditions are insufficiently described by a mechanistic model treating roots primarily as a sink for mineral nutrients supplied by mass flow or diffusion or both these processes. Also in view of the role of soil structure for nutrient uptake the root–soil contact zone requires separate treatment, particularly in soils of low nutrient availability and in ectomycorrhizal plants (Chapter 15).

13.2.2 Concentration of Nutrients in the Soil Solution

To meet the nutrient demand of soil-grown plants, nutrients must reach the root surface, and this is mainly mediated by movement in or transport with the soil solution, interrelated with root growth which decreases the length of transport pathways. The concentration of nutrients in the soil solution is therefore of primary importance for nutrient supply to roots. Soil solution mineral nutrient concentration varies widely, depending on such factors as soil moisture, soil depth, pH, cation-exchange capacity, redox potential, quantity of soil organic matter and microbial activity, season of the year, and fertilizer application (Asher, 1978). The concentrations of mineral nutrients in the soil solution, particularly nitrogen, are usually very low in many natural ecosystems, for example, of the tundra (Chapin, 1988) as compared with arable soils. An example of annual average concentrations of mineral nutrients in the soil solution of an arable soil is shown in Table 13.3. As a rule in aerobic soils of neutral pH the concentrations of calcium, magnesium and sulfate are fairly high, as are the concentrations of nitrate, whereas those of ammonium and particularly phosphate are very low. The potassium concentrations are mainly a function of clay content and clay mineral composition.

The concentration of mineral nutrients in the soil solution is an indicator of the mobility of nutrients both toward the root surface and in the vertical direction (i.e., in humid climates it indicates the potential for leaching). Compared with the concentration of other nutrients, that of phosphorus is extremely low (Table 13.3), leaching or transport by mass flow to root surfaces thus being generally of minor importance in

Table 13.3

Annual Average Concentrations of Mineral Nutrients in the Soil Solution (Topsoil, 0–20 cm) of an Arable Soil (Luvisol, pH 7.7)[a]

				Concentration in μM				
K	Ca	Mg	NH_4-N	NO_3-N	SO_4-S	PO_4-P	Zn	Mn
510	1650	490	48	3100	590	1.5	0.48	0.002

[a]Recalculated from Peters (1990).

mineral soils. In contrast to other anions such as nitrate and sulfate, phosphate strongly interacts with surface-active sesquioxides and oxidhydrates of clay minerals. In mineral soils, concentration and mobility of phosphate is enhanced by complexation of sesquioxides with organic ligands. Soil organic matter and microbial activity therefore increase concentration and mobility of phosphate (Seeling and Zasoski, 1993) and also lead to a higher percentage of organically bound forms (P_{org}) in the soil solution, of which in soils at higher pH (>6.0) more than 50% may exist as humic–Fe(Al)–P complexes (Gerke, 1992b). The percentage of P_{org} in the soil solution may vary between 20 and 70% (Welp et al., 1983; Ron Vaz et al., 1993) and reach 80–90% in the rhizosphere soil solution (Helal and Dressler, 1989).

In fertile, arable soils supporting high yielding crops the concentrations of mineral nutrients, particularly nitrogen, but also phosphorus, are usually far above average and fluctuate according to the time of fertilizer supply (Table 13.4). In such highly fertile, or fertilized soils, nutrient transport to the root surface does not limit uptake by the crop, even a concentration in the soil solution of 10 μM phosphorus and 87 μM nitrogen being adequate for the rape crop to ensure supply by diffusion (Barraclough, 1989).

The concentration of the micronutrients manganese, iron, zinc and copper in the soil solution mainly depends on the soil pH, redox potential, and soil organic matter

Table 13.4

Time Course of Nutrient Concentrations in the Soil Solution of the Topsoil (0–20 cm) of a High-Yielding Winter Rape (*Brassica napus*) Crop[a]

	Concentration (μM) at		
Nutrients	22 February	28 March[b]	15 May
NO_3-N	620	11 300	1843
NH_4-N	29	1100	<1
PO_4-P	14	14	10
K	91	202	133
Ca	1106	5258	1558
Mg	34	84	52

[a]Barraclough (1989). Reprinted by permission of Kluwer Academic Publishers.
[b]Split application of 265 kg N ha^{-1} as calcium ammonium nitrate on 25 February and 25 March.

content, and in a temperate climate may fluctuate very much throughout the season, with a maximum in early summer (Sinclair *et al.*, 1990). With a fall in the pH, or in redox potential, the concentration of the micronutrients manganese, iron, zinc and copper increase to various degrees (Sims and Patrick, 1978; Herms and Brümmer, 1980; Sanders, 1983; B. Marschner *et al.*, 1992).

Chelation by low-molecular-weight organic substances is another factor which exerts a dominant influence on the concentration of micronutrient cations in the soil solution and their transport to root surfaces by means of mass flow and diffusion. In soil solutions of calcareous soils, 40–75% of the zinc and 98–99% of the copper have been found in organic complexes (Hodgson *et al.*, 1966; Sanders, 1983). As a rule, dependent on soil organic matter content, the proportion of complexed manganese, zinc, and copper increases in the order manganese < zinc < copper, for example from 55%; 75%; 80% when organic matter was present in low amounts, and to 50%; 85% and >98% respectively, at high amounts of organic matter (McGrath *et al.* 1988). For the plants, the importance of complexed micronutrients in the soil solution is particularly evident in calcareous soils (Section 16.5). This is also indicated by the fact that soil extractions with synthetic chelators provide suitable soil tests for the estimation of available micronutrients (Sims and Johnson, 1991).

There is often a poor correlation, however, between the concentration of chelated micronutrients in the soil solution, on the one hand, and the availability of these micronutrients as indicated by plant uptake, on the other (Gilmour, 1977). This is because the metal–organic complexes in the soil solution differ both in electrical charge [they are negatively charged, uncharged, or positively charged (Sims and Patrick, 1978)] and in size. In general, the rate of uptake of metal cations from metal–organic complexes is lower than that of free cations (Section 2.5) and decreases with the size of the organic ligand, as has been demonstrated for copper (Jarvis, 1987). In contrast to such findings from experiments in nutrient solutions, however, chelation of micronutrient cations such as copper and nickel in soil does not give rise to a decrease but usually to an increase in their contents in plants. This results from an increase in concentration of these nutrients in soil solution and thus also in mobility and transport to the root surface (Table 13.5).

Table 13.5
Effect of a Metal Chelator Added to the Nutrient Solution
or Soil on the Trace Element Content of Bean Leaves[a,b]

	Content in leaves (μg g^{-1} dry wt)				
	Zn	Cu	Fe	Mn	Ni
Nutrient solution					
Control	34.0	37.3	125	132	32.7
$+10^{-4}$ M DTPA	19.2	3.8	149	118	0.0
Soil culture					
Control	23.4	7.6	124	108	2.0
$+10^{-3}$ M DTPA	26.8	18.6	230	136	12.8

[a]Based on Wallace (1980a,b).
[b]Chelator: diethylethylenetriamine pentaacetate (DTPA).

In view of the importance of concentration and binding forms of mineral nutrients in the soil solution for both transport of nutrients to the roots as well as for leaching of nutrients from the rooting zone, various new techniques have been developed and older methods modified in order to obtain representative samples of soil solution. Many of these methods are based on soil preparations such as drying, grinding, etc., prior to rewetting and collection of soil solution by displacement or centrifugation. However, to characterize soil solutions of relevance for field-grown plants collection by suction cups or from undisturbed soil cores by centrifugation or percolation is more appropriate (Section 13.5).

Not only the concentration of nutrients in the soil solution (the so-called *intensity*), but also the buffer power (the so-called *capacity*) of a soil is highly important for mineral nutrient supply to the plants (Section 13.2.4). Minor fluctuations in the concentrations of a given mineral nutrient in the soil solution throughout the season (e.g., phosphorus in Table 13.4) reflect on a macroscale the buffer power of the soil.

13.2.3 Role of Mass Flow

Mass flow is the convective transport of nutrients dissolved in the soil solution from the bulk of the soil to the root surface. Calculations of the contribution of mass flow to the nutrient supply of field-grown plants therefore rely on detailed data both on the concentration of nutrients in the soil solution throughout the season and on the water consumption of the plants in question. The results of such experiments are shown in Table 13.6. Expressed as average values for the whole growing season, the contribution made by mass flow to total supply differs not only between mineral nutrients but also between plant species (owing to differences in transpiration rate or uptake rate of a particular mineral nutrient or to both these factors). Mass flow is more than sufficient to supply calcium in both plant species and for supply of magnesium in spring wheat, but not in sugar beet. In contrast, mass flow is unimportant for potassium supply, and the potassium uptake was mainly achieved from supply via diffusion. It can therefore be concluded that the soil around the roots was depleted of potassium, whereas substantial accumulation of calcium and magnesium took place at the root surface which can indeed be demonstrated (Barber, 1984; Section 15.2).

Table 13.6
Plant Uptake and Estimates on Supply to the Roots by Mass Flow of Potassium, Magnesium, and Calcium in Spring Wheat and Sugar Beet Grown in a Silty Loam Soil (Luvisol Derived from Loess)[a]

| | Amount (kg ha^{-1}) | | | | | |
| | Spring wheat | | | Sugar beet | | |
	K	Mg	Ca	K	Mg	Ca
Plant uptake	215	13	35	326	44	104
Mass flow	5	17	272	3	10	236
(% of total uptake)	(2)	(131)	(777)	(1)	(23)	(227)

[a]From Strebel and Duynisveld (1989).

From a four-year field study growing different cereal crops and sugar beet at the same location (Table 13.6), the average contribution of nitrogen supply by mass flow was between 15 and 33% of the total nitrogen supply mainly being confined to the topsoil (Strebel and Duynisveld, 1989). No data are given in Table 13.6 for phosphorus, but a rough calculation can be made. The amount of water transpired by a crop varies in the range 2–4 million l ha^{-1} (Strebel *et al.*, 1983; Barber, 1984). Assuming a phosphorus concentration in the soil solution of 5 μM (0.15 mg l^{-1}) and a total water consumption of the crop by transpiration of 3 million liters, the amount of phosphorus supplied by mass flow, the product of the two values, can be calculated at about 0.45 kg. This value corresponds to about 2–3% of the total demand of a crop and agrees well with the 1–4% found in field experiments with winter wheat and sugar beet (Claassen, 1990).

The contribution of mass flow depends on the plant species (Table 13.6) and might, for example, be higher in onion than in maize since onion roots have a higher water uptake per unit length (Baligar and Barber, 1978). The relative contribution of mass flow also varies with the age of plants (Brewster and Tinker, 1970), and time of the day.

When the soil water content is high (e.g., at field capacity), mass flow is unrestricted and maintains a similar water content (potential) at the root surface. As the soil water content falls, the rate of water uptake by the roots may exceed the supply by mass flow and the soil may then dry out at the soil–root interface. This is observed around the roots, particularly when the transpiration rate is high (Nye and Tinker, 1977), and often occurs in the topsoil during the growing season. Under field conditions the rainfall pattern (or irrigation cycle) therefore has an important influence on the contribution of mass flow to the total nutrient supply as well as on the proportion of nutrients taken up from the subsoil (Section 13.4).

Since mass flow and diffusion to the root surface usually occur simultaneously, it is not possible strictly to separate these processes. The term 'apparent mass flow' has therefore been recommended to define the amount of solutes transported to the root by mass flow (Nye and Tinker, 1977). A principal limitation of these calculations by mechanistic models is the assumption that uptake rates of nutrients and water are uniform along the axis of individual roots, which is not the case for various reasons, particularly in plant species showing 'root dimorphism' (Section 2.6 and 2.7).

13.2.4   Role of Diffusion

Diffusion is the main mechanism for the movement of at least phosphorus and potassium to the root surface. The driving force of diffusion is a concentration gradient. In soil-grown plants a concentration gradient between the adjacent soil and the root surface is formed when uptake rate of ions exceeds the supply by mass flow. Depletion profiles develop with time and their shape depends mainly on the balance between uptake by roots, replenishment from soil, and mobility of ions by diffusion.

The mobility of ions is defined in terms of the diffusion coefficient. Diffusion coefficients in homogeneous media such as water (D_1) are fairly uniform for different ions and orders of magnitude higher than in nonhomogeneous porous media such as aerated soils (Table 13.7). This is true in particular for phosphate. In aerated soils ions diffuse only in pore spaces that are filled with water, and additionally ions in the soil solution interact with the solid phase of the soil. For describing the diffusion of ions in

Table 13.7
Estimates of Diffusion Coefficients ($m^2 s^{-1}$) of Ions in Water (D_1) and in
Soils (D_e), and of Movement per Day at Average Values of D_e[a]

Ion	Diffusion coefficient		Average D_e in soils	Movement in soils (mm per day)
	Water (D_1)	soil (D_e)		
NO_3^-	1.9×10^{-9}	10^{-10}–10^{-11}	5×10^{-11}	3.0
K^+	2.0×10^{-9}	10^{-11}–10^{-12}	5×10^{-12}	0.9
$H_2PO_4^-$	0.9×10^{-9}	10^{-12}–10^{-15}	1×10^{-13}	0.13

[a]From Jungk (1991). Reprinted by courtesy of Marcel Dekker Inc.

soils the term *effective diffusion coefficient* D_e has been introduced by Nye and Tinker (1977):

$$D_e = D_1 \Theta f \cdot dC_1/dC_s$$

where D_e is the effective diffusion coefficient in the soil ($m^2 s^{-1}$); D_1 is the diffusion coefficient in water ($m^2 s^{-1}$); Θ is the volumetric water content of the soil ($m^3 m^{-3}$); f is the impedance (or tortuosity) factor which takes into account the tortuous pathway of ions and other solutes through water-filled soil pores, increasing the path length and thus decreasing the concentration gradient. It is defined as the reciprocal of the impedance, i.e., becomes smaller when the soil water content falls; and dC_1/dC_s is the reciprocal of the soil buffer power for the ion concerned; C_1 is the concentration of the ion in the soil solution, and C_s the sum of both ions in the soil solution and those which can be released from the solid phase (e.g., exchangeable potassium). Soils with high adsorption capacity (e.g., clay soils for K^+) have therefore a high buffer power, and thus a low dC_1/dC_s value.

13.2.4.1 Soil Factors

As a rule, the concentration of potassium and phosphorus is much lower at the root surface than in the bulk soil, creating a typical depletion zone around the roots. As shown in Fig. 13.2 with increasing potassium content in the soil D_e increases. The nutrient-depleted area surrounding the plant roots also increases from ~4 mm in depleted soil (by previous intensive cropping) to 5.3 mm in unfertilized and 6.3 mm in fertilized soil. Hence, raising the content of exchangeable potassium by fertilizer application increases the amount of potassium supplied via diffusion by a factor of more than 20, i.e. much more than would be expected from the increase in the amount of exchangeable potassium per unit soil weight only. Application of NaCl or $MgCl_2$ also increased the extension of the depletion zone and thus the delivery of potassium to the root surface (Kuchenbuch and Jungk, 1984).

For potassium, shape and extension of the depletion zone in different soils strongly depend on their clay content (cation exchange capacity) which is an important parameter of the buffer power for potassium (Fig. 13.3). In soil A, with 21% clay and a

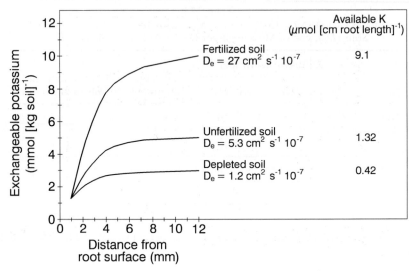

Fig. 13.2 Concentration gradient around roots of 7-day-old rape (*Brassica napus*) seedlings grown in a silt loam soil with different contents of exchangeable potassium. (Modified from Kuchenbuch and Jungk, 1984.)

correspondingly higher cation-exchange capacity, the equilibrium concentration of potassium in the soil solution was much lower than in soil B, with only 4% clay. In both soils, roots depleted the soil solution to about 2–3 μM K^+. The depletion zone in soil B was much wider than that in soil A, however, reflecting the much lower capacity of soil B to replenish the potassium in the soil solution.

Particularly in soils low in exchangeable potassium the plant demand might by far exceed the amount of potassium supply from this fraction, and a large proportion of the potassium taken up derives from the non-exchangeable fraction. In the experiment shown in Fig. 13.4 the proportion of potassium from the nonexchangeable fraction

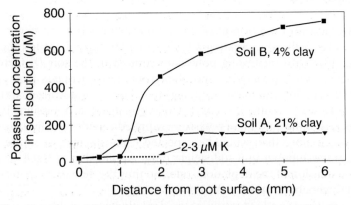

Fig. 13.3 Concentration gradient of potassium in the soil solution around maize roots growing in soils with different clay contents. Potassium concentration at the root surface, 2–3 μM. (Modified from Claassen and Jungk, 1982.)

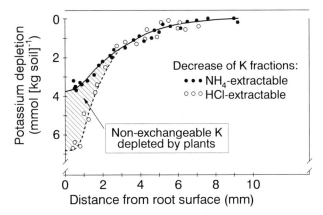

Fig. 13.4 Depletion of potassium in the rhizosphere of rape (*Brassica napus*) seedlings grown for 7 days in a silt loam soil. (From Jungk and Claassen, 1986.)

increased from 20% in the fertilized soil to 71% in the unfertilized and 83% in the depleted soil (Kuchenbuch and Jungk, 1984), i.e., in the latter two cases the rape seedlings received most of their potassium from a fraction which is either not, or only to a minor extent, characterized as plant available by soil testing methods. Similarly high proportions of potassium delivery in the rhizosphere from the nonexchangeable fraction have been found in ryegrass, and part of this potassium originated from the interlayer of clay minerals (Kong and Steffens, 1989; Hinsinger *et al.*, 1992).

Soil water content is another important factor affecting D_e. As shown for potassium in Table 13.8, for a given soil increasing the volumetric soil water content increases the cross-sectional area available for diffusion of ions and correspondingly raises the value of the impedance factor f (Section 13.2.4) and the buffer power of the soil. As a consequence D_e for potassium more than doubles. For phosphorus this effect of soil water content on D_e is even more pronounced. Increasing the volumetric soil water content in a Luvisol from 0.12 to 0.33 g cm^{-3} increased D_e from 0.10×10^{-13} to 4.45×10^{-13} m^2 s^{-1}, whereas changes in the bulk density of the soil had only a relatively small effect on D_e (Bhadoria *et al.*, 1991).

This pronounced effect of water content on D_e is also of importance in comparisons of water relations in soils of different textures (sandy soil, clay soil). Soils of different texture can have quite different water contents Θ at the same soil water potential (pF). As a rule, at the same pF the Θ increases with the clay content as shown by Cox and Barber (1992) by using four different soils in which at the same soil water potential (-33 kPa) the water content Θ varied between 0.13 and 0.40 g cm^{-3}. In order to achieve the same phosphorus uptake by maize plants, in the soil with the highest volumetric water content ($\Theta = 0.4$ g cm^{-3}) a concentration in the soil solution of only 10 μM P was necessary as compared with 200 μM in the soil with the lowest $\Theta = 0.13$ g cm^{-3} (Cox and Barber, 1992). These examples on the role of soil water content on ion supply to roots indicate the advantages particularly for phosphorus and potassium nutrition in plants adapted to submerged soils (e.g., lowland rice).

It is well known from field and greenhouse experiments that at low soil water contents

Table 13.8
Influence of Soil Volumetric Water Content (Θ) on Impedance Factor
(f), Buffer Power (b) and Effective Diffusion Coefficient (D_e) of
Potassium[a]

Water content (Θ, cm^3 cm^{-3})	Impedance factor (f)	Buffer power (b)	Effective diffusion coefficient (D_e, cm^2 s^{-1})
0.19	0.20	2.68	2.55×10^{-7}
0.26	0.30	3.09	4.91×10^{-7}
0.34	0.45	4.42	6.40×10^{-7}

[a]Based on Kuchenbuch *et al.* (1986). Reprinted by permission of Kluwer
Academic Publishers.

– or in dry years – the uptake of potassium and phosphorus are more impaired than that
of calcium and magnesium which may even be increased (Talha *et al.*, 1979). As has
been demonstrated by Zur *et al.* (1982) at low soil water contents, the water content at
the root surface is much lower than that of the bulk soil. Contact between root surface
and soil via the soil solution can thus be lost. Under these conditions it has been
postulated that long root hairs are of particular importance in preventing this loss of
contact (Nye and Tinker, 1977).

When soil water content is low, mechanical impedance of the soil increases and root
elongation growth is inhibited, which further limits nutrient supply to the root surface
by diffusion. However, root hair growth is strongly enhanced at low soil water content
(Mackay and Barber, 1985, 1987) and thus might in part compensate for any decrease in
surface area from impeded elongation growth of the root axis. At least for phosphorus
uptake, a similar compensatory effect, particularly for the decrease in D_e can be
expected, as a result of mycorrhizal infection by roots. At low soil water contents the
supply of boron to roots is depressed to a greater degree than that of any other
micronutrient (Kluge, 1971). As reliable data on boron concentrations in soil solutions
are lacking, it is not known whether it is mass flow or diffusion that is mainly responsible
for boron supply to roots.

13.2.4.2   Plant Factors

As a rule, in nonmycorrhizal plants the extension of the depletion zones of phosphorus
is closely related to root hair length. An example is given in Fig. 13.5 for maize. For
potassium this close relation between root hair length and depletion zone also holds
true for depleted soils (Fig. 13.2) or soils with a high buffer power for potassium (Fig.
13.3).

Root hair formation is modified by environmental factors (Chapter 14) and also
differs typically between plant species. Genotypical differences in root hair length are
particularly important for the concentration gradients of phosphorus and potassium
around roots. For example, in maize and rape the distance of the extent of maximum
phosphorus depletion in the rhizosphere is somewhat similar to the average root hair
length, which is 0.7 mm in maize and 1.3 mm in rape. The extension of the depletion

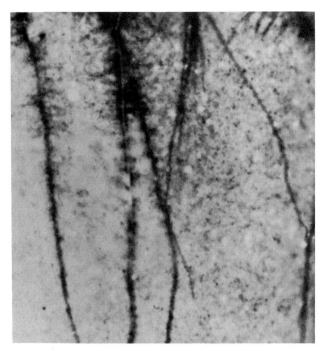

Fig. 13.5   Autoradiograph of maize roots in a soil labeled with ^{32}P showing zones of phosphorus depletion around the roots (removal of ^{32}P indicated by black zones).

zone is nearly identical with the maximal root hair length, namely 1.8 mm in maize and 2.6 mm in rape (Hendriks *et al.*, 1981).

When the uptake rates of phosphorus or potassium per unit root length of different plant species are compared, a close positive correlation can be demonstrated between the uptake rate per unit root length and volume of the root hair cylinder. This is shown for potassium in Fig. 13.6. In onion, root hairs were virtually absent, whereas of the plant species tested rape had the longest root hairs. Per unit length of onion root, the potassium of only 2–3 mm^3 soil was available, as compared with ~60 mm^3 for rape. Results similar to those shown in Fig. 13.6 were obtained for phosporus in comparative studies with plant species of different root hair lengths (Itoh and Barber, 1983a). In agreement with this, even within a given plant species (white clover) selection of genotypes results in more-efficient phosphorus uptake in genotypes with long root hairs than in those with short root hairs (Caradus, 1982).

The importance of root hairs for phosphorus uptake from soils is also reflected in simulation models for predicting phosphorus uptake by different plant species (Föhse *et al.*, 1988, 1991). In models in which root hairs are not taken into account uptake is only predicted well in species which are without root hairs or in which root hair formation is minimal. The inclusion of root hairs leads to a much better agreement with determined values. In soils low in extractable phosphorus the contribution of root hairs can account for up to 90% of the total uptake (Föhse *et al.*, 1991). Root hairs are more effective in absorbing phosphorus than the root cylinder when the influx per unit area of each are

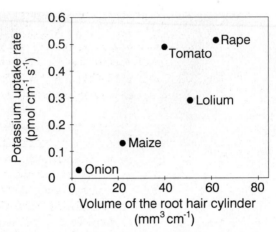

Fig. 13.6   Rate of potassium uptake per unit length of root in relation to the volume of the root hair cylinder. The plant species were grown in a silt loam with 21% clay. (Modified from Jungk *et al.*, 1982.)

compared because of the smaller diameter and geometric arrangement of the root hairs which maintain higher diffusion rates of phosphorus (Jungk and Claassen, 1989; Claassen, 1990). This advantage of root hairs is likely of particular importance for phosphorus nutrition not only at low soil water content (see above) but also in macrostructured soils with large proportions of voids (Misra *et al.*, 1988).

   However, close relationships between root hair length and extension of the depletion zone of phosphorus and potassium are not always found. For example, in a given soil, extension of the phosphorus depletion zone was confined mainly to the root hair cylinder in rape which possesses long root hairs (>0.5 mm), but considerably exceeded the root hair cylinder in species with short root hairs (~0.2 mm), for example cotton (Misra *et al.*, 1988). The extension of the depletion zone distinctly beyond the root hair cylinder in nonmycorrhizal plants can also indicate root-induced changes in the rhizosphere (e.g., release of root exudates, pH changes; Chapter 15) or higher efficiency in uptake parameters (K_m, I_{max}; Section 2.4.1) set by the plant demand per unit root length. In mycorrhizal plants the depletion zone of phosphorus by far exceeds the root hair cylinder (Jungk and Claassen, 1989) and in white clover, for example, can reach a value up to 11 cm (Li *et al.*, 1991a).

   Both the degree of depletion within the root hair cylinder and the extension of the depletion zone can also be affected by C_{min} values, i.e., the minimum concentration of nutrients to which the roots can deplete the external concentration. These differ between plant species and even genotypes within a given species (Section 2.4.1). In general compared with solution cultures C_{min} values are usually much higher in soil-grown plants, indicating the buffer power of the soil to replenish the removal of nutrients in the rhizosphere soil solution. Average C_{min} values for soil-grown plants are 2–3 μM potassium (Claassen and Jungk, 1982) and 1 μM for phosphorus (Hendriks *et al.*, 1981). For the rate of replenishment of phosphorus in the soil solution of the root hair cylinder the following calculation can be made: concentration in the soil solution, 5 μM = 0.15 mg phosphorus per liter; amount of soil solution in the topsoil (0–30 cm),

Table 13.9
Estimates of Proportions of Soil (Sandy Loam) Contribution to
the Phosphorus and Potassium Nutrition of Field Grown Maize[a]

Root length density	Proportion of soil contributing (%)	
	P	K
$\gg$2 cm cm^{-3}	20	50
<2 cm cm^{-3}	5	12

[a]Fusseder and Krauss (1986).

~500 000 liters = 75 g phosphorus per hectare; requirement during phase of rapid growth (e.g. in cereals between tillering and heading), ~300–500 g phosphorus per hectare and day. Since only ~25% of the topsoil is explored by roots during one growing season (Jungk, 1984) the rate of replenishment within the root hair cylinder has to be at least 10–20 times per day in order to meet the requirement of plants. For potassium, too, the rate of replenishment in the root hair cylinder has to be high. Within 2.5 days more than half of the potassium taken up by maize is derived from the so-called nonexchangeable fraction of the soil in the root hair cylinder (Claassen and Jungk, 1982), and in rape, within 7 days the contribution by the nonexchangeable fraction in an unfertilized and fertilized soil was 71% and 20%, respectively, of the total uptake (Section 13.2.4.1). From these data it can be concluded that field-grown plants do not uniformly deplete even the densely rooted topsoil; near the root surface a high proportion of the nonexchangeable potassium contributes to the total uptake, whereas at a distance from the root (that is, in the bulk soil) even the readily exchangeable potassium is not utilized. An example giving estimates of the proportion of soil delivering phosphorus and potassium to maize roots is shown in Table 13.9. Because of the typical differences in D_e values between potassium and phosphorus, the proportion of soil supplying nutrients is lower for phosphorus than for potassium.

Table 13.9 also clearly demonstrates the importance of the other main morphological component of nutrient acquisition of nonmycorrhizal roots from soils, the root length. This is expressed either as total root length per plant or total root length per unit soil volume as given in Table 13.9. Although the volume of the root hair cylinder of different plant species is often closely related to the uptake per unit root length of potassium (Claassen and Jungk, 1984) and phosphorus (Föhse et al., 1991), this is not necessarily true for the content of the two mineral nutrients in the shoots, or the demand of supply of nutrients required for shoot growth as shown for phosphorus in Table 13.10. If a plant species is highly efficient in phosphorus acquisition from soils this results either from high influx rates per unit root length (particularly root hair length), as well as from total root length, or root length–shoot dry weight ratio (Föhse et al. 1988; 1991), i.e., the ratio of 'source' size (root surface area) to 'sink' size (shoot weight). The relative importance of total root length versus uptake efficiency per unit root length also depends on the phosphorus level in the soil; for example, in wheat roots length becomes more important in low phosphorus soils and uptake efficiency in high phosphorus soils (Römer et al., 1988). In studies with young plants on phosphorus efficiency of different

Table 13.10
Relationship between Surface Area of Root Hairs (SAH) per Surface Area of Root Cylinder (SAC), and Demand of Phosphorus Supply in Soils for Obtaining 80% of Maximum Growth in Different Plant Species[a]

Plant species	SAH/SAC $(cm^2\,cm^{-2})$	Phosphorus demand	
		mg P $(100\,g)^{-1}$ soil	μM P in soil solution
Wheat (*Tr. aestivum*)	1.2	4	1.2
Rape (*Br. napus*)	1.3	5	1.4
Spinach (*Sp. oleracea*)	1.9	9	4.6
Bean (*P. vulgaris*)	0.4	9	4.6
Tomato (*L. esculentum*)	0.6	11	5.7
Onion (*A. cepa*)	<0.001	17	6.9

[a]Based on Föhse *et al.* (1988, 1991).

species which differ much in seed size, and thus in phosphorus reserves (e.g. rape and bean), some problems in interpretation can arise, as large-seeded plants can develop a larger root system from seed reserves only (Narayanan *et al.*, 1981).

Compared with red clover, ryegrass has a 2–4 times larger total root surface area and has a distinctly higher capacity for acquiring phosphorus (Steffens, 1984) and potassium (Mengel and Steffens, 1985) particularly when grown in soils low in phosphorus or potassium. As both plant species have a similar demand, for example of potassium, rates of potassium uptake per unit root length have to be at least three times higher in red clover compared with ryegrass (Mengel and Steffens, 1985). In mixed stands of both species, ryegrass is therefore a strong competitor for potassium and as a consequence the growth of red clover can rapidly be depressed in soils low in potassium (Steffens and Mengel, 1980).

13.3 Role of Root Density

Although a high root density and long root hairs are important factors in the uptake of nutrients supplied by diffusion, the relationship between root density and uptake rate is not linear, as shown in Fig. 13.7. When the root density is high, the uptake rate levels off. This is caused by overlapping of the depletion zones of individual roots and reflects interroot competition for nutrients (Fig. 13.8). For a given interroot distance the degree of competition mainly depends on D_e and is therefore usually much higher for nitrate than for potassium and is of minor importance for phosphorus, at least under field conditions for maize (Fusseder *et al.*, 1988). However, in poorly structured soils root aggregations are abundant and in those zones interroot competition for mineral nutrients can become important even for phosphorus at root length densities which on average are still relatively low (Fusseder and Kraus, 1986). This also holds true for root aggregations induced by localized fertilizer placement. In principle the same curvilinear relationship as shown in Fig. 13.7 can be expected between the rate of phosphorus uptake and root hair density, because of competition between individual root hairs

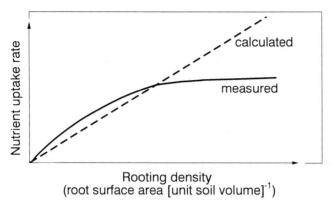

Rooting density
(root surface area [unit soil volume]$^{-1}$)

Fig. 13.7 Relationship between root density and uptake rate of nutrients supplied by diffusion.

(Itoh and Barber, 1983b). This competition has to be kept in mind when attempts are made to correlate root density, for example, to different soil layers or horizons, in relation to their contribution to nutrient supply.

In field-grown plants typical root density gradients occur between topsoil and subsoil (Table 13.11). The high rooting density in the topsoil is mainly caused by the usually more favorable physical, chemical and biological conditions in the topsoil as compared with the subsoil. As a first approximation and as an average of annual agricultural and horticultural crops, the logarithm of root density declines linearly with increasing depth (Greenwood *et al.*, 1982). However, at least in cereal species and maize this gradient becomes less steep during the growing season, and root density in the subsoil increases (Barber and Mackay, 1986; Vincent and Gregory, 1989). A representative example giving average root densities of cereal crops at heading is shown in Fig. 13.9.

Despite the lower root density in the subsoil, nutrient uptake from the subsoil can be considerable. The importance of subsoil nitrate for nitrogen nutrition of crop plants is

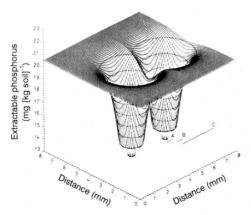

Fig. 13.8 Profile of extractable phosphorus around two individual maize roots with overlapping depletion zones. (Fusseder and Kraus, 1986). A = root cylinder; B = root hair cylinder; C = maximal depletion zone.

Table 13.11
Root Length Distribution of Maize at Flowering in a Luvisol[a]

Soil depth (cm)	Root length	
	Density (cm cm^{-3})	Total (km m^{-2})
0–15	6.19	9.3
15–30	3.07	4.6
30–45	1.12	1.7
45–60	0.48	0.7
60–75	0.41	0.6
75–90	0.26	0.4
90–135	0.17	0.3

[a]Horlacher (1991).

widely established (N_{min}; Section 13.6). For cereal crops such as winter wheat growing in deep loess soils, on average, 30% of the total nitrogen uptake by the crop can derive from the subsoil (Kuhlmann *et al.*, 1989). Uptake from the subsoil is important also for other mineral nutrients such as magnesium, potassium, and phosphorus. The relative importance of subsoil supply depends not only on root density in the subsoil (Fig. 13.9) but also on root density and nutrient availability in the topsoil (Barber and Mackay, 1986; Kuhlmann and Baumgärtel, 1991). Accessibility of mineral nutrients in the subsoil can also depend on the activity of the soil fauna, earthworms in particular; in barley and sugar beet between 20 and 40% of the roots in the subsoil (<65 cm) were found to follow earthworm channels (Meuser, 1991).

13.4 Nutrient Availability and Distribution of Water in Soils

Under many climatic conditions nutrient availability in the topsoil declines more or less steeply during the growing season because low soil water content becomes a limiting

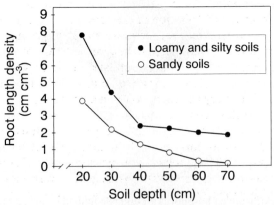

Fig. 13.9   Root length densities at heading of cereal crops in different soils as a function of soil depth. (Gäth *et al.*, 1989.)

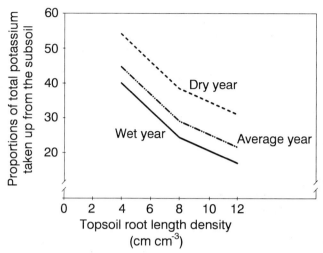

Fig. 13.10   Estimated potassium uptake of spring wheat from a Luvisol (derived from loess) at different rooting densities in the topsoil at flowering, and in years with different soil moisture contents. (Wessolek and Gäth, 1989.)

factor for nutrient delivery to the root surface. It has been shown by Fox and Lipps (1961) that under these conditions ~3% of the total root mass of alfalfa takes up from the subsoil more than 60% of the total nutrient uptake. Low soil water content in the top soil also impairs root elongation and, thus, additionally decreases nutrient uptake, although part of these negative effects on D_e and root elongation can be compensated for by enhancement of root hair development in dry soils (Mackay and Barber, 1985, Section 14.6). The effect of water supply on the root distribution, for example, of spring barley has been demonstrated in two successive years (Scott-Russell, 1977). In the first year, with high rainfall (82 mm) occurring a month after planting, more than 70% of the total root mass was found in the topsoil (2.5–12.5 cm) 2 months after planting and only ~10% of the roots had penetrated deeper than 22.5 cm; in contrast, in the following year, with inadequate rainfall (24 mm) during the first month after planting, the corresponding values for the distribution of the root mass were about 40% and 30%, respectively. This type of shift in root distribution has important consequences for nutrient uptake from various soil horizons. In spring wheat growing on Luvisols, on average ~50% of the total potassium taken up later in the growing season is derived from the subsoil (Grimme et al., 1981), but depending on the rainfall during the growing season the percentage may vary between ~60% in a dry year and ~30% in a wet year (Fleige et al., 1983). Within a given year (wet or dry) the relative importance of potassium uptake from the subsoil also depends very much on the root length density in the topsoil (Fig. 13.10). From large-scale field tests on root length distribution in various cereal crops (Fig. 13.9), combined with measurements of exchangeable potassium and consideration of the climatic water balance at a given location, models have been established which properly predict potassium delivery to roots of cereal crops (Gäth et al., 1989).

In principle, the uptake of phosphorus from different soil horizons should also

Table 13.12
Phosphorus Uptake by Spring Wheat and Phosphorus Delivery as a Function of
Soil Depth and Time[a]

		Phosphorus uptake (kg ha^{-1} d^{-1})		
Available P[b] (mg (100 g)$^{-1}$ soil)	Soil depth (cm)	Booting stage 0.345	Anthesis 0.265	Milky stage 0.145
		Delivery of P from different soil depths (%)		
11.5	0–30	83.3	58.8	67.4
4.5	30–50	8.1	17.8	15.5
2.5	50–75	5.9	16.3	12.0
2.0	75–90	2.7	7.1	5.1

[a]Luvisol derived from loess. Based on data from Fleige *et al.* (1981).
[b]Determined by extraction with calcium ammonium lactate.

respond similarly, or even more distinctly, to differences in soil moisture distribution within the soil profile. As shown in Table 13.12, despite the much higher level of extractable ('available') phosphorus in the topsoil, under low or erratic rainfall conditions during the growing season, in the later stages of growth between 30 and 40% of the total phosphorus taken up by the spring wheat crop comes from the subsoil.

For prediction of the nitrogen supply in the form of nitrate from the topsoil and subsoil the situation is somewhat different from potassium and phosphorus. Transport of nitrate by massflow can considerably contribute to the total delivery to the root surface (Strebel *et al.*, 1983; Strebel and Duynisveld, 1989). However, unless a high nitrate concentration is maintained in the soil solution by fertilizer application during the growing season, the relative proportions of nitrate supplied by mass flow and by diffusion shift most noticeably, as the buffer power for nitrate is usually low in soils, except in soils with high potential of readily mineralizable organic nitrogen. The concentration of nitrate is usually high in the topsoil early in the growing season but rapidly declines thereafter as a result of plant uptake. In spring wheat the change in nitrate concentration in the soil solution is correlated with a decline in the amount of nitrate supplied by mass flow and an increase in supply via diffusion, which then supplies more than 50% of the total nitrate (Strebel *et al.*, 1980).

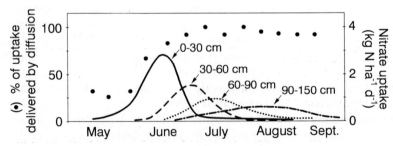

Fig. 13.11 Nitrate uptake rate and delivery by diffusion to sugar beet plants as a function of soil depth (cm) and time. Soil: Luvisol derived from loess. (Based on Strebel *et al.*, 1983.)

For sugar beet the supply by mass flow during the entire growing season is even less, on average 32 kg nitrate nitrogen, compared with 181 kg supplied by diffusion (Strebel *et al.*, 1983). A time course study (Fig. 13.11) demonstrates that the supply by mass flow is restricted to the early growing period; during this time, nitrate is taken up from the topsoil, which has a large nitrate concentration in the soil solution. Thereafter, the nitrate in the topsoil is depleted, and upon root proliferation into the subsoil nitrate is supplied exclusively by diffusion. This example illustrates that average data on the contribution of mass flow and diffusion (as well as of different soil horizons) to the total supply do not reflect reality. Simulation models for predicting nutrient uptake in field-grown plants have therefore to take into account nutrient dynamics in space and time as well as plant species and genotype within a species. In maize, for example, considerable genotypical differences exist in the extent of nitrate depletion in the subsoil; depletion was positively correlated with the root length densities in the subsoil at silking (Wiesler and Horst, 1993).

13.5 Role of Soil Structure

For various reasons, soil structure plays an important role in determining the amounts of mineral nutrients that are available for uptake by the roots. In structured soils not all roots have complete contact with the soil matrix, and in nonmycorrhizal plants the degree of root–soil contact varies between root segments (i.e., along the root axis) from 0 to 100% (Van Noordwijk *et al.*, 1992). In soils with higher bulk density contact is greater but root elongation growth is simultaneously impaired (Table 13.13). This impairment is in part compensated for by higher uptake rates per unit root length, at least of nitrate and water (Table 13.13), as well as of phosphate, particularly in soils high in available phosphorus (Cornish *et al.*, 1984). However, in experiments in which soil bulk density was increased and soil–root contact raised from 25 to 75%, root aggre-

Table 13.13
Relationship between Bulk Density of the Soil, Soil Porosity (Macropores > 30 μm), Root Length, Estimated Root-Soil Contact, and Uptake Rate of Nitrate and Water per Unit Root Length of Maize[a]

	Bulk density (g cm^{-3})		
Parameter	1.08	1.32	1.50
Soil porosity (%)	60	51	44
Root length (m per pot)	114	83	50
Root surface with soil contact (%)	60	72	87
Uptake (mmol m^{-1} root length)			
Nitrate	14	15	19
Water	18	21	24

[a]Compiled data from Van Noordwijk *et al.* (1992); Kooistra *et al.* (1992) and Veen *et al.* (1992)

gation was observed in certain zones and so too increased local O_2 demand (Asady and Smucker, 1989). For maintenance of root respiration at such sites of aggregation required external O_2 concentrations are more than triple. The optimal degree of soil–root contact and soil bulk density for nutrient uptake and plant growth thus depends on aeration and soil fertility (Van Noordwijk *et al.*, 1992).

In structured soils nutrient availability, particularly to perennial species, is also affected by other means. In arable soils at least in the plow layer established macro-structures are rare, these structures occur in the subsoil. However, in soils in which perennial species (pastures, forests) are growing established macrostructures are also present in the topsoil. The conventional soil testing methods for determination of available nutrients use soil samples which are homogenized prior to extraction. These methods have thus two major limitations. They not only ignore the importance of spatial availability of nutrients (as discussed above) but also destroy the soil structure and thereby, for example, gradients which occur in the cation exchange capacity and the base saturation between the external and internal surfaces of aggregates (Horn, 1987; Kaupenjohann and Hantschel, 1989). Whereas external surfaces of aggregates are usually in equilibrium with the soil solution in the macropores, this is not the case with the internal surfaces, as ion exchange processes between both surfaces are obviously limited and quite slow (Horn, 1989). These aspects are particularly important in acid forest soils where the structurally related gradients in microvariability of soil solution chemistry are very distinct, in addition to the spatial heterogenity related to the stem distance (Section 13.2.1).

Most of these limitations concerning soil structural aspects can be avoided by using soil solution collected from lysimeter or suction cups in the field, or from small soil cores. Soil solution can be obtained from these cores either by circulation of a percolating solution (Hildebrand, 1986) or by centrifuging after adjustment to field capacity. An example of the differences in cation concentration between so-called equilibrium soil solution from homogenized soil and of percolation solution of the same, but still structured acid soil, is shown in Table 13.14. As a rule, the concentrations of cations (except protons) are much higher in the homogenized samples as a result of the destruction of the aggregates and exposure of internal surfaces to the extractant. Accordingly, correlations were poor or absent between the contents of potassium and magnesium in the needles of Norway spruce, and the respective concentrations of the two nutrients in the extraction solution. For the same sites, however, correlations were

Table 13.14

Cation Concentrations in Soil Equilibrium Solutions (Homogenized Soil) and in Soil Percolation Solution (Structured Soil) of a Brown-Earth, pH (CaCl$_2$) 3.06[a]

Method employed	Concentration in soil solution (μM)				
	K	Ca	Mg	Al	Fe
Equilibr. sln.	55	41	39	104	39
Percolat. sln.	13	15	17	52	17

[a]Based on Hantschel *et al.* (1988) and Kaupenjohann and Hantschel (1989).

highly significant for both nutrients in the solutions of structured samples (Kaupenjohann and Hantschel, 1989).

13.6   Intensity/Quantity Ratio, Plant Factors, and Consequences for Soil Testing

In principle routine soil testing methods (Section 13.1, Table 13.1) determine the
fraction of 'chemically available' nutrients. In terms of an intensity/quantity concept,
depending on the extraction method, this mainly characterizes the intensity (e.g,. water
extraction) or a variable amount of the quantity, represented by the labile pool (Fig.
13.12). Soil testing for phosphorus in water extracts ($1 \, cm^3$ soil $(60 \, ml)^{-1} \, H_2O \, (22 \, h)^{-1}$)
is considered a reasonable compromise between a measurement of intensity and
capacity of phosphorus supply of the soil (Van Noordwijk *et al.*, 1990). Mild extractants
such as sodium bicarbonate (Olsen-method) primarily characterize the phosphate
fraction adsorbed to aluminum at clay surfaces (Kuo, 1990). More detailed information
concerning binding strengths, rate of replenishment, and the intensity/quantity ratios
for different mineral nutrients can be obtained with the electro-ultrafiltration method
(EUF), which involves the use of different electrical field strengths and temperatures in
an aqueous soil suspension (Nemeth, 1982; Nemeth *et al.*, 1987). However, it is not
possible to characterize adequately the buffer power of the soil for phosphate in terms
of predicting plant uptake of phosphate (Nair and Mengel, 1984). Also, for routine soil
testing the EUF method is not necessarily superior or technically simpler than
conventional extraction methods (e.g. with $CaCl_2$) in the prediction of fertilizer
requirement (Houba *et al.*, 1986).

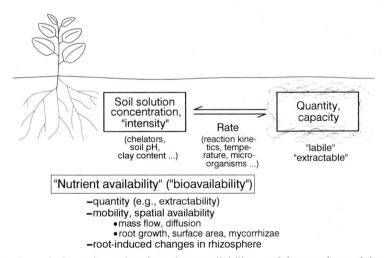

Fig. 13.12   Intensity/quantity ratio of nutrient availability, and factors determining the 'bioavailability' of mineral nutrients. (Marschner, 1993.) Reprinted by permission of Kluwer
Academic Publishers.

A large number of extraction methods are used in routine soil testing for micronutrients which, as a rule, mainly characterize the quantity component (Fig. 13.12) and predict fertilizer requirement reasonably well when the extracted amounts are considerably different from those considered as adequate (Sims and Johnson, 1991). Predictions can sometimes be improved by consideration of other soil properties such as pH, redox potential, clay and organic matter content (Moraghan and Mascogni, 1991; Brennan, 1992c). In terms of 'bioavailability' (Fig. 13.12) no systematic studies have been carried out on the relative importance of mass flow and diffusion for delivery of micronutrients to plant roots, nor on the subsoil as a potential contributor to micronutrient supply to roots. Only for zinc is there direct and indirect evidence to support the view that in aerated soils of high pH at least is diffusion the main component of delivery to the roots (Marschner, 1993).

Because of the importance of both ion concentrations in the soil solution and the rates of replenishment of these ions, the use of ion-exchange resins has attracted new interest as a possible means of characterizing the buffer power of soils, as for example for potassium (Gäth et al., 1989) or simultaneously for various cations and anions (Yang et al., 1991). In experiments with bean and maize under field conditions prediction of zinc uptake was more precise with ion exchange resins than with the routine DTPA extraction (Hamilton and Westermann, 1991).

Soil testing as the basis for recommendation of nitrogen fertilizer application for various agricultural and horticultural field crops has been very much improved by the N_{min} method. With this method the amount of mineralized nitrogen, mainly nitrate, in the soil profile is measured at the beginning of the growing season, thus taking into account components of 'bioavailability' (Fig. 13.12), namely the high mobility of nitrate in the soil profile (mass flow) and nitrogen uptake from the subsoil (root growth). Depending on plant species and rooting depth, N_{min} is determined up to a soil depth of 90 cm (Wehrmann and Scharpf, 1986; Schenk et al., 1991). The N_{min} method can also improve fertilizer recommendations in rainfed agriculture under dryland conditions (Soltanpour et al., 1989). Since in humid and semihumid climates most of the nitrate in the subsoil originates from mineralization of organically bound nitrogen (N_{org}) and nitrification of ammonium nitrogen in the topsoil, various attempts have been made to characterize this particular N_{org} fraction in the topsoil prior to nitrate leaching into the subsoil, as for example, by use of the EUF method (Nemeth, 1985; Mengel, 1991) or of $CaCl_2$ as an extractant (Appel and Steffens, 1988). For cereals, both EUF N_{org} and $CaCl_2$ extraction seem to be a suitable alternative to the N_{min} method (Appel and Mengel, 1992).

The principal limitation of soil testing methods is that they only characterize some of the factors which determine nutrient supply to the roots of field-grown plants. Improving the reliability of fertilizer recommendations based on chemical soil analysis does not depend primarily on extraction method used, but rather on the systematic consideration of the roots and environmental factors such as soil water content. Current models for predicting nutrient availability and nutrient uptake under field conditions are therefore based on both soil and plant factors (Fig. 13.12) in which root parameters are the key element (Greenwood, 1983; Barber, 1984; Van Noordwijk et al., 1990; Claassen, 1990). These models have been very much refined in recent years and predictions on mineral nutrient uptake, and actual uptake by crops, for example of

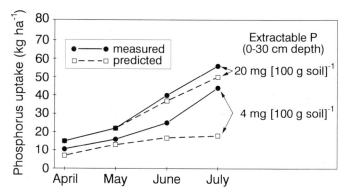

Fig. 13.13 Phosphorus uptake of winter wheat in a field experiment after long-term application of either 100 kg P ha^{-1} per year (high P soil) or without P fertilizer application (low P soil). Comparison of measured P uptake with P uptake predicted by the Claassen–Barber model. (Jungk and Claassen, 1989.)

phosphorus (Fig. 13.13) or potassium (Seward *et al.*, 1990) are often, but not always, in good agreement. As shown in Fig. 13.13 the predicted and measured uptake of phosphorus were closely related in the soil with high phosphorus content. However, in the soil with low phosphorus content predicted uptake was much lower than the measured uptake, indicating that the wheat plants in the low phosphorus soil had access to soil phosphorus sources which were not considered in the model. The acquisition of relatively large amounts of phosphorus by VA mycorrhiza to the wheat plants offers one possible explanation (Jungk and Claassen, 1989), and root-induced changes in the rhizosphere another (Silberbush and Barber, 1984). The role of mycorrhizae and of root-induced changes in the rhizosphere for the 'bioavailability' of nutrients (Fig. 13.12) is discussed in detail in Chapter 15.

In contrast to phosphorus, in the case of potassium the predictions were in close agreement with the measured uptake of the wheat crop only in potassium-deficient soils whereas in the potassium-sufficient soil the models overpredicted the potassium uptake by as much as four times (Seward *et al.*, 1990). This overprediction was obviously the result of poor characterization of the plant demand and thus an underestimation of the role of negative feedback regulation of potassium uptake by the roots at high internal content (Section 2.5.6).

In conclusion, at present, but presumably also at least in the near future, mechanistic simulation models cannot replace soil testing but can refine recommendations for different yield levels, and can make more accurate predictions for new crops or on the effects of changing soil water balance (Van Noordwijk *et al.*, 1990).

14

Effect of Internal and External Factors on Root Growth and Development

14.1 General

Plant root systems vary widely, both within and between plant species. Root dimorphism in C_4 species is an example of this variation (Section 2.7). Root systems are characterized by very high adaptability, and their growth and development involves complex interactions between both the soil environment and the shoots. Since the environment in which root systems develop is highly heterogeneous, both in space and time, the root system has to have the ability to react to heterogeneity and, thus, must possess high phenotypic plasticity (Fitter, 1991b). The most well-known responses of roots showing phenotypic plasticity are to localized supplies of nutrients, or to other localized changes involving chemical, physical and microbiological factors in the root environment. Changes in endogenous levels of phytohormones play an important role as transmitters of environmental signals which are either transformed in the roots into alterations of growth and developmental processes or transmitted as root signals to the shoot. In this chapter some examples are given of these aspects of root growth and development which are closely related to, or important for, the mineral nutrition of plants.

14.2 Carbohydrate Supply

Depending upon species and the developmental stage of a plant, on average 25–50% of the photosynthates produced per day in the shoot are allocated to the roots for growth, maintenance and other functions, for example, ion uptake (Section 2.4.1). About half of these carbohydrates allocated to the roots are used in respiration (Lambers *et al.*, 1991). In very young plants root carbohydrates can originate from seed reserves as well as from current photosynthesis, but the seed source becomes of lesser importance as the plant develops. After the seedling stage there is therefore a close relationship between root growth and photosynthesis as shown in Fig. 14.1 for coniferous plants. Growth of new roots in early spring depends on current photosynthesis and is proportional to light

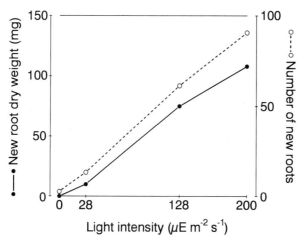

Fig. 14.1 Production of new roots in 2-year-old Douglas fir seedling transplants after 24 days of growth at different light intensities. (Van den Driessche, 1987.)

intensity and rate of photosynthesis. In contrast, there is only a very limited use of carbohydrates stored in older parts of roots for new growth (Van den Driessche, 1987).

Typically, limitation of photosynthesis by low light intensity inhibits root growth more than shoot growth as shown in an example for pea plants in Table 14.1. Root length per plant is particularly depressed under low light and, accordingly acquisition of mineral nutrients is impaired, and especially those of low mobility in soils.

In mycorrhizal plants the fungal symbiont competes effectively for carbohydrates with the host root, between 15 and 30% of the carbohydrates allocated to the roots might be used by the fungus (Lambers *et al.*, 1991). A similar proportion of carbohydrates is used in nodulated legumes for establishment and functioning of the symbiotic association. Accordingly, there is a marked competition between the symbiont and the roots for carbohydrates. Generally the symbionts are very effective in

Table 14.1

Shoot and Root Growth of Pea (*Pisum sativum* L.) Plants Grown for 35 Days at Different Light Intensities and without or with *Rhizobium* and VA Mycorrhiza (*Glomus mosseae*) under Non-Limited Phosphorus Supply[a]

Treatment Symbiont		Light intensity ($\mu E\ m^{-2}\ s^{-1}$)	Dry weight (mg per plant)		Root/shoot ratio	Root length (m per plant)
Rh.	VAM		Shoot	Root		
−	−	390	506	306	0.60	31.7
−	−	190	248	105	0.42	11.3
+	+	390	592	168	0.28	17.7
+	+	190	326	73	0.22	6.4

[a]Based on Reinhard *et al.* (1992).

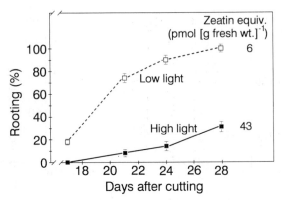

Fig. 14.2   Effect of high light (270 μE m^{-2} s^{-1}) and low light (30 μE m^{-2} s^{-1}) pretreatment of Norway spruce seedlings on CYT content in the hypocotyls and subsequent rooting percentage of cuttings. (Bollmark and Eliasson, 1990.)

competition for carbohydrates and there is a correspondingly sharp decline in root growth and in the root/shoot dry weight ratio, particularly under conditions of low light intensity (Table 14.1). Under these conditions part of the root functions may be compensated for by the symbiont (acquisition of nitrogen and phosphorus) but not all (e.g., uptake of water).

Shoot–root interactions are not confined to supply of carbohydrates or reduced nitrogen (Section 3.4). Recycling of mineral nutrients in the phloem from the shoot to the roots is important as a feedback signal in regulating ion uptake by the roots (Section 3.4.4). Root development and growth are also under the control of shoot-derived phytohormones and root export of phytohormones. Light intensity can therefore also affect root growth and development by factors other than carbohydrate supply. As shown in Fig. 14.2, root formation was severely inhibited in cuttings of Norway spruce seedlings which were grown under high light intensity. This inhibition of adventitious root formation is causally related to high endogenous levels of cytokinins in the high light seedlings. External application of cytokinin has a similar inhibitory effects on rooting as that of high light intensity.

14.3   Root Morphology, Hormonal Interactions

Despite large genotypical and phenotypical differences growing roots have some morphological features in common as shown schematically in Fig. 14.3. Attached to the root tip is the root cap which is a structural entity and its cells are continuously being renewed during root growth. The root cap not only protects the root meristem and facilitates penetration of the root through pores (by secretion of mucilage as 'lubricant'), it is also the site of perception of chemical and physical signals (Sievers and Hensel, 1991). Behind the root meristem is the elongation (extension) zone where the root cap-transmitted signals are rapidly transformed into differences in expansion rate of the cells and the root. Root elongation is correlated with stretching of the cell walls, particularly the outer tangential cell wall which, for example, in maize decreases in

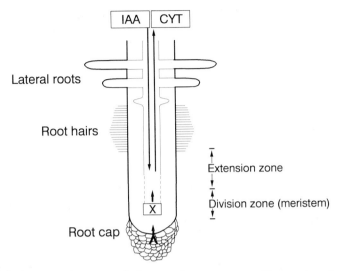

Fig. 14.3   Main features of root morphology, and some aspects of interactions in a growing root (IAA = indolylacetic acid, auxin; CYT   = cytokinins; X = unidentified growth modulating compound, ABA?).

thickness from $5.72\,\mu m$ near the apex to about $0.38\,\mu m$ at about 7.3 mm behind the apex (Hofer and Pilet, 1986).

Behind the elongation zone is the region of root hair development. Root hairs are elongated epidermal (rhizodermal) cells, of length usually between 80 and $1500\,\mu m$, and are normally short lived and collapse after a few days. However, depending on plant species and environmental factors, length and longevity of root hairs vary a great deal (Hofer, 1991). In peanut there are two types of root hairs one of which may be as long as 4 mm (Meisner and Karnock, 1991). In certain tomato strains with 'cottony roots' root hair density is exceptionally high and they are up to 3.1 mm long (Hochmuth *et al.*, 1985). There is much circumstantial evidence that ethylene plays a key role in root hair formation (Michael, 1990).

Lateral roots (Fig. 14.3) arise from the root pericycle, often adjacent to mature protoxylem strands. Lateral root development is strongly influenced by environmental factors (e.g., nutrient supply, rhizosphere microorganisms, soil strength) and under distinct hormonal control, particularly by IAA and CYT.

Auxins play an important role in the regulation of root growth and development. It is formed in the shoot (or the endosperm) and moves to the root tips (Martin and Elliott, 1984) where it can accumulate (Hasenstein and Evans, 1988). Roots are highly sensitive to IAA, concentrations as low as 10^{-9} M enhance cell extension; the receptor sites for IAA are rhizodermal cells of the apical root zone (Radermacher and Klämbt, 1993), similarly to the epidermal cells in coleoptiles (Section 5.6.3). High concentrations of endogenous IAA in root tips inhibit cell elongation, either directly or indirectly via enhanced production of ethylene (Eliasson *et al.*, 1989; Jackson, 1991b). On the other hand, auxins promote the formation of lateral roots, or at least their initiation (Table 14.2).

Table 14.2
Effect of Auxin (10^{-5} M) and Kinetin (2×10^{-5} M) on Formation
of Lateral Root Primordia (LRP) in the Primary Root of Lettuce
Seedlings after 72h[a]

Treatment	No. of LRP per root	Length of primary root (mm)
Control	5.7	29.6
+ Auxin	32.4	10.8
+ Kinetin	0	12.2

[a]Based on MacIsaac et al. (1989).

Root apical zones are the main sites of CYT synthesis in plants (Section 5.6.2) and they are exported in the xylem to the shoot. In contrast to auxins, a high concentration of CYT not only inhibits elongation of the main axis but also the formation of lateral roots (Table 14.2). In agreement with this, initiation of lateral roots can be dramatically increased by removal (decapitation) of the apex of the main axis (Forsyth and van Staden, 1981). Thus auxins and CYT play an important role in the regulation of both elongation of the main axis and the formation of lateral roots, in the latter case phytohormones have more or less an opposite effect.

The role of ABA in regulating root growth and development is less clear (Fig. 14.3). The apical zone (0–2 mm) contains high concentrations of ABA which decline basipetally (Ribaut and Pilet, 1991). ABA concentrations in apical zones are inversely related to the growth rates of the roots, but this does not necessarily imply a direct inhibitory effect of ABA on the rate of root elongation (Pilet, 1991). It is also not clear whether the enhancing effects of ABA on lateral root initiation and root hair formation (Biddington and Dearman, 1982) results directly from inhibitory effects on the extension of the apical root zone. Although ABA synthesis is high in the root cap (Pilet, 1981) ABA itself is probably not the growth-modifying signal which moves from the root cap basipetally to the elongation zone and modifies root curvature in response to mechanical impedance (Goss and Scott-Russell, 1980), aluminum (Bennet and Breen, 1991) or gravity (Moore et al., 1990; Sievers and Hensel, 1991). Recently calcium has been suggested as the factor responsible not only in signal perception of aluminum (Bennet and Breen, 1991) and gravity but also in the transduction of these signals and their conversion into growth responses in the elongation zone (Poovaiah and Reddy, 1991; Takahashi et al., 1992). The roles of calcium and calmodulin in signal transduction at the cellular level are well established (Section 8.6.7). It is not clear, however, how such a role of calcium in root growth responses, which require calcium transport from the root cap to the extension zone can be fulfilled without the involvement of hormones (e.g., IAA/Ca^{2+} counterflow, Section 3.4.3).

In contrast to the open questions as to the role of ABA in apical root meristems, it is well established that its accumulation in roots under drought stress (Ribaut and Pilet, 1991) and its transport to the shoot acts as a nonhydraulic root signal leading to inhibition in shoot and leaf elongation and a decrease in stomatal aperture (Section 5.6).

There are numerous and often contradictory reports on the effect of ethylene on root growth which appear to depend on applied concentrations. Low concentrations (<1 mg l^{-1}) may enhance root elongation, whereas high concentrations severely inhibit root elongation but simultaneously increase root diameter and root hair formation (Michael, 1990; Jackson, 1991b). Light strongly inhibits root elongation, the receptor again being the root cap, and ethylene is causally involved in the signal transduction (Eliasson and Bollmark, 1988). The most remarkable effect of elevated ethylene concentrations is the formation of aerenchyma in the root cortex which occurs in response to waterlogging as a mechanism of adaptation of roots to submerged conditions (Section 16.4.3).

14.4 Soil Chemical Factors

14.4.1 Mineral Nutrient Supply

Mineral nutrient supply can strongly affect root growth, morphology and distribution of root systems in the substrate (e.g., soil profile). This effect is particularly marked with nitrogen, less distinct with phosphorus, and usually absent with other nutrients, except for magnesium (Section 8.5). In the responsive zone (i.e., concentration range where nutrients limit plant growth), increasing nitrogen supply enhances both shoot and root growth, but usually the shoot growth more than the root growth, leading to a typical fall in root/shoot dry weight ratio with increase in nitrogen supply (Table 14.3). However, this comparison is somewhat misleading in terms of nutrient acquisition as the roots become finer (higher branching) and the surface area increases so that, despite the marked decrease in root/shoot dry weight ratio, the sink–source relationship, shoot weight (demand) and root surface (source, supply) for nutrient and water acquisition is less affected. In soil-grown plants this effect of nitrogen in increasing root surface area is usually more distinct with supply of ammonium as nitrogen source as compared with nitrate (Marschner et al., 1986b).

The reasons for the different effects of nitrogen forms on root morphogenesis are unknown; differences in both pathways of assimilation in the roots and in plant hormonal balance (Chapter 6) are probably involved. Hormonal effects are probably also responsible for the increase in the formation of aerenchyma in the cortex of maize

Table 14.3
Effect of Increasing Nitrate Concentrations in Nutrient Solution on Shoot and Root Growth of Potato Plants (cv. Astrid)[a]

N supply (mM)	Dry weight (g per plant)		Root/shoot ratio	Root surface area (dm^2 per plant)	Root length (m per plant)
	Shoot	Roots			
0.05	0.8	0.45	0.56	63	67
0.5	3.5	1.39	0.40	314	277
5.0	9.2	1.82	0.20	577	502

[a]From Sattelmacher et al. (1990a).

Table 14.4

Effect of Nitrogen Supply (NH_4NO_3) on Shoot and Root Growth of
15-Week-Old Field-Grown Maize[a]

N supply (kg ha^{-1})	Dry weight (g per plant)		Roots per plant		Root/shoot dry wt ratio
	Shoot	Grain	Length (m)	Dry wt (g)	
0	186	54	2189	42	0.23
180	352	138	2521	38	0.11

[a]Based on Anderson (1988).

roots even in well-aerated solutions when the nitrogen supply is low (Konings and Verschuren, 1980). An increase in aerenchyma formation is a typical root response to elevated levels of ethylene (Section 16.4).

Under field conditions the enhancing effect of nitrogen supply on root growth is usually less distinct (Table 14.4) but in principle the same pattern occurs as in nutrient solutions, namely an increase in total root length and a distinct decline in root/shoot dry weight ratio. In field-grown plants, however, data on root dry weight and root length in particular often much underestimate true values because of considerable losses of fine roots and leaching of solutes during collection and preparation (Grzebisz *et al.*, 1989). Furthermore, the turnover rate of roots is presumably much higher in plants adequately supplied with nitrogen than those that are nitrogen deficient. In broadleaf and coniferous forest stands the turnover rate of fine roots per year increased from 50% in low nitrogen soils to up to 200% in high nitrogen soils (Aber *et al.*, 1985).

Root growth is enhanced at sites of high as compared with low nutrient supply. This effect of increased root growth can be demonstrated in split-root experiments, by fertilizer placement in soils, or by localized nutrient supply to only one zone along the root axis. In longterm split-root experiments with Sitka spruce trees, root growth rate and total dry weight increased much more on the side of nutrient supply. Nitrogen was the most effective nutrient, with phosphorus giving some beneficial effects, potassium none (Coutts and Philipson, 1977; Philipson and Coutts, 1977). The growth enhancement on the nutrient-rich side was correlated with reduced root growth on the side without nutrient supply possibly but not necessarily, reflecting source limitation. The distribution of roots in soils can thus be modified by the placement of fertilizers. In annual species rooting density rapidly increases severalfold in zones of higher nutrient concentrations, especially of nitrogen (Fig. 14.4). This also demonstrates the potential risk of having high nitrogen availability only in the topsoil as roots concentrate there at the expense of subsoil penetration. Deep placement of fertilizer, nitrogen in particular, therefore enhances plant growth under drought stress conditions when the water potential of the topsoil decreases but ample water is available in the subsoil (Garwood and Williams, 1967).

Placement of phosphorus fertilizers is a common and effective practice in soils low in readily soluble phosphorus to ensure an adequate supply of the nutrient to the roots, especially in the early growing stages when soil temperatures are low. Rooting density

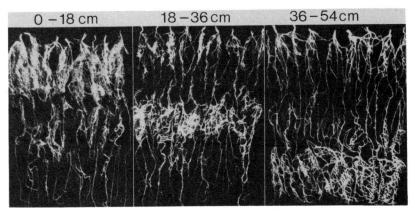

Fig. 14.4 Effect of nitrogen fertilizer placement in different soil depths on the root distribution of barley growing in a sandy soil. (From Gliemeroth, 1953.)

increases in zones of phosphorus fertilizer placement, although the effect is relatively small compared with that of combined nitrogen–phosphorus fertilizers (Böhm, 1974).

The effects of localized nutrient supply on root morphology can be demonstrated in more detail if the nutrients are supplied at a high concentration to only one particular zone of the root. As shown in Fig. 14.5, in the root zone supplied with a large amount of nitrate, lateral root growth was spectacularly enhanced.

Although the effect of a localized supply of phosphorus under field conditions is usually less dramatic than that of nitrate, in principle similar effects are obtained if phosphorus is supplied to one root zone only and the remainder of the root system is kept in a solution without phosphorus (Table 14.5). Over the 21-day period the total length of the lateral roots increased 15-fold over that of controls. The corresponding

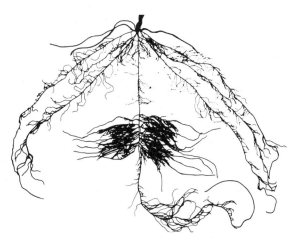

Fig. 14.5 Modification of the root system of barley by providing 1 mM nitrate to the midpart of one root axis for 15 days; the remainder of the root system received only 0.01 mM nitrate. (From Drew and Saker, 1975.)

Table 14.5
Effect of Localizing the Supply of Phosphorus to a Root Segment on the
Length of Lateral Roots and the Dry Weight of Barley Plants[a]

	Uniform supply		Localized supply[b]	
Root zone	Length of lateral roots (cm)	Dry weight (mg)	Length of lateral roots (cm)	Dry weight (mg)
A (basal)	40.0	8.9	14.3	3.5
B (middle)	27.2	3.7	332.0	37.8
C (apical)	17.5	10.2	11.1	4.9

[a]Based on Drew and Saker (1978).
[b]Phosphorus was supplied to a 4-cm segment (middle, or B zone) of a single seminal root axis. Duration of experiment: 21 days.

increase in dry weight (by a factor of 10) took place partially at the expense of the remaining root zones, A and C.

The question arises as to how mineral nutrients bring about this morphological change in the development of root systems. In maize roots with localized nitrate supply phloem unloading of photosynthates is enhanced at the zone of supply as soon as two days after the treatment begins, and after four days cell division rate is increased (Thoms and Sattelmacher, 1990). Respiration rates also increase at the sites of supply, but not of the total root system (Granato and Raper, 1989) suggesting alteration in photosynthate partitioning within the root system in favor of sites with high nutrient supply. The simultaneously enhanced initiation of lateral roots at sites of high nutrient supply is presumably not caused by higher unloading of photosynthates per se, or higher respiration rates, but by phloem unloading of IAA together with the photosynthates (Thoms and Sattelmacher, 1990).

Not only lateral root formation is influenced by mineral nutrient supply but also length and density of root hairs per unit root length. Again nitrogen and phosphorus seem to have the most pronounced effect. In rape, for example, with decreasing nitrate concentrations root hairs become much more frequent all along the root axis, and root hair length increases from 0.3 to about 3.0 mm (Bhat *et al.*, 1979). Increase in root hair length under nitrogen limitation, in principle, is also the case for grasses, but large differences in response to nitrogen occur between species (Robinson and Rorison, 1987).

The effect of phosphorus on root hair formation is similar to that of nitrate. In rape, spinach, and tomato grown at high phosphorus concentration ($>100 \mu M$), root hairs are absent or rudimentary, whereas at low concentrations ($<10 \mu M$) long root hairs are abundant (Föhse and Jungk, 1983). In soil-grown maize plants a fall in phosphorus availability has no effect on root hair length but distinctly increases the density of root hairs per unit root length (Fig. 14.6). This effect is relatively small, however, compared with the response in root hair formation to a decrease in soil water content. This increase in root hair growth at low soil water content can in part offset depressing effects of root elongation which occur in dry soils (Section 14.6.3).

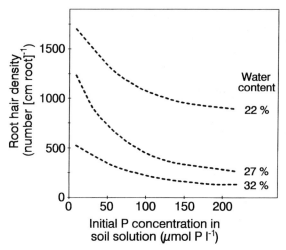

Fig. 14.6   Root hair density in 21-day-old maize plants grown in soils with different phosphorus supply and soil water contents. (Mackay and Barber, 1985.) Reprinted by permission of Kluwer Academic Publishers.

Phosphorus deficiency, like nitrogen deficiency, leads to an increase in the root/shoot dry weight ratio (Table 14.6). Increasing the duration of phosphorus starvation increases root dry weight, and root length in particular. The roots also become finer. In phosphorus-deficient plants roots become the dominant sink not only for photosynthates but also of phosphorus from the shoot (Section 8.4.6; Table 14.6). An increase in root surface area in deficient plants can be considered as a strategy for enhancing phosphorus acquisition from soils (Section 13.3).

In certain plant species formation of root clusters is another type of root response to phosphorus deficiency. The best-known root clusters are *proteoid roots* which are dense clusters of determinate lateral roots, characteristic of Proteaceae (Lamont, 1982; Vorster and Jooste, 1986), although they also occur in species of Myricaceae (Louis *et al.*, 1990), in leguminous trees such as Casuarinaceae (Racette *et al.*, 1990), and in annual legumes such as white lupin (Fig. 14.7). In infertile soils proteoid roots may

Table 14.6

Effects of Phosphorus Starvation on the Shoot and Root Growth of 12-Day-Old Maize Plants[a]

	Shoot		Root		
No. of days without P	Dry weight (g per pot)	P (%)	Dry weight (g per pot)	Length (m per pot)	Radius ($\times 10^2$ cm)
1	2.10	0.95	0.27	46.4	2.27
2	2.34	0.65	0.31	57.7	2.23
4	1.93	0.32	0.40	75.7	1.99
6	1.65	0.27	0.43	90.8	1.84

[a]Based on Anghinoni and Barber (1980).

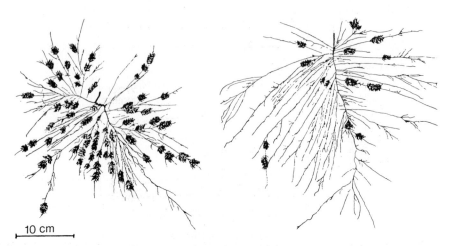

Fig. 14.7 Root system of soil-grown phosphorus-deficient (*left*) and phosphorus-sufficient (*right*) white lupin (*Lupinus albus* L.) plants (Marschner *et al.*, 1987). Number of proteoid roots per plant: P deficient: 53; P sufficient (foliar application of P): 15.

make up as much as 80% of the total root dry weight (Lamont, 1982), in white lupin (Fig. 14.7) it was about 50%.

In some plant species proteoid root formation is also enhanced under nitrogen deficiency (Lamont, 1972; Dennis and Prasad, 1986) and iron deficiency (White and Robson, 1989), but in all these species phosphorus deficiency induces the most pronounced effect. Proteoid roots are characterized by high respiration rates and thus, high oxygen demand (Vorster and Jooste, 1986). Their capacity to mobilize sparingly soluble phosphates is increased by excretion of organic acids or phenolics or both so that the limited soil volume in the immediate vicinity of the proteoid root zones is subjected to intense chemical extraction (Section 15.4.2).

14.4.2 Soil pH, Calcium/Total Cation Ratio

As a rule root growth is not much affected by external pH in the range of 5.0–7.5. In contrast to low pH (<5) stress relatively little information is available on inhibition of root growth at high pH. Root growth can possibly be inhibited by high pH either directly or indirectly. Direct effects of high pH are to be expected in relation to the functioning of transmembrane pH gradient, electropotential gradients, and the proton–anion cotransport at the plasma membrane (Section 2.4.2). The most well-known direct effect at high pH is ammonia toxicity. Root elongation is severely inhibited by ammonia concentrations of as low as 0.05 mM (Schenk and Wehrmann, 1979). Ammonia toxicity is probably also the reason for the inhibition of root growth in neutral or alkaline soils after the application of ammonium phosphate (Bennett and Adams, 1970) or the band application of urea (Creamer and Fox, 1980). At high soil pH root growth, particularly of calcifuge plant species (Lee and Woolhouse, 1969a) and of lowland rice (Dogar and van Hai, 1980) might be also inhibited by elevated bicarbonate concentrations.

Root growth of *Lupinus angustifolius* is exceptionally sensitive to high pH and is depressed even at pH 6.0 (Fig. 14.8). This inhibition in growth is caused by a decrease in

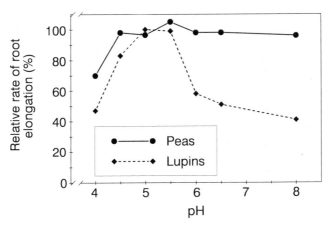

Fig. 14.8 Root elongation of *Lupinus angustifolius* and *Pisum sativum* grown in nutrient solutions of different pH for 60 hours. (Tang *et al.*, 1992b.)

the rate of cell elongation, which occurs within 1 h and leads to an increase in root diameter. The effect is reversible and on lowering the pH, the cell elongation rate is restored. This inhibitory effect at moderate to high pH might be related to the scavenging of protons from the apoplasm which are needed for growth ('acid growth theory'). These findings are in accord with the poor field performance of *L. angustifolius* in alkaline soils (Tang *et al.*, 1992b).

Inhibition of root elongation at pH values below 5 (Fig. 14.8) is a common feature in many plant species and is caused by various factors such as impairment of H^+ efflux (Schubert *et al.*, 1990b) and related processes as discussed in Section 2.4. In soil-grown plants inhibition of root elongation at these low pH values is often closely and causally related to high activities of monomeric aluminum and, thus, aluminum toxicity (Section 16.3.4). Calcium plays a key role in protecting root growth against low pH stress. For a given species, the calcium requirement for root growth is not fixed but is rather a function of both pH and the concentrations of other cations, including aluminum. For example, in cotton at pH 5.6, even ~1 μM calcium in the external solution is sufficient to obtain maximal root growth, compared with more than 50 μM calcium which is required at pH 4.5 (Lund, 1970). On average a molar ratio in the soil solution of calcium to total cations of ~0.15 is needed for maximal root growth (Fig. 14.9).

In acid mineral soils, ratios lower than 0.15 often occur and root growth is inhibited correspondingly. Liming these soils therefore enhances root extension and also root hair length (Haynes and Ludecke, 1981). This is particularly important for subsoil penetration by roots and thus the utilization of nutrients and water from the subsoil (Howard and Adams, 1965). In addition to poor aeration and high mechanical impedance (Sections 14.4.3 and 14.6.1), low subsoil pH is an important factor in restricted subsoil penetration by roots. Since calcium is phloem immobile, the calcium required for root growth must be taken up from the external solution by the apical zones. Roots are therefore severely inhibited in their capacity to penetrate acid subsoils even when an adequate amount of lime has been mixed with the topsoil only (Table

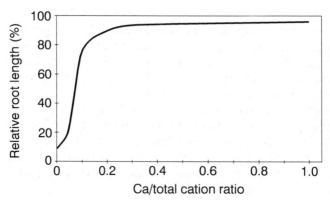

Fig. 14.9   Effect of calcium/total cation molar activity ratio of soil solutions on the growth of seminal roots of cotton seedlings. (Reproduced from Adams, 1966, by permission of the Soil Science Society of America.)

14.7). There is a close correlation between the increasing proportion of soil mass mixed with lime and root elongation in the soil.

In a comparison of aluminum-tolerant and -sensitive wheat genotypes the unexpected result was obtained that root penetration into acid, aluminum toxic subsoil was enhanced by phosphorus fertilization of the topsoil, although the plants were not phosphorus deficient (McLaughlin and Jones, 1991). The mechanisms of this phosphorus-mediated beneficial effect on root growth in acid subsoil is not known but might have practical implications.

Inhibition of root elongation is in many instances the most sensitive parameter of heavy metal toxicity, as for lead (Pb) for example (Breckle, 1991). Probably due to binding of heavy metal cations to plant constituents the order of toxicity usually conforms to the stability of the metal–organic complexes, for example, $Cu > Ni > Cd > Zn > Al > Fe$ (Wong and Bradshaw, 1982). In crop production aluminum toxicity is one of the major growth-limiting factors in acid mineral soils (Section 16.3). In sensitive

Table 14.7
Effect of Liming an Acid Subsoil (pH 4.6) on the Elongation of
Cotton Roots[a]

Percentage of subsoil mass limed[b]	Distance between limed layers (cm)	Relative root length
Unlimed	—	32
10	4.5	38
20	4.0	57
40	3.0	57
60	2.0	70
100	—	100

[a]Based on Pearson *et al.* (1973).
[b]Application of the same dose, but differently distributed.

species such as cotton root elongation rate may be severely inhibited with aluminum/calcium molar activity ratios greater than 0.02 (Lund, 1970). In soils this critical ratio may vary widely, depending on aluminum species and root-induced changes in the rhizosphere (Section 15.4). Chelation of aluminum most likely also contributes to the beneficial effects of mulching on plant growth, especially in highly permeable acid mineral soils of the humid tropics.

The calcium/total cation ratio is also of importance for root growth under saline conditions (Kafkafi, 1991; Section 16.6) and in relation to the application of ammonium phosphate fertilizer. In acid soils with a low cation-exchange capacity, ammonium phosphate can severely inhibit root growth by inducing calcium deficiency (Bennett and Adams, 1970) due to both a low calcium/total cation ratio and a decrease in rhizosphere pH (uptake rate cations > anions).

14.4.3   Aeration

Because of their high rates of respiration (Section 14.1) roots have a high demand of oxygen which in nonwetland species has to be provided externally, i.e., in soil-grown plants by gas exchange between atmosphere and soil. Additionally, soil aeration is essential for meeting the respiratory requirements of soil microorganisms. In a dense crop stand, oxygen consumption and CO_2 evolution may be as high as 17 l m^{-2} d^{-1} (Cannell, 1977). The transfer of gases between soil and atmosphere occurs mainly in air-filled pores since gas diffusion in air is about 10^4 times more rapid than in water. Although in many species adapted to waterlogging (e.g., wetland rice) sufficient internal diffusion of O_2 from leaves to roots takes place in the aerenchyma (Section 16.4), in mesophytic (nonwetland) species this internal transfer is either unimportant or at least insufficient to meet the requirement of large root systems.

In most mesophytic species O_2 concentrations can be lowered to about 15–10% in the gas phase without severely affecting root growth (Geisler, 1967). But some plant species may be more senstive than others. In maize, for example lowering the O_2 concentration from 21 to about 10%, leaves root respiration almost unaffected whereas root extension is severely impaired, indicating that at least in this concentration range processes other than respiration are responsible for the inhibitory effect of poor aeration on root growth (Saglio et al., 1984). It is difficult to define the critical O_2 concentrations in soils, for various reasons. For example, a decrease in O_2 is usually correlated with a simultaneous increase in CO_2, and in many instances also with increase in ethylene (Michael, 1990).

Compared with the ambient CO_2 concentration (~0.03 %) the soil atmosphere CO_2 concentrations increase with soil depth, and in temperate climates, a typical pattern occurs throughout the year with maximum values in summer. On average, CO_2 concentrations are in the range 2–4% at a depth of 10–20 cm and may increase to 10–15% at a depth of 40–60 cm in summer (Nakayama and Kimball, 1988). Depending on its concentration, CO_2 itself has either stimulatory (~1–2% CO_2) or inhibitory effects (>5%) on root growth (Geisler, 1968). In contrast to most other species, desert succulents like *Agave deserti* are extremely sensitive to elevated CO_2 concentrations. Even with values as low as 0.5% CO_2 in the soil atmosphere root growth rates are severely inhibited, explaining the confinement of these species to coarse textured, well aerated soils (Nobel, 1990). Also in forest stands with CO_2 concentrations of 0.5% in the soil atmosphere, enhancement of aeration markedly increases root growth of Norway spruce throughout the soil profile and at a depth of 40 cm in particular (Murach et al., 1993). It is not clear to what extent decreases in the concentration of gases other than CO_2 (e.g., ethylene) are responsible for the enhanced root growth.

In many instances, inhibition of root growth in poorly aerated soils is caused by elevated levels of ethylene (Jackson, 1991b). In poorly aerated soils, loss of ethylene from roots by radial diffusion is impaired as result of water around the roots. Additionally, ethylene produced by soil microorganisms is trapped within the soil, and ethylene production in the roots is often enhanced under oxygen deficiency. Low ethylene concentrations $(0.1–1.0\ \mu l\ l^{-1})$ in the soil atmosphere can stimulate root growth in some plant species, however, elevated ethylene concentrations which are typical for poorly aerated soils strongly depress root elongation and enhance root hair formation (Michael, 1990). Ethylene may or may not stimulate lateral root formation, but definitely induces aerenchyma formation in the cortex (Section 16.4.3).

In response to oxygen deficiency in the roots a signal is transmitted to the shoots leading to suppression of leaf elongation rate by lowering the cell wall extensibility (Smit et al., 1989). Phytohormones such as ABA and ethylene, or ethylene precursors, are presumably involved in this signal transduction (Section 16.4.3).

14.4.4   Low-Molecular-Weight Organic Solutes

Root growth is affected in various ways by the water-soluble fraction of soil organic matter. When present in low concentrations root initiation and elongation might be enhanced by the high-molecular-weight fraction, especially fulvic acids (Mylonas and McCants, 1980) and also by some phenols in the low-molecular-weight fraction (Pingel, 1976; Wilson and van Staden, 1990). However, at higher concentrations the low-molecular-weight fraction is quite effective in inhibiting root growth. This is particularly true of phenolic and short-chain fatty acids, which often accumulate in poorly aerated or waterlogged soils during decomposition of organic material (e.g., straw or green manure). In well-structured soils, uneven distribution of organic matter (clumps) may cause local anaerobiosis (the formation of anaerobic microhabitats). Poor germination and emergence of plants on these soils is often not caused by oxygen limitation or elevated ethylene levels but by high concentrations of phenolics and short-chain fatty acids. Evidence for this comes from the similar inhibitory effects on germination and emergence (Hicks et al. 1989) or root respiration, root growth, and root hair formation (Patrick, 1971) that can be achieved with water extracts from these soils, particularly 3–4 weeks after incorporation of organic matter (Patrick, 1971).

During decomposition of organic substances with a large lignin content (e.g., straw), phytotoxic substances, including phenolic acids such as p-coumaric and p-hydroxybenzoic acid, may accumulate and severely inhibit root elongation in sensitive species such as rye and wheat even at concentrations of between about 7 and $70\ \mu M$ (Börner, 1957) and in tolerant species such as sugarcane at about $750\ \mu M$ (Wang et al., 1967). In paddy soils after incorporation of straw, inhibition of rice root elongation is particularly caused by phenylpropionic acid which is very effective even at concentrations below $50\ \mu M$ (Tanaka et al., 1990). The critical toxicity concentrations of these

Table 14.8

Influence of Phenolic Acids on the Root Fresh Weight of Pea Plants Grown under Axenic Conditions in High (15 mM) and Low (1.5 mM) Nitrate-N Nutrient Solution for 10 Days[a]

Phenolic acids (1 μM)	Root fresh weight (g per plant)	
	High nitrogen	Low nitrogen
None (control)	0.52	0.39
Ferulic	0.56	0.14
Vanillic	0.54	0.22
p-Coumaric	0.54	0.11
p-Hydroxybenzoic	0.54	0.33

[a]From Vaughan and Ord (1990).

phenolic acids to plants seem to depend greatly on their nitrogen nutritional status (Table 14.8). In pea plants even a concentration of 1 μM of these phenolic acids was detrimental to root growth when the plants were inadequately supplied with nitrogen whereas for the nitrogen-sufficient plants to induce marked depression in root growth concentrations of at least 100 μM were required (Vaughan and Ord, 1990). The reasons for the much higher sensitivity of nitrogen-deficient plants are not known, but might be related to the endogenous IAA level. Phenolic acids either stimulate or inhibit IAA oxidase activity (Baranov, 1979). For example, p-coumaric acid induces changes in the morphology of the roots that are comparable to those of supraoptimal IAA supply (Svensson, 1972). Nitrogen-sufficient plants might possibly have a higher capacity to detoxify these phenolic acids enzymatically (e.g., higher polyphenoloxidase activity).

Leachates from oat hulls strongly inhibit root elongation rates of the seedlings, cause swelling of the roots and extensive root hair formation, effects which are known to arise from supraoptimal IAA treatments (Häggquist et al., 1988a,b). The compound responsible for this inhibitory effect has been identified as the amino acid tryptophan which either acts as inhibitor as such or after microbial conversion into IAA (Häggquist et al., 1988a,b). In soil-grown plants, changes in density and composition of the rhizosphere microflora by phenolic acids have also to be taken into account in interpretating the effects of phenolic acids on root growth (Shafer and Blum, 1991).

In waterlogged soils, other phytotoxic substances, primarily acetic acid and other volatile (short-chain) fatty acids may accumulate in concentrations that are phytotoxic (Harper and Lynch, 1982). These acids are detrimental to root elongation (Lynch, 1978), and inhibit root and shoot growth, even in plant species adapted to waterlogging (Table 14.9).

As a rule phytotoxicity increases with chain length (Table 14.9) and with decreasing substrate pH (Jackson and St. John, 1980). This pH effect is related to the high rates of permeation of the plasma membrane by the undissociated species of acids (Section 2.5). The inhibitory effect of decomposition products (e.g., of straw) on root growth therefore depends not only on the concentration of volatile fatty acids but also on the pH of the rooting medium, as demonstrated in Fig. 14.10. Between pH 4 and 8 the root

Table 14.9

Effect of Volatile Fatty Acids in Growth of Rice Plants 28 Days
after Transplanting[a]

Volatile fatty acid Each supplied at 1 mmol $(100\ g)^{-1}$ soil	Dry weight (g per plant)	
	Roots	Shoots
None	1.70	4.56
Acetic (C_2)	1.40	3.70
Propionic (C_3)	1.00	3.54
Butyric (C_4)	0.90	2.99
Valeric (C_5)	0.29	1.20

[a]From Chandrasekaran and Yoshida (1973).

elongation of the control plants was not much different. With the addition of the extracts, however, root elongation clearly became pH dependent. The phytotoxicity of organic acids produced by *Fusarium culmarum* on root extension and root hair formation in barley is also inversely related to the substrate pH (Katouli and Marchant, 1981).

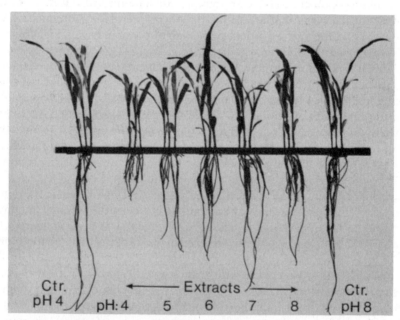

Fig. 14.10   Relationship between the pH of the nutrient solution and root growth of maize. Control plants (Ctr.) in nutrient solution only. (*Center*) Addition of 25 ml l^{-1} each of extracts from anaerobically fermented oat straw. (Courtesy of A. Wolf.)

14.5   Rhizosphere Microorganisms

14.5.1   General

Microorganisms can stimulate, inhibit, or be without effect on root growth, depending on the type of microorganism, plant species, and environmental conditions. Inoculation with a complex rhizosphere microflora, or growth in nonsterile compared with sterile culture, usually tends to inhibit elongation of the primary roots (Fusseder, 1984), and decrease root hair formation (Rovira *et al.*, 1983). An example of effects of such a complex rhizosphere microflora on root growth and morphology is shown in Table 14.10. It is also evident that different pure cultures of bacteria of the complex microflora have either a strong inhibitory or a strong stimulating effect on root growth indicating, that inoculation with a complex rhizosphere microflora might induce a variety of effects on root growth, depending on the proportions of detrimental and beneficial populations in the mixture.

According to Bowen and Rovira (1991), soil microorganisms can be classified into two categories in relation to their effects on plant growth:

1. *negative* (detrimental): root pathogens; subclinical pathogens; detrimental rhizo-bacteria; cyanide producers;
2. *positive* (beneficial): rhizobia, mycorrhizae; antagonists (biocontrol) of detrimental microorganisms; hormone producers; plant growth promoting bacteria (PGPB).

However, in many instances the distinctions are not clear within a category, for example, between detrimental rhizobacteria or cyanide producers, or between hormone producers and PGPB.

14.5.2   Pathogens and Pests

Traditionally, in studies of negative soil microbe–plant interactions the main interest has focused on soil-borne pathogens such as *G. graminis* or on cyst nematodes (Section 11.5) or on pathogens which impair specific root functions such as cytokinin production (Cahill *et al.*, 1986). However, in the last decade interest has increased very much in

Table 14.10

Influence of a Soil Inoculum with Rhizosphere Microflora and of Pure Cultures of Rhizosphere Bacteria on Root Morphology of Maize Seedlings[a]

Treatment	Dry weight (mg per plant) Shoot	Roots	Primary root length (mm)	Lateral roots (no. per plant)	Adventitious roots (no. per plant)
Sterile	33.4	28.2	122	31	1.0
Soil inoculum	33.3	21.4	97	31	1.0
Pure culture I[b]	19.3	16.4	86	17	4.9
Pure culture II[b]	42.4	37.8	137	25	5.0

[a]From Schönwitz and Ziegler (1986a).
[b]Bacterial colonies isolated from rhizosphere microflora.

Table 14.11

Effect of Deleterious Soil Microflora on Growth of Norway Spruce Seedlings in Soil of a Declining Spruce Stand (Grand-fontaine)[a]

Soil treatment	Dry weight (g per plant)		Mycorrhizal Short roots (%)
	Shoot	Roots	
Untreated	1.83	1.08	97
Pasteurized	2.80	2.72	98
Pasteurized + reinoculated	0.88	0.93	98

[a]From Devevre *et al.* (1993).

'minor pathogens' which inhibit root growth by production of toxins (Bolton *et al.*, 1989) and cyanide (Schippers *et al.*, 1990), and on beneficial microorganisms which might, for example, suppress these pathogens.

As a rule the harmful rhizosphere microorganisms belong to various genera of bacteria and fungi and they are most likely very often causally responsible for depression in growth and yield of crop plants in short rotations or in monocultures (de Weger *et al.*, 1987). In fruit trees or grapevine this situation is often described as 'soil sickness' or 'replant disease'. It might also play a role, at least in some sites, in forest decline in central Europe (Table 14.11). Typically, in case of soil sickness a range of different types of deleterious microorganisms (and also pests such as nematodes) is involved. These microorganisms can be eliminated by soil sterilization, and the harmful effect on plant growth again established by reinoculation (Table 14.11). In this case Norway spruce ectomycorrhizal infection was uniformly high and obviously not related to the growth differences brought about by the various treatments.

14.5.3 Beneficial Rhizosphere Bacteria

It has long been known that certain bacteria can stimulate root growth considerably (e.g., Table 14.10). Many of these are diazotrophic bacteria (e.g., *Azospirillum*, *Azotobacter*, or *Pseudomonas* ssp.). These beneficial bacteria are often classified as plant growth-stimulating bacteria (PGPB) which induce their effects by various means (Jagnow *et al.*, 1991). One of their major indirect effects on root growth is the suppression of pathogens, for example *Fusarium oxysporum* in potato (Beauchamp *et al.*, 1991), *Alternaria* in sunflower (Hebbar *et al.*, 1991) or cyst nematodes in soybean (Kloepper *et al.*, 1992a).

Many rhizosphere bacteria enhance root growth directly by production of phytohormones, IAA in particular. This is true for diazotrophic rhizosphere bacteria in particular (Section 7.6), and an example showing this effect is given in Fig. 14.11. Inoculation of soil-grown wheat plants with *A. brasilense* Cd stimulated root growth in general and the formation of lateral roots and root hairs in particular. Similar stimulating effects on root growth could be obtained by IAA application to soil-grown wheat plants (Martin *et al.*, 1989). These effects of *A. brasilense* and other diazotrophic

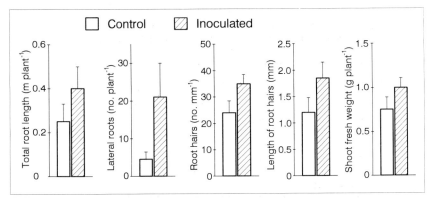

Fig. 14.11   Effect of inoculation with *Azospirillum brasilense* Cd on root and shoot growth of soil-grown wheat plants. (Based on Martin *et al.*, 1989.)

rhizosphere bacteria in enhancing root growth and development shall improve nutrient acquisition. This effect is especially important for phosphorus, and is presumably involved in the improvement in millet growth brought about by the incorporation of straw into the soil (Kretzschmar *et al.*, 1991).

Phytohormone production of rhizosphere microorganisms is not confined to IAA. Particularly depending on the substrate availability (precursors of phytohormones) the production, for example, of CYT by *Azotobacter* can be quite high and lead to remarkable increase in root and shoot growth of radish plants (Nieto and Frankenberger, 1990). Another example of such a specific substrate-induced enhancement of plant growth is shown in Table 14.12 for a leguminous tree. The amino acid L-methionine is a precursor of ethylene, and addition of this amino acid to the soil leads to marked increase in microbial production, and soil release of ethylene steeply increases (Arshad and Frankenberger, 1990). It is therefore most likely that the increase in shoot and root growth and especially that of nodulation was caused by microbial production

Table 14.12

Effect of Soil-Applied L-Methionine (L-MET) on Growth and Nodulation of *Albizia lebbeck* L. 32 Weeks after Treatment[a]

Treatment (g L-MET kg^{-1} soil)	Dry weight (g per plant)		Nodules (no. per plant)
	Shoot	Roots	
Control	64.5	42.5	19.2
10^{-8}	74.5	45.0	29.0
10^{-6}	81.5	48.8	36.0
10^{-4}	71.0	40.8	32.5
10^{-2}	67.2	35.8	19.2

[a]Based on Arshad *et al.* (1993).

of ethylene in the rhizosphere. However, rhizosphere microorganisms may affect root growth also by enhanced decomposition of ethylene. In maize and soybean under field conditions in well aerated soils on average ethylene concentrations were only 38 nl l^{-1} in the rhizosphere compared to 207 nl l^{-1} in the bulk soil (Otani and Ae, 1993).

Stimulation of root growth by PGPB might therefore very much depend on the availability of specific substrates as precursor of phytohormones (e.g., L-methionine for ethylene, L-tryptophan for IAA, and adenine for CYT). These substrates might be provided by plant root exudates (Section 15.5.3), soil organic matter, and incorporation of organic fertilizer.

14.6 Soil Physical Factors

14.6.1 Mechanical Impedance

Mechanical impedance refers to the resistance offered by the soil matrix against deformation by growing roots. Except for cracks and macropores (e.g., earthworm channels) that provide niches for roots to grow in (Passioura, 1991), root elongation in soils is possible only to the extent to which the root pressure exceeds the mechanical impedance (Bennie, 1991). When soils are compacted, the bulk density increases and the number of larger pores is reduced. The forces of the roots necessary for deformation and displacement of soil particles readily become limiting, and root elongation rates decrease (Fig. 14.12). There may be various reasons for differences in root response to soil strength between plant species, the difference in average diameter of the roots being one of them (Bennie, 1991; Materechera et al., 1992b).

The decrease which occurs in the rate of root elongation in response to increasing soil strength is correlated with an increase in root diameter, mainly because of the greater radial expansion of cortical cells under these conditions (Atwell, 1990). For a given soil bulk density the mechanical impedance increases as the soil dries out. This is because

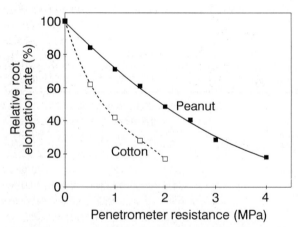

Fig. 14.12   Root elongation rates of peanut and cotton as a function of soil strength (resistance to root penetration). (Based on Taylor and Ratliff, 1969.)

particle mobility decreases, i.e. the forces required to displace and deform the soil particles increase, and accordingly the root elongation is suppressed. However, in compacted soils at a given bulk density this effect can be even greater in wet compared with dry soils, indicating that factors other than mechanical impedance are involved. Such factors include oxygen deficiency (Kirkegaard *et al.*, 1992) and, more likely elevated concentrations of ethylene. In maize roots subjected to an external pressure ethylene release was detected as soon as 1h after application of the treatment and rates of evolution increased linearly for at least 10h, but fell rapidly on release of pressure; exposure to ethylene (1 μl l^{-1}) caused similar morphological effects to roots as those occurring in response to mechanical impedance (Sarquis *et al.*, 1991).

Inhibition of root extension in compacted soils, particularly under field conditions is therefore brought about by various factors. In dry soils increase in mechanical impedance and decrease in soil water potential (and root pressure) may be most important, and in wet soils oxygen deficiency and accumulation of ethylene and phytotoxins are the primary factors (Section 14.4.3). In order to distinguish more clearly between the various factors Scott-Russell and Goss (1974) simulated high soil strength by applying radial pressure to roots growing in a solid substrate (glass ballotini) with unlimited supply of mineral nutrients, water, and oxygen. In these experiments it could be demonstrated that a pressure of only 0.5 bar (50 kPa), required to enlarge the pores of the substrate, was sufficient to inhibit root extension in barley by 80%. Despite the presence of aeration, accumulation of ethylene in the water-filled pores of the substrate was most likely the cause of this extremly high sensitivity of the roots to mechanical impedance.

Typical responses of roots to increased soil strength include inhibition of elongation rate of the main axis, enhanced formation of lateral roots which are initiated closer to the apex with a higher density per unit root length (Fig. 14.13). When only one compacted layer occurs in the soil (e.g., from tillage operations) a reduction in root

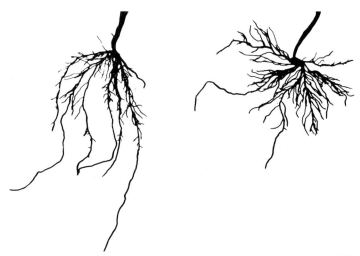

Fig. 14.13 Root system of young barley plants grown in the field in soils with different bulk densities. (*Left*) 1.35 g cm^{-3}; (*right*) 1.50 g cm^{-3}. (From Scott-Russell and Goss, 1974.)

growth in the zone of high soil strength is often compensated for by higher growth rates in loose soil above or below the compacted zone (Kirkegaard *et al.*, 1992). This occurs unless gas exchange (O_2/CO_2) becomes a limiting factor for root growth and activity because of a high rooting density in the loose soil zone (Asady and Smucker, 1989). The roots of a number of plant species can also readily penetrate calcic horizons (>15 cm thick) in 50–100 cm depth (Georgen *et al.*, 1991). In view of the soil strength in such horizons this is unexpected but might be related to penetration through continuous 'biopores', i.e., root channels of preceding crops or earthworm channels (Bennie, 1991; Meuser, 1991).

Inhibition of root elongation is not necessarily correlated with inhibited uptake of mineral nutrients (Shierlaw and Alston, 1984). In compacted soils the contact between roots and soil increases (Section 13.5) and thereby also the delivery rate for mineral nutrients (Silberbush *et al.*, 1983), as indicated, for example, by higher uptake rates per unit root length of nitrate (Veen *et al.*, 1992) and phosphorus (Cornish *et al.*, 1984). The increase in phosphorus uptake per unit root length can occur for various reasons, not only the higher buffer power of soils. It may be, for example, expression of a higher demand placed by the shoot on a smaller root system (Krannitz *et al.*, 1991) or brought about by root-induced changes in the rhizosphere. An increase in soil bulk density decreased the total root length of maize very considerably but did not alter the percentage of photosynthates allocated to the roots, thus leading to a twofold increase in consumption of photosynthates per unit root length (Sauerbeck and Helal, 1986).

Despite the various compensatory reactions of root systems in compacted soils, plants usually grow poorly in soils of high bulk density (Bennie, 1991). Insufficient water and nutrient supply might play a role, but often both shoot growth and transpiration are first reduced, regardless of the plant nutrient and water status. In compacted soils shoot growth is also often more depressed than root growth, suggesting root-derived hormonal signals in response to soil compaction (Masle and Passioura, 1987). Most likely the root cap is also the sensor of this stress factor (Section 14.2). In shoot responses to soil compaction considerable differences occur not only between plant species but also genotypes within species like barley and wheat. There is a tendency, for genotypes with lower growth rates in noncompacted soils to respond less to soil compaction than genotypes with high growth rates (Masle, 1992).

14.6.2 Root Volume Restriction

Restriction of the rooting volume, for example, by small pot size, has inhibitory effects on shoot growth similar to those of soil compaction. In soil-grown plants inhibited shoot growth by 'root restriction stress' (bonsai effect) might be caused at least in part by limited nutrient and water supply to the shoots. However, the same effect occurs in sand cultures percolated with nutrient solutions (Robbins and Pharr, 1988) or in containers with a flow-through hydroponic culture system (Fig. 14.14). Root activity, expressed as oxygen consumption per unit root weight is distinctly lower under root restriction stress, and some increase in ethylene formation occurs, but this is probably a secondary event (Peterson *et al.*, 1991b).

Under root-restriction stress there is a relatively similar decline in root and shoot dry weight with decrease in pot size (Robbins and Pharr, 1988). A reduction in leaf

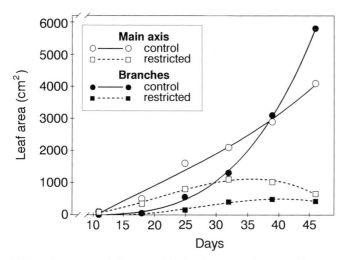

Fig. 14.14 Effect of 'root restriction stress' on leaf area development in tomato plants grown in a flow-through hydroponic culture system. Container volume: control = 1500 cm^3; restricted = 25 cm^3. (Based on Peterson *et al.*, 1991a.)

elongation rates is the most distinct response to root-restriction stress, and this reduction is primarily the result of reduced cell division and not so much of cell size (Cresswell and Causton, 1988). Thus, inhibited shoot growth under root-restriction stress is most likely a process regulated by root-derived hormonal signals in which nutritional factors or plant water relations may, or may not, play a secondary role. Root-restriction stress has practical implications for pot-grown plants in horticulture, but also deserves more attention in pot experiments.

14.6.3 Moisture

The effect of soil water content (or of the soil water potential) on root growth is expressed in the form of a typical optimum curve for which chemical and physical factors in the soil are the responsible parameters. In dry soils mechanical impedance and low soil water potential are the dominant stress factors. In saline soils the salt accumulation in general and at the root surface in particular (Section 15.2) further decrease soil water potential and, in combination with ion imbalance in the rhizosphere, impose an additional stress factor for root growth and functioning (Kafkafi, 1991). Increasing soil water content relieves these stress factors and there is a much lower impairment of extension growth by mechanical impedance. However, with further increase in soil water content aeration becomes a limiting factor for root growth and activity (Section 14.4.3) unless counterbalanced by morphological and physiological adaptation of the roots (Section 16.4.3).

In dry soils, despite increasing soil strength root growth is usually much less depressed than shoot growth, leading to a typical increase in root/shoot dry weight ratio in response to drought stress. For example, in maize seedlings without drought stress the root/shoot dry weight ratio is 1.45 compared with 5.79 under drought stress (Sharp

Table 14.13
Effect of Drought Stress on Shoot and Root Dry Weight of Maize Plants
Grown in Vermiculite[a]

Treatment		Dry weight (mg per plant)		Root/shoot ratio
		Shoot	Roots	
21 d	Control (watered)	114	40	0.39
	Drought stress	44	28	0.64
34 d	Control (watered)	230	106	0.46
	Drought stress	48	45	0.94

[a]From Stasovski and Peterson (1991).

et al., 1988), or in apple trees, prolonged drought stress decreases fruit and shoot dry weight to 21% and 26%, respectively, but is without effect on root dry weight (Buwalda and Lenz, 1992). A faster osmotic adjustment of the roots compared to the shoot might be involved (Schildwacht, 1988). However, soil drying inhibits leaf elongation rate without a measurable decline in leaf water potential (Saab and Sharp, 1989). In plants grown in substrates with low mechanical impedance, such as vermiculite, the root system of drought-stressed plants continues to grow by elongation of existing roots and initiation of new roots (Table 14.13). The apical root zones and the stele remain alive, despite severe collapse of the cortex in basal zones. On rehydration, the root system of stressed plants resumes growth by elongation of existing roots (Stasovski and Peterson, 1991).

As discussed in Section 5.6 ABA, or related compounds, play a central role in the shift in root/shoot growth under drought stress. In apical root zones ABA synthesis is enhanced under drought stress and transmitted as a nonhydraulic signal to the shoot which has a much higher sensitivity to ABA than the roots in terms of inhibition of extension growth (Creelman *et al.*, 1990). This can be considered to be an effective mechanism of adaptation to conserve water as the soil dries out, before the occurrence of shoot water deficit (Saab and Sharp, 1989).

The realization of this inherent potential of roots to continue growth at low soil water contents, however, depends very much on mechanical impedance which increases as soils dry out. On the other hand, in dry soils the root–soil contact decreases either by shrinking of the roots and, in many soils, also of the soil matrix (Faiz and Weatherby, 1982), or preferential root growth in soil cracks. Part of this loss in contact is probably compensated for by the increase in length and density of root hairs (Mackay and Barber, 1987; Fig. 14.6) and, in mycorrhizal plants by the external mycelium (Section 15.7).

14.6.4 Temperature

Root growth is often limited by low (suboptimal) or high (supraoptimal) soil temperatures. The temperature optimum varies among species and tends to be lower for root

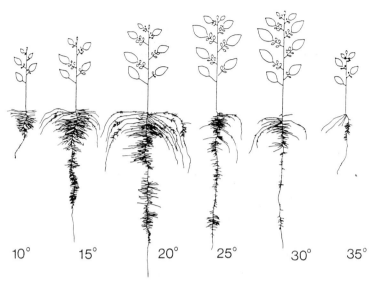

Fig. 14.15   Influence of root zone temperature on root morphology and shoot growth of potato seedlings. (Sattelmacher *et al.*, 1990c.)

growth than for shoot growth (Brouwer, 1981). Depending on their origin (C$_4$, C$_3$ species; tropical or temperate climate) temperature optima differ and are, for example, about 35°C for cassava (Sattelmacher *et al.*, 1990c), 30°C for cotton (Pearson *et al.*, 1970), 25°C for wheat (Huang *et al.*, 1991a,b) and 15–20°C for potato (Sattelmacher *et al.*, 1990b, c). A typical temperature response curve is shown for potato in Fig. 14.15.

Although in many parts of the lowland tropics soil temperatures of 40°C and above at 15 cm depth are common, relatively little information is available on root growth and functioning at supraoptimal temperatures (Bowen, 1991). Within a given species considerable genotypic differences exist in tolerance to supraoptimal root zone temperatures, for example in potato (Sattelmacher *et al.*, 1990b). Inhibition of root growth at supraoptimal root zone temperature is also dependent on other environmental factors, for example, high light intensities and high supply of nitrate nitrogen increase the sensitivity of roots to supraoptimal temperatures (Sattelmacher and Marschner, 1990). Although insufficient carbohydrate supply might be involved (Cumbus and Nye, 1982) it is not likely the primary reason for impaired root growth at supraoptimal temperatures. In potato early responses to supraoptimal root zone temperatures are inhibition of cell division in the root apical meristem and less geotropic response (Sattelmacher *et al.*, 1990c), and in sorghum a decrease in epidermal cell length suggesting accelerated maturation of the epidermal cells (Pardales *et al.*, 1992). As feedback response, senescence of the shoots is enhanced at supraoptimal root zone temperatures (Struik *et al.*, 1989).

Compared to supraoptimal root zone temperatures, much more detailed information is available on effects of low (suboptimal) temperatures on root growth and development. Typically, at low temperatures root growth is retarded, the roots become shorter and thicker, and particularly lateral root formation is depressed (Gregory, 1983). An example is shown in Table 14.14. In wheat the temperature optimum for root growth is

Table 14.14

Growth and Development of Roots of Winter Wheat Grown for 20 Days at Different Temperatures[a]

Temp. (°C)	Length of seminal roots (m per plant)	Primary lateral roots per plant		Specific root length (m g^{-1} dry wt)	Root/shoot dry wt ratio
		Number	Length (m)		
10	77	56	63	45	2.22
20	98	167	463	125	0.52
25	275	556	1536	160	0.57
30	138	389	352	125	0.58

[a]Compiled data from Huang et al. (1991a,b).

25°C, and at 20°C or 30°C root growth declines distinctly. This relatively narrow temperature optimum of root growth can also be observed when the temperature of only a certain zone of a root is altered, for example, in maize from 25 to 20°C, and vice versa (Sattelmacher et al., 1993). At low temperatures in wheat (Table 14.14) but also in many other plant species (Bowen, 1991) shoot growth is much more inhibited than root growth, leading to a high root/shoot dry weight ratio (Table 14.14). However, shortage of photosynthates is not likely the cause of poor root growth at low temperatures. This conclusion is supported by an increase in dry matter content in both roots and shoots in plants grown at low root zone temperatures (White et al., 1990a).

A fall in the elongation rates of roots at low temperatures is caused by a decrease in cell wall extensibility of the cells in the extension zone, not a loss in turgor. In maize, lowering the temperature from 30° to 15°C causes cell wall extensibility in the extension zone to decrease to 25% of its original value (Pritchard et al., 1990). Root cells arrested in growth by cold treatment did not resume growth on raising the temperature, although root growth did recover by the temperature effect on extension of newly formed cells. Low temperatures also distinctly alter root anatomy. In wheat lignification of late metaxylem vessels is delayed and the axial hydraulic conductivity much higher in roots grown at low temperatures compared with roots grown at higher temperatures (Huang et al., 1991c). However, this enhancement effect on axial hydraulic conductivity may be counteracted in many plant species where suberinization of roots occurs much closer to the growing apex at low temperatures (Bowen, 1991).

Cooling of the roots inhibits shoot and leaf elongation rates without affecting leaf water potential (Milligan and Dale, 1988) and is associated with increase in ABA concentration in the leaves (Smith and Dale, 1988). In maize a decrease in root zone temperature from 28°C to 8°C increases the ABA concentration in the xylem exudate about twofold (Atkin et al., 1973). This is another example of root to shoot communication upon stress in the root environment. This signal upon low temperature stress seems to be very much dependent on the nutritional status of the plants, it is particularly effective in nitrogen or phosphorus-deficient plants (Radin, 1990).

At low root zone temperatures also the production and root export of CYT is depressed, in maize at 18°C to about 15% of that at 28°C (Atkin et al., 1973). In grapevine roots at a temperature of 12°C the concentration of CYT in the xylem sap drops to nearly one-half of that at 25°C, and in addition qualitative changes occur in the CYT spectrum (Zelleke and Kliewer, 1981).

Low root zone temperatures might also inhibit shoot growth by insufficient uptake and supply of nutrients to the shoot. In general, at low root zone temperatures, particularly in combination with soil compaction (Al-Ani and Hay, 1983) root growth is more inhibited than uptake kinetics per unit root length of nutrients such as phosphorus (Mackay and Barber, 1984). After long-term exposure to low temperatures roots of cold-tolerant species also adapt by increasing the uptake rates of nutrients according to the demand (White *et al.*, 1987). Thus, despite lower root growth, shoot growth is not likely be limited by low root zone temperatures, when plants are grown in a nutrient-rich substrate. However, at low root zone temperatures nutrient supply might limit shoot growth in plants growing in nutrient-poor substrates, particularly at high shoot demand, for example, at high shoot temperatures (Engels and Marschner, 1990).

14.7 Shoot/Root Ratio

The ratio of shoot to root growth varies widely between plant species, during ontogenesis of plants, and is strongly modified by external factors. If considered on a broader scale there is a general tendency both between and within species to maintain a characteristic relationship between root and shoot dry weight (e.g., graminaceous species ≫ trees). When parts of the shoots are removed, plants tend to compensate this by lower root growth and returning to a ratio characteristic for the species. However, there is some controversy as to whether this reflects 'functional equilibrium' between roots and shoots (Klepper, 1991).

There are various, reasonably well-defined feedback regulation mechanisms, and some of them are under hormonal control. Examples are retardation or cessation of shoot growth when roots are exposed to drought stress (Section 14.6.3), soil compaction (Section 14.6.1) or poor soil aeration (Section 14.4.3). As with low root zone temperatures (Atkin *et al.*, 1973) waterlogging also drastically decreases root export of CYT (Burrows and Carr, 1969) and gibberellins (Reid *et al.*, 1969) within 1–2 days, and markedly depresses shoot elongation and enhances leaf senescence. Foliar sprays with CYT can counteract at least some of the negative effects of waterlogging on shoot growth (Reid and Railton, 1974). Inhibition of shoot growth but continuation or even enhancement of root growth under nutrient deficiency (Section 14.4.1) might mainly reflect alteration in photosynthate allocation, but is at least in some instances under direct hormonal control as has been shown in experiments in which shoot growth has been restored by CYT supply to nutrient-starved plants (Section 5.6.5).

In annual species competition for photosynthates between shoot and roots becomes the dominant factor during reproductive growth in limiting root growth and activity. This is of particular interest in perennial species. The proportion of photosynthates allocated belowground and used for fine root production can be up to 44% in a tropical broadleaf stand (Cuevas *et al.*, 1991) and more than 50% in Scotch pine stands in Sweden (Persson, 1979). However, despite this high allocation, after about 10–15 years of growth total fine root biomass in forest stands remains constant, which indicates a high turnover rate of fine root biomass (Olsthoorn and Tiktak, 1991), including their

Table 14.15
Effect of Fruit Load on Shoot and Root Dry Weight of 3-Year-Old Apple Trees[a]

	Treatment	
	Without fruits	With fruits
Leaf area (m^2 per plant)	443	400
Fruit dry weight (g per plant)	0	2994
Total shoot dry weight (g per plant)	1521	3522
Root dry weight (g per plant)	665	399

[a]Based on Buwalda and Lenz (1992).

ectomycorrhizal structures in boreal forest ecosystems (Söderström, 1992). This high rate of root replacement ensures a high proportion of young roots with corresponding advantages for water and mineral nutrient uptake, particularly in soils with low fertility. In fruit trees such as apples the turnover rate of fine root biomass seems to be even higher and maximum fine root biomass is already reached in four-year-old plants (Hughes and Gandar, 1993). This high turnover rate can become critical when fruit loads are high and there is a corresponding shoot/root competition for photosynthates (Table 14.15). It is usually the fine root biomass formation that is depressed by sink competition of the shoot.

The 'functional equilibrium' between root and shoot growth can therefore be very much affected by internal and external factors. In terms of root functions (supply of mineral nutrients, water, phytohormones) the size of the root system, and also the root/shoot dry weight ratio, required for these functions mainly depend on the concentration of nutrients in the soil or nutrient solution, and the physical, chemical and microbiological conditions in the substrate for root activity and formation of new roots. For example, within a given crop species in dryland areas the root/shoot ratio is considerably larger than in temperate climates (Gregory *et al.*, 1984). In the natural vegetation too the root/shoot dry weight ratio increases as soil fertility decreases (Chapin, 1988). On the other hand, when there is a large and continuous supply of water and nutrients only 10–20% of the root system may be required (Greenwood, 1983) as shown by plant production in water culture (e.g., the 'Nutrient Film Technique') in commercial horticulture.

15

The Soil–Root Interface (Rhizosphere) in Relation to Mineral Nutrition

15.1 General

The conditions in the rhizosphere differ in many respects from those in the soil some distance from the root, the so-called bulk soil. Roots not only act as a sink for mineral nutrients transported to the root surface by mass flow and diffusion. In addition, they take up either ions or water preferentially, which may lead to the depletion or accumulation of ions (Chapter 13). They also release H^+ or HCO_3^- (and CO_2) which changes the pH, and they consume or release O_2, which may cause alterations in the redox potential. Low-molecular-weight root exudates may mobilize mineral nutrients directly, or indirectly by providing the energy for microbial activity in the rhizosphere. These root-induced modifications are of crucial importance for the mineral nutrition of plants. Although the chemical properties of the bulk soil (e.g., the pH) are very important for root growth and mineral nutrient availability, the conditions in the rhizosphere and the extent to which roots can modify these conditions play a very decisive role in mineral nutrient uptake in general (Marschner *et al.*, 1986b), and in micronutrient uptake in particular (Marschner, 1991a). Conditions in the rhizosphere are also of importance for the adaptation of plants to adverse soil chemical conditions, as occur, for example, in acid mineral soils (Marschner, 1991b).

In soil-grown plants the rhizosphere is characterized by gradients which occur both in a radial and longitudinal direction along an individual root (Fig. 15.1). Gradients may exist for mineral nutrients, pH, redox potential and reducing processes, root exudates and microbial activity. These gradients are determined by soil chemical and physical factors, and by plant factors such as species and nutritional status of the plants, and by microbial activity in the rhizosphere. In the last decade much progress had been made to gain a better understanding of these rhizosphere processes (Darrah, 1993), and in the following discussion some examples of these findings will be presented.

15.2 Ion Concentration in the Rhizosphere

The concentration of a particular ion in the rhizosphere can be lower, higher, or similar to that in the bulk soil, depending on the concentration in the bulk soil solution, the rate

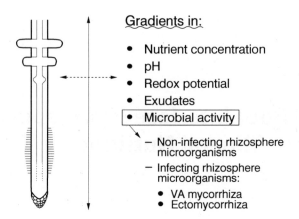

Fig. 15.1   Gradients at the root–soil interphase (rhizosphere).

of delivery of the ion to the root surface, and its rate of uptake by the root itself. Examples of depletion of phosphate and potassium have been given in Chapter 13. In soils low in available potassium this can lead to the disaggregation of polymineralytic shale particles and the accumulation of amorphous iron and aluminum oxyhydrates, indicative of enhanced weathering of soil material at the soil-root interface (Sarkar *et al.*, 1979; Kong and Steffens, 1989). In ryegrass (*Lolium multiflorum*) depletion of the potassium concentration below 80 μM in the rhizosphere soil solution enhances within a few days release of interlayer potassium and concomitant transformation of trioctahedric mica into vermiculite in the rhizosphere (Hinsinger and Jaillard, 1993). In the rhizosphere of rape (*Brassica napus* L.) depletion of both potassium and magnesium, associated with a decrease in pH to about 4, increases not only the release of interlayer potassium but also of octahedral magnesium and, thus, induces irreversible transformations of mica (Hinsinger *et al.*, 1993).

On the other hand, a greater uptake of water than of ions leads to ion accumulation in the rhizosphere. This can be predicted from calculations based on models of solute transport by diffusion and mass flow to the root surface for those ions which are present in high concentrations in the soil solution (Section 13.2). This accumulation can also be demonstrated by carefully separating roots from loosely adhering soil (*rhizosphere soil*) and closely adhering soil (*rhizoplane soil*). In a more elegant manner, root and bulk soil compartments can be separated by nets, allowing analysis of soil at defined distances from the root surface. Examples of this have been given for depletion of potassium and phosphorus in Chapter 13. Using a similar technique, the accumulation of ions in the rhizosphere can also be measured as shown in Fig. 15.2 for calcium and magnesium. After two months' growth in a sandy loam soil the concentrations of calcium and magnesium in the rhizosphere was increased 2–3-fold as compared with the bulk soil. Whether or not ions such as calcium accumulate in the rhizosphere depends on both transport by mass flow (transpiration) to the roots (Section 13.2) and rate of uptake by the roots. Plant species differ in both respects. For example, in ryegrass and lupin grown in the same soil, calcium supply per mass flow was 2.8 and 8 mg Ca, respectively, but calcium uptake was 0.8 mg in ryegrass and 9.0 mg in lupin. Thus, despite the much

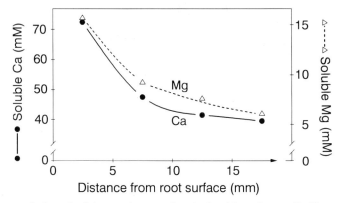

Fig. 15.2 Accumulation of calcium and magnesium in the rhizosphere soil of 2-month-old barley plants. (Redrawn from Youssef and Chino, 1987.)

higher supply, calcium was depleted in the rhizosphere of lupin but accumulated in the case of ryegrass (Barber and Ozanne, 1970).

At sufficiently high Ca^{2+} and SO_4^{2-} concentrations in the soil solution, the precipitation of $CaSO_4$ at the root surface can be demonstrated (Jungk, 1991). Over a long period in soil-grown plants, these precipitations occasionally may form a solid mantle around the roots (*pedotubules*) a few millimeters or even more than 1 cm in diameter (Barber, 1984).

In calcareous soils (e.g., Rendzinas) abundant quantities of calcified roots of herbaceous plants often occur in which the calcite elements retain the structure of the original cortex cells (Jaillard, 1985). Evidence has been presented that these cytomorphic calcite elements ($\sim$60–80 μm) are formed by root activities and cycles of rhizosphere acidification and precipitation of calcium carbonate within root cells. In agreement with this the calcified roots are surrounded by a decalcified rhizocylinder with a silico-aluminum matrix (Jaillard, 1985; Jaillard *et al.*, 1991). This is an interesting example of the role of root-induced changes in the rhizosphere which can be of importance in pedogenesis, since in certain locations the cytomorphic calcite fraction may represent up to a quarter of the soil mass (Jaillard *et al.*, 1991).

Accumulation of salts of low solubility in the rhizosphere (e.g., $CaCO_3$, $CaSO_4$) may not be very harmful to plants. This is different, however, in saline soils with high concentrations of water-soluble salts such as sodium chloride. As shown in Table 15.1 there is a concentration gradient for both chloride and sodium from the bulk soil to the root surface, and this gradient becomes steeper as the transpiration rate increases. Accordingly, the electrical conductivity of the soil increases near the root surface, especially at high transpiration rates.

Increasing the salt concentration and osmotic potential of the soil solution decreases water availability to plants and can severely impair plant water relations. For nonhalophytes ('salt excluders') grown in saline soils over a four-day period the salt concentrations in the rhizo-soil solution can rise from 50 to 300 mM (Schleiff, 1986). At high salt concentrations the relationships between transpiration rate and salt accumulation in the

Table 15.1

Relationship between Water Uptake per Unit Length of Root and Sodium and Chloride Accumulation around Maize Roots[a]

Water uptake (transpiration, 100 ml cm^{-1})	Chloride (mg (100 g)$^{-1}$ soil)			Sodium (mg (100 g)$^{-1}$ soil)			Elect. conductivity, close[d] (mmho cm^{-1})
	Bulk[b]	Loose[c]	Close[d]	Bulk[b]	Loose[c]	Close[d]	
0.38	31	41	58	22	34	41	1.38
0.46	36	43	65	28	33	45	2.28
0.82	43	66	97	36	49	68	3.79
0.95	44	64	128	38	57	90	5.02

[a]Based on Sinha and Singh (1974).
[b]Bulk soil.
[c]Loosely adhering soil (rhizosphere soil).
[d]Closely adhering soil (rhizoplane soil).

rhizosphere are not linear, indicating some back-diffusion of solutes from the root surface, counteracting in part the salt accumulation (Hamza and Aylmore, 1991).

Accumulation of soluble salts at the root surface is important for plant growth and irrigation in saline soils. Estimations of expected growth reduction of plants growing in saline soils are usually based on calculations of salt concentrations in soil-saturated extracts (Maas and Hoffman, 1977). The salt concentration in the soil solution under field conditions is estimated to be about two to four times higher than that in the saturation extract (soil paste). This, however, does not necessarily reflect the actual condition in the rhizosphere where water may be unavailable to plants long before the critical conductivity levels (see Section 16.6.3) are obtained in the bulk soil (Schleiff, 1986, 1987).

Average values for the rhizosphere as compared with the bulk soil also do not provide a true picture of nutrient relationships in the rhizosphere as they ignore gradients along the root axis, for example, in uptake rates of mineral elements and water (Fig. 15.1). Uptake rates of mineral nutrients can differ very much along the root axis (Section 2.6). The same holds true for uptake rates of water which are, for example, much higher in bare than in sheathed roots of maize and other C_4 plants (Section 2.7), or in perennial species where much higher rates occur in apical root zones where the formation of endodermis and exodermis is incomplete (Moon et al., 1986; Häussling et al., 1988).

Gradients in ion uptake rates along the root axis are also important for ion competition and selectivity in uptake (Fig. 15.3). The strong depression of magnesium uptake by potassium which can be readily demonstrated in nutrient solution culture (Section 2.5.3) occurs in soil-grown plants only as long as potassium is high in the rhizosphere. Depletion of potassium in the rhizosphere soil solution below 20 μM doubles the uptake rate of magnesium by ryegrass (Fig. 15.3). The radial extension in the depletion zone of potassium from apical to basal zones permits higher uptake rates of magnesium in basal zones. Thus, the spatial separation of ions in the rhizosphere along the root axis of soil-grown plants can overcome limitations in mineral nutrition of plants caused by ion competition for uptake sites. However, in saline soils with high concentrations of sodium, the preferential potassium uptake in apical root zones also

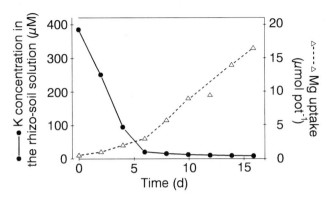

Fig. 15.3 Time course of magnesium uptake in ryegrass as affected by the potassium concentration in the rhizo-soil solution. (Seggewiss and Jungk, 1988.)

increases the chance of higher sodium uptake rates in basal zones and thus, decreases the overall selectivity in K^+/Na^+ uptake (Section 16.6).

15.3 Rhizosphere pH and Redox Potential

15.3.1 General

The rhizosphere pH may differ from the bulk soil pH by up to two units, depending on plant and soil factors. The most important factors for root-induced changes in rhizosphere pH are imbalance in cation/anion uptake ratio and corresponding differences in net release of H^+ and HCO_3^- (or OH^-), and the excretion of organic acids. Organic acids may also be produced by microbial activity stimulated by the release of organic carbon from roots, and CO_2 production by roots and rhizosphere microorganisms. In aerated soils, CO_2 per se is of minor importance for the rhizosphere pH since it rapidly diffuses away from the roots through air-filled pores (Nye, 1986). It is mainly the CO_2 dissolved in the soil solution (H^+, HCO_3^-) which affects rhizosphere pH as the mobility of H^+ and HCO_3^- are relatively low in the soil solution.

The pH buffering capacity of the soil and the initial soil pH (bulk soil) are the main factors determining the extent to which plant roots can change rhizosphere pH. The pH buffering capacity of soils depends not so much on clay content but primarily on initial pH and organic matter content; the pH buffering capacity is lowest at about pH 6, and increases to both lower and higher pH values (Schaller and Fischer, 1985; Nye, 1986).

Marked differences in redox potential in the rhizosphere occur between plants grown in aerated soils and those grown in submerged soils. In both cases, however, root-induced changes in redox potential of the rhizosphere can be substantial and therefore they also affect the availability and uptake of mineral nutrients.

15.3.2 Source of Nitrogen Supply and Rhizosphere pH

The form of nitrogen supply has the most prominent influence on the cation/anion uptake ratio (Section 2.5.3) and thus on rhizosphere pH both in annual species

Table 15.2
Nitrogen Supply, Rhizosphere pH, and Content of Mineral Nutrients in Shoots of Bean (*Phaseolus vulgaris* L.) Plants Grown in a Luvisol (pH 6.8)

Nitrogen supply	Rhizosphere pH	Contents in shoot dry matter				
		$(mg\ g^{-1})$		$(\mu g\ g^{-1})$		
		K	P	Fe	Mn	Zn
NO_3-N	7.3	13.6	1.5	130	60	34
NH_4-N	5.4	14.0	2.9	200	70	49

(Marschner and Römheld, 1983) and perennial species (Rollwagen and Zasoski, 1988). Nitrate supply is correlated more with a higher rate of HCO_3^- net release (or H^+ consumption) than of H^+ excretion, and with ammonium supply the reverse is the case. In neutral or alkaline soils rhizosphere acidification in plants fed with ammonium can enhance mobilization of sparingly soluble calcium phosphates and thereby favor the uptake of phosphorus (Gahoonia *et al.*, 1992), as well as the uptake of micronutrients such as boron (Reynolds *et al.*, 1987), iron, manganese and zinc (Table 15.2). Naturally, on acid soils a further pH decrease is not very effective in mobilizing these mineral nutrients. On acid soils the pH increase induced by nitrate supply enhances phosphorus uptake, presumably by exchange with HCO_3^- for phosphate adsorbed to iron and aluminum oxides (Gahoonia *et al.*, 1992). Thus for various pasture grasses grown in phosphorus-deficient acid soils depletion of phosphorus in the rhizosphere and increase in rhizosphere pH have been found to be closely associated (Armstrong and Helyar, 1992).

Average values of rhizosphere pH can be misleading and may result in erroneous conclusions being reached concerning nutrient relationships in the rhizosphere. For example, within the root system of an individual plant, distinct pH differences exceeding two pH units may sometimes occur between the primary and lateral roots or along the root axis (Marschner and Römheld, 1983; Marschner *et al.*, 1986b). As shown in an example for Norway spruce (Fig. 15.4) in acid soils pH is increased at the root apex and decreased in the subapical (extension) zone, irrespective of the form of nitrogen in the soil solution. In contrast in the more basal root zones the expected pH changes occur, namely a pH increase with nitrate supply only, and a pH decrease when ammonium is simultaneously present at concentrations of about 500 μM. As ammonium is usually taken up very much more preferentially than nitrate (Section 2.5.4; Arnold, 1992), rhizosphere acidification takes place despite the presence of higher nitrate concentrations, particularly at high soil water content which correspondingly facilitates diffusion of NH_4^+ (Gijsman, 1991). A higher pH at the root apex (Fig. 15.4) is a common feature of plants grown in acid soils and might be related to the release of root exudates (see below) or in nitrate-fed plants to high nitrate reductase activity in root apical zones (Klotz and Horst, 1988a).

Striking differences in the rhizosphere pH exist between plant species growing in the same soil and supplied with nitrate nitrogen. Buckwheat (Raij and van Diest, 1979) and

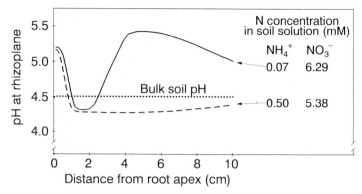

Fig. 15.4 Influence of nitrogen form in the soil solution on rhizoplane pH along roots of 4-year-old Norway spruce grown in a Luvisol of pH 4.5 (H_2O). (Leisen *et al.*, 1990.)

chickpea (Marschner and Römheld, 1983) have a very low rhizosphere pH compared, for example, with that of wheat or maize. These genotypical differences reflect differences in cation/anion uptake ratios (Bekele *et al.*, 1983).

Legumes and actinorhizal plants which meet their nitrogen requirement by symbiotic N_2 fixation rather than nitrate nutrition, take up more cations than anions since uncharged N_2 enters the roots. The cation/anion uptake ratio of plants fixing N_2 is thus quite large and so also is the net excretion of H^+, although per unit assimilated nitrogen it is less than in ammonium-fed plants (Raven *et al.*, 1991; Section 2.5.4). The consequences of the different cation-anion uptake ratio in alfalfa are reflected in differences in the acidity (net release of H^+) and alkalinity (net release of HCO_3^- or OH^-) and in the rhizosphere pH (Table 15.3). The capacity of plants to utilize phosphorus from rock phosphate is therefore higher in N_2-fixing plants than in nitrate-fed plants. In soybean, N_2-fixing plants were also higher in iron and manganese contents than nitrate-fed plants and did not show iron-deficiency symptoms (Wallace, 1982).

On severely phosphorus-deficient soils utilization of rock phosphate as a phosphorus source for legumes can be low when nodulation is limited by phosphorus deficiency (Section 7.4.5). Thus, a starter supply of soluble phosphorus can enhance nodulation,

Table 15.3
Effects of Nitrogen Sources on Acidity and Alkalinity Generated by Roots of Alfalfa Plants, on the Soil pH, and on the Utilization of Rock Phosphate[a]

Treatment		Acidity (meq g^{-1} dry wt)	Alkalinity (meq g^{-1} dry wt)	Soil pH (H_2O)	Phosphorus uptake (mg per pot)	Yield (g dry wt per pot)
Nitrogen source	Rock phosphate					
Nitrate	—	—	1.1	6.3	1	2.5
Nitrate	+	—	0.8	7.3	23	18.8
N_2	—	0.5	—	6.2	4	4.7
N_2	+	1.4	—	5.3	49	26.9

[a]From Aguilar S and van Diest (1981).

N_2 fixation and rhizosphere acidification and thereby utilization of rock phosphate (De Swart and van Diest, 1987). In interplantings of N_2-fixing legumes with nonlegumes, rhizosphere acidification of legumes can increase phosphorus uptake from rock phosphate by nonlegumes, for example, in black walnut tree seedlings by twofold when interplanted with alfalfa (Gillespie and Pope, 1989). Simulation models on predicting phosphorus uptake by symbiotically grown legumes, particularly when supplied with rock phosphate, or grown at alkaline soil pH, have therefore to consider this phosphorus mobilization by rhizosphere acidification, otherwise actual uptake by far exceeds the predicted uptake (Gillespie and Pope, 1990; Li and Barber, 1991).

In the long run, symbiotic nitrogen fixation also affects the acidification of the bulk soil and thus the lime requirement (Section 7.4). An alfalfa crop fixing N_2 and giving an annual shoot dry matter production of 10 t per hectare produces soil acidity equivalent to 600 kg $CaCO_3$ per hectare (Nyatsanaga and Pierre, 1973). In legume pastures which are not limed there is a distinct negative correlation between age of the pasture and soil pH (Haynes, 1983). In soils in which legumes are grown continuously exchangeable soil manganese can thus be released into soil solution to increase the risk of manganese toxicity to plants (Bromfield *et al.*, 1983a,b). In humid climates the loss of symbiotically fixed nitrogen from the system through leaching of nitrate and an equivalent amount of cations such as magnesium and calcium contribute to soil acidification under leguminous pastures. A similar impact on the long-term soil acidification by N_2 fixation can be observed in forest ecosystems when the pH under Red alder is compared with Douglas fir (Van Miegroet and Cole, 1984), and in crop rotations with a high proportion of legumes (Coventry and Slattery, 1991).

15.3.3 Nutritional Status of Plants and Rhizosphere pH

Root-induced changes in rhizosphere pH are also related to the nutritional status of plants. Examples are rhizosphere acidification in cotton and other dicots under zinc deficiency (Cakmak and Marschner, 1990), and in nongraminaceous species under iron deficiency (Römheld, 1987a,b; Section 2.5.6). In both instances the increase in net release of H^+ is closely related to increase in cation/anion uptake ratio. Under iron deficiency this acidification also takes place in nitrate-fed plants (Fig. 15.5), and based on the root system as a whole, the rates of net H^+ release per unit root weight are in a similar order of magnitude to the iron sufficient, ammonium-fed plants. Again, however, average values are misleading as under iron deficiency enhanced net release of H^+ is confined to the apical root zones where the actual rates are nearly eight times higher than in the ammonium-fed plants (Fig. 15.5). This highly localized acidification might enable the roots to decrease rhizosphere pH in apical zones even in calcareous soils to enhance iron mobilization (Section 16.5.3).

At least in dicots rhizosphere acidification is also a widespread phenomenon of root responses to phosphorus deficiency as shown in an example for rape plants in Table 15.4. When rape plants were grown in a phosphorus-deficient soil a decrease in the phosphorus concentration in rhizosphere soil solution was associated during the first 2 weeks with an increase in rhizosphere pH. Thereafter, this relationship was reversed. The changes in rhizosphere pH were related to the cation/anion uptake ratio, which increased in the older plants, most probably due to lower uptake rates of nitrate but also

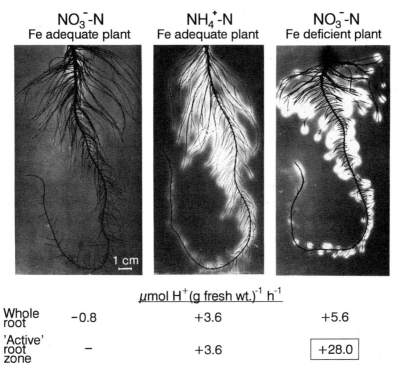

$NO_3^- \text{-} N$
Fe adequate plant

$NH_4^+ \text{-} N$
Fe adequate plant

$NO_3^- \text{-} N$
Fe deficient plant

1 cm

$\mu mol\ H^+ (g\ fresh\ wt.)^{-1}\ h^{-1}$

	NO_3^--N Fe adequate	NH_4^+-N Fe adequate	NO_3^--N Fe deficient
Whole root	−0.8	+3.6	+5.6
'Active' root zone	—	+3.6	+28.0

Fig. 15.5 Rhizosphere acidification (indicated by agar with bromocresol purple, *above*) and rates of net release of H^+ by roots of intact sunflower plants. (Modified from Römheld *et al.*, 1984.)

Table 15.4

Time Course of Dry Matter Production, Phosphorus Concentration, and pH in the Rhizosphere and Ion Uptake of Rape Plants Grown in a Low-Phosphorus Soil[a]

Age of plants (d)	Dry weight (g per vessel)	Phosphorus concentration in rhizosphere soil solution (μM)	Rhizosphere pH	Uptake of cations and anions[b]
0	—	5.17	6.1	—
7	0.16	2.56	6.3	Cat < An
14	0.89	0.82	6.5	Cat < An
20	1.89	1.40	5.3	Cat > An
28	3.69	2.47	4.3	Cat > An

[a]Based on Grinsted *et al.* (1982) and Hedley *et al.* (1982).
[b]Nitrogen supplied as $Ca(NO_3)_2$.

to higher uptake rates of calcium and magnesium (Moorby *et al.*, 1988). Similar responses in cation/anion uptake ratio and rhizosphere acidification have been found in phosphorus-deficient sunflower plants (Hoffland *et al.*, 1989a). However, there is increasing evidence that in many instances phosphorus deficiency-induced rhizosphere acidification is either exclusively, or at least to a high extent caused by excretion of organic acids (Section 15.4.2.3).

15.3.4 Redox Potential and Reducing Processes

As soil water content increases, redox potentials tend to decrease until in submerged soils negative values are obtained. The fall in redox potential is correlated with a range of changes in the solubility of mineral nutrients (e.g., manganese and iron, occasionally phosphorus) and also the accumulation of phytotoxic organic solutes (Section 14.4.4). Plants adapted to waterlogging and to submerged soils (e.g. lowland rice) maintain high redox potentials in the rhizosphere by the transport of O_2 from the shoot through the aerenchyma in the roots to release O_2 into the rhizosphere (Section 16.1). This oxidation of the rhizosphere (Fig. 15.6) is essential to decrease the phytotoxic concentrations of organic solutes (Section 14.4) and Fe^{2+} and Mn^{2+} present in the bulk soil solution of submerged soils. Both O_2 transport into the roots and the rate of O_2 consumption in the roots and particularly in the rhizosphere are strongly affected by mineral nutrition.

The distance that the oxidation zone extends from the rhizoplane into the bulk soil (Fig. 15.6) varies between 1 and 4 mm, depending on O_2 supply and O_2 consumption, and on the redox buffer capacity of the soil. The distance also varies along the axis of individual roots (Flessa and Fischer, 1992). In flooded rice, the redox potential increases steeply behind the root apex, for example, from -250 mV to about $+100$ mV, drops again in more basal zones, and rises again at sites where lateral roots penetrate the cortex. This pattern in redox potential along the root axis might be related to the pattern in population density of rhizosphere microorganisms (as the main O_2 consumers) which is low at the apex and steeply increases in basal zones prior to the emergence of lateral roots (Murakami *et al.*, 1990).

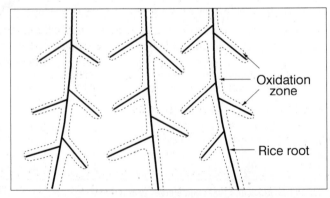

Fig. 15.6   Schematic representation of wetland rice roots in submerged soil.

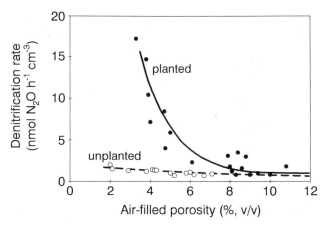

Fig. 15.7   Effect of air-filled porosity and wheat growth (planted) on denitrification in a chernozem soil (C_{org}: 1.8%). (According to Prade and Trolldenier, 1989.)

In aerated soils average redox potentials are in the range of $+500$–700 mV. However, aerated soils are not homogeneous and a mosaic of anaerobic microsites occurs which vary in location and time. Such microsites are most likely much more abundant in the rhizosphere than in the bulk soil (Fischer *et al.*, 1989), and are particularly important for the acquisition of manganese (Section 15.5) and iron, and for gaseous nitrogen losses (e.g., N_2, N_2O ...). As the redox potential decreases and the O_2 concentration falls nitrate is used by microorganisms as an alternative electron acceptor, followed by manganese oxides (Section 16.4.1). As O_2 consumption is higher in the rhizosphere compared with the bulk soil, in poorly aerated, compacted soils the risk of nitrogen losses by denitrification (or incomplete nitrification of ammonium nitrogen; Klemedtsson *et al.*, 1988; Papen *et al.*, 1989) is higher in a planted than in an unplanted soil (Fig. 15.7). Rates of denitrification rise with increasing inputs of organic carbon from the roots into the rhizosphere (Bakken, 1988), as is particularly the case in plants suffering from potassium deficiency (Rheinbaben and Trolldenier, 1984; Trolldenier, 1989).

Enhanced reducing activity at the root surface is a typical feature of root responses to iron deficiency in dicots and nongraminaceous monocots (Section 2.5.6). This root response can also lead to enhanced reduction of iron and manganese in the rhizosphere soil when reductants (e.g. phenolics) are released by the iron-deficient roots (Section 16.4).

15.4   Rhizodeposition and Root Exudates

15.4.1   Rhizodeposition

On average, 30–60% of the net photosynthetic carbon is allocated to the roots (Table 15.5) and of this carbon an appreciable proportion is released as organic carbon into the rhizosphere. This release of carbon, also termed as *rhizodeposition*, is highly variable,

Table 15.5
Percentage of Net Photosynthetic Carbon Allocated to, and Lost by, Roots of Annual Plant Species[a]

Percentage allocated to roots	Percentage of allocated carbon lost by roots		
	Respiration (A)	Rhizodeposition (B)	Total (A + B)
28–59	16–76	4–70	42–90

[a]Compiled data from the literature, based on Lynch and Whipps (1990).

but for annual species it is as much as 40% and for forest trees such as Douglas fir, values of more than 70% are quite common (Lynch and Whipps, 1990).

Rhizodeposition is increased by various forms of stress such as mechanical impedance, anaerobiosis, drought, and mineral nutrient deficiency (Whipps and Lynch, 1986; Lynch and Whipps, 1990). Therefore for a given plant species rates of rhizodeposition vary much and can be, for example, 2–4 times higher for soil-grown plants than for plants grown in nutrient solution (Trofymow *et al.*, 1987). Rhizosphere microorganisms increase rhizodeposition (Meharg and Killham, 1991), particularly the low-molecular-weight (LMW) fraction (Schönwitz and Ziegler, 1982; Kloss *et al.*, 1984). Over a 3-week period, roots of wheat seedlings released 7–13% of the net photosynthates when grown in the absence and 18–25% when grown in presence of microorganisms (Barber and Martin, 1976). Evidence has been presented that this enhancement effect of rhizosphere microorganisms is primarily caused not by increasing the efflux but by lowering the proportion of reabsorption ('retrieval') of LMW exudates by the roots (Jones and Darrah, 1993). In a comprehensive study using sealed systems over an entire growing period, Sauerbeck and Johnen (1976) found the highest release of organic carbon by the roots of soil-grown wheat during the period of rapid vegetative growth. At harvest the following amounts of carbon (in grams per pot) were measured: root dry weight, 3.0; root respiration, 1.9; root exudation and rhizodeposition, 7.6. In other words, more than twice as much organic carbon was released into the rhizosphere (rhizodeposition) as was retained in the root system at harvest. Similar data have been obtained with other annual species (Sauerbeck *et al.*, 1981).

The major components of rhizodeposition are shown in Fig. 15.8. The low-molecular-weight (LMW) exudates include organic acids, which can directly mobilize mineral nutrients in the rhizosphere. Some direct effects on mineral nutrient mobilization and binding can also be attributed to the mucilage and sloughed-off cells and tissues which are primarily a carbon substrate for rhizosphere microorganisms, but can become effective for mineral nutrient mobility as metabolites of microbial activity. Certain constituents of the LMW root exudates might also be transformed by rhizosphere microorganisms to highly physiological active compounds (e.g. phytohormones).

Rhizodeposition includes mineral nutrients previously taken up by the plant. In young wheat plants, for example, this contributes 1–5% of the phosphorus (McLaughlin *et al.*, 1987) and in wheat plants based on the whole growing period, it makes up 18%

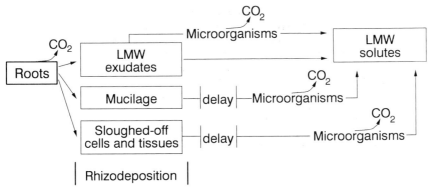

Fig. 15.8 Model of carbon flux in the rhizosphere. LMW = low-molecular-weight. (Modified from Warembourg and Billes, 1979.)

of the total nitrogen in plants of low nitrogen status and 33% in plants of high nitrogen status (Janzen, 1990). In principle, rhizodeposition by growing roots enhances the turnover rate of organic carbon in the rhizosphere ('priming effect'; Helal and Sauerbeck, 1989), particularly in plants well supplied with nitrogen (Liljeroth *et al.*, 1990).

15.4.2 Root Exudates

15.4.2.1 General

Root exudates comprise both high and low-molecular-weight solutes released or secreted by the roots. The most important components of the high-molecular-weight solutes are mucilage and ectoenzymes (Section 15.4.2.4) and those of the low-molecular-weight fraction are organic acids, sugars, phenolics, and amino acids (including phytosiderophores). Lysates from autolysis of epidermal and cortex cells are subsumed in the root exudate category (Lynch and Whipps, 1990; Bowen and Rovira, 1991). Root exudation is affected by various endogenous and exogenous factors (Uren and Reisenauer, 1988), and for nutrient dynamics in the rhizosphere and nutrient acquisition, the nutritional status of plants and the mechanical impedance of the substrate appear to be of particular importance (Fig. 15.9).

In cereal plants grown under sterile conditions 5% of the dry weight of the roots appeared as root exudates when plants were grown in nutrient solution but this value increased to 9% in a solid (glass ballotini) substrate (Barber and Gunn, 1974). Corresponding differences were found between the two substrates in root exudation of maize by a factor of three for sugars and vitamins (Schönwitz and Ziegler, 1982) and a factor of ten for phenolics in legumes (D'Arcy-Lameta, 1982). In principle, the same holds true for soil-grown plants (Fig. 15.9), and in particular when the mechanical impedance is increased by high soil bulk density. Increase in soil bulk density from 1.2 to 1.6 g cm^{-3} drastically depresses root length of maize, but the allocation of photosynthates to the roots remains similar (40% of net photosynthesis) leading to an

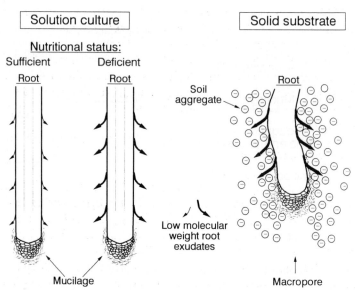

Fig. 15.9 Schematic presentation of root exudation as affected by mineral nutrient deficiency
and mechanical impedance.

increase in consumption of photosynthates per unit root length by a factor of two
(Sauerbeck and Helal, 1986).

Increased root exudation as a result of mechanical impedance has important
implications not only for the nutrient dynamics in the rhizosphere but also for the
tolerance of plants to high concentrations of aluminum (Table 15.6). Whereas in
nutrient solution a concentration of only 74 μM aluminum greatly inhibited root
elongation, the same concentration in the percolating nutrient solution in sand culture
was without effect (Horst *et al.*, 1990). Even increasing the aluminum concentration

Table 15.6
Influence of Aluminum on Root Length and Mineral Element Content in
Apical (0–5 mm) Root Zones of Soybean Grown in Nutrient Solution or in
Sand Culture Percolated with Nutrient Solution[a]

Substrate	Root length (cm per plant)	Mineral element content in root tips (mg g^{-1} dry wt)		
		Al	Ca	Mg
Nutrient solution				
Control (−Al)	189	<0.1	0.69	1.37
+74 μM Al	39	3.9	0.36	0.47
Sand culture				
Control (−Al)	114	<0.1	1.56	1.39
+741 μM Al	50	0.9	1.22	1.02

[a]Based on Horst *et al.* (1990).

about tenfold (741 μM), in plants grown in sand culture the detrimental effects on root growth were less severe than those at 74 μM in water culture (Table 15.6). As indicated by the mineral element content of the root apical zones, suppression of aluminum uptake and correspondingly the lesser depression in calcium and magnesium contents in the roots were presumably the responsible factors for higher aluminum tolerance of the roots growing in the solid substrate, an effect brought about by higher root exudation and a corresponding decrease in the concentration of the toxic monomeric aluminum species (Section 16.3).

15.4.2.2 Mucilage and Mucigel

Root surfaces, particularly apical zones, are covered by high-molecular-weight gelatinous material (*mucilage*), which consists mainly of polysaccharides which includes about 20–50% polyuronic acids, dependent upon plant species (Morel *et al.*, 1986; Ray *et al.*, 1988). This material is secreted by the root cap cells and is also released by epidermal cells (Vermeer and McCully, 1981). The production of mucilage is positively correlated with the root growth rate (Trolldenier and Hecht-Buchholz, 1984). In nonsterile media it also includes substances produced by bacterial degradation of the cell walls (Rovira *et al.*, 1983). In soil-grown plants the mucilage is usually invaded by microorganisms, and both organic and inorganic soil particles are embedded in it. This mixture of gelatinous material, microorganisms, and soil particles is termed *mucigel* (Bowen and Rovira, 1991).

Mucilage may have a diversity of biological functions (Ray *et al.*, 1988) including protection of root apical zones from desiccation, lubrication of the root as it moves through the soil (Fig. 15.9), ion uptake (facilitation or restriction), interaction with soil particles, and improving the soil–root contact, especially in dry soil, and causing aggregation of soil in the rhizosphere. Mucilage from maize may increase the proportion of water-stable soil aggregates from about 2% to nearly 40% (Morel *et al.*, 1991) and certainly contributes to the positive correlation between root length density and proportion of water-stable soil aggregates in field-grown plants (Materechera *et al.*, 1992a).

Under certain conditions the close contact between soil particles and root surface via mucilage (Fig. 15.9) can be of considerable importance for the uptake of mineral nutrients. This applies particularly to the micronutrients and phosphorus and also to toxic heavy metals and aluminum. In this ill-defined transition zone at the soil–root interface effects take place which are different from those occurring in the free solution ('two-phase-effect'; Matar *et al.*, 1967). These workers demonstrated that in phosphorus-deficient soil, plants take up phosphorus which is not in equilibrium with the soil solution but is mobilized at the root–soil interface presumably via phosphate desorption from clay surfaces by the polygalacturonic acid component of mucilage (Nagarajah *et al.*, 1970). Two-phase effects supply only a minor fraction of the total demand for macronutrients such as phosphorus, but this is not the case with micronutrients such as iron. As shown in Table 15.7 maize plants grown in quartz sand with FeOOH take up sufficient iron for normal growth and chlorophyll formation. This iron was mobilized at the sand–root interface and was not in equilibrium with the free solution. This is indicated by the extremely low iron content of the plants grown in a

Table 15.7

Rhizosphere Effect on the Utilization of Sparingly Soluble FeOOH by
Maize [a,b]

Treatment	Dry weight (g (6 plants)$^{-1}$)	Chlorophyll (mg g^{-1} dry wt)	^{59}Fe Content (μg g^{-1} dry wt)
Sand + ^{59}FeOOH	2.85	13.3	26
Nutrient solution	1.45	1.7^c	0.3

[a]Based on Azarabadi and Marschner (1979).
[b]The plants were grown in a sand and water culture system connected by
circulating an iron-free nutrient solution.
[c]Severe chlorosis.

nutrient solution in which iron was supplied only at the equilibrium concentration in the
quartz sand. Most probably, in the quartz sand iron was mobilized in the mucigel layer
of the sand–root interface by localized high concentrations of phytosiderophores in root
exudates of iron-deficient maize plants.

   Although in dry soils the role of root cap mucilage as a lubricant has been questioned
(Guinel and McCully, 1986) mucigel may be of importance for micronutrient uptake
from dry soils. Nambiar (1976a) presented evidence that roots growing through a layer
of soil drier than the wilting point can take up a significant amount of zinc, provided that
the roots have access to water elsewhere (e.g., in the subsoil). In dry soils more
mucilage is released in response to mechanical impedance, and this probably facilitates
zinc transport from the soil particles within the mucigel to the plasma membrane of root
cells (Nambiar, 1976b). Water transport in the roots from the subsoil and release of
water in the dry topsoil (*hydraulic lift*) might be involved in this enhancement effect
(Vetterlein and Marschner, 1993).

Table 15.8

Effect of Mucilage on the Growth and Aluminum Content of Roots of Cowpea
(cv. Tvu 354) Grown in Nutrient Solutions with or without Aluminum[a]

			Al Content of root tips (0–5 mm)			
Treatment	Mucilage	Root growth (cm d^{-1})	Roots (μg Al (25 tips)$^{-1}$)	Mucilage	Roots (mg Al g^{-1} dry wt)	Mucilage
−Al	+	6.3	—	—	—	—
	−^{b}	5.9	—	—	—	—
+Al^c	+	4.8	12.4	16.6	2.1	16.6
	−^{b}	2.1	20.6	3.6	3.2	14.5

[a]From Horst et al. (1982).
[b]Mucilage removed mechanically three times per day.
[c]+5 mg Al l^{-1}.

Mucilage has a high capacity to form complexes with heavy metal cations (Pb > Cu > Cd) mainly by exchange with Ca^{2+} (Morel *et al.*, 1986). The preferential binding of lead (Pb) might be an important factor in restricting uptake by roots. For aluminum uptake such a restriction by mucilage can be demonstrated (Table 15.8). In apical zones of growing roots, mucilage is continuously produced and may represent more than 10% of the dry weight of the apical 5 mm of cowpea roots (Horst *et al.*, 1982). In roots exposed to aluminum a high proportion of the aluminum is bound specifically to the mucilage. On a dry weight basis the mucilage contains about eight times more aluminum than the root tissue (Table 15.8). Accordingly, removal of the mucilage leads to an increase in the aluminum content of the root tissue and severe inhibition of root extension. The enhancement of mucilage production by mechanical impedance is therefore a major contributing factor to the much higher aluminum tolerance of roots growing in solid substrates compared with nutrient solutions (Table 15.6).

15.4.2.3 Low-Molecular-Weight (LMW) Root Exudates

The main constituents of the LMW root exudates (Fig. 15.8) are sugars, organic acids, amino acids, and phenolics. As a rule sugars and organic acids are the predominant compounds. However, not only the total amounts but also the proportions of these compounds vary considerably between plant species and the nutritional status of plants. Precise data on LMW root exudates are difficult to obtain as under nonsterile conditions, especially in nutrient solutions, microorganisms may utilize a major part of it as a carbon source (Bar-Ness *et al.*, 1992; Wirén *et al.*, 1993), and under sterile conditions the amounts released are considerably lower.

In general, root exudation of LMW compounds is higher in apical than in basal root zones (Uren and Reisenauer, 1988), in the case of sugars and amino acids this might reflect in part release by diffusion from cells and tissues with high internal concentrations. In apical root zones, phloem-derived amino acids lead to elevated concentrations of amino acids in the apoplasm and, despite an effective retrieval mechanism of reabsorption by plasma membrane-bound uptake systems (Schobert and Komor, 1987), release of amino acids into the external solution cannot be prevented. The same probably holds true for sugars (Jones and Darrah, 1993), whereas for organic acids, at high rates of exudation (e.g. under phosphorus deficiency) excretion by a H^+-coupled cotransport is more likely (Section 2.4).

Sugars have only minor direct effects on the mobilization of mineral nutrients. In this respect organic acids, amino acids and phenolics play a much more dominant role. Some of the principal reactions involved in mineral nutrient mobilization in the rhizosphere by these LMW compounds are shown in Fig. 15.10. Increased solubility of MnO_2 by root exudates seems to result mainly from organic acids (Uren and Reisenauer, 1988). For a given pH, root exudates from wheat dissolved 10–50 times more manganese from MnO_2 than did a buffer solution alone (Godo and Reisenauer, 1980). Malic acid is an important component of root exudates. During the oxidation of 1 mol of malic acid to CO_2 at the surface of MnO_2, 6 mol of Mn^{2+} are released (Jauregui and Reisenauer, 1982); chelation of Mn^{2+} prevents its reoxidization and increases the mobility of reduced manganese in the rhizosphere (Fig. 15.10, A). Phenolics contribute to the enhanced manganese reduction (Marschner, 1988). Organic acids are of general

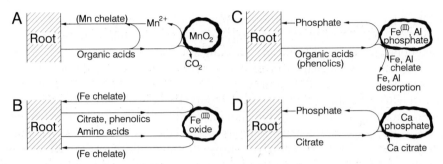

Fig. 15.10 Schematic presentation of various mechanisms in the rhizosphere for the solubilization of sparingly soluble inorganic compounds by root exudates in relation to the mineral nutrition of plants (see text).

importance in mobilizing sparingly soluble Fe(III) in the rhizosphere, and in response to iron deficiency increased rates of root exudation of phenolics and amino acids (phytosiderophores) play a particularly significant role (Section 2.5.6).

Organic acids, as well as phenolics, in root exudates are also important in bringing sparingly soluble inorganic phosphates into solution. The means by which organic acids mobilize phosphate are not confined to lowering the rhizosphere pH. Citrate, for example, desorbs phosphate from sesquioxide surfaces by anion (ligand) exchange (Parfitt, 1979; Gerke, 1992a). As a rule, however, a combination of both desorption and chelation of aluminum and iron is responsible for the phosphate mobilization from iron and/or aluminum phosphates (Fig. 15.10, C; Gerke, 1992a). Citric and malic acid, and phenolics, form relatively stable chelates with Fe(III) and aluminum, thereby increasing the solubility and rate of phosphorus uptake. As a side effect chelation of aluminum alleviates the harmful effects on root growth exerted by high concentrations of monomeric aluminum.

In certain plant species adapted to acid mineral soils with extremely low phosphorus availability, such as *Eucalyptus* ssp. (Mulette *et al.*, 1974) and tea plants (Jayman and Sivasubramaniam, 1975), this mechanism is of major importance for phosphorus nutrition. The high efficiency in these plant species is presumably a response to phosphorus deficiency (see below). Organic acids in root exudates are not only of importance for mobilizing soil phosphorus, but also for micronutrients. Iron, zinc and manganese in calcareous soils are increased in availability either by lowering the pH in the rhizosphere, or by chelation of these micronutrients, and by lowering the Ca^{2+} concentration by chelation and formation of sparingly soluble salts such as calcium citrate (Fig. 15.10, D).

Low-molecular-weight root exudates also mobilize heavy metals such as copper, lead, and cadmium by the formation of stable complexes (Mench *et al.*,1988). This can have important consequences for the heavy metal uptake rates. Root exudates of two tobacco species and of maize have been shown to mobilize cadmium from soils in an order (*N. tabacum* > *N. rustica* > *Z. mays*) which also reflected the differences in cadmium uptake ('bioavailability') between these three plant species (Mench and Martin, 1991).

Table 15.9
Root Exudation of Organic Acids in Phosphorus-Deficient Legume Species[a]

Species	Total[b]	Malonic	Fumaric	Malic	Citric
	Exudation (nmol g^{-1} root fresh wt $(12\ h)^{-1}$)				
Soybean	2.83	—	1.03	0.78	1.02
Chickpea	66.54	7.04	6.87	12.67	35.63
Peanut	47.21	—	24.44	12.84	9.17
Pigeon pea	6.17	0.34	0.73	4.31	0.79

[a]Based on Ohwaki and Hirata (1992).
[b]Including other organic acids.

15.4.2.4 Root Exudates and Nutritional Status of Plants

When plants are nutrient deficient the amounts of LMW root exudates often increase and the composition of the exudates is altered. For example, under potassium deficiency in maize the amounts of exudates increase and the proportions of sugars and organic acids are shifted in favor of the organic acids (Kraffczyk *et al.*, 1984). Under zinc deficiency in both dicots and grasses the amounts of amino acids, sugars, and phenolics in root exudates increase, but the increase in specific zinc-mobilizing root exudates is confined to grasses (Zhang *et al.*, 1991a). Examples are given below showing deficiency-induced enhancement of root exudates, or compounds in root exudates, which are rather specific for particular nutrient deficiencies and of special importance in enhancing the mobility of mineral nutrients in the rhizosphere.

Enhanced root exudation of organic acids is often observed under phosphorus deficiency in dicots in general and legumes in particular. In alfalfa (*Medicago sativa*) even under latent phosphorus deficiency when the total dry weight has not yet been depressed but the root/shoot dry weight ratio has begun to increase, root exudation of citric acid increases about twofold (Lipton *et al.*, 1987). Legume species respond quite differently to phosphorus deficiency in terms of increase in root exudation of organic acids (Table 15.9). Exudation is very high in chickpea and peanut, but low in soybean and moderate in pigeon pea. The main compounds are citric acid in chickpea and fumaric acid in peanut. Much smaller differences were observed between species in the organic acid contents of the roots than in the organic acid content of the root exudates (Ohwaki and Hirata, 1992).

The relatively low exudation of organic acids in pigeon pea (*Cajanus cajan*) is in some contrast to the outstanding capacity of this species for phosphorus acquisition from Alfisols (low pH; Fe(III)-P dominating). This high capacity is presumably caused by another organic acid in the root exudates, namely piscidic acid (*p*-hydroxybenzyl tartaric acid):

Piscidic acid

Table 15.10

Organic Acids in Exudates from Different Root Zones of Rape
(*Brassica napus* L.) Plants grown for Seven Days without or with
Phosphorus[a]

Phosphorus supply	Root zone	Organic acids in exudates (nmol cm^{-1} root (2 h)$^{-1}$)	
		Malic	Citric
−P	apical	0.87	0.27
	basal	0.20	0.13
+P	apical	0.15	0.06
	basal	0.03	0.03

[a]Based on Hoffland *et al.* (1989b). Reprinted by permission of Kluwer
Academic Publishers.

Piscidic acid is a strong chelator of Fe(III) and, thus, mobilizes sparingly soluble iron phosphates, but it is not very effective in bringing sparingly soluble calcium phosphates into solution (Ae *et al.*, 1990). Pigeon pea is therefore highly phosphorus efficient when grown on Alfisols but not on Vertisols (high pH, Ca-P dominating).

In symbiotically grown legumes rhizosphere acidification can thus be caused by both a high cation/anion uptake ratio (Table 15.3), and the excretion of organic acids, the relative importance of both processes depending on the phosphorus nutritional status of the plants.

In rape, rhizosphere acidification in phosphorus-deficient plants might be closely related to the cation/anion uptake ratio (Table 15.4), but also to enhanced net excretion of organic acids (Table 15.10). This enhanced excretion was mainly confined to the apical root zones and coincided with higher contents of malate and citrate in the apical root zones of the phosphorus-deficient plants (Hoffland *et al.*, 1989b, 1992). The reasons for the different results on the mechanisms of rhizosphere acidification in rape (Tables 15.4 and 15.10) are not clear.

A high local exudation rate of organic acids and other solutes (e.g. phytosidero-phores), as well as protons (Fig. 15.5) has ecological advantages. In well-buffered soils a pH decline can only be achieved by high flux densities of protons and organic acids. Furthermore, sites of localized low pH in the rhizosphere may inhibit the growth of microorganisms and thereby prevent or at least restrict microbial degradation of exudates (Hoffland *et al.*, 1989b; Römheld, 1991). This principle is nearly perfectly realized in cluster-rooted plants such as various tree species (e.g. *Banksia* ssp.) and also in annual legume species such as *Lupinus albus* (Section 14.4). Within the root clusters (*proteoid roots*) an intensive extraction of a limited soil volume is made possible by root exudate which would otherwise diffuse into a larger soil volume with corresponding dilution (Gardner *et al.*, 1983b). This spatial effect of root exudates within proteoid root zones is the opposite to that described for the acquisition of phosphorus, or potassium, in plants without cluster root formation where at high rooting density the overlapping of depletion zones decreases efficiency (Section 13.3).

Table 15.11

Soil pH and Contents of Citrate and Micronutrients in Bulk Soil and Rhizosphere Soil of White Lupin (*Lupinus albus* L.) Grown in a Phosphorus-Deficient Soil (23% $CaCO_3$)[a]

	Bulk soil	Rhizosphere soil (proteoid root zone)
pH (H_2O)	7.5	4.8
Citrate (μg g^{-1} soil)	ND	47.7
DTPA extractable (μmol kg^{-1} soil)		
Iron	34	251
Manganese	44	222
Zinc	2.8	16.8

[a]Dinkelaker *et al.* (1989), ND = not determined.

Citric acid is the dominant compound in the proteoid root exudates of white lupin (Gardner *et al.*, 1983a) and effective in the mobilization of phosphorus from both acid and calcareous soils. The high local citric acid excretion acidifies the rhizosphere even in calcareous soils (Table 15.11) and mobilizes sparingly soluble calcium phosphates by dissolution and subsequent formation of sparingly soluble calcium citrate in the rhizosphere (Fig. 15.11).

Localized acidification by citric acid not only mobilizes phosphorus but also iron, manganese, and zinc in the rhizosphere (Table 15.11) and increases their uptake rates and contents in the plants. When white lupin or other proteoid root-forming plant

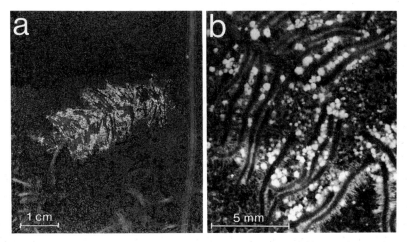

Fig. 15.11   Precipitation of calcium citrate in the rhizosphere of proteoid roots of 13-week-old *Lupinus albus* plants growing in a phosphorus-deficient Luvisol; (a) proteoid root zone; (b) details of (a) with calcium citrate particles. (Dinkelaker *et al.*, 1989.)

Table 15.12

Dry Weight and Phosphorus Uptake of Wheat (*T. aestivum*) and White
Lupin (*L. albus*) Grown in Mixed Culture in a Phosphorus-Deficient
Soil (pH 6.5) Supplied with Rock Phosphate and Nitrate Nitrogen[a]

Root systems of both species	Dry weight (g per pot)		Phosphorus uptake (mg per pot)	
	Wheat	Lupin	Wheat	Lupin
Separated[b]	23.5	27.3	23.6	36.6
Intertwining	39.9	24.5	46.5	40.2

[a]Based on Horst and Waschkies (1987).
[b]Root systems of both species separated by a stainless steel screen.

species are supplied with soluble phosphorus, proteoid root formation is depressed
(Section 14.4) and accordingly the excretion of citric acid. In plants such as white lupin
the contents of phosphorus can remain similar (Gardner *et al.*, 1982) but those of
micronutrients decrease, for example, manganese in the shoot dry weight decreases
from $4970\,\mu g\,g^{-1}$ to $833\,\mu g\,g^{-1}$ and zinc from $30\,\mu g\,g^{-1}$ to $16\,\mu g\,g^{-1}$.

The amounts of citric acid released into the rhizosphere of white lupin (Table 15.11)
accounted for 1 g per plant, or 23% of the net photosynthesis after 13 weeks' growth.
This appears to be a high cost for phosphorus acquisition. However, in view of the
benefits (mobilization not only of phosphorus but also other mineral nutrients), and the
costs of alternatives (allocation of 10–20% of the net photosynthates to the fungus in
VA mycorrhizal associations, Section 15.6.3); or increase in root surface area (Section
14.4), this strategy of proteoid root plants seems to be quite efficient.

When grown in mixed culture in a phosphorus-deficient soil supplied with rock
phosphate, wheat can profit from the phosphorus mobilized in the rhizosphere of the
white lupin (Table 15.12), provided the root systems of both plant species can
intertwine. Thus, white lupin is able not only to mobilize sparingly soluble phosphates
for its own demand but also to provide additional phosphorus for other plant species
such as wheat which, despite its much higher root length density (12.6 cm cm^{-3},
compared with 3.6 cm cm^{-3} in lupin) is much less phosphorus efficient.

Also in the proteoid root soil layer of mature stands of Proteaceae such as *Banksia
interifolia*, the amounts of organic acids in water leachates is about 10 times higher than
in the surrounding soil, and citric acid represents about 50% of the total organic acids
(Grierson, 1992).

A particular type of root exudation exists in graminaceous species (Section 2.5.6). In
response to iron deficiency, and also to withholding zinc supply (Zhang *et al.*, 1989,
1991b) the release of nonproteinogenic amino acids, the so-called phytosiderophores is
increased. Root exudates of iron-sufficient graminaceous species such as barley are able
to mobilize iron and other micronutrient cations from calcareous soils (Fig. 15.12). The
mobilization of manganese is much higher than of other cations, probably relating to
manganese reduction by organic acids in the exudates. Compared to iron-sufficient
plants, root exudates of iron-deficient barley plants have a much higher capacity for

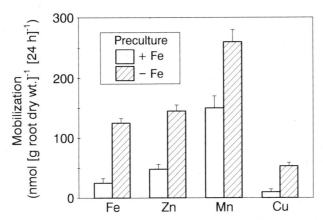

Fig. 15.12 Mobilization of micronutrients from a calcareous soil (Luvisol, 7% $CaCO_3$) by root exudates of iron-sufficient (+Fe) and iron-deficient (−Fe) barley plants. (Treeby *et al.*, 1989.)

micronutrient mobilization (Fig. 15.12), which for zinc and copper is in a similar range as for synthetic chelators (Treeby *et al.*, 1989). The much higher concentration of phytosiderophores in the exudate is responsible for the increase in micronutrient mobilization by the exudates of iron-deficient plants. Thus, in graminaceous species under iron deficiency, as a side-effect the solubility and mobility of other micronutrients in the rhizosphere is increased.

15.4.2.5 Ectoenzymes

In most agricultural soils between 30% and 70% of the total soil phosphorus is present in the soil organic matter (*organic phosphorus*, P_{org}). In forest soils the proportion of P_{org} may rise to 95% (Zech *et al.*, 1987). In the rhizosphere, part of this P_{org} is mobilized from, or incorporated into, this fraction by rhizosphere microorganisms (Helal and Dressler, 1989). Hydrolysis of P_{org} is mediated by root-borne acid phosphatase, fungal acid or alkaline phosphatase, and bacterial alkaline phosphatase. A distinct gradient in phosphatase activity therefore exists from the bulk soil to the root surface as shown in Fig. 15.13 for acid phosphatase.

Phosphatases are adaptive enzymes and accordingly activity of the root-borne acid phosphatase increases in response to phosphorus deficiency (Helal and Dressler, 1989; Tadano and Sakai, 1991). The root-borne acid phosphatase as determined *in vivo* is an ectoenzyme secreted, or released, by the roots, particularly in apical zones (Dinkelaker and Marschner, 1992). The differences in acid phosphatase activity between the three plant species shown in Fig. 15.13 are therefore not necessarily typical for the species but probably relate to differences in the phosphorus nutritional status of the plants (Tadano and Sakai, 1991). In view of the high proportion of P_{org} in the bulk soil solution (Section 13.2) and the high turnover of P_{org} in rhizosphere microorganisms, the importance of high activity of root-borne acid phosphatase for phosphorus acquisition of plants is evident, particularly when grown on low phosphorus soils. In agreement with this suggested role, of the total phosphorus depletion in the rhizosphere (reflecting uptake

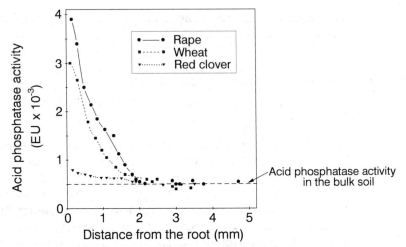

Fig. 15.13   Acid phosphatase activity in the rhizosphere of different plant species grown in a silt loam soil. (Tarafdar and Jungk, 1987).

by the roots) the proportion accounted for by the P_{org} fraction was about 50% (Helal and Dressler, 1989) or even more (Tarafdar and Jungk, 1987). In the rhizosphere of Norway spruce (Fig. 15.14) the depletion of phosphorus was even confined to the P_{org} fraction and closely correlated with the acid phosphatase activity which, in these ectomycorrhizal root systems might derive from both roots and fungi. In *Pinus rigida* the activity of acid phosphatase is much higher in nonmycorrhizal roots than in roots colonized with the ECM fungus *Pisolithus tinctorius* (Cumming, 1993).

Many other enzymes are located in the root apoplasm, particularly in epidermal cells of apical root zones. These enzymes include polyphenol oxidases as well as those

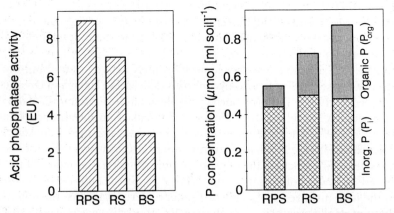

Fig. 15.14   Acid phosphatase activity and content of inorganic (P_i) and organic (P_{org}) phosphorus in the water-extractable fraction of rhizoplane soil (RPS), rhizosphere soil (RS), and of bulk soil (BS) in an 80-year-old Norway spruce stand. (Häussling and Marschner, 1989.)

needed for cell wall biosynthesis. Their role in nutrient dynamics in the rhizosphere and mineral nutrient acquisition, however, is not clear.

15.5 Noninfecting Rhizosphere Microorganisms

15.5.1 Root Colonization

Since roots act as a source of organic carbon, the population density of microorganisms, especially bacteria, is considerably higher in the rhizosphere than in the bulk soil. The relative increase in the number of microorganisms is expressed as an R/S ratio, R being the numbers per gram of soil in the rhizosphere and S in the bulk soil. The ratios vary greatly, between 5 and 50, depending for example on plant age, plant species, and nutritional status of plant. In general, all endogenous and exogenous factors which affect rhizodeposition and thus provision of organic carbon have a similar impact on population density on the rhizoplane and in the rhizosphere. The root colonization by noninfecting microorganisms is not confined to the rhizoplane but takes place to a varying degree also in the apoplasm of the cortex (e.g., of *A. brasilense*, Section 7.6). For such cases sometimes the term 'endorhizosphere' is used, but the term 'bacterial endophytes' is considered to be more appropriate (Kloepper *et al.*, 1992b).

As a rule, in soil-grown plants between 75% (Haller and Stolp, 1985) and more than 85% (Barber and Martin, 1976) of the total organic carbon supply for microbial activity in the rhizosphere is represented by sloughed-off cells and tissues (Fig. 15.8). Despite the high supply of organic carbon compounds the rhizosphere microorganisms can be nutrient limited, particularly of nitrogen. Therefore, in nonlegumes as a rule the number of rhizosphere bacteria increases with nitrogen fertilization (Liljeroth *et al.*, 1990a), as also does their activity and turnover rate. Limitation of nitrogen is probably also a main reason for the drastic decrease in bacterial turnover rates at the rhizoplane of rape from 9.2 h for 6-day-old to 160 h for 26-day-old plants (Baath and Johansson, 1990).

For growth and physiology of the roots and nutrient dynamics in the rhizosphere not only is the total number of rhizosphere microorganisms (bacteria, fungi) important but even more the types (species, strains) and their physiological characteristics, for example, producers of phytohormones, N_2 fixers, minor pathogens, and antagonists. Different plant species bear a different rhizosphere microflora both in number and physiological characteristics (Kloepper *et al.*, 1991). This is also true for a given plant species for different zones of roots, for example sheathed and bare root zones of C_4 species (Gochnauer *et al.*, 1989). Within a given plant species the amount and form of nitrogen fertilizer supply also alter the rhizosphere microflora. For example, as the nitrogen supply increases both the number and proportion of diazotrophic bacteria decrease at the rhizoplane of various grasses, whereas the total number of bacteria increases (Kolb and Martin, 1988). In wheat, depending on whether nitrogen is supplied as ammonium or nitrate, there is a considerable shift in the proportion of pathogens (*G. graminis*) and antagonists (*Pseudomonas* ssp.) in the rhizosphere (Sarniguet *et al.*, 1992a,b).

As shown schematically in Fig. 15.15, in fast growing roots there is usually a steep

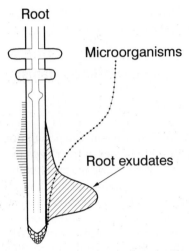

Fig. 15.15 Schematic presentation of spatial separation of LMW root exudates (e.g., phyto-siderophores, organic acids) and microbial activity in the rhizosphere of soil-grown plants.

gradient of rhizoplane and rhizosphere microorganisms from apical to basal zones along the root axis (Bowen and Rovira, 1991). In maize, for example, of the total root surface area, bacteria surface cover is about 4% in apical zones, 7% in the root hair zone, and might rise up to 20% in basal zones (Schönwitz and Ziegler, 1986b).

This gradient of microbial populations along the root axis has important implications for the efficiency of root exudates released in response to deficiency, for example, phytosiderophores under iron deficiency (Fig. 15.15). In addition, release of phytosiderophores is confined to a short period of 2–8 h which further increases their effectivity in iron acquisition in the rhizosphere (Römheld, 1991). All factors which favor a more uniform distribution of rhizosphere microorganisms along the roots therefore decrease effectivity of root-released phytosiderophores (Wirén et al., 1993). Model calculations of effectivity of root exudates in nutrient acquisition have to consider this spatial separation of root exudation and microbial activity (Darrah, 1991, 1993).

15.5.2 Role in Mineral Nutrition of Plants

Noninfecting rhizosphere microorganisms might affect mineral nutrition of plants through their influence on: (a) growth, morphology and physiology of roots; (b) the physiology and development of plants; (c) the availability of nutrients; and (d) nutrient uptake processes. Points (a) and (b) were discussed in Chapter 14, and some aspects of (c) and (d) in previous sections of this chapter. The role of N_2 fixing rhizosphere bacteria have been discussed extensively in Section 7.5. In the following discussion therefore the main emphasis is on the evaluation of the various aspects for the mineral nutrition of soil-grown plants.

It is generally agreed that rhizosphere microorganisms can influence the acquisition of phosphorus, potassium, and to some extent also nitrogen from soils mainly via their effects on root morphology and physiology. The relative importance of diazotrophic

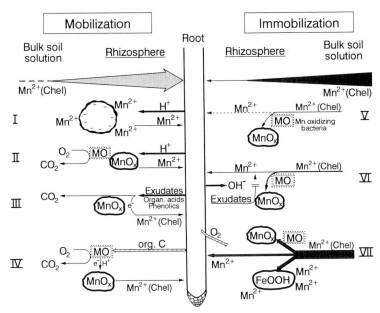

Fig. 15.16 Schematic presentation of various mechanisms for mobilization and immobilization of manganese in the rhizosphere. MO = microorganisms; MnO_x = Mn(III)+Mn(IV) oxides. (Marschner, 1988). Reprinted by permission of Kluwer Academic Publishers.

bacteria for nitrogen nutrition depends on nitrogen availability in soils and plant species; at high nitrogen supply their effects on nutrient acquisition by roots are indirect via modification of root growth. Rhizosphere microorganisms enhance turnover rate of organic carbon, nitrogen and phosphorus, and thus 'recycling' of organically bound nutrients, but it is difficult to draw firm conclusions on the net effects on nitrogen uptake (e.g. higher mineralization versus higher denitrification). Beneficial effects on the acquisition of phosphorus and various micronutrients by plants can be expected from rhizosphere microorganims which, for example, use sugars of the LMW root exudates or use sloughed-off cells and tissues as a carbon source for the production of chelators or organic acids which can act similarly to corresponding root exudates.

There is an extensive literature on so-called phosphate-dissolving bacteria, and there has been considerable speculation as to whether such bacteria might allow increased utilization of soil and fertilizer phosphorus. Although these bacteria are capable of dissolving sparingly soluble inorganic phosphates (e.g., rock phosphates), it is questionable whether this mechanism operates to any greater extent in the rhizosphere (Tinker, 1980). These bacteria have to compete with other rhizosphere microorganisms for organic carbon as an energy substrate. They are therefore difficult to establish and maintain in high quantity in the rhizosphere. The situation is different, however, for microorganisms such as mycorrhizas which receive photosynthates directly from the root cells (see Section 15.7).

When a large organic carbon supply from the roots is combined with a low O_2 partial pressure in the rhizosphere, high microbial activity might increase availability of manganese. On aerated soils with high pH such increase should be beneficial for manganese nutrition of plants, but in general becomes critical in poorly aerated, low pH soils. In submerged soils, microbial activity in the rhizosphere of lowland rice is enhanced, for example, by higher root exudation in plants suffering from potassium, phosphorus, or calcium deficiency, with corresponding increase in risk of iron toxicity ('bronzing') in the plants (Ottow et al., 1982).

In the case of manganese the role of rhizosphere microorganisms is particularly evident, and also complex (Fig. 15.16). Microbial activity is primarily responsible for oxidation of Mn^{2+} in the bulk soil and rhizosphere soil, and many soil-borne pathogens are effective manganese oxidizers (Section 11.4). Rhizosphere microorganisms not only immobilize (oxidize) but also mobilize manganese (reduce manganese oxides), depending on the conditions, and this might be favorable or unfavorable for the manganese nutrition of plants. Particularly in calcareous soils, mobilization of sparingly soluble manganese oxides in the rhizosphere can be quite high (Warden, 1991) being the result of both root and microbial activity (Fig. 15.16).

In aerated soils rhizosphere microorganisms may mobilize manganese by reduction, favoured by root excretion of protons (Mechanism II) or via decomposition of sloughed-off cells and tissues (Mechanism IV). In contrast, manganese oxidizing bacteria in the rhizosphere may decrease manganese availability and thereby either increase the risk of manganese deficiency in aerated, calcareous soils (Mechanisms V and VI), or decrease the risk of manganese toxicity in poorly aerated, or in submerged soils (Mechanism VII).

15.5.3 Root Exudates as Signals and Phytohormone Precursors

Recently in plant root–microbial interactions root exudates have attracted much interest not so much in terms of carbon source but as 'signals' for recognition, or as precursors for phytohormone production (Fig. 15.17). Particularly in functioning as signals for soil microorganisms only very low concentrations are required, and in many instances the active components are flavonoids released by the roots. As a signal for chemotaxis of rhizobia a concentration as low as 10^{-9} M luteolin is sufficient, and at 10^{-6} M concentration luteolin also stimulates nod gene expression (Bauer and Caetano-Anolles, 1990). Other flavonoids in root exudates of legumes may simultaneously act as suppressors for certain pathogenic fungi (Hartwig U. A. et al., 1991). Quercetin also acts as a signal for spore germination and hyphal growth of VA mycorrhizal fungi (Phillips and Tsai, 1992), although for hyphal growth the elevated CO_2 concentrations in the rhizosphere are obviously of much greater importance (Bécard and Piché, 1989). Root exudates of Pinus sylvestris enhance growth of ectomycorrhizal fungi, and one of the growth-stimulating compounds in the exudates has been identified as palmitic acid (Sun and Fries, 1992).

Specific root exudates, however, not only act as a signal for establishment of symbiotic interactions but also for parasitic flowering plants (Fig. 15.17). In root exudates of Sorghum bicolor a hydroquinone (sorgolactone) strongly stimulates germination of Striga asiatica, and thus the formation of the parasitic interactions (Fate et al., 1990; Hauck et al., 1992).

Root cap cells and root cap mucilage also seem to play a role in the establishment of

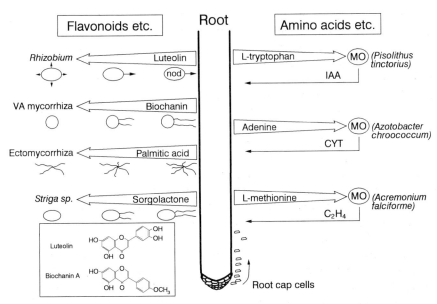

Fig. 15.17   Possible role of certain low-molecular-weight root exudates as 'signal' or as sources (precursors) for phytohormone production for microorganisms (MO) in the rhizosphere.

specific plant root–microbial interactions. Root cap mucilage of maize has a strongly chemotactic action on strains of *Azospirillum lipoferum* isolated from maize rhizoplane but not on strains isolated from rice rhizoplane (Mandimba *et al.*, 1986). During root penetration through the soil, detached root cap cells come into contact with more basal root zones, and root cap cells seem to carry host-specific traits into the rhizosphere for establishment of a characteristic rhizosphere bacterial flora, and to suppress certain soil-borne root pathogens (Hawes, 1990; Gochnauer *et al.*, 1990). In *Eucalyptus* the root cap cells are strongly chemotactic to the ectomycorrhizal fungus *Pisolithus tinctorius* (Horan and Chilvers, 1990).

A relatively large number of typical rhizosphere bacteria are effective producers of phytohormones such as IAA and CYT. In agreement with this, in field-grown maize the concentrations of IAA, CYT, and ABA, are several times higher in the rhizosphere compared with the bulk soil, and the concentrations in the rhizosphere are sufficiently high at least to produce morphogenetic effects on roots (Müller *et al.*, 1989). As sources for this phytohormone production in the rhizosphere certain compounds in the root exudates are presumably of key importance (Fig. 15.17). Examples are the enhancement of CYT and IAA production by *Azotobacter chroococcum* supplied either with maize root exudates (Gonzales-Lopez *et al.*, 1991) or with adenine (Nieto and Frankenberger, 1990). Production of ethylene (C_2H_4) by rhizosphere microorganisms, either fungi such as *Acremonium falciforme* (Arshad *et al.*, 1991) or many other soil fungi or bacteria (Arshad and Frankenberger, 1993) can be strongly enhanced by supplying L-methionine as a precursor. These various precursors for phytohormone production are components of root exudates or of lysates from decaying root tissues.

Some of these microorganisms also increase phytohormone production when supplied with organic acids and sugars.

In view of these specific effects of certain components in root exudates it is evident that the evaluation of the importance of root exudates in terms of amounts, or proportion of net photosynthates, is totally inadequate particularly with respect to their overall impact on the mineral nutrition of plants.

15.6 Mycorrhizas

15.6.1 General

Mycorrhizas are the most widespread associations between microorganisms and higher plants. The roots of most soil-grown plants are usual mycorrhizal. On a global basis mycorrhizas occur in 83% of dicotyledonous and 79% of monocotyledonous plants, and all Gymnosperms are mycorrhizal (Wilcox, 1991). Nonmycorrhizal plants occur in habitats where the soils are either very dry or saline, or waterlogged (submerged), severely disturbed (e.g., mining activities), or where soil fertility is extremely high or extremely low (Brundrett, 1991). Mycorrhizas are absent under all environmental conditions in the Cruciferae and Chenopodiaceae (Harley and Harley, 1987), and also quite rare or absent in many members of the Proteaceae or other typical root cluster-forming plant species (Brundrett and Abbott, 1991).

As a rule the fungus is strongly or wholly dependent on the higher plant, whereas the plant may or may not benefit. Only in some instances (orchids) are the mycorrhizas essential. Mycorrhizal associations are therefore either mutualistic, neutral, or parasitic, depending on the circumstances. As a rule mutualism dominates and therefore in the literature the term mycorrhizal symbiosis is often used. However, in this text the term association is preferred for two reasons: neutral and parasitic relationships are not rare between the fungus and the host plant, and in contrast, for example, to the *Rhizobium* symbiosis in legumes, in mycorrhizal associations the host plant can only regulate to a very limited extent the degree of root infection, growth and competition for carbohydrates of the fungus. For a comprehensive review on mycorrhizas in natural ecosystems the reader is referred to Alexander (1989), Brundrett (1991), Fitter (1991a), and Read (1991).

15.6.2 Mycorrhizal Groups, Morphology and Structures

There are two major mycorrhizal groups according to how the fungal mycelium relates to the root structure, the endomycorrhizas and the ectomycorrhizas (Fig. 15.18).
Endomycorrhizas. The fungi live within cortical cells and also grow intercellularly. There are several distinct types of endomycorrhizas, the best known being the *vesicular-arbuscular mycorrhizas* (VAM), the *ericoid*, and the *orchidaceous mycorrhizas*.

The VAM is by far the most abundant of endo- and ectomycorrhizas (Harley and Harley, 1987). The VAM is characterized by the formation of branched haustorial structures (arbuscules) within the cortex cells and by a mycelium which extends well

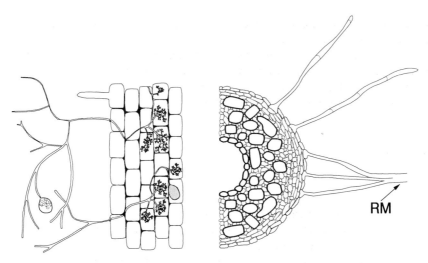

Fig. 15.18 Schematic presentation of the main structural features of the vesicular-arbuscular (VA) mycorrhizas (*left*) and of ecto- (EC) mycorrhizas (*right*). RM, rhizomorphs.

into the surrounding soil (external hyphae, extraradical mycelium; Fig. 15.19). The arbuscules are short lived, about 10–12 days, and are the main sites of solute exchange with the host. The VAM fungi belong mainly to four genera, *Acaulaspora*, *Gigaspora*, *Glomus*, and *Sclerocystis*. *Glomus* is thought to be the most abundant genus of the soil fungi (Lamont, 1982). Many, but not all of the endomycorrhizal fungi form vesicles as lipid-rich storage organs (Fig. 15.18). Therefore, in recent literature instead of VAM also the term AM (*arbuscular mycorrhiza*) is used for endomycorrhizas.

Ericoid mycorrhizas occur in Ericales, either as the endomycorrhizal type, or as the ectendomycorrhizal type. The endomycorrhizal type is characterized by coils of hyphae within infected rhizodermal (epidermal) cells. Each cell is invaded only through the outer cell wall, and the individual hyphae extend into the soil as in the case of VAM. In the ectendomycorrhizal type a thin layer of external hyphae may also surround the root.

Ectomycorrhizas occur mainly on roots of woody plants and only occasionally on herbaceous and graminaceous perennial plants (Wilcox, 1991). Ectomycorrhizas (ECM) are characterized by two main features (Fig. 15.18), an interwoven mantle of hyphae around the root surface (fungal sheath, or mantle) and hyphae that penetrate the root intercellular space of the cortex to form a network of fungal mycelium, the *Hartig net*, which envelops the cortex cells more or less totally and increases the surface area at the fungus–root interface similarly as in the case of transfer cells in higher plants. Most ECM fungi are Basidiomycetes, but relatively many are Ascomycetes. In most ECM types hyphal strands or *rhizomorphs* are produced which extend well into the surrounding soil (Fig. 15.18). The rhizomorphs are differentiated multihyphal organs with a diameter of up to 200 μm, and are important for solute transport over far distances, whereas the solute interchange between host and fungus takes place at the Hartig net. Sheath-forming ECM may totally enclose the mycorrhizal short roots (Fig. 15.19). During regrowth of the roots (e.g., in spring) both the host root and the fungal

Fig. 15.19 Mycorrhizal root systems. *Upper*: Root of soil-grown potato with extramatrical hyphae of the VAM fungus *Glomus mosseae*. *Lower*: Ectomycorrhizal short roots of soil-grown Norway spruce. (Courtesy of G. Hahn.)

tissue may grow slowly together, or with more rapid root growth the apex breaks through the hyphal sheath (Kottke and Oberwinkler, 1986).

Ectomycorrhizas are most common in the northern hemisphere, especially in Pinaceae, Betulaceae, Fagaceae and Salicaceae. However, ECM may also occur in some tropical and subtropical forests (Högberg, 1986).

In many forest tree species both VAM and ECM occur simultaneously, for example, in *Salix* and *Populus* (Lodge, 1989) or *Eucalyptus* (Gardner and Malajczuk, 1988), and the proportion of both types seem to depend on exogenous factors such as soil water content and aeration, and endogenous factors such as age of the trees. On a global scale, ECM is more abundant in boreal and temperate forests with a distinct surface humus horizon, and in nitrogen-limited ecosystems, whereas VAM is more abundant in warmer climates with drier soils, in pasture land and deciduous forests with high turnover of organic material, and where phosphorus supply is limited. VAM is usually the only form of mycorrhizas in crop plants and pastures and in fruit trees.

In addition to the differences in distribution, morphology and structure (Fig. 15.18) there is another principal difference between ECM and VAM. Whereas most ECM fungi can be grown in pure culture (*in vitro*) this is not possible for VAM fungi. Therefore, knowledge on physiology of VAM fungi is based on studies of fungal structures and fungal functions associated with the host roots (Smith and Gianninazzi-Pearson, 1988).

15.6.3   Root Infection, Photosynthate Demand, and Host Plant Growth

15.6.3.1   Root Infection

Root infection with mycorrhizas is initiated either from soil-borne propagules (spores, root residues) or from neighbouring roots of the same or different plants and plant species. Infection is enhanced by a preexisting network in the soil, and therefore severe soil disturbances (e.g., clear-cut logging or rigorous soil mixing (Jasper *et al.*, 1989b)), as well as tillage compared with no-tillage (Miller and McGonigle, 1992) severely depress and delay mycorrhizal infection.

Root exudates of host plants have a strongly chemotactic action on ECM (Horan and Chilvers, 1990) and VAM (Gianninazzi-Pearson *et al.*, 1989) fungi, and the effectivity of the responsible flavenoids in these root exudates (Section 15.5.3) is much enhanced by elevated CO_2 concentrations (Bécard *et al.*, 1992). Noninfecting rhizosphere bacteria may enhance or suppress mycorrhizal infection. Distinct stimulation of VAM infection has been obtained by inoculation with *Azospirillum* (Pacovsky *et al.*, 1985) and in case of the ECM fungus *Laccaria laccata* with so-called 'mycorrhization helper bacteria' (Duponnois and Garbaye, 1991).

In nonhost plants of VAM, for example members of the Chenopodiaceae and Cruciferae, incompatibility may be caused by the composition of root exudates, toxins, or enhanced defense reactions of the host against infection, as in response to pathogens (Anderson, 1988; Parra-Garcia *et al.*, 1992). Also in host plant species such as red clover, two SOD isoenzymes are induced by VAM infections, presumably as a response to infection and elevated O_2^- levels (Palma *et al.*, 1993). Different levels of host plant responses may be involved in the large varietal differences in root infection with VAM which has been found in wheat and cowpea to vary between zero and 18–30% (Mercy *et al.*, 1990; Vierheilig and Ocampo, 1991). There are some suspicions that selection of genotypes for high resistance to root pathogens might also involve the risk of simultaneous selection against high VAM infection (Toth *et al.*, 1990).

Mineral nutrient supply may enhance or suppress root infection and colonization with mycorrhizas. A distinct enhancement of root infection of shortleaf pine seedlings with the ECM fungus *Pisolithus tinctorius* has been achieved with boron supply at a level which markedly exceeded the host plant demand (Mitchell *et al.*, 1990), probably by suppression of the host plant response reaction against fungal invasion. At extremely low soil phosphorus level the root infections are low with VAM fungi (Bolan *et al.*, 1984) as well as with ECM fungi (see below), as phosphorus might limit the growth of the fungi itself. With increasing supply root growth and the proportion of infected root length increase until an optimum supply of phosphorus is approached and beyond this level infection rate is depressed to a different degree, depending on VAM species (Bolan *et al.*, 1984) or the ECM species (Jones *et al.*, 1990) and also the host species (Davis *et al.*, 1984). At high phosphorus supply decrease in infected root length and in soluble carbohydrates in the roots might be correlated (Thomson *et al.*, 1990) or not correlated (Amijee *et al.*, 1990). Negative relationships between root infection and phosphorus supply are probably more finely regulated by the host plant as indicated, for example, by an increase in necrotic root infection points (Amijee *et al.*, 1990) or a drastic decrease in the chemotaxis of the root exudates on hyphal growth in plants with high phosphorus supply (Elias and Safir, 1987).

High nitrogen supply also depresses VAM and ECM infection, particularly in combination with high phosphorus levels and when nitrogen is supplied as ammonium (Baath and Spokes, 1988). In ECM particularly the mass of the mycelium decreases at high nitrogen supply (Wallander and Nylund, 1991). Decrease in infected root percentage (VAM) or in the proportion of ECM root tips at high supply of phosphorus or nitrogen is, however, not necessarily an expression of a specific regulation mechanism but often the result of enhanced root growth whereas that of the associated fungus lags behind. Total VAM-infected root length or number of ECM root tips is often a more appropriate parameter, but for evaluation of effectivity in nutrient acquisition, quantification of the extraradical mycelium would be the most important parameter (see below).

15.6.3.2 Photosynthate Demand

In mycorrhizal roots a substantial proportion of the net photosynthates allocated to the roots is required for fungal growth and maintenance. In VAM plants 'root' respiration may be 20–30% higher than in nonmycorrhizal plants, and 87% of the higher respiration can be attributed to the fungus (Baas *et al.*, 1989). In cucumber, of the net photosynthates 20% were allocated below ground in nonmycorrhizal plants and 43% in VAM plants, and about half of it was respired (Jakobsen and Rosendahl, 1990). In highly infected plants the VAM fungal biomass may reach 20% of the root biomass, typically 10% can be assumed (Fitter, 1991a).

These costs in terms of photosynthates are not relevant in plants with sink limitation, for example when the source capacity exceeds the demand in nonmycorrhizal plants, and the mycorrhizal plants can compensate the higher demand by an increase in rate of photosynthesis per unit leaf area. Enhanced rates of photosynthesis in mycorrhizal plants are therefore often an expression of a higher sink activity (Dosskey *et al.*, 1990) and not of a specific stimulatory effect of the mycorrhizal association. The costs of photosynthates have also to be compared with the benefits such as enhanced uptake of nutrients like phosphorus when it limits photosynthesis and growth in nonmycorrhizal plants. However, despite this beneficial effect on overall plant growth, as a rule in mycorrhizal plants host root growth (dry weight) is less enhanced or even depressed (Dosskey *et al.*, 1991) compared with shoot growth and the root/shoot dry weight ratio decreases in a typical manner (Section 15.6.3.3).

In general, in ECM associations the proportion of photosynthate allocation to the fungus is considerably higher than in VAM associations. This is true in particular for ECM fungi with extended extramatrical mycelium (Fig. 15.20). An extended extramatrical mycelium is favorable for nutrient acquisition but also has a higher photosynthate demand. In studies with different ECM fungi in their effect on growth of *Pinus sylvestris* seedlings grown in a nutrient-poor substrate a close negative correlation was found between fungal biomass and host plant growth, which was attributed to both higher energy demand and also higher nitrogen incorporation into fungal structures and corresponding nitrogen limitation of the host plant (Colpaert *et al.*, 1992).

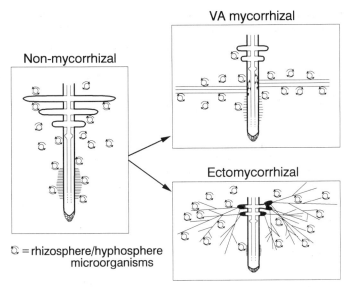

Fig. 15.20 Schematic presentation of effects of mycorrhizal colonization on root morphology and distribution of noninfecting rhizosphere microorganisms.

In ECM Douglas fir ecosystems about 60–70% of the net photosynthates are allocated below ground for growth of roots, mycorrhizas, and for respiration (Fogel, 1988). As in forest stands usually after 15–20 years the fine root biomass remains the same, and the increase in coarse root biomass is relatively small, a high proportion of the net photosynthates in forest stands is allocated to mycorrhizal structures and their turnover. Estimates on the proportions of carbon flow to the ECM in forest stands varies between 5 and 30% of the net photosynthates (Söderström, 1992). Thus ECM fungi play an important role in carbon import into the soil via the extramatrical mycelium, particularly in view of the high turnover rate of the fungal carbon which is about five times higher than that from litter fall (Fogel and Hunt, 1979).

This higher input in mycorrhizal plants of organic carbon into the soil also has important consequences on number, activity and distribution of soil microorganisms (Fig. 15.20). Most likely as a result of the higher carbon drain, the number of bacteria in the rhizosphere of VAM plants decrease (Paulitz and Linderman, 1989) whereas the number and particularly activity increase the greater the distance from the rhizoplane where more carbon is provided by the extraradical mycelium, especially in ECM plants. Thus, in mycorrhizal plants the soil–root interface is altered and an additional, or new, '*mycorrhizosphere*' is formed (Fogel, 1988; Linderman, 1988). As most soil-grown plants are mycorrhizal, this mycorrhizosphere might be the rule, not the exception, and around the external hyphae and mycelia a new interface with the soil is formed, the 'hyphosphere' (Linderman, 1988). The external mycelium may not only alter soil microbial activity but also provide substrate for soil fauna such as collembola feeding on mycorrhizal hyphae (McGonigle and Fitter, 1988). Also the soil structure may be changed by VAM extraradical hyphae by binding microaggregates into stable macro-

Table 15.13
Effect of Different VAM Species on Bacteria and Actinomycetes in
the Rhizosphere Soil of *Panicum maximum*[a]

Treatment	Rhizosphere populations (cfu g^{-1} soil)[b]		
	Bacteria ($\times 10^6$)	N$_2$ Fixer ($\times 10^5$)	Actinomycetes ($\times 10^4$)
Control (-VAM)	14.7	12.4	13.4
Glomus fasciculatum	41.9	42.0	26.1
Gigaspora margarita	34.0	87.9	17.6
Acaulospora laevis	8.1	10.6	28.6

[a]Based on Secilia and Bagyaraj (1987).
[b]cfu, colony forming units.

aggregates (Davies *et al.*, 1992), via intermeshing by hyphae or production of extra-cellular polysaccharides (Tisdall, 1991).

Mycorrhizal colonization not only alters the quantity of rhizosphere microorganisms (Fig. 15.20) but also their composition. In view of the role of rhizosphere micro-organisms on root morphology and activity (Section 14.5) this alteration has implications on mineral nutrient acquisition and growth of the roots and also shoots. But it is not only the mycorrhizal colonization per se which is important but also the mycorrhizal species which alter rhizosphere microflora differently (Table 15.13). Depending on the VAM species the total number of bacteria, diazotroph bacteria, and actiomycetes are affected to a different degree. The low numbers of rhizosphere bacteria in the nonmycorrhizal plants are caused by phosphorus limitation and correspondingly poor plant growth.

15.6.3.3 Host Plant Root and Shoot Growth

Mycorrhizal colonization affects root and shoot growth differently. In a nutrient-poor substrate the external mycelium adds surface area and compared with the nonmycorrhizal plants, the mycorrhizal plants have more access to the growth-limiting nutrients, for example phosphorus and nitrogen (Fig. 15.20). As a typical plant response to higher nutrient supply, shoot growth is more enhanced than root growth leading to a decrease in root/shoot dry weight ratio (Section 14.7). At a given nutritional status of the host plant, in mycorrhizal plants this shift is more pronounced (Bell *et al.*, 1989) as the fungus competes with the roots for photosynthates. This shift is particularly distinct in nodulated legumes (Piccini *et al.*, 1988) as the VAM fungi might represent a similar sink for photosynthates as the rhizobia in functioning nodules.

If the mycorrhizas are either ineffective in delivering nutrients, or nutrients are not growth-limiting factors in nonmycorrhizal plants – and other beneficial mycorrhizal effects (Section 15.10) are absent – mycorrhization depresses root growth primarily by sink competition. Unfavorable environmental conditions such as shading and defoliation also depress mycorrhizal growth (Same *et al.*, 1983), but to a lesser degree than that of the host root and, in nodulated legumes, the nodule weight (Bayne *et al.*, 1984). The

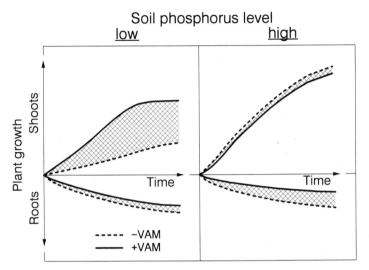

Fig. 15.21   Schematic presentation of effects of soil phosphorus level and root colonization with VAM fungi on root and shoot growth.

mycorrhizal fungi remain as a strong sink for photosynthates irrespectively of their contribution to host plant growth (Douds *et al.*, 1988). In most instances there is therefore an optimal level of mycorrhizal root colonization above which the plants receive no benefit from the fungus (Thomson *et al.*, 1992) and might even be substantially depressed in growth both by VAM (Amijee *et al.*, 1990) and ECM (Colpaert *et al.*, 1992). These relationships between root and shoot growth as affected by mycorrhization and phosphorus supply are summarized schematically in Fig. 15.21. In principle, growth depression can be predicted when root colonization remains high at high phosphorus supply and limited photosynthetic source capacity to compensate for the extra costs of mycorrhization (Sanders, 1993).

   Also other factors than competition for photosynthates are probably involved in impairment of root growth in mycorrhizal plants. In ECM plants most likely hormonal effects are involved, for example, inhibition of elongation growth of the short lateral roots by IAA production of the invading fungus (Section 15.10.1). In VAM plants, for example, in *Allium porum* infection with *Glomus mosseae* decreases total root length but simultaneously increases branching and number of lateral roots per unit root length, or per plant (Berta *et al.*, 1990). However, the newly formed roots are shorter and the activity of their apical meristems declines rapidly, which would explain the higher branching rate in mycorrhizal root systems. Higher rates of phloem unloading of photosynthates and IAA at the sink sites of VAM colonization in basal root zones could well explain such changes in rooting pattern as is known from localized phosphorus supply in nonmycorrhizal root systems (Section 14.4.1). In view of the role of root branching and of apical meristems in number and activity, effects of VAM colonization on balance of plant-borne phytohormones are to be expected (Section 15.10.1).

   Decrease in root surface area and root activity as well as in root/shoot dry weight ratios are, however, not necessarily harmful for shoot growth and plant yield as long as the external mycelium of the mycorrhizal fungi can fully compensate for the root's function in uptake of nutrients and water.

15.7 Role of Mycorrhizas in Mineral Nutrition of Their Host Plants

15.7.1 Vesicular-Arbuscular Mycorrhizas

The most distinct growth enhancement effect by VAM occurs by improved supply of mineral nutrients of low mobility in the soil solution, predominantly phosphorus. External hyphae can absorb and translocate phosphorus to the host from soil outside the root depletion zone of nonmycorrhizal roots (Tinker *et al.*, 1992). In view of the key importance, for example of the root hair length on the phosphorus depletion zone and phosphorus acquisition (Section 13.2) such an enhancing effect of VAM is to be expected. As a rule in mycorrhizal plants uptake rate of phosphorus per unit root length is 2–3 times higher than in nonmycorrhizal plants (Tinker *et al.*, 1992).

   An example of the different extension of depletion zones of phosphorus in mycorrhizal and nonmycorrhizal roots is shown in Fig. 15.22. By restriction of the host root extension by a net, and of the hyphal extension by a membrane, the phosphorus depletion could be measured at the root–soil interphase, in the hyphal compartment, and at the hyphae–soil interface. In nonmycorrhizal plants depletion zone extended not more than 1 cm from the rhizoplane but in the mycorrhizal plants phosphorus was uniformly depleted in the hyphal compartment (2 cm from the rhizoplane). At the hyphae–soil interface a new depletion zone was formed, extending several millimeters into the bulk soil. Of the total phosphorus the hyphae contributed between 70 and 80% in the mycorrhizal plants (Li *et al.*, 1991c). In mycorrhizal white clover with larger hyphal compartments phosphorus was uniformly depleted more than 11 cm from the rhizoplane (Li *et al.*, 1991a).

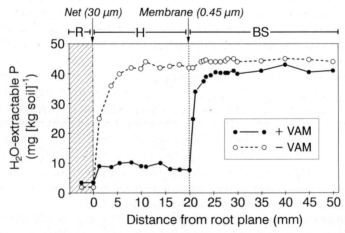

Fig. 15.22 Depletion profile of water-extractable phosphorus in the root (R), hyphal (H), and bulk soil (BS) compartment of nonmycorrhizal (−VAM) and mycorrhizal (*Glomus mosseae*, +VAM) white clover plants grown in a Luvisol. (Li *et al.*, 1991c.)

In VAM plants uptake of phosphorus is normally from the same labile pool from which the roots and thus, also nonmycorrhizal plants take up phosphorus (Bolan, 1991), and the greatest benefit for the host plant is achieved with supply of the least soluble source of phosphorus, for example, crystalline iron phosphate (Bolan *et al.*, 1987) or sparingly soluble organic phosphate sources like RNA (Jayachandran *et al.*, 1992) or phytate (Tarafdar and Marschner, 1994). In calcareous soils the higher rates of respiration (CO_2 production) of mycorrhizal compared to nonmycorrhizal roots increase the solubility of sparingly soluble calcium phosphates and might thereby also increase the effectivity in phosphorus acquisition (Knight *et al.*, 1989).

The high effectivity of VAM hyphae in phosphorus uptake is not only caused by their small diameter and large surface area, but also the accumulation of polyphosphates (Poly-P) in their vacuoles where it serves storage functions (Smith and Gianninazzi-Pearson, 1988) and in terms of energy as an alternative to ATP (Jennings, 1989). Polyphosphates are presumably also involved in transport of phosphate in the hyphae to the infected root where it is hydrolyzed in the arbuscules and most likely transported as inorganic phosphate (P_i) across the plasma membrane of the host root cell (Smith and Gianninazzi-Pearson, 1988). Polyphosphates are strongly negative polyanions and also serve important functions in the cation/anion balance of the fungus, and by binding cations such as magnesium, potassium and basic amino acids such as arginine and glutamine also as carriers for hyphal transport of these solutes to the host root cell (Jennings, 1989). The solute transport in the hyphae is bidirectional, carbohydrates versus phosphates and other mineral elements, and the plasma streaming is likely the driving force for this transport (Smith and Gianninazzi-Pearson, 1988), although in ectomycorrhizas other mechanisms might also be involved (Section 15.7.2).

The effectivity of VAM fungi in providing phosphorus to the host plants depends very much on the VAM species (Table 15.14). Compared with *Gl. macrocarpum*, the root colonization of the other two species was not only lower but they either did not provide any, or only small amounts of phosphorus to the host plant. This poor effectivity cannot properly be explained by lower root colonization but might be related to poor development and activity of the external hyphae, low hyphal transport rates, and poor solute interchange at the arbuscule–host root cell interface. Comparable differences in effectivity may even occur between ecotypes of the same VAM species *Glomus*

Table 15.14

Effect of Different VAM Species (*Glomus* sp.) on Root Colonization, Dry Weight and Phosphorus Uptake of *Sorghum bicolor* Grown for 48 Days at 25°C[a]

VAM species	Root colonization		Dry wt (g per plant)		P content
	Percentage	Length (m per plant)	Shoot	Roots	(mg per plant)
Control (-VAM)	0	0	0.46	0.25	0.29
Gl. macrocarpum	58	189.5	5.27	5.77	5.86
Gl. intraradices	27	7.1	0.45	0.32	0.30
Gl. fasciculatum	28	19.2	1.10	0.87	0.74

[a]Based on Raju *et al.* (1990).

mosseae, for example (Bethlenfalvay *et al.*, 1989). These examples demonstrate that percentage of root colonization only is a poor parameter of VAM effectivity in terms of acquisition of phosphorus, but also other mineral nutrients.

The diameter of external VAM hyphae is in the range 1–12 μm, and large differences in diameter exist between VAM species (Abbott and Robson, 1985). Also the total length of the external hyphae, which may be in the range 1–10 m cm^{-1} infected root length, differs much between VAM species as well as spread of the hyphae in the soil and their uptake rate of phosphorus per unit hyphal length (Jakobsen *et al.*, 1992). Only a proportion of the external hyphae is active, and this proportion often declines with distance from the root surface (Sylvia, 1988).

Similarly to the host roots the external hyphae of VAM fungi also possess acid phosphatase activity (Fig. 15.23) and, thus, also have access to organically bound phosphorus in their 'hyphosphere' (Tarafdar and Marschner, 1994).

In VAM plants the uptake and contents of zinc and copper are also usually distinctly higher than in nonmycorrhizal plants (Kothari *et al.*, 1991b; Lambert and Weidensaul, 1991). The capacity of the external hyphae for delivery of copper and zinc is high and might account for about 50–60% of the total uptake in white clover and 25% in maize (Fig. 15.24). By varying the phosphorus supply in the hyphal compartment the molar ratio of P/Cu transport in the hyphae could be varied by a factor of about 25, indicating that hyphal uptake and/or transport of both mineral nutrients are separately regulated (Li *et al.*, 1991b).

In agreement with the high capacity in hyphal delivery of zinc and copper, as a rule in VAM plants the shoot contents not only of phosphorus but also of zinc and copper are higher compared with nonmycorrhizal plants (Table 15.15). Increasing phosphorus in the soil is associated with a decrease in VAM colonization of the roots, or of hyphal length and activity, and is usually compensated for by higher root uptake of phosphorus. This is not necessarily so for zinc and copper in soils with low contents of these micronutrients. Accordingly, depressing effects of phosphorus fertilizer application on

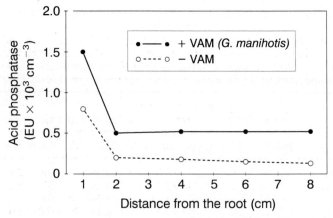

Fig. 15.23   Acid phosphatase activity in the rhizosphere of mycorrhizal and nonmycorrhizal wheat plants. (Tarafdar and Marschner, 1994.)

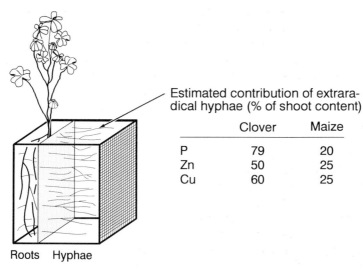

Estimated contribution of extraradical hyphae (% of shoot content)

	Clover	Maize
P	79	20
Zn	50	25
Cu	60	25

Roots   Hyphae

Fig. 15.24   Contribution of extraradical hyphae (*Glomus mosseae*) to the uptake of phosphorus, zinc and copper in white clover and in maize plants grown in a Luvisol in compartmented boxes. (Compiled data of Kothari *et al.*, 1991b and Li *et al.*, 1991b.)

plant content of zinc and copper, which are often reported in the literature and which by far exceed 'dilution effects' by growth indicate the importance of VAM in acquisition of zinc and copper from soils (Table 15.15).

In contrast to zinc and copper the shoot contents of manganese are often much lower in VAM plants (Table 15.15). In red clover there is a distinct negative correlation between percentage of root colonization with VAM and the manganese content in roots and shoots (Arines *et al.*, 1989).

Table 15.15
Effect of Increasing Phosphorus Fertilizer Supply on Shoot Growth and Shoot Contents of Mineral Nutrients in Nonmycorrhizal (NM) and Mycorrhizal (M; *Glomus fasciculatum*) Soybean[a]

P supply (mg kg^{-1} soil)	Shoot dry weight (g per plant)		Contents per g shoot dry matter							
			P (mg)		Cu (μg)		Zn (μg)		Mn (μg)	
	NM	M	NM	M	NM	M	NM	M	NM	M
0	1.25	2.80	0.61	1.73	3.3	10.3	21	44	366	111
60	1.61	3.21	0.75	2.09	3.7	7.9	27	35	515	109
150	1.85	3.42	0.81	2.08	2.9	6.3	30	36	412	115
270	2.78	3.83	1.40	1.79	3.5	4.6	29	33	556	123

[a]Based on Lambert and Weidensaul (1991).

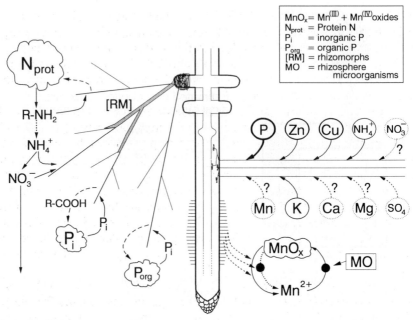

Fig. 15.25 Schematic presentation of components of the nutrient dynamics in and acquisition from the 'hyphosphere' of endo- (VA-) mycorrhizal roots and of additional components found in ectomycorrhizal roots. (Marschner and Dell, 1994.) Reprinted by permission of Kluwer Academic Publishers.

    The decrease in manganese uptake in mycorrhizal plants not only suggests a lack of substantial uptake and transport of manganese in the external hyphae but an additional effect of VAM on manganese acquisition by the roots. A direct inhibitory effect is, however, unlikely, the decrease in manganese contents is most probably caused by VAM-induced changes in rhizosphere microorganisms in general and decrease in population of manganese reducers in particular. These relationships are shown schematically in Fig. 15.25. In VAM maize plants not only the manganese contents in shoots and roots were lower compared with nonmycorrhizal plants but so also was the number of manganese reducing bacteria and amount of exchangeable manganese (Mn^{2+}) in the rhizosphere soil (Kothari *et al.*, 1991a). In red clover, lower manganese contents in roots and shoots of mycorrhizal plants were associated with higher numbers of manganese oxidizing bacteria (Arines *et al.*, 1992) which would cause a similar depression in manganese availability in the rhizosphere as a decrease in manganese-reducing bacteria.

    Not much is known about the role of VAM in uptake of potassium, magnesium and sulfur (Fig. 15.25). In *Agropyron repens* about 10% of the total potassium in mycorrhizal plants was attributable to hyphal uptake and delivery (George *et al.*, 1992). Although hyphal transport has been demonstrated for sulfur and calcium by using radioisotopes, the amounts transported are probably very small, at least for calcium as indicated by the frequently reported distinctly lower calcium contents in shoots of mycorrhizal plants (Kothari *et al.*, 1990a,b; Azcon and Barea, 1992). Lower calcium contents in mycorrhizal plants are probably related to changes in root morphology and differentiation, for example enhanced lignification and suberinization of the endodermis upon VAM infection (Dehne and Schönbeck, 1979a,b). In agreement with this in mycorrhizal maize plants the silicon contents in the shoots were also lower than in nonmycorrhizal plants (Kothari *et al.*, 1990b).

Compared with phosphorus, there is little information on the role of VAM in nitrogen acquisition – at least in nonlegumes – although both natural and agricultural ecosystems are often limited by nitrogen, rather than phosphorus. In celery about 20% of the total nitrogen uptake was attributable to hyphal uptake and delivery to the host plant (Ames *et al.*, 1983), and in *Agropyron repens* this proportion was about 31% (George *et al.*, 1992). High transport rates of reduced nitrogen in the hyphae in the form of arginine or glutamine together with polyphosphates are likely. But even at similar capacity for uptake and delivery – on a molar basis – of phosphorus, nitrogen and potassium by the hyphae, because of the much higher total demand by the host plant, the proportion of potassium and nitrogen contributed by external hyphae should remain relatively low compared with phosphorus.

A major problem in evaluation and quantification of the role of VAM in mineral nutrition of plants arises from the simultaneous changes in growth, and particularly root morphology and physiology, brought about by mycorrhizal colonization. As summarized in Table 15.16 for maize, at similar shoot and root dry weight the root surface area was much lower in mycorrhizal plants compared with nonmycorrhizal plants. As the mycorrhizal plants had a larger leaf blade area (Kothari *et al.*, 1990a) and also a higher photosynthate demand (and thus lower stomatal resistance) transpiration rates were

Table 15.16
Dry Weight, Water Relations and Nutrient Contents in Nonmycorrhizal and Mycorrhizal (*Gl. mosseae*) Maize (*Zea mays*) Grown in Calcareous Soil with Root and Hyphal Compartments[a]

Growth and water relations

	Dry weight (g per plant)		Root length (m per plant)	Root hair No. (per mm)	Root hair Length (μm)	Transpiration (l per plant (42 d)$^{-1}$)	Water uptake ((ml cm^{-1} root s^{-1}) $\times$ 10^7)
	Shoot	Root					
−VAM	20.0	4.8	619	35	347	3.40	0.61
+VAM	22.8	4.6	367	25	235	4.08	1.34

Mineral nutrients

Contents in shoot dry matter

	(mg g^{-1})				(μg g^{-1})					Mn reducer (10^5 g^{-1} soil)
	K	P	Mg	Ca	Zn	Cu	Mn	Fe	B	
−VAM	17	2.1	4.0	9.0	10	5.6	139	88	46	44.1
+VAM	12	3.7	4.1	5.3	36	7.1	95	58	35	1.7

[a]Compiled data of Kothari *et al.* (1990a,b, 1991a).

higher and also water uptake rates per unit root length and the rates of mass flow to the root surface. The lower potassium content in the shoots of mycorrhizal plants is in accordance with the reduction in root surface area and the relatively low hyphal transport of potassium. The contents of phosphorus, zinc and copper are much higher, that of magnesium not affected and of calcium much lower. The content of manganese is much lower in mycorrhizal plants and in accordance with the much lower number of manganese reducing bacteria in the rhizosphere. The amounts of iron and boron are lower in mycorrhizal plants suggesting that at least hyphal uptake and transport of these two micronutrients is small or absent.

In legumes growing in phosphorus-deficient soils, VAM enhances nodulation, N_2 fixation and host plant growth (Table 15.17). In view of the high phosphorus require-ment for nodulation a high VAM dependency is to be expected under these conditions. However, *Rhizobium* symbiosis imposes a strong sink for photosynthates, and mycor-rhizal colonization adds a new sink of similar size. This sink competition is reflected not only in a decrease in root/shoot dry weight ratio but also in a lower nitrogenase activity of the nodules compared with the phosphorus-sufficient, nonmycorrhizal plants.

The existence of extraradical VAM hyphal bridges between individual plants of the same, or different plant species in mixed stands is a potential pathway of nutrient transfer between plants. In principle, such transfer is also possible for nitrogen between legumes and nonlegumes in a mixed stand, or intercropped, as shown in Fig. 15.26 for soybean and maize. However, a substantial nitrogen transfer from the legume to maize took place only when the legume was supplied with mineral nitrogen, not when relying on N_2 fixation. In view of the high carbon costs for N_2 fixation, it is not surprising that legumes have mechanisms to prevent the drain of fixed nitrogen, via VAM hyphae, to the nonlegume. Also in field-grown soybean intercropped with maize, a direct transfer of fixed nitrogen from the soybean to maize via VAM hyphae was negligible (Hamel and Smith, 1992). Slightly higher nitrogen contents in mycorrhizal maize are presum-ably related to low microbial activity in the rhizosphere and thus less nitrogen is sequestered compared with nonmycorrhizal plants (Hamel *et al.*, 1991).

In communities of wild species transfer via VAM hyphae of phosphorus between different individual plants (e.g., large and small) exists but it is very slow and not of

Table 15.17

Plant Dry Weight and Phosphorus Content in Leaves, Number of Nodules and Nitrogenase Activity (ARA) in Nodules of Soybean Grown at Low and High Phosphorus Supply[a]

	Low P	High P	Low P+VAM[b]
Shoot dry weight (g)	2.8	3.8	5.6
Root dry weight (g)	1.7	1.9	2.0
P content (mg per plant)	2.9	6.0	5.8
Nodules (no. per plant)	33	30	97
ARA (μmol C_2H_4 per plant h^{-1})	4.6	22.8	9.0

[a]Brown *et al.* (1988).
[b]*Glomus mosseae.*

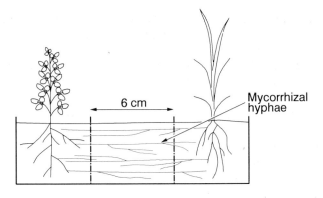

N supply to soybean	Dry weight (g plant^{-1})		N content (mg plant^{-1})	
	Soybean	Maize	Soybean	Maize
−N	3.9	7.2	30	33
+NH$_4$NO$_3$	21.8	8.6	351	55
+N$_2$ fix	25.1	6.9	419	40

Fig. 15.26   Dry weight and nitrogen uptake of soil-grown VA-mycorrhizal (*Glomus mosseae*) maize and soybean plants grown either without nitrogen (−N), with NH$_4$NO$_3$, or nodulated (N$_2$ fix). (Based on Bethlenfalvay *et al.*, 1991.)

ecological importance; this seems to be different for nitrogen, particularly from dying roots to living roots of neighboring plants (Newman *et al.*, 1992).

So far most of the enhancement effects of VAM on host plant growth and mineral nutrient uptake have been obtained under controlled environmental conditions, and usually optimized for the fungus. The results demonstrate the potential of the mycorrhizal association for host plant growth. Under field conditions the realization of this potential might be severely restricted, for example, by soil collembola grazing on the

Table 15.18

Effect of VA Mycorrhiza Inoculation on Shoot Growth, Seed Yield and Harvest Index of Chickpeas Grown on Fumigated Field Soil in Northern Syria[a]

Treatment	Flowering stage		Maturity		Harvest index (%)
	Shoot dry wt (g per plant)	Shoot content (mg P per plant)	Shoot dry wt (g per plant)	Seed yield (g per plant)	
Low VAM infection (fumigated soil)	2.7	4.4	6.0	2.5	41
High VAM infection (reinoculated)[b]	4.9	9.3	6.9	1.9	27

[a]Weber *et al.* (1993).
[b]Indigenous VAM.

external mycelium and thus decreasing the absorbing surface area (McGonigle and Fitter, 1988). Another limitation is shown in Table 15.18. Suppression of the indigenous VAM flora by soil fumigation limited phosphorus uptake and depressed shoot dry weight until flowering. At maturity, however, shoot dry weight was similar in both treatments but seed yield and therefore harvest index distinctly lower in the plants with high mycorrhizal colonization. Additional measurements suggested that in the plants with high mycorrhizal colonization, the more vigorous shoot growth in combination with a more shallow root system and a lower root length/shoot weight ratio caused a more severe drought stress during reproductive growth.

15.7.2   Ectomycorrhizas

With respect to their role in mineral nutrition of their host plant, ectomycorrhizal (ECM) fungi have many features in common with VAM. In addition, there are some principal differences in terms of structural arrangements with the roots and mechanism of nutrient acquisition (Fig. 15.25). In ECM plants such as Norway spruce, more than 90% of the root apical zones may be enclosed by a fungal sheath. On the other hand, in some broadleafed species, such as *Eucalyptus*, this proportion may not exceed 40–50%. Thus, depending on the tree species, and on root growth rate and season of the year, a varied proportion of the root supply of mineral nutrients from the soil may occur via the fungal hyphae of the extramatrical mycelium and the sheath. However, there is a wide variation in the thickness of the sheath (Agerer, 1987) and the hydraulic resistance to solute flow. The fungal sheath may be more or less sealed and prevent an apoplasmic route of solute and water flux into the root cortex, for example, in *Eucalyptus* with the ECM fungus *Pisolithus tinctorius* (Ashford *et al.*, 1989), whereas it provides a fairly unrestricted apolasmic route in *Pinus sylvestris* with the ECM fungi *Suillus bovinus* (Behrmann and Heyser, 1992).

   Also the extent of the extramatrical mycelium varies much between ECM species, an average of 1–3 m cm^{-1} infected root length has been found in *Salix* seedlings (Jones *et al.*, 1990). However, in contrast to VAM, in many ECM fungal species rhizomorphs are formed (Fig. 15.25) and act as main routes for the bidirectional solute transport. Their large diameter (~100 μm) and 'empty' sections in the centre may also allow rapid apoplasmic solute transport for some distances, although also in rhizomorphs solute transport driven by plasma streaming and concentration gradients is the rule (Jennings, 1987; Cairney, 1992). Similarly to VAM, in ECM hyphae fairly large amounts of polyphosphates are formed (Orlovich *et al.*, 1989), even at limited external phosphorus supply (McFall *et al.*, 1992). In ECM hyphae polyphosphates are present in the vacuoles mainly in soluble form (Orlovich and Ashford, 1993). In *Pisolithus tinctorius* the vacuoles are motile and interconnected with tubular elements, and transport of solutes (including polyphosphates and associated cations) seems to occur by peristaltic movements along the hyphae from cell to cell and independent of the plasma streaming (Shepherd *et al.*, 1993).

   Striking similarities exist between VAM and ECM not only in their importance for phosphorus nutrition of the host plant but also inflow rates of phosphorus and responses to increasing external phosphorus supply (Tinker *et al.*, 1992). At low external supply ECM infection strongly stimulates host plant growth by enhanced phosphorus acquisi-

Table 15.19

Growth, Phosphorus Uptake and Ectomycorrhizal Root Length in *Eucalyptus diversicolor* Seedlings Inoculated with *Laccaria laccata*[a]

Treatment (mg P kg^{-1} soil)	ECM +/−	Dry wt. (g per plant)	P content (mg per plant)	P uptake (mg g^{-1} fine root)	ECM root length (m per plant)
0	−	0.09	0.02	0.38	—
	+	0.16	0.07	0.74	0.25
8	−	0.32	1.73	0.58	—
	+	2.22	2.41	2.17	4.10
16	−	2.46	2.03	1.42	—
	+	3.46	4.26	2.14	4.71
32	−	8.58	10.56	3.75	—
	+	8.69	11.57	3.59	0.90

[a]Based on Bougher *et al.* (1990).

tion, and at high external supply this stimulation disappears (Table 15.19). In this case, high phosphorus supply strongly suppresses mycorrhizal colonization and therefore prevents growth depression in mycorrhizal plants at high phosphorus supply, which often occurs in other ECM–host plant associations (Jones *et al.*, 1990; Browning and Whitney, 1992).

Quantitative data on the delivery of mineral nutrients such as potassium, magnesium or of micronutrients via the external mycelium of ECM to the host plant are scarce, except for zinc (see below). However, it can be assumed that the potential is sufficient to provide all these mineral nutrients in amounts which can cover at least a large proportion of the host plant demand for growth.

The presence of acid phosphatase as ectoenzyme of ECM fungi (Fig. 15.25) is well established, its activity being high all along the external mycelium (Dinkelaker and Marschner, 1992) and at the surface of the sheathed roots (Gourp and Pargney, 1991). As VAM fungi also possess acid phosphatase as an ectoenzyme (Fig. 15.23), the capacity of utilization of P_{org} is therefore not unique for ECM but a common property of root systems of mycorrhizal as well as nonmycorrhizal plants. Some ericoid fungi such as *Hymenoscyphus ericae* produce siderophores (Schuler and Haselwandter, 1988) and thereby increase the iron acquisition and iron content in the shoot of the host plant (*Calluna vulgaris*) when grown on substrates with low iron contents (Shaw *et al.*, 1990) including calcareous soils. The tolerance of this calcifuge plant species to 'lime-induced chlorosis' is thereby increased (Leake *et al.*, 1990). Siderophore production is probably also involved in the enhanced weathering of goethite by the ECM fungus *Suillus granulus* (Watteau and Berthelin, 1990).

In contrast to VAM fungi, several ECM fungi have a considerable capacity for producing and excreting organic acids. These acids, and perhaps also siderophores, are presumably contributing factors in enhancing weathering of mica in the substrate of ECM as compared with nonmycorrhizal pine plants (Leyval and Berthelin, 1991). Some ECM fungi such as *Paxillus involutus* release large amounts of oxalic acid, particularly when supplied with nitrate nitrogen (Lapeyrie *et al.*, 1987). Oxalic acid dissolves sparingly soluble calcium phosphates, and for example when *Eucalyptus* seedlings are grown in calcareous soils, large amounts of calcium oxalate crystals cover the extramatrical mycelium and the hyphal sheath of mycorrhizal roots (Lapeyrie *et al.*, 1990). Production of siderophores for iron acquisition and of oxalic acid to bring calcium phosphates into solution, and precipitate calcium oxalate to restrict calcium uptake are suggested as coordinated mechanisms by which certain ECM fungi play a key role in adaptation of their host plant to calcareous soils (Lapeyrie, 1990). However, nonmycorrhizal roots, for example of Norway spruce, also form considerable amounts of calcium oxalate in the apoplasm of the cortex (Section 2.2.1).

Similarly to their host plant, ECM fungi prefer ammonium compared with nitrate as a source of nitrogen (Plassard *et al.*, 1991). Accordingly, when both ammonium and nitrate are supplied (e.g., NH_4NO_3) ECM fungi acidify their substrate similarly to the host roots (Section 8.2.4). There is not much storage capacity for nitrate in ECM fungi. However, a number of ECM fungi can efficiently reduce nitrate, and their nitrate reductase activity is in a similar order of magnitude to that in higher plants, and these fungi increase their substrate pH when supplied with nitrate nitrogen. Therefore, in ECM trees such as pine species there are no major differences in nitrogen uptake and assimilation between mycorrhizal and nonmycorrhizal parts of the root system (Plassard *et al.*, 1991).

After uptake of ammonium, or nitrate reduction, in the cells of the extramatrical mycelium and the fungal sheath, the ammonium is incorporated into glutamate and glutamine by the action of glutamate dehydrogenase (GDH) and glutamine synthase (GS), respectively (Fig. 15.27). This key role of GDH in ECM fungi is in contrast to higher plants where ammonium assimilation occurs via the glutamate synthase cycle involving the sequential action of GS and glutamate synthase (GOGAT; Fig. 8.9) and where GDH plays a minor role, except at very high ammonia concentrations (Section 8.2). In the extramatrical hyphae of ECM, after incorporation of the ammonia into glutamine, transport takes place to the sheath (Fig. 15.27). In the sheath and the Hartig net GOGAT may also become important in ammonia assimilation of some ECM (Martin *et al.*, 1992).

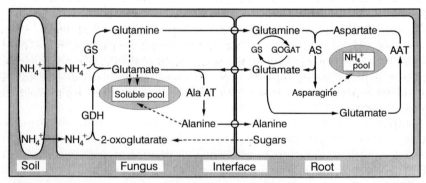

Fig. 15.27   Proposed scheme for nitrogen assimilation in Norway spruce ectomycorrhiza and of the localization of nitrogen-assimilating enzymes in the fungus and host cells. GDH, glutamate dehydrogenase; GS, glutamine synthase; GOGAT, glutamate synthase. (Chalot *et al.*, 1991.)

Table 15.20
Nitrogen Content in *Pinus contorta* Seedlings either Nonmycorrhizal or
Infected with *Suillus bovinus*, and *Pisolithus tinctorius* and Supplied
with Ammonium or Protein as Source of Nitrogen[a]

	NH$_4$-N	Protein-N
Treatment	Nitrogen content (mg per plant)[b]	
Nonmycorrhizal	3.66	1.14
Suillus bovinus	4.05	3.20
Pisolithus tinctorius	3.27	1.30

[a]Based on Abuzinadah *et al.* (1986).
[b]Seed content and starter nitrogen: 1.6 mg N per flask.

The extent to which inorganic nitrogen is assimilated in the fungal cells or passes the sheath to be assimilated in the host root cells is unclear and may depend on the relative enzyme activities, carbohydrate supply, and thickness of the sheath.

As a peculiarity, and of particular ecological importance, some ericoid and ECM fungi produce acid proteinases as ectoenzymes and thereby provide access for the host plant to complex organic sources of nitrogen such as protein (Fig. 15.25). Since the host plants themselves have little or no access to these resources their fungal associate may play a crucial role for host plant growth on substrates with complex organic nitrogen. As shown in Table 15.20, in contrast to nonmycorrhizal pine seedlings, seedlings in association with the ECM fungus *Suillus bovinus* can readily utilize nitrogen from protein sources, similarly to plants provided with ammonium nitrogen. However, Table 15.20 also shows that the capacity of the ECM fungi for utilization of protein nitrogen is confined to certain species. The number of those ECM fungal species which have access to protein as nitrogen source appears to be relatively small (Hutchinson, 1990), *Paxillus involutus* seems to be another candidate (Finlay *et al.*, 1992).

The higher proportion of ECM, compared with VAM, in deciduous woodlands poor in both phosphorus and nitrogen (Högberg, 1986), and the dominance of ECM in coniferous forest ecosystems of the northern hemisphere with nitrogen as the most limiting nutrient (unless atmospheric input is high, Section 4.4.2) might be causally related to the capability of some ECM to utilize complex organic nitrogen (Högberg, 1990). Such a capability is important to shorten and tighten the nitrogen cycle thereby minimizing nitrogen losses from the soil by leaching, and by gaseous losses, and simultaneously to decrease competition for nitrogen from other soil microorganisms (Vogt *et al.*, 1991).

15.8   Role of Mycorrhizas in Heavy Metal Tolerance

A relatively large number of ECM fungi are effective in increasing heavy metal tolerance of their host plant (Wilkins, 1991; Colpaert and van Assche, 1993). For

Table 15.21

Shoot and Root Contents of Zinc in Nonmycorrhizal and Ectomycorrhizal Seedlings of *Pinus sylvestris* Supplied with High Zinc Concentrations[a]

Treatment (fungus)	Shoot dry wt. (g per plant)	Shoot zinc (mg per plant)	Shoot zinc (μg g^{-1}dry wt)	Fungal biomass (% of short roots)	Short root Zn content (μg g^{-1} dry wt)
Nonmycorrhizal	16.2	3.19	197	—	273
Paxillus involutus	14.3	1.52	106	54	708
Thelephora terrestris	16.2	3.89	240	66	309

[a]Based on Colpaert and Van Assche (1992).

example, in birch seedlings tolerance to high nickel concentrations in the substrate is increased by inoculation with the ECM fungi *Lactarius rufus* or *Scleroderma flavidum* (Jones and Hutchinson, 1988), or the tolerance to high zinc concentrations in the substrate by inoculation with *Paxillus involutus* (Denny and Wilkins, 1987). In *Pinus banksiana* seedlings, tolerance to various heavy metals (lead, nickel, zinc) can be increased by *Suillus luteus* at low and intermediate, but not at high external concentrations which are directly harmful to the fungus (Dixon and Buschena, 1988). In all these examples an increase in heavy metal tolerance was brought about by sequestering of the heavy metals in the fungal structures, either of the extramatrical mycelium or in the sheath (Fig. 15.25) and thereby decreasing the concentration of the heavy metals in the soil solution around the host root, in the roots, and particularly in the shoot tissue. As most heavy metals, and also aluminum, exert their toxic influence by damaging root apical zones, the protecting effects of sheath-forming ECM is evident.

The preferential binding of heavy metals may occur in the mucilage of the hyphal surface, in the fungal cell walls, or in vacuoles presumably associated with poly-P granules (Kottke, 1992) or granules high in nitrogen but low in phosphorus (Turnau *et al.*, 1993). The specific heavy metal binding capacity of the extramatrical mycelium, and its mass, are therefore important components determining the effectiveness of heavy metal retention in the ECM (Colpaert and van Assche, 1992).

As shown in Table 15.21, *Paxillus involutus* has a high zinc retention capacity in its mycelium and thereby effectively decreases zinc content in the host plant, compared with the nonmycorrhizal plants. In contrast, despite a similar fungal biomass, *Thelephora terrestris* hardly retains zinc in its structures, and even further increases the zinc content in the host plant. This and many other examples in the literature demonstrate that it is not possible to generalize concerning increases in heavy metal tolerance in ECM plants. In ECM fungi with high zinc retention capacity this property is also retained in the low concentration range, i.e. not a specific mechanism according to the host plant demand of zinc (Colpaert and van Assche, 1992). In contrast, in some ericoid mycorrhizal fungi the transfer of iron to the host plant seems to be in some accordance with the host plant demand of iron (Shaw *et al.*, 1990).

Some ECM fungi may also increase the aluminum tolerance of the host plant. For example, in *Pinus rigida* even at 50 μM aluminum in the substrate growth is reduced and there is a marked increase in aluminum content in the needles, whereas in the mycorrhizal (*Pisolithus tinctorius*) plants even at 200 μM aluminum growth is not affected and aluminum content in the needles remains relatively low (Cumming and Weinstein, 1990). The ameliorating effect of ECM in *P. rigida* was brought about in part by improved phosphorus nutrition of the host plant. In other instances binding of aluminum to fungal structures, and particularly complexing the aluminum, and thereby detoxifying it at the fungus–host interface and in the root cortex may be involved. Therefore, lack of differences in aluminum distribution between nonmycorrhizal and ECM root tissues (Jentschke *et al.*, 1991) are not necessarily indications of ineffectiveness of the ECM in aluminum detoxification.

In contrast to ECM there are only a few reports on the effect of VAM on heavy metal tolerance of the host plant. In view of the principal differences in size and structural arrangement of the extraradical mycelium of the two types of mycorrhizae (Fig. 15.25), and the main mechanism of ECM in heavy metal tolerance, corresponding direct ameliorating effects of VAM are exceptions. Indirect effects may occur, however, for example, by improving phosphorus nutritional status and growth of the host plant on a phosphorus-deficient soil high in heavy metals or aluminum, i.e. by a dilution effect. A more specific effect is the alleviation of manganese toxicity by depression of manganese uptake (see above). Additional effects of VAM in increasing the tissue tolerance against high manganese concentrations may also play a role (Bethlenfalvay and Franson, 1989).

On the other hand, VAM may enhance zinc toxicity of the host plant. The high effectivity of VAM in the acquisition and delivery of zinc to the host plant is also retained when plants are grown at high external zinc supply (Schuepp *et al.*, 1987; Symeonidis, 1990). However, more cadmium was retained in roots of VAM plants leading to an increase in the Zn/Cd selectivity ratio in shoot transfer, compared with nonmycorrhizal plants (Schuepp *et al.*, 1987). The increase in copper tolerance in plants infected with VAM (Gildon and Tinker, 1983a,b) is probably related to a high retention of copper in the intraradical fungal mycelium in the host roots (Li *et al.*, 1991b).

15.9 Mycorrhizal Responsiveness

A major beneficial effect of mycorrhization on host plant growth is brought about by the increase in belowground combined surface area (roots and mycorrhizas) for acquisition of nutrients. The beneficial effect of mycorrhizas is therefore of special importance for those plants which have a coarse and poorly branched root system (Hetrick, 1991), and simultaneously lack specific root response mechanisms as, for example, cluster-rooted plants to phosphorus deficiency (Section 15.3 and 15.4). The beneficial effects of mycorrhizas on host plant growth is commonly regarded as *mycorrhizal dependency*, but in most instances the term *mycorrhizal responsiveness* would be more appropriate (Alexander, 1989).

In view of the abundance of low phosphorus soils and the particular role of VAM in enhancing phosphorus acquisition, most studies so far on mycorrhizal responsiveness have been focused on VAM and phosphorus. As in most soils roots are infected with indigenous VAM, studies on mycorrhizal responsiveness require soil sterilization (fumigation, steaming) and reinoculation with the indigenous soil microflora (MF), except VAM, and reinoculation with both MF and VAM. Growth response of plants is

Table 15.22
Effect of Suppressing and Reintroducing VAM Fungi on the Growth of
Plants[a]

Plant species	Soil treatment[b]		
	Nonfumigated	Fumigated	Fumigated–reinoculated
Carrot	8.5 (61)	0.4 (0)	7.4 (60)
Leek	4.4 (50)	0.4 (0)	4.0 (67)
Tomato	4.1 (61)	2.5 (0)	5.1 (90)
Wheat	2.0 (63)	1.7 (0)	2.1 (79)
Cabbage	11.9 (0)	14.2 (0)	13.6 (0)

[a]Plant growth expressed on a dry weight basis as grams per pot. From Plenchette *et al.* (1983).
[b]Values in parantheses indicate root colonization (percentage of total root) with VAM fungi.

then used as the parameter of mycorrhizal responsiveness. The results shown in Table 15.22 represent the range of mycorrhizal responsiveness of crop species grown in soils low in phosphorus. Elimination of VAM by soil fumigation elicited three kinds of growth responses. (a) Carrot and leek grew very poorly; growth was restored to about the level of growth in nonfumigated soil after reinoculation with VAM. (b) Tomato and wheat exhibited small or negligible growth responses despite high infection rates. (c) In cabbage, as a nonmycorrhizal plant species (member of Cruciferae) fumigation increased plant growth and inoculation with VAM had no further effect. The growth enhancement in cabbage by fumigation was presumably due to elimination of soil-borne pathogens, and omission of reinoculation with the indigenous MF together with VAM.

The results in Table 15.22 also suggest that one should not expect a large growth stimulation by inoculation of field-grown plants unless indigenous VAM fungi have been damaged, for example, by fungicides (Hale and Sanders, 1982; Khasa *et al.*, 1992). However, on severely phosphorus-deficient Oxisols in the tropics, growth enhancement in several crop species might be achieved by VAM inoculation even without soil sterilization (Howeler *et al.*, 1987; Sieverding, 1991).

Particularly coarse root systems are abundant in many woody species, and of the crop species in cassava (*Manihot esculentum*). Accordingly, in nonmycorrhizal cassava plants the critical deficiency level of extractable soil phosphorus is 190 mg, compared with only 15 mg in mycorrhizal plants (Fig. 15.28).

The importance of root morphology for mycorrhizal responsiveness in different plant species is shown in Fig. 15.29. In the grass with the large root surface area mycorrhizal responsiveness is absent even at extremely low soil phosphorus levels. In contrast in the legume with both short roots and short root hairs mycorrhizal responsiveness is high. By using five pasture species an inverse relationship has been found between root hair length and mycorrhizal dependency (Schweiger, 1994). In various legume species this inverse relationship between root hair length and mycorrhizal responsiveness is also indicated in the number of nodules (Crush, 1974).

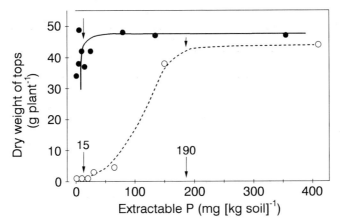

Fig. 15.28 Relationship between the dry weight of tops of inoculated (with VAM, •——•) and noninoculated (○– – –○) cassava and level of extractable phosphorus (soil testing method, Bray II) in sterilized soil. Arrows indicate the critical phosphorus levels corresponding to 95% of maximal yield. (Redrawn from Howeler *et al.*, 1982a.)

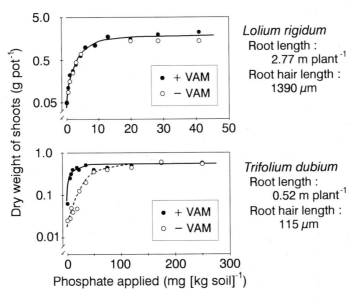

Fig. 15.29 Relationship between root morphology and mycorrhizal (*Glomus* sp.) benefits in phosphorus acquisition in two pasture species. (Based on Schweiger, 1994.)

There also seems to be a typical difference in VAM responsiveness between C_3 and C_4 grasses (Hetrick *et al.*, 1990). Cool-season C_3 grasses have highly fibrous root systems whereas warm-season C_4 grasses have coarser root systems. In C_4 grasses, but not in C_3 grasses, there was a strong positive growth response either to phosphorus fertilization or to VAM inoculation. However, in C_4 grasses large differences also exist between species in VAM responsiveness (Howeler *et al.*, 1987; Sieverding, 1991).

There is a tendency for a given plant species for wild ancestors to be less responsive to VAM than the cultivars, for example, in oat (Koide *et al.*, 1988) and tomato (Bryla and Koide, 1990a,b), partly due to differences in root morphology and root/shoot dry weight ratio, but also in growth rate and growth potential. Inherent differences in the latter parameters are often overlooked in comparisons between species in VAM responsiveness. Seed size, and thus seed reserves in phosphorus, as well as in other nutrients, is another important factor for VAM responsiveness. In a comparison of 15 wild species grown on a phosphorus-deficient soil a clear negative correlation has been found between VAM responsiveness and seed size (Allsopp and Stock, 1992).

There are some indications that VAM dependency of a given species differs whether grown in pure or mixed stands with other plant species. Legumes such as *Medicago sativa* and *Phaseolus vulgaris* appear to be more competitive in mixed stands when inoculated with VAM (Barea *et al.*, 1989; Reeves, 1992). This might be attributed to improved phosphorus nutritional status of the legume or, as observed in mixed stands of nonlegumes, an expression of general increase in 'fitness' by VAM (Carey *et al.*, 1992).

An absolute mycorrhizal dependency exists in orchids (Section 5.7.2), and many woody and forest tree species are highly responsive to mycorrhizas, an example for ECM has been given in Table 15.19. The same also holds true for ericoid mycorrhizal plants such as *Calluna*. In many natural ecosystems the dependency on ECM or ericoid mycorrhizas might mainly be a matter of nitrogen and not so much phosphorus availability, but systematic studies on this topic are scarce. In forest trees ECM dependency may also depend on plant age. In forest stands ECM hyphae can act as a conduit for photosynthate transfer from overstorey plants to those shaded by these plants, i.e. exert some kind of nursery influence for seedling establishment (Griffiths *et al.*, 1991).

15.10   Other Mycorrhizal Effects

15.10.1   Hormonal Effects, Plant Water Relations

Mycorrhizas may also alter host plant growth and development through direct and indirect effects on hormonal balance and plant water relations. In Section 14.5 hormonal effects of noninfecting rhizosphere microorganisms have been described, and some examples given for the production of hormones by ECM fungi. The typical morphological changes of ECM short roots (reduction or cessation of elongation growth, or branching) are brought about by IAA and C_2H_4 production by certain ECM fungi (Rupp and Mudge, 1985). Enhancement of shoot elongation growth of ECM (*Pisolithus tinctorius*) pine seedlings was closely related to the IAA production of the associated fungus (Frankenberger and Poth, 1987). However, not all ECM fungi produce hormones, and quite often poor or no correlations exist between the capability of an ECM fungus to produce a given hormone and the enhancing effect of the fungus on host plant growth. For example, in *Pinus taeda* inoculation with *Pisolithus tinctorius* enhances shoot growth and increases CYT content in the needles, although this fungus does not produce CYT, whereas inoculation with the CYT producer *Suillus punctipes* neither affects shoot growth nor CYT content in the needles (Wullschleger and Reid, 1990). Thus, depending on the ECM fungal species, hormones produced by these species, or their effects on synthesis of plant-borne hormones, altered hormonal balance in plants may, or may not, play a role in enhancing effects of ECM fungi on host plant growth.

There are also reports of higher CYT contents in plants inoculated with VAM, for example in flax (Drüge and Schönbeck, 1992) and in citrus (Dixon *et al.*, 1988). In maize inoculation with VAM increased the ABA contents twofold in roots and shoots compared with the nonmycorrhizal plants (Danneberg *et al.*, 1992). However, in all these cases the increase in hormone contents in the plants (per unit fresh or dry weight) was associated with distinct growth-enhancing effects by VAM inoculation, most probably via improved phosphorus nutritional status of the host plants. As the phosphorus nutritional status also affects plant hormone contents in nonmycorrhizal plants (Section 5.6.4), it is most likely that VAM inoculation affects host plant hormone contents indirectly.

Mycorrhizal colonization may also affect plant water relations directly or indirectly. Enhancement of water supply to the host plant is documented for ECM fungi which form extended extramatrical mycelium and rhizomorphs (Brownlee *et al.*, 1983). Rhizomorphs are very much suited to rapid and substantial water transport to the host plant (Lamhamedi and Fortin, 1991). In agreement with this in *Pinus pinaster* positive relationships have been found between the diameter of the rhizomorphs of the fungi, xylem water potential, and speed of recovery of the host plant after drought stress (Lamhamedi *et al.*, 1992).

An increase in drought stress tolerance has also been observed in VAM plants compared with nonmycorrhizal plants. Differences in phosphorus nutritional status of the plants (Section 8.4.6) might account in part for this effect. However, in plants with comparable phosphorus nutritional status and similar shoot size, plant water status may also differ between VAM and nonmycorrhizal plants (Augé and Stodola, 1990). In view of the effects of VAM on root morphology (root branching, number of apical meristems) and on root anatomy (e.g., lignification) changes in plant water balance may be indirectly the consequence of hormonal and structural changes in the host plant.

Water transport in the extraradical VAM hyphae to the host plant has been observed (Faber *et al.*, 1991). However, the small diameter of the hyphae ($\sim$2–14 μm) would require an unrealistically high mass flow of water towards the roots in order to make a substantial contribution to the transpirational demand of the host plant. Such mass flow of water can also not be reconciled with the necessity of bidirectional solute transport in the hyphae. It can also be shown experimentally (Fig. 15.30) that despite severe drought stress of the host plant there is no substantial water transport from the hyphal compartment through the hyphae to the host plant when a direct soil contact at the root–hyphal compartment interface is omitted. Despite this lack in significant water transport, hyphal uptake from the outer compartment accounted for 49% of the total

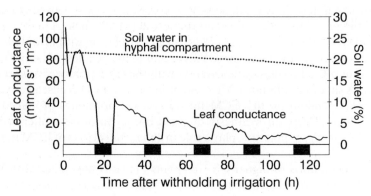

Fig. 15.30 Effect of withholding water supply on leaf stomatal conductance and soil water content in the hyphal compartment in a pot with mycorrhizal (*Glomus mosseae*) *Agropyron repens*. (E. George *et al.*, 1992.)

phosphorus and 35% of the total nitrogen taken up by the myorrhizal plants (George, E. *et al.*, 1992).

The more intensive soil water extraction often observed in VAM plants, compared with nonmycorrhizal plants, is most likely an indirect effect brought about by changes in soil structure by the extraradical hyphae (Fitter, 1985; Davies *et al.*, 1992) thereby increasing the unsaturated hydraulic conductivity of the soil. Such changes in soil structure presumably also contribute to the increase in water transport in ECM tree seedlings (Lamhamedi *et al.*, 1992).

15.10.2 Suppression of Root Pathogens

There is a long list of examples on suppression of soil-borne fungal and bacterial root pathogens by inoculation with mycorrhizas, VAM in particular. For example, inoculation with VAM fungi increases resistance of tomato to *Fusarium oxysporum* (Dehne and Schönbeck, 1979a), of casuarina to *Fusarium vesicubesum* (Gunjal and Paril, 1992), and of tomato to *Pseudomonas syringae* (Garcia-Garrido and Ocampo, 1989). This suppressing effect of VAM is also evident in cases of 'soil sickness', or 'replant disease', where minor pathogens or deleterious soil microorganisms may harm root growth and activity. An example of such a suppressing effect is shown in Table 15.23. The growth of grapevine seedlings was poor on soil with replant disease but could be considerably improved by inoculation with VAM which also elevated the level of root colonization with VAM to that in the control soil (without replant disease). Suppression of *Pseudomonas fluorescens* by VAM inoculation was presumably a main responsible factor for improvement of plant growth in the soil with replant disease. Soil sterilization was, however, more effective than VAM inoculation as it restored plant growth to the level in the control soil (Waschkies *et al.*, 1993).

In the northern wheat belt of Australia wheat root infection with common root rot (*Bipolaris sorokiniana*) has been found to be inversely related to root colonization with VAM (Thompson and Wildermuth, 1989). The differences in root colonization were

Table 15.23

Effect of Inoculation with *Glomus mosseae* on Growth and Number of Rhizoplane Bacteria in Grapevine (*Vitis vinifera*) Cuttings Growing in Soils without (Control) and with 'Replant Disease' (RPD)[a]

Soil treatment	Weight per plant		VAM root coloniz. (%)	Bacterial no. g^{-1} root fresh wt.	
	Shoot (g dry wt)	Root (g fresh wt)		Total × 10^7	Pseud.[b] × 10^5
Control					
−VAM	6.3	10.1	33	3.2	0.18
+VAM	6.2	12.5	39	3.7	0.16
RPD					
−VAM	1.3	3.6	21	4.4	5.88
+VAM	2.3	7.8	34	3.2	0.71

[a]Waschkies *et al.* (1994).
[b]*Pseudomonas fluorescens.*

caused by alterations in the VAM infection potential in the soils via the length of fallow periods, or relations between host/nonhost VAM plants in the crop rotation.

In several instances, suppression of root pathogens by VAM can be attributed to improved phosphorus nutritional status of the host plants (Perrin, 1990). This may also hold true for the zinc and copper nutritional status. However, in many cases specific interactions are involved, including changes in root exudation and rhizosphere microflora (see above), or in host root physiology (defense reactions) and anatomy such as lignification and suberization of the root tissue (Dehne and Schönbeck, 1979b).

In certain instances the VAM fungus can become parasitic, not only in terms of a carbon drain without a return of beneficial effects, but by directly damaging the roots. In tobacco, stunt disease is known to be controlled by soil fumigation. The same symptoms and growth depressions can be obtained by inoculation with *Glomus macrocarpum*, and the severity of the symptoms is correlated with the degree of root colonization with this VAM fungus (Modjo and Hendrix, 1986; Hendrix *et al.*, 1992). Stunt disease is less detrimental in plants inoculated with *Glomus microcarpum*, and absent in plants inoculated with *Glomus fasciculatum*. Less beneficial or even detrimental effects of certain *Glomus* species to crops in which they preferentially proliferate have also been found after long-term cropping history of maize or soybean monoculture (Johnson *et al.*, 1991, 1992).

Protection of the host plant from root pathogens is well documented for certain ECM fungi, an example being shown in Table 15.24 for *Paxillus involutus*. The ECM fungus can effectively suppress the harmful effects of *Fusarium oxysporum* on seedling growth. This suppression is achieved by production of oxalic acid by *Paxillus involutus*, and production of oxalic acid by the ECM fungus is enhanced by root exudates of *Pinus resinosa* (Duchesne *et al.*, 1989).

In ECM the mechanisms for protection of the host plants are more diverse than in VAM. Besides improving mineral nutritional status, changes in root exudation and noninfecting rhizosphere microflora, in ECM the fungal sheath can also act as a

Table 15.24
Suppression of Root Pathogens in *Pinus resinosa* Seedlings by *Paxillus involutus*[a]

Treatment	Seedling mortality (%)	Length (cm per plant)	
		Shoot	Root
Control	0	3.0	2.3
+ *Paxillus involutus*	0	3.0	2.5
+ *Fusarium oxysporum*	50	1.5	0.6
+ *P. involutus* + *F. oxysporum*	20	2.5	1.5

[a]Based on Chakravarty *et al.* (1991).

mechanical barrier (Perrin, 1990), or the fungi can produce phenolic compounds with strong inhibitory effects on various pathogenic fungi (Kope and Fortin, 1990).

15.11 Mycorrhizas: Practical Implications

Although higher plants may benefit from their associated mycorrhizas in most instances by improving their nutritional status, phosphorus in particular, other beneficial effects may occur and should not be overlooked. Diverse beneficial mycorrhizal effects can readily be demonstrated under controlled environmental conditions. Examples of the potential beneficial effects have been given in the preceding sections. Predictions on effects to be expected from inoculation with VAM or ECM require at least the consideration of the fungal species or strains. The major limitation for predictions, however, is our poor knowledge of the functioning of the associations under field conditions. More systematic studies are required, for example, on comparisons between mycorrhizal plant species and nonhost species under field conditions and in plant communities in order to evaluate more accurately the relative importance of the various potential beneficial effects of mycorrhizas on their host plants under given ecological conditions.

Despite the above limitations there are certain areas where it is feasible to inoculate with mycorrhizas on a commercial level. Many horticultural plants, and most of the fruit trees and forest trees are first established in seedbeds or maintained during early development in nurseries before transplanting to the field. Inoculation of forest trees with ECM in nurseries can substantially decrease transplantation shock and increase survival and growth rate in the field (Guehl and Garbaye, 1990; Villeneuve *et al.*, 1991). The same is true for VAM in reforestation of mining sites (Jasper *et al.*, 1989a) or for seedling production in fruit trees in nurseries (Menge, 1983).

For such technology use of ECM fungi is not very difficult as many of them can readily be multiplied in pure culture. This is different, however, for VAM fungi, particularly with respect to inoculation of field-grown plants. Inoculation of field-grown plants with VAM is rather limited for technical reasons, and also not very promising in most instances. The limitations are: lack or difficulties in the production of pathogen-free inoculum in sufficient quantities; poor knowledge on specific host plant (species, cultivar)/VAM fungi (species, strains) interactions; competition with indigenous VAM fungi; introduced VAM fungi have to be applied at high inoculum density because of competition (Sanders and Sheikh, 1983); soil sterilization on a large scale is neither possible nor to be recommended for obvious reasons.

So far, field results showing a clear yield response to inoculation with VAM fungi in nonsterilized soils are scanty (Kang *et al.*, 1980; Howeler *et al.*, 1987; Sieverding, 1991). To make use of the beneficial effects of VAM on crop plants it seems to be more promising under most circumstances to manipulate the infection potential of the indigenous VAM indirectly by soil management and crop rotation (Thompson and Wildermuth, 1989; Sattelmacher *et al.*, 1991). These efforts deserve more attention for economical reasons in low-input production systems, and for ecological reasons in both, low- and high-input production systems. For a comprehensive review of possibilities and limitations of using VAM in tropical agrosystems the reader is referred to Sieverding (1991).

Mycorrhizal colonization per se, and any marked change in VAM root colonization of field-grown plants have implications on soil testing and on simulation models for plant available mineral nutrients, and phosphorus in particular. The poor correlations between chemical soil analysis of phosphorus and plant response to phosphorus supply often found are in part due to differences in VAM root colonizations (Stribley *et al.*, 1981). Underestimation by simulation models of phosphorus uptake by plants from low phosphorus soils (Section 13.6) are certainly in part due to a high contribution of VAM to phosphorus delivery from outside the rhizo-cylinder of the host roots.

16

Adaptation of Plants to Adverse Chemical Soil Conditions

16.1 Natural Vegetation

Soil chemical factors such as pH, salinity, and nutrient availability determine the distribution of natural vegetation, i.e., of wild plants. Species and ecotypes can be classified in ecophysiological terms according to their distribution in soils. Some examples of species grouped in this way are acidophobes and acidophiles; calcifuges and calcicoles; halophytes and glycophytes; and metallophytes (adapted to metalliferous soils). Reviews have been published on ecophysiological aspects of plant responses, especially those of natural vegetation, to soil pH (Kinzel, 1983), salinity (Munns *et al.*, 1983; Läuchli and Schubert, 1989), heavy metals (Woolhouse, 1983; Baker, 1987; Baker and Walker, 1989b), and mineral nutrient availability (Chapin, 1980, 1983, 1988).

Crop species are usually selected for high soil fertility and their nutritional characteristics may be quite different from those of the natural vegetation grown on soils of low fertility, i.e., nutrient poor sites (Chapin, 1988). But also within the natural vegetation there are species adapted to high soil fertility (ruderal species) and with corresponding similarities in nutritional characteristics to crop species (Table 16.1).

Wild plants adapted to nutrient poor sites (Type I, slow growers) consistently have high root/shoot ratios. However, with the exception of some specialized plants (e.g., Proteaceae, Section 15.4.2) they are not more efficient in nutrient acquisition than ruderal species (Type II) or crop species. The most striking characteristic of Type I plants is their low maximum potential growth rate. They grow slowly even at high nutrient supply (Hommels *et al.*, 1989b) but store more nutrients which can be utilized when supply is limited (luxury consumption in periods of nutrient flushes). Plants adapted to high altitudes are also Type I plants (Körner, 1989). In contrast, in Type II plants growth rate is greatly reduced at periods of limited supply and visual deficiency symptoms can develop. These plants respond rapidly to high nutrient supply by enhanced rates of growth particularly of the shoots. Of the wild species the stinging nettle (*Urtica dioica* L.) is one such example of this type of plants (Fetene *et al.*, 1993; Fetene and Beck, 1993).

Wild plants of Type I usually have higher nutrient concentrations in their dry matter

Table 16.1

Characteristics of Strategies of Wild Plants in Adaptation to Soils with Low or High Nutrient Availability[a]

Nutrient availability	Type I (slow growers)	Type II (ruderal species)
Low (nutrient poor sites)	Low nutrient uptake rates Low growth rates of roots and shoots High root/shoot ratio High leaf longevity High nutrient concentrations in the tissue	Low growth rates Low nutrient storage High root/shoot ratio
High (nutrient rich sites)	Small growth response of roots and shoots High nutrient storage (luxury consumption) Low nutrient use efficiency	High nutrient uptake rate High growth rate High nutrient use efficiency Decrease in root/shoot ratio

[a]Based on Chapin (1980, 1988).

and thus, lower nutrient use efficiency (dry matter produced per unit nutrient in the dry matter) as in these species storage of nutrients is of key importance for survival and reproduction (Chapin, 1988). This strategy of wild plants is still retained in certain tropical root crop species, where in spite of large differences in yield brought about by variation in soil acidity (Section 16.3.6) the mineral element content of the leaves remains about the same (Abruna-Rodriguez *et al.*, 1982). Large seeds with correspondingly large storage of nutrients for seedling establishment on nutrient poor sites is another strategy found in certain Type I plants (i.e. Proteaceae; Stock *et al.*, 1990).

In adaptation to nitrogen limited ecosystems in cold climates specialized systems have been developed to utilize the organic soil nitrogen, either by hydrolysis of proteins in ectomycorrhizal forest trees (Section 15.7.2) or preferential uptake of amino acids as in the case of the nonmycorrhizal artic sedge (*Eriophorum vaginatum*). In its natural environment at least 60% of the nitrogen demand of this sedge may be met by amino acid uptake (Chapin *et al.*, 1993). This adaptation is also reflected in the much higher biomass production when amino acids rather than inorganic nitrogen are provided as nitrogen source (Table 16.2). This contrasts with the behavior of barley as a typical species adapted to nutrient rich sites.

Many natural ecosystems are also phosphorus-limited. Although VA mycorrhizas play a particularly important role in the acquisition of phosphorus at low soil phosphorus levels (Section 15.7.1) there are no principal differences between the wild and their related crop species in VAM dependency for phosphorus acquisition (Bryla and Koide, 1990a,b). The much higher phosphorus efficiency in wild oat (*Avena fatua* L.) compared with cultivated oat (*Avena sativa* L.) is caused by higher root mass, higher uptake rates both at low and high phosphorus supply, and also a higher phosphorus use efficiency in the shoots (Haynes *et al.*, 1991). As compared with Type I plants, in wild plants with high potential growth rates (Type II) not only are uptake and utilization

Table 16.2

Biomass Production of *Eriophorum vaginatum* and *Hordeum vulgare* Supplied with Different Forms of Nitrogen in Nutrient Solution Culture[a]

Nitrogen form	Biomass (mg per plant)	
	Eriophorum	*Hordeum*
−N	310	80
Nitrate	420	240
Ammonium	460	240
Amino acids	620	150

[a]Based on Chapin *et al.* (1993).

rates of nutrients higher but also the specific respiratory costs for nutrient uptake are lower (Poorter *et al.*, 1991).

16.2   High-Input versus Low-Input Approach

16.2.1   General

In the past, the approach to soil fertility problems in crop production emphasized the importance of changing the soil to fit the plant. Soil fertility factors, such as pH and nutrient availability, were adjusted to optimum levels for a given plant species. This high-input approach, coupled with the heavy use of chemical fertilizers, was very successful in the temperate zones in increasing yields of crops grown in soils that do not, as a rule, have extreme chemical properties. However, the high-input genotypes of crop plants often have a limited adaptability to the adverse chemical soil conditions that usually prevail in the tropics and subtropics. These conditions cannot easily be ameliorated because of their extent, the cost of improving the soils, or both (Vose, 1983). In tropical America, for example, ~70% of the soils are acid and infertile (Sanchez and Salinas, 1981). In subtropical and semiarid regions, soil salinity and alkalinity and related nutritional problems such as iron and zinc deficiency are widespread. About 25% of the world's area of cultivable soils has acute chemical problems (Vose, 1983).

The realization of the difficulties or even failure of the high-input approach in most tropical and subtropical soils led over the last twenty or thirty years to a shift in approach toward more emphasis on fitting plants to soils. This approach requires genotypes better adapted to given ecological conditions, as well as selection and breeding programs for high nutrient efficiency and high tolerance to such constraints as aluminum and manganese toxicity, waterlogging and salinity.

This low-input approach using adapted genotypes with a more efficient use of nutrients from soil reserves and fertilizers leads to yields that are only 80–90% of the maximum. For both economic and ecological reasons this approach addresses itself not only to extreme soil chemical conditions (e.g., salinity) but also to the selection and breeding of genotypes that are highly efficient in utilizing soil and fertilizer nutrients. In the past progress in selection and breeding for higher yield was achieved to a large extent by increasing harvest index, i.e., the proportion of biomass allocated to seeds and storage organs rather than by increasing total biomass (Dambroth and El Bassam, 1990). In recent years greater emphasis has been given to combining high crop yields with high efficiency in acquisition and utilization of mineral nutrients. Modern cultivars, for example of wheat or potato, tend to have lower root/shoot dry weight ratios than old, traditional cultivars, but the efficiency in nutrient acquisition is often similar, or even higher. This is presumably because of a finer and more active root system in the modern cultivars (Sattelmacher *et al.*, 1990d), often in combination with a more efficient internal utilization (retranslocation) as has been shown for phosphorus in wheat cultivars (Horst *et al.*, 1993).

16.2.2 Genetic Basis for Mineral Nutrition

Mineral nutrition of plants is under genetic control. In crop plants this is indicated by the numerous examples of nutritional differences between cultivars or strains. More specific evidence comes from inheritance studies involving cultivars and strains differing in nutritional requirements. Since the early 1960s there has been an enormous increase in interest in and research on genetically based mineral nutrition of plants, for example, by Brown and his group and Foy and his group, at the Plant Stress Laboratory in Beltsville, Maryland, and by Epstein and Läuchli at the University of California, Davis. Worldwide there has been impressive progress in both breeding programs for improving the adaptation of crop species to problem soils and in research on physiological mechanisms which are responsible for or at least involved in, adaptation. International meetings on genetic aspects of plant mineral nutrition reflect this development (El Bassam *et al.*, 1990; Randall *et al.*, 1993). In some instances major nutritional features are under the control of a single gene pair; in most instances more complex genetic systems are involved in acquisition and utilization of mineral nutrients (Gerloff and Gabelman, 1983; Graham, 1984; Hoan *et al.*, 1992) or in control of salt tolerance (Section 16.6.4). In aluminum tolerance of wheat there are different views as to whether it is a dominant character controlled by several genes (Aniol, 1990) or by a single gene with incomplete dominance (Wheeler *et al.*, 1992b).

The genotypical differences in copper and iron efficiency in cereals highlight the genetic control of mineral nutrition, and the progress which is possible in understanding the biochemical mechanisms involved by applying molecular biological approaches. Wheat and oat are generally sensitive to a low copper supply, whereas rye is relatively insensitive (Table 16.3). Important differences also exist in the copper efficiency of wheat cultivars. When the supply is suboptimal, cv. Gabo either fails totally or has only a very low grain yield compared with the relatively copper-efficient cv. Chinese Spring. Triticale, a hybrid of wheat and rye, also has a high copper efficiency similar to its parent rye, indicating that the specific mechanisms for copper uptake in rye are genetically controlled and transferable to triticale.

The genes controlling copper efficiency are carried on the long arm (L) of chromosome 5 R of rye (Graham *et al.*, 1987a). Chinese Spring wheat is carrying the 5 RL chromosome of rye and has therefore a high copper efficiency. Under field conditions

Table 16.3
Responses of Grain Yield of Various Genotypes to Copper[a,b]

Species, cultivars	Copper supply (mg $(11 \text{ kg})^{-1}$ soil)			
	0	0.1	0.4	4.0
Wheat				
cv. Gabo	0	0	10	100
cv. Halberd	2	7	52	100
cv. Chinese Spring	0	26	94	100
Rye				
cv. Imperial	100	114	114	100
Triticale				
cv. Beagle	99	95	99	100

[a]Data compiled from Nambiar (1976c) and Graham and Pearce (1979).
[b]Data represent grain yields expressed as relative values. Plants were grown in a copper-deficient soil and supplied with different levels of copper.

the presence of this chromosome in wheat cultivars increases wheat grain yield on copper-deficient soils by more than 100%. Copper uptake of 5 RL lines under copper deficiency is 50–100% greater than those of their recurrent parents (Graham *et al.*, 1987a).

The 5 R chromosome from rye is also the carrier of genes which encode enzymes operative in Strategy II plants in iron acquisition (Section 2.5.6). These enzymes regulate the synthesis of the phytosiderophore mugineic acid (MA) from deoxymugineic acid (DMA), and hydroxymugineic acid (HMA) from MA (Mori *et al.*, 1990). In barley the gene which encodes the synthesis of MA from DMA is carried by chromosome no. 4, and introducing this chromosome into wheat leads to synthesis of MA in wheat (Mori and Nishizawa, 1989). The identification of genes encoding certain biochemical reactions in a given genotype, and transferring these genes into other genotypes, also opens new approaches to the better adaptation of crop plants to adverse soil chemical properties, and low nutrient availability in particular.

16.2.3 Nutrient Efficiency

Genotypical differences in nutrient efficiency are related to differences in efficiency in acquisition by the roots, or in utilization by the plant, or both. Efficiency in acquisition is frequently defined in terms of total uptake per plant or specific uptake rate per unit root length, and efficiency in utilization (nutrient use efficiency, NUE) as dry matter production per unit nutrient in the dry matter. However, from the agronomical point of view and in an operational sense genotypical differences in the nutrient efficiency of crop plants are usually defined by differences in relative growth or in the yield when grown in a deficient soil. For a given genotype, nutrient efficiency is reflected by the ability to produce a high yield in a soil that is limiting in one or more mineral nutrients for a standard genotype (Graham, 1984). This definition can be applied to comparisons of both genotypes (cultivars or lines) within a species or between plant species.

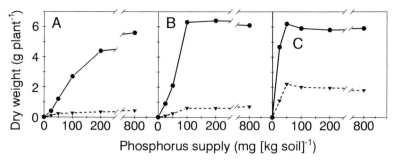

Fig. 16.1 Growth response of three pasture species to phosphorus fertilizer applied to a phosphorus-deficient soil. (A) *Trifolium cherleri*; (B) *Trifolium subterraneum*; (C) *Lolium rigidum* (●, total plants; ▼, roots). (Based on Ozanne *et al.*, 1969.)

There have been a large number of reports in recent years on nutrient efficiency, with the main emphasis on these agronomical aspects, comparing the yield, or the percentage of yield reduction, in genotypes supplied with insufficient amounts of mineral nutrients (Graham *et al.*, 1982; Gabelman and Gerloff, 1983; Dambroth and El Bassam, 1990; Randall *et al.*, 1993). A typical example of the phosphorus efficiency of three pasture species is given in Fig. 16.1. Despite a similar final dry weight at the highest phosphorus supply, the growth response of the three species to a given supply increased from *Trifolium cherleri* to *T. subterraneum* and *Lolium rigidum*. The minimum levels of applied phosphorus required in the three species to give 90% of maximum yield were 302, 87, and 26 mg P kg^{-1} of soil, respectively. As one would expect from the role of root growth and root surface area in phosphorus acquisition (Sections 13.2 and 13.3) there is a close correlation between the efficiency of fertilizer phosphorus utilization and root dry weight (Fig. 16.1).

In mixed pastures on phosphorus-deficient soils, with low levels of fertilizer phosphorus a shift in botanical composition can be expected, favoring the growth of grass, as also occurs in a mixed stand of legumes and grasses growing in potassium-deficient soils (Section 13.2.3.2).

Genotypical differences in nutrient efficiency occur for a number of reasons, these being related to uptake, transport, and utilization within plants (Fig. 16.2). There are typical differences in nitrogen requirement at the cellular level, for example, between C_3 and C_4 species (Section 12.3.6) or in calcium and boron requirements between monocots and dicots (Chapters 8 and 9). Within a given species the nutrient efficiency, for example for calcium, may differ between cultivars depending on binding stage of calcium (Section 12.3), transport rate to the apical meristem (Behling *et al.*, 1989), or differences in functional requirement within the tissue (English and Barker, 1987; Horst *et al.*, 1992a). For phosphorus higher use efficiency in certain genotypes may be related to better use of stored P_i (Caradus and Snaydon, 1987; Hart and Colville, 1988) either within a given tissue or by better retranslocation between shoot organs (Youngdahl, 1990). This latter factor is also mainly responsible for higher nitrogen use efficiency in sorghum cultivars (Alagarswamy *et al.*, 1988) or higher molybdenum use efficiency in bean (*Phaseolus vulgaris*) cultivars (Brodrick and Giller, 1991b). In the case of potassium use efficiency genotypical differences in sodium accumulation in the

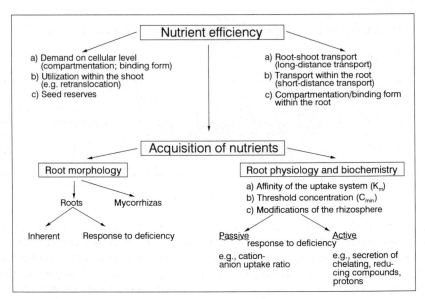

Fig. 16.2   Possible mechanisms of genotypical differences in nutrient efficiency.

shoots and thus, potential potassium replacement, play a major role (Section 10.2). Higher phosphorus contents in seeds, either per unit dry weight or per seed (large seeds) can produce a larger root surface in early growth and increase phosphorus acquisition of soil and fertilizer phosphorus even at higher soil phosphorus levels (Bolland and Baker, 1989; Riley *et al.*, 1993; Section 16.1).

There are several examples of genotypical differences between cultivars or mutants in the short-distance transport of mineral nutrients within the roots or in long-distance transport from the roots to the shoots (Läuchli, 1976b). The 'iron-inefficient' PI soybean cultivar represents a genotype in which a slow rate of transport of iron from roots to shoot seems to be responsible for the low nutrient efficiency (Brown *et al.*, 1967). Impaired root to shoot transport of iron may often be due in fact to sequestering of iron (precipitation) at the rhizoplane and in the apoplasm of rhizodermal cells. A combination of differences in the rates of both magnesium uptake and transport to the shoot has been reported by Clark (1975) in maize; in the efficient inbred line B 57 maximal dry weight was obtained with a magnesium concentration in the substrate of only 0.12 mM compared with 4.1 mM in the inefficient inbred line Oh 40 B. Genotypical differences in magnesium efficiency in sorghum seem to be related to differences in potassium uptake rates: there is a tendency for magnesium-efficient genotypes to have lower K/Mg ratios in the shoots (Keisling *et al.*, 1990).

As a rule in nutrient efficiency the acquisition of nutrients by the roots plays the most important role (Gutschick, 1993). Genotypes within a species can differ widely in both the affinity of the uptake system (K_m) and the threshold concentration (C_{min}), for example, for phosphorus in corn inbred lines (Schenk and Barber, 1979). Differences between plant species in phosphorus uptake rate per unit root length may be caused by either higher influx rates, longer root hairs or different root length/shoot weight ratios

Table 16.4

Root and Shoot Growth of Two Strains of Bean Receiving Adequate and Inadequate Phosphorus Supplies[a]

Strain no.	Adequate P supply			Inadequate P supply		
	Root dry wt (mg per plant)	Shoot dry wt (mg per plant)	Root/shoot ratio	Root dry wt (mg per plant)	Shoot dry wt (mg per plant)	Root/shoot ratio
6	242	1465	0.17	124	777	0.16
11	181	1233	0.15	365	1141	0.31

[a]Based on Whiteaker *et al.* (1976).

(Föhse *et al.*, 1988). Between genotypes not only the root–shoot ratio differs but also the response of this ratio to phosphorus deficiency (Table 16.4). Growth of strain 6 was greatly reduced, but the root/shoot ratio was approximately the same. In contrast, root growth and root/shoot ratio of strain 11 nearly doubled. The capacity to distribute a higher proportion of photosynthates to the roots is obviously under genetic control and is an important aspect of phosphorus efficiency for plants grown in deficient soils.

The efficiency in acquisition and internal utilization also depends on the level of nutrients supplied (Godwin and Blair, 1991; Smith *et al.*, 1990a) and on plant age (Schjorring and Nielsen, 1987; Brouder and Cassman, 1990). To evaluate genotypical differences in nutrient efficiency dose-response curves to increasing nutrient supply need to be obtained and sequential harvests taken. When grown in deficient soils, phosphorus-efficient genotypes in barley (Schjorring and Nielsen, 1987) and potassium-efficient genotypes in cotton (Brouder and Cassman, 1990) were characterized by higher uptake rates of these mineral nutrients after ear emergence and flowering due to maintenance of a high root growth and activity.

Certain important components of nutrient acquisition from soils and of genotypical differences in acquisition can not, or only incompletely, be assessed in experiments using nutrient solution cultures. Examples of this are root-induced changes in the rhizosphere by secretion of organic acids and chelating compounds (Itoh, 1987; Section 15.4.2), or associative N_2 fixation in C_4 species (Miranda *et al.*, 1990), and mycorrhizas in particular (Smith *et al.*, 1992; Section 15.7). An example for the role of VAM in genotypical differences in phosphorus acquisition by alfalfa is shown in Table 16.5. In the absence of VAM all three cultivars responded similarly to the increase in phosphorus supply. With VAM, however, growth increased nearly 100-fold in all three cultivars, but there were obvious differences in the efficiency of the VAM association with the host plants. In Du Puits, VAM could totally replace phosphorus application but in Buffalo this replacement was less than 50%. These differences in mycorrhizal response are probably related to differences in root infection with VAM which not only differs markedly between species but also between genotypes of a species as has been reviewed by Smith *et al.* (1992).

A particular example of the limitations in assessing phosphorus efficiency in nutrient solutions for prediction for field grown plants is pigeon pea (*Cajanus cajan* L.). This plant species readily uses sparingly soluble iron phosphates by secretion of the Fe(III)

Table 16.5

Shoot Dry Weight of Three Alfalfa Cultivars with or without Vesicular-Arbuscular Mycorrhizas (VAM) at Different Levels of Phosphorus Supply[a]

Soil treatment		Shoot dry weight (mg per plant)		
mg P kg^{-1}	± VAM	Buffalo	Cherokee	Du Puits
0	−	22	18	32
20	−	114	235	375
80	−	2389	2058	2115
0	+	1113	1740	2177

[a]Plants were grown in a phosphorus-deficient soil, pH 7.2. From Lambert *et al.* (1980).

complexing piscidic acid (Section 15.4.2) but simultaneously relies on VAM for efficient uptake rates of phosphorus mobilized by the root exudates (Ae *et al.*, 1993).

16.2.4 Tolerance to Excessive Supply of Mineral Elements

In many instances adaptation to adverse chemical soil conditions requires tolerance to excessive levels of mineral elements such as aluminum and manganese in acid mineral soils, manganese and iron in waterlogged soils, and sodium chloride in saline soils. Thus, *multiple stress tolerance* is often necessary for adaptation. In the case of rice impressive progress has been made in bringing about this adaptation as shown in Table 16.6. In a large-scale screening and breeding program adapted cultivars have been developed yielding up to 2.7 t ha^{-1} over traditional cultivars in farmers' fields on unamended soils with adverse chemical properties. This has allowed marginal land to be brought into production without the need to take costly reclamation measures.

Table 16.6

Grain Production of Rice Genotypes with Different Degrees of Adaptation to Adverse Chemical Soil Conditions in Farmers' Fields in the Philippines 1977–1988[a]

Soil conditions	No. of cultivars tested	Mean grain yield (tons ha^{-1})		
		Farmers cultivars	Selected cultivars	Advantage
Phosphorus deficiency	336	2.2	4.9	2.7
Zinc deficiency	411	1.8	4.4	2.6
Iron deficiency	89	0.9	2.8	1.9
Salinity	120	1.4	3.4	2.0
Alkalinity	103	0.8	3.4	2.6
Iron toxicity	104	2.2	4.1	1.9
Aluminum/manganese toxicity	44	1.2	3.0	1.8

[a]Neue *et al.* (1990). Reprinted by permission of Kluwer Academic Publishers.

Boron toxicity is another growth-limiting factor in certain dryland areas. In wheat and barley the mechanism of boron tolerance is based on reduced uptake and shoot transport of boron which is under control of several major additive genes, one of which is located on chromosome 4 A in wheat (Paull *et al.*, 1992a). However, boron-tolerant genotypes bear the risk of becoming boron deficient when grown on soils low in boron (Nable *et al.*, 1990a).

16.3 Acid Mineral Soils

16.3.1 Major Constraints

Soil acidity limits plant growth in many parts of the world. Plant growth inhibition results from a variety of specific chemical factors and interactions between these factors (Marschner, 1991b). In acid mineral soils the major constraints to plant growth are the following:

1. Increase in H^+ concentration: H^+ toxicity
2. Increase in aluminum concentration: aluminum toxicity
3. Increase in manganese concentration: manganese toxicity
4. Decrease in macronutrient cation concentration: magnesium, calcium (and potassium) deficiency
5. Decrease in phosphorus and molybdenum solubility: phosphorus and molybdenum deficiency
6. Inhibition in root growth and water uptake: nutrient deficiency, drought stress, increased nutrient leaching.

The relative importance of these constraints depends on plant species and genotype, soil type and horizon, parent material, soil pH value, concentration and species of aluminum, soil structure and aeration, and climate. The nitrogen levels in acid mineral soils are generally low unless there is high atmospheric input by air pollution (Schulze, 1989). Aluminum toxicity and calcium and magnesium deficiency occur in more than 70% of the acid soils of tropical America, and nearly all of these soils are phosphorus-deficient or have a high phosphorus-fixing capacity (Sanchez and Salinas, 1981). Subsoil acidity is a potential growth-limiting factor throughout many areas of the USA (Foy *et al.*, 1974) and of the tropics (Van Raij, 1991).

Forest soils in many regions of the world are typically acidic. In recent years concern has grown about the increasing acidification of forest soils by atmospheric emissions of SO_2 and nitrogen oxides ('acid rain') as causative factors of forest damage (forest decline) particularly in Europe and North America. There is a great deal of controversy as to the prime cause of forest decline by soil acidification. This has been thought to be an increase in aluminum solubility and thus aluminum toxicity (Murach and Ulrich, 1988), a decrease in concentration and uptake of nutrients, magnesium in particular, and thus magnesium deficiency (Zoettl and Huettl, 1986; Kaupenjohann *et al.*, 1987; Liu and Huettl, 1991), and an increase in magnesium and calcium deficiency because of high atmospheric nitrogen input in particular (Schulze, 1989; Aber *et al.*, 1989).

It is not possible to generalize, however, as to the prime factors of soil acidity stress without considering specific sites. The role of atmospheric nitrogen deposition depends not only on the amounts but also the cropping history of forest sites (Zöttl, 1990). In European beech (*Fagus sylvatica* L.) root growth is much more sensitive to high concentrations of H^+ than of aluminum whereas in Norway spruce (*Picea abies* (L.) Karst.) the reverse appears to be the case (Murach and Ulrich, 1988). For a given plant species the location and distribution of roots within the soil profile may be an important factor in determining the form of expression of soil acidity. In the topsoil where organic matter content is higher, H^+ toxicity may dominate, but in the subsoil root growth may be depressed by aluminum toxicity. Whether magnesium deficiency becomes a dominant factor in stress induced by soil acidity depends mainly on the parent material (Zoettl and Huettl, 1986) and the atmospheric input of magnesium (e.g. distance from the open sea). In soils high in manganese reserves and in exchangeable Mn^{2+}, as for example, brought about by continuous cultivation of N_2 fixing legumes, manganese toxicity may become a major factor in soil acidity stress (Bromfield *et al.*, 1983a,b).

In view of the different ways in which soil acidity can restrict plant growth, plants adapted to acid mineral soils require a variety of mechanisms to cope with the adverse soil chemical factors involved (Howeler, 1991). It is generally agreed that, as a rule and on a worldwide scale, high concentrations of aluminum, and in some instances also of manganese, are key factors of soil acidity stress, and correspondingly high tolerance to these two factors is required for adaptation particularly of crop plants to soils of pH <5.

16.3.2 Solubility of Aluminum and Manganese

In acid mineral soils below pH 5.5 an increasing proportion of the cation exchange sites of clay minerals is occupied by aluminum where it especially replaces other polyvalent cations (Mg^{2+}, Ca^{2+}) and simultaneously acts as a strong adsorber of phosphate and molybdate. The percentage of exchangeable aluminum in soils is thus correlated with soil pH (see Table 16.9), and inhibition of root growth of most plant species. However, these correlations are often not very close as not only the concentrations but particularly the aluminum species in soil solution (Section 16.3.4) determine the phytotoxicity of aluminum to roots. Crucial factors for the aluminum species in soil solution include the pH (see Fig. 16.6) and the concentrations of organic and inorganic compounds forming complexes with aluminum.

With decreasing pH the amount of exchangeable manganese increases in many soils (Table 16.7). Increase in exchangeable manganese (Mn^{2+}), however, is also a function of the redox potential ($MnO_2 + 4 H^+ + 2e^- \rightleftharpoons Mn^{2+} + 2 H_2O$). High levels of Mn^{2+} at the exchange sites and in the soil solution are therefore to be expected only in acid soils with large amounts of readily reducible manganese in combination with a large content of organic matter, high microbial activity, and anaerobiosis, either temporarily (e.g., short-term flooding) or permanently (Section 16.4). On soils with high levels of readily reducible manganese, soil acidification brought about by symbiotic legumes, can greatly increase the amount of exchangeable Mn^{2+} (Table 16.7) and the risk of manganese toxicity in permanent pastures. When a soil is acidified, amounts of exchangeable Mn^{2+} as well as concentrations of Mn^{2+} in the soil solution increase, for example, from 0.8 μM at pH 7.3 to 182 μM at pH 5.2, without much changes in the manganese species, i.e. the ratio Mn^{2+}/total Mn in the soil solution (Sanders, 1983).

Table 16.7

Relationship Between Age of *Trifolium subterraneum* Dominated Pastures, Soil pH and Exchangeable Manganese in Soils[a]

Age (years since establishment)	$pH_{(H_2O)}$	Exchangeable Mn (mg kg^{-1} soil)
0	6.1	4.6
25–30	5.6	22.7
30–35	5.3	33.3
35–40	5.1	37.3
50–55	4.8	46.1

[a]Based on Bromfield *et al.* (1983a).

Many acid soils in the tropics are highly weathered, and their total manganese content is often low. Thus, in these soils there is less risk of manganese toxicity than of aluminum toxicity.

16.3.3 Effects of Excessive Aluminum Levels

16.3.3.1 *Inhibited Mineral Nutrient Uptake and Induced Deficiency*

As the pH falls, i.e., the H$^+$ concentration increases, uptake of cations is inhibited for two reasons: (1) impairment of net extrusion of H$^+$ by the plasma membrane-bound ATPase (Section 2.4.2), and (2) decrease of loading of polyvalent cations (Mg^{2+}, Ca^{2+}, Zn^{2+}, Mn^{2+}) in the apoplasm of root cortical cells. Apoplasmic loading of polyvalent cations strongly enhances uptake rates of these cations into the symplasm (Section 2.2). Accordingly, at a given external concentration of these cations, lowering the pH, for example from 6 to 3 (i.e. 1 mM H$^+$), decreases their uptake, and the addition of aluminum as strongly competing polyvalent cation for binding sites in the apoplasm (Godbold *et al.*, 1988; Marschner, 1991b) accentuates this decrease (Table 16.8). Aluminum may also inhibit calcium uptake by blocking Ca^{2+} channels in the plasma membrane (Huang *et al.*, 1992b), and magnesium uptake by blocking binding sites of transport proteins (Rengel and Robinson, 1989a). High concentrations of Mn^{2+} also inhibit calcium and particularly magnesium uptake (Table 16.8).

In contrast to calcium and magnesium, the uptake of potassium is usually unaffected by manganese or aluminum (Jorns and Hecht-Buchholz, 1985; Wheeler *et al.*, 1992a), leading to an increase in K/(Ca + Mg) ratio in the shoots. This increases not only the risk of calcium or magnesium deficiency or both but also the potential risk of grass tetany in ruminants using grass as a forage (Rengel and Robinson, 1989b).

The strong competing effect of aluminum on calcium and magnesium uptake explains why the molar ratios of Ca/Al or Mg/Al in the soil or nutrient solutions are sometimes better parameters for predicting of the risk of aluminum-induced calcium and magnesium deficiency than the concentrations of any of the individual elements (Kruger and Sucoff, 1989). An example of this is given in Fig. 16.3 for aluminum-induced magnesium deficiency in soybean. Although increasing external concentrations of

Table 16.8

Effect of Substrate pH, Manganese and Aluminum on the Mineral Element Content in Roots and Needles of 2-year-old Norway Spruce (*Picea abies* (L.) Karst.) Grown in Sand Culture and Percolated with Nutrient Solution[a]

Treatment		Mineral element content (mmol kg^{-1} dry wt)							
		Fine roots				Needles			
pH	Mn, Al	Ca	Mg	Mn	Al	Ca	Mg	Mn	Al
6	—	132	115	0.9	ND[b]	205	74	ND	ND
3	—	100	82	ND	ND	77	37	ND	ND
3	1.5 mM Al	20	33	0.6	30	37	21	0.05	0.01
3	1.5 mM Mn	70	25	35	ND	67	16	25	ND

[a]Based on Stienen and Bauch (1988).
[b]ND = not determined.

aluminum decrease the magnesium content in shoot dry weight at both low and high levels of magnesium supply, the effect of aluminum in depressing growth was prevented at the high magnesium level (i.e., high magnesium/aluminum ratio) as the magnesium content in the plant tissue remained above the critical deficiency level.

Calcium deficiency in apical meristems is a well-documented manifestation of aluminum toxicity in soybean (Foy *et al.*, 1969) and snapbean (Foy *et al.*, 1972). In maize increasing the external calcium concentration, greatly reduces the aluminum-induced inhibition of root elongation in highly sensitive cultivars and completely prevents inhibition in less sensitive cultivars (Rhue and Grogan, 1977). In wheat,

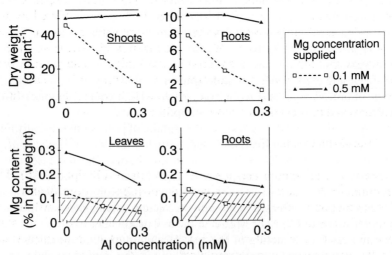

Fig. 16.3 Relationship between concentrations of magnesium and aluminum in the nutrient solution, dry weight and magnesium contents in leaves and roots of soybean. Hatched areas indicate critical deficiency concentrations. (Based on Grimme, 1984.)

Table 16.9

Effect of Liming a Highly Aluminum Saturated Soil on the Growth and Nodulation of Soybean[a]

Soil pH	Al Saturation[b] (%)	Dry wt (g per plant)		Nodulation		
		Shoot	Root	No. per plant	Dry wt (mg per nodule)	N Content (mg per shoot)
4.55	81	2.4	1.07	21	79	65
5.20	28	3.2	1.08	65	95	86
5.90	4	3.6	1.08	77	99	93

[a]Reproduced from Sartain and Kamprath (1975) by permission of the American Society of Agronomy.
[b]Percentage of cation-exchange capacity (CEC) saturated by aluminum.

however, the results on aluminum-induced calcium deficiency are inconsistent (Foy *et al.*, 1974), and in cowpea calcium deficiency is not a primary effect of aluminum toxicity (Horst *et al.*, 1983).

There is increasing awareness of the importance of magnesium deficiency as a secondary effect of aluminum toxicity and its amelioration by increasing the magnesium supply in forest stands on acid soils (Kaupenjohann *et al.*, 1987; Huettl, 1989), to sorghum genotypes grown in acid soils (Tan *et al.*, 1992a) or nutrient solution (Tan *et al.*, 1992b, 1993). Amelioration of acid soils by liming decreases the risk of aluminum-induced calcium and magnesium deficiency and direct aluminum toxicity on root growth. However, in soils or other substrates low in magnesium increasing calcium concentrations combined with enhanced growth may again increase the risk of induced magnesium deficiency (Edmeades *et al.*, 1991; Shamsuddin *et al.*, 1992).

Growth of legumes in acid soils can be limited by detrimental effects on host plant growth per se and delay in nodulation (Schubert *et al.*, 1990a) and a decrease in number of nodules in particular (Table 16.9). Nodulation is severely inhibited by high H^+ concentrations in combination with low calcium concentrations (Alva *et al.*, 1987; Section 7.4.5) and by high aluminum concentrations in particular. In soybean critical aluminum concentrations for host plant growth (10% growth reduction) were 5–9 μM compared with only 0.4 μM for nodulation (Alva *et al.*, 1987). Changes in root morphology and thus infection sites may be involved in the inhibition of nodulation by high aluminum concentrations (Carvalho *et al.*, 1982).

16.3.3.2 Inhibition of Root Growth

The toxic actions of aluminum are primarily root-related (Taylor, 1988a). The root system becomes stubby as a result of inhibition of elongation of the main axis and lateral roots (Klotz and Horst, 1988a). The severity of inhibition of root growth is a suitable indicator of genotypical differences in aluminum toxicity (Fig. 16.4). That aluminum toxicity was the cause of the acid soil injury of the root system of the cultivar Kearney was confirmed by nutrient solution experiments (Foy *et al.*, 1967).

With increasing soil acidification, root penetration is inhibited particularly into the subsoil (lower Ca/Al and higher Al^{n+}/Al_{tot} ratios) leading to a more shallow root system

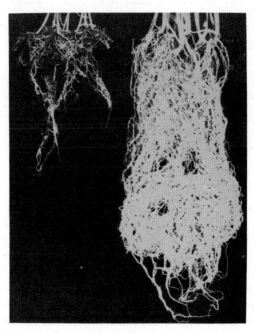

Fig. 16.4 Roots of Kearney (left) and Dayton (right) barley cultivars grown in acid soil of pH 4.6. (From Foy, 1974.)

(Marschner, 1991b) with a correspondingly lower utilization of mineral nutrients and water from the subsoil (Section 14.4.2). The risk of drought stress is increased (Goldman *et al.*, 1989) and also leaching of mineral nutrients. Aluminum toxicity is therefore often expressed simultaneously in two ways, namely induced deficiency of mineral nutrients, such as magnesium, and inhibition in root elongation (Tan *et al.*, 1992b). Inhibition of root growth by aluminum toxicity should further increase the risk of phosphorus deficiency on acid mineral soils, unless other growth-limiting factors (e.g., magnesium deficiency) dominate, or a high proportion of the phosphorus demand is provided by mycorrhizas to the host plant (Section 15.7). In acid mineral soils aluminum toxicity may inhibit shoot growth by limiting supply of nutrients and water by poorer subsoil penetration or lower root hydraulic conductivity (Kruger and Sucoff, 1989). Evidence that a limited supply of cytokinins from the roots may inhibit shoot growth of soybean grown either in acid soil or nutrient solutions with aluminum, has been obtained from the beneficial effect on growth by the application of cytokinins to the shoots of such plants (Pan *et al.*, 1989).

The physiological and biochemical mechanisms of the toxic effects of aluminum on root elongation are poorly understood and a matter of controversy. Inhibition of cell division in root apical meristems is a rapid response to aluminum treatment (Fig. 16.5). Cell division may resume after some time but remains at a lower level than in controls not exposed to aluminum. Genotypical differences in aluminum tolerance are reflected in the degree of recovery from primary aluminum stress. Although aluminum may bind

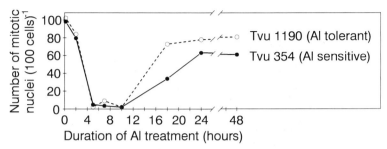

Fig. 16.5   Effect of aluminum (5 mg l^{-1}) on the rate of cell division in root tips of two cowpea genotypes. Relative values; control (no aluminum) = 100. (Based on Horst *et al.*, 1982.)

to DNA (Matsumoto, 1991; Lüttge and Clarkson, 1992), of the root cap cells in particular (Naidoo *et al.*, 1978), inhibition of cell division is presumably an indirect effect of aluminum (see below). Furthermore, inhibition in root elongation (mainly cell expansion) is an even faster response to aluminum. In wheat, root elongation resumes rapidly as soon as 30 min after treatment with citrate to chelate the aluminum, indicating that, at least in presence of high calcium concentrations (400 μM), aluminum remains in a compartment (apoplasm?) where it can readily be removed (Ownby and Popham, 1989). In agreement with these observations, in wheat plants root growth in sensitive genotypes was distinctly inhibited by aluminum within 4 h, although aluminum could be detected only in the rhizodermal layer and the cortical layer below the rhizodermis (Delhaize *et al.*, 1993a). In oat plants even after long-term treatment with aluminum and severe inhibition of root elongation aluminum was still confined to the apoplasm, mainly of peripheral root cells (Marienfeld and Stelzer, 1993).

Aluminum may bind to the outer surface of the plasma membrane of root rhizodermal and cortical cells and thereby impair plasma membrane functions (Taylor, 1988a; Haug and Shi, 1991). Aluminum (Al^{3+}) has a 560-fold higher affinity than Ca^{2+} for certain phospholipids in membranes (Akeson *et al.*, 1989). However, in wheat roots despite severe inhibition in elongation growth, the root cells retained their capacity to excrete H^+ indicating that the plasma membrane was intact (Kinraide, 1988; Ryan *et al.*, 1992). Nevertheless, plasma membrane properties might be altered as indicated, for example, by a decrease in potassium efflux (Cakmak and Horst, 1991a), increase in callose formation (Horst *et al.*, 1992b), or after long-term aluminum treatment, by increase in peroxidation of membrane lipids (Cakmak and Horst, 1991b).

In contrast to these suggestions as to the direct effects of aluminum on plasma membrane properties or cellular metabolism (Haug and Shi, 1991), Bennet and Breen (1991, 1992, 1993) put forward the hypothesis that in roots the primary target of aluminum is the root cap which perceives the 'aluminum signal', similarly to that of gravity or of mechanical impedance (Section 14.6). Aluminum reduces secretion of mucilage of the peripheral cap cells, and these cap cells are sources of endogenous regulators of extension growth. The mucilage seems to be required as a pathway for the apoplasmic transport of the signal substances (Moore *et al.*, 1990). In this model Ca^{2+} functions as a mediator of the signal conducting chain (Section 8.6.7), and secretion of mucilage in the peripheral cap cells is causally involved. Apoplasmic Ca^{2+} is required for the secretory functions of the cap cells, and aluminum changes Ca^{2+} homeostasis and thereby reduces secretion of mucilage. After transfer to aluminum-free solution resumption of elongation growth is preceded by increased secretory activity of the Golgi apparatus of root cap cells (Bennet and Breen, 1991). Accordingly, inhibition of root elongation would not be a direct effect of aluminum at the elongation zone but at the root cap. Arguments against such a key role of the root cap in aluminum toxicity have been provided by Ryan *et al.* (1993) who showed that in the presence of external calcium supply removal of the root cap had no effect on aluminum-induced inhibition of root elongation in maize. The particular role of calcium in aluminum toxicity has been recently reviewed by Rengel (1992a).

16.3.4 Aluminum Toxicity and Aluminum Species

The phytotoxictiy of aluminum also in terms of inhibition in root elongation clearly decreases with increasing ionic strength of the soil or nutrient solution (Blamey *et al.*, 1991) and with a decrease in the ratio of Al^{n+}/complexed Al. The relationships are less clear, however, with respect to soil or nutrient solution pH and phytotoxicity of aluminum. This is not only because of the increase in both total aluminum and H^+ in the soluble solution with decreasing pH but also the simultaneous shift in the aluminum species, assuming the absence of polynuclear aluminum and aluminum ligands other than OH^- (Fig. 16.6).

Aluminum released from soil minerals to the soil solution under acid conditions, or the aluminum in nutrient solutions of pH 4 and below, mainly appears as $Al(H_2O)_6^{3+}$ (or Al^{3+} for convenience). As the pH increases mononuclear hydrolysis products such as $Al(OH)^{2+}$ and $Al(OH)_2^+$ are formed (Fig. 16.6). At elevated OH^-/aluminum ratios in solution polynuclear hydroxyl aluminum species such as $AlO_4Al_{12}(OH)_{24}(H_2O)_{12}^{7+}$ (for convenience 'Al_{13}') may form, which are metastable intermediates in the precipitation

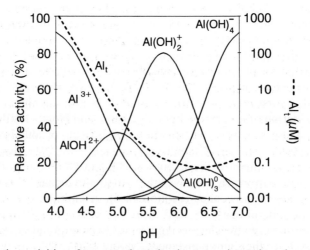

Fig. 16.6 Relative activities of mononuclear aluminum species and total concentration (Al_t) of soluble aluminum as a function of pH. (Kinraide, 1991.) Reprinted by permission of Kluwer Academic Publishers.

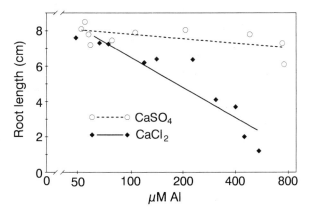

Fig. 16.7 Root length of wheat (*Triticum aestivum* L.) as a function of the aluminum concentration in soil solutions from soils treated with $CaSO_4$ or $CaCl_2$. (Modified from Wright *et al.*, 1989b.)

of solid $Al(OH)_3^0$. There are contradicting results as to the relative toxicity of the various mononuclear aluminum species (Bruce *et al.*, 1988; Kinraide and Parker, 1990), except for the nonphytotoxicity of the aluminate ion $Al(OH)_4^-$ (Kinraide, 1990). A particular high phytotoxicity is attributed to the polynuclear 'Al_{13}' which may form at pH 4.5 and may lead to unexpected maximum inhibition of root elongation rates at pH 4.5 (Parker *et al.*, 1988, 1989).

Contradicting results on the relative phytotoxicity of various aluminum species may arise for various reasons, a major one being the uncertainty of activities and identity of the aluminum species and of the H^+ concentrations at the rhizoplane and in the apoplasm of rhizodermal cells (Kinraide, 1991). Another main reason being that it is not possible to vary the relative distribution of the aluminum species without varying the pH, and vice versa (Fig. 16.6). Usually, results are interpreted exclusively in terms of aluminum speciation whereas the possible influence of varying H^+ activities is not considered. The ameliorating effect of high H^+ concentrations (i.e., very low pH) on aluminum toxicity has been stressed because of the much higher competitiveness with Al^{3+} of H^+ as compared with Ca^{2+} (Grauer and Horst, 1992; Grauer, 1993). Proton amelioration of aluminum toxicity is also assumed to be the responsible factor for the lower aluminum content in apical root zones and less severe inhibition in root elongation in plants supplied with NH_4^+ compared with NO_3^- (Klotz and Horst, 1988b; Grauer and Horst, 1990). The physiological relevance of H^+ amelioration of aluminum toxicity is, however, confined to plant species of high H^+ tolerance (Kinraide and Parker, 1990). Proton amelioration of aluminum toxicity is also not likely to be relevant for plants grown in acid soils because of the simultaneously enhanced release of Al^{3+} from the solid phase, and of the H^+ competition for Ca^{2+} and Mg^{2+} uptake.

Some of the mononuclear aluminum species associated with inorganic ligands such as AlF^{2+}, AlF_2^+ or $AlSO_4^+$ are nonphytotoxic (Kinraide, 1991). The nonphytotoxicity of $AlSO_4^+$ is of particular practical importance as, for example, the application of gypsum ($CaSO_4$) to acid soils may ameliorate phytotoxicity of aluminum (Fig. 16.7). Because of the sulfate component and higher water solubility of gypsum compared with lime ($CaCO_3$), gypsum, or gypsum-containing phosphate fertilizers (e.g., single superphosphate compared with triple superphosphate) are more suitable for amelioration of subsoil acidity than lime (Ritchey *et al.*, 1980; Alva and Sumner, 1990). Aluminum toxicity on plant growth can also be effectively decreased by supply of silicon in nutrient solutions, an aspect which deserves attention in screening for aluminum tolerance (Barcelo *et al.*, 1993).

The amelioration of aluminum toxicity by soil organic matter is well documented. As demonstrated by Adams and Moore (1983), root growth of soybean was depressed in the subsoil (low in organic matter) by 4 μM Al compared with 9–134 μM Al in the soil solution of the top soil (high in organic matter). Application of mulch (Duong and Diep, 1986) or green manure (Hue and Amien, 1989) are therefore quite effective in the amelioration of aluminum toxicity in acid soils. Fulvic acid is one of the compounds which effectively complexes aluminum and thereby ameliorates the phytotoxicity of monomeric (and polymeric) aluminum species on root growth (Suthipradit *et al.*, 1990). Organic acids also play a major role in aluminum detoxification. Their detoxifying capacity decreases in the order citric, oxalic > malic > succinic acid, and depends upon the orientation of their OH/COOH groups and the corresponding possibility of forming a stable 5- or 6-bond ring structure with aluminum (Hue *et al.*, 1986). An example of detoxification of aluminum by citric acid is shown in Table 16.10. Between pH 3.5 and 6 the predominant citrate (L) complexes are $AlL°$ and $AlLH^{-1}$ (Martin, 1988). As with soil organic matter extracts, citric acid completely protected roots from the harmful effects of free aluminum. With aluminum citrate, aluminum accumulation in roots was much lower than with inorganic aluminum, and there was only a minor reduction in the phosphorus and calcium contents in roots and shoots. The results also clearly demon-

Table 16.10

Effect of Aluminum Treatments on the Dry Weight and Mineral Element Content of Maize[a]

Treatment[b]	Dry weight (g per plant)	Mineral element content (μmol g^{-1} dry wt)		
		Al	P	Ca
Shoots				
Control (−Al)	1.96	0.5	58	55
Al(OH)$_2$Cl	1.05	1.9	24	28
Al Citrate	2.09	0.9	54	55
Al-Soil organic matter extract	1.85	0.5	68	63
Roots				
Control (−Al)	1.14	0.2	96	103
Al(OH)$_2$Cl	0.51	276.0	91	37
Al Citrate	1.17	62.0	99	89
Al-Soil organic matter extract	1.03	14.0	77	71

[a]From Bartlett and Riego (1972).
[b]The concentration of aluminum in aluminum-containing treatments was 0.33 mM.

strate the limited transport of aluminum into the shoots when inorganic aluminum was supplied.

16.3.5 Effects of Excessive Manganese Levels

In contrast to aluminum, manganese is readily transported from the roots to shoots and therefore, as a rule, symptoms of manganese toxicity first occur on the shoots. The effects of an excessive manganese supply on the uptake of other mineral nutrients, on metabolism, and on phytohormone balance have been summarized by Horst (1988) and were discussed in Section 9.2. Of particular importance for plant growth in acid mineral soils is the inhibition of calcium and magnesium uptake by high manganese concentrations. Crinkle leaf in young leaves and chlorotic speckling in mature leaves are well-known symptoms of manganese toxicity in dicotyledonous species grown in acid soils and are probably expressions, respectively, of induced deficiency of calcium and perhaps magnesium. Under these conditions visible symptoms of manganese toxicity are observed even at levels which may decrease growth only slightly, in contrast to aluminum toxicity, which severely inhibits growth without producing clearly identifiable symptoms in the shoot. Hence in acid mineral soils with high exchangeable levels of both aluminum and manganese the growth depression observed may be erroneously attributed to manganese toxicity when in fact aluminum toxicity is the more important of the two factors (Foy et al., 1978). On the other hand, in many instances manganese toxicity may in fact be induced magnesium deficiency. Manganese depresses magnesium uptake by blocking binding sites of magnesium in the roots (Le Bot et al., 1990; Section 2.5.3). Therefore, high manganese concentrations in the rooting medium may inhibit root and shoot growth by induced magnesium deficiency (Langheinrich et al., 1992). In manganese toxic soils growth inhibition may therefore be overcome by higher magnesium supply (Goss and Carvalho, 1992).

As a rule, with soil acidification manganese concentrations in the soil solution increase much more than the corresponding manganese uptake and contents in the shoots (Marschner, 1988). This effect can mainly be attributed to the strong inhibitory effects of high H^+ concentrations on uptake of manganese (Mn^{2+}). The occurrence of manganese toxicity is not only a function of soil pH, concentrations of manganese and other polyvalent cations in the soil solution, plant species and genotype and microbial activity in the rhizosphere (Section 15.3.4) but also of the availability of silicon. Silicon strongly increases the tolerance of the shoot tissue to high manganese contents (Section 10.3). Thus, on acid mineral soils the harmful effects of excessive manganese levels on plant growth may also depend on solubility and uptake of silicon.

In legumes, manganese toxicity also depends on the form of nitrogen nutrition. When the supply of manganese to bean (*Phaseolus vulgaris* L.) is large, the nitrogen content of the shoots decreases to a much greater extent in plants depending on N_2 fixation than in plants fed with mineral nitrogen (Döbereiner, 1966). Similar negative effects of high manganese levels on nodulation have been observed in other legume species (Foy et al., 1978; Evans et al., 1987), although, at least in isolated culture, most *Rhizobium* strains are more sensitive to aluminum than to manganese (Keyser and Munns, 1979). In conclusion, nodulation is a very critical step for legumes in acid mineral soils; it is adversely affected by a combination of high aluminum or manganese or both and low calcium concentrations. Low phosphorus availability on acid mineral soils is another effect impairing nodulation (Section 7.4.5).

16.3.6   Mechanisms of Adaptation to Acid Mineral Soils

16.3.6.1   General

Plants adapted to acid mineral soils utilize a variety of mechanisms to cope with the adverse chemical soil factors. These mechanisms are regulated separately (e.g., those of aluminum and manganese tolerance) or are interrelated (e.g., those of aluminum tolerance and efficiency in phosphorus acquisition). From the agronomic view-point, for crop plants the sum of the individual mechanisms is of importance because it determines the requirement of inputs for amelioration of acid soils (fertilizers and lime in particular). In large areas of the tropics and subtropics, phosphorus deficiency is the most important nutritional factor limiting the growth of crop plants (Sanchez and Salinas, 1981).

Large differences occur between the various crop species in their tolerance to acid soils. For example, the annual root crop species, cassava (*Manihot esculenta* Crantz) is known for its high tolerance to acid soils, compared, for example, to sweet potatoes, taniers, and yams (Fig. 16.8). Other acid soil-tolerant crop species are cowpea, peanut, and potato, whereas maize, soybean, and wheat are nontolerant species (Sanchez and Salinas, 1981). A fairly large input of nutrients is necessary to adjust the soil chemical properties, mainly by liming, to the requirements of nontolerant species (Fig. 16.8). It is noteworthy that despite the large yield differences brought about by an alteration in soil pH by liming in three of the four root crops, the macronutrient and manganese content of the leaves was hardly affected, except that of calcium (Abruna-Rodriguez *et al.*, 1982). Here, foliar analyis would be of limited value in determining both the mechanisms of adaptation and nutritional status of plants.

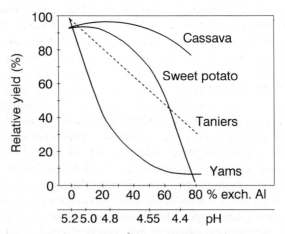

Fig. 16.8   Relationship between exchangeable aluminum (percentage of total cation-exchange capacity), soil pH and yield of four tropical root crops. (Redrawn from Abruna-Rodriguez *et al.*, 1982, by permission of the Soil Science Society of America.)

Differences in acid soil tolerance between cultivars of a given species can be quite large. For example, in an unlimed soil of pH 4.5 and 80% aluminum saturation, a traditional, adapted dryland rice cultivar produced ~2.3 t of grain per hectare, compared with an introduced nonadapted cultivar, which produced only 1 t; the latter required ~6 t of lime ha^{-1} and a corresponding decrease in aluminum saturation to 15% to achieve the grain yield of the traditional, adapted cultivar in the unlimed soil (Spain *et al.*, 1975).

As a rule aluminum tolerance is the most important individual factor required for adaptation of species and cultivars to acid mineral soils. As a first approximation growth inhibition by increasing aluminum concentration in a nutrient solution is therefore a suitable parameter for such adaptation of contrasting plant species or genotypes within a species, if the various limitations of this approach are taken into account (e.g., role of root exudates; Horst *et al.*, 1990). In a large-scale screening program on aluminum tolerance of 34 plant species (including 87 cultivars) grown in diluted nutrient solutions, aluminum concentrations needed for a reduction of shoot dry weight to 50% varied between species and cultivars from less than 1 μM Al^{3+} in the most sensitive genotypes to more than 30 μM Al^{3+} in tolerant genotypes (Wheeler *et al.*, 1992c). By using inhibition of root elongation as a parameter, the critical aluminum concentrations in the nutrient solution varied between 1.8 μM in barley and 150 μM in rye and yellow lupin (Horst and Göppel, 1986a). Large differences in aluminum tolerance also exist within a given species (e.g., Fig. 16.4), and in crop plants some of this genetic variability appears to have been introduced unintentionally by breeding the same species in different regions with high or low soil pH, as in the case of wheat (Foy *et al.*, 1974; Mugwira *et al.*, 1981) or soybean (Lafever *et al.*, 1977).

16.3.6.2   Tolerance versus Avoidance

According to Levitt's general stress concept adaptation can be achieved by tolerating the stress or avoidance of the stress factor(s) or both of these means. Because of the various stress factors to which plants are exposed when growing on acid mineral soils, as a rule both strategies are most probably required simultaneously, although to varying degrees. The most important components of the tolerance and avoidance strategy of plant adaptation are summarized in Fig. 16.9.

16.3.6.3   Aluminum Tolerance

Aluminum tolerance achieved by accumulation (includer, Fig. 16.9) seems to be especially common in those plant families that were present in the early fossil history. The Proteaceae belong to this group (Chenery and Sporne, 1976). In tropical rain forests includers and excluders which coexist at the same sites vary in aluminum concentrations in the leaf press sap between less than 10 mg l^{-1} and 4780 mg l^{-1} (Cuenca *et al.*, 1990). Only a few cultivated species are aluminum includers, such as tea plants, the old leaves of which may contain up to 30 mg aluminum per gram dry weight (Matsumoto *et al.*, 1976b). Tea plants not only tolerate high aluminum contents but also their growth is strongly enhanced by aluminum supply (Konishi *et al.*, 1985). Growth stimulation by aluminum supply (up to 100 μM Al) has also been observed in calcifuge

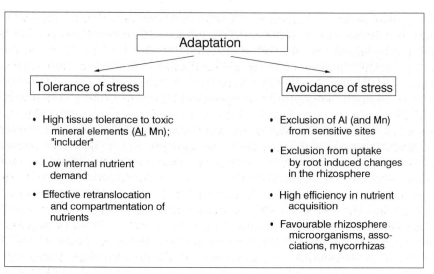

Fig. 16.9   Strategies of plant adaptation to acid mineral soils.

species such as *Deschampsia flexuosa* and *Arnica montana* L. (Pegtel, 1987), but information is lacking as to whether these species also belong to the includer types. There are also several reports on stimulatory effects of low aluminum concentrations on growth of tolerant crop species and cultivars (Foy, 1983) but the mechanisms of this stimulation is not clear (Section 10.6). In turnip (*Brassica campestris* L.) which is highly sensitive to low pH, stimulation of root elongation by low aluminum concentrations (0.6–1.2 μM) in nutrient solution was confined to the low pH (4.6), suggesting amelioration of proton toxicity by aluminum (Kinraide and Parker, 1990).

Most plant species, and crop species in particular, achieve aluminum tolerance mainly by exclusion, at least from the shoots. In some tolerant species such as rye and yellow lupin, between 3–4 mg Al g^{-1} dry weight might accumulate in root apical zones before elongation growth is inhibited (Horst and Göppel, 1986b), but how this aluminum is compartmented between apoplasm, cytoplasm, and vacuole is not known. In principle in excluder types aluminum tolerance may be achieved by exclusion from sensitive sites in roots, or from uptake in general by root-induced changes in the rhizosphere (Fig. 16.9). The following discussion considers factors which may be involved in exclusion mechanisms.

Exclusion from Sensitive Sites. In terms of aluminum toxicity sensitive sites are the cytoplasm (e.g. interfering with calmodulin; Suhayda and Haug, 1985), the plasma membrane–apoplasm interface, the apoplasm, and the peripheral root cap cells (Section 16.3.3.2). In wheat, genotypical differences in aluminum tolerance were not reflected in differences in aluminum distribution between apoplasm and symplasm (Tice *et al.*, 1991). Speculations concerning the induction of aluminum tolerance in wheat by formation of particular proteins to inactivate aluminum in the symplasm (Aniol, 1984) deserve further attention, as in wheat cultivars aluminum treatment induces or increases the level of several proteins, and three of the cytoplasmic proteins were exclusive to the tolerant cultivar (Ownby and Hruschka, 1991). Differences in the plasma membrane surface potential (Wagatsuma and Akiba, 1989) and binding of aluminum at the plasma membrane (Caldwell, 1989) have also been suggested as possible causative factors for differences in aluminum tolerance, but the evidence so far is weak. For a review of these aspects see Haug and Shi (1991).

For many years differences between species and cultivars in cation exchange capacity (CEC) of the root apoplasm have been discussed in relation to respective differences in aluminum tolerance. As the cation-exchange capacity (CEC) of root tissue increases, there is also a greater exchange adsorption of polyvalent cations in the apoplasm. In acid soils this could lead to a higher aluminum accumulation as indicated, for example, by a close positive correlation between the CEC of different plant species and the aluminum content in their roots (Wagatsuma, 1983). There is also some evidence for a negative correlation between CEC and aluminum tolerance in wheat and barley cultivars (Foy *et al.*, 1967), two *Lotus* species (Blamey *et al.*, 1990) and in populations of certain species of wild plants (Büscher and Koedam, 1983). A general role of CEC in this respect is unlikely, however, as for example dicots, which have a high CEC, are generally not less aluminum tolerant than monocots, which have a low CEC. In a study of 12 monocots and dicots differing in CEC by a factor of 17, the aluminum tolerance of these species was not closely associated with the root CEC (Blamey *et al.*, 1992). However, to characterize a possible causal role of aluminum binding in the apoplasm for aluminum tolerance these quantitative methods for CEC determination are inadequate and have to be complemented by qualitative methods providing information on binding strength and affinities for aluminum species and other cations, H^+, Ca^{2+} and Mg^{2+} in particular (Grauer and Horst, 1992; Grauer, 1993).

Rhizosphere pH. In the pH range between 4 and 5 an increase in rhizosphere pH should greatly decrease the concentration of Al^{3+} (Fig. 16.6), but may simultaneously increase the phytotoxicity of aluminum (Section 16.3.4). Although in nutrient solutions aluminum-tolerant cultivars of certain crop species tend to increase solution pH more than do aluminum-sensitive cultivars (Mugwira and Patel, 1977; Foy and Fleming, 1982), studies with different forms of nitrogen (NH_4^+ versus NO_3^-) indicate that an increase in solution pH is of minor importance for higher aluminum tolerance in cultivars of wheat (Taylor, 1988b) and soybean (Klotz and Horst, 1988a). When the pH of the nutrient solution is strictly controlled, yellow lupin was even more tolerant at pH 4.1 than at 4.5 (Grauer and Horst, 1990), probably reflecting proton amelioration. However, the findings of nutrient solution experiments have to be interpreted with care in relation to the aluminum tolerance of soil-grown plants (Haug and Feger, 1990). In plants grown in, and adapted to, acid soils the pH in apical root zones is often considerably higher than in basal zones and in the bulk soil (Fig. 16.10), and such an increase in rhizosphere pH may decrease both the exchangeable aluminum (Gahoonia, 1993) and the release of aluminum from the solid phase into the rhizosoil solution. For example, in the pH range between 4.0 and 4.5 a pH increase of 0.2 units depresses the aluminum concentration in the soil solution by a factor of 2–3 (Tyler *et al.*, 1987). Furthermore, polynuclear aluminum species like 'Al_{13}' have a high affinity to negatively charged surfaces of the solid phase of the soil (Wright, 1989).

For plant adaptation to acid mineral soils (Fig. 16.9) root-induced changes in

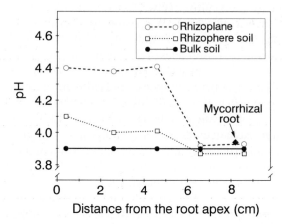

Fig. 16.10 pH pattern in bulk and rhizosphere soil and at the rhizoplane along nonmycorrhizal roots of 80-year-old Norway spruce. (Marschner, 1991b; courtesy of M. Häussling.)

rhizosphere pH cannot only be evaluated in terms of aluminum concentrations and species. With increasing rhizosphere pH, H^+ toxicity can be eliminated and the binding of Ca^{2+} and Mg^{2+} in the root apoplasm can be increased. Also the function of calcium in secretion of cell wall material in expanding tissues (Steer, 1988b) is presumably supported by an increase in rhizosphere pH.

Root exudates. Root exudates such as mucilage and organic acids play a key role in the avoidance strategy. For a given genotype an increase in root exudation by mechanical impedance increases aluminum tolerance more than 10-fold (Section 15.4.2). Mucilage is mainly secreted at the root cap and root apical zones and has a high capacity for binding (Section 15.4.2.2) and complexing aluminum (Fig. 16.11). The role of mucilage for genotypical differences in aluminum tolerance has been shown in wheat (Puthota *et*

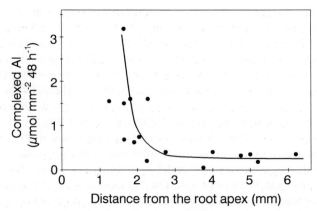

Fig. 16.11 Complexation of monomeric aluminum in the external solution along apical zones of intact roots of 4-year-old Norway spruce seedlings. (Häussling *et al.*, 1990.)

Table 16.11
Effect of Aluminum on Dry Weight and Root Exudation of Citric Acid
(Concentration in Medium) in Axenic Culture of two Cultivars of Bean
(*Phaseolus vulgaris* L.) Differing in Aluminium Tolerance[a]

Cultivar	Al (μM)	Dry weight (g per plant)		Citric acid concn. (μM)	Final pH
		Shoot	Roots		
Dade	0	0.58	0.20	0.52	6.48
(Al tolerant)	148	0.46	0.13	38.41	4.99
Romano	0	0.86	0.35	2.76	6.04
(Al sensitive)	148	0.54	0.16	3.07	4.74

[a]Based on Miyasaka *et al.* (1991). Reprinted by permission of the American Society of Plant Physiologists.

al., 1991). In the absence of aluminum mucilage production was three times higher in the tolerant (Atlas) than the sensitive (Victor) cultivar, and mucilage production ceased at 20 μM aluminum supply in the sensitive and at 400 μM in the tolerant cultivar. In natural grassland on acid soils the dominance of the unpalatable grass *Aristida juniformis* is most likely related to its high aluminum tolerance brought about by an unusually high production of root cap mucilage (Johnson and Bennet, 1991).

The release of relatively large amounts of organic acids is a typical feature of plant species adapted to highly acid mineral soils, as for example, by certain lupin species (Section 15.4.2) and tea plants (Jayman and Sivasubramaniam, 1975). Complexing aluminum by organic acids such as citric acid (Table 16.10) not only provides protection against the harmful effects of free aluminum on root growth, it is also important for the acquisition of phosphorus (Section 15.4). In acid forest soils concentrations of low-molecular-weight organic acids (especially oxalic acid) in the soil solution may vary between 25 and 1000 μM and their concentrations are distinctly higher in the rhizosphere as compared with the bulk soil solution (Fox and Comerford, 1990).

In wheat cultivars differing in aluminum tolerance the release of organic acids was about three times higher in the tolerant cultivar which also had a much higher capacity for associative N_2 fixation in the rhizosphere (Christiansen-Weniger *et al.*, 1992), thus combining tolerance and nutrient efficiency. By using axenic cultures convincing evidence has been presented on the crucial role of excretion of organic acids for aluminum tolerance in isogenic wheat lines. In sensitive genotypes (lines) aluminum accumulation in the apical root zones was 5–10 times higher than in tolerant genotypes (Delhaize *et al.*, 1993a) which excreted 5–10 times more malic acid than the sensitive genotypes (Delhaize *et al.*, 1993b). Excretion was enhanced by as little as 10 μM Al and continued linearly over 24 h. Malic acid was excreted from the apical 3–4 mm. When exposed to 50 μM Al excretion rate of malic acid per seedling (nmol h^{-1}) increased from 0.08 to 0.33 in sensitive and from 0.08 to 3.57 in tolerant genotypes. In bean too one of the mechanisms of aluminum tolerance is most probably enhanced release of organic acids, citric acid in particular (Table 16.11). Citric acid release in the tolerant cultivar may have been induced either by toxic effects of aluminum, or by lowering phosphorus availability, i.e., as a root response to phosphorus deficiency. In this experiment only high aluminum concentrations which severely inhibited root growth of the sensitive cultivar were tested, and it is possible that the sensitive cultivar may also have the capacity to release organic acids, but only at a lower aluminum level.

Enhanced release of organic acids under phosphorus deficiency is a typical feature in many dicots (Section 15.4.2.4) and may be an important component in the strategies of plant adaptation to acid mineral soils for both increasing efficiency in mineral nutrient acquisition and avoidance of aluminum toxicity. A similar mechanism is assumed to operate in certain *Eucalyptus* species adapted to extremely acid, phosphorus-deficient soils (Mulette *et al.*, 1974). The formation of complexes of aluminum with polyphenols or organic acids leached from leaves or litter might offer an indirect way for certain *Eucalyptus* species to achieve both high aluminum tolerance and acquisition of phosphorus from extremely deficient soils (Ellis, 1971).

Mycorrhizas. Root colonization with mycorrhizas is another important component in the adaptation to acid mineral soils with inherent low phosphorus availability and high aluminum concentrations. The role of VA mycorrhiza is particularly evident for plant species with coarse root systems such as cassava (Howeler *et al.*, 1987; Sieverding, 1991). It is also important in plant species or genotypes where phosphorus deficiency-induced root responses such as enlargement of the root system are impaired by aluminum toxicity (low or moderate aluminum tolerance), or where an enhanced release of organic acids by the roots does not occur. The dependency on VA mycorrhiza for plant growth on aluminum-toxic tropical soils has been shown for many arable tropical crop species (Howeler *et al.*, 1987; Sieverding, 1991) and may also hold true for tropical pasture species such as *Stylosanthes guianensis* (Lambais and Cardoso, 1990).

In temperate climates in forest trees grown on acid soils, ectomycorrhiza is important not only for phosphorus acquisition (Section 15.4) but may also play a role in the protection of roots from aluminum toxicity. In *Pinus rigida* a concentration as low as 50 μM aluminum in the nutrient solution reduced root and shoot growth and phosphorus content in the needles of nonmycorrhizal plants (Cumming and Weinstein, 1990). At 200 μM aluminum growth inhibition was severe in nonmycorrhizal plants, but growth inhibition was prevented by mycorrhizal inoculation. Mycorrhizal plants also maintained a higher phosphorus but lower aluminum content in the needles compared with the nonmycorrhizal plants. Also in Norway spruce seedlings grown in sand culture root colonization with ectomycorrhiza increased the tolerance of these plants to aluminum (Hentschel *et al.*, 1993); shoot growth and chlorophyll contents in particular were higher in the mycorrhizal plants, despite the lower calcium and magnesium contents of the needles. Protection of root hormonal balance from the detrimental effects of aluminum (Pan *et al.*, 1989; Rengel, 1992a) may be involved in these beneficial effects of mycorrhizas. Generalizations on the increase in aluminum tolerance of trees by ectomycorrhizas are, however, not possible. This varies between fungal species as has been shown in jack pine seedlings (Jones *et al.*, 1986).

Screening for Aluminum Tolerance. Field screening of genotypes in acid soils is a labour-intensive process, requiring several months for completion, and is often influenced by secondary factors such as differences in resistance to diseases and pests which occur between genotypes. To avoid these problems, rapid screening methods have been developed, mainly in water culture systems. In most crop species the relative root length of plants exposed compared to those not exposed to aluminum provides a suitable growth parameter, and the formation of callose a more physiological parameter (Wissemeier et al., 1992). Because aluminum toxicity to roots is affected by a number of factors (e.g., concentration of aluminum, calcium, and magnesium, solution pH) the biggest problem in developing rapid screening techniques is finding an appropriate combination of these factors to use (Rhue and Grogan, 1977; Blamey et al., 1991). Tissue cultures exposed to high aluminum concentrations, and regeneration of plants from selected calli of high aluminum tolerance may offer another approach in the use of screening techniques for high aluminum tolerance in plants (Arihara et al., 1991).

In some cases the classification of genotypes (cultivars) based on their aluminum tolerance in rapid screening methods correlates well with the growth response of these genotypes in acid soils, for example, with wheat and barley (Foy et al., 1967, 1974) and dryland rice (Howeler and Cadavid, 1976). Quite often, however, the correlations are poor, for example, with dryland rice (Nelson, 1983), or the results differ between various acid soils, for example, with wheat and triticale (Mugwira et al., 1981). These discrepancies are not surprising in view of the various stress factors, in addition to aluminum toxicity, to which plants are exposed on acid mineral soils, and in view of the various strategies plants have developed to adapt to these soil conditions (Fig. 16.9). Short-term bioassay techniques by using root elongation of seedlings in acid mineral soils (Horst, 1985; Wright et al., 1989a) may serve as useful practical orientated compromise to overcome at least some of these problems.

Selection of aluminum-tolerant species and cultivars in nutrient solutions for improved root development in acid subsoils under field conditions may also be inadequate for another reason (Hairiah et al., 1992). Highly aluminum tolerant species show aluminum avoidance in a heterogeneous medium and proliferate roots into phosphorus-rich zones (e.g., top soil).

In legume genotypes impaired growth in acid mineral soils can be due to failure of one of the symbiotic partners, or the symbiosis is more sensitive than either host or bacterium independently (Munns, 1986). At least in alfalfa there is promising progress showing that it is possible to select tolerant host plant germplasm and combine it with acid and aluminum tolerant *Rhizobium meliloti* strains to increase plant growth under field conditions in acid soils (Hartel and Bouton, 1991). The possibilities and limitations of the various screening techniques have been summarized by Howeler (1991).

16.3.6.4 Manganese Tolerance

Mechanisms of manganese tolerance are located in the shoots, as indicated, for example, by reciprocal root stock–scion grafts of tolerant and nontolerant genotypes (Heenan et al., 1981). It is the genotype of the scion that determines the tolerance of the plant to high manganese concentrations in the substrate and within the leaf tissue.

The critical toxicity concentrations of manganese in leaf tissue differ to a great extent between plant species (Section 9.2), as well as cultivars of a species such as soybean (Table 16.12) and cowpea (Horst, 1983, 1988). In mature leaves of tolerant cultivars manganese is uniformly distributed, whereas in nontolerant cultivars with the same

Table 16.12

Effect of the Manganese Concentration in the Nutrient Solution on the Dry
Weight and Manganese Content of Soybean Cultivars[a]

Cultivar	Mn Supply (mg l⁻¹)	Dry weight (g per plant)		Mn content of shoots (μg g⁻¹ dry wt)
		Shoot	Roots	
T 203	1.5	5.4	0.61	208
	4.5	6.6	0.55	403
	6.5	7.0	0.55	527
Bragg	1.5	5.7	0.59	297
	4.5	5.3	0.64	438
	6.5	4.5	0.68	532

[a]Based on Brown and Devine (1980).

manganese content, a local spotlike accumulation of manganese occurs, which is
correlated with chlorosis and necrosis at the sites of accumulation (Horst, 1983, 1988).
The distribution of manganese in leaf blades of nontolerant and tolerant cultivars is
similar to that obtained when a plant is supplied with an excessive amount of manganese
in the absence or presence of silicon. In the absence of silicon in nontolerant cultivars
the manganese is concentrated in localized areas whereas in tolerant cultivars it is
evenly distributed (Section 10.3). Restricted translocation of manganese to young
leaves might also be involved in the high manganese tolerance of certain lettuce
cultivars (Blatt and Diest, 1981). An alternative explanation for higher manganese
tolerance might be that calcium transport to apical meristems and young leaves is less
impaired (Horst, 1988). It might be that in tolerant perennial species the formation of
manganese complexes of high stability, for example with oxalic acid (Memon and
Yatazawa, 1984) or polyphenols (Horst, 1988) are also of importance for genotypical
differences in manganese tolerance. Such mechanisms may play a role in the high
manganese tolerance in needles of coniferous trees such as Norway spruce.

Tolerance to excessive levels of manganese is not necessarily correlated with
tolerance to excessive levels of aluminum (Nelson, 1983) and vice versa, as has been
shown for wheat cultivars (Foy *et al.*, 1973). Separate screening for manganese
tolerance and aluminum tolerance is therefore necessary; a combination is required
only under certain soil conditions. The application of manganese to the petioles of
leaves provides a simple, rapid, and nondestructive method for screening cowpea for
manganese tolerance during vegetative growth (Horst, 1982). Induction of callose
formation is inversely related to the manganese tolerance of a tissue and thus a suitable
physiological parameter for screening genotypes of high manganese tolerance (Wissemeier *et al.*, 1992).

Manganese tolerance during vegetative growth is not necessarily correlated with
tolerance during reproductive growth (Horst, 1982, 1988). At least in legumes (soybean, cowpea) manganese toxicity may reduce grain yield more than vegetative growth.
For example, under field conditions some cowpea genotypes growing in a soil with an
excessive manganese level produce little or no grain despite vigorous vegetative growth

Table 16.13

Dry Matter Production and Calcium Contents in Roots and Shoots of Two Cowpea (*Vigna unguiculata* L. Welp) Cultivars Supplied with Different Calcium
Concentrations[a]

Supply (μM Ca^{2+})	Cultivar	Dry wt (g per plant)	Calcium content (μmol g^{-1} dry wt)		
			Roots	Stem	Leaves
10	Solojo	0.75	34	40	35
	TVu 354	1.75	25	46	34
50	Solojo	2.10	37	58	62
	TVu 354	1.80	32	70	57

[a]Based on Horst *et al.* (1992a).

(Kang and Fox, 1980). The application of manganese to the peduncle seems promising
as a technique for screening for manganese tolerance during reproductive growth
(Horst, 1982).

16.3.6.5 Nutrient Efficiency

Adaptation of plants to acid soils requires highly efficient uptake or utilization of
nutrients or both, especially of phosphorus, calcium, and magnesium. Many of the
plant species considered to be adapted to acid mineral soils are usually heavily infected
with mycorrhizas (Sanchez and Salinas, 1981; Howeler *et al.*, 1987). Tolerance to
aluminum and high phosphorus efficiency coexist in tropical root crops such as cassava
(Howeler *et al.*, 1987) and in certain cultivars of wheat and dryland rice (Sanchez and
Salinas, 1981).

In contrast to phosphorus efficiency, a high calcium efficiency is usually related to
better utilization within plants. In this respect genotypical differences can be quite
distinct, as shown for cowpea in Table 16.13. With low calcium supply the inefficient
cultivar Solojo died whereas the efficient cultivar TVu 354 was hardly affected in dry
matter production. Differences could not be explained by differences in calcium uptake
and concentrations in the plant tissue. The main reason for cultivar differences is
obviously a lower functional calcium requirement in the efficient cultivar (Horst *et al.*,
1992a). In various cowpea cultivars a positive relationship has been observed between
aluminum tolerance and calcium efficiency (Horst, 1987). Selection for aluminum
tolerance should therefore take into account that these genotypes are also adapted to
low calcium supply.

In sorghum genotypes marked differences in the inhibition of magnesium uptake by
aluminum have been found, suggesting that genotypical differences in binding of
magnesium in the root apoplasm in the presence of aluminum might be a contributing
factor to aluminum tolerance of this species (Tan *et al.*, 1993) as has also been suggested
for ryegrass (Rengel and Robinson, 1989a).

In acid mineral soils molybdenum availability is very low (Section 9.5). Thus
molybdenum efficiency may sometimes be involved in adaptation to acid mineral soils.
This was demonstrated by Brown and Clark (1974) in a comparison of two maize inbred

lines, Pa 36 and WH, grown in an acid soil (pH 4.3). The poorer growth of Pa 36 compared with that of WH was caused by insufficient molybdenum uptake. Thus, low molybdenum efficiency may limit the overall adaptation of Pa 36 to certain acid mineral soils, despite the high efficiency of this inbred line to take up phosphorus from low phosphorus soils, even in the presence of aluminum (Brown and Clark, 1974).

16.4 Waterlogged and Flooded Soils

16.4.1 Soil Chemical Factors

In waterlogged* soils, air is displaced from the pore spaces either to different depths of subsoil (which results in a high water table) or in the topsoil. This often occurs in temperate climates during the winter and spring and also, temporarily, during summer following heavy rainfall or excessive irrigation on slowly draining or poorly structured soils. Flooded or submerged soils are permanently below the water table, or at least under water for several months every year. Paddy soils are the most well-known agricultural example of such soils (Ponnamperuma, 1972). Since oxygen (and other gases) diffuse in air about 10^3–10^4 times more rapidly than in water or water-saturated soils (Armstrong, 1979), oxygen is depleted more or less rapidly by the respiration of soil microorganisms and plant roots in waterlogged soils. Various degrees of oxygen depletion (*hypoxia*) and *anoxia* (the absence of molecular oxygen) occur. Once molecular oxygen has been consumed in respiration, various populations of micro-organisms utilize other terminal electron acceptors for respiration. The sequence of reduction takes place at specific redox potentials and is shown for mineral nutrients in Table 16.14. As soils are extremely heterogeneous and characterized by microsites differing in pore size, water content and microbial activity, redox potentials vary widely over short distances. The change from oxygen sufficiency to deficiency can occur within a few millimeters, and even in aerobic soils the interior of soil aggregates may be anaerobic (Bohn *et al.*, 1985).

Table 16.14
Redox Reactions in Relation to the Soil Redox Potential[a]

Redox reaction	Redox potential E_0' (mV) at pH 7
Onset of nitrate reduction (denitrification)	450–550
Onset of Mn^{2+} formation	350–450
Absence of free O_2	330
Absence of nitrate	220
Onset of Fe^{2+} formation	150
Onset of sulfate reduction (H_2S formation)	−50
Absence of sulfate	−180

[a]Based on Brümmer (1974).

*The terms *waterlogged* and *flooded* are, as a rule, used synonymously to describe soils with excessive water levels.

As soon as free oxygen is depleted nitrate is used by soil microorganisms as an alternative electron acceptor in respiration. Nitrate is reduced to nitrite (NO_2^-), various nitrous oxides (e.g., N_2O, NO), and molecular nitrogen (N_2) in the process of denitrification. Nitrite can accumulate temporarily during this process, especially in soils that are alternatively wet and dry. Particularly in acid soils subjected to heterotrophic nitrification (Killham, 1990), nitrite and various nitrous oxides are also formed during nitrification ($NH_4^+ \rightarrow NO_3^-$). Denitrification may be enhanced in the rhizosphere, for example, in wheat when soil water contents are high, suggesting lower redox potentials through oxygen consumption by roots and rhizosphere microorganisms (Prade and Trolldenier, 1990a,b). Manganese oxides [mainly Mn(VI)] are the next electron acceptors. In acid soils high in manganese oxides and organic matter but low in nitrate, very high levels of water-soluble and exchangeable Mn^{2+} can build up within a few days. After prolonged waterlogging the reduction of Fe(III) occurs. Iron reduction is associated with a marked increase in soil pH (Kirk *et al.*, 1990). For plant species adapted to flooded soils (wetland species, e.g., rice) enhanced iron reduction increases the availability of iron, and quite often plants may suffer from excessive iron levels. Iron reduction increases phosphate solubility and availability to wetland species if Fe(III)PO$_4$ is present in the soil in sufficient amounts. The reduction of sulfate to H_2S in submerged soils may decrease the solubility of iron, zinc, copper, and cadmium by the formation of sparingly soluble sulfides (Ponnamperuma, 1972; Herms and Brümmer, 1979). Zinc deficiency is widespread in rice, not through formation of zinc sulfide but because of the formation of sparingly soluble zinc compounds in the oxidized rhizosphere (Sajwan and Lindsay, 1988).

Various products of microbial carbon metabolism, such as ethylene, accumulate in waterlogged soils. During prolonged waterlogging, volatile fatty acids and phenolics accumulate in soils high in readily decomposable organic matter (e.g., after application of green manure, straw), which has a detrimental effect on root metabolism and growth (Section 14.4.4). In submerged soils at very low redox potential large amounts of methane (CH_4) may also be formed, for example, from acetic acid. Wetland rice fields are a major source of CH_4 emission and therefore of concern in relation to global warming. Methane emissions are higher in planted than in unplanted paddy fields, as root exudates may increase methane production in the rhizosphere (Kimura *et al.*, 1991) and the porous rice roots enhance its transport in the aerenchyma to the shoots and subsequent release into the atmosphere (Nouchi *et al.*, 1990).

16.4.2 Plant Responses

16.4.2.1 Waterlogging Injury

Most plant species not adapted to waterlogging (nonwetland, mesophytic species) develop injury symptoms sequentially over a period of several days if waterlogging continues. Wilting, leaf senescence and, in herbaceous species, epinasty (downward bending of leaves) are likely to be the first symptoms (Drew, 1990). A decrease in the hydraulic conductivity of the roots and an accumulation of ethylene in the shoots are responsible, respectively, for wilting and epinasty (Bradford *et al.*, 1982; Drew, 1990). A rapid decline in or cessation of shoot extension growth is another typical symptom, followed after several days of waterlogging by enhanced senescence of the lower leaves, indicating nitrogen deficiency or lack of root-borne cytokinins. Waterlogging can impair the nitrogen nutrition of legumes by interfering with nodulation (Tran Dang Hong *et al.*, 1977).

The severity of the effects of waterlogging on growth and yield depends on the plant species, developmental stage of the plants, soil properties (e.g., pH, organic matter content), and soil temperature in particular. Peas are especially sensitive to waterlogging. After only one day or several days, enhanced senescence occurs, and shortly after flowering 24 h of waterlogging can depress yields appreciably (Cannell *et al.*, 1979). Because low soil temperatures decrease the oxygen requirement for respiration, the drop in redox potential, and thus injury from waterlogging is less severe at low than at high soil temperatures (Trought and Drew, 1982). Flooding wheat for 30 days during grain filling at soil temperatures of 15 and 25°C reduced grain yield by about 20 and 70%, respectively (Luxmoore *et al.*, 1973). In wheat nitrogen losses by denitrification in the rhizosphere become substantial at an air-filled porosity of the soil of 12–10% and increase steeply below 9–8%, particularly under potassium deficiency. This is presumably because of enhanced root exudation and oxygen consumption by rhizosphere microorganisms in potassium-deficient plants (Prade and Trolldenier, 1990a,b).

Whether the principal cause of waterlogging injury to the shoots is related directly to oxygen deficiency in the roots or more indirectly to the production of toxic substances in the soil depends on circumstances. These causes of injury are discussed in the following section, with emphasis on plant metabolism and mineral nutrition. Comprehensive reviews of the subject are those of Cannell (1977) and Drew (1988, 1990).

16.4.2.2 Growth and Nutrient Uptake

Short-term responses of plants to anaerobic soil conditions can readily be demonstrated by waterlogging of a previously well-aerated soil. The growth of existing roots ceases immediately (Fig. 16.12), and they may die within a few days. In contrast, shoot growth

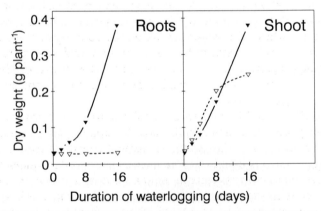

Fig. 16.12   Effect of waterlogging on the dry weight of seminal roots and shoots of winter wheat seedlings (▼, control; ▽, waterlogged). (Based on Trought and Drew, 1980a.)

Table 16.15

Effect of Waterlogging on the Growth and Mineral Nutrient Content of the Shoots of Barley Seedlings[a]

	2 d of waterlogging		6 d of waterlogging		Net uptake, 2–6 d (% of control)
	Control	Waterlogged	Control	Waterlogged	
Leaf extension[b] (cm)	6.4	4.2	12.3	5.2	—
Shoot dry weight (mg per plant)	170	170	380	360	—
Content of the shoots (μmol g^{-1} dry wt)					
Nitrate	390	139	470	14.3	9.9
Phosphorus	217	149	210	71	2.9
Potassium	1540	1190	1420	615	9.6

[a]From Drew and Sisworo (1979).
[b]Leaf no. 3 (youngest).

in terms of dry weight increase continues for several days at a similar or an even somewhat higher rate, although visible symptoms of waterlogging injury (transient wilting, inhibition of leaf extension, and chlorosis) are observed within a few days (Trought and Drew, 1980a).

Cessation of root growth and root respiration leads to a drastic drop in the uptake and transport of mineral nutrients to the shoot within a few days of waterlogging (Table 16.15). Because the shoot dry weight continues to increase, the nutrient concentration in the shoot declines by 'dilution'. There is evidence that at least in several instances inhibited nutrient uptake and thus nutrient deficiency are causally involved in enhanced leaf senescence and cessation of shoot growth in plants subjected to waterlogging (Trought and Drew, 1980a). In maize a lack of root aeration leads to distinctly lower concentrations of nitrogen, potassium and phosphorus in the shoot elongation zone and a decline in shoot elongation growth (Atwell and Steer, 1990). In agreement with this role of nutrient deprivation, in wheat symptoms of enhanced leaf senescence induced by waterlogging could be prevented by the daily application of nitrogen (nitrate or ammonium) to the surface of a waterlogged soil where new roots were developing (Trought and Drew, 1980b). Also, in long-term experiments, high nitrogen fertilizer application alleviated waterlogging injury to cereal crops (Watson et al., 1976) due both to compensation for losses by denitrification and to impaired uptake from poorly aerated soils.

The beneficial effect of additional nitrogen application on shoot growth should not be overestimated and generalized, however, because the uptake of other mineral nutrients is also impaired (Table 16.15). Furthermore, nutrient deficiency is only one aspect of waterlogging injury. In soils high in organic matter and nitrate, sudden waterlogging might lead to an accumulation of nitrite in the soil solution to concentrations which are toxic to the roots of sensitive plant species. Tobacco, for example, is injured by nitrite concentrations as low as 5 mg l^{-1}, but values 10 times higher than this are often found in waterlogged soils high in organic matter (Hamilton and Lowe, 1981). Waterlogging

Table 16.16
Effect of a 3-Day Period of Waterlogging on Shoot Dry Weight and Manganese Content in the Shoots of Alfalfa Grown in a Soil with a Large Organic Matter Content[a]

Soil application (g lime kg^{-1} soil)	Waterlogging	Soil pH	Shoot dry weight (g per pot)	Mn content of shoots (mg kg^{-1} dry wt)
0	−	4.8	3.1	426
0	+	5.2	1.2	6067
2.5	−	6.4	5.7	99
2.5	+	6.7	3.0	957

[a]Based on Graven *et al.* (1965).

injury caused primarily by manganese toxicity occurs in plant species inherently low in tolerance to manganese such as alfalfa (Table 16.16), especially in acid soils containing high levels of manganese oxides. At pH ~5, with a lack of a substantial amount of nitrate to act as an alternative electron acceptor, even a 3-day period of waterlogging leads to a manganese content in the leaves which is extremely toxic to alfalfa. Although liming cannot prevent manganese toxicity induced by short-term waterlogging it can at least considerably lower the manganese content and thus the detrimental effects on growth.

In wetland species excessive iron uptake may cause iron toxicity. 'Bronzing' of leaves is a typical nutritional disorder in wetland rice caused by iron toxicity and leaf contents of 700 mg Fe kg^{-1} dry matter and higher (Yamauchi, 1989). With iron toxicity the activity of polyphenol oxidases is increased and oxidized polyphenols are the cause of 'bronzing' (Peng and Yamauchi, 1993) as is the case with 'brown speckles' with manganese toxicity (Section 9.2.7). In principle, high amounts of iron in plants can give rise to oxygen radical formation as shown below

$$Fe^{2+} + O_2 \rightarrow Fe^{3+} + O_2^-$$
$$O_2^- + O_2^- \xrightarrow{SOD} H_2O_2$$
$$Fe^{2+} + H_2O_2 \rightarrow Fe^{3+} + HO^{\cdot} + HO^-$$

Hydroxyl radicals in particular are highly phytotoxic and responsible for peroxidation of membrane lipids (Sections 2.3) and protein degradation as shown in Table 16.17 for *E. hirsutum*, an iron-intolerant wetland species. Linked with enhanced iron uptake and superoxide radical formation is an increased activity of superoxide dismutase (SOD) which leads to the formation of extra H_2O_2 which has to be detoxified by peroxidases or catalase. However, in the roots of *E. hirsutum* catalase activity was not detectable and the rise in peroxidase activity was small (Hendry and Brocklebank, 1985). In leaves the activities of the H_2O_2 scavenging enzymes are usually higher than in roots (Section 5.2.2), and 'bronzing' may be considered as an expression of enhanced activity of peroxidase and polyphenol oxidase induced by excessive iron levels (Peng and Yamauchi, 1993).

A low concentration of oxygen in the rooting medium decreases the selectivity of K^+/Na^+ uptake by roots in favor of Na^+ and retards the transport of K^+ to the shoots

Table 16.17

Effect of Increasing Iron Concentrations in Nutrient Solution, Gassed for 5 Days with or without O_2, on 6-Week-Old *Epilobium hirsutum* Plants[a]

Iron supply (μM)	Protein (mg g^{-1} fresh wt)		Lipid peroxidation[b]		Root SOD activity (EU mg^{-1} protein)	
	$+O_2$	$-O_2$	$+O_2$	$-O_2$	$+O_2$	$-O_2$
50	3.39	3.41	0.71	0.63	447	543
180	4.07	3.45	0.54	0.67	513	931
450	2.98	1.44	0.78	1.71	412	2725

[a]Compiled data from Hendry and Brocklebank (1985).
[b]μmol malonyldialdehyde oxidized mg^{-1} protein.

(Thomson *et al.*, 1989b; Section 2.5). The enhanced shoot transport of sodium even after short-term (1 h) oxygen deficiency may remain as a 'memory-effect' for several days and is associated with increased synthesis of a membrane-bound protein in roots (Brauer *et al.*, 1987). In accordance with this effect of oxygen deficiency on selectivity of ion uptake, a particular type of waterlogging injury may occur in semiarid and arid regions after heavy irrigation with saline water. In most crop species salt tolerance is based on mechanisms which prevent or at least restrict salt accumulation in the shoots (exclusion mechanisms, Section 16.6). These mechanisms rely on a high metabolic activity in the roots and thus, in nonwetland species, on soil aeration. For a given salinity level in the substrate contents of sodium and chloride in the leaves are therefore increased by both an increase in temperature and waterlogging (Table 16.18). Because of the higher oxygen requirement at 20°C than at 10°C, waterlogging is much more effective at 20°C in increasing the levels of sodium and chloride in the leaves. Salt injury is therefore more likely to take place at 20°C than at 10°C. Similar results have been found in apple trees (West, 1978). In sunflower, waterlogging increased the sodium content in the leaves much more than the chloride content, indicating the impairment of

Table 16.18

Sodium and Chloride Content of Leaves of Tomato Plants Grown at Different Temperatures in Soil and Treated for 15 Days with Saline Solution (90 mM NaCl)[a]

Temperature	Root zone conditions	Leaf content (% dry wt)	
		Sodium	Chloride
10°C	Drained	1.53	2.61
	Waterlogged	1.84	3.04
20°C	Drained	2.96	3.25
	Waterlogged	5.82	9.49

[a]From West and Taylor (1980).

an exclusion mechanism (Na^+ efflux pump, Section 16.6.5) in the roots not adapted to O_2 deficiency in their environment (Kriedemann and Sands, 1984). This interaction between salinity and soil aeration should be considered in the practice of irrigation when saline water has to be used, especially on poorly structured soils.

16.4.2.3 Ethanol, Phytotoxic Metabolites

Even in aerated substrates, because of a high oxygen consumption in root apical meristems in particular, hypoxic conditions may prevail and a high proportion of cells be involved in anaerobic metabolism (Drew, 1990). Under oxygen deficiency in the rooting medium (e.g., waterlogging) anaerobic metabolism is enhanced in the roots of most plant species, regardless of their flooding tolerance, and ethanol and lactate formation is increased at the expense of the carbohydrate degradation in the tricarboxylic acid cycle (TCA):

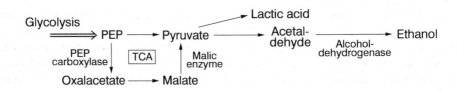

Fermentation to ethanol is inefficient in carbon utilization, 2 mole ATP per mol hexose are produced compared with 36 mole ATP per mole hexose in the TCA cycle and aerobic respiration. The enhanced rate of glycolysis under anaerobiosis (Pasteur effect) may be considered as compensatory reaction in terms of energy charge. In pea plants after 3 to 4 days of flooding the concentration of ethanol in the roots increased sharply and the release of ethanol into the xylem was raised to 2.1 mM compared with 0.07 mM in the non-flooded control (Jackson et al., 1982). Poor soil aeration induced by excessive irrigation of apple trees may even increase the ethanol content of the fruits at harvest time and reduce the keeping quality during storage (Gur and Meir, 1987). Accumulation of ethanol in roots had been suggested as the main factor responsible for flooding injury to nonwetland species (Crawford and Zochowski, 1984) but this seems unlikely for a number of reasons. In the first place plant cells and tissues are able to tolerate rather high concentrations of ethanol, and although close correlations have often been found between ethanol concentration and flooding injury the detrimental effects on plant growth are probably caused by acetaldehyde which is an intermediate in ethanol formation (see above) and is highly toxic (Perata and Alpi, 1991). It is questionable, however, whether there is a universal cause of anoxia injury in all plant species and tissues (Crawford and Wollenweber-Ratzer, 1992; Kennedy et al., 1992). The spread of injury from anoxia varies greatly between species, tissues and experimental conditions, and the time taken for tissues or plants to die can take from a few hours to months. When death is rapid cellular malfunctions may be due to decline in ATP level leading to impairment of the H^+ efflux pump and acidification of the cytoplasm. In tissues and plants where damage is less rapid and ATP levels are maintained, carbohydrate shortage may limit survival under anaerobiosis (Crawford, 1993). In agreement with this under anaerobic conditions an exogenous supply of glucose prolongs root tip viability (Webb and Armstrong, 1983) and delays the loss of elongation potential of the roots (Waters *et al.*, 1991).

A possible aggravating factor in anoxia injury is the resupply of air, which occurs typically under temporary flooding conditions. This 'post-anoxia stress' depends on the accumulation of ethanol and activities of SOD and peroxidases formed under anaerobiosis (Crawford and Wollenweber-Ratzer, 1992; Crawford, 1993). Resupplying air enhances the conversion of ethanol into acetaldehyde and the generation of oxygen radicals. Pretreatment with ascorbic acid as an antioxidant is therefore an effective means of raising tolerance of such plants to 'post-anoxia stress'.

In soils of high organic matter content and with simultaneously high microbial activity prolonged periods of waterlogging lead to the accumulation of volatile fatty acids and phenolics in the soil, which are additional stress factors affecting root metabolism and growth, especially at low soil pH (Section 14.4).

The relative importance of toxic substances that accumulate in the soil, and of root-borne toxins from waterlogging depends on particular circumstances. For example, in nonwetland species sudden waterlogging at a high soil temperature primarily affects root metabolism via anaerobiosis. After prolonged periods of waterlogging of soils high in organic matter, however, accumulation of soil-borne toxins may become an increasingly important cause of injury (Section 14.4).

16.4.2.4 *Phytohormones, Root–Shoot Signals*

Waterlogging (or anaerobiosis) inhibits root growth and the synthesis and shoot export of cytokinins (Smit *et al.*, 1990) and gibberellins (Section 14.4). Correspondingly, foliar application of cytokinins and gibberellins may counteract at least temporarily the inhibition of shoot elongation and enhanced leaf senescence imposed by waterlogging (Jackson and Campbell, 1979). The rapid decline in leaf elongation rate and stomatal aperture in response to flooding, however, is not caused by lower cytokinin export from the roots (Neuman *et al.*, 1990) but by elevated ABA levels in the leaves. In *Phaseolus vulgaris* under hypoxia within 3–4 h leaf elongation rate decreases from 0.94 to 0.18 mm h^{-1}, stomatal conductance from 0.94 to 0.25 cm s^{-1}, and ABA increases from 0.77 to 3.99 nmol g^{-1} leaf dry weight (Neuman and Smit, 1991). These authors attributed at least part of the increase in ABA levels in the leaves to xylem import from the roots. On the other hand the likelihood of an increase in root export of ABA under flooding has been questioned by Jackson *et al.* (1988) and Jackson (1991a).

The accumulation of ethylene in soils as well as in roots under waterlogging is well documented. Accumulation of ethylene in soils becomes increasingly important at oxygen concentrations in the soil atmosphere below 9% (Hunt *et al.*, 1981). Because of the much lower rates of diffusion of gases in water as compared with air, the water film around roots entraps ethylene in the root tissue ('waterjacket effect'). The resulting increase in ethylene concentration in root tissue has a number of effects on root growth and morphology (Chapter 14), simultaneously triggering anatomical changes in the

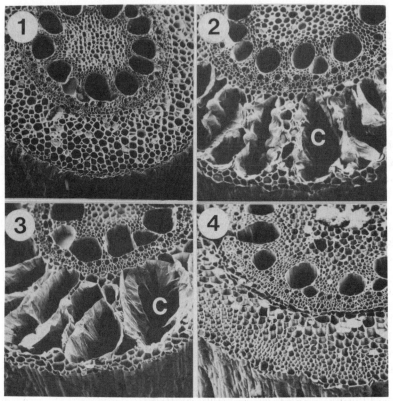

Fig. 16.13 Transverse sections of maize roots under a scanning electron microscope. (1) Control grown in well-aerated solution; (2) root receiving $5\,\mu$l ethylene l^{-1} in air; (3) root from nonaerated solution; (4) root receiving nitrogen gas (anoxic treatment). C, Cortical air space. (From Drew *et al.*, 1979.)

root tissue and the export of ethylene, or its precursor 1-aminocyclopropane-1-carboxylate (ACC, Section 5.6). It also acts as a root signal inducing epinastic responses to flooding in the leaves of herbaceous plants (Jackson, 1990b; Wang and Arteca, 1992).

The formation of aerenchyma, the prominent air spaces in the root cortex which are produced by waterlogging or growth of roots in nonaerated solutions (Fig. 16.13) is brought about by the accumulation of ethylene. Supplying ethylene externally to roots via a nutrient solution induces a similar effect as waterlogging on aerenchyma formation. Preventing ethylene accumulation by bubbling N_2 gas through the solution prevents aerenchyma formation. This response to elevated ethylene levels is not restricted to the roots but is also observed at the shoot base and basal parts of the stem (Kuznetsova *et al.*, 1981; Jackson, 1990a,b) and is essential for the internal ventilation of roots growing in anaerobic environments (Section 16.4.3.2).

Enhanced aerenchyma formation in roots is not confined to waterlogging or anaerobiosis but also occurs in response to nutrient deficiency, and nitrogen deficiency in particular, despite only low rates of ethylene formation (Table 16.19). Increase in

Table 16.19
Effect of Nutrient Deprivation (Withholding Supply for 4 Days) on
Ethylene Production and Aerenchyma Formation in Roots of Maize
Seedlings Grown in Oxygenated Nutrient Solution[a]

Treatment	Ethylene production (pmol g^{-1} fresh wt h^{-1})	Aerenchyma (% area of root cortex)
Control	~200	6
−N	~165	34
+N	~120	10

[a]Compiled and recalculated data from Drew *et al.* (1989).

aerenchyma formation under nutrient deficiency may be brought about by a higher
tissue sensitivity to ethylene.

16.4.3   Mechanisms of Adaptation

16.4.3.1   Tolerance versus Avoidance

Plants differ widely in their capacity to adapt to waterlogging, as is apparent from the
differences between nonwetland and wetland species. Extreme examples of crop plants
are barley and wetland rice. Other well-known genotypical differences exist among
forage species (Cannell, 1977). The principal stress factor for plants in waterlogged soils
is oxygen deficit. According to the general stress concept of Levitt (1980) adaptation
can be achieved by avoidance of the stress factor or tolerance of the stress or both these
strategies:

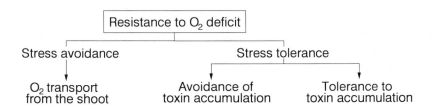

The relative importance of the two components tolerance versus avoidance depends
on various factors, the duration of oxygen deficit in particular. For short-term oxygen
deficit (e.g., after heavy rains) stress tolerance is required, whereas for long-term
oxygen deficit in the soil stress avoidance is also needed. Stress avoidance becomes the
key factor in the adaptation of plants to permanently waterlogged soils. In order to cope
with the high concentrations of Fe^{2+} in submerged soils, wetland species require
particular mechanisms to detoxify iron either by exclusion from uptake (oxygenated
rhizosphere) or uptake (includer) and detoxification by high activities of SOD and
peroxidase and catalase:

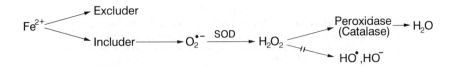

16.4.3.2 Metabolic Adaptation

Compared with flooding-sensitive species, flooding-tolerant species are better able to regulate their rate of glycolysis and fermentation to ethanol (Drew, 1990). Evidence for the importance of this regulation is shown in Table 16.20 for barley and wetland rice. In barley under hypoxia the activity of alcohol dehydrogenase (ADH) increased about tenfold whereas cytochrome oxidase activity decreased markedly. Despite the high carbon flow to ethanol, barley was not able to maintain energy charge (Pearson and Havill, 1988). In contrast, rice retained a high cytochrome oxidase activity. This enzyme has a high affinity for O_2 and may help to scavenge any available oxygen in the roots to maintain a high ATP level. According to the Davies-Roberts pH stat hypothesis, prevention of acidification of the cytosol (e.g., by transport of malate and lactate into the vacuole) is a key factor in resistance to anaerobiosis (Drew, 1990; Kennedy et al., 1992).

There is some evidence that in flooding-tolerant species, increased SOD activity under anoxia is an important protection mechanism in preventing oxidative damage during recovery from anoxia stress, e.g., after transient flooding (Monk et al., 1987; Crawford, 1993).

Under natural conditions concentrations of oxygen in the root tissue are usually lower than in the soil. Thus, roots already experience hypoxia before the onset of anoxia under waterlogging (Drew, 1990). Roots of soil-grown plants therefore have more time to adapt than in conditions imposed in many laboratory studies. Accordingly, pretreatment of roots of wheat or maize for several hours under hypoxia (1–6% oxygen) results in severalfold increases in ADH activity and in the subsequent period of anoxia corresponding increases in ATP levels and survival rates (Johnson et al., 1989; Waters et al., 1991). Under oxygen deficiency special proteins are synthesized, the

Table 16.20
Effect of Hypoxia on Metabolic Changes in Roots of Barley (*Hordeum vulgare* L.) and Rice (*Oryza sativa* L.)[a]

Plant species	Cytochrome oxidase (nmol min^{-1} mg^{-1} protein)		Alcohol dehydrogenase (nmol min^{-1} mg^{-1} protein)		Malate (μmol g^{-1} fresh wt)		Ethanol (μmol g^{-1} fresh wt)	
	$+O_2$	$-O_2$	$+O_2$	$-O_2$	$+O_2$	$-O_2$	$+O_2$	$-O_2$
H. vulgare	193	47	125	1290	1.56	0.60	2.81	20.87
O. sativa	131	129	32	167	1.45	1.45	1.29	3.68

[a]Based on Pearson and Havill (1988).

Table 16.21
Effect of Manganese Supply on the Growth and Manganese
Content of Mature Leaves of Barley and Wetland Rice[a]

Mn supply ($mg\ l^{-1}$)	Shoot dry weight (g per plant)		Mn content ($mg\ kg^{-1}$ dry wt)	
	Barley	Rice	Barley	Rice
0.2	14.0	14.5	70	100
0.5	12.1	15.5	190	400
2.0	6.5	15.0	310	2200
5.0	6.0	12.1	960	5300

[a]Data recalculated from Vlamis and Williams (1964).

'anaerobic' polypeptides (ANP_s), and in maize several of these proteins have been identified as enzymes of glycolysis and fermentation (Bailey-Serres et al., 1988). In plant species not known for formation of root nodules the presence of genes encoding for hemoglobin has been demonstrated, and in these species deoxygenation of oxyhemoglobin under oxygen deficiency may serve as a cellular message to initiate anaerobic metabolism (Appleby et al., 1988).

For adaptation to waterlogging a high manganese tolerance of shoot tissue is important, barley and wetland rice represent extremes in this respect (Table 16.21). Whereas even less than 200 mg manganese per kilogram leaf dry weight is toxic to barley, more than 10 times this amount is tolerated by the leaf tissue of rice without any growth depression. In barley grown in soils high in manganese, sensitivity to waterlogging may well be related in part to the low manganese tolerance of the shoots. Mechanisms by which shoots become tolerant of high manganese content in the tissue (e.g., complexing manganese with oxalic acid) have been discussed in Section 16.3.6. In wetland species such as rice, tissue tolerance to iron is also considerably higher than in nonwetland species. However, many wetland species are intolerant to high iron levels in the tissue (Table 16.17) and regardless of the flooding tolerance iron toxicity may occur at similar iron levels in the leaves ($\sim$1100–1600 mg Fe kg^{-1} dry weight) of flooding tolerant and intolerant species as has been shown for Rumex (Laan et al., 1991b).

16.4.3.3 Anatomical and Morphological Adaptation

Transport of oxygen from the shoot to the roots and the rhizosphere is readily demonstrated in both wetland and nonwetland species (Greenwood, 1967), and may provide a substantial proportion of the oxygen demand of roots also of nonwetland species grown in aerated soils (De Willigen and van Noordwijk, 1989). Oxygen transport takes place to a limited extent in air-filled intercellular spaces; the main pathway, however, is the aerenchyma in the root cortex (Fig. 16.13). The proportion of air-filled intercellular spaces of an aerenchyma is an expression of root porosity (air-filled spaces as a proportion of the total root volume). Root porosity differs between

Table 16.22
Rooting Depth and Root Porosity of Nonwetland Plants Grown
under Drained and Flooded Conditions in a Loam Soil[a]

Plant species, cultivars	Root porosity (%)		Root depth (cm)	
	Drained	Flooded	Drained	Flooded
Maize	6.5	15.5	47	17
Sunflower	5	11	33	15
Wheat				
cv. Pato	5.5	14.5	10	5
cv. Inia	3.0	7.5	23	10
Barley	3.5	2.0	32	15

[a]Data recalculated from Yu et al. (1969).

plant species and is also adaptive (Fig. 16.13; Table 16.19). For wetland rice, maize, and barley grown in an aerated nutrient solution, the relative values for root porosity are 1.0, 0.25, and 0.10, respectively (Jensen et al., 1967). To a certain extent, the root system of nonwetland species has the capacity to adapt to waterlogging conditions (Table 16.22). In this study plants were grown in well-drained soils for 2 weeks, and thereafter were or were not flooded. With the exception of barley, the root porosity of all plants tested was higher under conditions of flooding than of nonflooding. Maize showed the greatest degree of adaptation; the wheat cultivar Pato behaved similarly. The differences in the root porosity of wheat cultivars corresponded well with the higher tolerance of Pato than of Inia to waterlogging under field conditions (Yu et al., 1969).

As a rule flooding-tolerant species develop an extended aerenchyma not only in the roots but also the rhizomes (Laan et al., 1989), and in most instances close correlations occur between flooding-tolerance and size of the aerenchyma (Laan et al., 1990). In *Rumex* species root porosity was 10% in intolerant, 35% in the intermediate, and 50% in the highly flooding-tolerant species (Laan et al., 1989). Aerenchymatous tissues are formed either by cell wall separation and cell wall collapse (*lysigeny*) or by separation without collapse (*schizogeny*). Although in wetland species such as rice or the sedge *Scirpus americanus* Pers. aerenchyma formation is constitutional, flooding enhances arenchyma formation and ethylene is also involved in this enhanced effect (Seliskar, 1988; Justin and Armstrong, 1991). Long distance transport of oxygen in the aerenchyma to apical zones of roots growing in flooded soils requires restriction of oxygen loss by diffusion into the rhizosphere in basal zones (Armstrong and Beckett, 1987). The cell wall permeability of the rhizodermis is presumably drastically decreased by the formation of the exodermis (Fig. 16.14).

Changes in root anatomy in response to flooding are closely correlated with changes in root morphology. After waterlogging many old roots die, but numerous adventitious roots with well-developed aerenchyma emerge from the base of the stem and grow to a limited extent into the anaerobic soil. Whether existing roots die on sudden waterlogging (anaerobiosis) and new (adapted) roots have to be formed, or whether the development of aerenchyma is enhanced in the existing roots, depends mainly on the

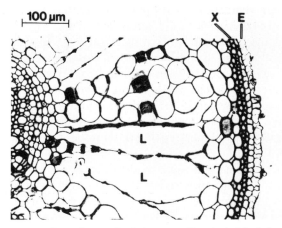

Fig. 16.14 Transverse section of a seminal rice root 20 mm behind the apex. E = epidermis (rhizodermis); X = exodermis; L = lacunae = aerenchyma. (Prade and Trolldenier, 1990a.)

plant species and the flooding tolerance of the species (Laan *et al.*, 1991a). The principal differences between species are shown schematically in Fig. 16.15.

In nonwetland species the internal ventilation (Armstrong, 1979) is insufficient to prevent an oxygen deficit from occurring in the roots, the apical meristems in particular, when the external supply of oxygen is cut off by flooding. New roots have to be formed in the basal zones or at the stem. Roots of wetland species, such as rice, with a

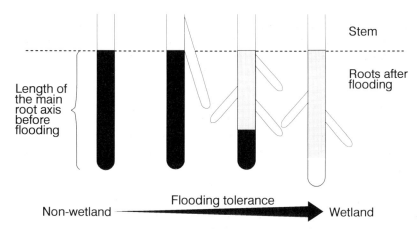

Fig. 16.15 Suggested relationship between the responses of roots of nonwetland and wetland species to a limited period of soil flooding. Black areas, dead tissue; stippled areas, surviving tissue; white areas, regrowth. (Based on Armstrong, 1979.)

constitutionally higher ventilation are capable of adapting to anaerobiosis even in aerated substrates by enlargement of the aerenchyma. Air spaces are formed behind the apical meristem, and the aerenchyma in the basal zones may occupy 20 and up to 50% of the total root volume in well-drained and flooded soils, respectively (Armstrong, 1979). Aerenchyma formation in the basal part of the stem also effectively connects the root aerenchyma with the leaf aerenchyma. In deep-water rice, where most of the leaves are submerged in the early growing stages, the main pathway of oxygen diffusion in the shoot tissue seems to be along a continuous air layer between the hydrophobic leaf surface and the surrounding water rather than in the leaf aerenchyma (Raskin and Kende, 1983).

Oxygen transport to submerged roots by diffusion is not very effective over long distances. For efficient long-distance transport in the aerenchyma from shoots to roots other mechanisms are required such as pressurized gas transport, driven by temperature gradients between aerated parts (leaves, stem) and submerged parts (Schröder et al., 1986). Light-induced increase in shoot temperature and thus in internal gas pressure acts as the driving force of root ventilation in flooding-tolerant species like Alnus, but such an effect is absent in flooding-intolerant species like beech (Table 16.23) or Acer pseudoplatanus (Grosse et al., 1992).

In common reed (Phragmites australis) in addition to pressurized gas transport from shoot to roots, the gas flow rate is enhanced by Venturi-convection driven by wind which sucks air into the belowground system via dead culms cut off near the ground level (Armstrong et al., 1992). Accordingly high wind speed can considerably enhance rhizosphere aeration in common reed.

In wetland species oxygenation of the rhizosphere through the release of molecular oxygen by the roots is vital for plants growing in anaerobic soils (Yoshida and Tadano, 1978). Differences in the capacity of cultivars of wetland rice to release oxygen from the roots and thus oxidize Fe^{2+} in the rhizosphere are correlated with a resistance to iron toxicity (bronzing disease, or Akagare disease, Armstrong, 1969). These differences in the 'oxidation power' of roots are readily apparent from differences in the amount of $Fe(OOH)_x$ precipitated on the roots when grown under field conditions in flooded soils

Table 16.23
Light-Enhanced Gas Transport to the Roots of 6-Month-Old
Seedlings of Various Tree Species without and with Leaves[a]

Leaves	Species	Gas transport (μl air min^{-1})	
		Dark	Light
−	Fagus sylvatica	3.3	3.3
−	Aesculus hippocastanum	1.1	2.6
−	Alnus glutinosa	1.4	5.8
+	Alnus glutinosa	1.8	3.2
+	Alnus glutinosa (12 month)	15.0	22.7

[a]Based on Grosse and Schroeder (1985).

Table 16.24
Relationship between the Rate of Oxygen Release from the Roots and FeOOH
Precipitation on the Roots of Wetland Rice Cultivars[a]

	Brazos	Bluebelle	Lebonnet	Labelle
Oxygen release[b]	0.947	0.692	0.528	0.359
FeOOH (% of total root dry wt)	14.10	8.45	7.15	5.09

[a]Cultivars were grown in Katy fine sandy loam soil. Based on Chen et al. (1980).
[b]Increase in O_2 saturation in 10 ml H_2O (5 min)$^{-1}$; laboratory studies.

(Table 16.24). The largest amount of oxygen was released from the roots of the cultivar Brazos. This cultivar also had the highest grain yield capacity when grown in different locations. It has been calculated that at maturity ~500 kg $Fe(OOH)_x$ per hectare is present as root coating ('plaque') each season (Chen et al., 1980). Microbial iron oxidation may play a direct role in plaque formation (Trolldenier, 1988), or indirectly by enhancing oxygen release into the rhizosphere as has been shown for *Azospirillum brasilense* Cd in the rhizosphere of rice or Kallar grass (Ueckert et al., 1990). A high oxidation power of roots and plaque formation may also lower the zinc concentration in the rhizosphere through formation of sparingly soluble $ZnFe_2O_4$ and thus increase the risk of zinc deficiency in rice (Sajwan and Lindsay, 1988).

In rice and probably other wetland species, the formation and stability of the aerenchyma are dependent on silicon supply. Uptake of iron and particularly manganese are distinctly lower in rice plants supplied with silicon (Ma and Takahashi, 1990a). When the silicon content of rice plants increased from 0.2 to 5.7% dry wt, the iron uptake from flooded soil decreased to about one-third (Okuda and Takahashi, 1965). Mineral nutrition also affects 'oxidation power' indirectly: nutrient deficiencies that increase the exudation of photosynthates from the roots simultaneously enhance microbial activity and oxygen consumption in the rhizosphere (Chapter 15) and thus increase the risk of iron toxicity (Yoshida and Tadano, 1978; Ottow et al., 1982; Yamauchi, 1989).

Permanently flooded soils (e.g., mangrove swamps) may contain high concentrations of both Fe^{2+} and hydrogen sulfide (H_2S). In addition to internal ventilation, tannins at the rhizoplane may play a role in oxidation of Fe^{2+} and H_2S and in the formation of the sparingly soluble FeS (Kimura and Wada, 1989).

16.5 Alkaline Soils

16.5.1 Soil Characteristics

Alkaline soils (pH > 7) are very common in semiarid and arid climates. Calcareous soils cover more than 30% of the earth's surface (Chen and Barak, 1982). Their content of $CaCO_3$ in the upper horizon varies from a few per cent to 95%. The predominant alkaline soil units are the Rendzinas, which are shallow soils with high organic matter content overlying calcareous material. Less common are the Chernozems and Xero-

Table 16.25
Relative Abundance of Alkaline Soils and Major Constraints on Plant Growth in These Soils

	Soil pH		
	7	8	9
Dominant soil associations	Rendzinas (calcareous soils) (Chernozems, Xerosols)		Solonetz (sodic soils) Solonchaks (saline soils)
Relative abundance			
Major nutritional constraints	Deficiency: Fe, Zn, P (Mn)[a]		Toxicity: Na, B Deficiency: Zn, Fe, P (Ca, K, Mg)[a]
Other constraints	Excess of HCO_3^- Water deficit Mechanical impedance		Poor aeration Excess of HCO_3^- Water deficit Mechanical impedance

[a]Parentheses indicate less frequently, or only in certain situations.

sols. The pH of calcareous soils is determined by the presence of $CaCO_3$, which buffers the soils in the pH range 7.5–8.5 (Table 16.25). At pH > 8, Solonetz is the predominant soil unit. Solonetz or sodic (alkali) soils are characterized by a sodium adsorption ratio (SAR)* of the soil matrix greater than 15, and they often contain sodium carbonate. Sodic soils usually occur in association with saline soils (Solonchaks), and saline–sodic soils are more abundant than purely sodic soils. In the context of constraints on plant growth it is necessary, however, to make a clear distinction between salinity and sodicity. Saline soils are not necessarily alkaline, and plant growth on saline soils is affected mainly by high levels of sodium chloride (ion toxicity, ion imbalance) and impairment of water balance (Section 16.6). Sodic soils, on the other hand, are alkaline and plant growth is impaired mainly by high pH, high bicarbonate, and often poor aeration.

The major nutritional constraints in calcareous soils differ from those in sodic soils (Table 16.25). The differences are directly related either to other soil chemical factors such as bicarbonate concentration or to soil physical factors. In sodic soils, poor physical conditions and correspondingly poor soil aeration are the major constraints and are often correlated with sodium and boron toxicity.

Nitrogen is a growth-limiting factor for most crop species (other than legumes) growing in alkaline soils. More than 90% of the soil nitrogen is organically bound nitrogen (humus) which becomes available to plants after mineralization by soil

*The *sodium adsorption ratio* or SAR describes a relationship between soluble sodium and soluble divalent cations which can be used to predict the exchangeable sodium percentage of soil equilibrated with a given solution:

$$\frac{sodium}{(calcium + magnesium)^{1/2}}$$

microorganisms. Both the total amount of soil nitrogen and its availability to plants are therefore closely related to the soil organic matter content and conditions of mineralization (soil moisture, temperature, aeration). The soil pH is only of minor importance for the level and turnover of nitrogen in alkaline soils.

16.5.1.1 Iron

Mineral soils have, on average, a total iron content of $\sim2\%$. Most crop species remove only between 1 and 2 kg iron per hectare annually. In well-aerated soils with a high pH, however, the concentration of Fe^{2+} and Fe^{3+} in the soil solution is extremly low, and the total concentration of inorganic iron species [between pH 7 and 9 mainly $Fe(OH)_2^+$, $Fe(OH)_3$, and $Fe(OH)_4^-$] in the soil solution is only around 10^{-10} M (Fig. 16.16). Although it is not possible to give precise information on the forms or concentrations of inorganic iron at the root surface, the concentration of chelated iron required for optimal growth is of the order of 10^{-6} to 10^{-5} M (Fig. 16.16). These values have to be interpreted with reservation, however, as they are based on the supply of synthetic iron chelates (e.g., Fe EDTA) which are utilized relatively poorly, at least by grasses (Section 16.5.3).

In aerated soils the solubility of iron is largely controlled by Fe(III) oxides (group name), especially ferrihydrite ($5Fe_2O_3 \cdot 9H_2O$) and amorphous ferric hydroxide ($Fe(OH)_3$), and the formation of chelated iron derived from soil organic matter or microbial production of siderophores (Lindsay, 1991). For example, for a soil of pH 7.9 more than 35 000 times the concentration of soluble iron was found than predicted from inorganic equilibrium constants (O'Connor et al., 1971). In alkaline soils with a high organic matter content the concentration of organic iron chelates in the soil solution can reach values of 10^{-4} to 10^{-3} M (Mashhady and Rowell, 1978). The application of farmyard manure to calcareous soils low in organic matter may therefore be an effective

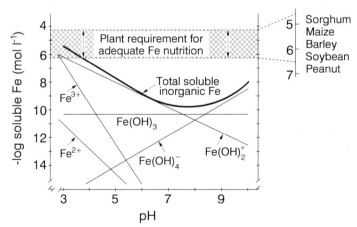

Fig. 16.16 Solubility of inorganic iron species in equilibrium with iron oxides (synthetic Fe(III) chelates) in well-aerated soils in comparison to the requirement of soluble iron at the root surface of various plant species. (Römheld and Marschner, 1986.)

procedure for increasing iron solubility (Bar-Ness and Chen, 1991), and iron uptake by crop species with low iron efficiency such as sorghum (Mathers *et al.*, 1980).

Besides humic acids, organic acids, phenolics and siderophores are major Fe(III) chelating compounds. Siderophores are produced by almost all microorganisms subjected to iron deficiency. There are two major types: hydroxamate siderophores produced by both fungi and bacteria, and catecholate siderophores produced by bacteria (Crowley *et al.*, 1987), and the more recently discovered rhizoferrin, a low-molecular-weight polycarboxylate siderophore produced by fungi (Drechsel *et al.*, 1991). The concentrations of siderophores are considerably higher in the rhizosphere than in the bulk soil (Reid *et al.*, 1984), and it is supposed that the corresponding iron–siderophore chelates may exceed the concentrations of soluble nonchelated iron by several orders of magnitude (Crowley *et al.*, 1987). However, the actual concentrations of soluble iron in the rhizosphere of soil-grown plants is difficult to estimate as siderophores are strongly adsorbed by the soil matrix at higher pH (Cline *et al.*, 1983) and on the other hand iron solubility in the rhizosphere may be governed primarily by root-induced changes in the rhizosphere, under iron deficiency in particular (Section 16.5.3.2).

In sodic soils (pH > 8.5), sodium carbonate disperses organic matter, and low-molecular-weight organic substances (mainly sodium humates) form soluble complexes with iron (and manganese). Increasing the $NaHCO_3$ concentration from 12 to 75 mM (pH $8.0 \rightarrow 8.8$) increased the concentration of iron and manganese in the soil solution by a factor of 18 and 2.3, respectively (Mashhady and Rowell, 1978). This humate effect was also demonstrated in solution culture, where the addition of sodium humates prevented iron deficiency in tomato plants grown at high pH in the presence of high bicarbonate concentrations (Badurowa *et al.*, 1967).

16.5.1.2 Zinc

The solubility of inorganic zinc, like that of inorganic iron, decreases with increasing pH. In the pH range 5.5–7.0 the equilibrium concentration of zinc may decrease 30 to 45 times for each unit increase in soil pH (Moraghan and Mascagni, 1991). Diffusion coefficients for zinc in calcareous soils are therefore about 50 times lower than in acid soils (Melton *et al.*, 1973). As the zinc concentration in the soil solution is also determined by adsorption and desorption processes occurring in the soil matrix, the concentration of zinc at a given soil pH may also depend on other solute components (e.g., Ca^{2+}; Brümmer, 1981), and soil organic matter and microbial activity. The application of farmyard manure to alkaline soils low in organic matter may therefore increase the solubility and plant uptake of zinc (Srivastava and Sethi, 1981). Making soils more alkaline by increase in $NaHCO_3$ concentration, however, enhances the risk of zinc deficiency in plants because of both, a decrease in zinc extractability in the soil (Mehrotra *et al.*, 1986) and direct impairment of root growth (see below).

16.5.1.3 Manganese

The chemistry of manganese in soils and soil solutions is governed by pH and redox (Section 16.4.1; Marschner, 1988). Although manganese may form organic and inorganic complexes, in soil solutions Mn^{2+} is the major species (Norvell, 1988). In well-aerated calcareous soils (Rendzinas), the solubility of manganese decreases with increasing levels of both $CaCO_3$ and MnO_2 due to the adsorption of manganese on $CaCO_3$ and its oxidation on MnO_2 surfaces and probably to the precipitation of manganese calcite (Jauregui and Reisenauer, 1982). The effect of soil pH and $CaCO_3$ on the availability of manganese is well known in temperate zones as a result of the manganese deficiency induced by overliming of acid soils low in manganese. On calcareous soils manganese availability to plants is mainly determined by soil structure and aeration (anaerobic microsites) and root-induced changes in the rhizosphere (Sections 15.3, 15.4 and 15.5).

16.5.1.4 Boron and Phosphorus

Boron adsorption to clay minerals increases sharply above pH 6.5 and is maximal at pH ~9 (Kluge and Beer, 1979; Keren *et al.*, 1981). In alkaline soils, the low boron solubility dictated by boron adsorption to clay minerals is usually compensated for, however, by a lack of leaching or by boron supplied by irrigation water. Boron toxicity is thus much more likely, particularly in sodic soils, than is boron deficiency.

In alkaline soils (except Chernozems) the availability of phosphorus is generally low. As with zinc the concentration of phosphorus in the soil solution is determined primarily not by dissolution or precipitation of definite inorganic compounds such as tricalcium phosphate but by the desorption and adsorption of phosphorus, particularly in soils with more than 1% organic matter. In these soils, at least in the pH range 6–8, the phosphorus concentration in the soil solution may not decline but rather increase with pH (Welp *et al.*, 1983). In alkaline soils of increasing pH and decreasing soil organic matter content, the equilibrium constants of inorganic phosphates become increasingly important for the concentration of phosphorus in the soil solution. In general, however, phosphorus deficiency in crop plants growing in alkaline soils is caused primarily by very low levels of total phosphorus and low soil moisture levels, that is, when the mobility of phosphorus is limited and root growth is restricted (Chapters 13 and 14).

16.5.2 Major Chemical Constraints to Plant Growth

16.5.2.1 Iron Deficiency

The most prominent nutritional disorder of crop plants grown in soils with more than 20% $CaCO_3$ is iron deficiency or so-called 'lime-induced chlorosis' (Schinas and Rowell, 1977). Plant species that are mainly affected include apple, peaches, citrus, grapevine, peanut, soybean, sorghum and upland rice. It is the major problem in sorghum and soybean production in the Great Plains of the United States (Clark, 1982a). For reviews on this topic the reader is referred to Chen and Barak (1982) and Vose (1982).

Iron deficiency on calcareous soils is often enhanced by poor soil aeration caused by soil compaction or high water content, and low soil temperatures which keep the soil wet for longer (Römheld, 1985). It is, however, not oxygen deficiency, which enhances

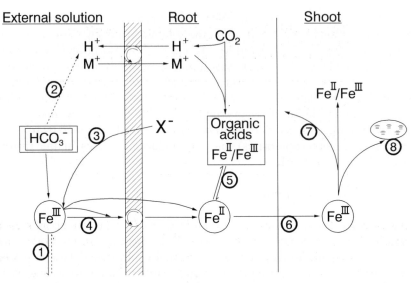

Fig. 16.17   Eight possible effects of a high bicarbonate concentration in the substrate on the uptake, transport, and availability of iron for chlorophyll formation in leaves. For a description of mechanisms (1)–(8), see text.

chlorosis but elevated concentrations of bicarbonate (HCO_3^-). In soils with free $CaCO_3$ increase in CO_2 concentration (e.g. by impaired gas exchange or amendments of organic matter) leads to formation of $Ca(HCO_3)_2$. Levels of 4–20 mM HCO_3^- are predicted based on 1–5% CO_2 (Chaney, 1984). Whether or not high HCO_3^- concentrations in the soil play the key role in iron deficiency on calcareous soils depends on the plant species and the type of root response to iron deficiency. In graminaceous species (Strategy II) such as sorghum and maize, iron deficiency is either not or only poorly correlated with elevated HCO_3^- concentrations in the soil (Chaney, 1984; Yen *et al.*, 1988), but closely correlated with decreasing levels of poorly crystalline or amorphous iron oxides (Loeppert and Hallmark, 1985). In contrast, in nongraminaceous species (Strategy I) close correlations exist between levels of HCO_3^- in soils and severity of iron deficiency, for example, in soybean (Chaney, 1984; Inskeep and Bloom, 1986, 1987), grapevine (Römheld, 1985) and apple (Ao *et al.*, 1987). Based on this key role of bicarbonate in Strategy I plants, bicarbonate-buffered nutrient solutions containing low levels of soluble iron allow screening for susceptibility to iron deficiency chlorosis, for example in soybean, chickpea and citrus (Chaney *et al.*, 1992a). The relative susceptibility observed in these solutions has a high correlation with the relative susceptibility observed in wet calcareous soils which induce iron deficiency chlorosis.

In leaves of Strategy I plants suffering from lime-induced chlorosis the content of iron in the dry matter is often lower (Häussling *et al.*, 1985; Ao *et al.*, 1987; Dockendorf and Höfner, 1990), but might also be similar or even higher (Mengel *et al.*, 1979, 1984a,b; Mengel and Bübl, 1983) than in green leaves. Similar or higher iron contents in chlorotic leaves indicate 'physiological inactivation' of iron (Section 9.1.5).

Fig. 16.17 summarizes some of the major means by which high HCO_3^- concentration

Table 16.26
Rate of Fe(III) Reduction and ^{59}Fe Uptake (^{59}Fe EDDHA
Supply) in Iron Deficient Peanut (*Arachis hypogaea* L.)
Plants as Affected by the pH Buffering Capacity of the
Nutrient Solution pH 8.5[a]

Treatment (Nutrient solution)	Fe(III) reduction (nmol g^{-1} root dry wt h^{-1})	^{59}Fe uptake
Unbuffered	4208	658
+10 mM HCO$_3^-$	1592	95
+10 mM HEPES	1513	72

[a]Marschner *et al.* (1989).

may affect the uptake, translocation, and utilization of iron in plants. A high HCO$_3^-$ concentration in the soil solution both raises and buffers the pH and thus further lowers the concentration of soluble inorganic iron [mechanism (1)]. Simultaneously in Strategy I plants, root responses to iron deficiency are severely inhibited by high pH; these responses include impairment of the effectivity of the H$^+$-efflux pump by neutralization of the H$^+$ [mechanism (2)], lowering the release of phenolics [mechanism (3)] and Fe(III) reduction at the plasma membrane [mechanism (4)] (Römheld and Marschner, 1986; see also Section 2.5.6). In agreement with this high HCO$_3^-$ concentrations lead to a sharp decrease in the uptake and transport of iron to the shoot (Kolesch *et al.*, 1984; Dockendorf and Höfner, 1990). At least in short-term studies this inhibitory effect of high HCO$_3^-$ concentrations can be simulated by an organic pH buffer (HEPES) which demonstrates the importance of acidification at least of the rhizoplane and the apoplasm of the rhizodermis cells for iron acquisition of Strategy I plants (Table 16.26).

In roots supplied with a high HCO$_3^-$ concentration, CO$_2$ fixation and organic acid synthesis increase, particularly in so-called calcifuge species (Lee and Woolhouse, 1969b). It is not clear to what extent sequestering of iron in vacuoles by certain organic acids [mechanism (5)] contributes to the inhibition of iron transport to the shoot [mechanism (6)]. The transport of iron to expanding leaves might be especially impaired (Rutland and Bukovac, 1971), and the distribution of iron within the leaf tissue uneven [mechanism (7)] (Rutland, 1971). These effects are discussed in relation to the alkalinization of the tissue (Mengel and Malissiovas, 1981) and of the cytoplasm in particular (Kolesch *et al.*, 1984). However, inactivation of iron in leaves by high HCO$_3^-$ concentrations could not be confirmed by Dockendorf and Höfner (1990).

High HCO$_3^-$ concentrations in the rooting medium may indirectly affect content and utilization of iron in leaves. In many plant species, the chlorosis-sensitive species in particular, high HCO$_3^-$ concentrations inhibit root growth (Lee and Woolhouse, 1969a), root respiration, root pressure-driven solute export into the xylem (Wallace *et al.*, 1971) and the rate of cytokinin export to the shoot. Cytokinins are necessary for protein synthesis and chloroplast development (Parthier, 1979). In agreement with this, high HCO$_3^-$ concentrations might inhibit shoot growth considerably prior to the occurrence of iron deficiency chlorosis, for example, in sorghum (McCray and Matocha, 1992) and peach trees (Shi *et al.*, 1993). High iron contents in leaves of plants suffering from lime-induced chlorosis may therefore also be the consequence of a limitation on other factors required for leaf expansion growth, chloroplast development and chlorophyll formation [mechanism (8)].

The role of phosphorus in lime-induced chlorosis is complex. It may impair iron nutrition at various levels, for example, decreasing the rate of dissolution of iron from iron oxides in the bulk soil and in the rhizosphere (root exudates). In many instances, increase in phosphorus supply suppresses phosphorus deficiency-induced root exudation of organic acids (Section 15.4) and, thus also a mechanism that enhances iron solubility in the rhizosphere. Such an example has been reported for *Banksia ericifolia* (Proteaceae) where an increase in phosphorus supply suppressed proteoid root formation and simultaneously induced iron-deficiency chlorosis (Handreck, 1991, 1992). Inactivation of iron in plants by high phosphorus contents (Cumbus *et al.*, 1977) may occur, but is unlikely to play an ecological role. Elevated phosphorus levels in chlorotic leaves are probably the result of growth inhibition (concentration effect) and are thus the consequence and not the cause of iron chlorosis (Mengel *et al.*, 1984b). Although iron deficiency can be induced in crop plants growing in calcareous soils supplied with very high levels of fertilizer phosphorus, there is substantial doubt that phosphorus is of major importance for lime-induced chlorosis under field conditions (Kovanci *et al.*, 1978; Mengel *et al.*, 1979). Many laboratory and greenhouse studies on phosphorus–iron interactions have been conducted with phosphorus concentrations that are orders of magnitude higher than those typical of soil solutions in calcareous soils.

Lime-induced chlorosis is of minor importance in sodic soils, mainly for two reasons: the increase in iron solubility by elevated concentrations of low-molecular-weight organic compounds and growth inhibition by soil constraints other than iron deficiency (Mashhady and Rowell, 1978).

16.5.2.2   Zinc and Manganese Deficiency

Increasing the pH of a soil by liming usually decreases the plant availability of zinc and manganese much more than of any other mineral nutrient, including phosphorus (Table 16.27). The risk of zinc deficiency therefore is high in soils after liming, or on calcareous soils in general. In cereals zinc deficiency is probably the most widespread micronutrient deficiency on calcareous soils (Graham *et al.*, 1992). Besides high soil pH, high

Table 16.27

Effect of pH Increase by Liming a Sandy Soil on Mineral Element Content in Leaves of Peanut[a]

	Content in dry matter				
	$(mg\ kg^{-1})$		$(g\ kg^{-1})$		
Soil pH	Zn	Mn	P	K	Mg
5.2	200	310	1.8	18.5	4.5
6.0	54	66	1.9	17.5	3.8
6.8	20	19	1.9	19.0	3.9

[a]Parker and Walker (1986).

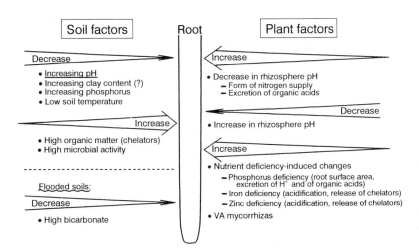

Fig. 16.18 Schematic presentation of main soil and plant factors decreasing or increasing zinc availability and uptake by soil-grown plants. (Marschner, 1993). Reprinted by permission of Kluwer Academic Publishers.

clay content, high phosphorus supply, and low soil temperatures also enhance the risk of zinc deficiency (Fig. 16.18).

The fall in zinc contents in the shoot dry matter in plants grown with high levels of phosphorus fertilizers may be the result of growth enhancement and thus 'dilution' by growth, inhibition of VA mycorrhizas, and suppression of phosphorus deficiency-induced changes in the rhizosphere. The particular role of VA mycorrhizas in zinc acquisition had been discussed in Section 15.7.1 and shown in two examples (Table 15.16; Fig. 15.24). Low soil temperatures often enhance the incidence and severity of zinc deficiency symptoms, and increasing phosphorus supply increases the likelihood of low temperature-induced zinc deficiency on calcareous soils (Moraghan, 1980). At low temperatures zinc uptake from soils is not specifically impaired, but root activity and root colonization with VA mycorrhizas are lower. Of the plant factors, rhizosphere acidification brought about either by enhanced net excretion of protons or of organic acids (Fig. 16.18) are of particular importance for acquisition of zinc and manganese on calcareous soils (Sections 15.3 and 15.4).

In white lupin grown on alkaline soils either manganese deficiency or toxicity may occur, depending on the source of nitrogen supply and the phosphorus level (Moraghan, 1992). Manganese contents in the shoot dry matter were high in plants depending on N_2 fixation compared with nitrate fed plants. Application of phosphorus drastically decreased manganese contents in the shoots towards the deficiency range, presumably by inhibition of proteoid root formation.

Zinc deficiency is widespread in flooded rice in soils with high pH and with high organic matter content (Moraghan and Mascagni, 1991). Thus, besides salinity and iron toxicity, the next most important nutritional limitation on yield in flooded rice is zinc deficiency (Ikehashi and Ponnamperuma, 1978). In neutral and alkaline soils there is a

Table 16.28
Influence of Soil pH and Zinc Supply on Grain Yield and Leaf
Content of Zinc in Flooded Rice[a]

Treatment		Grain yield (kg ha^{-1})	Zinc content in leaves (mid tillering) (mg kg^{-1} dry wt)
Soil pH	kg Zn ha^{-1}		
6.8	0	5934	9
	1.9	7212	17
7.3	0	5265	9
	1.9	6171	18
7.7	0	2788	10
	1.9	6637	14

[a]Based on Sedberry *et al.* (1988).

strong negative correlation between soil pH and rice yield when no zinc fertilizer is applied (Table 16.28). As in the pH range 6.5–8.0 the content of DTPA-extractable zinc in paddy soils only slightly decreases, the drastic decrease in zinc uptake in plants without zinc fertilizer supply is mainly caused by elevated HCO_3^- concentrations. In paddy soils 3–6 weeks after planting HCO_3^- concentrations over 10 mM are common; these concentrations of HCO_3^- not only inhibit zinc transport to the shoots (Forno *et al.*, 1975) but also uptake into the roots and seem to injure rice roots directly (Dogar and van Hai, 1980). This response of flooded rice to high HCO_3^- concentrations resembles that typical of many calcifuge species, but there are marked cultivar differences in rice in this respect (Section 16.5.3.3).

16.5.2.3 Other Constraints

In many calcareous upland soils high pH and low soil moisture contents are the main environmental factors, impairing nutrient mobility in the soil and root extension growth. Boron availability is particularly impaired by low soil moisture (Moraghan and Mascogni, 1991). The risk of boron deficiency in crop plants is therefore increased in calcareous soils in dryland areas (Section 9.7). In contrast to boron, for phosphorus acquisition plants have developed a whole range of mechanisms (Chapter 14 and 15).

16.5.3 Mechanisms of Adaptation

16.5.3.1 Calcicoles versus Calcifuges

Plant species and populations within species (ecotypes) of the natural vegetation that preferentially grow on calcareous soils (*calcicoles*) possess adaptive mechanisms for coping with constraints on growth and mineral nutrition such as low iron and zinc availability and often high calcium and bicarbonate concentrations in the soil solution. For example, calcicoles have a higher capacity for iron acquisition than *calcifuges*, i.e., species and ecotypes which have adapted to acid soils (De Neeling and Ernst, 1986; Gries and Runge, 1992). Calcicoles are often highly efficient in phosphorus uptake (Musick, 1978), at least in some instances because roots are highly infected with VA mycorrhizas (Kianmehr, 1978). In agreement with this, inability to mobilize phosphorus in calcareous soils is considered as a key factor in the calcifuge behavior of many plant species such as *Rumex acetosella* and *Silene rupestris* (Tyler, 1992). In addition, and in contrast to calcifuges, in calcicoles high HCO_3^- concentrations have only negligible inhibitory effects on root extension growth (Lee and Woolhouse, 1969a).

The role of high calcium concentrations in adaptation of plants to calcareous soils seems to be rather complex. In calcicoles both avoidance and tolerance of high calcium concentrations in the plant tissue seem to occur. For example, many members of the Brassicaceae accumulate large amounts of soluble calcium in their vacuoles (calciotrophic types) which may have advantages in terms of osmoregulation on dry limestone habitats (Kinzel and Lechner, 1992). On the other hand, in certain calcicole species the uptake of calcium is more restricted than in calcifuge species (Bousquet *et al.*, 1981), presumably due to a lower affinity of root plasma membranes for Ca^{2+} (Monestiez *et al.*, 1982). The mechanisms of calcium toxicity in general and in calcifuges in particular is poorly understood. In the calcifuge species *Lupinus luteus* growth is severely depressed even at a concentration of 8 mM calcium in the external solution. This depression could not be explained by interference of high Ca^{2+} concentrations in the leaf apoplasm with the role of ABA in stomatal regulation (Section 8.6.6) and thus, control of plant water relations (Atkinson, 1991). Probably in calcifuges the strict compartmentation of Ca^{2+} at cellular level and maintenance of low Ca^{2+} concentrations in the cytosol (Section 8.6.2) is less effective than in calcicoles. A higher level of calcium-binding proteins in the cytpolasm of calcicole species (Le Gales *et al.*, 1980) would support this suggestion.

In many herbaceous plants grown in calcareous soils there are abundant calcified roots which are formed by solubilization of $CaCO_3$ in the rhizosphere and precipitation of $CaCO_3$ in their cortex cells (Jaillard, 1985; Jaillard *et al.*, 1991; Section 15.2). This might reflect enhanced mobilization of sparingly soluble mineral nutrients (phosphorus, iron, zinc) in the rhizosphere and simultaneously protection of the shoot tissue from excessive calcium concentrations by precipitation of $CaCO_3$ in the root tissue.

There is substantial evidence (Section 15.7.2) that certain ecto- and ericoid-mycorrhizal fungi play a role in adaptation of perennial plant species to calcareous soils. This may be achieved by the release of siderophores and enhanced iron acquisition and thereby increasing the tolerance of the host plant to 'lime-induced-chlorosis'. It may also be brought about by the production of oxalic acid which dissolves sparingly soluble calcium phosphates and simultaneously protects the host plant from excessive calcium uptake by precipitation of calcium oxalate around the fungal mycelium. The ectomycorrhizal fungus *Paxillus involutus* is particularly effective in dissolving calcium phosphates by oxalic acid excretion, particularly when supplied with nitrate as source of nitrogen (Lapeyrie *et al.*, 1991). However, there are marked differences between strains: those from calcareous soils accumulate much less calcium in their hyphae than those from acid soils (Lapeyrie and Bruchet, 1986).

16.5.3.2 Iron Efficiency and Chlorosis Resistance

In terms of an ecological classification crop species or cultivars within species which grow in alkaline soils without developing symptoms of chlorosis are called iron efficient;

those that become chlorotic are called iron inefficient (Brown and Jones, 1976). The large differences which occur between crop species and genotypes within a species in iron efficiency are basically related to differences in iron acquisition by the roots. The responsible mechanisms have been reviewed elsewhere (Römheld and Marschner, 1986; Römheld, 1987a,b).

Mobilization of iron in the rhizosphere of plants grown on calcareous soils can be brought about by both nonspecific and specific mechanisms.

Nonspecific Mechanisms. These are not related to the iron nutritional status of plants and have been discussed in detail in Sections 15.3–15.5.

1. Root-induced decrease in pH as a consequence of preferential cation uptake (e.g., induced by ammonium sulfate; Kafkafi and Ganmore-Neumann, 1985), or N_2 fixation in legumes (Soerensen *et al.*, 1989).

2. Release of organic acids by the roots (e.g., in response to phosphorus deficiency; Hoffland, 1992).

3. Release by roots of photosynthates as substrate for rhizosphere microorganisms, which in turn affect pH, redox potential, and chelator concentration (e.g., siderophores) in the rhizosphere.

Specific mechanisms. There is general agreement as to the occurrence of at least two distinct root response mechanisms (strategies) to iron deficiency in higher plants, Strategy I in dicots and monocots except of grasses, and Strategy II in grasses (Section 2.5.6). The root responses of Strategy I are not restricted to crop plants but are also typical of indigenous shrubs and forbs of alkaline soils (Nelson, 1992). In principle there is a close positive correlation between the extent to which iron deficiency induces enhanced reducing capacity of roots and net excretion of protons on the one hand, and the resistance of plants to iron deficiency on calcareous soils (chlorosis resistance) on the other. This is also true for different genotypes within a species such as tomato (Olsen and Brown, 1980), sunflower (Alcantara and de la Guardia, 1991) or grapevine (Bavaresco *et al.*, 1991). An example of such genotypical differences is shown in Fig. 16.19 for soybean cultivars.

The differences obtained in resistance to chlorosis between soybean cultivars when grown on calcareous soils provide a classical example of genetically controlled mineral nutrition in general and iron nutrition in particular (Weiss, 1943). In a given species there is a large genetic potential from which to select for high resistance to chlorosis. In soybean yield reduction in calcareous soils may vary between 6.4 and 81.9% for adapted and nonadapted cultivars, respectively (Froehlich and Fehr, 1981). Another example of the genetic potential is shown in Table 16.29 for peanut. The nonadapted cultivar Congo Red, originating from acid soils, became severely chlorotic when grown in a calcareous soil and iron chelates had to be applied to overcome chlorosis and to obtain a reasonable yield. In contrast, in the adapted cultivar 71–238 chlorosis was absent, the yield was much higher, and iron application had only a slightly beneficial effect.

As the main factors responsible for high resistance to chlorosis of Strategy I plants are known (high reducing capacity and proton excretion, less impairment by high HCO_3^-

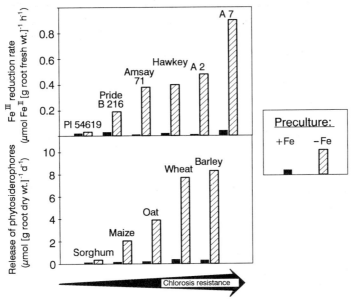

Fig. 16.19 Relationship between resistance against 'lime-induced chlorosis' under field conditions and reducing capacity of roots of soybean cultivars (*upper*) and release of phytosiderophores of graminaceous species (*lower*) under controlled conditions. (Compiled from Römheld, 1987a,b and Römheld and Marschner, 1990.)

concentrations), effective screening programs are possible based on nutrient solutions using either intact plants (Hintz *et al.*, 1987) or tissue cultures (Graham *et al.*, 1992), allowing genotypes to develop with exceptionally high levels of chlorosis resistance by recurrent selection.

When different species of Strategy I plants are compared which differ in chlorosis resistance, it is sometimes more difficult to attribute chlorosis resistance to a single component such as root-reducing capacity, for example in citrus species (Treeby and

Table 16.29
Effect of Iron Chelate Application on the Pod Yield of Peanut Grown in a Calcareous Soil (23% $CaCO_3$) of pH $7.8^{a,b}$

Cultivar	Iron chelate application	Pod yield (kg ha^{-1})	Yield increase (%)
Congo Red	−	833	—
	+	2583	210
Shulamit	−	3305	—
	+	4749	44
71–238	−	4388	—
	+	4777	9

[a]Based on Hartzook *et al.* (1974).
[b]10 kg Fe ha^{-1} (as Fe-EDDHA).

Uren, 1993), or effects of bicarbonate on iron uptake in *Lupinus* and *Pisum* (White and Robson, 1990). Qualitative differences in root-reducing capacity, in release of phenolics, or in seed reserves and redistribution of iron in the plants may account for these difficulties.

In Strategy II plants there is a close positive correlation between the extent of enhanced release of phytosiderophores and the resistance of plants to iron deficiency when grown on calcareous soils as shown in Fig. 16.19 for different cereal species. In a study on genotypical variation between Indian cereal species, Singh *et al.* (1993) obtained a similar sequence in iron deficiency-induced enhancement of phytosiderophore release: wheat > barley > rye, oat >> maize >>> sorghum. The release of phytosiderophores under iron deficiency is particularly low in flooded rice (Mori *et al.*, 1991) which reflects another feature of the calcifuge behavior of this species. Because of microbial degradation of phytosiderophores (Wirén *et al.*, 1993) however, genotypical differences in the amounts of phytosiderophores recovered in nutrient solutions have to be interpreted with care (Section 15.5; Römheld, 1991). Nevertheless, the low recovery rates of phytosiderophores in maize and sorghum raised the question as to the classification of these species as Strategy II plants (Brown *et al.*, 1991; Lytle and Jolley, 1991).

Root responses and the pattern of iron uptake by Strategy II plants under iron deficiency has many features in common with the microbial siderophore system (Crowley *et al.*, 1987). The existence of a specific transport system for microbial siderophores in the plasma membrane of Strategy II plants has therefore been suggested (Crowley *et al.*, 1991). Despite their similar capacity to mobilize iron from a calcareous soil, however, the uptake rates of iron from Fe(III) siderophores (ferrioxamine B = Desferal) are very low as compared to the plant-borne phytosiderophores such as hydroxymugineic acid (Table 16.30).

In contrast to the low effectivity of microbial siderophores of the hydroxamate and catecholate type in providing iron to higher plants in short-term studies, in long-term studies they may be of considerable ecological importance by providing soluble iron to the root surface and the plasma membrane of root cortical cells of plants growing on calcareous soils (Jurkevitch *et al.*, 1986, 1988). Higher concentrations of siderophores in the rhizosphere soil compared with the bulk soil (Nelson *et al.*, 1988) might indicate

Table 16.30

Iron Mobilization from a Calcareous Soil (Luvisol) by Fe(III) Chelators and Uptake of Iron Supplied as ^{59}Fe(III) Chelates by Iron-Deficient Barley Plants[a]

Chelator $(10^{-5}$ M)	Mobilization (nmol Fe g^{-1} soil (12 h)$^{-1}$)	Uptake (nmol Fe g^{-1} root dry wt (4 h)$^{-1}$)
Phytosiderophore (HMA)	23.6	3456.0
Siderophore (Desferal)	19.2	1.21
Synth. Chelate (DTPA)	2.0	0.51

[a]Römheld and Marschner (1990).

Table 16.31
Zinc Content of Leaves at Maturity and Grain Yield of Pigeon Pea Cultivars Grown in a Zinc-Deficient Soil (pH 7.8)[a]

Cultivar	Zn content of leaves (mg kg^{-1} dry wt)			Grain yield (g per pot)		
	0	5 mg Zn	50 mg Zn	0	5 mg Zn	50 mg Zn
T 21	15.0	19.7	37.1	3.8	8.5	10.4
Plant A-3	21.2	30.8	90.8	6.7	10.1	10.0

[a]The cultivars were supplied with 5 and 50 mg zinc (as $ZnSO_4$) per kilogram soil. From Shukla and Raj (1980).

such an ecological role. However, as yet no evidence exists either in Strategy I or Strategy II plants that genotypical differences in efficiency of iron acquisition and in chlorosis resistance when grown on calcareous soils are related to differences in siderophore production by rhizosphere microorganisms. This lack of evidence also holds true for relationships between VA mycorrhizal infection of roots and chlorosis resistance.

16.5.3.3 Zinc Efficiency

Differences in zinc efficiency of crop species are well documented. When grown on alkaline soils, sensitivity to zinc deficiency is high in bean, maize, cotton and apple compared, for example, to wheat, oat or pea. These differences between species in zinc acquisition from soils are probably related to inherent differences in rhizosphere pH, root exudation, or root colonization with VA mycorrhizas (Thompson, 1990; Marschner, 1993). Also within a given species there are distinct differences between cultivars in zinc efficiency, and these differences are closely related to higher uptake rates of zinc by the efficient cultivar when grown in zinc-deficient alkaline soils. Examples for this have been presented for maize (Shukla and Raj, 1976), wheat (Shukla and Raj, 1974; Graham *et al.*, 1992), barley (Graham *et al.*, 1992) and soybean (Hartwig, E. E. *et al.*, 1991) and are shown for pigeon pea in Table 16.31. The zinc content in the leaves of the efficient cultivar was distinctly higher, and only small amounts of zinc fertilizer were necessary to obtain maximum grain yields.

The mechanisms responsible for higher zinc acquisition in efficient genotypes of a species are poorly understood. In soybean only a few genes seem to control uptake efficiency, or inefficiency (Hartwig, E. E. *et al.*, 1991). In bean (*Phaseolus vulgaris*) in the zinc-inefficient cultivar Sanilac, typical iron deficiency-induced root responses were enhanced under zinc deficiency and the iron content in the leaves increased (Jolley and Brown, 1991b). In the zinc-efficient cultivar Saginaw this behavior was not observed. In graminaceous species such as wheat and barley release of phytosiderophores may also be enhanced under zinc deficiency (Zhang *et al.*, 1989, 1991b). It is not clear whether this enhanced release is an expression of a separate regulation of phytosiderophore biosynthesis by iron and zinc, or more likely, of impaired iron metabolism under zinc deficiency. In wheat cultivars differences in zinc efficiency found on calcareous soils are related to differences in release of phytosiderophores as occur under zinc deficiency

Table 16.32
Grain Yield and Phytosiderophore (PS) Release of Two Wheat Cultivars in
Response to Zinc Deficiency[a]

Cultivar	Grain yield (t ha^{-1})		PS release (μmol (60 plants)$^{-1}$ (4 h)$^{-1}$)	
	$-$Zn	$+$Zn	$-$Zn	$+$Zn
Aroona	1.21	1.42	6.9	0.5
Durati	0.45	1.12	1.8	0.5

[a]Compiled data from field experiments (1988) on calcareous soils (Graham *et al.*, 1992) and from laboratory studies with 18-day-old plants (Cakmak *et al.*, 1994).

under controlled conditions (Table 16.32). The zinc-efficient Aroona released much higher quantities of phytosiderophores under zinc deficiency as compared with the zinc-inefficient Durati. In contrast, under iron deficiency both cultivars release similar amounts of phytosiderophores (Cakmak *et al.*, 1994). It well might be that in the zinc efficient Aroona iron metabolism is more severely impaired under zinc deficiency than in the zinc-inefficient Durati.

Lowland rice cultivars also differ very much in zinc efficiency, particularly when grown in soils of high pH (Section 16.2.4). In zinc-efficient cultivars high bicarbonate concentrations as well as low root zone temperatures have only small effects on growth and shoot contents of zinc, iron, and manganese, but large depressing effects in zinc-inefficient cultivars (Yang *et al.*, 1993). Thus zinc efficiency in rice seems to be causally related to high bicarbonate tolerance of the roots, which is also reflected in even slight increases in root growth by bicarbonate and a better control of organic acid accumulation in the roots as compared with the zinc-inefficient cultivar (Yang *et al.*, 1994).

16.5.3.4 Manganese Efficiency

Plant species differ considerably in susceptibility to manganese deficiency when grown on soils low in available manganese (Section 9.2.7). Genotypes within a species differ also in manganese efficiency which is highly heritable, and major dominant genes appear to be involved (Graham, 1988). In barley grown on manganese-deficient calcareous soils the manganese-efficient cultivar Weeah, which is derived from old English landraces, achieved grain yields of 3.3 t ha^{-1} both without and with manganese fertilization whereas in the inefficient cultivar Galleon the grain yield dropped from 3.2 to 1.8 t ha^{-1} without manganese fertilization (Ralph, 1986). In wheat differences between cultivars in manganese efficiency when grown on a manganese-deficient calcareous soil were related to differences in manganese acquisition but not internal utilization (Marcar and Graham, 1987).

The mechanisms responsible for cultivar differences in manganese efficiency are not known. It well might be that the particular dynamics of manganese in the rhizosphere and the role of manganese reducing and oxidizing rhizosphere microorganisms (Section 15.5.2) are responsible for the difficulties in identifying mechanisms.

Table 16.33
Iron Deficiency-Induced Manganese Toxicity in Flax (*Linum usitatissimum* L.)
Grown in a Calcareous Soil of pH 8.0[a]

Treatment	Shoot dry wt (g per pot)	Contents in shoot dry weight		
		Manganese (mg kg^{-1})	Iron (mg kg^{-1})	Phosphorus (%)
Control (−Fe)	3.60	881	83	0.32
2 mg Fe per pot (Fe-EDDHA)	5.55	64	174	0.32

[a]From Moraghan (1979).

In Strategy I plants manganese acquisition from calcareous soils is largely dependent on the iron nutritional status of plants. In flax manganese content in the shoot dry matter was poorly related to the amount of manganese extracted from the soils, but was inversely related to the amount of extractable iron (Moraghan and Freeman, 1978). In agreement with this, manganese toxicity could be eliminated by the application of Fe-EDDHA, which drastically decreased the manganese content of the plants (Table 16.33). High iron efficiency may in many instances therefore prevent manganese deficiency in Strategy I plants growing in calcareous soils; it may even increase the risk of manganese toxicity, as shown for flax (Table 16.33) and for an iron-efficient genotype of soybean (Brown and Jones, 1977).

In proteoid root-forming species like white lupin (*Lupinus albus*) manganese acquisition and contents in the shoots are primarily related to the proteoid root formation and not the iron nutritional status (Moraghan, 1991b). Phosphorus deficiency increases proteoid root formation, and raises the manganese contents in the shoots and may even lead to manganese toxicity when grown on calcareous soils. In contrast, the iron contents in the shoots are generally low in white lupin. The sensitivity of white lupin to iron deficiency when grown on calcareous soils probably reflects an inherent restricted iron transport from roots to the shoot in this species in order to prevent iron toxicity, despite the very high iron contents in the proteoid roots.

16.6 Saline Soils

16.6.1 General

The saline areas of the world consist of salt marshes of the temperate zones, mangrove swamps of the subtropics and tropics, and their interior salt marshes adjacent to salt lakes. Saline soils are abundant in semiarid and arid regions, where the amount of rainfall is insufficient for substantial leaching. Salinity problems occur in nonirrigated croplands and rangelands either as a result of evaporation and transpiration of saline underground water or due to salt input from rainfall. They are particularly critical in irrigated areas. Salinity has been an important historical factor and has influenced the life spans of agricultural systems. It frequently destroyed ancient agrarian societies, and more recently large areas of the Indian subcontinent have been rendered unproductive by salt accumulation and poor water management. In Pakistan, for example, about 10 million of 15 million hectares of canal-irrigated land are becoming saline (Wyn Jones, 1981). Worldwide about 33% of the irrigated land is affected by salinity, and presumably more land is going out of irrigation due to salinity than there is new land coming into irrigation (Vose, 1983). Salinity is the major nutritional constraint on the growth of wetland rice.

Even water of good quality may contain 100–1000 g salt m^{-3}. With an annual application of $10\,000\,m^3\,ha^{-1}$, between 1 and 10 t of salt are added to the soil. As a result of transpiration and evaporation of water, soluble salts further accumulate in the soil and must be removed periodically by leaching and drainage. But even when proper technology is applied to these soils, they contain salt concentrations which often impair the growth of crop plants of low salt tolerance.

Although salt tolerance is relatively low in most crop species and cultured woody species, it is encouraging that genetic variability exists not only between species but also between cultivars within a species. Therefore, not only are selection and breeding for salt tolerance an important issue for traditional agricultural production in semiarid and arid regions, they may also offer the potential for utilizing the unlimited resource of seawater for irrigation. Examples of this have been highlighted by Epstein *et al.* (1980), who showed that, with certain barley strains, grain yields of up to 1 t ha^{-1} can be obtained even when undiluted seawater, supplemented with nitrogen and phosphorus, is used for irrigation.

16.6.2 Soil Characteristics and Classification

Soils are considered saline if they contain soluble salts in quantities sufficient to interfere with the growth of most crop species. This, however, is not a fixed amount of salt but depends on the plant species, the texture and water capacity of the soil, and the composition of the salts. Thus the criterion for distinguishing saline from nonsaline soils is arbitrary. According to the definition of the US Salinity Laboratory the saturation extract (the solution extracted from a soil at its saturation water content) of a saline soil has an electrical conductivity (EC) greater than 4 mmho cm^{-1} or 4 deciSiemens m^{-1} (equivalent to ~40 mM NaCl l^{-1}) and an exchangeable sodium percentage (ESP) of less than 15. Although the pH of saline soils can vary over a wide range, it is usually around neutrality, with a tendency toward slight alkalinity. The dominant soil association consists of Solonchaks (Table 16.25). Saline soils with an ESP of greater than 15 are termed saline–alkali soils (or saline–sodic soils), have high pH values, and tend to become rather impermeable to both water and aeration when the soluble salts are removed by leaching. These distinctions between saline and alkali (sodic) soils are often insufficiently appreciated in nutrient solution studies by adding high concentrations of single salts (mainly NaCl) but maintaining calcium concentrations low. Such wide Na^+/Ca^{2+} ratios in the substrate are typical for sodic soils but not saline agricultural soils (Maas and Grieve, 1987).

In evaluations of the suitability of saline soils for crop production, the measurement of EC offers a rapid and simple method for characterizing the salt content (Fig. 16.20).

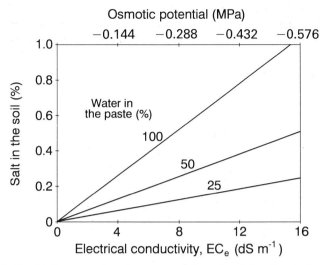

Fig. 16.20 Relationship between the salt content of the soil and the electrical conductivity (dS = deci Siemens) of the saturation extract at 25°C. Water in the paste (%); relative values; saturation extract = 100; 50 and 25 = calculated changes in EC and osmotic potential in the soil solution when the water content drops to 50 and 25% respectively of that at saturation (50% is about field capacity of soils). (Based on US Salinity Laboratory Staff, 1954.)

From the EC the osmotic potential of the saturation extract can also be calculated: osmotic potential (MPa) = EC × 0.036.

Since the EC is measured in the saturation extract (the value 100 in Fig. 16.20), the salt concentration in the soil solution at field capacity will be about twice that of the saturation extract (i.e., 50 in Fig. 16.20) and even higher when the soil moisture level declines below field capacity. For comparison, the EC of seawater is in the range of 44–55 dS m^{-1}, and as a rule irrigation water of good quality must have an EC_e below 2 dS m^{-1}. For plant growth in saline soils, however, the EC of the saturation extract only is an insufficient indicator, mainly for two reasons: (a) The actual salt concentration at the root surface can be much higher than in the bulk soil (Section 15.2) and (b) the EC characterizes only the total salt content, not its composition. Although NaCl is

Table 16.34
Solubility of Salts at 25°C

Salt	Solubility (mmol l^{-1} H$_2$O)
Calcium chloride (CaCl$_2$ · 6H$_2$O)	12 735
Magnesium chloride (MgCl$_2$ · 6H$_2$O)	97 478
Sodium chloride (NaCl)	6 108
Magnesium sulfate (MgSO$_4$ · 7H$_2$O)	2 880
Sodium bicarbonate (NaHCO$_3$)	821
Sodium sulfate (Na$_2$SO$_4$ · 10H$_2$O)	341
Calcium sulfate (CaSO$_4$ · 2H$_2$O)	15

usually the dominant salt, others may be abundant in various combinations, depending on the origin of the saline water and the solubility of the salts (Table 16.34).

Furthermore, in both saline soils and irrigation waters high boron concentrations might become more critical for plant growth than salt concentrations *per se*. Irrigation water with more than 0.5 mg boron l^{-1} might injure sensitive species such as citrus and walnut, and more than 2.0 mg l^{-1} is harmful to most crop species.

16.6.3 Salinity and Plant Growth

16.6.3.1 *Genotypical Differences in Growth Response to Salinity*

Plant species differ greatly in their growth response to salinity, as shown schematically in Fig. 16.21. The growth of halophytes (group I) is optimal at relatively high levels of NaCl, a response which can be explained only in part by the role of sodium as a mineral nutrient in these species (Section 10.2). Only a few crop species are slightly stimulated by low salinity levels (group II). Most are nonhalophytes (glycophytes), and either their salt tolerance is relatively low (group III) or their growth is severely inhibited even at low substrate salinity levels (group IV).

Generally, classification of the salt tolerance (or sensitivity) of crop species, forage species and fruit trees is based on two parameters: the threshold EC and the slope, i.e., percentage of yield decrease beyond threshold. Examples taken from an extensive study are given in Table 16.35. It is evident that barley tolerates relatively high salinity levels in comparison, for example, with bean or grapevine.

It has been known for many years that there are large differences in salt tolerance between cultivars within a crop species. Some of the literature data have been summarized by Duvick *et al*. (1981) and Vose (1983). A few examples of the differences are given in Table 16.36. The genetic variability within a species is not only a valuable

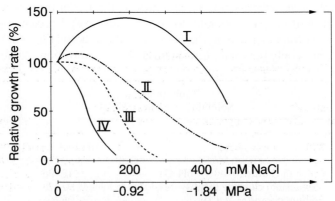

Fig. 16.21   Growth response of various plant species to increasing substrate salinity and related osmotic potential. I, Halophytes; II, halophilic crop species related to I (e.g., sugar beet); III, salt-tolerant crop species (e.g., barley); IV, salt-sensitive crop species (e.g., bean). (Modified from Greenway and Munns, 1980, with permission from the *Annual Review of Plant Physiology*. Copyright 1980 by Annual Review Inc.)

T   (25°C)
=   ~~~~~~ ~~~~~~~ ~~~~ ~~~~ Does Not Reduce Yield; Slope =
Yield Reduction Per Unit Increase in EC Beyond Threshold[a]

	EC saturation soil extract	
Crop species	Threshold ($dS\ m^{-1}$)	Slope (% per $dS\ m^{-1}$)
Barley (*H. vulgare*)	8.0	5.0
Sugar beet (*B. vulgaris*)	7.0	5.9
Bermuda grass (*C. dactylon*)	6.9	6.4
Wheat (*T. aestivum*)	6.0	7.1
Soybean (*G. max*)	5.0	20.0
Tomato (*L. esculentum*)	2.5	9.9
Maize (*Z. mays*)	1.7	12.9
Orange (*C. sinensis*)	1.7	16.0
Grapevine (*Vitis* sp.)	1.5	22.0
Bean (*P. vulgaris*)	1.0	19.0

[a]Based on Maas (1985).

tool for studying mechanisms of salt tolerance but also an important basis for screening and breeding for higher salt tolerance. Progress in this field has been particularly impressive with wetland rice (Table 16.6) and barley (Epstein *et al.*, 1980).

The sensitivity to salinity of a given species or cultivar may change during ontogeny. It may decrease or increase, depending on the plant species, cultivar, or environmental factors. Sugar beet, for example, is highly tolerant during most of its life cycle but sensitive during germination. In contrast, the salt sensitivity of rice, tomato, wheat, and

Table 16.36
Effect of Salinity on the Growth Depressions of Cultivars within Crop Species[a]

Species	Treatment (mM NaCl)	Growth parameter	Growth depression of cultivars	Reference
Barley	125	Grain yield	45 to <5	Greenway (1962)
Wheat	~50	Grain yield	90–50	Bernal *et al.* (1974)
Sugar beet	150	Total dry weight	93–49	Marschner *et al.* (1981a)
Soybean	50	Total dry weight	75–44	Läuchli and Wienecke (1979)
Tobacco	500	Surviving plants	100–15	Nabors *et al.* (1980)
Bean	Saline–sodic soil	Surviving plants	79–1	Ayoub (1974)

[a]Differences in the salt tolerance of the cultivars are indicated by the range of relative values; control (without salt) = 100.

barley usually increases after germination (Maas and Hoffman, 1977). In maize, salt sensitivity is particularly high at tasseling and low at grain filling (Maas *et al.*, 1983). Reports are often contradictory in respect to changes in salt sensitivity during ontogenesis. The difficulties of generalizing even for a given species such as barley have been demonstrated by Lynch *et al.* (1982), who found that the most sensitive cultivar at the seedling stage was rather tolerant at maturity, whereas another cultivar showed the opposite pattern.

16.6.3.2 Major Constraints

There are three major constraints for plant growth on saline substrates (Fig. 16.22): (1) water deficit ('drought stress') arising from the low (more negative) water potential of the rooting medium; (2) ion toxicity associated with the excessive uptake mainly of Cl^- and Na^+; (3) nutrient imbalance by depression in uptake and/or shoot transport and impaired internal distribution of mineral nutrients, and calcium in particular. It is often not possible to assess the relative contribution of these three major constraints to growth inhibition at high substrate salinity, as many factors are involved. These include ion concentrations and relations in the substrate, duration of exposure, plant species, cultivar and root stock (excluder, includer, Fig. 16.22), stage of plant development, plant organ, and environmental conditions. The long-term exposure of a plant to salinity may for example result mainly in ion toxicity in the older leaves and water deficit and shortage of carbohydrates in the younger leaves. The following examples illustrate

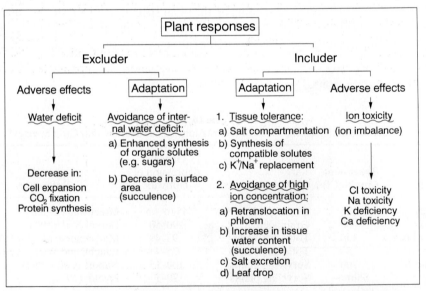

Fig. 16.22   Adverse effects of salinity and possible mechanisms of adaptation. (Reproduced from Greenway and Munns, 1980, with permission from the *Annual Review of Plant Physiology*. Copyright 1980 by Annual Review Inc.)

the possible role of the three major constraints, and also the difficulties of generalizing about the effects of salinity.

16.6.3.3   Water Deficit

As a rule, in saline substrates shoot growth is more depressed than root growth (Termaat and Munns, 1986) although root elongation can also be immediately depressed by the application of a high concentration of NaCl together with a low Ca^{2+} concentration (Cramer et al., 1988). Most of the rapid responses in leaf elongation rate to substrate salinity are attributable to changes in leaf water status (Fig. 16.23). At removal of root-zone salinity, leaf extension rate immediately reverts to the presalinized level, suggesting that water deficit was the main reason for growth reduction by root zone salinity rather than salt toxicity.

In bean plants too (Neumann et al., 1988) a decrease in turgor of the leaf cells is considered to be the main factor responsible for inhibited leaf elongation growth at mild root zone salinization and not a decrease in cell wall extensibility as has been found in leaves of barley plants exposed to high NaCl concentrations (Lynch et al., 1988). As an artificial increase in root pressure of wheat plants was not found to ameliorate the impaired leaf elongation rate in saline substrates the explanation of leaf water deficit as the decisive factor has been questioned and a hormonal message from the roots been suggested as an alternative explanation (Munns and Termaat, 1986).

Substrate salinization decreases water availablity and water uptake and thus reduces root pressure-driven xylem transport of water and solutes. In tomato and pepper plants even 27 days after employment of the salt stress (50 mM NaCl) the xylem exudation flow decreased by a factor of 17–20 compared with the nonsalinized plants, and ion concentrations in the xylem sap increased only by a factor of 2–3 (Kafkafi, 1991). Thus, both the rate of supply of water and mineral nutrients to the shoot becomes restricted in saline substrates. It is only a matter of controversy as to whether even in the short-term a decrease also in amounts of nutrients or unfavorable nutrient ratios (e.g. Na^+/Ca^{2+}) in the leaf elongation zones are important causative factors for impaired leaf elongation rate (Lynch et al., 1988; Munns et al., 1989).

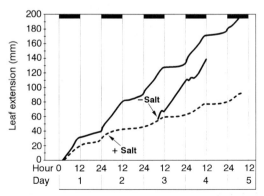

Fig. 16.23   Time course of leaf extension in sugar beet (*Beta vulgaris* L.) grown under 12 h photoperiod without or with salinization (100 mM NaCl). (Modified from Waldron et al., 1985.)

16.6.3.4 Ion Toxicity and Ion Imbalance

In saline substrates Na^+ and Cl^- are usually the dominant ions. Despite the essentiality of chloride as a micronutrient for all higher plants (Section 9.8) and of sodium as mineral nutrient for many halophytes and some C_4 species (Section 10.2), the concentrations of both ions in saline substrates by far exceed this demand and lead to toxicity in non-salt-tolerant plants. In many herbaceous crop species, grapevine, and many fruit trees growth inhibition and injury of the foliage (marginal chlorosis and necrosis on mature leaves) occur even at low levels of NaCl salinization (Sykes, 1992; Maas, 1993). Under such conditions water deficit is not a constraint (Greenway and Munns, 1980) and, at least in *Citrus* species, high chloride sensitivity and thus chloride toxicity is the major constraint (Maas, 1993). Decidious woody trees such as *Tilia* and *Aesculus*, as well as coniferous trees such as *Picea omorika* also suffer from chloride toxicity in the foliage when grown on substrates high in NaCl (Alt *et al.*, 1982; Mekdaschi *et al.*, 1988).

Many leguminous species are also very sensitive to high chloride levels. Isoosmotic concentrations of Cl^- compared with SO_4^{2-} salinity are therefore much more toxic on growth, for example of peanut (Chavan and Karadge, 1980), or cause severe leaf chlorosis and depression in photosynthesis as shown for bean in Table 16.37. Certain soybean cultivars provide outstanding examples of high chloride sensitivity. In poorly drained coastal soils the application of potash fertilizer containing KCl can raise the chloride contents of some cultivars to 1% of the leaf dry weight, causing leaf scorch and severe reduction in grain yield (Parker *et al.*, 1983, 1987). Although chloride toxicity is very common in many plant species grown on saline substrates, in some plant species such as *Sorghum*, compared with chloride (NaCl), sulfate salinization (Na_2SO_4) can decrease growth similar as NaCl at low and moderate salinity and even more at high substrate salinity. This particular decrease in growth is in part due to depression in the

Table 16.37
Effect of Salt Stress on Chlorophyll Content and $^{14}CO_2$ Assimilation of
Phaseolus vulgaris L.[a]

Treatment EC (dS m^{-1})	Chlorophyll (mg g^{-1} fresh wt)	$^{14}CO_2$ Assimilation[b] (cpm mg^{-1} fresh wt)
Control	2.44	248
NaCl + CaCl$_2$		
2.5	1.88	172
5.0	1.53	159
10.0	0.96	147
Na$_2$SO$_4$		
2.5	2.67	283
5.0	3.41	273
10.0	3.38	264

[a]Based on Bhivare *et al.* (1988).
[b]Provided for one hour.

shoot contents of potassium and magnesium at high sulfate salinization (Boursier and Läuchli, 1990).

With the exception for example of graminaceous species such as wheat (Gorham, 1993), sorghum or rice (Flowers *et al.*, 1991; Yeo, 1993), sodium toxicity seems not be as widespread as chloride toxicity, and is mainly related to low Ca^{2+} concentrations in the substrate (Section 16.6.3.5) or poor soil aeration. Many crop species with relatively low salt tolerance are typical Na^+ excluders (Section 10.2), and capable at low and moderate salinity levels of restricting the transport of Na^+ into the leaves where it is highly toxic in salt sensitive species. Under conditions of poor soil aeration, however, the restriction in uptake and shoot transport of Na^+ (Drew and Dikumwin, 1985) and also Cl^- (Barrett-Lennard, 1986) is nullified leading to massive accumulation in the leaves and corresponding salt toxicity (Section 16.4.2).

In describing mechanisms of salt toxicity in leaves, in the past most emphasis has been put on the inhibition of enzyme reactions and inadequate compartmentation between cytoplasm and vacuoles. There is increasing support, however, of the hypothesis of Oertli (1968) of salt accumulation in the leaf apoplasm as an important component of salt toxicity, leading to dehydration and turgor loss and death of leaf cells and tissues (Munns, 1988; Flowers, 1988). In rice, for example, an early symptom of root exposure to only 50 mm NaCl is wilting, although accumulation of ions in the leaves is more than sufficient for osmotic adjustment. In the leaf apoplasm, however, the Na^+ concentrations may reach 500 mm which cause dehydration of the leaf cells (Flowers *et al.*, 1991). The higher salt sensitivity of pea as compared with spinach also seems to be causally related not only to a more regulated control of Na^+ and Cl^- transport to the shoots in spinach but also to the maintenance of lower concentrations of both these ions in the leaf apoplasm (Speer and Kaiser, 1991). In pea leaves sodium chloride toxicity is associated with higher generation of superoxide radicals (O_2^-) and lipid peroxidation which suggests that at the level of mitochondria an oxidative stress is involved (Hernandez *et al.*, 1993).

At high substrate salinity growth depression may also originate from inhibited nutrient uptake, transport and utilization in the plants. For example, in barley supplied with low concentrations of manganese high NaCl concentrations in the substrate depressed growth mainly by inhibited manganese uptake and induced manganese deficiency (Cramer and Nowak, 1992). Although high Cl^- concentrations may inhibit NO_3^- uptake (Section 2.5.3), induced nitrogen deficiency is not likely to be an important factor in the growth depression caused by soil salinity.

In substrates with high phosphorus availability which are at least an order of magnitude higher than those typical for soil solutions, NaCl salinity may enhance phosphorus uptake and depress plant growth by phosphorus toxicity (Roberts *et al.*, 1984). In contrast, for example in cotton at low phosphorus concentrations (10–30 μm) high NaCl concentrations (150 mm) depress uptake and translocation of phosphorus (Martinez and Läuchli, 1991). In tomato also the utilization efficiency of phosphorus in the leaves is depressed, with increasing NaCl concentrations in the substrate from 10 to 50 and 100 mm the corresponding phosphorus contents in the youngest mature leaf required to obtain 50% yield were increased from 1.8 to 2.4 and 3.0 g P kg^{-1} dry matter (Awad *et al.*, 1990). In principle, a similar increase in mineral nutrient demand for potassium in the leaf tissue of spinach plants was found in saline substrates; for optimal photosynthetic capacity the potassium demand in the leaf dry matter was at least twice as high as in nonsaline substrates (Chow et al., 1990).

16.6.3.5 Role of Calcium, Calcium-Related Disorders

The particular role of calcium in increasing salt tolerance of plants is well documented, as is the induction of calcium deficiency in plants grown in saline substrates. Application of gypsum is a common practice in reclamation of saline–sodic and sodic soils. Under these conditions the application of gypsum can considerably increase salt tolerance. In potato, for example, addition of gypsum markedly increases tuber yield (Table 16.38). Even at high salinity levels, the yield is only slightly lower than that of the control, although 1.2% salt corresponds to an EC of ~20 dS m^{-1} and the average threshold EC for the species potato is ~2 dS m^{-1} (Maas, 1993). As has been shown in soybean growing in saline-sodic soil (Coale et al., 1984) addition of gypsum has a dual effect: (a) it improves soil structure and thus soil aeration and (b) increases the Ca^{2+}/Na^+ ratio and thus supports the capacity of roots to restrict Na^+ influx.

The importance of calcium for salt tolerance of plants has been highlighted by LaHaye and Epstein (1971) who demonstrated that in bean (*Phaseolus vulgaris*) with 50 mM NaCl in the substrate increasing the supply of $CaSO_4$ from 0.1 to 10.0 mM enhanced shoot dry weight from 0.46 to 0.74 g per plant, and this was associated with a drastic decline in sodium content of the leaves from ~1.4 to less than 0.1 mmol g^{-1} dry weight. Similarly, in *Citrus* species salt tolerance can be very much increased by high Ca^{2+} concentrations in the substrate which suppress both Na^+ and Cl^- transport to the leaves (Banuls et al., 1991). This ameliorating effect of Ca^{2+} is in accordance with its functions in membrane integrity and control of selectivity in ion uptake and transport (Section 2.5). On the other hand high Na^+ concentrations in the substrate inhibit uptake and transport of Ca^{2+} and may therefore induce calcium deficiency in plants growing in substrates with low Ca^{2+} concentrations or high Na^+/Ca^{2+} ratios (Lynch and Läuchli, 1985). In coastal saline–sodic soils, rolling and bleaching of young leaves of rice may occur and can be prevented by soil amelioration with gypsum (Muhammed et al., 1987). These symptoms resemble calcium deficiency caused by extremly wide Na^+/Ca^{2+} ratios of ~150 in these soils. Accordingly at a given high NaCl concentration, increasing Ca^{2+}

Table 16.38

Effect of Salinization and of Gypsum on the Growth of Potato (cv. Red Lasoda) in a Sandy Loam Soil[a]

	Tuber yield (g fresh wt per plant)	
Treatment	Without gypsum	With gypsum (2%)
Control (no salt)	221	226
0.6% salt	183	280
1.2% salt	149	207

[a]Salination was achieved with a mixture of sodium, magnesium, and calcium salts ($Cl^- > SO_4^{2-} > HCO_3^-$). Based on Abdullah and Ahmad (1982).

Table 16.39
Shoot Dry Weight and Mineral Nutrient Content in Shoots of
Rice (*Oryza sativa* L.) as Affected by the Na^+/Ca^{2+} Ratio in the
Nutrient Solution[a]

Treatment (mM)		Shoot dry wt (g (10 plants)$^{-1}$)	Content in shoot dry matter (%)		
Na	Ca		Na	K	Ca
99.9	0.1	9.0	2.55	2.21	0.07
99.5	0.5	25.0	1.06	2.18	0.11
95.2	4.8	33.4	0.45	2.42	0.14

[a]Based on Muhammed *et al.* (1987).

concentrations can very much enhance growth and prevent Na^+ induced calcium deficiency (Table 16.39). However, plant species differ considerably in their sensitivity to Na^+-induced calcium deficiency. In *Brassica* species supply of 80 mM NaCl severely inhibited growth, and growth could not be improved by increase in the Ca^{2+} concentrations from 1 to 10 mM (Schmidt *et al.*, 1993).

In saline soils of the Canadian prairies sodium and magnesium sulfates are often present in high amounts. In these soils poor growth of wheat and barley may be caused by calcium deficiency as indicated by very much improved growth resulting from calcium amendments (Ehret *et al.*, 1990). High overnight humidity dramatically improved wheat growth under these saline conditions, and increasing calcium supply was without effect on growth at high humidity. Such results are in accordance with the role of root pressure for calcium transport to the shoot apex (Section 3.4.3), supporting the assumption of calcium deficiency as a dominant factor of growth depression of wheat and barley in saline soils.

In vegetable crops soil salinity increases the incidence of calcium-related physiological disorders such as tipburn in lettuce and blossom end rot in tomato (Sonneveld and Ende, 1975) and may thereby limit the use of saline irrigation water in vegetable production. For example, in Chinese cabbage (*Brassica pekinensis*) an increase in salinity of the irrigation water from 1.2 to 10 dS m^{-1} had little effect on plant fresh weight but increased the percentage of tipburn from zero to 80% (Mizrahi and Pasternak, 1985). Increase in the overnight relative humidity in part ameliorated this negative effect of substrate salinity on tipburn (Mizrahi and Pasternak, 1985; Berkel, 1988). Under field conditions in dryland areas with high transpiration rates, however, even an increase in Ca^{2+} concentration in the irrigation water cannot compensate for the decrease in root pressure-driven Ca^{2+} transport in the xylem to the shoot apex (Table 16.40). In contrast, the mainly transpiration-mediated Ca^{2+} transport into the leaf blades was not impaired and even slightly increased with salinity.

More recently salinity-induced calcium deficiency has also attracted interest in terms of Ca^{2+} homeostasis at cellular level in the roots and in transmitting a signal of salinity stress in the roots to the shoots (Lazof and Läuchli, 1991; Rengel, 1992b). It has been suggested that high external Na^+ concentrations may displace Ca^{2+} from the binding

Table 16.40

Effect on Increased Root Zone Salinity Achieved by $NaCl + CaCl_2$ (Equal Amounts) in the Irrigation Water on Calcium Content and Calcium Deficiency Symptoms in Artichokes (*Cynara scolymus* L.)[a]

Root zone salinity (dS m^{-1})	Ca content (mmol kg^{-1} dry wt)		Percentage of moderately–extremely damaged buds
	Leaf blades	Inner bracts	
4.6 (control)	306	25.1	11
7.4	336	16.4	22
10.6	362	13.7	42

[a]Compiled data from Francois *et al.* (1991).

sites on the outer surface of the plasma membrane of the root cells (Lynch *et al.*, 1987), or more likely, from intracellular membranes (Lynch and Läuchli, 1988) and thereby impair Ca^{2+} homeostasis in cells (Rengel, 1992b) and its role as secondary messenger (Section 8.6.7). Inhibition in shoot elongation by high NaCl salinity is suggested as a combination of increase in Na^+/Ca^{2+} in the leaf apoplasm, together with, if not preceded by, limited water supply; hormonal signals from the roots are considered to be secondary in nature (Rengel, 1992b).

16.6.3.6 Photosynthesis and Respiration

Salinity level and leaf area are usually inversely related. With increasing salinity water loss per plant by transpiration may therefore also decrease for this reason. Not only the total leaf area, however, but also net CO_2 fixation per unit leaf area may decline, whereas respiration (dark respiration) increases, leading to a drastic reduction in net CO_2 assimilation per unit leaf area per day (Table 16.41). Lower rates of net CO_2 fixation during the light period may be caused by water deficit and partial stomatal closure, loss of turgor of mesophyll cells through salt accumulation in the apoplasm, or direct toxic effects of ions.

Table 16.41

Effect of NaCl Salinity (Osmotic Potential, MPa) on the CO_2 Balance of Cotton (cv. Alcala SJ-1)[a]

Salinity (MPa)	Leaf area (dm^2 per plant)	CO_2 Balance (mg CO_2 dm^{-2} (24 hr)$^{-1}$)		
		Net fixation light period	Evolution dark period	Net assimilation
−0.04	30	57	11	46
−0.64	24	44	16	29
−1.24	18	41	19	23

[a]Based on Hoffman and Phene (1971).

Photosynthesis of the whole shoot is often not informative for causal interpretation of the mechanisms of salt injury, as salts primarily accumulate in mature leaves. In rice at low substrate salinity net photosynthesis in the whole shoot was not affected but it was in the older leaves, where decrease in net photosynthesis was inversely related to the Na^+ concentration in the leaves, presumably mainly in the apoplasm (Yeo *et al.*, 1985). In grapevine rates of CO_2 fixation were inversely related to the Cl^- concentrations in the leaves and to an increase in mesophyll resistance but not stomatal resistance, as would be expected from water deficit (Downton, 1977). In the long-term, growth responses to salinity shall be determined by the maximum salt concentration tolerated by fully expanded leaves. If the rate of leaf death approaches the rate of new leaf expansion the photosynthetic area will become too low to support continuous growth (Munns and Termaat, 1986).

Salinity may also increase the respiration rate of the roots, which have a higher carbohydrate requirement for maintenance respiration in saline substrates (Schwarz and Gale, 1981). This higher requirement presumably results from the compartmentation of ions, ion secretion (e.g., Na^+-efflux pump), or the repair of cellular damage. Increasing atmospheric CO_2 concentrations above normal level may increase the rate of photosynthesis and thus may play an important role under saline conditions. For plants growing in saline substrates it may compensate for decreased stomatal aperture, leaf area and higher respiration rates and thereby increase the salt tolerance markedly, for example of tomato (Meiri and Plaut, 1985). Similarly, high irradiation may also increase salt tolerance, for example of faba bean (Helal and Mengel, 1981) and muskmelon (Meiri *et al.*, 1982). Under field conditions, however, it is not feasible to increase atmospheric CO_2 concentrations, and the beneficial effect of high irradiation on salt tolerance might be counteracted by negative effects of low relative humidity and correspondingly high transpiration rates, leading to salt accumulation at the root surface (Section 15.2) and in the leaves.

16.6.3.7 Protein Synthesis

Protein synthesis in the leaves of plants growing in saline substrates may decline in response either to a water deficit or to a specific ion excess. When there is a low substrate water potential imposed by Carbowax (a high-molecular-weight organic solute) or NaCl, protein synthesis in the leaves of bean is inhibited by both substrates, but inhibition is more severe with salinity stress than with water deficit alone (Frota and Tucker, 1978). The effects of NaCl salinity on protein synthesis might be due to chloride toxicity in sensitive species (e.g., soybean), whereas in the more salt tolerant barley, Na^+/K^+ imbalance in the leaves is probably the responsible factor (Table 16.42). The adverse effect of high NaCl concentrations on both potassium content and protein synthesis in barley can be counterbalanced by KCl, despite the further decrease in the osmotic potential and increase in the Cl^- concentration of the substrate (Table 16.42). In the barley cultivar, replacement of K^+ by Na^+ may allow osmotic adjustment in expanded leaves but not the maintenance of protein synthesis. Except in the case of a few halophytes, Na^+ cannot replace K^+ in its function in protein synthesis, irrespectively of the salt tolerance of cultivars within a given species, as has been shown in wheat (Gibson *et al.*, 1984).

Table 16.42
Effect of Substrate Salinization on Growth, Mineral Element Content, and Protein Synthesis in Barley (cv. Miura)[a]

Treatment	Shoot dry weight (mg per plant)	Content (mmol $(100 \text{ g})^{-1}$ dry wt)		^{15}N Content (% of total ^{15}N)[b]	
		K	Na	Protein N	Inorganic N
Control	371	126	14	44	3
+80 mM NaCl	286	80	208	29	20
+80 mM NaCl + 10 mM KCl	323	136	160	49	1

[a]Based on Helal and Mengel (1979).
[b]After supply of ^{15}NH$_4$^{15}NO$_3$ for 24 h.

16.6.3.8 Phytohormones

Changes in phytohormone levels are associated with a response of plants to salinity. Typically levels of CYT decrease (Kuiper *et al.*, 1990) whereas levels of ABA increase in response to salinity, in a similar way as to drought stress (Section 5.6). Elevated ABA levels are important for rapid osmotic adjustment to salinity, both of individual cells (La Rosa *et al.*, 1985) and of intact plants. Application of ABA may therefore increase salt tolerance by enhancement of mechanisms which are important for rapid adaptation to salinity (Table 16.43), for example by increasing leaf PEP carboxylase activity which may enhance CO_2 fixation rate despite reduced stomatal aperture (Amzallag *et al.*, 1990). There are several reports of applications of CYT counteracting salinity-induced leaf senescence (Katz *et al.*, 1978; Bejaoui, 1985) and the same is true of polyamines like putrescine (Krishnamurthy, 1991). In *Sorghum* growth at high NaCl salinity could be very much improved by supplying CYT, particularly combined with GA (Amzallag *et al.*, 1992). A similar improvement of growth in the saline substrate could be achieved by doubling the nutrient concentration in the solution. It is therefore suggested that at least part of the growth depression by high salinity was caused by inadequate phytohormone production due to impaired nutrient uptake or utilization, or both.

Table 16.43
Effect of Foliar Sprays of ABA (40 μM) on Shoot Dry Weight of Sorghum (*S. bicolor* (L.) Moench) Plants Grown at 150 mM NaCl Salinity[a]

Days after salinization	Shoot dry weight (mg per plant)	
	150 mM NaCl	150 mM NaCl + ABA
3	76	127
8	156	297
15	508	1135
25	2240	3340

[a]Based on Amzallag *et al.* (1990).

Polyamines like putrescine accumulate in plants under salinity stress. They are known for their stabilizing effect on plant cell membranes and their enhancing effect on protein synthesis (Section 5.6). Application of polyamines may increase plant growth under salinity. For example, in rice foliar application of putrescine was ineffective in control (nonsalinized) plants but ameliorated the salt-induced depression in growth rate, chlorophyll, RNA and DNA content (Krishnamurthy, 1991).

Despite these examples showing interactions between salinity and phytohormones, the role of endogenous phytohormone levels in salt tolerance is still not clear. In barley cultivars differing in salt tolerance application of CYT made the salt-tolerant cultivars behave like salt-sensitive cultivars (Kuiper *et al.*, 1990), and in soybean cultivars differences in salt tolerance were not related to differences in endogenous levels of ABA or CYT (Roeb *et al.*, 1982).

16.6.4 Mechanisms of Adaptation to Saline Substrates

16.6.4.1 Salt Exclusion versus Salt Inclusion

In principle, salt tolerance can be achieved by salt exclusion or salt inclusion (Fig. 16.22). Adaptation by salt exclusion requires mechanisms for avoidance of an internal water deficit. Adaptation by salt inclusion requires either high tissue tolerance to Na^+ and Cl^-, or avoidance of high tissue concentrations. Although a clear distinction is often made between salt excluders and salt includers, there is, in reality, a continuous spectrum of different degrees of exclusion and inclusion, both between Na^+ and Cl^-, and between different parts and organs of plants. For comprehensive reviews the reader is referred to Greenway and Munns (1980) and Gorham *et al.* (1985).

In terrestrial halophytes of the Chenopodiaceae, high salt tolerance is based mainly on the inclusion of salts and their utilization for turgor maintenance or for the replacement of K^+ in various metabolic functions by Na^+. The crop species sugar beet is included in this group. Of the monocotyledons the highly salt tolerant kallar grass (*Leptochloa fusca*) is also a salt includer (Gorham, 1987), although components of excluders, such as intensive retranslocation from the shoot to the roots and root release of Na^+ and Cl^- can also be observed (Bhatti and Wieneke, 1984). In highly salt tolerant *Casuarina* species (Aswathappa and Bachelard, 1986) and in the halophylic monocotyledons *Puccinellia peisonis* (Stelzer and Läuchli, 1977) and *Festuca rubra* (Khan and Marshall, 1981) exclusion is also an important contributory factor to high salt tolerance, but these species suffer from adverse effects on water balance and very much reduced growth rates on saline substrates (Gorham *et al.*, 1985).

In glycophytes, which comprise most crop species, there is generally an inverse relationship between salt uptake and salt tolerance; that is, exclusion is the predominant strategy (Greenway and Munns, 1980; Gorham *et al.*, 1985). However, the classification of glycophytes as excluders is only a relative term, that is, it means a much lower salt uptake in comparison with includers. It usually applies only to salt transport from the roots to the leaves in general, and expanding leaves and the shoot apex in particular.

Typical differences between crop species in response to NaCl salinity in terms of

Table 16.44
Effect of Increasing NaCl Concentrations in the Substrate on Three Crop Species[a]

Species	Concentration of NaCl (mM)	Dry weight (relative)	Content (meq g^{-1} dry wt)			
			Na	Cl	K	Ca
Sugar beet	0	100	0.1	0.05	3.3	1.6
(cv. Monohill)	25	108	1.7	1.0	2.2	0.5
	50	115	2.1	1.2	2.0	0.4
	100	101	2.6	1.5	1.9	0.3
Maize	0	100	0.02	0.01	1.6	0.5
(cv. DC 790)	25	90	0.2	0.5	1.8	0.3
	50	70	0.2	0.6	2.0	0.3
	100	62	0.3	0.8	2.0	0.3
Bean	0	100	0.02	0.01	1.7	2.9
(cv. Contender)	25	64	0.04	1.0	2.2	3.7
	50	47	0.2	1.4	1.9	3.4
	100	37	0.4	1.5	2.2	3.6

[a]From Lessani and Marschner (1978) and H. Marschner (unpublished data).

growth and the mineral element content of the shoots are shown in Table 16.44. Sugar beet has the typical features of a salt-tolerant halophytic includer. Growth is enhanced by NaCl salinity and the levels of Cl$^-$ and especially Na$^+$ in the shoot increase with increasing external supply. On the other hand, the potassium and calcium levels decline due to cation competition. Maize is much less salt tolerant than sugar beet; growth is inhibited, although the levels of chloride and especially sodium in the shoot remain relatively low. Induced potassium or calcium deficiency is an unlikely cause of growth depression; impaired osmoregulation is probably the main factor. Of the three species shown in Table 16.44 bean has the lowest salt tolerance, and chloride toxicity is the main reason for growth depression at the low salinity level. In contrast to Cl$^-$, the shoot transport of Na$^+$ is effectively restricted in bean. Thus bean, like many other salt-sensitive crop species, is an effective excluder of Na$^+$ but not of Cl$^-$.

Differences in the capacity for Na$^+$ and Cl$^-$ exclusion also exist between cultivars of species. For example, the higher salt tolerance of certain cultivars of wheat (Bernal *et al.*, 1974), barley (Greenway and Munns, 1980) and citrus (Maas, 1993) is related to a more effective restriction of shoot transport of both Na$^+$ and Cl$^-$. In grapevine differences in salt tolerance are closely related to the capacity of rootstocks for Na$^+$ and particularly Cl$^-$ exclusion from the shoots (Downton, 1985). The capacity of Cl$^-$ exclusion seems to be the effect of a major dominant gene and appears to be independent of the ability of Na$^+$ exclusion from the shoot (Sykes, 1992).

Salt tolerance in soybean is also primarily related to the restriction of Cl$^-$ transport from roots to shoots (Parker *et al.*, 1987). The differences between soybean cultivars are particularly impressive. When grown in saline soil, the chloride content in the leaf dry matter of the salt-sensitive Jackson was ~0.9% compared with only 0.05% in the salt-tolerant Lee (Läuchli and Wieneke, 1979). Inheritance of the capacity for Cl$^-$ exclusion in these soybean cultivars is controlled by a single gene pair (Abel, 1969).

Mechanisms which restrict excessive Na$^+$ and Cl$^-$ transport to the shoots of plants grown in saline substrates operate at root level (such as membrane properties, anatomical features) and along the pathway from roots to the shoot (Sections 2.5 and 3.2). Retranslocation of Na^+ from the shoots to the roots may contribute to low sodium contents in the shoots of both salt-sensitive species such as bean, *Phaseolus vulgaris* (Section 3.4) or salt-tolerant species such as common reed, *Phragmites communis* (Matsushika and Matoh, 1991) and berseem, *Trifolium alexandrinum* (Winter, 1982). In halophytes such as the mangrove *Avicennia marina* exposed to saline seawater, about 80% of the salts delivered by mass flow to the root surface are excluded from uptake (Waisel *et al.*, 1986). In many halophytes barriers are particularly developed in the roots against passive influx of salts. For example, the width of the Casparian band is 2–3 times greater than of glycophytes (Poljakoff-Mayber, 1975), and the inner cortex cell layer may be differentiated into a second endodermis (Stelzer and Läuchli, 1977).

In glycophytes differences in passive membrane permeability to Na^+ and Cl^-, or efflux pumps at the plasma membrane are the main mechanisms at the root level for restricting uptake and root to shoot transfer. In different rootstocks of grapevine a striking negative correlation was observed between certain phospholipids in the membranes of roots and the chloride content of the leaves (Kuiper, 1968). In *Citrus* species genotypical differences in degree of Cl^- exclusion from the shoots were inversely correlated with the ratio phospholipids/free sterols in the plasma membrane of root cells (Douglas and Sykes, 1985).

No such correlations between lipid composition of root cell membranes and uptake and transport of Na^+ have been found so far (Hajibagheri *et al.*, 1989). The operation of an Na^+/H^+ antiport (Section 2.4) at the plasma membrane and tonoplast of root cells may be enhanced under salinity (Blumwald and Poole, 1987) but the functioning of such an antiport seems not to be related to the salt tolerance of different plant species (Mennen *et al.*, 1990). In wheat species (*T. aestivum*, *T. turgidum*) there is also no evidence that the mechanisms of salt tolerance, or salt sensitivity, are based on differences in K^+/Na^+ selectivity of a cation channel in the plasma membrane (Schachtman *et al.*, 1991). In maize cultivars differences in degree of Na^+ exclusion seem to be related to differences in passive Na^+ permeability of the root cell membranes (Schubert and Läuchli, 1990), and cultivars with higher salt tolerance have a higher capacity to maintain low Na^+ concentrations in the cytoplasm (Hajibagheri *et al.*, 1989). In response to high NaCl salinity enhancement of plasma membrane and tonoplast-bound H^+ pumping ATPase activity can be observed in barley roots (Matsumoto and Chung, 1988), and this enhancement is also much more distinct in *Sorghum* as an Na^+ excluder than in *Spartina* as an Na^+ includer (Koyro *et al.*, 1993).

In wheat where salt tolerance seems to be positively correlated with restriction of uptake and shoot transport of Na^+ in particular, much progress has been made in breeding salt tolerant genotypes by hybridization and introducing the D-genome from the salt tolerant diploid *Aegilops squarrosa* and producing a synthetic hexaploid type (Shah *et al.*, 1987; Gorham, 1993). The D-genome strongly increased the K^+/Na^+ selectivity in both roots and shoot and the salt tolerance of the synthetic hexaploid plants. In contrast to cation selectivity the D-genome had no effect on Cl^- accumulation in the leaves.

In contrast to wheat, in barley cultivars high salt tolerance is correlated with even higher xylem transport of Na^+ to the shoot, i.e., properties typical for salt includers

(Wolf and Jeschke, 1986). Even within a species such as tomato, salt tolerance of cultivars can be based on quite different mechanisms: in the most tolerant cultivars, one effectively restricted translocation of Na^+ and Cl^- to the shoots, the other accumulated Na^+ and Cl^- in the shoots and simultaneously had lower potassium contents than the other cultivar (Alfocea *et al.*, 1993).

16.6.4.2   Salt Distribution in Shoot Tissue

In includer types partitioning of Na^+ and Cl^- into various shoot organs and tissues is of principal importance. This holds true for partitioning between old and young leaves, leaf sheath and leaf blades, cell types within leaf blades, and vegetative and reproductive organs. Restricted import of Na^+ and Cl^- into young leaves is characteristic for salt tolerant species. In *Kosteletzkya virginica*, a dicotyledonous halophyte, at substrate concentrations of 85 mm NaCl, the optimum concentration for growth, from the oldest to the youngest leaf the Na^+ concentration in the leaf water decreased from 230 to 25 mm whereas the K^+ concentration increased from 100 to 320 mm (Blits and Gallagher, 1990). A much more effective restriction of Na^+ und Cl^- import into young leaves compared to old leaves was also typical for a clone of *Agrostis stolonifera* from salt marshes compared with a clone from inland (Robertson and Wainwright, 1987).

For salt tolerance of crop species it is also not so much the total salt content in the shoot dry matter that is of importance but the capacity to maintain steep concentration gradients of Na^+ and Cl^- between old and young leaves by restricting the import into young leaves, inflorescences and seeds as has been shown for wheat (Gorham *et al.*, 1986) and maize (Hajibagheri *et al.*, 1987). In sugar beet as a salt-tolerant crop species similar steep inverse Na^+/K^+ gradients between old and young leaves are maintained (Section 10.2.4) as is typical for halophytes. High K^+ but low Na^+ concentrations in young leaves and reproductive organs are achieved by a general low xylem import of both K^+ and Na^+, but high phloem import of K^+ from mature leaves (Wolf *et al.*, 1991). An example of this had been given in Section 3.5 for barley grown at high substrate salinity.

The importance of Cl^- partitioning within individual leaves for salt tolerance has been demonstrated for sorghum (Boursier and Läuchli, 1989) and barley (Table 16.45). In both species Cl^- is particularly accumulated in the leaf sheath and in the epidermal cells of the blades, and present at much lower concentrations in the mesophyll (barley, sorghum) and bundle sheath cells (sorghum). In principle, similar differences in partitioning as shown in Table 16.45 at 100 mm NaCl were achieved after 4 days at 50 mm NaCl (Huang and van Steveninck, 1989c). The maintenance of a lower Cl^- concentration in the mesophyll cells of leaf blades of the salt-tolerant barley cultivar California Mariout might be considered an important factor for protecting photosynthetic tissues from salt stress. These examples again demonstrate how misleading average values for the shoots are in terms of interpreting mechanisms of salt tolerance.

16.6.4.3   Osmotic Adjustment

With a sudden increase in salinity, osmotic adjustment is achieved first by a decrease in tissue water content (partial dehydration). Salt tolerance and further growth in a saline

Table 16.45
Accumulation of Na$^+$ and Cl$^-$ in Leaves of Two Barley Cultivars after 1 Day Exposure to 100 mM NaCl[a]

Cultivar	Organ	Tissue	Concentration in vacuoles (mM)	
			Na$^+$	Cl$^-$
California Mariout	Blade	Epidermis	35	110
(salt tolerant)		1st Mesophyll	42	4
	Sheath	Epidermis	134	204
		1st Mesophyll	72	223
Clipper	Blade	Epidermis	41	170
(salt sensitive)		1stMesophyll	58	44
	Sheath	Epidermis	171	238
		1st Mesophyll	157	191

[a]Huang and van Steveninck (1989c). Reprinted by permission of the American Society of Plant Physiologists.

substrate, however, require a net increase in the quantity of osmotically active solutes in the tissue (Gorham *et al.*, 1985). In genotypes in which salt exclusion is the principal mechanism of salt tolerance, either the synthesis of organic solutes such as sugars and amino acids or the uptake rate of, for example, K$^+$, Ca^{2+}, or NO$_3^-$ must be increased. In terms of energy demand this is most expensive and growth rates of such genotypes is naturally very low.

In genotypes in which salt inclusion is the predominant strategy (Fig. 16.22) osmotic adjustment is achieved by the accumulation of salts (mainly NaCl) in the leaf tissue (Flowers, 1988). In natrophilic species Na$^+$ can replace K$^+$ not only in its function as an osmotically active solute in the vacuoles but to some extent also in specific functions in cell metabolism (Section 10.2). In salt-tolerant includer species, such as members of Chenopodiaceae, and also in tobacco, at high substrate salinity in the leaves the capacity of the mesophyll vacuoles to accumulate Na$^+$ and Cl$^-$ may be vastly increased. This occurs by increasing cell size, mainly the vacuole (succulence) and thereby 'diluting' the accumulated salts and preventing, or at least delaying, accumulation of Na$^+$ and Cl$^-$ in the leaf apoplasm and in the cytoplasm (Gorham *et al.*, 1985; Flowers *et al.*, 1986).

A shift from C$_3$ photosynthesis to CAM (Section 5.2.4) is another adaptive response to substrate salinity in facultative halophytes such as *Mesembryanthemum crystallinum* (Cushman *et al.*, 1990), thereby drastically decreasing transpiration and thus salt transport to the shoots. The rapidity of this shift depends on plant development and is very much increased with plant age.

In saline substrates osmotic adjustment in includers requires accumulation of high salt concentrations in the leaf cells in the order of 300–500 mM of both Cl$^-$ and Na$^+$. In the cytoplasm the concentration of inorganic ions has to be kept in the range 100–200 mM (Gorham *et al.*, 1985). In nonhalophytes protein synthesis, studied *in vitro* as a translation step by polysomes, is optimal between 100 and 150 mM K$^+$ and 2–4 mM Mg^{2+}. Higher concentrations of K$^+$, and even Na$^+$ and Cl$^-$ concentrations below 100 mM, are inhibitory. However, in certain salt-tolerant species these optimal concentrations are higher and in the range of 200 mM K$^+$ and 6–8 mM Mg^{2+} (Flowers and

Table 16.46
Solute Concentration in Isolated Chloroplasts and in Leaf Extracts of
Spinach Plants Grown without (Control) or with 300 mM NaCl[a]

Solute	Control		+300 mM NaCl	
	Chloroplast	Leaf	Chloroplast	Leaf
meq l^{-1}				
Na^+	7	2	22	405
K^+	180	318	108	191
Mg^{2+}	18	32	13	39
Cl^-	1	21	25	335
HPO_4^{2-}	30	31	16	51
mmol l^{-1}				
QAC_s[b]	57	21	181	47

[a]Based on Schröppel-Meier and Kaiser (1988).
[b]Quarternary ammonium compounds (e.g. glycine betaine).

Dalmond, 1992). Furthermore, Na^+ is better able to substitute for the role of K^+ in the translation step of polysomes from halophytes than from glycophytes. In certain halophytes high NaCl concentrations also stimulate photosynthetic O_2 evolution in PS II whereas the same concentrations of KCl are inhibitory (Preston and Critchley, 1986). Thus, salt tolerance of certain halophytes is achieved also by inclusion of Na^+ in the cytoplasm and chloroplasts.

16.6.4.4 Compartmentation and Compatible Solutes

Certain enzymes, such as membrane-bound ATPase, in the roots are either activated or inhibited *in vitro* by high salt concentrations, depending on the salt tolerance of the intact plants; that is, membrane-bound ATPases of halophytes may be less sensitive to salt than those of glycophytes (Lerner *et al.*, 1983). However, *in vitro* in many instances enzymes such as malate dehydrogenase and aspartate transaminase of halophytes such as *Atriplex spongiosa* and of glycophytes like *Phaseolus vulgaris* are similarly sensitive to high NaCl concentrations (Greenway and Osmond, 1972). Thus, as a rule, includer species which accumulate large quantities of Na^+ and Cl^- or other inorganic ions in their leaves for osmotic adjustment must protect their enzymes in the cytoplasm and in the chloroplasts from high concentrations of Na^+ and Cl^-. Such strict compartmentation of Na^+ and Cl^- in the vacuoles has been demonstrated, for example, in spinach growing in saline substrates and containing high NaCl levels in the leaf tissue (Coughlan and Wyn Jones, 1980).

For osmotic adjustment of the cytoplasm and its organelles 'non-toxic' or 'compatible' organic solutes such as gylcine betaine (Section 8.2.2) accumulate. In the leaves of spinach exposed to 300 mM NaCl, Na^+ and Cl^- are mainly sequestered in the vacuoles and the K^+ concentrations in the chloroplasts are still sufficiently high (Table 16.46) in order to maintain photosynthesis. For osmotic adjustment high concentrations of compatible solutes accumulated in the chloroplasts of salt-stressed plants whereas for the whole leaf tissue this increase was much less distinct. In a comparable study with

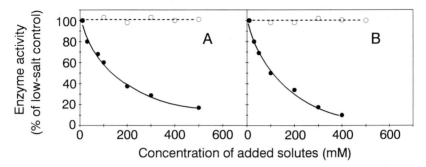

Fig. 16.24 Comparative effects of NaCl (●) and glycine betaine (○) on soluble enzymes of barley. A, malate dehydrogenase; B, pyruvate kinase. (Based on Pollard and Wyn Jones, 1979.)

spinach exposed to 200 mM NaCl the concentrations of glycine betaine in the leaf tissue increased from 2.5 to 16.4 mM and in the chloroplasts from 26 to 289 mM (Robinson and Jones, 1986). At least 30–40% of the total leaf glycine betaine was located in the chloroplasts of the salt-stressed plants. Most of the remaining glycine betaine is usually located in the cytosol whereas the concentrations in the vacuoles are very low. In chloroplasts high chloride concentrations seem to enhance dissociation of intrinsic proteins from the O_2 evolving system, and glycine betaine prevents this effect of chloride (Papageorgiou *et al.*, 1991).

Glycine betaine is a very effective cytosolute because it is highly water soluble and does not carry a net charge, which could affect the charge balance of the cytoplasm. As shown in Fig. 16.24 enzyme activities are not inhibited even at extremely high concentrations of glycine betaine, whereas even much lower concentrations of NaCl severely inhibit these enzymes. In pyruvate kinase isolated from halophytes such as *Atriplex gmelini*, glycine betaine not only protects activity but simultaneously reduces the K^+ requirement for enzyme activation. The K_m value for K^+ decreased from 5.6 mM in the absence of glycine betaine to 3.2 and 1.3 mM in presence of 500 and 1000 mM glycine betaine, respectively (Matoh *et al.*, 1988). Thus, in halophytes glycine betaine may reduce the demand of K^+ in the cytosol by more than half, at least for the pyruvate kinase.

The chemical nature of compatible solutes involved in osmotic adaptation varies between taxonomic groups (Table 16.47). Glycine betaine is the most well studied

Table 16.47
Examples of Taxonomic Distribution of Compatible Solutes[a]

Solute	Distribution
D-Sorbitol	Rosaceae, Plantaginaceae
D-Pinitol	Leguminoseae, Caryophyllaceae
Glycine betaine	Chenopondiaceae, Gramineae, Solanaceae
Proline	Asteraceae, Gramineae
3-dimethylsulfonioipropionate	Asteraceae, Gramineae

[a]Based on Gorham *et al.* (1985).

compatible solute in relation to plant adaptation to saline substrates. D-Pinitol (an inositol derivate) is probably also important as a compatible solute in *Mesembryanthemum crystallinum* where it is located usually in the cytosol and chloroplasts (Paul and Cockburn, 1989), whereas in *Viscum album* it may also contribute to the osmotic potential in the vacuoles (Richter and Popp, 1992). Proline accumulation is a well-known response to water deficit and to salt stress and it has a protective effect on seed germination in saline substrates (Bar-Nun and Poljakoff-Mayber, 1977).

Osmotic adjustment in plants via salt inclusion or exclusion has important implications for energy balance. Since NaCl and other soluble salts are abundant in saline substrates, they can be regarded as potentially 'cheap', although dangerous, osmotica. According to Wyn Jones (1981) the approximate energy cost of accumulating 1 mole solute for osmoregulation is as follows:

	Energy requirement (mole ATP mole^{-1} solute)
NaCl uptake	0.54
Synthesis of K^+-malate (CO_2 fixation)	13
Accumulation of C_6 sugars	54

A concentration of 300 mM C_6 sugars in the cell sap would account for 20–30% of the tissue dry weight. Thus, species with effective salt exclusion from the shoot tissue do not have much chance of achieving substantial growth rates in highly saline substrates. However, for osmotic adjustment average values for the whole shoots are also an inadequate parameter. An instructive example of the diversity of osmotic adjustment in various organs of the same plant is shown in Table 16.48 for *Aster tipolium* growing in a saline substrate. Osmotic adjustment in the leaves is mediated by Na^+ and Cl^-, whereas that in the flowers is mediated by K^+, glycine betaine, and sugars. But even within a given leaf the role of solutes may vary; in young leaves of sorghum, glycine betaine is important for osmotic adjustment only in the leaf blades, not in the leaf sheaths (Grieve and Maas, 1984). In barley grown on saline substrates, sugars play an insignificant role, compared with Na^+ and Cl^- in osmotic adjustment in mature leaves, whereas they

Table 16.48
Chemical Composition of *Aster tipolium* Leaves and Florets[a]

Component	Leaves (mM)[b]	Florets (mM)[b]
Sodium	360	56
Chloride	320	51
Potassium	72	133
Glycine betaine	18	82
Total soluble sugars	53	493

[a]Based on Gorham *et al.* (1980).
[b]Plant water basis.

contribute more than 20% to the osmotic adjustment in expanding leaves (De Lane *et al.*, 1982).

16.6.4.5   Salt Excretion

Halophytes may reduce the salt concentrations of the active photosynthetic tissue by various mechanisms: accumulation in bladder hairs, secretion by salt glands, shedding of old leaves, and retranslocation to other organs (Waisel *et al.*, 1986). Salt glands vary much in their anatomy and effectivity. They may be multicellular organs of highly specialized cells, for example in *Avicennia marina*, or are simple-type glands comprising only two cells, for example in kallar grass, *Leptochloa fusca* (Wieneke *et al.*, 1987). Salt glands are highly selective and may eliminate relatively large quantities of salt by secretion to the leaf surface, where it can be washed off by rain or dew. Secretion as an active process is highly temperature-dependent (Gorham, 1987), and usually Na^+ and Cl^- are secreted in equivalent amounts (Gorham 1987; Ball, 1988). The importance of salt excretion to the salt tolerance of many halophytes is indicated by the fact that the salt tolerance of intact plants (*Sueda* or *Atriplex*) cannot be reproduced in callus cultures (Smith and McComb, 1981). In the mangrove *A. marina*, of the salts transported in the xylem to the shoot between 40% (Waisel *et al.*, 1986) and 90% (Ball, 1988) are excreted by salt glands. However, salt excretion in *A. marina* or *L. fusca* is considered to be a secondary mechanism of salt tolerance, the exclusion at the roots (i.e., avoidance) being the major mechanism (Waisel *et al.*, 1986; Gorham, 1987).

16.6.5   Outlook

Selection and breeding programs designed to improve the adaptation of crop plants to saline soils have to consider the various mechanisms responsible for salt tolerance and sensitivity. In Cl^--sensitive species such as soybean and grapevine, efforts have to be focused on the exclusion of Cl^- from leaf tissue. However, effective excluders of both Na^+ and Cl^- will probably not become very productive crops in highly saline soils because water deficit and large photosynthate requirements for osmotic adjustment seriously restrict their growth. Crop plants of the includer type have a much greater potential for better adaptation coupled with sufficient productivity when grown in highly saline soils. Includers of both Na^+ and Cl^- rely on a strict compartmentation of salts within individual leaf cells and on their capacity to maintain a high K^+/Na^+ ratio in growing tissue. Salt deposition in nonphotosynthetic tissue or excretion to the leaf surface or both these processes, are also important for adaptation.

Efforts to develop new crop cultivars with improved salt tolerance have been intensified over the past 15–20 years. However, with the exception of rice (Section 16.1) there is only a limited number of cultivars that have been developed with improved salt tolerance, and for all of them selection has been based on agronomic characters such as yield or survival, thus characters that integrate various physiological mechanisms (e.g., K^+/Na^+ selectivity) responsible for tolerance (Noble and Rogers, 1992).

In vitro selection in cell and tissue culture for improving salt tolerance of plants is frequently suggested (Nabors *et al.*, 1980; Smith and McComb, 1981). However, plants regenerated from selected cells have not shown an unequivocal increase in salt tolerance (Dracup, 1991, 1993). It is also a matter of question whether cell cultures are appropriate for selection because salt tolerance is in most instances multigenic and depends on the structural and physiological integrity of plants as it has been shown in this section. *In vitro* selection should therefore be confined to traits where a large proportion of the whole plant tolerance may be cell-based, for example Na^+/Ca^{2+} interactions and Cl^- exclusion (Dracup, 1991; Noble and Rogers, 1992).

In addition to using existing genetic variability within crop species for breeding (Epstein *et al.*, 1980; Yeo, 1993) it seems promising to introduce important traits of salt tolerance into crop species from their wild relations through interspecific hybridization, as has been shown, for example in tomato (Tal, 1985) and wheat (Shah *et al.*, 1987; Gorham, 1993). However, in these approaches it is often not sufficiently appreciated that these traits in wild plants are favorable mainly for survival under natural conditions. Furthermore, for crop production in dry saline soils, high water use efficiency (WUE) is of key importance, and the already existing wheat and barley cultivars have a higher WUE and total biomass production on saline soils than most of the salt-tolerant wild species including C_4 halophytes (Richards, 1991).

References

Abadia, J. (1992). Leaf responses to Fe deficiency: a review. *J. Plant Nutr.* **15**, 1699–1713.

Abadia, A., Sanz, M., de las Rivas, J. and Abadia, J. (1989). Photosynthetic pigments and mineral composition of iron deficient pear leaves. *J. Plant Nutr.* **12**, 827–838.

Abadia, A., Poc, A. and Abadia, J. (1991). Could iron nutrition status be evaluated through photosynthetic pigment changes? *J. Plant Nutr.* **14**, 987–999.

Abbott, A. J. (1967). Physiological effects of micronutrient deficiencies in isolated roots of *Lycopersicon esculentum*. *New Phytol.* **66**, 419–437.

Abbott, L. K. and Robson, A. D. (1985). Formation of external hyphae in soil by four species of vesicular-arbuscular mycorrhizal fungi. *New Phytol.* **99**, 245–255.

Abdullah, Z. and Ahmad, R. (1982). Salt tolerance of *Solanum tuberosum* L. growing on saline soils amended with gypsum. *Z. Acker- Pflanzenbau* **151**, 409–416.

Abel, G. H. (1969). Inheritance of the capacity for chloride inclusion and chloride exclusion by soybeans. *Crop Sci.* **9**, 697–698.

Aber, J. D., Melillo, J. M., Nadelhoffer, K. J., McClaugherty, C. A. and Pastor, J. (1985). Fine root turnover in forest ecosystems in relation to quantity and form of nitrogen availability: a comparison of two methods. *Oecologia* **66**, 317–321.

Aber, J. D., Nadelhoffer, K. J., Steudler, P. and Melillo, J. M. (1989). Nitrogen saturation in northern forest ecosystems. *Bioscience* **39**, 378–386.

Abou, A. A. and Volk, O. H. (1971). Nachweis von Cytokinin-Aktivität in rost-infizierten Pelargonium-Blättern. *Z. Pflanzenphysiol.* **65**, 240–247.

Abrams, M. M., Shennan, C., Zasoski, J. and Burau, R. G. (1990). Selenomethionine uptake by wheat seedlings. *Agron. J.* **82**, 1127–1130.

Abruna-Rodriguez, F., Vicente-Chandler, J., Rivera, E. and Rodriguez, J. (1982). Effect of soil acidity factors on yield and foliar composition of tropical root crops. *Soil Sci. Soc. Am. J.* **46**, 1004–1007.

Abuzinadah, R. A., Finlay, R. D. and Read, D. J. (1986). The role of proteins in the nitrogen nutrition of ectomycorrhizal plants. II. Utilization of proteins by mycorrhizal plants of *Pinus contorta*. *New Phytol.* **103**, 495–506.

Adam, G. and Marquardt, V. (1986). Brassinosteroids; Review. *Phytochemistry* **25**, 1787–1799.

Adams, F. (1966). Calcium deficiency as a causal agent of ammonium phosphate injury to cotton seedlings. *Soil Sci. Soc. Am. Proc.* **30**, 485–488.

Adams, F. and Moore, B. L. (1983). Chemical factors affecting root growth in subsoil horizons of coastal plain soils. *Soil Sci. Soc. Am. J.* **47**, 99–102.

Adams, J. F., Burmester, C. H. and Mitchell, C. C. (1990). Long-term fertility treatments and molybdenum availability. *Fert. Res.* **21**, 167–170.

Adatia, M. H. and Besford, R. T. (1986). The effect of silicon on cucumber plants grown in recirculating nutrient solution. *Ann. Bot.* **56**, 343–351.

Adriaanse, F. G. and Human, J. J. (1990). The effect of nitrate: ammonium ratios and nitrapyrin on the nitrogen response of *Zea mays* L. under field conditions. *Plant Soil* **122**, 287–293.

Adu-Gyamfi, J. J., Fujita, K. and Ogata, S. (1989). Phosphorus absorption and utilization efficiency of pigeon pea (*Cajanus cajan* (L) Millsp.) in relation to dry matter production and dinitrogen fixation. *Plant Soil* **119**, 315–324.

Ae, N., Arihara, J., Okada, K., Yoshihara, T. and Johansen, C. (1990). Phosphorus uptake by pigeon pea and its role in cropping systems of the Indian subcontinent. *Science* **248**, 477–480.

Ae, N., Arihara, J., Okada, K., Yoshihara, T., Otani, T. and Johansen, C. (1993). The role of piscidic acid secreted by pigeon pea roots grown in an Alfisol with low-P fertility. *In* 'Genetic Aspects of Plant Mineral Nutrition' (P. J. Randall, E. Delhaize, R. A. Richards and R. Munns, eds.), pp. 279–288. Kluwer Academic, Dordrecht.

Agarwala, S. C., Sharma, C. P., Farooq, S. and Chatterjee, C. (1978). Effect of molybdenum deficiency on the growth and metabolism of corn plants raised in sand culture. *Can. J. Bot.* **56**, 1905–1908.

Agarwala, S. C., Chatterjee, C., Sharma, P. N., Sharma, C. P. and Nautiyal, N. (1979). Pollen development in maize plants subjected to molybdenum deficiency. *Can. J. Bot.* **57**, 1946–1950.

Agarwala, S. C., Sharma, P. N., Chatterjee, C. and Sharma, C. P. (1981). Development and enzymatic changes during pollen development in boron deficient maize plants. *J. Plant Nutr.* **3**, 329–336.

Agerer, R. (1987). 'Colour Atlas of Ectomycorrhizae'. Einhorn, Schwäb. Gmünd.

Aguilar S. A. and van Diest, A. (1981). Rock-phosphate mobilization induced by the alkaline uptake pattern of legumes utilizing symbiotically fixed nitrogen. *Plant Soil* **61**, 27–42.

Ahmad, R., Zaheer, S. H. and Ismail, S. (1992). Role of silicon in salt tolerance of wheat (*Triticum aestivum* L.). *Plant Sci.* **85**, 43–50.

Ahmed, C. M. S. and Sagar, G. R. (1981). Effects of a mixture of NAA + BA on numbers and growth rates of tubers of *Solanum tuberosum* L. *Potato Res.* **24**, 267–278.

Ahmed, S. and Evans, H. J. (1960). Cobalt: a micronutrient element for the growth of soybean plants under symbiotic conditions. *Soil Sci.* **90**, 205–210.

Akeson, M., Munns, D. and Burau, R. G. (1989). Adsorption of Al^{3+} to phosphatidyl-choline vesicles. *Biochim. Biophys. Acta* **986**, 33–40.

Alagarswamy, G., Gardner, J. C., Maranville, J. W. and Clark, R. B. (1988). Measurement of instantaneous nitrogen use efficiency among pearl millet genotypes. *Crop Sci.* **28**, 681–685.

Alagna, L., Hasnain, S. S., Pigott, B. and Williams, D. L. (1984). The nickel ion environment in jack bean urease. *Biochem. J.* **220**, 591–595.

Al-Ani, M. K. A. and Hay, R. K. M. (1983). The influence of growing temperature on the growth and morphology of cereal seedling root systems. *J. Exp. Bot.* **34**, 1720–1730.

Alazard, D., Ndoye, I. and Dreyfus, B. (1988). *Sesbania rostrata* and other stem-nodulated legumes. *In* 'Nitrogen Fixation: Hundred Years After' (H. Bothe, F. de Bruijn and W. E. Newton, eds.), pp. 765–769. Fischer, Stuttgart.

Albert, L. S. (1965). Ribonucleic acid content, boron deficiency systems, and elongation of tomato root tips. *Plant Physiol.* **40**, 649–652.

Alcantara, E. and de la Guardia, M. D. (1991). Variability of sunflower inbred lines to iron deficiency stress. *Plant Soil* **130**, 93–96.

Alcantara, E., Fernandez, M. and de la Guardia, M. D. (1990). Genetic studies on the acidification capacity of sunflower roots induced under iron stress. *Plant Soil* **123**, 239–241.

Alcantara, E., de la Guardia, M. D. and Romer, F. J. (1991). Plasmalemma redox activity and H^+ extrusion in roots of Fe-deficient cucumber plants. *Plant Physiol.* **96**, 1034–1037.

Alcaraz, C. F., Martinez-Sánchez, F., Sevilla, F. and Hellin, E. (1986). Influence of ferredoxin levels on nitrate reductase activity in iron deficient lemon leaves. *J. Plant Nutr.* **9**, 1405–1413.

Alcubilla, M., Diaz-Palacio, M. P., Kreutzer, K., Laatsch, W., Rehfuess, K. E. and Wenzel, G. (1971). Beziehungen zwischen dem Ernährungszustand der Fichte (*Picea abies* (L.) Karst.), ihrem Kernfäulebefall und der Pilzhemmwirkung ihres Basts. *Eur. J. For. Pathol.* **1**, 100–114.

Alexander, I. (1989). Mycorrhizas in tropical forests. *In* 'Mineral Nutrients in Tropical Forests and Savannah Ecosystems' (J. Protector, ed.), pp. 169–188. Blackwell, Oxford.

Alexander, D. B. and Zuberer, D.A. (1989). $^{15}N_2$ fixation by bacteria associated with maize roots at a low partial O_2 pressure. *Appl. Environ. Microbiol.* **55**, 1748–1753.

Alfocea, F. P., Estan, M. T., Caro, M. and Bolarin, M. C. (1993). Response of tomato cultivars to salinity. *Plant Soil* **150**, 203–211.

Ali, A. H. N. and Jarvis, B. C. (1988). Effects of auxin and boron on nucleic acid metabolism and cell division during adventitious root regeneration. *New Phytol.* **108**, 383–391.

Allan, A. C. and Rubery, P. H. (1991). Calcium deficiency and auxin transport in *Cucurbita pepo* L. seedlings. *Planta* **183**, 604–612.

Allan, D. L. and Jarrell, W. M. (1989). Proton and copper adsorption to maize and soybean root cell walls. *Plant Physiol.* **89**, 823–832.

Allen, M. (1960). The uptake of metallic ions by leaves of apple trees. II. The influence of certain anions on uptake from magnesium salts. *J. Hortic. Sci.* **35**, 127–135.

Allen, R. D. (1969). Mechanism of the seismonastic reaction in *Mimosa pudica. Plant Physiol.* **44**, 1101–1107.

Allen, S., Raven, J. A. and Sprent, J. I. (1988). The role of long-distance transport in intracellular pH regulation in *Phaseolus vulgaris* grown with ammonium or nitrate as nitrogen source, or nodulated. *J. Exp. Bot.* **39**, 513–528.

Alleweldt, G., Düring, H. and Waitz, G. (1975). Untersuchungen zum Mechanismus der Zuckereinlagerung in wachsende Weinbeeren. *Angew. Bot.* **49**, 65–73.

Allinger, P., Michael, G. and Martin, P. (1969). Einfluß von Cytokininen und anderen Wuchsstoffen auf die Stoffverlagerung in abgeschnittenen Blättern. *Flora (Jena), Abt. A* **160**, 538–551.

Alloush, G. A., Le Bot, J., Sanders, F. E. and Kirkby, E. A. (1990). Mineral nutrition of chickpea plants supplied with NO_3^-- or NH_4^-- N. I. Ionic balance in relation to iron stress. *J. Plant Nutr.* **13**, 1575–1590.

Alloway, B. J. and Tills, A. R. (1984). Copper deficiency in world crops. *Outlook Agric.* **13**, 32–42.

Allsopp, N. and Stock, W. D. (1992). Mycorrhizas, seed size and seedling establishment in a low nutrient environment. *In* 'Mycorrhizas in Ecosystems' (D. J. Read, D.

H. Lewis, A. H. Fitter and I. J. Alexander, eds.), pp. 59–64. CAB International, Wallingford, UK.

Aloni, R. and Griffith, M. (1991). Functional xylem anatomy in root–shoot functions of six cereal species. *Planta* **184**, 123–129.

Alt, D. and Stüwe, S. (1982). Decline of the nitrate content in lettuce (*Lactuca sativa* var. Capitata L.) by means of monitoring the nitrogen content of the nutrient solution in hydroponic systems. *In* 'Proceedings of the Ninth International Plant Nutrition Colloquium, Warwick, England' (A. Scaife, ed.), pp. 17–21. Commonwealth Agricultural Bureau, Farnham Royal, Bucks, UK.

Alt, D., Zimmer, R., Stock, M., Peters, I. and Krupp, J. (1982). Erhebungsuntersuchungen zur Nährstoffversorgung von *Picea omorika* im Zusammenhang mit dem Omorikasterben. *Z. Pflanzenernähr. Bodenk.* **145**, 117–127.

Altman, A., Levin, N., Cohen, P., Schneider, M. and Nadel, B. (1989). Polyamines in growth and differentiation of plant cell cultures: The effect of nitrogen nutrition, salt stress and embryogenic media. *In* 'Polyamines in Biochemical and Clinical Sciences' (V. Zappia, ed.). Raven Press, New York.

Alva, A. K. and Sumner, M. E. (1990). Amelioration of acid soil infertility by phosphogypsum. *Plant Soil* **128**, 127–134.

Alva, A. K., Edwards, D. G., Asher, C. J. and Suthipradit, S. (1987). Effects of acid soil infertility factors on growth and nodulation of soybean. *Agron. J.* **79**, 302–306.

Alva, A. K., Asher, C. J. and Edwards, D. G. (1990). Effect of solution pH, external calcium concentration, and aluminium activity on nodulation and early growth of cowpea. *Aust. J. Agric. Res.* **41**, 359–365.

Ambak, K. and Tadano, T. (1991). Effect of micronutrient application on the growth and occurrence of sterility in barley and rice in a Malaysian deep peat soil. *Soil Sci. Plant Nutr.* **37**, 715–724.

Amberger, A. (1989). Research on dicyandiamide as a nitrification inhibitor and future outlook. *Commun. Soil Sci. Plant Anal.* **20**, 1933–1955.

Ames, R. N., Reid, C. P. P., Porter, L. K. and Cambardella, C. (1983). Hyphal uptake and transport of nitrogen from two ^{15}N-labelled sources by *Glomus mosseae*, a vesicular-arbuscular mycorrhizal fungi. *New Phytol.* **95**, 381–396.

Amijee, F., Stribley, D. P. and Tinker, P. B. (1990). Soluble carbohydrates in roots of leek (*Allium porrum*) plants in relation to phosphorus supply and VA mycorrhiza. *Plant Soil* **124**, 195–198.

Amzallag, G. N., Lerner, H. R. and Poljakoff-Mayber, A. (1990). Exogenous ABA as a modulator of the response of *Sorghum* to high salinity. *J. Exp. Bot.* **41**, 1529–1534.

Amzallag, G. N., Lerner, H. R. and Poljakoff-Mayber, A. (1992). Interaction between mineral nutrients, cytokinin and gibberellic acid during growth of *Sorghum* at high NaCl salinity. *J. Exp. Bot.* **43**, 81–87.

Anderson, A. J. (1988). Mycorrhizae-host specifity and recognition. *Phytopathology* **78**, 375–378.

Anderson, A. J. and Spencer, D. (1950). Sulphur in nitrogen metabolism of legumes and non-legumes. *Aust. J. Sci. Res.*, *Ser. B* **3**, 431–449.

Anderson, D. L. (1991). Soil and leaf nutrient interactions following application of calcium silicate slag to sugarcane. *Fert. Res.* **30**, 9–18.

Anderson, D. L., Jones, D. B. and Snyder, G. H. (1987). Response of a rice-sugarcane rotation to calcium silicate slag on everglade histosols. *Agron. J.* **79**, 531–535.

Anderson, E. L. (1988). Tillage and N fertilization effects on maize root growth and root:shoot ratio. *Plant Soil* **108**, 245–251.

Anderson, J. V., Hess, J. L. and Chevone, B. I. (1990). Purification, characterization, and immunological properties for two isoforms of glutathione reductase from eastern white pine needles. *Plant Physiol.* **94**, 1402–1409.

Andersson, B., Critchley, C., Ryrie, I. J., Jansson, C., Larsson, C. and Anderson, J. M. (1984). Modification of the chloride requirement for photosynthetic O_2 evolution. The role of the 23 kDa polypeptide. *FEBS Lett.* **168**, 113–117.

Andrews, M. (1986a). The partitioning of nitrate assimilation between root and shoot of higher plants. *Plant Cell Environ.* **9**, 511–519

Andrews, M. (1986b). Nitrate and reduced-N concentrations in the xylem sap of *Stellaria media*, *Xanthium strumarinum* and six legume species. *Plant Cell Environ.* **9**, 605–608.

Angelini, R., Manes, F. and Federico, R. (1990). Spatial and functional correlation between diamine-oxidase and peroxidase activities and their dependence upon de-etiolation and wounding in chickpea stems. *Planta* **182**, 89–96.

Anghinoni, I. and Barber, S. A. (1980). Phosphorus influx and growth characteristics of corn roots as influenced by phosphorus supply. *Agron. J.* **72**, 685–688.

Aniol, A. (1984). Induction of aluminium tolerance in wheat seedlings by low doses of aluminium in the nutrient solution. *Plant Physiol.* **75**, 551–555.

Aniol, A. (1990). Genetics of tolerance to aluminium in wheat (*Triticum aestivum* L. Thell). *Plant Soil* **123**, 223–227.

Ankel-Fuchs, D. and Thauer, R. K. (1988). Nickel in biology: Nickel as an essential trace element. *In* 'The Bioinorganic Chemistry of Nickel' (J. R. Lancaster jr, ed.), pp. 93–110. Verlag Chemie, Weinheim.

Antkowiak, B., Engelmann, W., Herbjornsen, R. and Johnsson, A. (1992). Effect of vanadate, N_2, and light on the membrane potential of motor cells and the lateral leaflet movements of *Desmodium motorium*. *Physiol. Plant.* **86**, 551–558.

Anuradha, M. and Narayanan, A. (1991). Promotion of root elongation by phosphorus deficiency. *Plant Soil* **136**, 273–275.

Ao, T. Y., Chaney, R. L., Korcak, R. F. and Faust, M. (1987). Influence of soil moisture level on apple iron chlorosis development in a calcareous soil. *Plant Soil* **104**, 85–92.

App, A., Santiago, T., Daez, C., Menguito, C., Ventura, W., Tirol, A., Po, J., Watanabe, I., De Datta, S. K. and Roger, P. (1984). Estimation of the nitrogen balance for irrigated rice and the contribution of phototropic nitrogen fixation. *Field Crops Res.* **9**, 17–27.

Appel, T. and Mengel, K. (1992). Nitrogen uptake of cereals grown on sandy soils as related to nitrogen fertilizer application and soil nitrogen fractions obtained by electro-ultrafiltration (EUF) and $CaCl_2$ extraction. *Eur. J. Agron.* **1**, 1–9.

Appel, T. and Steffens, D. (1988). Vergleich von Elektro-Ultrafiltration (EUF) und Extraktion mit 0,01 molarer $CaCl_2$-Lösung zur Bestimmung des pflanzenverfügbaren Stickstoffs im Boden. *Z. Pflanzenernähr. Bodenk.* **151**, 127–130.

Appleby, C.A., Bogusz, D., Dennis, E. S. and Peacock, W. J. (1988). A role for haemoglobin in all plant roots? *Plant Cell Environ.* **11**, 359–367.

Arihara, A., Kumagai, R., Koyama, H. and Ojima, K. (1991). Aluminium-tolerance of carrot (*Daucus carota* L.) plants regenerated from selected cell cultures. *Soil Sci. Plant Nutr.* **37**, 699–705.

Arines, J., Vilarino, A. and Sainz, M. (1989). Effect of different inocula of vesicular-arbuscular mycorrhizal fungi on manganese content and concentration in red clover (*Trifolium pratense* L.) plants. *New Phytol.* **112**, 215–219.

Arines, J., Porto, M. E. and Vilarino, A. (1992). Effect of manganese on vesicular-arbuscular mycorrhizal development in red clover plants and on soil Mn-oxidizing bacteria. *Mycorrhiza* **1**, 127–131.

Armstrong, J., Armstrong, W. and Beckett, P. M. (1992). Venturi- and humidity-induced pressure flows enhance rhizome aeration and rhizosphere oxidation. *New Phytol.* **120**, 197–207.

Armstrong, M. J. and Kirkby, E. A. (1979a). Estimation of potassium recirculation in tomato plants by comparison of the rates of potassium and calcium accumulation in the tops with their fluxes in the xylem stream. *Plant Physiol.* **63**, 1143–1148.

Armstrong, M. J. and Kirkby, E. A. (1979b). The influence of humidity on the mineral composition of tomato plants with special reference to calcium distribution. *Plant Soil* **52**, 427–435.

Armstrong, R. D. and Helyar, K. R. (1992). Changes in soil phosphate fractions in the rhizosphere of semi-arid pasture grasses. *Aust. J. Soil Res.* **30**, 131–143.

Armstrong, W. (1969). Rhizosphere oxidation in rice: an analysis of intervarietal differences in oxygen flux from the roots. *Physiol. Plant.* **22**, 296–303.

Armstrong, W. (1979). Aeration in higher plants. *Adv. Bot. Res.* **7**, 225–232.

Armstrong, W. and Beckett, P. M. (1987). Internal aeration and the development of stelar anoxia in submerged roots. A multishelled mathematical model combining axial diffusion of oxygen in the cortex with radial losses to the stele, the wall layers and the rhizosphere. *New Phytol.* **105**, 221–245.

Arnold, G. (1992). Soil acidification as caused by the nitrogen uptake pattern of Scots pine (*Pinus sylvestris*). *Plant Soil* **142**, 41–51.

Arnon, D. I. and Stout, P. R. (1939). The essentiality of certain elements in minute quantity for plants with special reference to copper. *Plant Physiol.* **14**, 371–375.

Arnozis, P. A. and Findenegg, G. R. (1986). Electrical charge balance in the xylem sap of beet and *Sorghum* plants grown with either NO_3 or NH_4 nitrogen. *J. Plant Physiol.* **125**, 441–449.

Arnozis, P. A., Nelemans, J. A. and Findenegg, G. R. (1988). Phosphoenolpyruvate carboxylase activity in plants grown with either NO_3^- or NH_4^+ as inorganic nitrogen source. *J. Plant Physiol.* **132**, 23–27.

Arora, R. and Palta, J. P. (1988). *In vivo* perturbation of membrane-associated calcium by freeze–thaw stress in onion bulb cells. Simulation of this perturbation in extracellular KCl and alleviation by calcium. *Plant Physiol.* **85**, 622–628.

Arora, S. K. and Luchra, Y. P. (1970). Metabolism of sulphur containing amino acids in *Phaseolus aureus* Linn. *Z. Pflanzenernähr. Bodenk.* **126**, 151–158.

Arrese-Igor, C., Garcia-Plazaola, J. I., Hernandez, A. and Aparicio-Tejo, P. M. (1990). Effect of low nitrate supply to nodulated lucerne on time course of activities of enzymes involved in inorganic nitrogen metabolism. *Physiol. Plant.* **80**, 185–190.

Arshad, M. and Frankenberger jr, W. T. (1990). Response of *Zea mays* and *Lycopersicon esculentum* to the ethylene precursors, l-methionine and l-ethionine applied to soil. *Plant Soil* **122**, 219–227.

Arshad, M. and Frankenberger jr, W. T. (1991). Microbial production of plant hormones. *Plant Soil* **133**, 1–8.

Arshad, M., Hussain, A., Javed, M. and Frankenberger jr, W. T. (1993). Effect of soil applied l-methionine on growth, nodulation and chemical composition of *Albizia lebbeck* L. *Plant Soil* **148**, 129–135.

Arulanathan, A. R., Rao, I. M. and Terry, N. (1990). Limiting factors in photosynthesis. VI. Regeneration of ribulose 1,5-bisphosphate limits photosynthesis at low photochemical capacity. *Plant Physiol.* **93**, 1466–1475.

Asada, K. (1992). Ascorbate peroxidase – a hydrogen peroxide-scavenging enzyme in plants. *Physiol. Plant.* **85**, 235–241.

Asady, G. H. and Smucker, A. J. M. (1989). Compaction and root modifications of soil aeration. *Soil Sci. Soc. Am. J.* **53**, 251–254.

Asher, C. J. (1978). Natural and synthetic culture media for Spermatophytes. *CRC Handb. Ser. Nutr. Food, Sect. G* **3**, 575–609.

Asher, C. J. (1991). Beneficial elements, functional nutrients, and possible new essential elements. *In* 'Micronutrients in Agriculture', 2nd edition (J. J. Mortvedt, F. R. Cox, L. M. Shuman and R. M. Welch, eds.), pp. 703–723. Soil Sci. Soc. Amer. Book Series No. 4, Madison, WI, USA.

Asher, C. J. and Edwards, D. G. (1983). Modern solution culture techniques. *In* 'Encyclopedia of Plant Physiology, New Series' Vol. 15A (A. Läuchli and R. L. Bieleski, eds.), pp. 94–119. Springer-Verlag, Berlin and New York.

Asher, C. J., Butler, G. W. and Peterson, P. J. (1977). Selenium transport in root systems of tomato. *J. Exp. Bot.* **28**, 279–291.

Ashford, A. E., Allaway, W. G., Peterson, C. A. and Cairney, J. W. G. (1989). Nutrient transfer and the fungus-root interface. *Aust. J. Plant Physiol.* **16**, 85–97.

Aslam, M. and Huffaker, R. C. (1982). *In vivo* nitrate reduction in roots and shoots of barley (*Hordeum vulgare* L.) seedlings in light and darkness. *Plant Physiol.* **70**, 1009–1013.

Aslam, M. and Huffaker, R. C. (1984). Dependency of nitrate reduction on soluble carbohydrates in primary leaves of barley under aerobic conditions. *Plant Physiol.* **75**, 623–628.

Aslam, M. and Huffaker, R. C. (1989). Role of nitrate and nitrite in the induction of nitrate reductase in leaves of barley seedlings. *Plant Physiol.* **91**, 1152–1156.

Aswathappa, N. and Bachelard, E. P. (1986). Ion regulation in the organs of *Casuarina* species differing in salt tolerance. *Aust. J. Plant Physiol.* **13**, 533–545.

Atalay, A., Garrett, H. E., Mawhinney, T. P. and Mitchell, R. J. (1988). Boron fertilization and carbohydrate relations in mycorrhizal and nonmycorrhizal short-leaf pine. *Tree Physiol.* **4**, 275–280.

Atkin, R. K., Barton, G. E. and Robinson, D. K. (1973). Effect of root-growing temperature on growth substances in xylem exudate of *Zea mays*. *J. Exp. Bot.* **24**, 475–487.

Atkins, C. A. (1987). Metabolism and translocation of fixed nitrogen in the nodulated legume. *Plant Soil* **100**, 157–169.

Atkins, C. A., Kuo, J., Pate, J. S., Flynn, A. M. and Steele, T. W. (1977). Photosynthetic pod wall of pea (*Pisum sativum* L.). Distribution of carbon dioxide-fixing enzymes in relation to pod structure. *Plant Physiol.* **60**, 779–786.

Atkinson, C. J. (1991). The influence of increasing rhizosphere calcium on the ability of *Lupinus luteus* L. to control water use efficiency. *New Phytol.* **119**, 207–215.

Atkinson, C. J., Mansfield, T. A., McAinsh, M. R., Brownlee, C. and Hetherington, A. M. (1990). Interactions of calcium with abscisic acid in the control of stomatal aperture. *Biochem. Physiol. Pflanzen* **186**, 333–339.

Atkinson, M. M., Baker, C. J. and Collmer, A. (1986). Transient activation of plasmalemma K^+ efflux and H^+ influx in tobacco by a pectate lyase isoenzyme from *Erwinia chrysanthemi*. *Plant Physiol.* **82**, 142–146.

Atkinson, M. M., Koppler, L. D., Orlandi, E. W., Baker, C. J. and Mischke, C. F. (1990). Involvement of plasma membrane calcium influx in bacterial induction of the K^+/H^+ and hypersensitive responses in tobacco. *Plant Physiol.* **92**, 215–221.

Atwell, B. J. (1990). The effect of soil compaction on wheat during early tillering. I. Growth, development and root structure. *New Phytol.* **115**, 29–35.

Atwell, B. J. and Steer, B. T. (1990). The effect of oxygen deficiency on uptake and distribution of nutrients in maize plants. *Plant Soil* **122**, 1–8.

Auderset, G., Sandelius, A. S., Penel, C., Brightman, A., Greppin, H. and Morré, D. J. (1986). Isolation of plasma membrane and tonoplast fractions from spinach leaves by preparative free-flow electrophoresis and effect of photoinduction. *Physiol. Plant.* **68**, 1–12.

Augé, R. M. and Stodola, A. J. W. (1990). An apparent increase in symplastic water contributes to greater turgor in mycorrhizal roots of droughted *Rosa* plants. *New Phytol.* **115**, 285–295.

Austin, R. B. (1989). Genetic variation in photosynthesis. *J. Agric. Sci. (Camb.)* **112**, 287–294.

Avivi, Y. and Feldman, M (1982). The response of wheat to bacteria of the genus Azospirillum. *Isr. J. Bot.* **31**, 237–245.

Awad, A. S., Edwards, D. G. and Campbell, L. C. (1990). Phosphorus enhancement of salt tolerance of tomato. *Crop Sci.* **30**, 123–128.

Ayala, M. B. and Sandmann, G. (1989a). Activities of Cu-containing proteins in Cu-depleted pea leaves. *Physiol. Plant.* **72**, 801–806.

Ayala, M. B. and Sandmann, G. (1988b). The role of Cu in respiration of pea plants and heterotrophically growing *Scenedesmus* cells. *Z. Naturforsch.* **43c**. 438–442.

Ayala, M. B., Gorgé, J. L., Lachica, M. and Sandmann, G. (1992). Changes in carotenoids and fatty acids in photosystem II of Cu-deficient pea plants. *Physiol. Plant.* **84**, 1–5.

Ayoub, A. T. (1974). Causes of inter-varietal differences in susceptibility to sodium toxicity injury in *Phaseolus vulgaris*. *Agric. Sci.* **83**, 539–543.

Azarabadi, S. and Marschner, H. (1979). Role of the rhizosphere in utilization of inorganic iron III compounds by corn plants. *Z. Pflanzenernähr. Bodenk.* **142**, 751–764.

Azcon, R. and Barea, J. M. (1992). The effect of vesicular-arbuscular mycorrhizae in decreasing Ca acquisition by alfalfa plants in calcareous soils. *Biol. Fertil. Soils* **13**, 155–159.

Baas, R., van der Werf, A. and Lambers, H. (1989). Root respiration and growth in *Plantago major* as affected by vesicular-arbuscular mycorrhizal infection. *Plant Physiol.* **91**, 227–232.

Baath, E. and Johansson, T. (1990). Measurement of bacterial growth rates on the rhizoplane using ^{3}H-thymidine incorporation into DNA. *Plant Soil* **126**, 133–139.

Baath, E. and Spokes, J. (1988). The effect of added nitrogen and phosphorus on mycorrhizal growth response and infection in *Allium schoenoprasum*. *Can. J. Bot.* **67**, 3227–3232.

Bachthaler, G. and Stritesky, A. (1973). Wachstumsuntersuchungen an Kulturpflanzen auf einem mit Kupfer überversorgten Mineralboden. *Bayer. Landwirtsch. Jahrb.* **50**, 73–81.

Badurowa, M., Guminski, S. and Suder-Morav, A. (1967). Die Wirkung steigender Konzentrationen von Natriumhydrogenkarbonat in Wasserkulturen und die Gegenwirkung des Na-Humats. *Biol. Plant* **9**, 92–101.

Bailey-Serres, J., Kloeckener-Gruissem, B. and Freeling, M. (1988). Genetic and molecular approaches to the study of the anaerobic response and tissue specific genetic expression in maize. *Plant Cell Environ.* **11**, 351–357.

Baker, A. J. M. (1987). Metal tolerance. *New Phytol.* **106**, 93–111.

Baker, A. J. M. and Walker, P. L. (1989a). Physiological responses of plants to heavy metals and the quantification of tolerance and toxicity. *Chem. Speciation Bioavail.* **1**, 7–17.

Baker, A. J. M. and Walker, P. L. (1989b). Ecophysiology of metal uptake by tolerant plants. *In* 'Heavy Metal Tolerance in Plants: Evolutionary Aspects' (A. J. Shaw, ed.), pp. 155–177. CRC Press, Boca Raton, FL.

Baker, D. A., Malek, F. and Dehvar, F. D. (1980). Phloem loading of amino acids from the petioles of ricinus leaves. *Ber. Dtsch. Bot. Ges.* **93**, 203–209.

Bakken, L. R. (1988). Denitrification under different cultivated plants: effects of soil moisture tension, nitrate concentration, and photosynthetic activity. *Biol. Fertil. Soils* **6**, 271–278.

Balandreau, J., Rinaudo, G., Fares-Hamad, I. and Dommergues, Y. (1975). Nitrogen fixation in the rhizosphere of rice plants. *In* 'Nitrogen Fixation by Free-living Micro-organisms' (W. D. P. Setwart, ed.), pp. 57–70. Cambridge University Press, Cambridge.

Balasta, M. L. F. C., Perez, C. M., Juliano, B. O., Villareal, C. P., Lott, J. N. A. and Roxas, D. B. (1989). Effect of silica level on some properties of *Oriza sativa* straw and hull. *Can. J. Bot.* **67**, 2356–2363.

Baldi, B. G., Franceschi, V. R. and Loewus, F. A. (1987). Localization of phosphorus and cation reserves in *Lilium longiflorum* pollen. *Plant Physiol.* **83**, 1018–1021.

Baligar, V. C. and Barber, S. A. (1978). Potassium uptake by onion roots characterized by potassium/rubidium ratio. *Soil Sci. Soc. Am. J.* **42**, 618–622.

Balke, N. E. and Hodges, T. K. (1975). Plasma membrane adenosine triphosphatase of oat roots. *Plant Physiol.* **55**, 83–86.

Ball, M. C. (1988). Salinity tolerance in the mangroves *Aegiceras corniculatum* and *Avicennia marina*. I. Water use in relation to growth, carbon partitioning, and salt balance. *Aust. J. Plant Physiol.* **15**, 447–464.

Ball, M. C., Taylor, S. E. and Terry, N. (1984). Properties of thylakoid membranes of the mangrove *Avicennia germinans* and *Avicennia marina*, and the sugar beet, *Beta vulgaris*, grown under different salinity conditions. *Plant Physiol.* **76**, 531–535.

Bamji, M. S. and Jagendorf, A. T. (1966). Amino acid incorporation by wheat chloroplasts. *Plant Physiol.* **41**, 764–770.

Bangerth, F. (1979). Calcium-related physiological disorders of plants. *Annu. Rev. Phytopathol.* **17**, 97–122.

Bangerth, F. (1989). Dominance among fruits/sinks and the search for a correlative signal. *Physiol. Plant.* **76**, 608–614.

Bangerth, F., Dilley, D. R. and Dewey, D. H. (1972). Effect of calcium infusion on internal break-down and respiration of apple fruits. *J. Am. Soc. Hortic. Sci.* **97**, 679–682.

Bansal, K. N., Motiramani, D. P. and Pal, A. R. (1983). Studies on sulphur in vertisols. I. Soil and plant tests for diagnosing sulphur deficiency in soybean (*Glycine max* (L.) Merr.). *Plant Soil* **70**, 133–140.

Banuelos, G. S. and Meek, D. W. (1989). Selenium accumulation in selected vegetables. *J. Plant Nutr.* **12**, 1255–1272.

Banuelos, G. S., Bangerth, F. and Marschner, H. (1987). Relationship between polar basipetal auxin transport and acropetal Ca^{2+} transport into tomato fruits. *Physiol. Plant.* **71**, 321–327.

Banuelos, G. S., Bangerth, F. and Marschner, H. (1988). Basipetal auxin transport in lettuce and its possible involvement in acropetal calcium transport and incidence of tipburn. *J. Plant Nutr.* **11**, 525–533.

Banuls, J., Legaz, F. and Primo-Millo, E. (1991). Salinity–calcium interactions on growth and ionic concentration of Citrus plants. *Plant Soil* **133**, 39–46.

Bar-Akiva, A. (1984). Substitutes for benzidine as H-donor in the peroxidase assay, for rapid diagnosis of iron deficiency in plants. *Commun. Soil Sci. Plant Anal.* **15**, 929–934.

Bar-Akiva, A., Sagiv, J. and Leshem, J. (1970). Nitrate reductase activity as an indicator for assessing the nitrogen requirement of grass crops. *J. Sci. Food Agric.* **21**, 405–407.

Bar-Akiva, A., Maynard, D. N. and English, J. E. (1978). A rapid tissue test for diagnosing iron deficiencies in vegetable crops. *HortScience* **13**, 284–285.

Baranov, V. I. (1979). Biological activity of oxidized phenolic compounds and their role in break of indolyl-3-acetic acid. *Sov. Plant Physiol.* (*Engl. Transl.*) **26**, 688–695.

Barber, D. A. and Gunn, K. B. (1974). The effect of mechanical forces on the exudation of organic substances by the roots of cereal plants grown under sterile conditions. *New Phytol.* **73**, 30–45.

Barber, D. A. and Lee, R. B. (1974). The effect of micro-organisms on the absorption of manganese by plants. *New Phytol.* **73**, 97–106.

Barber, D. A. and Martin, J. K. (1976). The release of organic substances by cereal roots into soil. *New Phytol.* **76**, 69–80.

Barber, S. A. (1962). A diffusion and massflow concept of soil nutrient availability. *Soil Sci.* **93**, 39–49.

Barber, S. A. (1979). Growth requirement for nutrients in relation to demand at the root surface. *In* 'The Soil–Root Interface' (J. L. Harley and R. Scott-Russell, eds.), pp. 5–20. Academic Press, London.

Barber, S. A. (1984). 'Soil Nutrient Bioavailability. A Mechanistic Approach'. John Wiley, New York.

Barber, S. A. and Mackay, A. D. (1986). Root growth and phosphorus and potassium uptake by two corn genotypes in the field. *Fert. Res.* **10**, 217–230.

Barber, S. A. and Ozanne, P. G. (1970). Autoradiographic evidence for the differential effect of four plant species in altering the calcium content of the rhizosphere soil. *Soil Sci. Soc. Am. Proc.* **34**, 635–637.

Barcelo, J., Guevara, P. and Poschenrieder, Ch. (1993). Silicon amelioration of aluminium toxicity in teosinte (*Zea mays* L. ssp. *mexicana*). *Plant Soil* **154**, 249–255.

Barea, J. M., El-Atrach, F. and Azcon, R. (1989). Mycorrhiza and phosphate interactions as affecting plant development, N_2-fixation, N-transfer and N-uptake from soil in legume–grass mixtures by using a ^{15}N dilution technique. *Soil Biol. Biochem.* **21**, 581–589.

Barel, D. and Black, C.A. (1979). Effect of neutralization and addition of urea, sucrose and various glycols on phosphorus absorption and leaf damage from foliar-applied phosphate. *Plant Soil* **52**, 515–525.

Barker, A. V. (1979). Nutritional factors in photosynthesis of higher plants. *J. Plant Nutr.* **1**, 309–342.

Bartlett, R. J. and Riego, D. C. (1972). Effect of chelation on the toxicity of aluminium. *Plant Soil* **37**, 419–423.

Barneix, A. J., Breteler, H. and van de Geijn, S. C. (1984). Gas and ion exchanges in wheat roots after nitrogen supply. *Physiol. Plant.* **61**, 357–362.

Bar-Ness, E. and Chen, Y. (1991). Manure and peat based iron–argeno complexes. II. Transport in soils. *Plant Soil* **130**, 45–50.

Bar-Ness, E., Chen, Y., Hadar, Y., Marschner, H. and Römheld, V. (1991). Siderophores of *Pseudomonas putida* as an iron source for dicot and monocot plants. *Plant Soil* **130**, 231–241.

Bar-Ness, E., Hadar, Y., Chen, Y., Römheld, V. and Marschner, H. (1992). Short-term effects of rhizosphere microorganisms on Fe uptake from microbial sidero-phores by maize and oat. *Plant Physiol.* **100**, 451–456.

Barnett, K. H. and Pearce, R. B. (1983). Source–sink ratio alteration and its effect on physiological parameters in maize. *Crop Sci.* **23**, 294–299.

Bar-Nun, N. and Poljakoff-Mayber, A. (1977). Salinity stress and the content of proline in roots of *Pisum sativum* and *Tamarix tetragyna*. *Ann. Bot. (London) [N.S.].* **41**, 173–179.

Barraclough, P. B. (1989). Root growth, macronutrient uptake dynamics and soil fertility requirements of a high-yielding winter oilseed rape crop. *Plant Soil* **119**, 59–70.

Barrett-Lennard, E. G. (1986). Effects of waterlogging on the growth and CaCl uptake by vascular plants under saline conditions. *Reclam. Reveg. Res.* **5**, 245–261.

Barrett-Lennard, E. G. and Greenway, H. (1982). Partial separation and characteriz-ation of soluble phosphatases from leaves of wheat grown under phosphorus deficiency and water deficit. *J. Exp. Bot.* **33**, 694–704.

Barry, D. A. J. and Miller, M. H. (1989). Phosphorus nutritional requirement of maize seedlings for maximum yield. *Agron. J.* **81**, 95–99.

Bartels, U. (1990). Organischer Kohlenstoff im Niederschlag nordrhein-westfälischer Fichten- und Buchenbestände. *Z. Pflanzenernähr. Bodenk.* **153**, 125–127.

Bartloth, W. (1990). Scanning electron microscopy of the epidermal surface in plants. *In* 'Scanning Electron Microscopy in Taxonomy and Functional Morphology' (D. Claugher, ed.), pp. 69–83. Clarendon Press, Oxford.

Barz, W. (1977). Degradation of polyphenols in plants and cell suspension cultures. *Physiol. Veg.* **134**, 37–52.

Bashan, Y. (1990). Short exposure to *Azospirillum brasilense* Cd inoculation enhanced proton efflux of intact wheat roots. *Can. J. Microbiol.* **36**, 419–425.

Basiouny, F. M. (1984). Distribution of vanadium and its influence on chlorophyll formation and iron metabolism in tomato plants. *J. Plant Nutr.* **7**, 1059–1073.

Baszynski, T., Ruszkowska, M., Krol, M., Tukendorf, A. and Wolinska, D. (1978). The effect of copper deficiency on the photosynthetic apparatus of higher plants. *Z. Pflanzenphysiol.* **89**, 207–216.

Baszynski, T., Warcholowa, M., Krupa, Z., Tukendorf, A., Krol, M. and Wolinska, D. (1980). The effect of magnesium deficiency on photochemical activities of rape and buckwheat chloroplasts. *Z. Pflanzenphysiol.* **99**, 295–303.

Bateman, D. F. and Lumsden, R. D. (1965). Relation between calcium content and nature of the peptic substances in bean hypocotyls of different ages to susceptibility to an isolate of *Rhizoctonia solani*. *Phytopathology* **55**, 734–738.

Bates, T. E. (1971). Factors affecting critical nutrient concentrations in plants and their evaluation: a review. *Soil Sci.* **112**, 116–126.

Batten, G. D. and Wardlaw, I. F. (1987a). Senescence and grain development in wheat plants grown with contrasting phosphorus regimes. *Aust. J. Plant Physiol.* **14**, 253–265.

Batten, G. D. and Wardlaw, I. F. (1987b). Senescence of the flag leaf and grain yield following late foliar and root application of phosphate on plants of differing phosphorus status. *J. Plant Nutr.* **10**, 735–748.

Batten, G. D., Wardlaw, I. F. and Aston, M. J. (1986). Growth and the distribution of phosphorus in wheat developed under various phosphorus and temperature regimes. *Aust. J. Agric. Res.* **37**, 459–469.

Baudet, J., Huet, J.-C., Lesaint, C., Mosse, J. and Pernollet, J.-C. (1986). Changes in accumulation of seed nitrogen compounds in maize under conditions of sulphur deficiency. *Physiol. Plant.* **68**, 608–614.

Bauer, W. D. and Caetano-Anolles, G. (1990). Chemotaxis, induced gene expression and competitiveness in the rhizosphere. *Plant Soil* **129**, 45–52.

Bavaresco, L. and Eibach, R. (1987). Investigations on the influence of N fertilizer on resistance to powdery mildew (*Oidium tuckeri*) downy mildew (*Plasmopara viticola*) and on phytoalexin synthesis in different grapevine varieties. *Vitis* **26**, 192–200.

Bavaresco, L., Fregoni, M. and Fraschini, P. (1991). Investigations on iron uptake and reduction by excised roots of different grapevine rootstocks and a *V. vinifera* cultivar. *Plant Soil* **130**, 109–113.

Bayne, H. G., Brown, M. S. and Bethlenfalway, G. J. (1984). Defoliation effects on mycorrhizal colonization, nitrogen fixation and photosynthesis in the *Glycine–Glomus–Rhizobium* symbiosis. *Physiol. Plant.* **62**, 576–580.

Beauchamp, C. J., Dion, P., Kloepper, J. W. and Antoun, H. (1991). Physiological characterization of opine-utilizing rhizobacteria for traits related to plant growth-promoting activity. *Plant Soil* **132**, 273–279.

Becana, M. and Rodriguez-Barrueco, C. (1989). Protective mechanisms of nitrogenase against oxygen excess and partially-reduced oxygen intermediates. *Physiol. Plant.* **75**, 429–438.

Becana, M. and Salim, M. L. (1989). Superoxide dismutases in nodules of leguminous plants. *Can. J. Bot.* **67**, 415–421.

Becana, M., Aparicio-Tejo, P. M. and Sánchez-Diaz, M. (1985). Nitrate and nitrite reduction by alfalfa root nodules: accumulation of nitrite in *Rhizobium meliloti* bacteroids and senescence of nodules. *Physiol. Plant.* **64**, 353–358.

Bécard, G. and Piché, Y. (1989). Fungal growth stimulation by CO_2 and root exudates in vesicular-arbuscular mycorrhizal symbiosis. *Appl. Environ. Microbiol.* **55**, 2320–2325.

Bécard, G., Douds, D. D. and Pfeffer, P. E. (1992). Extensive *in vitro* hyphal growth of vesicular-arbuscular mycorrhizal fungi in presence of CO_2 and flavenols. *Appl. Environ. Microbiol.* **58**, 821–825.

Beck, S. T. (1965). Resistance of plants to insects. *Annu. Rev. Entomol.* **10**, 207–232.

Becker, M., Ladha, J. K. and Ottow, J. C. G. (1990). Einfluss von NPK auf die Biomasseproduktion und Stickstoffbindung der stengelknöllchenbildenden Gründüngungsleguminosen *Sesbania rostrata* und *Aeschynomene afraspera* im Nassreisanbau. *Z. Pflanzenernähr. Bodenk.* **153**, 333–339.

Becking, J. H. (1961). A requirement of molybdenum for the symbiotic nitrogen fixation in alder. *Plant Soil* **15**, 217–227.

Beebe, D. U. and Evert, R. F. (1992). Photoassimilate pathways(s) and phloem loading in the leaf of *Moricandia arvensis* (L.) DC. (Brassicaceae). *Int. J. Plant Sci.* **153**, 61–77.

Beebe, D. U. and Turgeon, R. (1991). Current perspectives on plasmodesmata: structure and function. *Physiol. Plant.* **83**, 194–199.

Beevers, L. and Hageman, R. H. (1983). Uptake and reduction of nitrate: bacteria and higher plants. *In* 'Encyclopedia of Plant Physiology, New Series' (A. Läuchli and R. L. Bieleski, eds.), Vol. 15A, pp. 351–375. Springer-Verlag, Berlin and New York.

Behl, R. and Jeschke, W. D. (1982). Potassium fluxes in excised barley roots. *J. Exp. Bot.* **33**, 584–600.

Behl, R., Tischner, R. and Raschke, K. (1988). Induction of a high-capacity nitrate uptake mechanism in barley roots prompted by nitrate uptake through a constitutive low-capacity mechanism. *Planta* **176**, 235–240.

Behling, J. P., Gabelman, W. H. and Gerloff, G. C. (1989). The distribution and utilization of calcium by two tomato (*Lycopersicon esculentum* Mill.) lines differing in calcium efficiency when grown under low-Ca stress. *Plant Soil* **113**, 189–196.

Behrmann, P. and Heyser, W. (1992). Apoplastic transport through the fungal sheath of *Pinus sylvestris/Suillus bovimus* ectomycorrhizae. *Bot. Acta* **105**, 427–434.

Beinert, H. and Kennedy, M. C. (1989). Engineering of protein bound iron-sulfur clusters. A tool for the study of protein and cluster chemistry and mechanisms of iron–sulfur enzymes. *Eur. J. Biochem.* **186**, 5–15.

Bejaoui, M. (1985). Interactions entre NaCl et quelques phytohormones sur la croissance du soja. *J. Plant Physiol.* **120**, 95–110.

Bekele, T., Cino, B. J., Ehlert, P. A. I., van der Mass, A. A. and van Diest, A. (1983). An evaluation of plant-borne factors promoting the solubilization of alkaline rock phosphates. *Plant Soil* **75**, 361–378.

Bell, P. F., Chaney, R. L. and Angle, J. S. (1991). Free metal activity and total metal concentrations as indices of micronutrient availability to barley (*Hordeum vulgare* (L.) Klages). *Plant Soil* **130**, 51–62.

Bell, R. W., Edwards, D. G. and Asher, C. J. (1990). Growth and nodulation of tropical food legumes in dilute solution culture. *Plant Soil* **122**, 249–258.

Bell, M. J., Middleton, K. J. and Thompson, J. P. (1989). Effects of vesicular-arbuscular mycorrhizae on growth and phosphorus and zinc nutrition of peanut (*Arachis hypogaea* L.) in an oxisol from subtropical Australia. *Plant Soil* **117**, 49–57.

Bell, R. W., Brady, D., Plaskett, D. and Loneragan, J. F. (1987). Diagnosis of potassium deficiency in soybean. *J. Plant Nutr.* **10**, 1947–1953.

Bell, R. W., McLay, L., Plaskett, D., Dell, B. and Loneragan, J. F. (1989). Germination and vigour of black gram (*Vigna mungo* (L.) Hepper) seed from plants grown with and without boron. *Aust. J. Agric. Res.* **40**, 273–279.

Belucci, S., Keller, E. R. and Schwendimann, F. (1982). Einfluss von Wachstumsregulatoren auf die Entwicklung und den Ertragsaufbau der Ackerbohne (*Vicia faba* L.). I. Wirkung von Gibberellinsäure (GA$_3$) auf die Ertragskomponenten und die Versorgung der jungen Früchte mit ^{14}C. *Angew. Bot.* **56**, 35–53.

Benepal, P. S. and Hall, C. V. (1967). The influence of mineral nutrition of varieties of *Cucurbita pepo* L. on the feeding response of Squash bug *Anasa tristis* De Geer. *Proc. Am. Soc. Hortic. Sci.* **90**, 304–312

Bennet, R. J. and Breen, C. M. (1989). Towards understanding root growth responses to environmental signals. The effect of aluminium on maize. *S. Afr. J. Sci.* **85**, 9–12.

Bennet, R. J. and Breen, C. M. (1991). The recovery of the roots of *Zea mays* L. from various aluminum treatments: Towards elucidating the regulatory processes that underlie root growth control. *Environ. Exp. Bot.* **31**, 153–163.

Bennet, R. J. and Breen, C. M. (1992). The use of lanthanum to delineate the aluminium signalling mechanisms functioning in the roots of *Zea mays* L. *Environ. Exp. Bot.* **32**, 365–376.

Bennet, R. J. and Breen, C. M. (1993). Aluminium toxicity: towards an understanding of how plant roots react to the physical environment. *In* 'Genetic Aspects of Plant Mineral Nutrition' (P. J. Randall, E. Delhaize, R. A. Richards and R. Munns, eds.), pp. 103–116. Kluwer Academic Publ., Dordrecht.

Bennet, R. J., Breen, C. M. and Bandu, V. H. (1990). A role for Ca^{2+} in the cellular differentiation of root cap cells: a re-examination of root growth control mechanisms. *Environ. Exp. Bot.* **30**, 515–523.

Bennett, A. B., O'Neill, S. D. and Spanswick, R. M. (1984). H^+-ATPase activity from storge tissue of *Beta vulgaris*. I. Identification and characterization of an anion-sensitive H^+-ATPase. *Plant Physiol.* **74**, 538–544.

Bennett, A. C. and Adams, F. (1970). Calcium deficiency and ammonia toxicity as separate causal factors of $(NH_4)_2HPO_4$ injury to seedlings. *Soil Sci. Soc. Am Proc.* **34**, 255–259.

Bennett, D. M. (1982). Silicon deposition in the roots of *Hordeum sativum* Jess. *Avena sativa* L. and *Triticum aestivum* L. *Ann. Bot.* (*London*) [N.S.] **50**, 239–245.

Bennie, A. T. P. (1991). Growth and mechanical impedance. *In* 'The Plant Root, the Hidden Half' (Y. Waisel, A. Eshel and U. Kafkafi, eds.), pp. 393–414. Marcel Dekker, New York.

Bentley, B. L. and Carpenter, E. J. (1984). Direct transfer of newly-fixed nitrogen from free-living epiphyllous microorganisms. *Oecologia* **63**, 52–56.

Bentrup, F.-W. (1989). Cell electrophysiology and membrane transport. *Progr. Bot.* **51**, 70–79.

Ben-Zioni, A., Vaadia, Y. and Lips, S. H. (1971). Nitrate uptake by roots as regulated by nitrate reduction products of the shoot. *Physiol. Plant.* **24**, 288–290.

Beraud, J., Brun, A., Feray, A., Hourmant, A. and Penot, M. (1992). Long distance transport of ^{14}C-putrescine in potato plantlets (*Solanum tuberosum* cv. Bintje). *Biochem. Physiol. Pflanzen* **188**, 169–176.

Bergersen, F. J. (1991). Physiological control of nitrogenase and uptake hydrogenase. *In* 'Biology and Biochemistry of Nitrogen Fixation' (M. J. Dilworth and A. R. Glenn, eds.), pp. 76–102. Elsevier, Amsterdam.

Bergmann, L. and Rennenberg, H. (1993). Glutathione metabolism in plants. *In* 'Sulfur Nutrition and Assimilation in Higher Plants' (L. J. De Kok, I. Stulen, H. Rennenberg, C. Brunold and W. E. Rauser, eds.) , pp. 109–123. SPB Academic Publishing, The Hague, The Netherlands.

Bergmann, W. (1988). 'Ernährungsstörungen bei Kulturpflanzen. Entstehung, visuelle und analytische Diagnose'. Fischer Verlag, Jena.

Bergmann, W. (1992). 'Nutritional Disorders of Plants – Development, Visual and Analytical Diagnosis'. Fischer Verlag, Jena.

Bergmann, W. and Neubert, P. (1976). 'Pflanzendiagnose und Pflanzenanalyse'. Fischer Verlag, Jena.

Beringer, H. (1966). Einfluss von Reifegrad und N-Düngung auf Fettbildung und Fettsäurezusammensetzung in Haferkörnern. *Z. Pflanzenernähr. Bodenk.* **114**, 117–127.

Beringer, H. and Forster, H. (1981). Einfluss variierter Mg-Ernährung auf Tausendkorngewicht und P-Fraktionen des Gerstenkorns. *Z. Pflanzenernähr. Bodenkd.* **144**, 8–15.

Beringer, H., Haeder, H. E. and Lindhauer, M. G. (1983). Water relationships and incorporation of ^{14}C assimilates in tubers of potato plants differing in potassium nutrition. *Plant Physiol.* **73**, 956–960.

Beringer, H., Koch, K. and Lindhauer, M. G. (1986). Sucrose accumulation and osmotic potentials in sugar beet at increasing levels of potassium nutrition. *J. Sci. Food Agric.* **37**, 211–218.

Berkel, N. van (1988). Preventing tipburn in chinese cabbage by high relative humidity during the night. *Neth. J. Agric. Sci.* **36**, 301–308.

Bernal, C. T., Bingham, F. T. and Oertli, J. (1974). Salt tolerance of Mexican wheat.

II. Relation to variable sodium chloride and length of growing season. *Soil Sci. Soc. Am. Proc.* **38**, 777–780.

Bernhard-Reversat, F. (1975). Nutrients in through fall and their quantitative importance in rain forest mineral cycle. *Ecol. Stud.* **11**, 153–159.

Bernstein, L. and Francois, L. E. (1975). Effects of frequency of sprinkling with saline waters compared with daily drip irrigation. *Agron. J.* **67**, 185–190.

Berry, S. Z., Madumadu, G. G. and Uddin, M. R. (1988). Effect of calcium and nitrogen nutrition on bacterial cancer disease of tomato. *Plant Soil* **112**, 113–120.

Berta, G., Fusconi, A., Trotta, A. and Scannerini, S. (1990). Morphogenetic modifications induced by the mycorrhizal fungus *Glomus* strain E_3 in the root system of *Allium porrum* L. *New Phytol.* **114**, 207–215.

Bertelsen, F. and Jensen, E. S. (1992). Gaseous nitrogen losses from field plots grown with pea (*Pisum sativum* L.) and spring barley (*Hordeum vulgare* L.) estimated by ^{15}N mass balance and acetylene inhibition techniques. *Plant Soil* **142**, 287–295.

Bertl, A., Felle, H. and Bentrup, F.-W. (1984). Amine transport in *Riccia fluitans*. Cytoplasmic and vacuolar pH recorded by a pH-sensitive microelectrode. *Plant Physiol.* **76**, 75–78.

Besford, R. T. (1978a). Effect of replacing nutrient potassium by sodium on uptake and distribution of sodium in tomato plants. *Plant Soil* **50**, 399–409.

Besford, R. T. (1978b). Use of pyruvate kinase activity of leaf extracts for the quantitative assessment of potassium and magnesium status of tomato plants. *Ann. Bot.* (*London*) [*N.S.*] **42**, 317–324.

Besford, R. T. and Syred, A. D. (1979). Effect of phosphorus nutrition on the cellular distribution of acid phosphatase in the leaves of *Lycopersicon esculentum* L. *Ann. Bot.* (*London*) [*N.S.*] **43**, 431–435.

Bethlenfalvay, G. J. and Franson, R. L. (1989). Manganese toxicity alleviated by mycorrhizae in soybean. *J. Plant Nutr.* **12**, 953–970.

Bethlenfalvay, G. J., Abu-Shakra, S. S., Fishbeck, K. and Phillips, D. A. (1978). The effect of source–sink manipulations on nitrogen fixation in peas. *Physiol. Plant.* **43**, 31–34.

Bethlenfalvay, G. J., Franson, R. L., Brown, M. S. and Mihara, K. L. (1989). The *Glycine–Glomus–Bradyrhizobium* symbiosis. IX. Nutritional, morphological and physiological responses of nodulated soybean to geographic isolates of the mycorrhizal fungus *Glomus mosseae*. *Physiol. Plant.* **76**, 226–232.

Bethlenfalvay, G. J., Reyes-Solis, M. G., Camel, S. B. and Ferrera-Cerrato, R. (1991). Nutrient transfer between the root zones of soybean and maize plants connected by a common mycorrhizal mycelium. *Physiol. Plant.* **82**, 423–432.

Bhadoria, P. B. S., Kaselowsky, J., Claassen, N. and Jungk, A. (1991). Phosphate diffusion coefficients in soil as affected by bulk density and water content. *Z. Pflanzenernähr. Bodenk.* **154**, 53–57.

Bhat, K. K. S., Nye, P. H. and Brereton, A. J. (1979). The possibility of predicting solute uptake and plant growth response from independently measured soil and plant characteristics. VI. The growth and uptake of rape in solutions of constant nitrate concentration. *Plant Soil* **53**, 137–167.

Bhattarai, T. and Hess, D. (1993). Yield responses of Nepalese spring wheat (*Triticum aestivum* L.) cultivars to inoculation with *Azospirillum* spp. of Nepalese origin. *Plant Soil* **151**, 67–76.

Bhatti, A.S. and Wieneke, J. (1984). Na^+- and Cl^--leaf extrusion, retranslocation and root efflux in *Diplachne fusca* (Kallar grass) grown in NaCl. *J. Plant Nutr.* **7**, 1233–1250.

Bhivare, V. N., Nimbalkar, J. D. and Chavan, P. D. (1988). Photosynthetic carbon metabolism in french bean leaves under saline conditions. *Environ. Exp. Bot.* **28**, 117–121.

Biddington, N. L. and Dearman, A. S. (1982). The effect of abscisic acid on root and shoot growth of cauliflower plants. *Plant Growth Regul.* **1**, 15–24.

Bieleski, R. L. and Ferguson, I. B. (1983). Physiology and metabolism of phosphate and its compounds. *In* 'Encyclopedia of Plant Physiology, New Series' (A. Läuchli and R. L. Bieleski, eds.), Vol. 15A, pp. 422–449. Springer-Verlag, Berlin and New York.

Bienfait, H. F. (1985). Regulated redox processes at the plasmalemma of plant root cells and their function in iron uptake. *J. Bioenerg. Biomembr.* **17**, 73–83.

Bienfait, H. F. (1989). Preventation of stress in iron metabolism of plants. *Acta Bot. Neerl.* **38**, 105–129.

Bienfait, F. and Lüttge, U. (1988). On the function of two systems that can transfer electrons across the plasma membrane. *Plant Physiol. Biochem.* **26**, 665–671.

Bienfait, H. F., Lubberding, H. J., Heutink, P., Lindner, L., Visser, J., Kaptain, R. and Dijkstra, K. (1989). Rhizosphere acidification by iron deficient bean plants: the role of trace amounts of divalent metal ions. *Plant Physiol.* **90**, 359–364.

Biles, C. L. and Abeles, F. B. (1991). Xylem sap proteins. *Plant Physiol.* **96**, 597–601.

Birnbaum, E. H., Beasley, C. A. and Dugger, W. M. (1974). Boron deficiency in unfertilized cotton (*Gossypium hirsutum*) ovules grown in vitro. *Plant Physiol.* **54**, 931–935.

Birnbaum, E. H., Dugger, W. M. and Beasley, B. C. A. (1977). Interaction of boron with components of nucleic acid metabolism in cotton ovules cultered in vitro. *Plant Physiol.* **59**, 1034–1038.

Björkman, C., Larsson, S. and Gref, R. (1991). Effects of nitrogen fertilization on pine needle chemistry and sawfly performance. *Oecologia* **86**, 202–209.

Bjorkman, T. and Cleland, R. E. (1991). The role of extracellular free-calcium gradients in grovitropic signalling in maize roots. *Planta* (*Berlin*) **185**, 379–384.

Blamey, F. P. C., Edmeades, D. C. and Wheeler, D. M. (1990). Role of root cation-exchange capacity in differential aluminium tolerance of *Lotus* species. *J. Plant Nutr.* **13**, 729–744.

Blamey, F. P. C., Robinson, N. J. and Asher, C. J. (1992). Interspecific differences in aluminium tolerance in relation to root cation-exchange capacity. *Plant Soil* **146**, 77–82.

Blamey, F. P. C., Joyce, D.C., Edwards, D. G. and Asher, C. J. (1986). Role of trichomes in sunflower tolerance to manganese toxicity. *Plant Soil* **91**, 171–180.

Blamey, F. P. C., Edmeades, D. C., Asher, C. J., Edwards, D. G. and Wheeler, D. M. (1991). Evaluation of solution culture techniques for studying aluminium toxicity in plants. *In* 'Plant–Soil Interactions at Low pH' (R. J. Wright, V. C. Baligar and R. P. Murrmann, eds.), pp. 905–912. Kluwer Academic Publ., Dordrecht, The Netherlands.

Blanke, M. M., Hucklesby, D. P. and Notton, B. A. (1987). Distribution and physiological significance of photosynthetic phosphoenolpyruvate carboxylase in developing apple fruit. *J. Plant Physiol.* **129**, 319–325.

Blaser-Grill, J., Knoppik, D., Amberger, A. and Goldbach, H. (1989). Influence of boron on the membrane potential in *Elodea densa* and *Helianthus annuus* roots and H^+ extrusion of suspension cultured *Daucus carota* cells. *Plant Physiol.* **90**, 280–284.

Blatt, C. R. and van Diest, A. (1981). Evaluation of a screening technique for manganese toxicity in relation to leaf manganese distribution and interaction with silicon. *Neth. J. Agric. Sci.* **29**, 297–304.

Blatt, M. R. and Thiel, G. (1993). Hormonal control of ion channel gating. *Annu. Rev. Plant Physiol. Plant Mol. Biol.* **44**, 543–567.

Blevins, D. G. (1989). An overview of nitrogen metabolism in higher plants. *In* 'Plant Nitrogen Metabolism' (J. E. Poulton *et al.*, eds.), pp. 1–41. Plenum, New York.

Bligny, R. and Douce, R. (1977). Mitochondria of isolated plant cells (*Acer pseudoplatanus* L.). II. Copper deficiency effects on cytochrome *c* oxidase and oxygen uptake. *Plant Physiol.* **60**, 675–679.

Blits, K. C. and Gallagher, J. L. (1990). Salinity tolerance of *Kosteletzkya virginica*. I. Shoot growth, ion and water relations. *Plant Cell Environ.* **13**, 409–418.

Blom-Zandstra, G. and Lampe, J. E. M. (1983). The effect of chloride and sulphate salts on the nitrate content in lettuce plants. (*Lactuca sativa* L.). *J. Plant Nutr.* **6**, 611–628.

Bloom, A. J. and Caldwell, R. M. (1988). Root excision decreases nutrient absorption and gas fluxes. *Plant Physiol.* **87**, 794–796.

Bloom, A. J., Sukrapanna, S. S. and Warner, R. L. (1992). Root respiration associated with ammonium and nitrate absorption and assimilation by barley. *Plant Physiol.* **99**, 1294–1301.

Blum, A., Johnson, J. W., Ramseur, E. L. and Tollner, E.W. (1991). The effect of a drying top soil and a possible non-hydraulic root signal on wheat growth and yield. *J. Exp. Bot.* **42**, 1225–1231.

Blum, A., Poiarkova, H., Golan, G. and Mayer, J. (1983). Chemical desiccation of wheat plants as a simulator of post-anthesis stress. I. Effects on translocation on kernel growth. *Field Crops Res.* **6**, 51–58.

Blumwald, E. and Poole, R. J. (1987). Salt tolerance in suspension cultures of sugar beet. Induction of Na^+/H^+ antiport activity at the tonoplast by growth in salt. *Plant Physiol.* **83**, 884–887.

Boardman, R. and McGuire, D. O. (1990). The role of zinc in forestry. I. Zinc in forest environments, ecosystems and tree nutrition. *Forest Ecol. Mgmt*, **37**, 167–205.

Boddey, R. M. and Döbereiner, J. (1988). Nitrogen fixation associated with grasses and cereals: recent results and perspectives for future research. *Plant Soil* **108**, 53–65.

Boddey, R. M., Urquiaga, S., Reis, V. and Döbereiner, J. (1991). Biological nitrogen fixation associated with sugar cane. *Plant Soil* **137**, 111–117.

Boguslawski, E. von (1958). Das Ertragsgesetz. *In* 'Encyclopedia of Plant Physiology' (W. Ruhland, ed.), Vol. 4, pp. 943–976. Springer-Verlag, Berlin and New York.

Bohlool, B. B., Ladha, J.K., Garrity, D. P. and George, T. (1992). Biological nitrogen fixation for sustainable agriculture: A perspective. *Plant Soil* **141**, 1–11.

Böhm, W. (1974). Phosphatdüngung und Wurzelwachstum. *Phosphorsaure* **30**, 141–157.

Bohn, H. L., McNeal, B. L. and O'Connor, G.-A. (1985). 'Soil Chemistry', pp. 268–271. Wiley, New York.

Bohnsack, C. W. and Albert, L. S. (1977). Early effects of boron deficiency on indoleacetic acid oxidase levels of squash root tips. *Plant Physiol.* **59**, 1047–1050.

Bolan, N. S. (1991). A critical review on the role of mycorrhizal fungi in the uptake of phosphorus by plants. *Plant Soil* **134**, 189–207.

Bolan, N. S., Robson, A. D. and Barrow, N. J. (1984). Increasing phosphorus supply can increase the infection of plant roots by vesicular-arbuscular mycorrhizal fungi. *Soil Biol. Biochem.* **16**, 419–420.

Bolan, N. S., Robson, A. D. and Barrow, N. J. (1987). Effects of vesicular-arbuscular mycorrhiza on the availability of iron phosphates to plants. *Plant Soil* **99**, 401–410.

Bolland, M. D. A. and Baker, M. J. (1989). High phosphorus concentration in *Trifolium balansae* and *Medicago polymorpha* seeds increases herbage and seed yields in the field. *Aust. J. Exp. Agric.* **29**, 791–795.

Bolland, M. D. A. and Gilkes, R. J. (1992). Evaluation of the Bray 1, calcium acetate lactate (CAL), Truog and Colwell soil test as predictors of triticale grain production on soil fertilized with superphosphate and rock phosphate. *Fert. Res.* **31**, 363–372.

Bollard, E. G. (1983). Involvement of unusual elements in plant growth and nutrition. *In* 'Encyclopedia of Plant Physiology, New Series' (A. Läuchli and R. L. Bieleski, eds.), Vol. 15B, pp. 695–755. Springer-Verlag, Berlin and New York.

Bolle-Jones, E. W. and Hilton, R. N. (1956). Zinc-deficiency of *Hevea brasiliensis* as a predisposing factor to *Oidium* infection. *Nature* (*London*) **177**, 619–620.

Bolle-Jones, E. W. and Mallikarjuneswara, V. R. A. (1957). A beneficial effect of cobalt on the growth of *Hevea brasiliensis*. *Nature* (*London*) **179**, 738–739.

Boller, B. C. and Nösberger, J. (1987). Symbiotically fixed nitrogen from field-grown white and red clover mixed with ryegrasses at low levels of ^{15}N-fertilization. *Plant Soil* **104**, 219–226.

Bollmark, M. and Eliasson, L. (1990). A rooting inhibitor present in Norway spruce seedlings grown at high irradiance: a putative cytokinin. *Physiol. Plant.* **80**, 527–533.

Bolton jr, H., Elliott, L. F., Gurusiddaiah, S. and Fredrickson, J. K. (1989). Characterization of a toxin produced by a rhizobacterial *Pseudomonas* sp. that inhibits wheat growth. *Plant Soil* **114**, 279–287.

Bonilla, I., Garcia-González, M. and Mateo, P. (1990). Boron requirement in cyanobacteria. Its possible role in the early evolution of photosynthetic organisms. *Plant Physiol.* **94**, 1554–1560.

Borchert, R. (1990). Ca^{2+} as developmental signal in the formation of Ca-oxalate crystal spacing patterns during leaf development in *Carya ovata*. *Planta* **182**, 339–347.

Börner, H. (1957). Die Abgabe organischer Verbindungen aus Karyopsen, Wurzeln und Ernterückständen von Roggen, Weizen und Gerste und ihre Bedeutung bei der gegenseitigen Beeinflussung der höheren Pflanzen. *Beitr. Biol. Pflanz.* **33**, 33–83.

Bors, W., Langebartels, C., Michel, C. and Sandermann, H. (1989). Polyamines as radical scavengers and protectants against ozone damage. *Phytochemistry* **28**, 1589–1595.

Bosch, C. (1983). Ernährungskundliche Untersuchung über die Erkrankung der Fichte (*Picea abies* Karst.) in den Hochlagen des Bayrischen Waldes. Diplomarbeit, Universität München.

Bothe, H., Yates, M. G. and Cannon, F. C. (1983). Physiology, biochemistry and genetic dinitrogen fixation. *In* 'Encyclopedia of Plant Physiology, New Series' (A. Läuchli and R. L. Bieleski, eds.), Vol. 15A, pp. 241–285. Springer-Verlag, Berlin and New York.

Bottrill, D. E., Possingham, J. V. and Kriedemann, P. E. (1970). The effect of nutrient deficiencies on photosynthesis and respiration in spinach. *Plant Soil* **32**, 424–438.

Bougher, N. L., Grove, T. S. and Malajczuk, N. (1990). Growth and phosphorus acquisition of karri (*Eucalyptus diversicolor* F. Muell.) seedlings inoculated with ectomycorrhizal fungi in relation to phosphorus supply. *New Phytol.* **114**, 77–85.

Bould, C. (1966). Leaf analysis of deciduous fruits. *In* 'Temperate to Tropical Fruit Nutrition' (N. F. Childers, ed.), pp. 651–684. Horticultural Publications, Rutgers University, New Brunswick, NJ.

Bould, C. and Parfitt, R. I. (1973). Leaf analysis as a guide to the nutrition of fruit crops. X. Magnesium and phosphorus sand culture experiments with apple. *J. Sci. Food Agric.* **24**, 175–185.

Bouma, D. (1983). Diagnosis of mineral deficiencies using plant tests. *In* 'Encyclopedia of Plant Physiology, New Series' (A. Läuchli and R. L. Bieleski, eds.), Vol. 15A, pp. 120–146. Springer-Verlag, Berlin and New York.

Bouma, T. J. and De Visser, R. (1993). Energy requirements for maintenance of ion concentrations in roots. *Physiol. Plant.* **89**, 133–142.

Bourquin, S., Bonnemain, J.-L. and Delroth, S. (1990). Inhibition of loading of ^{14}C assimilates by *p*-chlormercuribenzene-sulfonic acid. Localization of the apoplastic pathway in *Vicia faba*. *Plant Physiol.* **92**, 97–102.

Boursier, P. and Läuchli, A. (1989). Mechanism of chloride partitioning in leaves of salt-stressed *Sorghum bicolor* L. *Physiol. Plant.* **77**, 537–544.

Boursier, P. and Läuchli, A. (1990). Growth responses and mineral nutrient relations of salt-stressed sorghum. *Crop Sci.* **30**, 1226–1233.

Bousquet, U., Scheidecker, D. and Heller, R. (1981). Effect des conditions de culture sur la nutrition calcique de plantules calcifuge ou calcicoles. *Physiol. Veg.* **19**, 253–262.

Bouzayen, M., Felix, G., Latché, A., Pech, J.-C. and Boller, T. (1991). Iron: an essential cofactor for the conversion of 1-aminocyclopropane-1-carboxylic acid to ethylene. *Planta* **184**, 244–247.

Bowen, G. D. (1991). Soil temperature, root growth, and plant function. *In* 'Plant Roots: the Hidden Half' (Y. Waisel, A. Eshgel and U. Kafkafi, eds.), pp. 309–330. Marcel Dekker, New York.

Bowen, G. D. and Rovira, A. D. (1991). The rhizosphere, the hidden half of the hidden half. *In* 'Plant Roots: the Hidden Half' (Y. Waisel, A. Eshel and U. Kafkafi, eds.), pp. 641–669. Marcel Dekker, New York.

Bowen, P., Menzies, J., Ehret, D., Samuels, L. and Glass, A. D. M. (1992). Soluble silicon sprays inhibit powdery mildew development on grape leaves. *J. Am. Soc. Hortic. Sci.* **117**, 906–912.

Bowler, C., Slooten, L., Vandenbranden, S., De Rycke, R., Botterman, J., Sybesma, C., Van Montagu, M. and Inzé, D. (1991). Manganese superoxide dismutase can reduce cellular damage mediated by oxygen radicals in transgenic plants. *EMBO* **10**, 1723–1732.

Bowling, D. J. F. (1981). Release of ions to the xylem in roots. *Physiol. Plant.* **53**, 392–397.

Bowling, D. J. F. (1987). Measurement of the apoplastic activity of K^+ and Cl^- in the leaf epidermis of *Commelina communis* in relation to stomatal activity. *J. Exp. Bot.* **38**, 1351–1355.

Bowman, D. C. and Paul, J. L. (1992). Foliar absorption of urea, ammonium, and nitrate by perennial ryegrass turf. *J. Am. Soc. Hortic. Sci.* **117**, 75–79.

Bowsher, C. G., Hucklesby, D. P. and Emes, M.J. (1989). Nitrite reduction and carbohydrate metabolism in plastids purified from roots of *Pisum sativum* L. *Planta* **177**, 359–366.

Boyer, R. F. and Van der Ploeg, J. R. (1986). Iron metabolism in higher plants. The influence of nutrient iron on bean leaf lipoxygenase. *J. Plant Nutr.* **9**, 1585–1600.

Boyle, M. G., Boyer, J. S. and Morgan, P. W. (1991). Stem infusion of liquid culture medium prevents reproductive failure of maize at low water potential. *Crop Sci.* **31**, 1246–1252.

Boylston, E. K. (1988). Presence of silicon in developing cotton fibers. *J. Plant Nutr.* **11**, 1739–1747.

Boylston, E. K., Hebert, J. J., Hensarling, T. P., Bradow, J. M. and Thibodeaux, D. P. (1990). Role of silicon in developing cotton fibres. *J. Plant Nutr.* **13**, 131–148.

Braconnier, S. and d'Auzac, J. (1989). Effet d'une carence en chlorure au champ chez le cocotier hybride PB 121. *Oléagineux* **44**, 467–474.

Braconnier, S. and d'Auzac, J. (1990a). Chloride and stomatal conductance in coconut. *Plant Physiol. Biochem.* **28**, 105–112.

Bradfield, E. G. and Guttridge, C. G. (1984). Effects of night-time humidity and nutrient solution concentration on the calcium-content of tomato fruit. *Sci. Hortic. (Amsterdam)* **22**, 207–217.

Bradford, K. J., Hsiao, T. C. and Yang, S. F. (1982). Inhibition of ethylene synthesis in tomato plants subjected to anaerobic root stress. *Plant Physiol.* 70, 1503–1507.

Brauer, D., Leggett, J. E. and Egli, D. B. (1987). Changes in K, Rb, and Na transport to shoots after anoxia. *Plant Physiol.* **83**, 219–224.

Brauer, M., Sanders, D. and Stitt, M. (1990). Regulation of photosynthetic sucrose synthesis: a role for calcium? *Planta* **182**, 236–243.

Bravo, F. P. and Uribe, E. G. (1981). Temperature dependence of the concentration kinetics of absorption of phosphate and potassium in corn roots. *Plant Physiol.* **67**, 815–819.

Breckle, S.-W. (1991). Growth under stress. Heavy metals. *In* 'The Plant Root, the Hidden Half' (Y. Waisel, A. Eshel and U. Kafkafi, eds.), pp. 351–373. Marcel Dekker, New York.

Breeze, V. G., Wild, A., Hopper, M. J. and Jones, L. H. P. (1984). The uptake of phosphate by plants from flowing nutrient solution. II. Growth of *Lolium perenne* L. at constant phosphate concentrations. *J. Exp. Bot.* **35**, 1210–1221.

Breimer, T. (1982). Environmental factors and cultural measures affecting the nitrate content of spinach. *Fert. Res.* **3**, 191–292.

Brennan, R. F. (1989). Effect of nitrogen and phosphorus deficiency in wheat on the infection of roots by *Gaeumannomyces graminis* var. *tritici. Aust. J. Agric. Res.* **40**, 489–495.

Brennan, R. F. (1992a). The role of manganese and nitrogen nutrition in the susceptibility of wheat plants to take-all in Western Australia. *Fert. Res.* **31**, 35–41.

Brennan, R. F. (1992b). Effect of superphosphate and nitrogen on yield and take-all of wheat. *Fert. Res.* **31**, 43–49.

Brennan, R. F. (1992c). The relationship between critical concentrations of DTPA-extractable zinc from the soil for wheat production and properties of south-western Australian soils responsive to applied zinc. *Commun. Soil Sci. Plant Anal.* **23**, 747–759.

Breteler, H. and Nissen, P. (1982). Effect of exogenous and endogenous nitrate concentration on nitrate utilization by dwarf bean. *Plant Physiol.* **70**, 754–759.

Breteler, H. and Siegerist, M. (1984). Effect of ammonium on nitrate utilization by roots of dwarf bean. *Plant Physiol.* **75**, 1099–1103.

Breteler, H. and Smit, A. L. (1974). Effect of ammonium nutrition on uptake and metabolism of nitrate in wheat. *Neth. J. Agric. Sci.* **22**, 73–81.

Brevedan, R. E., Egli, D. B. and Leggett, J. E. (1978). Influence of N nutrition on flower and pod abortion and yield of soybeans. *Agron. J.* **70**, 81–84.

Brewster, J. L. and Tinker, P. B. (1970). Nutrient cation flows in soil around plant roots. *Soil Sci. Soc. Am. Proc.* **34**, 421–426.

Brinckmann, E., Hartung, W. and Wartinger, M. (1990). Abscisic acid levels of individual leaf cells. *Physiol. Plant.* **80**, 51–54.

Briskin, D. P. (1990). Ca^{2+}-translocating ATPase of the plant plasma membrane. *Plant Physiol.* **94**, 397–400.

Briskin, D. P. and Poole, R. J. (1983). Characterization of a K^{+}-stimulated adenosine triphosphatase associated with the plasma membrane of red beet. *Plant Physiol.* **71**, 350–355.

Brodrick, S. J. and Giller, K. E. (1991a). Root nodules of *Phaseolus*: efficient scavengers of molybdenum for N$_2$-fixation. *J. Exp. Bot.* **42**, 679–686.

Brodrick, S. J. and Giller, K.E. (1991b). Genotypic difference in molybdenum accumulation affects N$_2$-fixation in tropical *Phaseolus vulgaris*. J. Exp. Bot. **42**, 1339–1343.

Bromfield, S. M., Cumming, R. W., David, D. J. and Williams, C. H. (1983a). Change in soil pH, manganese and aluminium under subterranean clover pasture. *Aust. J. Exp. Agric. Anim. Husb.* **23**, 181–191.

Bromfield, S. M., Cumming, R. W., David, D. J. and Williams, C. H. (1983b). The assessment of available manganese and aluminium status in acid soils from subterranean clover pastures of various ages. *Aust. J. Exp. Agric. Anim. Husb.* **23**, 192–200.

Brookes, A., Collins, J. C. and Thurman, D. A. (1981). The mechanism of zinc tolerance in grasses. *J. Plant Nutr.* **3**, 695–705.

Brookes, R. R. and Malaisse, F. (1989). Metal-enriched sites in South Central Africa. *In* 'Heavy Metal Tolerance in Plants: Evolutionary Aspects' (A. J. Shaw, ed.), pp. 53–73. CRC Press, Boca Raton, FL.

Brouder, S. M. and Cassman, K. G. (1990). Root development of 2 cotton cultivars in relation to potassium uptake and plant growth in a vermiculitic soil. *Field Crops Res.* **23**, 187–204.

Brouquisse,R., Gaillard, J. and Douce, R. (1986). Electron paramagnetic resonance characterization of membrane bound iron-sulfur clusters and aconitase in plant mitochondria. *Plant Physiol.* **81**, 247–252.

Brouquisse, R., James, F., Raymond, P. and Pradet, A. (1991). Study of glucose starvation in excised maize root tips. *Plant Physiol.* **96**, 619–626.

Brouwer, R. (1981). Co-ordination of growth phenomena within a root system of intact maize plants. *Plant Soil* **63**, 65–72.

Brown, D. J. and DuPont, F. M. (1989). Lipid composition of plasma membranes and endomembranes prepared from roots of barley (*Hordeum vulgare* L.). *Plant Physiol.* **90**, 955–961.

Brown, J. C. (1979). Role of calcium in micronutrients stresses of plants. *Commun. Soil Sci. Plant Anal.* **10**, 459–472.

Brown, J. C. and Clark, R. B. (1974). Differential response of two maize inbreds to molybdenum stress. *Soil Sci. Soc. Am. Proc.* **38**, 331–333.

Brown, J. C. and Clark, R. B. (1977). Copper as essential to wheat reproduction. *Plant Soil* **48**, 509–523.

Brown, J. C. and Devine, T. E. (1980). Inheritance of tolerance or resistance to manganese toxicity in soybeans. *Agron. J.* **72**, 898–904.

Brown, J. C. and Jones, W. E. (1971). Differential transport of boron in tomato (*Lycopersicon esculentum* Mill.). *Physiol. Plant.* **25**, 279–282.

Brown, J. C. and Jones, W. E. (1972). Effect of germanium on utilization of boron in tomato (*Lycopersicon esculentum* Mill.). *Plant Physiol.* **49**, 651–653.

Brown, J. C. and Jones, W. E. (1976). A technique to determine iron efficiency in plants. *Soil Sci. Soc. Am. J.* **40**, 398–405.

Brown, J. C. and Jones, W. E. (1977). Manganese and iron toxicities dependent on soybean variety. *Commun. Soil Sci. Plant Anal.* **8**, 1–15.

Brown, J. C., Weber, C. R. and Caldwell, B. E. (1967). Efficient and inefficient use of iron by two soybean genotypes and their isolines. *Agron. J.* **59**, 459–462.

Brown, J. C., Jolley, V. D. and Lytle, C. M. (1991). Comparative evaluation of iron solubilizing substances (phytosiderophores) released by oats and corn: Iron-efficient and iron-inefficient plants. *Plant Soil* **130**, 157–163.

Brown, M. S., Thamsurakul, S., Bethlenfalvay, G. J. (1988). The *Glycine-Glomus-Bradyrhizobium* symbiosis. IX. Phosphorus-use efficiency of CO_2 and N_2 fixation in mycorrhizal soybean. *Physiol. Plant.* **74**, 159–163.

Brown, P. H., Graham, R. D. and Nicholas, D. J. D. (1984). The effects of manganese and nitrate supply on the level of phenolics and lignin in young wheat plants. *Plant Soil* **81**, 437–440.

Brown, P. H., Welch, R. M. and Cary, E. E. (1987a). Nickel: a micronutrient essential for higher plants. *Plant Physiol.* **85**, 801–803.

Brown, P. H., Welch, R. M., Cary, E. E. and Checkai, R. T. (1987b). Beneficial effects of nickel on plant growth. *J. Plant Nutr.* **10**, 2125–2135.

Brown, P. H., Dunemann, L., Schulz, R. and Marschner, H. (1989). Influence of redox potential and plant species on the uptake of nickel and cadmium from soils. *Z. Pflanzenernähr. Bodenk.* **152**, 85–91.

Brown, P. H., Welch, R. M. and Madison, J. T. (1990). Effect of nickel deficiency on soluble anion, amino acid, and nitrogen levels in barley. *Plant Soil* **125**, 19–27.

Brown, R. H. (1985). Growth of C_3 and C_4 grasses under low N levels. *Crop Sci.* **25**, 954–957.

Brown, T. A. and Shrift, A. (1982). Selenium: toxicity and tolerance in higher plants. *Biol. Rev. Camb. Philos. Soc.* **57.**, 59–84.

Brownell, P. F. (1979). Sodium as an essential micronutrient element for plants and its possible role in metabolism. *Adv. Bot. Res.* **7**, 117–224.

Brownell, P. F. (1965). Sodium as an essential micronutrient element for a higher plant (*Atriplex vesicaria*). *Plant Physiol.* **40**, 460–468.

Brownell, P. F. and Crossland, C. J. (1972). The requirement for sodium as a micronutrient by species having the C_4 dicarboxylic photosynthetic pathway. *Plant Physiol.* **49**, 794–797.

Brownell, P. F. and Crossland, C. J. (1974). Growth responses to sodium by *Bryophyllum tubiflorum* under conditions inducing crassulacean acid metabolism. *Plant Physiol.* **54**, 416–417.

Browning, M. H. R. and Whitney, R. D. (1992). The influence of phosphorus concentration and frequency of fertilization on ectomycorrhizal development in containerized black spruce and jack pine seedlings. *Can. J. For. Res.* **22**, 1263–1270.

Brownlee, C., Duddridge, J. A., Malibari, A. and Read, D. J. (1983). The structure and function of mycelial systems of ectomycorrhizal roots with special reference to their role in forming inter-plant connections and providing pathways for assimilate and water transport. *Plant Soil* **71**, 433–443.

Broyer, T. C. (1966). Chlorine nutrition of tomato: observations on inadvertment accretion and loss and their implications. *Physiol. Plant.* **19**, 925–936.

Broyer, T. C., Carlton, A. B., Johnson, C. M. and Stout, P. R. (1954). Chlorine – a micronutrient element for higher plants. *Plant Physiol.* **29**, 526–532.

Broyer, T. C., Johnson, C. M. and Houston, R. P. (1972). Selenium and nutrition of *Astragalus*. I. Effect of selenite or selenate supply on growth and selenium content. *Plant Soil* **36**, 635–649.

Bruce, R. C., Warrell, L. A., Edwards, D. G. and Bell, L. C. (1988). Effects of aluminium and calcium in the soil solution of acid soils on root elongation of *Glycine max* cv. Forrest. *Aust. J. Agric. Res.* **39**, 319–338.

Brüggemann, W., Moog, P. R., Nakagawa, H., Janiesch, P. and Kuiper, P. J. C. (1991). Plasma membrane-bound NADH:Fe^{3+}-EDTA reductase and iron deficiency in tomato (*Lycopersicon esculentum*). Is there a Turbo reductase? *Physiol. Plant.* **79**, 339–346.

Brümmer, G. (1974). Redoxpotentiale und Redoxprozesse von Mangan-, Eisen- und Schwefelverbindungen in hydromorphen Böden und Sedimenten. *Geoderma* **12**, 207–222.

Brümmer, G. (1981). Ad- und Desorption oder Ausfällung und Auflösung als Lösungskonzentration bestimmende Faktoren in Böden. *Mitt. Dtsch. Bodenkd. Ges.* **30**, 7–18.

Brüning, D. (1967). Befall mit *Eulecanium corni* Bché. f. *robinarium* Dgl. und *Eulecanium rufulum* Ckll. in Düngungsversuchen zu Laubgehölzen. *Arch. Pflanzenschutz* **3**, 193–200.

Bruinsma, J. (1977). Rolle der Cytokinine bei Blüten- und Fruchtentwicklung. *Z. Pflanzenernähr. Bodenk.* **140**, 15–23.

Brumagen, D. M. and Hiatt, A. J. (1966). The relationship of oxalic acid to the translocation and utilization of calcium in *Nicotiana tabacum*. *Plant Soil* **24**, 239–249.

Brumme, R., Leimcke, U. and Matzner, E. (1992). Interception and uptake of NH$_4$ and NO$_3$ from wet deposition by aboveground parts of young beech (*Fagus silvatica* L.) trees. *Plant Soil* **142**, 273–279.

Brummell, D.A. and Hall, J. L. (1987). Rapid cellular responses to auxin and the regulation of growth. *Plant, Cell Environ.* **10**, 523–543.

Brundrett, M. C. (1991). Mycorrhizas in natural ecosystems. *Adv. Ecol. Res.* **21**, 171–313.

Brundrett, M. C. and Abbott, L. K. (1991). Roots of jarrah forest plants. I. Mycorrhizal associations of shrubs and herbaceous plants. *Aust. J. Bot.* **39**, 445–457.

Brunold, C. (1993). Regulatory interactions between sulfate and nitrate assimilation. *In* 'Sulfur Nutrition and Assimilation in Higher Plants' (L. J. De Kok, I. Stulen, H. Rennenberg, C. Brunold and W. E. Rauser, eds.), pp. 62–75. SPB Academic Publishing, The Hague, The Netherlands.

Brunold, C. and Schmidt, A. (1978). Regulation of sulfate assimilation in plants. 7. Cysteine inactivation of adenosine 5-phosphosulfate sulfotransferase in *Lemna minor* L. *Plant Physiol.* **61**, 342–347.

Brunold, C. and Suter, M. (1984). Regulation of sulfate assimilation by nitrogen nutrition in the duckweed *Lemna minor* L. *Plant Physiol.* **76**, 579–583.

Bryla, D. R. and Koide, R. T. (1990a). Regulation of reproduction in wild and cultivated *Lycopersicon esculentum* Mill. by vesicular-arbuscular mycorrhizal infection. *Oecologia* **84**, 74–81.

Bryla, D. R. and Koide, R. T. (1990b). Role of mycorrhizal infection in the growth and reproduction of wild vs. cultivated plants. II. Eight wild accessions and two cultivars of *Lycopersicon esculentum* Mill. *Oecologia* **84**, 82–92.

Buban, T., Varga, A., Tromp, J., Knegt, E. and Bruinsma, J. (1978). Effects of ammonium and nitrate nutrition on the level of zeatin and amino nitrogen in xylem sap of apple rootstocks. *Z. Pflanzenphysiol.* **89**, 289–295.

Budde, R. J. A. and Chollet, R. (1988). Regulation of enzyme activity in plants by reversible phosphorylation. *Physiol. Plant.* **72**, 435–439.

Budde, R. J. A. and Randall, D. D. (1990). Protein kinases in higher plants. *In* 'Inositol

Metabolism in Plants' (D. J. Morré, W. F. Boss and F. A. Loewus, eds.), pp. 351–367, Wiley–Liss, New York.

Buerkert, A., Cassman, K. G., de la Piedra, R., and Munns, D. N. (1990). Soil acidity and liming effects on stand, nodulation, and yield of common bean. *Agron. J.* **82**, 749–754.

Bulder, H. A. M., Speek, E. J., van Hasselt, P. R. and Kuiper, P. J. C. (1991). Growth temperature and lipid composition of cucumber genotypes differing in adaptation to low energy conditions. *J. Plant Physiol.* **138**, 655–660.

Bunce, J. A. (1990). Abscisic acid mimics effects of dehydration on area expansion and photosynthetic partitioning in young soybean leaves. *Plant, Cell Environ.* **13**, 295–298.

Buntje, G. (1979). Untersuchungen zum Einfluss der Mangan- und Kupferversorgung auf die Kälteresistenz von Winterweizen, Hafer und Mais anhand von Gefässversuchen. PhD Thesis, Universität Kiel.

Burke, J. J., Holloway, P. and Dalling, M. J. (1986). The effect of sulfur deficiency on the organization and photosynthetic capability of wheat leaves. J. Plant Physiol. **125**, 371–375.

Burnell, J. N. (1981). Selenium metabolism in *Neptunia amplexicauli*. Plant Physiol. **67**, 316–324.

Burnell, J. N. (1986). Purification and properties of phosphoenolpyruvte carboxykinase from C_4 plants. *Aust. J. Plant Physiol.* **13**, 577–587.

Burnell, J. N. (1988). The biochemistry of manganese in plants. *In* 'Manganese in Soils and Plants' (R. D. Graham, R. J. Hannam and N. C. Uren, eds.), pp. 125–137. Kluwer Academic,Dordrecht.

Burnell, J. N. and Hatch, M. D. (1988). Low bundle sheath carbonic anhydrase is apparent by essential for effective C_4 pathway operation. *Plant Physiol.* **86**, 1252–1256.

Burnell, J. N., Suzuki, I. and Sugiyama, T. (1990). Light induction and the effect of nitrogen status upon the activity of carbonic anhydrase in maize leaves. *Plant Physiol.* **94**, 384–387.

Burns, I. G. (1992). Influence of plant nutrient concentration on growth rate: Use of a nutrient interruption technique to determine critical concentrations of N, P, and K in young plants. *Plant Soil* 142, 221–233.

Burns, J. K. and Pressey, R. (1987). Ca^{2+} in cell walls of ripening tomato and peach. *J. Am. Soc. Hortic. Sci.* **112**, 783–787.

Burris, R. H., Hartmann, A., Zhang, Y. and Fu, H. (1991). Control of nitrogenase in *Azospirillum* sp. *Plant Soil* **137**, 127–134.

Burrows, W. J. and Carr, D. J. (1969). Effects of flooding the root system of sunflower plants on the cytokinin content in the xylem sap. *Physiol. Plant.* **22**, 1105–1112.

Büscher, P. and Koedam, N. (1983). Soil preference of populations of genotypes of *Asplenium trichomanes* L. and *Polypodium vulgare* L. in Belgium as related to cation exchange capacity. *Plant Soil* **72**, 275–282.

Bush, D. S., Biswas, A. K. and Jones, R. J. (1993). Hormonal regulation of Ca^{2+} transport in the endomembrane system of the barley aleurone. *Planta* **189**, 507–515.

Bush, D. S., Cornejo, M.-J. Huang, C.-N. and Jones, R. L. (1986). Ca^{2+}-stimulated secretion of α-amylase during development in barley aleurone protoplasts. *Plant Physiol.* **82**, 566–574.

Bussler, W. (1963). Die Entwicklung von Calcium-Mangelsymptomen. *Z. Pflanzenernaehr., Dueng. Bodenkd.* **100**, 53–58.

Bussler, W. (1964). Die Bormangelsymptome und ihre Entwicklung. *Z. Pflanzener-naehr. Dueng. Bodenkd.* **105**, 113–136.

Bussler, W. (1970). Die Entwicklung der Mo-Mangelsymptome an Blumenkohl. *Z. Pflanzenernähr. Bodenk.* **125**, 36–50.

Bussler, W. (1981a). Microscopic possibilities for the diagnosis of trace element stress in plants. *J. Plant Nutr.* **3**, 115–128.

Bussler, W. (1981b). Physiological functions and utilization of copper. *In* 'Copper in Soils and Plants' (J. F. Loneragan, A. D. Robson and R. D. Graham, eds.), pp. 213–234. Academic Press, London.

Buwalda, J. G. and Lenz, F. (1992). Effects of cropping, nutrition and water supply on accumulation and distribution of biomass and nutrients for apple trees on 'M 9' root systems. *Physiol. Plant.* **84**, 21–28.

Buwalda, J. G. and Smith, G. S. (1991). Influence of anions on the potassium status and productivity of kiwifruit (*Actinidia deliciosa*) vines. *Plant Soil* **133**, 209–218.

Cacciari, I., Lippi, D., Pietrosanti, T. and Pietrosanti, W. (1989). Phytohormone-like substances produced by single and mixed diazotrophic cultures of Azospirillum and Arthrobacter. *Plant Soil* **115**, 151–153.

Caetano-Anollés, G. and Gresshoff, P. M. (1991). Alfalfa controls nodulation during the onset of *Rhizobium*-induced cortical cell division. Plant Physiol. **95**, 366–373.

Caetano-Anollés, G., Lagares, A. and Favelukes, G. (1989). Adsorption of *Rhizobium meliloti* to alfalfa roots: Dependence on divalent cations and pH. *Plant Soil* **117**, 67–74.

Cahill, D. M., Weste, G. M. and Grant, B. R. (1986). Changes in cytokinin concentrations in xylem exudate following infection of *Eucalyptus marginata* Donn ex Sm with *Phytophthora cinnamomi* Rands. *Plant Physiol.* **81**, 1103–1109.

Cairney, J. W. G. (1992). Translocation of solutes in ectomycorrhizal and saprophytic rhizomorphs. *Mycol. Res.* **96**, 135–141.

Cairns, A. L. P. and Kritzinger, J. H. (1992). The effect of molybdenum on seed dormancy in wheat. *Plant Soil* **145**, 295–297.

Cakmak, I. and Horst, W. J. (1991a). Effect of aluminium on net efflux of nitrate and potassium from root tips of soybean (*Glycine max.* L.). *J. Plant Physiol.* **138**, 400–403.

Cakmak, I. and Horst, W. J. (1991b). Effect of aluminium on lipid peroxidation, superoxide dismutase, catalase, and peroxidase activities in root tips of soybean (*Glycine max*). *Physiol. Plant.* **83**, 463–468.

Cakmak, I. and Marschner, H. (1986). Mechanism of phosphorus-induced zinc deficiency in cotton. I. Zinc deficiency-enhanced uptake rate of phosphorus. *Physiol. Plant.* **68**, 483–490.

Cakmak, I. and Marschner, H. (1987). Mechanism of phosphorus-induced zinc deficiency in cotton. III. Changes in physiological availability of zinc in plants. *Physiol. Plant.* **70**, 13–20.

Cakmak, I. and Marschner, H. (1988a). Enhanced superoxide radical production in roots of zinc-deficient plants. *J. Exp. Bot.* **39**, 1449–1460.

Cakmak, I. and Marschner, H. (1988b). Zinc-dependent changes in ESR signals, NADPH oxidase and plasma membrane permeability in cotton roots. *Physiol. Plant.* **73**, 182–186.

Cakmak, I. and Marschner, H. (1988c). Increase in membrane permeability and exsudation of roots of zinc deficient plants. *J. Plant Physiol.* **132**, 356–361.

Cakmak, I. and Marschner, H. (1990). Decrease in nitrate uptake and increase in proton release in zinc deficient cotton, sunflower and buckwheat plants. *Plant Soil* **129**, 261–268.

Cakmak, I. and Marschner, H. (1992). Magnesium deficiency and high light intensity enhance activities of superoxide dismutase, ascorbate peroxidase and glutathione reductase in bean leaves. *Plant Physiol.* **98**, 1222–1227.

Cakmak, I., Marschner, H. and Bangerth, F. (1989). Effect of zinc nutritional status on growth, protein metabolism and levels of indole-3-acetic acid and other phytohormones in bean (*Phaseolus vulgaris* L.). *J. Exp.Bot.* **40**, 405–412.

Cakmak, I., Gülüt, K. Y., Marschner, H. and Graham, R. D. (1994). Effect of zinc and iron deficiency on phytosiderophore release in wheat genotypes differing in zinc efficiency. *J. Plant Nutr.* **17**, 1–17.

Caldwell, C. R. (1989). Analysis of aluminum and divalent cation binding to wheat root plasma membrane proteins using terbium phosphorescence. *Plant Physiol.* **91**, 233–241.

Caldwell, C. R. and Haug, A.(1981). Temperature dependence of the barley root plasma membrane-bound Ca^{2+} and Mg^{2+}-dependent ATPase. *Physiol. Plant.* **53**, 117–124.

Cammack, R., Fernandez, V. M. and Schneider, K. (1988). Nickel in hydrogenases from sulfate-reducing, photosynthetic, and hydrogen-oxidizing bacteria. *In* 'The Bioorganic Chemistry of Nickel' (J. R. Lancaster jr, ed.), pp. 167–190. Verlag Chemie, Weinheim.

Cammarano, P., Felsani, A., Gentile, M., Gualerzi, C., Romeo, C. and Wolf, G. (1972). Formation of active hybrid 80-S particles from subunits of pea seedlings and mammalian liver ribosomes. *Biochim. Biophys. Acta* **281**, 625–642.

Campbell, L. C. and Nable, R. O. (1988). Physiological functions of manganese in plants. *In* 'Manganese in Soils and Plants' (R. D. Graham, R. J. Hannan and N. C. Uren, eds.), pp. 139–154. Kluwer Academic, Dordrecht.

Campbell, L. C., Miller, M. H. and Loneragan, J. F. (1975). Translocation of boron to plant fruits. *Aust. J. Plant Physiol.* **2**, 481–487.

Campbell, M., Dunn, R., Ditterline, R., Pickett, S. and Raboy, V. (1991). Phytic acid represents 10 to 15% of total phosphorus in alfalfa root and crown. *J. Plant Nutr.* **14**, 925–937.

Campbell, N. A. and Thomson, W. W. (1977). Effects of lanthanum and ethylenediaminetetraacetate on leaf movements of *Mimosa*. *Plant Physiol.* **60**, 635–639.

Campbell, W. H. and Gowri, G. (1990). Codon usage in higher plants, green algae, and cyanobacteria. *Plant Physiol.* **92**, 1–11.

Campbell, W. H. and Redinbaugh, M. G. (1984). Ferric-citrate reductase activity of nitrate reductase and its role in iron assimilation by plants. *J. Plant Nutr.* **7**, 799–806.

Cannell, R. Q. (1977). Soil aeration and compaction in relation to root growth and soil management. *Appl. Biol.* **2**, 1–86.

Cannell, R. Q., Gales, K., Snaydon, R. W. and Suhail, B. A. (1979). Effects of short-term water logging on the growth and yield of peas (*Pisum sativum*). *Ann. Appl.Biol.* **93**, 327–335.

Canny, M. J. (1988). Bundle sheath tissues of legume leaves as a site of recovery of solutes from the transpiration stream. *Physiol. Plant.* **73**, 457–464.

Canny, M. J. (1990). Transley Review No. 22: What becomes of the transpiration stream? *New Phytol.* **114**, 341–368.

Canny, M. J. and McCully, M. E. (1989). The xylem sap of maize roots: its collection, composition and formation. *Aust. J. Plant Physiol.* **15**, 557–566.

Caradus, J. R. (1982). Genetic differences in the length of root hairs in white clover and their effect on phosphorus uptake. *In* 'Proceedings of the Ninth International Plant Nutrition Colloquium, Warwick, England' (A.Scaife, ed.), pp. 84–88. Commonwealth Agricultural Bureau, Farnham Royal, Bucks, UK.

Caradus, J. R. and Snaydon, R. W. (1987). Aspects of the phosphorus nutrition of white clover populations. I. Inorganic phosphorus content of leaf tissue. *J. Plant Nutr.* **10**, 273–285.

Carey, P. D., Fitter, A. H. and Watkinson, A. R. (1992). A field study using the fungicide benomyl to investigate the effect of mycorrhizal fungi on plant fitness. *Oecologia* **90**, 550–555.

Carmi, A. and Koller, D. (1979). Regulation of photosynthetic activity in the primary leaves of bean (*Phaseolus vulgaris* L.) by materials moving in the water-conducting system. *Plant Physiol.* **64**, 285–288.

Carpita, N., Sabularse, D., Montezinos, D. and Delmer, D. P. (1979). Determination of the pore size of cell walls of living plant cells. *Science* **205**, 1144–1147.

Carroll, B. J., Hansen, A. P., McNeil, D. L. and Gresshoff, P. M. (1987). Effect of oxygen supply on nitrogenase activity of nitrate- and dark-stressed soybean (*Glycine max.* (L.) Merr.) plants. *Aust. J. Plant Physiol.* **14**, 679–687.

Cartwright, B. and Hallsworth, E. G. (1970). Effects of copper deficiency on root nodules of subterranean clover. *Plant Soil* **33**, 685–698.

Carvalho, M. M. de, Edwards, D. G. and Asher, C. J. (1982). Effects of aluminium on nodulation of two stylosanthes species grown in nutrient solution. *Plant Soil* **64**, 141–152.

Carver, T. L. W., Zeyen, R. J. and Ahlstrand, G. G. (1987). The relationship between insoluble silicon and success or failure of attempted primary penetration by powdery mildew (*Erysiphe graminis*) germlings of barley. *Physiol. Mol. Plant Pathol.* **31**, 133–148.

Casimiro, A., Barroso, J. and Pais, M. S. (1990). Effect of copper deficiency on photosynthetic electron transport in wheat plants. *Physiol. Plant.* **79**, 459–464.

Cassab, G. I. and Varner, J. E. (1988). Cell wall proteins. *Annu. Rev. Plant Physiol. Plant Mol. Biol.* **39**, 321–353.

Cassells, A. L. and Barlass, M. (1976). Environmentally induced changes in the cell walls of tomato leaves in relation to cell and protoplast release. *Physiol. Plant.* **37**, 239–246.

Cassman, K. G., Whitney, A. S. and Stockinger, K. R. (1980). Root growth and dry matter distribution of soybean as affected by phosphorus stress, nodulation, and nitrogen source. *Crop Sci.* **20**, 239–244.

Castle, S. L. and Randall, P. J. (1987). Effects of sulfur deficiency on the synthesis and accumulation of proteins in the developing wheat seed. *Aust. J. Plant Physiol.* **14**, 503–516.

Cataldo, D. A., McFadden, K. M., Garland, T. R. and Wildung, R. E. (1988). Organic constituents and complexation of nickel (II), iron (III), cadmium (II), and plutonium (IV) in soybean xylem exudates. *Plant Physiol.* **86**, 734–739.

Causin, H. F. and Barneix, A. J. (1993). Regulation of NH_4^+ uptake in wheat plants: Effect of root ammonium concentration and amino acids. *Plant Soil* **151**, 211–218.

Chaboussou, F. (1976). Cultural factors and the resistance of citrus plants to scale insects and mites. *Proc. 12th Colloq. Int. Potash Inst. Bern*, pp. 259–280.

Chakravarty, C., Peterson, R. L. and Ellis, B. E. (1991). Interaction between the ectomycorrhizal fungus *Paxillus involutus*, damping-off fungi and *Pinus resinosa* seedlings. *J. Phytopathol.* **132**, 207–218.

Chalk, P. M. (1991). The contribution of associative and symbiotic nitrogen fixation to the nitrogen nutrition of non-legumes. *Plant Soil* **132**, 29–39.

Chalot, M., Stewart, G. R., Brun, A., Martin, F. and Botton, B. (1991). Ammonium assimilation by spruce-*Hebeloma* sp. ectomycorrhizas. *New Phytol.* **119**, 541–550.

Chamel, A. (1988). Foliar uptake of chemicals studied with whole plants and isolated cuticles. *In* 'Plant Growth and Leaf-Applied Chemicals' (P. M. Neumann, ed.), pp. 27–50. CRC Press, Boca Raton, FL.

Chamel, A., Andréani, A. M. and Elroy, J. F. (1981). Distribution of foliar applied boron measured by spark-source mass spectrometry and laser-probe mass spectrography. *Plant Physiol.* **67**, 457–459.

Champigny, M. L. and Foyer, C. (1992). Nitrate activation of cytosolic protein kinases diverts photosynthetic carbon from sucrose to amino acid biosynthesis. *Plant Physiol.* **100**, 7–12.

Chandrasekaran, S. and Yoshida, T. (1973). Effects of organic acid transformation in submerged soils on growth of the rice plant. *Soil Sci. Plant Nutr.* **19**, 39–45.

Chaney, R. L. (1984). Diagnostic practices to identify iron deficiency in higher plants. *J. Plant Nutr.* **7**, 47–67.

Chaney, R. L. (1988). Recent progress and needed research in plant Fe nutrition. *J. Plant Nutr.* **11**, 1589–1603.

Chaney, R. L., Coulombe, B. A., Bell, P. F. and Angle, J. S. (1992a). Detailed method to screen dicot cultivars for resistance to Fe-chlorosis using FeDTPA and bicarbonate in nutrient solutions. *J. Plant Nutr.* **15**, 2063–2083.

Chaney, R. L., Chen, Y., Green, C. E., Holden, M. J., Bell, P. F., Luster, D. G. and Angle, J. S. (1992b). Root hairs on chlorotic tomatoes are an effect of chlorosis rather than part of the adaptive Fe-stress response. *J. Plant Nutr.* **15**, 1857–1875.

Chang, K. and Roberts, J. K. M. (1992). Quantitation of rates of transport, metabolic fluxes, and cytoplasmic levels of inorganic carbon in maize root tips during K^+ ion uptake. *Plant Physiol.* **99**, 291–297.

Changzhi, L., Hongmin, D., Hechen, J., Guang-Yong, Y. and Zhongxi, C. (1990). Effects of ^{10}B application on the distribution characteristics of boron in rape leaves. *Sci. Agric. Sin.* **23**, 67–72.

Chanson, A. (1991). A Ca^{2+}/H^+ antiport system driven by the tonoplast pyrophosphate-dependent proton pump from maize roots. *J. Plant Physiol.* **137**, 471–476.

Chanson, A., Fichmann, J., Spear, D. and Taiz, L. (1985). Pyrophosphate-driven proton transport by microsomal membranes of corn coleoptiles. *Plant Physiol.* **79**, 159–164.

Chapin, F. S. III (1980). The mineral nutrition of wild plants. *Annu. Rev. Ecol. Syst.* **11**, 233–260.

Chapin, F. S. III (1983). Adaptation of selected trees and grasses to low availability of phosphorus. *Plant Soil* **72**, 283–297.

Chapin, F. S. III (1988). Ecological aspects of plant mineral nutrition. *In* 'Advances in Plant Nutrition' Vol. 3 (B. Tinker and A. Läuchli, eds.), pp. 161–191. Praeger, New York.

Chapin, F. S. III and Wardlaw, I. F. (1988). Effect of phosphorus deficiency on source–sink interactions between the flag leaf and developing grain in barley. *J. Exp. Bot.* **39**, 165–177.

Chapin, F. S. III, Bloom, A. J., Field, C. B. and Waring, R. H. (1987). Plant responses to multiple environmental factors. *BioScience* **37**, 49–57.

Chapin, F. S. III, Walter, C. H. S. and Clarkson, D. T. (1988). Growth response of barley and tomato to nitrogen stress and its control by abscisic acid, water relations and photosynthesis. *Planta* **173**, 352–366.

Chapin, F. S. III, Moilanen, L. and Kielland, K. (1993). Preferential use of organic nitrogen for growth by a non-mycorrhizal artic sedge. *Nature* **361**, 150–153.

Chapman, H. D. (1966) 'Diagnostic Criteria for Plants and Soils.' Riverside Div. Agric. Sci., University of California.

Chatel, D. L., Robson, A. D., Gartrell, J. W. and Dilworth, M. J. (1978). The effect of inoculation and cobalt application on the growth of and nitrogen fixation by sweet lupinus. *Aust. J. Agric. Res.* **29**, 1191–1202.

Chatterjee, C., Nautiyal, N. and Agarwala, S. C. (1985). Metabolic changes in mustard plant associated with molybdenum deficiency. *New Phytol.* **100**, 511–518.

Chatterjee, C., Nautiyal, N. and Agarwala, S. C. (1992). Excess sulphur partially alleviates copper deficiency effects in mustard. *Soil Sci. Plant Nutr.* **38**, 57–64.

Chavan, P. D. and Karadge, B. A. (1980). Influence of sodium chloride and sodium sulfate salinization on photosynthetic carbon assimilation in peanut. *Plant Soil* **56**, 201–207.

Cheeseman, J. M. and Hanson, J. B. (1979). Energy-linked potassium influx as related to cell potential in corn roots. *Plant Physiol.* **64**, 842–845.

Chen, C. H. and Lewin, J. (1969). Silicon as a nutrient element for *Equisetum arvense*. *Can. J. Bot.* **47**, 125–131.

Chen, C. C., Dixon, J. B. and Turner, F. T. (1980). Iron coatings on rice roots: mineralogy and quantity influencing factors. *Soil Sci. Soc. Am. J.* **44**, 635–639.

Chen, Y. and Barak, P. (1982). Iron nutrition of plants in calcareous soils. *Adv. Agron.* **35**, 217–240.

Chenery, E. M. and Sporne, K. R. (1976). A note on the evolutionary status of aluminium-accumulators among dicotyledons. *New Phytol.* **76**, 551–554.

Chereskin, B. M. and Castelfranco, P. A. (1982). Effects of iron and oxygen on chlorophyll biosynthesis. II. Observations on the biosynthetic pathway in isolated etio-chloroplasts. *Plant Physiol.* **68**, 112–116.

Chhabra, R., Ringoet, A., Lamberts, D. and Scheys, I. (1977). Chloride losses from tomato plants (*Lycopersicon esculentum* Mill.). *Z. Pflanzenphysiol.* **81**, 89–94.

Chino, M., Fukumorita, T., Kawabe, S. and Ando, Y. (1982). Chemical composition of rice phloem sap collected by 'insect technique'. *In* 'Proceedings of the Ninth International Plant Nutrition Colloquium , Warwick, England' (A. Scaife, ed.), pp. 105–110. Commonwealth Agricultural Bureau, Farnham Royal, Bucks, UK.

Chisholm, R. H. and Blair, G. J. (1981). Phosphorus uptake and dry weight of stylo and white clover as affected by chlorine. *Agron. J.* **73**, 767–771.

Chow, W. S., Ball, M. C. and Naderson, J. M. (1990). Growth and photosynthetic response of spinach to salinity: implications of K^+ nutrition for salt tolerance. *Aust. J. Plant Physiol.* **17**, 563–578.

Christensen, N. W., Powelson, R. L. and Brett, M. (1987). Epidemiology of wheat take-all as influenced by soil pH and temporal changes in inorganic soil N. *Plant Soil* **98**, 221–230.

Christiansen, M. N., Carns, H. R. and Slyter, D. J. (1970). Stimulation of solute loss from radicles of *Gossypium hirsutum* L. by chilling, anaerobiosis, and low pH. *Plant Physiol.* **46**, 53–56.

Christiansen-Weniger, C. and van Veen, J. A. (1991). Nitrogen fixation by *Azospirillum brasilense* in soil and the rhizosphere under controlled environmental conditions. *Biol. Fertil. Soils* **12**, 100–106.

Christiansen-Weniger, C., Groneman, A. F. and van Veen, J. A. (1992). Associative

N_2 fixation and root exudation of organic acids from wheat cultivars of different aluminium tolerance. *Plant Soil* **139**, 167–174.

Churchill, K. A. and Sze, H. (1984). Anion-sensitive, H^+-pumping ATPase of oat roots. Direct effects of Cl^-, NO_3^-, and a disulfonic stilbene. *Plant Physiol.* **76**, 490–497.

Citernesi, U., Neglia, R., Seritti, A., Lepidi, A. A., Filippi, C., Bagnoli, G., Nuti, M. P. and Galluzzi, R. (1977). Nitrogen fixation in the gastroenteric cavity of soil animals. *Soil Biol. Biochem.* **9**, 71–72.

Claassen, N. (1990). Nährstoffaufnahme höherer Pflanzen aus dem Boden – Ergebnis von Verfügbarkeit und Aneignungsvermögen. Severin Verlag Göttingen.

Claassen, N. and Jungk, A. (1982). Kaliumdynamik im wurzelnahen Boden in Beziehung zur Kaliumaufnahme von Maispflanzen. *Z. Pflanzenernärh. Bodenk.* **145**, 513–525.

Claassen, N. and Jungk, A. (1984). Bedeutung von Kaliumaufnahmerate, Wurzelwachstum und Wurzelhaaren für das Kaliumaneignungsvermögen verschiedener Pflanzenarten. *Z. Pflanzenernähr. Bodenk.* **147**, 276–289.

Clark, R. B. (1975). Differential magnesium efficiency in corn inbreds. I. Dry-matter yields and mineral element composition. *Soil Sci. Soc. Am. Proc.* **39**, 488–491.

Clark, R.B. (1977). Effect of aluminium on growth and mineral elements of Al-tolerant and Al-intolernt corn. *Plant Soil* **47**, 653–662.

Clark, R. B. (1982a). Iron deficiency in plants grown in the great plains of the U.S. *J. Plant Nutr.* **5**, 251–268.

Clark, R. B. (1982b). Nutrient solution growth of sorghum and corn in mineral nutrition studies. *J. Plant Nutr.* **5**, 1039–1057.

Clark, R. B., Römheld, V. and Marschner, H. (1988). Iron uptake and phytosiderophore release by roots of sorghum genotypes. *J. Plant Nutr.* **11**, 663–676.

Clarkson, D. T. (1977). Membrane structure and transport. *In* 'The Molecular Biology of Plant Cells' (H. Smith, ed.), pp. 24–63. Blackwell, Oxford.

Clarkson, D. T. (1984). Calcium transport between tissues and its distribution in the plant. *Plant, Cell Environ.* **7**, 449–456.

Clarkson, D. T. (1988). The uptake and translocation of manganese by plant root. *In* 'Manganese in Soils and Plants' (R. D. Graham, R. J. Hannan and N. C. Uren, eds.), pp. 101–111. Kluwer Academic, Dordrecht.

Clarkson, D. T. and Hanson, J. B (1980). The mineral nutrition of higher plants. *Annu. Rev. Plant Physiol.* **31**, 239–298.

Clarkson, D. T. and Saker, L. R. (1989). Sulphate influx in wheat and barley roots becomes more sensitive to specific protein-binding reagents when plants are sulphate-deficient. *Planta* **178**, 249–257.

Clarkson, D. T. and Scattergood, C. B. (1982). Growth and phosphate transport in barley and tomato plants during the development of, and recovery from, phosphate-stress. *J. Exp. Bot.* **33**, 865–875.

Clarkson, D. T. and Warner, A. J. (1979). Relationships between root temperature and the transport of ammonium and nitrate ions by Italian and perennial ryegrass (*Lolium multiflorum* and *Lolium perenne*). *Plant Physiol.* **64**, 557–561.

Clarkson, D. T., Robards, A. W. and Sanderson, J. (1971). The tertiary endodermis in barley roots: Fine structure in relation to radial transport of ions and water. *Planta* **96**, 292–305.

Clarkson, D. T., Sanderson, J. and Scattergood, C. B. (1978). Influence of phosphate-stress and phosphate absorption and translocation by various parts of the root system of *Hordeum vulgare* L. (Barley). *Planta* **139**, 47–53.

Clarkson, D. T., Hopper, M. J. and Jones, L. H. P. (1986). The effect of root temperature on the uptake of nitrogen and the relative size of the root system in *Lolium perenne*. I. Solutions containing both NH_4^+ and NO_3^-. *Plant, Cell Environ.* **9**, 535–545.

Clarkson, D. T., Robards, A. W., Stephens, J. E. and Stark, M. (1987). Suberin lamellae in the hypodermis of maize (*Zea mays*) roots; development and factors affecting the permeability of hypodermal layers. *Plant, Cell Environ.* **10**, 83–93.

Clarkson, D. T., Jones, L. H. P. and Purves, J. V. (1992). Absorption of nitrate and ammonium ions by *Lolium perenne* from flowing solution cultures at low root temperatures. *Plant, Cell Environ.* **15**, 99–106.

Clasen, K., Grosse, F. and Diepenbrock, W. (1991). Die Stoffbildung und -verteilung bei Winterraps (*Brassica napus* L.). *Kali-Briefe* **20**, 685–713.

Cleland, R. E., Virk, S. S., Taylor, D. and Björkman, T. (1990). Calcium, cell walls and growth. *In* 'Calcium in Plant Growth and Development' (R. T. Leonard and P. K. Hepler, eds.), pp. 9–16. The American Society of Plant Physiology, Symposium Series, Vol. 4.

Clement, C. R., Hopper, M. J. and Jones, L. H. P. (1978a). The uptake of nitrate by *Lolium perenne* from flowing nutrient solution. I. Effect of NO_3-concentration. *J. Exp. Bot.* **29**, 453–464.

Clement, C. R., Hopper, M. J., Jones, L. H. P. and Leafe, E. L. (1978b).The uptake of nitrate by *Lolium perenne* from flowing nutrient solution. II. Effect of light, defoliation, and relationship to CO_2 flux. *J. Exp. Bot.* **29**, 1173–1183.

Clement, C. R., Jones, L. H. P. and Hopper, M. J. (1979). Uptake of nitrogen from flowing nutrient solution: effect of terminated and intermittent nitrate supplies. *In* 'Nitrogen Assimilation in Plants' (E. J. Hewitt and C. V. Cutting, eds.), pp. 123–133. Academic Press, London.

Cline, G. R., Powell, P. E., Szaniszlo, P. J. and Reid, C. P. P. (1983). Comparisons of the abilities of hydroxamic and other organic acids to chelate iron and other ions in soils. *Soil Sci.* **136**, 145–157.

Clutterbuck, B. J. and Simpson, K. (1978). The interactions of water and fertilizer nitrogen in effects on growth pattern and yield of potatoes. *J. Agric. Sci.* 91, 161–172.

Coale, F. J., Evangelou, V. P. and Grove, J. H. (1984). Effects of saline-sodic soil chemistry on soybean mineral composition and stomatal resistance. *J. Environ. Qual.* **13**, 635–639.

Codignola, A., Verotta, L., Panu, P., Maffei, M., Scannerini, S. and Bonfante-Fasolo, P. (1989). Cell wall-bound phenols in roots of vesicular-arbuscular mycorrhizal plants. *New Phytol.* **112**, 221–228.

Cohen, E., Okon, Y., Kigel, J., Nur, I. and Henis, Y. (1980). Increase in dry weight and total nitrogen content in *Zea mays* and *Setaria italica* associated with nitrogen-fixing *Azospirillum* ssp. *Plant Physiol.* **66**, 746–749.

Coke, L. and Whittington, W. J. (1968). The role of boron in plant growth, IV. Interrelationships between boron and indol-3-yl acetic acid in the metabolism of bean radicles. *J. Exp. Bot.*19, 295–308.

Coker, G. T. III and Schaefer, J. (1985). ^{15}N and ^{13}C NMR determination of allantoin metabolism in developing soybean cotyledons. *Plant Physiol.* **77**, 129–135.

Coleman, J. E. (1992). Zinc proteins: enzymes, storage proteins, transcription factors, and replication proteins. *Annu. Rev. Biochem.* **61**, 897–946.

Coleman, J., Evans, D. and Hawes, C. (1988). Plant coated vesicles. *Plant, Cell Environ.* **11**, 669–684.

Coleman, W. J., Govindjee and Gutowsky, H. S. (1987). The location of the chloride binding sites in the oxygen-evolving complex of spinach Photosystem II. *Biochim. Biophys. Acta* **894**, 453–459.

Collier, G. F. and Tibbits, T. W. (1984). Effects of relative humidity and root temperature on calcium concentration and tipburn development in lettuce. *J. Am. Soc. Hortic. Sci.* **109**, 128–131.

Collins, M. and Duke, S. H. (1981). Influence of potassium-fertilization rate and form on photosynthesis and N_2 fixation of alfalfa. *Crop Sci.* **21**, 481–485.

Colman, R. L. and Lazemby, A. (1970). Factors affecting the response of tropical and temperate grasses to fertilizer nitrogen. *Proc. 11th, Int. Grassl. Conf. Surf. Paradise*, pp. 393–397.

Colpaert, J. V. and van Assche, J. A. (1992). Zinc toxicity in ectomycorrhizal *Pinus sylvestris*. *Plant Soil* **143**, 201–211.

Colpaert, J. V. and van Assche, J. A. (1993). The effect of cadmium on ectomycorrhizal *Pinus sylvestris* L. *New Phytol.* **123**, 325–333.

Colpaert, J. V., van Assche, J. A. and Luijtens, K. (1992). The growth of the extramatrical mycelium of ectomycorrhizal fungi and the growth response of *Pinus sylvestris* L. *New Phytol.* **120**, 127–135.

Conroy, J. P. (1992). Influence of elevated atmospheric CO_2 concentrations on plant nutrition. *Aust. J. Bot.* **40**, 445–456.

Constantopoulus, G. (1970). Lipid metabolism of manganese-deficient algae. I. Effect of manganese deficiency on the greening and the lipid composition of *Euglena gracilis* Z. *Plant Physiol.* **45**, 76–80.

Conway, B. E. (1981). 'Ionic Hydration in Chemistry and Biophysics'. Elsevier, Amsterdam.

Coombes, A. J., Lepp, N. W. and Phipps, D. A. (1976). The effect of copper on IAA-oxidase activity in root tissue of barley (*Hordeum vulgare*, cv. Zephyr). *Z. Pflanzenphysiol.* **80**, 236–242.

Cooper, H. D. and Clarkson, D. T. (1989). Cycling of amino-nitrogen and other nutrients between shoots and roots in cereals – a possible mechanism integrating shoot and root in the regulation of nutrient uptake. *J. Exp. Bot.* **40**, 753–762.

Cooper, T. and Bangerth, F. (1976). The effect of Ca and Mg treatment on the physiology, chemical composition and bitter-pit development of 'Cox orange' apples. *Sci. Hortic. (Amsterdam)* **5**, 49–57.

Corden, M. E. (1965). Influence of calcium nutrition on Fusarium wilt of tomato and polygalacturonase activity. *Phytopathology* **55**, 222–224.

Cornish, P. S., So, H. B. and McWilliam, J. R. (1984). Effects of soil bulk density and water regime on root growth and uptake of phosphorus by ryegrass. *Aust. J. Agric. Res.* **35**, 631–644.

Corzo, A., Plasa, R. and Ulrich, W. R. (1991). Extracellular ferricyanide reduction and nitrate reductase activity in the green alga *Monoraphidium braunii*. *Plant Science* **75**, 221–228.

Costigan, P. A. and Mead, G. P. (1987). The requirements of cabbage and lettuce seedlings for potassium in the presence and absence of sodium. *J. Plant Nutr.* **10**, 385–401.

Coughlan, S. J. and Wyn Jones, R. G. (1980). Some responses of *Spinacea oleracea* to salt stress. *J. Exp. Bot.* **31**, 883–893.

Coutts, M. P. and Philipson, J. J. (1977). The influence of mineral nutrition on the root development of trees. III. Plasticity of the root growth in response to changes in the nutrient environment. *J. Exp. Bot.* **28**, 1071–1075.

Coventry, D. R. and Slattery, W. J. (1991). Acidification of soil associated with lupins grown in a crop rotation in north-eastern Victoria. *Aust. J. Agric. Res.* **42**, 391–397.

Cowan, I. R., Raven, J. A., Hartung, W. and Farquhar, G. D. (1982). A possible role for abscisic acid in coupling stomatal conductance and photosynthetic carbon metabolism in leaves. *Aust. J. Plant Physiol.* **9**, 489–498.

Cowling, D. W. and Lockyer, D. R. (1981). Increased growth of ryegrass exposed to ammonia. *Nature (London)* **292**, 337–338.

Cox, F. R. and Reid, P. H. (1964). Calcium–boron nutrition as related to concealed damage in peanuts. *Agron. J.* **56**, 173–176.

Cox, M. S. and Barber, S. A. (1992). Soil phosphorus levels needed for equal P uptake from four soils with different water contents at the same water potential. *Plant Soil* **143**, 93–98.

Crafts, A. S. and Broyer, T. C. (1938). Migration of salts and water into xylem of roots of higher plants. *Am. J. Bot.* **24**, 415–431.

Cram, W. J. (1973). Internal factors regulating nitrate and chloride influx in plant cells. *J. Exp. Bot.* **24**, 328–341.

Cram, W. J. (1983). Characteristics of sulfate transport across plasmalemma and tonoplast of carrot root cells. *Plant Physiol.* **72**, 204–211.

Cramer, G. R. and Nowak, R. S. (1992). Supplemental manganese improves the relative growth, net assimilation and photosynthetic rates of salt-stressed barley. *Physiol. Plant.* **84**, 600–605.

Cramer, G. R., Epstein, E. and Läuchli, A. (1988). Kinetics of root elongation of maize in response to short-term exposure to NaCl and elevated calcium concentrations. *J. Exp. Bot.* **39**, 1513–1522.

Cramer, M. D., Lewis, O. A. M. and Lips, S. H. (1993). Inorganic carbon fixation and metabolism in maize roots as affected by nitrate and ammonium nutrition. *Physiol. Plant.* **89**, 632–639.

Crawford, R. M. M. (1993). Plant survival without oxygen. *Biologist* **40**, 110–114.

Crawford, R. M. M. and Wollenweber-Ratzer, B. (1992). Influence of L-ascorbic acid on post-anoxic growth and survival of chickpea seedlings (*Cicer arietinum* L.). *J. Exp. Bot.* **43**, 703–708.

Crawford, R. M. M. and Zochowski, Z. M. (1984). Tolerance of anoxia and ethanol toxicity in chickpea seedlings (*Cicer arietinum* L.). *J. Exp. Bot.* **35**, 1472–1480.

Creamer, F. L. and Fox, R. H. (1980). The toxicity of banded urea or diammonium phosphate to corn as influenced by soil temperature, moisture and pH. *Soil Sci. Soc. Am. J.* **44**, 296–300.

Creelman, R. A., Mason, H. S., Bensen, R. J., Boyer, J. S. and Mullet, J. E. (1990). Water deficit and abscisic acid cause differential inhibition of shoot *versus* root growth in soybean seedlings. Analysis of growth, sugar accumulation, and gene expression. *Plant Physiol.* **92**, 105–214.

Cresswell, A. and Causton, D. R. (1988). The effect of rooting space on whole plant and leaf growth in Brussels sprouts (*Brassica oleracea* var. *gemmifera*). *Ann. Bot.* **62**, 549–558.

Crisp, P., Collier, G. F. and Thomas, T. H. (1976). The effect of boron on tipburn and auxin activity in lettuce. *Sci. Hortic. (Amsterdam)* **5**, 215–226.

Critchley, C. (1985). The role of chloride in photosystem II. *Biochim. Biophys. Acta* **811**, 33–46.

Crittenden, H. W. and Svec, C. V. (1974). Effect of potassium on the incidence of *Diaporthe sojae* in soybean. *Agron. J.* **66**, 696–698.

Crooke, W. M. and Knight, A. H. (1962). An evaluation of published data on the mineral composition of plants in the light of cation exchange capacities of their roots. *Soil Sci.* **93**, 365–373.

Crossett, R. N. (1968). Effect of light upon the translocation of phosphorus by seedlings of *Hordeum vulgare* (L.). *Aust. J. Biol. Sci.* **21**, 225–233.

Crowley, D. E., Reid, C. P. P. and Szaniszlo, P. J. (1987). Microbial siderophores as iron source for plants. *In* 'Iron Transport in Microbes, Plants and Animals' (G. Winkelmann, D. van der Helm and J. B. Neilands, eds.), pp. 375–386. Verlag Chemie, Weinheim.

Crowley, D. E., Wang, Y. C., Reid, C. P. P. and Szaniszlo, P. J. (1991). Mechanisms of iron acquisition from siderophores by microorganisms and plants. *Plant Soil* **130**, 179–198.

Crowley, D. E., Römheld, V., Marschner, H. and Szaniszlo, P. J. (1992). Root-microbial effects on plant iron uptake from siderophores and phytosiderophores. *Plant Soil* **142**, 1–7.

Crush, J. R. (1974). Plant growth responses to vesicular-arbuscular mycorrhiza. VII. Growth and nodulation of some herbage legumes. *New Phytol.* **73**, 743–752.

Cruz, R. T., Jordan, W. R. and Drew, M. C. (1992). Structural changes and associated reduction of hydraulic conductance in roots of *Sorghum bicolor* L. following exposure to water deficit. *Plant Physiol.* **99**, 203–212.

Cuenca, G., Herrera, R. and Medina, E. (1990). Aluminium tolerance in trees of a tropical cloud forest. *Plant Soil* **125**, 169–175.

Cuevas, E., Brown, S. and Lugo, A. E. (1991). Above- and belowground organic matters storage and production in a tropical pine plantation and a paired broadleaf secondary forest. *Plant Soil* **135**, 257–268.

Cumbus, I. P. (1985). Development of wheat roots under zinc deficiency. *Plant Soil* **83**, 313–316.

Cumbus, I. P. and Nye, P. H. (1982). Root zone temperature effects on growth and nitrate absorption in rape (*Brassica napus* cv. Emerald). *J. Exp. Bot.* **33**, 1138–1146.

Cumbus, I. P., Hornsey, D. J. and Robinson, L. W. (1977). The influence of phosphorus, zinc and manganese on absorption and translocation of iron in water cress. *Plant Soil* **48**, 651–660.

Cumming, J. R. (1993). Growth and nutrition of nonmycorrhizal and mycorrhizal pitch pine (*Pinus rigida*) seedlings under phosphorus limitation. *Tree Physiol.* **13**, 173–187.

Cumming, J. R. and Weinstein, L. H. (1990). Aluminum–mycorrhizal interactions in the physiology of pitch pine seedlings. *Plant Soil* **125**, 7–18.

Curvetto, N. R. and Rauser, W. E. (1979). Isolation and characterization of copper binding proteins from roots of *Agrostis gigantea* tolerant to excess copper. *Plant Physiol.* **63**, Suppl., 59.

Cushman, J. C., Michalowski, C. B. and Bohnert, H. J. (1990). Developmental control of Crassulacean acid metabolism inducibility by salt stress in the common ice plant. *Plant Physiol.* **94**, 1137–1142.

Cutcliffe, J. A. and Gupta, U. C. (1987). Effects of foliar sprays of boron applied at different stages of growth on incidence of brown-heart of rutabagas. *Can. J. Soil Sci.* **67**, 705–708.

Cutting, J. G. M. and Bower, J. P. (1989). The active control of calcium allocation in avocado trees. Progress report. *SA Avocado Grower's Assoc. Yrb.* **12**, 50–52.

Dadson, R. B. and Acquaah, G. (1984). *Rhizobium japonicum*, nitrogen and phosphorus effects on nodulation, symbiotic nitrogen fixation and yield of soybean (*Glycine max* (L.) Merill) in the southern savanna of Ghana. *Field Crops Res.* **9**, 101–108.

Dahse, J., Bernstein, M., Müller, E. and Petzold, U. (1989). On possible functions of electron transport in the plasmalemma of plant cells. *Biochem. Physiol. Pflanzen* **185**, 145–180.

Dalling, M. J., Halloran, G. M. and Wilson, J. H. (1975). The relationship between nitrate reductase activity and grain nitrogen productivity in wheat. *Aust. J. Agric. Res.* **26**, 1–10.

Dalton, D. A., Evans, H. J. and Hanus, F. J. (1985). Stimulation by nickel of soil microbial urease activity and urease and hydrogenase activity in soybeans grown in a low-nickel soil. *Plant Soil* **88**, 245–258.

Dalton, D. A., Post, C. J. and Langeberg, L. (1991). Effects of ambient oxygen and of fixed nitrogen on concentrations of glutathione, ascorbate, and associated enzymes in soybean root nodules. *Plant Physiol.* **96**, 812–818.

Dalton, D. A., Langeberg, L. and Treneman, N. C. (1993). Correlation between the ascorbate–glutathione pathway and effectiveness in legume root nodules. *Physiol. Plant.* **87**, 365–370.

Dambroth, M. and El Bassam, N. (1990). Genotypic variation in plant productivity and consequences for breeding of 'low-input cultivars'. *In* 'Genetic Aspects of Plant Mineral Nutrition' (N. El Bassam, M. Dambroth and B. C. Loughman, eds.), pp. 1–7. Kluwer Academic, Dordrecht.

Danneberg, G., Latus, C., Zimmer, W., Hundeshagen, B., Schneider-Poetsch, Hj. and Bothe, H. (1992). Influence of vesicular-arbuscular mycorrhiza on phytohormone balances in maize (*Zea mays* L.). *J. Plant Physiol.* **141**, 33–39.

Danso, S. K. A., Bowen, G. D. and Sanginga, N. (1992). Biological nitrogen fixation in trees in agro-ecosystems. *Plant Soil* **141**, 177–196.

Danso, S. K. A., Hardarson, G. and Zapata,F. (1993). Misconceptions and practical problems in the use of ^{15}N soil enrichment techniques for estimating N_2 fixation. *Plant Soil* **152**, 25–52.

D'Arcy-Lameta, A. (1982). Etude des exudats racinaires de soja et de lentille. I. Cinetique d'exsudation des composés phenoliques, des amino acides et des sucres, au cours de premiers jours de la vie des plantules. *Plant Soil* **68**, 399–403.

Darrah, P. R. (1991). Models of the rhizosphere. I. Microbial population dynamics around a root releasing soluble and insoluble carbon. *Plant Soil* **133**, 187–199.

Darrah, P. R. (1993). The rhizosphere and plant nutrition: a quantitative approach. *In* 'Plant Nutrition – from Genetic Engineering to Field Practice' (N. J. Barrow, ed.), pp. 3–22. Kluwer Academic, Dordrecht.

Darwinkel, A. (1980a). Grain production of winter wheat in relation to nitrogen and diseases. I. Relationship between nitrogen dressing and yellow rust infection. *Z. Acker-Pflanzenbau* **149**, 299–308.

Darwinkel, A. (1980b). Grain production of winter wheat in relation to nitrogen and diseases. II. Relationship between nitrogen dressing and mildew infection. *Z. Acker-Pflanzenbau* **149**, 309–317.

Da Silva, M. C. and Shelp, B. J. (1990). Xylem-to-phloem transfer of organic nitrogen in young soybean plants. *Plant Physiol.* **92**, 797–801.

Dave, I. C. and Kannan, S. (1980). Boron deficiency and its associated enhancement of RNAase activity in bean plants. *Z. Pflanzenphysiol.* **97**, 261–264.

Davies, D. D. (1986). The fine control of cytosolic pH. *Physiol. Plant.* **67**, 702–706.

Davies, E. (1987). Action potentials as multifunctional signals in plants: a unifiying hypothesis to explain apparently disparate wound responses. *Plant Cell Environ.* **10**, 623–631.

Davies, F. T. jr, Potter, J. R. and Linderman, R. G. (1992). Mycorrhiza and repeated drought exposure affect drought resistance and extraradical hyphae development of pepper plants independent of plant size and nutrient content. *J. Plant Physiol.* **139**, 289–294.

Davies, J. M., Rea, P. A. and Sanders, D. (1991b). Vacuolar proton-pumping pyrophosphatase in *Beta vulgaris* shows vectorial activation by potassium. *FEBS Lett.* **278**, 66–68.

Davies, J. N., Adams, P. and Winsor, G. W. (1978). Bud development and flowering of *Chrysanthemum morifolium* in relation to some enzyme activities and to the copper, iron and manganese status. *Commun. Soil Sci. Plant Anal.* **9**, 249–264.

Davies, K. L., Davies, M. S. and Francis, D. (1991a). The influence of an inhibitor of phytochelatin synthesis on root growth and root meristem activity in *Festuca rubra* L. in response to zinc. *New Phytol* **118**, 565–570.

Davies, W. J. and Meinzer, F. C. (1990). Stomatal responses of plants in drying soil. *Biochem. Physiol. Pflanzen* **186**, 357–366.

Davis, A. M. (1986). Selenium uptake in *Astragalus* and *Lupinus* species. *Agron. J.* **78**, 727–729.

Davis, E. A., Young, J. L. and Rose, S. L. (1984). Detection of high-phosphorus tolerant VAM-fungi colonizing hops and peppermint. *Plant Soil* **81**, 29–36

Day, D. A., Lambers, H., Bateman, J., Carroll, B. J. and Gresshoff, P. M. (1986). Growth comparisons of a super-nodulating soybean (*Glycine max.*) mutant and its wild-type parent. *Physiol. Plant.* **68**, 375–382.

Day, D. A., Carroll, B. J., Delves, A. C. and Gresshoff, P. M. (1989). Relationship between autoregulation and nitrate inhibition of nodulation on soybeans. *Physiol. Plant.* **75**, 37–42.

Deane-Drummond, C. E. (1987). The regulation of sulphate uptake following growth of *Pisum sativum* L. seedlings in S nutrient limiting conditions. Interaction between nitrate and sulphate transport. *Plant Science* **50**, 27–35.

Deane-Drummond, C. E. and Chaffey, N. J. (1985). Characteristics of nitrate uptake into seedlings of pea (*Pisum sativum* L. cv. Feltham First). Changes in net NO_3^- uptake following inoculation with *Rhizobium* and growth in low nitrate concentrations. *Plant Cell Environ.* **8**, 517–523.

Deane-Drummond, C. E. and Gates, P. (1987). A novel technique for identification of sites of anion transport in intact cells and tissues using a fluorescent probe. *Plant Cell Environ.* **10**, 221–227.

DeBoer, D. L. and Duke, S. H. (1982). Effects of sulphur nutrition on nitrogen and carbon metabolism in lucerne (*Medicago sativa* L.). *Physiol. Plant.* **54**, 343–350.

DeBoer, A. H., Prius, H. B. A. and Zanstra, P. E. (1983). Biphasic composition of transroot electrical potential in roots of *Plantago* species: involvement of spatially separated electrogenic pumps. *Planta* **157**, 259–266.

DeBoer, A. H., Katou, K., Mizuno, A., Kojima, H. and Okamoto, H. (1985). The role of electrogenic xylem pump in K^+ absorption from the xylem of *Vigna unguiculata*: the effect of auxin and fusicoccin. *Plant, Cell Environ.* **8**, 579–586.

DeGuzman, C. C. and De la Fuente, R. K. (1984). Polar calcium flux in sunflower hypocotyl segments. I. The effect of auxin. *Plant Physiol.* **76**, 347–352.

DeGuzman, C. C. and De la Fuente, R. K. (1986). Polar calcium flux in sunflower hypocotyl segments. II. The effect of segment orientation, growth, and respiration. *Plant Physiol.* **81**, 408–412.

Dehne, H.-W. und Schönbeck, F. (1979a). Untersuchungen zum Einfluss der endotrophen Mykorrhiza auf Pflanzenkrankheiten. I. Ausbreitung von *Fusarium oxysporum* f. sp. *lycopersici* in Tomaten. *Phytopathol. Z.* **95**, 105–110.

Dehne, H.-W. und Schönbeck, F. (1979b). Untersuchungen zum Einfluss der endotrophen Mykorrhiza auf Pflanzenkrankheiten. II. Phenolstoffwechsel und Lignifizierung. *Phytopathol. Z.* **95**, 210–216.

DeKok, L. J. (1990). Sulfur metabolism in plants exposed to atmospheric sulfur. *In* 'Sulfur Nutrition and Sulfur Assimilation in Higher Plants' (H. Rennenberg, Ch. Brunold, L. J. DeKok and I. Stulen, eds.), pp. 111–130, SPB Academic Publishing, The Hague, The Netherlands.

DeKok, L. J. and Stulen, I. (1993). Role of glutathione in plants under oxidative stress. *In* 'Sulfur Nutrition and Assimilation in Higher Plants' (L. J. De Kok, I. Stulen, H. Rennenberg, C. Brunold and W. E. Rauser, eds.), pp. 125–138. SPB Academic Publishing, The Hague, The Netherlands.

DeKok, L. J., Stahl, K. and Rennenberg, H. (1989). Fluxes of atmospheric hydrogen sulphide to plant shoots. *New Phytol.* **112**, 533–542.

De Kreij, C., Janse, J., van Goor, B. J. and van Doesburg, J. D. J. (1992). The incidence of calcium oxalate crystals in fruit walls of tomato (*Lycopersicon esculentum* Mill.) as affected by humidity, phosphate and calcium supply. *J. Hortic. Sci.* **67**, 45–50.

De Lane, R., Greenway, H., Munns, R. and Gibbs, J. (1982). Ion concentration and carbohydrate status of the elongating leaf tissue of *Hordeum vulgare* growing at high external NaCl. I. Relationship between solute concentration and growth. *J. Exp. Bot.* **33**, 557–573.

Del Gallo, M. and Fabbri, P. (1991). Effect of soil organic matter on chickpea inoculated with *Azospirillum brasilense* and *Rhizobium leguminosarum* bv. *ciceri*. *Plant Soil* **137**, 171–175.

Delhaize, E., Loneragan, J. F. and Webb, J. (1982). Enzymic diagnosis of copper deficiency in subterranean clover. II. A simple field test. *Aust. J. Agric. Res.* **33**, 981–987.

Delhaize, E., Loneragan, J. F. and Webb, J. (1985). Development of three copper metalloenzymes in clover leaves. *Plant Physiol.* **78**, 4–7.

Delhaize, E., Dilworth, M. J. and Webb, J. (1986). The effect of copper nutrition and developmental state on the biosynthesis of diamine oxidase in clover leaves. *Plant Physiol.* **82**, 1126–1131.

Delhaize, E., Craig, S., Beaton, C. D., Bennet, R. J., Jagadish, V. C. and Randall, P. J. (1993a). Aluminum tolerance in wheat (*Triticum aestivum* L.). I. Uptake and distribution of aluminum in root apices. *Plant Physiol.* **103**, 685–693.

Delhaize, E., Ryan, P. R. and Randall, P. J. (1993b). Aluminum tolerance in wheat (*Triticum aestivum* L.). II. Aluminum-stimulated excretion of malic acid from root apices. *Plant Physiol.* **103**, 695–702.

Dell, B. (1981). Male sterility and outer wall structure in copper-deficient plants. *Ann. Bot.* (*London*) **48**, 599–608.

Dell, B. and Wilson, S. A. (1985). Effect of zinc supply on growth of three species of Eucalyptus seedlings and wheat. *Plant Soil* **88**, 377–384.

Deloch, H. W. (1960). Über die analytische Bestimmung des Schwefels in biochemischen Substanzen und die Schwefelaufnahme durch landwirtschaftliche Kulturpflanzen in Abhängigkeit von der Düngung. PhD Thesis, Universität Giessen.

Delrot, S. (1987). Phloem loading: apoplastic or symplastic? *Plant Physiol. Biochem.* **25**, 667–676.

Delrot, S. and Bonnemain, J.-L. (1981). Involvement of protons as a substrate for the sucrose carrier during phloem loading in *Vicia faba* leaves. *Plant Physiol.* **67**, 560–564.

Delwiche, C. C., Johnson, C. M. and Reisenauer, H. M. (1961). Influence of cobalt on nitrogen fixation by Medicago. *Plant Physiol.* **36**, 73–78.

Demming, B. and Winter, K. (1988). Light response of CO_2 assimilation, reduction state of Q and radiationless energy dissipation in intact leaves. *Aust. J. Plant Physiol.* **15**, 151–162.

Demming-Adams, B. and Adams III, W. W. (1992). Photoprotection and other responses of plants to high light stress. *Annu. Rev. Plant Physiol. Plant. Mol. Biol.* **43**, 599–626.

De Neeling, A. J. and Ernst, W. H. O. (1986). Response of an acidic and a calcareous population of *Chamaenerion angustifolium* (L.) Scop. to iron, manganese and aluminium. *Flora (Jena)* **178**, 85–92.

Deng, M., Moureaux, T. and Caboche, M. (1989). Tungstate, a molybdate analog inactivating nitrate reductase, deregulates the expression of the nitrate reductase structural gene. *Plant Physiol.* **91**, 304–309.

Dennis, D. J. and Prasad, M. (1986). The effect of container media on the growth and establishment of *Leucadendrou* 'Safari sunset'. *Acta Hortic.* **185**, 253–257.

Denny, H. J. and Wilkins, D. A. (1987). Zinc tolerance in *Betula* spp. IV. The mechanism of ectomycorrhizal amelioration of zinc toxicity. *New Phytol.* **106**, 545–553.

De-Polli, H., Boyer, C. D. and Neyra, C. A. (1982). Nitrogenase activity associated with roots and stems of field-grown corn (*Zea mays* L.) plants. *Plant Physiol.* **70**, 1609–1613.

Deroche, M.-E. and Carrayol, E. (1988). Nodule phosphoenolpyruvate carboxylase: a review. *Physiol. Plant.* **74**, 775–782.

Desai, N. and Chism, G. W. (1978). Changes in cytokinin activity in the ripening tomato fruit. *J. Food Sci.* **43**, 1324–1326.

Deshaies, R. J., Fish, L. E. and Jagendorf, A. T. (1984). Permeability of chloroplast envelopes to Mg^{2+} – Effects on protein synthesis. *Plant Physiol.* **74**, 956–961.

De Swart, P. H. and van Diest, A. (1987). The rock-phosphate solubilizing capacity of *Pueraria javanica* as affected by soil pH, superphosphate priming effect and symbiotic N_2 fixation. *Plant Soil* **100**, 135–147.

De Veau, E. J. and Burris, J. E. (1989). Photorespiratory rates in wheat and maize as determined by ^{18}O-labelling. *Plant Physiol.* **90**, 500–511.

Devevre, O., Garbaye, J. and Perrin, R. (1993). Experimental evidence of a deleterious soil microflora associated with Norway spruce decline in France and Germany. *Plant Soil* **148**, 145–153.

De Vos, C. R., Lubberding, H. J. and Bienfait, H. F. (1986). Rhizosphere acidification as a response to iron deficiency in bean plants. *Plant Physiol.* **81**, 842–846.

De Vos, Ch. H. R., Vonk, M. J., Vooijs, R. and Schat, H. (1992). Glutathione depletion due to copper-induced phytochelatin synthesis causes oxidative stress in *Silene cucubalus*. *Plant Physiol.* **98**, 853–858.

De Vos, C. H. R., Schat, H., de Waal, M. A. M., Vooijs, R. and Ernst, W. H. O. (1991). Increased resistance to copper-induced damage of the root cell plasmalemma in copper tolerant *Silene cucubalus*. *Physiol. Plant.* **82**, 523–528.

De Weger, L. A., Schippers, B. and Lugtenberg, B. (1987). Plant growth stimulation by biological interference in iron metabolism in the rhizosphere. *In* 'Iron Transport in Microbes, Plants and Animals' (G. Winkelmann *et al.*, eds.), pp. 387–400. Verlag Chemie, Weinheim.

De Wet, E., Robbertse, P. J. and Groeneveld, H. T. (1989). The influence of temperature and boron on pollen germination in *Mangifera indica* L. *S. Afr. Tydskr. Plant Grond* **6**, 228–234.

De Willigen, P. and van Noordwijk, M. (1989). Model calculations on the relative importance of internal longitudinal diffusion for aeration of roots of non-wetland plants. *Plant Soil* **113**, 111–119.

Dhugga, K. S., Waines, J. G. and Leonard, R. T. (1988). Nitrate absorption by corn roots. Inhibition by phenylglyoxal. *Plant Physiol.* **86**, 759–763.

Dias, M. A. and Oliveira, M. M. (1987). Nitrate reductase and petiole nitrate as indicator of the nitrogen nutrition status of field grown sugar beet. *Agron. Lusit.* **42**, 275–284.

Dickinson, C. D., Altabella, T. and Chrispeels, M. J. (1991). Slow-growth phenotype of transgenic tomato expressing apoplastic invertase. *Plant Physiol.* **95**, 420–425.

Dickinson, D. B. (1978). Influence of borate and pentaerythriol concentrations on germination and tube growth of *Lilium longiflorum* pollen. *J. Am. Soc. Hortic. Sci.* **103**, 413–416.

Diem, H. G. and Dommergues, Y. R. (1990). Current and potential uses and management of Casuarinaceae in the tropics and subtropics. *In* 'The Biology of Frankia and Actinorrhizal plants' (C. R. Schwintzer and J. D.Tjepkema, eds.), pp. 317–342. Academic Press, New York.

Dietz, K.-J. (1989). Recovery of spinach leaves from sulfate and phosphate deficiency. *J. Plant Physiol.* **134**, 551–557.

Dijkshoorn, W. and van Wijk, A. L. (1967). The sulphur requirement of plants as evidenced by the sulphur–nitrogen ratio in the organic matter. A review of published data. *Plant Soil* **26**, 129–157.

Dilworth, M. J., and Bisseling, T. (1984). Cobalt and nitrogen fixation in *Lupinus angustifolius* L. III. DNA and methionine in bacteroids. *New Phytol.* **98**, 311–316.

Dilworth, M. J., Robson, A. D. and Chatel, D. L. (1979). Cobalt and nitrogen fixation in *Lupinus angustifolius* L.II. Nodule formation and functions. *New Phytol.* **83**, 63–79.

Dilworth, M. J., Eady, R. R. and Eldridge, M. E. (1988). The vanadium nitrogenase of *Azotobacter chroococcum*. Reduction of acetylene and ethylene to ethane. *Biochem. J.* **249**, 745–751.

Dinkelaker, B. and Marschner, H. (1992). In vivo demonstration of acid phosphatase activity in the rhizosphere of soil-grown plants. *Plant Soil* **144**, 199–205.

Dinkelaker, B., Römheld, V. and Marschner, H. (1989). Citric acid excretion and precipitation of calcium citrate in the rhizosphere of white lupin (*Lupinus albus* L.). *Plant Cell Environ.* **12**, 285–292.

Dionisio-Sese, M. L., Fukuzawa, H. and Miyachi, S. (1990). Light-induced carbonic anhydrase expression in *Chlamydomonas reinhardtii*. *Plant Physiol.* **94**, 1103–1110.

Dixon, N. E., Gazola, C., Blakeley, R. L. and Zerner, B. (1975). Jack bean urease (EC 3.5.1.5), a metalloenzyme. A simple biological role for nickel? *J. Am. Chem. Soc.* **97**, 4131–4133.

Dixon, N. E., Hinds, J. A., Fihelly, A. K., Gazola, C., Winzor, D. J., Blakeley, R. L. and Zerner, B. (1980). Jack bean urease (EC 3.5.1.5). IV. The molecular size and the mechanism of inhibition by hydroxamic acids. Spectrophotometric tiration of enzymes with reversible inhibitors. *Can. J. Biochem.* **58**, 1323–1334.

Dixon, R. K. and Buschena, C. A. (1988). Response of ectomycorrhizhal *Pinus banksiana* and *Picea glauca* to heavy metals in soil. *Plant Soil* **105**, 265–271.

Dixon, R. K., Garrett, H. E. and Cox, G. S. (1988). Cytokinins in the root pressure exudate of *Citrus jambhiri* Lush. colonized by vesicular-arbuscular mycorrhizae. *Tree Physiol.* **4**, 9–18.

Dixon, R. K., Garrett, H. E. and Cox, G. S. (1989). Boron fertilization, vesicular-arbuscular mycorrhizal colonization and growth of *Citrus jambhiri* Lush. *J. Plant Nutr.* **12**, 687–700.

Djordjevic, M. A. and Weinman, J. J. (1991). Factors determining host recognition in the clover–*Rhizobium* symbiosis. *Aust. J. Plant Physiol.* **18**, 543–557.

Döbereiner, J. (1966). Manganese toxicity effects on nodulation and nitrogen fixation of beans (*Phaseolus vulgaris* L.) in acid soils. *Plant Soil* **24**, 153–166.

Döbereiner, J. (1983). Dinitrogen fixation in rhizosphere and phyllosphere associations. *In* 'Encyclopedia of Plant Physiology, New Series' (A. Läuchli and R. L. Bieleski, eds.), Vol. 15A, pp. 330–350. Springer-Verlag, Berlin and New York.

Dockendorf, H. and Höfner, W. (1990). Einfluss von Bikarbonat auf die subzelluläre Verteilung von blatt- und wurzelappliziertem Eisen bei Sonnenblumen (*Helianthus annuus* L.). *Z. Pflanzenernähr. Bodenk.* **153**, 313–317.

Dogar, M. A. and van Hai, T. (1980). Effect of P, N and HCO_3^- levels in the nutrient solution on rate of Zn absorption by rice roots and Zn content in plants. *Z. Pflanzenphysiol.* **98**, 203–212.

Doll, S., Rodier, F. and Willenbrink, J. (1979). Accumulation of sucrose in vacuoles isolated from red beet tissue. *Planta* **144**, 407–411.

Domingo, A. L., Nagatomo, Y., Tamai, M. and Takaki, H. (1992). Free-tryptophan and indolacetic acid in zinc-deficient radish shoots. *Soil Sci. Plant Nutr.* **38**, 261–267.

Dorenstouter, H., Pieters, G. A. and Findenegg, G. R. (1985). Distribution of magnesium between chlorophyll and other photosynthetic functions in magnesium deficient 'sun' and 'shade' leaves of poplar. *J. Plant Nutr.* **8**, 1088–1101.

Dosskey, M. G., Linderman, R. G. and Boersma, L. (1990). Carbon-sink stimulation of photosynthesis in Douglas-fir seedlings by some ectomycorrhizas. *New Phytol.* **115**, 269–274.

Dosskey, M. G., Boersma, L. and Linderman, R. G. (1991). Role for the photosynthate demand of ectomycorrhizas in the response of Douglas fir seedlings to drying soil. *New Phytol.* **117**, 327–334.

Douds jr, D. D., Johnson, C. R. and Koch, K. E. (1988). Carbon cost of the fungal symbiont relative to net leaf P accumulation in a split-root VA mycorrhizal symbiosis. *Plant Physiol.* **86**, 491–496.

Douglas, T. J. and Sykes, S. R. (1985). Phospholipid, galactolipid and free sterol composition of fibrous roots from citrus genotypes differing in chloride exclusion ability. *Plant Cell Environ.* **8**, 693–699.

Douglas, T. J. and Walker, R. R. (1983). 4-Desmethylsterol composition of citrus rootstocks of different salt exclusion capacity. *Physiol. Plant.* **58**, 69–74.

Downton, W. J. S. (1977). Photosynthesis in salt-stressed grapevines. *Aust. J. Plant Physiol.* **4**, 183–192.

Downton, W. J. S. (1985). Growth and mineral composition of the sultana grapevine as influenced by salinity and rootstock. *Aust. J. Agric. Res.* **36**, 425–434.

Dracup, M. (1991). Increasing salt tolerance of plants through cell culture requires greater understanding of tolerance mechanisms. *Aust. J. Plant Physiol.* **18**, 1–15.

Dracup, M. (1993). Why does in vitro cell selection not improve the salt tolerance of plants? *In* 'Genetic Aspects of Plant Mineral Nutrition' (P. J. Randall, E. Delhaize, R. A. Richards and R. Munns, eds.), pp. 137–142. Kluwer Academic, Dordrecht.

Drechsel, P. and Zech, W. (1991). Foliar nutrient of broad-leaved tropical trees: a tabular review. *Plant Soil* **131**, 29–46.

Drechsel, H., Metzger, J., Freund, S., Jung, G., Boelaert, J. R. and Winkelmann, G. (1991). Rhizoferrin – a novel siderophore from the fungus *Rhizopus microsporus* var. *rhizopodiformis*. *Biol. Metals* **4**, 238–243.

Dressel, J. and Jung, J. (1979). Gehaltsniveau an Vitaminen des B-Komplexes in Abhängigkeit von Stickstoffzufuhr und Standort. *Landwirtsch. Forsch., Sonderh.* **35**, 261–270.

Drew, M. C. (1988). Effect of flooding and oxygen deficiency on plant mineral nutrition. *In* 'Advances in Plant Nutrition' Vol. 3 (B. Tinker and A. Läuchli, eds.), pp. 115–159. Praeger Publishers, New York.

Drew, M. C. (1990). Sensing soil oxygen. *Plant, Cell Environ.* **13**, 681–693.

Drew, M. C. and Dikumwin, E. (1985). Sodium exclusion from the shoots by roots of *Zea mays* (cv. LG 11) and its break-down with oxygen deficiency. *J. Exp. Bot.* **36**, 55–62.

Drew, M. C. and Fourcy, A. (1986). Radial movement of cations across aerenchyma-tous roots of *Zea mays* measured by electron probe X-ray microanalysis. *J. Exp. Bot.* **37**, 823–831.

Drew, M. C. and Läuchli, A. (1987). The role of the mesocotyl in sodium exclusion from the shoot of *Zea mays* L. (cv. Pioneer 3906). *J. Exp. Bot.* **38**, 409–418.

Drew, M. C. and Saker, L. R. (1975). Nutrient supply and the growth of the seminal root system in barley. II. Localized compensatory increases in lateral root growth and rates of nitrate uptake when nitrate supply is restricted to only part of the root system. *J. Exp. Bot.* **26**, 79–90.

Drew, M. C. and Saker, L. R. (1978). Nutrient supply and the growth of the seminal root system in barley. III. Compensatory increase in growth of lateral roots, and in rates of phosphate uptake, in response to a localized supply of phosphate. *J. Exp. Bot.* **29**, 435–451.

Drew, M. C. and Saker, L. R. (1984). Uptake and long-distance transport of phosphate, potassium and chloride in relation to internal ion concentration in barley: evidence for non-allosteric regulation. *Planta* **160**, 500–507.

Drew, M. C. and Saker, L. R. (1986). Ion transport to the xylem in aerenchymatous roots of *Zea mays* L. *J. Exp. Bot.* **37**, 22–33.

Drew, M. C. and Sisworo, E. J. (1979). The development of waterlogging damage in young barley plants in relation to plant nutrient status and changes in soil properties. *New Phytol.* **82**, 301–314.

Drew, M. C., Jackson, M. B. and Giffard, S. (1979). Ethylene-promoted adventitious rooting and development of cortical air spaces (aerenchyma) in roots may be adaptive responses to flooding in *Zea mays* L. *Planta* **147**, 83–88.

Drew, M. C., Saker, L. R., Barber, S. A. and Jenkins, W. (1984). Changes in the kinetics of phosphate and potassium absorption in nutrient-deficient barley roots measured by a solution-depletion technique. *Planta* **160**, 490–499.

Drew, M. C., He, C.-J. and Morgan, P. W. (1989). Decreased ethylene biosynthesis, and induction of aerenchyma, by nitrogen- or phosphate-starvation in adventitious roots of *Zea mays* L. *Plant Physiol.* **91**, 266–271.

Drew, M. C., Webb, J. and Saker, L. R. (1990). Regulation of K^+ uptake and transport to the xylem in barley roots: K^+ distribution determined by electron probe X-ray microanalysis of frozen-hydrated cells. *J. Exp. Bot.* **41**, 815–825.

Dreyer, D. L. and Campbell, B. C. (1987). Chemical basis of host-plant resistance to aphids. *Plant Cell Environ.* **10**, 353–361.

Droillard, M. J. and Paulin, A. (1990). Isoenzymes of superoxide dismutase in mitochondria and peroxisomes isolated from petals of carnation (*Dianthus caryophyllus*) during senescence. *Plant Physiol*. **94**, 1187–1192.

Droppa, M., Terry, N. and Horvath, G. (1984). Effects of Cu deficiency on photosynthetic electron transport. *Proc. Natl. Acad. Sci*. **81**, 2369–2373.

Drüge, U. and Schönbeck, F. (1992). Effect of vesicular-arbuscular mycorrhizal infection on transpiration, photosynthesis and growth of flax (*Linum usitatissimum* L.) in relation to cytokinin levels. *J. Plant Physiol*. **141**, 40–48.

Dubois, D., Winzeler, M. and Nösberger, J. (1990). Fructan accumulation and sucrose: sucrose fructosyltransferase activity in stems of spring wheat genotypes. *Crop Sci*. **30**, 315–319.

Duchesne, L. C., Ellis, B. E. and Peterson, R. L. (1989). Disease suppression by the ectomycorrhizal fungus *Paxillus involutus*: contribution of oxalic acid. *Can. J. Bot*. **67**, 2726–2730.

Dudel, G. and Kohl, G. (1974). Über die Verteilung der Nitratreduktaseaktivität in Wurzel und Blatt bei *Hordeum vulgare* L. und ihre Abhängigkeit vom exogenen Nitratangebot. *Arch. Acker-Pflanzenbau Bodenkd*. **18**, 233–242.

Duff, S. M. G., Moorhead, G. B. G., Lefebvre, D. D. and Plaxton, W. C. (1989). Phosphate starvation inducible 'bypasses' of adenylate and phosphate dependent glycolytic enzymes in *Brassica nigra* suspension cells. *Plant Physiol*. **90**, 1275–1278.

Dugger, W. M. (1983). Boron in plant metabolism. *In* 'Encyclopedia of Plant Physiology, New Series' (A. Läuchli and R. L. Bieleski, eds.), Vol.15B, pp. 626–650. Springer-Verlag, Berlin.

Dugger, W. M. and Palmer, R. L. (1985). Effect of boron on the incorporation of glucose by cotton fibers grown *in vitro*. *J. Plant Nutr*. **8**, 311–325.

Duke, M. V. and Salin, M. L. (1985). Purification and characterization of an iron-containing superoxide dismutase from a eucaryote, *Ginkgo biloba*. *Arch. Biochem. Biophys*. **243**, 305–314.

Dumon, J. C. and Ernst, W. H. O. (1988). Titanium in plants. *J. Plant Physiol*. **133**, 203–209.

Dunlap, J. R. and Robacker, K. M. (1990). Abscisic acid alters the metabolism of indole-3-acetic acid in senescing flowers of *Cucumis melo* L. *Plant Physiol*. **94**, 870–874.

Dunlop, J. (1989). Phosphate and membrane electropotentials in *Trifolium repens* L. *J. Exp. Bot*. **40**, 803–807.

Dunlop, J. and Bowling, D. J. F. (1978). Uptake of phosphate by white clover. II. The effect of pH on the electrogenic phosphate pump. *J. Exp. Bot*. **29**, 1147–1153.

Duong, T. P. and Diep, C. N. (1986). An inexpensive cultural system using ash for cultivation of soybean (*Glycine max*.(L.) Merrill) on acid clay soils. *Plant Soil* **96**, 225–237.

Duponnois, R. and Garbaye, J. (1991). Effect of dual inoculation of Douglas fir with the ectomycorrhizal fungus *Laccaria laccata* and mycorrhization helper bacteria (MHB) in two bare-root forest nurseries. *Plant Soil* **138**, 169–176.

Durrant, M. J., Draycott, A. P. and Milford, G. F. J. (1978). Effect of sodium fertilizer on water status and yield of sugar beet. *Ann. Appl. Biol*. **88**, 321–328.

Duvick, D. N., Kleese, R. A. and Frey, N. M. (1981). Breeding for tolerance of nutrient imbalance and constraints to growth in acid, alkaline and saline soils. *J. Plant Nutr*. **4**, 111–129.

Duyzer, J. H., Verhagen, H. L. M., Weststrate, J. H. and Bosveld, F. C. (1992). Measurement of the dry deposition flux of NH_3 on coniferous forest. *Environ. Pollut*. **75**, 3–13.

Dwelle, R. B., Kleinkopf, G. E., Steinhorst, R. K., Pavek, J. J. and Hurley, P. J. (1981). The influence of physiological processes on tuber yield of potato clones (*Solanum tuberosum* L.). Stomatal diffusive resistance, stomatal conductance, gross photosynthetic rate, leaf canopy, tissue nutrient levels, and tuber enzyme activities. *Potato Res.* **24**, 33–47.

Dwivedi, R. W. and Takkar, P. N. (1974). Ribonuclease activity as an index of hidden hunger of zinc in crops. *Plant Soil* **40**, 173–181.

Dwyer, L. M., Tollenaar, M. and Stewart, D. A. (1991). Changes in plant density dependence of leaf photosynthesis of maize (*Zea mays* L.) hybrids, 1959–1988. *Can. J. Plant Sci.* **71**, 1–11.

Eady, R. R., Robson, R. L., Richardson, T. H., Miller, R. W. and Hawkins, M. (1987). The vanadium nitrogenase of *Azotobacter chroococcum*. Purification and properties of the VFe protein. *Biochem. J.* **244**, 197–207.

Eastman, P. A. K., Peterson, C. A. and Dengler, N. G. (1988). Suberized bundle sheaths in grasses (*Poaceae*) of different photosynthetic types. II. Apoplastic permeability. *Protoplasma* **142**, 112–126.

Eberl, D., Preissler, M., Steingraber, M. and Hampp, R. (1992). Subcellular compartmentation of pyrophosphate and dark/light kinetics in comparison to fructose 2,6-bisphosphate. *Physiol. Plant.* **84**, 13–20.

Edelbauer, A. (1980). Auswirkung von abgestuftem Schwefelmangel auf Wachstum, Substanzbildung und Mineralstoffgehalt von Tomate (*Lycopersicon esculentum* Mill.) in Nährlösungskultur. *Die Bodenkultur* **31**, 229–241.

Edelmann, H. G. and Kutschera, U. (1993). Tissue pressure and cell-wall metabolism in auxin-mediated growth of sunflower hypocotyls. *J. Plant Physiol.* **142**, 467–473.

Edmeades, D. C., Wheeler, D. M., Blamey, F. P. C. and Christie, R. A. (1991). Calcium and magnesium amelioration of aluminium toxicity in Al-sensitive and Al-tolerant wheat. *In* 'Plant–Soil Interactions at Low pH' (R. J. Wright, V. C. Baligar and R. P. Murrmann, eds.), pp. 755–761. Kluwer Academic, Dordrecht, The Netherlands.

Edwards, D. G. and Asher, C. J. (1982). Tolerance of crop and pasture species to manganese toxicity. *In* 'Proceedings of the Ninth Plant Nutrition Colloquium, Warwick, England' (A. Scaife, ed.), pp. 145–150. Commonwealth Agricultural Bureau, Farnham Royal, Bucks, UK.

Edwards, G. and Walker, D. (1983). 'C$_3$, C$_4$: Mechanisms, and Cellular and Environmental Regulation, of Photosynthesis'. Blackwell, Oxford.

Edwards, M. C., Smith, G. N. and Bowling, D. J. F. (1988). Guard cells extrude protons prior to stomatal opening – a study using fluorescence microscopy and pH microelectrodes. *J. Exp. Bot.* **39**, 1541–1547.

Egmond, F. van and Breteler, H. (1972). Nitrate reductase activity and oxalate content of sugar-beet leaves. *Neth. J. Agric. Sci.* **20**, 193–198.

Ehret, D. L., Redmann, R. E., Harvey, B. L. and Cipywnyk, A. (1990). Salinity-induced calcium deficiency in wheat and barley. *Plant Soil* **28**, 143–151.

Elawad, S. H., Gascho, G. J. and Street, J. J. (1982a). Response of sugarcane to silicate source and rate. I. Growth and yield, *Agron. J.* **74**, 481–484.

Elawad, S. H., Stret, J. J. and Gascho, G. J. (1982b). Response of sugarcane to silicate source and rate. II. Leaf freckling and nutrient content. *Agron. J.* **74**, 484–487.

El Bassam, N., Dambroth, M. and Loughman, B. C. (eds.) (1990). 'Genetic Aspects of Plant Mineral Nutrition'. Kluwer Academic, Dordrecht.

El-Baz, F. K., Maier, P., Wissemeier, A. H. and Horst, W. J. (1990). Uptake and distribution of manganese applied to leaves of *Vicia faba* (cv. Herzfreya) and *Zea mays* (cv. Regent) plants. *Z. Pflanzenernähr. Bodenk*. **153**, 279–282.

Elias, K. S. and Safir, G. R. (1987). Hyphal elongation of *Glomus fasciculatus* in response to root exudates. *Appl. Environ. Microbiol*. **53**, 1928–1933.

Eliasson, L. and Bollmark, M. (1988). Ethylene as a possible mediator of light-induced inhibition of root growth. *Physiol. Plant*. **72**, 605–609.

Eliasson, L., Bertell, G. and Bolander, E. (1989). Inhibitory action of auxin and root elongation not mediated by ethylene. *Plant Physiol*. **91**, 310–314.

Elliott, G. C. and Läuchli, A. (1985). Phosphorus efficiency and phosphate-iron interaction in maize. *Agron. J*. **77**, 399–403.

Elliott, G. C. and Läuchli, A. (1986). Evaluation of an acid phosphatase assay for detection of phosphorus deficiency in leaves of maize (*Zea mays* L.). *J. Plant Nutr*. **9**, 1469–1477.

Elliott, G. C., Lynch, J. and Läuchli, A. (1984). Influx and efflux of P in roots of intact maize plants. *Plant Physiol*. **76**, 336–341.

Ellis, R. C. (1971). The mobilization of iron by extracts of Eucalyptus leaf litter. *J. Soil Sci*. **22**, 8–22.

El-Sheikh, A. M., Ulrich, A., Awad, S. K. and Mawardy, A. E. (1971). Boron tolerance of squash, melon, cucumber and corn. *J. Am. Soc. Hortic. Sci*. **96**, 536–537.

Elstner, E. F. (1982). Oxygen activation and oxygen toxicity. *Annu. Rev. Plant Physiol*. **33**, 73–96.

Elwali, A. M. O. and Gascho, G. J. (1984). Soil testing, foliar analysis, and DRIS as guides for sugarcane fertilization. *Agron. J*. **76**, 466–470.

Emadian, S. F. and Newton, R. J. (1989). Growth enhancement of loblolly pine (*Pinus taeda* L.) seedlings by silicon. *J. Plant Physiol*. **134**, 98–103.

Emmert, F. H. (1972). Effect of time, water flow, and pH on centripetal passage of radio-phosphorus across roots of intact plants. *Plant Physiol*. **50**, 332–335.

Ender, Ch., Li, M. Q., Martin, B., Povh, B., Nobiling, R., Reiss, H.-D. and Traxel, K. (1983). Demonstration of polar zinc distribution in pollen tubes of *Lilium longiflorum* with the Heidelberg proton microprobe. *Protoplasma* **116**, 201–203.

Engels, C. and Marschner, H. (1986). Allocation of photosynthates to individual tubers of *Solanum tuberosum* L. III. Relationship between growth rate of individual tubers, tuber weight and stolon growth prior to tuber initiation. *J. Exp. Bot*. **37**, 1813–1822.

Engels, C. and Marschner, H. (1987). Effects of reducing leaf area and tuber number on the growth rates of tubers on individual potato plants. *Potato Res*. **30**, 177–186.

Engels, C. and Marschner, H. (1990). Effect of suboptimal root zone temperatures at varied nutrient supply and shoot meristem temperature on growth and nutrient concentrations in maize seedlings (*Zea mays* L.). *Plant Soil* **126**, 215–225.

Engels, C. and Marschner, H. (1992a). Root to shoot translocation of macronutrients in relation to shoot demand in maize (*Zea mays* L.) grown at different root zone temperatures. *Z. Pflanzenernähr. Bodenk*. **155**, 121–128.

Engels, C. and Marschner, H. (1992b). Adaptation of potassium translocation into the shoot of maize (*Zea mays*) to shoot demand: evidence for xylem loading as a regulating step. *Physiol. Plant* **86**, 263–268.

Engels, C. and Marschner, H. (1993). Influence of the form of nitrogen supply on root uptake and translocation of cations in the xylem exudate of maize (*Zea mays* L.). *J. Exp. Bot*. **44**, 1695–1701.

Engels, C., Münkle, L. and Marschner, H. (1992). Effect of root zone temperature and shoot demand on uptake and xylem transport of macronutrients in maize (*Zea mays* L.). *J. Exp. Bot*. **43**, 537–547.

English, J. E. and Barker, A. V. (1987). Ion interactions in calcium-efficient and calcium-inefficient tomato lines. *J. Plant Nutr.* **10**, 857–869.

English, J. E. and Maynard, D. N. (1981). Calcium efficiency among tomato strains. *J. Am. Soc. Hortic. Sci.* **106**, 552–557.

Engvild, K. C. (1986). Chlorine-containing natural compounds in higher plants. *Phytochemistry* **25**, 781–791.

Engvild, K. C. (1989). The death hormone hypothesis. *Physiol. Plant.* **77**, 282–285.

Enstone, D. E. and Peterson, C. A. (1992). The apoplastic permeability of root apices. *Can. J. Bot.* **70**, 1502–1512.

Epstein, E. (1965). Mineral metabolism. *In* 'Plant Biochemistry' (J. Bonner and J. E. Varner, eds.), pp. 438–466. Academic Press, London.

Epstein, E. (1972). 'Mineral Nutrition of Plants: Principles and Perspectives'. Wiley, New York.

Epstein, E. (1994). The anomaly of silicon in plant biology. *Proc. Natl. Acad. Sci. U.S.A.* **91**, 11–17.

Epstein, E., Rains, D. W. and Elzam, O. E. (1963). Resolution of dual mechanisms of potassium absorption by barley roots. *Proc. Natl. Acad. Sci. U.S.A.* **49**, 684–692.

Epstein, E., Norlyn, J. D., Rush, D. W., Kingsbury, R. W., Kelley, D. B., Cunningham, G. A. and Wrona, A. F. (1980). Saline culture of crops: a genetic approach. *Science* **210**, 399–404.

Erdei, L., Stuiver, B. and Kuiper, P. J. C. (1980). The effect of salinity on lipid composition and on activity of Ca^{2+} and Mg^{2+}-stimulated ATPases in salt-sensitive and salt-tolerant *Plantago* species. *Physiol. Plant.* **49**, 315–319.

Erdei, L. and Trivedi, S. (1991). Caesium/potassium selectivity in wheat and lettuce of different K^+ status. *J. Plant Physiol.* **138**, 696–699.

Ergle, D. R. and Eaton, F. M. (1951). Sulfur nutrition of cotton. *Plant Physiol.* **26**, 639–654.

Ericsson, A., Norden, L.-G., Näsholm, T. and Walheim, M. (1993). Mineral nutrient imbalances and arginine concentration in needles of *Picea abies* (L.) Karst. from two areas with different levels of airborne deposition. *Trees* **8**, 67–74.

Eriksson, M. (1979). The effect of boron on nectar production and seed setting of red clover (*Trifolium pratense* L.). *Swed. J. Agric. Res.* **9**, 37–41.

Ernst, M., Römheld, V. and Marschner, H. (1989). Estimation of phosphorus uptake capacity by different zones of the primary root of soil-grown maize (*Zea mays* L.). *Z. Pflanzenernähr. Bodenk.* **152**, 21–25.

Ernst, W. H. O. (1982). Schwermetallpflanzen. *In* 'Pflanzenökologie und Mineralstoffwechsel' (H. Kinzel, ed.), pp. 472–506. Ulmer, Stuttgart.

Ernst, W. H. O. (1993). Ecological aspect of sulfur in higher plants: the impact of SO_2 and the evolution of the biosynthesis of organic sulfur compounds on populations and ecosystems. *In* 'Sulfur Nutrition and Assimilation in Higher Plants' (J. L. De Kok, I. Stulen, H. Rennenberg, C. Brunold and W. E. Rauser, eds.), pp. 295–313. SPB Academic Publishing, The Hague, The Netherlands.

Ernst, W. H. O. and Joosse-van Damme, E. N. G. (1983). 'Umweltbelastung durch Mineralstoffe – Biologische Effekte.' Fischer, Stuttgart.

Eschrich, W. (1976). 'Strasburger's Kleines Botanisches Praktikum für Anfänger.' Fischer, Stuttgart.

Eschrich, W. (1984). Untersuchungen zur Regulation des Assimilattransports. *Ber. Dtsch. Bot. Ges.* **97**, 5–14.

Eshel, A. (1985). Response of *Sueda aegyptiaca* to KCl, NaCl and Na_2SO_4 treatments. *Physiol. Plant.* **64**, 308–315.

Eskew, D. L., Welch, R. M. and Norwell, W. A. (1984). Nickel in higher plants. Further evidence for an essential role. *Plant Physiol.* **76**, 691–693.

Eustice, D. C., Kull, F. J. and Shrift, A. (1981). In vitro incorporation of selenomethionine into protein by *Astragalus polysomes*. *Plant Physiol.* **67**, 1059–1060.

Evans, D. E., Briars, S.-A. and Williams, L. E. (1991). Active calcium transport by plant cell membranes. *J. Exp. Bot.* **42**, 285–303.

Evans, H. J. and Barber, L. E. (1977). Biological nitrogen fixation for food and fiber production. *Science*, **197**, 332–339.

Evans, H. J. and Wildes, R. A. (1971). Potassium and its role in enzyme activation. *Proc. 8th Colloq. Int. Potash Inst. Bern*, pp. 13–39.

Evans, J., Scott, B. J. and Lill, W. J. (1987). Manganese tolerance in subterranean clover (*Trifolium subterraneum* L.) genotypes grown with ammonium nitrate or symbiotic nitrogen. *Plant Soil* **97**, 207–215.

Evans, P. T. and Malmberg, R. L. (1989). Do polyamines have roles in plant development? *Annu. Rev. Plant Physiol. Plant Mol. Biol.* **40**, 235–269.

Ewald, E. (1964). Die Wirkung unterschiedlicher Stickstoffdüngung auf Sommerweizen unter besonderer Berücksichtigung der Kornproteine und der Backqualität. PhD Thesis, Universität Hohenheim.

Ewart, J. A. D. (1978). Glutamin and dough tenacity. *J. Sci. Food Agric.* **29**, 551–556.

Ewens, M. and Leigh, R. A. (1985). The effect of nutrient solution composition on the length of root hairs of wheat (*Triticum aestivum* L.). *J. Exp. Bot.* **36**,713–724.

Eyster, C., Brown, T. E., Tanner, H. A. and Hood, S. L. (1958). Manganese requirement with respect to growth, Hill reaction and photosynthesis. *Plant Physiol.* **33**, 235–241.

Faber, B. A., Zasoski, R. J., Munns, D. N. and Shackel, K. (1991). A method for measuring hyphal nutrient and water uptake in mycorrhizal plants. *Can. J. Bot.* **69**, 87–94.

Fackler, U., Goldbach, H., Weiler, E. W. and Amberger, A. (1985). Influence of boron deficiency on indol-3yl-acetic acid and abscisic acid levels in root and shoot tips. *J. Plant Physiol.* **119**, 295–299.

Faiz, S. M. A. and Weatherley, P. E. (1982). Root contraction in transpiring plants. *New Phytol.* **92**, 333–344.

Falchuk, K. H., Ulpino, L., Mazus, B. and Valee, B. L. (1977). *E. gracilis* RNA polymerase. I. A zinc metalloenzyme. *Biochem. Biophys. Res. Commun.* **74**, 1206–1212.

Faller, N. (1972). Schwefeldioxid, Schwefelwasserstoff, nitrose Gase und Ammoniak als ausschliessliche S- bzw. N-Quellen der höheren Pflanzen. *Z. Pflanzenernähr. Bodenk.* **131**, 120–130.

Fankhauser, H. and Brunold, C. (1978). Localization of adenosine 5′-phosphosulfate sulfotransferase in spinach leaves. *Planta* **143**, 285–289.

Fankhauser, H. and Brunold, C. (1979). Localization of *O*-acetyl-L-serine sulfhydroylase in *Spinacia oleracea* L. *Plant Sci. Lett.* **14**, 185–192.

Farley, R. F. and Draycott, A. P. (1973). Manganese deficiency of sugar beet in organic soil. *Plant Soil* **38**, 235–244.

Farrar, J. F. and Minchin, P. E. H. (1991). Carbon partitioning in split root systems of barley: relation to metabolism. *J. Exp. Bot.* **42**, 1261–1268

Farrow, R. P., Johnson, J. H., Gould, W. A. and Charbonneu, J. E. (1971). Detinning in canned tomatoes caused by accumulations of nitrate in the fruit. *J. Food Sci.* **36**, 341–345.

Farwell, A. J., Farina, M. P. W. and Channon, P. (1991). Soil acidity effects on premature germination in immature maize grain. *In* 'Plant-Soil Interactions at Low pH' (V. C. Baligar and R. P. Murrmann, eds.), pp. 355–361. Kluwer Academic, Dordrecht, The Netherlands.

Fate, G., Chang, M. and Lynn, D. G. (1990). Control of germination in *Striga asiatica*: chemistry of spatial definition. *Plant Physiol.* **93**, 201–207.

Faust, M. and Klein, J. D. (1974). Levels and sites of metabolically active calcium in apple fruit. *J. Am. Soc. Hortic. Sci.* **99**, 93–94.

Favilli, F. and Messini, A.(1990). Nitrogen fixation at phyllospheric level in coniferous plants in Italy. *Plant Soil* **128**, 91–95.

Federico, R., Cona, A., Angelini, R., Schinina, M. E. and Giartosio, A. (1990). Characterization of maize polyamine oxidase. *Phytochemistry* **29**, 2411–2414.

Felle, H. (1982). Effects of fusicoccin upon membrane potential, resistance and current-voltage characteristics in root hairs of *Sinapis alba*. *Plant Sci. Lett.* **25**, 219–225.

Felle, H. (1988). Cytoplasmic free calcium in *Riccia fluitans* L. and *Zea mays* L.: Interaction of Ca^{2+} and pH? *Planta* **176**, 248–255.

Fensom, D. S., Thompson, R. G. and Alexander, K. G. (1984). Stem anoxia temporarily interrupts translocation of ^{11}C-photosynthate in sunflower. *J. Exp. Bot.* **35**, 1582–1594.

Ferguson, I. B. and Bollard, E. G. (1976). The movement of calcium in woody stems. *Ann. Bot.* (*London*) [N.S.] **40**, 1057–1065.

Ferguson, I. B. and Clarkson, D. T. (1975). Ion transport and endothermal suberization in the roots of *Zea mays*. *New Phytol.* **75**, 69–79.

Ferguson, I. B. and Clarkson, D. T. (1976). Simultaneous uptake and translocation of magnesium and calcium in barley (*Hordeum vulgare* L.) roots. *Planta* **128**, 267–269.

Fernando, M., Kulpa, J., Siddiqi, M. Y. and Glass, A. D. M. (1990). Potassium-dependent changes in the expression of membrane-associated proteins in barley roots. *Plant Physiol.* **92**, 1128–1132.

Ferrario, S., Agius, I. and Morisot, A. (1992a). Daily variations of xylemic exudation rate in tomato. *J. Plant Nutr.* **15**, 69–83.

Ferrario, S., Agius, I. and Morisot, A. (1992b). Daily variations of the mineral composition of xylemic exudates in tomato. *J. Plant Nutr.* **15**, 85–98.

Ferrol, N., Belver, A., Roldan, M., Rodriguez-Rosales, M. P. and Donaire, J. P. (1993). Effects of boron on proton transport and membrane properties of sunflower (*Helianthus annuus* L.) cell microsomes. *Plant Physiol.* **103**, 763–769.

Fetene, M. and Beck, E. (1993). Reversal of the direction of photosynthate allocation in *Urtica dioica* L. plants by increasing cytokinin import into the shoot. *Bot. Acta* **106**, 235–240.

Fetene, M., Möller, I. and Beck, E. (1993).The effect of nitrogen supply to *Urtica dioica* L. plants on the distribution of assimilate between shoot and roots. *Bot. Acta* **106**, 228–234.

Fido, R. J., Gundry, C. S., Hewitt, E. J. and Notton, B. A. (1977). Ultrastructural features of molybdenum deficiency and whiptail of cauliflower leaves. Effect of nitrogen source and tungsten substitution for molybdenum. *Aust. J. Plant Physiol.* **4**, 675–689.

Fieuw, S. and Willenbrink, J. (1990). Sugar transport and sugar-metabolizing enzymes in sugar beet storage roots (*Beta vulgaris* ssp. *altissima*). *J. Plant Physiol.* **137**, 216–223.

Figdore, S. S., Gabelman, W. H. and Gerloff, G. C. (1987). The accumulation and distribution of sodium in tomato strains differing in potassium efficiency when grown under low-K stress. *Plant Soil* **99**, 85–92.

Figdore, S. S., Gerloff, G. C. and Gabelmann, W. H. (1989). The effect of increasing NaCl levels on the potassium utilization efficiency of tomatoes grown under low-K stress. *Plant Soil* **119**, 295–303.

Findenegg, G. R., Salihu, M. and Ali, N. A. (1982). Internal self-regulation of H^+-ion concentration in acid damaged and healthy plants of *Sorghum bicolor* (L.) Moench. *In 'Proc. 9th Plant Nutrition Colloquium. Warwick'* (A. Scaife, ed.), pp. 174–179. Commonwealth Agricultural Bureaux, Farnham Royal, Bucks, UK.

Findenegg, G. R., Nelemans, J. A. and Arnozis, P. A. (1989). Effect of external pH and Cl on the accumulation of NH_4-ions in the leaves of sugar beet. *J. Plant Nutr.* **12**, 593–601.

Fink, S. (1991a). Comparative microscopical studies on the patterns of calcium oxalate distribution in the needles of various conifer species. *Bot. Acta* **104**, 306–315.

Fink, S. (1991b). The micromorphological distribution of bound calcium in needles of Norway spruce (*Picea abies* (L.) Karst.). *New Phytol.* **119**, 33–40.

Fink, S. (1991c). Unusual patterns in the distribution of calcium oxalate in spruce needles and their possible relationships to the impact of pollutants. *New Phytol.* **119**, 41–51.

Fink, S. (1991d). Structural changes in conifer needles due to Mg and K deficiency. *Fert. Res.* **27**, 23–27.

Fink, S. (1992a). Physiologische und strukturelle Veränderungen an Bäumen unter Magnesiummangel. *In 'Magnesiummangel in Mitteleuropäischen Waldökosystemen'* (G. Glatzel, R. Jandel, M. Sieghardt, H. Hager, eds.), S. 16–26. Forstliche Schriftenreihe Band 5, Universität für Bodenkultur Wien.

Fink, S. (1992b). Occurrence of calcium oxalate cristals in non-mycorrhizal fine roots of *Picea abies* (L.) Karst. *J. Plant Physiol.* **140**, 137–140.

Finlay, R. D., Frostegard, A. and Sonnerfeldt, A. M. (1992). Utilization of organic and inorganic nitrogen sources by ectomycorrhizal fungi in pure culture and in symbiosis with *Pinus contorta* Dougl. ex Loud. *New Phytol.* **120**, 105–115.

Firman, D. M. and Allen, E. J. (1988). Field measurements of the photosynthetic rate of potatoes grown with different amounts of nitrogen fertilizer. *J. Agric. Sci.Camb.* **111**, 85–90.

Fischer, E. S. and Bussler, W. (1988). Effects of magnesium deficiency on carbohydrates in *Phaseolus vulgaris*. *Z. Pflanzenernähr. Bodenk.* **151**, 295–298.

Fischer, G. and Hecht-Buchholz, Ch. (1985). The influence of boron deficiency on glandular scale development and structure in *Mentha piperita*. *Planta Medica* **5**, 371–377.

Fischer, W., Felssa, H. and Schaller, G. (1989). pH values and redox potentials in microsites of the rhizosphere. *Z. Pflanzenernähr. Bodenk.* **152**, 191–195.

Fiscus, E. L. and Kramer, P. J. (1970). Radial movement of oxygen in plant roots. *Plant Physiol.* **45**, 667–669.

Fisher, D. (1978). An evaluation of the Münch hypothesis for phloem transport in soybean. *Planta* **139**, 25–28.

Fisher, D. B. (1987). Changes in the concentration and composition of peduncle sieve tube sap during grain filling in normal and phosphate-deficient wheat plants. *Aust.J. Plant Physiol.* **14**, 147–156.

Fisher, J. D., Hausen, D. and Hodges, T. K. (1970). Correlation between ion fluxes and ion stimulated adenosine triphosphatase activity of plant roots. *Plant Physiol.* **46**, 812–814.

Fitter, A. H. (1985). Functioning of vesicular-arbuscular mycorrhizas under field conditions. *New Phytol.* **99**, 257–265.

Fitter, A. H. (1991a). Costs and benefits of mycorrhizas: Implications for functioning under natural conditions. *Experientia* **47**, 350–355.

Fitter, A. H. (1991b). Characteristics and functions of root systems. *In* 'Plant Roots: the Hidden Half' (Y. Waisel, A. Eshel and U. Kafkafi, eds.), pp. 3–25. Marcel Dekker, New York.

Fitzgerald, M. A. and Allaway, W. G. (1991). Apoplastic and symplastic pathways in the leaf of the grey mangrove *Avicennia marina* (Forsk.) Vierh. *New Phytol.* **119**, 217–226.

Fixen, P. E., Gelderman, R. H., Gerwing, J. and Cholick, F. A. (1986a). Response of spring wheat, barley, and oats to chloride in potassium chloride fertilizers. *Agron. J.* **78**, 664–668.

Fixen, P. E., Buchenau, G. W., Gelderman, R. H., Schumacher, T. E., Gerwing, J. R., Cholik, F. A. and Farber, B. G. (1986b). Influence of soil and applied chloride on several wheat parameters. *Agron. J.* **78**, 736–740.

Fleige, H., Strebel, O., Renger, M. and Grimme, H. (1981). Die potentielle P-Anlieferung durch Diffusion als Funktion von Tiefe, Zeit und Durchwurzelung bei einer Parabraunerde aus Löss. *Mitt. Dtsch. Bodenkd. Ges.* **32**, 305–310.

Fleige, H., Grimme, H., Renger, M. and Strebel, O. (1983). Zur Erfassung der Nährstoffanlieferung durch Diffusion im effektiven Wurzelraum. *Mitt. Dtsch. Bodenkd. Ges.* **38**, 381–386.

Flessa, H. and Fischer, W. R. (1992). Plant-induced changes in the redox potential of rice rhizospheres. *Plant Soil* **143**, 55–60.

Florijn, P. J. and van Beusichem, M. L. (1993). Cadmium distribution in maize inbred lines: Effect of pH and level of Cd supply. *Plant Soil* **153**, 79–84.

Flowers, T. J. (1988). Chloride as a nutrient and as an osmoticum. *In* 'Advances in Plant Nutrition' Vol. 3 (B. Tinker and A. Läuchli, eds.), pp. 55–78. Praeger, New York.

Flowers, T. J. and Dalmond, D. (1992). Protein synthesis in halophytes: The influence of potassium, sodium and magnesium in vitro. *Plant Soil* **146**, 153–161.

Flowers, T. J. and Läuchli, A. (1983). Sodium versus potassium: Substitution and compartmentation. *In* 'Encyclopedia of Plant Physiology, New Series' (A. Läuchli and R. L. Bieleski, eds.), Vol. 15B, pp. 651–681. Springer-Verlag, Berlin.

Flowers, T. J., Troke, P. F. and Yeo, A. R. (1977). The mechanism of salt tolerance in halophytes. *Annu. Rev. Plant Physiol.* **28**, 89–121.

Flowers, T. J., Flowers, S. A. and Greenway, H. (1986). Effects of sodium chloride on tobacco plants. *Plant Cell Environ.* **9**, 645–651.

Flowers, T. J., Hajibagheri, M. A. and Yeo, A. R. (1991). Ion accumulation in the cell walls of rice plants growing under saline conditions: evidence for the Oertli hypothesis. *Plant Cell Environ.* **14**, 319–325.

Flügge, U. I., Freisl, M. and Heldt, H. W. (1980). Balance between metabolite accumulation and transport in relation to photosynthesis by isolated spinach chloroplasts. *Plant Physiol.* **65**, 574–577.

Fogel, R. (1988). Interactions among soil biota in coniferous ecosystems. *Agric. Ecos. Environ.* **24**, 69–85.

Fogel, R. and Hunt, G. (1979). Fungal and arboreal biomass in a western Oregon Douglas-fir ecosystem: distribution patterns and turnover. *Can. J. For. Res.* **9**, 245–256.

Föhse, D. and Jungk, A. (1983). Influence of phosphate and nitrate supply on root hair formation of rape, spinach and tomato plants. *Plant Soil* **74**, 359–368.

Föhse, D., Claassen, N. and Jungk, A. (1988). Phosphorus efficiency of plants. I. External and internal P requirement and P uptake efficiency of different plant species. *Plant Soil* **110**, 101–109.

Föhse, D., Claassen, N. and Jungk, A. (1991). Phosphorus efficiency of plants. II. Significance of root hairs and cation–anion balance for phosphorus influx in seven plant species. *Plant Soil* **132**, 261–272.

Forno, D. A., Yoshida, S. and Asher, C. J. (1975). Zinc deficiency in rice. I. Soil factors associated with the deficiency. *Plant Soil* **42**, 537–550.

Foroughi, M., Marschner, H. and Döring, H.-W. (1973). Auftreten von Bormangel bei *Citrus aurantium* L. (Bitterorangen) am Kaspischen Meer (Iran). *Z. Pflanzenernähr. Bodenk.* **136**, 220–228.

Forster, H. (1970). Der Einfluss einiger Ernährungsunterbrechungen auf die Ausbildung von Ertrags- und Qualitätsmerkmalen der Zuckerrübe. *Landwirtsch. Forsch. Sonderh.* **25**(II), 99–105.

Forster, H. (1980). Einfluss von unterschiedlich starkem Magnesiummangel bei Gerste auf den Kornertrag und seine Komponenten. *Z. Pflanzenernähr. Bodenk.* **143**, 627–637.

Förster, J. C. and Jeschke, W. D. (1993). Effects of potassium withdrawal on nitrate transport and on the contribution of the root to nitrate reduction in the whole plant. *J. Plant Physiol.* **141**, 322–328.

Forsyth, C. and Van Staden, J. (1981). The effect of root decapitation on lateral root formation and cytokinin production in *Pisum sativum*. *Physiol. Plant.* **51**, 375–379.

Fournier, J. M., Benlloch, M. and de la Guardia, M. D. (1987). Effect of abscisic acid on exudation of sunflower roots as affected by nutrient status, glucose level and aeration. *Physiol. Plant.* **69**, 675–679.

Fownes, J. H. and Anderson, D. G. (1991). Changes in nodule and root biomass of *Sesbania sesban* and *Leucaena leucocephala* following coppicing. *Plant Soil* **138**, 9–16.

Fox, R. L. and Lipps, R. C. (1961). Distribution and activity of roots in relation to soil properties. *Trans. Int. Congr. Soil Sci.*, *7th*, 1960, pp. 260–267.

Fox, R. H., Roth, G. W., Iversen, K. V. and Piekielek, W. P. (1989). Soil and tissue nitrate tests compared for predicting soil nitrogen availability to corn. *Agron. J.* **81**, 971–974.

Fox, T. R. and Comerford, N. B. (1990). Low-molecular-weight organic acids in selected forest soils of the southwestern USA. *Soil Sci. Soc. Am. J.* **54**, 1139–1144.

Foy, C. D. (1983). The physiology of plant adaptation to mineral stress. *Iowa State J. Res.* **57**, 355–391.

Foy, C. D. (1974). Effect of aluminium on plant growth. *In* 'The Plant Root and its Environment' (E. W. Carson, ed.), pp. 601–642. University Press of Virginia, Charlottesville.

Foy, C. D. and Fleming, A. L. (1982). Aluminium tolerances of two wheat genotypes related to nitrate reductase activities. *J. Plant Nutr.* **5**, 1313–1333.

Foy, C. D., Fleming, A. L., Burns, G. R. and Armiger, W. H. (1967). Characterization of differential aluminium tolerance among varieties of wheat and barley. *Soil Sci. Soc. Am. Proc.* **31**, 513–521.

Foy, C. D., Fleming, A. L. and Armiger, W. H. (1969). Aluminium tolerance of soybean varieties in relation to calcium nutrition. *Agron. J.* **61**, 505–511.

Foy, C. D., Fleming, A. L. and Gerloff, G. C. (1972). Differential aluminium tolerance on two snapbean varieties. *Agron. J.* **64**, 815–818.

Foy, C. D., Fleming, A. L. and Schwartz, J. W. (1973). Opposite aluminium and manganese tolerances of two wheat varieties. *Agron. J.* **65**, 123–126.

Foy, C. D., Lafever, H. N., Schwartz, J. W. and Fleming, A. L. (1974). Aluminium tolerance of wheat cultivars related to region of origin. *Agron. J.* **66**, 751–758.

Foy, C. D., Chaney, R. L. and White, M. C. (1978). The physiology of metal toxicity in plants. *Annu. Rev. Plant Physiol.* **29**, 511–566.

Foy, C. D., Webb, H. W. and Jones, J. E. (1981). Adaptation of cotton genotypes to an acid, manganese toxic soil. *Agron. J.* **73**, 107–111.

Foyer, C. and Spencer, C. (1986). The relationship between phosphate status and photosynthesis in leaves. Effects of intracellular orthophosphate distribution, photosynthesis and assimilate partitioning. *Planta* **167**, 369–375.

Foyer, C. H. (1988). Feedback inhibition of photosynthesis through source–sink regulation in leaves. *Plant Physiol. Biochem.* **26**, 483–492.

Franck, E. von and Finck, A. (1980). Ermittlung von Zink-Ertragsgrenzwerten für Hafer und Weizen. *Z. Pflanzenernähr. Bodenk.* **143**, 38–46.

Franco, A. A. and Munns, D. N. (1981). Response of *Phaseolus vulgaris* L. to molybdenum under acid conditions. *Soil Sci. Soc. Am. J.* **45**, 1144–1148.

Franco, A. A. and Munns, D. N. (1982). Acidity and aluminium restraints on nodulation, nitrogen fixation, and growth of *Phaseolus vulgaris* in solution culture. *Soil Sci. Soc. Am. J.* **46**, 296–301.

Francois, L. E. (1989). Boron tolerance of snap bean and cowpea. *J. Amer. Soc. Hort. Sci.* **144**, 615–619.

Francois, L. E. and Clark, R. A. (1979). Accumulation of sodium and chloride in leaves of sprinkler-irrigated grapes. *J. Am. Soc. Hortic. Sci.* **104**, 11–13.

Francois, L. E., Donovan, T. J. and Maas, E. V. (1991). Calcium deficiency of artichoke buds in relation to salinity. *HortScience* **26**, 549–553.

Franke, W. (1967). Mechanism of foliar penetration of solutions. *Annu. Rev. Plant Physiol.* **18**, 281–300.

Frankenberger, W. T. jr. and Poth, M. (1987). Biosynthesis of indole-3-acetic acid by the pine ectomycorrhizal fungus *Pisolithus tinctorius*. *Appl. Environ. Microbiol.* **53**, 2908–2913.

Fredeen, A. L., Rao, I. M. and Terry, N. (1989). Influence of phosphorus nutrition on growth and carbon partitioning in *Glycine max*. *Plant Physiol.* **89**, 225–230.

Fredeen, A. L., Raab, T. K., Rao, I. M. and Terry, N. (1990). Effects of phosphorus nutrition on photosynthesis of *Glycine max* (L.) Merr. *Planta* **181**, 399–405.

Freney, J. R., Delwiche, C. C. and Johnson, C. M. (1959). The effect of chloride on the free amino acids of cabbage and cauliflower plants. *Aust. J. Soil Sci.* **12**, 160–167.

Freney, J. R., Spencer, K. and Jones, M. B. (1978). The diagnosis of sulphur deficiency in wheat. *Aust. J. Agric. Res.* **29**, 727–738.

Fridovich, I. (1983). Superoxide radical: an endogenous toxicant. *Annu. Rev. Pharmacol. Toxicol.* **23**, 239–257.

Friedman, R., Levin, N. and Altman, A. (1986). Presence and identification of polyamines in xylem and phloem exudates of plants. *Plant Physiol.* **82** 1154–1157.

Fritzsche, R. and Thiele, S. (1984). Erste Hinweise auf mögliche Wechselbeziehungen zwischen der Mangan-Versorgung von Böden und Pflanzen und dem Auftreten bzw. der Intensität der Symptomausbildung bei Zuckerrüben durch Infektion mit dem Milden Rübenvergilbungsvirus (beet mild yellowing virus). *Arch. Phytophathol. Pflanzenschutz, Berlin* **20**, 449–451.

Froehlich, D. M. and Fehr, W. R. (1981). Agronomic performance of soybeans with differing levels of iron deficiency chlorosis on calcareous soil. *Crop Sci.* **21**, 438–440.

Fromm, J. (1991). Control of phloem unloading by action potentials in *Mimosa*. *Physiol. Plant.* **83**, 529–533.

Fromm, J. and Eschrich, W. (1988). Transport processes in stimulated and non-stimulated leaves of *Mimosa pudica*. III. Displacement of ions during seismonastic leaf movements. *Trees* **2**, 65–72.

Fromm, J. and Eschrich, W. (1989). Correlation of ionic movements with phloem unloading and loading in barley leaves. *Plant Physiol. Biochem.* **27**, 577–585.

Fromm, J. and Eschrich, W. (1993). Electrical signals released from roots of willow (*Salix viminalis* L.) change transpiration and photosynthesis. *J. Plant Physiol.* **141**, 673–680.

Frota, J. N. E. and Tucker, T. C. (1978). Salt and water stress influences nitrogen metabolism in red kidney beans. *Soil Sci. Soc. Am. J.* **42**, 743–746.

Fuchs, W. H. and Grossmann, F. (1972). Ernährung und Resistenz von Kulturpflanzen gegenüber Krankheitserregern und Schädlingen. *In* 'Handbuch der Pflanzenernährung und Düngung' (H. Linser, ed.), Vol. 1, part 2, pp. 1007–1107. Springer-Verlag, Berlin.

Fujita, K., Masudo, T. and Ogata, S. (1988). Dinitrogen fixation, ureide concentration in xylem exudate and translocation of photosynthates in soybean as influenced by pod removal and defoliation. *Soil Sci. Plant Nutr.* (*Tokyo*) **34**, 265–275.

Furbank, R. T., Jenkins, C. L. D. and Hatch, M. D. (1989). CO_2 concentrating mechanism of C_4 photosynthesis. Permeability of isolated bundle sheath cells to inorganic carbon. *Plant Physiol.* **91**, 1364–1371.

Furlani, A. M. C., Clark, R. B., Sullivan, C. Y. and Maranville, J. W. (1986). Sorghum genotype differences to leaf 'red-speckling' induced by phosphorus. *J. Plant Nutr.* **9**, 1435–1451.

Fusseder, A. (1984). Der Einfluss von Bodenart, Durchlüftung des Bodens, N-Ernährung und Rhizosphärenflora auf die Morphologie des seminalen Wurzelsystems von Mais. *Z. Pflanzenernähr. Bodenk.* **147**, 553–565.

Fusseder, A. und Kraus, M. (1986). Individuelle Wurzelkonkurrenz und Ausnutzung der immobilen Makronährstoffe im Wurzelraum von Mais. *Flora* **178**, 11–18.

Fusseder, A., Kraus, M. and Beck, E. (1988). Reassessment of root competition for P of field-grown maize in pure and mixed cropping. *Plant Soil* **106**, 299–301.

Gabbrielli, R., Pandolfini, T., Vergnano, O. and Palandri, M. R. (1990). Comparison of two serpentine species with different nickel tolerance strategies. *Plant Soil* **122**, 271–277.

Gabelman, W. H. and Gerloff, G. C. (1983). The search for an interpretation of genetic controls that enhance plant growth under deficiency levels of a macronutrient. *Plant Soil* **72**, 335–350.

Gahoonia, T. S. (1993). Influence of root-induced pH on the solubility of soil aluminium in the rhizosphere. *Plant Soil* **149**, 289–291.

Gahoonia, T. S., Claassen, N. and Jungk, A. (1992). Mobilization of phosphate in different soils by ryegrass supplied with ammonium or nitrate. *Plant Soil* **140**, 241–248.

Galamay, T. O., Yamauchi, A., Kono, Y. and Hioki, M. (1992). Specific colonization of the hypodermis of sorghum roots by an endophyte, *Polymyxa* sp. *Soil Sci. Plant Nutr.* (*Tokyo*) **38**, 573–578.

Galling, G. (1963). Analyse des Magnesium-Mangels bei synchronisierten Chlorellen. *Arch. Mikrobiol.* **46**, 150–184.

Galston, A. W. and Sawhney, R. K. (1990). Polyamines in plant physiology. *Plant Physiol.* **94**, 406–410.

Gantar, M., Kerby, N. W., Rowell, P. and Obreht, Z. (1991). Colonization of wheat (*Triticum vulgare* L.) by N_2-fixing cyanobacteria: I. A survey of soil cyanobacterial isolates forming associations with roots. *New Phytol.* **118**, 477–484.

Gao, Y.-P., Motosugi, H. and Sugiura, A. (1992). Rootstock effects on growth and flowering in young apple trees grown with ammonium and nitrate nitrogen. *J. Amer. Soc. Hort. Sci.* **117**, 446–452.

Garbarino, J. and DuPont, F.M. (1989). Rapid induction of Na^+/H^+ exchange activity in barley root tonoplast. *Plant Physiol.* **89**, 1–4.

Garcia-Garrido, J. M. and Ocampo, J. A. (1989). Effect of VA mycorrhizal infection of tomato on damage caused by *Pseudomonas syringae*. *Soil Biol. Biochem.* **21**, 165–167.

Garcia-González, M., Mateo, P. and Bonilla, I. (1988). Boron protection for O_2 diffusion in heterocysts of *Anabaena* sp. PCC 7119. *Plant Physiol.* **87**, 785–789.

Garcia-González, M., Mateo, P. and Bonilla, I. (1991). Boron requirement for envelope structure and function in *Anabaena* PCC 7119 heterocysts. *J. Exp. Bot.* **42**, 925–929.

Gardner, B. R. and Roth, R. L. (1989). Midrib nitrate concentration as a means for determining nitrogen needs of cabbage. *J. Plant Nutr.* **12**, 1073–1088.

Gardner, B. R. and Roth, R. L. (1990). Midrib nitrate concentration as a means for determing nitrogen needs of cauliflower. *J. Plant Nutr.* **13**, 1435–1451.

Gardner, J. H. and Malajczuk, N. (1988). Recolonization of rehabilitated bauxite mine sites in Western Australia by mycorrhizal fungi. *For. Ecol. and Mgmt* **24**, 27–42.

Gardner, W. K. and Flynn, A. (1988). The effect of gypsum on copper nutrition of wheat grown in marginally deficient soil. *J. Plant Nutr.* **11**, 475–493.

Gardner, W. K., Parbery, D. G. and Barber, D. A. (1982). The acquisition of phosphorus by *Lupinus albus* L. II. The effect of varying phosphorus supply and soil type on some characteristics of the soil/root interface. *Plant Soil* **68**, 33–41.

Gardner, W. K., Barber, D. A. and Parbery, D. G (1983a). The acquisition of phosphorus by *Lupinus albus* L. III. The probable mechanism by which phosphorus movement in the soil/root interface is enhanced. *Plant Soil* **70**, 107–114.

Gardner, W. K., Parbery, D. G., Barber, D. A. and Swinden, L. (1983b). The acquisition of phosphorus by *Lupinus albus* L. V. The diffusion of exudates away from roots: a computer simulation. *Plant Soil* **72**, 13–29.

Garg, O. K., Sharma, A. N. and Kona, G. R. S. S. (1979). Effect of boron on the pollen vitality and yield of rice plant (*Oryza sativa* L. var Jaya). *Plant Soil* **52**, 591–594.

Gärtel, W. (1974). Die Mikronährstoffe – ihre Bedeutung für die Rebenernährung unter besonderer Berücksichtigung der Mangel- und Überschusserscheinungen. *Weinberg Keller* **21**, 435–507.

Garten, C. T. jr and Hanson, P. J. (1990). Foliar retention of ^{15}N-nitrate and ^{15}N-ammonium by red maple (*Acer rubrum*) and white oak (*Quercus alba*) leaves from simulated rain. *Environ. Exp. Bot.* **30**, 333–342.

Garwood, E. A. and Williams, T. E. (1967). Growth, water use and nutrient uptake from the subsoil by grass swards. *J. Agric. Sci.* **69**, 125–130.

Garz, J. (1966). Menge, Verteilung und Bindungsform der Mineralstoffe (P, K, Mg und Ca) in den Leguminosensamen in Abhängigkeit von der Mineralstoffumlagerung innerhalb der Pflanze und den Ernährungsbedingungen. *Kuehn-Arch.* **80**, 137–194.

Gascho, G. J. (1977). Response of sugarcane to calcium silicate slag. I. Mechanisms of response in Florida. *Proc. Soil Crop Sci. Soc. Fla.* **37**, 55–58.

Gäth, S., Meuser, H., Abitz, C.-A., Wessolek, G. and Renger. M. (1989). Determination of potassium delivery to the roots of cereal plants. *Z. Pflanzenernähr. Bodenk.* **152**, 143–149.

Gebauer, G., Schubert, B., Schuhmacher, M. I., Rehder, H. and Ziegler, H. (1987). Biomass production and nitrogen content of C_3- and C_4-grasses in pure and mixed culture with different nitrogen supply. *Oecologia (Berlin)* **71**, 613–617.

Geisler, G. (1967). Interactive effects of CO_2 and O_2 in soil on root and top growth of barley and peas. *Plant Physiol.* 42, 305–307.

Geisler, G. (1968). Über den Einfluss von Unterbodenverdichtungen auf den Luft- und Wasserhaushalt des Bodens und das Wurzelwachstum. *Landwirtsch. Forsch.*, *Sonderh.* **22**, 61–69.

George, E., Häussler, K.-U., Vetterlein, D., Gorgus, E. and Marschner, H. (1992). Water and nutrient translocation by hyphae of *Glomus mosseae*. *Can. J. Bot.* **70**, 2130–2137.

George, T., Ladha, J. K., Buresh, R. J. and Garrity, D. P. (1992). Managing native and legume-fixed nitrogen in lowland rice-based cropping systems. *Plant Soil* **141**, 69–91.

Georgen, P. G., Davis-Carter, J. and Taylor, H. M. (1991). Root growth and water extraction patterns from a calcic horizon. *Soil Sci. Soc. Am. J.* **55**, 210–215.

Gerath, H., Borchmann, W. and Zajonc, I. (1975). Zur Wirkung des Mikronährstoffs Bor auf die Ertragsbildung von Winterraps (*Brassica napus* L. ssp. *oleifera*). *Arch. Acker- Pflanzenbau Bodenkd.* **19**, 781–792.

Gerendás, J. and Sattelmacher, B. (1990). Influence of nitrogen form and concentration on growth and ionic balance of tomato (*Lycopersicon esculentum*) and potato (*Solanum tuberosum*). *In* 'Plant Nutrition-Physiology and Application' (M. L. van Beusichem, ed.), pp. 33–37. Kluwer Academic, Dordrecht.

Gerendás, J., Ratcliffe, R. G. and Sattelmacher, B. (1990). [31]P nuclear magnetic resonance evidence for differences in intracellular pH in the roots of maize seedlings grown with nitrate or ammonium. *J. Plant Physiol.* **137**, 125–128.

Gerhardt, R., Stitt, M. and Heldt, H. W. (1987). Subcellular metabolite levels in spinach leaves. Regulation of sucrose synthesis during diurnal alterations in photosynthetic partitioning. *Plant Physiol.* **83**, 399–407.

Gerke, J. (1992a). Phosphate, aluminium and iron in the soil solution of three different soils in relation to varying concentrations of citric acid. *Z. Pflanzenernähr. Bodenk.* **155**, 339–343.

Gerke, J. (1992b). Orthophosphate and organic phosphate in the soil solution of four sandy soils in relation to pH-evidence for humic-Fe(Al-)phosphate complexes. *Commun. Soil Sci. Plant Anal.* **23**, 601–612.

Gerloff, G. C. and Gabelman, W. H. (1983). Genetic basis of inorganic plant nutrition. *In* 'Encyclopedia of Plant Physiology, New Series' (A. Läuchli and R. L. Bieleski, eds.), Vol. 15B, pp. 453–480. Springer-Verlag, Berlin.

Gettier, S. W., Martens, D. C. and Brumback jr, T. B. (1985). Timing of foliar manganese application for correction of manganese deficiency in soybean. *Agron. J.* **77**, 627–629.

Geyer, B. and Marschner, H. (1990). Charakterisierung des Stickstoffversogungsgrades bei Mais mit Hilfe des Nitrat-Schnelltests. *Z. Pflanzenernähr. Bodenk.* **153**, 341–348.

Gianinazzi-Pearson, V., Branzanti, B. and Gianinazzi, S. (1989). *In vitro* enhancement of spore germination and early hyphal growth of a vesicular-arbuscular mycorrhizal fungus by host root exudates and plant flavonoids. *Symbiosis* **7**, 243–255.

Giaquinta, R. T. (1977). Phloem loading of sucrose. *Plant Physiol.* **59**, 750–755.

Giaquinta, R. T. (1978). Source and sink leaf metabolism in relation to phloem translocation. *Plant Physiol.* **61**, 380–385.

Giaquinta, R. T. and Geiger, D. R. (1977). Mechanisms of cyanide inhibition of phloem translocation. *Plant Physiol.* **59**, 178–180.

Giaquinta, R. T. and Quebedeaux, B. (1980). Phosphate-induced changes in assimilate partitioning in soybean leaves during pod filling. *Plant Physiol.* **65**, Suppl., 119.

Gibson, T. S., Speirs, J. and Brady, C. J. (1984). Salt tolerance in plants. II. In vitro translation of m-RNA from salt-tolerant and salt-sensitive plants on wheat germ ribosomes. Responses to ions and compatible organic solutes. *Plant Cell Environ.* **7**, 579–587.

Gijsman, A. J. (1991). Soil water content as a key factor determining the source of nitrogen (NH_4^+ or NO_3^-) absorbed by Douglas-fir (*Pseudotsuga menziesii*) along its roots. *Can. J. For. Res.* **21**, 616–625.

Gildon, A. and Tinker, P. B. (1983a). Interactions of vesicular-arbuscular mycorrhizal infection and heavy metals in plants. I. The effects of heavy metals on the development of vesicular-arbuscular mycorrhizas. *New Phytol.* **94**, 247–261.

Gildon, A. and Tinker, P. B. (1983b). Interactions of vesicular-arbuscular mycorrhizal infections and heavy metals in plants. II. The effects of infection on uptake of copper. *New Phytol.* **95**, 263–268.

Gillespie, A. R. and Pope, P. E. (1989). Alfalfa N_2-fixation enhances the phosphorus uptake of walnut in interplantings. *Plant Soil* **113**, 291–293.

Gillespie, A. R. and Pope, P. E. (1990). Rhizosphere acidification increases phosphorus recovery of black locust: II. Model predictions and measured recovery. *Soil Sci. Soc. Am. J.* **54**, 338–341.

Gillis, M., Kersters, K., Hoste, B., Janssens, D., Kroppenstedt, R. M., Stephan, M. P., Teixeira, K. R. S., Döbereiner, J. and De Ley, J. (1989). *Acetobacter diazotrophicus* sp. nov., a nitrogen-fixing acetic acid bacterium associated with sugarcane. *Int. J. Syst. Bact.* **39**, 361–364.

Gilmour, J. T. (1977). Micronutrient status of the rice plant. I. Plant and soil solution concentrations as function of time. *Plant Soil* **46**, 549–557.

Gissel-Nielsen, G. (1987). Fractionation of selenium in barley and rye-grass. *J. Plant Nutr.* **10**, 2147–2152.

Gladstones, J. S., Loneragan, J. F. and Goodchild, N. A. (1977). Field responses to cobalt and molybdenum by different legume species, with interferences on the role of cobalt in legume growth. *Aust. J. Agric. Res.* **28**, 619–628.

Glass, A. D. M. (1983). Regulation of ion transport. *Annu. Rev. Plant Physiol.* **34**, 311–326.

Glass, A. D. M. and Dunlop, J. (1979). The regulation of K^+ influx in excised barley roots. Relationship between K^+ influx and electrochemical potential differences. *Planta* **145**, 395–397.

Glass, A. D. M. and Siddiqi, M. Y. (1985). Nitrate inhibition of chloride influx in barley: Implications for a proposed chloride homeostat. *J. Exp. Bot.* **36**, 556–566.

Glass, A. D. M., Siddiqi, M. Y., Ruth, T. J. and Rufty jr, T. W. (1990). Studies on the uptake of nitrate in barley. II. Energetics. *Plant Physiol.* **93**, 1585–1589.

Glavac, V., Koenies, H., Ebben, U. and Avenhaus, U. (1991). Jahreszeitliche Veränderung der NO_3^--Konzentrationen im Xylemsaft des unteren Stammteiles von Buchen (*Fagus sylvatica* L.). *Z. Pflanzenernähr. Bodenk.* **154**, 121–125.

Gliemeroth, G. (1953). Bearbeitung und Düngung des Unterbodens in ihrer Wirkung auf Wurzelentwicklung, Stoffaufnahme und Pflanzenwachstum. *Z. Acker- Pflanzenbau.* **96**, 1–44.

Glinka, Z. and Reinhold, L. (1971). Abscisic acid raises the permeability of plant cells to water. *Plant Physiol.* **48**, 103–105.

Gochnauer, M. B., McCully, M. E. and Labbé, H. (1989). Different populations of bacteria associated with sheathed and bare regions of roots of field-grown maize. *Plant Soil* **114**, 107–120.

Gochnauer, M. B., Sealey, L. J. and McCully, M. E. (1990). Do detached root-cap cells influence bacteria associated with maize roots? *Plant, Cell Environ.* **13**, 793–801.

Godbold, D. L., Horst, W. J., Marschner, H., Collins, J. C. and Thurman, D. A. (1983). Root growth and Zn uptake by two ecotypes of *Deschampsia caespitosa* as affected by high Zn concentrations. *Z. Pflanzenphysiol.* **112**, 315–324.

Godbold, D. L., Horst, W. J., Collins, J. C., Thurman, D.A. and Marschner, H. (1984). Accumulation of zinc and organic acids in roots of zinc tolerant and non-tolerant ecotypes of *Deschampsia caespitosa*. *J. Plant Physiol.* **116**, 59–69.

Godbold, D. L., Fritz, E. and Hüttermann, A. (1988). Aluminium toxicity and forest decline. *Proc. Natl. Acad. Sci. USA* **85**, 3888–3892.

Godo, G. H. and Reisenauer, H. M. (1980). Plant effects on soil mangnese availability. *Soil Sci. Soc. Am. J.* **44**, 993–995.

Godwin, D. C. and Blair, G. J. (1991). Phosphorus efficiency in pasture species. V. A comparison of white clover accessions. *Aust. J. Agric. Res.* **42**, 531–540.

Goeschl, J. D., Magnuson, C. E., Fares, Y., Jaeger, C. H., Nelson, C. E. and Strain, B. R. (1984). Spontaneous and induced blocking and unblocking of phloem transport. *Plant, Cell Environ.* **7**, 607–613.

Gojon, A., Bussi, C., Grignon, C. and Salsac, L. (1991a). Distribution of NO_3^- reduction between roots and shoots of peach-tree seedlings as affected by NO_3^- uptake rate. *Physiol. Plant.* **82**, 505–512.

Gojon, A., Wakrim, R., Passama, L. and Robin, P. (1991b). Regulation of NO_3^- assimilation by anion availability in excised soybean leaves. *Plant Physiol.* **96**, 396–405.

Goldbach, E., Goldbach, H., Wagner, H. and Michael, G. (1975). Influence of N-deficiency on the abscisic acid content of sunflower plants. *Physiol. Plant.* **34**, 138–140.

Goldbach, H. (1985). Influence of boron nutrition on net uptake and efflux of ^{32}P and ^{14}C glucose in *Helianthus annuus* roots and cell cultures of *Daucus carota*. *J. Plant Physiol.* **118**, 431–438.

Goldbach, H. and Michael, G. (1976). Abscisic acid content of barley grains during ripening as affected by temperature and variety. *Crop Sci.* **16**, 797–799.

Goldbach, H., Goldbach, E. and Michael, G. (1977). Transport of abscisic acid from leaves to grains in wheat and barley plants. *Naturwissenschaften* **64**, 488.

Goldbach, H. E., Hartmann, D. and Rötzer, T. (1990). Boron is required for the stimulation of the ferric cyanide-induced proton release by auxins in suspension-cultured cells of *Daucus carota* and *Lycopersicon esculentum*. *Physiol. Plant.* **80**, 114–118.

Goldman, I. L., Carter jr, T. E. and Patterson, R. P. (1989). A detrimental interaction of subsoil aluminium and drought stress on the leaf water status of soybean. *Agron. J.* **81**, 461–463.

Goldstein, A. H., Baertlein, D. A. and McDaniel, R. G. (1988). Phosphate starvation inducible metabolism in *Lycopersicon esculentum*. I. Excretion of acid phosphatase by tomato plants and suspension-cultured cells. *Plant Physiol.* **87**, 711–715.

Gollan, T., Schurr, U. and Schulze, E.-D. (1992). Stomatal response to drying soil in relation to changes in the xylem sap composition of *Helianthus annuus*. I. The concentration of cations, anions, amino acids in, and pH of, the xylem sap. *Plant Cell Environ*. **15**, 551–559.

Gonzalez-Lopez, J., Martinez-Toledo, M. V., Reina, S. and Salmeron, V. (1991). Root exudates of maize and production of auxins, gibberellins, cytokinins, amino acids and vitamins by *Azotobacter chloococcum* in chemically-defined media and dialised-soil media. *Technol. and Environ. Chem*. **33**, 69–78.

Gonzalez-Reyes, J. A., Döring, O., Navas, P., Obst, G. and Böttger, M. (1992). The effect of ascorbate free radical on the energy state of the plasma membrane of onion (*Allium cepa* L.) root cells: alteration of K^+ efflux by ascorbate? *Biochim. Biophys. Acta* **1098**, 177–183.

Gooding, M. J. and Davies, W. P. (1992). Foliar urea fertilization of cereals: a review. *Fert. Res*. **32**, 209–222.

Goor, B. J. van (1966). The role of calcium and cell permeability in the disease blossom end rot of tomatoes. *Physiol. Plant*. **21**, 1110–1121.

Goor, B. J. van and Lune, P. van (1980). Redistribution of potassium, boron, iron, magnesium and calcium in apple trees determined by an indirect method. *Physiol. Plant*. **48**, 21–26.

Gordon, W. R., Schwemmer, S. S. and Hillman, W. S. (1978). Nickel and the metabolism of urea by *Lemna paucicostata* Hegelm. 6746. *Planta* **140**, 265–268.

Gorham, J. (1987). Photosynthesis, transpiration and salt fluxes through leaves of *Leptochloa fusca* L. Kunth. *Plant Cell Environ*. **10**, 191–196.

Gorham, J. (1993). Genetics and physiology of enhanced K/Na discrimination. *In* 'Genetic Aspects of Plant Mineral Nutrition' (P. J. Randall, E. Delhaize, R. A. Richards and R. Munns, eds.), pp. 151–158. Kluwer Academic, Dordrecht, The Netherlands.

Gorham, J., Hughes, L. and Wyn Jones, R. G. (1980). Chemical composition of salt-marsh plants from Ynys Mon (Anglesey): the concept of physiotypes. *Plant Cell Environ*. **3**, 309–318.

Gorham, J., Wyn Jones, R. G. and McDonnell, E. (1985). Some mechanisms of salt tolerance in crop plants. *Plant Soil* **89**, 15–40.

Gorham, J., Forster, B. P., Budrewicz, E., Wyn Jones, R. G., Miller, T. E. and Law, C. N. (1986). Salt tolerance in the Triticeae: solute accumulation and distribution in an amphidiploid derived from *Triticum aestivum* cv. Chinese Spring and *Thinopyrum bessarabicum*. *J. Exp. Bot*. **37**, 1435–1449.

Goss, M. J. and Carvalho, M. J. G. P. R. (1992). Manganese toxicity: the significance of magnesium for the sensitivity of wheat plants. *Plant Soil* **139**, 91–98.

Goss, M. J. and Scott-Russell, R. (1980). Effects of mechanical impedance on root growth in barley (*Hordeum vulgare* L.). III. Observations on the mechanism of response. *J. Exp. Bot*. **31**, 577–588.

Goulding, K. W. T. (1990). Nitrogen deposition to land from the atmosphere. *Soil Use Mgmt* **6**, 61–63.

Gourp, V. and Pargney, J.-C. (1991). Immuno-cytolocalisation des phosphatases acides de *Pisolithus tinctorius* L. lors de sa confrontation avec le système racinaire de *Pinus sylvestris* (Pers.) Desv. *Cryptogamie Mycol*. **12**, 293–304.

Gowing, D. J. G., Davies, W. J. and Jones, H. G. (1990). A positive root-source signal as an indicator of soil drying in apple, *Malus* × *domestica* Berkh. *J. Exp. Bot*. **41**, 1535–1540.

Grabau, L. J., Blevins, D. G. and Minor, H. C. (1986a). P nutrition during seed development. Leaf senescence, pod retention, and seed weight of soybean. *Plant Physiol*. **82**, 1008–1012.

Grabau, L. J., Blevins, D. G. and Minor, H. C. (1986b). Stem infusions enhanced methionine content of soybean storage protein. *Plant Physiol.* **82**, 1013–1018.

Graham, M. J., Stephens, P. A., Widholm, J. M. and Nickell, C. D. (1992). Soybean genotype evaluation for iron deficiency chlorosis using sodium bicarbonate and tissue culture. *J. Plant Nutr.* **15**, 1215–1225.

Graham, M. Y. and Graham, T. L. (1991). Rapid accumulation of anionic peroxidase and phenolic polymers in soybean cotyledon tissues following treatment with *Phytophthora megasperma* f. sp. *Glycinea* woll glucan. *Plant Physiol.* **97**, 1445–1455.

Graham, R. D. (1975). Male sterility in wheat plants deficient in copper. *Nature (London)* **254**, 514–515.

Graham, R. D. (1976). Anomalous water relations in copper-deficient wheat plants. *Aust. J. Plant Physiol.* **3**, 229–236.

Graham, R. D. (1979). Transport of copper and manganese to the xylem exudate of sunflower. *Plant Cell Environ.* **2**, 139–143.

Graham, R. D. (1980a). The distribution of copper and soluble carbohydrates in wheat plants grown at high and low levels of copper supply. *Z. Pflanzenernähr. Bodenk.* **143**, 161–169.

Graham, R. D. (1980b). Susceptibility to powdery mildew of wheat plants deficient in copper. *Plant Soil* **56**, 181–185.

Graham, R. D. (1983). Effects of nutrient stress on susceptibility of plants to disease with particular reference to the trace elements. *Adv. Bot. Res.* **10**, 221–276.

Graham, R. D. (1984). Breeding for nutritional characteristics in cereals. *In* 'Advances in Plant Nutrition' (P. B. Tinker and A. Läuchli, eds.), Vol. 1, pp. 57–102. Praeger, New York.

Graham, R. D. (1988). Genotypic differences in tolerance to manganese deficiency. *In* 'Manganese in Soils and Plants' (R. D. Graham, R. J. Hannam and N. C. Uren, eds.), pp. 261–276. Kluwer Academic, Dordrecht, The Netherlands.

Graham, R. D. and Pearce, D. T. (1979). The sensitivity of hexaploid and octaploid triticales and their parent species to copper deficiency. *Aust. J. Agric. Res.* **30**, 791–799.

Graham, R. D. and Webb, M. J.(1991). Micronutrients and plant disease resistance and tolerance in plants. *In* 'Micronutrients in Agriculture' (J. J. Mortvedt, F. R. Cox, L. M. Shuman, R. M. Welch, eds.), pp. 329–370. SSSA Book Series No. 4, Madison, WI.

Graham, R. D., Davies, W. J., Sparrow, D. H. B. and Ascher, J. S. (1982). Tolerance of barley and other cereals to manganese-deficient calcareous soils of South Australia. *In* 'Genetic Specificity of Mineral Nutrition of Plants' (M. R. Saric, ed.), Vol. 13, pp. 277–283. Serb. Acad. Sci. Arts, Beograd.

Graham, R. D., Ascher, J. S., Ellis, P. A. E. and Shepherd, K. W. (1987a). Transfer to wheat of the copper efficiency factor carried on rye chromosome arm 5RL. *Plant Soil* **99**, 107–114.

Graham, R. D., Welch, R. M., Grunes, D. L., Cary, E. E. and Norvell, W. A. (1987b). Effect of zinc deficiency on the accumulation of boron and other mineral nutrients in barley. *Soil Sci. Soc. Am. J.* **51**, 652–657.

Graham, R. D., Ascher, J. S. and Hynes, S. C. (1992). Selecting zinc-efficient cereal genotypes for soils low in zinc status. *Plant Soil* **146**, 241–250.

Granato, T. C. and Raper jr, C. D. (1989). Proliferation of maize (*Zea mays* L.) roots in response to localized supply of nitrate. *J. Exp. Bot.* **40**, 263–275.

Grasmanis, V. O. and Edwards, G. E. (1974). Promotion on flower initiation in apple trees by short exposure to the ammonium ion. *Aust. J. Plant Physiol.* **1**, 99–105.

Grauer, U. E. (1993). Modelling anion amelioration of aluminium phytoxicity. *Plant Soil* **157**, 319–331.

Grauer, U. E. and Horst, W. J. (1990). Effect of pH and nitrogen source on aluminium tolerance of rye (*Secale cereale* L.) and yellow lupin (*Lupinus luteus* L.). *Plant Soil* **127**, 13–21.

Grauer, U. E. and Horst, W. J. (1992). Modeling cation amelioration of aluminum phytotoxicity. *Soil Sci. Soc. Am. J.* **56**, 166–172.

Graven, E. H., Attoe, O. J. and Smith, D. (1965). Effect of liming and flooding on manganese toxicity in alfalfa. *Soil Sci. Soc. Am. Proc.* **29**, 702–706.

Green, D. G. and Warder, F. G. (1973). Accumulation of damaging concentrations of phosphorus by leaves of Selkirk wheat. *Plant Soil* **38**, 567–572.

Green, J. F. and Muir, R. M. (1979). Analysis of the role of potassium in the growth effects of cytokinin, light and abscisic acid on cotyledon expansion. *Physiol. Plant.* **46**, 19–24.

Greene, R. M., Geider, R. J., Kilber, Z. and Falkowski, P. G. (1992). Iron-induced changes in light harvesting and photochemical energy conversion processes in eukaryotic marine algae. *Plant Physiol.* **100**, 565–575.

Greenway, H. (1962). Plant response to saline substrates. I. Growth and ion uptake of several varieties of Hordeum during and after sodium chlorine treatment. *Aust. J. Biol. Sci.* **15**, 16–38.

Greenway, H. and Gunn, A. (1966). Phosphorus retranslocation in *Hordeum vulgare* during early tillering. *Planta* **71**, 43–67.

Greenway, H. and Munns, R. (1980). Mechanism of salt tolerance in nonhalophytes. *Annu. Rev. Plant Physiol.* **31**, 149–190.

Greenway, H. and Osmond, C. B. (1972). Salt responses of enzymes from species differing in salt tolerance. *Plant Physiol.* **49**, 256–259.

Greenway, H. and Pitman, M. G. (1965). Potassium retranslocation in seedlings of *Hordeum vulgare*. *Aust. J. Biol. Sci.* **18**, 235–247.

Greenwood, D. J. (1967). Studies on the transport of oxygen through the stems and roots vegetable seedlings. *New Phytol.* **66**, 337–347.

Greenwood, D. J. (1983). Quantitative theory and the control of soil fertility. *New Phytol.* **94**, 1–18.

Greenwood, D. J., Gerwitz, A., Stone, D. A. and Barnes, A. (1982). Root development of vegetable crops. *Plant Soil* **68**, 75–96.

Gregory, P. J. (1983). Response to temperature in a stand of pearl millet (*Pennisetum thypoides* S. & H). VIII. Root development. *J. Exp. Bot.* **37**, 379–388.

Gregory, P., Shepherd, K. and Cooper, P. (1984). Effects of fertilizer on root growth and water use of barley in N.-Syria. *J. Agric. Res., Cambridge* **103**, 429–438.

Grewal, J. S. and Singh, S. N. (1980). Effect of potassium nutrition on frost damage and yield of potato plants on alluvial soils of the Punjab (India). *Plant Soil* **57**, 105–110.

Grierson, P. F. (1992). Organic acids in the rhizosphere of *Banksia integrifolia* L. f. *Plant Soil* **144**, 259–265.

Gries, D. and Runge, M. (1992). The ecological significance of iron mobilization in wild grasses. *J. Plant Nutr.* **15**, 1727–1737.

Grieve, C. M. and Maas, E. V. (1984). Betaine accumulation in salt stressed sorghum. *Physiol. Plant.* **61**, 167–171.

Griffiths, R. P., Castellano, M. A. and Caldwell, B. A. (1991). Hyphal mats formed by two ectomycorrhizal fungi and their association with Douglas-fir seedlings: A case study. *Plant Soil* **134**, 255–259.

Grill, E., Winnacker, E.-L. and Zenk, M. H. (1987). Phytochelatins, a class of heavy metal binding peptides from plants are functionally analogeous to metallothioneins. *Proc. Natl. Acad. Sci. (U.S.A.)* **84**, 439–443.

Grill, E., Winnacker, E.-L. and Zenk, M. H.(1988). Occurrence of heavy metal binding phytochelatins in plants growing in a mining refuse area. *Experientia* **44**, 539–540.

Grimes, H. D. and Hodges, T. K. (1990). The inorganic NO_3^- : NH_4^+ ratio influences plant regeneration and auxin sensitivity in primary callus derived from immature embryo of indica rice (*Oryza sativa* L.). *J. Plant Physiol.* **136**, 362–367.

Grimm, E., Bernhardt, G., Rothe, K. E. and Jacob, F. (1990). Mechanism of sucrose retrieval along the phloem path – a kinetic approach. *Planta* **182**, 480–485.

Grimme, H. (1984). Aluminium tolerance of soybean plants as related to magnesium nutrition. *In* 'Proceedings of the Sixth International Colloquium Optimizing of Plant Nutrition Montpellier, France' (P. Martin-Prevel, ed.), pp. 243–249.

Grimme, H., Strebel, O., Renger, M. and Fleige, H. (1981). Die potentielle K-Anlieferung an die Pflanzenwurzel durch Diffusion. *Mitt. Dtsch. Bodenkd. Ges.* **36**, 367–374.

Grinsted, M. J., Hedley, M. J., White, R. E. and Nye, P. H. (1982). Plant-induced changes in the rhizosphere of rape (*Brassica napus* var. Emerald) seedlings. I. pH change and the increase in P concentration in the soil solution. *New Phytol.* **91**, 19–29.

Grof, C. P. L., Johnston, M. and Brownell, P. F. (1989). Effect of sodium nutrition on the ultrastructure of chloroplasts of C_4 plants. *Plant Physiol.* **89**, 539–543.

Grosse, W. and Schröder, P. (1985). Aeration of the roots and chloroplast-free tissues of trees. *Ber. Dtsch. Bot. Ges.* **98**, 311–318.

Grosse, W., Frye, J. and Lattermann, S. (1992). Root aeration in wetland trees by pressurized gas transport. *Tree Physiol.* **10**, 285–295.

Grosse-Brauckmann, E. (1957). Über den Einfluss der Kieselsäure anf den Mehltaubefall von Getreide bei unterschiedlicher Stickstoffdüngung. *Phytopathol. Z.* **30**, 112–115.

Grossmann, F. (1976). Outlines of host-parasite interactions in bacterial diseases in relation to plant nutrition. *Proc. 12th Colloq. Int. Potash Inst. Bern*, pp. 221–224.

Grossmann, K. (1990). Plant growth retardants as tools in physiological research. *Physiol. Plant.* **78**, 640–648.

Gruhn, K. (1961). Einfluss einer Molybdän-Düngung auf einige Stickstoff-Fraktionen von Luzerne und Rotklee. *Z. Pflanzenernaehr., Dueng., Bodenkd.* **95**, 110–118.

Grundon, N. J. (1980). Effectiveness of soil-dressing and foliar sprays of copper sulphate in correcting copper deficiency of wheat (*Triticum aestivum*) in Queensland. *Aust. J. Exp. Agric. Anim. Husb.* **20**, 717–723.

Grundon, N. J. and Asher, C. J. (1986). Volatile losses of sulfur by intact alfalfa plants. *J. Plant Nutr.* **9**, 1519–1532.

Grundon, N. J. and Asher, C. J. (1988). Volatile losses of sulfur from intact plants. *J. Plant Nutr.* **11**, 563–576.

Grunes, D. L., Stout, P. R. and Brownell, J. R. (1970). Grass tetany of ruminants. *Adv. Agron.* **22**, 332–374.

Grunwald, G., Ehwald, R., Pietzsch, W. and Göring, H. (1979). A special role of the rhizodermis in nutrient uptake by plant roots. *Biochem. Physiol. Pflanz.* **174**, 831–837.

Grusak, M. A., Welch, R. M. and Kochian, L. V. (1990). Does iron deficiency in *Pisum sativum* enhance the activity of the root plasmalemma iron transport protein? *Plant Physiol.* **94**, 1353–1357.

Grzebisz, W., Floris, J. and van Noordwijk, M. (1989). Loss of dry matter and cell contents from fibrous roots of sugar beet due to sampling, storage and washing. *Plant Soil* **113**, 53–57.

Guardia, M. D. de la and Benlloch, M. (1980). Effects of potassium and gibberellic acid on stem growth of whole sunflower plants. *Physiol. Plant.* **49**, 443–448.

Gubler, W. D., Gorgan, R. G. and Osterli, P. P. (1982). Yellows of melons caused by molybdenum deficiency in acid soil. *Plant Dis.* **66**, 449–451.

Guehl, J. M. and Garbaye, J. (1990). The effects of ectomycorrhizal status on carbon dioxide assimilation capacity, water-use efficiency and response to transplanting in seedlings of *Pseudotsuga menziesii* (Mirb) Franco. *Ann. Sci. For.* **21**, 551–563.

Guinel, F. C. and McCully, M. E. (1986). Some water-related physical properties of maize root-cap mucilage. *Plant Cell Environ.* **9**, 657–666.

Gunasena, H. P. M. and Harris, P. M. (1971). The effect of CCC, nitrogen and potassium on the growth and yield of two varieties of potatoes. *J. Agric. Sci.* **76**, 33–52.

Gunjal, S. S. and Paril, P. L. (1992). Mycorrhizal control of wilt in casuarina. *Agrofor. Today* **14–15**, April–June issue.

Gupta, G. and Narayanan, R. (1992). Nitrogen fixation in soybean treated with nitrogen dioxide and molybdenum. *J. Environ. Qual.* **21**, 46–49.

Gupta, U. C. (1979). Boron nutrition of crops. *Adv. Agron.* **31**, 273–307.

Gupta, U. C. and Lipsett, J. (1981). Molybdenum in soils, plants and animals. *Adv. Agron.* **34**, 73–115.

Gur, A. and Meir, S. (1987). Root hypoxia and storage breakdown of 'Jonathan' apples. *J. Am. Soc. Hortic. Sci.* **112**, 777–783.

Gurley, W. H. and Giddens, J. (1969). Factors affecting uptake, yield response, and carry over of molybdenum on soybean seed. *Agron. J.* **61**, 7–9.

Gutschick, V. P. (1993). Nutrient-limited growth rates: roles of nutrient-use efficiency and of adaptations to increase uptake rates. *J. Exp. Bot.* **44**, 41–52.

Guttridge, C. G., Bradfield, E. G. and Holder, R. (1981). Dependence of calcium transport into strawberry leaves on positive pressure in the xylem. *Ann. Bot. (London)* [N.S.] **48**, 473–480.

Guyette, R. P., Cutter, B. E. and Henderson, G. S. (1989). Long-term relationships between molybdenum and sulfur concentrations in redcedar tree rings. *J. Environ. Qual.* **18**, 385–389.

Haeder, H.-E. and Beringer, H. (1981). Influence of potassium nutrition and water stress on the abscisic acid content in grains and flag leaves during grain development. *J. Sci. Food Agric.* **32**, 552–556.

Haeder, H.-E. and Beringer, H. (1984a). Long distance transport of potassium in cereals during grain filling in detached ears. *Physiol. Plant.* **62**, 433–438.

Haeder, H.-E. and Beringer, H. (1984b). Long distance transport of potassium in cereals during grain filling in intact plants. *Physiol. Plant.* **62**, 439–444.

Hafner, H., Ndunguru, B. J., Bationo, A. and Marschner, H. (1992). Effect of nitrogen, phosphorus and molybdenum application on growth and symbiotic N$_2$-fixation of groundnut in an acid sandy soil in Niger. *Fert. Res.* **31**, 69–77.

Hafner, H., Bley, J., Bationo, A., Martin, P. and Marschner, H. (1993). Long-term nitrogen balance for pearl millet (*Pennisetum glaucum* L.) in an acid sandy soil of Niger. *Z. Pflanzenernähr. Bodenk.* **156**, 169–176.

Hager, A. and Helmle, M. (1981). Properties of an ATP-fueled, Cl$^-$-dependent proton pump localized in membranes of microsomal vesicles from maize coleoptiles. *Z. Naturforsch., Biosci. C:* **36**, 997–1008.

Häggquist, M.-L., Stird, L., Widell, K.-O. and Liljenberg, C. (1988a). Identification of tryptophan in leachate of oat hulls (*Avena sativa*) as a mediator of root growth regulation. *Physiol. Plant.* **72**, 423–427.

Häggquist, M.-L., Widell, K. O., Fredriksson, M. and Liljenberg, C. (1988b). Growth inhibitors in oat grains. II. Bioassays for characterization of a new substance in leachate of oat hulls (*Avena sativa*) regulating root growth. *Physiol. Plant.* **72**, 414–422.

Hähndel, R. and Wehrmann, J. (1986). Einfluss der NO₃-bzw. NH₄-Ernährung auf Ertrag und Nitratgehalt von Spinat und Kopfsalat. *Z. Pflanzenernähr. Bodenk.* **149**, 290–302.

Hairiah, K., van Noordwijk, M., Stulen, I. and Kuiper, P. J. C. (1992). Aluminium avoidance by *Mucuna pruriens*. *Physiol. Plant.* **86**, 17–24.

Hajibagheri, M. A., Harvey, D. M. R. and Flowers, T. J. (1987). Quantitative ion distribution within maize root cells in salt-sensitive and salt-tolerant varieties. *New Phytol.* **105**, 367–379.

Hajibagheri, M. A., Yeo, A. R., Flowers, T. J. and Collins, J. C. (1989). Salinity resistance in *Zea mays*: fluxes of potassium, sodium, and chloride, cytoplasmic concentrations and microsomal membrane lipids. *Plant Cell Environ.* **12**, 753–757.

Hale, K. A. and Sanders, F. E. (1982). Effects of benomyl on vesicular-arbuscular mycorrhizal infection of red clover (*Trifolium pratense* L.) and consequences for phosphorus inflow. *J. Plant Nutr.* **5**, 1355–1367.

Hall, A. J. and Milthorpe, F. L. (1978). Assimilation source-sink relationship in *Capsicum annuum* L. III. The effect of fruit excision on photosynthesis and leaf and stem carbohydrates. *Aust. J. Plant Physiol.* **5**, 1–13.

Haller, T. and Stolp, H. (1985). Quantitative estimation of root exudation of maize plants. *Plant Soil* **86**, 207–216.

Halliwell, B. (1978). Biochemical mechanisms accounting for the toxic action of oxygen on living organisms: The key role of superoxide dismutase. *Cell Biol. Int. Rep.* **2**, 113–128.

Halliwell, B. and Gutteridge, J. M. C. (1986). Iron and free radical reactions: two aspects of antioxidant protection. *Trends Biochem. Sci.* **11**, 372–375.

Hallock, D. L. and Garren, K. H. (1968). Pod breakdown, yield and grade of Virginia type peanuts as affected by Ca, Mg, and K sulfates. *Agron. J.* **60**, 253–357.

Hambidge, K. M. and Walravens, P. A. (1976). Zinc deficiency in infants and preadolescent children. *In* 'Trace Elements in Human Health and Disease' (A. S. Prasad and D. Overleas, eds.), Vol. 1, Chapter 2, pp. 21–32. Academic Press, New York.

Hamel, C. and Smith, D. L. (1992). Mycorrhizae-mediated ¹⁵N transfer from soybean to corn in field-grown intercrops: effect of component crop spatial relationships. *Soil Biol. Biochem.* **24**, 499–501.

Hamel, C., Nesser, C., Barrautes-Cartin, U. and Smith, D. L. (1991). Endomycorrhizal fungal species mediate ¹⁵N transfer from soybean to maize in non-fumigated soil. *Plant Soil* **138**, 41–47.

Hamilton, D. A. and Davies, P. J. (1988). Mechanism of export of organic material from the developing fruits of pea. *Plant Physiol.* **86**, 956–959.

Hamilton, J. L. and Lowe, R. H. (1981). Organic matter and N effects on soil nitrite accumulation and resultant nitrite toxicity to tobacco transplants. *Agron. J.* **73**, 787–790.

Hamilton, M. A. and Westermann, D. T. (1991). Comparison of DTPA and resin extractable soil Zn to plant zinc uptake. *Commun. Soil Sci. Plant Anal.* **22**, 517–528.

Hammes, P. S. and Beyers, E. A (1973). Localization of the photoperiodic perception in potatoes. *Potato Res.* **16**, 68–72.

Hampe, T. and Marschner, H. (1982). Effect of sodium on morphology, water relations and net photosynthesis in sugar beet leaves. *Z. Pflanzenphysiol.* **108**, 151–162.

Hamza, M. and Aylmore, L. A. G. (1991). Liquid ion exchanger microelectrodes used to study soil solute concentrations near plant roots. *Soil Sci. Soc. Am. J.* **55**, 954–958.

Hancock, J. G. and Huisman, O. C. (1981). Nutrient movement in host–pathogen systems. *Annu. Rev. Phytopathol.* **19**, 309–331.

Handreck, K. A. (1991). Interactions between iron and phosphorus in the nutrition of *Banksia ericifolia* L. f. var. *ericifolia* (Proteaceae) in soil-less potting media. *Aust. J. Bot.* **39**, 373–384.

Handreck, K. A. (1992). Relative effectiveness of iron sources for an iron-inefficient species growing in a soilless medium. *J. Plant Nutr.* **15**, 179–189.

Handreck, K. K. and Riceman, D. S. (1969). Cobalt distribution in several pasture species grown in culture solutions. *Aust. J. Agric. Res.* **20**, 213–226.

Hanisch, H.-C. (1980). Zum Einfluss der Stickstoffdüngung und vorbeugender Spritzung von Natronwasserglas zu Weizenpflanzen auf deren Widerstandsfähigkeit gegen Getreideblattläuse. *Kali-Briefe* **15**, 287–296.

Hannam, R. J. and Ohki, K. (1988). Detection of manganese deficiency and toxicity in plants. *In* 'Manganese in Soils and Plants' (R. D. Graham, R. J. Hannam and N. C. Uren, eds.), pp. 243–259. Kluwer Academic, Dordrecht.

Hannam, R. J., Riggs, J. L. and Graham, R. D. (1987). The critical concentration of manganese in barley. *J. Plant Nutr.* **10**, 2039–2048.

Hansen, A. P., Pate, J. S. and Atkins, C. A. (1987). Relationships between acetylene reduction activity, hydrogen evolution and nitrogen fixation in nodules of *Acacia* ssp: Experimental background to assaying fixation by acetylene reduction under field conditions. *J. Exp. Bot.* **38**, 1–12.

Hansen, A. P., Yoneyama, T. and Kouchi, H. (1992). Short term nitrate effects on hydroponically grown soybean cv. Bragg and its supernodulating mutant. II. Distribution and respiration of recently fixed ^{13}C-labelled photosynthate. *J. Exp. Bot.* **43**, 9–14.

Hansen, A. P., Yoneyama, T., Kouchi, H. and Martin, P. (1993). Respiration and nitrogen fixation of hydroponically cultured *Phaseolus vulgaris* L. cv. OAC Rico and a supernodulating mutant.I. Growth, mineral composition and effect of sink removal. *Planta* **189**, 538–545.

Hanson, E. J. (1991a). Movement of boron out of tree fruit leaves. *HortScience* **26**, 271–273.

Hanson, E. J. (1991b). Sour cherry trees respond to foliar boron applications. *HortScience* **26**, 1142–1145.

Hanson, E. J., Chaplin, M. H. and Breen, P. J. (1985). Movement of foliar applied boron out of leaves and accumulation in flower buds and flower parts of 'Italian' prune. *HortScience* **20**, 747–748.

Hanson, J. B. (1984). The function of calcium in plant nutrition. *In* 'Advances in Plant Nutrition' (P. B. Tinker and A. Läuchli, eds.), pp. 149–208. Praeger, New York.

Hanson, P. J., Sucoff, E. I. and Markhardt III, A. H. (1985). Quantifying apoplastic flux through red pine root systems using trisodium, 3-hydroxy-5,8,10-pyrenetrisulfonate. *Plant Physiol.* **77**, 21–24.

Hantschel, R., Kaupenjohann, M., Horn, R., Gradl, J. and Zech, W. (1988). Ecologically important differences between equilibrium and percolation soil extracts. *Geoderma* **43**, 213–227.

Haque, M. Z. (1988). Effect of nitrogen, phosphorus and potassium un spikelet sterility induced by low temperature at the reproductive stage of rice. *Plant Soil* **109**, 31–36.

Harley, J. L. and Harley, E. L. (1987). A check-list of mycorrhiza in the British flora. *New Phytol.* (*Suppl.*) **105**, 1–102.

Harper, J. E. and Gibson, A. H. (1984). Differential nodulation tolerance to nitrate among legume species. *Crop Sci.* **24**, 797–801.

Harper, L. A., Sharpe, R. R., Langdale, G. W. and Giddens, J. E. (1987). Nitrogen cycling in a wheat crop: soil, plant, and aerial nitrogen transport. *Agron. J.* **79**, 965–973.

Harper, S. H. T. and Lynch, J. M. (1982). The role of water-soluble components in phytotoxicity from decomposing straw. *Plant Soil* **65**, 11–17.

Harris, J. M., Lucas, J. A., Davey, M. R., Lethbridge, G. and Powell, K. A. (1989). Establishment of *Azospirillum* inoculant in the rhizosphere of winter wheat. *Soil Biol. Biochem.* **21**, 59–64.

Harrison, S. J., Lepp, N. W. and Phipps, D. A. (1979). Uptake of copper by excised roots. II. Copper desorption from the free space. *Z. Pflanzenphysiol.* **94**, 27–34.

Hart, A. L. (1989). Distribution of phosphorus in nodulated white clover plants. *J. Plant Nutr.* **12**. 159–171.

Hart, A. L. and Colville, C. (1988). Differences among attributes of white clover genotypes at various levels of phosphorus supply. *J. Plant Nutr.* **11**, 189–207.

Hartel, H. (1977). Wirkung einer Harnstofffernährung auf Harnstoffumsatz und N-Stoffwechsel von Mais und Sojabohnen. PhD Thesis, Technische Universität, München.

Hartel, P. G. and Bouton, J. H. (1991). *Rhizobium meliloti* inoculation of alfalfa selected for tolerance to acid, aluminium-rich soils. *In* 'Plant–Soil Interactions at low pH' (R. J. Wright, V. C. Baligar, R. P. Murrmann, eds.), pp. 597–601. Kluwer Academic, Dordrecht, The Netherlands.

Hartt, C. E. (1969). Effect of potassium deficiency upon translocation of ^{14}C in attached blades and entire plants of sugarcane. *Plant Physiol.* **44**, 1461–1469.

Hartung, W. and Slovik, S. (1991). Translay Review No. 35. Physicochemical properties of plant growth regulators and plant tissues determine their distribution and redistribution: stomatal regulation by abscisic acid in leaves. *New Phytol.* **119**, 361–382.

Hartung, W. J., Radin, J. W. and Hendrix, D. L. (1988). Abscisic acid movement into the apoplasmic solution of water stressed cotton leaves: Role of apoplastic pH. *Plant Physiol.* **86**, 908–913.

Hartung, W., Weiler, E. W. and Radin, J. W. (1992). Auxin and cytokinins in the apoplastic solution of dehydrated cotton leaves. *J. Plant Physiol.* **140**, 324–327.

Hartwig, E. E., Jones, W. F. and Kiolen, T. C. (1991a). Identification and inheritance of inefficient zinc absorption in soybean. *Crop Sci.* **31**, 61–63.

Hartwig, U. A., Joseph, C. M. and Phillips, D. A. (1991b). Flavenoids released naturally from alfalfa seeds enhance growth rate of *Rhizobium meliloti*. *Plant Physiol.* **95**, 797–803.

Hartwig, V., Boller, B. and Nösberger, H. P. (1987). Oxygen supply limits nitrogenase activity of clover nodules after defoliation. *Ann. Bot.* **59**, 285–291.

Hartzook, A., Karstadt, D., Naveh, M. and Feldman, S. (1974). Differential iron absorption efficiency of peanut (*Arachis hypogaea* L.) cultivars grown on calcareous soils. *Agron. J.* **66**, 114–115.

Haschke, H. P. and Lüttge, K. (1975). Interactions between IAA, potassium, and malate accumulation on growth in *Avena* coleoptile segments. *Z. Pflanzenpysiol.* **76**, 450–455.

Hasenstein, K. H. and Evans, M. L. (1988). Effects of cations on hormone transport in primary roots of *Zea mays*. *Plant Physiol.* **86**, 890–894.

Hatch, M. D. and Burnell, J. N. (1990). Carbonic anhydrase activity in leaves and its role in the first stop of C_4 photosynthesis. *Plant Physiol.* **93**, 825–828.

Hauck, C., Müller, S. and Schildknecht, H. (1992). A germination stimulant from parasitic flowering plants from *Sorghum bicolor*, a genuine host plant. *J. Plant Physiol.* **139**, 474–478.

Haug, A. and Shi, B. (1991). Biochemical basis of aluminium tolerance in plant cells. *In* 'Plant–Soil Interactions at Low pH' (R. J. Wright, V. C. Baligar and R. P. Murrmann, eds.), pp.839–850. Kluwer Academic, Dordrecht, The Netherlands.

Haug, I. and Feger, K. H. (1990). Effects of fertilization with $MgSO_4$ and $(NH_4)_2SO_4$ on soil solution chemistry, mycorrhiza and nutrient content of fine roots in a Norway spruce stand. *Water, Air and Soil Pollution* **54**, 453–468.

Häussling, M. and Marschner, H. (1989). Organic and inorganic soil phosphates and acid phosphatase activity in the rhizosphere of 80-year-old Norway spruce (*Picea abies* (L.) Karst.) trees. *Biol. Fertil. Soils* **8**, 128–133.

Häussling, M., Leisen, E. and Marschner, H. (1990). Gradienten von pH-Werten und Nährstoffaufnahmeraten bei Langwurzeln von Fichten (*Picea abies* (L.) Karst.) unter kontrollierten Bedingungen und auf Standorten in Baden-Württemberg. *Kali-Briefe* **20**, 431–439.

Häussling, M., Römheld, V. und Marschner, H. (1985). Beziehungen zwischen Chlorosegrad, Eisengehalten und Blattwachstum von Weinreben auf verschiedenen Standorten. *Vitis* **24**, 158–168.

Häussling, M., Jorns, C. A., Lehmbecker, G., Hecht-Buchholz, Ch. and Marschner, H. (1988). Ion and water uptake in relation to root development in Norway spruce (*Picea abies* (L.) Karst.). *J. Plant Physiol.* **133**, 486–491.

Hawes, M. C. (1990). Living plant cells released from the root cap: a regulator of microbial populations in the rhizosphere. *Plant Soil* **129**, 19–27.

Hawker, J. S., Jenner, C. F. and Niemietz, C. M. (1991). Sugar metabolism and compartmentation. *Aust. J. Plant Physiol.* **18**, 227–237.

Hawker, J. S., Marschner, H. and Downton, W. J. S. (1974). Effect of sodium and potassium on starch synthesis in leaves. *Aust. J. Plant Physiol.* **1**, 491–501.

Hawkesford, M. J. and Belcher, A. R. (1991). Differential protein synthesis in response to sulphate and phosphate deprivation: Identification of possible components of plasma-membrane transport systems in cultured tomato roots. *Planta* **185**, 323–329.

Hayashi, H. and Chino, M. (1990). Chemical composition of phloem sap from the uppermost internode of the rice plant. *Plant Cell Physiol.* **31**, 247–251.

Haynes, B., Koide, R. T. and Elliott, G. (1991). Phosphorus uptake and utilization in wild and cultivated oats (*Avena* ssp.). *J. Plant Nutr.* **14**, 1105–1118.

Haynes, R. J. (1980). Ion exchange properties of roots and ionic interactions within the root apoplasm: their role in ion accumulation by plants. *Bot. Rev.* **46**, 75–99.

Haynes, R. J. (1983). Soil acidification induced by leguminous crops. *Grass Forage Sci. (Oxford)* **38**, 1–11.

Haynes, R. J. (1990). Active ion uptake and maintenance of cation–anion balance: a cricital examination of their role in regulating rhizosphere pH. *Plant Soil* **126**, 247–264.

Haynes, R. J. and Ludecke, T. E. (1981). Yield, root morphology and chemical composition of two pasture legumes as affected by lime and phosphorus applications to an acid soil. *Plant Soil* **62**, 241–254.

Haystead, A. and Sprent, J. I. (1981). Symbiotic nitrogen fixation. *In* 'Physiological Processes Limiting Plant Productivity' (C. B. Johnson, ed.), pp. 345–364.

He, C.-J., Morgan, P. W. and Drew, M. C. (1992). Enhanced sensitivity to ethylene in nitrogen- or phosphate-starved roots of *Zea mays* L. during aerenchyma formation. *Plant Physiol.* **98**, 137–142.

Heath, M. C. and Stumpf, M. A. (1986). Ultrastructural observations of penetration sites of the cowpea rust fungus in untreated and silicon-depleted French bean cells. *Physiol. Mol. Plant Pathol.* **29**, 27–39.

Hebbar, P., Berge, O., Heulin, T. and Singh, S. P. (1991). Bacterial antagonists of sunflower (*Helianthus annuus* L.) fungal pathogens. *Plant Soil* **133**, 131–140.

Heber, U., Kirk, M. R., Gimmler, H. and Schäfer, G. (1974). Uptake and reduction of glycerate by isolated chloroplasts. *Planta* **120**, 32–46.

Heber, U., Viil, J., Neimanis, S., Mimura, T. and Dietz, K.-J. (1989). Photoinhibitory damage to chloroplasts under phosphate deficiency and alleviation of deficiency and damage by photorespiratory reactions. *Z. Naturforsch.* **44c**, 524–536.

Hecht-Buchholz, C. (1967). Über die Dunkelfärbung des Blattgrüns bei Phosphormangel. *Z. Pflanzenernähr. Bodenk.* **118**, 12–22.

Hecht-Buchholz, C. (1972). Wirkung der Mineralstoffernährung auf die Feinstruktur der Pflanzenzelle. *Z. Pflanzenernähr. Bodenk.* **132**, 45–68.

Hecht-Buchholz, C. (1973). Molybdänverteilung und -verträglichkeit bei Tomate, Sonnenblume und Bohne. *Z. Pflanzenernähr. Bodenk.* **136**, 110–119.

Hecht-Buchholz, C. (1979). Calcium deficiency and plant ultrastructure. *Commun. Soil. Sci. Plant Anal.* **10**, 67–81.

Hecht-Buchholz, C., Pflüger, R. and Marschner, H. (1971). Einfluss von Natriumchlorid auf Mitochondrienzahl und Atmung von Maiswurzelspitzen. *Z. Pflanzenphysiol.* **65**, 410–417.

Heckmann, M.-O., Drevon, J.-J., Saglio, P. and Salsac, L. (1989). Effect of oxygen and malate on NO_3^- inhibition of nitrogenase in soybean nodules. *Plant Physiol.* **90**, 224–229.

Hedley, M. J., Nye, P. H. and White, R. E. (1982). Plant-induced changes in the rhizosphere of rape (*Brassica napus* var. Emerald) seedlings. II. Origin of the pH change. *New Phytol.* 91, 31–44.

Hedrich, R. and Schroeder, J. I. (1989). The physiology of ion channels and electrogenic pumps in higher plants. *Annu. Rev. Plant Physiol. Plant Mol. Biol.* **40**, 539–569.

Hedrich, R., Busch, H. and Raschke, K. (1990). Ca^{2+} and nucleotide dependent regulation of voltage dependent anion channels in the plasma membrane of guard cells. *EMBO J.* **9**, 3889–3892.

Hedrich, R., Schroeder, J. I. and Fernandez, J. M.(1986). Patch-clamp studies on higher plant cells: a perspective. *Trends Biochem. Sci.* **12**, 49–52.

Heenan, D. P. and Campbell, L. C. (1981). Influence of potassium and manganese on growth and uptake of magnesium by soybeans (*Glycine max* (L.) Merr. cv Bragg). *Plant Soil* **61**, 447–456.

Heenan, D. P. and Carter, O. G. (1977). Influence of temperature on the expression of manganese toxicity by two soybean varieties. *Plant Soil* **47**, 219–227.

Heenan, D. P., Campbell, L. C. and Carter, O. G (1981). Inheritance of tolerance to high manganese supply in soybean. *Crop Sci.* **21**, 625–627.

Hefler, S. K. and Averill, B. A. (1987). The 'manganese (III)-containing' purple acid phosphatase from sweet potatoes is an iron enzyme. *Biochem. Biophys. Res. Commun.* **146**, 1173–1177.

Hehl, G. and Mengel, K. (1972). Der Einfluss einer variierten Kalium- und Stickstoffdüngung auf den Kohlenhydratgehalt verschiedener Futterpflanzen. *Landwirtsch. Forsch. Sonderh.* **27(2)**, 117–129.

Hein, M. B., Brenner, M. L. and Brun, W. A. (1984). Concentrations of abscisic acid and indole-3-acetic acid in soybean seeds during development. *Plant Physiol.* **76**, 951–954.

Heineke, D. and Heldt, H. W. (1988). Measurement of light-dependent changes of the stromal pH in wheat leaf protoplasts. *Bot. Acta* **101**, 45–47.

Heineke, D., Sonnewald, U., Büssis, D., Günter, G., Leidreiter, K., Wilke, I., Raschke, K., Willmitzer, L. and Heldt, H. W. (1992). Apoplastic expression of yeast-derived invertase in potato. Effects on photosynthesis, leaf solute composition, water relations, and tuber composition. *Plant Physiol.* **100**, 301–308.

Heinze, M. and Fiedler, H. J. (1992). Ernährung der Gehölze. *In* 'Physiologie und Ökologie der Gehölze' (H. Lyr, H. J. Fiedler and W. Tranquillini, eds.), pp. 43–115. Fischer Verlag, Stuttgart.

Helal, H. M. and Dressler, A. (1989). Mobilization and turnover of soil phosphorus in the rhizosphere. *Z. Pflanzenernähr. Bodenk.* **152**, 175–180.

Helal, H. M. and Mengel, K. (1979). Nitrogen metabolism of young barley plants as affected by NaCl-salinity and potassium. *Plant Soil* **51**, 457–462.

Helal, H. M. and Mengel, K. (1981). Interaction between light intensity and NaCl salinity and their effects on growth, CO_2 assimilation, and photosynthate conversion in young broad beans. *Plant Physiol.* 67, 999–1002.

Helal, H. M. and Sauerbeck, D. (1989). Input and turnover of plant carbon in the rhizosphere. *Z. Pflanzenernähr. Bodenk.* **152**, 211–216.

Helder, R. J. and Boerma, J. (1969). An electron-microscopical study of the plasmodesmata in the roots of young barley seedlings. *Acta Bot. Neerl.* **18**, 99–107.

Heldt, H. W., Chon, C. J., Maronde, D., Herold, A., Stankovic, Z. S., Walker, D. A., Kraminer, A., Kirk, M. R. and Heber, U. (1977). Role of orthophosphate and other factors in the regulation of starch formation in leaves and isolated chloroplasts. *Plant Physiol.* **59**, 1146–1155.

Heldt, H. W., Flügge, U.-I. and Borchert, S. (1991). Diversity of specificity and function of phosphate translocators in various plastids. *Plant Physiol.* **95**, 341–343.

Hell, R. and Bergmann, L. (1988). Glutathione synthetase in tobacco suspension cultures: catalytic properties and localization. *Physiol. Plant.* **72**, 70–76.

Hendricks, T. and van Loon, L. C. (1990). Petunia peroxidase a is localized in the epidermis of aerial plant organs. *J. Plant Physiol.* **136**, 519–525.

Hendriks, L., Claassen, N. and Jungk, A. (1981). Phosphatverarmung des wurzelnahen Bodens und Phosphataufnahme von Mais und Raps. *Z. Pflanzenernähr. Bodenk.* **144**, 486–499.

Hendrix, J. E. (1967). The effect of pH on the uptake and accumulation of phosphate and sulfate ions by bean plants. *Am. J. Bot.* **54**, 560–564.

Hendrix, J. W., Jones, J. J. and Nesmith, W. C. (1992). Control of pathogenic mycorrhizal fungi in maintenance of soil productivity of crop rotation. *J. Prod. Agric.* **5**, 383–386.

Hendry, G. A. F. and Brocklebank, K. J. (1985). Iron-induced oxygen radical metabolism in waterlogged plants. *New Phytol.* **101**, 199–206.

Henriques, F. S. (1989). Effects of copper deficiency on the photosynthetic apparatus of sugar beet (*Beta vulgaris* L.). *J. Plant Physiol.* **135**, 453–458.

Hentschel, E., Godbold, D. L., Marschner, P., Schlegel, H. and Jentschke, G. (1993). The effect of *Paxillus involutus* Fr. on aluminum sensitivity of Norway spruce seedlings. *Tree Physiol.* **12**, 379–390.

Herms, U. and Brümmer, G. (1979). Einfluss der Redoxbedingungen auf die Löslichkeit von Schwermetallen in Böden und Sedimenten. *Mitt. Dtsch. Bodenkd. Ges.* **29**, 533–544.

Herms, U. and Brümmer, G. (1980). Einfluss der Bodenreaktion auf Löslichkeit und tolerierbare Gesamtgehalte an Nickel, Kupfer, Zink, Cadmium und Blei in Böden und kompostierten Siedlungsabfällen. *Landwirtsch. Forsch.* **33**, 408–423.

Hernandez, J. A., Corpas, F. J., Gomez, M., del Rio, L. A. and Sevilla, F. (1993). Salt-induced oxidative stress mediated by activated oxygen species in pea leaf mitochondria. *Physiol. Plant.* **89**, 103–110.

Herridge, D. F. and Pate, J. S. (1977). Utilization of net photosynthate for nitrogen fixation and protein production in an annual legume. *Plant Physiol.* **60**, 759–764.

Herzog, H. (1981). Wirkung von zeitlich begrenzten Stickstoff- und Cytokiningaben auf die Fahnenblatt- und Kornentwickung von Weizen. *Z. Pflanzenernähr. Bodenk.* **144**, 241–253.

Herzog, H. and Geisler, G. (1977). Der Einfluss von Cytokininapplikation auf die Assimilateinlagerung und die endogene Cytokininaktivität der Karyopsen bei zwei Sommerweizensorten. *Z. Acker- Pflanzenbau* **144**, 230–242.

Heslop-Harrison, J. S. and Reger, B. J. (1986). Chloride and potassium ions and turgidity in the grass ştigma. *J. Plant Physiol.* **124**, 55–60.

Hether, N. H., Olsen, R. A. and Jackson, L. L. (1984). Chemical identification of iron reductants exuded by plant roots. *J. Plant Nutr.* **7**, 667–676.

Hetrick, B. A. D. (1991). Mycorrhizas and root architecture. *Experientia* **47**, 355–362.

Hetrick, B. A. D., Wilson, G. W. T. and Todd, T. C. (1990). Differential responses of C_3 and C_4 grasses to mycorrhizal symbiosis, phosphorus fertilization, and soil microorganisms. *Can. J. Bot.* **68**, 461–467.

Hewitt, E. J. (1983). Essential and functional methods in plants. *In* 'Metals and Micronutrients: Uptake and Utilization by Plants' (D. A. Robb and W. S. Pierpoint, eds.), pp. 313–315. Academic Press, New York.

Hewitt, E. J. and Gundry, C. S. (1970). The molybdenum requirement of plants in relation to nitrogen supply. *J. Hortic. Sci.* **45**, 351–358.

Hewitt, E. J. and McCready, C. C. (1956). Molybdenum as a plant nutrient. VII. The effects of different molybdenum and nitrogen supplies on yields and composition of tomato plants grown in sand culture. *J. Hortic. Sci.* **31**, 284–290.

Heyns, K. (1979). Über die endogene Nitrosamin-Entstehung beim Menschen. *Landwirtsch. Forsch. Sonderh.* **39**, 145–162.

Heyser, W., Evert, F. R., Fritz, E. and Eschrich, W. (1978). Sucrose in the free space of translocating maize leaf bundles. *Plant Physiol.* **62**, 491–494.

Hiatt, A. J. (1967a). Relationship of cell sap pH to organic acid change during ion uptake. *Plant Physiol.* **42**, 294–298.

Hiatt, A. J. (1967b). Reactions *in vitro* of enzymes involved in CO_2-fixation accompanying salt uptake by barley roots. *Z. Pflanzenphysiol.* **56**, 233–245.

Hiatt, A. J. and Hendricks, S. B (1967). The role of CO_2-fixation in accumulation of ions by barley roots. *Z. Pflanzenphysiol.* **56**, 220–232.

Hicks, S. K., Wendt, C. W., Gannaway, J. R. and Baker, R. B. (1989). Allelopathic effects of wheat straw on cotton germination, emergence, and yield. *Crop Sci.* **29**, 1057–1061.

Higinbotham, N., Etherton, B. and Foster, R. J. (1967). Mineral ion contents and cell transmembrane electropotentials of pea and oat seedling tissue. *Plant Physiol.* **42**, 37–46.

Hildebrand, D. F. (1989). Lipoxygenases. *Physiol. Plant.* **76**, 249–253.

Hildebrand, E. E. (1986). Ein Verfahren zur Gewinnung der Gleichgewichts-Bodenporenlösung. *Z. Pflanzenernähr. Bodenk.* **149**, 340–346.

Hill, J., Robson, A. D. and Loneragan, J. F. (1978). The effect of copper and nitrogen supply on the retranslocation of copper in four cultivars of wheat. *Aust. J. Agric. Res.* **29**, 925–939.

Hill, J., Robson, A. D. and Loneragan, J. F. (1979a). The effects of Cu supply and shading on Cu retranslocation from old wheat leaves. *Ann. Bot. (London)* [*N.S.*] **43**, 449–457.

Hill, J., Robson, A. D. and Loneragan, J. F. (1979b). The effect of copper supply on the senescence and the retranslocation of nutrients of the oldest leaf of wheat. *Ann. Bot. (London)* [*N.S.*] **44**, 279–287.

Hill, J., Robson, A. D. and Loneragan, J. F. (1979c). The effect of copper and nitrogen supply on the distribution of copper in dissected wheat grain. *Aust. J. Agric. Res.* **30**, 233–237.

Hills, M. J. and Beevers, H. (1987). Ca^{2+} stimulated neutral lipase activity in castor bean lipid bodies. *Plant Physiol.* **84**, 272–276.

Hinsinger, P. and Jaillard, B. (1993). Root-induced release of interlayer potassium and vermiculitization of phlogopite as related to potassium depletion in the rhizosphere of ryegrass. *J. Soil Sci.* **44**, 525–534.

Hinsinger, P., Jaillard, B. and Dufey, J. E. (1992). Rapid weathering of a trioctahedral mica by the roots of ryegrass. *Soil Sci. Soc. Am. J.* **56**, 977–982.

Hinsinger, P., Elsass, F., Jaillard, B. and Robert, M. (1993). Root-induced irreversible transformation of a trioctahedral mica in the rhizosphere of rape. *J. Soil Sci.* **44**, 535–545.

Hintz, R. W., Fehr, W. R. and Cianzio, S. R. (1987). Population development for the selection of high-yielding soybean cultivars with resistance to iron-deficiency chlorosis. *Crop Sci.* **27**, 707–710.

Hippeli, S. and Elstner, E. F. (1991). Oxygen radicals and air pollution. *In* 'Oxidative Stress: Oxidants and Antioxidants' (H. Sies, ed.), pp. 1–55. Academic Press.

Hirsch, A. M. and Torrey, J. G. (1980). Ultrastructural changes in sunflower root cells in relation to boron deficiency and added auxin. *Can. J. Bot.* **58**, 856–866.

Hirsch, A. M., Pengelly, W. L. and Torrey, J. G. (1982). Endogenous IAA levels in boron-deficient and control root tips of sunflower. *Bot. Gaz. (Chicago)* **143**, 15–19.

Hoagland, D. R. (1948). 'Lectures on the Inorganic Nutrition of Plants', pp. 48–71. Chronica Botanica, Waltham, MA.

Hoan, N. T., Prasado Rao, U. and Siddiq, E. A. (1992). Genetics of tolerance to iron chlorosis in rice. *Plant Soil* **146**, 233–239.

Hochmuth, G. J., Gabelman, W. H. and Gerloff, G. C. (1985). A gene affecting tomato root morpholgy. *HortScience* **20**, 1099–1101.

Hocking, P. J. (1980a). Redistribution of nutrient elements from cotyledons of two species of annual legumes during germination and seedling growth. *Ann. of Bot.* **45**, 383–396.

Hocking, P. J. (1980b). The composition of phloem exudate and xylem sap from tree tobacco (*Nicotiana glauca* Groh). *Ann. Bot. (London)* [*N.S.*] **45**, 633–643.

Hocking, P. J. and Meyer, C. P. (1991). Effects of enrichment and nitrogen stress on growth, and partitioning of dry matter and nitrogen in wheat and maize. *Aust. J. Plant Physiol.* **18**, 339–356.

Hocking, P. J. and Pate, J. S. (1978). Accumulation and distribution of mineral elements in annual lupins *Lupinus albus* and *Lupinus angustifolius* L. *Aust. J. Agric. Res.* **29**, 267–280.

Hodenberg, A. von and Finck, A. (1975). Ermittlung von Toxizitäts-Grenzwerten für Zink, Kupfer und Blei in Hafer und Rotklee. *Z. Planzenernähr. Bodenk.* **138**, 489–503.

Hodgson, J. F., Lindsay, W. L. and Trierweiler, J. T. (1966). Micronutrient cation complexing in soil solution. II. Complexing of zinc and copper in displaced solution from calcareous soils. *Soil Sci. Soc. Am. Proc.* **30**, 723–726.

Hodgson, R. A. J. and Raison, J. K. (1991). Superoxide production by thylakoids during chilling and its implication in the susceptibility of plants to chilling induced photoinhibition. *Planta* **183**, 222–228.

Hodson, M. J. and Parry, W. D. (1982). The ultrastructure and analytical microscopy of silicon deposition in the aleurone layer of the caryopsis of *Setaria italica* (L.) Beauv. *Ann. Bot. (London)* [*N.S.*] **50**, 221–228.

Hodson, M. J. and Sangster, A. G. (1988). Observations on the distribution of mineral elements in the leaf of wheat (*Triticum aestivum* L.), with particular reference to silicon. *Ann. Bot.* **62**, 463–471.

Hodson, M. J. and Sangster, A. G. (1989a). X-ray microanalysis of the seminal root of *Sorghum bicolor* with particular reference to silicon. *Ann. Bot.* **64**, 659–667.

Hodson, M. J. and Sangster, A. G. (1989b). Silica deposition in the inflorescence bracts of wheat (*Triticum aestivum*). II. X-ray microanalysis and backscattered electron imaging. *Can. J. Bot.* **67**, 281–287.

Hodson, M. J. and Sangster, A. G. (1989c). Subcellular localization of mineral deposits in the roots of wheat (*Triticum aestivum* L.). *Protoplasma* **151**, 19–32.

Hofer, R.-M. (1991). Root hairs. *In* 'Plant Roots: the Hidden Half' (Y. Waisel, A. Eshel and U. Kafkafi, eds.), pp. 129–148. Marcel Dekker, New York.

Hofer, R. M. and Pilet, P. E. (1986). Structural and cytochemical analysis of the cell walls in growing maize roots. *J. Plant Physiol.* **122**, 395–402.

Hoffland, E. (1992). Quantitative evaluation of the role of organic acid exudation in the mobilization of rock phosphate by rape. *Plant Soil* **140**, 279–289.

Hoffland, E., Findenegg, G. R. and Nelemans, J. A. (1989a). Solubilization of rock phosphate by rape. I. Evaluation of the role of the nutrient uptake pattern. *Plant Soil* **113**, 155–160.

Hoffland, E., Findenegg, G. R. and Nelemans, J. A. (1989b). Solubilization of rock phosphate by rape. II. Local root exudation of organic acids as a response to P-starvation. *Plant Soil* **113**, 161–165.

Hoffland, E., van den Boogaard, R., Nelemans, J. A. and Findenegg, G. R. (1992). Biosynthesis and root exudation of citric and malic acid in phosphate-starved rape plants. *New Phytol.* **122**, 675–680.

Hoffman, G. J. and Phene, C. J. (1971). Effect of constant salinity levels on water use efficiency of bean and cotton. *Trans. ASAE* **14**, 1103–1106.

Hoffmann, B. and Bentrup, F.-W. (1989). Two proton pumps operate in parallel across the tonoplast of vacuoles isolated from suspension cells of *Chenopodium rubrum* L. *Bot. Acta* **102**, 297–301.

Hofinger, M. and Böttger, M. (1979). Identification by GC-MS of 4-chloroindolylacetic acid and its methyl ester in immature *Vicia faba* seeds. *Phytochemistry* **18**, 653–654.

Höfner, W. and Grieb, R. (1979). Einfluss von Fe- und Mo-Mangel auf den Ionengehalt mono- und dikotyler Pflanzen unterschiedlicher Chloroseanfälligkeit. *Z. Pflanzenernähr. Bodenk.* **142**, 626–638.

Höfte, M., Seong, K. Y., Jurkevitch, E. and Verstraete, W. (1991). Pyoverdin production by the plant growth beneficial *Pseudomonas* strain 7NSK₂: ecological significance in soil. *Plant Soil* **130**, 249–257.

Högberg, P. (1986). Soil nutrient availability, root symbioses and tree species composition in tropical Africa: a review. *J. Trop. Ecol.* **2**, 359–372.

Högberg, P. (1990). ¹⁵N natural abundance as a possible marker of the ectomycorrhizal habit of trees in mixed African woodlands. *New Phytol.* **115**, 483–486.

Höglund, A.-S., Lenman, M., Falk, A. and Rask, L. (1991). Distribution of myrosinase in rape-seed tissues. *Plant Physiol.* **95**, 213–221.

Holden, M. J., Luster, D. G., Chaney, R. L., Buckhout, T. J. and Robinson, C. (1991). Fe^{3+}-chelate reductase activity of plasma membranes isolated from tomato (*Lycopersicon esculentum* Mill.) roots. Comparison of enzymes from Fe-deficient and Fe-sufficient roots. *Plant Physiol.* **97**, 537–544.

Holobradá, M. and Kubica, S. (1988). The role of maize root tissues in sulphate absorption and radial transport. *Plant Soil* **111**, 177–181.

Homann, P. H. (1988). Structural effects of Cl⁻ and other anions on the water oxidizing complex of chloroplast photosystem II. *Plant Physiol.* **88**, 194–199.

Homer, F. A., Reeves, R. D., Brooks, R. R. and Baker, A. J. M. (1991). Characterization of the nickel-rich extract from the nickel hyperaccumulator *Dichapetalum gelonioides*. *Phytochemistry* **30**, 2141–2145.

Hommels, C. H., Kuiper, P. J. C. and de Haan, A. (1989a). Responses to internal potassium ion concentrations of two *Taraxacum* microspecies of contrasting mineral ecology: the role of inorganic ions in growth. *Physiol. Plant.* **77**, 562–568.

Hommels, C. H., Kuiper, P. J. C. and Tanczos, O. G. (1989b). Luxury consumption and specific utilization rates of three macro-elements in two *Taraxacum* microspecies of contrasting mineral ecology. *Physiol. Plant.* **77**, 569–578.

Hope, A. B. and Stevens, P. G. (1952). Electrical potential differences in bean roots on their relation to salt uptake. *Aust. J. Sci. Res.*, *Ser. B* **5**, 335–343.

Hopkins, H. T., Specht, A.W. and Hendricks, S.B. (1950). Growth and nutrient accumulation as controlled by oxygen supply to plant roots. *Plant Physiol.* **25**, 193–208.

Hopmans, P. (1990). Stem deformity in *Pinus radiata* plantations in south-eastern Australia: I. Response to copper fertiliser. *Plant Soil* **122**, 97–104.

Horak, O. (1985a). Zur Bedeutung des Nickels für *Fabaceae*. I. Vergleichende Untersuchungen über den Gehalt vegetativer Teile und Samen an Nickel und anderen Elementen. *Phyton (Austria)* **25**, 135–146.

Horak, O. (1985b). Zur Bedeutung des Nickels für *Fabaceae*. II. Nickelaufnahme und Nickelbedarf von *Pisum sativum* L. *Phyton (Austria)* **25**, 301–307.

Horan, D. P. and Chilvers, G. A (1990). Chemotropism – the key to ectomycorrhizal formation? *New Phytol.* **116**, 297–302.

Horesh, I. and Levy, Y. (1981). Response of iron-deficient citrus trees to foliar iron sprays with a low-surface-tension surfactant. *Sci. Hortic.* (*Amsterdam*) **15**, 227–233.

Horgan, J. M. and Wareing, P. F. (1980). Cytokinins and the growth response of seedlings of *Betula pendula* Roth. and *Acer pseudoplatanus* L. to nitrogen and phosphorus deficiency. *J. Exp. Bot.* **31**, 525–532.

Horiguchi, T. (1988a). Mechanism of mangnese toxicity and tolerance of plants. IV. Effects of silicon on alleviation of manganese toxicity of rice plants. *Soil Sci. Plant Nutr.* **34**, 65–73.

Horiguchi, T. (1988b). Mechanism of manganese toxicity and tolerance of plants. VII. Effect of light intensity on manganese-induced chlorosis. *J. Plant Nutr.* **11**, 235–246.

Horiguchi, T. and Morita, S. (1987). Mechanism of manganese toxicity and tolerance of plants. VI. Effect of silicon on alleviation of manganese toxicity of barley. *J. Plant Nutr.* **10**, 2299–2310.

Horlacher, D. (1991). Einfluss organischer und mineralischer N-Dünger auf Sprosswachstum und Nitratauswaschung bei Silomais sowie Quantifizierung der Ammoniakverluste nach Ausbringung von Flüssigmist. PhD. Thesis, University Hohenheim.

Horn, M. A., Heinstein, P. F. and Low, P. S. (1990). Biotin-mediated delivery of exogenous macromolecules into soybean cells. *Plant Physiol.* **93**, 1492–1496.

Horn, R. (1989). Die Bedeutung der Bodenstruktur für die Nährstoffverfügbarkeit. *Kali-Briefe* **19**, 505–515.

Horn, R. (1987). Die Bedeutung der Aggregierung für die Nährstoffsorption in Böden. *Z. Pflanzenernähr. Bodenk.* **150**, 13–16.

Horst, W. J. (1982). Quick screening of cowpea genotypes for manganese tolerance during vegetative and reproductive growth. *Z. Pflanzenernähr. Bodenk.* **145**, 423–435.

Horst, W. J. (1983). Factors responsible for genotypic manganese tolerance in cowpea (*Vigna unguiculata*). *Plant Soil* **72**, 213–218.

Horst, W. J. (1985). Quick screening of cowpea (*Vigna unguiculata*) genotypes for aluminium tolerance in an aluminium-treated acid soil. *Z. Pflanzenernähr. Bodenk.* **148**, 335–348.

Horst, W. J. (1987). Aluminium tolerance and calcium efficiency of cowpea genotypes. *J. Plant Nutr.* **10**, 1121–1129.

Horst, W. J. (1988). The physiology of manganese toxicity. *In* 'Manganese in Soils and Plants' (R. D. Graham, R. J Hannam and N. C. Uren, eds.), pp. 175–188. Kluwer Academic, Dordrecht.

Horst, W. J. and Göppel, H. (1986a). Aluminium-Toleranz von Ackerbohne (*Vicia faba*), Lupine (*Lupinus luteus*), Gerste (*Hordeum vulgare*) und Roggen (*Secale cereale*). I. Spross- und Wurzelwachstum in Abhängigkeit vom Aluminium-Angebot. *Z. Pflanzenernähr. Bodenk.* **149**, 83–93.

Horst, W. J. and Göppel, H. (1986b). Aluminium-Toleranz von Ackerbohne (*Vicia faba*), Lupine (*Lupinus luteus*), Gerste (*Hordeum vulgare*) und Roggen (*Secale cereale*). II. Mineralstoffgehalte in Spross und Wurzeln in Abhängigkeit vom Aluminium-Angebot. *Z. Pflanzenernähr. Bodenk.* **149**, 94–109.

Horst, W. J. and Marschner, H. (1978a). Effect of silicon on manganese tolerance of beanplants (*Phaseolus vulgaris* L.). *Plant Soil* **50**, 287–303.

Horst, W. J. and Marschner, H. (1978b). Effect of excessive manganese supply on uptake and translocation of calcium in bean plants (*Phaseolus vulgaris* L.). *Z. Pflanzenphysiol.* **87**, 137–148.

Horst, W. J. and Waschkies, Ch. (1987). Phosphatversorgung von Sommerweizen (*Triticum aestivum* L.) in Mischkultur mit Weisser Lupine (*Lupinus albus* L.). *Z. Pflanzenernähr. Bodenk.* **150**, 1–8.

Horst, W. J., Wagner, A. and Marschner, H. (1982). Mucilage protects root meristems from aluminium injury. *Z. Pflanzenphysiol.* **105**, 435–444.

Horst, W. J., Wagner, A. and Marschner, H. (1983). Effect of aluminium on root growth, cell division rate and mineral element contents in roots of *Vigna unguiculata* genotypes. *Z. Pflanzenphysiol.* **109**, 95–103.

Horst, W. J., Klotz, F. and Szulkiewicz, P. (1990). Mechanical impedance increases aluminium tolerance of soybean (*Glycine max*) roots. *Plant Soil* **124**, 227–231.

Horst, W. J., Currle, C. and Wissemeier, A. H. (1992a). Differences in calcium efficiency between cowpea (*Vigna unguiculata* (L.) Walp.) cultivars. *Plant Soil* **146**, 45–54.

Horst, W. J., Asher, C. J., Cakmak, I., Szulkiewicz, P. and Wissemeier, A. H. (1992b). Short-term responses of soybean roots to aluminium. *J. Plant Physiol.* **140**, 174–178.

Horst, W. J., Abdou, M. and Wiesler, F. (1993). Genotypic differences in phosphorus efficiency of wheat. *In* 'Plant Nutrition – from Genetic Engineering to Field Practice' (N. J. Barrow, ed.), pp. 367–370. Kluwer Academic, Dordrecht.

Houba, V. J. G., Novozamsky, I., Huybregts, A. W. M. and van der Lee, J. J. (1986). Comparison of soil extractions by 0.01 $CaCl_2$, by EUF and by some conventional extraction procedures. *Plant Soil* **96**, 433–437.

Houman, F., Godbold, D. L., Majcherczyk, A., Shasheng, W. and Hüttermann, A. (1991). Polyamines in leaves and roots of *Populus maximoviczii* grown in differing levels of potassium and phosphorus. *Can. J. For. Res.* **21**, 1748–1751.

Houtz, R. L., Nable, R. O. and Cheniae, G. M. (1988). Evidence for effects on the *in vivo* activity of ribulose-bisphosphate carboxylase/oxygenase during development of Mn toxicity in tobacco. *Plant Physiol.* **86**, 1143–1149.

Howard, D. D. and Adams, R. (1965). Calcium requirement for penetration of subsoil by primary cotton roots. *Soil Sci. Soc. Am. Proc.* **29**, 558–562.

Howeler, R. H. (1991). Identifying plants adaptable to low pH conditions. *In* 'Plant–Soil Interactions at Low pH' (R. J. Wright, V. C. Baligar and R. P. Murrmann, eds.), pp. 885–904. Kluwer Academic, Dordrecht, The Netherlands.

Howeler, R. H. and Cadavid, L. F. (1976). Screening of rice cultivars for tolerance to Al-toxicity in nutrient solutions as compared with field screening methods. *Agron. J.* **68**, 554–555.

Howeler, R. H., Cadavid, L. F. and Burckhardt, E. (1982a). Response of cassava to VA mycorrhizal inoculation and phosphorus application in greenhouse and field experiments. *Plant Soil* **69**, 327–339.

Howeler, R. H., Edwards, D. G. and Asher, C. J. (1982b). Micronutrient deficiencies and toxicities of cassava plants grown in nutrient solutions. I. Critical tissue concentrations. *J. Plant Nutr.* **5**, 1059–1076.

Howeler, R. H., Sieverding, E. and Saif, S. (1987). Practical aspects of mycorrhizal technology in some tropical crops and pastures. *Plant Soil* **100**, 249–283.

Howieson, J. G., Ewing, M. A. and D'Antuono, M. F. (1988). Selection for acid tolerance in *Rhizobium meliloti*. *Plant Soil* **105**, 179–188.

Howieson, J. G., Robson, A. D. and Abbott, L. K. (1992). Acid-tolerant species of *Medicago* produce root exudates at low pH which induce the expression of nodulation genes in *Rhizobium meliloti*. *Aust. J. Plant Physiol.* **19**, 287–296.

Hsiao, T. C. and Läuchli, A. (1986). Role of potassium in plant-water relations. *In* 'Advances in Plant Nutrition' (B. Tinker and A. Läuchli, eds.) Vol. 2, pp. 281–312. Praeger Scientific, New York.

Hsu, W. and Miller, G. W. (1968). Iron in relation to aconitate hydratase activity in *Glycine max*. Merr. *Biochim. Biophys. Acta* **151**, 711–713.

Huang, B.-R., Taylor, H. M. and McMichael, B. L. (1991a). Behaviour of lateral roots in winter wheat as affected by temperature. *Environ. Exp. Bot.* **31**, 187–192.

Huang, B.-R., Taylor, H. M. and McMichael, B. L. (1991b). Growth and development of seminal and crown roots of wheat seedlings as affected by temperature. *Environ. and Exp. Bot.* **31**, 471–477.

Huang, B. R., Taylor, H. M. and McMichael, B. L. (1991c). Effects of temperature on the development of metaxylem in primary wheat roots and its hydraulic consequence. *Ann. Bot. (London)* [*N.S.*] **67**, 163–166.

Huang, C. and Graham, R. D. (1990). Resistance of wheat genotypes to boron toxicity is expressed at the cellular level. *Plant Soil* **126**, 295–300.

Huang, C. X. and Van Steveninck, R. F. M. (1989a). Longitudinal and transverse profiles of K^+ and Cl^- concentration in 'low-' and 'high-salt' barley roots. *New Phytol.* **112**, 475–480.

Huang, C. X. and Van Steveninck, R. F. M. (1989b). The role of particular pericycle cells in the apoplastic transport in root meristems of barley. *J. Plant Physiol.* **135**, 554–558.

Huang, C. X. and Van Steveninck, R. F. M. (1989c). Maintenance of low Cl^- concentrations in mesophyll cells of leaf blades of barley seedlings exposed to salt stress. *Plant Physiol.* **90**, 1440–1443.

Huang, I. J., Welkie, G. W. and Miller, G. W. (1992a). Ferredoxin and flavodoxin analysis in tobacco in response to iron stress. *J. Plant Nutr.* **15**, 1765–1782.

Huang, J. W., Shaff, J. E., Grunes, D. L. and Kochian, L. V. (1992b). Aluminum effects on calcium fluxes at the root apex of aluminum-tolerant and aluminum-sensitive wheat cultivars. *Plant Physiol.* **98**, 230–237.

Huang, W. Z., Schoenau, J. J. and Elmy, K. (1992c). Leaf analysis as a guide to sulfur fertilization of legumes. *Commun. Soil Sci. Plant Anal.* **23**, 1031–1042.

Huber, D. M. (1980). The role of mineral nutrition in defense. *In* 'Plant Disease' Vol. V (J. G. Harsfall and E. B. Cowling, eds.), pp. 381–406. Academic Press, New York.

Huber, D. M. (1989). The role of nutrition in the take-all disease of wheat and other small grains. *In* 'Soilborne Plant Pathogens: Management of Diseases with Macro- and Microelements' (A. W. Engelhard, ed.), pp. 46–75. APS Press. The American Phytopathological Society St Paul, Minnesota.

Huber, D. M. and Watson, R. D. (1974). Nitrogen form and plant disease. *Annu. Rev. Phytophath.* **12**, 139–165.

Huber, D. M. and Wilhelm, N. S. (1988). The role of manganese in resistance to plant diseases. *In* 'Manganese in Soils and Plants' (R. D. Graham, R. J. Hannan, N. C. Uren, eds.), pp. 155–173. Kluwer Academic, Dordrecht.

Hucklesby, D. P. and Blanke, M. M. (1992). Limitation of nitrogen assimilation in plants. IV. Effect of defruiting on nitrate assimilation, transpiration, and photosynthesis in tomato leaf. *Gartenbauwiss.* **57**, 53–56.

Hue, N. V. and Amien, I. (1989). Aluminium detoxification with green manure. *Commun. Soil Sci. Plant Anal.* **20**, 1499–1511.

Hue, N. V., Craddock, G. R. and Adams, F. (1986). Effect of organic acids on Al toxicity in subsoils. *Soil Sci. Soc. Am. J.* **50**, 28–34.

Huettl, R. F. (1989). Liming and fertilization as migitation tools in declining forest ecosystems. *Water Air Soil Pollut.* **44**, 93–118.

Hughes, J. C. and Evans, J. L. (1969). Studies on after-cooking blackening. V. Changes in after-cooking blackening and the chemistry of Magestic and Ulster Beacon tubers during the growing season. *Eur. Potato J.* **12**, 26–40.

Hughes, K. A. and Gandar, P. W. (1993). Length density, occupancies and weights of apple root systems. *Plant Soil* **148**, 211–221.

Hughes, N. P. and Williams, R. J. P. (1988). An introduction to manganese biological chemistry. *In* 'Manganese in Soils and Plants' (R. D. Graham, R. J. Hannam and N. C. Uren, eds.), pp. 7–19. Kluwer Academic, Dordrecht.

Humble, G. D. and Raschke, K. (1971). Stomatal opening quantitatively related to potassium transport. *Plant Physiol.* **48**, 447–453.

Hunt, P. G., Campbell, R. B., Sojka, R. E. and Parsons, J. E. (1981). Flooding-induced soil and plant ethylene accumulation and water status response of field-grown tobacco. *Plant Soil* **59**, 427–439.

Hunt, S. and Layzell, D. B. (1993). Gas exchange of legume nodules and the regulation of nitrogenase activity. *Annu. Rev. Plant Physiol. Plant Mol. Biol.* **44**, 483–511.

Hunter, W. J., Fahring, C. J., Olsen, S. R. and Porter, L. K. (1982). Location of nitrate reduction in different soybean cultivars. *Crop Sci.* **22**, 944–948.

Hutchison, L. J. (1990). Studies on the systematics of ectomycorrhizal fungi in axenic culture. II. The enzymatic degradation of selected carbon and nitrogen compounds. *Can. J. Bot.* **68**, 1522–1530.

Hutton, J. T. and Norrish, K. (1974). Silicon content of wheat husks in relation to water transpired. *Aust. J. Agric. Res.* **25**, 203–212.

Hwang, B. K., Ibenthal, W.-D. and Heitefuss, R. (1983). Age, rate of growth, carbohydrate and amino acid content of spring barley plants in relation to their resistance to powdery mildew (*Erysiphe graminis* f. sp. *hordei*). *Physiol. Plant Pathol.* **22**, 1–14.

Hylton, L.-O., Jr., Ulrich, A. and Cornelius, D. R. (1967). Potassium and sodium interrelations in growth and mineral content of Italian ryegrass. *Agron. J.* **59**, 311–314.

Ichioka, P. S. and Arnon, D. I. (1955). Molybdenum in relation to nitrogen metabolism. II. Assimilation of ammonia and urea without molybdenum by Scenedesmus. *Plant Physiol.* **69**, 1040–1045.

Idris, M., Hossain, M. M. and Choudhury, F. A. (1975). The effect of silicon on lodging of rice in presence of added nitrogen. *Plant Soil* **43**, 691–695.

Idris, M., Vinther, F. P. and Jensen, V. (1981). Biological nitrogen fixation associated with roots of field-grown barley (*Hordeum vulgare* L.). *Z. Pflanzenernähr. Bodenk.* **144**, 385–394.

Ikeda, M., Choi, W. K. and Yamada, Y. (1991). Sucrose fatty acid esters enhance efficiency of foliar-applied urea nitrogen to soybeans. *Fert. Res.* **29**, 127–131.

Ikeda, M., Mizoguchi, K. and Yamakawa, T. (1992). Stimulation of dark carbon fixation in rice and tomato roots by application of ammonium nitrogen. *Soil Sci. Plant Nutr.* (*Tokyo*) **38**, 315–322.

Ikehashi, H. and Ponnamperuma, F. N. (1978). Varietal tolerance or rice for adverse soils. *In* 'Soils and Rice', pp. 801–823, International Rice Research Institute, Los Baños, Philippines.

Ingestad, T. and Ågren, G. I. (1992). Theories and methods on plant nutrition and growth. *Physiol. Plant.* **84**, 177–184.

Inskeep, W. P. and Bloom, P. R. (1986). Effect of soil moisture on soil pCO_2, soil solution bicarbonate, and iron chlorosis in soybeans. *Soil Sci. Soc. Am. J.* **50**, 946–952.

Inskeep, W. P. and Bloom, P. R. (1987). Soil chemical factors associated with soybean chlorosis in calciaquolls of Western Minnesota. *Agron. J.* **79**, 779–786.

Isegawar, Y., Nakano, Y. and Kitaoka, S. (1984). Conversion and distribution of cobalamin in *Euglena gracilis* Z, with special reference to its location and probable function within chloroplasts. *Plant Physiol.* **76**, 814–818.

Ishizuka, J. (1982). Characterization of molybdenum absorption and translocation in soybean plants. *Soil Sci. Plant Nutr.* (*Tokyo*) **28**, 63–78.

Islam, A. K. M. S., Asher, C. J. and Edwards, D. G (1987). Response of plants to calcium concentration in flowing solution culture with chloride or sulphate as the counter-ion. *Plant Soil* **98**, 377–395.

Islam, A. K. M. S., Edwards, D. G. and Asher, C. J. (1980). pH optima for crop growth. Results of a flowing solution culture experiment with six species. *Plant Soil* **54**, 339–357.

Islam, M. M. and Ponnamperuma, F. N. (1982). Soil and plant tests for available sulfur in wetland rice soils. *Plant Soil* **68**, 97–113.

Ismunadji, M. (1976). Rice diseases and physiological disorders related to potassium deficiency. *Proc. 12th Colloq. Int. Potash Inst. Bern*, 47–60.

Ismunadji, M. and Dijkshoorn, W. (1971). Nitrogen nutrition of rice plants measured by growth and nutrient content in pot experiments. Ionic balance and selective uptake. *Neth. J. Agric. Sci.* **19**, 223–236.

Israel, D. W. (1987). Investigation of the role of phosphorus in symbiotic dinitrogen fixation. *Plant Physiol.* **84**, 835–840.

Itoh, S. (1987). Characteristics of phosphorus uptake of chickpea in comparison with pigeonpea, soybean, and maize. *Soil Sci. Plant Nutr. (Tokyo)* **33**, 417–422.

Itoh, S. and Barber, S. A. (1983a). Phosphorus uptake by six plant species as related to root hairs. *Agron. J.* **75**, 457–461.

Itoh, S. and Barber, S. A. (1983b). A numerical solution of whole plant nutrient uptake for soil–root systems with root hairs. *Plant Soil* **70**, 403–413.

Itoh, S. and Uwano, S. (1986). Characteristics of the Cl^- action site in the O_2 evolving reaction in PSII particles: Electrostatic interaction with ions. *Plant Cell Physiol.* **27**, 25–36.

Ivins, J. D. and Bremner, P. M. (1964). Growth, development and yield in the potato. *Outlook Agric.* **4**, 211–217.

Iwasaki, K., Sakurai, K. and Takahashi, E. (1990). Copper binding by the root cell walls of italian ryegrass and red clover. *Soil Sci. Plant Nutr. (Tokyo)* **36**, 431–440.

Jackson, C., Dench, J., Moore, A. L., Halliwell, B., Foyer, C. H. and Hall, D. O. (1978). Subcellular localization and identification of superoxide dismutase in the leaves of higher plants. *Eur. J. Biochem.* **91**, 339–344.

Jackson, M. B. (1990a). Communication between the root and shoots of flooded plants. *In* 'Importance of Root to Shoot Communication in the Response to Environmental Stress', pp. 115–133. British Secondary Plant Growth Regulation Monograph 21.

Jackson, M. B. (1990b). Hormones and developmental change in plants subjected to submergence or soil waterlogging. *Aquatic Bot.* **38**, 49–72.

Jackson, M. B. (1991a). Regulation of water relationships in flooded plants by ABA from leaves, roots and xylem sap. *In* 'Abscisic Acid, Physiology and Biochemistry' (W. J. Davies and H. G. Jones, eds.), pp. 217–226. Bios. Scientific, Oxford.

Jackson, M. B. (1991b). Ethylene in root growth and development. *In* 'The Plant Hormone Ethylene' (A. K. Mattoo and J. C. Suttle, eds.), pp. 159–181. CRC Press, Boca Raton, Fl.

Jackson, M. B. and Campbell, D. L. (1979). Effects of benzyladenine and gibberellic acid on the responses of tomato plants to anaerobic root environments and to ethylene. *New Phytol.* **82**, 331–340.

Jackson, M. B., Hermann, B. and Goodenough, A. (1982). An examination of the importance of ethanol in causing injury to flooded plants. *Plant Cell Environ.* **5**, 163–172.

Jackson, M. B., Young, S. F. and Hall, K. C. (1988). Are roots the source of abscisic acid for the shoots of flooded pea plants? *J. Exp. Bot.* **39**, 1631–1637.

Jackson, P. C. and St. John, J. B. (1980). Changes in membrane lipids of roots associated with changes in permeability. I. Effect of undissociated organic acids. *Plant Physiol.* **66**, 801–804.

Jackson, W. A., Chaillou, S., Morot-Gaudry, J.-F. and Volk, R. J. (1993). Endogenous ammonium generation in maize roots and its relationship to other ammonium fluxes. *J. Exp. Bot.* **44**, 731–739.

Jacobs, M. and Rubery, P. H. (1988). Naturally occuring auxin transport regulators. *Science* **241**, 346–349.

Jacobsen, K. R., Fisher, D. G., Maretzki, A. and Moore, P. H. (1992). Developmental changes in the anatomy of the sugarcane stem in relation to phloem unloading and sucrose storage. *Bot. Acta* **105**, 70–80.

Jacobson, L., Moore, D. P. and Hannapel, R. J. (1960). Role of calcium in absorption of monovalent cations. *Plant Physiol.* 35, 352–358.

Jacoby, B. (1967). The effect of the roots on calcium ascent in bean stems. *Ann. Bot. (London) [N.S.]* **31**, 725–730.

Jacoby, B. and Rudich, B. (1985). Sodium fluxes in corn roots: comparison to Cl and K fluxes, and to Na-fluxes in barley. *Plant Cell Environ.* **8**, 235–238.

Jaffe, M. J., Huberman, M., Johnson, J. and Telewski, F. W. (1985). Thigmomorphogenesis: the induction of callose formation and ethylene evolution by mechanical perturbation in bean stem. *Physiol. Plant.* **64**, 271–279.

Jagnow, G. (1987). Inoculation of cereal crops and forage grasses with nitrogen-fixing rhizosphere bacteria: possible causes of success and failure with regard to yield response – a review. *Z. Pflanzenernähr. Bodenk.* **150**, 361–368.

Jagnow, G. (1990). Differences between cereal crop cultivars in root-associated nitrogen fixation, possible causes of variable yield response to seed inoculation. *Plant Soil* **123**, 255–259.

Jagnow, G., Höflich, G. and Hoffmann, K.-H. (1991). Inoculation of non-symbiotic rhizosphere bacteria: possibilities of increasing and stabilizing yields. *Angew. Botanik* **65**, 97–126.

Jaillard, B. (1985). Activité racinaire et rhizostructures en milieu carbonate. *Pédologie* **35**, 297–313.

Jaillard, B., Guyon, A. and Maurin, A. F. (1991). Structure and composition of calcified roots, and their identification in calcareous soils. *Geoderma* **50**, 197–210.

Jain, D. K., Beyer, D. and Rennie, R. J. (1987). Dinitrogen fixation (C_2H_2 reduction) by bacterial strains at various temperatures. *Plant Soil* **103**, 233–237.

Jakobsen, I. (1985). The role of phosphorus in nitrogen fixation by young pea plants (*Pisum sativum*). *Physiol. Plant.* **64**, 190–196.

Jakobsen, I. and Rosendahl, L. (1990). Carbon flow into soil and external hyphae from roots of mycorrhizal cucumber plants. *New Phytol.* **115**, 77–83.

Jakobsen, I., Abbott, L. K. and Robson, A. D. (1992). External hyphae of vesicular–arbuscular mycorrhizal fungi associated with *Trifolium subterraneum* L. I. Spread of hyphae and phosphorus inflow into roots. *New Phytol.* **120**, 371–380.

Jameson, P. E., McWha, J. A. and Wright, G. J. (1982). Cytokinins and changes in their activity during development of grains of wheat (*Triticum aestivum* L.). *Z. Pflanzenphysiol.* **106**, 27–36.

Janzen, H. H. (1990). Deposition of nitrogen into the rhizosphere by wheat roots. *Soil Biol. Biochem.* **22**, 1155–1160.

Jarrell, W. M. and Beverly, R. B. (1981). The dilution effect in plant nutrition studies. *Adv. Agron.* **34**, 197–224.

Jarvis, S. C. (1981). Copper concentrations in plants and their relationship to soil properties. *In* 'Copper in Soils and Plants' (J. F. Loneragan, A. D. Robson and R. D. Graham, eds.), pp. 265–285. Academic Press, London.

Jarvis, S. C. (1982). Sodium absorption and distribution in forage grasses of different potassium status. *Ann. Bot. (London) [N.S.]* **49**, 199–206.

Jarvis, S. C. (1987). The uptake and transport of silicon by perennial ryegrass and wheat. *Plant Soil* **97**, 429–437.

Jarvis, S. C. and Hatch, D. J. (1985). Rates of hydrogen ion efflux by nodulated legumes grown in flowing solution culture with continuous pH monitoring and adjustment. *Ann. Bot.* **55**, 41–51.

Jarvis, S. C. and Robson, A. D. (1982). Absorption and distribution of copper in plants with sufficient or deficient supplies. *Ann. Bot. (London)* [*N.S.*] **50**, 151–160.

Jasper, D. A., Abbott, L. K. and Robson, A. D. (1989a). The loss of VA mycorrhizal infectivity during bauxite mining may limit the growth of *Acacia pulchella* R. Br. *Aust. J. Bot.* **37**, 33–42.

Jasper, D. A., Abbott, L. K. and Robson, A. D. (1989b). Hyphae of a vesicular–arbuscular mycorrhizal fungus maintain infectivity in dry soil, except when the soil is disturbed. *New Phytol.* **112**, 101–107.

Jauregui, M. A. and Reisenauer, H. M. (1982). Dissolution of oxides of manganese and iron by root exudate components. *Soil Sci. Soc. Am. J.* **46**, 314–317.

Jayachandran, K., Schwab, A. P. and Hetrick. B. A. D. (1992). Mineralization of organic phosphorus by vesicular-arbuscular mycorrhizal fungi. *Soil Biol. Biochem.* **24**, 897–903.

Jayman, T. C. Z. and Sivasubramaniam, S. (1975). Release of bound iron and aluminium from soils by the root exudates of tea (*Camellia sinensis*) plants. *J. Sci. Food Agric.* **26**, 1895–1898.

Jeffrey, D. W. (1968). Phosphate nutrition of Australian heath plants. II. The formation of polyphosphate by five heath species. *Aust. J. Bot.* **16**, 603–613.

Jenkyn, J. F. (1976). Nitrogen and leaf diseases of spring barley. *Proc. 12th Colloq. Int. Potash Inst. Bern*, pp. 119–128.

Jennings, D. H. (1976). The effect of sodium chloride on higher plants. *Biol. Rev. Cambridge Philos. Soc.* **51**, 453–486.

Jennings, D. H. (1987). Translocation of solutes in fungi. *Biol. Rev.* **62**, 215–243.

Jennings, D. H. (1989). Some perspectives on nitrogen and phosphorus metabolism in fungi. *In* 'Nitrogen, Phosphorus and Sulphur Utilization by Fungi' (L. Boddy, R. Marchant and D. J. Read, eds.). Cambridge University Press, Cambridge.

Jensen, C. R., Stolzy, L. H. and Letey, J. (1967). Tracer studies of oxygen diffusion through roots of barley corn, and rice. *Soil Sci.* **103**, 23–29.

Jensén, P., Erdei, L. and Moller, I. M. (1987). K^+ uptake in plant roots: experimental approach and influx models. *Physiol. Plant.* **70**, 743–748.

Jentschke, G., Schlegel, H. and Godbold, D. L. (1991). The effect of aluminium on uptake and distribution of magnesium and calcium in roots of mycorrhizal Norway spruce seedlings. *Physiol. Plant.* **82**, 266–270.

Jeschke, W. D. (1977). K^+-Na^+-exchange and selectivity in barley root cells: effect of Na^+ on the Na^+ fluxes. *J. Exp. Bot.* **28**, 1289–1305.

Jeschke, W. D. and Jambor, W. (1981). Determination of unidirectional sodium fluxes in roots of intact sunflower seedlings. *J. Exp. Bot.* **32**, 1257–1272.

Jeschke, W. D. and Pate, J. S. (1991a). Modelling of the partitioning, assimilation and storage of nitrate within root and shoot organs of castor bean (*Ricinus communis* L.). *J. Exp. Bot.* **42**, 1091–1103.

Jeschke, W. D. and Pate, J. S. (1991b). Cation and chloride partitioning through xylem and phloem within the whole plant of *Ricinus communis* L. under conditions of salt stress. *J. Exp. Bot.* **42**, 1105–1116.

Jeschke, W. D. and Pate, J. S. (1992). Temporal patterns of uptake, flow and utilization of nitrate, reduced nitrogen and carbon in a leaf of salt-treated castor bean (*Ricinus communis* L.). *J. Exp. Bot.* **43**, 393–402.

Jeschke, W. D. and Stelter, W. (1976). Measurement of longitudinal ion profiles in single roots of *Hordeum* and *Atriplex* by use of flameless atomic absorption spectroscopy. *Planta* **128**, 107–112.

Jeschke, W. D., Atkins, C. A. and Pate, J. S. (1985). Ion circulation via phloem and xylem between root and shoot of nodulated white lupin. *J. Plant Physiol.* **117**, 319–330.

Jeschke, W. D., Pate, J. S. and Atkins, C. A. (1986). Effects of NaCl salinity on growth, development, ion transport and ion storge in white lupin (*Lupinus albus* L. vc. Ultra). *J. Plant Physiol.* **124**, 257–274.

Jeschke, W. D., Pate, J. S. and Atkins, C. A. (1987). Partitioning of K^+, Na^+, Mg^{2+}, and Ca^{2+} through xylem and phloem to component organs of nodulated white lupin under mild salinity. *J. Plant Physiol.* **128**, 77–93.

Jewell, A. W., Murray, B. G. and Alloway, B. J. (1988). Light and electron microscope studies on pollen development in barley (*Hordeum vulgare* L.) grown under copper-sufficient and deficient conditions. *Plant Cell Environ.* **11**, 273–281.

Jiao, J.-A., Vidal, J., Echevarria, C. and Chollet, R. (1991). *In vivo* regulatory phosphorylation site in C_4-leaf phosphoenolpyruvate carboxylase from maize and sorghum. *Plant Physiol.* **96**, 297–301.

Joachim, S. and Robinson, D. G. (1984). Endocytosis of cationic ferritin by bean leaf protoplasts. *Eur. J. Cell Biol.* **34**, 212–216.

Johnson, A. D. and Simons, J. G. (1979). Diagnostic indices of zinc deficiency in tropical legumes. *J. Plant Nutr.* **1**, 123–149.

Johnson, C. M., Stout, P. R., Broyer, T. C. and Carlton, A. B. (1957). Comparative chlorine requirements of different plant species. *Plant Soil* **8**, 337–353.

Johnson, J., Cobb, B. G. and Drew, M. C. (1989). Hypoxic induction of anoxia tolerance in root tips of *Zea mays*. *Plant Physiol.* **91**, 837–841.

Johnson, N. C., Pfleger, F. L., Crookston, R. K., Simmons, S. R. and Copeland, P. J. (1991). Vesicular-arbuscular mycorrhizas respond to corn and soybean cropping history. *New Phytol.* **117**, 657–663.

Johnson, N. C., Copeland, P. J., Crookston, R. K. and Pfleger, F. L. (1992). Mycorrhizae: possible explanation for yield decline with continuous corn and soybean. *Agron. J.* **84**, 387–390.

Johnson, P. A. and Bennet, R. J. (1991). Aluminium tolerance of root cap cells. *J. Plant Physiol.* **137**, 760–762.

Johnson, S. L. and Smith, K. W. (1976). The interaction of borate and sulfite with pyridine nucleotides. *Biochemistry* **15**, 553–559.

Johnson, W. C. and Wear, J. I. (1967). Effect of boron on white clover (*Trifolium repens* L.) seed production. *Agron. J.* **59**, 205–206.

Johnston, M., Grof, C. P. L. and Brownell, P. F. (1984). Responses to ambient CO_2 concentration by sodium-deficient C_4 plants. *Aust. J. Plant Physiol.* **11**, 137–141.

Johnston, M., Grof, C. P. L. and Brownell, P. F. (1988). The effect of sodium nutrition on the pool sizes of intermediates of the C_4 photosynthetic pathway. *Aust. J. Plant Physiol.* **15**, 749–760.

Johnston, M., Grof, C. P. L. and Brownell, P. F. (1989). Chlorophyll a/b ratios and photosystem activity of mesophyll and bundle sheath fractions from sodium-deficient C_4 plants. *Aust. J. Plant Physiol.* **16**, 449–457.

Jolivet, Y., Larher, F. and Hamelin, J. (1982). Osmoregulation in halophytic higher plants: the protective effect of glycine betaine against the heat destabilization of membranes. *Plant Sci. Lett.* **25**, 193–201.

Jolley, D. and Brown, J. C. (1991a). Differential response of Fe-efficient corn and Fe-inefficient corn and oat to phytosiderophore released by Fe-efficient Coker 227 oat. *J. Plant Nutr.* **14**, 45–58.

Jolley, V. D. and Brown, J. C. (1991b). Factors in iron-stress response mechanism enhanced by Zn deficiency stress in Sanilac, but not Saginaw navy bean. *J. Plant Nutr.* **14**, 257–265.

Jones, A. M. (1990). Do we have the auxin receptor yet? *Physiol. Plant.* **80**, 154–158.

Jones, D. L. and Darrah, P. R (1993). Re-absorption of organic compounds by roots of *Zea mays* L. and its consequences in the rhizosphere. II. Experimental and model evidence for simultaneous exudation and re-absorption of soluble C compounds. *Plant Soil* **153**, 47–59.

Jones, J. B. jr (1967). Interpretation of plant analysis for several agronomic crops. *In* 'Soil Testing and Plant Analysis, Part II: Plant Analysis', pp. 49–58. Publisher SSSA, Madison, WI.

Jones, J. B. jr (1991). Plant tissue analysis in micronutrients. *In* 'Micronutrients in Agriculture', 2nd Ed (J. J. Mortvedt, F. R. Cox, L. M. Shuman and R. M. Welch, eds.), pp. 477–521. Soil Science Society of America Book Series No. 4, Madison WI.

Jones, L. H.-P. (1978). Mineral components of plant cell walls. *Am. J. Clin. Nutr.* **31**, 94–98.

Jones, L. H.-P. and Handreck, K. A. (1965). Studies of silica in the oat plant. III. Uptake of silica from soils by the plant. *Plant Soil* **23**, 79–96.

Jones, L. H.-P. and Handreck, K. H. (1967). Silica in soils, plants and animals. *Adv. Agron.* **19**, 107–149.

Jones, L. H.-P. and Handreck, K. A. (1969). Uptake of silica by *Trifolium incarnatum* in relation to the concentration in the external solution and to transpiration. *Plant Soil* **30**, 71–80.

Jones, L. H.-P., Hartley, R. D. and Jarvis, S. C. (1978). Mineral content of forage plants in relation to nutritional quality. Silicon. *In* 'Annual Report of Grassland Research Institute, Hurley', pp. 25–26.

Jones, M., Browning, M. H. R. and Hutchinson, T. C. (1986). The influence of mycorrhizal associations on paper birch and jack pine seedlings when exposed to elevated copper, nickel or aluminum. *Water Air Soil Pollut.* **31**, 441–448.

Jones, M. D. and Hutchinson, T. C. (1988). Nickel toxicity in mycorrhizal birch seedlings infected with *Lactarius rufus* or *Sleroderma flavidum*: II. Uptake of nickel, calcium, magnesium, phosphorus and iron. *New Phytol.* **108**, 461–470.

Jones, M. D., Durall, D. M. and Tinker, P. B. (1990). Phosphorus relationships and production of extramatrical hyphae by two types of willow ectomycorrhizas at different soil phosphorus levels. *New Phytol.* **115**, 259–267.

Jones, R. L., Gilroy, S. and Hillmer, S. (1993). The role of calcium in the hormonal regulation of enzyme synthesis and secretion in barley aleurone. *J. Exp. Bot.* **44** (Suppl.), 207–212.

Jongruaysup, S., Dell, B. and Bell, R. B. (1994). Distribution and redistribution of molybdenum in black gram (*Vigna mungo* L. Hepper) in relation to molybdenum supply. *Annals of Botany* **73**, 161–167.

Jorns, A. C. and Hecht-Buchholz, C. (1985). Aluminium induzierter Magnesium- und Calciummangel im Laborversuch bei Fichtensämlingen. *Allgem. Forstzeitschr.* **46**, 1248–1252.

Jorns, A. C., Hecht-Buchholz, C. and Wissemeier, A. H. (1991). Aluminium-induced callose formation in root tips of Norway spruce (*Picea abies* (L.) Karst). *Z. Pflanzenernähr. Bodenk.* **154**, 349–353.

Judel, G. K. (1972). Änderungen in der Aktivität der Peroxidase und der Katalase und im Gehalt an Gesamtphenolen in den Blättern der Sonnenblume unter dem Einfluss von Kupfer- und Stickstoffmangel. *Z. Pflanzenernähr. Bodenk.* **133**, 81–92.

Jung, J. (1980). Zur praktischen Anwendung pflanzlicher Bioregulatoren. *Arzneim.- Forsch.* **30**, 1974–1980.

Jung, J. and Sturm, H. (1966). 'Der Wachstumsregulator CCC', Rep. pp. 257–280. Landwirtsch. Vers. Stn. Limburgerhof der BASF.

Jungk, A. (1984). Phosphatdynamik in der Rhizosphäre und Phosphatverfügbarkeit für Pflanzen. *Die Bodenkultur (Wien)* **35**, 99–107.

Jungk, A. (1991). Dynamics of nutrient movement at the soil–root interface. *In* 'Plant Roots, The Hidden Half' (J. Waisel, A. Eshel and U. Kafkafi, eds.), pp. 455–481. Marcel Dekker, New York.

Jungk, A. and Claassen, N. (1986). Availability of phosphate and potassium as the result of interactions between root and soil in the rhizosphere. *Z. Pflanzenernähr. Bodenk.* **149**, 411–427.

Jungk, A. and Claassen, N. (1989). Availability in soil and acquisition by plants as the basis for phosphorus and potassium supply to plants. *Z. Pflanzenernähr. Bodenk.***152**, 151–157.

Jungk, A., Claassen, N. and Kuchenbuch, R. (1982). Potassium depletion of the soil–root interface in relation to soil parameters and root properties. *In* 'Proceedings of the Ninth International Plant Nutrition Colloquium, Warwick, England' (A. Scaife, ed.), pp. 250–255. Commonwealth Agricultural Bureau, Farnham Royal, Bucks.

Jungk, A., Asher, C. J., Edwards, D. G. and Meyer, D. (1990). Influence of phosphate status on phosphate uptake kinetics of maize (*Zea mays*) and soybean (*Glycine max*). *Plant Soil* **124**, 175–182.

Jurkevitch, E., Hadar, Y. and Chen, Y. (1986). The remedy of lime-induced chlorosis in peanuts by *Pseudomonas* sp. siderophores. *J. Plant Nutr.* **9**, 535–545.

Jurkevitch, E., Hadar, Y. and Chen, Y. (1988). Involvement of bacterial siderophores in the remedy of lime-induced chlorosis in peanut. *Soil Sci. Soc. Am. J.* **52**, 1032–1037.

Justin, S. H. F. W. and Armstrong, W. (1991). Evidence for the involvement of ethene in aerenchyma formation in adventitious roots of rice (*Oryza sativa* L.). *New Phytol.* **118**, 49–62.

Kafkafi, U. (1990). Root temperature, concentration and the ratio NO_3^-/NH_4^+ effect on plant development. *J. Plant Nutr.* **13**, 1291–1306.

Kafkafi, U. (1991). Root growth under stress. Salinity. *In* 'Plant Roots: the Hidden Half' (E. Waisel, A. Eshel and U. Kafkafi, eds.), pp. 375–391. Marcel Dekker, New York.

Kafkafi, U. and Ganmore-Neumann, R. (1985). Correction of iron chlorosis in peanut (*Arachis hypogea*, Shulamit) by ammonium sulfate and nitrification inhibitor. *J. Plant Nutr.* **8**, 303–309.

Kaiser, W. M. (1987). Effects of water deficit on photosynthetic capacity. *Physiol. Plant.* **71**, 142–149.

Kaiser, W. M. and Hartung, W. (1981). Uptake and release of abscisic acid by isolated photoautotrophic mesophyll cells, depending on pH gradients. *Plant Physiol.* **68**, 202–206.

Kaiser, W. M. and Spill, D. (1991). Rapid modulation of spinach leaf nitrate reductase by photosynthesis. II. *In vitro* modulation by ATP and AMP. *Plant Physiol.* **96**, 368–375.

Kaiser, W., Dittrich, A. and Heber, U. (1993). Sulfate concentrations in Norway spruce needles in relation to atmospheric SO_2: a comparison of trees from various forests in Germany with trees fumigated with SO_2 in growth chambers. *Tree Physiol.* **12**, 1–13.

Kakie, T. (1969). Phosphorus fractions in tobacco plants as affected by phosphate application. *Soil Sci. Plant Nutr.* (*Tokyo*) **15**, 81–85.

Kaltofen, H. (1988). Warum steigert Stickstoffdüngung oft den Ligningehalt von Gräsern? *Arch. Acker- Pflanzenbau Bodenkd. Berlin* **21**, 255–260.

Kanayama, Y. and Yamamoto, Y. (1991). Formation of nitrosylleghemoglobin in nodules of nitrate-treated cowpea and pea plants. *Plant Cell Physiol.* **32**, 19–23.

Kang, B. T. and Fox, R. L. (1980). A methodology for evaluating the manganese tolerance of cowpea (*Vigna unguiculata*) and some preliminary results of field trials. *Field Crops Res.* **3**, 199–210.

Kang, B. T., Islam, R., Sanders, F. E. and Ayanaba, A. (1980). Effect of phosphate fertilization and inoculation with VA-mycorrhizal fungi on performance of cassava (*Manihot esculenta* Crantz) grown on an alfisol. *Field Crops Res.* **3**, 83–94.

Kannan, S. and Ramani, S. (1978). Studies on molybdenum absorption and transport in bean and rice. *Plant Physiol.* **62**, 179–181.

Kapoor, A. C. and Li, P. H. (1982). Effects of age and variety on nitrate reductase and nitrogen fractions in potato plants. *J. Sci. Food Agric.* **33**, 401–406.

Karmoker, J. L., Clarkson, D. L., Saker, L. R., Rooney, J. M. and Purves, J. V. (1991). Sulphate deprivation depresses the transport of nitrogen to the xylem and the hydraulic conductivity of barley (*Hordeum vulgare* L.) roots. *Planta* **185**, 269–278.

Kasai, M. and Muto, S. (1990). Ca^{2+} pump and Ca^{2+}/H^+ antiporter in plasma membrane vesicles isolated by aqueous two phase partitioning from corn leaves. *J. Membr. Biol.* **114**, 133–142.

Katouli, M. and Marchant, R. (1981). Effect of phytotoxic metabolites of *Fusarium culmorum* on barley root and root-hair development. *Plant Soil* **60**, 385–397.

Katz, A., Dehan, K. and Itai, C. (1978). Kinetin reversal of NaCl effects. *Plant Physiol.* **62**, 836–837.

Kaupenjohann, M. and Hantschel, R. (1989). Nährstofffreisetzung aus homogenen und in situ Bodenproben: Bedeutung für die Waldernährung und Gewässerversauerung. *Kali-Briefe* **19**, 557–572.

Kaupenjohann, M., Zech, W., Hantschel, R. and Horn, R. (1987). Ergebnisse von Düngungsversuchen mit Magnesium an vermutlich immissionsgeschädigten Fichten (*Picea abies* L. Karst.) im Fichtelgebirge. *Forstw. Cbl.* **106**, 78–84.

Kaupenjohann, M., Schneider, B. U., Hantschel, R., Zech, W. and Horn, R. (1988). Sulfuric acid rain treatment of *Picea abies* (L.) Karst: effects on nutrient solution, throughfall chemistry, and tree nutrition. *Z. Pflanzenernähr. Bodenk.* **151**, 123–126.

Kauss, H. (1987). Some aspects of calcium-dependent regulation in plant metabolism. *Annu. Rev. Plant Physiol.* **38**, 47–72.

Kauss, H., Waldmann, T., Jeblick, W. and Takemoto, J. Y. (1991). The phytotoxin syringomycin elicits Ca^{2+}-dependent callose synthesis in suspension-cultured cells of *Catharanthus roseus*. *Physiol. Plant.* **81**, 134–138.

Keeling, P. L., Wood, J. R., Tyson, R. H. and Lang, I. (1988). Starch biosynthesis in developing wheat grain. Evidence against the direct involvement of triose phosphates in the metabolic pathway. *Plant Physiol.* **87**, 311–319.

Keisling, T. C., Hanna, W. and Walker, M. E. (1990). Genetic variation for Mg tissue concentration in pearl millet lines grown under Mg stress conditions. *J. Plant Nutr.* **13**, 1371–1379.

Keller, P. and Deuel, H. (1957). Kationenaustauschkapazität und Pektingehalt von Pflanzenwurzeln. *Z. Pflanzenernähr. Düng. Bodenk.* **79**, 119–131.

Keller, T. (1981). Auswirkungen von Luftverunreinigungen auf Pflanzen. *HLH Heizl Lueft. Klimatech. Haustech.* **48**, 22–24.

Kelman, A., McGuire, R. G. and Tzeng, K.-C. (1989). Reducing the severity of bacterial soft rot by increasing the concentration of calcium in potato tubers. *In* 'Soilborne Plant Pathogens: Managements' (A. W. Engelhard, ed.), pp. 102–123. APS Press, The American Phytopathological Society, St. Paul, Minnesota.

Keltjens, W. G. and Nijenstein, J. H. (1987). Diurnal variations in uptake, transport and assimilation of NO_3^- and efflux on OH^- in maize plants. *J. Plant Nutr.* **10**, 887–900.

Kenis, J. D., Silvente, S. T., Luna, C. M. and Campbell, W. H. (1992). Induction of nitrate reductase in detached corn leaves: the effect of the age of the leaves. *Physiol. Plant.* **85**, 49–56.

Kennedy, C. D. and Gonsalves, F. A. N. (1988). H^+ efflux and trans-root potential measured while increasing the temperature of solutions bathing excised roots of *Zea mays*. *J. Exp. Bot.* **39**, 37–49.

Kennedy, I. R. and Tchan, Y.-T. (1992). Biological nitrogen fixation in non-leguminous field crops: recent advances. *Plant Soil* **141**, 93–118.

Kennedy, R. A., Rumpho, M. E. and Fox, T. C. (1992). Anaerobic metabolism in plants. *Plant Physiol.* **100**, 1–6.

Keren, R., Gast, R. G. and Bar-Josef, B. (1981). pH dependent boron adsorption by Na-montmorillonite. *Soil Sci. Soc. Am. J.* **45**, 45–48.

Kerridge, P. C., Cook, B. G. and Everett, M. L. (1973). Application of molybdenum trioxide in the seed pellet for sub-tropical pasture legumes. *Trop. Grassl.* **7**, 229–232.

Kevekordes, K. G., McCully, M. E. and Canny, M. J. (1988). Late maturation of large metaxylem vessels in soybean roots: significance for water and nutrient supply to the shoot. *Ann. Bot.* **62**, 105–117.

Keyser, H. H. and Munns, D. N. (1979). Effects of calcium, manganese, and aluminium on growth of rhizobia on acid media. *Soil Sci. Soc. Am. J.* **43**, 500–503.

Khamis, S., Chaillou, S. and Lamaze, T. (1990a). CO_2 assimilation and partitioning of carbon in maize plants deprived of orthophosphate. *J. Exp. Bot.* **41**, 1619–1625.

Khamis, S., Lamaze, T., Lemoine, Y. and Foyer, C. (1990b). Adaptation of the photosynthetic apparatus in maize leaves as a result of nitrogen limitation. Relationship between electron transport and carbon assimilation. *Plant Physiol.* **94**, 1436–1443.

Khamis, S., Lamaze, T. and Farineau, J. (1992). Effect of nitrate limitation on the photosynthetically active pools of aspartate and malate in maize, a NADP malic enzyme C_4 plant. *Physiol. Plant.* **85**, 223–229.

Khan, A. H. and Marshall, C. (1981). Salt tolerance within populations of chewing fescue (*Festuca rubra* L.). *Commun. Soil Sci. Plant Anal.* **12**, 1271–1281.

Khasa, P., Furlan, V. and Fortin, J. A. (1992). Response of some tropical plant species to endomycorrhizal fungi under field conditions. *Trop. Agric. (Trinidad)* **69**, 279–283.

Kianmehr, H. (1978). The response of *Helianthemum chamaecistus* Mill. to mycorrhizal infection in two different types of soil. *Plant Soil* **50**, 719–722.

Killham, K. (1990). Nitrification in coniferous forest soils. *Plant Soil* **128**, 31–44.

Kimura, M. and Wada, H. (1989). Tannins in mangrove tree roots and their role in the root environment. *Soil Sci. Plant Nutr.* (*Tokyo*) **35**, 101–108.

Kimura, M., Murakami, H. and Wada, H. (1991). CO_2, H_2, and CH_4 production in rice rhizosphere. *Soil Sci. Plant Nutr.* (*Tokyo*) **37**, 55–60.

Kinney, A. J., Clarkson, D. T. and Loughman, B. C. (1987). Phospholipid metabolism and plasma membrane morphology of warm and cool rye roots. *Plant Physiol. Biochem.* **25**, 769–774.

Kinraide, T. B. (1988). Proton extrusion by wheat roots exhibiting severe aluminum toxicity symptoms. *Plant Physiol.* **88**, 418–423.

Kinraide, T. B. (1990). Assessing the rhizotoxicity of the aluminate ion, $Al(OH)_4^-$. *Plant Physiol.* **94**, 1620–1625.

Kinraide, T. B. (1991). Identity of the rhizotoxic aluminium species. *Plant Soil* **134**, 167–178.

Kinraide, T. B. (1993). Aluminum enhancement of plant growth in acid rooting media. A case of reciprocal alleviation of toxicity by two toxic cations. *Physiol. Plant.* **88**, 619–625.

Kinraide, T. B. and Parker, D. R. (1990). Apparent phytotoxicity of mononuclear hydroxy-aluminum to four dicotyledonous species. *Physiol. Plant.* **79**, 283–288.

Kinzel, H. (1983). Influence of limestone, silicates and soil pH on vegetation. *In* 'Encyclopedia of Plant Physiology, New Series' (O. L. Lange, P. S. Nobel, C. B. Osmond and H. Ziegler, eds.), Vol. 12C, pp. 201–244. Springer-Verlag, Berlin and New York.

Kinzel, H. (1989). Calcium in the vacuoles and cell walls of plant tissue. Forms of deposition and their physiological and ecological significance. *Flora* **182**, 99–125.

Kinzel, H. and Lechner, I. (1992). The specific mineral metabolism of selected plant species and its ecological implications. *Bot. Acta* **105**, 355–361.

Kiraly, Z. (1964). Effect of nitrogen fertilization on phenol metabolism and stem rust susceptibility of wheat. *Phytopathol. Z.* **51**, 252–261.

Kiraly, Z. (1976). Plant disease resistance as influenced by biochemical effects of nutrients in fertilizers. *Proc. 12th Colloq. Int. Potash Inst. Bern*, pp. 33–46.

Kirk, G. J. and Loneragan, J. F. (1988). Functional boron requirement for leaf expansion and its use as a critical value for diagnosis of boron deficiency in soybean. *Agron. J.* **80**, 758–762.

Kirk, G. J. D., Ahmad, A. R. and Nye, P. H. (1990). Coupled diffusion and oxidation of ferrous iron in soils. II. A model of the diffusion and reaction of O_2, Fe^{2+}, H^+ and HCO_3^- in soils and a sensitivity analysis of the model. *J. Soil Sci.* **41**, 411–431.

Kirkby, E. A. (1967). A note on the utilization of nitrate, urea, and ammonium nitrogen by *Chenopodium album*. *Z. Pflanzenernähr. Bodenkd.* **117**, 204–209.

Kirkby, E. A. (1979). Maximizing calcium uptake by plants. *Commun. Soil Sci. Plant Anal.* **10**, 89–113.

Kirkby, E. A. (1981). Plant growth in relation to nitrogen supply. *In* 'Terrestrial Nitrogen Cycles, Processes, Ecosystem Strategies and Management Impacts' (F. E. Clarke and T. Rosswall, eds.), pp. 249–267. Ecol. Bull., Stockholm.

Kirkby, E. A. and Knight, A. H. (1977). Influence of the level of nitrate nutrition on ion uptake and assimilation, organic acid accumulation, and cation-anion balance in whole tomato plants. *Plant Physiol.* **60**, 349–353.

Kirkby, E. A. and Mengel, K. (1967). Ionic balance in different tissues of the tomato plant in relation to nitrate, urea, or ammonium nutrition. *Plant Physiol.* **42**, 6–14.

Kirkby, E. A. and Mengel, K. (1970). Preliminary observations on the effect of urea nutrition on the growth and nitrogen metabolism of sunflower plants. *In* 'Nitrogen Nutrition of the Plant' (E. A. Kirkby, ed.), pp. 35–38. The University of Leeds.

Kirkby, E. A. and Pilbeam, D. J. (1984). Calcium as a plant nutrient. *Plant Cell Environ.* **7**, 397–405.

Kirkegaard, J. A., So, H. B. and Troedson, R. J. (1992). The effect of soil strength on the growth of pigeonpea radicles and seedlings. *Plant Soil* **140**, 65–74.

Kirkham, D. S. (1954). Significance of the ratio of the water soluble aromatic and nitrogen constituents of apple and pear in the host–parasite relationship of *Venturia* sp. *Nature* (*London*) **173**, 690–691.

Kitagishi, K. and Obata, H. (1986). Effects of zinc deficiency on the nitrogen metabolism of meristematic tissues of rice plants with reference to protein synthesis. *Soil Sci. Plant Nutr.* (*Tokyo*) **32**, 397–405.

Kitagishi, K., Obata, H. and Kondo, T. (1987). Effect of zinc deficiency on 80S ribosome content of meristematic tissues of rice plant. *Soil Sci. Plant Nutr.* (*Tokyo*) **33**, 423–430.

Kivilaan, A. and Scheffer, R. P. (1958). Factors affecting development of bacterial stem rot of *Pelargonium*. *Phytopathology* **48**, 185–191.

Klämbt, D. (1990). A view about the function of auxin-binding proteins at plasma membranes. *Plant Mol. Biol.* **14**, 1045–1050.

Klauer, S. F., Franceschi, V. R. and Ku, M. S. B. (1991). Protein composition of mesophyll and paraveinal mesophyll of soybean leaves at various developmental stages. *Plant Physiol.* **97**, 1306–1316.

Klein, H., Priebe, A. and Jäger, H.-J. (1979). Putrescine and spermidine in peas: effects of nitrogen source and potassium supply. *Physiol. Plant.* **45**, 497–499.

Kleinkopf, G. E., Westermann, D. T. and Dwelle, R. B. (1981). Dry matter production and nitrogen utilization by six potato cultivars. *Agron. J.* **73**, 799–802.

Klemedtsson, L., Svensson, B. H. and Rosswall, T. (1988). Relationships between soil moisture content and nitrous oxide production during nitrification and denitrification. *Biol. Fertil. Soils* **6**, 106–111.

Klemm, K. (1966). Der Einfluss der N-Form auf die Ertragsbildung verschiedener Kulturpflanzen. *Bodenkultur* **17**, 265–284.

Klepper, B. (1991). Root–shoot relationships. *In* 'Plant Roots: the Hidden Half' (Y. Waisel, A. Eshel and U. Kafkafi, eds.), pp. 265–286. Marcel Dekker, New York.

Klepper, B. and Kaufmann, M. R. (1966). Removal of salt from xylem sap by leaves and stems of guttating plants. *Plant Physiol.* **41**, 1743–1747.

Kliewer, M. and Evans, H. J (1963a). Identification of cobamide coenzyme in nodules of symbionts and isolation of the B_{12} coenzyme from *Rhizobium meliloti*. *Plant Physiol.* **38**, 55–59.

Kliewer, M. and Evans, H. J. (1963b). Cobamide coenzyme contents of soybean nodules and nitrogen fixing bacteria in relation to physiological conditions. *Plant Physiol.* **38**, 99–104.

Kloepper, J. W., Leong, J., Teintze, M. and Schroth, M. N. (1980). *Pseudomonas* siderophores: a mechanism explaining disease-suppressive soils. *Curr. Microbiol.* **4**, 317–320.

Kloepper, J. W., Rodriguez-Kabana, R., McInroy, J. A. and Collins, D. J. (1991). Analysis of populations and physiological characterisation of microorganisms in rhizosphere of plants with antagonistic properties to phytopathogenic nematodes. *Plant Soil* **136**, 95–102.

Kloepper, J. W., Rodriguez-Kabana, R., McInroy, J. A. and Young, R. W. (1992a). Rhizosphere bacteria antagonistic to soybean cyst (*Heterodere glycine*) and root-knot

(*Meloidogyne incognita*)nematodes: Identification by fatty acid analysis and frequency of biological control activity. *Plant Soil* **139**, 75–84.

Kloepper, J. W., Schippers, B. and Bakker, P. A. H. M. (1992b). Proposed elimination of the term *Endorhizosphere*. *Phytopathology* **82**, 726–727.

Kloss, M., Iwannek, K.-H., Fendrik, I. and Niemann, E. G. (1984). Organic acids in the root exudates of *Diplachne fusca* (Linn.) Beauv. *Environ. Exp. Bot.* **24**, 179–188.

Klotz, F. and Horst, W. J. (1988a). Genotypic differences in aluminium tolerance of soybean (*Glycine max.* L.) as affected by ammonium and nitrate-nitrogen nutrition. *J. Plant Physiol.* **132**, 702–707.

Klotz, F. and Horst, W. J. (1988b). Effect of ammonium- and nitrate-nitrogen nutrition on aluminium tolerance of soybean (*Glycine max.* L.). *Plant Soil* **111**, 59–65.

Klucas, R. V. (1991). Associative nitrogen fixation in plants. *In* 'Biology and Biochemistry of Nitrogen Fixation' (M. J. Dilworth and A. R. Glenn, eds.), pp. 187–198. Elsevier, Amsterdam.

Klucas, R. V., Hanus, F. J., Russell, S. A. and Evans, H. J. (1983). Nickel: a micronutrient element for hydrogen-dependent growth of *Rhizobium japonicum* and for expression of urease activity in soybean leaves. *Proc. Natl. Acad. Sci. (USA)* **80**, 2253–2257.

Kluge, R. (1971). Beitrag zum Problem des B-Mangels bei landwirtschaftlichen Kulturen als Folge der Bodentrockenheit. *Arch. Acker Pflanzenbau Bodenkd.* **15**, 749–754.

Kluge, R. (1990). Symptombezogene toxische Pflanzengrenzwerte zur Beurteilung von Bor(B)-Überschuss bei ausgewählten landwirtschaftlichen Nutzpflanzen. *Agribiol. Res.* **43**, 234–243.

Kluge, R. and Beer, K. H. (1979). Einfluss des pH-Wertes auf die B-Adsorption von Aluminiumhydroxigel, Tonmineralen und Böden. *Arch. Acker Pflanzenbau Bodenkd.* **23**, 279–287.

Knight, A. H., Crooke, W. M. and Burridge, J. C. (1973). Cation exchange capacity, chemical composition and the balance of carboxylic acids in the floral parts of various plant species. *Ann. Bot. (London)* [N.S.] **37**, 159–166.

Knight, W. G., Allen, M. F., Jurinak, J. J. and Dudley, L. M. (1989). Elevated carbon dioxide and solution phosphorus in soil with vesicular-arbuscular mycorrhizal western wheatgrass. *Soil Sci. Soc. Am. J.* **53**, 1075–1082.

Knowles, T. C., Doerge, T. A. and Ottman, M. J. (1991). Improved nitrogen management in irrigated durum wheat using stem nitrate analysis: II. Interpretation of nitrate-nitrogen concentrations. *Agron. J.* **83**, 353–356.

Ko, M. P., Huang, P.-Y., Huang, J.-S. and Barker, K. R. (1987). The occurrence of phytoferritin and its relationship to effectiveness of soybean nodules. *Plant Physiol.* **83**, 299–305.

Koch, A. S. and Matzner, E. (1993). Heterogeneity of soil and soil solution chemistry under Norway spruce (*Picea abies* Karst.) and European beech (*Fagus silvatica* L.) as influenced by distance from the stem basis. *Plant Soil* **151**, 227–237.

Koch, K. and Mengel, K. (1974). The influence of the level of potassium supply to young tobacco plants (*Nicotiana tabacum* L.) on short-term uptake and utilisation of nitrate nitrogen. *J. Sci. Food Agric.* **25**, 465–471.

Kochian, L. V. (1991). Mechanism of micronutrient uptake and translocation in plants. *In* 'Micronutrients in Agriculture' (J. J. Mortvedt, ed.), pp. 229–296. *Soil Sci. Soc. Am. Book Series* **No.4**. Madison, WI.

Kochian, L. V. and Lucas, W. J. (1982). Potassium transport in corn roots. I.

Resolution of kinetics into a saturable and linear component. *Plant Physiol.* **70**, 1723–1731.

Kochian, L. V., Shaff, J. E. and Lucas, W. J. (1989). High affinity K^+ uptake in maize roots. A lack of coupling with H^+ efflux. *Plant Physiol.* **91**, 1202–1211.

Koda, Y. (1982). Changes in levels of butanol- and water-soluble cytokinins during the life cycle of potato tubers. *Plant Cell Physiol.* **23**, 843–850.

Kohls, S. J. and Barker, D. D. (1989). Effects of substrate nitrate concentration on symbiotic nodule formation in actinorhizal plants. *Plant Soil* **118**, 171–179.

Koide, R., Li, M., Lewis, J. and Irby, C. (1988). Role of mycorrhizal infection in the growth and reproduction of wild vs. cultivated oats. *Oecologia* **77**, 537–543.

Kojima, M. and Conn, E. E. (1982). Tissue distribution of chlorogenic acid and of enzymes involved in its metabolism in leaves of *Sorghum bicolor*. *Plant Physiol.* **70**, 922–925.

Kolb, W. and Martin, P. (1988). Influence of nitrogen on the number of N_2-fixing and total bacteria in the rhizosphere. *Soil Biol. Biochem.* **20**, 221–225.

Kolesch, H., Oktay, M. and Höfner, W. (1984). Effect of iron chlorosis-inducing factors on the pH of the cytoplasm of sunflower (*Helianthus annuus*). *Plant Soil* **82**, 215–221.

Koller, D. (1990). Light-driven leaf movements. *Plant Cell Environ.* **13**, 615–632.

Komor, E., Rotter, M. and Tanner, W. (1977). A proton-cotransport system in a higher plant: sucrose transport in *Ricinus communis*. *Plant Sci. Lett.* **9**, 153–162.

Kong, T. and Steffens, D. (1989). Bedeutung der Kalium-Verarmung in der Rhizos-phäre und der Tonminerale für die Freisetzung von nichtaustauschbarem Kalium und dessen Bestimmungmit HCl. *Z. Pflanzenernähr. Bodenk.* **152**, 337–343.

Konings, H. and Verschuren, G. (1980). Formation of aerenchyma in roots of *Zea mays* in aerated solutions and its relation to nutrient supply. *Physiol. Plant.* **49**, 265–270.

Konishi, S., Miyamoto, S. and Taki, T. (1985). Stimulatory effects of aluminum on tea plants grown under low and high phosphorus supply. *Soil Sci. Plant Nutr. (Tokyo)* **31**, 361–368.

Konno, H., Yamaya, T., Yamasaki, Y. and Matsumoto, H. (1984). Pectic polysacchar-ide break-down of cell walls in cucumber roots grown with calcium starvation. *Plant Physiol.* **76**, 633–637.

Kooistra, M. J., Schoonderbeek, D., Boone, F. R., Veen, B. W. and Van Noordwijk, M.(1992). Root–soil contact of maize, as measured by a thin-section technique. II. Effects of soil compaction. *Plant Soil* **139**, 119–129.

Kope, H. H. and Fortin, J. A. (1990). Antifungal activity in culture filtrates of the ectomycorrhizal fungus *Pisolithus tinctorius*. *Can. J. Bot.* **68**, 1254–1259.

Koritsas, V. M. (1988). Effect of ethylene and ethylene precursors on protein phosphorylation and xylogenesis in tuber explants of *Helianthus tuberosus* (L.). *J. Exp. Bot.* **39**, 375–386.

Körner, C. (1989). The nutritional status of plants from high altitudes – A worldwide comparison. *Oecologia* **81**, 379–391.

Kothari, S. K., Marschner, H. and George, E. (1990a). Effect of VA mycorrhizal fungi and rhizosphere microorganisms on root and shoot morphology, growth and water relations in maize. *New Phytol.* **116**, 303–311.

Kothari, S. K., Marschner, H. and Römheld, V. (1990b). Direct and indirect effects of VA mycorrhiza and rhizosphere microorganisms on mineral nutrient acquisition by maize (*Zea mays* L.) in a calcareous soil. *New Phytol.* **116**, 637–645.

Kothari, S. K., Marschner, H. and Römheld, V. (1991a). Effect of a vesicular-

arbuscular mycorrhizal fungus and rhizosphere microorganisms on manganese reduction in the rhizosphere and mangnese concentrations in maize (*Zea mays* L.).
New Phytol. **117**, 649–655.

Kothari, S. K., Marschner, H. and Römheld, V. (1991b). Contribution of the VA
mycorrhizal hyphae in acquisition of phosphorus and zinc by maize grown in a
calcareous soil. *Plant Soil* **131**, 177–185.

Kottke, I. (1992). Ectomycorrhizas – organs for uptake and filtering of cations. *In*
'Mycorrhizas in Ecosystems' (D. J. Read, D. H. Lewis, A H. Fitter and I. J.
Alexander, eds.), pp. 316–322. CAB International, Wallingford, UK.

Kottke, I. and Oberwinkler, F. (1986). Mycorrhiza of forest trees – structure and
function. *Trees* **1**, 1–24.

Kouchi, H. (1977). Rapid cessation of mitosis and elongation of root tip cells of *Vicia
faba* as affected by boron deficiency. *Soil Sci. Plant Nutr.* (*Tokyo*) **23**, 113.

Kouchi, H. and Kumazawa, K. (1976). Anatomical responses of root tips to boron
deficiency. III. Effect of boron deficiency on sub-cellular structure of root tips,
particularly on morphology of cell wall and its related organelles. *Soil Sci. Plant Nutr.*
(*Tokyo*) **22**, 53–71.

Kouno, K. and Ogata, S. (1988). Sulfur-supplying capacity of soils and critical sulfur
values of forge crops. *Soil Sci. Plant Nutr.* (*Tokyo*) **34**, 327–339.

Kourie, J. and Goldsmith, M. H. M. (1992). K^+ channels are responsible for an
inwardly rectifying current in the plasma membrane of mesophyll protoplasts of
Avena sativa. *Plant Physiol.* **98**, 1087–1097.

Kovanci, I. and Colakoglu, H. (1976). The effect of varying K level on yield components and susceptibility of young wheat plants to attack by *Puccinia striiformis*
West. *Proc. 12th Colloq. Int. Potash Inst. Bern*, pp. 177–182.

Kovanci, I., Hakerlerler, H. and Höfner, W. (1978). Ursachen der Chlorosen an
Mandarinen (*Citrus reticulata* Blanco) der ägäischen Region. *Plant Soil* **50**, 193–205.

Koyro, H.-W., Stelzer, R. and Huchzermeyer, B. (1993). ATPase activities and
membrane fine structure of rhizodermal cells from *Sorghum* and *Spartina* roots
grown under mild salt stress. *Bot. Acta* **106**, 110–119.

Kraffczyk, I., Trolldenier, G. and Beringer, H. (1984). Soluble root exudates of maize:
influence of potassium supply and rhizosphere microorganisms. *Soil Biol. Biochem.*
16, 315–322.

Kramer, D., Läuchli, A., Yeo, A. R. and Gullasch, J. (1977). Transfer cells in roots of
Phaseolus coccineus: Ultrastructure and possible function in exclusion of sodium
from the shoot. *Ann. Bot.* (*London*). [N.S.] **41**, 1031–1040.

Kramer, D., Römheld, V., Landsberg, E. and Marschner, H. (1980). Induction of
transfer-cell formation by iron deficiency in the root epidermis of *Helianthus annuus*.
Planta **147**, 335–339.

Krannitz, P. G., Aarssen, L. W. and Lefebvre, D. D. (1991). Correction for non-linear
relationships between root size and short term P_i uptake in genotype comparisons.
Plant Soil **133**, 157–167.

Krauss, A. (1971). Einfluss der Ernährung des Salats mit Massennährstoffen auf den
Befall mit *Botrytis cinera* Pers. *Z. Pflanzenernähr. Bodenk.* **128**, 12–23.

Krauss, A. (1978a). Tuberization and abscisic acid content in *Solanum tuberosum* as
affected by nitrogen nutrition. *Potato Res.* **21**, 183–193.

Krauss, A. (1978b). Endogenous regulation mechanisms in tuberization of potato
plants in relation to environmental factors. *EAPR Abstr. Conf. Pap.* **7**, 47–48.

Krauss, A. (1980). Influence of nitrogen nutrition on tuber initiation of potatoes. *Proc.
15th Colloq. Int. Potash Inst. Bern*, pp. 175–184.

Krauss, A. and Marschner, H.(1971). Einfluss der Stickstoffernährung der Kartoffeln auf Induktion und Wachstumsrate der Knolle. *Z. Pflanzenernähr. Bodenk.* **128**, 153–168.

Krauss, A. and Marschner, H. (1975). Einfluss des Calcium-Angebotes auf Wachstumsrte und Calcium-Gehalt von Kartoffelknollen.*Z. Pflanzenernähr. Bodenk.* **138**, 317–326.

Krauss, A. and Marschner, H. (1976). Einfluss von Stickstoffernährung und Wuchsstoffapplikation auf die Knolleninduktion bei Kartoffelpflanzen. *Z. Pflanzenernähr. Bodenk.* **139**, 143–155.

Krauss, A. and Marschner, H. (1982). Influence of nitrogen nutrition, daylength and temperature on contents of gibberellic and abscisic acid and on tuberization in potato plants. *Potato Res.* **25**, 13–21.

Krauss, A. and Marschner, H. (1984). Growth rate and carbohydrate metabolism of potato tubers exposed to high temperatures. *Potato Res.* **27**, 297–303.

Kreimer, G., Melkonian, M., Holtum, J. A. M. and Latzko, E. (1988). Stromal free calcium concentration and light-mediated activation of chloroplast fructose-1,6-bisphosphatase. *Plant Physiol.* **86**, 423–428.

Kretzschmar, R. M., Hafner, H., Bationo, A. and Marschner, H. (1991). Long- and short-term effects of crop residues on aluminum toxicity, phosphorus availability and growth of pearl millet in an acid sandy soil. *Plant Soil* **136**, 215–223.

Kriedemann, P. E. and Anderson, J. E. (1988). Growth and photosynthetic response to manganese and copper deficiencies in wheat (*Triticum aestivum*) and barley grass (*Hordeum glaucum* and *H. leporinum*). *Aust. J. Plant Physiol.* **15**, 429–446.

Kriedemann, P. E. and Sands, R. (1984). Salt resistance and adaptation to root-zone hypoxia in sunflower. *Aust. J. Plant Physiol.* **11**, 287–301.

Kriedemann, P. E., Graham, R. D. and Wiskich, J. T. (1985). Photosynthetic disfunction and *in vivo* changes in chlorophyll *a* fluorescence from manganese-deficient wheat leaves. *Aust. J. Agric. Res.* **36**, 157–169.

Krishnamurthy, R. (1991). Amelioration of salinity effect in salt tolerant rice (*Oryza sativa* L.) by foliar application of putrescine. *Plant Cell Physiol.* **32**, 699–703.

Krogmann, D. W., Jagendorf, A. T. and Avron, M. (1959). Uncouplers of spinach chloroplast photosynthetic phosphorylation. *Plant Physiol.* **34**, 272–277.

Krogmeier, M. J., McCarty, G. W. and Bremner, J. M. (1989). Phytotoxicity of foliar-applied urea. *Proc. Natl. Acad. Sci. (USA)* **86**, 8189–8191.

Krogmeier, M. J., McCarty, G. W., Shogren, D. R. and Bremner, J. M. (1991). Effect of nickel deficiency in soybeans on the phytotoxicity of foliar-applied urea. *Plant Soil* **135**, 283–286.

Kröniger, W., Rennenberg, H. and Polle, A. (1992). Purification of two superoxide dismutase isoenzymes and their subcellular localization in needles and roots of Norway spruce (*Picea abies* L.) trees. *Plant Physiol.* **100**, 334–340.

Krosing, M. (1978). Der Einfluss von Bormangel und von mechanischer Zerstörung des Spitzenmeristems auf die Zellteilung bei Sonnenblumen. *Z. Pflanzenernähr. Bodenk.* **141**, 641–654.

Krueger, R. W., Lovatt, C. J. and Albert, L. S. (1987). Metabolic requirement of *Cucurbita pepo* for boron. *Plant Physiol.* **83**, 254–258.

Krug, H., Wiebe, H.-J. and Jungk, A. (1972). Calciummangel an Blumenkohl unter konstanten Klimabedingungen. *Z. Pflanzenernähr. Bodenk.* **133**, 213–226.

Kruger, E. and Sucoff, E. (1989). Aluminium and the hydraulic conductivity of *Quercus rubra* L. root systems. *J. Exp. Bot.* **40**, 659–665.

Krumm, M. (1991). Regulation der Kornzahl in der Weizenähre: Rolle von nichtstrukturellen Kohlenhydraten, insbesondere von Fructanen. PhD Thesis, University Hohenheim.

Kubota, J. and Allaway, W. H. (1972). Geographic distribution of trace element problems. *In* 'Micronutrients in Agriculture' (J. J. Mortvedt, P. M. Giordano and W. L. Lindsay, eds.), pp. 525–554. Soil Sci. Soc. Am., Madison, Wisconsin.

Kubota, J., Welch, R. M. and Van Campen, D. R. (1987). Soil-related nutritional problem areas for grazing animals. *Adv. Soil Sci.* **6**, 189–215.

Kuchenbuch, R. and Jungk, A. (1984). Wirkung der Kaliumdüngung auf die Kalium-verfügbarkeit in der Rhizosphäre von Raps. *Z. Pflanzenernähr. Bodenk.* **147**, 435–448.

Kuchenbuch, R., Claassen, N. and Jungk, A. (1986). Potassium availability in relation to soil moisture. I. Effect of soil moisture on potassium diffusion, root growth and potassium uptake of anion plants. *Plant Soil* **95**, 221–231.

Kuehn, G. D., Rodriguez-Garay, B., Bagga, S. and Phillips, G. C. (1990). Novel occurrence of uncommon polyamines in higher plants. *Plant Physiol.* **94**, 855–875.

Kuehny, J. S., Peet, M. M., Nelson, P. V. and Willits, D. H. (1991). Nutrient dilution by starch in CO_2-enriched *chrysanthemum*. *J. Exp. Bot.* **42**, 711–716.

Kuhlmann, H. and Baumgärtel, G. (1991). Potential importance of the subsoil for the P and Mg nutrition of wheat. *Plant Soil* **137**, 259–266.

Kuhlmann, H., Barraclough, P. B. and Weir, A. H. (1989). Utilization of mineral nitrogen in the subsoil by winter wheat. *Z. Pflanzenernähr. Bodenk.* **152**, 291–295.

Kuiper, D. (1988). Growth responses of *Plantago major* L. ssp. *pleiosperma* (Pilger) to changes in mineral supply. *Plant Physiol.* **87**, 555–557.

Kuiper, D. and Kuiper, P. J. C. (1979). Ca^{2+} and Mg^{2+} stimulated ATPases from roots of *Plantago lanceolata*, *Plantago media* and *Plantago coronopus*: Response to alterations of the level of mineral nutrition and ecological significance. *Physiol. Plant.* **45**, 240–244.

Kuiper, D., Schuit, J. and Kuiper, P. J. C. (1988). Effect of internal and external cytokinin concentrations on root growth and shoot to root ratio of *Plantago major* ssp. *pleiosperma* at different nutrient concentrations. *Plant Soil* **111**, 231–236.

Kuiper, D., Kuiper, P. J. C., Lambers, H., Schuit, J. and Staal, M.(1989). Cytokinin concentration in relation to mineral nutrition and benzyladenine treatment in *Plantago major* ssp. *pleiosperma*. *Physiol. Plant.* **75**, 511–517.

Kuiper, D., Schuit, J. and Kuiper, P. J. C. (1990). Actual cytokinin concentrations in plant tissue as an indicator for salt resistance in cereals. *Plant Soil* **123**, 243–250.

Kuiper, D., Sommarin, M. and Kylin, A. (1991). The effects of mineral nutrition and benzyladenine on the plasmalemma ATPase activity from roots of wheat and *Plantago major* ssp. *pleiosperma*. *Physiol. Plant.* **81**, 169–174.

Kuiper, P. J. C. (1968). Lipids in grape roots in relation to chloride transport. *Plant Physiol.* **43**, 1367–1371.

Kuiper, P. J. C. (1980). Lipid metabolism as a factor in environmental adaptation. *In* 'Biogenesis and Function of Plant Lipids' (P. Mazliak *et al.*, eds.), pp. 169–196. Elsevier/North-Holland Biomedical Press, Amsterdam.

Kumar, S., Patil, B. C. and Singh, S. K. (1990). Cyanide resistant respiration is involved in temperature rise in ripening mangoes. *Biochem. Biophys. Res. Commun.* **168**, 818–822.

Kuo, J., Pate, J. S., Rainbird, R. M. and Atkins, C. A.(1980). Internodes of grain legumes – New location of xylem parenchyma transfer cells. *Protoplasma* **104**, 181–185.

Kuo, S. (1990). Phosphate sorption implications on phosphate soil tests and uptake by corn. *Soil Sci. Soc. Am. J.* **54**, 131–135.

Kurdjian, A. and Guern, J. (1989). Intracellular pH: measurement and importance in cell activity. *Annu. Rev. Plant Physiol. Plant Mol. Biol.* **40**, 271–303.

Kurvits, A. and Kirkby, E. A. (1980). The uptake of nutrients by sunflower plants (*Helianthus annuus*) growing in a continuous flowing culture system, supplied with nitrate or ammonium as nitrogen source. *Z. Pflanzenernähr. Bodenk.* **143**, 140–149.

Kutschera, U. (1989). Tissue stresses in growing plant organs. *Physiol. Plant.* **77**, 157–163.

Kutschera, U., Bergfeld, R. and Schopfer, P. (1987). Cooperation of epidermis and inner tissues in auxin-mediated growth of maize coleoptiles. *Planta* **170**, 168–180.

Kuznetsova, G. A., Kuznetsova, M. G. and Grineva, G. M. (1981). Characteristics of water exchange and anatomical-morphological structure in corn plants under conditions of flooding. *Sov. Plant Physiol.* (*Engl. Transl.*) **28**, 241–248.

Kwiatowsky, J., Safianowska, A. and Kaniuga,Z. (1985). Isolation and characterization of an iron-containing superoxide dismutase from tomato leaves, *Lycopersicon esculentum*. *Eur. J. Biochem.* **146**, 459–466.

Kylin, A. and Hansson, G. (1971). Transport of sodium and potassium, and properties of (sodium+potassium) activated adenosine triphosphatase: Possible connection with salt tolerance in plants. *Proc. 8th Colloq.Int.Potash Inst. Bern*, pp. 64–68.

Laan, P., Berrevoets, M. J., Lythe, S., Armstrong, W. and Blom, C. W. P. M. (1989). Root morphology and aerenchyma formation as indicators of the flood-tolerance of *Rumex* species. *J. Ecol.* **77**, 693–703.

Laan, P., Tosserams, M., Blom, C. W. P. M. and Veen, B. W. (1990). Internal oxygen transport in *Rumex* species and its significance for respiration under hypoxic conditions. *Plant Soil* **122**, 39–46.

Laan, P., Clement, J. M. A. M. and Blom, C. W. P. M. (1991a). Growth and development of *Rumex* roots as affected by hypoxic and anoxic conditions. *Plant Soil* **136**, 145–151.

Laan, P., Smolders, A. and Blom, C. W. P. M. (1991b). The relative importance of anaerobiosis and high iron levels in the flood tolerance of *Rumex* species. *Plant Soil* **136**, 153–161.

Lafever, H. N., Campbell, L. G. and Foy, C. D. (1977). Differential response of wheat cultivars to Al. *Agron. J.* **69**, 563–568.

LaHaye, P. A. and Epstein, E. (1971). Calcium and salt tolerance by bean plants. *Physiol. Plant.* **25**, 213–218.

Lamattina, L., Anchoverri, V., Conde, R. D. and Pont Lezia, R. (1987). Quantification of the kinetin effect on protein synthesis and degradation in senescing wheat leaves. *Plant Physiol.* **83**, 497–499.

Lamaze, T., Sentenac, H. and Grignon, C. (1987). Orthophosphate relations of root: NO_3^- effects on orthophosphate influx, accumulation and secretion into the xylem. *J. Exp. Bot.* **38**, 923–934.

Lambais, M. R. and Cardoso, E. J. B. N. (1990). Response of *Stylosanthes guianensis* to endomycorrhizal fungi inoculation as affected by lime and phosphorus application. I. Plant growth and development. *Plant Soil* **129**, 283–289.

Lambers, H. (1982). Cyanide resistant respiration: a non-phosphorylating electron transport pathway acting as an energy overflow. *Physiol. Plant.* **55**, 478–485.

Lambers, H., Day, D. A. and Azcón-Bieto, J. (1983). Cyanide-resistant respiration in roots and leaves. Measurements with intact tissues and isolated mitochondria. *Physiol. Plant.* **58**, 148–154.

Lambers, H., van der Werf, A. and Konings, H. (1991). Respiratory patterns in roots in relation to their functioning. *In* 'Plant Roots, the Hidden Half' (Y. Waisel, A. Eshel and U. Kafkafi, eds.), pp. 229–263. Marcel Dekker, Inc. New York.

Lambers, H., Posthumus, F., Stulen, I., Lantin, L., van de Dijk, S. J. and Hostra, R. (1981). Energy metabolism of *Plantago lanceolata* as dependent on the supply of mineral nutrients. *Physiol. Plant.* **51**, 85–92.

Lambert, D. H. and Weidensaul, T. C. (1991). Element uptake by mycorrhizal soybean from sewage-sludge-treated soil. *Soil Sci. Soc. Am. J.* **55**, 393–397.

Lambert, D. H., Cole jr, H. and Baker, D. E. (1980). Variation in the response of alfalfa clones and cultivars to mycorrhizae and phosphorus. *Crop Sci.* **20**, 615–618.

Lamhamedi, M. S. and Fortin, J. A. (1991). Genetic variations of ectomycorrhizal fungi: extramatrical phase of *Pisolithus* sp. *Can. J. Bot.* **69**, 1927–1934.

Lamhamedi, M. S., Bernier, P. Y. and Fortin, J. A. (1992). Hydraulic conductance and soil water potential at the soil-root interface of *Pinus pinaster* seedlings inoculated with different dikaryons of *Pisolithus* sp. *Tree Physiol.* **10**, 231–244.

Lamont, B. (1972). The effect of soil nutrients on the production of proteoid roots by *Hakea* species. *Aust. J. Bot.* **20**, 27–40.

Lamont, B. (1982). Mechanisms for enhancing nutrient uptake in plants with particular reference to mediterranean South Africa and Western Australia. *Bot. Rev.* **48**, 597–689.

Landsberg, E.-C. (1981). Organic acid synthesis and release of hydrogen ions in response to Fe deficiency stress of mono- and dicotyledonous plant species. *J. Plant Nutr.* **3**, 579–591.

Landsberg, E.-C. (1989). Proton efflux and transfer cell formation as response to Fe deficiency of soybean in nutrient solution culture. *Plant Soil* **114**, 53–61.

Lang, A. (1983). Turgor regulated translocation. *Plant Cell Environ.* **6**, 683–689.

Lang, A. and Thorpe, M. R. (1989). Xylem, phloem and transpiration flows in a grape: application of a technique for measuring the volume of attached fruits to high resolution using archimedes' principle. *J. Exp. Bot.* **40**, 1069–1078.

Lange, O. L., Zellner, H., Gebel, J., Schrameli, P., Köstner, B. and Czygan, F.-C. (1987). Photosynthetic capacity, chloroplast pigments, and mineral content of the previous year's needles with and without the new flush: analysis of the forest-decline phenomenon of needle bleaching. *Oecologia* (*Berlin*) **73**, 351–357.

Langheinrich, U., Tischner, R. and Godbold, D. L. (1992). Influence of a high Mn supply on Norway spruce (*Picea abies* (L.) Karst.) seedlings in relation to the nitrogen source. *Tree Physiol.* **10**, 259–271.

Langmeier, M., Ginsburg, S. and Matile, P. (1993). Chlorophyll breakdown in senescent leaves: demonstration of Mg-chelatase activity. *Physiol. Plant.* **89**, 347–353.

Lanning, F. C. and Eleuterius, L. N. (1989). Silica deposition in some C_3 and C_4 species of grasses, sedges and composites in the USA. *Ann. of Bot.* **63**, 395–410.

Lantzsch, H. J., Marschner, H., Wilberg, E. and Scheuermann, S. (1980). The improvement of the bioavailability of zinc in wheat and barley grains following application of zinc fertilizer. *Proc. Miner. Elements*, *Helsinki* 1980, Part I. pp. 323–328.

Lapeyrie, F. (1990). The role of ectomycorrhizal fungi in calcareous soil tolerance by 'symbiocalcicole' woody plants. *Ann. Sci. For.* **21**, 579–589.

Lapeyrie, F. F. and Bruchet, G. (1986). Calcium accumulation by two strains, calcicole and calcifuge, of the mycorrhizal fungus *Paxillus involutus*. *New Phytol.* **103**, 133–141.

Lapeyrie, F., Chilvers, G. A. and Behm, C. A. (1987). Oxalic acid synthesis by the mycorrhizal fungus *Paxillus involutus*. *New Phytol.* **106**, 139–146.

Lapeyrie, F., Picatto, C., Gerard, J. and Dexheimer, J. (1990). T. E. M. study of intracellular and extracellular calcium oxalate accumulation by ectomycorrhizal fungi in pure culture or in association with *Eucalyptus* seedlings. *Symbiosis* **9**, 163–166.

Lapeyrie, F., Ranger, J. and Vairelles, D. (1991). Phosphate-solubilizing activity of ectomycorrhizal fungi *in vitro*. *Can. J. Bot.* **69**, 342–346.

Larcher, W. (1980). 'Ökologie der Pflanzen'. Ulmer Verlag, Stuttgart.

LaRosa, P. C., Handa, A. K., Hasegawa, P. M. and Bressan, R. A. (1985). Abscisic acid accelerates adaptation of cultured tobacco cells to salts. *Plant Physiol.* **79**, 138–142.

Larsen, J. B. (1976). Untersuchungen über die Frostempfindlichkeit von Douglasien-herkünften und über den Einfluss der Nährstoffversorgung auf die Frostresistenz der Douglasie. *Forst- und Holzwirt.* **15**, 299–302.

Larsson, C.-M., Larsson, M., Purves, J. V. and Clarkson, D. T. (1991). Translocation and cycling through roots of recently absorbed nitrogen and sulphur in wheat (*Triticum aestivum*) during vegetative and generative growth. *Physiol. Plant.* **82**, 345–352.

Larsson, S., Wiren, A., Lundgren, L. and Ericsson, T. (1986). Effects of light and nutrient stress on leaf phenolic chemistry in *Salix dasyclados* and susceptibility to *Garlerucella lineola* (COL. Chrysomelidea). *Oikos* **47**, 205–210.

Lascaris, D. and Deacon, J. W. (1991a). Comparison of methods to assess senescence of the cortex of wheat and tomato roots. *Soil Biol. Biochem.* **23**, 979–986.

Lascaris, D. and Deacan, J. W. (1991b). Relationship between root cortical senescence and growth of wheat as influenced by mineral nutrition, *Idriella bolleyi* (Sprague) von Arx and pruming of leaves. *New Phytol.* **118**, 391–396.

Lass, B. and Ullrich-Eberius, C. I. (1984). Evidence for proton/sulfate co-transport and its kinetics in *Lemna gibba* G 1. *Planta* **161**, 53–60.

Läuchli, A. (1976a). Symplasmic transport and ion release to the xylem. *In* 'Transport and Transfer Processes in Plants' (I. F. Wardlaw and J. B. Passioura, eds.), Chapter 9, pp. 101–112. Academic Press, New York.

Läuchli, A. (1976b). Genotypic variation in transport. *In* 'Transport in Plants 2, Part A' (U. Lüttge and M. G. Pitman, eds.), pp. 372–393. Springer-Verlag, Berlin.

Läuchli, A. (1993). Selenium in plants: uptake functions, and environmental toxicity. *Bot. Acta* **106**, 455–468.

Läuchli, A. and Pflüger, R. (1978). Potassium transport through plant cell membranes and metabolic role of potassium in plants. *Proc. 11th Congr. Int. Potash Inst. Bern*, pp. 111–163.

Läuchli, A. and Schubert, S. (1989). The role of calcium in the regulation of membrane and cellular growth processes under salt stress. *In* 'NATO ASI Series Vol.G 19, Environmental Stress in Plants' (J. H. Cherry, ed.), pp. 131–138. Springer-Verlag Berlin.

Läuchli, A. and Wieneke, J. (1979). Studies on growth and distribution of Na^+, K^+ and Cl^- in soybean varieties differing in salt tolerance. *Z. Pflanzenernähr. Bodenk.* **142**, 3–13.

Läuchli, A., Pitman, M. G., Kramer, D. and Ball, E. (1978). Are developing xylem vessels the sites of ion exudation from root to shoot? *Plant Cell Environ.* **1**, 217–222.

Lauer, M. J. and Blevins, D. G. (1989). Flowering and podding characteristics on the main stem of soybean grown on varying levels of phosphate nutrition. *J. Plant Nutr.* **12**, 1061–1072.

Lauer, M. J., Pallardy, S. G., Blevins, D. G. and Randall, D. D. (1989a). Whole leaf carbon exchange characteristics of phosphate deficient soybeans (*Glycine max* L.). *Plant Physiol.* **91**, 848–854.

Lauer, M. J., Blevins, D. G. and Sierzputowska-Gracz, H. (1989b). ^{31}P-nuclear magnetic resonance determination of phosphate compartmentation in leaves of reproductive soybeans (*Glycine max* L.) as affected by phosphate nutrition. *Plant Physiol.* **89**, 1331–1336.

Laurie, S. H., Tancock, N. P., McGrath, S. P. and Sanders, J. R. (1991). Influence of complexation on the uptake by plants of iron, manganese, copper and zinc. I. Effect of EDTA in a multi-metal and computer simulation study. *J. Exp. Bot.* **42**, 509–513.

Lavoie, N., Vézina, L.-P. and Margolis, H. A. (1992). Absorption and assimilation of nitrate and ammonium ions by jack pine seedlings. *Tree Physiol.* **11**, 171–183.

Lawlor, D. W. and Milford, G. F. J. (1973). The effect of sodium on growth of water-stressed sugar-beet. *Ann. Bot.* (*London*) [N.S.] **37**, 597–604.

Layzell, D. B., Gaito, S. T. and Hunt, S. (1988). Model of gas exchange and diffusion in legume nodules. I. Calculation of gas exchange rates and the energy cost of N_2 fixation. *Planta* **173**, 117–127.

Lazof, D. and Läuchli, A. (1991). The nutritional status of the apical meristem of *Lactuca sativa* as affected by NaCl salinization: An electron-probe microanalytic study. *Planta* **184**, 334–342.

Lazzaro, M. D. and Thomson, W. W. (1989). Ultrastructure of organic acid secreting trichomes of chickpea (*Cicer arietinum*). *Can. J. Bot.* **67**, 2669–2677.

Leake, J. R., Shaw, G. and Read, D. J. (1990). The biology of mycorrhiza in the Ericaceae. XVI. Mycorrhiza and iron uptake in *Calluna vulgaris* (L.) Hull in the presence of two calcium salts. *New Phytol.* **114**, 651–657.

Le Bot, J. and Kirkby, E. A. (1992). Diurnal uptake of nitrate and potassium during the vegetative growth of tomato plants. *J. Plant Nutr.* **15**, 247–264.

Le Bot, J., Kirkby, E. A. and van Beusichem, M. L.(1990). Manganese toxicity in tomato plants: effects on cation uptake and distribution. *J. Plant Nutr.* **13**, 513–525.

Ledgard, S. F. (1991). Transfer of fixed nitrogen from white clover to associated grasses in swards grazed by dairy cows, estimated using ^{15}N methods. *Plant Soil* **131**, 215–223.

Lee, B., Martin, P. and Bangerth, F. (1989). The effect of sucrose in the levels of abscisic acid, indoleacetic acid and zeatin/zeatin ribose in wheat ears growing in liquid culture. *Physiol. Plant.* **77**, 73–80.

Lee, J. A. and Woolhouse, H. W. (1969a). A comparative study of bicarbonate inhibitions of root growth in calcicole and calcifuge grasses. *New Phytol.* 68, 1–11.

Lee, J. A. and Woolhouse, H. W. (1969b). Root growth and dark fixation of carbon dioxide in calcicoles and calcifuges. *New Phytol.* **68**, 247–255.

Lee, J. S., Mulkey, T. J. and Evans, M. L. (1984). Inhibition of polar calcium movement and gravitropism in roots treated with auxin-transport inhibitors. *Planta* **160**, 536–543.

Lee, R. B (1977). Effects of organic acids on the loss of ions from barley roots. *J. Exp. Bot.* **28**, 578–587.

Lee, R. B. (1982). Selectivity and kinetics of ion uptake of barley plants following nutrient deficiency. *Ann. Bot.* (*London*) [N.S.] **50**, 429–449.

Lee, R. B. (1988). Phosphate influx and extracellular phosphatase activity in barley roots and rose cells. *New Phytol.* **109**, 141–148.

Lee, R. B. and Clarkson, D. T. (1986). Nitrogen-13 studies of nitrate fluxes in barley roots. I. Compartmental analysis from measurements of ^{13}N efflux. *J. Exp. Bot.* **37**, 1753–1767.

Lee, R. B. and Drew, M. C. (1989). Rapid, reversible inhibition of nitrate influx in barley by ammonium. *J. Exp. Bot.* **40**, 741–752.

Lee, R. B. and Ratcliffe, R. G. (1983). Phosphorus nutrition and the intracellular distribution of inorganic phosphate in pea root tips: A quantitative study using ^{31}P-NMR. *J. Exp. Bot.* **34**, 1222–1244.

Lee, R. B. and Ratcliffe, R. G. (1993). Subcellular distribution of inorganic phosphate, and levels of nucleoside triphosphate, in mature maize roots at low external phosphate concentrations: Measurements with ^{31}P-NMR. *J. Exp. Bot.* **44**, 587–598.

Lee, R. B. and Rudge, K. A. (1986). Effect of nitrogen deficiency on the absorption of nitrate and ammonium by barley plants. *Ann. Bot.* **57**, 471–486.

Lee, R. B., Ratcliffe, R. G. and Southon, T. E. (1990). ^{31}P NMR mesurements of the cytoplasmic and vacuolar P_i content of mature maize roots: relationship with phosphorus status and phosphate fluxes. *J. Exp. Bot.* **41**, 1063–1078.

Lee, R. B., Purves, J. V., Ratcliffe, R. G. and Saker, L.R. (1992). Nitrogen assimilation and the control of ammonium and nitrate absorption by maize roots. *J. Exp. Bot.* **43**, 1385–1396.

Lefebvre, D. D. and Glass, A. D. M. (1982). Regulation of phosphate influx in barley roots; effects of phosphate deprivation and reduction of influx with provision of orthophosphate. *Physiol. Plant.* **54**, 199–206.

Lefebvre, D. D., Duffi, S. M. G., Fife, C. A., Julien-Inalsingh, C. and Plaxton, W. C. (1990). Response to phosphate deprivation in *Brassica nigra* suspension cells. *Plant Physiol.* **93**, 504–511.

Le Gales, Y., Lamant, A. and Heller, R. (1980). Fixation du calcium par des fractions macromoleculaires solubles isolées a partir de végétaux supérieurs. *Physiol. Vég.* **18**, 431–441.

Legge, R. L., Thompson, E., Baker, J. E. and Lieberman, M. (1982). The effect of calcium on the fluidity and phase properties of microsomal membranes isolated from postclimacteric Golden Delicious apples. *Plant Cell Physiol.* **23**, 161–169.

Lehle, L. (1990). Phosphatidyl inositol metabolism and its role in signal transducing in growing plants. *Plant Mol. Biol.* **15**, 647–658.

Lehr, J. J. (1953). Sodium as a plant nutrient. *J. Sci. Food Agric.* **4**, 460–468.

Leidi, E. O. and Gomes, M. (1985). A role for manganese in the regulation of soybean nitrate reductase activity? *J. Plant Physiol.* **118**, 335–342.

Leidi, E. O., Gómez, M. and del Rio, L. A. (1987). Evaluation of biochemical indicators of Fe and Mn nutrition for soybean plants. II. Superoxide dismutase, chlorophyll contents and photosystem II activity. *J. Plant Nutr.* **10**, 261–271.

Leigh, R. A. and Wyn Jones, R. G. (1984). A hypothesis relating critical potassium concentrations for growth to the distribution and functions of this ion in the plant cell. *New Phytol.* **97**, 1–13.

Leigh, R. A. and Wyn Jones, R. G. (1986). Cellular compartmentation in plant nutrition: the selective cytoplasm and the promiscuous vacuole. In 'Advances in Plant Nutrition 2' (B. Tinker and A. Läuchli, eds.), pp. 249–279. Praeger Scientific, New York.

Leigh, R. A., Stribley, D. P. and Jonston, A. E. (1982). How should tissue nutrient concentrations be expressed? In 'Proceedings of the Ninth International Plant Nutrition Colloquium, Warwick, England' (A. Scaife, ed.), pp. 39–44. Commonwealth Agricultural Bureau, Farnham Royal, Bucks.

Leigh, R. A., Chater, M., Storey, R. and Johnston, E. A. (1986). Accumulation and subcellular distribution of cations in relation to the growth of potassium-deficient barley. *Plant Cell Environ.* **9**, 595–604.

Leisen, E. und Marschner, H. (1990). Einfluss von Düngung und saurer Benebelung auf Nadelverluste sowie Auswaschung und Gehalte an Mineralstoffen und Kohlenhydraten in Nadeln von Fichten (*Picea abies* (L.) Karst.). *Forstwiss. Cbl.* **109**, 253–263.

Leisen, E., Häussling, M. und Marschner, H. (1990). Einfluss von Stickstoff-Form und -Konzentration und saurer Benebelung auf pH-Veränderungen in der Rhizosphäre von Fichten (*Picea abies* (L.) Karst.). *Forstwiss. Cbl.* **109**, 275–286.

Lemoine, R., Daie, J. and Wyse, R. (1988). Evidence for the presence of a sucrose carrier in immature sugar beet tap roots. *Plant Physiol.* **86**, 575–580.

Lemon, E. and van Houtte, R. (1980). Ammonia exchange at the land surface. *Agron. J.* **72**, 876–883.

Lenz, F. (1970). Einfluss der Früchte auf das Wachstum, den Wasserverbrauch und die Nährstoffaufnahme von Auberginen. *Gartenbauwissenschaft* **35**, 281–292.

Lenz, F. and Döring, H. W. (1975). Fruit effects on growth and water consumption in Citrus. *Gartenbauwissenschaft* **6**, 257–260.

Leonard, R. T. and Hotchkiss, C. W. (1976). Cation-stimulated adenosine triphosphatase activity and cation transport in corn roots. *Plant Physiol.* **58**, 331–335.

Leonardi, S. and Flückiger, W. (1989). Effects of cation leaching on mineral cycling and transpiration: Investigations with beech seedlings, *Fagus sylvatica* L. *New Phytol.* **111**, 173–179.

Lerchl, D., Hillmer, S., Grotha, R. and Robinson, D. G. (1989). Ultrastructural observations on CTC-induced callose formation in *Riella helicophylla*. *Bot. Acta* **102**, 62–72.

Lerer, M. and Bar-Akiva, A. (1976). Nitrogen constituents in manganese-deficient lemon leaves. *Physiol. Plant.* **38**, 13–18.

Lerner, H. R., Reinhold, L., Guy, R., Braun, Y., Hasidim, M. and Poljakoff-Mayber, A. (1983). Salt activation and inhibition of membrane ATPase from roots of the halophyte *Atriplex nummularia*. *Plant Cell Environ.* **6**, 501–506.

Lessani, H. and Marschner, H. (1978). Relation between salt tolerance and long distance transport of sodium and chloride in various crop species. *Aust. J. Plant Physiol.* **5**, 27–37.

Leusch, H.-J. and Buchenauer, H. (1988a). Si-Gehalte und Si-Lokalisation im Weizenblatt und deren Bedeutung für die Abwehr einer Mehltauinfektion. *Kali-Briefe* **19**, 13–24.

Leusch, H.-J. and Buchenauer, H. (1988b). Einfluss von Bodenbehandlung mit siliziumreichen Kalken und Natriumsilikat auf den Befall des Weizens mit *Erysiphe graminis* und *Septoria nodorum* in Abhängigkeit von der Form der N-Dünger. *J. Plant Dis. Protect.* **96**, 154–172.

Levin, S. A., Mooney, H. A. and Field, C. (1989). The dependence of plant root:shoot ratios on internal nitrogen concentration. *Ann. Bot. (London)* **64**, 71–75.

Levitt, J. (1980). 'Responses of Plants to Environmental Stresses', 2nd. edn., Vol. 2. Academic Press, New York.

Levy, Y. and Horesh, I. (1984). Importance of penetration through stomata in the correction of chlorosis with iron salts and low-surface-tension surfactants. *J. Plant Nutr.* **7**, 279–281.

Lewin, J. and Reimann, B. E. F. (1969). Silicon and plant growth. *Annu. Rev. Plant Physiol.* **20**, 289–304.

Lewis, D. C. (1992). Effect of plant age on the critical inorganic and total phosphorus concentrations in selected tissues of subterranean clover (cv. Trikkala). *Aust. J. Agric. Res.* **43**, 215–223.

Lewis, D. H. (1980a). Boron, lignification and the origin of vascular plants – a unified hypothesis. *New Phytol.* **84**, 209–229.

Lewis, D. H. (1980b). Are there inter-relations between the metabolic role of boron, synthesis of phenolic phytoalexins and the germination of pollen? *New Phytol.* **84**, 261–270.

Lexmond, T. M. and Vorm, P. D. J. van der (1981). The effect of pH on copper toxicity to hydroponically grown maize. *Neth. J. Agric. Sci.* **29**, 217–238.

Leyval, C. and Berthelin, J. (1991). Weathering of a mica by roots and rhizosphere microorganisms of pine. *Soil Sci. Soc. Am. J.* **55**, 1009–1016.

Li, C. J. and Bangerth, F. (1992). The possible role of cytokinins, ethylene and indoleacetic acid in apical dominance. *In* 'Progress in Plant Growth Regulation' (C. M. Karsten, L. C. van Loon and D. Vreugdenhil, eds.), pp. 431–436. Kluwer Academic, Dordrecht.

Li, X.-L., George, E. and Marschner, H. (1991a). Extension of the phosphorus depletion zone in VA-mycorrhizal white clover in a calcareous soil. *Plant Soil* **136**, 41–48.

Li, X.-L., Marschner, H. and George, E. (1991b). Acquisition of phosphorus and copper by VA-mycorrhizal hyphae and root-to-shoot transport in white clover. *Plant Soil* **136**, 49–57.

Li, X.-L., George, E. and Marschner, H. (1991c). Phosphorus depletion and pH decrease at the root–soil and hyphae–soil interfaces of VA mycorrhizal white clover fertilized with ammonium. *New Phytol.* **119**, 397–404.

Li, Y. and Barber, S. A. (1991). Calculating changes of legume rhizosphere soil pH and soil solution phosphorus from phosphorus uptake. *Commun. Soil Sci. Plant Anal.* **22**, 955–973.

Li, Z.-C. and Bush, D. R. (1991). pH-dependent amino acid transport into plasma membrane vesicles isolated from sugar beet (*Beta vulgaris* L.) leaves. II. Evidence for multiple olipathic, neutral amino acid symport. *Plant Physiol.* **96**, 1338–1344.

Li, Z. Z. and Gresshoff, P. M. (1990). Developmental and biochemical regulation of 'constitutive' nitrate reductase activity in leaves of nodulating soybean. *J. Exp. Bot.* **41**, 1231–1238.

Liegel, W. (1970). Calciumoxalat-Abscheidung in Fruchtstielen einiger Apfelvarietäten. *Angew. Bot.* **44**, 223–232.

Liljeroth, E., Schelling, G. C. and Van Veen, J. A. (1990a). Influence of different application rates of nitrogen to soil on rhizosphere bacterial. *Neth. J. Agric. Sci.* **38**, 355–264.

Liljeroth, E., Van Veen, J. A. and Miller, H. J. (1990b). Assimilate translocation to the rhizosphere of two wheat lines and subsequent utilization by rhizosphere microorganisms at two soil nitrogen concentrations. *Soil Biol. Biochem.* **22**, 1015–1021.

Lin, C. H. and Stocking, C. R. (1978). Influence of leaf age, light, dark and iron deficiency on polyribosome levels in maize leaves. *Plant Cell Physiol.* **19**, 461–470.

Lin, D. C. and Nobel, P. S. (1971). Control of photosynthesis by Mg^{2+}. *Arch. Biochem. Biophys.* **145**, 622–632.

Lin, M. S. and Kao, C. H. (1990). Senescence of rice leaves. XIII. Changes of Zn^{2+}-dependent acid inorganic pyrophosphatase. *J. Plant Physiol.* **137**, 41–45.

Lin, P. P. C., Egli, D. B., Li, G. M. and Meckel, L. (1984). Polyamine titer in the embryonic axis and the cotyledons of *Glycine max* (L.) during seed growth and maturation. *Plant Physiol.* **76**, 366–371.

Linderman, R. G. (1988). Mycorrhizal interactions with the rhizosphere microflora: the mycorrhizosphere effect. *Phytopathology* **78**, 366–371.

Lindhauer, M. G. (1985). Influence of K nutrition and drought on water relations and growth of sunflower (*Helianthus annuus* L.). *Z. Pflanzenernähr. Bodenk.* **148**, 654–669.

Lindhauer, M. G., Haeder, H. E. and Beringer, H. (1990). Osmotic potentials and solute concentrations in sugar beet plants cultivated with varying potassium/sodium ratios. *Z. Pflanzenernähr. Bodenk.* **153**, 25–32.

Lindsay, W. L. (1991). Iron oxide solubilization by organic matter and its effect on iron availability. *In* 'Iron Nutrition and Interactions in Plants' (Y. Chen and Y. Hadar, eds.), pp. 29–36. Kluwer Academic, Dordrecht, The Netherlands.

Lingle, J. C. and Lorenz, O. A. (1969). Potassium nutrition of tomatoes. *J. Am. Soc. Hortic. Sci.* **94**, 679–683.

Linser, H., Raafat, A. and Zeid, F. A. (1974). Reinprotein und Chlorophyll bei *Daucus carota* im Verlauf der Vegetationsperiode des ersten Jahres unter dem Einfluss von Wachstumsregulatoren. *Z. Pflanzenernähr. Bodenk.* **137**, 36–48.

Lips, S. H., Leidi, E. O., Silberbush, M., Soares, M. I. M. and Lewis, O. E. M. (1990). Physiological aspects of ammonium and nitrate fertilization. *J. Plant Nutr.* **13**, 1271–1289.

Lipton, D. S., Blanchar, R. W. and Blevins, D. G. (1987). Citrate, malate, and succinate concentration in exudates from P-sufficient and P-stressed *Medicago sativa* L. seedlings. *Plant Physiol.* **85**, 315–317.

Liu, J. C. and Huettl, R. F. (1991). Relations between damage symptoms and nutritional status of Norway spruce stands (*Picea abies* Karst.) in southwestern Germany. *Fert. Res.* **27**, 9–22.

Liu, W. (1979). Potassium and phosphate uptake in corn roots. Further evidence for an electrogenic H^+/K^+ exchanger and an OH^-/P_i antiporter. *Plant Physiol.* **63**, 952–955.

Lobreaux, S. and Briat, J. F. (1991). Ferritin accumulation and degradation in different organs of pea (*Pisum sativum*) during development. *Biochem. J.* **274**, 601–606.

Lobreaux, S., Massenet, O. and Briat, J.-F. (1992). Iron induces ferritin synthesis in maize plantlets. *Plant Mol. Biol.* **19**, 563–575.

Lodge, D. J. (1989). The influence of soil moisture and flooding on formation of VA-endo- and ectomycorrhizae in *Populus* and *Salix*. *Plant Soil* **117**, 243–253.

Loeppert, R. H. and Hallmark, C. T. (1985). Indigenous soil properties influencing the availability of iron in calcareous soils. *Soil Sci. Soc. Am. J.* **49**, 597–603.

Loescher, W. H. (1987). Physiology and metabolism of sugar alcohols in higher plants. *Physiol. Plant.* **70**, 553–557.

Löhnis, M. P. (1960). Effect of magnesium and calcium supply on the uptake of manganese by various crop plants. *Plant Soil* **12**, 339–376.

Lolas, G. M., Palamidis, N. and Markakis, P. (1976). The phytic-acid total phosphorus relationship in barley, oats, soybeans and wheat. *Cereal Chem.* **53**, 867–870.

Loneragan, J. F. and Asher, C. J. (1967). Response of plants to phosphate concentration in solution culture. II. Role of phosphate absorption and its relation to growth. *Soil Sci.* **103**, 311–318.

Loneragan, J. F. and Dowling, E. J. (1958). The interaction of calcium and hydrogen ions in the nodulation of subterranean clover. *Aust. J. Agric. Res.* **9**, 464–472.

Loneragan, J. F., Snowball, K. (1969). Calcium requirements of plants. *Aust. J. Agric. Res.* **20**, 465–478.

Loneragan, J. F., Snowball, K. and Simmons, W. J. (1968). Response of plants to calcium concentration in solution culture. *Aust. J. Agric. Res.* **19**, 845–857.

Loneragan, J. F., Snowball, K. and Robson, A. D. (1976). Remobilization of nutrients and its significance in plant nutrition. *In* 'Transport and Transfer Process in Plants' (I. F. Wardlaw and J. B. Passioura, eds.), pp. 463–469. Academic Press, London.

Loneragan, J. F., Grove, T. S., Robson, A. D. and Snowball, K. (1979). Phosphorus toxicity as a factor in zinc-phosphorus interactions in plants. *Soil Sci. Soc. Am. J.* **43**, 966–972.

Loneragan, J. F., Delhaize, E. and Webb, J. (1982a). Enzymic diagnosis of copper deficiency in subterranean clover. I. Relationship of ascorbate oxidase activity in leaves to plant copper status. *Aust. J. Agric. Res.* **33**, 967–979.

Loneragan, J. F., Grunes, D. L., Welch, R. M., Aduayi, E. A., Tengah, A., Lazar, V. A. and Cary, E. E. (1982b). Phosphorus accumulation and toxicity in leaves in relation to zinc supply. *Soil Sci. Soc. Am. J.* **46**, 345–352.

Loneragan, J. F., Kirk, G. J. and Webb, M. J. (1987). Translocation and function of zinc in roots. *J. Plant Nutr.* **10**, 1247–1254.

Long, J. M. and Widders, I. E. (1990). Quantification of apoplastic potassium content by elution analysis of leaf lamina tissue from pea (*Pisum sativum* L. cv. Argenteum). *Plant Physiol.* **94**, 1040–1047.

Longnecker, N. and Welch, R. M. (1990). Accumulation of apoplastic iron in plant roots. A factor in the resistance of soybeans to iron-deficiency induced chlorosis? *Plant Physiol.* **92**, 17–22.

Longnecker, N. E., Graham, R. D. and Card, G. (1991b). Effects of manganese deficiency on the pattern of tillering and development of barley (*Hordeum vulgare* cv. Galleon). *Field Crops Res.* **28**, 85–102.

Longnecker, N. E., Marcar, N. E. and Graham, R. D. (1991a). Increased manganese content of barley seeds can increase grain yield in manganese-deficient conditions. *Aust. J. Agric. Res.* **42**, 1065–1074.

Loomis, W. D. and Durst, R. W. (1991). Boron and cell walls. *Curr. Top. in Plant Biochem. Physiol.* **10**, 149–178.

Loomis, W. D. and Durst, R. W. (1992). Chemistry and biology of boron. *BioFactors* **3**. 229–239.

Lott, J. N. A. and Buttrose, M. S. (1978). Globoids in protein bodies of legume seed cotyledons. *Aust. J. Plant Physiol.* **5**, 89–111.

Lott, J. N. A. and Vollmer, C. M. (1973). Changes in the cotyledons of *Cucurbita maxima* during germination. IV. Protein bodies. *Protoplasma* **78**, 255–271.

Loughman, B. C., Webb, M. J. and Loneragan, J. F. (1982). Zinc and the utilization of phosphate in wheat plants. *In* 'Proceedings of the Ninth International Plant Nutrition Colloquium, Warwick, England' (A. Scaife, ed.), pp. 335–340. Commonwealth Agricultural Bureau, Farnham Royal, Bucks.

Louis, I., Racette, S. and Torrey, J. G. (1990). Occurrence of cluster roots on *Myrica cerifera* L. (Myricaceae) in water culture in relation to phosphorus nutrition. *New Phytol.* **115**, 311–317.

Lovatt, C. J. (1985). Evolution of xylem resulted in a requirement for boron in the apical meristems of vascular plants. *New Phytol.* **99**, 509–522.

Lowther, W. L. and Loneragan, J. F. (1968). Calcium and nodulation in subterranean clover. (*Trifolium subterraneum* L.). *Plant Physiol.* **43**, 1362–1366.

Lu, J. L., Ertl, J. R. and Chen, C. M. (1992). Transcriptional regulation of nitrate reductase mRNA levels by cytokinin-abscisic acid interactions in etiolated barley leaves. *Plant Physiol.* **98**, 1255–1260.

Lubberding, H. J., de Graaf, F. H. J. M. and Bienfait, H. F. (1988). Ferric reducing activity in roots of Fe-deficient *Phaseolus vulgaris*: Source of reducing equivalents. *Biochem. Physiol. Pflan.(BPP)* **183**, 271–276.

Lucena, J. J., Garate, A., Ramon, A. M. and Manzanares, M. (1990). Iron nutrition of a hydroponic strawberry culture (*Fragaria vesca* L.) supplied with different Fe chelates. *Plant Soil* **123**, 9–15.

Lui, W. C., Lund, L. J. and Page, A. L. (1989). Acidity produced by leguminous plants through symbiotic dinitrogen fixation. *J. Environ. Qual.* **18**, 529–534.

Lukaszewski, K. M., Blevins, D. G. and Randall, D. D. (1992). Asparagine and boric acid cause allantoate accumulation in soybean leaves by inhibiting manganese-dependent allantoate amidohydrolase. *Plant Physiol.* **99**, 1670–1676.

Lund, Z. F. (1970). The effect of calcium and its relation to several cations in soybean root growth. *Soil Sci. Soc. Am. Proc.* **34**, 456–459.

Lune, P. van and van Goor, B. J. (1977). Ripening disorders of tomato as affected by the K/Ca ratio in the culture solution. *J. Hortic. Sci.* **52**, 173–180.

Luster, D. G. and Buckhout, T. J. (1988). Characterization and partial purification of multiple electron transport activities in plasma membranes from maize (*Zea mays*) roots. *Physiol. Plant.* **73**, 339–347.

Lüttge, U. (1988). Day–night changes of citric-acid levels in crassulacean acid metabolism: phenomenon and ecophysiological significance. *Plant Cell Environ.* **11**, 445–451.

Lüttge, U. and Clarkson, D. T. (1992). Mineral nutrition: aluminium. *In* 'Progress in Botany Vol. 53', pp. 63–77. Springer Verlag Berlin

Lüttge, U. and Laties, G. G. (1966). Dual mechanism of ion absorption in relation to long distance transport in plants. *Plant Physiol.* **41**, 1531–1539.

Luxmoore, R. J., Fischer, R. A. and Stolzy, L. H. (1973). Flooding and soil temperature effects on wheat during grain filling. *Agron. J.* **65**, 361–364.

Lynch, J. and Läuchli, A. (1985). Salt stress disturbs the calcium nutrition of barley (*Hordeum vulgare* L.). *New Phytol.* **99**, 345–354.

Lynch, J. and Läuchli, A. (1988). Salinity affects intracellular calcium in corn root protoplasts. *Plant Physiol.* **87**, 351–356.

Lynch, J. and White, J. W. (1992). Shoot nitrogen dynamics in tropical common bean. *Crop Sci.* **32**, 392–397.

Lynch, J., Epstein, E. and Läuchli, A. (1982). Na$^+$-K$^+$ relationship in salt-stressed barley. *In* 'Proceedings of the Ninth International Plant Nutrition Colloquium, Warwick, England' (A. Scaife, ed.), pp. 347–352. Commonwealth Agricultural Bureau, Farnham Royal, Bucks.

Lynch, J., Cramer, G. R. and Läuchli, A. (1987). Salinity reduces membrane-associated calcium in corn root protoplasts. *Plant Physiol.* **83**, 390–394.

Lynch, J., Thiel, G. and Läuchli, A. (1988). Effects of salinity on the extensibility and Ca availability in the expanding region of growing barley leaves. *Bot. Acta* **101**, 355–361.

Lynch, J., Läuchli, A. and Epstein, E. (1991). Vegetative growth of the common bean in response to phosphorus nutrition. *Crop Sci.* **31**, 380–387.

Lynch, J. M. (1978). Production and phytotoxicity of acetic acid in anaerobic soils containing plant residues. *Soil Biol. Biochem.* **10**, 131–135.

Lynch, J. M. and Whipps, J. M. (1990). Substrate flow in the rhizosphere. *Plant Soil* **129**, 1–10.

Lyshede, O. B. (1982). Structure of the outer epidermal wall in xerophytes. *In* 'The Plant Cuticle' (D. F. Cutler, K. L. Alvin and C. E. Price, eds.), pp. 87–98. Academic Press London.

Lytle, C. M. and Jolley, V. D. (1991). Iron deficiency stress response of various C-3 and C-4 grain crop genotypes: strategy II mechanism evaluated. *J. Plant Nutr.* **14**, 341–362.

Ma, J. and Takahashi, E. (1990a). Effect of silicon on the growth and phosphorus uptake of rice. *Plant Soil* **126**, 115–119.

Ma, J. and Takahashi, E. (1990b). The effect of silicic acid on rice in a P-deficient soil. *Plant Soil* **126**, 121–125.

Ma, J. and Takahashi, E. (1991). Availability of rice straw to rice plants. *Soil Sci. Plant Nutr. (Tokyo)* **37**, 111–116.

Ma, J. F. and Takahashi, E. (1993). Interaction between calcium and silicon in water-cultured rice plants. *Plant Soil* **148**, 107–113.

Ma, J., Nishimura, K. and Takahashi, E. (1989). Effect of silicon on the growth of rice plant at different growth stages. *Soil Sci. Plant Nutr. (Tokyo)* **35**, 347–356.

Ma, J. F., Kusano, G., Kimura, S. and Nomoto, K. (1993). Specific recognition of mugineic acid–ferric complex by barley roots. *Phytochemistry* **34**, 599–603.

Maas, E. V. (1985). Crop tolerance to saline sprinkling water. *Plant Soil* **89**, 372–284.

Maas, E. V. (1993). Salinity and citriculture. *Tree Physiol.* **12**, 195–216.

Maas, E. V. and Grieve, C. M. (1987). Sodium-induced calcium deficiency in salt-stressed corn. *Plant Cell Environ.* **10**, 559–564.

Maas, E. V. and Hoffman, G. J. (1977). Crop salt tolerance – current assessment. *J. Irrig. Drin. Div. Am. Soc. Civ. Eng.* **103**, 115–134.

Maas, E. V., Hoffman, G. J., Chaba, G. D., Poss, J. A. and Shannon, M. C. (1983). Salt sensitivity of corn at various growth stages. *Irrig. Sci.* **4**, 45–57.

Maas, F. M., van de Wetering, D. A. M., van Beusichem, M. L. and Bienfait, H. F. (1988). Characterization of phloem iron and its possible role in the regulation of Fe-efficiency reactions. *Plant Physiol.* **87**, 167–171.

Maathuis, F. J. M. and Sanders, D. (1993). Energization of potassium uptake in *Arabidopsis thaliona. Planta* **191**, 302–307.

MacAdam, J. W., Volenec, J. J. and Nelson, C. J. (1989). Effect of nitrogen on mesophyll cell division and epidermal cell elongation in tall fescue leaf blades. *Plant Physiol.* **89**, 549–556.

MacDonald, E. M. S., Powell, G. K., Regier, D. A., Glass, N. L., Roberto, F., Kosuge, T. and Morris, R. O. (1986). Secretion of zeatin, ribosylzeatin, and ribosyl-1'-methylzeatin by *Pseudomonas savastanoi. Plant Physiol.* **82**, 742–747.

MacDonald, I. R., Macklon, A. E. S. and MacLeod, R. W. G. (1975). Energy supply and light-enhanced chloride uptake in wheat laminae. *Plant Physiol.* **56**, 699–702.

MacDuff, J. H. and Jackson, S. B. (1991). Growth and preference for ammonium or nitrate uptake by barley in relation to root temperature. *J. Exp. Bot.* **42**, 521–530.

MacFall, J. S., Johnson, G. A. and Kramer, P. J. (1991). Comparative water uptake by roots of different ages in seedlings of loblolly pine (*Pinus taeda* L.). *New Phytol.* **119**, 551–560.

MacFall, J. S., Slack, S. A. and Wehrli, S. (1992). Phosphorus distribution in red pine roots and the ectomycorrhizal fungus *Hebeloma arenosa. Plant Physiol.* **100**, 713–717.

Machold, O. (1967). Untersuchungen an stoffwechseldefekten Mutanten der Kulturto-mate. III. Die Wirkung von Ammonium- und Nitratstickstoff auf den Chlorophyllge-halt. *Flora (Jena), Abt. A* **157**, 536–551.

Machold, O. (1968). Einfluss der Ernährungsbedingungen auf den Zustand des Eisens in den Blättern, den Chlorophyllgehalt und die Katalase- sowie Peroxydaseaktivität. *Flora (Jena), Abt. A* **159**, 1–25.

Machold, O., Meisel, W. and Schnorr, H. (1968). Bestimmung der Bindungsformen des Eisens in Blättern durch Mössbauer-Spektrometrie. *Naturwissenschaften* **55**, 499–500.

MacInnes, C. B. and Albert, L. S. (1969). Effect of light intensity and plant size on rate of development of early boron deficiency symptoms in tomato root tips. *Plant Physiol.* **44**, 965–976.

MacIsaac, S. M., Sawhney, V. K. and Pohorecky, Y. (1989). Regulation of lateral root formation in lettuce (*Lactuca sativa*) seedling roots. I. Interacting effects of α-naphthaleneacetic acid and kinetin. *Physiol. Plant.* **77**, 287–293.

Mäck, G. and Tischner, R. (1990). The effect of endogenous and externally supplied nitrate on nitrate uptake and reduction in sugarbeet seedlings. *Planta* **182**, 169–173.

Mackay, A. D. and Barber, S. A. (1984). Soil temperature effects on root growth and phosphorus uptake by corn. *Soil Sci. Soc. Am. J.* **48**, 818–823.

Mackay, A. D. and Barber, S. A. (1985). Effect of soil moisture and phosphate level on root hair growth of corn roots. *Plant Soil* **86**, 321–331.

Mackay, A. D. and Barber, S. A. (1987). Effect of cyclic wetting and drying of a soil on root hair growth of maize roots. *Plant Soil* **104**, 291–293.

Macklon, A. E. S., Ron, M. M. and Sim, A. (1990). Cortical cell fluxes of ammonium and nitrate in excised root segments of *Allium cepa* L.; studies using [15]N. *J. Exp. Bot.* **41**, 359–370.

MacLeod, L. B. (1969). Effects of N, P and K and their interactions on the yield and kernel weight of barley in hydroponic culture. *Agron. J.* **61**, 26–29.

MacNish, G. C. (1988). Changes in take-all (*Gaeumannomyces graminis* var. *tritici*) rhizoctonia root rot (*Rhizoctonia solani*) and soil pH in continuous wheat with annual applications of nitrogenous fertilizer in Western Australia. *Aust. J. Exp. Agric.* **28**, 333–341.

Mäder, M. and Füssl, R. (1982). Role of peroxidase in lignification of tobacco cells. II. Regulation by phenolic compounds. *Plant Physiol.* **70**, 1132–1134.

Magalhäes, J. R. and Huber, D. M. (1989). Ammonium assimilation in different plant species as affected by nitrogen form and pH control in solution culture. *Fert. Res.* **21**, 1–6.

Mahli, S. S., Piening, L. J. and MacPherson, D. J. (1989). Effect of copper on stem melanosis and yield of wheat: sources, rates and methods of application. *Plant Soil* **119**, 199–204.

Maier, R. J., Phil, T. D., Stults, L. and Sray, W. (1990). Nickel accumulation and storage in *Bradyrhizobium japonicum*. *Appl. Environ. Microbiol.* **56**, 1905–1911.

Maier-Maercker, U. (1979). 'Peristomatal transpiration' and stomatal movement: a controversial view. I. Additional proof of peristomatal transpiration by photography and a comprehensive discussion in the light of recent results. *Z. Pflanzenphysiol.* **91**, 25–43.

Maiti, R. K., Ramaiah, K. V., Bisen, S. S. and Chidley, V. L. (1984). A comparative study of the haustorial development of *Striga asiatica* (L.) Kuntze on *Sorghum* cultivars. *Ann. Bot. (London)* [*N.S.*] **54**, 447–457.

Maki, H., Yamagishi, K., Sato,T., Ogura, N. and Nakagawa, H. (1986). Regulation of nitrate reductase activity in cultured spinach cells as studied by an enzyme-linked immunosorbent assay. *Plant Physiol.* **82**, 739–741.

Mallarino, A. P., Wedin, W. F., Goyenola, R. S., Perdomo, C. H. and West, C. P.

(1990a). Legume species and proportion effects on symbiotic dinitrogen fixation in legume–grass mixtures. *Agron. J.* **82**, 785–789.

Mallarino, A. P., Wedin, W. F., Perdomo, C. H., Goyenola, R. S. and West, C. P. (1990b). Nitrogen transfer from white clover, red clover, and birdsfoot trefoil to associated grass. *Agron. J.* **82**, 790–795.

Mandava, N. B. (1988). Plant growth-promoting brassinosteroids. *Annu. Rev. Plant Physiol.* **39**, 23–52.

Mandimba, G., Heulin, T., Bally, R., Guckert, A. and Balandreau, J. (1986). Chemotaxis of free-living nitrogen-fixing bacteria towards maize mucilage. *Plant Soil* **90**, 129–139.

Manrique, L. A. and Bartholomew, D. P. (1991). Growth and yield performance of potato grown at three elevations in Hawaii. II. Dry matter production and efficiency of partitioning. *Crop Sci.* **31**. 367–371.

Mansfield, T. A., Hetherington, A. M. and Atkinson, C. J. (1990). Some aspects of stomatal physiology. *Annu. Rev. Plant Physiol. Mol. Biol.* **41**, 55–75.

Marcar, N. E. and Graham, R. D. (1987). Genotypic variation for manganese efficiency in wheat. *J. Plant Nutr.* **10**, 2049–2055.

Marder, J. B. and Barber, J. (1989). The molecular anatomy and function of thylakoid proteins. *Plant Cell Environ.* **12**, 595–614.

Marienfeld, S. and Stelzer, R. (1993). X-ray microanalyses in roots of Al-treated *Avena sativa* plants. *J. Plant Physiol.* **141**, 569–573.

Mark, F. van der, Lange, T. de and Bienfait, H. F. (1981). The role of ferritin in developing primary bean leaves under various light conditions. *Planta* **153**, 338–342.

Marmé, D. (1983). Calcium transport and function. *In* 'Encyclopedia of Plant Physiology, New Series' (A. Läuchli and R. L. Bieleski, eds.), Vol. 15B, pp. 599–625. Springer-Verlag, Berlin and New York.

Marquard, R., Kühn, H. and Linser, H. (1968). Der Einfluss der Schwefelernährung auf die Senfölbildung. *Z. Pflanzenernähr. Bodenk.* **121**, 221–230.

Marquardt, G. and Lüttge, U. (1987). Proton transporting enzymes at the tonoplast of leaf cells of the CAM plant *Kalenchoë daigremontiana*. II. The pyrophosphatase. *J. Plant Physiol.* **129**, 269–286.

Marschner, B., Stahr, K. and Renger, M. (1991). Element inputs and canopy interactions in two pine forest ecosystems in Berlin, Germany. *Z. Pflanzenernähr. Bodenk.* **145**, 147–151.

Marschner, B., Stahr, K. and Renger, M. (1992). Lime effects on pine forest floor leachate chemistry and element fluxes. *J. Environ. Qual.* **21**, 410–419.

Marschner, H. (1971). Why can sodium replace potassium in plants? *Proc. 8th Colloq. Int. Potash Inst. Bern*, pp. 50–63.

Marschner, H. (1983). General introduction to the mineral nutrition of plants. *In* 'Encyclopedia of Plant Physiology, New Series' (A. Läuchli and R. L. Bieleski, eds.), Vol. 15A, pp. 5–60. Springer-Verlag, Berlin.

Marschner, H. (1988). Mechanism of manganese acquisition by roots from soils. *In* 'Manganese in Soils and Plants' (R. D. Graham, R. J. Hannam and N. C. Uren, eds.), pp. 191–204. Kluwer Academic, Dordrecht, The Netherlands.

Marschner, H. (1991a). Root-induced changes in the availability of micronutrients in the rhizosphere. *In* 'Plant Roots: the Hidden Half' (Y. Waisel, A. Eshel and U. Kafkafi, eds.), pp. 503–528. Marcel Dekker, New York.

Marschner, H. (1991b). Mechanism of adaptation of plants to acid soils. *Plant Soil* **134**, 1–20.

Marschner, H. (1992). Bodenversauerung und Magnesiumernährung der Pflanzen. *In*

'Magnesiummangel in Mitteleuropäischen Waldökosystemen (G. Glatzel, R. Jandl, M. Sieghardt und H. Hager, eds.), pp. 1–15. Forstliche Schriftenreihe, Band 5, Universität für Bodenkultur, Wien.

Marschner, H. (1993). Zinc uptake from soils. In 'Zinc in Soils and Plants' (A. D. Robson, ed.), pp. 59–77. Kluwer Academic, Dordrecht, The Netherlands.

Marschner, H. and Cakmak, I. (1986). Mechanism of phosphorus-induced zinc deficiency in cotton. II. Evidence for impaired shoot control of phosphorus uptake and translocation under zinc deficiency. *Physiol. Plant.* **68**, 491–496.

Marschner, H. and Cakmak, I. (1989). High light intensity enhances chlorosis and necrosis in leaves of zinc, potassium, and magnesium deficient bean (*Phaseolus vulgaris*) plants. *J. Plant Physiol.* **134**, 308–315.

Marschner, H. and Dell, B. (1994). Nutrient uptake in mycorrhizal symbiosis. *Plant Soil* **159**, 89–102.

Marschner, H. and Ossenberg-Neuhaus, H. (1977). Wirkung von 2,3,5-Trijodbenzoesäure (TIBA) auf den Calciumtransport und die Kationenaustauschkapazität in Sonnenblumen. *Z. Pflanzenphysiol.* **85**, 29–44.

Marschner, H. and Possingham, J. V. (1975). Effect of K^+ and Na^+ on growth of leaf discs of sugar beet and spinach. *Z. Pflanzenphysiol.* **75**, 6–16.

Marschner, H. and Richter, C. (1973). Akkumulation und Translokation von K^+, Na^+ und Ca^{2+} bei Angebot zu einzelnen Wurzelzonen von Maiskeimpflanzen. *Z. Pflanzenernähr. Bodenk.* **135**, 1–15.

Marschner, H. and Richter, C. (1974). Calcium-Transport in Wurzeln von Mais- und Bohnenkeimpflanzen. *Plant Soil* **40**, 193–210.

Marschner, H. and Römheld, V. (1983). *In vivo* measurement of root-induced pH changes at the soil-root interface: Effect of plant species and nitrogen source. *Z. Pflanzenphysiol.* **111**, 241–251.

Marschner, H. and Schafarczyk, W. (1967). Vergleich der Nettoaufnahme von Natrium und Kalium bei Mais- und Zuckerrübenpflanzen. *Z. Pflanzenernähr. Bodenk.* **118**, 172–187.

Marschner, H. and Schropp, A. (1977). Vergleichende Untersuchungen über die Empfindlichkeit von 6 Unterlagensorten der Weinrebe gegenüber Phosphat-induziertem Zink-Mangel. *Vitis* **16**, 79–88.

Marschner, H., Kylin, A. and Kuiper, P. J. C. (1981a). Differences in salt tolerance of three sugar beet genotypes. *Physiol. Plant.* **51**, 234–238.

Marschner, H., Kylin, A. and Kuiper, P. J. C. (1981b). Genotypic differences in the response of sugar beet plants to replacement of potassium by sodium. *Physiol. Plant.* **51**, 239–244.

Marschner, H., Römheld, V. and Kissel, M. (1986a). Different strategies in higher plants in mobilization and uptake of iron. *J. Plant Nutr.* **9**, 695–713.

Marschner, H., Römheld, V., Horst, W. J. and Martin, P. (1986b). Root-induced changes in the rhizosphere: importance for the mineral nutrition of plants. *Z. Pflanzenernähr. Bodenk.* **149**, 441–456.

Marschner, H., Römheld, V. and Cakmak, I. (1987). Root-induced changes of nutrient availability in the rhizosphere. *J. Plant Nutr.* **10**, 1175–1184.

Marschner, H., Treeby, M. and Römheld, V. (1989). Role of root-induced changes in the rhizosphere for iron acquisition in higher plants. *Z. Pflanzenernähr. Bodenk.* **152**, 197–204.

Marschner, H., Oberle, H., Cakmak, I. and Römheld, V. (1990). Growth enhancement by silicon in cucumber (*Cucumis sativus*) plants depends on imbalance on phosphorus and zinc supply. *In* 'Plant Nutrition–Physiology and Applications' (M. L. van Beusichem, ed.), pp. 241–249, Kluwer Academic, Dordrecht.

Marschner, H., Häussling, M. and George, E. (1991). Ammonium and nitrate uptake rates and rhizosphere-pH in non-mycorrhizal roots of Norway spruce (*Picea abies* (L.) Karst.). *Trees* **5**, 14–21.

Marschner, P., Asher, J. S. and Graham, R. D. (1991). Effect of manganese-reducing rhizosphere bacteria on the growth of *Gaeumannomyces graminis* var. *tritici* and on manganese uptake by wheat (*Triticum aestivum* L.). *Biol. Fertil. Soils* **12**, 33–38.

Martin, F., Chalot, M., Brun, A., Lorrilou, S., Botton, B. and Dell, B. (1992). Spatial distribution of nitrogen assimilation pathways in ectomycorrhizas. *In* 'Mycorrhizas in Ecosystems' (D. J. Read, D. H. Lewis, A. H. Fitter and I. J. Alexander, eds.), pp. 311–315. CAB International, Wallingford, UK.

Martin, H. V. and Elliott, M. C. (1984). Ontogenetic changes in the transport of indol-3yl-acetic acid into maize roots from the shoot and caryopsis. *Plant Physiol.* **74**, 971–974.

Martin, P. (1971). Wanderwege des Stickstoffs in Buschbohnenpflanzen beim Aufwärtstransport nach der Aufnahme durch die Wurzel. *Z. Pflanzenphysiol.* **64**, 206–222.

Martin, P. (1982). Stem xylem as a possible pathway for mineral retranslocation from senescing leaves to the ear in wheat. *Aust. J. Plant Physiol.* **9**, 197–207.

Martin, P. (1989). Long-distance transport and distribution of potassium in crop plants. *Proc. 21st Colloq. Int. Potash Inst. Bern*, pp. 83–100.

Martin, P. (1990). Einfluss von Mineralstoffen auf das symbiontische N_2-Bindungssystem bei Leguminosen. *Kali-Briefe* **20**, 93–110.

Martin, P., Glatzle, A., Kolb, W., Omay, H. and Schmidt, W. (1989). N_2-fixing bacteria in the rhizosphere: Quantification and hormonal effects on root development. *Z. Pflanzenernähr. Bodenk.* **152**, 237–245.

Martin, R. B. (1988). Bioinorganic chemistry of aluminum. *In* 'Metal Ions in Biological Systems Vol. 24: Aluminum and its Role in Biology' (H. Sigel, ed.). pp. 1–57. Marcel Dekker, New York, NY.

Martinez, V. and Läuchli, A. (1991). Phosphorus translocation in salt-stressed cotton. *Physiol. Plant.* **83**, 627–632.

Martinoia, E., Heck, U. and Wienecken, A. (1981). Vacuoles as storage compartments for nitrate in barley leaves. *Nature (London)* **289**, 292–294.

Martin-Prével, P., Gagnard, J. and Gautier, P. (1987). Plant analysis as a guide to the nutrient requirements of temperate and tropical crops. Lavoisier, New York, Paris.

Marziah, M. and Lam, C. H. (1987). Polyphenol oxidase from soybeans (*Glycine max.* cv. Palmetto) and its response to copper and other micronutrients. *J. Plant Nutr.* **10**, 2089–2094.

Mascagni, H. J. jr and Cox, F. R. (1985). Effective rates of fertilization for correcting manganese deficiency in soybeans. *Agron. J.* **77**, 363–366.

Mascarenhas, J. P. and Machlis, L. (1964). Chemotropic response of the pollen of *Antirrhinum majus* to calcium. *Plant Physiol.* **39**, 70–77.

Mashhady, A. S. and Rowell, D. L. (1978). Soil alkalinity. II. The effect of Na_2CO_3 on iron and manganese supply to tomatoes. *J. Soil Sci.* **29**, 367–372.

Masle, J. (1992). Genetic variation in the effects of root impedance on growth and transpiration rates of wheat and barley. *Aust. J. Plant Physiol.* **19**, 109–125.

Masle, J. and Passioura, J. B. (1987). The effect of soil strength on the growth of young wheat plants. *Aust. J. Plant Physiol.* **14**, 643–656.

Massey, H. F. and Loeffel, A. (1967). Species specific variations in zinc content of corn kernels. *Agron. J.* **59**, 214–217.

Matar, A. E., Paul, J. L. and Jenny, H. (1967). Two phase experiments with plants growing in phosphate-treated soil. *Soil Sci. Soc. Am. Proc.* **31**, 235–237.

Mateo, P., Bonilla, I., Fernández-Valienta, E. and Sanchez-Maseo, E. (1986). Essentiality of boron for dinitrogen fixation in *Anabaena* sp. PCC 7119. *Plant Physiol.* **81**, 430–433.

Materechera, S. A., Dexter, A. R. and Alston, A. M. (1992a). Formation of aggregates by plant roots in homogenised soils. *Plant Soil* **142**, 69–79.

Materechera, S. A., Alston, A. M., Kirby, J. M. and Dexter, A. R. (1992b). Influence of root diameter on the penetration of seminal roots into a compacted subsoil. *Plant Soil* **144**, 297–303.

Mathers, A. C., Thomas, J. D., Steward, B. A. and Herring, J. E. (1980). Manure and inorganic fertilizer effects on sorghum and sunflower growth on iron-deficient soil. *Agron. J.* **72**, 1025–1029.

Matocha, J. E. and Smith, L. (1980). Influence of potassium on *Helminthosporium cynodontis* and dry matter yields of coastal Bermudagrass. *Agron. J.* **72**, 565–567.

Matoh, T., Yasuoka, S., Ishikawa, T. and Takahashi, E. (1988). Potassium requirement of pyruvate kinase extracted from leaves of halophytes. *Physiol. Plant.* **74**, 675–678.

Matsumoto, H. (1988). Repression of proton extrusion from intact cucumber roots and the proton transport rate of microsomal membrane vesicles of the roots due to Ca^{2+} starvation. *Plant Cell Physiol.* **29**, 79–84.

Matsumoto, H. (1991). Biochemical mechanism of the toxicity of aluminium and the sequestration of aluminium in plant cells. *In* 'Plant–Soil Interactions at Low pH' (R. J. Wright, V. C. Baligar and R. P. Murrmann, eds.), pp. 825–838. Kluwer Academic, Dordrecht, The Netherlands.

Matsumoto, H. and Chung, G. C. (1988). Increase in proton-transport activity of tonoplast vesicles as an adaptive response of barley roots to NaCl stress. *Plant Cell Physiol.* **29**, 1133–1140.

Matsumoto, H. and Tamura, K. (1981). Respiratory stress in cucumber roots treated with ammonium or nitrate nitrogen. *Plant Soil* **60**, 195–204.

Matsumoto, H., Hirasawa, F., Torikai, H. and Takahashi, E. (1976a). Localization of absorbed aluminium in pea root and its binding to nucleic acid. *Plant Cell Physiol.* **17**, 127–137.

Matsumoto, H., Hirasawa, E., Morimura, S. and Takahashi, E. (1976b). Localization of aluminium in tea leaves. *Plant Cell Physiol.* **17**, 627–631.

Matsushita, N. and Matoh, T. (1991). Characterization of Na^+ exclusion mechanisms of salt-tolerant reed plants in comparision with salt-sensitive rice plants. *Physiol. Plant.* **83**, 170–176.

Matsushita, N. and Matoh, T. (1992). Function of the shoot base of salt-tolerant reed (*Phragmites communis* Trinius) plants for Na^+ exclusion from the shoots. *Soil Sci. Plant Nutr.* **38**, 565–571.

Matsuyama, N. (1975). The effect of ample nitrogen fertilizer on cell wall materials and its significance to rice blast disease. *Ann. Phytopathol. Soc. Jpn* **4**, 56–61.

Matsuyama, N. and Dimond, A. E. (1973). Effect of nitrogenous fertilizer on biochemical processes that could affect lesion size of rice blast. *Phytopathology* **63**, 1202–1203.

Mattoo, A. K., Baker, J. E. and Moline, H. E. (1986). Induction by copper ions of ethylene production in *Spirodela oligorrhiza*: evidence for a pathway independent of 1-amminocyclopropane-1-carboxylic acid. *J. Plant Physiol.* **123**, 193–202.

Mauk, C. S. and Noodén, L. D. (1992). Regulation of mineral redistribution in pod-bearing soybean explants. *J. Exp. Bot.* **43**, 1429–1440.

Mauk, C. S., Brinker, A. M. and Noodén, L. D. (1990). Probing monocarpic senescence and pod development through manipulation of cytokinin and mineral supplies in soybean explants. *Ann. Bot.* **66**, 191–201.

Mayland, H. F., Wright, J. L. and Sojka, R. E. (1991). Silicon accumulation and water uptake by wheat. *Plant Soil* **137**, 191–199.

Mayland, H. F., James, L. F., Panter, K. E. and Sonderegger, J. L. (1989). Selenium in seleniferous environments. *In* 'Selenium in Agriculture and the Environment' (L. W. Jacobs, ed.), pp. 15–50. SSSA Special Publication no. 23, Madison, WI.

Mazzolini, A. P., Pallaghy, C. K. and Legge, G. J. F. (1985). Quantitative microanalysis of Mn, Zn and other elements in mature wheat seed. *New Phytol.* **100**, 483–509.

McCain, D. C. and Markley, J. L. (1989). More manganese accumulates in maple sun leaves than in shade leaves. *Plant Physiol.* **90**, 1417–1421.

McClendon, J. H. (1976). Elemental abundance as a factor on the origins of mineral nutrient requirments. *J. Mol. Evol.* **8**, 175–195.

McClure, J. M. (1976). Physiology and functions of flavanoids. *In* 'The Flavanoids' (J. B. Harborne, T. Mabry and H. Mabry, eds.), pp. 970–1055. Chapman and Hall, London.

McClure, P. R., Kochian, L. V., Spanswick, R. M. and Shaff, J.E. (1990a). Evidence for cotransport of nitrate and protons in maize roots. I. Effects of nitrate on the membrane potential. *Plant Physiol.* **93**, 281–289.

McClure, P., Kochian, L. V., Spanswick, R. M. and Shaff, J. E. (1990b). Evidence for cotransport of nitrate and protons in maize roots. II. Measurement of NO_3^- and H^+ fluxes with ion-selective microelectrodes. *Plant Physiol.* **93**, 290–294.

McCray, J. M. and Matocha, J. E. (1992). Effect of soil water levels on solution bicarbonate, chlorosis and growth of sorghum. *J. Plant Nutr.* **15**, 1877–1890.

McCully, M. E. and Canny, M. J. (1988). Pathways and processes of water and nutrient movement in roots. *Plant Soil* **111**, 159–170.

McCully, M. E. and Mallett, J. E. (1993). The branch roots of *Zea*. 3. Vascular connections and bridges for nutrient recycling. *Ann. Bot.* **71**, 327–341.

McCully, M. E., Canny, M. J. and Van Steveninck, R. F. M. (1987). Accumulation of potassium by differentiating metaxylem elements of maize roots. *Physiol.Plant.* **69**, 73–80.

McGonigle, T. P. and Fitter, A. H. (1988). Ecological consequences of arthropod grazing on VA mycorrhizal fungi. *Proc. Roy. Soc. Edin.* **94B**, 25–32.

McGrath, J. F. and Robson, A. D. (1984). The movement of zinc through excised stems of seedlings of *Pinus radiata* D. Don. *Ann. Bot.* **54**, 231–242.

McGrath, S. P., Sanders, J. R. and Shalaby, M. H. (1988). The effect of soil organic matter levels on soil solution concentrations and extractabilities of manganese, zinc and copper. *Geoderma* **42**, 177–188.

McGregor, A. J. and Wilson, G. C. S. (1964). The effect of applications of manganese sulphate to a neutral soil upon the yield of tubers and the incidence of common scab in potatoes. *Plant Soil* **20**, 59–64.

McIlrath, W. J. and Skok, J. (1966). Substitution of germanium for boron in plant growth. *Plant Physiol.* **41**, 1209–1212.

McKay, I. A., Dilworth, M. J. and Glenn, A. R. (1988). C_4-dicarboxylate metabolism in free-living and bacteroid forms of *Rhizobium leguminosarum* MNF 3841. *J. Gen. Microbiol.* **134**, 1433–1440.

McLaughlin, M. J. and Jones. T. R. (1991). Effect of phosphorus supply to the surface roots of wheat on root extension and rhizosphere chemistry in an acid subsoil. *Plant Soil* **134**, 73–82.

McLaughlin, M. J., Alston, A. M. and Martin, J. K. (1987). Transformations and movement of P in the rhizosphere. *Plant Soil* **97**, 391–399.

McNeil, D. L. (1980). The role of the stem in phloem loading of minerals in *Lupinus albus* L. cv. Ultra *Ann. Bot. (London)* [N.S.] **45**, 329–338.

McPharlin, I. R. and Bieleski, R. L. (1987). Phosphate uptake by *Spirodela* and *Lemna* during early phosphorus deficiency. *Aust. J. Plant Physiol.* **14**, 561–572.

McPharlin, I. R. and Bieleski, R. L. (1989). P_i efflux and influx in P-adequate and P-deficient *Spirodela* and *Lemna*. *Aust. J. Plant Physiol.* **16**, 391–399.

McSwain, B. D., Tsujimoto, H. Y. and Arnon, D. I. (1976). Effects of magnesium and chloride ions on light-induced electron transport in membrane fragments from a blue–green alga. *Biochim. Biophys. Acta* **423**, 313–322.

Meharg, A. A. and Macnair, M. R. (1992). Suppression of the high affinity phosphate uptake system: a mechanism of arsenate tolerance in *Holcus lanatus* L. *J. Exp. Bot.* **43**, 519–524.

Meharg, A. A. and Killham, K. (1991). A novel method of quantifying root exudation in the presence of soil microflora. *Plant Soil* **133**, 111–116.

Mehlhorn, H. (1990). Ethylene-promoted ascorbate peroxidase activity protects plants against hydrogen peroxide, ozone and paraquat. *Plant Cell Environ.* **13**, 971–976.

Mehrotra, N. K., Khana, V. K. and Agarwala, S. C. (1986). Soil sodicity-induced zinc deficiency in maize. *Plant Soil* **92**, 63–71.

Meinzer, F. C. and Moore, P. H. (1988). Effect of apoplastic solutes on water potential in elongating sugarcane leaves. *Plant Physiol.* **86**, 873–879.

Meinzer, F. C., Grautz, D. A. and Smit, B. (1991). Root signals mediate coordination of stomatal and hydraulic conductance in growing sugarcane. *Aust. J. Plant Physiol.* **18**, 329–338.

Meiri, A. and Plaut, Z. (1985). Crop production and management under saline conditions. *Plant Soil* **89**, 253–271.

Meiri, A., Hofmann, G. J., Shannon, M. C. and Poss, J. A. (1982). Salt tolerance of 3 muskmelon cultivars under 2 radiation levels. *J. Am. Soc. Hortic. Sci.* **107**, 1168–1172.

Meisner, C. A. and Karnok, K. J. (1991). Root hair occurrence and variation with environment. *Agron. J.* **83**, 814–818.

Mekdaschi, R., Horlacher, D., Schulz, R. and Marschner, H. (1988). Streusalzschäden und Sanierungsmassnahmen zur Verminderung der Streusalzbelastung von Strassenbäumen in Stuttgart. *Angew. Bot.* **62**, 355–371.

Melton, J. R., Mahtab, S. K. and Swoboda, A. R. (1973). Diffusion of zinc in soils as a function of applied zinc, phosphorus and soil pH. *Soil Sci. Soc. Am. Proc.* **37**, 379–381.

Memon, A. R. and Yatazawa, M. (1984). Nature of manganese complexes in manganese accumulator plant – *Acanthopanax sciadophylloides*. *J. Plant Nutr.* **7**, 961–974.

Menary, R. C. and Van Staden, J. (1976). Effect of phosphorus nutrient and cytokinins on flowering in the tomato, *Lycopersicon esculentum* Mill. *Aust. J. Plant Physiol.* **3**, 201–205.

Mench, M. and Martin, E. (1991). Mobilization of cadmium and other metals from two soils by root exudates of *Zea mays* L., *Nicotiana tabacum* L. and *Nicotiana rustica* L. *Plant Soil* **132**, 187–196.

Mench, M., Morel, J. L., Guckert, A. and Guillet, B. (1988). Metal binding with root exudates of low molecular weight. *J. Soil Sci.* **39**, 521–527.

Menge, J. A. (1983). Utilization of vesicular-arbuscular mycorrhizal fungi in agriculture. *Can. J. Bot.* **61**, 1015–1024.

Mengel, K. (1962). Die K- und Ca-Aufnahme der Pflanze in Abhängigkeit vom Kohlenhydratgehalt ihrer Wurzel. *Z. Pflanzenernähr. Dueng. Bodenkd.* **98**, 44–54.

Mengel, K. (1991). Available nitrogen in soils and its determination by the 'N$_{min}$-method' and by electroultrafiltration (EUF). *Fert. Res.* **28**, 251–262.

Mengel, K. and Arneke, W. W. (1982). Effect of potassium on the water potential, the osmotic potential, and cell elongation in leaves of *Phaseolus vulgaris*. *Physiol. Plant.* **54**, 402–408.

Mengel, K. and Bübl, W. (1983). Verteilung von Eisen in Blättern von Weinreben mit HCO$_3^-$ induzierter Chlorose. *Z. Pflanzenernähr. Bodenk.* **146**, 560–571.

Mengel, K. and Geurtzen, G. (1988). Relationship between iron chlorosis and alkalinity in *Zea mays*. *Physiol. Plant.* **72**, 460–465.

Mengel, K. and Haeder, H. E. (1977). Effect of potassium supply on the rate of phloem sap exudation and the composition of phloem sap of *Ricinus communis*. *Plant Physiol.* **59**, 282–284.

Mengel, K. and Helal, M. (1968). Der Einfluss einer variierten N- und K-Ernährung auf den Gehalt an löslichen Aminoverbindungen in der oberirdischen Pflanzenmasse von Hafer. *Z. Pflanzenernähr. Bodenk.* **120**, 12–20.

Mengel, K. and Malissiovas, N. (1981). Bicarbonat als auslösender Faktor der Eisenchlorose bei der Weinrebe (*Vitis vinifera*). *Vitis* **20**, 235–243.

Mengel, K. and Steffens, D. (1985). Potassium uptake of rye-grass (*Lolium perenne*) and red clover (*Trifolium pratense*) as related to root parameters. *Biol. Fertil. Soils* **1**, 53–58.

Mengel, K., Haghparast, M. and Koch, K. (1974). The effect of potassium on the fixation of molecular nitrogen by root nodules of *Vicia faba*. *Plant Physiol.* **54**, 535–538.

Mengel, K., Viro, M. and Hehl, G. (1976). Effect of potassium on uptake and incorporation of ammonium-nitrogen of rice plants. *Plant Soil* **44**, 547–558.

Mengel, K., Scherer, H. W. and Malissiovas, N. (1979). Die Chlorose aus der Sicht der Bodenchemie und Rebenernährung. *Mitt. Klosterneuburg* **29**, 151–156.

Mengel, K., Breiniger, M. T. and Bübl, W. (1984a). Bicarbonate, the most important factor inducing iron chlorosis in vine grapes on calcareous soil. *Plant Soil* **81**, 333–344.

Mengel, K., Bübl, W. and Scherer, H. W. (1984b). Iron distribution in vine leaves with HCO$_3^-$ induced chlorosis. *J. Plant Nutr.* **7**, 715–724.

Mengel, K., Lutz, H.-J. and Breininger, M. Th. (1987). Auswaschung von Nährstoffen durch sauren Nebel aus jungen intakten Fichten (*Picea abies*). *Z. Pflanzenernähr. Bodenk.* **150**, 61–68.

Mennen, H., Jacoby, B. and Marschner, H. (1990). Is sodium proton antiport ubiquitous in plant cells? *J. Plant Physiol.* **137**, 180–183.

Menzies, J. G., Ehret, D. L., Glass, A. D. M. and Samuels, A. L. (1991). The influence of silicon on cytological interactions between *Sphaerotheca fuliginea* and *Cucumis sativus*. *Physiol. Mol. Plant Pathol.* **39**, 403–414.

Mercy, M. A., Shivshankar, G. and Bagyaraj, D. J. (1990). Mycorrhizal colonization on cowpea is host dependent and heritable. *Plant Soil* **121**, 292–294.

Merker, E. (1961). Welche Ursachen hat die Schädigung der Insekten durch die Düngung im Walde? *Allg. Forst- Jagdzt.* **132**, 73–82.

Merwin, I. A. and Stiles, W. C. (1989). Root-lesion nematodes, potassium deficiency, and prior cover crops as factors in apple replant disease. *J. Am. Soc. Hort. Sci.* **114**, 724–728.

Mettler, I. J., Mandala, S. and Taiz, L. (1982). Characterization of in vitro proton pumping by microsomal vesicles isolated from corn coleoptiles. *Plant Physiol.* **70**, 1738–1742.

Meuser, H. (1991). Bodenkundliche Aspekte bei Wurzeluntersuchungen an Kulturpflanzen. *Die Geowissenschaft* **9**, 247–250.

Michael, B., Zink, F. and Lantzsch, H. J. (1980). Effect of phosphate application of phytin-P and other phosphate fractions in developing wheat grains. *Z. Pflanzenernähr. Bodenk.* **143**, 369–376.

Michael, G. (1941). Über die Aufnahme und Verteilung des Magnesiums und dessen Rolle in der höheren grünen Pflanze. *Z. Pflanzenernaehr. Dueng. Bodenkd.* **25**, 65–120.

Michael, G. (1990). Vorstellungen über die Regulation der Wurzelhaarbildung. *Kali-Briefe* **20**, 411–429.

Michael, G. and Beringer, H. (1980). The role of hormones in yield formation. *Proc. 15th Colloq. Int. Potash Inst. Bern*, pp. 85–116.

Michael, G., Faust, H. and Blume, B. (1960). Die Verteilung von spät gedüngtem ^{15}N in der reifenden Gerstenpflanze unter besonderer Berücksichtigung der Korneiweisse. *Z. Pflanzenernaehr. Dueng. Bodenkd.* **91**, 158–169.

Mietkowski, K. and Beringer, H. (1990). Schwefeldüngebedarf von Raps – Einfluss einer K_2SO_4 und KCl-Düngung auf Samenertrag und -qualität zweier Sorten. *Kali-Briefe* **20**, 287–292.

Mikkelsen, R. L. and Wan, H. F. (1990). The effect of selenium on sulfur uptake by barley and rice. *Plant Soil* **121**, 151–153.

Mikkelsen, R. L., Page, A. L. and Bingham, F. T. (1989). Factors affecting selenium accumulation by agricultural crop species. *In* 'Selenium in Agriculture and the Environment' (L. W. Jacobs, ed.), pp. 65–94. SSSA Special Publication No. 23, Madison, WI.

Mikus, M., Bobák, M. and Lux, A. (1992). Structure of protein bodies and elemental composition of phytin from dry germ of maize (*Zea mays* L.). *Bot. Acta* **105**, 26–33.

Milford, G. F. J., Cormack, W. F. and Durrant, M. J. (1977). Effects of sodium chloride on water status and growth of sugar beet. *J. Exp. Bot.* **28**, 1380–1388.

Millard, P. (1988). The accumulation and storage of nitrogen by herbaceous plants. *Plant Cell Environ.* **11**, 1–8.

Miller, E. R., Lei, X. and Ullrey, D. E (1991). Trace elements in animal nutrition. *In* 'Micronutrients in Agriculture' 2nd edn. (J. J. Mortvedt, F. R. Cox, L. M. Shuman and R. M. Welch, eds.), pp. 593–662. SSSA Book Series No. 4, Madison, WI.

Miller, G. W., Shigematsu, A., Welkie, G. W., Motoji, N. and Szlek, M. (1990). Potassium effect on iron stress in tomato. II. The effects on root CO_2-fixation and organic acid formation. *J. Plant Nutr.* **13**, 1355–1370.

Miller, M. H. and McGonigle, T. P. (1992). Soil disturbance and the effectiveness of arbuscular mycorrhizas in an agricultural ecosystem. *In* 'Mycorrhizas in Ecosystems' (D. J. Read, D. H. Lewis, A. H. Fitter and I. J. Alexander, eds.), pp. 156–163. CAB International, Wallingford, UK.

Milligan, S. P. and Dale, J. E. (1988). The effects of root treatments on growth of the primary leaves of *Phaseolus vulgaris* L.: biophysical analysis. *New Phytol.* **109**, 35–40.

Mills, D. and Hodges, T. K. (1988). Characterization of plasma membrane ATPase from roots of *Atriplex nummularia*. *J. Plant Physiol.* **132**, 513–519.

Mimura, T., Dietz, K.-J., Kaiser, W., Schramm, M. J., Kaiser, G. and Heber, U. (1990). Phosphate transport across biomembranes and cytosolic phosphate homeostasis in barley leaves. *Planta* **180**, 139–146.

Minchin, P. E. H. and Thorpe, M. R. (1982). Evidence of a flow of water into sieve tubes associated with phloem loading. *J. Exp. Bot.* **33**, 233–240.

Minchin, P. E. H. and Thorpe, M. R (1987). Measurement of unloading and reloading of photoassimilate within the stem of bean. *J. Exp. Bot.* **38**, 211–220.

Minorsky, P. V. and Spanswick, R. M. (1989). Electrophysiological evidence for a role for calcium in temperature sensing by roots of cucumber seedlings. *Plant Cell Environ.* **12**, 137–143.

Miranda, C. H. B., Urquiaga, S. and Boddey, R. M. (1990). Selection of ecotypes of *Panicum maximum* for associated biological nitrogen fixation using the ^{15}N isotope dilution technique. *Soil Biol. Biochem.* **22**, 657–663.

Mirswa, W. and Ansorge, H. (1981). Einfluss der K-Düngung auf Ertrag und Qualität der Kartoffel. *Arch. Acker- Pflanzenbau Bodenkd.* **25**, 165–171.

Mishra, D. and Kar, M. (1974). Nickel in plant growth and metabolism. *Bot. Rev.* **40**, 395–452.

Misra, R. K., Alston, A. M. and Dexter, A. R. (1988). Role of root hairs in phosphorus depletion from a macrostructured soil. *Plant Soil* **107**, 11–18.

Mitchell, R. J., Garrett, H. E., Cox, G. S. and Atalay, A. (1986). Boron and ectomycorrhizal influences on indole-3-acetic acid levels and indole-3-acetic acid oxidase and peroxidase activities of *Pinus echinata* Mill. roots. *Tree Physiol.* **1**, 1–8.

Mitchell, R. J., Garrett, H. E., Cox, G. S. and Atalay, A. (1990). Boron and ectomycorrhizal influences on mineral nutrition of container-grown *Pinus ehinata* Mill. *J. Plant Nutr.* **13**, 1555–1574.

Mitscherlich, E. A. (1954). 'Bodenkunde für Landwirte, Förster und Gärtner', 7th edn. Parey, Berlin.

Mitsui, T., Christeller, J. T., Hara-Nishimura, I. and Akazawa, T. (1984). Possible roles of calcium and calmodulin in the biosynthesis and secretion of α-amylase in rice seed scutellar epithelium. *Plant Physiol.* **75**, 21–25.

Mix, G. P. and Marschner, H. (1976a). Calciumgehalte in Früchten von Paprika, Bohnen, Quitte und Hagebutte im Verlauf des Fruchtwachstums. *Z. Pflanzernähr. Bodenk.* **139**, 537–549.

Mix, G. P. and Marschner, H. (1976b). Einfluss exogener und endogener Faktoren auf den Calciumgehalt von Paprika- und Bohnenfrüchten. *Z. Pflanzenernähr. Bodenk.* **139**, 551–563.

Mix, G. P. and Marschner, H. (1976c). Calcium-Umlagerung in Bohnenfrüchten während des Samenwachstums. *Z. Pflanzenphysiol.* **80**, 354–366.

Miyake, Y. and Takahashi, E. (1978). Silicon deficiency of tomato plant. *Soil Sci. Plant Nutr.* (*Tokyo*) **24**, 175–189.

Miyake, Y. and Takahashi, E. (1983). Effect of silicon on the growth of solution-cultured cucumber plant. *Soil Sci. Plant Nutr.* (*Tokyo*) **29**, 71–83.

Miyake, Y. and Takahashi, E. (1985). Effect of silicon on the growth of soybean plants in a solution culture. *Soil Sci. Plant Nutr.* (*Tokyo*) **31**, 625–636.

Miyake, Y. and Takahashi, E. (1986). Effect of silicon on the growth and fruit production of strawberry plants in a solution culture. *Soil Sci. Plant Nutr.* (*Tokyo*) **32**, 321–326.

Miyasaka, S. C. and Grunes, D. L. (1990). Root temperature and calcium level effects in winter wheat forage: II. Nutrient composition and tetany potential. *Agron. J.* **82**, 242–249.

Miyasaka, S. C., Buta, J. G., Howell, R. K. and Foy, C. D. (1991). Mechanism of aluminum tolerance in snapbeans. Root exudation of citric acid. *Plant Physiol.* **96**, 737–743.

Miyazaki, J. H. and Yang, S. F. (1987). The methionine salvage pathway in relation to ethylene and polyamine synthesis. *Physiol. Plant.* **69**, 366–370.

Mizrahi, Y. and Pasternak, D. (1985). Effect of salinity on quality of various agricultural crops. *Plant Soil* **89**, 301–307.

Mizuno, A., Kojima, H., Katou, K. and Okamoto, H. (1985). The electrogenic proton pumping from parenchyma symplast into xylem – direct demonstration by xylem perfusion. *Plant Cell Environ.* **8**, 525–529.

Mizuno, N., Inazu, O. and Kamada, K. (1982). Characteristics of concentrations of copper, iron and carbohydrates in copper deficient wheat plants. *In* 'Proceedings of the Ninth International Plant Nutrition Colloquium, Warwick, England' (A. Scaife, ed.), pp. 396–399. Commonwealth Agricultural Bureau, Farnham Royal, Bucks.

Modjo, H. S. and Hendrix, J. W. (1986). The mycorrhizal fungi *Glomus macrocarpum* as a cause of tobacco stunt disease. *Phytopathology* **76**, 688–690.

Mohabir, G. and John, P. (1988). Effect of temperature on starch synthesis in potato tuber tissue and in amyloplasts. *Plant Physiol.* **88**, 1222–1228.

Mohapatra, S. S., Poole, R. J. and Dhindsa, R. S. (1988). Alterations in membrane protein-profile during cold treatment of alfalfa. *Plant Physiol.* **86**, 1005–1007.

Möller, I. and Beck, E. (1992). The fate of apoplastic sucrose in sink and source leaves of *Urtica dioica*. *Physiol. Plant.* **85**, 618–624.

Monestiez, M., Lamant, A. and Heller, R. (1982). Endocellular distribution of calcium and Ca-ATPases in horse-bean roots: Possible relation to the ecological status of the plant. *Physiol. Plant.* **55**, 445–452.

Monk, L. S., Fagerstedt, K. V. and Crawford, R. M. M. (1987). Superoxide dismutase as an anaerobic polypeptide. A key factor in recovery from oxygen deprivation in *Iris pseudodacorus*? *Plant Physiol.* **85**, 1016–1020.

Monson, R. K. (1989). The relative contributions of reduced photorespiration, and improved water- and nitrogen-use efficiency, to the advantages of C_3-C_4 intermediate photosynthesis in *Flaveria*. *Oecologia* (*Berlin*) **80**, 215–221.

Moon, G. J., Clough, B. F., Peterson, C. A. and Allaway, W. G. (1986). Apoplastic and symplastic pathway in *Avicennia marina* (Forsk.) Vierh. roots revealed by fluorescent tracer dyes. *Aust. J. Plant Physiol.* **13**, 637–648.

Moorby, H., White, R. E. and Nye, P. H. (1988). The influence of phosphate nutrition on H ion efflux from the roots of young rape plants. *Plant Soil* **105**, 247–256.

Moore, H. M. and Hirsch, A. M. (1983). Effects of boron deficiency on mitosis and incorporation of tritiated thymidine into nuclei of sunflower root tips. *Am. J. Bot.* **70**, 165–172.

Moore jr, P. A. and Patrick jr, W. H. (1988). Effect of zinc deficiency on alcohol dehydrogenase activity and nutrient uptake in rice. *Agron. J.* **80**, 882–885.

Moore, R. and Black jr, C. C. (1979). Nitrogen assimilation pathways in leaf mesophyll and bundle sheath cells of C_4 photosynthetic plants formulated from comparative studies with *Digitaria sanguinalis* (L.). Scap. *Plant Physiol.* **64**, 309–313.

Moore, R., Evans, M. L. and Fondreu, W. M. (1990). Inducing gravitropic curvature of primary roots of *Zea mays* cv. Agrotropic. *Plant Physiol.* **92**, 310–315.

Moraghan, J. T. (1980). Effect of soil temperature on response of flux to P and Zn fertilizers. *Soil Sci.* **129**, 290–296.

Moraghan, J. T. (1979). Manganese toxicity in flax growing on certain calcareous soils low in available iron. *Soil Sci. Soc. Am. J.* **43**, 1177–1180.

Moraghan, J. T. (1991a). Removal of endogenous iron, manganese and zinc during plant washing. *Commun. Soil Sci. Plant Anal.* **22**, 323–330.

Moraghan, J. T. (1991b). The growth of white lupin on a Calcaquoll. *Soil Sci. Soc. Am. J.* **55**, 1353–1357.

Moraghan, J. T. (1992). Iron-manganese relationships in white lupin grown on a calciaquoll. *Soil Sci. Soc. Am. J.* **56**, 471–475.

Moraghan, J. T. and Freeman, T. J. (1978). Influence of Fe EDDHA on growth and manganese accumulation in flax. *Soil Sci. Soc. Am. J.* **42**, 445–460.

Moraghan, J. T. and Mascagni jr, H. J.(1991). Environmental and soil factors affecting micronutrient deficiencies and toxicities. *In* 'Micronutrients in Agriculture' (J. J. Mortvedt, F. R. Cox, L. M. Shumann and R. M. Welch, eds.), pp. 371–425. SSSA Book Series No. 4, Madison, WI.

Morales, F., Abadia, A. and Abadia, J. (1990). Characterization of the xanthophyll cycle and other photosynthetic pigment changes induced by iron deficiency in sugar beet (*Beta vulgaris* L.). *Plant Physiol.* **94**, 607–613.

Morales, F., Abadia, A. and Abadia, J. (1991). Chlorophyll fluorescence and photon yield of oxygen evolution in iron-deficient sugar beet (*Beta vulgaris* **L.**) leaves. *Plant Physiol.* **97**, 886–893.

Morel, J. L., Mench, M. and Guckert, A. (1986). Measurement of Pb^{2+}, Cu^{2+} and Cd^{2+} binding with mucilage exudates from maize (*Zea mays* L.) roots. *Biol. Fertil. Soils* **2**, 29–34.

Morel, J. L., Habib, L., Plantureux, S. and Guckert, A. (1991). Influence of maize root mucilage on soil aggregate stability. *Plant Soil* **136**, 111–119.

Morgan, J. M. (1980). Possible role of abscisic acid in reducing seed set in water stressed plants. *Nature* (*London*) **285**, 655–657.

Morgan, M. A. and Jackson, W. A. (1988). Suppression of ammonium uptake by nitrogen supply and its relief during nitrogen limitation. *Physiol. Plant.* **73**, 38–45.

Morgan, M. A., Volk, R. J. and Jackson, W. A. (1985). *p*-Fluorophenylalanine-induced restriction of ion uptake and assimilation by maize roots. *Plant Physiol.* **77**, 718–721.

Morgan, P. W., Taylor, D. M. and Joham, H. E. (1976). Manipulation of IAA-oxidase activity and auxine-deficiency symptoms in intact cotton plants with manganese nutrition. *Plant Physiol.* **37**, 149–156.

Mori, S. and Nishizawa, N. (1987). Methionine as a dominant precursor of phytosidero-phores in *Graminaceae* plants. *Plant Cell Physiol.* **28**, 1081–1092.

Mori, S. and Nishizawa, N. (1989). Identification of barley chromosome No. 4, possible encoder of genes of mugineic acid synthesis from 2'-deoxymugineic acid using wheat–barley addition lines. *Plant Cell Physiol.* **30**, 1057–1061.

Mori, S., Kishi-Nishizawa, N. and Fujigaki, J. (1990). Identification of rye chromosome 5R as a carrier of the gene for mugineic acid synthase and 3-hydroxymugineic acid synthase using wheat-rye addition lines. *Jpn J. Genet.* **65**, 343–352.

Mori, S., Nishizawa, N., Hayashi, H., Chino, M., Yoshimura, E. and Ishihara, J. (1991). Why are young rice plants highly susceptible to iron deficiency? *Plant Soil* **130**, 143–156.

Morré, D. J., Brightman, A. O., Wu, L.-Y., Barr, R., Leak, B. and Crane, F. L. (1988). Role of plasma membrane redox activities in elongation growth in plants. *Physiol. Plant.* **73**, 187–193.

Morris, D. R., Weaver, R. W., Smith, G. R. and Rouquette, F. M. (1990). Nitrogen transfer from arrowleaf clover to ryegrass in field plantings. *Plant Soil* **128**, 293–297.

Morrison, R. S., Brooks, R. D., Reeves, R. D., Malaise, F., Horowitz, P., Aronson, M. and Merriam, G. R. (1981). The diverse chemical forms of heavy-metals in tissue extracts of some metallophytes from Shaba province, Zaire. *Phytochemistry* **20**, 455–458.

Mortvedt, J. J. (1981). Nitrogen and molybdenum uptake and dry matter relationship in soybeans and forage legumes in response to applied molybdenum on acid soil. *J. Plant Nutr.* **3**, 245–256.

Mortvedt, J. J., Fleischfresser, M. H., Berger, K. C. and Darling, H. M. (1961). The relation of some soluble manganese to the incidence of common scab in potatoes. *Am. Potato J.* **38**, 95–100.

Mortvedt, J. J., Berger, K. C. and Darling, H. M. (1963). Effects of manganese and copper on the growth of *Streptomyces scabies* and the incidence of potato scab. *Am. Potato J.* **40**, 96–102.

Moseley, G. and Baker, D. H. (1991). The efficacy of a high magnesium grass cultivar in controlling hypomagnesaemia in grazing animals. *Grass Forage Sci.* **46**, 375–380.

Mostafa, M. A. E. and Ulrich, A. (1976). Absorption, distribution and form of Ca in relation to Ca deficiency (tip burn) of sugarbeets. *Crop Sci.* **16**, 27–30.

Mounla, M. A. K., Bangerth, F. and Stoy, V. (1980). Gibberellin-like substances and indole type auxins in developing grains of normal- and high-lysine genotypes of barley. *Physiol. Plant.* **48**, 568–573.

Muchovej, R. M. C. and Muchovej, J. J. (1982). Calcium suppression of *Sclerotium*-induced twin stem abnormality of soybean. *Soil Sci.* **134**, 181–184.

Mugwira, L. M. and Patel, S. U. (1977). Root zone pH changes and ion uptake imbalances by triticale, wheat and rye. *Agron. J.* **69**, 719–722.

Mugwira, L. M., Sapra, V. T., Patel, S. U. and Choudry, M. A. (1981). Aluminium tolerance of triticale and wheat cultivars developed in different regions. *Agron. J.* **73**, 470–475.

Muhammed, S., Akbar, M. and Neue, H. U. (1987). Effect of Na/Ca and Na/K ratios in saline culture solution on the growth and mineral nutrition of rice (*Oryza sativa* L.). *Plant Soil* **104**, 57–62.

Mukherji, S., Dey, B., Paul, A. K. and Sircar, S. M. (1971). Changes in phosphorus fractions and phytase activity of rice seeds during germination. *Physiol. Plant.* **25**, 94–97.

Mulette, K. L., Hannon, N. J. and Elliott, A. G. L. (1974). Insoluble phosphorus usage by *Eucalyptus*. *Plant Soil* **41**, 199–205.

Müller, M., Deigele, C. and Ziegler, H. (1989). Hormonal interactions in the rhizosphere of maize (*Zea mays* L.) and their effects on plant development. *Z. Pflanzenernähr. Bodenk.* **152**, 247–254.

Münch, E. (1930). 'Die Stoffbewegungen in der Pflanze'. Fischer, Jena.

Munk, H. (1982). Zur Bedeutung silikatischer Stoffe bei der Düngung landwirtschaftlicher Kulturpflanzen. *Landwirtsch. Forsch., Sonderh.* **38**, 264–277.

Munns, D. N. (1986). Acid tolerance in legumes and rhizobia. *In* 'Advances in Plant Nutrition' Vol. **2** (B. Tinker and A. Läuchli, eds.), pp. 63–91. Praeger Scientific, New York.

Munns, R. (1988). Effect of high external NaCl concentrations on ion transport within the shoot of *Lupinus albus*. I. Ions in xylem sap. *Plant Cell Environ.* **11**, 283–289.

Munns, R. (1992). A leaf elongation assay detects an unknown growth inhibitor in xylem sap from wheat and barley. *Aust. J. Plant Physiol.* **19**, 127–135.

Munns, R. and Termaat, A. (1986). Whole-plant responses to salinity. *Aust. J. Plant Physiol.* **13**, 143–160.

Munns, R., Greenway, H. and Kirst, G. O. (1983). Halotolerant Eukaryotes. *In* 'Encyclopedia of Plant Physiology' (O. L. Lange, P. S. Nobel, C. B. Osmond and H. Ziegler, eds.), Vol. 12C, pp. 59–135. Springer-Verlag, Berlin.

Munns, R., Fisher, D. B. and Tonnet, M. L. (1987). Na^+ and Cl^- transport in the phloem from leaves of NaCl-treated barley. *Aust. J. Plant Physiol.* **13**, 757–766.

Munns, R., Gardner, P. A., Tonnet, M. L. and Rawson, H. M (1989). Growth and development in NaCl-treated plants. II. Do Na^+ or Cl^- concentrations in dividing or expanding tissues determine growth in barley? *Aust. J. Plant Physiol.* **15**, 529–540.

Murach, D. and Ulrich, B. (1988). Destabilization of forest ecosystems by acid deposition. *Geo J.* **17.2**, 253–260.

Murach, D., Ilse, L., Klaproth, F., Parth, A. and Wiedemann, H. (1993). Rhizotron-Experimente zur Wurzelverteilung der Fichte. *Forstarchiv* **64**, 191–194.

Murakami, H., Kimura, M. and Wada, H. (1990). Microbial colonization and decomposition processes in rice rhizoplane. II. Decomposition of young and old roots. *Soil Sci. Plant Nutr. (Tokyo)* **36**, 441–450.

Murakami, T., Ise, K., Hayakawa, M., Kamei, S. and Takagi, S. (1989). Stabilities of metal complexes of mugineic acids and their specific affinities for iron (III). *Chem. Lett.*, pp. 2137–2140.

Murphy, M. D. and Boggan, J. M. (1988). Sulphur deficiency in herbage in Ireland. 1. Causes and extent. *Irish J. Agric. Res.* **27**, 83–90.

Murty, K. S., Smith, T. A. and Bould, C. (1971). The relation between the putrescine content and potassium status of black current leaves. *Ann. Bot. (London)* [N.S.] **35**, 687–695.

Musick, H. B. (1978). Phosphorus toxicity in seedlings of *Larrea divaricata* grown in solution culture. *Bot. Gaz. (Chicago)* **139**, 108–111.

Myers, R. J. K., Foale, M. A., Smith, F. W. and Ratcliff, D. (1987). Tissue concentration of nitrogen and phosphorus in grain sorghum. *Field Crops Res.* **17**, 289–303.

Myers, P. N., Setter, T. L., Madison, J. T. and Thompson, J. F. (1990). Abscisic acid inhibition of endosperm cell division in cultured maize kernels. *Plant Physiol.* **94**, 1330–1336.

Mylonas, V. A. and McCants, C. B. (1980). Effects of humic and fulvic acids on growth of tobacco. I. Root initiation and elongation. *Plant Soil* **54**, 485–490.

Nable, R. O. (1991). Distribution of boron within barley genotypes with differing susceptibilities to boron toxicity. *J. Plant Nutr.* **14**, 453–461.

Nable, R. O. and Loneragan, J. F. (1984). Translocation of manganese in subterranean clover (*Trifolium subterraneum* L. cv. Seaton Park). II. Effects of leaf senescence and of restricting supply of manganese to part of a split root system. *Aust. J. Plant Physiol.* **11**, 113–118.

Nable, R. O. and Paull, J. G. (1990). Effect of excess grain boron concentrations on early seedling development and growth of several wheat (*Triticum aestivum*) genotypes with different susceptibility to boron toxicity. *In* 'Plant Nutrition – Physiology and Application' (M. L. van Beusichem, ed.), pp. 291–295. Kluwer Academic, Dordrecht, The Netherlands.

Nable, R. O. and Paull, J. G. (1991). Mechanism of genetics of tolerance to boron toxicity in plants. *Curr. Top. Plant Biochem. Physiol.* **10**, 257–273.

Nable, R. O., Bar-Akiva, A. and Loneragan, J. F. (1984). Functional manganese requirement and its use as a critical value for diagnosis of manganese deficiency in subterranean clover (*Trifolium subterraneum* L. cv. Seaton Park). *Ann. Bot.* **54**, 39–49.

Nable, R. O., Houtz, R. L. and Cheniae, G. M. (1988). Early inhibition of photosynthesis during development of Mn toxicity in tobacco. *Plant Physiol.* **86**, 1136–1142.

Nable, R. O., Cartwright, B. and Lance, R. C. M. (1990a). Genotypic differences in boron accumulation in barley: relative susceptibilities to boron deficiency and toxicity. *In* 'Genetic Aspects of Plant Mineral Nutrition' (N. El Bassam *et al.*, eds.), pp. 243–251. Kluwer Academic, Dordrecht, TheNetherlands.

Nable, R. O., Lance, R. C. M. and Cartwright, B. (1990b). Uptake of boron and silicon by barley genotypes with differing susceptibilities to boron toxicity. *Ann. Bot.* **66**, 83–90.

Nable, R. O., Paull, J. G. and Cartwright, B. (1990c). Problems associated with the use of foliar analysis for diagnosing boron toxicity in barley. *Plant Soil* **128**, 225–232.

Nabors, M. W., Gibbs, S.-E., Bernstein, C. S. and Mais, M. E. (1980). NaCl-tolerant tobacco plants from cultured cells. *Z. Pflanzenphysiol.* **97**, 13–17.

Nagarajah, S., Posner, A. M. and Quirk, J. P. (1970). Competitive adsorptions of phosphate with polygalacturonate and other organic anions on kaolinite and oxide surfaces. *Nature (London)* **228**, 83–84.

Nagarathna, K. C., Shetty, A., Bhat, S. G. and Shetty, H. S. (1992). The possible involvement of lipoxygenase in downy mildew resistance in pearl millet. *J. Exp. Bot.* **43**, 1283–1287.

Naidoo, G., Steward, J. McD. and Lewis, R. J. (1978). Accumulation sites of Al in snapbean and cotton roots. *Agron. J.* **70**, 489–492.

Nair, K. P. P. and Mengel, K. (1984). Importance of phosphate buffer power for phosphate uptake by rye. *Soil Sci. Soc. Am. J.* **48**, 92–95.

Naito, K., Nagumo, S., Furuya, K. and Suzuki, H. (1981). Effect of benzyladenine on RNA and protein synthesis in intact bean leaves at various stages of ageing. *Physiol. Plant.* **52**, 343–348.

Nakayama, F. S. and Kimball, B. A. (1988). Soil carbon dioxide distribution and flux within the open-top chamber. *Agron. J.* **80**, 394–398.

Nambiar, E. K. S. (1976a). Uptake of Zn^{65} from dry soil by plants. *Plant Soil* **44**, 267–271.

Nambiar, E. K. S. (1976b). The uptake of zinc-65 by roots in relation to soil water content and root growth. *Aust. J. Soil Res.* **14**, 67–74.

Nambiar, E. K. S. (1976c). Genetic differences in the copper nutrition of cereals. I. Differential responses of genotypes to copper. *Aust. J. Agric. Res.* **27**, 453–463.

Nandi, A. S. and Sen, S. P. (1981). Utility of some nitrogen fixing microorganism in the phyllosphere of crop plants. *Plant Soil* **63**, 465–476.

Narayanan, A., Saxena, N. P. and Sheldrake, A. R. (1981). Cultivar differences in seed size and seedling growth of pigeonpea and chickpea. *Indian J. Agric. Sci.* **51**, 389–393.

Nátr, L. (1975). Influence of mineral nutrition on photosynthesis and the use of assimilates. *Photosynth. Prod. Differ. Environ.* [*Proc. IBP Synth. Meet.*], 1973, Vol. 2, pp. 537–555.

Nayyar, V. K. and Takkar, P. N. (1980). Evaluation of various zinc sources for rice grown on alkali soil. *Z. Pflanzenernähr. Bodenk.* **143**, 489–493.

Neilsen, G. H. and Hogue, E. J. (1986). Some factors affecting leaf zinc concentration of apple seedlings grown in nutrient solution. *HortScience* **21**, 434–436.

Nelson, L. E. (1983). Tolerance of 20 rice cultivars to excess Al and Mn. *Agron. J.* **75**, 134–138.

Nelson, M., Cooper, C. R., Crowley, D. E., Reid, C. P. P. and Szaniszlo, P. J. (1988). An *Escherichia coli* bioassay of individual siderophores in soil. *J. Plant Nutr.* **11**, 915–924.

Nelson, S. D. (1992). Response of several wildland shrubs and forbs of arid regions to iron-deficiency stress. *J. Plant Nutr.* **15**, 2015–2023.

Nemeth, K. (1982). Electro-ultrafiltration of aqueous soil suspension with simultaneously varying temperature and voltage. *Plant Soil* **64**, 7–23.

Nemeth, K. (1985). Recent advances in EUF research (1980–1983). *Plant Soil* **83**, 1–19.

Nemeth, K., Irion, H. and Maier, J. (1987). Einfluss der EUF-K-, EUF-Na- und EUF-Ca-Fraktionen auf die K-Aufnahme sowie den Ertrag von Zuckerrüben. *Kali-Briefe* **18**, 777–790.

Ness, P. J. and Woolhouse, H. W. (1980). RNA synthesis in *Phaseolus* chloroplasts. I. Ribonucleic acid synthesis and senescing leaves. *J. Exp. Bot.* **31**, 223–233.

Neue, H. U., Lantin, R. S., Cayton, M. T. C. and Autor, N. U. (1990). Screening of rices for adverse soil tolerance. *In* 'Genetic Aspects of Plant Mineral Nutrition' (N. El Bassam, M. Dambroth and B. C. Loughman, eds.), pp. 523–531. Kluwer Academic, Dordrecht.

Neuman, D. S. and Smit, B. A. (1991). The influence of leaf water status and ABA on leaf growth and stomata of *Phaseolus* seedlings with hypoxic roots. *J. Exp. Bot.* **42**, 1499–1506.

Neuman, D. S., Rood, S. B. and Smit, B. A. (1990). Does cytokinin transport from root-to-shoot in the xylem sap regulate leaf responses to root hypoxia? *J. Exp. Bot.* **41**, 1325–1333.

Neumann, K. H. and Steward, F. C. (1968). Investigations on the growth and metabolism of cultured explants of *Daucus carota*. I. Effects of iron, molybdenum and manganese on growth. *Planta* **81**, 333–350.

Neumann, P. M. (1987). Sequential leaf senescence and correlatively controlled increase in xylem flow resistance. *Plant Physiol.* **83**, 941–944.

Neumann, P. M. (1982). Late-season foliar fertilization with macronutrients – Is there a theoretical basis for increased seed yields? *J. Plant Nutr.* **5**, 1209–1215.

Neumann, P. M., Ehrenreich, Y. and Golab, Z. (1983). Foliar fertilizer damage to corn leaves: relation to cuticular penetration. *Agron. J.* **73**, 979–982.

Neumann, P. M., van Volkenburgh, E. and Cleland, R. E. (1988). Salinity stress inhibits bean leaf expansion by reducing turgor, not wall extensibility. *Plant Physiol.* **88**, 233–237.

Nevins, D. J. and Loomis, R. S. (1970). Nitrogen nutrition and photosynthesis in sugar beet (*Beta vulgaris* L.). *Crop Sci.* **10**, 21–25.

Newman, E. I., Eason, W. R., Eissenstat, D. M. and Ramos, M. I. R. F. (1992). Interactions between plants: the role of mycorrhizae. *Mykorrhiza* **1**, 47–53.

Nguyen, J. and Feierabend, J. (1978). Some properties and subcellular localization of xanthine dehydrogenase in pea leaves. *Plant Sci. Lett.* **13**, 125–132.

Ni, M. and Beevers, L. (1990). Essential arginine residues in the nitrate uptake system from corn seedling roots. *Plant Physiol.* **94**, 745–751.

Nicholson III, C., Stein, J. and Wilson, K. A. (1980). Identification of the low molecular weight copper protein from copper-intoxicated mung bean plants. *Plant Physiol.* **66**, 272–275.

Niegengerd, E. und Hecht-Buchholz, Ch. (1983). Elektronenmikroskopische Untersuchungen einer Virusinfektion (BYMV) von *Vicia faba* bei gleichzeitigem Mineralstoffmangel. *Z. Pflanzenernähr. Bodenk.* **146**, 589–603.

Nielsen, F. H. (1984). Ultratrace elements in nutrition. *Annu. Rev. Nutr.* **4**, 21–41.

Nieto, K. F. and Frankenberger, jr, W. T.(1990). Influence of adenine, isopentyl alcohol and *Azotobacter chroococcum* on the growth of *Raphanus sativus*. *Plant Soil* **127**, 147–157.

Nieto, K. F. and Frankenberger, jr, W. T. (1991). Influence of adenine, isopentyl alcohol and *Azotobacter chroococcum* on the vegetative growth of *Zea mays*. *Plant Soil* **135**, 213–221.

Nieto-Sotelo, J. and Ho, T.-H. D. (1986). Effects of heat shock on the metabolism of glutathione in maize roots. *Plant Physiol.* **82**, 1031–1035.

Nishikuza, Y. (1986). Studies and perspectives of protein kinase C. *Science* **233**, 305–312.

Nishizawa, N. and Mori, S. (1987). The particular vesicle appearing in the barley root cells and its relation to mugineic acid secretion. *J. Plant Nutr.* **10**, 1013–1020.

Nissen, P. (1991). Multiphasic uptake mechanisms in plants. *Int. Rev. Cytol.* **126**, 89–134.

Nitsche, K., Grossmann, K., Sauerbrey, E. and Jung, J. (1985). Influence of the growth retardant tetcyclacis on cell division and cell elongation in plants and cell cultures of sunflower, soybean, and maize. *J. Plant Physiol.* **118**, 209–218.

Nitsos, R. E. and Evans, H. J. (1969). Effects of univalent cations on the activity of particulate starch synthetase. *Plant Physiol.* **44**, 1260–1266.

Noble, C. L. and Rogers, M. E. (1992). Arguments for the use of physiological criteria for improving the salt tolerance in crops. *Plant Soil* **146**, 99–107.

Nobel, P. S. (1990). Soil O_2 and CO_2 effects on apparent cell viability for roots of desert succulents. *J. Exp. Bot.* **41**, 1031–1038.

Nomoto, K., Sugiura, Y. and Takagi, S. (1987). Mugineic acids, studies on phytosiderophores. *In* 'Iron Transport in Microbes, Plants and Animals' (G.Winkelmann *et al.*, eds.), pp. 401–425. Verlag Chemie, Weinheim.

Noodén, L. D. and Letham, D. S. (1986). Cytokinin control of monocarpic senescence in soybean. *In* 'Plant Growth Substances' (M. Bopp, ed.), pp. 324–332. Springer-Verlag, Berlin.

Noodén, L. D. and Mauk, C. S. (1987). Changes in the mineral composition of soybean xylem sap during monocarpic senescence and alterations by depodding. *Physiol. Plant.* **70**, 735–742.

Noodén, L. D., Singh, S. and Letham, D. S. (1990a). Correlation of xylem sap cytokinin levels with monocarpic senescence in soybean. *Plant Physiol.* **93**, 33–39.

Noodén, L. D., Guiamet, J. J., Singh, S., Lethani, D. S., Tsuji, J. and Schneider, M. J. (1990b). Hormonal control of senescence. *In* 'Plant Growth Substances' (R. P. Pharis and S. B. Rood, eds.), pp. 537–547. Springer-Verlag, Berlin.

Noordwijk, M. van, de Willigen, P., Ehlert, P. A. I. and Chardon, W. J. (1990). A simple model of P uptake by crops as a possible basis for P fertilizer recommendations. *Neth. J. Agric.Sci.* **38**, 317–332.

Noordwijk, M. van, Kooistra, J. J., Boone, F. R., Veen, B. W. and Schoonderbeek, D. (1992). Root–soil contact of maize, as measured by a thin-section technique. I. Validity of the method. *Plant Soil* **139**, 108–118.

Norvell, W. A. (1988). Inorganic reactions of manganese in soils. *In* 'Manganese in Soils and Plants' (R. D. Graham, R. J. Hannam and N. C. Uren, eds.), pp. 37–58. Kluwer Academic, Dordrecht, The Netherlands.

Notton, B. A. and Hewitt, E. J. (1979). Structure and properties of higher plant nitrate reductase especially *Spinacia oleracea*. In 'Nitrogen Assimilation of Plants' (E. J. Hewitt and C. V. Cutting, eds.), pp. 227–244. Academic Press, London.

Nouchi, I., Mariko, S. and Aoki, K. (1990). Mechanisms of methane transport from the rhizosphere to the atmosphere through rice plants. *Plant Physiol.* **94**, 59–66.

Nunes, M. A., Dias, M. A., Correia, M. and Oliveira, M. M. (1984). Further studies on growth and osmoregulation of sugar beet leaves under low salinity conditions. *J. Exp. Bot.* **35**, 322–331.

Nyatsanaga, T. and Pierre, W. H. (1973). Effect of nitrogen fixation by legumes on soil acidity. *Agron. J.* **65**, 936–940.

Nye, P. H. (1986). Acid–base changes in the rhizosphere. In 'Advances in Plant Nutrition Vol. 2' (B. Tinker and A. Läuchli, eds.), pp. 129–153. Praeger Scientific, New York.

Nye, P. H. and Greenland, D. J. (1960). 'The Soil under Shifting Cultivation'. Commonwealth Agricultural Bureau, Farnham Royal, Bucks.

Nye, P. H. and Tinker, P. B. (1977). 'Solute Movements in the Root–Soil System'. Blackwell, Oxford.

Oaks, A. (1991). Nitrogen assimilation in roots: a re-evaluation. *BioScience* **42**, 103–111.

Oaks, A. and Hirel, B. (1985). Nitrogen metabolism in roots. *Annu. Rev. Plant Physiol.* **36**, 345–365.

Oaks, A., Wallace, W. and Stevens, D. (1972). Synthesis and turnover of nitrate reductase in corn roots. *Plant Physiol.* **50**, 649–654.

Obata, H. and Umebayashi, M. (1988). Effect of zinc deficiency on protein synthesis in cultured tobacco plant cells. *Soil Sci. Plant Nutr.* (*Tokyo*) **34**, 351–357.

Ockenden, I. and Lott, J. N. A. (1988a). Mineral storage in *Cucurbita* embryos. III. Calcium storage as compared with storage of magnesium, potassium, and phosphorus. *Can. J. Bot.* **66**, 1486–1489.

Ockenden, I. and Lott, J. N. A. (1988b). Changes in the distribution of magnesium, potassium, calcium and phosphorus during growth of *Cucurbita* seedlings. *J. Exp. Bot.* **39**, 973–980.

O'Connell, A. M. and Grove, T. S. (1985). Acid phosphatase activity in Karri (*Eucalyptus diversicolor* F. Muell.) in relation to soil phosphate and nitrogen supply. *J. Exp. Bot.* **36**, 1359–1372.

O'Conner, G. A., Lindsay, W. L. and Olsen, S. R. (1971). Diffusion of iron and iron chelates in soil. *Soil Sci. Soc. Am. Proc.* **35**, 407–410.

Oertli, J. J. (1962). Loss of boron from plants through guttation. *Soil Sci.* **94**, 214–219.

Oertli, J. J. (1968). Extracellular salt accumulation, a possible mechanism of salt injury in plants. *Agrochimica* **12**, 461–469.

Oertli, J. J. and Grgurevic, E. (1975). Effect of pH on the absorption of boron by excised barley roots. *Agron. J.* **67**, 278–280.

Oertli, J. J. and Roth, J. A. (1969). Boron supply of sugar beet, cotton and soybean. *Agron. J.* **61**, 191–195.

Ogawa, M., Tanaka, K. and Kasai, Z. (1979a). Accumulation of phosphorus, magnesium and potassium in developing rice grains: followed by electron microprobe X-ray analysis focusing on the aleurone layer. *Plant Cell Physiol.* **20**, 19–27.

Ogawa, M., Tanaka, K. and Kasai, Z. (1979b). Energy-dispersive X-ray analysis of phytin globoids in aleurone particles of developing rice grains. *Soil Sci. Plant Nutr.* (*Tokyo*) **25**, 437–448.

O'Hara, G. W., Boonkerd, N. and Dilworth, M. J. (1988a). Mineral constraints to nitrogen fixation. *Plant Soil* **108**, 93–110.

O'Hara, G. W., Dilworth, M. J., Boonkerd, N. and Parkpian, P. (1988b). Iron-deficiency specifically limits nodule development in peanut inoculated with *Bradyrhizobium* sp. *New Phytol.* **108**, 51–57.

O'Hara, G. W., Franklin, M. and Dilworth, M. J. (1987). Effect of sulfur supply on sulfate uptake, and alkaline sulfatase activity in free-living and symbiotic bradyrhizobia. *Arch. Microbiol.* **149**, 163–167.

Ohki, K. (1976). Effect of zinc nutrition on photosynthesis and carbonic anhydrase activity in cotton. *Physiol. Plant.* **38**, 300–304.

Ohki, K. (1984). Zinc nutrition related to critical deficiency and toxicity levels for sorghum. *Agron. J.* **76**, 253–256.

Ohki, K., Boswell, F. C., Parker, M. B., Shuman, L. M. and Wilson, D. O. (1979). Critical manganese deficiency level of soybean related to leaf position. *Agron. J.* **71**, 233–234.

Ohki, K., Wilson, D. O. and Anderson, O. E. (1981). Manganese deficiency and toxicity sensitivities of soybean cultivar. *Agron. J.* **72**, 713–716.

Ohnishi, J. and Kanai, R. (1987). Na^+-induced uptake of pyruvate into mesophyll chloroplasts of a C_4 plant, *Panicum miliaceum*. *FEBS Lett.* **219**, 347–350.

Ohnishi, J., Flügge, U.-I., Heldt, H. W. and Kanai, R. (1990). Involvement of Na^+ in active uptake of pyruvate in mesophyll chloroplasts of some C_4 plants. *Plant Physiol.* **94**, 950–959.

Ohta, D., Matoh, T. and Takahashi, E. (1987). Early responses of sodium-deficient *Amaranthus tricolor* L. plants to sodium application. *Plant Physiol.* **84**, 112–117.

Ohta, D., Matsui, J., Matoh, T. and Takahashi, E. (1988). Sodium requirement of monocotyledonous C_4 plants for growth and nitrate reductase activity. *Plant Cell Physiol.* **29**, 1429–1432.

Ohta, D., Yasuoka, S., Matoh, T. and Takahashi, E. (1989). Sodium stimulates growth of *Amaranthus tricolor* L. plants through enhanced nitrate assimilation. *Plant Physiol.* **89**, 1102–1105.

Ohwaki, Y. and Hirata, H. (1992). Differences in carboxylic acid exudation among P-starved leguminous crops in relation to carboxylic acid contents in plant tissues and phospholipid level in roots. *Soil Sci. Plant Nutr.* (*Tokyo*) **38**, 235–243.

Oja, V., Laisk, A. and Heber, U. (1986). Light-induced alkalization of the chloroplast stroma *in vivo* as estimated from the CO_2 capacity of intact sunflower leaves. *Biochim. Biophys. Acta* **849**, 355–365.

O'Kelley, J. C. (1969). Mineral nutrition of algae. *Annu. Rev. Plant Physiol.* **19**, 89–112.

Okon, Y., Fallik, S., Yahalom, E. and Tal, S. (1988). Plant growth promoting effects of *Azospirillum*. *In* 'Nitrogen Fixation: Hundred Years After' (H. Bothe, F. J. de Bruijn and W. E. Newton, eds.), pp. 741–746. Fischer Verlag, Stuttgart.

Okuda, A. and Takahashi, E. (1965). The role of silicon. *Min. Nutr. Rice Plant, Proc. Symp. Int. Rice Res. Inst. Manila* 1964, pp. 123–146.

Okumura, N., Nishizawa, N.-K., Umehara, Y. and Mori, S. (1991). An iron deficiency-specific CDNA from barley roots having two homologous cysteine-rich MT domains. *Plant Mol. Biol.* **17**, 531–533.

Olbe, M. and Sommarin, M. (1991). ATP-dependent Ca^{2+} transport in wheat root plasma membrane vesicles. *Physiol. Plant.* **83**, 535–543.

Oldenkamp, L. and Smilde, K. W. (1966). Copper deficiency in douglas fir (*Pseudotsuga menziesii* Mirb. Franco). *Plant Soil* **25**, 150–152.

Ollagnier, M. and Renard, J.-L. (1976). The influence of potassium on the resistance of oil palms to *Fusarium*. *Proc. 12th Colloq. Int. Potash Inst. Bern*, pp. 157–166.

Ollagnier, M. and Wahyuni, M. (1986). Die Ernährung und Düngung mit Kalium und Chlor der Kokospalme, Hybride Nain de Malaisie × Grand Quest. Africain. *Kali-Briefe, Int. Kali-Institut Bern*. Fachgeb. 27, Nr. 2, pp. 1–8.

Olsen, R. A. and Brown, J. C. (1980). Factors related to iron uptake by dicotyledonous and monocotyledonous plants. I. pH and reductant. *J. Plant Nutr.* **2**, 629–645.

Olsen, R. A., Bennett, J. H., Blume, D. and Brown, J. C. (1981). Chemical aspects of the Fe stress response mechanism in tomatoes. *J. Plant Nutr.* **3**, 905–921.

Olsthoorn, A. F. M. and Tiktak, A. (1991). Fine root density and root biomass of two Douglas-fir stands on sandy soils in the Netherlands. 2. Periodicity of fine root growth and estimation of belowground carbon allocation. *Neth. J. Agric. Sci.* **39**, 61–77.

Olsthoorn, A. F. M., Keltjens, W. G., van Baren, B. and Hopman, M. C. G. (1991). Influence of ammonium on fine root development and rhizosphere pH of Douglas-fir seedlings in sand. *Plant Soil* **133**, 75–81.

O'Neal, D. and Joy, K. W. (1974). Glutamine synthetase of pea leaves. Divalent cation effects, substrate specificity, and other properties. *Plant Physiol.* **54**, 775–779.

O'Neill, S. D. and Spanswick, R. M (1984). Characterization of native and reconstituted plasma membrane H^+-ATPase from the plasma membrane of *Beta vulgaris*. *J. Membr. Biol.* **79**, 245–256.

O'Neill, S. D., Bennett, A. B. and Spanswick, R. M (1983). Characterization of a NO_3^- sensitive H^+-ATPase from corn roots. *Plant Physiol.* **72**, 837–846.

Oparka, K. J. (1986). Phloem unloading in the potato tuber. Pathways and sites of ATPase. *Protoplasma* **131**, 201–210.

Oparka, K. J. (1990). What is phloem unloading? *Plant Physiol.* **94**, 393–396.

Ordentlich, A., Linzer, R. A. and Raskin, I. (1991). Alternative respiration and heat evolution in plants. *Plant Physiol.* **97**, 1545–1550.

Orlovich, D. A. and Ashford, A. E. (1993). Polyphosphate granules are an artefact of specimen preparation in the ectomycorrhizal fungus *Pisolithus tinctorius*. *Protoplasma* **173**, 91–102.

Orlovich, D. A., Ashford, A. E. and Cox, G. C. (1989). A reassessment of polyphosphate granule composition in the ecto-mycorrhizal fungus *Pisolithus tinctorius*. *Aust. J. Plant Physiol.* **16**, 107–115.

Osmond, C. B. (1967). Acid metabolism in *Atriplex*. I. Regulation in oxalate synthesis by the apparent excess cation absorption. *Aust. J. Biol. Sci.* **20**, 575–587.

Osswald, W. F. and Elstner, E. F. (1986). Fichtenerkrankungen in den Hochlagen der Bayerischen Mittelgebirge. *Ber. Deutsch. Bot. Ges.* **99**, 313–339.

Osswald, W., Schütz, W. and Elstner, E. F. (1989). Indole-3-acetic acid and *p*-hydroxyacetophenone driven ethylen formation from 2-aminocyclopropane-1-carboxylic acid catalized by horseradish peroxidase. *J. Plant Physiol.* **134**, 510–513.

O'Sullivan, M. (1970). Aldolase activity in plants as an indicator of zinc deficiency. *J. Sci. Food Agric.* **21**, 607–609.

Osuna-Canizalez, F. J., Datta, S. K. and Bonman, J. M. (1991). Nitrogen form and silicon nutrition effects on resistance to blast disease of rice. *Plant Soil* **135**, 223–231.

Otani, T. and Ae, N. (1993). Ethylene and carbon dioxide concentrations of soils as influenced by rhizosphere of crops under field and pot conditions. *Plant Soil* **150**, 255–262.

Oteifa, B. A. and Elgindi, A. Y. (1976). Potassium nutrition of cotton, *Gossypium barbadense*, in relation to nematode infection by *Meloidogyne incognita* and *Rotylenchulus reniformis*. *Proc. 12th Colloq. Int. Potash Inst. Bern*, pp. 301–306.

Ottow, J. C. G., Benckiser, G., Santiago, S. and Watanabe, I. (1982). Iron toxicity of wetland rice (*Oriza sativa* L.) as a multiple nutritional stress. *In* 'Proceedings of the Ninth International Plant Nutrition Colloquium, Warwick, England' (A. Scaife, ed.), pp. 454–460.

Outlaw jr, W. H. (1983). Current concepts on the role of potassium in stomatal movements. *Physiol. Plant.* **49**, 302–311.

Overlach, S., Diekmann, W. and Raschke, K. (1993). Phosphate translocator of isolated guard-cell chloroplasts from *Pisum sativum* L. transport glucose-6-phosphate. *Plant Physiol.* **101**, 1201–1207.

Ownby, J. D. and Hruschka, W. R. (1991). Quantitative changes in cytoplasmic and microsomal proteins associated with aluminium toxicity in two cultivars of winter wheat. *Plant Cell Environ.* **14**, 303–309.

Ownby, J. D. and Popham, H. R. (1989). Citrate reverses the inhibition of wheat root growth caused by aluminium. *J. Plant Physiol.* **135**, 588–591.

Ozanne, P. G. (1958). Chlorine deficiency in soils. *Nature (London)* **182**, 1172–1173.

Ozanne, P. G., Greenwood, E. A. N. and Shaw, T. C. (1963). The cobalt requirement of subterranean clover in the field. *Aust. J. Agric. Res.* **14**, 39–50.

Ozanne, P. G., Keay, J. and Biddiscombe, E. F. (1969). The comparative applied phosphate requirement of eight annual pasture species. *Aust. J. Biol. Sci.* **20**, 809–818.

Pacovsky, R. S., Fuller, G. and Paul, E. A. (1985). Influence of soil on the interactions between endomycorrhizae and *Azospirillum* in sorghum. *Soil Biol. Biochem.* **17**, 525–531.

Pais, I. (1983). The biological importance of titanium. *J. Plant Nutr.* **6**, 3–131.

Palavan, N. and Galston, A. W. (1982). Polyamine biosynthesis and titer during various development stages of *Phaseolus vulgaris*. *Physiol. Plant.* **55**, 438–444.

Paliyath, G. and Thompson, J. E. (1987). Calcium- and calmodulin-regulated break-down for phospholipid by microsomal membranes from bean cotyledons. *Plant Physiol.* **83**, 63–68.

Palma, J. M., Sandalio, L. M. and del Rio, L. A. (1986). Manganese superoxide dismutase and higher plant chloroplats: a reappraisal of a controverted cellular localization. *J. Plant Physiol.* **125**, 427–439.

Palma, J. M., Yanez, J., Gomez, M. and del Rio, L. A.(1990). Copper-binding proteins and copper tolerance in *Pisum sativum* L. Characterization of low-molecular-weight metalloproteins from plants with different sensitivity to copper. *Planta* **18**, 487–495.

Palma, J. M., Longa, M. A., del Rio, L. A. and Arines, J. (1993). Superoxide dismutase in vesicular arbuscular-mycorrhizal red clover plants. *Physiol. Plant.* **87**, 77–83.

Palmgren, M. G. (1991). Regulation of plant plasma membrane H^+-ATPase activity. *Physiol. Plant.* **83**, 314–323.

Pan, W. L., Hopkins, A. G. and Jackson, W. A. (1989). Aluminum inhibition of shoot lateral branches of *Glycine max* and reversal by exogenous cytokinin. *Plant Soil* **120**, 1–9.

Pandita, M. L. and Andrew, W. T. (1967). A correlation between phosphorus content of leaf tissue and days to maturity in tomato and lettuce. *Proc. Am. Soc. Hortic. Sci.* **91**, 544–549.

Papageorgiou, G. C., Fujimura, Y. and Murat, N. (1991). Protection of the oxygen-evolving Photosystem II complex by glycine-betaine. *Biochim. Biophys. Acta (B)* **1057**, 361–366.

Papen, H., von Berg, R., Hinkel, I., Thoene, B. and Rennenberg, H. (1989). Heterotrophic nitrification by *Alcaligenes faecalis*; NO_2^-, NO_3^-, N_2O, and NO production in exponentially growing cultures. *Appl. Environ. Microbiol.* **55**, 2068–2072.

Pardales, J. R. jr, Kono, Y. and Yamauchi (1992). Epidermal cell elongation in sorghumn seminal roots exposed to high root-zone temperature. *Plant Sci.* **81**, 143–146.

Pardee, A. B. (1967). Crystallization of a sulfate binding protein (permease) from *Salmonella typhimurium. Science* **156**, 1627–1628.

Parets-Soler, A., Pardo, J. M. and Serrano, R. (1990). Immunocytolocalization of plasma membrane H^+-ATPase. *Plant Physiol.* **93**, 1654–1658.

Parfitt, R. L. (1979). The availability of P from phosphate-goethite bridging complexes. Description and uptake by ryegrass. *Plant Soil* **53**, 55–65.

Parker, D. R., Kinraide, T. B. and Zelazny, L. W. (1988). Aluminum speciation and phytotoxicity in dilute hydroxy-aluminum solutions. *Soil Sci. Soc. Am. J.* **52**, 438–444.

Parker, D. R., Kinraide, T. B. and Zelazny, L. W. (1989). On the phytotoxicity of polynuclear hydroxy-aluminum complexes. *Soil Sci. Soc. Am. J.* **53**, 789–796.

Parker, M. B. and Harris, H. B. (1977). Yield and leaf nitrogen of nodulating soybeans as affected by nitrogen and molybdenum. *Agron. J.* **69**, 551–554.

Parker, M. B. and Walker, M. E. (1986). Soil pH and manganese effects on manganese nutrition of peanut. *Agron. J.* **78**, 614–620.

Parker, M. B., Gascho, G. J. and Gaines, T. P. (1983). Chloride toxicity of soybeans grown on Atlantic coast flatwoods soils. *Agron. J.* **75**, 439–443.

Parker, M. B., Gaines, T. P., Hook, J. E., Gascho, G. J. and Maw, B. W. (1987). Chloride and water stress effects on soybean in pot culture. *J. Plant Nutr.* **10**, 517–538.

Parr, A. J. and Loughman, B. C. (1983). Boron and membrane functions in plants. *In* 'Metals and Micronutrients: Uptake and Utilization by Plants' (Annu. Proc. Phytochem. Soc. Eur. No. 21; D. A. Robb and W. S. Pierpoint, eds.), pp. 87–107. Academic Press London.

Parra-Garcia, M. D., Lo Giudice, V. and Ocampo, J. A. (1992). Absence of VA colonization in *Oxalis pes-caprae* inoculated with *Glomus mosseae. Plant Soil* **145**, 298–300.

Parry, A. D. and Horgan, R. (1991). Carotenoids and abscisic acid (ABA) biosynthesis in higher plants. *Physiol. Plant.* **82**, 320–326.

Parry, D. W. and Hodson, M. J. (1982). Silica distribution in the caryopsis and inflorescence bracts of foxtail millet (*Setaria italica* L. Beauv.) and its possible significance in carcinogenesis. *Ann. Bot. (London)* [*N.S.*] **49**, 531–540.

Parry, D. W. and Kelso, M. (1975). The distribution of silicon deposits in the root of *Molinia caerulea* (L.) Moench and *Sorghum bicolor* (L.) Moench. *Ann. Bot. (London)* [*N.S.*] **39**, 995–1001.

Parry, D. W. and Smithson, F. (1964). Types of opaline silica deposition in the leaves of British grasses. *Ann. Bot. (London)* [*N.S.*] **28**, 169–185.

Parsons, R., Silvester, W. B., Harris, S., Gruijters, W. I. M. and Bullivant, S. (1987). Frankia vesicles provide inducible and absolute oxygen protection for nitrogenase. *Plant Physiol.* **83**, 728–731.

Parthier, B. (1979). The role of phytohormones (cytokinin) in chloroplast development. *Biochem. Physiol. Pflanz.* **144**, 173–214.

Parthier, B. (1991). Jasmonates, new regulators of plant growth and development: many facts and few hypotheses on their actions. *Bot. Acta* **104**, 446–454.

Parton, W. J., Morgan, J. A., Altenhofer, J. M. and Harper, L. A. (1988). Ammonia volatilization from spring wheat plants. *Agron. J.* **80**, 419–425.

Pasricha, N. S., Nayyar, V. K., Randhawa, N. S. and Sinha, M. K. (1977). Influence of sulphur fertilization on suppression of molybdenum uptake by berseem (*Trifolium alexandrinum*) and oats (*Avena sativa*) grown on a molybdenum-toxic soil. *Plant Soil* **46**, 245–250.

Pasricha, N. S., Baddesha, H. S., AuLakh, M. S. and Nayyar, V. R. (1987). The quantity–intensity relationships in four different soils as influenced by phosphorus. *Soil Sci.* **143**, 1–4.

Passioura, J. B. (1991). Soil structure and plant growth. *Aust. J. Soil Res.* **29**, 717–728.

Passioura, J. B. and Gardner, P. A. (1990). Control of leaf expansion in wheat seedlings growing in drying soil. *Aust. J. Plant Physiol.* **17**, 149–157.

Pate, J. S. (1975). Exchange of solutes between phloem and xylem and circulation in the whole plant. *In* 'Encyclopedia of Plant Physiology, New Series' (M. H. Zimmermann and J. A. Milburn, eds.), Vol. 1, pp. 451–468. Springer-Verlag, Berlin.

Pate, J. S. (1983). Patterns of nitrogen metabolism in higher plants and their ecological significance. *In* 'Nitrogen as an Ecological Factor' (J. A. Lee, S. McNeill and I. H. Rorison, eds.), pp. 225–255. Blackwell Scientific Publications, Oxford.

Pate, J. S. and Atkins, C. A. (1983). Xylem and phloem transport and the functional economy of carbon and nitrogen of a legume leaf. *Plant Physiol.* **71**, 835–840.

Pate, J. S. and Gunning, B. E. S. (1972). Transfer cells. *Annu. Rev. Plant Physiol.* **23**, 173–196.

Pate, J. S. and Herridge, D. F. (1978). Partitioning and utilization of net photosynthate in nodulated annual legumes. *J. Exp. Bot.* **29**, 401–412.

Pate, J. S., Wallace, W. and Die, van J. (1964). Petiole bleeding sap in the examination of the circulation of nitrogenous substances in plants. *Nature* (*London*) **204**, 1073–1074.

Pate, J. S., Sharkey, P. J. and Lewis, O. A. M. (1974). Xylem to phloem transfer of solutes in fruiting shoots of legumes, studied by a phloem bleeding technique. *Planta* **122**, 11–26.

Pate, J. S., Kuo, J. and Hocking, P. J. (1978). Functioning of conducting elements of phloem and xylem in the stalk of the developing fruit of *Lupinus albus* L. *Aust. J. Plant Physiol.* **5**, 321–326.

Pate, J. S., Layzell, D. B. and Atkins, C. A. (1979). Economy of carbon and nitrogen in a nodulated and nonnodulated (NO_3-grown) legume. *Plant Physiol.* **64**, 1083–1088.

Pate, J. S., Lindblad, P. and Atkins, C. A. (1988). Pathways of assimilation and transfer of fixed nitrogen in coralloid roots of cycad–*Nostoc* symbioses. *Planta* **176**, 461–471.

Patrick, J. W. (1990). Sieve element unloading: cellular pathway, mechanism and control. *Physiol. Plant.* **78**, 298–308.

Patrick, J. W. (1993). Osmotic regulation of assimilate unloading from seed coats of *Vicia faba*. Role of turgor and identification of turgor-dependent fluxes. *Physiol. Plant.* **89**, 87–96.

Patrick, Z. A. (1971). Phytotoxic substances associated with the decomposition in soil of plant residues. *Soil Sci.* **111**, 13–18.

Paul, M. J. and Cockburn, W. (1989). Pinitol, a compatible solute in *Mesembryanthemum crystallinum* L.? *J. Exp. Bot.* **40**, 1093–1098.

Paulitz, T. C. and Linderman, R. G. (1989). Interactions between fluorescent pseudo-monads and VA mycorrhizal fungi. *New Phytol.* **113**, 37–45.

Paull, J. G., Cartwright, B. and Rathjen, A. J. (1988). Responses of wheat and barley genotypes toxic concentrations of soil boron. *Euphytica* **39**, 137–144.

Paull, J. G., Rathjen, A. J. and Cartwright, B. (1988). Genetic control of tolerance to high concentrations of soil boron in wheat. *In* Proc. 7th International Wheat Genetics Symposium, Cambridge, UK (T. E. Miller and R. M. D. Loebuer, eds.), pp. 871–877.

Paull, J. G., Rathjen, A. J. and Cartwright, B. (1991). Major gene control of tolerance of bread wheat (*Triticum aestivum* L.) to high concentrations of boron. *Euphytica* **55**, 217–228.

Paull, J. G., Nable, R. O. and Rathjen, A. J. (1992a). Physiological and genetic control of the tolerance of wheat to high concentrations of boron and implications for plant breeding. *Plant Soil* **146**, 251–260.

Paull, J. G., Nable, R. O., Lake, A. W. H., Materne, M. A. and Rathjen, A. J. (1992b). Response of annual medics (*Medicago* spp.) and field peas (*Pisum sativum*) to high concentration of boron: genetic variation and the mechanism of tolerance. *Aust. J. Agric. Res.* **43**, 203–213.

Pearson, C. J., Volk, R. J. and Jackson, W. A. (1981). Daily changes in nitrate influx, efflux and metabolism in maize and pearl millet. *Planta* **152**, 319–324.

Pearson, J. and Havill, D. C. (1988). The effect of hypoxia and sulphide on culture-grown wetland and non-wetland plants. II. Metabolic and physiological changes. *J. Exp. Bot.* **39**, 432–439.

Pearson, R. W., Radcliffe, L. F. and Taylor, H. M. (1970). Effect of soil temperature, strength and pH on cotton seedlings root elongation. *Agron. J.* **62**, 243–246.

Pearson, R. W., Childs, J. and Lund, Z. F. (1973). Uniformity of limestone mixing in acid subsoil as a factor in cotton root penetration. *Soil Sci. Soc. Am. Proc.* **37**, 727–732.

Peck, N. H., Grunes, D. L., Welch, R. M. and MacDonald, G. E. (1980). Nutritional quality of vegetable crops as affected by phosphorus and zinc fertilizer. *Agron. J.* **72**, 528–534.

Peel, A. J. and Rogers, S. (1982). Stimulation of sugar loading into sieve elements of willow by potassium and sodium salts. *Planta* **154**, 94–96.

Pegtel, D. M. (1987). Effect of ionic Al in culture solutions on the growth of *Arnica montana* L. and *Deschampsia flexuosa* (L.) Trin. *Plant Soil* **102**, 85–92.

Peirson, D. R. and Elliot, J. R. (1981). In vivo nitrite reduction in leaf tissue of *Phaseolus vulgaris*. *Plant Physiol.* **68**, 1068–1072.

Pelacho, A. M. and Mingo-Castel, A. M. (1991). Jasmonic acid induces tuberization of potato stolons cultured *in vitro*. *Plant Physiol.* **97**, 1253–1255.

Peng, X. X. and Yamauchi, M. (1993). Ethylene production in rice bronzing leaves induced by ferrous iron. *Plant Soil* **149**, 227–234.

Peoples, M. B. and Craswell, E. T. (1992). Biological nitrogen fixation: investments, expectations and actual contributions to agriculture. *Plant Soil* **141**, 13–39.

Peoples, M. B., Pate, J. S. and Atkins, C. A. (1983). Mobilization of nitrogen in fruiting plants of a cultivar of cowpea. *J. Exp. Bot.* **34**, 562–578.

Peoples, M. B., Hebb, D. M., Gibson, A. H. and Herridge, D. H. (1989). Development of the xylem ureide assay for the measurement of nitrogen fixation by pigeonpea (*Cajanus cajan* (L.) Millsp.). *J. Exp. Bot.* **40**, 535–542.

Peoples, T. R. and Koch, D. W. (1979). Role of potassium in carbon dioxide assimilation in *Medicago sativa* L. *Plant Physiol.* **63**, 878–881.

Perata, P. and Alpi, A. (1991). Ethanol-induced injuries to carrot cells. The role of acetaldehyde. *Plant Physiol.* **95**, 748–752.

Perrenoud, S. (1977). Potassium and plant health. *In* 'Research Topics', No. 3, pp. 1–118. International Potash Institute, Bern, Switzerland.

Perrin, R. (1990). Interactions between mycorrhizae and diseases caused by soil-borne fungi. *Soil Use Mgmt* **6**, 189–195.

Perry, C. C., Williams, R. J. P. and Fry, S. C. (1987). Cell wall biosynthesis during silification of grass hairs. *J. Plant Physiol.* **126**, 437–448.

Persson, H. (1979). Fine root production, mortality, and decomposition in forest ecosystems. *Vegetatio* **41**, 101–109.

Persson, L. (1969). Labile-bound sulfate in wheat-roots: Localization, nature and possible connection to active absorption mechanism. *Physiol. Plant.* **22**, 959–977.

Perumalla, C. J. and Peterson, C. A. (1986). Deposition of Casparian bands and suberin lamellae in the exodermis and endodermis of young corn and onion roots. *Can. J. Bot.* **64**, 1873–1878.

Perur, N. G., Smith, R. L. and Wiebe, H. H. (1961). Effect of iron chlorosis on protein fraction on corn leaf tissue. *Plant Physiol.* **36**, 736–739.

Peters, M. (1990). Nutzungseinfluss auf die Stoffdynamik schleswig-holsteinischer Böden – Wasser-, Luft-, Nähr- und Schadstoffdynamik. *Schriftenreihe Inst. für Pflanzenernährung und Bodenkunde, Universität Kiel* (H. P. Blume *et al.*, eds.) Nr. **8**.

Peterson, C. A. (1988). Exodermal Casparian bands: their significance for ion uptake by roots. *Physiol. Plant.* **72**, 204–208.

Peterson, T. A., Reinsel, M. D. and Krizek, D. T. (1991a). Tomato (*Lycopersicon esculentum* Mill., cv. 'Better Bush') plant response to root restriction. II. Root respiration and ethylene generation. *J. Exp. Bot.* **42**, 1241–1249.

Peterson, T. A., Reinsel, M. D. and Krizek, D. T. (1991b). Tomato (*Lycopersicon esculentum* Mill., cv. 'Better Bush') plant response to root restriction. I. Alteration of plant morphology. *J. Exp. Bot.* **42**, 1233–1240.

Peterson, C. A., Murrmann, M. and Steudle, E. (1993). Location of the major barriers to water and ion movement in young roots of *zea mays* L. *Planta* **190**, 127–136.

Petit, C. M., Ringoet, A. and Myttenaere, C. (1978). Stimulation of cadmium uptake in relation to the cadmium content of plants. *Plant Physiol.* **62**, 554–557.

Petzold, U., Neumann, St. and Dahse, I. (1989). Amino acid and sucrose uptake into cotyledons and roots of *Sinapis alba* L. *Biochem. Physiol. Pflanzen* **185**, 27–40.

Peverley, J. H., Adamec, J. and Parthasarathy, M. V. (1978). Association of potassium and some other monovalent cations with occurrence of polyphosphate. *Plant Physiol.* **62**, 120–126.

Pfeffer, P. E., Tu, S.-I., Gerasimowicz, W. V. and Cavanaughk J. R. (1986). *In vivo* ^{31}PNMR studies of corn root tissue and its uptake of toxic metals. *Plant Physiol.* **80**, 77–84.

Pfeiffenschneider, Y. and Beringer, H. (1989). Measurement of turgor potential in carrots of different K-nutrition by using the cell pressure probe. *Proc. 21st Colloq. Int. Potash Inst. Bern* pp. 203–217.

Pfirrmann, T., Runkel, K. H., Schramel, P. and Eisenmann, T. (1990). Mineral and nutrient supply, content and leaching in Norway spruce exposed for 14 month to ozone and acid mist. *Environ. Pollut.* **64**, 229–254.

Pflüger, R. and Cassier, A. (1977). Influence of monovalent cations on photosynthetic CO_2 fixation. *Proc. 13th Colloq. Int. Potash Inst. Bern* pp. 95–100.

Pflüger, R. and Wiedemann, R. (1977). Der Einfluss monovalenter Kationen auf die Nitratreduktion von *Spinacia oleracea* L. *Z. Pflanzenphysiol.* **85**, 125–133.

Philipson, J. J. and Coutts, M. P. (1977). The influence of mineral nutrition on the development of trees. II. The effect of specific nutrient elements on the growth of individual roots of Sitka spruce. *J. Exp. Bot.* **28**, 864–871.

Phillips, D. A. and Tsai, S. M. (1992). Flavonoids as signals to rhizosphere microbes. *Mycorrhiza* **1**, 55–58.

Piccini, D., Ocampo, J. A. and Bedmar, E. J. (1988). Possible influence of *Rhizobium* on VA mycorrhiza metabolic activity in double symbiosis of alfalfa plants (*Medicago sativa* L.) grown in a pot experiment. *Biol. Fertil. Soils* **6**, 65–67.

Pich, A. and Scholz, G. (1991). Nicotianamine and the distribution of iron into apoplast and symplast of tomato (*Lycopersicon esculentum* Mill.). II. Uptake of iron by protoplasts from the variety Bonner Beste and its nicotianamine-less mutant *chloronerva* and the compartmentation of iron in leaves. *J. Exp. Bot.* **42**, 1517–1523.

Pich, A., Scholz, G. and Seifert, K.(1991). Effect of nicotianamine on iron uptake and citrate accumulation in two genotypes of tomato, *Lycopersicon esculentum* Mill. *J. Plant Physiol.* **137**, 323–326.

Pier, P. A. and Berkowitz, G. A. (1987). Modulation of water stress affects on photosynthesis by altered leaf K^+. *Plant Physiol.* **85**, 655–661.

Pierce, J. (1986). Determinants of substrate specificity and the role of metal in the reaction of ribolosebisphosphate carboxylase/oxygenase. *Plant Physiol.* **81**, 943–945.

Pijnenborg, J. W. M. and Lie, T. A. (1990). Effect of lime-pelleting on the nodulation of lucerne (*Medicago sativa* L.) in acid soil: a comparative study carried out in the field, in pots and in rhizotrons. *Plant Soil* **121**, 225–234.

Pilbeam, D. J. and Kirkby, E. A. (1983). The physiological role of boron in plants. *J. Plant Nutr.* **6**, 563–582.

Pilet, P. E. (1981). Root growth and gravireaction: Endogenous hormone balance. *In* 'Structure and Function of Plant Roots' (R. Brouwer, O. Gasparikova, J. Kolek and B. E. Loughman, eds.), pp. 89–93. Martin Nijhoff/Junk Publ., The Hague.

Pilet, P.-E. (1991). Root growth and gravireaction. Implications of hormones and other regulators. *In* 'Plant Roots: the Hidden Half' (Y. Waisel, A. Eshel and U. Kafkafi, eds.), pp. 179–204. Marcel Dekker, Inc. New York.

Pingel, U. (1976). Der Einfluss phenolischer Aktivatoren und Inhibitoren der IES-Oxidase-Aktivität auf die Adventivbewurzelung bei *Tradescantia albiflora*. *Z. Pflanzenphysiol.* **79**, 109–120.

Pinton, R., Cakmak, I. and Marschner, H. (1993). Effect of zinc deficiency on proton fluxes in plasma membrane-enriched vesicles isolated from bean roots. *J. Exp. Bot.* **44**, 623–630.

Pirson, A. (1937). Ernährungs- und stoffwechselphysiologische Untersuchungen an *Frontalis* und *Chlorella*. *Z. Bot.* **31**, 193–267.

Pissarek, H. P. (1973). Zur Entwicklung der Kalium-Mangelsymptome von Sommerraps. *Z. Pflanzenernähr. Bodenk.* **136**, 1–19.

Pissarek, H. P. (1974). Untersuchungen der durch Kupfermangel bedingten anatomischen Veränderungen bei Hafer- und Sonnenblumen. *Z. Pflanzenernähr. Bodenk.* **137**, 224–234.

Pissarek, H. P. (1979). Der Einfluss von Grad und Dauer des Mg-Mangels auf den Kornertrag von Hafer. *Z. Acker- Pflanzenbau* **148**, 62–71.

Pissarek, H. P. (1980). Makro- und Mikrosymptome des Bormangels bei Sonnenblumen, Chinakohl und Mais. *Z. Pflanzenernähr. Bodenk.* **143**, 150–160.

Pitcher, L. H. and Daie, J. (1991). Growth and sink to source transition in developing leaves of sugar-beet. *Plant Cell Physiol.* **32**, 335–342.

Pitman, M. G. (1972a). Uptake and transport of ions in barley seedlings. II. Evidence for two active stages in transport to the shoot. *Aust. J. Biol. Sci.* **25**, 243–257.

Pitman, M. G. (1972b). Uptake and transport of ions in barley seedlings. III. Correlation between transport to the shoot and relative growth rate. *Aust. J. Biol. Sci.* **25**, 905–919.

Pitman, M. G., Mowat, J. and Nair, H. (1971). Interactions of processes for accumulation of salt and sugar in barley plants. *Aust. J. Biol. Sci.* **24**, 619–631.

Pitman, M. G., Wellfare, D. and Carter, C. (1981). Reduction of hydraulic conductivity during inhibition of exudation from excised maize and barley roots. *Plant Physiol.* **61**, 802–808.

Plassard, C., Scheromm, P., Mousain, D. and Salsac, L. (1991). Assimilation of mineral nitrogen and ion balance in the two partners of ectomycorrhizal symbiosis: Data and hypothesis. *Experientia* **47**, 340–349.

Platero, M. and Tejerina, G. (1976). Calcium nutrition in *Phaseolus vulgaris* in relation to its resistance to *Erwinia carotavora*. *Phytopathol. Z.* **85**, 314–319.

Platt-Aloia, K. A., Thomson, W. W. and Terry, N. (1983). Changes in plastid ultrastructure during iron nutrition-mediated chloroplast development. *Protoplasma* **114**, 85–92.

Plaxton, W. C. and Preiss, J. (1987). Purification and properties of nonproteolytic degraded ADPglucose pyrophosphorylase from maize endosperm. *Plant Physiol.* **83**, 105–112.

Plenchette, C., Fortin, J. A. and Furlan, V. (1983). Growth response of several plant species to mycorrhizae in a soil of moderate P-fertility. II. Soil fumigation induced stunting of plants corrected by reintroduction of the wild endomycorrhizal flora. *Plant Soil* **70**, 211–217.

Poffenroth, M., Green, D. B. and Tallman, G. (1992). Sugar concentrations in guard cells of *Vicia faba* illuminated with red or blue light. *Plant Physiol.* **98**, 1460–1471.

Pohlman, A. A. and McColl, J. G. (1982). Nitrogen fixation in the rhizosphere and rhizoplane of barley. *Plant Soil* **69**, 341–352.

Poirier, Y., Thoma, S., Somerville, C. and Schiefelbein, J.(1991). A mutant of *Arabidopsis* deficient in xylem loading of phosphate. *Plant Physiol.* **97**, 1087–1093.

Polacco, J. C. (1977). Nitrogen metabolism in soybean tissue culture. II. Urea utilization and urease synthesis require Ni^{2+}. *Plant Physiol.* **59**, 827–830.

Poljakoff-Mayber, A. (1975). Morphological and anatomical changes in plants as a response to salinity stress. *In* 'Plants in Saline Environments' (A. Poljakoff-Mayber and J. Gale, eds.), pp. 97–117. Springer-Verlag, Berlin.

Pollard, A. S. and Wyn Jones, R. G. (1979). Enzyme activities in concentrated solutions of glycinebetaine and other solutes. *Planta* **144**, 291–298.

Pollard, A.S., Parr, A. J. and Loughman, B. C. (1977). Boron in relation to membrane function in higher plants. *J. Exp. Bot.* **28**, 831–841.

Polle, A., Chakrabarti, K., Schürmann, W. and Rennenberg, H. (1990). Composition and properties of hydrogen peroxide decomposing systems in extracellular and total extracts from needles of Norway spruce (*Picea abies* L. Karst.). *Plant Physiol.* **94**, 312–319.

Polle, A., Chakrabarti, K., Chakrabarti, S., Seifert, F., Schramel, P. and Rennenberg, H. (1992). Antioxidants and manganese deficiency in needles of Norway spruce (*Picea abies* L.) trees. *Plant Physiol.* **99**, 1084–1089.

Ponnamperuma, F. N. (1972). The chemistry of submerged soils. *Adv. Agron.* **24**, 29–96.

Poole, R. J. (1978). Energy coupling for membrane transport. *Annu. Rev. Plant Physiol.* **29**, 437–460.

Poorter, H., van der Werf, A., Atkin, O. K. and Lambers, H. (1991). Respiratory energy requirements of root vary with the potential growth rate of a plant species. *Physiol. Plant.* **83**, 469–475.

Poovaiah, B. W. (1979). Role of calcium in ripening and senescence. *Commun. Soil Sci. Plant Anal.* **10**, 83–88.

Poovaiah, B. W. and Reddy, A. S. N. (1991). Calcium and root development. Importance of calcium in signal transduction. *In* 'Plant Roots: the Hidden Half' (Y. Waisel, A. Eshel and U. Kafkafi, eds.), pp. 205–227. Marcel Dekker, New York.

Poovaiah, B. W., McFadden, J. J. and Reddy, A. S. N. (1987). The role of calcium ions in gravity signal perception and translocation. *Physiol. Plant.* **71**, 401–407.

Pope, A. J. and Leigh, R. A. (1990). Characterization of chloride transport at the tonoplast of higher plants using a chloride-sensitive fluorescent probe. Effects of other anions, membrane potential, and transport inhibitors. *Planta* (Berlin) **181**, 406–413.

Portis jr, A. R. (1981). Evidence of a low stromal Mg^{2+} concentration in intact chloroplasts in the dark. I. Studies with the ionophore A 23187. *Plant Physiol.* **67**, 985–989.

Portis jr, A. R. (1982). Effects of the relative extrachloroplastic concentrations of inorganic phosphate, 3-phosphoglycerate and dihydroxyacetone phosphate on the rate of starch synthesis in isolated spinach chloroplasts. *Plant Physiol.* **70**, 393–396.

Portis jr, A. R. and Heldt, H. W. (1976). Light-dependent changes of the Mg^{2+} concentration in the stroma in relation to the Mg^{2+} depending of CO_2 fixation in intact chloroplasts. *Biochim. Biophys. Acta* **449**, 434–446.

Poston, J. M. (1978). Coenzyme B_{12}-dependent enzymes in potato: leucine 2,3-aminomutase and methylmalonyl-CoA mutase. *Phytochemistry* **17**, 401–402.

Pourmohseni, H. and Ibenthal, W.-D. (1991). Novel β-cyanoglucosides in the epidermal tissue of barley and their possible role in the barley-powdery mildew interaction. *Angew. Bot.* **65**, 341–350.

Powlson, D. S., Poulton, P. R., Møller, N. E., Hewitt, M. V., Penny, A. and Jenkinson, D. S. (1989). Uptake of foliar-applied urea by winter wheat (*Triticum aestivum*). The influence of application time and the use of a new ^{15}N technique. *J. Sci. Food Agric.* **48**, 429–440.

Powrie, J. K. (1964). The effect of cobalt on the growth of young lucerne on a silicous sand. *Plant Soil* **21**, 81–93.

Prade, K. and Trolldenier, G. (1989). Further evidence concerning the importance of soil air-filled porosity, soil organic matter and plants for denitrification. *Z. Pflanzenernähr. Bodenk.* **152**, 391–393.

Prade, K. and Trolldenier, G. (1990a). Denitrification in the rhizosphere of rice and wheat seedlings as influenced by plant K status, air-filled porosity and substrate organic matter. *Soil Biol. Biochem.* **22**, 769–773.

Prade, K. and Trolldenier, G. (1990b). Denitrification in the rhizosphere of plants with inherently different aerenchyma formation: wheat (*Triticum aestivum*) and rice (*Oryza sativa*). *Biol. Fertil. Soils* **9**, 215–219.

Pradet, A. and Raymond, P. (1983). Adenine nucleotide ratios and adenylate energy charge in energy metabolism. *Annu. Rev. Plant Physiol.* **34**, 199–224.

Prask, J. A. and Plocke, D. J. (1971). A role of zinc in the structural integrity of the cytoplasmic ribosomes of *Euglena gracilis*. *Plant Physiol.* **48**, 150–155.

Prattley, C. A., Stanley, D. W., Smith, T. K. and van de Voort, F. R. (1982). Protein-phytate interaction in soybeans. III. The effect of protein-phytate complexes on zinc bioavailability. *J. Food Biochem.* **6**, 273–282.

Preston, C. and Critchley, C. (1986). Differential effects of K^+ and Na^+ on oxygen evolution activity of photosynthetic membranes from two halophytes and spinach. *Aust. J. Plant Physiol.* **13**, 491–498.

Preusser, E., Khalil, F. A. and Göring, H. (1981). Regulation of activity of the granule-bond starch synthetase by monovalent cations. *Biochem. Physiol. Pflanz.* **176**, 744–752.

Price, A. H. and Hendry, G. A. F. (1991). Iron-catalysed oxygen radical formation and its possible contribution to drought damage in nine native grasses and three cereals. *Plant Cell Environ.* **14**, 477–484.

Prins, W. H. (1983). Effect of a wide range of nitrogen applications in the herbage nitrate content in long-term fertilizer trials on all-grass swards. *Fert. Res.* **4**, 101–113.

Pritchard, J., Barlow, P. W., Adam, J. S. and Tomos, A. D. (1990). Biophysics of the inhibition of the growth of maize roots by lowered temperature. *Plant Physiol.* **93**, 222–230.

Pugliarello, M. C., Rasi-Caldogno, F., De Michelis, M. I. and Olivari, C. (1991). The tonoplast H^+-pyrophosphatase of radish seedlings: biochemical characteristics. *Physiol. Plant.* **83**, 339–345.

Pukacka, S. and Kuiper, P. J. C. (1988). Phospholipid composition and fatty acid peroxidation during ageing of *Acer platanoides* seeds. *Physiol. Plant.* **72**, 89–93.

Purves, D. and Mackenzie, E. J. (1974). Phytotoxicity due to boron in municipal compost. *Plant Soil* **40**, 231–235.

Pushnik, J. C. and Miller, G. W. (1989). Iron regulation of chloroplast photosynthetic function: mediation of PS I development. *J. Plant Nutr.* **12**, 407–421.

Pushnik, J. C., Miller, G. W. and Manwaring, J. H. (1984). The role of iron in higher plant chlorophyll biosynthesis, maintenance and chloroplast biogenesis. *J. Plant Nutr.* **7**, 733–758.

Puthota, V., Cruz-Ortega, R., Johnson, J. and Ownby, J. (1991). An ultrastructural study of the inhibition of mucilage reaction in the wheat root cap by aluminium. *In* 'Plant-Soil Interactions at Low pH' (R. J. Wright, V. C. Baligar, R. P. Murrmann, eds.), pp. 779–787. Kluwer Academic, Dordrecht, The Netherlands.

Qiu, J., Israel, D. W. (1992). Diurnal starch accumulation and utilization in phosphorus-deficient soybean plants. *Plant Physiol.* **98**, 316–323.

Quilez, R., Abadia, A. and Abadia, J. (1992). Characteristics of thylakoids and photosystem II membrane preparations from iron deficient and iron sufficient sugar beet (*Beta vulgaris* L.). *J. Plant Nutr.* **15**, 1809–1819.

Quispel, A. (1991). A critical evaluation of the prospects for nitrogen fixation in non-legumes. *Plant Soil* **137**, 1–11.

Qureshi, J. A., Thurman, D. A., Hardwick, K. and Collin, H. A. (1985). Uptake and accumulation of zinc, lead and copper in zinc and lead tolerant *Anthoxanthum odoratum* L. *New Phytol.* **100**, 429–434.

Raboy, V., Noaman, M. W., Taylor, G. A. and Pickett, S. G. (1991). Grain phytic acid and protein are highly correlated in winter wheat. *Crop Sci.* **31**, 631–635.

Racette, S., Louis, I. and Torrey, J. G. (1990). Cluster root formation by *Gymnostoma papuanum* (Casuarinaceae) in relation to aeration and mineral nutrient availability in water culture. *Can. J. Bot.* **68**, 2564–2570.

Rademacher, W. (1978). Gaschromatographische Analyse der Veränderungen im Hormongehalt des wachsenden Weizenkorns. PhD Thesis, Universität Göttingen.

Radermacher, E. and Klämbt, D. (1993). Auxin dependent growth and auxin binding proteins in primary roots and root hairs of corn (*Zea mays* L.). *J. Plant Physiol.* **141**, 698–703.

Radin, J. W. (1983). Control of plant growth by nitrogen: Differences between cereals and broadleaf species. *Plant Cell Environ.* **6**, 65–68.

Radin, J. W. (1984). Stomatal responses to water stress and to abscisic acid in phosphorus-deficient cotton plants. *Plant Physiol.* **76**, 392–394.

Radin, J. W. (1990). Responses of transpiration and hydraulic conductance to root temperature in nitrogen- and phosphorus-deficient cotton seedlings. *Plant Physiol.* **92**, 855–857.

Radin, J. W. and Ackerson, R. C. (1981). Water relations of cotton plants under nitrogen deficiency. III. Stomatal conductance. *Plant Physiol.* **67**, 115–119.

Radin, J. W. and Boyer, J. S. (1982). Control of leaf expansion by nitrogen nutrition in sunflower plants: role of hydraulic conductivity and turgor. *Plant Physiol.* **69**, 771–775.

Radin, J. W. and Eidenbock, M. P. (1984). Hydraulic conductance as a factor limiting leaf expansion of phosphorus-deficient cotton plants. *Plant Physiol.* **75**, 372–377.

Radin, J. W. and Hendrix, D. L. (1988). The apoplastic pool of abscisic acid in cotton leaves in relation to stomatal closure. *Planta* **174**, 180–186.

Radin, J. W. and Matthews, M. A. (1989). Water transport properties of cortical cells in roots of nitrogen- and phosphorus-deficient cotton seedlings. *Plant Physiol.* **89**, 264–268.

Radin, J. W. and Parker, L. L. (1979). Water relation of cotton plants under nitrogen deficiency. I. Dependence upon leaf structure. *Plant Physiol.* **64**, 495–498.

Radin, J. W., Parker, L. L. and Guinn, G. (1982). Water relations of cotton plants under nitrogen deficiency. V. Environmental control of abscisic acid accumulation and stomatal sensitivity to abscisic acid. *Plant Physiol.* **70**, 1066–1070.

Radley, M. (1978). Factors affecting grain enlargement in wheat. *J. Exp. Bot.* **29**, 919–934.

Raghavendra, A. A., Rao, J. M. and Das, V. S. R. (1976). Replaceability of potassium by sodium for stomatal opening in epidermal strips of *Commelina benghalensis*. *Z. Pflanzenphysiol.* **80**, 36–42.

Rahimi, A. (1970). Kupfermangel bei höheren Pflanzen. *Landwirtsch. Forsch. Sonderh.* **25**(I), 42–47.

Rahimi, A. and Bussler, W. (1973a). Die Diagnose des Kupfermangels mittels sichtbarer Symptome an höheren Pflanzen. *Landwirtsch. Forsch. Sonderh.* **25 (I)**, 42–47.

Rahimi, A. and Bussler, W. (1973b). Physiologische Voraussetzungen für die Bildung der Kupfermangelsymptome. *Z. Pflanzenernähr. Bodenk.* **136**, 25–32.

Rahimi, A. and Bussler, W. (1974). Kupfermangel bei höheren Pflanzen und sein histochemischer Nachweis. *Landwirtsch. Forsch. Sonderh.* **30 (II)**, 101–111.

Rahimi, A. and Schropp, A.(1984). Carboanhydraseaktivität und extrahierbares Zink als Maßstab für die Zink-Versorgung von Pflanzen. *Z. Pflanzenernähr. Bodenk.* **147**, 572–583.

Rahman, M. S. and Wilson, J. H. (1977). Effect of phosphorus applied as superphosphate on rate of development and spikelet number per ear in different cultivars of wheat. *Aust. J. Agric. Res.* **28**, 183–186.

Raij, B. van and van Diest, A. (1979). Utilization of phosphate from different sources by six plant species. *Plant Soil* **51**, 577–589.

Rainbird, R. M., Atkins, C. A. and Pate, J. S. (1983). Diurnal variation in the functioning of cowpea nodules. *Plant Physiol.* **72**, 308–312.

Rains, D. W. (1968). Kinetics and energetics of light-enhanced potassium absorption by corn leaf tissue. *Plant Physiol.* **43**, 394–400.

Rains, D. W. (1969). Cation absorption by slices of stem tissues of bean and cotton. *Experientia* **25**, 215–216.

Rajaratnam, J. A. and Hock, L. I. (1975). Effect of boron nutrition on intensity of red spider mite attack on oil- palm seedlings. *Exp. Agric.* **11**, 59–63.

Rajaratnam, J. A. and Lowry, J. B. (1974). The role of boron in the oil-palm (*Elaeis guineensis*). *Ann. Bot. (London)* [*N.S.*] **38**, 193–200.

Rajaratnam, J. A., Lowry, J. B., Avadhani, P. N. and Corley, P. H. V. (1971). Boron: possible role in plant metabolism. *Science* **172**, 1142–1143.

Raju, P. S., Clark, R. B., Ellis, J. R. and Maranville, J. W. (1990). Effects of species of VA-mycorrhizal fungi on growth and mineral uptake of sorghum at different temperatures. *Plant Soil* **121**, 165–170.

Ralph, W. (1986). Managing manganese deficiency (Report). *Rural Res. CSIRO* **130**, 18–22.

Ramón, A. M., Carpena-Ruiz, R. O. and Gárate, A. (1990). The effects of short-term deficiency of boron on potassium, calcium, and magnesium distribution in leaves and roots of tomato (*Lycopersicon esculentum*) plants. *In* 'Plant Nutrition – Physiology and Application (M. L. van Beusichem, ed.), pp. 287–290. Kluwer Academic, Dordrecht.

Randall, H. C. and Sinclair, T. R. (1988). Sensitivity of soybean leaf development to water deficits. *Plant Cell Environ.* **11**, 835–839.

Randall, P. J. (1969). Changes in nitrate and nitrate reductase levels on restoration of molybdenum to molybdenum-deficient plants. *Aust. J. Agric. Res.* **20**, 635–642.

Randall, P. J. and Wrigley, C. W. (1986). Effects of sulfur supply on the yield, composition, and quality of grain from cereals, oilseeds and legumes. *Adv. Cereal Sci. Technol.* **8**, 171–206.

Randall, P. J., Delhaize, E., Richards, R. A. and Munns, R. (eds.) (1993). 'Genetic Aspects of Plant Mineral Nutrition'. Kluwer Academic, Dordrecht.

Ranjeva, R. and Boudet, A. M. (1987). Phosphorylation of proteins in plants: Regulatory effects and potential involvement in stimulus/response coupling. *Annu. Rev. Plant Physiol.* **38**, 73–93.

Rao, I. M. and Terry, N. (1989). Leaf phosphate status, photosynthesis, and carbon partitioning in sugar beet. I. Changes in growth, gas exchange, and Calvin cycle enzymes. *Plant Physiol.* **90**, 814–819.

Rao, I. M., Fredeen, A. L. and Terry, N. (1990). Leaf phosphate status, photosynthesis, and carbon partitioning in sugar beet. III. Diurnal changes in carbon partitioning and carbon export. *Plant Physiol.* **92**, 29–36.

Rao, I. M., Sharp, R. E. and Boyer, J. S. (1987). Leaf magnesium alters photosynthetic response to low water potentials in sunflower. *Plant Physiol.* **84**, 1214–1219.

Raper jr, C. D., Vessey, J. K. and Henry, L. T.(1991). Increase in nitrate uptake by soybean plants during interruption of the dark period with low light intensity. *Physiol. Plant.* **81**, 183–189.

Raschke, K., Hedrich, R., Beckmann, U. and Schroeder, J. L. (1988). Exploring biophysical and biochemical components of the osmotic motor that drives stomatal movement. *Bot. Acta* **101**, 283–294.

Rashid, A. and Fox, R. L.(1992). Evaluating internal zinc requirements of grain crops by seed analysis. *Agron. J.* **84**, 469–474.

Raskin, J. and Kende, H. (1983). How does deep water rice solve its aeration problem? *Plant Physiol.* **72**, 447–454.

Rasmussen, H. P. (1968). Entry and distribution of aluminium in *Zea mays*. The mode of entry and distribution of aluminium in *Zea mays*: electron microprobe-X-ray analysis. *Planta* **81**, 28–37.

Rasmussen, J. B., Hammerschmidt, R. and Zook, M. N. (1991). Systemic induction of salicylic acid accumulation in cucumber after inoculation with *Pseudomonas syringae* pv. *syringae*. *Plant Physiol.* **97**, 1342–1347.

Rasmussen, P. E., Ramig, R. E., Ekin, L. G. and Rhode, C. R. (1977). Tissue analyses guidelines for diagnosing sulfur deficiency in white wheat. *Plant Soil* **46**, 153–163.

Ratnayake, M., Leonard, R. T. and Menge, A. (1978). Root exudation in relation to supply of phosphorus and its possible relevance to mycorrhizal infection. *New Phytol.* **81**, 543–552.

Rauser, W. E. (1990). Phytochelatins. *Annu. Rev. Biochem.* **59**, 61–86.

Raven, J. A. (1983). The transport and function of silicon in plants. *Biol. Rev. Camb. Philos. Soc.* **58**, 179–207.

Raven, J. A. (1985). Regulation of pH and generation of osmolarity in vascular land plants: costs and benefits in relation to efficiency of use of water, energy and nitrogen. *New Phytol.* **101**, 25–77.

Raven, J. A. (1986). Biochemical disposal of excess H^+ in growing plants? *New Phytol.* **104**, 175–206.

Raven, J. A. and Smith, F. A. (1976). Nitrogen assimilation and transport in vascular land plants in relation to intracellular pH regulation. *New Phytol.* **76**, 415–431.

Raven, J. A., Rothemund, C. and Wollenweber, B. (1991). Acid–base regulation by *Azolla* spp. with N_2 as sole N source and with supplementation by NH_4^+ or NO_3^-. *Acta Bot.* **104**, 132–138.

Ray, T. B. and Black, C. C. (1979). The C4 pathway and its regulation. *In* 'Photosynthesis II' Encycl. Plant Physiol. New Series Vol. 6 (M. Gibbs and E. Latzko, eds.), pp. 77–101. Springer-Verlag, Berlin.

Ray, T. C., Callow, J. A. and Kennedy, J. F. (1988). Composition of root mucilage polysaccharides from *Lepidium sativum*. *J. Exp. Bot.* **39**, 1249–1261.

Rea, P. A. and Sanders, D. (1987). Tonoplast energization: two H^+ pumps, one membrane. *Physiol. Plant.* **71**, 131–141.

Read, D. J. (1991). Mycorrhizas in ecosytems. *Experientia* **47**, 376–391.

Rebafka, F.-P. (1993). Deficiency of phosphorus and molybdenum as major growth limiting factors of pearl millet and groundnut on an acid sandy soil in Niger, West Africa. PhD Thesis University Hohenheim.

Rebafka, F.-P., Ndunguru, B. J. and Marschner, H. (1993). Single superphosphate depresses molybdenum uptake and limits yield response to phosphorus in groundnut (*Arachis hypogaea* L.) grown on an acid sandy soil in Niger, West Africa. *Fert. Res.* **34**, 233–242.

Rebafka, F.-P., Schulz, R. and Marschner, H. (1990). Erhebungsuntersuchungen zur Pflanzenverfügbarkeit von Nickel auf Böden mit hohen geogenen Nickelgehalten. *Angew. Bot.* **64**, 317–328.

Reckmann, U., Scheibe, R. and Raschke, K. (1990). Rubisco activity in guard cells compared with the solute requirement for stomatal opening. *Plant Physiol.* **92**, 246–253.

Reddy, D. T. and Raj, A. S. (1975). Cobalt nutrition of groundnut in relation to growth and yield. *Plant Soil* **42**, 145–152.

Reddy, V. R., Baker, D. N. and Hodges, H. F. (1991). Temperature effects on cotton canopy growth, photosynthesis, and respiration. *Agron. J.* **83**, 699–704.

Redinbaugh, M. G. and Campbell, W. H. (1991). Higher plant responses to environmental nitrate. *Physiol.Plant.* **82**, 640–650.

Reeves, M. (1992). The role of VAM fungi in nitrogen dynamics in maize–bean intercrops. *Plant Soil* **144**, 85–92.

Reggiani, R., Aurisano, N., Mattana, M. and Bertani, A. (1993). Influence of K^+ ions on polyamine level in wheat seedlings. *J. Plant Physiol.* **141**, 136–140.

Reid, D. M. and Railton, I. D. (1974). Effect of flooding on the growth of tomato plants: involvement of cytokinins and gibberellins. *In* 'Mechanisms of Regulation of Plant Growth' (R. L. Bieleski *et al.*, eds.), Cresswell Bull. No. 12, pp. 789–792. Royal Society of New Zealand, Wellington.

Reid, D. M., Crozier, A. and Harvey, B. M. R. (1969). The effects of flooding on the export of gibberellins from the root to the shoot. *Planta* **89**, 376–379.

Reid, R. K., Reid, C. P. P., Powell, P. E. and Szaniszlo, P. J. (1984). Comparison of siderophore concentrations in aqueous extracts of rhizosphere and adjacent bulk soils. *Pedobiologia* **26**, 263–266.

Reinhard, S., Martin, P. and Marschner, H. (1992). Interactions in the tripartite symbiosis of pea (*Pisum sativum* L.), *Glomus* and *Rhizobium* under non-limiting phosphorus supply. *J. Plant Physiol.* **141**, 7–11.

Reinhold, B., Hurek, T. and Fendrik, I. (1988). Plant–bacteria interactions with special emphasis on the kallar grass association. *Plant Soil* **110**, 249–257.

Reinhold, J. G., Nasr, K., Lahimgarzadeh, A. and Hedayati, H. (1973). Effects of purified phytate and phytate-rich bread upon metabolism of zinc, calcium, phosphorus and nitrogen in man. *Lancet* **1**, 283–291.

Rengel, Z. (1990). Net Mg^{2+} uptake in relation to the amount of exchangeable Mg^{2+} in the Donnan free space of ryegrass roots. *Plant Soil* **128**, 185–189.

Rengel, Z. (1992a). Role of calcium in aluminium toxicity. *New Phytol.* **121**, 499–514.

Rengel, Z. (1992b). The role of calcium in salt toxicity. *Plant Cell Environ.* **15**, 625–632.

Rengel, Z. and Elliott, D. C. (1992). Mechanism of aluminum inhibition of net $^{45}Ca^{2+}$ uptake by *Amaranthus* protoplasts. *Plant Physiol.* **98**, 632–638.

Rengel, Z. and Robinson, D. L. (1989a). Competitive Al^{3+} inhibition of net Mg^{2+} uptake by intact *Lolium multiflorum* roots. I. Kinetics. *Plant Physiol.* **91**, 1407–1413.

Rengel, Z. and Robinson, D. L. (1989b). Aluminum effects on growth and macronutrient uptake by annual ryegrass. *Agron. J.* **81**, 208–215.

Renger, G. and Wydrzynski, T. (1991). The role of manganese in photosynthetic water oxidation *Biol. Metals* **4**, 73–80.

Rennenberg, H. (1989). Synthesis and emission of hydrogensulfide by higher plants *In* 'Biogenic Sulfur in the Environment' (E. S. Saltzman and W. J. Cooper, eds.), pp. 44–57. ACS Symp. Series 3296, Washington, DC.

Rennenberg, H. and Lamoureux, G. L. (1990). Physiological processes that modulate the concentration of glutathione in plant cells. *In* 'Sulfur Nutrition and Sulfur Assimilation in Higher Plants' (H. Rennenberg *et al.* eds.), pp. 53–65. XPB Academic Publishers bv, The Hague, The Netherlands.

Rennenberg, H., Schmitz, K. and Bergmann, L. (1979). Long-distance transport of sulfur in *Nicotiana tabacum*. *Planta* **147**, 57–62.

Rennenberg, H., Kemper, O. and Thoene, B. (1989). Recovery of sulfate transport into heterotrophic tobacco cells from inhibition by reduced glutathione. *Physiol. Plant.* **76**, 271–276.

Rennenberg, H., Huber, B., Schröder, P., Stahl, K., Haunold, W., Georgii, H.-W., Slovik, S. and Pfanz, H. (1990). Emission of volatile sulfur compounds from spruce trees. *Plant Physiol.* **92**, 560–564.

Rensing, L. and Cornelius, G. (1980). Biologische Membranen als Komponenten oszillierender Systeme. *Biol. Rundsch.* **18**, 197–209.

Rerkasem, B., Netsangtip, R., Bell, R. W., Loneragan, J. F. and Hiranburana, N. (1988). Comparative species responses to boron on a typic Tropaqualf in Northern Thailand. *Plant Soil* **106**, 15–21.

Reuter, D. J. and Robinson, J. B. (1986). Plant Analysis: An Interpretation Manual. Inkata Press Ltd., Melbourne, Australia.

Reuter, D. J., Alston, A. M. and McFarlane, J. D. (1988). Occurrence and correction of manganese deficiency in plants. *In* 'Manganese in Soils and Plants' (R. D. Graham, R. J. Hannan and N. C. Uren, eds.), pp. 205–224. Kluwer Academic, Dordrecht.

Reuter, D. J., Robson, A.D., Loneragan, J. F. and Tranthim-Fryer, D. J. (1981). Copper nutrition of subterranean clover (*Trifolium subterraneum* L. cv. Seaton Park). II. Effects of copper supply on distribution of copper and the diagnosis of copper deficiency by plant analysis. *Aust. J. Agric. Res.* **32**, 267–282.

Reynolds, S. B., Scaife, A. and Turner, M. K. (1987). Effect of nitrogen form on boron uptake by cauliflower. *Commun. Soil Sci. Plant Anal.* **18**, 1143–1154.

Rheinbaben, W. von and Trolldenier, G. (1984). Influence of plant growth on denitrification in relation to soil moisture and potassium nutrition. *Z. Pflanzenernähr. Bodenk.* **147**, 730–738.

Rhue, R. D. and Grogan, C. O. (1977). Screening corn for Al tolerance using different Ca and Mg concentrations. *Agron. J.* **69**, 755–760.

Ribaut, J.-M. and Pilet, P.-E. (1991). Effect of water stress on growth, osmotic potential and abscisic acid content of maize roots. *Physiol. Plant.* **81**, 156–162.

Rice, C. F., Lukaszewski, K. M., Walker, S., Blevins, D. G., Winkler, R. G. and Randall, D. D. (1990). Changes in ureide synthesis, transport and assimilation following ammonium nitrate fertilization of nodulated soybeans. *J. Plant Nutr.* **13**, 1539–1553.

Richards, B. N. and Bevege, D. I. (1969). Critical foliage concentrations of nitrogen and phosphorus as a guide to the nutrient status of *Araucaria* underplanted to *Pinus*. *Plant Soil* **31**, 328–336.

Richards, D. (1981). Root–shoot interactions in fruiting tomato plants. *In* 'Structure and Function of Plant Roots' (P. Brouwer, O. Gasparikova, J. Kolek and B. C. Loughman, eds.), pp. 373–380. Martinus Nijhoff/Junk, The Hague.

Richards, R. A. (1992). Increasing salinity tolerance of grain crop: is it worthwhile? *Plant Soil* **146**, 89–98.

Richards, R. L. (1991). The chemistry of dinitrogen reduction. *In* 'Biology and Biochemistry of Nitrogen Fixation' (M. J. Dilworth and A. R. Glenn, eds.), pp. 58–75. Elsevier, Amsterdam.

Richardson, A. E., Djordjevic, M. A., Rolfe, B. G. and Simpson, R. J. (1988). Effect of pH, Ca and Al on the exudation from clover seedlings of compounds that induce the expression of nodulation genes in *Rhizobium trifolii*. *Plant Soil* **109**, 37–47.

Richardson, M. D. and Croughan, S. S. (1989). Potassium influence on susceptibility of bermudagrass to *Helminthosporium cynodontis* toxin. *Crop Sci.* **29**, 1280–1282.

Richter, A. and Popp, M. (1992). The physiological importance of accumulation of cyclitols in *Viscum album* L. *New Phytol.* **121**, 431–438.

Riens, B. and Heldt, H. W. (1992). Decrease of nitrate reductase activity in spinach leaves during a light–dark transition. *Plant Physiol.* **89**, 573–577.

Rigney, C. J. and Wills, R. B. H. (1981). Calcium movement, a regulating factor in the initiation of tomato fruit ripening. *HortScience* **16**, 550–551.

Rijven, A. H. G. C. and Gifford, R. M. (1983). Accumulation and conversion of sugars by developing wheat grains. IV. Effects of phosphate and potassium ions in endosperm slices. *Plant Cell Environ.* **6**, 625–631.

Riley, I. T. and Dilworth, M. J. (1985a). Cobalt requirement for nodule development and function in *Lupinus angustifolius* L. *New Phytol.* **100**, 347–359.

Riley, I. T. and Dilworth, M. J. (1985b). Recovery of cobalt-deficient root nodules in *Lupinus angustifolius* L. *New Phytol.* **100**, 361–365.

Riley, I. T. and Dilworth, M. J. (1985c). Cobalt status and its effects on soil populations of *Rhizobium lupini*, rhizosphere colonization and nodule initiation. *Soil Biol. Biochem.* **17**, 81–85.

Riley, M. M., Adcock, K. G. and Bolland, M. D. A. (1993). A small increase in the concentration of phosphorus in the sawn seed increased the early growth of wheat. *J. Plant Nutr.* **16**, 851–864.

Rincon, M. and Hanson, J. B. (1986). Controls on calcium ion fluxes in injured or shocked corn root cells: importance of proton pumping and cell membrane potential. *Physiol. Plant.* **67**, 576–583.

Ritchey, K. D., Souza, D. M. G., Lobato, E. and Correa, O. (1980). Calcium leaching to increase rooting depth in a Brasilian savanah Oxisol. *Agron. J.* **72**, 40–44.

Rivera, C. M. and Penner, D. (1978). Effect of calcium and nitrogen on soybean (*Glycine max*) root fatty acid composition and uptake of linuron. *Weed Sci.* **26**, 647–650.

Robbins, N. S. and Pharr, D. M. (1988). Effect of restricted root growth on carbohydrate metabolism and whole plant growth of *Cucumis sativus* L. *Plant Physiol.* **87**, 409–413.

Roberts, D. M. and Harmon, A. C. (1992). Calcium-modulated proteins: targets of intracellular calcium signals in higher plants. *Annu. Rev. Plant Physiol. Plant Mol. Biol.* **43**, 375–414.

Roberts, J. K. M. and Pang, M. K. L. (1992). Estimation of ammonium ion distribution between cytoplasm and vacuole using nuclear magnetic resonance spectroscopy. *Plant Physiol.* **100**, 1571–1574.

Roberts, J. K. M., Linker, C. S., Benoit, A. G., Jardetzky, O. and Nieman, R. H. (1984). Salt stimulation of phosphate uptake in maize root tips studied by ^{31}P nuclear magnetic resonance. *Plant Physiol.* **75**, 947–950.

Roberts, T. M., Skeffington, R. A. and Blank, L. W.(1989). Causes of type I spruce decline in Europe. *Forestry* **62**, 179–222.

Robertson, G. A. and Loughman, B. C. (1974). Response to boron deficiency: A comparison with responses produced by chemical methods of retarding root elongation. *New Phytol.* **73**, 821–832.

Robertson, K. P. and Wainwright, S. J. (1987). Photosynthetic responses to salinity in two clones of *Agrostis stolonifera*. *Plant Cell Environ.* **10**, 45–52.

Robinson, D. and Rorison, I. H. (1987). Root hairs and plant growth at low nitrogen availabilities. *New Phytol.* **107**, 681–693.

Robinson, N. J. (1990). Metal binding polypeptides in plants. *In* 'Heavy Metal Tolerance in Plants: Evolutionary Aspects' (A. J. Shaw, ed.), pp. 195–214. CRC Press Inc., Boca Raton, FL.

Robinson, P. W. and Hodges, C. F. (1981). Nitrogen-induced changes in the sugars and amino acids of sequentially senescing leaves of *Poa pratensis* and pathogenesis by *Drechslera sorokiniana*. *Phytopathol. Z.* **101**, 348–361.

Robinson, S. P. and Downton, W. J. S. (1984). Potassium, sodium, and chloride content of isolated intact chloroplasts in relation to ionic compartmentation in leaves. *Arch. Biochem. Biophys.* **228**, 197–206.

Robinson, S. P. and Downton, W. J. S. (1985). Potassium, sodium, and chloride ion concentration in leaves and isolated chloroplasts of the halophyte *Sueda australis* R. Br. *Aust. J. Plant Physiol.* **12**, 471–479.

Robinson, S. P. and Giersch, C. (1987). Inorganic phosphate concentration in the stroma of isolated chloroplasts and its influence on photosynthesis. *Aust. J. Plant Physiol.* **14**, 451–462.

Robinson, S. P. and Jones, G. P. (1986). Accumulation of glycinebetaine in chloroplasts provide osmotic adjustment during salt stress. *Aust. J. Plant Physiol.* **13**, 659–668.

Robinson-Beers, K., Sharkey, Th. D. and Evert, R. F. (1990). Import of ^{14}C-photosynthate by developing leaves of sugarcane. *Bot. Acta* **103**, 424–429.

Roblin, G., Fleurat-Lessard, P. and Bonmort, J. (1989). Effects of compounds affecting calcium channels on phytochrome- and blue pigment-mediated pulvinar movements of *Cassia fasciculata*. *Plant Physiol.* **90**, 697–701.

Robson, A. D. and Bottomley, P. J. (1991). Limitations in the use of legumes in agriculture and forestry. *In* 'Biology and Biochemistry of Nitrogen Fixation' (M. J. Dilworth and A. R. Glenn, eds.), pp. 320–349. Elsevier, Amsterdam.

Robson, A. D. and Mead, G. R. (1980). Seed cobalt in *Lupinus angustifolius*. *Aust. J. Agric. Res.* **31**, 109–116.

Robson, A. D. and Pitman, M. G. (1983). Interactions between nutrients in higher plants. *In* 'Encyclopedia of Plant Physiology, New Series' (A. Läuchli and R. L. Bieleski, eds.), Vol. 15A, pp. 147–180. Springer-Verlag, Berlin and New York.

Robson, A. D. and Reuter, D. J. (1981). Diagnosis of copper deficiency and toxicity. *In* 'Copper in Soils and Plants' (J. F. Loneragan, A. D. Robson and R. D. Graham, eds.), pp. 287–312. Academic Press, London.

Robson, A. D. and Snowball, K. (1987). Response of narrow-leafed lupins to cobalt application in relation to cobalt concentration in seed. *Aust. J. Exp. Agric.* **27**, 657–660.

Robson, A. D., Dilworth, M. J. and Chatel, D. L. (1979). Cobalt and nitrogen fixation in *Lupinus angustifolius* L. I. Growth nitrogen concentrations and cobalt distribution. *New Phytol.* **83**, 53–62.

Robson, A. D., Hartley, R. D. and Jarvis, S. C. (1981). Effect of copper deficiency on phenolic and other constituents of wheat cell walls. *New Phytol.* **89**, 361–373.

Rodriguez-Barrueco, C., Cervantes, E. and Rodriguez-Caceres, E. (1991). Growth promoting effect of *Azospirillum brasilense* on *Casuarina cunninghamiana* Miq. seedlings. *Plant Soil* **135**, 121–124.

Roeb, G. W., Wieneke, J. and Führ, F. (1982). Auswirkungen hoher NaCl-Konzentrationen im Nährmedium auf die Transpiration, den Abscisinsäure-, Cytokinin- und Prolingehalt zweier Sojabohnensorten. *Z. Pflanzenernär. Bodenk.* **145**, 103–116.

Roger, P. A. and Ladha, J. K. (1992). Biological N_2 fixation in wetland rice fields: Estimation and contribution to nitrogen balance. *Plant Soil* **141**, 41–55.

Rognes, S. E. (1980). Anion regulation of lupin asparagine synthetase: Chloride activation of the glutamine-utilizing reaction. *Phytochemistry* **19**, 2287–2293.

Rohozinski, J., Edwards, G. R. and Hoskyns, P. (1986). Effects of brief exposure to nitrogenous compounds on floral initiation in apple trees. *Physiol. Veg.* **24**, 673–677.

Roldán, M., Belver, A., Rodriguez-Rosales, P., Ferrol, N. and Donaire, J. P. (1992).

In vivo and *in vitro* effects of boron on the plasma membrane proton pump of sunflower roots. *Physiol. Plant.* **84**, 49–54.

Rolfe, B. G. and Gresshoff, P. M. (1988). Genetic analysis of legume nodule initiation. *Ann. Rev. Plant Physiol. Plant Mol. Biol.* **39**, 297–319.

Rollwagen, B. A. and Zasoski, R. J. (1988). Nitrogen source effects on rhizosphere pH and nutrient accumulation by Pacific Northwest conifers. *Plant Soil* **105**, 79–86.

Römer, W. (1971). Untersuchungen über die Auslastung des Photosyntheseapparates bei Gerste (*Hordeum distichon* L.) und weissem Senf (*Sinapis alba* L.) in Abhängigkeit von den Umweltbedingungen. *Arch. Bodenfruchtbarkeit Pflanzenprod.* **15**, 414–423.

Römer, W. and Schilling, G. (1986). Phosphorus requirements of the wheat plant in various stages of its life cycle. *Plant Soil* **91**, 221–229.

Römer, W., Augustin, J. and Schilling, G. (1988). The relationship between phosphate absorption and root length in nine wheat cultivars. *Plant Soil* **111**, 199–201.

Romera, F. J., Alcantara, E. and de la Guardia, M. D. (1991). Characterization of the tolerance to iron chlorosis in different peach rootstocks grown in nutrient solution. *Plant Soil* **130**, 121–125.

Romera, F. J., Alcantara, E. and de la Guardia, M. D. (1992). Role of roots and shoots in the regulation of the Fe efficiency responses in sunflower and cucumber. *Physiol. Plant.* **85**, 141–146.

Römheld, V. (1985). Schlechtwetterchlorose der Rebe: Einfluss von Bikarbonat und niedrigen Bodentemperaturen auf die Aufnahme und Verlagerung von Eisen und das Auftreten von Chlorose. VDLUFA-Schriftenreihe **16**, Kongressband. S. 211–217.

Römheld, V. (1987a). Different strategies for iron acquisition in higher plants. *Physiol. Plant.* **70**, 231–234.

Römheld, V. (1987b). Existence of two different strategies for the acquisition of iron in higher plants. *In* 'Iron Transport in Microbes, Plant and Animals' (G. Winkelmann, D. van der Helm and J. B. Neilands, eds.), pp. 353–374, VCH Verlag Weinheim, FRG.

Römheld, V. (1991). The role of phytosiderophores in acquisition of iron and other micronutrients in graminaceous species: An ecological approach. *Plant Soil* **130**, 127–134.

Römheld, V. and Kramer, D. (1983). Relationship between proton efflux and rhizodermal transfer cells induced by iron deficiency. *Z. Pflanzenphysiol.* **113**, 73–83.

Römheld, V. and Marschner, H. (1981a). Rhythmic iron stress reactions in sunflower at suboptimal iron supply. *Physiol. Plant.* **53**, 347–353.

Römheld, V. and Marschner, H. (1981b). Iron deficiency stress induced morphological and physiological changes in root tips of sunflower. *Physiol. Plant.* **53**, 354–360.

Römheld, V. and Marschner, H. (1986). Mobilization of iron in the rhizosphere of different plant species. *In* 'Advances in Plant Nutrition' Vol. 2 (B. Tinker and A. Läuchli, eds.), pp. 155–204. Praeger Scientific, New York.

Römheld, V. and Marschner, H. (1990). Genotypical differences among graminaceous species in release of phytosiderophores and uptake of iron phytosiderophores. *Plant Soil* **123**, 147–153.

Römheld, V. and Marschner, H. (1991). Functions of micronutrients in plants. *In* 'Micronutrients in Agriculture' 2nd edn (J. J. Mordvedt, F. R. Cox, L. M. Shuman and R. M. Welch, eds.), pp. 297–328. SSSA Book Series, No., 4, Madison, WI, USA.

Römheld, V., Müller, Ch. and Marschner, H. (1984). Localization and capacity of proton pumps in roots of intact sunflower plants. *Plant Physiol.* **76**, 603–606.

Ron Vaz, M. D., Edwards, A. C., Shand, C. A. and Cresser, M. S. (1993). Phosphorus fractions in soil solution: influence of soil acidity and fertilizer addition. *Plant Soil* **148**, 175–183.

Roper, M. M., Marschke, G. W. and Smith, N. A. (1989). Nitrogenase activity (C_2H_2 reduction) in soils following wheat straw retention: Effects of straw management. *Aust. J. Agric. Res.* **40**, 241–253.

Rorison, I. H. (1987). Mineral nutrition in time and space. *In* 'Frontiers of comparative plant ecology' (I. H. Rorison, J. P. Grime, R. Hunt, G. A. F. Hendry and D. H. Lewis, eds.). *New Phytol. (Suppl.)* **106**, 79–92.

Rosendahl, L., Glenn, A. R. and Dilworth, M. J. (1991). Organic and inorganic inputs into legume root nodule nitrogen fixation. *In* 'Biology and Biochemistry of Nitrogen Fixation' (M. J. Dilworth and A. R. Glenn, eds.), pp. 259–291. Elsevier, Amsterdam.

Rosenfeld, I. and Beath, O. A. (1964). Seleniuim: Geobotany, biochemistry, toxicity, and nutrition. Academic Press, New York.

Rosenfield, C.-L., Reed, D. W. and Kent, M. W. (1991). Dependency of iron reduction on development of a unique root morphology in *Ficus benjamina* L. *Plant Physiol.* **95**, 1120–1124.

Ross, G. S., Minchin, P. E. H., and McWha, J. A. (1987). Direct evidence of abscisic acid affecting phloem unloading within seed coat of peas. *J. Plant Physiol.* **129**, 435–441.

Rossiter, R. C. (1978). Phosphorus deficiency and flowering in subterranean clover (*Tr. subterraneum* L.). *Ann. Bot. (London)* [*N.S.*] **42**, 325–329.

Roth-Bejerano, N. and Itai, C. (1981). Effect of boron on stomatal opening of epidermal strips of *Commelina communis*. *Physiol. Plant.* **52**, 302–304.

Rovira, A. D., Bowen, G. D. and Foster, R. C. (1983). The significance of rhizosphere microflora and mycorrhizas in plant nutrition. *In* 'Encyclopedia of Plant Physiology, New Series' (A. Läuchli and R. L. Bieleski, eds.), Vol. 15A, pp. 61–89. Springer-Verlag, Berlin.

Rozema, J., De Bruin, J. and Broekman, R. A. (1992). Effect of boron on the growth and mineral ecomony of some halophytes and nonhalophytes. *New Phytol.* **121**, 249–256.

Ruano, A., Barceló, J. and Poschenrieder, Ch. (1987). Zinc toxicity-induced variation of mineral element composition in hydroponically grown bush bean plants. *J. Plant Nutr.* **10**, 373–384.

Ruano, A., Poschenrieder, Ch. and Barcelo, J. (1988). Growth and biomass partitioning in zinc-toxic bush beans. *J. Plant Nutr.* **11**, 577–588.

Rufty, T. W. jr, Jackson, W. A. and Raper, C. D. jr. (1981). Nitrate reduction in roots as affected by presence of potassium and by flux of nitrate through the roots. *Plant Physiol.* **68**, 605–609.

Rufty, T. W. jr, Miner, W. S. and Raper, C. D. jr (1979). Temperature effects on growth and manganese tolerance in tobacco. *Agron. J.* **71**, 638–644.

Rufty, T. W. jr, Jackson, W. A. and Raper, C. D. jr. (1982a). inhibition of nitrate assimilation in roots in the presence of ammonium: the moderating influence of potassium. *J. Exp. Bot.* **33**, 1122–1137.

Rufty, T. W. jr, Raper, C. D. and Jackson, W. A. (1982b). Nitrate uptake, root and shoot growth, and ion balance of soybean plants during acclimation to root-zone acidity. *Bot. Gaz. (Chicago)* **143**, 5–14.

Rufty, T. W. jr, Volk, R. J., McClure, R. R., Israel, D. W. and Raper, C. D. jr (1982c). Relative content of NO_3^- and reduced N in xylem exudate as an indicator of root reduction of concurrently absorbed $^{15}NO_3$. *Plant Physiol.* **69**, 166–170.

Ruinen, J. (1975). Nitrogen fixation in the phyllosphere. *In* 'Nitrogen Fixation by Free-living Micro-organisms' (W. D. P. Stewart, ed.), pp. 85–100. Cambridge University Press, Cambridge.

Rupp, L. A. and Mudge, K. W. (1985). Ethephon and auxin induce mycorrhiza-like changes in the morphology of root organ cultures of Mugo pine. *Physiol. Plant.* **64**, 316–322.

Ruppel, S., Hecht-Buchholz, C., Remus, R., Ortmann, U. and Schmelzer, R. (1992). Settlement of the diazotrophic, phytoeffective bacterial strain *Pantoea agglomerans* on and within winter wheat. An investigation using ELISA and transmission electron microscopy. *Plant Soil* **145**, 261–273.

Rutherford, A. W. (1989). Photosystem II, the water-splitting enzyme. *Trends Biochem. Sci.* **14**, 227–232.

Rutland, R. B. (1971). Radioisotopic evidence of immobilization of iron in *Azalea* by excess calcium carbonate. *J. Am. Soc. Hortic. Sci.* **96**, 653–655.

Rutland, R. B. and Bukovac, M. J. (1971). The effect of calcium bicarbonate on iron absorption and distribution by *Chrysanthemum morifolium* (Ram.). *Plant Soil* **35**, 225–236.

Ryan, P. R., Shaff, J. E. and Kochian, L. V. (1992). Aluminum toxicity in roots. Correlation among ionic currents, ion fluxes, and root elongation in aluminum-sensitive and aluminum-tolerant wheat cultivars. *Plant Physiol.* **99**, 1193–1200.

Ryan, P. R., Ditomaso, J. M. and Kochian, L. V. (1993). Aluminium toxicity in roots: an investigation of spatial sensitivity and the role of the root cap. *J. Exp. Bot.* **44**, 437–446.

Rychter, A. M. and Mikulska, M. (1990). The relationship between phosphate status and cyanide-resistant respiration in bean roots. *Physiol. Plant.* **79**, 663–667.

Ryle, G. J. A., Powell, C. E. and Gordon, A. J. (1979). The respiratory costs of nitrogen fixations in soybean, cowpea, and white clover. II. Comparisons of the costs of nitrogen fixation and the utilization of combined nitrogen. *J. Exp. Bot.* **30**, 145–153.

Rynders, L. and Vlassak, K. (1982). Use of *Azospirillum brasilense* as biofertilizer in intensive wheat cropping. *Plant Soil* **66**, 217–223.

Saab, I. N. and Sharp, R. E. (1989). Non-hydraulic signals from maize roots in drying soil: inhibition of leaf elongation but not stomatal conductance. *Planta* **179**, 466–474.

Saalbach, E. and Aigner, H. (1970). Über die Wirkung einer Natriumdüngung auf Natriumgehalt, Ertrag und Trockensubstanzgehalt einiger Gras- und Kleearten. *Landwirtsch. Forsch.* **23**, 264–274.

Saftner, R. A. and Wyse, R. E. (1980). Alkali cation/sucrose co-transport in the root sink of sugar beet. *Plant Physiol.* **66**, 884–889.

Saftner, R. A., Daie, J. and Wyse, R. E. (1983). Sucrose uptake and compartmentation in sugar beet transport tissue. *Plant Physiol.* **72**, 1–6.

Sage, R. F., Pearcy, R. W. and Seemann, J. R. (1987). The nitrogenase efficiency of C_3 and C_4 plants. III. Leaf nitrogen effects on the activity of carboxylating enzymes in *Chenopodium album* (L.) and *Amaranthus retroflexus* (L). *Plant Physiol.* **85**, 355–359.

Saglio, P. H., Rancillac, M., Bruzan, F. and Pradet, A. (1984). Critical oxygen pressure for growth and respiration of excised and intact roots. *Plant Physiol.* **76**, 151–154.

Saini, H. S. and Aspinall, D. (1982). Sterility in wheat (*Triticum aestivum* L.) induced by water deficit or high temperature: Possible mediation by abscisic acid. *Aust. J. Plant Physiol.* **9**, 529–537.

Sajwan, K. S. and Lindsay, W. L. (1988). Effect of redox, zinc fertilization and incubation time on DTPA-extractable zinc, iron and manganese. *Commun. Soil Sci. Plant Anal.* **19**, 1–11.

Salama, A. M. S. El-D. A. and Wareing, P. F (1979). Effects of mineral nutrition on endogenous cytokinins in plants of sunflower (*Helianthus annuus* L.). *J. Exp. Bot.* **30**, 971–981.

Salami, A. U. and Kenefick, D. G. (1970). Stimulation of growth in zinc-deficient corn seedlings by the addition of tryptophan. *Crop Sci.* **10**, 291–294.

Salim, M. and Pitman, M. G. (1984). Pressure-induced water and solute flow through plant roots. *J. Exp. Bot.* **35**, 869–881.

Salim, M. and Saxena, R. C. (1991). Nutritional stresses and varietal resistance in rice: Effects on whitebacked plant-hopper. *Crop Sci.* **31**, 797–805.

Salin, M. L. (1988). Toxic oxygen species and protective systems of the chloroplasts. *Physiol. Plant.* **72**, 681–689.

Salsac, L., Chaillou, S., Morot-Gaudry, J.-F., Lesaint, C. and Jolivet, E. (1987). Nitrate and ammonium nutrition in plants. *Plant Physiol. Biochem.* **25**, 805–812.

Same, B. I., Robson, A. D. and Abbott, L. K. (1983). Phosphorus, soluble carbohydrates and endomycorrhizal infection. *Soil Biol. Biochem.* **15**, 593–597.

Samimy, C. (1978). Influence of cobalt on soybean hypocotyl growth and its ethylene evolution. *Plant Physiol.* **62**, 1005–1006.

Samuels, A. L., Glass, A. D. M., Ehret, D. L. and Menzies, J. G. (1991). Mobility and deposition of silicon in cucumber plants. *Plant Cell Environ.* **14**, 485–492.

Sanchez, P. A. and Salinas, G. (1981). Low input technology for managing Oxisols and Ultisols in tropical America. *Adv. Agron.* **34**, 280–406.

Sánchez-Alonso, F. and Lachica, M. (1987a). Seasonal trends in the mineral content of sweet cherry leaves. *Commun. Soil Sci. Plant Anal.* **18**, 17–29.

Sánchez-Alonso, F. and Lachica, M. (1987b). Seasonal trends in the elemental content of plum leaves. *Commun. Soil Sci. Plant Anal.* **18**, 31–43.

Sandalio, L. M. and Del Rio, L. A. (1987). Localization of superoxide dismutase in glyoxysomes from *Citrullus vulgaris*. Functional implications in cellular metabolism. *J. Plant Physiol.* **127**, 395–409.

Sanders, F. E. (1993). Modelling plant growth responses to vesicular arbuscular mycorrhizal infection. *Adv. Plant Pathol.* **9**, 135–166.

Sanders, F. E. and Sheikh, N. A (1983). The development of vesicular-arbuscular mycorrhizal infection in plant root system. *Plant Soil* **71**, 223–246.

Sanders, J. R. (1983). The effect of pH on the total and free ionic concentrations of manganese, zinc and cobalt in soil solutions. *J. Soil Sci.* **34**, 315–323.

Sanderson, J. (1983). Water uptake by different regions of barley root. Pathway of radial flow in relation to development of the endodermis. *J. Exp. Bot.* **34**, 240–253.

Sandmann, G. and Böger, P. (1983). The enzymatological function of heavy metals and their role in electron transfer processes of plants. *In* 'Encyclopedia of Plant Physiology, New Series' (A. Läuchli and R. L. Bieleski, eds.), Vol. 15A, pp. 563–596. Springer-Verlag, Berlin.

Sandstrom, R. P. and Cleland, R. E. (1989). Comparison of the lipid composition of oat root and coleoptile plasma membranes. Lack of short-term change in response to auxin. *Plant Physiol.* **90**, 1207–1213.

Sangster, A. G. (1970). Intracellular silica deposition in immature leaves in three species of the *Gramineae. Ann. Bot. (London) [N.S.]* **34**, 245–257.

Sangster, A. G., Hodson, M. J. and Wynn Parry, D. (1983). Silicon deposition and anatomical studies in the inflorescence with their possible relevance to carcinogenesis. *New Phytol.* **93**, 105–122.

Sano, Y., Fujii, T., Iyama, S., Hirota, Y. and Komagata, K. (1981). Nitrogen fixation in the rhizosphere of cultivated and wild rice strains. *Crop Sci.* **21**, 758–761.

Santoro, L. G. and Magalhaes, A. C. N. (1983). Changes in nitrate reductase activity during development of soybean leaf. *Z. Pflanzenphysiol.* **112**, 113–121.

Saric, S., Okon, Y. and Blum, A. (1990). Promotion of leaf area development and yield in *Sorghum bicolor* inoculated with *Azospirillum brasilense.* Symbiosis **9**, 235–245.

Sarkar, A. N., Jenkins, D. A. and Wyn Jones, R. G. (1979). Modification to mechanical and mineralogical composition of soil within the rhizosphere. *In* 'The Soil–Root Interface' (J. L. Harley and R. Scott-Russell, eds.), pp. 125–136. Academic Press, London.

Sarniguet, A., Lucas, P. and Lucas, M. (1992a). Relationships between take-all, soil conduciveness to the disease, populations of fluorescent pseudomonads and nitrogen fertilizer. *Plant Soil* **145**, 17–27.

Sarniguet, A., Lucas, P., Lucas, M. and Samson, R. (1992b). Soil conduciveness to take-all of wheat: influence of the nitrogen fertilizers on the structure of populations of fluorescent pseudomonads. *Plant Soil* **145**, 29–26.

Sarquis, J. I., Jordan, W. R. and Morgan, P. W. (1991). Ethylene evolution from maize (*Zea mays* L.) seedling roots and shoots in response to chemical impedance. *Plant Physiol.* **96**, 1171–1177.

Sartain, J. B. and Kamprath, E. J. (1975). Effect of liming a highly Al-saturated soil on the top and root growth and soybean nodulation. *Agron. J.* **67**, 507–510.

Sartirana, M. L. and Bianchetti, R. (1967). The effect of phosphate on the development of phytase in the wheat embryo. *Physiol. Plant.* **20**, 1066–1075.

Sasakawa, H. and La Rue, T. A. (1986). Root respiration associated with nitrate assimilation by cowpea. *Plant Physiol.* **81**, 972–975.

Sasaki, Y., Okubo, A., Murakami, T., Arima, Y. and Kumazawa, K. (1987). Radial transport of phosphate in corn roots (II). *J. Plant Nutr.* **10**, 1263–1271.

Sattelmacher, B. and Marschner, H. (1978a). Nitrogen nutrition and cytokinin activity in *Solanum tuberosum. Physiol. Plant.* **42**, 185–189.

Sattelmacher, B. and Marschner, H. (1978b). Relation between nitrogen nutrition, cytokinin and tuberization in *Solanum tuberosum. Physiol. Plant.* **44**, 65–68.

Sattelmacher, B. and Marschner, H. (1979). Tuberization in potato plants as affected by application of nitrogen to the roots and leaves. *Potato Res.* **22**, 39–47.

Sattelmacher, B. and Marschner, H. (1990). Effects of root-zone temperature on growth and development of roots of two potato (*Solanum tuberosum* L.) clones as influenced by plant age, nutrient supply, and light intensity. *J. Agron. Crop. Sci.* **165**, 190–197.

Sattelmacher, B., Klotz, F. and Marschner, H. (1990a). Influence of the nitrogen level on root growth and morphology of two potato varieties differing in nitrogen acquisition. *Plant Soil* **123**, 131–137.

Sattelmacher, B., Marschner, H. and Kühne, R. (1990b). Effects of root zone temperature on root activity of two potato (*Solanum tuberosum* L.) clones with different adaptation to high temperatures. *J. Agron.Crop Sci.* **165**, 131–137.

Sattelmacher, B., Marschner, H. and Kühne, R. (1990c). Effects of the temperature of the rooting zone on the growth and development of roots of potato (*Solanum tuberosum* L.). *Ann. Bot.* (*London*) [*N.S.*] **65**, 27–36.

Sattelmacher, B., Kühne, R., Malagamba, P. and Moreno, U. (1990d). Evaluation of tuber bearing *Solanum* species belonging to different ploidy levels for its yielding potential at low soil fertility. *Plant Soil* **129**, 227–233.

Sattelmacher, B., Reinhard, S. and Pomilkalko, A. (1991). Differences in mycorrhizal colonization of rye (*Secale cereale* L.) grown in conventional or organic (biological-dynamic) farming systems. *J. Agron. Crop Sci.* **167**, 350–355.

Sattelmacher, B., Gerendàs, J., Thoms, K., Brück, H. and Bagdady, N. H. (1993). Interaction between root growth and mineral nutrition. *Environ. Exp. Bot.* **33**, 63–73.

Satter, R., Morse, M. J., Lee, Y., Crain, R. C., Coté, G. G. and Moran, N. (1988). Light and clock-controlled leaflet movements in *Samanea saman*: a physiological, biophysical and biochemical analysis. *Bot. Acta* **101**, 205–213.

Sauerbeck, D. and Helal, H. M. (1986). Plant root development and photosynthate consumption depending on soil compaction. *Trans. XIII Congr. Int. Soil Sci. Soc. (Hamburg)*, Vol. *3*, 948–949.

Sauerbeck, D. and Johnen, B. (1976). Der Umsatz von Pflanzenwurzeln im Laufe der Vegetationsperiode und dessen Beitrag zur 'Bodenatmung'. *Z. Pflanzenernähr. Bodenk.* **139**, 315–328.

Sauerbeck, D., Nonnen, S. and Allard, J. L. (1981). Assimilateverbrauch und -umsatz im Wurzelraum in Abhängigkeit von Pflanzenart und -anzucht. *Landwirtsch. Forsch. Sonderh.* **37**, 207–216.

Savage, W., Berry, W. L. and Reed, C. A. (1981). Effects of trace element stress on the morphology of developing seedlings of lettuce (*Lactuca sativa* L. Grand Rapids) as shown by scanning electron microscopy. *J. Plant Nutr.* **3**, 129–138.

Saxena, M. C., Malhotra, R. S. and Singh, K. B. (1990). Iron deficiency in chickpea in the Mediterranean region and its control through resistant genotypes and nutrient application. *Plant Soil* **123**, 251–254.

Scaife, A. (1988). Derivation of critical nutrient concentrations for growth rate from data from field experiments. *Plant Soil* **109**, 159–169.

Scalet, M., Federico, R. and Angelini, R. (1991). Time course of diamine oxidase and peroxidase activities, and polyamine changes after mechanical injury of chick-pea seedlings. *J. Plant Physiol.* **137**, 571–575.

Schacherer, A. and Beringer, H. (1984). Zahl und Grösse von Endospermzellen im wachsenden Getreidekorn als Indikator der Speicherkapazität. *Ber. Dtsch. Bot. Ges.* **97**, 182–195.

Schachtman, D. P., Tyerman, S. D. and Terry, B. R. (1991). The K^+/Na^+ selectivity of a cation channel in the plasma membrane of root cells does not differ in salt-tolerant and salt-sensitive wheat species. *Plant Physiol.* **97**, 598–605.

Schachtschabel, P. and Beyme, B. (1980). Löslichkeit des anorganischen Bodenphosphors und Phosphatdüngung. *Z. Pflanzenernähr. Bodenk.* **143**, 306–316.

Schaffer, A. A., Nerson, H. and Zamski, E. (1991). Premature leaf chlorosis in cucumber associated with high starch accumulation. *J. Plant Physiol.* **138**, 186–190.

Schaller, G. and Fischer, W. R. (1985). Kurzfristige pH-Pufferung von Böden. *Z. Pflanzenernähr. Bodenk.* **148**, 471–480.

Schaller, G. E. and Sussman, M. R. (1988). Phosphorylation of the plasma membrane H^+-ATPase of oat roots by calcium-stimulated protein kinase. *Planta* **173**, 509–518.

Schauf, C. L. and Wilson, J. J. (1987). Effects of abscisic acid on K^+ channels in *Vicia faba* L. guard cell protoplasts. *Biochem. Biophys. Res. Commun.* **145**, 284–290.

Scheffer, K., Schreiber, A. and Kickuth, R. (1982). Die sorptive Bindung von Düngerphosphaten im Boden und die phosphatmobilisierende Wirkung der Kieselsäure. 2. Mitt. Die phosphatmobilisierende Wirkung der Kieselsäure. *Arch. Acker-Pflanzenbau Bodenkd.* **26**, 143–152.

Schenk, M. K. and Barber, S. A. (1979). Root characteristics of corn genotypes as related to phosphorus uptake. *Agron. J.* **71**, 921–924.

Schenk, M. K. and Wehrmann, J. (1979). The influence of ammonia in nutrient solution on growth and metabolism of cucumber plants. *Plant Soil* **52**, 403–414.

Schenk, M. K., Heins, B. and Steingrobe, B. (1991). The significance of root development of spinach and kohlrabi for N fertilization. *Plant Soil* **135**, 197–203.

Scheromm, P. and Plassard, C. (1988). Nitrogen nutrition of non-mycorrhized pine (*Pinus pinaster*) grown on nitrate or ammonium. *Plant Physiol. Biochem.* **26**, 261–269.

Schiff, J. A. (1983). Reduction and other metabolic reactions of sulfate. *In* 'Encyclopedia of Plant Physiology, New Series' (A. Läuchli and R. L. Bieleski, eds.), Vol. 15A, pp. 401–421. Springer-Verlag,Berlin and New York.

Schildwacht, P.M. (1988). Changes in the osmotic potential of the root as factor in the decrease in the root–shoot ratio of *Zea mays* plants under water stress. *Plant Soil* **111**, 271–275.

Schilling, G. (1983). Genetic specificity of nitrogen nutrition in leguminous plants. *Plant Soil* **72**, 321–334.

Schilling, G. and Trobisch, S. (1970). Einfluss zusätzlich später Stickstoffgaben auf die Ertragsbildung von Kruziferen in Gefäss- und Feldversuchen. *Albrecht-Thaer-Arch.* **14**, 739–750.

Schilling, G. and Trobisch, S. (1971). Untersuchungen über die Verlagerung ^{15}N-markierter Stickstoffverbindungen in Abhängigkeit von der Proteinsynthese am Zielort bei *Sinapis alba*. *Arch. Acker- Pflanzenbau Bodenkd.* **15**, 671–682.

Schimansky, C. (1981). Der Einfluss einiger Versuchsparameter auf das Fluxverhalten von ^{28}Mg bei Gerstenkeimpflanzen in Hydrokulturversuchen. *Landwirtsch. Forsch.* **34**, 154–165.

Schinas, S. and Rowell, D. L. (1977). Lime-induced chlorosis. *J. Soil Sci.* **28**, 351–368.

Schippers, B., Bakker, A. W., Bakker, P. A. H. M. and Van Peer, R. (1990). Beneficial and deleterious effects of HCN-producing pseudomonads on rhizosphere interactions. *Plant Soil* **129**, 75–83.

Schjorring, J. K. (1986). Nitrate and ammonium absorption by plants growing at a sufficient or insufficient level of phosphorus in nutrient solution. *Plant Soil* **91**, 313–318.

Schjorring, J. K. and Jensen, P. (1984). Phosphorus nutrition of barley, buckwheat and rape seedlings. II. Influx and efflux of phosphorus by intact roots of different P status. *Physiol. Plant.* **61**, 584–590.

Schjorring, J. K. and Nielsen, N. E. (1987). Root length and phosphorus uptake by four barley cultivars grown under moderate deficiency of phosphorus in field experiments. *J. Plant Nutr.* **10**, 1289–1295.

Schlee, D., Reinbothe, D. and Fritsche, W. (1968). Der Einfluss von Eisen auf den Purinstoffwechsel und die Riboflavinbildung von *Candida guilliermondii* (Cast.) Lang et G. *Allg. Mikrobiol.* **8**, 127–138.

Schlegel, H. G., Cosson, J.-P. and Baker, A. J. M. (1991). Nickel-hyperaccumulating plants provide a niche for nickel-resistant bacteria. *Bot. Acta* **104**, 18–25.

Schleiff, U. (1986). Water uptake by barley roots as affected by the osmotic and matric potential in the rhizosphere. *Plant Soil* **94**, 354–360.

Schleiff, U. (1987). Eine Vegetationstechnik zur quantitativen Bestimmung der Wasseraufnahme durch Wurzeln aus versalzten Rhizoböden. *Z. Pflanzenernähr. Bodenk.* **150**, 139–146.

Schlichting, E. (1976). Pflanzen- und Bodenanalysen zur Charakterisierung des Nähr-stoffzustandes von Standorten. *Landwirtsch. Forsch.* **29**, 317–321.

Schmid, K. (1967). Zur Stickstoffdüngung im Tabakbau. *Dtsch. Tabakbau* **14**, 129–133.

Schmidt, A. (1986). Regulation of sulfur metabolism in plants. *Prog. Bot.* **48**, 133–150.

Schmidt, A. and Jäger, K. (1992). Open questions about sulfur metabolism in plants. *Annu. Rev. Plant Physiol. Plant Mol. Biol.* **43**, 325–349.

Schmidt, C., He, T. and Cramer, G. R. (1993). Supplemental calcium does not improve growth of salt-stressed Brassicas. *In* 'Plant Nutrition – from Genetics Engineering to Field Practice' (N. J. Barrow, ed.), pp. 617–620. Kluwer Academic, Dordrecht.

Schmidt, H. E., Wrazidlo, W., Bergmann, W. and Schmelzer, K. (1972). Nachweis von Zinkmangel als Ursache der Kräuselkrankheit des Hopfens. *Biol. Zentralbl.* **91**, 729–742.

Schmidt, S. and Buban, T. (1971). Beziehungen zwischen dem ^{32}P-Einbau in anorga-nische Polyphosphate der Blätter und dem Beginn der Blütendifferenzierung bei *Malus domestica. Biochem. Physiol. Pflanze* **162**, 265–271.

Schmidt, W. and Janiesch, P. (1991). Specificity of the electron donor for transmem-brane redox systems in bean (*Phaseolus vulgaris* L.) roots. *J. Plant Physiol.* **138**, 450–453.

Schmit, J.-N. (1981). Le calcium dans le cellule génératrice en mitose. Etude dans le tube pollinique en germination du *Clivia nobilis* Lindl. (*Amaryllidacee*) *C. R. Acad. Sci. Ser. [III]* **293**, 755–760.

Schmucker, T. (1934). Über den Einfluss von Borsäure auf Pflanzen, insbesondere keimende Pollenkörner. *Planta* **23**, 264–283.

Schmutz, D. and Brunold, C. (1982). Regulation of sulfate assimilation in plants. XIII. Assimilatory sulfate reduction during ontogenesis of primary leaves of *Phaseolus vulgaris* L. *Plant Physiol.* **70**, 524–527.

Schmutz, D. and Brunold, C. (1984). Intercellular localization of assimilatory sulfate reduction in leaves of *Zea mays* and *Triticum aestivum. Plant Physiol.* **74**, 866–870.

Schnabl, H. (1980). Der Anionenmetabolismus in stärkehaltigen und stärkefreien Schließzellenprotoplasten. *Ber. Dtsch. Bot. Ges.* **93**, 595–605.

Schnabl, H. and Ziegler, H. (1977). The mechanism of stomatal movement in *Allium cepa* L. *Planta* **136**, 37–43.

Schnug, E. (1993). Physiological functions and environmental relevance of sulfur-containing secondary metabolites. *In* 'Sulfur Nutrition and Assimilation in Higher Plants' (L. J. De Kok, I. Stulen, H. Rennenberg, C. Brunold and W. E. Rauser, eds.), pp. 179–190. SPB Academic Publishing, The Hague, The Netherlands.

Schnyder, H., Nelson, C. J. and Spollen, W.G. (1988). Diurnal growth of tall fescue leaf blades. II. Dry matter partitioning and carbohydrate metabolism in the elongation zone and adjacent expanded tissue. *Plant Physiol.* **86**, 1077–1083.

Schobert, C. and Komor, E. (1987). Amino acid uptake by *Ricinus communis* roots: characterization and physiological significance. *Plant Cell Environ.* **10**, 493–500.

Schobert, C. and Komor, E. (1989). The differential transport of amino acids into the phloem of *Ricinus communis* L. seedlings as shown by the analysis of sieve-tube sap. *Planta* **177**, 342–349.

Schobert, C. and Komor, E. (1990). Transfer of amino acids and nitrate from the roots into the xylem of *Ricinus communis* seedlings. *Planta* **181**, 85–90.

Scholz, G. (1958). Über die Bedeutung des Bors für die Alkaloidproduktion von *Nicotiana rustica*. *Z. Pflanzenernaehr*. *Dueng*. *Bodenk*. **80**, 149–155.

Scholz, G., Becker, R., Stephan, U. W., Rudolph, A. and Pich, A. (1988). The regulation of iron uptake and possible functions of nicotianamine in higher plants. *Biochem. Physiol. Pflanzen* **183**, 257–269.

Scholz, G., Becker, R., Pich, A. and Stephan, U. W. (1992). Nicotianamine – a common constituent of strategies I and II of iron aquisition by plants: a review. *J. Plant Nutr*. **15**, 1647–1665.

Schon, M. K. and Blevins, D. G. (1990). Foliar boron applications increase the final number of branches and pods on branches of field-grown soybeans. *Plant Physiol*. **92**, 602–607.

Schon, M. K., Novacky, A. and Blevins, D. G. (1990). Boron induces hyperpolarization on sunflower root cell membranes and increases membrane permeability to K^+. *Plant Physiol*. **93**, 566–571.

Schönherr, J. (1976). Water permeability of isolated cuticular membranes: the effect of cuticular waxes on diffusion of water. *Planta* **131**, 159–164.

Schönherr, J. and Bukovac, M. J. (1978). Penetration of succinic acid-2,2-dimethylhydrazide: mechanism and rate limiting step. *Physiol. Plant*. **42**, 243–251.

Schönwitz, R. and Ziegler, H. (1982). Exudation of water-soluble vitamins and of some carbohydrates by intact roots of maize seedlings (*Zea mays* L.) into a mineral nutrient solution. *Z. Pflanzenphysiol*. **107**, 7–14.

Schönwitz, R. and Ziegler, H. (1986a). Influence of rhizosphere bacteria on morphological characteristics of maize seedlings (*Zea mays* L.). *Z. Pflanzenernähr*. *Bodenk*. **149**, 614–622.

Schönwitz, R. and Ziegler, H. (1986b). Quantitative and qualitative aspects of a developing rhizosphere microflora and hydroponically grown maize seedlings. *Z. Pflanzenernähr*. *Bodenk*. **149**, 623–634.

Schreiber, L., Breiner, H.-W., Riederer, M., Düggelin, M. and Guggenheim, R. (1994). Casparian strip of *Clivia miniata* Reg. roots: Isolation, fine structure and chemical nature. *Bot. Acta* **107**, 353–361.

Schröder, J. I. and Hedrich, R. (1989). Involvement of ion channels and active transport in osmoregulation and signalling of higher plant cells. *Trends in Biochm. Sci*. **14**, 187–192.

Schröder, P., Grosse, W. and Woermann, D. (1986). Localization of thermo-osmotically active partitions in young leaves of *Nuphar lutea*. *J. Exp. Bot*. **37**, 1450–1461.

Schröder, P., Rusness, D. G. and Lamoureux, G. L. (1990). Detoxification of xenobiotics in spruce trees is mediated by glutathione-*S*-transferases. *In* 'Sulfur Nutrition and Sulfur Assimilation in Higher Plants' (H. Rennenberg *et al*., eds.), pp. 145–248. SPB Academic, The Hague, The Netherlands.

Schropp, A. and Marschner, H. (1977). Wirkung hoher Phosphatdüngung auf die Wachstumsrate, den Zn-Gehalt und das P/Zn-Verhältnis in Weinreben (*Vitis vinifera*). *Z. Pflanzenernähr*. *Bodenk*. **140**, 525–529.

Schröppel-Meier, G. and Kaiser, W. M. (1988). Ion homeostasis in chloroplasts under salinity and mineral deficiency. I. Solute concentrations in leaves and chloroplasts from spinach plants under NaCl or $NaNO_3$ salinity. *Plant Physiol*. **87**, 822–827.

Schubert, E., Mengel, K. and Schubert, S. (1990a). Soil pH and calcium effect on nitrogen fixation and growth of broad bean. *Agron. J*. **82**, 969–972.

Schubert, K. R., Jennings, N. T. and Evans, H. J. (1978). Hydrogen reactions of nodulated leguminous plants. *Plant Physiol*. **61**, 398–401.

Schubert, S. and Läuchli, A. (1988). Metabolic dependence of Na$^+$ efflux from roots of intact maize seedlings. *J. Plant Physiol.* **133**, 193–198.

Schubert, S. and Läuchli, A. (1990). Sodium exclusion mechanisms at the root surface of two maize cultivars. *Plant Soil* **123**, 205–209.

Schubert, S. and Mengel, K. (1986). Effect of light intensity on proton extrusion by roots of intact maize plants. *Physiol. Plant.* **67**, 614–619.

Schubert, S., Schubert, E. and Mengel, K. (1990b). Effect of low pH of the root medium on proton release, growth, and nutrient uptake of field beans (*Vicia faba*). *Plant Soil* **124**, 239–244.

Schüepp, H., Dehn, B. und Sticher, H. (1987). Interaktionen zwischen VA-Mykorrhizen und Schwermetallbelastungen. *Angew. Bot.* **61**, 85–96.

Schuepp, P. H. and Hendershot, W. H. (1989). Nutrient leaching from dormant trees at an elevated site. *Water Air Soil Pollut.* **45**, 253–264.

Schuler, R. and Haselwandter, K. (1988). Hydroxamate siderophore production by ricoid mycorrhizal fungi. *J. Plant Nutr.* **11**, 907–913.

Schulze, E.-D. (1989). Air pollution and forest decline in a spruce (*Picea abies*) forest. *Science* **244**, 776–783.

Schulze, E. D., Turner, N. C. and Glatzle, G. (1984). Carbon, water and nutrient relations of two mistletoes and their hosts: a hypothesis. *Plant Cell Environ.* **7**, 293–299.

Schulze, W. (1957). Über den Einfluss der Düngung auf die Bildung der Chloroplastenpigmente. *Z. Pflanzenernähr. Düng. Bodenkd.* **76**, 1–19.

Schum, A., Forchthammer, L. und Fischer, P. (1988). Einfluss des Kupferernährungszustandes der Mutterpflanzen auf die Polyphenoloxidaseaktivität und die in vitro-Sprossregeneration bei Infloreszensexplantaten von *Gerbera jamesonii*. *Gartenbauwissensch.* **53**, 263–269.

Schumacher, R. and Frankenhauser, F. (1968). Fight against bitter pit. *Schweiz. Z. Obst- Weinbau* **104**, 424.

Schumaker, K. S. and Sze, H. (1990). Solubilization and reconstitution of the oat root vacuolar H$^+$/Ca^{2+} exchanger. *Plant Physiol.* **92**, 340–345.

Schupp, R. and Rennenberg, H. (1988). Diurnal changes in the glutathione content of spruce needles (*Picea abies* L.). *Plant Sci.* **57**, 113–117.

Schupp, R., Glavac, V. and Rennenberg, H. (1991). Thiol composition of xylem sap of beech trees. *Phytochemistry* **30**, 1415–1418.

Schussler, J. R., Brenner, M. I. and Brun, W. A. (1984). Abscisic acid and its relationship to seed filling in soybeans. *Plant Physiol.* **76**, 301–306.

Schürmann, P. (1993). Plant thioredoxins. *In* 'Sulfur Nutrition and Assimilation in Higher Plants' (L. J. De Kok, I. Stulen, H. Rennenberg, C. Brunold and W. E. Rauser, eds.), pp. 153–162. SPB Academic, The Hague, The Netherlands.

Schütte, K. H. (1967). The influence of boron and copper deficiency upon infection by *Erysiphe graminis* D. C. the powdery mildew in wheat var. Kenya. *Plant Soil* **27**, 450–452.

Schütz, B., De Kok, L. J. and Rennenberg, H. (1991). Thiol accumulation and cysteine desulfurylase activity in H$_2$S-fumigated leaves and leaf homogenates of cucurbit plants. *Plant Cell Physiol.* **32**, 733–736.

Schwarz, M. and Gale, J. (1981). Maintenance respiration and carbon balance of plants at low levels of sodium chloride salinity. *J. Exp. Bot.* **32**, 933–941.

Schweiger, P. (1994). Factors affecting VA mycorrhizal uptake of phosphorus. PhD Thesis, University of Western Australia, Nedlands, Australia.

Scott, B. J. and Robson, A. D (1990a). Changes in the content and form of magnesium in the first trifoliate leaf of subterranean clover under altered or constant root supply. *Aust. J. Agric. Res.* **41**, 511–519.

Scott, B. J. and Robson, A. D. (1990b). Distribution of magnesium in subterranean clover (*Trifolium subterraneum* L.) in relation to supply. *Aust. J. Agric. Res.* **41**, 499–510.

Scott, B. J. and Robson, A. D. (1991). Distribution of magnesium in wheat (*Triticum aestivum* L.) in relation to supply. *Plant Soil* **136**, 183–193.

Scott, J. J. and Loewus, F. A. (1986). A calcium-activated phytase from pollen of *Lilium longiflorum*. *Plant Physiol.* **82**, 333–335.

Scott-Russell, R. (1977). 'Plant Root Systems: Their Function and Interaction with the Soil'. McGraw-Hill, New York.

Scott-Russell, R. and Goss, M. J. (1974). Physical aspects of soil fertility – The response of root to mechanical impedance. *Neth. J. Agric. Sci.* **22**, 305–318.

Scriber, J. M. and Slansky, F. (1981). The nutritional ecology of immature insects. *Annu. Rev. Entomol.* **26**, 183–211.

Secilia, J. and Bagyaraj, D. J. (1987). Bacteria and actinomycetes associated with pot cultures of vesicular-arbuscular mycorrhizas. *Can. J. Bot.* **33**, 1069–1073.

Seckbach, J. (1982). Ferreting out the secrets of plant ferritin – a review. *J. Plant Nutr.* **5**, 369–394.

Sedberry jr, J. E., Bligh, D. P., Peterson, F. J. and Amacher, M. C. (1988). Influence of soil pH and application of zinc on the yield and uptake of selected nutrient elements by rice. *Commun. Soil Sci. Plant Anal.* **19**, 597–615.

Seeling, B. and Claassen, N. (1990). A method for determining Michaelis–Menten kinetic parameters of nutrient uptake for plants growing in soil. *Z. Pflanzenernähr. Bodenk.* **153**, 301–303.

Seeling, B. and Zasoski, R. J. (1993). Microbial effects in maintaining organic and inorganic solution phosphorus concentrations in a grassland topsoil. *Plant Soil* **148**, 277–284.

Seggewiss, B. and Jungk, A. (1988). Einfluss der Kaliumdynamik im wurzelnahen Boden auf die Magnesiumaufnahme von Pflanzen. *Z. Pflanzenernähr. Bodenk.* **151**, 91–96.

Sekiya, J., Schmidt, A., Wilson, L.G. and Filner, P. (1982a). Emission of hydrogen sulfide by leaf-tissue in response to L-cysteine. *Plant Physiol.* **70**, 430–436.

Sekiya, J., Wilson, L. G. and Filner, P. (1982b). Resistance to injury by sulfur dioxide. Correlation with its reduction to and emission of hydrogen sulfide in *Cucurbitaceae*. *Plant Physiol.* **70**, 437–441.

Seliskar, D. M. (1988). Waterlogging stress and ethylene production in the dune slack plant, *Scirpus americanus*. *J. Exp. Bot.* **39**, 1639–1648.

Sen Gupta, A., Berkowitz, G. A. and Pier, P. A. (1989). Maintenance of photosynthesis at low leaf water potential in wheat. *Plant Physiol.* **89**, 1358–1365.

Sen Gupta, B., Nandi, A. S. and Sen, S. P. (1982). Utility of phyllosphere N_2-fixing micro-organisms in the improvement of crop growth. I. Rice. *Plant Soil* **68**, 55–67.

Senden, M. H. M. N. and Wolterbeek, H. T. (1990). Effect of citric acid on the transport of cadmium through xylem vessels of excised tomato stem–leaf systems. *Acta Bot. Neerl.* **39**, 297–303.

Sentenac, H. and Grignon, C. (1985). Effect of pH on orthophosphate uptake by corn roots. *Plant Physiol.* **77**, 136–141.

Seresinhe, P. S. J. W. and Oertli, J. J. (1991). Effects of boron on growth of tomato cell suspensions. *Physiol. Plant.* **81**, 31–36.

Serrano, R. (1989). Structure and function of plasma membrane ATPase. *Annu. Rev. Plant Physiol. Plant Mol. Biol.* **40**, 61–94.

Serrano, R. (1990). Recent molecular approaches to the physiology of the plasma membrane proton pump. *Bot. Acta.* **103**, 230–234.

Seth, A. K. and Wareing, P. F. (1967). Hormone-directed transport of metabolites and its possible role in plant senescence. *J. Exp. Bot.* **18**, 65–77.

Setter, T. L. and Meller, V. H. (1984). Reserve carbohydrate in maize stem. [14]C glucose and ^{14}C sucrose uptake characteristics. *Plant Physiol.* **75**, 617–622.

Setter, T. L., Brun, W. A. and Brenner, M. L. (1980). Effect of obstructed translocation on leaf abscisic acid, and associated stomatal closure and photosynthesis decline. *Plant Physiol.* **65**, 1111–1115.

Sevilla, F., del Rio, L. A. and Hellin, E. (1984). Superoxide dismutases from a citrus plant: presence of two iron-containing isoenzymes in leaves of lemon trees (*Citrus limonum* L.). *J. Plant Physiol.* **116**, 381–387.

Seward, P., Barraclough, P. B. and Gregory, P. J. (1990). Modelling potassium uptake by wheat (*Triticum aestivum*) crops. *Plant Soil* **124**, 303–307.

Shafer, S. R. and Blum, U. (1991). Influence of phenolic acids on microbial populations in the rhizosphere of cucumber. *J. Chem. Ecol.* **17**, 369–389.

Shah, S. H., Gorham, J., Forter, B. P. and Wyn Jones, R. G. (1987). Salt tolerance in the Triticeae: the contribution of the D genome to cation selectivity in hexaploid wheat. *J. Exp. Bot.* **38**, 254–269.

Shaked, A. and Bar-Akiva, A. (1967). Nitrate reductase activity as an indication of molybdenum level and requirement of citrus plants. *Phytochemistry* **6**, 347–350.

Shamsuddin, Z. H., Kasrand, R., Edwards, D. G. and Blamey, F. P. C. (1992). Effects of calcium and aluminium on nodulation nitrogen fixation and growth of groundnut in solution culture. *Plant Soil* **144**, 273–279.

Shaner, D. L. and Boyer, J. S. (1976). Nitrate reductase activity in maize (*Zea mays* L.) leaves. II. Regulation by nitrate flux at low leaf water potential. *Plant Physiol.* **58**, 505–509.

Shanmugam, K. T., O'Gara, F., Andersen, K. and Valentine, R. C. (1978). Biological nitrogen fixation. *Annu. Rev. Plant Physiol.* **29**, 263–276.

Sharkey, T. D. (1988). Estimating the rate of photorespiration in leaves. *Physiol. Plant.* **73**, 147–152.

Sharma, S. and Sanwal, G. G. (1992). Effect of Fe deficiency on the photosynthetic system of maize. *J. Plant Physiol.* **140**, 527–530.

Sharma, C. P. and Singh, S. (1990). Sodium helps overcome potassium deficiency effects on water relations of cauliflower. *HortScience* **25**, 458–459.

Sharma, C. P., Sharma, P. N., Bisht, S. S. and Nautiyal, B. D. (1982). Zinc deficiency induced changes in cabbage. *In* 'Proceedings of the Ninth Plant Nutrition Colloquium, Warwick, England' (A. Scaife, ed.), pp. 601–606. Commonwealth Agricultural Bureau, Farnham Royal, Bucks.

Sharma, C. P., Sharma, P. N., Chatterjee, C. and Agarwala, S C. (1991). Manganese deficiency in maize effects pollen viability. *Plant Soil* **138**, 139–142.

Sharma, P. N., Chatterjee, C., Agarwala, S. C. and Sharma, C. P. (1990). Zinc deficiency and pollen fertility in maize (*Zea mays*). *Plant Soil* **124**, 221–225.

Sharp, R. E., Silk, W. K. and Hsiao, T. C. (1988). Growth of the maize primary root at low water potentials. I. Spatial distribution of expansive growth. *Plant Physiol.* **87**, 50–57.

Sharpless, R. O. and Johnson, D. S. (1977). The influence of calcium on senescense changes in apple. *Ann. Appl.Biol.* **85**, 450–453.

Shaviv, A., Hagin, J. and Neumann, P. M. (1987). Effects of a nitrification inhibitor on efficiency of nitrogen utilization by wheat and millet. *Commun. Soil Sci. Plant Anal.* **18**, 815–833.

Shaw, G., Leake, J. R., Baker, A. J. M. and Read, D. J. (1990). The biology of mycorrhiza in the Ericaceae. XVII. The role of mycorrhizal infection in the regulation of iron uptake by ericaceous plants. *New Phytol.* **115**, 251–258.

Shea, P. F., Gabelman, W. H. and Gerloff, G. C. (1967). The inheritance of efficiency in potassium utilization in snap beans (*Phaseolus vulgaris* L.) *Proc. Am. Soc. Hortic. Sci.* **91**, 286–293.

Shear, C. B. (1975). Calcium related disorders of fruits and vegetables. *HortScience* **10**, 361–365.

Shelp, B. J. (1987). The composition of phloem exudate and xylem sap from broccoli (*Brassica oleracea* var. *italica*) supplied with NH_4^+, NO_3^- or NH_4NO_3. *J. Exp. Bot.* **38**, 1619–1636.

Shelp, B. J. (1988). Boron mobility and nutrition in broccoli (*Brassica oleracea* var. *italica*). *Ann. Bot.* **61**, 83–91.

Shelp, B. J. (1993). Physiology and biochemistry of boron in plants. *In* 'Boron and its Role in Crop Production' (U. C. Gupta, ed.), pp. 53–85. CRC Press, Boca Raton, FL.

Shepherd, V. A., Orlovich, D. A. and Ashford, A. E. (1993). A dynamic continuum of pleiomorphic tubules and vacuoles in growing hyphae of a fungus. *J. Cell Sci.* **104**, 495–507.

Sherwood, R. T. and Vance, C. P. (1980). Resistance to fungal penetration in *Gramineae*. *Phytopathology* **70**, 273–279.

Shetty, A. S. and Miller, G. W. (1966). Influence of iron chlorosis on pigment and protein metabolism in leaves of *Nicotiana tabacum* L. *Plant Physiol.* **41**, 415–421.

Shi, Y., Byrne, D. H., Reed, D. W. and Loeppert, R. H. (1993). Iron chlorosis development and growth response of peach root-stocks to bicarbonate. *J. Plant Nutr.* **16**, 1039–1046.

Shierlaw, J. and Alston, A.M. (1984). Effect of soil compaction on root growth and uptake of phosphorus. *Plant Soil* **77**, 15–28.

Shih, L.-M., Kaur-Sawhney, R., Führer, J., Samat, S. and Galston, A. W. (1982). Effect of exogenous 1,3-diaminopropane and spermidine on senescence of oat leaves. I. Inhibition of protease activity, ethylene production and chlorophyll loss as related to polyamine content. *Plant Physiol.* **70**, 1592–1596.

Shimshi, D. (1969). Interaction between irrigation and plant nutrition. *Proc. 7th Colloq. Int. Potash Inst. Bern*, pp. 111–120.

Shiv Raj, A. (1987). Cobalt nutrition of pigeon-pea and peanut in relation to growth and yield. *J. Plant Nutr.* **10**, 2137–2145.

Shkol'nik, M. Y. (1974). General conception of the physiological role of boron in plants. *Sov. Plant Physiol.* (*Engl. Transl.*) **21**, 140–150.

Shkol'nik, M. Y., Krupnikova, T. A. and Smirnov, Y. S. (1981). Activity of polyphenol oxidase and sensitivity to boron deficiency in monocots and dicots. *Sov. Plant Physiol.* (*Engl.Transl.*) **28**, 279–283.

Shojima, S., Nishizawa, N.-K. and Mori, S. (1989). Establishment of a cell-free system for the biosynthesis of nicotianamine. *Plant Cell Physiol.* **30**, 673–677.

Shojima, S., Nishizawa, N.-K., Fushiya, S., Nozoe, S., Irifune, T. and Mori, S. (1990). Biosynthesis of phytosiderophores. In vitro biosynthesis of 2'-deoxymugineic acid from L-methionine and nicotianamine. *Plant Physiol.* **93**, 1497–1503.

Shone, M. G. T. and Flood, A. V. (1985). Measurement of free space and sorption of large molecules by cereal roots. *Plant Cell Environ.* **8**, 309–315.

Shone, M. G. T., Clarkson, D. T., Sanderson, J. and Wood, A.V. (1973). A comparison of the uptake and translocation of some organic molecules and ions in higher plants. *In* 'Ion Transport in Plants' (W. P. Anderson, ed.), pp. 571–582. Academic Press, London.

Shrift, A. (1969). Aspects of selenium metabolism in higher plants. *Annu. Rev. Plant Physiol.* **20**, 475–494.

Shrotri, C. K., Mohanty, P., Rathore, V. C. and Tewari, M. N. (1983). Zinc deficiency limits the photosynthetic enzyme activation in *Zea mays* L. *Biochem. Physiol. Pflanz.* **178**, 213–217.

Shu, Z.-H., Wu, W. Y. and Oberly, G. H. (1991). Boron uptake by peach leaf slices. *J. Plant Nutr.* **14**, 867–881.

Shukla, U. C. and Raj, H. (1974). Influence of genetic variability on zinc response in wheat. *Proc. Soil Sci. Soc. Am.* **38**, 477–479.

Shukla, U. C. and Raj, H. (1976). Zinc response in corn as influenced by genetic variability. *Agron. J.* **68**, 20–22.

Shukla, U. C. and Raj, H. (1980). Zinc response in pigeon pea as influenced by genotypic variability. *Plant Soil* **57**, 323–333.

Sicher, R. C. and Kremer, D. F. (1988). Effects of phosphate deficiency on assimilate partitioning in barley seedlings. *Plant Sci.* **57**, 9–17.

Siddiqi, M. Y., Glass, A. D. M., Ruth, T. J. and Fernando, M. (1989). Studies of the regulation of nitrate influx by barley seedlings using $^{13}NO_3^{-1}$. *Plant Physiol.* **90**, 806–813.

Siddiqi, M. Y., Glass, A. D. M., Ruith, T. J. and Rufty jr, T. W. (1990). Studies of the uptake of nitrate in barley. I. Kinetics of $^{13}NO_3^-$ influx. *Plant Physiol.* **93**, 1426–1432.

Siddique, A. M. and Bal, A. K. (1991). Nitrogen fixation in peanut nodules during dark periods and detopped conditions with special reference to lipid bodies. *Plant Physiol.* **95**, 896–899.

Siedow, J. N. and Berthold, D. A. (1986). The alternative oxidase: a cyanide-resistant respiratory pathway in higher plants. *Physiol. Plant.* **66**, 569–573.

Sieverding, E. (1991). 'Vesicular-arbuscular mycorrhiza management in tropical agro-systems'. Technical Cooperation (GTZ), TZ-Verlagsgesellschaft Rossdorf, Germany.

Sievers, A. and Hensel, W. (1991). Root cap, structure and function. *In* 'Plant Roots: the Hidden Half' (Y. Waisel, A. Eshel, U. Kafkafi, eds.), pp. 53–74. Marcel Dekker, New York.

Sijmons, P. C., Kolattukudy, P. E. and Bienfait, H. F. (1985). Iron-deficiency decreases suberization in bean roots through a decrease in suberin-specific peroxidase activity. *Plant Physiol.* **78**, 115–120.

Silberbush, M. and Barber, S. A. (1984). Phosphorus and potassium uptake of field-grown soybean cultivars predicted by a simulation model. *Soil Sci. Soc. Am. J.* **48**, 592–596.

Silberbush, M., Hallmark, W. B. and Barber, S. A. (1983). Simulation of effects of soil bulk density and P addition on K uptake of soybean. *Comm. Soil Sci. Plant Anal.* **14**, 287–296.

Silva, P. R. F. da and Stutte, C. A. (1981). Nitrogen loss in conjunction with transpiration from rice leaves as influenced by growth stage, leaf position, and N supply. *Agron. J.* **73**, 38–42.

Silvius, J. E., Kremer, D. F. and Lee, D. R. (1978). Carbon assimilation and translocation in soybean leaves at different stages of development. *Plant Physiol.* **62**, 54–58.

Simán, G. and Jansson, S. L. (1976). Sulphur exchange between soil and atmosphere with special attention to sulphur release directly to the atmosphere. 2. The role of vegetation in sulphur exchange between soil and atmosphere. *Swedish J. Agric. Res.* **6**, 135–144.

Simpson, D. J. and Robinson, S. R. (1984). Freeze-fracture ultrastructure of thylakoid membranes in chloroplasts from manganese-deficient plants. *Plant Physiol.* **74**, 735–741.

Simpson, F. B. (1987). The hydrogen reactions of nitrogenase. *Physiol. Plant.* **69**, 187–190.

Simpson, R. J., Lambers, H. and Dalling, M. J. (1982). Translocation of nitrogen in a vegetative wheat plant (*Triticum aestivum*). *Physiol. Plant.* **56**, 11–17.

Sims, G. K. and Dunigan, E. P. (1984). Diurnal and seasonal variation in nitrogenase activity (C_2H_2 reduction) of rice roots. *Soil Biol. Biochem.* **16**, 15–18.

Sims, J. L. and Patrick jr, W. H. (1978). The distribution of micronutrient cations in soil under conditions of varying redox potential and pH. *Soil Sci. Soc. Am. J.* **42**, 258–262.

Sims, J. T. and Johnson, G. V. (1991). Micronutrient soil tests. *In* 'Micronutrients in Agriculture' 2nd. edn. (J. J. Mortvedt, F. R. Cox, L. M. Shuman, R. M.Welch, eds.), pp. 427–476. SSSA Book Series No. 4, Madison, WI.

Sinclair, A. H., Mackie-Dawson, L. A. and Linehan, D. L.(1990). Micronutrient inflow rates and mobilization into soil solution in the root zone of winter wheat (*Triticum aestivum* L.). *Plant Soil* **122**, 143–146.

Sinclair, T. R. and Horie, T. (1989). Leaf nitrogen, photosynthesis, and crop radiation use efficiency: a review. *Crop Sci.* **29**, 90–98.

Singh, A. and Paolillo jr, D. J. (1990). Role of calcium in the callose response of self-pollinated brassica stigmas. *Am. J. Bot.* **77**, 128–133.

Singh, B. and Song, P.-S. (1990). Phytochrome and protein phosphorylation. *Phytochem. Photobiol.* **52**, 249–254.

Singh, B., Dang, Y. P. and Mehta, S. C. (1990a). Influence of nitrogen on the behaviour of nickel in wheat. *Plant Soil* **127**, 213–218.

Singh, B. K. and Jenner, C. F. (1982). Association between concentrations of organic nutrients in the grain, endosperm cell number and grain dry weight within the ear of wheat. *Aust. J. Plant Physiol.* **9**, 83–95.

Singh, J. P., Dahiya, D. J. and Narwal, R. P. (1990b). Boron uptake and toxicity in wheat in relation to zinc supply. *Fert. Res.* **24**, 105–110.

Singh, K., Chino, M., Nishizawa, N. K., Ohata, T. and Mori, S. (1993). Genotypic variation among Indian graminaceous species with respect to phytosiderophore secretion. *In* 'Genetic Aspects of Plant Mineral Nutrition' (P. J. Randall, E. Delhaize, R. A. Richards and R. Munns, eds.), pp. 335–339. Kluwer Academic, Dordrecht.

Singh, M. (1981). Effect of zinc, phosphorus and nitrogen on tryptophan concentration in rice grains grown on limed and unlimed soils. *Plant Soil* **62**, 305–308.

Singh, S. P. and Paleg, L. G. (1984). Low temperature-induced GA_3 sensitivity of isolated aleurone of kite. *Plant Physiol.* **76**, 143–147.

Sinha, B. K. and Singh, N. T. (1974). Effect of transpiration rate on salt accumulation around corn roots in a saline soil. *Agron. J.* **66**, 557–560.

Skinner, P. W., Matthews, M. A. and Carlson, R. M. (1987). Phosphorus requirements of wine grapes: extractable phosphate of leaves indicates phosphorus status. *J. Am. Soc. Hortic. Sci.* **112**, 449–454.

Skirver, K. and Mundy, J. (1990). Gene expression in response to abscisic acid and osmotic stress. *Plant Cell* **2**, 505–512.

Skubatz, H., Williamson, P. S., Schneider, E. L. and Meeuse, B. J. D. (1990). Cyanide-insensitive respiration in thermogenic flowers of *Victoria* and *Nelumbo*. *J. Exp. Bot.* **41**, 1335–1339.

Slocum, R. C. and Roux, S. J. (1983). Cellular and subcellular localization of calcium in gravistimulated oat coleoptiles and its possible significance in the establishment of tropic curvature. *Planta* **157**, 481–492.

Slone, J. H. and Buckhout, T. J. (1991). Sucrose-dependent H^+ transport in plasma-membrane vesicles isolated from sugarbeet leaves (*Beta vulgaris* L.). Evidence in support of the H^+-symport model for sucrose transport. *Planta* **183**, 584–589.

Smeulders, F. and van de Geijn, S. C. (1983). In situ immobilization of heavy metals with tetraethylenepentamine(tetren) in natural soils and its effect on toxicity and plant growth. III. Uptake and mobility of copper and its tetren-complex in corn plants. *Plant Soil* **70**, 59–68.

Smiciklas, K. D. and Below, F. E. (1992). Role of cytokinin in enhanced productivity of maize supplied with NH_4^+ and NO_3^-. *Plant Soil* **142**, 307–313.

Smirnoff, N. and Stewart, G. R. (1985). Nitrate assimilation and translocation by higher plants: comparative physiology and ecological consequences. *Physiol. Plant.* **64**, 133–140.

Smirnoff, N. and Stewart, G. R (1987). Nitrogen assimilation and zinc toxicity to zinc-tolerant and non-tolerant clones of *Deschampsia caespitosa* (L.) Beauv. *New Phytol.* **107**, 671–680.

Smirnov, Y. S., Krupnikova, T. A. and Shkol'nik, M. Y. (1977). Content of IAA in plants with different sensitivity to boron deficits. *Sov. Plant Physiol.(Engl. Transl.)* **24**, 270–276.

Smit, B. A., Neumann, D. S. and Stachowiak, M. L. (1990). Root hypoxia reduces leaf growth. Role of factors in the transpiration stream. *Plant Physiol.* **92**, 1021–1028.

Smit, B., Stachowiak, M. and van Volkenburgh, E. (1989). Cellular processes limiting leaf growth in plants under hypoxic root stress. *J. Exp. Bot.* **40**, 89–94.

Smith, B. N. (1984). Iron in higher plants: storage and metabolic rate. *J. Plant Nutr.* **7**, 759–766.

Smith, C. J., Freney, J. R., Sherlock, R. R. and Galbally, I. E. (1991). The fate of urea nitrogen applied in a foliar spray to wheat at heading. *Fert. Res.* **28**, 129–138.

Smith, F. W. (1974). The effect of sodium on potassium nutrition and ionic relations in Rhodes grass. *Aust. J. Agric. Res.* **25**, 407–414.

Smith, F. W., Jackson, W. A. and van den Berg, P. J. (1990a). Internal phosphorus flows during development of phosphorus stress in *Stylosanthes hamata*. *Aust. J. Plant Physiol.* **17**, 451–464.

Smith, G. S. and Watkinson, J. H. (1984). Selenium toxicity in perennial ryegrass and white clover. *New Phytol.* **97**, 557–564.

Smith, G. S., Middleton, K. R. and Edmonds, A. S. (1978). Sodium and potassium contents of top-dressed pastures in New Zealand in relation to plant and animal nutrition. *N. Z. J. Exp. Agric.* **6**, 217–225.

Smith, G. S., Middleton, K. R. and Edmonds, A. S. (1980). Sodium nutrition of pasture plants.II. Effects of sodium chloride on growth, chemical composition and reduction of nitrate nitrogen. *New Phytol.* **84**, 613–622.

Smith, G. S., Lauren, D. R., Cornforth, I. S. and Agnew, M. P. (1982). Evaluation of putrescine as a biochemical indicator of the potassium requirments of lucerne. *New Phytol.* **91**, 419–428.

Smith, G. S., Cornforth, I. S. and Henderson, H. V. (1984). Iron requirements of C_3 and C_4 plants. *New Phytol.* **97**, 543–556.

Smith, G. S., Clark, C. J. and Holland, P. T. (1987). Chloride requirement of kiwi-fruit (*Actinidia deliciosa*). *New Phytol.* **106**, 71–80.

Smith, I. K. and Lang, A. L. (1988). Translocation of sulfate in soybean (*Glycine max* L. Merr). *Plant Physiol.* **86**, 798–802.

Smith, I. K., Polle, A. and Rennenberg, H. (1990b). Glutathione. *In* 'Stress Responses in Plants: Adaptation and Acclimation Mechanisms', pp. 201–215. Wiley-Liss, New York.

Smith, J. A. C. (1991). Ion transport in the transpiration stream. *Bot. Acta* **104**, 416–421.

Smith, J. A. C. and Milburn, J. A. (1980). Water stress and phloem loading. *Ber. Dtsch. Bot. Ges.* **93**, 269–280.

Smith, M. K. and McComb, J. A. (1981). Effect of NaCl on the growth of whole plants and their corresponding callus cultures. *Aust. J. Plant Physiol.* **8**, 267–275.

Smith, P. G. and Dale, J. E. (1988). The effect of root cooling and excision treatments on the growth of primary leaves of *Phaseolus vulgaris* L. Rapid and reversible increases in abscisic acid content. *New Phytol.* **110**, 293–300.

Smith, R. H. and Johnson, W. C. (1969). Effect of boron on white clover nectar production. *Crop Sci.* **9**, 75–76.

Smith, S. and Stewart, G. R. (1990). Effect of potassium levels on the stomatal behavior of the hemi-parasite *Striga hermonthica*. *Plant Physiol.* **94**, 1472–1476.

Smith, S. E. and Gianinazzi-Pearson, V. (1988). Physiological interactions between symbionts in vesicular-arbuscular mycorrhizal plants. *Annu. Rev. Plant Physiol. Plant Biol.* **39**, 221–244.

Smith, S. E., Robson, A. D. and Abbott, L. K. (1992). The involvement of mycorrhizas in assessment of genetically dependent efficiency of nutrient uptake and use. *Plant Soil* **146**, 169–179.

Smith, T. A. (1988). Symposium report: amines in plants. *Phytochemistry* **27**, 1233–1234.

Smith, T. A. and Sinclair, C. (1967). The effect of acid feeding on amine formation in barley. *Ann. Bot.* (*London*) [*N.S.*] **31**, 103–111.

Smith, T. A. and Wilshire, G. (1975). Distribution of cadaverine and other amines in higher plants. *Phytochemistry* **14**, 2341–2346.

Smyth, D. A. and Chevalier, P. (1984). Increases in phosphatase and β-glucosidase activities in wheat seedlings in response to phosphorus-deficient growth. *J. Plant Nutr.* **7**, 1221–1231.

Smyth, T. J. and Cravo, M. S. (1990). Critical phosphorus levels for corn and cowpea in a Brazilian Amazon Oxisol. *Agron. J.* **82**, 309–312.

Söderbäck, E. and Bergman, B. (1992). The *Nostoc–Gumera magellanica* symbiosis: phycobiliproteins, carboxysomes and Rubisco in the cyanobiont. *Physiol. Plant.* **84**, 425–432.

Söderström, B. (1992). The ecological potential of the ectomycorrhizal mycelium. *In* 'Mycorrhizas in Ecosystems' (D. J. Read, D. H. Lewis, A. H. Fitter and I. J. Alexander, eds.), pp. 77–83. CAB International, Wallingford, UK.

Soerensen, K. U., Terry, R. E., Jolley, V. D. and Brown, J. C. (1989). Iron-stress response of inoculated and non-inoculated roots of an iron inefficient soybean cultivar in a split-root system. *J. Plant Nutr.* **12**, 437–447.

Sogawa, K. (1982). The rice brown plant hopper: Feeding physiology and host plant interactions. *Annu. Rev. Entomol.* **27**, 49–73.

Solomonson, L. P. and Barber, M. J.(1990). Assimilatory nitrate reductase: Functional properties and regulation. *Annu. Rev. Plant Physiol. Plant Mol. Biol.* **41**, 225–253.

Soltanpour, P. N., El Gharous, M., Azzaouri, A. and Abdelmonum, M. (1989). A soil test based N recommendation model for dryland wheat. *Commun. Soil Sci. Plant Anal.* **20**, 1053–1068.

Sommer, K. and Six, R. (1982). Ammonium als Stickstoffquelle beim Anbau von Futtergerste. *Landw. Forsch.* **38**, 151–161.

Sonneveld, C. and Ende, J. van den (1975). The effect of some salts on head weight and tipburn of lettuce and on fruit production and blossom-end rot of tomatoes. *Neth. J. Agric. Sci.* **23**, 192–201.

Sovonick, S. A., Geiger, D. R. and Fellows, R. J. (1974). Evidence for active phloem loading in the minor veins of sugar beet. *Plant Physiol.* **54**, 886–891.

Spaeth, S. C. and Sinclair, T. R. (1983). Variation in nitrogen accumulation among soybean cultivars. *Field Crops Res.* **7**, 1–12.

Spain, J. M., Francis, C. A., Howeler, R. H. and Calvo, F. (1975). Differential species and varietal tolerances to soil acidity in tropical crops and pastures. *In* 'Soil Management in Tropical America' (E. Bornemisza and A. Alvarado, eds.), pp. 308–329. North Carolina State University, Raleigh.

Spanner, D. C. (1975). Electroosmotic flow. *In* 'Encyclopedia of Plant Physiology, New Series' (M. H. Zimmermann and J. A. Milburn, eds.), Vol. 1, pp. 301–327. Springer-Verlag, Berlin.

Speer, M. and Kaiser, W. M. (1991). Ion relations of symplastic and apoplastic space in leaves from *Spinacia oleracea* L. and *Pisum sativum* L. under salinity. *Plant Physiol.* **97**, 990–997.

Spencer, D. and Possingham, J. V. (1960). The effect of nutrient deficiencies on the Hill reaction of isolated chloroplasts from tomato. *Aust. J. Biol. Sci.* **13**, 441–445.

Sperrazza, J. M. and Spremulli, L. L. (1983). Quantitation of cation binding to wheat germ ribosomes: influences on subunit association equilibria and ribosome activity. *Nucleic Acids Res.* **11**, 2665–2679.

Spilatro, S. R. and Preiss, J. (1987). Regulation of starch synthesis in the bundle sheath and mesophyll of *Zea mays* L. *Plant Physiol.* **83**, 621–627.

Spiller, S. C., Castelfranco, A. M. and Castelfranco, P. A. (1982). Effects of iron and oxygen on chlorophyll biosynthesis. I. *In vivo* observations on iron and oxygen-deficient plants. *Plant Physiol.* **69**, 107–111.

Spiller, S. C., Kaufman, L. S., Thompson, W. F. and Briggs, W. R. (1987). Specific mRNA and rRNA level in greening pea leaves during recovery from iron stress. *Plant Physiol.* **84**, 409–414.

Spiteri, A., Viratelle, O. M., Raymond, P., Rancillac, M., Labouesse, J. and Pradet, A. (1989). Artefactual origins of cyclic AMP in higher plant tissues. *Plant Physiol.* **91**, 624–628.

Sprent, J. I. and Raven, J. A. (1985). Evolution of nitrogen-fixing symbioses. *Proc. R. Soc. Edin.* **85B**, 215–237.

Spurr, A. R. (1957). The effect of boron on cell-wall structure. *Am. J. Bot.* **44**, 637–650.

Srivastava, O. P. and Sethi, B. C. (1981). Contribution of farm yard manure on the build up of available zinc in an aridisol. *Commun. Soil Sci. Plant Anal.* **12**, 355–361.

Starck, Z., Choluj, D. and Szczepanska, B. (1980). Photosynthesis and photosynthates distribution in potassium-deficient radish plants treated with indolyl-3-acetic acid or gibberellic acid. *Photosynthetica* **14**, 497–505.

Stark, J. M. and Redente, E. F. (1990). Copper fertilization to prevent molybdenosis on retorted oil shale disposal piles. *J. Environ. Qual.* **19**, 502–504.

Starrach, N. and Mayer, W.-E. (1989). Changes of the apoplastic pH and K^+ concentration in the *Phaseolus* pulvinus in situ in relation to rhythmic leaf movements. *J. Exp. Bot.* **40**, 865–873.

Starrach, N., Flach, D. and Mayer, W. E. (1985). Activity of fixed negative charges of isolates extensor cell walls of the laminar pulvinus of primary leaves of *Phaseolus*. *J. Plant Physiol.* **120**, 441–455.

Stasovski, E. and Peterson, C. A. (1991). The effects of drought and subsequent rehydration on the structure and vitality of *Zea mays* seedling roots. *Can. J. Bot.* **69**, 1170–1178.

Steen, E. and Wünsche, U. (1990). Root growth dynamics of barley and wheat in field trials after CCC application. *Swedish J. Agric. Res.* **20**, 57–62.

Steer, B. T., Hocking, P. J., Kortt, A. A. and Roxburgh, C. M. (1984). Nitrogen nutrition of sunflower (*Helianthus annuus* L.) yield components, the timing of their establishment and seed characteristics in response to nitrogen supply. *Field Crops Res.* **9**, 219–236.

Steer, M. W. (1988a). Plasma membrane turnover in plant cells. *J. Exp. Bot.* **39**, 987–996.

Steer, M. W. (1988b). The role of calcium in exocytosis and endocytosis in plant cells. *Physiol. Plant.* **72**, 213–220.

Steffens, D. (1984). Wurzelstudien und Phosphat-Aufnahme von Weidelgras und Rotklee unter Feldbedingungen. *Z. Pflanzenernähr. Bodenk.* **147**, 85–97.

Steffens, D. and Mengel, K.(1980). Das Aneignungsvermögen von *Lolium perenne* im Vergleich zu *Trifolium pratense* für Zwischenschicht-Kalium der Tonminerale. *Landwirtsch. Forsch.* **36**, 120–127.

Steingröver, E. (1981). The relationship between cyanide-resistant root respiration and the storage of sugars in the transport in *Daucus carota* L. *J. Exp. Bot.* **32**, 911–919.

Steingröver, E. (1983). Storage of osmotically active compounds in the taproot of *Daucus carota* L. *J. Exp. Bot.* **34**, 425–433.

Steingröver, E., Oosterhuis, R. and Wieringa, F. (1982). Effect of light treatment and nutrition on nitrate accumulation in spinach (*Spinacea oleracea* L.). *Z. Pflanzenphysiol.* **107**, 97–102.

Stelzer, R. and Läuchli, A. (1977). Salz- und Überflutungstoleranz von *Puccinellia peisonis*. II. Strukturelle Differenzierung der Wurzel in Beziehung zur Funktion. *Z. Pflanzenphysiol.* **84**, 95–108.

Stelzer, R., Lehmann, H., Kramer, D. and Lüttge, U. (1990). X-ray microprobe analysis of vacuoles of spruce needle mesophyll, endodermis and transfusion parenchyma cells at different seasons of the year. *Bot. Acta* **103**, 415–423.

Stephan, U. W. and Grün, M. (1989). Physiological disorders of the nicotianamine-auxotroph tomato mutant *chloronerva* at different levels of iron nutrition. II. Iron deficiency response and heavy metal metabolism. *Biochem. Physiol. Pflanzen* **185**, 189–200.

Stephan, U. W. and Scholz, G. (1993). Nicotianamine: mediator of transport of iron and heavy metals in the phloem? *Physiol. Plant.* **88**, 522–529.

Stewart, G. R., Joly, C. A. and Smirnoff, N. (1992). Partitioning of inorganic nitrogen assimilation between roots and shoots of cerrado and forest trees of contrasting plant communities of South East Brazil. *Oecologia* **91**, 511–517.

Stewart, G. R., Gracia, C. A., Hegarty, E. E. and Specht, R. L. (1990). Nitrate reductase activity and chlorophyll content in sun leaves of subtropical Australian closed-forest (rainforest) and open-forest communities. *Oecologia* **82**, 544–551.

Stienen, H. and Bauch, J. (1988). Element content in tissues of spruce seedlings from hydroponic cultures simulating acidification and deacidification. *Plant Soil* **106**, 231–238.

Stock, W. D., Pate, J. S. and Delfs, J. (1990). Influence of seed size and quality on seedling development under low nutrient conditions in five Australian and South African members of the Proteaceae. *J. Ecol.* **78**, 1005–1020.

Stockman, Y. M., Fischer, R. A. and Brittain, E. G. (1983). Assimilate supply and floret development within the spike of wheat (*Triticum aestivum* L.). *Aust. J. Plant Physiol.* **10**, 585–594.

Storey, R. and Walker, R. R. (1987). Some effects of root anatomy on K, Na and Cl loading of citrus roots and leaves. *J. Exp. Bot.* **38**, 1769–1780.

Stout, P. R., Meager, W. R., Pearson, G. A. and Johnson, C. M. (1951). Molybdenum nutrition of crop plants. I. The influence of phosphate and sulfate on the absorption of molybdenum from soils and solution cultures. *Plant Soil* **3**, 51–87.

Strebel, O. and Duynisveld, W. H. M. (1989). Nitrogen supply to cereals and sugar beet by mass flow and diffusion on a silty loam soil. *Z. Pflanzenernähr. Bodenk.* **152**, 135–141.

Strebel, O., Grimme, H., Renger, M. and Fleige, H. (1980). A field study with nitrogen-15 of soil and fertilizer nitrate uptake and of water withdrawal by spring wheat. *Soil Sci.* **130**, 205–210.

Strebel, O., Duynisveld, W. H. M., Grimme, H., Renger, M. and Fleige, H. (1983). Wasserentzug durch Wurzeln und Nitratanlieferung (Massenfluss, Diffusion) als Funktion von Bodentiefe und Zeit bei einem Zuckerrübenbestand. *Mitt. Dtsch. Bodenkd. Ges.* **38**, 153–158.

Streeter, J. G. (1978). Effect of N starvation on soybean plants at various stages of growth on seed yield on N-concentration in plant parts at maturity. *Agron. J.* **70**, 74–76.

Streeter, J. G. (1979). Allantoin and allantoic acid in tissues and stem exudate from field-grown soybean plants. *Plant Physiol.* **63**, 478–480.

Streeter, J. G. (1985). Carbon metabolism in legume nodules. *In* 'Nitrogen Fixation Research Progress' (H. J. Evans, P. J. Bottomley and W. E. Newton, eds.), pp. 277–291. Martinus Nijhoff, Dordrecht.

Streeter, J. G. (1993). Translocation – a key factor limiting the efficiency of nitrogen fixation in legume nodules. *Physiol. Plant.* **87**, 616–623.

Stribley, D. P., Tinker, P. B. and Snellgrove, R. C. (1981). Effect of vesicular-arbuscular mycorrhizal fungi on the relations of plant growth, internal phosphorus concentration and soil phosphate analyses. *J. Soil Sci.* **31**, 655–672.

Strother, S. (1988). The role of free radicals in leaf senescence. *Gerontology* **34**, 151–156.

Struik, P. C., Geertsema, J. and Custers, C. H. M. G. (1989). Effect of shoot, root and stolon temperature on the development of the potato (*Solanum tuberosum* L.) plant. I. Development of the haulm. *Potato Res.* **32**, 133–141.

Strullu, D. G., Harley, L., Gourret, J. P. and Garrec, J. P. (1982). Ultra-structure and microanalysis of the polyphosphate granules of the endomycorrhizas of *Fagus sylvatica*. *New Phytol.* **92**, 417–424.

Stryker, R. B., Gilliam, J. W. and Jackson, W. A. (1974). Nonuniform transport of phosphorus from single roots to the leaves of *Zea mays*. *Physiol. Plant.* **30**, 231–239.

Stuiver, C. E. E., Kuiper, P. J. C. and Marschner, H. (1978). Lipids from bean, barley and sugar beet in relation to salt resistance. *Physiol. Plant.* **42**, 124–128.

Stuiver, C. E. E., Kuiper, P. J. C., Marschner, H. and Kylin, A. (1981). Effects of salinity and replacement of K^+ by Na^+ on lipid composition in two sugar beet inbred lines. *Physiol. Plant.* **52**, 77–82.

Stulen, I. and De Kok, L. J. (1993). Whole plant regulation of sulfur metabolism – a theoretical approach and comparison with current ideas on regulation of nitrogen metabolism. *In* 'Sulfur Nutrition and Assimilation in Higher Plants' (L. J. De Kok, I. Stulen, H. Rennenberg, C. Brunold and W. E. Rauser, eds.), pp. 77–91. SPB Academic, The Hague, The Netherlands.

Sturtevant, D. B. and Taller, B. J. (1989). Cytokinin production by *Bradyrhizobium japonicum*. *Plant Physiol.* **89**, 1247–1252.

Suelter, C. H. (1970). Enzymes activated by monovalent cations. *Science* **168**, 789–795.

Sugiyama, T., Nakyama, N. and Akazawa, T. (1968). Structure and function of chloroplast proteins. V. Homotropic effect of bicarbonate in RuBP carboxylase relation and the mechanism of activation by magnesium ions. *Arch. Biochem. Biophys.* **126**, 734–745.

Sugiyama, T., Matsumoto, C., Akazawa, T. and Miyachi, S. (1969). Structure and function of chloroplast proteins. VII. Ribulose-1,5-diphosphate carboxylase of *Chlorella ellipsoida*. *Arch. Biochem. Biophys.* **129**, 597–602.

Suhayda, C. G. and Haug, A. (1985). Citrate chelation as a potential mechanism against aluminum toxicity: The role of calmodulin. *Can. J. Biochem. Cell Biol.* **63**, 1167–1175.

Sumner, M. E. (1977). Application of Beaufils' diagnostic indices to maize data published in the literature irrespectively of age and conditions. *Plant Soil* **46**, 350–360.

Sumner, M. E. and Farina, P. M. W. (1986). Phosphorus interactions with other nutrients and lime in field cropping systems. *Adv. Soil Sci.* **5**, 210–236.

Sun, Y.-P. and Fries, N. (1992). The effect of tree-root exudates on the growth rate of ectomycorrhizal and saprotrophic fungi. *Mycorrhiza* **1**, 63–69.

Sundstrom, F. J., Morse, R. D. and Neal, J. L. (1982). Nodulation and nitrogen fixation of *Phaseolus vulgaris* L. grown in minesoil as affected by soil compaction and N fertilization. *Commun. Soil Sci. Plant Anal.* **13**, 231–242.

Sutherland, M. W. (1991). The generation of oxygen radicals during host plant responses to infection. *Physiol. Mol. Plant Pathol.* **39**, 79–93.

Sutherland, T. D., Bassam, B. J., Schuller, L. J. and Gresshoff, P. M. (1990). Early nodulation signals of the wild type and symbiotic mutants of soyben (*Glycine max*). *Mol. Plant–Microbe Interact.* **3**, 122–128.

Suthipradit, S. (1991). Effects of aluminium on growth and nodulation of some tropical crop legumes. PhD Thesis, University of Queensland, Australia.

Suthipradit, S., Edwards, D. G. and Asher, C. J. (1990). Effects of aluminium on tap-root elongation of soybean (*Glycine max*), cowpea (*Vigna unguiculata*) and green gram (*Vigna radiata*) grown in the presence of organic acids. *Plant Soil* **124**, 233–237.

Svensson, S.-B. (1971). The effect of coumarin on root growth and root histology. *Physiol. Plant.* **24**, 446–470.

Svensson, S.-B. (1972). The effect of coumarin on growth, production of dry matter, protein and nucleic acids in roots of maize and wheat and the interaction of coumarin with metabolic inhibitors. *Physiol. Plant.* **27**, 13–24.

Swamy, P. M. and Suguna, P. (1992). Influence of calcium chloride and benzyladenine on lipoxygenase of *Vigna unguiculata* leaf discs during senescence. *Physiol. Plant.* **84**, 467–471.

Sykes, S. R. (1992). The inheritance of salt exclusion in woody perennial fruit species. *Plant Soil* **146**, 123–129.

Sylvia, D. (1988). Activity of external hyphae of vesicular-arbuscular mycorrhizal fungi. *Soil Biol. Biochem.* **20**, 39–43.

Symeonidis, L. (1990). Tolerance of *Festuca rubra* L. to zinc in relation to mycorrhizal infection. *Biol. Metals* **3**, 204–207.

Ta, C. T. (1991). Nitrogen metabolism in the stalk tissue of maize. *Plant Physiol.* **97**, 1375–1380.

Tachibana, S. (1987). Effect of root temperature on the rate of water and nutrient absorption in cucumber cultivars and figleafgourd. *J. Jpn. Soc. Hortic. Sci.* **55**, 461–467.

Tachibana, S. (1988). Cytokinin concentrations in roots and root xylem exudate of cucumber and figleafgourd as affected by root temperature. *J. Jpn. Soc. Hortic. Sci.* **56**, 417–425.

Tachibana, S. (1991). Import of calcium by tomato fruit in relation to the day-night periodicity. *Sci. Hortic.* **45**, 235–243.

Tachimoto, M., Fukutomi, M., Matsushiro, H., Kobayashi, M. and Takahashi, E. (1992). Role of putrescine in *Lemna* plants under potassium deficiency. *Soil Sci. Plant Nutr. (Tokyo)* **38**, 307–313.

Tadano, T. and Sakai, H. (1991). Secretion of acid phosphatase by roots of several crop species under phosphorus-deficient conditions. *Soil Sci. Plant Nutr. (Tokyo)* **37**, 129–140.

Takagi, S., Nomoto, K. and Takemoto, T. (1984). Physiological aspect of mugineic acid, a possible phytosiderophore of graminaceous plants. *J. Plant Nutr.* **7**, 469–477.

Takahashi, E. and Miyake, Y. (1977). Silica and plant growth. *Proc. Int. Semin. Soil Environ. Fert. Manage. Intensive Agric.*, pp. 603–611.

Takahashi, H., Scott, T. K. and Suge, H. (1992). Stimulation of root elongation and curvature by calcium. *Plant Physiol.* **98**, 246–252.

Takeoka, Y., Kondo, K. and Kaufman, P. B. (1983). Leaf surface fine-structures in rice plants cultured under shaded, and non-shaded conditions. *Jpn. J. Crop Sci.* **52**, 534–543.

Takeoka, Y., Wada, T., Naito, K. and Kaufman, P. B. (1984). Studies on silification of epidermal tissues of grasses as investigated by soft X-ray image analysis. II. Differences in frequency of silica bodies in bulliform cells at different positions in the leaves of rice plants. *Jpn. J. Crop Sci.* **53**, 197–203.

Tal, M. (1985). Genetics of salt tolerance in higher plants: theoretical and practical considerations. *Plant Soil* **89**, 199–226.

Talha, M., Amberger, A., and Burkart, N. (1979). Effect of soil compaction and soil moisture level on plant growth and potassium uptake, *Z. Acker- Pflanzenbau* **148**, 156–164.

Tallman, G. and Zeiger, E. (1988). Light quality and osmoregulation in *Vicia* guard cells. Evidence for involvement of three metabolic pathways. *Plant Physiol.* **88**, 887–895.

Tan, K., Keltjens, W. G. and Findenegg, G. R. (1991). Role of magnesium in combination with liming in alleviating acid-soil stress with the aluminium-sensitive sorghum genotype CV 323. *Plant Soil* **136**, 65–71.

Tan, K., Keltjens, W. G. and Findenegg, G. R. (1992a). Acid soil damage in sorghum genotypes: Role of magnesium deficiency and root impairment. *Plant Soil* **139**, 149–155.

Tan, K., Keltjens, W. G. and Findenegg, G. R. (1992b). Aluminium toxicity with sorghum genotypes in nutrient solutions and its amelioration by magnesium. *Z. Pflanzenernähr. Bodenk.* **155**, 81–86.

Tan, K., Keltjens, W. G. and Findenegg, G. R. (1993). Evaluating the contribution of magnesium deficiency in the aluminium toxicity syndrome in twelve sorghum genotypes. *Plant Soil* **149**, 255–261.

Tanada, T. (1978). Boron-key element in the actions of phytochrome and gravity. *Planta* **143**, 109–111.

Tanada, T. (1982). Role of boron in the far-red delay of nyctinastic closure of *Albizzia* pinnules. *Plant Physiol.* **70**, 320–321.

Tanada, T. (1983). Localization of boron in membranes. *J. Plant Nutr.* **6**, 743–749.

Tanaka, A. and Navasero, S. A. (1964). Loss of nitrogen from the rice plant through rain or dew. *Soil Sci. Plant Nutr.* (*Tokyo*) **10**, 36–39.

Tanaka, H. (1966). Response of *Lemna pausicostata* to boron as affected by light intensity. *Plant Soil* **25**, 425–434.

Tanaka, H. (1967). Boron adsorption by plant roots. *Plant Soil* **27**, 300–302.

Tanaka, F., Ono, S. and Hayasaka, T. (1990). Identification and evaluation of toxicity of rice root elongation inhibitors in flooded soils with added wheat straw. *Soil Sci. Plant Nutr.* (*Tokyo*) **36**, 97–104.

Tang, C., Robson, A. D. and Dilworth, M. J. (1990). A split-root experiment shows that iron is required for nodule initiation in *Lupinus angustifolius* L. *New Phytol.* **115**, 61–67.

Tang, C., Robson, A. D. and Dilworth, M. J. (1991). Inadequate iron supply and high bicarbonate impair the symbiosis of peanut (*Arachis hypogaea* L.) with different *Bradyrhizobium* strains. *Plant Soil* **138**, 159–168.

Tang, C., Robson, A. D., Dilworth, M. J. and Kuo, J. (1992a). Microscopic evidence on how iron deficiency limits nodule initiation in *Lupinus angustifolius* L. *New Phytol.* **121**, 457–467.

Tang, C., Longnecker, N. E., Thomson, C. J., Greenway, H. and Robson, A. D. (1992b). Lupin (*Lupinus angustifolius* L.) and pea (*Pisum sativum* L.) roots differ in their sensitivity to pH above 6.0. *J. Plant Physiol.* **140**, 715–719.

Tang, P. M. and de la Fuente, R. K. (1986). Boron and calcium sites involved in indole-3-acetic acid transport in sunflower hypocotyl segments. *Plant Physiol.* **81**, 651–655.

Tanji, K., Läuchli, A. and Meyer, J. (1986). Selenium in the San Joaquin Valley. *Environment* **28**, No. 6-11, 34–39.

Tannenbaum, S. R., Fett, D., Young, V. R., Lan, P. D. and Bruce, W. R. (1978). Nitrite and nitrate are formed by endogenous synthesis in the human intestine. *Science* **200**, 1487–1488.

Tanner, P. D. (1978). A relationship between premature sprouting on the cob and the molybdenum and nitrogen status of maize grain. *Plant Soil* **49**, 427–432.

Tanner, P. D. (1982). The molybdenum requirements of maize in Zimbabwe. *Zimbabwe Agric. J.* **79**, 61–64.

Tanner, W. and Beevers, H. (1990). Does transpiration have an essential function in long-distance ion transport in plants? *Plant Cell Environ.* **13**, 745–750.

Tarafdar, J. C. and Jungk, A. (1987). Phosphatase activity in the rhizosphere and its relation to the depletion of soil organic phosphorus. *Biol. Fertil. Soils* **3**, 199–204.

Tarafdar, J. C. and Marschner, H.(1994). Phosphatase activity in the rhizosphere of VA-mycorrhizal wheat supplied with inorganic and organic phosphorus. *Soil Biology Biochem.* **26**, 387–395.

Tardieu, F., Zhang, J. and Davies, W. J. (1992). What information is conveyed by an ABA signal from maize roots in drying field soil? *Plant Cell Environ.* **15**, 185–191.

Taylor, G. J. (1988a). The physiology of aluminum phytotoxicity. *In* 'Metal Ions in

Biological Systems' (H. Sigel and A. Sigel, eds.) Vol. **24**, pp. 123–163. Marcel Dekker Inc. New York.

Taylor, G. J. (1988b). Mechanism of aluminum tolerance in *Triticum aestivum* (wheat). V. Nitrogen nutrition, plant-induced pH, and tolerance to aluminum; correlation without causality. *Can. J. Bot.* **66**, 694–699.

Taylor, G. J. and Foy, C. D. (1985). Differential uptake and toxicity of ionic and chelated copper in *Triticum aestivum*. *Can. J. Bot.* **63**, 1271–1275.

Taylor, H. M. and Ratliff, L. F. (1969). Root elongation rates of cotton and peanuts as a function of soil strength and soil water content. *Soil Sci.* **108**, 113–119.

Taylor, J. S, Thompson, B., Pate, J. S., Atkins, C. A. and Pharis, R. P. (1990). Cytokinins in the phloem sap of white lupin (*Lupinus albus* L.). *Plant Physiol.* **94**, 1714–1720.

Teasdale, R. D. and Richards, D. K. (1990). Boron deficiency in cultured pine cells. Quantitative studies of the interaction with Ca and Mg. *Plant Physiol.* **93**, 1071–1077.

Terashima, I. and Evans, J. R. (1988). Effects of light and nitrogen nutrition on the organization of the photosynthetic apparatus in spinach. *Plant Cell Physiol.* **29**, 143–155.

Termaat, A. and Munns, R. (1986). Use of concentrated macronutrient solutions to separate osmotic from NaCl-specific effects on plant growth. *Aust. J. Plant Physiol.* **13**, 509–522.

Terry, N. (1977). Photosynthesis, growth, and the role of chloride. *Plant Physiol.* **60**, 69–75.

Terry, N. (1980). Limiting factors in photosynthesis. I. Use of iron stress to control photochemical capacity in vivo. *Plant Physiol.* **65**, 114–120.

Terry, N. and Abadia, J. (1986). Function of iron in chloroplasts. *J. Plant Nutr.* **9**, 609–646.

Terry, N. and Low, G. (1982). Leaf chlorophyll content and its relation to the intracellular location of iron. *J. Plant Nutr.* **5**, 301–310.

Terry, N., Carlson, C., Raab, T. K. and Zayed, A. M. (1992). Rates of selenium volatilization among crop species. *J. Environ. Qual.* **21**, 341–344.

Tester, M. (1990). Plant ion channels: whole-cell and single-channel studies. *New Phytol.* **114**, 305–340.

Tester, M. and Blatt, M. R. (1989). Direct measurement of K^+ channels in thylakoid membranes by incorporation of vesicles into planar lipid bilayers. *Plant Physiol.* **91**, 249–252.

Thauer, R. K., Diekert, G. and Schönheit, P. (1980). Biological role of nickel. *Trends Biochem. Sci.* **5**, 304–306.

Thellier, M., Duval, Y. and Demarty, M. (1979). Borate exchanges of *Lemna minor* L. as studied with the help of the enriched stable isotope and of a (n, α) nuclear reaction. *Plant Physiol.* **63**, 283–288.

Theodorides, T. N. and Pearson, C. J. (1982). Effect of temperature on nitrate uptake, translocation and metabolism in *Pennisetum americanum*. *Aust. J. Plant Physiol.* **9**, 309–320.

Theodorou, M. E. and Plaxton, W. C. (1993). Metabolic adaptations of plant respiration to nutritional phosphate deprivation. *Plant Physiol.* **101**, 339–344.

Theologis, A., Zarembinski, T. I., Oeller, P. W., Liang, X. and Abel, S. (1992). Modification of fruit ripening by suppressing gene expression. *Plant Physiol.* **100**, 549–551.

Thiel, H. and Finck, A. (1973). Ermittlung von Grenzwerten optimaler Kupfer-Versorgung für Hafer und Sommergerste. *Z. Pflanzenernähr. Bodenk.* **134**, 107–125.

Thoene, B., Schröder, P., Papen, H., Egger, A. and Rennenberg, H. (1991). Absorption of atmospheric NO_2 by spruce (*Picea abies* L. Karst.) trees. I. NO_2 influx and its correlation with nitrate reduction. *New Phytol.* **117**, 575–585.

Thomas, J. and Prasad, R. (1983). Mineralization of urea, coated urea and nitrification inhibitor treated urea in different rice growing soils. *Z. Pflanzenernähr. Bodenk.* **146**, 341–347.

Thomas, R. B. and Strain, B. R.(1991). Root restriction as a factor in photosynthetic acclimation of cotton seedlings grain in elevated carbon dioxide. *Plant Physiol.* **96**, 627–634.

Thomas, W. A. (1967). Dye and calcium ascent in dogwood trees. *Plant Physiol.* **42**, 1800–1802.

Thompson, J. P. (1990). Soil sterilization methods to show VA-mycorrhizae aid P and Zn nutrition of wheat in vertisols. *Soil Biol. Biochem.* **22**, 229–240.

Thompson, J. P. and Wildermuth, G. B. (1989). Colonization of crop and pasture species with vesicular-arbuscular mycorrhizal fungi and a negative correlation with root infection by *Bipolaris sorokiniana*. *Can. J. Bot.* **69**, 687–693.

Thoms, K. and Sattelmacher, B. (1990). Influence of nitrate placement on morphology and physiology of maize (*Zea mays*) root systems. *In* 'Plant nutrition-physiology and applications' (M. L. van Beusichem, ed.), pp. 29–32. Kluwer Academic, Dordrecht, The Netherlands.

Thomson, B. D., Robson, A. D. and Abbott, L. K. (1990). Mycorrhizas formed by *Gigaspora calospara* and *Glomus fasciculatum* on subterranean clover in relation to soluble carbohydrate concentrations in roots. *New Phytol.* **114**, 217–225.

Thomson, B. D., Robson, A. D. and Abbott, L. K. (1992). The effect of long-term applications of phosphorus fertilizer on populations of vesicular-arbuscular mycorrhizal fungi in pastures. *Aust. J. Agric. Res.* **43**, 1131–1142.

Thomson, C. J., Atwell, B. J. and Greenway, H. (1989a). Response of wheat seedlings to low O_2 concentrations in nutrient solution. I. Growth, O_2 uptake and synthesis of fermentative end-products by root segments. *J. Exp. Bot.* **40**, 985–991.

Thomson, C. J., Atwell, B. J. and Greenway, H. (1989b). Response of wheat seedlings to low O_2 concentrations in nutrient solution. II. K^+/Na^+ selectivity of root tissues of different age. *J. Exp. Bot.* **40**, 993–999.

Thongbai, P., Graham, R. D., Neate, S. M. and Webb, M. J. (1993). Interaction between zinc nutritional status of cereals and *Rhizoctonia* root rot severity. II. Effect of Zn on disease severity of wheat under controlled conditions. *Plant Soil* **153**, 215–222.

Thorneley, R. N. F. (1992). Nitrogen fixation – a new light on nitrogenase. *Nature* **360**, 532–533.

Tice, K. R., Parker, D. R. and De Mason, D. A. (1991). Operational defined apoplastic and symplastic aluminum fractions in root tips of aluminum-intoxicated wheat. *Plant Physiol.* **100**, 309–318.

Tiller, K. and Merry, R. H. (1981). Copper pollution of agricultural soils. *In* 'Copper in Soils and Plants' (J. F. Loneragan, A. D. Robson and R. D. Graham, eds.), pp. 119–137. Academic Press, London.

Timm, C. A., Goos, R. J., Johnson, B. E., Siobolik, F. J. and Stack, R. W. (1986). Effect of potassium fertilizers on malting barley infected with common root rot. *Agron. J.* **78**, 197–200.

Tinker, P. B. (1980). Role of rhizosphere microorganisms in phosphorus uptake by plants. In *Role Phosphorus Agric.*, *Proc. Symp.*, 1976, pp. 617–654. ASA-CSSA-SSSA, Madison, WI.

Tinker, P. B., Jones, M. D. and Durall, D. M. (1992). A functional comparison of ecto- and endomycorrhizas. In 'Mycorrhizas in Ecosystems' (D. J. Read, D. H. Lewis, A. H. Fitter and I. J. Alexander, eds.), pp. 303–310. CAB International, Wellingford, UK.

Tisdall, J. M. (1991). Fungal hyphae and structural stability of soil. Aust. J. Soil Res. 29, 729–743.

Tollenaar, M. (1991). Physiological basis of genetic improvement of maize hybrids in Ontario from 1959 to 1988. Crop Sci. 31. 119–124.

Tomati, U. and Galli, E. (1979). Water stress and -SH-dependent physiological activities in young maize plants. J. Exp. Bot. 30, 557–563.

Tomlinson, J. A. and Hunt, J. (1987). Studies on watercress chlorotic leaf spot virus and on the control of the fungus vector (Spongospora subterranea) with zinc. Ann. Appl. Biol. 110, 75–88.

Tomsett, A. B. and Thurman, D. A. (1988). Molecular biology of metal tolerances of plants. Plant Cell Environ. 11, 383–394.

Torrey, J. G. and Racette, S. (1989). Specificity among the Casuarinaceae in root nodulation by Frankia. Plant Soil 118, 157–164.

Toth, R., Toth, D., Starke, D. and Smith, D. R. (1990). Vesicular-arbuscular mycorrhizal colonization in Zea mays affected by breeding for resistance to fungal pathogens. Can. J. Bot. 68, 1039–1044.

Toulon, V., Sentenac, H., Thibaud, J.-B., Soler, A., Clarkson, D. T. and Grignon, C. (1989). Effect of HCO_3^- concentration in the absorption solution on the energetic coupling of H^+-cotransport in roots of Zea mays L. Planta 179, 235–241.

Touraine, B., Grignon, N. and Grignon, C. (1990). Interaction between nitrate assimilation in shoots and nitrate uptake by roots of soybean (Glycine max) plants: Role of carboxylate. Plant Soil 124, 169–174.

Tran Dang Hong, Minchin, F. R. and Summerfield, R. J. (1977). Recovery of nodulated cowpea plants (Vigna unguiculata (L.) Walp.) from waterlogging during vegetative growth. Plant Soil 48, 661–672.

Treeby, M. and Uren, N. (1993). Iron deficiency stress response amongst citrus rootstocks. Z. Pflanzenernähr. Bodenk. 156, 75–81.

Treeby, M., Marschner, H. and Römheld, V. (1989). Mobilization of iron and other micronutrients from a calcareous soil by plant-borne, microbial, and synthetic metal chelators. Plant Soil 114, 217–226.

Treeby, M. T., Van Steveninck, R. F. M. and de Vries, H. M. (1987). Quantitative estimates of phosphorus concentrations within Lupinus luteus leaflets by means of electron probe X-ray microanalysis. Plant Physiol. 85, 331–334.

Trehan, S. P. and Sekhon, G. S. (1977). Effect of clay, organic matter and $CaCO_3$ content of zinc adsorption by soils. Plant Soil 46, 329–336.

Treharne, K. J. and Cooper, J. P. (1969). Effect of temperature on the activity of carboxylase in tropical and temperate Gramineae. J. Exp. Bot. 20, 170–175.

Trewavas, A. (1981). How do plant growth substances work? Plant Cell Environ. 4, 203–208.

Tricot, F., Crozat, Y., Tardieu, F. and Sebillotte, M. (1990). Establishment and distribution of pea primary root nodules (Pisum sativum L.) as affected by shading. Symbiosis 9, 97–103.

Triplett, E. W., Barnett, N. M. and Blevins, D. G. (1980). Organic acids and ionic balance in xylem exudate of wheat during nitrate or sulfate absorption. Plant Physiol. 65, 610–613.

Trobisch, S. and Schilling, G. (1969). Untersuchungen über Zusammenhänge zwischen

Massenentwicklung und N-Umsatz während der generativen Phase bei *Sinapis alba* L. *Albrecht-Thaer-Arch.* **13**, 867–878.

Trobisch, S. and Schilling, G. (1970). Beitrag zur Klärung der physiologischen Grundlage der Samenbildung bei einjährigen Pflanzen und zur Wirkung später zusätzlicher N-Gaben auf diesen Prozess am Beispiel von *Sinapis alba* L. *Albrecht-Thaer-Arch.* **14**, 253–265.

Trofymow, J. A., Coleman, D. C. and Cambardella, C. (1987). Rates of rhizodeposition and ammonium depletion in the rhizosphere of axenic oat roots. *Plant Soil* **97**, 333–344.

Trolldenier, G. (1977). Influence of some environmental factors on nitrogen fixation in the rhizosphere of rice. *Plant Soil* **47**, 203–317.

Trolldenier, G. (1981). Influence of soil moisture, soil acidity and nitrogen source on take-all of wheat. *Phytopathol. Z.* **102**, 163–177.

Trolldenier, G. (1988). Visualisation of oxidizing power of rice roots and of possible participation of bacteria in iron deposition. *Z. Pflanzenernähr. Bodenk.* **151**, 117–121.

Trolldenier, G. (1989). Plant nutritional and soil factors in relation to microbial activity in the rhizosphere, with particular emphasis on denitrification. *Z. Pflanzenernähr. Bodenk.* **152**, 223–230.

Trolldenier, G. and Hecht-Buchholz, C. (1984). Effect of aeration status of nutrient solution on microorganisms, mucilage and ultrastructure of wheat roots. *Plant Soil* **80**, 381–390.

Trought, M. C. T. and Drew, M. C. (1980a). The development of waterlogging damage in wheat seedlings (*Triticum aestivum* L.). I. Shoot and root growth in relation to changes in the concentration of dissolved gases and solutes in the soil solution. *Plant Soil* **54**, 77–94.

Trought, M. C. T. and Drew, M. C. (1980b). The development of waterlogging damage in wheat seedlings (*Triticum aestivum* L.). II. Accumulation and redistribution of nutrients by the shoot. *Plant Soil* **56**, 187–199.

Trought, M. C. T. and Drew, M. C. (1982). Effects of waterlogging on young wheat plants (*Triticum aestivum* L.) and on soil solutes at different soil temperatures. *Plant Soil* **69**, 311–326.

Trüby, P. and Lindner, M. (1990). Mangan-Verteilung in Fichten (*Picea abies* Karst.). *Angew. Botanik* **64**, 1–12.

Tsai, S. M., Da Silva, P. M., Cabezas, W. L. and Bonetti, R. (1993). Variability in nitrogen fixation of common bean (*Phaseolus vulgaris* L.) intercropped with maize. *Plant Soil* **152**, 93–101.

Tschaplinski, T. J. and Blake, T. J. (1989). The role of sink demand in carbon partitioning and photosynthetic reinvigoration following shoot decapitation. *Physiol. Plant.* **75**, 166–173.

Tsui, C. (1948). The role of zinc in auxin synthesis in the tomato plant. *Am. J. Bot.* **35**, 172–179.

Tucker, E. B. (1990). Calcium-loaded 1,2-bis(2-aminophenoxy)ethane-N, N, N, N', N'-tetraacetic acid blocks cell-to-cell diffusion of carbofluorescein in staminal hairs of *Setcreasea purpurea. Planta* **182**, 34–38.

Tukendorf, A. and Rauser, W. E. (1990). Changes in glutathione and phytochelatins in roots of maize seedlings exposed to cadmium. *Plant Sci.* **70**, 155–166.

Tukey jr, H. B. and Morgan, J. V. (1963). Injury to foliage and its effect upon the leaching of nutrients from above-ground plant parts. *Physiol. Plant.* **16**, 557–564.

Turgeon, R. (1989). The sink–source transition in leaves. *Annu. Rev. Plant Physiol. Plant Mol. Biol.* **40**, 119–138.

Turgeon, R. and Beebe, D. U. (1991). The evidence for symplastic phloem loading. *Plant Physiol.* **96**, 349–354.

Turnau, K., Kottke, I. and Oberwinkler, F. (1993). *Paxillus involutus – Pinus sylvestris* mycorrhizae from heavily polluted forest. I. Elemental localization using electron energy loss spectroscopy and imaging. *Bot. Acta* **106**, 213–219.

Turner, D. P. and Tingey, D. T. (1990). Foliar leaching and root uptake of Ca, Mg, and K in relation to acid fog effects on Douglas-fir. *Water Air Soil Pollut.* **49**, 205–214.

Turner, R. G. (1970). The subcellular distribution of zinc and copper within the roots of metal-tolerant clones of *Agrostis tenuis* Sibth. *New Phytol.* **69**, 725–731.

Twary, S. N. and Heichel, G. H. (1991). Carbon costs of dinitrogen fixation associated with dry matter accumulation in alfalfa. *Crop Sci.* **31**, 985–992.

Tyagi, V. K. and Chauhan, S. K. (1982). The effect of leaf exudates on the spore germination of phylloplane mycoflora of chilli (*Capsicum annuum* L.) cultivars. *Plant Soil* **65**, 249–256.

Tyerman, S. D. (1992). Anion channels in plants. *Annu. Rev. Plant Physiol. Plant Mol. Biol.* **43**. 351–371.

Tyler, G. (1992). Inability to solubilize phosphate in limestone soils – key factor controlling calcifuge habit of plants. *Plant Soil* **145**, 65–70.

Tyler, G., Berggren, D., Bergkvist, B., Falkengren-Grerup, U., Folkeson, L. and Rühling, Å. (1987). Soil acidification and metal solubility in forests of South Sweden. *In* 'Effects of Atmospheric Pollutants on Forests, Wetlands and Agricultural Ecosystems' (T. C. Hutchinson and K. M. Meema, eds.), pp. 347–359. NATO ASI Series, Vol G. 16. Springer-Verlag Berlin.

Tyree, M. T. (1970). The symplast concept. A general theory of symplastic transport according to the thermodynamics of irreversible processes. *J. Theor. Biol.* **26**, 181–224.

Tyree, M. T. and Ewers, F. W. (1991). The hydraulic architecture of trees and other woody plants. *New Phytol.* **119**, 345–360.

Tyree, M. T., Scherbatskoy, T. D. and Tabor, C. A. (1990). Leaf cuticles behave as asymetric membranes. Evidence from the measurement of diffusion potentials. *Plant Physiol.* **92**, 103–109.

Udvardi, M. K. and Day, D. A. (1990). Ammonia (^{14}C-methylamine) transport across the bacteroid and peribacteroid membranes of soybean root nodules. *Plant Physiol.* **94**, 71–76.

Ueckert, J., Hurek, T., Fendrik, I. and Niemann, E.-G. (1990). Radial gas diffusion from roots of rice (*Oryza sativa* L.) and Kallar grass (*Leptochloa fusca* L. Kunth), and effects of inoculation with *Azospirillum brasilense* Cd. *Plant Soil* **122**, 59–65.

Uehara, K., Fujimoto, S. and Taniguchi, T. (1974). Studies on violet-colored acid phosphatase of sweet potato. II. Enzymatic properties and amino acid composition. *J. Biochem. (Tokyo)* **75**, 639–649.

Uexküll, H. R. von (1985). Chlorine in the nutrition of palm trees. *Oleagineux* **40**, 67–72.

Ullrich, C. I. and Novacky, A. J. (1990). Extra- and intracellular pH and membrane potential changes induced by K^+, Cl^-, $H_2PO_4^-$, and NO_3^- uptake and fusicoccin in root hairs of *Limnobium stoloniferum*. *Plant Physiol.* **94**, 1561–1567.

Ulrich, A. and Ohki, K. (1956). Chlorine, bromine and sodium as nutrients for sugar beet plants. *Plant Physiol.* **31**, 171–181.

Uren, N. C. and Reisenauer, H. M. (1988). *In* 'The role of root exudation in nutrient acquisition', Advances Plant Nutrition, Vol. 3 (B. Tinker and A. Läuchli, eds.), pp.79–114. Praeger, New York.

Urquiaga, S., Cruz, K. H. S. and Boddey, R. M. (1992). Contribution of nitrogen fixation to sugar cane: nitrogen-15 and nitrogen-balance estimates. *Soil Sci. Soc. Am. J.* **56**, 105–114.

U.S. Salinity Laboratory Staff (1954). Diagnosis and improvement of saline and alkali soils. *U.S. Dept. Agric. Agric. Handb.* **60**.

Utsunomiya, E. and Muto, S. (1993). Carbonic anhydrase in the plasma membranes from leaves of C_3 and C_4 plants. *Physiol. Plant.* **88**, 413–419.

Vakhmistrov, D. B. (1967). On the function of the apparent free space in plant roots. A comparative study of the absorption power of epidermal and cortex cells in barley roots. *Sov. Plant Physiol.* (*Engl. Transl.*) **14**, 123–129.

Vakhmistrov, D. B. (1981). Specialization of root tissues in ion transport. *Plant Soil* **63**, 33–38.

Valenti, V., Scalorbi, M. and Guerrini, F. (1991). Induction of plasma membrane NADH-ferricyanide reductase following iron stress in tomato roots. *Plant Physiol. Biochem.* **29**, 249–255.

Vallee, B. L. and Auld, D. S. (1990). Zinc coordination, function, and structure of zinc enzymes and other proteins. *Biochemistry* **29**, 5647–5659.

Vallee, B. L. and Falchuk, K. H. (1993). The biochemical basis of zinc physiology. *Physiol. Rev.* **73**, 79–118.

Van Assche, F. and Clijsters, H. (1986a). Inhibition of photosynthesis in *Phaseolus vulgaris* by treatment with toxic concentration of zinc: effect on ribulose-1,5-bisphosphate carboxylase/oxygenase. *J. Plant Physiol.* **125**, 355–360.

Van Assche, F. and Clijsters, H. (1986b). Inhibition of photosynthesis in *Phaseolus vulgaris* by treatment with toxic concentrations of zinc: effects on electron transport and photophosphorylation. *Physiol. Plant.* **66**, 717–721.

Van Bel, A. J. E. (1984). Quantification of the xylem-to-phloem transfer of amino acids by use of inulin (^{14}C) carboxylic acid as xylem transport marker. *Plant Sci. Lett.* **35**, 81–85.

Van Bel, A. J. E. (1989). The challenge of symplastic phloem loading. *Bot. Acta* **102**, 183–185.

Van Bel, A. J. E. (1993). Strategies of phloem loading. *Annu. Rev. Plant Physiol. Plant Mol. Biol.* **44**, 253–281.

Van Bel, A. J. E., van Kesteren, W. J. P. and Papenhuiijzen, C. (1988). Ultrastructural indications for coexistence of symplastic and apoplastic phloem loading in *Commelina benghalensis* leaves. *Planta* **176**, 159–172.

Van Bel, A. J. E. and van Patrick, J. W. (1985). Proton extrusion in seed coats of *Phaseolus vulgaris* L. *Plant Cell Environ.* **8**, 1–6.

Van Beusichem, M. L., Kirkby, E. A. and Baas, R. (1988). Influence of nitrate and ammonium nutrition and the uptake, assimilation, and distribution of nutrients in *Ricinus communis*. *Plant Physiol.* **86**, 914–921.

Van Campen, D. R. (1991). Trace elements in human nutrition. *In* 'Micronutrients in Agriculture' 2nd edn. (J. J. Mortvedt, F. R. Cox, L. M. Shuman and R. M. Welch, eds.), pp. 663–701. SSSA Book Series No. 4, Madison, WI.

Vance, C. P. and Gantt, J. S. (1992). Control of nitrogen and carbon metabolism in root nodules. *Physiol. Plant.* **85**, 266–274.

Vance, C. P. and Heichel, G. H. (1991). Carbon in N_2 fixation: limitation or exquisite adaptation. *Annu. Rev. Plant Physiol. Plant Mol. Biol.* **42**, 373–392.

Van Cutsem, P. and Gillet, C. (1982). Activity coefficient and selectivity values of

Cu^{2+}, Zn^{2+} and Ca^{2+} ions adsorbed in the *Nitella flexilis* L. cell wall during triangular ion exchanges. *J. Exp. Bot.* **33**, 847–853.

Van den Berg, H. J., Vreugdenhil, D., Ludford, P. M., Hillman, L. L. and Ewing, E. E. (1991). Changes in starch, sugar, and abscisic acid content associated with second growth in tubers of potato (*Solanum tuberosum* L.) one-leaf cuttings. *J. Plant Physiol.* **139**, 86–89.

Van den Driessche, R. (1987). Importance of current photosynthate to new growth in planted conifer seedlings. *Can. J. For. Res.* **17**, 776–782.

Van der Boon, J., Steenhuizen, J. W. and Steingrover, E. G. (1990). Growth and nitrate concentration of lettuce as affected by total nitrogen and chloride concentrations, NH_4/NO_3 ratio and temperature of the circulating nutrient solution. *J. Hortic. Sci.* **65**, 309–321.

Van der Mark, F., van den Briel, M. L., van Oers, J. W. A. M. and Bienfait, H. F. (1982). Ferritin in bean leaves with constant and changing iron status. *Planta* **156**, 341–344.

Van Miegroet, H. and Cole, D. W. (1984). The impact of nitrification on soil acidification and cation leaching in a red alder ecosystem. *J. Environ. Qual.* **13**, 586–590.

Van Raij, B. (1991). Fertility of acid soils. *In* 'Plant–Soil Interactions at Low pH' (R. J. Wright, V. C. Baligar and R. P. Murrmann, eds.), pp. 159–167. Kluwer Academic, Dordrecht, The Netherlands.

Van Sanford, D. A. and MacKown, C. T. (1987). Cultivar differences in nitrogen remobilization during grain fill in soft red winter wheat. *Crop Sci.* **27**, 295–300.

Van Staden, J. and Davey, J. E. (1979). The synthesis, transport and metabolism of endogenous cytokinins. *Plant Cell Environ.* **2**, 93–106.

Van Steveninck, R. F. M. (1965). The significance of calcium on the apparent permeability of cell membranes and the effects of substitution with other divalent ions. *Physiol. Plant.* **18**, 54–69.

Van Steveninck, R. F. M., Van Steveninck, M. E., Fernando, D. R., Horst, W. J. and Marschner, H. (1987a). Deposition of zinc phytate in globular bodies in roots of *Deschampsia caespitosa* ecotypes; a detoxification mechanism? *J. Plant Physiol.* **131**, 247–257.

Van Steveninck, R. F. M., Van Steveninck, M. E., Fernando, D. R., Godbold, D. L., Horst, W. J. and Marschner, H. (1987b). Identification of zinc-containing globules in roots of a zinc-tolerant ecotype of *Deschampsia caespitosa*. *J. Plant Nutr.* **10**, 1239–1246.

Van Steveninck, R. F. M., Fernando, D. R., Anderson, C. A., Edwards, L. B. and Van Steveninck, M. E. (1988). Chloride and sulphur concentrations in chloroplasts of spinach. *Physiol. Plant.* **74**, 651–658.

Vaughan, A. K. F. (1977). The relation between the concentration of boron in the reproductive and vegetative organs of maize plants and their development. *Rhod. J. Agric. Res.* **15**, 163–170.

Vaughan, D. and Ord, B. (1990). Influence of phenolic acids on morphological changes in roots of *Pisum sativum*. *J. Sci. Food Agric.* **52**, 289–299.

Vaughan, D., DeKock, P. C. and Ord, B. G. (1982). The nature and localization of superoxide dismutase in fronds of *Lemna gibba* L. and the effect of copper and zinc deficiency on its activity. *Physiol. Plant.* **54**, 253–257.

Vaughn, K. C. and Campbell, W. H. (1988). Immunogold localization of nitrate reductase in maize leaves. *Plant Physiol.* **88**, 1354–1357.

Vazquez, M. D., Barcelo, J., Poschenrieder, Ch., Madico, J., Hatton, P., Baker, A. J.

M. and Cope, G. H. (1992). Localization of zinc and cadmium in *Thlaspi caerulescens* (Brassicaceae), a metallophyte that can hyperaccumulate both metals. *J. Plant Physiol.* **140**, 350–355.

Veen, B. W. and Kleinendorst, A. (1985). Nitrate accumulation and osmotic regulation in Italian ryegrass (*Lolium multiflorum* Lam.). *J. Exp. Bot.* **36**, 211–218.

Veen, B. W., Van Noordwijk, M., De Willigen, P., Boone, F. R. and Kooistra, M. J. (1992). Root–soil contact of maize, as measured by a thin-section technique. III. Effects on shoot growth, nitrate and water uptake efficiency. *Plant Soil* **139**, 131–138.

Venkatarayappa, T., Tsujita, M. J. and Murr, D. P. (1980). Influence of cobaltous ion (Co^{2+}) on the postharvest behaviour of 'Samanta' roses. *J. Am. Soc. Hortic. Sci.* **105**, 148–151.

Venter, H. A. van de and Currier, H. B. (1977). The effect of boron deficiency on callose formation and ^{14}C translocation in bean (*Phaseolus vulgaris*) and cotton (*Gossypium hirsutum* L.). *Am. J. Bot.* **64**, 861–865.

Verkleij, J. A. C. and Schat, H. (1989). Mechanism of metal tolerance in higher plants. *In* 'Heavy Metal Tolerance in Plants: Evolutionary Aspects' (A. J. Shaw, ed.), pp. 179–193. CRC Press, Boca Raton, FL.

Verkleij, J. A. C., Koevoets, P., van't Riet, J., Bank, R., Nijdam, Y. and Ernst, W. H. O. (1990). Poly (γ-glutamylcysteinyl) glycines or phytochelatins and their role in cadmium tolerance of *Silene vulgaris*. *Plant Cell Environ.* **13**, 413–421.

Vermeer, J. and McCully, M. E. (1981). Fucose in the surface deposits of axenic and field grown roots of *Zea mays* L. *Protoplasma* **109**, 233–248.

Vertregt, N. (1968). Relation between black spot and composition of the potato tuber. *Eur. Potato J.* **11**, 34–44.

Vesk, M., Possingham, V. and Mercer, F. V. (1966). The effect of mineral nutrient deficiency on the structure of the leaf cells of tomato, spinach and maize. *Aust. J. Bot.* **14**, 1–18.

Vessey, J. K. and Waterer, J. (1992). In search of the mechanism of nitrate inhibition of nitrogenase activity in legume nodules: recent developments. *Physiol. Plant.* **84**, 171–176.

Vessey, J. K., Walsh, K. B. and Layzell, D. B. (1988). Oxygen limitation of N_2 fixation in stem-girdled and nitrate-treated soybean. *Physiol. Plant.* **73**, 113–121.

Vetter, H. and Teichmann, W. (1968). Feldversuche mit gestaffelten Kupfer- und Stickstoff-Düngergaben in Weser-Ems. *Z. Pflanzenernähr. Bodenk.* **121**, 97–111.

Vetter, H., Früchtenicht, K. and Mählhop, R.(1978). Untersuchungen über den Aussagewert verschiedener Bodenuntersuchungsmethoden für die Ermittlung des Phosphatdüngerbedarfs. *Landwirtsch. Forsch. Sonderh.* **34**, 121–132.

Vetterlein, D. and Marschner, H. (1993). Use of a microtensiometer technique to study hydraulic lift in a sandy soil planted with pearl millet (*Pennisetum americanum* L. Leeke). *Plant Soil* **149**, 275–282.

Vianello, A. and Macri, F. (1991). Generation of superoxide anion and hydrogen peroxide at the surface of plant cells. *J. Bioenerg. Biomembr.* **23**, 409–423.

Vielemeyer, H. P., Fischer, F. and Bergmann, W. (1969). Untersuchungen über den Einfluss der Mikronährstoffe Eisen und Mangan auf den Stickstoff-Stoffwechsel landwirtschaftlicher Kulturpflanzen. 2. Mitt.: Untersuchungen über die Wirkung des Mangans auf die Nitratreduktion und den Gehalt an freien Aminosäuren in jungen Buschbohnenpflanzen. *Albrecht-Thaer-Arch.* **13**, 393–404.

Vierheilig, H. and Ocampo, J. A. (1991). Receptivity of various wheat cultivars to infection by VA-mycorrhizal fungi as influenced by inoculum potential and the relation of VAM-effectiveness to succinic dehydrogenase activity of the mycelium in the roots. *Plant Soil* **133**, 291–296.

Viets jr, F. G. (1944). Calcium and other polyvalent cations as accelerators of ion accumulation by excised barley roots. *Plant Physiol.* **19**, 466–480.

Villeneuve, N., Le Tacon, F. and Bouchard, D. (1991). Survival of inoculated *Laccaria bicolor* in competition with native ectomycorrhizal fungi and effects on the growth of outplanted Douglas fir seedlings. *Plant Soil* **135**, 95–107.

Vincent, C. D. and Gregory, P. J. (1989). Effects of temperature on the development and growth of winter wheat roots. II. Field studies of temperature, nitrogen and irradiance. *Plant Soil* **199**, 99–110.

Vinther, F. P. (1982). Nitrogenase activity (acetylene reduction) during the growth cycle of spring barley (*Hordeum vulgare* L.). *Z. Pflanzenernähr. Bodenk.* **145**, 356–362.

Vlamis, J. and Williams, D. E. (1964). Iron and manganese relations in rice and barley. *Plant Soil* **20**, 221–231.

Vlamis, J. and Williams, D. E. (1967). Manganese and silicon interaction in the *Gramineae*. *Plant Soil* **27**, 131–140.

Vogt, H., Holtum, J., Bücker, J. and Latzko, E. (1987). Daily pattern of proton extrusion by roots of *Zea mays* cv. Limac. *J. Plant Physiol.* **128**, 405–415.

Vogt, K. A., Publicover, D. A. and Vogt, D. J. (1991). A critique of the role of ectomycorrhizas in forest ecology. *Agric. Ecosyst. Environ.* **35**, 171–190.

Volk, R. J., Kahn, R. P. and Weintraub, R. L. (1958). Silicon content of the rice plant as a factor influencing its resistance to infection by the blast fungus *Pyricularia oryzae*. *Phytopathology* **48**, 179–184.

Von Schaewen, A., Stitt, M., Schmidt, R., Sonnewald, U. and Willmitzer, L. (1990). Expression of yeast-derived invertase in the cell wall of tobacco and *Arabidopsis* plants leads to accumulation of carbohydrate and inhibition of photosynthesis and strongly influences growth and phenotype of transgenic tobacco plants. *EMBO J.* **9**, 3033–3044.

Vorm, P. D. J. van der (1980). Uptake of Si by five plant species as influenced by variations in Si-supply. *Plant Soil* **56**, 153–156.

Vorster, P. W. and Jooste, J. H. (1986). Translocation of potassium and phosphate from ordinary and proteoid roots to shoots in the Proteaceae. *S. Afr. J. Bot.* **52**, 282–285.

Vose, P. B. (1982). Iron nutrition in plants: a world overview. *J. Plant Nutr.* **5**, 233–249.

Vose, P. B. (1983). Rationale of selection for specific nutritional characters in crop improvement with *Phaseolus vulgaris* L. as a case of study. *Plant Soil* **72**, 351–364.

Vreugdenhil, D. (1983). Characterization of absorption sites for aluminium in the roots. *Soil Sci. Plant Nutr.* (*Tokyo*) **29**, 499–515.

Vreugdenhil, D. (1991). Hormonal regulation of tuber formation. *Kali-Briefe* **20**, 605–611.

Vunkova-Radeva, R., Schiemann, J., Mendel, R.-R., Salcheva, G. and Georgieva, D. (1988). Stress and activity of molybdenum-containing complex (molybdenum cofactor) in winter wheat seeds. *Plant Physiol.* **87**, 533–535.

Wagatsuma, T. (1983). Characterization of absorption sites for aluminium in the roots. *Soil Sci. Plant Nutr.* (*Tokyo*) **29**, 499–515.

Wagatsuma, T. and Akiba, R. (1989). Low surface negativity of root protoplasts from aluminum-tolerant plant species. *Soil Sci. Plant Nutr.* (*Tokyo*) **35**, 443–452.

Wagner, H. and Michael, G. (1971). Der Einfluss unterschiedlicher Stickstoffversorgung auf die Cytokininbildung in Wurzeln von Sonnenblumenpflanzen. *Biochem. Physiol. Pflanz.* **162**, 147–158.

Wahle, K. W. J. and Davies, N. T. (1977). Involvement of copper in microsomal mixed-function oxidase reactions: a review. *J. Sci. Food Agric.* **28**, 93–97.

Waisel, Y., Eshel, A. and Agami, M. (1986). Salt balance of leaves of the mangrove *Avicennia marina*. *Physiol. Plant.* **67**, 67–72.

Wakabayashi, K., Sakurai, M. and Kuraishi, S. (1991). Differential effect of auxin on molecular weight distributions of xyloglucans in cell walls of outer and inner tissues form segments of dark grown squash (*Cucurbita maxima* Duch.) hypocotyls. *Plant Physiol.* **95**, 1070–1076.

Wakhloo, J. L. (1975a). Studies of the growth, flowering, and production of female sterile flowers as affected by different levels of foliar potassium in *Solanum sisymbrifolium* Lam. I. Effect of potassium content of the plant on vegetative growth and flowering. *J. Exp. Bot.* **26**, 425–433.

Wakhloo, J. L. (1975b). Studies on the growth, flowering and production of female sterile flowers as affected by different levels of foliar potassium in *Solanum sisymbrifolium* Lam. II. Interaction between foliar potassium and applied gibberellic acid and 6-furfurylaminopurine. *J. Exp. Bot.* **26**, 433–440.

Waldron, L. J., Terry, N. and Nemson, J. A. (1985). Diurnal cycles of leaf extension in unsalinized and salinized *Beta vulgaris*. *Plant Cell Environ.* **8**, 207–211.

Walker, C. D. and Lance, R. C. M. (1991). Silicon accumulation and ^{13}C composition as indices of water-use efficiency in barley cultivars. *Aust. J. Plant Physiol.* **18**, 427–434.

Walker, C. D. and Welch, R. M. (1987). Low molecular weight complexes of zinc and other trace metals in lettuce leaf. *J. Agric. Food Chem.* **35**, 721–726.

Walker, C. D., Graham, R. D., Madison, J. T., Cary, E. E. and Welch, R. M (1985). Effects of Ni deficiency on some nitrogen metabolites in cowpea (*Vigna unguiculata* L. Walp). *Plant Physiol.* **79**, 474–479.

Walker, C. J. and Weinstein, J. D. (1991). Further characterization of the magnesium chelatase in isolated developing cucumber chloroplasts. *Plant Physiol.* **95**, 1189–1196.

Walker, D. (1992). Excited leaves. *New Phytol.* **121**, 325–346.

Walker, D. A. (1980). Regulation of starch synthesis in leaves – the role of orthophosphate. *Proc. 15th Colloq. Int. Potash Inst. Bern*, pp. 195–207.

Walker, J. M. (1969). One-degree increments in soil temperature affect maize seedling behaviour. *Soil Sci. Soc. Am. Proc.* **33**, 729–736.

Wallace, A. (1980a). Effect of excess chelating agent on micronutrient concentrations in bush beans grown in solution culture. *J. Plant Nutr.* **2**, 163–170.

Wallace, A. (1980b). Effect of chelating agents on uptake of trace metals when chelating agents are supplied to soil in contrast to when they are applied to solution culture. *J. Plant Nutr.* **2**, 171–175.

Wallace, A. (1982). Effect of nitrogen fertilizer and nodulation on lime-induced chlorosis in soybean. *J. Plant Nutr.* **5**, 363–368.

Wallace, A., Frolich, E. and Lunt, O. R. (1966). Calcium requirements of higher plants. *Nature (London)* **209**, 634.

Wallace, A., Abou-Zamzan, A. M. and Motoyama, E. (1971). Cation and anion balance in the xylem exudate of tobacco roots. *Plant Soil* **35**, 433–438.

Wallace, T. (1961). 'The Diagnosis of Mineral Deficiencies in Plants by Visual Symptoms. A Colour Atlas and Guide'. HMSO, London.

Wallace, W. and Pate, J. S. (1965). Nitrate reductase in the field pea (*Pisum arvense* L.). *Ann. Bot.* (*London*) [*N.S.*] **29**, 655–671.

Wallander, H. and Nylund, J.-E. (1991). Effects of excess nitrogen on carbohydrate concentration and mycorrhizal development of *Pinus sylvestris* L. seedlings. *New Phytol.* **119**, 405–411.

Wallsgrove, R. M., Keys, A. J., Lea, P. J. and Miflin, B. J. (1983). Photosynthesis, photorespiration and nitrogen metabolism. *Plant Cell Environ.* **6**, 301–309.

Walsh, K. B. (1990). Vascular transport and soybean nodule function. III. Implications of a continual phloem supply of carbon and water. *Plant Cell Environ.* **13**, 893–901.

Walters, D. R. and Wylie, M. A.(1986). Polyamines in discrete regions of barley leaves infected with the powdery mildew fungus, *Erysiphe graminis*. *Physiol. Plant.* **67**, 630–633.

Walworth, J. L. and Sumner, M. E. (1988). Foliar diagnosis: a review. *In* 'Advances in Plant Nutrition' Vol. 3 (B. Tinker and A. Läuchli, eds.), pp. 193–241. Praeger, New York.

Wang, C. H., Liem, T. H. and Mikkelsen, D. S. (1976). Sulfur deficiency – a limiting factor in rice production in the lower Amazon basin. II.Sulfur requirement for rice production. *IRI Res. Inst.* [*Rep.*] **48**, 9–30.

Wang, T. S. C., Yang, T. K. and Chuang, Z. T. (1967). Soil phenolic acids as plant growth inhibitors. *Soil Sci.* **103**, 239–246.

Wang, T.-W. and Arteca, R. N. (1992). Effects of low O_2 root stress on ethylene biosynthesis in tomato plants (*Lycopersicon esculentum* Mill. cv. Heinz 1350). *Plant Physiol.* **98**, 97–100.

Wang,, X.-L., Canny, M. J. and McCully, M. E. (1991). The water status of the roots of soil-grown maize in relation to the maturity of their xylem. *Physiol. Plant.* **82**, 157–162.

Wani, S. P., Chandrapalaih, S., Zambre, M. A. and Lee, K. K. (1988). Association between N_2-fixing bacteria and pearl millet plants: responses, mechanisms and persistence. *Plant Soil* **110**, 289–302.

Warburg, O. and Lüttgens, W. (1946). Photochemical reduction of quinone in green cells and granules. *Biochimia* **11**, 303–322.

Warden, B. T. (1991). Manganese extracted from different chemical fractions of bulk and rhizosphere soil as affected by method of sample preparation. *Commun. Soil Sci. Plant Anal.* **22**, 169–176.

Warembourg, F. R. and Billes, G. (1979). Estimation carbon transfers in the plant rhizosphere. *In* 'The Soil–Root Interface' (J. L. Harley and R. Scott-Russell, eds.), pp.183–196. Academic Press, London.

Warembourg, F. R. and Roumet, C. (1989). Why and how to estimate the cost of symbiotic N_2 fixation? A progressive approach based on the use of ^{14}C and ^{15}N isotypes. *Plant Soil* **115**, 167–177.

Warner, R. L. and Kleinhofs, A. (1992). Genetics and molecular biology of nitrate metabolism in higher plants. *Physiol. Plant.* **85**, 245–252.

Waschkies, C., Schropp, A. and Marschner, H. (1993). Relations between replant disease, growth parameters and mineral nutrition status of grapevines (*Vitis* sp.). *Vitis* **32**, 69–76.

Waschkies, C., Schropp, A. and Marschner, H. (1994). Relations between grapevine replant disease and root colonization of grapevine (*Vitis* sp.) by fluorescent pseudomonads and endomycorrhizal fungi. *Plant Soil* **162**, 219–227.

Watanabe, I. (1986). Nitrogen fixation by non-legumes in tropical agriculture with special reference to wetland rice. *Plant Soil* **90**, 343–357.

Waterer, J. G., Vessey, J. K. and Raper jr, C. D. (1992). Stimulation of nodulation in field peas (*Pisum sativum*) by low concentrations of ammonium in hydroponic culture. *Physiol. Plant.* **86**, 215–220.

Waters, I., Morrell, S., Greenway, H. and Colmer, T. D. (1991). Effects of anoxia on wheat seedlings. II. Influence of O_2 supply prior to anoxia on tolerance to anoxia, alcoholic fermentation and sugar levels. *J. Exp. Bot.* **42**, 1437–1447.

Waters, S. P., Martin, P. and Lee, B. T. (1984). The influence of sucrose and abscisic acid on the determination of grain number in wheat. *J. Exp. Bot.* **35**, 829–840.

Watson, E. R., Lapins, P. and Barron, R. J. W. (1976). Effect of waterlogging on the growth, grain and straw yield of wheat, barley and oats. *Aust. J. Exp. Agric. Anim. Husb.* **16**, 114–122.

Watteau, F. and Berthelin, J. (1990). Iron solubilization by mycorrhizal fungi producing siderophores. *Symbiosis* **9**, 59–67.

Webb, M. J. and Loneragan, J. F. (1988). Effect of zinc deficiency on growth, phosphorus concentration, and phosphorus toxicity of wheat plants. *Soil Sci. Soc. Am.J.* **52**, 1676–1680.

Webb, M. J. and Loneragan, J. F. (1990). Zinc translocation to wheat roots and its implications for a phosphorus/zinc interaction in wheat plants. *J. Plant Nutr.* **13**, 1499–1512.

Webb, T. and Armstrong, W. (1983). The effect of anoxia and carbohydrates on the growth and viability of rice, pea and pumpkin roots. *J. Exp. Bot.* **34**, 579–603.

Weber, E., Saxena, M. C., George, E. and Marschner, H. (1993). Effect of vesicular-arbuscular mycorrhiza on vegetative growth and harvest index of chickpea grown in northern Syria. *Field Crops Res.* **32**, 115–128.

Wedding, R. T. and Black, M. K. (1988). Role of magnesium in the binding of substrate and effectors to phosphoenolpyruvate carboxylase from a CAM plant. *Plant Physiol.* **87**, 443–446.

Wedler, A. (1980). Untersuchungen über Nitratgehalte in einigen ausgewählten Gemüsearten. *Landwirtsch. Forsch. Sonderh.* **36**, 128–137.

Wegner, L. H. and Raschke, K. (1994). Ion channels in the xylem parenchyma of barley roots: a procedure to isolate protoplasts from this tissue and a patch-clamp exploration of salt passageways into xylem vessels. *Plant Physiol.* **105**, 799–813.

Wehrmann, J. and Scharpf, H. J. (1986). The N_{min}-method – an aid to integrating various objectives of nitrogen fertilization. *Z. Pflanzenernähr. Bodenk.* **149**, 428–440.

Weiler, E. W. (1993). Octadecanoid-derived signaling molecules involved in touch perception in a higher plant. *Bot. Acta* **106**, 2–4.

Weiler, E. W. and Ziegler, H. (1981). Determination of phytohormones in phloem exudate from species by radioimunoassay. *Planta* **152**, 168–170.

Weiner, H., Blechschmidt-Schneider, S., Mohme, H., Eschrich, W. and Heldt, H. W. (1991). Phloem transport of amino acids. Comparison of amino acid content of maize leaves and of the sieve tube exudate. *Plant Physiol. Biochem.* **29**, 19–23.

Weir, R. C. and Hudson, A. (1966). Molybdenum deficiency in maize in relation to seed reserves. *Aust. J. Exp. Agric. Anim. Husb.* **6**, 35–41.

Weiser, C. J., Blaney, L. T. and Li, P. (1964). The question of boron and sugar translocation. *Physiol. Plant.* **17**, 589–599.

Weiss, A. and Herzog, A. (1978). Isolation and characterization of a silicon-organic complex from plants. In 'Biochemistry of Silicon and Related Problems' (G. Bendz and I. Lindqvist, eds.), pp. 109–127. Plenum, New York.

Weiss, M. G. (1943). Inheritance and physiology of efficiency in iron utilization in soybeans. *Genetics* **28**, 253–268.

Weisz, P. R. and Sinclair, T. R. (1988). Soybean nodule gas permeability, nitrogen fixation and diurnal cycles in soil temperature. *Plant Soil* **109**, 227–234.

Welch, R. M. (1986). Effects of nutrient deficiencies on seed production and quality. *In* 'Advances in Plant Nutrition' (B. Tinker and A. Läuchli, eds.), pp. 205–247. Praeger Scientific, New York.

Welch, R. M. and House, W. A. (1984). Factors affecting the bioavailability of mineral nutrients in plant foods. *In* 'Crops as Sources of Nutrients for Humans' (R. M. Welch and W. H. Gabelman, eds.), pp. 37–54. ASA Special Publ. No. 48, Madison, WI, USA.

Welch, R. M., House, W. A and Allaway, W. H. (1974). Availability of zinc from pea seed to rats. *J. Nutr.* **104**, 733–740.

Welch, R. M., Webb, M. J. and Loneragan, J. F. (1982). Zinc in membrane function and its role in phosphorus toxicity. *In* 'Proceedings of the Ninth Plant Nutrition Colloquium, Warwick, England' (A. Scaife, ed.), pp. 710–715. Commonwealth Agricultural Bureau, Farnham Royal, Bucks.

Welch, R. M., Allaway, W. H., House, W. A. and Kubota, J. (1991). Geographic distribution of trace element problems. *In* 'Micronutrients in Agriculture' 2nd edn. (J. J. Mortvedt, F. R. Cox, L. M. Shuman and R. M. Welch, eds.), pp. 31–57. SSSA Book Series No. 4, Madison, WI, USA.

Welkie, G. W. and Miller, G. W. (1989). Sugar beet responses to iron nutrition and stress. *J. Plant Nutr.* **12**, 1041–1054.

Welp, G., Herms, U. and Brümmer, G. (1983). Einfluss von Bodenreaktion, Redox-bedingungen und organischer Substanz auf die Phosphatgehalte der Bodenlösung. *Z. Pflanzenernähr. Bodenk.* **146**, 38–52.

Welte, E. and Müller, K. (1966). Über den Einfluss der Kalidüngung auf die Dunkelung von rohem Kartoffelbrei. *Eur. Potato J.* **9**, 36–45.

Wenzel, C. L. and McCully, M. E. (1991). Early senescence of cortical cells in the roots of cereals. How good is the evidence? *Am. J. Bot.* **78**, 1528–1541.

Wenzel, C. L., McCully, M. E. and Canny, M. J. (1989). Development of water conducting capacity in the root system of young plants of corn and some other C_4 species. *Plant Physiol.* **89**, 1094–1101.

Wenzel, G. and Kreutzer, K. (1971). Der Einfluss des Manganmangels auf die Resistenz der Fichten (*Picea abies* Karst.) gegen *Fomes annosus* (Fr) Cooke. *Z. Pflanzenernähr. Bodenk.* **128**, 123–129.

Werf, A. van der, Kooijman, A., Welschen, R. and Lambers, H. (1988). Respiratory energy costs for the maintenance of biomass, for growth and for iron uptake in roots of *Carex diandra* and *Carex acutiformis*. *Physiol. Plant.* **72**, 483–491.

Werf, A. van der, Raaimakers, D., Poot, P. and Lambers, H. (1991). Evidence for a significant contribution by peroxidase-mediated O_2 uptake to root respiration of *Brachypodium pinnatum*. *Planta* **183**, 347–352.

Werner, D. (1967). Untersuchungen über die Rolle der Kieselsäure in der Entwicklung höherer Pflanzen.I. Analyse der Hemmung durch Germaniumsäure. *Planta* **76**, 25–36.

Werner, D. (1980). Stickstoff(N_2)-Fixierung und Produktionsbiologie. *Angew. Bot.* **54**, 67–75.

Werner, D. (1987). 'Pflanzliche und mikrobielle Symbiosen'. Georg Thieme Verlag, Stuttgart.

Werner, D. and Roth, R. (1983). Silica metabolism. *In* 'Encyclopedia of Plant

Physiology, New Series' (A. Läuchli and R. L. Bieleski, eds.), Vol. 15B, pp. 682–694. Springer-Verlag, Berlin and New York.

Werner, D., Wilcockson, J., Tripf, R., Mörschel, E. and Papen, H. (1981). Limitations of symbiotic and associated nitrogen fixation by developmental stages in the system *Rhizobium japonicum* with *Glycine max* and *Azospirillum brasilense* with grasses, e. g. *Triticum aestivum*. In 'Biology of Inorganic Nitrogen and Sulfur' (H. Bothe and A. Trebst, eds.), pp. 299–308. Springer-Verlag, Berlin.

Werner, D., Berggold, R., Jaeger, D., Krotzky, A., Papen, H., Schenk, S. and Thierfelder, H. (1989). Plant, microbial and soil factors, determining nitrogen fixation in the rhizosphere. *Z. Pflanzenernähr. Bodenk.* **152**, 231–236.

Werner, W. (1959). Die Wirkung einer Magnesiumdüngung zu Kartoffeln in Abhängigkeit von Bodenreaktion und Stickstofform. *Kartoffelbau* **10**, 13–14.

Wernicke, W. and Milkovits, L. (1987). Rates of uptake and metabolism of indole-3-acetic acid and 2,4-dichlorophenoxyacetic acid by cultured leaf segments at different stages of development in wheat. *Physiol. Plant.* **69**, 23–28.

Wesely, R. W., Shearman, R. C., Kinbacher, E. J. and Lowry, S. R. (1987). Ammonia volatilization from foliar-applied urea on field-grown Kentucky bluegrass. *HortScience* **22**, 1278–1280.

Wessolek, G. und Gäth, S. (1989). Integration der Wurzellängendichte in Wasserhaushalts- und Kaliumanlieferungsmodellen. *Kali-Briefe* **19**, 491–503.

West, D. W. (1978). Water use and sodium chloride uptake by apple trees. II. The response to soil oxygen deficiency. *Plant Soil* **50**, 51–65.

West, D. W. and Taylor, J. A. (1980). The effect of temperature on salt uptake by tomato plants with diurnal and nocturnal waterlogging of salinized root zones. *Plant Soil* **56**, 113–121.

Westcott, M. P., Stewart, V. R. and Lund, R. E. (1991). Critical petiole nitrate levels in potato. *Agron. J.* **83**, 844–850.

Westerman, L. and Roddick, J. G. (1981). Annual variation in sterol levels in leaves of *Taraxacum officinale* Weber. *Plant Physiol.* **68**, 872–875.

Wetselaar, R. and Farquhar, G. D. (1980). Nitrogen losses from tops of plants. *Adv. Agron.* **33**, 263–302.

Wheeler, D. M., Edmeades, D. C. and Christie, R. A. (1992a). Effect of aluminium on relative yield and plant chemical concentrations of cereals grown in solution culture at low ionic strength. *J. Plant Nutr.* **15**, 403–418.

Wheeler, D. M., Edmeades, D. C., Christie, R. A. and Gardner, R. (1992b). Comparison of techniques for determining the effect of aluminium on the growth of, and the inheritance of aluminium tolerance in wheat. *Plant Soil* **146**, 1–8.

Wheeler, D. M., Edmeades, D. C.,Christie, R. A. and Gardner, R. (1992c). Effect of aluminium on the growth of 34 plant species: a summary of results obtained in low ionic strength solution culture. *Plant Soil* **146**, 61–66.

Whipps, J. M. and Lynch, J. M. (1986). The influence of the rhizosphere on crop productivity. *Adv. Microbial Ecol.* **6**, 187–244.

White, P. J. (1993). Characterization of a high-conductance, voltage-dependent cation channel from the plasma membrane of rye roots in planar lipid bilayers. *Planta* **191**, 541–551.

White, M. C., Decker, A. M. and Chaney, R. L. (1979). Differential cultivar tolerance in soybean to phytotoxic levels of soil Zn. I. Range of cultivar response. *Agron. J.* **71**, 121–126.

White, M. C., Decker, A. M. and Chaney, R. L. (1981a). Metal complexation in xylem fluid. I. Chemical composition of tomato and soybean stem exudate. *Plant Physiol.* **67**, 292–300.

White, M. C., Decker, A. M. and Chaney, R. L. (1981b). Metal complexation in xylem fluid. II. Theoretical equilibrium model and computational computer program. *Plant Physiol.* **67**, 301–310.

White, P. F. and Robson, A.D. (1989). Rhizosphere acidification and Fe^{3+} reduction in lupins and peas: iron deficiency in lupins is not due to a poor ability to reduce Fe^{3+}. *Plant Soil* **119**, 163–175.

White, P. F. and Robson, A. D. (1990). Response of lupins (*Lupinus angustifolius* L.) and peas (*Pisum sativum* L.) to Fe deficiency induced by low concentrations of Fe in solution or by addition of HCO_3H-. *Plant Soil* **125**, 39–47.

White, P. J., Clarkson, D. T. and Earnshaw, M. J. (1987). Acclimation of potassium influx in rye (*Secale cereale*) to low root temperatures. *Planta* **171**, 377–385.

White, P. J., Cooper, H. D., Earnschaw, M. J. and Clarkson, D. T. (1990a). Effects of low temperature on the development and morphology of rye (*Secale cereale*) and wheat (*Triticum aestivum*). *Ann. Bot. (London)* **66**, 559–566.

White, P. J., Cooke, D. T., Earnshaw, M. J., Clarkson, D. T. and Burden, R. S. (1990b). Does plant growth temperature modulate the membrane composition and ATPase activities of tonoplast and plasmamembrane fractions from rye roots. *Phytochemistry* **29**, 3385–3393.

Whiteaker, G., Gerloff, G. C., Gabelman, W. H. and Lindgren, D. (1976). Intraspecific differences in growth of beans at stress levels of phosphorus. *J.Am. Soc. Hortic. Sci.* **101**, 472–475.

Whitehead, D. C. (1985). Chlorine deficiency in red clover grown in solution culture. *J. Plant Nutr.* **8**, 193–198.

Whitehead, D. C. and Lockyer, D. R. (1987). The influence of the concentration of gaseous ammonia on its uptake by the leaves of Italian ryegrass with and without an adequate supply of nitrogen to the roots. *J. Exp. Bot.* **38**, 818–827.

Wiebe, H. J., Schätzler, H. P. and Kühn, W. (1977). On the movement and distribution of calcium in white cabbage in dependence of the water status. *Plant Soil* **48**, 409–416.

Wieneke, J. (1992). Nitrate fluxes in squash seedlings measured with ^{13}N. *J. Plant Nutr.* **15**, 99–124.

Wieneke, J., Sarwar, G. and Roeb, M. (1987). Existence of salt glands on leaves of Kallar grass (*Leptochloa fusca* L.Kunth.). *J. Plant Nutr.* **10**, 805–820.

Wiersma, D. and van Goor, B. J. (1979). Chemical forms of nickel and cobalt in phloem of *Ricinus communis*. *Physiol. Plant.* **45**, 440–442.

Wiesler, F. and Horst, W. J. (1993). Differences among maize cultivars in the utilization of soil nitrate and the related losses of nitrate through leaching. *Plant Soil* **151**, 193–203.

Wilcox, H. E. (1991). Mycorrhizae. *In* 'Plant Roots: the Hidden Half' (Y. Waisel, A. Eshel and U. Kafkafi, eds.), pp. 731–765. Marcel Dekker, New York.

Wild, A., Woodhouse, P. J. and Hopper, M. J. (1979). A comparison between uptake of potassium by plants from solutions of constant potassium concentration and during depletion. *J. Exp. Bot.* **30**, 697–704.

Wilhelm, M. S., Fisher, J. M. and Graham, R. D. (1985). The effect of manganese deficiency and cereal cyst nematode infection on the growth of barley. *Plant Soil* **85**, 23–32.

Wilhelm, N. S., Graham, R. D. and Rovira, A. D. (1990). Control of Mn status and infection rate by genotype of both host and pathogen in the wheat take-all interaction. *Plant Soil* **123**, 267–275.

Wilkins, D. A. (1991). The influence of sheathing (ecto-) mycorrhizas of trees on the uptake and toxicity of metals. *Agric. Ecosyst. Environ.* **35**, 245–260.

Wilkinson, R. E. and Ohki, K. (1988). Influence of manganese deficiency and toxicity on isoprenoid synthesis. *Plant Physiol.* **87**, 841–846.

Willenbrink, J. (1964). Lichtabhängiger ^{35}S-Einbau in organische Bindung in Tomatenpflanzen. *Z. Naturforsch.* **19**, 356–357.

Willenbrink, J. (1967). Über Beziehungen zwischen Proteinumsatz und Schwefelversorgung der Chloroplasten. *Z. Pflanzenphysiol.* **56**, 427–438.

Willenbrink, J. (1983). Mechanismen des Zuckertransports durch Membranen bei *Beta vulgaris*. *Kali-Briefe* **16**, 585–594.

Willenbrink, J., Doll, S., Getz, H.-P. and Meyer, S. (1984). Zuckeraufnahme in isolierten Vakuolen und Protoplasten aus dem Speichergewebe von Beta-Rüben. *Ber. Dtsch. Bot. Ges.* **97**, 27–39.

Williams, C. M. J. and Maier, N. A. (1990). Determination of the nitrogen status of irrigated potato crops. I. Critical nutrient ranges for nitrate-nitrogen in petioles. *J. Plant Nutr.* **13**, 971–984.

Williams, D. E. and Vlamis, J. (1957). The effect of silicon on yield and manganese-54 uptake and distribution in the leaves of barley plants grown in culture solutions. *Plant Physiol.* **32**, 404–409.

Williams, E. G. and Knight, A. H. (1963). Evaluations of soil phosphate status by pot experiments, conventional extraction methods and labile phosphate values estimated with the aid of phosphorus-32. *J. Sci. Food Agric.* **14**, 555–563.

Williams, J. H. H. and Farrar, J. F. (1988). Endogenous control of photosynthesis in leaf blades of barley. *Plant Physiol. Biochem.* **26**, 503–509.

Williams, J. H., Dutta, M. and Manbiar, P. T. C. (1990). Light interception as a source of variation for nitrogen fixation in groundnut genotypes. *Plant Soil* **121**, 83–88.

Williams, L. and Hall, J. L. (1987). ATPase and proton pumping activities in cotyledons and other phloem-containing tissues of *Ricinus communis*. *J. Exp. Bot.* **38**, 185–202.

Wills, R. B. H., Tirmazi, S.I. H. and Scott, K. J. (1977). Use of calcium to delay ripening of tomatoes. *HortScience* **12**, 551–552.

Wilson, D. O., Boswell, F. C., Ohki, K., Parker, M. B., Shuman, L. M. and Jellum, M. D. (1982). Changes in soybean seed oil and protein as influenced by manganese nutrition. *Crop Sci.* **22**, 948–952.

Wilson, E. J. (1992). Foliar uptake and release of inorganic nitrogen compounds in *Pinus sylvestris* L. and *Picea abies* (L.) Karst. *New Phytol.* **120**, 407–416.

Wilson, P. J. and Van Staden, J. (1990). Rhizocaline, rooting co-factors, and the concept of promotors and inhibitors of adventitious rooting – A review. *Ann. Bot. (London)* **66**, 479–490.

Wilson, S. B. and Hallsworth, E. G. (1965). Studies on the nutrition of the forage legumes. IV. The effect of cobalt on the growth of nodulated and non-nodulated *Trifolium subterraneum* L. *Plant Soil* **22**, 260–279.

Wilson, S. B. and Nicholas, D. J. D. (1967). A cobalt requirement for non-nodulated legumes and for wheat. *Phytochemistry* **6**, 1057–1066.

Wilson, T. P., Canny, M. J. and McCully, M. E. (1988). Proton pump activity in bundle sheath tissues of broad-leaved trees in relation to leaf age. *Physiol. Plant.* **73**, 465–470.

Win, K., Berkowitz, G. A. and Henninger, M. (1991). Antitranspirant-induced increases in leaf water potential increase tuber calcium and decrease tuber necrosis in water-stressed potato plants. *Physiol. Plant.* **96**, 116–120.

Winer, L. and Apelbaum, A. (1986). Involvement of polyamines in the development and ripening of avocado fruits. *J. Plant Physiol.* **126**, 223–233.

Wink, M. (1993). The plant vacuole: a multifunctional compartment. *J. Exp. Bot.* **44** Suppl., 231–246.

Winkelmann, G. (1986). Iron complex products (siderophores). *In* 'Biotechnology' Vol. 4 (H. J. Rehm and G. Reed, eds.), pp. 216–243. VCH Verlagsgesellschaft, Weinheim.

Winkler, R. G., Blevins, D. G., Polacco, J. C. and Randall, D. D. (1987). Ureide catabolism of soybeans. II. Pathway of catabolism in intact leaf tissue. *Plant Physiol.* **83**, 585–591.

Winkler, R. G., Blevins, D. G., Polacco, J. C. and Randall, D. D. (1988). Ureide catabolism in nitrogen-fixing legumes. *Trends Biochem. Sci.* **11**, 97–100.

Winkler, R. G., Polacco, J. C., Blevins, D. G. and Randall, D. D. (1985). Enzymatic degradation of allantoate in developing soybeans. *Plant Physiol.* **79**, 878–793.

Winkler, R. G., Polacco, J. C., Eskew, D. L. and Welch, R. M. (1983). Nickel is not required for apo-urease synthesis in soybean seeds. *Plant Physiol.* **72**, 262–263.

Winter, E. (1982). Salt tolerance of *Trifolium alexandrinum* L. III. Effects of salt on ultrastructure of phloem and xylem transfer cells in petioles and leaves. *Aust. J. Plant Physiol.* **9**, 239–250.

Winter, H., Lohaus, G. and Heldt, H. W. (1992). Phloem transport of amino acids in relation to their cytosolic levels in barley leaves. *Plant Physiol.* **99**, 996–1004.

Wirén, N. von, Römheld, V., Morel, J. L., Guckert, A. and Marschner, H. (1993). Influence of microorganisms on iron acquisition in maize. *Soil Biol. Biochem.* **25**, 371–376.

Wirth, E., Kelly, G. J., Fischbeck, G. and Latzko, E. (1977). Enzyme activities and products of CO_2-fixation in various photosynthetic organs of wheat and oat. *Z. Pflanzenphysiol.* **82**, 78–87.

Wissemeier, A. H. and Horst, W. J. (1987). Callose deposition in leaves of cowpea (*Vigna unguiculata* L. Walp.) as a sensitive response to high Mn supply. *Plant Soil* **102**, 283–286.

Wissemeier, A. H. and Horst, W. J. (1991). Simplified methods for screening cowpea cultivars for manganese leaf-tissue tolerance. *Crop Sci.* **31**, 435–439.

Wissemeier, A. H. and Horst, W. J. (1992). Effect of light intensity on manganese toxicity symptoms and callose formation in cowpea (*Vigna unguiculata* (L.) Walp.). *Plant Soil* **143**, 299–309.

Wissemeier, A. H., Klotz, F. and Horst, W. J. (1987). Aluminium induced callose synthesis in roots of soybean (*Glycine max* L.). *J. Plant Physiol.* **129**, 487–492.

Wissemeier, A. H., Diening, A., Hergenröder, A., Horst, W. J. and Mix-Wagner, G. (1992). Callose formation as parameter for assessing genotypical plant tolerance of aluminium and manganese. *Plant Soil* **146**, 67–75.

Witt, H. H. and Jungk, A. (1974). The nitrate inducible nitrate reductase activity in relation to nitrogen nutritional status of plants. In (J. Wehrmann, ed.) *Proc. 7th Int. Colloq. Plant Anal. Fert. Probl.*, pp. 519–527. Hannover.

Witt, H. H. and Jungk, A. (1977). Beurteilung der Molybdänversorgung von Pflanzen mit Hilfe der Mo-induzierbaren Nitratreduktase-Aktivität. *Z. Pflanzenernähr. Bodenk.* **140**, 209–222.

Witty, J. F., Keay, P. J., Frogatt, P. J. and Dart, P. J. (1979). Algal nitrogen fixation on temperate arable fields. The Broadbalk experiment. *Plant Soil* **52**, 151–164.

Wolf, O. and Jeschke, W. D. (1986). Sodium fluxes, xylem transport of sodium, and K/

Na selectivity in roots of seedlings of *Hordeum vulgare*, cv. California Mariout and *H. distichon*, cv. Villa. *J. Plant Physiol.* **125**, 243–256.

Wolf, O., Jeschke, W. D. and Hartung, W. (1990a). Long distance transport of abscisic acid in NaCl-treated intact plants of *Lupinus albus*. *J. Exp. Bot.* **41**, 593–600.

Wolf, O., Munns, R., Tonnet, M. L. and Jeschke, W. D. (1990b). Concentrations and transport of solutes in xylem and phloem along the leaf axis of NaCl-treated *Hordeum vulgare*. *J. Exp. Bot.* **41**, 1133–1141.

Wolf, O., Munns, R., Tonnet, M. L. and Jeschke, W. D. (1991). The role of the stem in the partitioning of Na^+ and K^+ in salt-treated barley. *J. Exp. Bot.* **42**, 697–704.

Wollring, J. and Wehrmann, J. (1990). Der Nitratgehalt in der Halmbasis als Massstab für den Stickstoffdüngerbedarf bei Wintergetreide. *Z. Pflanzenernähr. Bodenk.* **153**, 47–53.

Wolswinkel, P., Ammerlaan, A. and Peters, F. C. (1984). Phloem unloading of amino acids at the site of attachment of *Cuscuta europaea*. *Plant Physiol.* **75**, 13–20.

Wolterbeek, H. T., van Luipen, J. and de Bruin, M. (1984). Non-steady state xylem transport of fifteen elements into the tomato leaf as measured by gamma-ray spectroscopy: a model. *Physiol. Plant.* **61**, 599–606.

Wong, M. H. and Bradshaw, A. D. (1982). A comparison of the toxicity of heavy metals, using root elongation of rye grass, *Lolium perenne*. *New Phytol.* **91**, 255–261.

Wong You Cheong, Y. and Chan, P. Y. (1973). Incorporation of ^{32}P in phosphate esters of the sugar cane plant and the effect of Si and Al on the distribution of these esters. *Plant Soil* **38**, 113–123.

Woo, K. C., Flügge, U. I. and Heldt, H. W. (1987). A two-translocator model for the transport of 2-oxoglutarate and glutamate in chloroplasts during ammonia assimilation in the light. *Plant Physiol.* **84**, 624–632.

Wood, L. J., Murray, B. J., Okatan, Y. and Noodén, L. D. (1986). Effect of petiole phloem distribution on starch and mineral distribution in senescing soybean leaves. *Am. J. Bot.* **73**, 1377–1383.

Woodrow, I. E. and Rowan, K. S. (1979). Change of flux of orthophosphate between cellular compartments in ripening tomato fruits in relation to the climacteric rise in respiration. *Aust. J. Plant Physiol.***6**, 39–46.

Woolhouse, H. W. (1983). Toxicity and tolerance in response of plants to metals. *In* 'Encyclopedia of Plant Physiology, New Series' (O. L. Lange *et al.*, eds.), Vol. 12C, pp. 245–300. Springer-Verlag, Berlin.

Wright, J. P. and Fisher, D. B. (1981). Measurement of the sieve tube membrane potential. *Plant Physiol.* **67**, 845–848.

Wright, R. J. (1989). Soil aluminum toxicity and plant growth. *Commun. Soil Sci. Plant Anal.* **20**, 1479–1497.

Wright, R. J., Baligar, V. C. and Ahlrichs, J. L. (1989a). The influence of extractable and soil solution aluminum on root growth of wheat seedlings. *Soil Sci.* **148**, 293–302.

Wright, R. J., Baligar, V. C., Ritchey, K. D. and Wright, S. F. (1989b). Influence of soil solution aluminum on root elongation of wheat seedlings. *Plant Soil* **113**, 294–298.

Wrigley, C. W., du Cros, D. L., Archer, M. J., Downie, P. G. and Roxburgh, C. M. (1980). The sulfur content of wheat endosperm and its relevance to grain quality. *Aust. J. Plant Physiol.* **7**, 755–766.

Wu, J., Neimanis, S. and Heber, U. (1990). Photorespiration is more effective than the Mchler reaction to protect the photosynthetic apparatus against photoinhibition. *Bot. Acta* **104**, 283–291.

Wu, L. and Huang, Z.-Z. (1992). Selenium assimilation and nutrient element uptake in white clover and tall fescue under the influence of sulphate concentration and selenium tolerance of the plants. *J. Exp. Bot.* **43**, 549–555.

Wu, L., Thurman, D. A. and Bradshaw, A. D. (1975). The uptake of copper and its effect upon respiratory processes of roots of copper-tolerant and non-tolerant clones of *Agrostis stolonifera*. *New Phytol.* **75**, 225–229.

Wu, W., Peters, J. and Berkowitz, G. A. (1991). Surface charge-mediated effects of Mg^{2+} on K^+ flux across the chloroplast envelope are associated with regulation of stromal pH and photosynthesis. *Plant Physiol.* **97**, 580–587.

Wullschleger, S. D. and Reid, C. P. P. (1990). Implication of ectomycorrhizal fungi in the cytokinin relations of loblolly pine (*Pinus taeda* L.). *New Phytol.* **116**, 681–688.

Wunderlich, F. (1978). Die Kernmatrix: Dynamisches Protein-Gerüst in Zellkernen. *Naturwiss. Rundsch.* **31**, 282–288.

Wydrzynski, T., Baumgart, F., MacMillan, F. and Renger, G. (1990). Is there a direct chloride cofactor requirement in the oxygen-evolving reactions of photosystem II? *Photosynthesis Res.* **25**, 59–72.

Wyn Jones, R. G. (1981). Salt tolerance. *In* 'Physiological Processes Limiting Plant Productivity' (C. B. Johnson, ed.), pp. 271–292. Butterworth, London.

Wyn Jones, R. G. and Pollard, A. (1983). Proteins, enzymes and inorganic ions. *In* 'Encyclopedia of Plant Physiology, New Series' (A. Läuchli and R. L. Bieleski, eds.), Vol. 15B, pp. 528–562. Springer-Verlag, Berlin.

Wyn Jones, R. G., Brady, C. J. and Speirs, J. (1979). Ionic and osmotic relations in plant cells. *In* 'Recent Advances in the Biochemistry of Cereals' (D. L. Laidman and R. G. Wyn Jones, eds.), pp. 63–103. Academic Press, London.

Wyttenbach, A., Tobler, L. and Bajo, S. (1991). Silicon concentration in spruce needles. *Z. Pflanzenernähr. Bodenk.* **154**, 253–258.

Xu, Q. F., Tsai, C. L. and Tsai, C. Y. (1992). Interaction of potassium with the form and amount of nitrogen nutrition on growth and nitrogen uptake of maize. *J. Plant Nutr.* **15**, 23–33.

Yamauchi, M. (1989). Rice bronzing in Nigeria caused by nutrient imbalances and its control by potassium sulfate application. *Plant Soil* **117**, 275–286.

Yang, J. E., Skogley, E. O. and Schaff, B. E. (1991). Nutrient flux to mixed-bed ion-exchange resin: temperature effects. *Soil Sci. Soc. Am. J.* **55**, 762–767.

Yang, X., Römheld, V. and Marschner, H. (1993). Effect of bicarbonate and root zone temperature on uptake of Zn, Fe, Mn and Cu in different rice varieties (*Oryza sativa* L.) growing in calcareous soil. *Plant Soil* **155/156**, 441–444.

Yang, X., Römheld, V. and Marschner, H. (1994). Effect of bicarbonate on root growth and accumulation of organic acid in Zn-inefficient and Zn-efficient rice varieties (*Oryza sativa* L.). *Plant Soil* **164**, 1–7.

Yazaki, Y., Asukagawa, N., Ishikawa, Y., Ohta, E. and Sakata, M. (1988). Estimation of cytoplasmic free Mg^{2+} levels and phosphorylation potentials in mung bean root tips *in vivo* ^{31}P NMR spectroscopy. *Plant Cell Physiol.* **29**, 919–924.

Yen, P. Y., Inskeep, W. P., Westerman, R. L. (1988). Effects of soil moisture and phosphorus fertilization on iron chlorosis of sorghum. *J. Plant Nutr.* **11**, 1517–1531.

Yeo, A. R. (1993). Variation and inheritance of sodium transport in rice. *In* 'Genetic Aspects of Plant Mineral Nutrition' (P. J. Randall, E. Delhaize, R. A. Richards and R. Munns, eds.), pp. 143–150. Kluwer Academic, Dordrecht, The Netherlands.

Yeo, A. R., Caporn, S. J. M. and Flowers, T. J. (1985). The effect of salinity upon photosynthesis in rice (*Oryza sativa* L.): gas exchange by individual leaves in relation to their salt content. *J. Exp. Bot.* **36**, 1240–1248.

Yeo, A. R., Yeo, M. E. and Flowers, T. J. (1987). The contribution of an apoplastic pathway to sodium uptake by rice roots in saline conditions. *J. Exp.Bot.* **38**, 1141–1153.

Yih, R. Y. and Clark, H. E. (1965). Carbohydrate and protein content of boron-deficient tomato root tips in relation to anatomy and growth. *Plant Physiol.* **40**, 312–315.

Yin, Z.-H., Neimanis, S. and Heber, U. (1990). Light-dependent pH changes in leaves of C_3 plants. II. Effect of CO_2 and O_2 on the cytosolic and the vacuolar pH. *Planta* **182**, 253–261.

Yokota, A., Shigeoka, S., Onishi, T. and Kitaoka, S. (1988). Selenium as inducer of glutathione peroxidase in low-CO_2-grown *Chlamydomonas reinhardtii*. *Plant Physiol.* **86**, 649–651.

Yokota, H. and Konishi, S. (1990). Effect of the formation of a sugar–borate complex on the growth inhibition of pollen tubes of *Camellia sinensis* and cultured cells of *Nicotiana tabacum* by toxic levels of borate. *Soil Sci. Plant Nutr.* **36**, 275–282.

Yoshida, S. and Tadano, T. (1978). Adaptation of plants to submerged soils. *ASA Spec. Publ.* **32**, 233–256.

Yoshida, S. and Uemura, M. (1986). Lipid composition of plasma membranes and tonoplasts isolated from etiolated seedlings of mung bean (*Vigna radiata* L.). *Plant Physiol.* **82**, 807–812.

Yoshida, S., Navasero, S. A. and Ramirez, E. A. (1969). Effects of silica and nitrogen supply on some leaf characters of the rice plant.*Plant Soil* **31**, 48–56.

Young, A. J. (1991). The photoprotective role of carotenoids in higher plants. *Physiol. Plant.* **83**, 702–708.

Young, T. F. and Terry, N. (1982). Transport of iron into leaves following iron resupply to iron-stressed sugar beet plants. *J. Plant. Nutr.* **5**, 1273–1283.

Youngdahl, L. J. (1990). Differences in phosphorus efficiency in bean genotypes. *J. Plant Nutr.* **13**, 1381–1392.

Youssef, R. A. and Chino, M. (1987). Studies on the behavior of nutrients in the rhizosphere. I. Establishment of a new rhizobox system to study nutrient status in the rhizosphere. *J. Plant Nutr.* **10**, 1185–1195.

Yu, J. and Wo, K. C. (1991). Correlation between the development of photorespiration and the change in activities of NH_3 assimilation enzymes in greening oat leaves. *Aust. J. Plant Physiol.* **18**, 583–588.

Yu, P. T., Stolzy, L. H. and Letey, J. (1969). Survival of plants under prolonged flooded conditions. *Agron. J.* **61**, 844–847.

Zayed, A. M. and Terry, N. (1992). Selenium volatilization in broccoli as influenced by sulfate supply. *J. Plant Physiol.* **140**, 646–652.

Zech, W., Alt, H. G., Haumaier, L. and Blasek, R. (1987). Characterisation of phosphorus fractions in mountain soils of the Bavarian alps by ^{32}P-NMR spectroscopy. *Z. Pflanzenernähr. Bodenk.* **150**, 119–123.

Zeevaart, J. A. D. and Boyer, G. L. (1984). Accumulation and transport of abscisic acid and its metabolites in *Ricinus* and *Xanthium*. *Plant Physiol.* **74**, 934–939.

Zehler, E. (1981). Die Natrium-Versorgung von Mensch, Tier und Pflanze. *Kali-Briefe* **15**, 773–792.

Zelleke, A. and Kliewer, W. M. (1981). Factors affecting the qualitative and quantitative levels of cytokinins in xylem sap of grapevine. *Vitis* **20**, 93–104.

Zeng, Z. R. and King, R. W. (1986). Regulation of grain number in wheat: changes in endogenous levels of abscisic acid. *Aust. J. Plant Physiol.* **13**, 347–352.

Zhang, F., Römheld, V. and Marschner, H. (1989). Effect of zinc deficiency in wheat on the release of zinc and iron mobilizing root exudates. *Z. Pflanzenernähr. Bodenk.* **152**, 205–210.

Zhang, F., Römheld, V. and Marschner, H. (1991a). Release of zinc mobilizing root exudates in different plant species as affected by zinc nutritional status. *J. Plant Nutr.* **14**, 675–686.

Zhang, F., Römheld, V. and Marschner, H. (1991b). Diurnal rhythm of release of phytosiderophores and uptake rate of zinc in iron-deficient wheat. *Soil Sci. Plant Nutr. (Tokyo)* **37**, 671–678.

Zhang, F., Römheld, V. and Marschner, H. (1991c). Role of the root apoplasm for iron acquisition by wheat plants. *Plant Physiol.* **97**, 1302–1305.

Zhang, J. and Davies, W. J. (1989). Sequential response of whole plant water relations to prolonged soil drying and the involvement of xylem sap ABA in the regulation of stomatal behaviour of sunflower plants. *New Phytol.* **113**, 167–174.

Zhang, J. and Davies, W. J. (1990). Changes in the concentration of ABA in xylem sap as a function of changing soil water status can account for changes in leaf conductance and growth. *Plant Cell Environ.* **13**, 277–285.

Zhang, J. and Davies, W. J. (1991). Antitranspirant activity in xylem sap of maize plants. *J. Exp. Bot.* **42**, 317–321.

Zhang, W. H. and Tyerman, S. D. (1991). Effect of low O_2 concentration and azide on hydraulic conductivity and osmotic volume of the cortical cells of wheat roots. *Aust. J. Plant Physiol.* **18**, 603–613.

Zhou, J. R., Fordyce, E. J., Raboy, V., Dickinson, D. B., Wong, M.-S., Burns, R. A. and Erdman jr, J. W. (1992). Reduction of phytic acid in soybean products improves zinc bioavailability in rats. *J. Nutr.* **122**, 2466–2473.

Zhu, G. L. and Steudle, E. (1991). Water transport across maize roots. Simultaneous measurement of flows at the cell and root level by double pressure probe technique. *Plant Physiol.* **95**, 305–315.

Ziegler, H. (1975). Nature of transported substances. *In* 'Encyclopedia of Plant Physiology, New Series' (M. H. Zimmermann and J. A. Milburn, eds.), Vol. 1, pp. 59–100. Springer-Verlag, Berlin.

Zimmermann, U. and Steudle, E. (1970). Bestimmung von Reflexionskoeffizienten an der Membran der Alge *Valonia utricularis*. *Z. Naturforsch. B: Anorg. Chem. Org. Chem. Biochem. Biophys. Biol.* **25B**, 500–504.

Zöttl, H. W. (1990). Remarks on the effects of nitrogen deposition to forest ecosystems. *Plant Soil* **128**, 83–89.

Zoettl, H. W. and Huettl, R. F. (1986). Nutrient supply and forest decline in southwest-Germany. *Water Air Soil Pollut.* **31**, 449–462.

Zsoldos, F. and Haunold, E. (1982). Influence of 2,4-D and low pH on potassium, ammonium and nitrate uptake by rice roots. *Physiol. Plant.* **54**, 63–68.

Zsoldos, F. and Karvaly, B. (1978). Effects of Ca^{2+} and temperature on potassium uptake along roots of wheat, rice and cucumber. *Physiol. Plant.* **43**, 326–330.

Zur, B., Jones, J. W., Boote, K. J. and Hammond, L. C. (1982). Total resistance to water flow in field soybeans. II. Limiting soil moisture. *Agron. J.* **74**, 99–105.

Index